MW01518036

A TEXTBOOK OF

MICROBIOLOGY

[For University and College Students in India & Abroad]

R.C. Dubey
M.Sc., Ph.D., F.B.S.,
F.P.S.I., F.N.R.S.

D.K. Maheshwari
M.Sc., Ph.D. F.B.S.,
F.P.S.I., A.I.S.I. (Szeged)

Department of Botany & Microbiology
Gurukul Kangri University
HARIDWAR- 249 404
(INDIA)

S. CHAND & COMPANY LTD.

(AN ISO 9001 : 2000 COMPANY)

RAM NAGAR, NEW DELHI - 110 055

S. CHAND & COMPANY LTD.

(An ISO 9001 : 2000 Company)

Head Office : 7361, RAM NAGAR, NEW DELHI - 110 055
Phones : 23672080-81-82, 9899107446, 9911310888;
Fax : 91-11-23677446

Shop at: **schandgroup.com**; E-mail: **schand@vsnl.com**

Branches :

- 1st Floor, Heritage, Near Gujarat Vidhyapeeth, Ashram Road,
  Ahmedabad-380 014. Ph. 27541965, 27542369, ahmedabad@schandgroup.com
- No. 6, Ahuja Chambers, 1st Cross, Kumara Krupa Road,
  Bangalore-560 001. Ph : 22268048, 22354008, bangalore@schandgroup.com
- 238-A M.P. Nagar, Zone 1, **Bhopal** - 462 011. Ph : 4274723. bhopal@schandgroup.com
- 152, Anna Salai, **Chennai**-600 002. Ph : 28460026, chennai@schandgroup.com
- S.C.O. 2419-20, First Floor, **Sector- 22**-C (Near Aroma Hotel), **Chandigarh**-160022,
  Ph-2725443, 2725446, chandigarh@schandgroup.com
- 1st Floor, Bhartia Tower, Badambadi, **Cuttack**-753 009, Ph-2332580; 2332581,
  cuttack@schandgroup.com
- 1st Floor, 52-A, Rajpur Road, **Dehradun**-248 001. Ph : 2740889, 2740861,
  dehradun@schandgroup.com
- Pan Bazar, **Guwahati**-781 001. Ph : 2738811, guwahati@schandgroup.com
- Sultan Bazar, **Hyderabad**-500 195. Ph : 24651135, 24744815, hyderabad@schandgroup.com
- Mai Hiran Gate, **Jalandhar** - 144008 . Ph. 2401630, 5000630, jalandhar@schandgroup.com
- A-14 Janta Store Shopping Complex, University Marg, Bapu Nagar, **Jaipur** - 302 015,
  Phone : 2719126, jaipur@schandgroup.com
- 613-7, M.G. Road, Ernakulam, **Kochi**-682 035. Ph : 2381740, cochin@schandgroup.com
- 285/J, Bipin Bihari Ganguli Street, **Kolkata**-700 012. Ph : 22367459, 22373914,
  kolkata@schandgroup.com
- Mahabeer Market, 25 Gwynne Road, Aminabad, **Lucknow**-226 018. Ph : 2626801, 2284815,
  lucknow@schandgroup.com
- Blackie House, 103/5, Walchand Hirachand Marg , Opp. G.P.O., **Mumbai**-400 001.
  Ph : 22690881, 22610885, mumbai@schandgroup.com
- Karnal Bag, Model Mill Chowk, Umrer Road, **Nagpur**-440 032 Ph : 2723901, 2777666
  nagpur@schandgroup.com
- 104, Citicentre Ashok, Govind Mitra Road, **Patna**-800 004. Ph : 2300489, 2302100,
  patna@schandgroup.com

Multicolour edition conceptualized by R.K. Gupta, CMD

First Edition 1999
Subsequent Editions and Reprints 2000, 2001, 2002, 2003, 2004

First Multicolour Illustrative Revised Edition 2005, *Reprint 2007*
Reprint 2008

ISBN : 81-219-2559-2
Code : 03 333

PRINTED IN INDIA

By Rajendra Ravindra Printers (Pvt.) Ltd., 7361, Ram Nagar, New Delhi-110 055
and published by S. Chand & Company Ltd. 7361, Ram Nagar, New Delhi-110 055

Preface to the Second Revised Edition

We are delighted to update the second edition of the book by weaving the current informations as and where required, besides writing two new chapters. Chapter 24 Cellular Microbiology and Chapter 26 Microbial Ecology III : Extremophiles, have been written as per **U.G.C. syllabus** provided to all University/Colleges where microbiology is being taught to the UG and/or PG students. Due to evolving the fast knowledge in different areas of microbiology, it became necessary to provide these new materials to the readers. However, even more are required to nurse them with current knowledge.

We thank the readers especially the students who have been demanding for a long time for the second edition of this book. Hopefully, the second edition will meet out their demand.

Helps and critical suggestions rendered by the students, colleagues and friends are gratefully acknowledged. We appreciate the constant availability of comments from the personnel of S. Chand & Company. Decision for bringing out **Multicolour Illustrative Edition** of this book by Mr. Ravindra K. Gupta (Managing Director) will certainly ease to grasp the matter to the readers. Helps and suggestions in multifarious ways done by Mr. Navin Joshi (General Manager, Sales & Marketing), Mr. R.S. Saxena (Advisor), Mr. Amit Goel (Manager, Sales & Promotion) and Mr. Shishir Bhatnagar (Editor), S. Chand & Co. for betterment of the book are gratefully acknowledged.

Any constructive suggestion from the readers shall always be welcome.

AUTHORS

Preface to the First Edition

Until recently, Microbiology has been a least concerned branch of biology. But at present it has revolutionised the field of biology throughout the world. Needless to say that the highest number of Nobel prize awardee are from microbiology.

This book contains 30 chapters covering a wide range of areas such as general microbiology, agriculture, medical, fermentation, food and dairy including the advanced fields as well. A detailed account of defense mechanism, working of immune system and antigen-antibody interactions have been given. Moreover, epidemiology, parasitism, antimicrobial chemotherapy and important human diseases (protozoan, fungal, bacterial and viral) are described in detail.

Ecology of microorganisms especially soil, water and air, and microbial interactions has been discussed. The present need of the country is the enhanced crop production and crop protection from pathogens and pests in field and storage conditions. These aspects have also been covered. The knowledge of handling the instruments is very necessary for the graduate and post-graduate students of microbiology. Therefore, for the interest of students a separate chapter has been written on Instrumentations in Microbiology.

The authors have received helps, encouragement and suggestions from their well-wishers, colleagues and friends throughout India and abroad while compiling this book. We gratefully acknowledge the helps and suggestions accorded by Dr. H. Uchiyama, N.I.E.S., Ibaraki, Japan for a SEM photograph of *Methylocystis* sp. strain ML, Prof. Y. Nishimura (Japan), Prof. Ajit Verma (JNU, New Delhi), Dr. Y.D. Gaur and Dr. S.K. Goel (IARI) , Prof. Subhash Chand (IIT, New Delhi), Prof. T. Satyanarayan (South Campus, Delhi), Dr. D.S. Arora (Amritsar), Prof. J.K. Gupta (Chandigarh), Dr. Swaranjit Singh (IMTECH, Chandigarh), Prof. B.S. Kundu (Kurukshetra), Dr. Geeta Sumbali (Jammu), Prof. S.K. Mahana (Ajmer), Prof. H.C. Dube (Bhavnagar), Dr. P.C. Trivedi (Jaipur), Prof. Shashi Chauhan (Gwalior), Prof. P.S. Bisen (Bhopal), Prof. R.C. Rajak (Jabalpur), Prof. S.C. Agrawal (Sagar), Prof. R.D. Khulbe and Dr. S.C. Sati (Nainital), Prof. K.N. Pandey and Dr. R.C. Gupta (Almora), Prof. B.N. Johri (Pantnagar), Prof. S.K. Garg and Dr. Rajeev Gaur (Faizabad), Dr. A.P. Garg (Meerut), Dr. R.K.S. Kushwaha (Kanpur), Prof. H.N. Verma (Lucknow), Prof. Sudhir Chandra (Allahabad), Prof. Bharat Rai (BHU, Varanasi), Prof. J.S. Singh (BHU), Dr. N.N. Tripathi (Gorakhpur), Dr. D.B. Singh (Sonbhadra), Prof. R.S. Bilgrami (Muzaffarpur), Prof. K.B. Mishra (Patna), Prof. R.P. Purkayastha (Calcutta), Prof. R.R. Mishra (Shillong), Dr. M.S. Singh and Dr. R.R. Pandey (Manipur), Prof. R.K.S. Chauhan (Ujjain), Prof. B.P. Kapadnis (Pune), Prof. C. Manoharachary (Hyderabad), Prof. C. Laxminarsinhan (Thanjavur), Dr. R. Saravanamuthu (Mayiladuthurai).

We thank to our colleagues Dr. G.P. Gupta and Dr. Navneet for helps and critical suggestions time to time. Assistance rendered by our research scholars Messrs V.K. Sharma, Roshan Lal, Vineet Kumar, Chandra Prakash, Navin K. Arora and Promen Sharma is gratefully acknowledged. Our sincere thanks are also due to our publisher Mr. R.K. Gupta of S. Chand & Co. Ltd., for timely bringing out the book.

We shall welcome the constructive suggestion, if any, from the readers.

AUTHORS

CONTENTS

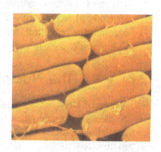

Artificial synthesis of human leucocyte interferon gene; Gene machine; Gene synthesis by using mRNA; The PCR technology.

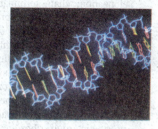
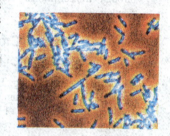
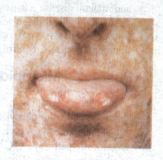

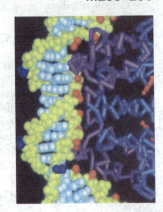
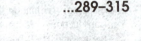

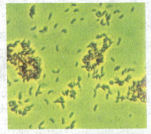

sulphur bacteria, purple non-sulphur bacteria, green bacteria, green sulphur bacteria, green non-sulphur bacteria); Heliobacteria; Oxygenic photosynthetic bacteria. members of prochlorophyta; Unclassified bacteria; Photosynthetic pigments – bactero chlorophylls; Carotenoids, Bacteriorhodopsin, Phycobilins; Metabolism in photosynthetic bacteria; Photosynthetic electron transport systems- purple bacteria, green bacteria, heliobacteria; Mechanism of photosynthesis; Dark reaction (Calvin-Banson cycle).

CHAPTER 14 : Nitrogen Fixation

Symbiotic nitrogen fixing systems – root nodulating symbiotic bacteria (process of root nodule formation – curling and deformation of root hairs, formation of infection – threads and nodule formation, development of nodule); Leghaemoglobin; Metabolism of nitrogen fixation (free living and symbiotic microorganisms); Genetics of nitrogen fixing microorganisms; Bacterial nodulation genes and regulation of nod gene expression; *nif* genes and their regulation; Nitrogen fixation mechanisms; nitrogenase types – structure and function; alternative nitrogenase; substrates for nitrogenase; electron proteins; hydrogen evolution.

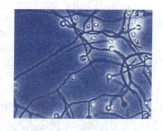

CHAPTER 15 : Viruses-I : Plant and Animal Viruses, Viroids and Prions

Historical account; General concepts – What is a virus?; Occurrence; Morphology of viruses : Shape, Size, Structure – helical viruses (naked and enveloped viruses), icosahedral (naked and enveloped) viruses, complex viruses; Viral envelope; Nucleic acids; Proteins, Carbohydrates; Classification of viruses; Isolation and cultivation of viruses (Cultivation of bacteriophages, cultivation of animal viruses, Identification of viruses); Plant viruses : Tobamovirus group (TMV - symptoms, viral structure, protein synthesis, transmission); Potex virus group (Potato virus X – PVX); Potyvirus group (Potato virus Y); Tymovirus group (Cuccumber mosaic virus); Tomato spotted wilt virus; Cauliflower mosaic virus; Potato leaf roll virus; Rice tungro virus; Mosaic disease of sugarcane, Transmission of plant viruses - mechanical transmission, vegetative and graft transmission, pollen, seed, nematode, fungal, insect vectors, dodder transmission; Effect of viruses on plants (symptoms of virus infection); External symptoms (mosaic, chlorosis, vein banding, ring spots, necrosis, leaf abnormalities); Internal symptoms histological and cytological abnormalities; Serological tests for diagnosis of plant viruses, preparation of viral antigens, histological test, agar gel double diffusion test, DAC-ELISA, DAS-ELISA, I-DAS-ELISA, NASH using probes, Dot-ELISA; Animal viruses : Classification of animal viruses; Multiplication of animal viruses (attachment, penetration, uncoating, replication, assembly and release); Examples of animal viruses. DNA containing viruses (Papovaviruses; SV40 virus; Adenovirus; Herpes virus; Poxviruses); RNA containing viruses (Picorna viruses; Togaviruses; Rabdoviruses; Orthomyxoviruses; Reoviruses; Retroviruses – HIV *i.e.*, AIDS virus- working of immune system in the presence of HIV, replication if HIV in target cell); Viroids – Host range, genome and origin of viroids; Virusoids; Prions (spread of prions, artificial prions).

improvement for lysine production, L-glutamic acid; Strain improvement for glutamic acid production. Pectolytic enzymes: Pectinases (production, harvest and recovery, uses); Invertase; Proteases; Lipases Cellulases- (production, recovery, uses); Vitamins: Vitamin B12 (cyanocobalamine); Riboflavin (vitamin B2); Steroid biotransformations.

Microbiology of milk and dairy; Microbiology of milk (Milk and starter culture; Measures to minimize contamination); Milk products – yoghurt, kefir, koumiss, butter milk, butter cheese – microbiology of cheese; Renin (milk coagulating enzyme); Biotechnology of dairy foods (plasmid concept in butter starter culture, gene manipulation). Microbial contamination and spoilage of poultry, fish and sea foods; Microbial contamination of meat (Microbiology of meat curing brines, general types of meat spoilage, spoilage of meat under anaerobic conditions, spoilage under aerobic conditions; Spoilage of different kinds of meat – fresh meat, fresh beef, hamburger, cured meat, sausage; Growth of microorganisms in meat; Fish and other seafoods: microbiology of fish brines, spoilage, factors influencing kind and rate of spoilage, control of spoiling microorganisms (use of heat, freezing of frozen storage, irradiation, drying; Food preservation method; Physical preservation methods, High temperature; Canning, Heat process for canning, cooling process, Chemical preservation method, wood smoke; use of preservatives; Oriental foods : Mycoprotein, tempeh, soy sauce, idli, natto, mirchin, poi, Food-feed source-bakers yeast, mushroom neutraceuticals; Aflatoxins- structures, function, aflatoxin– producing potential of fungi.

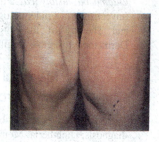

Historical perspectives Non-specific defense; Skin and mucous membranes (Mechanical factors; lacrimal apparatus, other glands; Chemical factors); Phagocytosis (Types of phagocytes; Mechanism of phagocytosis : chemotaxis, attachment, ingestion, digestion); Inflammation (Vasodilation and increased permeability of blood vessels; Phagocyte migration; Repair); Fever; Antimicrobial substances (Complements and properdin : cytolysis, inflammation, opsonization); Interferon Immunity : Types of immunity (Naturally acquired immunity: active and passive immunity; Artificially acquired immunity : active and passive immunity); Types of immune system (Humoral immune system : the antigens, the antibodies – their structure, immunoglobin isotypes : IgG, IgM, IgA, IgD and IgE, mechanism of humoral immunity interaction of antigen with B-cells, production of antibodies and antigen-antibody binding; Monoclonal antibodies; The cell mediated immune system : types of effector T cells, mechanism of cell mediated immunity; Genetics of antibodies : Dryer and Bennett's two gene model, multigene organisation of immunoglobulin gene : multigene family of (chain and heavy chain DNA, and in light and heavy chain DNA, antibody diversity; Major histocompatibility complex (MHC)- classes, structure of MHC molecules, Function of MHC molecules, gene regulation; Antigen-antibody reactions – Precipitation reaction; Immunodiffusion test; Counter current immunoelectrophoresis test;

Agglutination reaction (Direct agglutination test; Indirect agglutination test; Haemagglutination; Opsonization; Complement fixation reaction); Neutralization reaction (Diagnosis of viral infection, Schick test); Radioimmunoassay (RIA); Enzyme-linked immunosorbent assay (ELISA): indirect ELISA and double antibody sandwich ELISA; Fluorescent antibody (FA) technique.

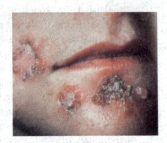

Epidemiology, Parasitism and Antimicrobial; Chemotherapy; Epidemiology; Terminology; Frequency of diseases; Characteristics of infectious diseases; Herd Immunity; Disease cycle (sources of diseases, reservoirs, carriers); Transmission of pathogens (air-borne transmission, contact transmission, Vehicle transmission, vector-borne transmission); Control of infectious diseases (breaking the links in disease cycle, elimination of sources of infection, immunization); Parasitism : Infection (attachment and colonization of pathogens, entry of pathogens, growth and multiplication of pathogens); Intoxications (exotoxins and endotoxins, and their mechanism of action); Antimicrobial chemotherapy : History of development of antibiotics; General features of antimicrobial drugs; Mechanism of action of antimicrobial drugs (inhibition of synthesis of bacterial cell wall, disruption of cell wall, alteration in membrane function, inhibition in synthesis of proteins, purines and pyrimidines and respiratory processes, antagonism of metabolic pathways); Drug resistance –origin of resistance.

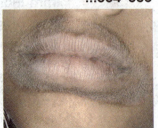

Protozoan disease : *Taxoplasma gondii*; *Plasmodium* (Life cycle, chemotherapy, malarial vaccines, malaria eradication movement); *Balantidium coli*; *Trichomonas vaginalis*; *Giardia*; *Trypanosoma*; *Entamoeba histolytica*. Fungal diseases : Mycoses; Mycotoxicoses (mycetismus); Epidemiology, Clinical types, Culture characteristics and Therapy of the Phycomycosis, Candidiasis, Actinomycosis, Dermatophytosis, Aspergillosis, Otomycosis and Penicillinosis.

Bacterial diseases : Air-borne diseases (Tuberculosis : immunity, diagnosis; Diphtheria; Meningitis; Pertussis; Streptococcal pneumonia); Food-borne and water-borne diseases (Cholera; Botulism; Shigellosis; Typhoid fever); Soil-borne diseases (Tetanus : immunization and control measures; Anthrax); Sexually transmitted and contact diseases (Gonorrhea; Syphilis : primary, secondary and tertiary; Leprosy : tuberculoid, lymphomatous); Viral diseases: Air-borne viral diseases (Influenza; Measles; Mumps; Rubella; Small pox); Insect-borne Diseases (Yellow fever; Dengue fever); Food and water-borne diseases (Polio); Direct contact diseases (Viral hepatides : hepatitis B; Rabies; Cold sores; AIDS transmission, AIDS in India, prevention and control of AIDS).

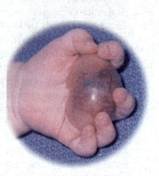

Based on carbon sources (autotrophs and heterotrophs); Based on temperature (psychrophiles, mesophiles, thermophiles, hyperthermophiles and super-hyperthermophiles); Based on habitat (soil microorganisms, aquatic microorganisms, air microorganisms); The extremophiles (acidophiles, alkaliphiles, halophiles); On the basis of nutrition (saprophytism, parasitism and symbiosis); Microbial interactions : Clay-humus-microbe interaction; Plant-microbe interaction (Interaction of above ground parts : destructive and beneficial associations; Interactions on below ground part 1. destructive association, 2. beneficial association (*i*) cyanobacterial, (*ii*) bacterial (associative symbiont, PGPR, *Rhizobium*), (*iii*) actinomycetes, and (*iv*) fungal symbiosis (mycorrhiza , Types : ectomycorrhiza, ectendomycorrhiza, VAM, ericoid mycorrhiza, arbutoid mycorrhiza, monotropoid mycorrhiza, orchid mycorrhiza), Effects of mycorrhizal fungi on their hosts, Works done on mycorrhiza in India); Animal-microbe interactions: Destructive association; Neutralism - normal microbiota of human body; Symbiotic association (ectosymbiosis of protozoa, bacteria and fungi with insects and birds, endosymbiosis of bacteria and fungi with birds and insects, ruminant symbiosis); Microbe-microbe interactions (Symbiosis between algae and fungi : lichens); Antagonistic interactions : amensalism, competition, and parasitism and predation (mycoparasitism, mycophagy, nematophagy : predaceous fungi/nematophagous fungi – nematode-trapping fungi, endoparasites, egg parasites).

Acidophiles- physiology, molecular adaptation, applications; Alkalophiles- physiology, molecular adaptation, applications (bacteriorhodopsin, physiology, polysaccharides, microbially enhanced oil recovery, cancer detection, drug screening, liposomes, enzymes, bioremediation, gas vacuoles, other products): Psychrophiles-physiology, molecular adaptations, applications (source of pharmaceuticals, bacterial ice nucleating agents, fermentation in industry, in microbial leaching, in bioremediation, denitrification of drinking water sources, anaerobic digestion of organic wasters); Thermophiles and hyperthermophiles- physiology molecular adaptations, applications (enzyme, chaperons); Barophiles- physiology, molecular adaptation, applications.

Soil as a habitat for microorganisms – Soil quality; Physico-chemical properties of soil (Organic matter; Soil water and air; Soil microbes: algae, bacteria, actinomycetes, bacteriophages, protozoa, nematode and fungi); Is habitat a better term for microorganisms; Microbial balance; Rhizosphere and rhizoplane microorganisms : reasons for increased microbial activity in rhizosphere; Composition of root exudates; Factors affecting exudation; Rhizosphere microorganisms : the rhizosphere effect; Rhizosphere engineering; Effects of microflora on host plants; Factors affecting microbial community in soil - soil moisture, organic and inorganic chemicals, soil organic matter, types of vegetation and its growth stages, different seasons; Organic matter decomposition : Composition of litter (cellulose, hemicellulose,

lignin, water soluble components, ether- and alcohol-soluble components , and proteins); Carbon assimilation and immobilization; Organic matter dynamics in soil; Microorganisms associated with organic matter decomposition (cellulose decomposers, hemicellulose decomposers, lignin decomposers); Factors affecting organic matter decomposition (litter quality, temperature, aeration, soil pH, inorganic chemicals, moisture); Microbial biomass as an index of soil fertility, Soil fertility Biogeochemical cycling : Carbon cycling; Nitrogen cycling (nitrogen fixation; ammonification : microbiology; nitrification-microbiology; denitrification-microbiology); Phosphorus cycling; Sulphur cycling.

Types of water : Atmospheric water; Surface water; Stored water (sedimentation, interaction of other microbes, light rays, temperature, food supply); Ground water; Water microorganisms : Marine microbiology, Fresh water mirobiology; Microbial analysis of water: Sanitary tests for coliforms (presumptive test, confirmed test, completed test); The MPN of coliforms (the membrane filter technique : advantages and disadvantages; defined substrate test; IMViC test). Purification of water : Sedimentation; Filtration (slow sand filtration, rapid sand filtration); Disinfection.

Early concept of air; Vedic technology for air purification; Works on aeromicrobiology in India (Fungi : occurrence and epidemiology, aflatoxin by aerofungi, seasonal occurrence of aflatoxin producing aerofungi; Algae); Indoor aeromicrobiology : aeromicroflora of pharmacy, aeromicroflora of hospitals and other houses, aeromicroflora of storage materials (library, wall paintings); Aeroallergens and aeroallergy; House dust allergens, Pollen grains, Cosmetics; Phylloplane microflora Phylloplane pathogens, microflora of floral parts, Characteristics of phylloplane microflora (morphological characters, physiological characters : nutrition, radiation, RH, temperature), Microbial interactions on leaf surfaces.

Waste as a resource : Organic compost (Definition, Process of composting, Factors affecting the composting - microorganisms, soil and organic matter, Role of compost), Vermicomposting - process of vermicomposting Biogas; (Benefits from biogas plants, Feed stock materials, Biogas production : solubilization, acetogenesis and methanogenesis), Mechanism of methane formation, Factors affecting methane formation; Sewage (wastewater) treatment : Sewage microorganisms, BOD and COD; Small scale sewage treatment (Cesspools, septic tanks), Large scale sewage treatment (primary treatment, secondary treatment (oxidation pond, trickling filter, the activated sludge, anaerobic digesters- fermentation, acetogenic reactions, methanogenesis), tertiary treatment; Microbial leaching : Copper leaching, Uranium leaching; Biodegradation : Biodegradation of petroleum, microbial degradation of xenobiotics : characteristics of microbial metabolism (enzymatic process, non-enzymatic process), common process of insecticidal metabolism (hydrolytic process, reductive process, oxidation); Microorganisms in abatement of heavy metal pollution : Heavy metal tolerance in microbes (algae, bacteria, actinomycetes, fungi, higher plants, cyanobacteria), mechanism of heavy metal resistance; Water pollution management; Use of commercial blends of microorganisms/enzymes in wastewater treatment-bioaugmentation, use of enzymes in wastewater treatment; Use of immobilised cells in wastewater treatment. Role of microorganisms in metal removal-mechanism of metal removal; Application of recombinant DNA technology in waste treatment; Biofiltration, Biofilters, microorganisms, biofilters media, mechanism of biofiltration; Biodeterioration-biodeterioration of stored plant food material, biodeterioration of leather, biodeterioration of store and building materials, biodeterioration of paper and other cellulosic materials, biodeterioration of fuel and lubricants, biodeterioration of metals, biodeterioration of plastics, biodeterioration of cosmetics and pharmaceutical products, safe practices. Microbial plastics.

Microbial inoculants, microbial consortium or mixed inoculants, Crop production : Production of bacterial biofertilizer, Production of biofertilizer: Criteria for strain selection, steps for preparing biofertilizers (seed pelleting, inoculant carriers, quality standard for inoculants), Green manuring; Algae and other biofertilizers : Mass cultivation of cyanobacteria biofertilizer, mass cultivation of *Azolla*; Endophytic nitrogen fixers, – facultative endophytic diazotrophs, obligate endophytic diazotrophs, other bacteria (isolation and identification of endophytes, applications in agriculture), Biofertilizers aiding phosphorus nutrients; Production of mycorrhizal biofertilizers : ectomycorrhizal fungi and VAM fungi; Crop protection : Microbial herbicides; Bacterial insecticides : *Pseudomonas* as a bacterial insecticide, *Bacillus* species as bacterial insecticide (*B. thuringiensis*, toxins produced by *B. thuringiensis*); Virus insecticide; Entomopathogenic fungi : *Metarhizium anisopliae, Verticillium lecanii, Hirsutella thompsonii, Nomuraea rileyi,* and other fungi; Microbiology of lignocellulose degradation in the rumen, use of recombinant DNA technology in rumen bacteria, rumen microflora.

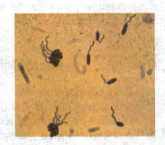

Microscopy : beginning of microscopy, types of microscope (light or compound microscope : parts of compounds microscope, Bright-field microscopy, Dark field microscopy, Phase contrast microscopy, Fluorescent microscopy, Electron microscopy, Transmission electron microscopy, Scanning electron microscopy); Colorimetry - Beer-Lamberts' Law and its limitations, Photoelectric calorimeter; Spectrophotometry : absorption spectra, absorption spectra and extinction coefficient, absorption spectra of different compounds, some practical points; Autoradiography, tracer technique; Chromatography : absorption column chromatography, thin layer chromatography, gas chromatography; Centrifugation (Centrifuge, Zonal centrifugation, Density gradient centrifugation, Differential centrifugation); Electrophoresis and electrofocussing; Hot air oven; Autoclave; Laminar air flow; Incubator.

Introduction

1

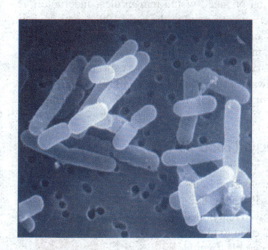

A. History of Microbiology

In the poem, De rerum nature, Lucretious (967-55B.C.) mentioned the existence of "seeds" of disease. India can also take pride in contributing to the development of ancient microbiology in the form of septic tanks in Mohenjodaro and Harappa regions (3000 B.C.). But the existence of microbes was not established until Antony van Leeuwenhoek (1677) could see them in simple (one-lens) microscope. Leeuwenhoek, a cloth merchant in Delft, Holland spent much of his time in grinding tiny lenses of high magnification (300 x or so). He took the scurf from the root of decayed tooth and mixed it with clean rain water, and saw the mobility in animalcules. He discovered major classes of bacteria (spheres, rods, and spirals), protozoa, algae, yeasts, erythrocytes, spermatozoa, and the capillary circulation. Leeuwenhoek's discoveries were described in a flow of letters to the Royal Society of London.

Leeuwenhoek's Microscope.

Antony van Leeuwenhoek
(1632—1723)

Aristotle (384-322 B.C.) emphasized that animals might evolve spontaneously from the soil, plants or other unlike animals. Virgil (70-19 B.C.) also gave opinion for the artificial propogation of bees. Discoveries about spontaneous generation persisted till 17th century.

1.   Spontaneous Generation of Organisms

The theory of spontaneous generation states that the microbes arise automatically in decomposing organic matter. In the 17th century Francesco Redi worked out the appearence of maggots in decomposing meat depended on the decomposition of eggs by flies, but the idea of spontaneous generation persisted for the new world of microbes. T. Needham in 1748 experienced the appearance of organisms not present previously and concluded that these organisms appeared from the decomposition of the vegetables and meat. Later Spallanzani (1729-1799) introduced the use of sterile culture media; he showed that infusion of meat, would remain clear indefinitely if boiled and properly sealed. This discovery was later on confirmed in the early 19th century, when a French confectioner, in the Appert, competing for a prize by Napolean, developed the art of preserving food by canning. In 1837, Schwann obtained similar results even when air was allowed to reenter the cooling flask before sealing. Preliminary reports on experiments concerning alcoholic fermentation and putrifection were given by Schwann (1837) and Liebig (1839). To give more weightage, Schroder and von Dusch applied the use of cotton plug, to exclude air borne contaminants which is still in use.

2.   Golden Era of Microbiology (1860-1910)

Louis Pasteur

Golden era of microbiology started with the work of Louis Pasteur (France) and Robert Koch (Germany). Louis Pasteur (1822-1895) investigated number of aspects such as he showed that boiled medium could remain clear in a "swan-neck" flask, open to the air through a sinuous horizontal tube in which dust particles would settle as air reentered the cooling vessel (Fig. 1.1). Pasteur also demonstrated that in the relatively dust free atmosphere of a quiet cellar, or of a mountain top, sealed flasks could be opened and then resealed so as to avoid contamination. Pasteur delivered a lecture in 1864 at Sorbonne and made sensation by discovering life is a germ and germ is a life. F. Cohn (1876) studied the biology of the bacili. John Tyndall (1820-1893) showed that the hay had contaminated his lab with an incredible kind of living organism. Ferdinand John (1877) demonstrated the resistant forms as small, refractile endospores, a special stage in the life cycle of hay bacillus (*Bacillus subtilis*). Since spores are readily sterilized in the presence of moisture at 120°C, the autoclave, which uses steam under pressure, became hallmark of the bacteriology.

Pasteur (1857) became interested in fermentation products and observed different kind of microbes associated with different kind of fermentation: spheres of variable size (now known as yeast cells) in the alcoholic fermentation, and smaller rods (lactobacilli) in the lactic fermentation. During this

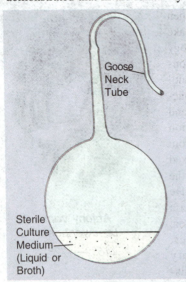

Goose Neck Tube

Sterile Culture Medium (Liquid or Broth)

Fig1.1 : Goose neck flask.

experiment, Pasteur established the study of microbial metabolism and in particular he showed that life is possible without air. Pasteur explained that in grape juice the high sugar concentration and the low protein content (i.e. low buffering power) lead to a low pH, which allows the outgrowth of acid-resistant yeasts and thus yields an alcoholic fermentation. In milk, in contrast, the much higher protein and lower sugar content favour the outgrowth of fast growing but more acid-sensitive bacteria, which cause a lactic fermentation. This finding led Pasteur to state that specific microbes might also be causes of specific disease in man.

Pasteur developed the procedure of gentle heating (i.e. pasteurization) to prevent the spoilage of beer and wine by undesired microbes. This process was later used to prevent milk borne diseases of man. Of the great economic importance was the extension of industrial fermentations from the production of foods and beverages to that of valuable chemicals, such as glycerol, acetone, and later vitamins, antibiotics and alkaloids.

The unity of biology at a molecular level concept was developed when it was discovered that the carbohydrate metabolism pathways are similar in some microbes and in mammals. This discovery was made towards the end of the Pasteurian era notably by Winogradsky in Russia and Beijerinck in Holland who discovered variety of metabolic patterns by different kinds of bacteria adapted to different ecological niches. The ecological niche is defined as 'the physical space occupied by an organism, but also its functional role in the community'. These organisms were isolated by using Pasteur's principle of selective cultivation: enrichment culture in which only a particular energy source is provided, and growth is restricted to those organisms that can use that source.

3. Germ Theory of Disease

The 'germ theory of disease' has presented a great stimulus in Microbiology and Medicine. Louis Pasteur and Robert Koch (1843-1910) were the national heroes. Preventive measures also supported the germ theory. Edward Jenner (1796) introduced vaccination (L. *vacce, cow*) against small pox, using material from lesions of a similar disease of cattle (cowpox). In 1860s Joseph Lister introduced antiseptic surgery, on the basis of Pasteur's evidence for the ubiquity of airbone microbes.

Robert Koch

Recognition of agents of infection first to be recognised were fungi: Agostinod Bassi (1836) demostrated that a fungus was the cause of disease (of silk worm), the etiologic role of bacteria was established by Koch (1876) for anthrax. The pure culture preparation is the key to the identification. Koch perfected the technique of identification including the use of solid media and the use of stain. After identifying the tubercle bacillus Koch formalized the criteria, introduced by Henle in 1840 but known as Koch's postulates, for distinguishing a pathogenic from an adventitious microbe:

1. The organism is regularly found in the lesion of the disease.
2. It can be isolated in pure culture.
3. Inoculation of this culture produces a similar disease in experiments on animals.

These criteria have proceeded invaluable in identifying pathogens, but they cannot be met: some organism such as viruses cannot grow on artificial media and some are pathogenic only for man.

Golden era of microbiology was established between 1860 and 1910 because of development of powerful methodology. Moreover, various members of the German school isolated (in addition to the tubercle bacillus), the Cholera *Vibrio*, Typhoid *Bacillus, Diphtheria, Bacillus, Pneumococcus, Staphylococcus, Streptococcus, Meningococcus, Gonococcus* and Tetanus bacillus.

4.   Viral Diseases and Immunization

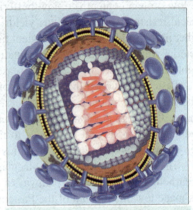

The discovery of viruses and their role in disease was made possible when Charles Chamberland (1851-1908), one of Pasteur's associates constructed a porcelain bacteria filter in 1884. The first virus to be recognized as filterable was tobacco mosaic virus, discovered by Russian, named lvanovski (1882) and by Bejerinck (1899) in Holland. On the otherhand, filterable animal viruses were discovered for foot and mouth disease of cattle by Loffler and Frosch (1898), and for a human disease, yellow fever by the US army commission. Twort in England and d' Herelle in France in 1916-1917 discovered viruses that infect bacteria i.e. bacteriophages. The first crystallization of virus was made by Stanley (1935).

Structure of HIV Virus.

After this discovery, it was a matter of great surprise to the scientist that how animals resisted disease. Pasteur observed that old cultures of the bacterium attenuated i.e. lost their disease causing ability, they remain healthy but developed the ability to resist disease. He called the attenuated culture as **Vaccine** in respect of Edward Jenner because, several years back Jenner had used vaccination with material from cowpox lesion to protects people against smallpox. Emil Von Boehring (1854-1917) and Shibasaburo Kitasato (1852-1931) used inactivated toxin into rabbit, inducing them for antitoxin production. This is how a tetanus antitoxin was prepared and now used in the treatment.

5.   Microbiology in 20th Century

The discovery of microbial effects on organic and inorganic matter started with the discovery of Theodore Schwann and others (1937) who observed that yeast cells are able to convert sugar to alcohol i.e. alcoholic fermentation. It was Pasteur's observations that revealed about anaerobic and aerobic microorganisms. Role of microorganisms in the carbon, nitrogen and sulphur cycles in soil and aquatic habitats were discussed by Sergei N.Winogradsky (1956-1953) and Martinus Beijerinck (1851-1931), The Russian microbiologist Winogradsky also discovered that (*i*) soil bacteria oxidize Iron, Sulphur and Ammonia to obtain energy, (*ii*) isolated anaerobic N_2 fixers and (*iii*) studied the decomposition of cellulosic organic matter. On the other hand, Beijerinck, contributed a lot in the area of microbial ecology. *Azotobacter*, a free living nitrogen fixer was isolated. Later a root nodulating bacterium named as *Rhizobium* and sulphate reducers were also isolated. Both these microbiologists developed the enrichment culture techniques and the use of selective media in the microbiology.

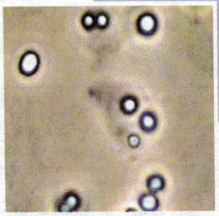

In 20th century, microbiology developed from the angle of other disciplines of biological sciences in such a way so that problems of cell structure to the evolution are solved. Although, more emphasis were laid down on the agents of infectious disease, the immune response, chemotherapeutic agents and bacterial metabolism.

Azotobacter is a free living nitrogen fixer.

Beadle and Tautam (1941) used mutants of the bread mold, *Neurospora* while Salvadore Luria and Max Delbruck (1943) used bacterial mutants to show that gene mutations were truly spontaneous and not directed by the environment. Avery, Macleod, and Mc Carty (1944) evidenced that DNA was the genetic material carried genetic information. Such discoveries made microbiology, genetics and biochemistry as modern molecularly oriented genetics. Microbiology contributed maximum in molecular biology which deals with the physical and chemical aspects of living matter and its function. The genetic code and the mechanism of DNA, RNA and protein synthesis were also studied by using several microorganisms. Regulation of gene expression and the control of enzymes activity were also discussed in the light of microbiology. In 1970's new discovery such as recombinant DNA technology and genetic engineering were also led to development of microbiology which gave the service of microbial biotechnology.

Scientists of West Cjester University, Pennsylvania have revived a microbe that had been in suspended animation for 250 million years, a remarkable feat which boosts theories that the ancient seeds for life arrived on Earth from space. Russell Vreeland (2003) isolated a spore forming *Bacillus* sp. from 250 years old sample of salt crystal found below ground (1850 ft.) in New Mexico. The bacterium seems to be similar to *Bacillus marismortui*. Earlier, there were reports of oldest living creatures of 254-40 million years.

The importance of this branch lies due to the fact that about 30% of the total Nobel prizes given in the physiology and medicine are awarded to those working on problems related to microbiology as shown in the Table 1.1.

Table: 1.1 : Nobel prizes awarded in the subject related in Microbiology Research (1945 onwards)

Name of Scientist	Area of research	Year
A. Fleming E.B.Chain H.W.Florey	Discovery of Penicillin & its therapeutic value	1945
M.Theiler	Development of vaccine against yellow fever	1951
S.A.Waksman	Discovery of streptomycin	1952
J.F.Enders T.H.Weller F.Robbins	Cultivations of poliovirus in tissue culture	1954
D.Bovet	Discovery of the first antihistamine	1957
G.W.Beadle E.I.Tatum J.Lederberg	Microbial genetics	1958
S.Ochoa A.Kornberg	Discovery of enzyme catalyzing nucleic acid synthesis	1959
F.H.C.Crick J.D.Watson M.Wilkins	Discoveries related to DNA	1962
F.Jacob A.Lwoft J.Monod	Discoveries about the regulation of genes	1965
F.P.Rous	Discovery of cancer virus	1966

Name of Scientist	Area of research	Year
R.W.Holley H.G.Khorana M.W.Nirenberg	Deciphering of the genetic code	1968
M.Delbruck A.D.Hershey S.E.Luria	Discoveries concerning viruses & viral infection of cells	1969
G.Edelman R.Porter	Research on the structure of antibodies	1972
H.Temin D.Baltimore R.Dulbecco	Discovery of RNA dependent DNA synthesis by RNA tumour viruses; reproduction of DNA tumour virus	1975
B.Blumberg D.C.Gajdusek	Mechanism and dissemination of hepatitis B virus; research on slow virus infection.	1976
R.Yalow	Development of the radioimmuno assay technique	1977
H.O.Smith D.Nathans W.Arber	Discovery of restriction enzymes and their application to the problem of molecular genetices	1978 1981
B.Benaclavaf G.Snell J.Dausset	Discovery of the histo- compatibilty antigens	1980
P.Berg W.Gilberg F.Sanger	Development of DNA technique (Berg); specially of DNA sequencing techniques	1981
A.Klug	Development of crystallogra- phics electron microscopy	1982
C.Milstein G.J.F.Kohler N.K.Jerne	Development of the technique for formation of monoclonal antibodies; theoretical work in immunology.	1984
E.Ruska	Development of the transmission electron microscope.	1986
S.Tonegawa	The genetic principle for generation of antibody diversity	1987
J.Deisenhofer R.Huber and H.Nichel	Crystallization and study of the photosynthetic reaction center from a bacterial membrane.	1988
J.M.Bishop H.E.Varmus	Discovery of oncogenes	1989
S.Altman T.R.Cech	Discovery of ribozyme	1989

Name of Scientist	Area of research	Year
K.B.Mullis M.Smith	For discovery of PCR technique and development of site directed mutagenesis	1993
E. Lewis C. Nusslein E. Wieschans	Physiology of Genetics of microbes	1995
S.B.Prussiner	Discovery of prions	1997

Noteworthy Events in the Development of Industrial Microbiology

Year	Events
1857	L.Pasteur showed that Lactic acid formation is due to microorganisms.
1881	L.Pasteur developed anthrax vaccine.
1885	Gave rise rabies vaccine.
1887	Buchner discovered that yeast extract ferment sugar.
1921	Fleming discovered lysozyme.
1923	First edition of Bergey's manual.
1929	Fleming discovered penicillin.
1933	Ruska developed electron microscope.
1935	Domagk discovered sulfa drugs.
1937	Chatton divided living organisms into prokaryotes and eukoryotes.
1941	Beadle and Tatum gave one gene one enzyme theory.
1944	Waksman discovered streptomycin.
1982	Recombinant hepatitis B vaccine developed.
1986	First vaccine (hepatitis B vaccine) produced by genetic engineering approved for human use.

Industrial microbiology involves the use of microorganism to produce organic chemicals, antibiotics and other pharmaceuticals and supplements. Microorganisms are also involved in insects and pests control, for the recovery of metals, and for the improvement and maintenance of environmental quality. In the beginning selection and mutation were the major means of improving cultures for use in industrial microbiology.

The earlier development of industrial microbiology began after the primary method of selecting and improving microbes which Pontecarvo has never claimed but described as a "Prehistoric" technique, mutation and screening for selection from the available gene pool. Once a suitable microbe is available, the microbiologists are able to manipulate in such a way so as to give products. Even the use of microorganisms in modern biotechnology is still depends on the principles of their mass culture developed over decades by industrial microbiologists.

Ninth edition of Bergey's Manual of Determinative Bacteriology lists only 754 named genera, the diversity of habitats makes them difficult above to non-availabiltiy of single identification scheme for the whole microbiota.

For a species level identification, it is to be considered that all identification have certain elements in common. Such data is to be compared with defined taxa, and on the basis of certain characters, allocation of the unknown to a defined taxon is drawn and requires a full taxonomic study.

B. Microbiology In India

There are number of institutes engaged in microbiological research in our country. The Indian Institutes of Petroleum, Dehradun; Tata Energy Research Institute, Delhi and National Chemical Laboratory, Pune have worked on microbial dewaxing of heavier petroleum fractions. The institutes has also played a vital role on the area of microbial enhanced oil recovery and production of biosurfactants.

National Institute of Nutrition, Hyderabad, ITRC, Lucknow, and National Institute of Occupational Health, Ahmedabad have already completed a long time plan on monitoring and surveillance of food contaminants hazards in India while genome analysis and synthetic gene design for modulation of genome expression *in vivo* was carried out by the scientists of Indian Institute of Science, Bangalore.

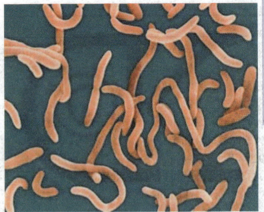

Vibrio cholerae, a curved shape bacillus causes cholera in human beings.

Anaerobic recycling of lignocellulosic wastes to fuels and feedstock chemicals have been successfully carried out at Indian Institute of Technology, Delhi and Madras. Molecular characterization of naturally occurring allergens for identification of immunopotential moieties have been worked at Indian Association for Cultivation of Science, Calcutta. The molecular biology of human enteric pathogens, bacteriophage typing techniques for identifying cholera infection, construction of physical and genetic map of *Vibrio cholerae* 569B, oral cholera vaccine besides establishment of a *Leichmania* parasite bank have been achieved at Indian Institute of Chemical Biology, Calcutta besides worked on structural anatomy of *Leishmania donovani* enzyme adenosine which has a structural role. A group at Central Salt and Marine Chemicals Research Institute, Bhavnagar has observed certain economically important plants with special reference to the marine algal flora occurring in the mangrove forests. Technology for bacterial grade agar agar, biotoxins from marine algae and bacteria. Manipulated strains, enzymatic conversions of rifomycinB to rifamysinS, development of cholera vaccine are major achievements of Institute of Microbial Technology, Chandigarh. The basic biochemical engineering studies on development of stable plasmid vectors in *Corynebacteria* and fusion hybrids between *Praerthes* and *Corynebacter* carried out at Indian Institute of Technology, New Delhi. The industrial waste treatment by anaerobic digestion with special reference to ultra structure of granular sludge and the cell immobilisation was studied at regional sophisticated instrumentation centre, I.I.T., Bombay.

Microbial degradation of polycyclic aromatic hydrocarbons with reference to construction of genetically mainpulated strain was completed at Institute of Microbial Technology, Chandigarh, Similar to that recombinant DNA application to anaerobic fixed film systems for methane biosynthesis construction of genetically engineered strains for microbial desulfurisation of petroleum crude, production of surfactants and bioplastics, distillates of cow urine for antimicrobial therapy (US patent) are few achievements of scientists at National Environmental Engineering Research Institute (NEERI), Nagpur.

National facility for disease control in fishes through quarantine and health certification was carried out at Central Institute of fresh water Aquaculture, Bhubneshwar. Monitoring and surveillance of food contaminants hazards in India, and production of various enzymes such as xylanase and B-amylase was developed at Bose Institute, Calcutta.

Microbial xylanases at semipilot scale fermentation and down stream processing and xylose metabolism in *Neurospora crassa*, ethanol biotechnology and yeast strain improvements were carried out at National Chemical Laboratory, Pune. Solid state fermentation for production of glucoamylase was studied at Regonial Research laboratory, Thiruvanathapuram (Kerala), while application of natural and recombinant microorganisms to bio-surfactant production, oil spill degradation and pollution control was studied at C.S.I.R. laboratories. Regional Research Laboratory, Jammu has designed technologies for the production of free gluconic acid, construction of genetically engineered strains for microbial desulfurization of petroleum crude, production of surfactants and biopolymers etc.

Biotechnology on mass production of natural enemies of *Spodoptera litura* was studied at Central Tobacco Research Institute, Rajamundry (A.P.), while the genetics of *Thiobacillus ferroxidans* and strain improvement by advanced genetical technique was explored at Bose Institute, Calcutta. The removal of heavy metal ions from industrial waste by microorganisms was carried out at Regional Research Laboratory, Bhubaneshwar. The improvement of strain efficiency, quality and mass production technology for heterotrophic microbial inoculants studied at SPIC Science foundation, Madras. The protein engineering and Biophysical and chemical approaches to the study of protein folding was carried out at TIFR, Bombay. The germplasm collection, quality improvement by genetic engineering, and down stream processing of *Spirulina* and its biotechnology was developed under an All India Coordinated Project involving several institutes including CFTRI. Mysore. The molecular and genetic approaches for the analysis of pre-mRNA splicing in yeast was the major contribution of Indian Institute of Science, Bangalore.

C. Applications of Microbiology

The microorganisms influence the man in several ways. They are ubiquitous in our environment i.e. they are found in the soil, mud, water, air, in animals, plants, food products, dead wood, cloths, jams and shoes, optical instruments, nails, skin even in space and at antarctica. Literature survey reveal that biosphere contains a variety of microorganisms that proliferate in extreme environments. The diversity of their activities varies from causing diseases in human and other animals and plants to the production of various useful products, recovery of metals, increasing in soil fertility and the deterioration of aeroplanes. The modern development of sanitation and public health has resulted in the reduction of the incidence of many diseases. The use of mineral water contributed a safer environment for man.

Making red wine: yeast produces alcohol from the sugar in grapes.

The major fields of applied microbiology are described below:

1.   Microbes in Food and Dairy Industries

(*i*) Molds : Food microbiology not only includes the study of those microbes which provide food due to their high protein value (such as yeast), but on the otherhand, those microbes also which use our food supply as a source of nutient for their growth and result in deterioration of the food by increasing their numbers, utilizing nutrients, producing enzymatic changes, and contributing off flavours by means of break down products. Microorganisms, such as molds (*Mucor, Rhizopus, Botrytis, Aspergillus, Penicillium* etc.) lead to deterioration of food. Special molds are useful in the manufacture of

Bread Mold causes food spoilage.

certain foods or ingredients of foods. Some cheese are mold ripened e.q. blue, Roquefort, camembert etc, molds are also used in production of oriental foods, e.g. soy sauce, miso, sonti, etc, used as food or feed and are involved in making enzymes such as amylase for bread making or citric acid used is soft drinks. Some molds are harmful (*Aspergillus flavus*) and some molds (*A. parasiticus*) produce toxic metabolites (mycotoxins).

(*ii*) **Yeasts :** Yeast refers to those fungi which are generally not filamentous but unicellular and void or spherical reproduce by budding or fission, and may be useful or harmful in food. Yeast fermentations are involved in the manufacture of foods such as bread, beer, wines, vinegar and surface ripened cheese and yeasts are grown for enzymes and for food. Yeasts are undesirable when spoil fruit juices, syrups, molasses, jam pickles wine, beer and other foods. Example of some of the genera are *Saccharomyces*,

Yeasts.

*Schizosaccharomyces, Candida, Kluyveromyces, Zygosaccharomyces, Pichia, Hansenula, Debaromyces, Hanseniaspora,*etc.

(*iii*) **Bacteria :** Bacteria in a food may be of special significance. Pigmented bacteria cause changes in colour on the surfaces of foods, form film over the surfaces of liquid food, etc. which result in undesirable cloudiness or sediment. Some genera, such as *Acetobacter* oxidises ethyl alcohol to acetic acid, *Aeromonas*, a facultative anaerobe also pathogenic not only to human beings but to fish, frogs and other mammals. *Alcaligenes* as the name indicates produce an alkaline reaction in the medium for growth,

Cheese manufacture.

causes ropiness in milk, gives slimy growth on cottage cheese. *Bacillus coagulans,*a proteolytic species, curdle milk. *Bacillus purimilus* is recommended as test organism in sterility testing. *B. stearothermophilus* is used for testing procedures involving steam sterilization. *B. subtilis* var. *niger* (ATCC 9372) is recommended for ethylene oxide sterilization testing. *Brochothrix* can spoil a wide variety of meats and their products. The genus *Microbacterium lacticum* used in production of vitamins while *Micrococcus luteus* and *M. rosens* help in meat-curing brines.

Lactobacillus viridescens causes greening of sausage. *Proteus* is involved in the spoilage of meats, seafood, and eggs. Its presence in food suspect as a cause of food poisoning. *Pseudomonas* can cause food spoilage. The greenish fluorescence is developed due to pyoverdin formation and white, cream-coloured reddish, brown, or even black colours are formed due to *P. nigrifaciens*. Different species such as *C. thermosaccharolyticum,* saccharolytic obligate thermophile, causes gaseous spoilage of canned vegetables. Putrefaction i.e. deterioration of food under anaerobic condition takes place due to proteolytic action of *C. putrefaciens* while *C. perfrigens* causes violent disruption of the curd in milk. *Desulfotmaculum* is responsible for sulfide stinker spoilage in canned food. *Erwinia carotovora* is associated with the bacterial rot in vegetables (carrot, etc.). *Flavobac-*

terium involved in the spoilage of shellfish, poultry, eggs, butter and milk, while *Halobacterium* causes discolouration in salted fish. *Lactobacillus* ferments sugars with the production of lactic acid, but also resulting in the deterioration of wine or bear. They synthesize most of the vitamins they require. *Leuconostoc cremoris*, ferments citric acid of milk and also stimulate lactic streptococci so called 'lactic starter' for butter milk, butter and cheese. *L. mesenteroides* produces high sugar concentration in syrups, ice-cream mixes, etc. which is a desirable characteristic.

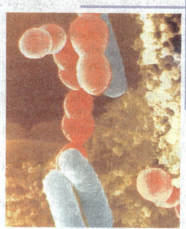

Bacteria of yoghurt 'starter culture'.

Pediococcus is salt tolerant acid producer and psychrophilic in nature. Pediococci have been found growing during the fermentation of brined vegetables and have been found responsible for the spoilage of alcoholic beverages, e.g. beer. *Photobacterium* causes phosphorescence of meats and fish, while *Propionibacterium freudeureichii* ferments the lactates to produce the gas that helps, forms the holes or eyes and also contributes flavours in cheese.

Salmonella, enteric pathogens, may grow in foods and cause food infections. Usually they are only transported by foods. *Staphylococcus aureus* gives yellow to orange growth and some produce an enterotoxin which causes food poisoning. Streptococci are homofermentative; *S. thermophilus*, a coccus important in cheese made by cooking the curd at high temperature and in certain cases femented milk such as yoghurt, and *S. boris* comes from cow manure and saliva. *S.lacti*, an important dairy bacterium used as starter for cheese, cultured butter milk and some types of butter alongwith *Leuconostoc* spp., *S.lactis* often causes souring of raw milk while *S. faecalis* and *S. faecium* are commonly found in raw foods. *Vibrio cholerae* and *V. vulrificus* both are pathogenic to human being and associated with the sea food. *Shigella sonnei* and *S. dysenteriae* are the bacteria found in macaroni salads, beans, potato etc. and cause bacillary dysentery (*Shigellosis*) in human beings. *Yersinia pestis*, causes food-borne diseases and also is causal organism of plague in humans and in rats (this bacteria caused disease out break in Gujarat in 1994-95).

2. Microbes in the Production of Industrial Products

Enzymes, amino acids, vitamins, antibiotics, organic, acids and alcohols are commercially produced by microorganisms. Some of them are used by microbes during their growth and therefore, are called primary microbial products or primary metabolites such as enzymes, amino acids and vitamins, while some are not used by the cell for their growth such as antibiotics, alcohol and organic acids. These are called secondary microbial products or metabolites. Commercially important microbial enzymes are usually extracellular and marketed in crude form. Lipases are supplemented in detergents such as surf and proteases are also used in detergents and in leather and food industries; pectinases used in clarifying fruit juices and penicillin acylases in the production of semi synthetic penicillin.

Citric acid produced by *Aspergillus niger* is used in food preservation.

Amylases are used in the preparation of starch hydrolysates which in turn is used in various product formation such as beer, vinegar, etc. Various small quantities of other enzymes of microbial origin such as asparaginase is used against leukemia (blood cancer) and lyphomas (cancer) which acts upon high percentage of the amino acid asparagin present in cancer cells.

Microbes produce some important amino acids such as glutamic acid, lysine and methionine. The microbes produce only L-isomers of the amino acids. *Monosodium glutamate* (MSG) imparts flavours, while lysine and methionine use essential amino acids which are obtained in human beings by their diets. On the otherhand, citric acid, a Kreb's cycle (TCA cycle) intermediate is produced mainly by *Aspergillus niger* (Koji) and used in preservation of food by microorganisms and as animal feed additives since they stimulate growth.

3.   Microbes in Genetic Engineering and Biotechnology

Molecular techniques for manipulating genetic information called *Genetic engineering* and the use of such genetically engineered microbes in industrial processes called biotechnology. Such microorganisms now being used in producing mammalian proteins such as insulin and human growth factors, make vaccines from microbial and viral genes, and induce the new strains of microbe. Many proteins of pharmacological importance can be produced by genetically engineered microorganisms. Vaccines and diagnostic kits also depend upon the use of improved strains of microorganisms. Human growth hormones have been cloned in bacteria, which can provide it in sufficient quantities required for clinical trials. Interferons are produced in animal cells if induced by viral infection. These are used in testing interleukins

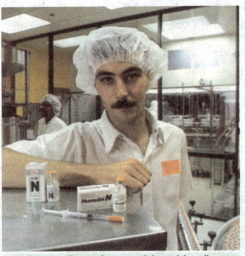

Production of recombinant insulin.

(which stimulate T-lymphocytes). The production of viral, bacterial or protozoan antigens for protecting human against dysentery, typhoid, cholera etc. have been outlined and viral vaccines against flu or influenza, chicken pox and human immunodeficiency are found to be cloned fully.

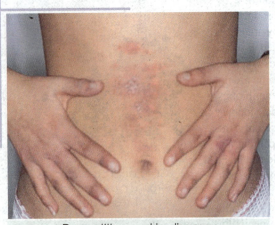

Dermatitis — a skin disease.

The viral vaccines that consist of cloned polypeptide antigens and produced by genetic engineering have been introduced to prevent foot and mouth disease and hepatitis B. The partial success with malaria initiated to switch, over to the genetic engineering of vaccines for controlling Schizstomiasis. The hepatitis B vaccine is the first genetically engineered vaccine approved for human use in U.S.A. It is an effective agent against all viral infection. Flavouring agent of candy, ice cream, fruit/juices and other confectionaries, as a preservative for stored blood, and in ointments and cosmetics microbes are being used. Microorganisms make some of the important vitamins such as vitamin B_{12} which human body cannot make. It is therefore, used to supplement in food besides the use in the pharmaceutical industry. Riboflavin (vit B_2) is also produced microbiologically by the yeast *Ashbya gossypii*. The deficiency of the vit B_2 causes dermatitis (skin disease).

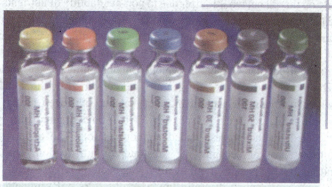

Vaccines and diagnostics Kits.

Lactic acid and acetic acid are commerically significant microbial by-products. Lactic acid is used as food preservative in tanning animal hides and in textile industry while acetic acid is a major component of vineger. On the otherhand ethanol is used as fuel and chemical feed stock.

Antibiotics such as b-lactam (penicillins and cephalosporins) are used in various treatment purposes. Cephalosporins are widely used in the treatment of noscomial Gram-negative sepsis. Penicillins are the good choice for treatment of gonoccocal, treponemal and streptococcal infection. Penicillins generally kill Gram-positive bacteria. Tetracyclines produced by various species of *Streptomyces*. Chlorotetracycline and Oxytetracycline are naturally produced. Transformants producing VP_3 protein were selected and used to make a vaccine that confers protection against foot and mouth disease in animals. In future some more new vaccines would be required for the check of protozoan diseases that affect people.

4. Microbes in Environmental Microbiology

Microbes are distributed everywhere i.e. soil, water and air even they are present deep in side the earth such as deep sea vents. The microbes play an important role in recycling of biological elements such as oxygen, carbon, nitrogen, sulphur and phosphorus.

(*i*) **Microbes in Biogeochemical Cycles :** Oxygen comprises 70 percent of cell's weight and above 23% oxygen is present in atmosphere which is available to all the microbes. On the otherhand, oxygen occurs in mineral deposits as salts of carbonates, silicate, aluminate and other oxides. About 80% of oxygen is unavailable to biological organisms. Cyanobacteria, heterotrophs and chemolithotrophs use it. Water is a product of aerobic processes hence again remains available for photosynthesis.

About 20% percent earth's carbon is an organic compound, and less that 1 % is in fossil fuel such as petroleum, coal, etc. Carbon is present in the form of CO_2 in the atmosphere. CO_2 is produced during the burning of fossil fuel, biological preparations and microbial decomposition.

Photosynthetic and chemosynthetic microgranisms convert CO_2 into organic carbon. Methane (CH_4) is generated anaerobically from CO_2 and H_2 by methanogenic archaeobacteria. Soil bacteria and fungi mainly use organic matter present in soil. Without the cycling action of these microbes

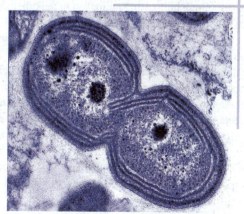

Nitrosomonas is a nitrifying bacteria.

life on this planet would suffer due to non-availabiltiy of nutrients essential for life. Microbes through their enzymatic machinery are able to release soluble products, making them available to microbes and plants. The dead remains of plants and animals are decomposed by microbes.

About 78% of N is present in atmosphere while 9-15 percent of a cell's dry weight contains essential cellular element N which contains amino acids, nucleic acid and in some coenzymes. The inorganic forms of nitrogen are interconverted by the metabolic activities of microorganisms, which maintain the natural nitrogen balance. Nitrogen fixing bacteria reduce nitrogen gas (N_2) to ammonia required for plant growth. Some organic nitrogen (amino acid, nucleic acid) is recycled by the process called ammonification. The ammonia produced is either incorporated into biomass or becomes the substrate for nitrification. The aerobic oxidation of ammonia to nitrate (NO_3^-) is carried out by nitrifying bacteria represented by *Nitrosomonas* and *Nitrosococcus*.

(ii) **Microbes in Pollution Microbiology :** Examination of aquatic environments where eutrophication is actively under way of several fascinating variety of microorganism. Among these would be protozoa, algae and less obvious visually but no less important functionally are non-motile organisms and smaller forms, including fungi, and the bacteria, certain viruses also show their presence if special procedures are adopted for the detection. Viruses that infect all the biological grouping have a bearing on water pollution problems. Plant and algal viruses may prevent the unrestricted development of the host. It has been suggested that excessive growth of blue-green algae might be controlled by seeding with specific viruses (LPP and AS viruses). Similarly, bacterial viruses (bacteriophage) may help in controlling the size of bacterial population through the lysis of susceptible groups. Of special interest, the possible role of bacteriophage is the destruction of bacteria pathogenic to man, and bacterial indicators of faecal pollution. On the other hand, bacteria are especially fit to exploit a wide range of environmental opportunities. Bacteria play an important role in waste decomposition. Some actinomycetes are inhabitant of lake muds, but in general actinomycetes do not seem responsible for the earthy odour so noticeable after ploughing.

Biological sewage treatment and self purification have much in common. Both result in the mineralization of organic pollutants and in the utilisation of dissolved oxygen. Complex microbial associations play an important role in self-purification, and such communities dominate the ecology of treatment plants. Some organic molecules present in industrial effluents are decomposed readily by a variety of microorganisms, while some compounds appear to resist biological attack completely.

5.   Microbes and Medical Microbiology

Microbes cause infections resulting in diseases among human and animals. On the other side, they help in creating a "disease-free world", where people are saved from the pain of being born with physical or mental deformities. The control of our infectious disease has been the greatest achievement of medical science. Vaccination (the use of attenuated or killed microbes in the preparation of vaccines) reduced the incidence of several epidemic diseases (e.g. smallpox, diphtheria, whooping cough, poliomyelitis) but for many organisms vaccination is not effective or feasible. It has been especially important for diseases transmitted by respiratory droplets.

Living cells infected with viruses produce

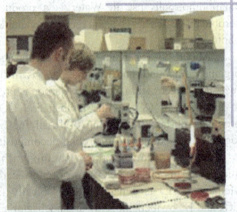

Preparation of vaccines from killed microbes.

viral proteins (glycoproteins) having a broad spectrum antiviral action called interferon. In addition to the antiviral action, interferon has a number of biological effects, and causes inhibition of parasitic infections due to chlamydia, rickettsiae, protozoa and bacteria.

The powerful methodology was developed and adopted by Robert Koch during "Golden Era" of medical bacteriology (1879-1889) which helped in isolation of tubercle bacillus, cholera vibrio typhoid bacillus, diphtheriae bacillus, *Pneumococcus*, *Staphylococus*, *Streptococcus*, *Meningococcus*, *Gonococcus* and tetanus bacillus. Mechanism of pathogenicity of these organisms the host responses, and the methods of prevention and treatment have already been evolved.

6. Microbes and Agriculture

Organic composting, increasing soil fertility, reclaimation of alkaline user land and use of biofertilizer and microbial pesticides in agriculture are some of the important areas in which different groups of microorganisms participate.

Incomplete and artificially composting in its broadest sense means the organic matter by mixed microbial population in moist, warm and aerobic conditions and the preparation and application of organic anaerobic environment. Compost or organic manure in the agricultural fields is a traditional system. But in the traditional system of composting generally the mature crop residues (wheat straw, rice straw, sugarcane bagasse, etc.) are being highly resistant lignified tissues that are difficult to degrade. Thus, breakdown or microbial conversion of these complex organic materials favour nutrient rich compost within a short span. Compost is the store house of major plant nutrients (NPK) as well as micronutrients. Different processes for the recovery of synthetic fuel from various sources of biomass i.e. animal production manures and organic wastes have been developed.

Among these processes, process of biogas production through anaerobic fermentation is most relevant for satisfying the energy needs of rural population. Microbiological conversion of organic materials into methane gas through anaerobic digestion has been quite successful in recent years. In India, the mass popularization of biogas technology in rural areas is being carried out under the ageis of Ministry of Non-conventional Energy Sources.

Compost or organic manure increases soil fertility.

The productivity of leguminous crops largely depends on an efficient and sustainable management of the ecosystems, involving specific rhizobial associations. The symbiotic association of rhizobia (*Rhizobium, Bradyrhizobum, Azorhizobuim, Sinorhizobium and Mesorhizobium*) with legumes and free living nitrogen fixing microorganisms (*Azotobacter, Azospirillum*) and other category (see Chapter 14, *Nitrogen Fixation*) of bacteria offer an advantage in the sense that molecular nitrogen is converted into assimilable form of ammonia thus enabling these crops to grow purely on biological sources of nitrogen. Thus, legume and other cereals are benefitted and such association mimicks the expensive, energy-intensive chemical fertilizer factories.

Bacteria are often associated with human disease and ill health. However, there are several bacteria that are benevolent as well as harmless to human beings. In addition there are certain bacteria which can be of immense benefit to mankind. *Bacillus thuringiensis* (Bt), *B. Papillae, B. sphaericus,*

Xenohrabdus nematophilus, X. fluorescence, etc. are such which kill a wide range of insects like moths, beetles, mosquitoes flies, aphids, ants, termites, midges and butterflies depending upon the host strain of the bacterium. Several *Pseudomonas* such as *Pseudomonas cepacia, P. fluorescencse, P. alcaligenes, P. acidovora* isolated from insects and pathogenic to them. Many entomopathogenic fungi overcome their hosts to cause their death. *Metarhizium anisopliae, Beauveria brassiana, Paecilomyces fumosporoseus, Verticilluim lecanii* are also being used as microbial insecticides. *Hirsutella thompsonii* and *Nomuraea rileyi* control lepidopterean insects while most viruses such as occluded viruses namely NPV, cytoplasmic polyhedrosis (CPV), granulosis (GV) and entomopox viruses (EPV) considered for control of insects (usually sauflies and lepidoptera). Certain plant pathogenic fungi causing diseases (charcoal rot, collar rot, damping off, wilt) can be checked by using various bacteria (*Pseudomonas cepacia* and *P. fluorescens*) and fungi (*Gliocladium virens* and *Trichoderma harzianum*). There are few laboratories on microbial insecticides. The commercialization of their technology is still under infancy. Large scale efforts are required in controlling agricultural pests, nematodes and mosquitoes from Indian soils.

7.   Microbes in Bio-terrorism

Bio-terrorism has been defined as the deliberate release of disease causing germs microorganisms with the intent of killing large number of people and panicking many more. Accordingly microorganisms used as weapons of mass destruction of people. Many microorganisms are included but important are those, which affect human, animal and its management. These microorganisms can be classified according to the different modes of transmission. Some of these microorganisms cause smallpox, plague and cholera but like anthrax is such that kill millions.

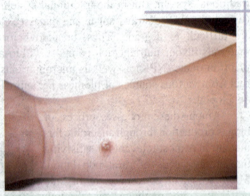

Skin anthrax caused by *Bacillus anthracis.*

Bacillus anthracis forms spores, which can survive in adverse conditions for several years. When anthrax spores get inside the body, they grow rapidly and produce anthrax toxin in the body that kills cells for the immune system. It can occur as skin anthrax, intestinal anthrax and inhalation anthrax. In Scotland during World War II, the British Army in an experiment at one of the islands found that the spores persisted for 36 years in the environment. Generally, it is not so easy to grow deadly anthrax. The spores have to be turned into a powder based carrier material, which could be sprayed over a large area. Another infectious agents is *Yersinia pestis*, transmitted by the infected rat flea (*Xenophilus cheopsis*) to human beings and animals and cause Plague. The bacteria is a Gram-negative, cocco-bacillus, appears like bipolar safety pin. It produces both endotoxins and exotoxins. A large number of other microorganisms create terror to both man and animals.

These microbes in the form of biological weapons intended to cause disease to death in human beings, depends for their effects on their multiplication in the body. The U. S. Govt. in their plan on Centres for Disease Control (CDC) Hazardous Biological Agent Regulation (1997) identified the infective agents that poses potential risk to public health.

This include a large number of bacteria (*Bacillus anthracis, Brucella abortus, Burkholderia mallei, Clostridium botulinum, Francissella turlarensis, Yersinia pestis* etc.) viruses (Crimean congo haemorrhagic fever virus, Eastern equine encephalitis virus, Ebola virus, Equina morbillivirus, Lassa fever virus, Masburg virus, Rift velley fever virus, South America haemorrhagic fever virus, Tick-borne encephalitis complex virus, Variola (small pox) major virus, Venezuelan equine encephalitis virus, Hantavirus, Yellow fever virus), *rickettsiae* such as *Coxiaell burnetii*, *Rickettsia prownzekii* and *R. rickettsii*.

Similarly, smallpox (*Variola*) has been affecting mankind for centuries. During 20th century, about 300 million people died of small pox. This virus spread from person to person, usually by droplet infection as shown in Table 1.2. There is no natural resistance to small pox. Recently, rDNA technology has been used for constructing novel, pathogenic microorganisms to yield significantly more effective or usable than conventional microorganisms.

Table 1.2 : Transmission mode of different microorganisms.

Mode of transmission	Microorganisms
Air droplets and dusts	Small pox, Plague, *Cl. Botulinum, Francisella tularensis, Legionella, Bacillus anthracis, influenza virus, Coccidiomycosis, Coxiella burnetii.*
Food, fruit and vegetables	*Salmonella. Shigella. E. coli, Bordetella, Hepatitis A, Clostridium botulinum and Cl perfringens*
Drinking Water	*Vibrio cholerae* and other entericpathogens, Hepatitis A & E.
Formities	*Bacillus anthracis* and acute enteric infections
Biological vector	*Aedes aegypti* (mosquito biter), Dengue virus, Ebola virus, Masburg virus and Rickettsia.

8. Computer Applications in Microbiology

Computers can serve a variety of functions in fermentation process control and analysis.

(*i*) **Optimisation via Computer :** Computers are used in scale up, to store and evaluate fermentation parameters and to measure the effects of individual parameters on the metabolic behaviour of cultures.

(*ii*) **Control via Computer :** Computers can control fermentation processes. On-line fermentation control is widely used in the production scale in many companies.

Computer applications in microbiology not yet as widespread as in the chemical industry for several reasons. Sensors suitable for use in sterile systems are not yet reliable

Microbiologist uses computer for data analysis.

enough to take advantage of computer capacity, biosynthesis. Regulation of metabolite formation are not yet fully understood. The fermentation cost reduction by using computers is difficult to calculate. Thus, in microbiology, computers are used primarily for data acquisition, data analysis and development of fermentation models.

(*a*) **Data acquisition :** Data can be acquired directly at the fermenter with on line-sensors. The informations acquired can be such as pH, temperature, pressure, viscosity, fermenter weight, power uptake, aeration rate and O_2 and CO_2 content in the gas stream. Other data can be obtained from laboratory measurements and fed into the computer off-line, e.g. biomass concentration nutrient content, metabolite formation. This information can be entered as raw data and can be converted by the computer to standard units, for example to adjust volumes for a standard temperature, temperature correction data can be used to calculate the true aeration rate for a production system. An alarm system can be looked upto the data-acquisition system to inform the attendant when deviations from standard value occur. Data about the course of fermentation can be stored, retrieved and printed out and product calculations can be documented.

(*b*) **Data Analysis :** The data entered or measured is used in calculations such as CO_2 formation rate, O_2 uptake rate, respiratory quotient, specific substrate uptake rate, yield coefficient, heat balance, productivity, volume-specific energy uptake, and Reyonld's number. When biomass is not continuously measured, the biomass concentration can be calculated through the O_2 uptake rate. It is assumed that the yield constant and the proportion of O_2 needed for maintenance metabolism are known. The calculations must be adjusted if, for example secondary metabolites are formed or the yield constant changes during the fermentation.

After fermentation at different pH, and temperature levels, "isoproduction and isotime curves" are computed. The fields are given as percentage of the minor production compared with the max-erythromycin titre. Optimal productivity (production/fermentation time) for a given set of operating conditions can be ascertained by this graph.

9. Development of Fermentation Models

By using mathematical models it is possible to better understand the fermentation process and to calculate the effects of process variables on the fermentation results, and thus optimize a process faster and more cost-effectively. The use of models can also help the development of better control strategies for fermentation. A large number of models exists for batch and continuous fermentations. However, each model is only applicable to a specific process and cannot be used for other processes.

D. Basic Concepts of Intellectual Property Rights (IPR)

1. Definition of IPR

In the common sense intellectual property is a product of mind. It is similar to the property (consisting of movable or immovable things) like a house or a car where in the property or owner may use his property as he wishes and nobody else can use his property without his permission as per Indian laws.

2. Function

World intellectual property organization (1967) one of the specialised agencies of the United Nations system provided that intellectual property shall include rights relating to the following.

(*a*) Literary, artistic and scientific works, performance of artists, phonograms and broadcast; innovation in all fields of human endeavor; scientific discoveries; trade marks, service marks and commercial names; industrial designs; protection against unfair competition and all other rights resulting from intellectual activity in the area of industrial, scientific, literary or artistic fields.

(*b*) The intellectual property is protected by and governed by appropriate national legislations. The national legislation specifically described the inventions, which are the subject matter of protection and those which are excluded from a protection, for example methods of the treatment of humans or therapy and invention whose use would be contrary to law or invention which are injurious to public health are excluded from patentability in the Indian legislation.

3. Forms of Protection

The forms of protection are as follows :

(i) **Patents :** A patent is a government granted and secured legal right to prevent others from making, using or selling the inventions covered by the patent. A patent is a personal property which can be licenced or sold by the person/organisation like any other property. For example Alexander Grahm Bell obtained patent for his telephone. This gave him the power to prevent any one else from making or using or selling a telephone.

It has been reported that the first patent was granted to Filippo Brunelleschi in the Republic of Florence of the Italian city states in 1421 on the discovery of special hoisting gear used on barges. An ordinance issued in a vential law in 1474 on patents. Later, in England, during 1533-1603, minister, Lord Burghley (1520-1598) in the ministry of Elizabeth, granted a series of patents with a view to inculcate and encourage inventors working in England. In India, the basics of intellectual property rights were first introduced by enacting the Act on protection of inventors in 1856 which was based on British Patent Law of 1852. Later, series of patent legislation established as shown below.

International and national agreements and treaties were founded as given below:

Year	Act/Law
1856	Act on Protection of Invention
1859	Patent monopolies of making, selling and using invention in India and authorizing others to do so for 14 years from date of filling application
1872	Patents & Design of protection
1883	Protection of Invention Act
1888	Consolidation as the Invention & Designs Act
1911	Indian patents and designs act w.e.f. August 15, 1947
1914	British Copyright Act 1911 modified for British India
1940	Legislation for protection of the Trade mark act w.e.f. June 1, 1942
1957	Adopted many principles for British Copyright Act
1959	Indian Trade & Merchandise Marks Act
1967	Patent bill introduced in Parliament
1970	Indian Patents Act
1970	Patents Act (Act 39 of 1970) introduced w.e.f. April 20, 1972
1983	Amendments to avail the benefits arising from the revision of the Berne convention and the Universal Copyright convention
1983	Amendments to discourage and prevent piracy.
1992	Amendment to increase protection time to Author's Life time + 60 years
1992	Amendment proposed for a "New Act"; debates continuing
1993	Ordinance framed to amend the Patents Act, 1970 and the concept of exclusive marketing rights introduced as pipeline protection
1994	Amendment due to obligation arising from the General Agreement of Trade & Tariff (GATT), copyright protection extended to computer industry.

As far as International and Regional agreements/treatises in Intellectual Property Rights are concerned, it begin from 1883 with Paris convention for the protection of Industrial Property; Berne convention for the protection of Literacy and Artistic works (1886); Madrid agreement for

Repression of false or deceptive indications of source of goods (1891); Hauge agreement concerning the International Deposit of Industrial designs (1925); Nice agreement concerning the International classification of Goods and services for the purposes of the Registration of Marks (1957); Lisbon agreement for the protection of appellation of origin and their International Registration (1958); Rome convention for the protection of performers, producers of phonograms and broadcasting organizations (1961); Locarno Agreement establishing an International classification for industrial designs (1968); Patent cooperation treaty (PCT in the year 1970); Strasbourg agreement concerning the International Patent Classification and Geneva convention for the protection of producers of phonograms against unauthorized duplication of phonograms (1971); Vienna agreement establishing an international classification of the figurative elements of marks (1973); Brussels convention relating to the distribution of programme carrying signals transmitted by satellites (1974); Budapest treaty on the International Recognition of the Deposit of Microorganisms for the purposes of patent procedures (1977); Nairobi treaty on the protection of Olympic symbol (1981); Protocol relating to the Madrid agreement concerning the International Registration of Marks (1989); Trademark law treaty and trademark related Intellectual property Rights (TRIPS) 1994; Community Trademark (1996), Documents for the diplomatic conference on certain copyrights and Neighbouring Rights (1996), WIPO Copyright Treaty WCT) and WIPO Performance and Phonograms Treaty (WPPT).

In India, the Controller General of Patents Designs and Trademarks (CGPDT) functioning under the Department of Industrial Development Control grant the patents, designs and trademarks. The Ministry of Human Resources and Development is in-charge of copyright board.

(*a*) **Conditions for patentability :** An invention or process is patentable if it is new, involves an inventive step (i.e. it is not obvious) and is industrially applicable.

(*b*) **Test of novelty of patents :** Patents specifications should be made before the date of filling of the application with complete information.

Any other document published in India or elsewhere before the date of the filling of the applicants complete specifications. This will cover forcing specifications whether publishing in India or not and text books and periodicals published any where related to the art in question. The only limitations being that they should be published before the date of the filling of the applicants complete specifications.

The economic and competitive position of a fermentation process depends on several factors such as yields, research costs, size of the market, profit potential, and patent or secret, process position of the forementation process or product. Patents are granted to inventors in return for a public disclosure of their inventions. This disclosure and the knowledge of the respective art help to advance the state of that art. The patent in terms give the inventor the right to exclude others from making, using or selling his particular invention as disclosed in the "claims" of the patent. Obviously, in case of certain inventions secrecy is difficult to maintain, for example in the process of fermentation.

The individuals working in an industrial research laboratory or any laboratory in which fermentation process of potential economic value are under study should know about how to read a patent in order to be able to determine the points of the invention which are actually protected by the patent. He should also understand the types of informations that are required for filling a patent application so that research can be directed towards obtaining information. As we shall see claiming too little or too much about the process or product can be disastrous. Guidance in these problems can be obtained from a qualified patent attorney.

(*c*) **Composition of a patent :** A patent consists of three parts, the grant, specifications and claims. The grant is filled at the patent office and is not published. It is signed document and is the agreement that grant patents right to the inventor. The specifications and claims are published as a single document which is available to the public at a minimum charge from the patent office. The specification section is narrative description of the subject matter of the inventiion and of how the invention is carried out, the claims section specifically defines the scope of the invention to be protected by the patent, that which other may not practice. Thus, a patent stands of falls depending upon the statements included in the claims section.

(ii) **Copyrights :** Copyrights broadly include literacy works, musical works, including any accompanying works, dramatics works, including any accompanying music pantomimes and choreographic works, pictorial graphics and sculptural works.

Recently an expression called neighboring rights has been added to the concept of copyrights. The expression neighboring rights is the abbreviated form of the rights neighboring on copyright. The following three types of rights are covered by the concept of neighboring rights.

(a) The rights of performing artists in their performance

(b) The rights of producers of phonograms in their phonograms

(c) The rights of broadcasting organizations in their radio and television broadcasts.

(iii) **Trade Mark :** A trade mark is an identification symbol which is used in the course of trade to enable the public to distinguish on trader's good from the similar goods of other traders. The public makes use of these trade works in order to choose whose goods they will have to buy. If they are satisfied with the purchase, they can simply repeat their order by using the trade mark, for example KODAK for photography goods and IBM for computers, Zodiac for readymade clothes, etc.

(iv) **Design :** Design means only the features of shape, configuration, pattern or ornament applied to any article in any industrial process or means whether manual, mechanical, chemical, separate or combined, which is the finished form appeal to end or judged solely by the eye. By registration under the designs act, the features are protected as design. Genetic informations can also be used to cure a disease, for example using the technology of gene therapy with a specific gene vector. The direct use of proteins as therapy is well established, and these products may be patented, though we should note, in general, that medical procedures have not been patented for ethical and practical reasons.

A patented product that reaches the commercial market gives the inventors some compensation for the time they spent in research for the development. In the USA, the average time required for biotechnology medicine to be approved for commercial scale of food & drug administration is 21.4 month after requests of trial based on chemical tests and it should be ten years after identifying the substance. Once a product is patented the sales can bring about high income for the company that produces it and this includes return for the inventors. The system is self sustaining; if patents are awarded, companies will invest time in research and if not these will be less incentive for companies to do research. Without patents it may be easy for other companies to copy the techniques soon after introducing and take a share of the commercial market, especially because they do not need to bear the cost of the long period of research for product development. Some system of reward is required to encourage commercial research, which is responsible for a significant number of biotechnology applications. The international recognition of intellectural properly rights (patents variety rights) is thus a basic concern.

The ethical principle of beneficence can be applied here. Does commercialization of biotechnology leads to more benefits than a bar on it? The benefits should be in terms of general, medical or agriculture developments, rather than the economic prosperity of one company or country over another. Patenting is not permitted useful information otherwise becomes trade secrets, or if plant variety rights are not recognized seeds may not be made widely available. However, property rights are not absolutely protected in any society because of the principle of justice, and for the sake of "public interest", "social need,"; and "public-utility", societies can confiscate intellectual property.

People arguing for patenting claim that patent laws regulates inventiveness not commercial uses of inventions. However, there was recent controversy regarding the commercial monopoly held by the company which was able to protect the first HIV/AIDS treatment, which enable to obtain large projects while it held a monopoly. It also meant that the drug was prohibitively expensive for developing countries. Another arrangement is that if other countries support patents, our country needs to if our biotechnology company is to compete; however, the reverse arrangement that some countries do not permit similar patents, is also used to justify exclusions.

(*v*) **Know-how :** Know-how is another important form of intellectual property generated by R&D institution that does not have the benefit of patent protection. This could be in the form of an aggregations of known procedures and accumulation of data. A secret formulation or a combination of any of these know-how is often transferred together with licensing of patent.

4.   Patenting of Biotechnological Discoveries

The question arises what properties must a product have in order for it to be patentable? Actually, there is some debate about whether living organisms should be patented. Folllowing characters are to be considered to qualify for a patent; (*a*) novel invention, (*b*) non-obvious, (*c*) usefulness

In the case of natural products, details of four molecules or coding nucleic acid sequence may have lost its novelty and non-obviousness. Patents are granted on molecules which have medical uses if the chemical structure or the useful activity was novel when the patent was applied for methods of gene-sequencing, mapping or expression, can be invented or patented.

The process for making 'oncomouse', a mouse that contains activated oncogenes sequences that is, therefore, sensitive to mutagens or carcinogens was patented in 1998 in the USA. Protected by copy right law is creativity in the choice and arrangement of words against those who 'copy'; who take and use the form in which the original work has expressed by the author e.g. such as books, photographical works, paintings etc.

5.   Biopiracy

There has been growing discontent amongst developing countries about this biopiracy i.e. unfair exploitation and monopolization of public domain knowledge and resources. Most of the industries of developed nations recent losses come from counterfeit goods and pirated technology in the developing countries. A new drug generally takes a decade or so involving several lakhs of rupees to develop. Intellectual Property Rights (IPRs) are justified to protect this enormous amount of investment. On the other hand, developed countries freely acquired most of its live material including crops from its neighbours. Developed countries also looted medicinal plants, dyes, spices, etc. from developing countries leading to discovery and conquest of India. S. America and S.E. Asia. For example new drug like resperine used against hypertension. Derived from an Indian Plant sarapagandha (*Rauwolfia serpentina*) has enormously enriched foreign drug companies. Similarly, genes from the Pattambi rice variety in Kerala (S. India) have been used to

A plant of Sarapagandha.

introduce to pest resistance in paddy. Now seed companies have started exploiting this character without giving any advantage to farmers. The developed countries have never paid for the benefits obtained from the developing countries. On the other hand, they get raw material of basic knowledge from developing countries but selling back on a very high price. Since, IPRs protect only the commercial inventions. The domestic and ongoing use of bioresources is not prevented. Thus, grand parents or ayurvedic practitioners can continue to use of market churan or decoctions as usual. However, they are not entitled for the claim from the profits out of these items.

Industrial Property : Industrial property includes inventions (process, products, apparatus); industrial designs (shapes & ornamentation); and Marks and Trade names to distinguish goods. Recently, the scope of industrial property has been expanded to include among others, the protection of distinctive geographical indications (in particular appellation of origin), plant varieties and the layout designs (topographies) or integrated circuits as well as the repression of unfair competition, including the protection of trade secrets.

6. Importance in Indian Scenario

A US patent granted for use of turmeric powder (*haldi*) as a wound healing agent to the University of Missisippi, Medical Centre, US, has been revoked following objections by the Council of Scientific and Indian Research (CSIR), New Delhi. The revocation order was passed on August 13, 1997, two years after the grant of patent in March 1995. It was a legal battle on turmeric powder which is used in India for ages now as a wound healing agent among other things, and was not a discovery of the US patentee.

Now patents have already been granted for food stuffs like idli, dosa, vada, churan, pickle, halwa and pizza topping. The Indian patents Act 1970 stresses that any patentable commodity must possess novelty. Apparently, the Chennai Patent office believes that South Indian delicacies like "medu vadai". "rava-uppuma", "badam halwa", "rice idli", "rice pongal", and even Green Peas Masala are novel process patent rights for these popular preparations were granted to the Dasaprakash hotel chain in 1973. The Mumbai office of the Patents has granted one Dilip Shantaram Dahanunkar, a process patent for the preparation of tomato rasam and a custard chilli jam spread used as a pizza topping. The same person has been given a patent for an improved process for preparation of vitaminised sweet and sour lemon pickle rice and a process for manufacturing banana sauce.

The purpose of patenting common products seems to improve their marketablility rather than to protect "inventions". Accordingly inventors blindly exploit legal loopholes to patent age-old products.

According to Calcutta based patent and trademark attorney D.P. Ahuja and company the Patents Act, 1970 states that a patent can be given for a novel article or a process even if it results in an old products.

On 30 September 1997 the European Patent Office (EPO) delivered a favourable interim judgment on the challenge of a European patent on the fungicide effects of neem oil (Patent No. 436 257 BY) owned by W.R. Grace & Co. The opposition division of the EPO issued a provisional statement on the basis of the European Patent convention (EPO) and delivered favourable interim judgment on opposition to Neem Patent in favour of Dr. Vandana Shiva, Ms. Magda Alvoet (MP of the European parliament) and other NGOs of Neem campaign. Recently, another controversy has arisen regarding patenting of 'Basmati Rice' by U.S.A. Indian Govt. filed reexamination request for the patent on Basmati rice lines and grains (US Patent no. 5666484) granted by US PTO , and Ricetech Co. from Texas has decided to withdraw the specific claims challenged by India.

There is problem on the grant of such patents linked to the indigenous knowledge of the developing world that needs to be addresses jointly by the developing and the developed world. Actually, the available data bases of different items available in patent international offices are to be considered by patent examiners while granting the patents. They search non-patent literature database that deal with traditional knowledge, captured electronically and placed in the appropriate classification within the international patent classification systems so that it can be easily searched and retrieved in the international patent office. This would help prevent the patenting of the products that have been based on the traditional knowledge of the developing world.

The Indian Govt. has taken a step to create a Traditional Knowledge Digital Library (TKDL) on traditional medicinal plants and systems, which will also lead to a Traditional Knowledge Resource Classification (TKRC). Such information shall fill up the gap between the knowledge contained in an old sanskrit shloka and the computer screen data of a patent examiner in Washington. This will eliminate the possibility of granting wrong patents since the Indian rights to that knowledge will be known to the examiners. Some of the countries are in process of securing patents (Table 1.2).

Table 1.3 : Patents secured abroad and country of origins.

Country of origin	Number of patents Secured
India	78
USA	1,09,146
Japan	80,905
Israel	933
Luxembourg	613
Brazil	275

Above data for the year 1995 is based on The Express Magazine, May 3, 1998.

7.   Forthcoming Laws

New laws to be enacted by government of India so as to bring India closer to the international patents regine. New laws relating to intellectual property rights that the government proposes to enact, are given below.

(i) **Trade Marks :** This Act will allow the registration of service marks and collective marks. The service mark will allow the entire service industry to register its logos that identify a firm while the collective mark will allow entrepreneurs from a certain region that is famous for a particular region. For example, all makers of footwear from Kolhapur will be able to register the name Kolhapuri so, anyone manufacturing chappals outside the town won't be allowed to use its name.

(ii) **Geographical Indicators :** Basmati is the suitable example which allows a country to register all products whose quality, reputation or other characteristics are essentially attributable to their geographial regions.

(iii) **Industrial Designs :** India's industrial design law dates back to 1911. It badly needs to be updated.

(iv) **Layout / Designs of Integrated Circuits :** India has a role to play in world's electronic market. The protection of integrated circuits is crucial to the development of the electronic industry. This is essential so that efficiency and the capability of each circuit is maintained.

(v) **Trade secrets :** This is something great that India does not have trade secrets. It will allow a company to register and protect formula details or processes. A patent usually runs out in 10-20 years, but under this law a company will have no obligation to reveal its secret. Coca-Cola, for example has covered its best-kept secret of its formula under this law.

QUESTIONS

1. What do you mean by Microbiology ? Discuss about the contribution of some pioneer of this field.

2. Introduce the following:

   (*a*) Robert Koch, (*b*) Louis Pasteur, (*c*) Anton von Leeuwenhoek,

   (*d*) Edward Jenner, (*e*) Lister, (*f*) Ehrlich.

3. How will you prove that microorganisms can cause diseases?

4. What circumstances provoked the idea of spontaneous generation ?

5. Write in detail about the germ theory of disease.

6. Write in short on the followings :

   (*a*) Vaccination, (*b*) Koch's postulates, (*c*) Golden age of Microbiology,

   (*d*) Pasteur's effect, (*e*) Bacterial nomenclature.

7. Write at least name and scientific contribution of three noble laureates of last two decades.

8. Briefly state the characters of the following:

   (*a*) Archaeobacteria (*b*) Cyanobacteria

9. What is the basic concept of intellectual property rights? Write its function and importance in Indian scenario.

10. Describe the basic concept of Bio-terrorism. What is its significance? Name some microoganisms associated with this act.

11. Write short notes on the following:

   (*a*) Patents (*b*) Trademark (*c*) Copyright (*d*) Designs (*e*) Biopiracy

Classification of Microorganisms

2

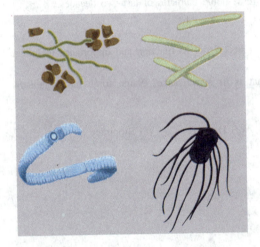

A. Microbial Diversity

The term 'biological diversity' or biodiversity has become so well known that a public servant is also aware about it. Biodiversity is defined as the variability among living organisms. The main key of biodiversity on Earth is due to evolution. The structural and functional diversity of any cell represents its evolutionary event which occurred through Darwinian theory of natural selection. Natural selection and survival of fittest theory is involved on the microorganisms. This includes diversity within species, between species and of ecosystems (convention at Rio de Janiero, 1992). This was first used in the title of a scientific meeting in Washington, D.C. in 1986 (Wilson, 1988).

The current list of the world's biodiversity is quite incomplete (Table 2.1) and that of viruses, microorganisms, and invertebrates is especially deficient. The fungal diversity indicate the total number of species in a particular taxonomic group. The estimates of 1.5 million fungal species is based principally on a ratio of vascular plants of fungi to about 1:6 (Fig. 2.1).

Attempts to estimate total numbers of bacteria, archaea, and viruses even more problematical because of difficulties such as

Charles Darwin
(1809–1882)

detection and recovery from the environment, incomplete knowledge of obligate microbial associations e.g. incomplete knowledge of *Symbiobacterium thermophilum*, and the problem of species concept in these groups. Take the case of mycoplasmas, which are prokaryotes having obligate associations with eukaryotic organisms, frequently have traditional nutritional requirements or are monoculturable and appear to have remarkable diversity. On the otherhand, there is one group *Spiroplasma*, which was discovered in 1972, may be the largest genus on earth. *Spiroplasma* species are principally associated with insects, and the overall rate of new species isolation from such sources of 6% annually indicates species richness. Similarly, marine ecosystems likely support a luxuriant microbial diversity. Further, microbial diversity can be seen on cell size, morphology, metabolism, motility, cell division developmental biology, adaptation to extreme conditions, etc.

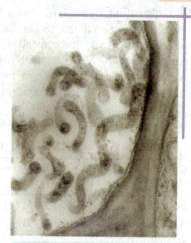

Spiroplasma seen under microscope.

The microbial diversity, therefore, appears in large measure to reflect obligate or facultative associations with higher organisms and to be determined by the spatio-temporal diversity of their hosts or associates.

1. Revealing Microbial Diversity

The perception of microbial diversity is being radically altered by DNA techniques such as DNA-DNA hybridization, nucleic acid fingerprinting and methods of assessing the outcome of DNA probing, and perhaps most important at present, is 16S rRNA sequencing (Olsen *et al*, 1986; Pace *et al*, 1986; Stahl, 1988 and Stahl *et al*, 1988).

The 16S rRNA has radically changed the classification of microbes into 3 domains, the Bacteria, Archaea and Eukarya (Woese *et al*, 1990). While DNA-based analysis (DNA finger printing by restriction fragment length polymorphism i.e. RFLP analysis) is another accepted technique for evaluation of relationships between organisms, especially if they are closely related. Holben *et al.* (1988) detected *Bradyrhizobium japonicum* selectively at densities as low as 4.3×10^3 organisms/gram dry soil.

2. The Concept of Microbial Species

Biological diversity or biodiversity is actually evolved as part of the evolution of organisms, and the smallest unit of microbial diversity is a species. Bacteria, due to lack of sexuality, fossil records etc., are defined as a group of similar strains distinguished sufficiently

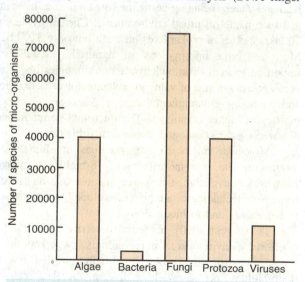

Fig. 2.1 : The number of known species of microorganisms in the world (based on Hawksworth, 1991).

from other similar groups of strains by genotypic, phenotypic, and ecological characteristics. The adhoc committee on the reconcillation of approach to international committee on systematic bacteriology (ICSB) recommended in 1987 that bacterial species would include strains with approximately 70% or more DNA-DNA relatedness and with 5% or less in thermal stability (Wayne *et al*, 1987). Hence, a bacterial species is a genomic species based on DNA-DNA relatedness and the modern concept of bacterial species differs from those of other organisms.

To date, more than 69,000 species in 5100 genera of fungi and about 4,760 species of about 700 genera of bacteria have been described in the literature as given in Table 2.1.

Table 2.1 : Estimation of biological species*

Group	Known species	Estimated total species	Percentage of known species
Viruses	5,000	1,30,000	4
Bacteria	4,760	40,000	12
Fungi	69,000	15,00,000	5
Algae	40,000	60,000	67
Bryophytes	17,000	25,000	68
Gymnosperms	750		
Angiosperms	2,50,000	2,70,000	93
Protozoa	30,800	1,00,000	31

* Based On DiCastri and Younes (1990), McNeely *et al* (1990), Bull and Hardman (1991), Hawks worth (1991).

3.   Significance of Study of Microbial Diversity

As quoted by American Society of Microbiology under Microbial Diversity Research Priority, "microbial diversity encompasses the spectrum of variability among all types of microorganisms in the natural world and as altered by human intervention". The role of microorganisms both on land and water, including being the first colonizer, have ameliorating effects of naturally occurring and man-made disturbed environments. Current evidence suggests there exist perhaps 3 lakh to 10 lakh species of prokaryotes on earth but only 3100 bacteria are described in Bergey's Manual. More and more informations are required and will be of value because microorganisms are important sources of knowledge about strategies and limits of life. There are resources for new genes and organisms of value to biotechnology, there diversity patterns can be used for monitoring and predicting environmental change. Microorganisms play role in conservation and restoration biology of higher organisms. The microbial communities are excellent model for understanding biological interactions and evolutionary history.

Molecular microbiological methods involving DNA-DNA hybridization and 16S rRNA sequencing, etc. now more helpful in establishing microbial diversity. Data bases are becoming more widely available as a source of molecular and macromolecular information on microorganisms. New technologies are being developed that are based on diverse organisms from diagnostics to biosensors and to biocatalysts.

In the year 1990s' microbial diversity has burst forward in a new and exciting form due to efforts of environmental microbiologists, who kept the diversity flame alive during the paradigm organism years. The molecular revolution that has been sweeping through environmental microbiology, has shown how diverse microbes really are. It has also leashed new waves of creativity in the from of RNA sequence analysis to prove the metabolic activities and gene regulation of microbes *in situ*.

Applications : The gainful advantages may occur by enriching microbial diversity. Microbial genomes can be used for recombinant DNA technology and genetic engineering of organisms with environmental and energy related applications. Emergence of new human pathogen such as SARS is becoming quite important due to threat to public health can be solved by analyzing the genomes of such pathogen. Culture collections can play a vital role in preserving the genetic diversity of microorganisms. Microbial informations including molecular, phenotypic, chemical, taxonomic, metabolic, and ecological information can be deposited on databases. A large number of yet unexplored microorganisms may lead to beneficial informa-

Genetic engineering enables scientists to use microorganisms like *E. coli* to produce an almost infinite number of replicas of any given piece of DNA.

tions. This can be further strengthened by multidisciplinary involvement of experts. There is a compelling need for discovery and identification of microbial biocontrol agents, an assessment of their efficacy etc,. The molecular nature of genomes of some important pathogens is necessary to understand the pathogenesis, biocontrol, and bioremediation of pollution etc., besides helping in rapid detection and diagnosis and in identification of genes for transfer of desirable properties. Microorganisms are sensitive indicators of environmental quality. Thus, microbial diversity may be helpful in determining the environmental state of a given habitat of ecosystem. The diverse microorganisms can cause disease and could potentially be used as biological weapons. Knowing what is likely to be present can help in rapid diagnosis and treatment. Biodegradation and bioremediation are potentially important to cleanup and destruction of unwanted materials. Microbial diversity of marine microorganisms is equally important. Sometimes, it is helpful to solve the contamination of sea-food by pathogenic microorganisms e.g. *Vibrio vulnificus* contaminated oysters. Blue green algae and cyanophages are another dangerous organisms to aquaculture industries.

One way to obtain impure enzyme is to grow bacteria (the fine strands) and then kill them using a treatment that does not harm their enzymes.

4. Microbial Evolution

The microbial evolution has entered a new era with the use of molecular phylogenies to determine relatedness. Certainly this type of phylogenetic analysis remains controversial, but it has opened up possibility of comparing very diverse microbes with a single yardstick and attempting to deduce their history. Some scientists have opined that the 'failure' of molecular methods of find a single unambiguous evolutionary progression from a single ancestor to the present panoply of microorganisms. The increasing appreciation of the ubiquity and frequency of gene transfer events open the possibility of learning quite essential prokaryotes is by establishing a central core of genes that has not participated in the general orgy of gene transfer. The increasing number of genome sequences may also contribute to a better understanding of the evolutionary history of microbe.

B. Classification of Microorganism

E. Haeckel (1866) separated living organism into three groups, plants, animals and protista. The primitive organisms were included in protista. This was 12 years before the Sedillot used of "microbe". Therefore, a substantial concept of microbes probably did not exist at that time, and microbes in a modern sense were scattered into plants and protista.

(i) Five-kingdom System of Classifications : Later, prokaryotic and eukaryotic organisms distinguished on the basis of cell anatomy, and the concept of a bacterium as a prokaryotic organism was established in microbiology in 1962 by Stamir and Van Niel. In 1969, Whittaker proposed a five kingdom system consisting of kingdom of plantae, fungi, animalia, protista and monera (Fig. 2.2) for all organisms on the basis of their energy-yielding systems and

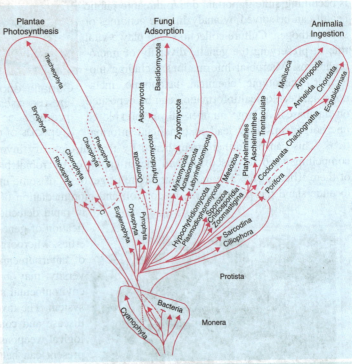

Fig. 2.2 : Five kingdom system of classification by R.H. Whittaker (1969).

cell anatomy. Microorganism with the common characterstics described above are distributed in the kingdoms of monera, protista, fungi and a part of plants. Recently, evolutionary relationships of living organisms have been defined on the basis of ribosomal RNA sequences and other data.

The kingdom **Monera** of prokaryoteae includes all prokaryotic microorganisms. **Protista** consists of unicellular or multicellular eukaryotic organisms but true tissues are lacking. The kingdom **Fungi** contains eukaryotic and multinucleate organisms. The members have absorptive mode of nutrition. **Animalia** contains multicellular animals devoid of cell wall. Ingestion is the mode of nutrition. The kingdom **Plantae** includes multicellular eukaryotes. Their mode of nutrition is photosynthesis.

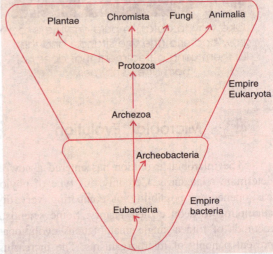

Fig. 2.3 : Classification of protists into eight kingdom as proposed by Cavalier-Smith (1987)

(ii) Eight Kingdom System of Classification : Cavalier-Smith (1987, 1993) classified protists into eight kingdom on the basis of ultrastructure of cell and genetic organisations (rRNA sequencing and other data). He divided all organisms into two **empires** (Bacteria and Eukaryota) (Fig. 2.3). The empire Bacteria includes two kingdoms (Eubacteria and Archaeobacteria). The empire Eukaryota contains six kingdoms (Archezoa, Protozoa, Plantae, Chromista, Fungi and Animalia).

The kingdom Chromista includes diatoms, brown algae, cryptomonads and oomycetes. The members of Chromista are photosynthetic and have their chromoplast within the lumen of endoplasmic reticulum but not in cytoplasm.

(iii) Three Domain System of Classification : Woese *et al.* (1990) noted that bacteria are distant from plants and animals and, by contrast, plants and animals are not so far from each other. Therefore, they established a new superior concept of domains over the kingdom, and proposed three domains, Bacteria, Archaea and Eukarya in 1991 (Fig. 2.4).

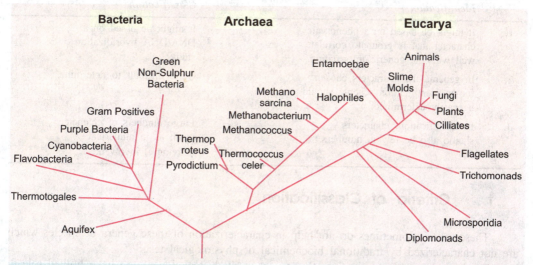

Fig. 2.4 : Universal phylogenetic tree derived from comparative sequencing of 16S or 18S ribosomal RNA (*Source* : data from Olsen, G.R. and Woese, 1993.)

The domains **Archaea** and **Eukarya** are distinctly related to each other. Eukarya includes the organisms which possess glycerol fatty acyldiester as membrane lipids and eukaryotic rRNA. The domain Bacteria consists of such members which have membrane lipids as diacyl glycerol diesters and eubacterial rRNA. The member of domain Archea consists of isoprenoid glycerol diester (or diglycerol tetraether) lipids in their membrane and archebacterial rRNA.

In a modern sense, bacteria, cyanobacteria, actinomycetes, etc. are distributed in the domain bacteria; methanogens extremely thermophilic organisms, extremely halophilic organisms, etc. are in the domain Archaea; and molds, yeasts,

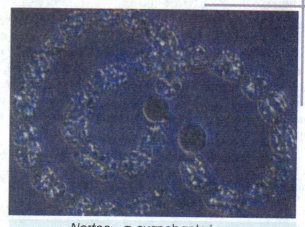

Nostoc - a cyanobacterium.

basidiomycetes, algae and protozoa, etc. in the domain Eukarya. Microorganisms are regarding as collections of evolutionary different organisms.

I.   What is Microbial Taxonomy?

Classification, nomenclature and identification comprise taxonomy of microorganisms. On the basis of common characters or properties, a set of organism is considered into groups (taxa). There are no formal rules to define the taxa. On the otherhand, nomenclature is the name given for defined taxa as also governed by bacteriological code of nomenclature. The features that are used to differentiate various organisms often have little to do with the fundamental basis for arranging the organisms into taxonomic groups.

Table 2.2 : Differences between identification and classification

	Identification	Classification
1.	It might be based on a phenotypic character that is group to correlate well with the generic information.	It might be based on a DNA/DNA hybridization study.
2.	In general, the characters chosen for an identification scheme should be easily determined.	Quite difficult to determine.
3.	In identification, characters should also be few in numbers.	Large numbers of characters (such as in numerical taxonomy) involved.

1.   Criteria of Classification

These criteria sometimes do not help in characterization of these genera or species which are not characterized by traditional biochemical or physiological tests.

Serological tests (agglutination, fluorescent antibody techniques etc.) have limited role in classification but have enormous value in identification. It is important to note that identification may not be based on only a few tests, but rather on the whole battery of test. Genetic tools are the modern one for identification of bacteria, based on the detection of a specific portion of an organism's genetic material.

2.   General Methods of Classifying Bacteria

(*i*) **The Intuitive Method** : Since a large array of microbiologists study the characterstics of organisms (morphological, physiological, biochemical, genetical, molecular), sometimes, it is difficult to assign an organism based on all the characters because a character may be important to a particular microbiologist may not be that important to another, hence, different taxonomists may arrive at very different groupings. Sometimes, this approach i.e. intuitive method is found useful.

(*ii*) **Numerical Taxonomy** : This is based on several characterstics for each strain and each character is given equal weightage. The % similarity of each strain may determine by the following formula.

$$\% \ S = NS \ / \ (NS + ND)$$

Where

NS = Number of characterstics for each strains which are similar or disimilar.

ND = Number of characterstics that are disimilar or different.

On the basis of % S, S = Similarity if it is high to each other, placed into groups larger and so on.

3. Genetic Relatedness

This classification is based on genetic relatedness (DNA and RNA) between organisms.

The % G + C determines the organism whether it is the same or different species. If two organisms have quite different mol % G + C, then the species are different and seem to be not related to each other. Further, characterization is based on the principles given below.

(*a*) **DNA-DNA hybridization or DNA - homology** : After annealing i.e. separation of two strands and converting them into single strand, the later one is mixed with those obtained from other organism. If the two organism are similar the pairing will occur in strands of both the organism, and will form heteroduplexes otherwise not.

(*b*) **16S rRNA sequencing** : Ribosomal RNA homology experiments and ribosomal RNA oligonucleotide catalogueing determine the molecular characteristics just to demonstrate degree of relatedness. Due to the highly conserve nature of rRNA genes, the 16S rRNA nucleotide sequence give better informations than DNA molecule. It is important to mention that if there is no DNA-DNA relatedness (homology) between two organisms, still nucleotide sequence of their rRNA cistrons may be studied so as to observe the similarity. Hence, the degree of similarity may be used as a measure of relatedness between organisms, but at a level beyond that of species.

II. Classification and identification of bacteria

Although the bacterial structures are quite simple but their classification is quite difficult. However, due to readily available devises in the field of molecular biology it make the taxonomy much easy. The use of old staining techniques such as Gram's staining for differentiation of bacteria on the basis of cell wall structures, their morphological, biochemical characters for proper characterization were known earlier. The use of molecular biological techniques such as restriction fragment length polymorphism (RFLP), DNA-DNA hybridization and 16S rRNA sequencing make the studies more easy as far as the phylogeny of the bacterial strains are concerned. As a result, there has been a considerable revival of interest in bacterial taxonomy.

1. Bacterial Nomenclature

After classifying and identification, it is important to assign name to the bacterium. The name should be such so that it is accepted internationally.

If we see the history of bacteria and its systematic studies, it was Robert Koch isolate of tuberculosis which was called 'Koch's bacillus'. Later, it was quickly discarded due to inadequate informations. There is still a tendency to honour eminent bacteriologists by assigning bacteria in Latinised forms of their names. For example Border, Bruce, Erwin, Escherich, King, Lister, Neisser, Pasteure, Ricketts, Salmon and Shiga, etc. each have a bacterial genus named after them. On the otherhand, some of the bacterial species are after the names of Byod, Stewart and Jenner. It is only Morgan who had both genus and species named.

The another system of nomenclature was based on disease caused. For example, 'Koch's bacillus was named later as 'The *tubercle bacillus*'. Recently, the casual organism of the legionnaire's disease was assigned in the genus *Legionella*. This system was also discarded due to the reason that some bacteria cause several disease processes. For instance, *Escherichia coli* causes diarrhea, hemorrhagic colitis, dysentery like illness similar to *Shigella* infection. Another problem with this system is that it does not consider those bacteria which do not cause any disease. There are certain evidences where bacteria was assigned according to their habitat. Thus, *Escherichia coli* is so called as it is found in colon. This system also did not work due to the fact that majority of bacteria are found in a diversity of locations.

It was Carl Linnaeus who gave hierachical classifi-cation system which was already attempted to assign both plants and animals. According to this, orders of bacteria divided into families and in each family, there are number of genera, each of which comprises one or more species.

In this process each bacteria was described according to bionomial nomenclature (genus and species). The genus name is always capitalised, often to just the first letter is in capital letter and species name is written with a lower initial letter. The care must be taken while abbreviating the common genera in share the same initial letter then the first few letters are written e.g. *Staphylococcus aureus* abbreviated as *Staph aureus*, *Streptococcus pyrogenes* abbreviated as *Strepto pyrogenes*. Generally bacterial

Carl Linnaeus.

names should be printed in italic letters but if it is not possible then bacterial names are usually underlined. When a group of bacteria is written, like staphylococci, the name is neither italicised nor underlined and it is not written with a initial capital letter. The unnamed species if it is one written as sp. or if many, the spp. should be used. It (sp. or spp.) should also not be underlined or italicised. Most of the families end with - *aceae* as in Entererobacteriaceae and order end in - *ales* e.g. Spirochaetales. Latinised names may be of either male (*Staphylococcus*) or female (*Salmonella*) or may be neuter gender (*Clostridium*). The plural forms of these genera are *staphylococci*, *salmonellae*, and *clostridia*, following strictly linguistic rules.

Currently, the bacterial nomenclature is regulating by a committee on systematic nomen-clature. List of approved names are published in International Journal of Systematic Bacteriology. It is kept that most of the bacterial names should often be descriptive e.g. *Mycobacterium tuberculosis* causes tubeculosis. But sometimes, they reflect the microscopic appearance as in the case of *Staphylococcus aureus*, which is golden (aureus) berry (coccus) that form clusters like, grapes (staphule).

2.   Modern Trends of Bacterial Taxonomy

The bacterial taxonomy has a rapid and radical change due to involvement of modern molecular biological techniques. This is why it has now refined. Much care has also been taken in ascertaining the history of a species so that species names reflect the original description, rather than a popularly used name. As a result of such activity, the species names assigned to bacteria may change.

The basic constitution of bacterial species is different from higher organisms due to the reason that in later forms the species is defined as those which shares many common features and its members breed to produce fertile offspring. While in case of bacteria, they reproduce only asexually by binary fission, hence the above concept does not suit in bacteriology.

The use of computers in bacteriology and in particular taxonomy gave an advantage to resolve the impact of the techniques of numerical taxonomy. This is based on comparing the strains of bacteria for the large numbers of characters. The more closely related organisms are to have the more characteristics in common. A similarity matrix can be drawn up depicting the degree of relationship between all the species examined and it is then presented in the form of dendogram.

Numerical taxonomy (taxonometrics) considers the phenotypes of organisms, and is sometimes described as phenetic system. Since, Charles Darwin gave the theory of evolution by natural selection, biologists have derived phyletic systems of classification in which the true evolutionary relationship of organism is represented. While the philosophy derives from the 18th century from a botanist Adanson, hence it is also referred to as Adansonian Taxonomy. This system determines the degree of relationship between strains by a statistical coefficient that later accounts for their similarities and differences in widest possible range of characters (morphological, physiological, antigenic i.e. all phenetic), all of which are of equal importance. It also defines taxospecies (a group of strains of high mutual phenetic similarity) and produces polythetic groups (consist of assemblages whose members share on high proportions of common attributes). The strains, species, genera for which no term is available, hence are called operational taxonomic units (OTUs). Now, taxonomists are paying attention to a direct analysis of their genetic material.

The similar G + C content of two or more bacterial strains indicate that they are closely related but it is always not the case because one may evolve from an ancestor with low G + C content. On the otherhand, DNA hybridization is also helpful by determining the degree of hybridization by measuring the temperature at which a mixture of melted DNA strands isolated from different bacteria re-annealed. This is performed by observing changes in the ultraviolet absorbance of the DNA mixture at various temperature.

The evolutionary relationship between organisms is a direct comparision of the nucleotide sequences of bacterial genes. The 16S rRNA is well

Charles Darwin

conserved, hence it is considered for such studies. The more divergent the nucleotide sequence of the genome, the more distantly related are the organisms. On the basis of above studies a phyletic classification of bacteria has become possible.

3. Ribosomal RNA and its Sequencing

Ribosomal RNA proved to be a universal tool for the phylogenetic analysis and interrelatonship among the organisms. It is ancient, universally distributed and most conserve region in the genome of the microorganisms. Although in prokaryotic ribosomes, there are three different ribosomal RNAs i.e. 5S, 16S and 23S but only 16S sequence is used because nucleotides are neither less nor more in length and easy to sequence. About 1,500 nucleotides are present in 16S rRNA which are determined by using polymerase chain reaction (PCR) and gene sequencer. The 5S rRNA contains only ~ 120 nucleotides that are too small to conclude any fruitful relationship, whereas 23S contains approximately ~2,900 nucleotides which are quite high in number; therefore, it is not found suitable for experiments. In eukaryotes, 18S rRNA is used for phylogenetic studies. The database of rRNA sequences of both prokaryotic and eukaryotic microorganisms can be accessed on internet.

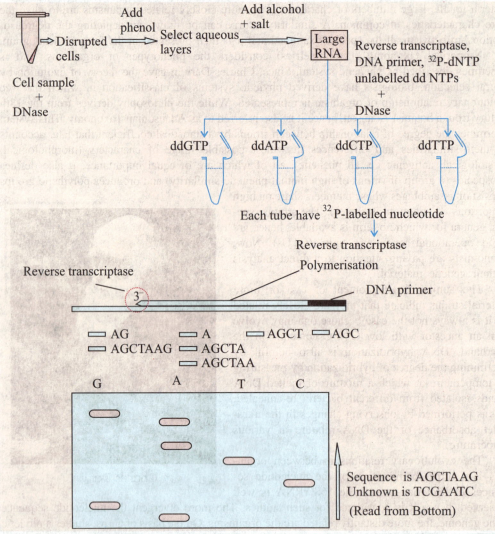

Fig. 2.5 : Methodology for the determination of sequences (diagrammatic).

Ribosomal RNase is easily extracted from cells. About 300-500 mg cells are taken in a centrifuge tube, mixed with appropriate quantity of DNAse to breakdown and isolate RNA for sequencing. The heated cells are mixed with phenol (0.3-0.5 g) for the removal of proteins and carbohydrates/polysaccharides, etc. to release the RNA which was precipitated by using alcohol and salt. The RNA so obtained is mixed with DNA primer (oligonucleotide) complementary to region in 16S rRNA. The enzyme reverse trancriptase is added along with ^{32}P- labelled deoxynucleotide triphosphate (dATP) and other unlabelled dideoxynucleotide triphosphate (ddNTP). These are mixed and taken equally in four tubes one for each 32P-labelled deoxynucleotide triphosphate. In each tube, a small amount of different ddNTP is added. Reverse transcriptase determines the 16S rRNA template and terminates the DNA copy. The fragments are then separated by gel-electrophoresis. The sequence of 16S rRNA is then determined by using standard cDNA sequence.

Now-a-days, ribosomal RNA sequencing using the polymerase chain reaction (PCR) to yield many copies of 16S ribosomal gene is amplified and the methodology given by Sanger (1977) is followed to sequence shown in Fig. 2.5.

4. Construction of the Phylogenetic Tree

The rRNA sequence is used to construct phylogenetic tree by applying distance-matrix method. The evolutionary distance (ED) is determined by recording the differences in the sequences of two or more organism(s) by software computer analysis. A statistical correction factor is applied due to the reason the some changes might have taken place in the genome which would lead back to the same sequence. This is analysed using computer software.

The construction of phylogenetic tree after ED measurements is shown in Fig. (2.6). The different ED of two organisms is directly proportional to the total length of the branches separating them. Depending on software/computer programme and the number of microorganisms, different formats of phylogenetic trees are possibly constructed. Fan-like (as in case of Carl Woese three Domain system of classification) or dendograms (highly branched) are some of the example of phylogenetic trees.

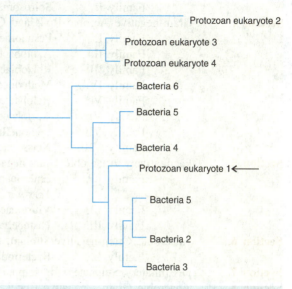

Fig. 2.6 : Phylogenetic tree based on genes that do not match progeny of organisms.

III. Bergey's System of Bacterial Classification

The 'Bergey's *Manual of Systematic Bacteriology'* has four volumes, that contain the internationally recognised names and descriptions of bacterial species. The details of the informations in the above volumes are summarised below.

Vol. I: (Sections 1-11) 1984: Gram – negative bacteria
Vol II: (Sections 12-17) 1986: Gram – positive bacteria, Phototrophic and other specialised bacteria including gliding bacteria.
Vol. III: (Sections 18-25) 1989: Archaeobacteria
Vol. IV: (Sections 26-33) 1991: Actinomycetes and other filamentous bacteria.

KINGDOM PROKARYOTAE
(Prokaryotic organism with primordial nucleus)

Division I. Gracilicutes
(Prokaryotes with thinner cell walls, with a Gram-negative type of cell wall)
Division II. Fimicutes
(Prokaryotes with thick and strong skin, indicating a Gram - positive type of cell wall)
Division III. Tenericutes
(Prokaryotes of pliable soft nature, indicative of lack of rigid cell wall)
Division IV. Mendosicutes
(Prokarotes having faulty cell walls, suggesting the lack of conventional peptidoglycan)
Section 1. The Spirochetes
        Order I.   Spirochaetales
        Family I.   Spirochaetaceae e.g. *Spirochaeta*
        Family II.   Leptospiraceae e.g. *Leptospira*

Section 2. Aerobic/Microaerophilic, Motile, Helical/Vibroid Gram-negative bacteria.

Section 3. Nonmotile (or rarely motile), Gram - negative, curved bacteria.

      Family I.   Spirosomaceae, e.g. *Spirosoma*

Section 4. Gram-negative Aerobic Rods and Cocci.

      Family I.   Pesudomonadacae e.g. *Pseudomonas*

      Family II.   Azotobacteriaceae e.g. *Azotobacter*

      Family III.   Rhizobiaceae e.g. *Rhizobium*

      Family IV.   Methylococcaeae, e.g. *Methylococcus*

      Family V.   Halobacteriaceae e.g. *Halobacterium*

      Family VI.   Acetobacteriaceae e.g. *Acetobacter*

      Family VII.   Legionellaceae e.g. *Legionella*

      Family VIII.   Neisseriaceae e.g. *Neisseria, Beijerinckia*

Section 5. Facultatively Anaerobic Gram-negative Rods.

      Family I.   Enterobacteriaceae e.g. *Escherichia, Shigella Yersinia*

      Family II.   Vibrionaceae e.g. *Vibrio*

      Family III.   Pasteuellaceae e.g. *Actinobacillus, Haemophilus.*

Section 6. Anaerobic Gram-negative straight, curved and Helical Rods.

      Family I.   Bacteriodaceae e.g. *Bacteroides*

Section 7. Dissmilatery Sulphate - or sulphur-reducing bacteria.

Section 8. Anaerobic Gram - negative Cocci

      Family I.   Veillonellacae e.g. *Veillonella*

Section 9. The Rickettsias and Chlamydias

Order I. Rickettsiales

      Family I.   Rickettsiaceae e.g. *Rickettsia*

      Family II.   Bartonellaceae e.g. *Bartonella*

      Family III.   Anaplasmataceae e.g. *Anaplasma*

Order II. Chlamydiales e.g. *Chlamydia*

      Family I.   Chlamydiaceae

Section 10. The Mycoplasmas

      Division   Tenericutes

      Class I.   Mollicutes

      Order I.   Mycoplasmatales

      Family I.   Mycoplasmataceae e.g. *Mycoplasma*

      Family II.   Acholeplasmataceae e.g. *Acholeplasma*

      Family III.   Spiroplasmataceae e.g. *Spiroplasma*

Section 11. Endosymbionts.

A. Endosymbionts of protozoa, ciliates, flagellates, amoebae

B. Endosymbionts of Insects

C. Endosymbionts of fungi and invertebrates other than Arthropods.

Section 12. Gram-postive cocci

      Family I.   Micrococcaceae e.g. *Micrococcus*

      Family II.   Deinococcaceae e.g. *Deinococcus*

Section 13. Endospore forming Gram-positive rods and cocci

For example, *Bacillus, Clostridium* etc.

Section 14. Regular, nonsporing, Gram-positive rods

For example, *Lactobacillus, Renibacterium* etc.

Section 15. Irregular, non-sporing, Gram-positive

For example, *Corynebacterium, Microbacterium* etc.

Section 16. The mycobacteria

Family: Mycobacteriaceae e.g. *Mycobacterium*

Section 17.   Nocardioforms

For example, *Nocardia, Rhodococcus* etc.

Section 18.   An oxygenic, phototrophic bacteria

I.   Purple bacteria

        Family I.   Chromatiaceae e.g. *Chromatium*

        Family II.   Ectothiorhodospiraceae e.g. *Ectothiorhodospira*

Purple non-Sulphur bacteria

e.g. *Rhodospirillum, Rhodobacter* etc.

II.   Green bacteria

      Green Sulphur bacteria

      e.g. *Chlorobium, Chloroherpeton* etc.

      Multicellular, filamentous, Green bacteria

      For example, *Chloroflexus, Heliothrix*, etc.

III.   General Incertae Sedis

      For example, *Heliobacterium* and *Erythrobacter*

Section 19.   Oxygenic Photosynthetic Bacteria

Group I: Cyanobacteria

Subsection I   Order: Chroococcales

Subsection II   Order: Pleurocapsales

Subsection III   Order: Oscillatoriales

Subsection IV   Order: Nostocales

Subsection V   Order: Stigonematales

Group II: Order: Prochlorales

Family   Prochloraceae

        For example *Prochloron* and *Prochlorothrix*

Section 20.   Aerobic Chemolithotrophic Bacteria and associated organisms

A.   Nitrifying Bacteria

      Family Nitrobacteriaceae

      —   Nitrite-oxidizing bacteria

            e.g. *Nitrobacter, Nitrospira, Nitrococcus* etc.

      —   Ammonia-Oxidizing, bacteria

            e.g. *Nitrosomonas, Nitrosococcus, Nitrosospira*

            Nitrosolobus, Nitrosovibrio.

B.   Colourless Sulphur Bacteria

      For example, *Thiobacterium, Macromonas, Thiospira* etc.

C.   Obligately Chemolithotrophic Hydrogen bacteria

      e.g. *Hydrogenobacter*

D.   Iron and manganese-Oxidizing and/or depositing bacteria

      Family Sidero capsaceae (e.g. *Siderocapsa*)

E.   Magnetotactic bacteria

      e.g. *Aquaspirillum magnetotacticum* and *Bilophococcus*

Section 21.   Budding and/or Appendaged Bacteria

I. Prosthecate Bacteria

A. Budding bacteria

    1. Buds produced at tip of prostheca e.g. *Hyphomonas*

    2. Buds produced on cell surface e.g. Prosthecomicrobium

B. Bacteria that divide by binary transverse fission

   e.g. *Caulobacter, Prosthecobacter.*

II.   Non-Prosthecate Bacteria

A. Budding bacteria

    1. Lack Peptidoglycan e.g. *Planctomyces*

      2. Contain peptidoglycan e.g. *Ensifer, Blastobacter*

  B. Non-budding stalked bacteria e.g. *Gallionella, Nevskia*

  C. Other bacteria

      1. Nonspinate bacteria e.g. *Seliberia, Thiodendron*

      2. Spinate bacteria

Section 22.	Sheathed Bacteria
	For example, *Sphaerotilus, Leptothrix, Clonothrix*
Section 23.	Non photosynthetic, Non fruiting Gliding Bacteria
	Order I. Cytophagales
	Family I. Cytophagaceae
	e.g. *Cytophaga, Capnocytophaga*
	Order II. Lysobacteriales
	Family I. Lysobacteriaceae
	e.g. *Lysobacter*
	Order III. Beggiatoales
	Family I. Beggiatoaceae
	e.g. *Beggiatoa, Thiothrix,Thioploca*
	Other families
	Family Simonsiellaceae
	Family Pelonemataceae
Section 24.	Fruiting Gliding Bacteria (The Myxobacteria)
	Order Myxococcales
	Family I. Myxococcaceae e.g. *Myxococcus*
	Family II. Arthangiaceae e.g. *Archangium*
	Family III. Cystobacteriaceae e.g. *Cystobacter*
	Family IV. Polyangiaceae e.g. *Polyangium*
Section 25.	Archaeobacteria
	Group I. Methanogenic Archacobacteria
	Order I. Methanobacteriales
	Family I. Methanobacteriaceae e.g. *Methanobacterium*
	Family II. Methanothermaceae e.g. *Methanothermus*
	Order II. Methanococcales
	Family Methanococcaceae e.g. *Methanococcus*
	Order III. Methanomicrobiales
	Family I. Methanomicrobiaceae e.g. *Methanomicrobium*
	Family II. Methanosarcinaceae e.g. *Methanosarcina,*
	Methanolobus

Group II. Archaeobacterial   Sulphate Reducers

                 Order   Archaeoglobales

                 Family   Archaeoglobaceae e.g. *Archaeoglobus*

Group III. Extremely Halophilic   Archaeobacteria

                 Order   Halobacteriales

                 Family   Halobacteriaceae e.g. *Halobacterium,*

                                          Halococcus, Haloferax etc.

Group IV. Cell Wall-less Archaeobacteria e.g. *Thermoplasma*

Group V. Extremely Thermophilic sulphate-Metabolizers

                 Order I.   Thermococcales

                 Family   Thermococcaceae e.g. *Thermococcus*

                 Order II.   Thermoproteales

Family I.	Thermoproteaceae e.g. *Thermoproteus*
Family II.	Desulfurococcaceae e.g. *Desulfurococcus*
OrderIII.	Sulfolobales
Family	Sulfolobaceae e.g. *Sulfolobus*

Section 26. Nocardioform Actinomycetes
 e.g. *Nocardia, Rhodococcus, Saccharomonospora*

Section 27. Actinomycetes with Multilocular Sporangia
 e.g. *Frankia, Dermatophilus*

Section 28. Actinoplanetes
 For example, *Actinoplanes, Micromonospora*

Section 29. Streptomycetes and related genera
 e.g. *Streptomyces, Kineosporia*

Section 30. Maduromycetes
 e.g. *Actinomadura, Microbispora*
Microtetraspora, Streptosporangium

Section 31. Thermomonospora and Related Genera
 e.g. *Thermomonospora, Nocardiopsis*

Section 32. Thermoactinomycetes
 e.g. *Thermoactinomyces*

Section 33. Othera Genera
 e.g. *Pasteuria, Saccharothrix, Kibdelosporangium.*

C. General Features and Classification of Some Groups of Microorganisms

I. The Rickettsias

This group include two subgroups, Rickettsias and chlamydias

1. Rickettsias

Most live in an obligate intracellular association parasitic or mutualistic, with eukaryotic hosts (vertebrates or arthopods); a few can be grown on moderately complex bacteriological media containing blood.

There cell wall contain muramic acid. While glutamate is oxidised with ATP generations. Mainly there are rod-shaped, coccoid, and often pleomorphic that stain gram-negative and lack flagella. The parasitic species are associated with the reticuloendothelial and vascular endothelial cells or erythrocytes of vertebrates and often with various organs of arthropods, which may acts as vectors or primary hosts. The mutualistic species are found in insects. Important Genera are: *Bartronella, Grahamella* and *Rochalimaea.*

It is named after Howard T. Ricketts who identified this microorganisms as the

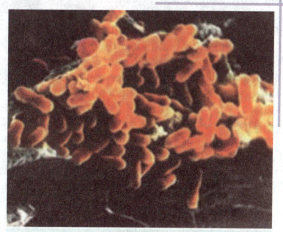

Rickettsias.

causal organism of typhus and Rocky Mountain spotted fever. Unfortunately he along with his co-worker, Baron von Prowazek died from this infection. They are small, Gram-negative coccobacilli which are obligate intracellular parasites, grow in the cytoplasm of infected cells of mammals and arthropods. Rickettias can be cultured only in embryonated eggs and grow at 35°C.

Genera of Rickettsias in infected persons cause Typhus fever in of three different forms that includes *(i)* Epidemic typhus by *R. prowzekii*. Brill-Zinsser disease, or recrudescent typhus by the type of antibodies formed. Epidemic typhus elicits IgM and then IgG antibodies. *R. tsutsugarmushi* causes scrub typhus and grows mostly in nuclei. *Coxiella rochalimaea quintana* causes trench fever and it is the only genus that grows on artificial medium. *Bartonella hemselae* causes bacillary angiomatosis and lesser the immune system. *Coxiella burnetii* causes Q fever which is a pneumonia-like disease.

2.   Chlamydias

These are nonmotile, obligate parasite, coccoid bacteria multiply with in membrane bounded vacuoles in the cytoplasm of the cells of humans, other mammals, and birds (Arthropods do not act as host). The multiplication occurs by means of a unique development cycle, which divide by fission. They are pathogenic. The specialised techniques are involved for their cultivation.

These are small (0.2–1.0 µm) rod or cocoid, Gram-negative having both DNA and RNA. They are non-motile, obligate para-site, cocoid bacteria live inside the host cells. They are pathogenic. The cell walls are devoid of mu-ramic acid and glucose is not oxi-dized with ATP generation, hence parasite depends upon the host ATP cells. The specialized techniques are involved for their cultivation. They are present in two different forms during their life cycle. A large, metabolically active reticu-late body which is non-infectious with flexible cell wall and a smaller form are present which is called elementary body. This body causes infection and can attack other cells of the same host or transmitted to

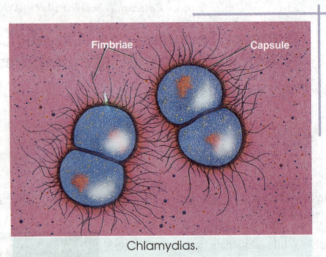

Fimbriae   Capsule

Chlamydias.

new hosts. They multiply by a process of attachment of elementary body to the host cell wall followed by phagocytosis, conversion, reproduction, condensation and release of the elementary and reticulate bodies by description of the wall material. The reticulate body divides by binary fission.

There are three species of *Chlamydia psittaci* which cause severe pneumonia in humans and parrot fever in birds, whereas *P. pneumonia* causes human respiratory infection. *C. trachomatis* is a major cause of the eye infection trachoma. Some strains of *C. trachomaits* also causes sexually transmitted disease called *lymphogranuloma venereum*. The Chlamydias are transmitted by air. The generation time is 2-3 hours which is faster than rickettsias and chlamydias are classified as below:

Order I Rickettsiaceae
Family I: Rickettsiaceae
Tribe I: Rickettsieae

Genera: *Rickettsia, Rochalimaea, Coxiella*
Tribe II: Ehrlichieae
Genera *Ehrlichina, Cowdria, Neorickttesia*
Tribe III: Wolbachieae
Genera *Wolbacia, Rickettsiella*
Family II: Bartonellaceae
Genera *Bartonella, Grahamella*
Family III: Anaplasmataceae
Genera *Anaplasma, Aegyptianella, Haemobartonella, Eperythrozoon*
Order II Chlamydiales
Family I: Chlamydiaceae
Genus *Chlamydia*

II. The Mycoplasmas (or Mollicutes) : The Cell Wall-Less Bacteria

This group consists of very small prokaryotes totally devoid of cell walls, consist of only a plasma membrane and incapable of synthesis of peptidoglycan and its precursors. These are resistant to penicillins and sensitive to lysis by osmotic shock, detergents and alcohols. These are pleomorphic, varying in shape from spherical or pear shaped structures (0.3-0.8 m in dia) to branched or helical filaments. Usually nonmotile, but some show gliding motility in liquid surfaces. They are Gram-negative. Most species require sterols and fatty acids in their growth media for growth. mostly these are facultatively anaerobic, but few are obligate anaerobes, and under favourable conditions, most species form colonies that have a characteristic of "fried egg" appearence. All are the parasites, commensals or saprophytes, but several are pathogenic to human, animals, plants and insects. The mol% G+C of the DNA ranges from 23-40.

Based on sterol requirement, mycoplasmas formed two separate groups. Genera *Mycoplasma, Anaeroplasma* and *Ureaplasma* require sterols to grow but *Acholeplasma, Asteroplasma* and *Thermoplasma* do not require sterols. *Mycoplasma* and *Anaeroplasma* contain lipoglycans long chain heteropolysaccharides covalently linked to membrane lipids. But other genus such as *Spiroplasma* and *Ureaplasma* do not contain lipoglycans.

Classification :

Division :	*Tenericutes*	
Class I :	*Mollicutes*	
Order I :	*Mycoplasmatales*	
Family I :	*Mycoplasmataceae*	**e.g.** *Mycoplama, Ureaplasma*
Family II :	*Ascholeplasmataceae*	**e.g.** *Acholeplasma*
Family III :	*Spiroplasmataceae*	**e.g.** *Spiroplasma*
Other genera		**e.g.** *Anaeroplasma, Thermoplasma*

Genus : *Mycoplasama*

It is pleomorphic, sometimes slightly ovoid to slender branched filaments. The cells lack a cell wall and are bounded by a three layered membrane only. They are Gram-negative and usually nonmotile, but gliding motiltiy has been described in some species. These are facultatively anaerobic, possessing a truncated flavin-terminated electron transport chain devoid of quinones and cytochromes. The typical colony has a "fried egg" appearance. They are chemoorganotrophic and require cholesterol or related sterols for growth. They are parasites and pathogens of mammals and avian hosts. But the mol% G+C of the DNA ranges from 23 to 40%. UGA codon is read as tryptophan signal.

Type species: *Mycoplasma mycoides.*

Genus: *Spiroplasma*

Cells are pleomorphic and shape varies from helical and branched to spherical or ovoid. The helical filaments are motile, with twitching movements, and often show an apparent rotatory mobility. Flagella, periplasmic fibrils, or other organelles of locomotion are not present, but intracellular fibrils have been demonstrated. The colonies are frequently diffuse, showing the mobility of the cells during active growth. Cholesterol is required for growth. Most of them isolated from ticks, from the haemolymph and guts of insects, from vascular plant fluids and insects that feed on the fluids and from the surfaces of flowers and other plant parts. The mol% G+C of the DNA is 25-31.

Type species: *Spiroplasma citri.*

III. Algae

A famous botanist F.E. Fritsch (1935) classified algae into following 11 classes, based on pigmentation, reserve food material, flagellation and reproduction.

Class 1. Chlorophyceae (= Isokontae)

Generally algae are fresh water and chlorophyllous thallophytes. Chlorophyll *b* and carotenoides are present in chloroplasts. The cell wall is made up of cellulose and food is synthesized in the form of starch. Motile spores and cilia are found. The sexual reproduction is isogamous, anisogamous and oogamous types.

Important genera are: *Chlamydomonas, Volvox, Chlorella, Ulothrix, Spirogyra*

Class 2. Xanthophyceae (= Heterokontae)

These are green-yellow in colour due to the presene of xanthophyll. The pyrenoids are absent and food is in the form of fat. Chlorophyll *e* is found in place of chlorophyll b. The sexual reproduction occurs by fission of two gametes having cilia of different length.

Important genera are: *Microspora, Vaucheria, Protosiphon*

Class 3. Chrysophyceae

In these organisms, besides chlorophyll, yellow-green pigments are present. Phycocyanin is the colouring material. Plants are unicellular, multicellular or colonial. The cell wall is present in the form of two overlapping halves and stored food is in the form of oil or insoluble carbohydrates, leuosin.

Example : *Chrysosphaera*

Class 4. Bacillariophyceae (= diatoms)

These are yellow-green-brown or olive green in colour. Diatomin is the colouring material which is found in chloroplast. Pyrenoids are also present. These are unicellular and non-motile Chlorophyll *c* is present in place of chlorophyll b.

Example : *Pinularia, Navicula, Fragilaria*

Class 5. Cryptophyceae

These are red, green-blue, olive-green or green coloured algae. Each cell consists of two large chloroplasts in which pyrenoids are present. They occur in fresh water and sea.

Examples: *Cryptomonas.*

Class 6. Dinophyceae

These are dark yellow or brown or red coloured algae. Stored food is oil or starch. Large nucleus and many disc like chromatophores are present.
Example: *Peridinium.*

Class 7. Chloromonadineae

These algae are bright green or olive green colour. Xanthophyll is in abundance. Fatty compounds acts as food. Reproduction takes place by longitudinal division.
Examples: *Vacuolaria*

Class 8. Eugleninae

They resemble microscopic animal due to presence of naked ciliated reproductive organs. Chlorophyll is present.
Example: *Euglena.*

Class 9. Phaeophyceae

These are yellow-brown coloured marine algae. Fucoxanthin pigment is the main colouring material. chlorophyll a and carotene are also found and chlorophyll c is found in place of Chlorophyll b. Storage food materials are laminarian, mannitol and fats. Zoospores are biciliated and one cilium is larger. There is no resting period in zygote.
Example: *Fucus, Sargassum.*

Class 10. Rhodophyceae

These are red in colour due to phycoerythrin pigment. Phycocyanin, chlorophyll a, carotene and xanthophyll are also present in small quantities. Storage food is floridean starch. Non-motile cells are found during reproduction. These are commonly found in sea water. Sexual reproduction is oogamous type. Chlorphyll d is present in place of chlorophyll b.
Example: *Polysiphonia* and *Batrachospermum.*

Class 11. Myxophyceae (=Cyanophyceae)

The nucleus is of prokaryotic type. The blue colour is due to the presence of phycocyanin pigment but phycoerythrin, chlorophyll b and carotene are also present in small quantity. The accessory pigment (*i.e.,* phycocyanin, phycoerythrin and allophycocyanin) conain structure is called phycobilisomes. The chlorophylls are found in thyllakoids. Storage food is myxophycean starch and protein granules. There is no motile stage in these algae. Sexual reproduction is absent. Mainly these algae are unicellular or filamentous.
Examples: *Nostoc, Oscillatoria, Anabaena, Lyngbya, Plectonema.*

IV. Archaeobacteria (Archaea)

1. Characterstics

All organism of this group are assumed to be descendent from a common ancestor, the so-called progenote and the organisms are primitive, cellular with a rudimentary not well coordinated translation apparatus. The unique feature is in the cell membrane structure. The membrane lipids are glycerol isopronyl ethers due to the unusual chemical composition. The cell envelopes show resistance against cell wall antibiotics and lytic agents. These bacteria exhibit unique cell envelope. All lack murein but exceptionally it occurs also in some eukaryotes such as *Plantomyces.* The absence of ribothymidine in the "common" arm of the tRNA is unusual. It is replaced by 1- methylpse-udouridine or pseudouridone.

All Gram-negative bacteria composed of single layered or more complex crystalline protein or glycoprotein subunits while all Gram-positive bacteria have pseudomurein, methanochondrin or heteropolysaccharide. Single S-layers (three dimensional structure) are found in most organisms which helps in withstanding extreme environmental condition such as high salt, low pH and high temperature. The size of the genome of archaeobacteria varies from 0.8×10^9 dalton (*Hermoplasma*) to 2.4×10^9 dalton (*Halobacterium*) and is generally smaller than *E.coli*. The G+C contents of the DNA exhibits a broad range from 25-68 mol%.

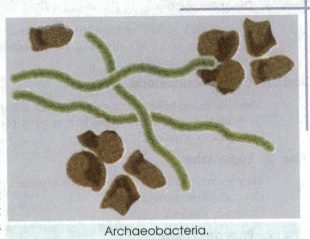

Archaeobacteria.

The archaeobacteria are distinguished from the other two kingdoms by the sequence signature of the 16S rRNA.

The unique properties of living conditions (archaeobacteria) are of extreme pH, temperature, salinity, and pressure to establish a "new biotechnology" which helps in production of biogas, in ultrafiltration production of thermophilic enzymes, restriction enzymes, leaching and as biosensors.

2. Classification : Archaeobacteria have 5 orders and 9 families

Classification of Methanogenic Bacteria

KINGDOM: ARCHAEOBACTERIA

Orders	Family	Genus
I. Methanobacteriales	I. Methanobacteriaceae	*Methanobrevibacter, Methanothermobacter, Methanosphaera, Methanobacterium.*
	II. Methanothermaceae	*Methanothermus*
II. Methanococcales	I. Methanococcaceae	*Methanococcus Methanothermococcus*
III. Methanomicrobiales	I. Methanomicrobiaceae	*Methanomicrobium, Methanolacinia, Methanogenium, Methanoculleus Methanofollis*
	II. Methanocorpusculaceae	*Methanocorpuculum*
	III. Methanospirillaceae	*Methanospirillum*
IV. Methanosarcinales	I. Methanosarcinaceae	*Mtehanosarcina Methanohalophilus Methanolobus, Methanohalobium Methanosalsus Methanococoides*
	II. Methanosetaceae	*Methanosaeta*
V. Methanopyrales	Methanopyraceae	*Methanopyrus*

Classification of Extremely Halophillic Bacteria

Orders	Family	Genus
I. Halobacteriales	I. Halobacteriaceae	1. *Halobacterium* 2. *Haloarcula* 3. *Haloferax* 4. *Halococcus* 5. *Natronobacterium* 6. *Natromococcus*

Classification of Thermophilic Sulfur-metabolizing Bacteria

Orders	Family	Genus
I. Sulfolobales	I. Sulfolobaceae	1. *Sulfolobus* 2. *Acidianus* 3. *Desulfurolobus*
II. Thermoproteales	I. Thermoproteaceae	1. *Thermoproteus* 2. *Pyrobaculum* 3. *Thermofilum*
	II. Desulfurococcaceae	*Desulfurococcus*
	III. Staphylothermaceae	*Staphylothermus*
	IV. Pyrodictiaceae	*Pyridictium*
	V. Thermodiscaceae	*Thermodiscus*

Classification of Intermediate Groups

Orders	Family	Genus
I. Thermoplasmales	I. Thermoplasmaceae	*Thermoplasma*
II. Archaeoglobales	II. Archaeoglobaceae	*Archaeoglobus*
III. Thermococcales	III. Thermococcaceae	*Pyrococcus* *Thermococcus*

The characters and classifications of viruses, viroids and prions are given in the chapter 15.

V. Actinomycetes

1. General Features

Actinomycetes are called actinobacteria or filamentous bacteria due to their mycelium like (slender and branched) structures. These filaments are long and it may fragment into much smaller units and less broad than that of the fungal mycelium usually 0.5 to 1.0 μm in diameter but sometimes reaches to 2.0 μm in few cases. A chain of sexual spores called conidia are produced on their hyphae, and few of the actinomycete (genera) found in soil bear the sporangium containing spores. The colonies are powdery mass over the surface of culture media, often these are pigmented when the aerial spores are produced.

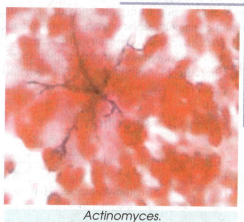

Actinomyces.

They are Gram-positive rod-shaped to filamentous, aerobic and generally non-notile in the vegetative phase. Due to their rod-shaped arrangement, they are called filamentous bacteria. Most of the genera have high mol (%) G+C. They are morphologically quite different with each other. *Actinomyces* is anaerobic to facultatively aerobic, filamentous microcolony, whereas *Streptomyces* has intact and abundant aerial mycelium with long chain spores. *Nocardia* a common soil organism is an obligate aerobe. Nitrogen-fixing actinomycetes, *Frankia* produces true mycelium, and they are microaerophilic and live symbiotically with non-leguminous plants. The characteristics of some important genera are given below.

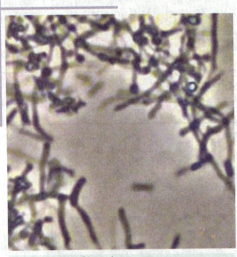

Streptomyces.

Micrococcus, is a non-motile, Gram-positive cocci, aerobic with a high DNA base composition (66 to 72 mol % G + C). It is generally found in soil and produce correlated pigments yellow to red in colour. This helps in protection against UV light.

Arthrobacter, is also a soil-inhabitant. They are coccoid but when growing actively, they form irregular rods. They are gram-positive but some strains are gram-negative. Some are motile in nature.

Frankia, is a plant symbiont that produces root-nodules fixes nitrogen in non-leguminous plants such as *Alnus* (alder), *Ceanothus* (wild lilae) and *Casuarina* (Australian pine or She-wood). It is difficult to cultivate on artificial culture medium. The aerial mycelium develops multicellular sporangium.

Actinomyces is a typical mycelial actinobacteria. They are anaerobic or facultative anaerobic microorganism. They do not produce an aerial mycelium. *A. bovis* grows in oral cavities and can cause serious infection such as lumpy jaw.

Bifidobacterium, is a mycelial form but recently, its phylogeny places it as a similarity to *Actinomyces*. It is anaerobic, irregular rod-shaped bacteria, found in the intestinal tracts of animal e.g. *B. bifidus*. It also grows in human breast milk which contains amino sugar not found in cow's milk.

Streptomyces, is mycelial forming actinobacteria that lives in soil, produces aerial as well as substrate mycelium. The aerial hyphae differentiate to form asexual conidiospores in chains. They are blue, gray, green, red, violet or yellow colour. They impart "earthy' odor to soil after rain which is due to the presence of geosmines (volatile organic compound), some important antibiotics namely, streptomycin, chloramphenicol and tetracycline are produced from this genus.

Helicobacteria, is recently discovered actinobacteria. This is the only phototrophic gram-negative bacteria, required organic carbon sources. They have a unique type of bacteriochlorophyll called bacteriochlorophyll *g* e.g. *Helicobacillus* and *Helicobacter*. They are anaerobes carrying out anoxygenic photosynthesis.

2.   Classification

These are divided into 7 families. The classification is based on hyphal and reproductive structures.

Family 1: Streptomycetaceae: Hyphae non-fragmented, aerial myceliium with chains of spores with 5 to 50 or more conidia per chain e.g. *Streptomyces, Microdlobaspone* and *Sporictilhya*.

Family 2: Nocardiaceae: Hyphae typically fragmented e.g. *Nocardia, Pseudonocardia*.

Family 3: Micromononsporaceae: Hyphae non-fragmented conidia borne singly or in pairs or in short chains. e.g. *Micromonospora, Thermonospora, Thermoactinomycetes, Actinobifida*.

Family 4: Actinoplanaceae: Sporangia bear the spores. The hyphal diameter varies from 0.2 to 2.0 μm e.g. *Streptosporangium, Actinoplanes, Plasmobispora* and *Dactylosporangium*.

Family 5: Dermatophilaceae: Hyphal fragments divide to form large numbers of round, motile structures e.g. *Geodermatophilus*.

Family 6: Frankiaceae: It is strictly associated with root of non-leguminous plant and form root nodules e.g. *Frankia*.

Family 7: Actinomycetaceae: No true myceluim is produced, usually strictly to facultative anaerobic e.g. *Actinomyces*.

Due to the development of modern techniques of molecular biology such as 16S rRNA sequencing, phylogeny and relationship, mycelium has the following criteria.

The actinomycetes line of descent (including the *Atopobium, Sphaerobacter* and the lineages) includes the following:

(*a*) The *Actinomycete, Bifidobacterium, Arthrobacter* and Relatives

(*b*) The family actinomycetaceae and relatives, *Propionibacterium* and relatives
The mycolic acid containing genera (formerly the CMN Cluster)

(*c*) The family Actinoplanaceae and relatives.

(*d*) The family Pseudonocardiaceae and relatives.

(*e*) The family Frankiaceae and relatives.

(*f*) The family Streptomycetaceae and Streptosporangiaceae and the genera *Actinomadura* and *Nocardiopsis*.

(*g*) Actinomycetes of uncertain phylogenetic applications.

VI. Fungi

1. General Characteristics

Representatives of two phyla of fungi: Ascomycetes and Basidiomycetes.

The fungi are cosmopolitan in distribution, some are aquatic, others are terrestrial and still others are air borne. Many are parasitic on plants, animals and human beings. The plant body typically consists of branched and filamentous hyphal form, a net like structures called as mycelium. The hyphae are aseptate (coenocytic) or septate and uni, bi or multi nucleate. The septa usually have simple pores, but the septa in basidiomycetous genera has dolipore septum. The cell wall mainly consists of fungal cellulose or chitin. These are devoid of chlorophyll but carotenoids are normally present. The fungi are heterotrophic i.e. cannot manufacture their own food, hence these are either parasites, saprophytes or symbionts.

There are three primary divisions of microbial world, based on rRNA sequences. Among them archaea and eubacteria are prokaryotes, while the third group is eukaryotic which contains a true nucleus. They are morphologically complex. Three multicellular eukaryotic kingdoms: fungi, the green plants and animals are grouped in this category but we shall remain confined to fungi.

Fungal cell wall consists of chitin but in some fungi mannans, galactosans and chitisans are also present. All the fungi are chemoorganotrophs, lacking chlorophyll but have simple nutritional requirements.

Slime molds are another category of fungi which have morphological similarity of both fungi and protozoa. They are of two types; cellular slime molds and acellular slime molds. The former are composed of single amoeba-like cells, while the later are naked massess of protoplasm of indefinite size and shape called plasmodia. They live on decaying plant matter and soil and take their food by engulfing bacteria (phagocytosis). The cellular slime mold in *Dictyostelin,* while acellular is *Plasmodium.*

Molds are filamentous fungi associated with bread, cheese of fruits. The mycelium is coenocytic (non-septate), branched. exo- or endogenously borne conidia are produced on the aerial branches. They are asexual reproductive structures of various colours such as black, blue-green, red, yellow or brown. Gametangia develop during sexual reproduction by the fission of specialized hyphae or unicellular gametes to form spores. When these spores are endogenous, they are called ascospores. They are found in a sac like structure called ascus as in ascomycetes, whereas if exogenous i.e. produced out side, they are called basidiospores which are present at the ends of a club-shaped structure known as basidium as in basidiomycetes. Yeasts belonging to ascomycetes are unicellular usually spherical, oval or cylindrical and cell division occurs by budding. Generally, yeasts do not form mycelium and remain present in unicellular form. But under certain conditions, baker's yeasts *Saccharomyces cerevisiae* forms pseudomycelium. Yeast exhibits sexual reproduction i.e. mating by cell fusion to form zygote which froms ascospores. They grow on sugary habitat.

Another major group is basidiomycetes. They are filamentous forming fruiting bodies called basidiocarp. They are also found on dead organic matter in the soil or on the trunks of trees. Sexual spores called basidiospores are produced by them. The fusion of two basidiospores or haploid mycelium resulting in the formation of dikaryotic mycelium. The mycelium grow further to give rise thallus.

2.   Classification

They are classfied into 7 divisions, distinguished by the presence or absence of a plasmodium or pseudoplasmodium. Further named as Division Myxomycota (fungi with plasmodia or pseudoplasmodia) while the majority of fungi consist of filamentous structure and named as Eumycota.

Division I: Myxomycota

It consists of four classes:

Class I: Acrasiomycetes: Assimilative phase is free living amoebae which unite as a pseudoplasmodium before reproduction.

Class II: Hydromyxomycetes: Plasmodium forming a network (net plasmodium)

Class III: **Myxomycetes:** Plasmodium is saprobic, free-living.

Class IV: **Plasmodiophoromycetes:** Plasmodium is parasitic within cell of the host plant.

Division II: Eumycota

Divided into five major groups, distinguished from each other as shown below:

Subdivision 1-*Mastigomycotina:* Motile cells (zoospores) present. Perfect stage spores is typically oospores.

Subdivision 2-*Zygomycotina:* Perfect-state spores are zygospores.

Subdivision 3-*Ascomycotina:* Perfect-state spores are ascospores.

Subdivision 4-*Basidiomycotina:* Perfect-state spores are basidiospores.

Subdivision 5-*Deuteromycotina:* Perfect-state is absent.

Subdivision I : Mastigomycotina

Mastigomycotina is divided into three classes:

Class I-**Chytridiomycetes:** Zoospores are posteriorly uniflagellate (flagella whiplash type).

Class II-**Hyphochytridiomycetes:** Zoospores are anteriorly uniflagellate (flagella-tinsel type).

Class III-**Oomycetes:** Zoospores biflagellate (posterior flagellum whiplash type anterior tinsel type); cell wall consists of cellulose.

Subdivision II-Zygomycotina

Zygomycotina is divided into two classes.

Class I-**Zygomycetes:** This class comprises of two orders: the Mucorales and Entomophthorales. Mucorales are ubiquitous in soil and dung (mostly saprophytes), entomophthorales include a number of insect parasites.

Class II-**Trichomycetes:** It is a group of uncertain affinity and are mostly parasitic in the guts of arthopods e.g. insect larvae, and millipedes.

Subdivision III-Ascomycotina

Ascomycotina consists of six classes.

Class I-**Hemiascomycetes:** In this asococarps and ascogenous hyphae lacking; thallus yeast like or mycelial.

Class II-**Loculoascomycetes:** Ascocarps and ascogenous hyphae present; thallus mycelial, asci are bitunicate; ascocarp an ascostroma.

Class III-**Plectomycetes:** Asci typically unitunicate; if bitunicate, ascocarp an apothecium. Asci evanescent, scatterd within the astomous ascocarp which is typically a cleistothecium; ascospores aseptate.

Class IV-**Laboulbenomycetes:** Asci regularly arranged within the ascocarp as a basal or peripheral layer. Exparasites of arthopods; thallus reduced; ascocarp a perithecium; asci inoperculate.

Class V-**Pyrenomycetes:** Not exoparasites on arthopods, ascocarp typically a perithecium which is usually ostiolate; asci inoperculate with an apical pore or slit.

Class VI-**Discomycetes:** Ascocarp an apothecium or a modified apothecium, frequently macrocarpic, epigean or hypogean; asci inoperculate or operculate.

Subdivision IV- Basidiomycotina

Basidiomycotina consists of three classes.

Class I-**Teliomycetes:** In these basidiocarp is lacking and replaced by teliospores or chlamydospores (encysted probasidia) grouped in sori or scattered within the host tissue; parasitic on vascular plants.

Class II- Hymenomycetes: In such cases basidiocarp usually well developed, basidia typically organised as a hymenium; saprobic or rarely parasitic. Basidiocarp typically gymnocarpous or semiangiocarpous; basidia (phragmbasidicy) or holobasidia; basidiospores balistospores type.

Class III-Gasteromycetes: Basidiocarp is typically angiocarpous; basidia holobasidia, basidiospores not ballistospores.

Subdivision V- Deuteromycotina

Ainsworth (1973) classified into three classes.

Class I-Blastomycetes: Budding (yeast or yeast like) cells with or without pseudomycelium; true mycelium lacking or not well developed.

Class II-Hyphomycetes: Mycelium well developed, assimilative budding cells absent. Mycelium sterile or bearing spores directly or on special branches (sporophores) which may be variously aggregated but not in pycnidia or acervuli.

Class III-Coelomycetes: Spores are found in acervuli or pycnidia.

VII. Protozoa

Protozoa are minute animalcules visible only under the microscope. About 50,000 species are known so far. A protozoan is an independent eukaryotic cell *i.e.* a complete unicellular organism. The branch of study is called Protozoology. It was Leeuwenhoek (1677) who first of all studied "protozoa" but its name was given by Goldfuss (1817). Locomotion takes place through pseudopodia, flagella or cilia. Nutrition principally heterotrophic, some are autotrophic.

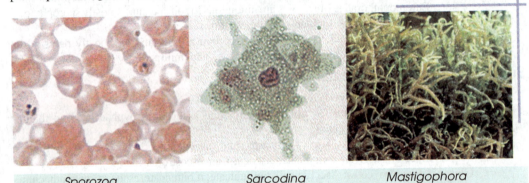

| *Sporozoa* | *Sarcodina* | *Mastigophora* |

They are unicellular eukaryotic microorganisms, lack cell walls, colourless and motile. They are larger than prokaryotes but lack chlorophyll. They obtain food by ingesting other organisms found in fresh water of ponds, pools, ditches, etc. Some are marine but majority of them are parasitic in other animals, including humans. Some are found in soil. The uptake of nutrients (macromolecules) occurs by a process called pinocytosis. Most of them also ingest the food by phagocytosis, but some form special structure called gullet. The mechanisms of their motility divides into taxonomic groups e.g. *Sarcodina* in which movement occurs by amoeboid motion, those using flagella i.e. *Mastigophora*. The protozoans use cilia belong to *Ciliophora* while non-motile and parasitic to higher animals, belong to *Sporozoa*.

Classification

Phylum Protozoa is divided into four subphyla based on locomotory organs.

Subphylum A: Sarcomastigophora

In these cases, the locomotory organelles are pseudopodia or flagella or both. There is presence of one nucleus or more nuclei in organism. There are three classes.

Class - 1: Mastigophora or Flagellata

Class - 2: Sarcodina or Rhizopoda

Class - 3: Opalinata

Subphylum B: Sporozoa (=Apicomplexa)

All the organisms are endoparasites. No specialized locomotory organelles and contractile vacuoles are present but possess an apical complex of ringlike, tubular, filamentous organelles at apical end at some stage of their life-cycle. The life cycle of such organisms is complicated, spores usually formed but have no polar filaments.

The sporozoans are divided into two classes on the basis of sporozoites (contained spores).

Class - 1: Telosporea

Class - 2: Piroplasmea

Subphylum C: Cnidospora (=Microspora)

They are intracellular endoparasites, especially on insects. No locomotory organelles and contractile vacuoles are present. The spores are formed but these differ from sporozoams in possessing one to four polar filaments.

Cnidiospora is divided into two classes on the basis of their mode of spore formation.

Class - 1: Myxosporea

Class - 2: Microsporea

Subphylum D: Ciliophora

This is the largest group of protozoans which are animal-like and structurally complex. The locomotory organelles are cilia or ciliary organelles; cilia are replaced in the adult by sucking tentacles for feeding. A single large-macronucleus of trophic function and one to several small micronuclei of reproductive function are present. The reproduction is mainly by asexual means (transverse fission) and occurs sexually by conjugation.

Class 1: Ciliata: There is only one class in this subphylum.

QUESTIONS

1. Define biological diversity. Discuss about microbial diversity in the context of future prospects.

2. Classify microorganism write in detail about 3-kingdom classification of microrganisms.

3. What a is the difference between identification and classification. Mention for points of modern molecular biological parameters which help in classifying bacteria.

4. Write in detail about the classification of bacteria adopted by Bergey's.

5. Describe the significance of microbial diversity. What are its major applications in the field of Microbiology ?

6. How ribosomal RNA sequences are determined and used in construction of phylogenetic tree?

7. Differentiate between growth rate and generation time. Discuss them in brief in relation to microbial growth.

8. Write short notes on:
   (i) Taxospecies   (ii) Intuitive method   (iii) Bacterial nomenclature
   (iv) 16s rRNA   (v) Numerical taxonomy

9. Discuss in detail about modern systems of bacterial taxonomy,

10. Why is it essential to study biodiversity or microbial diversity? List different diversified groups of microorganisms.

11. Mention the basis of outline classification of any three groups of microorganisms.

REFERENCES

Bull, A.T. and Hardman, D.J. (1991). Microbial diversity. *Curr. Opin. Biotechnol.* 2: 421-28.

Bull, T.A., Michael Goodfellow and J. Howard Slater (1992). Biodiversity as a source of innovation in Biotechnology. *Annu. Rev. Microbiol.* 46: 219-52.

Brock, T.D., 1995. *Biology of microorganisms.**Dicastri, F. and Youness, T. (1990). Ecosystem function of biological diversity. *Biol. Int. Spec. Issue.* 22: 1-20.

Dobell, C. 1960. Antony van Lecuwerhoek and his "Little animals". New York, Dover.

Dubos, R.J. 1950. Louis Pasteur Free Lance of Science, Proston Little, Brown.

Haeckel, E. (1866). *Generelle Morphologic der Organism* Vol.2. Reimer, Berlin. (Cited from M.A. Ragan and D.J. Chapman. 1978, A biochemical phylogeny of the protists, pp. 15, Academic Press, New York, San Francisco, London)

Hawksworth, D.L. (1991). The fungal dimension of biodiversity: Magnitude, significance, and conservation. *Mycol. Res.* 95, 641-645.

Hawksworth, D.L. and Mound, L.A. (1991). Biodiversity databases: The crucial significance of collections. In *Biodiversity of Microorganisms and Invertebrates and its role in Sustainable Agriculture.* (ed D.L. Hawksworth), CAB International: Walling ford, U.K.

Holben, W.E., Jansson, J.K., Chelm, B.K. and Tiedije, J.M. (1988). DNA probe method for the detection of specific microorganisms in soil

Ketchum, Paul A 1988 *Microbiology: Concepts and Applications:* John Wilay and sons, New York.

Mc Neely, G.A., Miller, K.R., Reid, W.V., Mittermeir, R.A., Werner, T.R. (1990). Conserving the World's Biological Diversity,. Gland: Int. Union for Conservation of Nature and Natural resources.

Olsen, G.J., Lane, D.L., Govannoni, S.J., Pace, N.R., and Stahl, D.A. (1986): Microbial ecology and evolution: A ribosomal RNA approach, *Annu. Rev. Microbiol.* 40: 337-365.

Pace, N.R., Stahl, D.A., Lane, D.J. and Olsen, G.J. (1986). The analysis of natural microbial populations by ribosomal RNA sequence *Adv. Microbial. Ecol.* 9: 1-55.

Stahl, D.A. (1988). Phylogenetically based studies of microbial ecosystem perturbation. In *Biotechnology for crop protection.* ed. P.A. Hedin, J.J. Menn, R.M. Hollingworth, pp. 373-390. Washington DC: Am. Chem. Soc.

Stahl, D.A., Flesher, B., Mansfield, H.R., Monotomery, L. (1988). Use of phylogenetically based hybridisation probes for studies of ruminant microbial ecology. *App. Environ. Microbiol* 54: 1079-84.

Wayne, L.G., D.J. Brenner, R.R. Colwell, et al (1987). Report of the ad hoc committee on reconcilation of approaches to bacterial systematic. *Int. J. Syst. Bacteriol.* 37, 463-464.

Whittaker, R.H. (1969). New concepts of kingdoms of organisms. Science, 163, 150-160.

Woese, C.R., O. Kandler, and M.L. Wheelis (1990). Towards a natural system of organisms: Proposals for the domains Archaea, Bacteria and Eukarya. *Proc. Natl. Acad. Sci. USA,* 87, 4576-4579.

Methods in
Microbiology

3

A. Microbial Cultures
B. Microbial Growth
C. Culture Media
D. Staining and
   Smearing

When a particular species of microbe is present in a very small number in comparison to the total number of microorganisms, such culture is called as mixed culture. A culture containing only one species of microbe is called pure culture. The process of screening a pure culture by separating one type of microbe from a mixed population, is called isolation. Although methods of isolation and study of microorganisms are given in a separate book (see *Practical Microbiology* by Dubey and Maheshwari, 2002) yet a brief description of such methods has been given in this chapter.

A. Microbial Cultures

In order to get a pure culture. It is necessary to first achieve an increase in the relative number of species, preferably to the point where the species becomes the numerically dominant component of the population. This can be achieved by using the growth of desired species, while killing or inhibiting others.

There are three categories of methods in order to obtain particular kind of organism, physical, chemical and biological methods.

Microbes of desired species can be cultured in a lab.

1. Physical Conditions for Growth

(*i*) **Heat Treatment :** This is applied in case of bacteria which form spores for example *Bacillus* spp. *Lactobacillus* spp. etc. To select the endospore-forming bacteria, a mixed culture can be heated at 80°C for 10 min before being used to inoculate culture medium. In such case spore forming bacteria will survive and all the bacteria (vegetative cells) will be killed.

Bacteria growing on a medium.

(*ii*) **Incubation Temperature :** Most of the microorganisms require a particular temperature to grow efficiently. Such a temperature is called *optimum temperature*. To select psychrophilic or psychrotrophic bacteria, cultures are incubated at extreme low temperature varying zero to 5°C. The mesophilic (20-40°C) and thermophilic (45°C and above) microbes will be killed, and, only psychrophiles will survive.

(*iii*) **pH :** It is an important factor which influence the growth of microorganisms. Most of the *lactobacilli* spp. grow at acidic range of pH (below 7) while some bacteria grow at alkaline medium e.g. *Vibrio cholerae*.

(*iv*) **Cell Size and Motility :** Most of the microorganisms has a definite size and are motile in nature. Due to 0.15 mm size *Treponema* can be isolated by placing gingival scrapping on the surface of an agar plate. It will penetrate from the filter and will swim through solid agar medium The hazy appearance in the agar indicates the presence of *Treponema*.

2. Requirement of Gases

Oxygen and carbon dioxide are the gases which affect bacterial growth. Aerobic bacteria require about 21% oxygen while anaerobic bacteria do not require oxygen as it is toxic to them. *Facultative anaerobic bacteria* may use it for growth. They are not inhibited by oxygen. *Microaerophilic bacteria* require low levels of oxygen for growth but cannot tolerate the level of oxygen present in an air atmosphere. This is due to inactivation of enzymes, reduced thiol (-SH) group. On the other hand, various cellular enzymes catalyze chemical reactions, involving molecular oxygen. Sometimes addition of a single electron to an oxygen molecule forms super oxide radical (O_2).

$$O_2 + e^- \longrightarrow O_2^-$$

The inactivation of cell component is also observed due to production of even more toxic chemical substances such as hydrogen peroxide (H_2O_2) and hydroxyl radicals means of the reaction given below :

$$2O_2 + 2H^+ \longrightarrow O_2 + H_2O_2$$
$$O_2^- + H_2O_2 \longrightarrow O_2 + OH^- + OH$$

Aerobic and facultative microbes have protective mechanism against oxygen due to super oxide dismutase. The H_2O_2 produced can in form be dissipated by catalase and peroxidase enzyme.

$$2H_2O_2 \xrightarrow{\text{Catalase}} 2H_2O + O_2$$
$$H_2O_2 + \text{reduce substrate} \xrightarrow{\text{Peroxidase}} 2H_2O + \text{oxidised substrate}$$

3. Chemical Methods

The chemicals select a particular microorganism. If it is supplemented in a medium only those microorganisms will grow well which can utilize the supplemented chemicals. This particular kind of selection is based on addition of either carbon or nitrogen source. Such method is often called *enrichment* culture and medium is called *enrichment medium*.

In addition certain dyes such as rose bengal are often used for the isolation of fast growing fungi, while certain antibiotics such as nystatin is added to avoid bacterial contaminants in the medium. Similarly, yeast mannitol agar (YEM) medium supplemented with congo red allows to differentiate *rhizobia* with that of *Agrobacterium*. Rhizobia do not take colour while *Agrobacterium* colonies become pink. Many Gram-positive bacteria do not grow in the presence of dyes due to their inhibitory action while Gram-negative bacteria grow well in the presence of dyes.

4. Biological Methods (Natural Selection)

The presence of particular microorganisms is specific in a habitat. Certain microorganisms would like to live in dung only (coprophilous fungi and methanogenic bacteria), while others live in citrus fruit e.g. penicillia. Nature itself select the microorganism. *Streptococcus pneumoniae* in sputum sample is contaminated by many other organisms. However, laboratory mice are extremely susceptible to infection by *S. pneumoniae* and if the sputum sample is injected into lab mice, the pathogen will multiply rapidly and rest of the microorganisms will be killed due to defensive mechanism of animal.

Penicillia growing on an orange.

I. Methods of Culturing Aerobic Bacteria

Aerobic or facultative bacteria are to be grown in an atmosphere rich in oxygen. For this a large surface of medium in such condition increases the exposure. There are several devices such as Fernbach flask, kolle flask and Roux bottle which are used for providing increased aeration (Fig. 3.1)

II. Methods of Culturing Anaerobes

The anaerobic cultures differ in their requirements and senstivity to

Fernhbach flask.

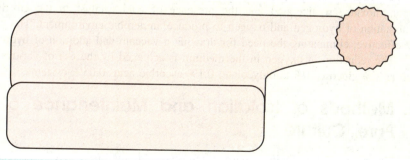

Fig. 3.1 : A Roux Bottle

oxygen. Some bacteria such as *Clostridium histolyticum* is aerotolerant and may show growth on the surface of anaerobic culture plates, while others such as *C. tetani* is strictly anaerobic and form surface growth if the oxygen is less than 2mm Hg. The obligate anaerobic forms can be grown in anaerobic growth cabins where oxygen is replaced by inert gas (Helium or CO_2). Such chambers or cabins are expensive. Hence, a less expensive device is the use of 'candle jar'. The inoculated cultures of the petri dishes are placed in the jar, a candle placed on the top of the petri dishes is lit, and the jar is sealed with an air tight lid. As the candle burns indicating the utilization of oxygen and production of carbon dioxide, which supports bacterial growth. The anaerobes grow in agar stab cultures, whereas facultative anaerobes are able to grow in either the absence or presence of the oxygen. The microtolerants or microaerophiles are those which grow in the presence of low oxygen concentration. Their growth stops in the presence of high oxygen concentration (Fig. 3.2)

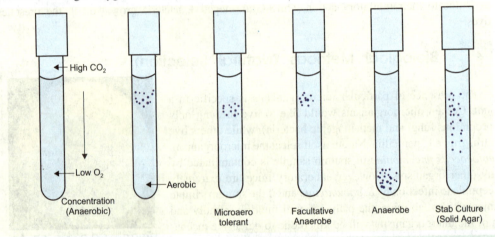

Fig. 3.2 : Growth response of different microbes to oxygen when grown in a semisolid medium or in a solid culture. The highest concentration of oxygen is present at the top of the tube, whereas the lowest concentration of oxygen is at the bottom of the tube.

Now-a-days commercial anerobic jar are available which has an indicator strip which change the colour from blue to colourless showed total consumption of oxygen (Fig. 3.3).

1.   Use of 'Gas Pak' for Preparing Anaerobic Jar

The gas pak is commercially available as a disposable envelope, containing chemicals which generate hydrogen and CO_2 on the addition of water. After the inoculated plates are kept on the jar, the gas pak envelope with H_2O added is placed inside and lid screwed tight. Hydrogen and carbon dioxide are liberated and the presence of cold catalyst in the envelope permits the combination of hydrogen and oxygen to produce an aerobic environment. The gas pak is simple and effective, eliminating the need for drawing a vaccum and addition of hydrogen.

The reduction of oxygen in the medium is achieved by the use of various reducing agents such as 7% glucose, 1% thioglycolate, 0.1% ascorbic acid, 0.05% cysteine.

III. Methods of Isolation and Maintenance of Pure Culture

The process of screening a pure culture by separating one type of microbes from a mixture is called *isolation*. A culture containing only one species of microbe is called *pure culture*. In a

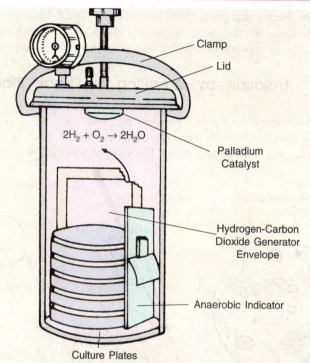

$$2H_2 + O_2 \rightarrow 2H_2O$$

Clamp

Lid

Palladium Catalyst

Hydrogen-Carbon Dioxide Generator Envelope

Anaerobic Indicator

Culture Plates

Fig. 3.3 : Anaerobic jars are used to incubate plates of anaerobic bacteria. Hydrogen and carbon dioxide gases are generated when water is added to the gas generator. Palladium catalyses the reaction of oxygen and hydrogen to form water. The methylene blue strips turns colourless when the oxygen is removed.

mixed culture, a particular species is present in small numbers in comparison to the numbers of others. Various techniques are developed for the isolation of pure culture from mixed culture. Some of which are given below:

1. Use of Micromanipulator

A single viable cell may be transferred on the culture medium to develop axenic pure culture by using micromanipulator which is used with a microscope for picking up a single colony from a mixed population.

2. Isolation by Exposure to Air

The nutrient agar slide or culture medium containing plate is exposed to the atmosphere for few minutes. After incubation (overnight or more), small colonies appear on the surface of medium which may be transferred on a fresh medium aseptically to obtain

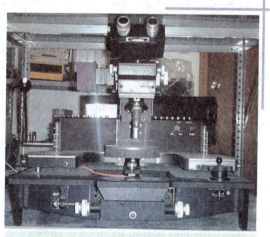

Micromanipulator.

pure culture. Such technique is called sub-culturing. When the transfer is from solid medium (agar) to liquid medium (broth), the term 'picking off' is used. In such cases the colour of the colony, their size, shape, appearance, form, consistency and optical properties are recorded.

3. Isolation by Streaking or Streak Plate Technique

In this method the tip of a fine structure wire loop called *inoculation needle* consists of a wooden or glass handle with a nichrome wire the end of which is bend to form a loop is used to transfer microbes from culture broth. The straight wires are similar to wire loop except they do

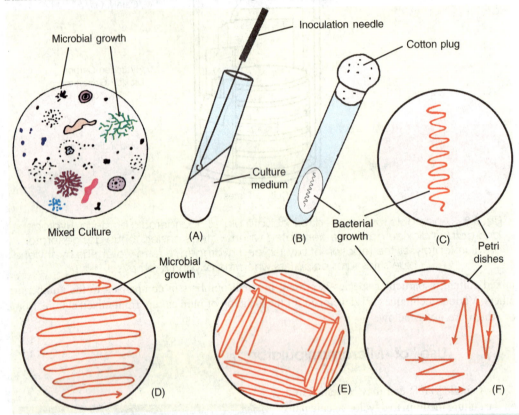

Fig. 3.4 : Isolation of bacteria from a mixed culture. A, isolation of pure culture and inoculation needle containing a loop; B, pure culture, C, D, E and F, streaking of bacterial culture by different methods.

not have loop. These are used to transfer culture in colony formed on solid culture medium. In such cases, the colony from solid medium is streaked on the surface of nutrient agar medium in a sterile Petri dish. This technique consists of the following steps:

(*i*) Hold the broth culture containing tube in left hand and shake it.

(*ii*) Sterilize the wire loop of the inoculation needle on burner flame.

(*iii*) Remove the cotton plug of the broth culture tube by little finger of right hand.

(*iv*) Flame the mouth of the test tube immediately

(*v*) Keep the test tube in such a way as given in figure; insert the wire loop to form a thin film and replace the cotton plug

(*vi*) The thin film in the loop is streaked in either a zig-zag manner by removing the loop backwards and forwards firmly. Care should be taken that loop should not be firmly pressed against the agar surface.

(*vii*) Incubate the Petri dish in incubator at a required temperature.

(*viii*) Growth of the bacteria will be visible (after an overnight incubation) on the streaked marks.

4. Isolation by Inoculating In Animals

If disease causing microbe is unable to grow on artificial culture media, the impure culture is injected into the susceptible animals such as guinea pigs or rabbits. These animals allow disease causing organism to grow while rest of the microorganism fail to grow. The etiologic microbe can be isolated in pure form from blood or affected tissue.

Guinea pigs are used as a medium to grow disease causing organism.

5. Isolation by Using Selective or Enrichment Media

Certain media are selective because these have chemicals or dyes which enrich the media. They have growth inhibitory effect on some microorganims. Such media separate the dominant species. Chemical dyes, such as malachite green and crystal violet, are used to inhibit the growth of bacteria and yeast. Sodium azide is a metal-binding agent that inhibits the growth of anaerobic bacteria, but does not affect the anaerobic lactic acid bacteria.

(a) Enrichment Medium : This medium is prepared by adding any number of growth factors. This addition enhances the growth and recovery of the desired microorganism. The differential media by virtue of their ingredients distinguish organism growing together. Thus, in EMB (eosin-methylene blue complex), *Escherichia coli* imparts metallic sheen which is due to precipitation of eosin-methylene blue complex while other enteric bacteria generally do not show metallic sheen. Similarly, on MacConkey agar medium, *E.coli* colonies are brick red in colour due to fermentation of lactose. On the other hand, *Salmonella typhi* does not give this appearence due to non-fermenter of lactose. The other media are selective media such as those which are of their special composition promote the growth of one organism and inhibit the growth of others. The minimal medium lack certain growth factors. The medium supports the growth of those microorganisms whose nutritional requirements do not exceed those of the corresponding wild type of strain. Such media are helpful when auxotrophs are required (in case of those microorganism which are having growth requirement of specific substance not necessary to the parent strain).

6. Other Methods

The other methods to isolate microorganisms are: (a) by controlling physical environment (especially temperature and pH), (b) by culturing highly diluted microbial suspesion by pour plate method in which isolated population of microorganisms growing from a single isolate can then be easily separated. In pour plate method, the sample is mixed with known quantity of distilled water. The suspension is plated (1 ml) on a presterilized Petri dish, spread the sample with the help of loop, and then pour the melted agar medium after cooling it, over the sample containing plate. Incubate the Petri dish in an incubator at optimum temperature ($27 \pm 1°C$). If counting is not possible, then sample dilutions are made as given in Fig. 3.5.

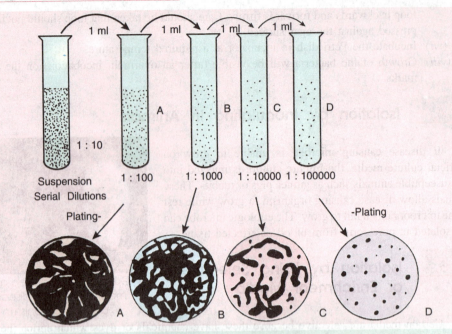

Fig. 3.5 : Dilution plate method for isolaton of microorganisms from soil, from different dilutions (A-D), and respective microbial colonies (A-D).

IV. Culture Characteristics

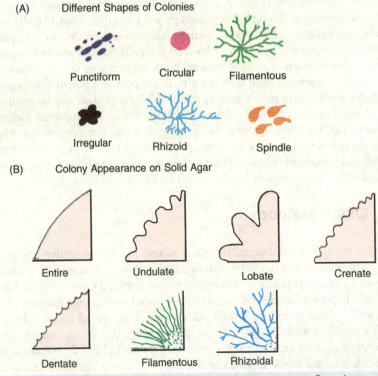

(A)   Different Shapes of Colonies

Punctiform   Circular   Filamentous

Irregular   Rhizoid   Spindle

(B)   Colony Appearance on Solid Agar

Entire   Undulate   Lobate   Crenate

Dentate   Filamentous   Rhizoidal

Fig. 3.6 : Colonies showing appearance; A, colony appearance; B, colony margins.

1.   Colony Appearence

Many fungi produce colonies with a fluffy appearance similar to cotton wool. The molds produce colonies which on aging develop a dry chalky appearaence.The colony characteristics are given in Fig. 3.6.

2.   Colony Forms

The colony shape may be circular, filamentous, rhizoidal, punctiform (dot like), irregular or spindle shape.

3.   Colony Elevation

This form is used to describe the depth of the colony developed by microbes. A colony may be flat (thin film over the agar surface), raised, convex or umbonate or with papillate surface.

4.   Colony Margins

The margins may be entire, undulate (wavy), crenate, dentate, lobate, rhizoidal or filamentous (Fig. 3.6).

5.   Optical Density

The colony may be transparent or translucent (foggy in appearance) or opaque (not permitting light to pass through) or irrediscent (rainbow colour).

6.   Colour

Many microbes develop colonies which are pigmented. Such coloured substances are either water soluble or insoluble. The soluble pigments diffuse out in the medium.

7.   Colony Odour

Some microbes produce a characteristic smell which sometimes helps in identifying the microbe. The actinomycetes produce an earthy odour which is quite often experienced after a first shower or rain. Many fungi produce fruity smell while *E. coli* produces a faecal odour.

8.   Colony Consistency

The degree of thickness, solidity or firmness of the medium should be examined as given below:

(*a*) If the entire broth appears milky and cloudy, it is called *turbid.*

(*b*) If a deposit of cells is present at the bottom of the tube, the term *sediment* is used.

(*c*) When microbial or bacterial growth forms a continuous or interrupted sheet over the broth, it is called *pellicle.*

(*d*) If the growth of microbe is similar in appearance to that of butter, it is called *butyrous*.

B. Microbial Growth

Growth is defined as an orderly increase in cellular components. Microorganisms grow in a variety of physical and chemical environments. To distinguish orderly growth with that of inorderly growth, in recent years the term "balanced growth" has been used. " Balanced growth" as defined by Campbell, is doubling of every biochemical unit of the cell within the time duration of a single division without a change in the rate of growth.

1. Growth Measurement

A number of methods are available for measuring microbial growth. The choice depends upon the measurement, objectives and on available techniques usefulness. In some cases of industrial fermentations which contain complex media, indirect methods for estimation need to be used however, no matter what method is used, considerable care is required in interpreting the results.

Bacterial growth can be measured either by (*i*) colony counting or cell counting, (*ii*) by weighing the cell i.e. cell mass measurement or (*iii*) by cell activity (turbidity method) measurement.

Microbiologist monitoring microbial growth.

2. Parameters of Growth

(*i*) **Cell Counting by Direct Microscopic Count Method** : Bacterial cell can be accurately counted by using Petroff-Hausser counting chamber (the chamber includes a glass slide, a cover slip which is framed and kept 1/50 mm above the slide so that bacterial suspension is present in each ruled square of the slide.

The area of square is $1/400$ mm^2; glass cover slip rests 1/50 mm above the slide hence volume over a square is $1/20,000$ mm^3 or $1/20,000,000$ cm^3, for example - if in one square, an average of five bacteria is present, then these are $5 \times 20,000,000$, or 10^8 bacteria per ml.

(*ii*) **Colony Counting (Plate-counting Technique)** : This method is based on the fact that one viable cell gives rise to one colony. Therefore, a colony count on an agar plate reveals the viable microbial population. For carrying out this, a measured amount of the sample of bacterial suspension is mixed in the agar medium (when it is in liquid form at 40-45±°C). It is plated after mixing thoroughly. Each organism grows, reproduces and forms a visible mass in the form of colony. These are counted either by using colony counter or with the aid of large magnifying lens. If too many colonies are appearing and overlapping each other, the sample is diluted so that the colonies are accurately counted. This method is called *pour plate method*. The plate count is also performed by *spread plate method*. In this method 0.1 ml sample containing bacteria is spread over the surface of an agar plate using a sterile glass spreader.

In both pour plate and spread plate methods the plates containing bacterial suspension are incubated until the colonies appear, and the colonies are counted. As stated earlier, to obtain the appropiate colony number, the sample must be diluted. Serial dilutions of the sample are usually adopted. To make a 10-fold (10^{-1}) dilution, 10 ml sample is mixed with 90 ml diluent. *Serial dilution of soil sample is shown in Fig. 3.5.* In most cases, serial dilutions are needed to obtain final dilution.

Demerits of the above methods are not only the suitability of the culture medium and incubation conditions but sometimes bacterial cells are deposited on the plate, does not show their visibility in the form of colony if incubation period is short. Further, viable counts, preparations of dilutions of the sample also give wrong informations. To get correct informations, viable counts are often expressed as the number of colony forming units (CFU) per millilitre rather than number of viable cells. This method is adopted in counting microorganisms in soil.

(*iii*) **Measurement of Cell mass and Turbidity** : Cell mass is directly propotional to cell number. This can be obtained after centrifugation of a known volume of culture and weighing the pellet obtained. This is called fresh weight but dry weight of cells is obtained by drying the pelleted cells at 90-110°C overnight.

(*iv*) **Turbidity Measurement by Optical Density Method** : The cell mass and number are also obtained by using optical density method. *Turbidity* is developed in the liquid medium due to the presence of cells which make cloudy appearance to the eyes. Measurement of microbial growth is given in Fig. 3.7.

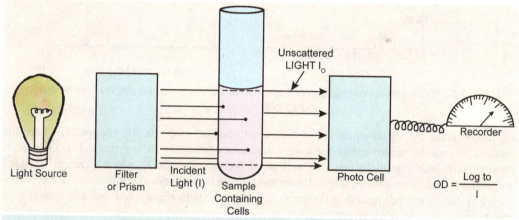

Fig. 3.7 : Measurement of microbial growth (turbidity) by optical method.

When sample is more turbid it means that more cells are present. Hence more light is scattered. *Turbidity* can be measured with a photometer or a spectrophotometer device that detect the amount of unscattered light recorded in photometer unit (for example "klett units" or optical density (OD) as shown in Fig. 3.8.

The method is indirect, hence some direct measurements of cell number should also be determined. Since OD is propotional to cell mass and thus also to cell number, therefore, turbidity reading acts as an estimate of cell number or cell mass. The plotting between semilogarthmic versus time, growth rate of microbial cultures is obtained and used to calculate the *generation time* of the growing culture. Such a curve can contain data for both cell number and cell mass, allowing for an estimate of both parameters from a single turbidity reading.

Demerits : If concentration of the cell in the sample is high, light scattered away from the detecting unit by one to one cell can be rescattered back by another. Hence, the one to one correspondance between cell number and turbidity does not follow linearity as shown. Secondly, dead cells also interfere during measurement. Hence. This method is reasonably accurate only for measurement of microbial growth till early log phase.

3. Growth in Continuous Culture

When growth occurs in a fixed volume of a culture medium, it is called *batch culture*. When

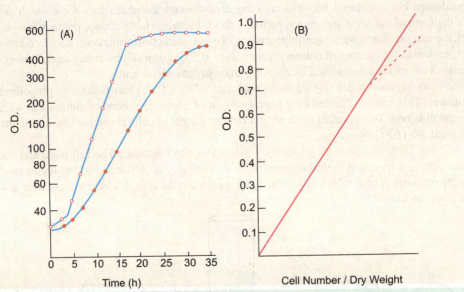

Fig. 3.8 : Enumeration of micro-organisms based on optical density measurement and cell number.

culture becomes old, the composition of the culture medium is drastically changed. To keep the cultures in constant environments for longer durations *continuous culture* method is adopted. A continuous culture essentially requires a flow of constant volume to which medium is added continuously and from which continuous removal of medium alongwith culture can occur. When such a system is in equilibirium, cell number and nutrient status remain constant and system is in *steady state.*

The fresh culture medium flows into a growth chamber at a carefully controlled rate, the volume of culture is maintained by controlling the rates of inflow and out flow, the rate of growth is then controlled by regulating the inflows rates. The rate of loss of cells through overflow can be expressed as

$$\frac{dM}{dt} = \frac{F}{V} = DM$$

Where flow rate F is measured in the culture volume V/hr. The expression F/VM is called the dilution rate (D). In a continuous culture, KM = DM

K = D, i.e. growth rate (K), will equal the dilution rate in a stabilised continuous culture. Under such conditions growth of the culture is linear rather than exponential, since the amount of mass/ unit volume (growth rate) remains constant. The continuous culture does not represent synchronous growth since these do not contain cells that are physiologically identical.

(i) Chemostat for Measurement of Growth : Chemostat (kee-mo-stat, kem. o. stat) means a device for maintaining organisms in continuous culture; it regulates the growth rate of the organism by regulating the concentration of an essential nutrient. The growth rate and growth yield can be controlled independently on each other by adjusting the dilution rate and by varying the concentration of nutrients present in a limiting amount. The high dilution rate does not allow the organism to grow fast enough to keep up with its dilution rates, a large fraction of the cells may die from starvation due to non-availability of nutrients required for maintenance of cell metabolism (Fig. 3.9). Hence, a minimum amount of energy is required to maintain cell structure. Such energy is called *maintenance energy.*

The number of cells/ml is called cell density which is controlled in the chemostat by the nutrients. If the concentration of the nutrient in the incoming medium is raised with the dilution rate remaining constant, the cell density will increase although growth rate will remain the same.

(iii) **Water Availability :** The water availability is one of the factors affecting the growth of microorganisms in nature. Water availability depends upon the water content present in the environment and water soluble salts, sugars and other substances. Actually, dissolved substances have an ability for water which make the water associated with solutes unavailable to organisms.

Water availability in physical sense is called as water activity denoted by a_w, which is a ratio between vapour pressure of the air in equilibrium with a substance or solution to the vapour pressure at the same temperature of pure water. Most of the agricultural soil has 0.9-1.0 water activities. If the soil is saline, it makes the dry environment. It means cells are in the

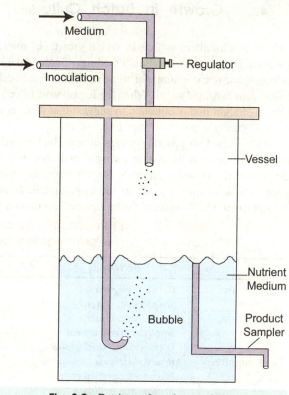

Fig. 3.9 : Design of a chemostat (continuous culture).

habitat of high salt concentration. Such organism are called *halophiles* such as *Pseudomonas* spp. and *Vibrio sp.*

Heterotolerant organisms can tolerate some reduction in water activity (aw) and grow best in the absence of added solute. The organisms capable to grow at high concentration of salt is called extreme halophiles for example *Halobacterium* and *Halococcus* spp.. These organisms generally require 15-30% sodium chloride for their optimum growth. Similar to these, there are certain organisms, that require high sugar percentage, are called osmophiles such as *Saccharomyces, Bacilli* and *Penicillium* spp. Lack of water supports the *Xerophiles* for example *Xeromyces bisporus*.

When the dissolved salt concentration increases in water it becomes unavailable to the microorganism. To overcome this situation, organisms start producing intracellular compatible solutes (solute present in the cell adjust cytoplasmic water activity which is non-inhibitory to biochemical processes of the cell), which make the cell in equilbrium state and make positive water balance (Table 3.1).

Table 3.1: Some of the compatible solutes and their derivatives.

Use	*Solute derivative*
Glycine betaine present in halophilic bacteria and Cyanobacteria	*Betaine*-derivative of glycine in which protons on the amino group are replaced by three methyl groups.
Non-photosynthetic extreme halophilic grown in the absence of glycine-betaine	*Ectoine*-derivative of the cyclic amino acid proline.

4.   Growth in Batch Culture

A batch culture is that in which growth of microbes occurs in a limited volume of liquid medium. During growth in liquid medium of unicellular microorganisms, the increase in cell number is logarithmic (exponential) for some time. Since each cell gives rise to two cells which in turn gives rise to four cells and so on. When the logarithmic of cell number in the population is plotted against time, a straight line is obtained to suggest that there is an equal percentage increase in the number of cells during a constant time interval.

There are four general patterns of microbial growth. Bacteria divides by binary fission. During the growth cycle cells double its mass and also the amount of all the cell constituents. Prior to division, cell wall and membrane is synthesized and the parent cell divides into two identical cells, each of which grows exactly as the original cell. *Doubling time* for most bacteria is reported to be as fast as 15-20 minutes under optimal conditions.

Table 3.2 : Generation time of some Gram-positive and Gram-negative bacteria.

	Bacteria	*Generation Time (minutes)*
Gram-positive	*Bacillus megaterium*	25
	Streptococcus lactis	26
	Streptococcus lactis	48
	Staphylococcus aureus	27-30
	Lactobacillus acidophilus	66-87
Gram-negative	*Bradyrrhizobium japonicum*	344-461
	Mycobacterium tuberculosis	792-932
	Treponema pallidum	1980

Yeasts divides by budding (exceptions to include yeast that grow by fission or by forming hyphae). Budding involves the formation of a bud on the mother cell. The bud grows until it becomes same in the size of the mother. At this point the daughter cell separates. Under optimal conditions, yeast may divide in as little as 45 minutes However, 90-120 minutes is optimum.

Molds grow by chain elongation and branching. Growth proceeds from the tip of the mycelium by forming septa between the cells. Depending upon the physiological environment, the mycelium may be long or diffuse, short and highly branched, or a mixture of two. Typically molds have optimal mass doubling time of 4-8 h although, some are reported to double as little as 60-90 min. *Actinomyces* and *Streptomyces* are classes of organisms closely related to bacteria that are extremely important industrially. Both classes also grow as mycelium organisms.

Microbial viruses or phages do not follow the normal growth patterns. They require a host to multiply. So they must infect a cell and utilised the cell protein and nucleic acid synthesizing machinery to produce new viral material. A virus inside the cell can replicate itself 50-300 times before the cell bursts. The growth is exponential but the exponent is much greater than two.

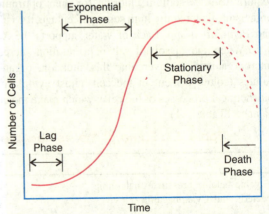

Fig. 3.10 : A typical bacterial growth curve for liquid culture. The dashed line is the total number of cells, and the solid line is the number of viable cells.

When bacteria are transferred from a slant culture to a known volume of liquid medium the population undergoes a characteristic sequence in its rate of increase in cell number. Four recognizable phases are seen when the increase in cell number is determined in relation to time. In *lag phase,* there is no increase in the number of viable cells. However, cell growth occurs as indicated by the increase in cell mass. During this period the cells increase in size as a result of extensive macromolecule synthesis. This stage represents a period of active growth without cell division and the cells prepare for division. The length of the lag phase depends upon a variety of factors such as the age of the inoculum, the composition of the growth medium and the environmental factors such as temperature, pH, aeration, etc. The lag phase is then followed by the phase of exponential growth *(log phase)* (Fig. 3.10).

In the *log* phase, the cell population increases logarithmically and the cells divide at the maximum rate permitted by the composition of medium and environmental conditions. The cell number during this period increases as a function of the exponent ($2^1, 2^2, 2^3, 2^4, \ldots\ldots 2^n$). Growth rates are measured during this period since growth occurs at a maximal rate. The growth rate can be expressed in time. For example, a culture which has a double time of 60 minutes will give a growth rate of one.

The rate of growth is influenced by a variety of factors such as the composition of medium, environmental and the inherent properties of the culture. The length of log period is determined mainly by the composition of the medium and the rate of accumulation of inhibitory products. Although in the log phase, the growth is maximum, a culture in logarthmic phase represents a population of cells of different ages, some have just divided and others are in the intermediate stages of the division cycle. Therefore, in physiological terms the population is not homogeneous.

The period of exponential growth is followed by *stationary phase* in which the total number of viable cells remain constant. In fact, we can say that the stationary phase is reached when the viable cell number does not increase. The duration of stationary phase varies with the organism and environmental conditions. Some organisms may display long stationary phase lasting for several days while others may show a very short stationary phase of only few hours before the next phase begins.

The *death phase* (the phase of decline) is characterized by an exponential decrease in number of the viable cells. A cell is considered to have died or to have to become non-viable when it is no longer capable of multiplying. The phase of decline is seen generally in bacteria, which also be rapid if cell lysis occurs.

Some bacteria such as the sporulating bacteria may form endospores as they reach the stationary phase of growth and these would be resistant to lysis or death. In such cases the number of viable cells will remain constant after attaining the stationary phase and the phase of decline may not be seen.

When a population of cell from a stationary phase or death phase is used to inoculate fresh growth medium, the cells will not continue to die but re-enter the lag phase and initiate new growth. Cells from the death phase may, however, show a longer lag in contrast to cells transferred from either the stationary phase or the logarithmic phase.

All the four phases are applied for population growth.

Mathematics of bacterial growth: Bacterial cells divide by binary fission, hence their increase in cell number is a function of the exponent ($2^1, 2^2, 2^3, 2^4, \text{------}2^n$). where n = exponent, number of cell division if M_1 is the number of cells at T_1 time and M_2 .... at T_2 time

Then,

$M_2 = M_1\ 2^n$

By taking log on both sides

$\log M_2 = \log M_1\ n\ \log 2$

if equation is simplified:

$\log M_2/M_1 = n \log 2;$

or $n = \dfrac{\log M_2 - \log M_1}{\log 2}$ This is growth equation

If t is the mean generation time. Then,

$n \times t = T_2 - T_1$ or $n = \dfrac{T_2 - T_1}{t}$

where n = number of cell division or generations that the population has undergone during an interval of time $(T_2 - T_1)$.

Hence, $\log \dfrac{M_2}{M_1} = \dfrac{T_2 - T_1}{t} \times \log 2$

The plot of log cell number against time will therefore be a straight line.

5.   Synchronous growth

When the cultures in log phase are analyzed, cells are present in various stages and division cycle. Analysis of such population therefore, yields only average value of any parameter. To understand the properties of individual cell, during the course of its division cycle, it is necessary to analyze each cell, which is practically not possible. A system that closely resembles and amplifies the behaviour of single cell is a synchronous culture, which contains cells that are physiologically identical and are in the stage of division cycle.

A synchronous population can be generated either by physically separating the cells in the same stage of division or by forcing a cell population to attain an identical physiological condition by a change in the environment. In the synchronized culture, the cells are physiologically identical, cell division occur periodically at constant intervals. The dry mass of the cell, optical density, total proteins, or RNA contents per cell increases at a constant rate. The amount per cell will increase in population to the cell number. On the other hand, the pattern of DNA synthesis can be either periodic or continuous depending upon how fast the culture is growing. E. coli grows in a medium with a generation time greater than 40 min, will show a period when no DNA synthesis occurs. Recently, to obtain synchronous culture, the exponentially grown culture is centrifuged either in sucrose, glycerol or sorbitol gradients in order to separate cells based on their densities which is directly related to their age. Such fractions provided the same results as by using synchronous culture.

6.   Diauxic Growth

In a culture medium containing two carbon sources, bacteria such as E.coli displays a growth curve, called diauxic (fig. 3.11). Under this conditions, if glucose and lactose are supplemented in medium having E. coli. First E. coli will utilize glucose and after it is exhausted lactose will be utilized. In between a short lag period is there. This led us to conclude that E. coli preferentially utilizes certain carbon sources.

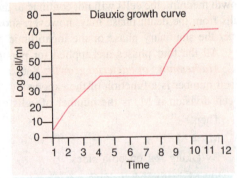

Fig. 3.11 : Diauxic growth curve.

C. Culture Media

The culture media (nutrients) consist of

chemicals which supports the growth of culture or microorganisms. Microbes can use the nutrients of culture media as their food is necessary for cultivating them *in vitro*.

1. Characteristics of Medium

The medium should neither be acidic nor alkaline. It should contain all kinds of nutrients in suitable amounts. It must be sterilized (free from microbes) before use.

2. Types of Media

The first medium prepared was meat-infusion broth. As most pathogenic microbes require complex food similar in composition to the fluids of the animal body. It was Robert Koch and his colleagues who used meat infusion and meat extracts as basic ingredients in their culture media for the isolation of pathogenic microbes, while one of his assistant named Petri designed and developed glass dishes known today as Petri dishes are used in microbiological work.

On the basis of chemical composition, the media are classified into two types:

(*i*) **Synthetic or chemically defined medium** : These media are prepared by mixing all the pure chemicals of known composition for e.g. Czapek Dox medium.

(*ii*) **Semi-synthetic or undefined medium** : Such are those media, where exact chemical composition is unknown e.g. potato dextrose agar or MacConkey agar medium.

On the basis of consistancy: This demonstrates the physical characteristics of the media. Such media are of three types:

(*a*) **Solid medium or synthetic medium** : When 5-7% agar agar or 10-20% gelatin is added the liquid broth becomes solidified. Such media are used for making agar slants or slopes and agar stab.

(*b*) **Liquid medium or broth** : In such cases no agar is added or used while preparing the medium. After inoculation and later incubation, the growth of cells become visible in the form of small mass on the top of the broth.

(*c*) **Semi-solid or floppy agar medium** : Such media are prepared by adding half quantity of agar (1/2 than required for solid medium) i.e. about 0.5% in the medium. This type of medium may be selective which promote the growth of one organism and retards the growth of the other organism. On the other hand, there are differential media which serve to differentiate organisms growing together.

3. Preparation of Medium

The liquid medium or broth is prepared by dissolving the known amounts of chemicals in distilled water, the pH is adjusted by adding N/10 HCl or 1N NaOH.

The liquid medium is dissolved into either Erlenmeyer flasks or rimless clean test tubes. In 15 ml capacity of test tube, 5 ml medium should be poured while in flask of 250 ml capacity, the amount of the medium should be 100 ml. These are then plugged with non-adsorbent cotton plugs. The plugged tubes or flasks should be wrapped by brown paper and placed for sterilization by autoclaving at a pressure of 15 lbs/inch^2 (at temperature 121°C), for 15 min.

The heat sensitive substances (protein or enzymes etc.) should be sterilized by using membrane filters (millipore). The agar agar is to be dissolved separately and dispensed after dissolving all ingredients of the medium. It is first to be noted that all the glassware in use should be sterilized in oven at 170°C for 3 h before using them. Such sterilized glassware are needed for pouring the medium used for culturing the microorganisms.

Each and every biological process requires energy for their vital activities. The basic cell building requirements are supplied by the nutrition, which is manipulated according to its requirement. Nutrition not only provides energy but also acts as precursors for growth of microorganisms. The nutritional requirement of an organism depends upon the biochemical capacity. If an organism is capable of synthesizing its own food using various inorganic components, requires a simple nutritional diet whereas organism unable to meet such synthesis requires complex organic substances.

A culture medium prepared for microbial testing.

4. Minimal Requirements

Every microbe has its own specific minimal nutritional requirement. If it is not provided, they do not grow. This minimal requirement consists of a carbon source, nitrogen source, sulphur source, phosphorus source besides energy source. They grown better in the presence of particular amino acids or vitamins or other compounds, so that the species could grow or develop better. Microbes can utilize a wide range of substrates from complex form of compounds (lignin etc.) that are generally not used by other forms of life.

Carbon source (glucose etc.) is essential for the basic cell structure because each and every biomolecule is made up of carbon along with other compounds. Nitrogen source is required for the biosynthesis of amino acids, nucleic acids, enzymes etc. Sulphur and phosphorous required for synthesizing nucleic acids, vitamins, and certain amino acids.

A photosynthetic microorganism eg. *Cyanobacteria* do not require a energy source. They use sunlight and trap the form of chemical energy, used frequently. With the help of CO_2 and water, they synthesize food in the form of carbohydrate. But many microorganisms need some energy sources. This is met out by organic compounds. Some microbes have special capacity. They can harvest energy from redox potential for their vital activities.

5. Nutritional Types of Microorganisms

Based on the way of harvesting energy, they are classified into two major groups.

Those organisms that can make use of external energy sources and assimilate inorganic carbon are called as autotrophs. Blue green algae and some chemosynthetic bacteria belong to this group. They can make use of sunlight/ redox potential as their energy source. CO_2 is the main and sole carbon source. Nitrogen is assimilated in the form of NH_4^+, sulphur as SO_4^{--} and phosphorus in PO_4^{--} from their surroundings. Further, autotrophs may be of two types:

Photoautotrophs are bacteriochlorophyll containing microorganisms, while chemoautotrophs, utilize various oxidation-reduction reactions as their energy source. During oxidation, energy is released hence, the microbes oxidize the reduced traditional compounds and make use of the released electrons i.e. energy in case of sulphur bacteria (*Thiobacillus* spp.) and nitrifying bacteria (*Nitrosomonas* spp.). The phototrophs utilize solar energy to oxidizes from O^- (singlet) stage to O_2 stage and thus utilizes the electrons released (Table 3.3).

Table 3.3 : Properties of different nutritional groups of microorganisms

Property	Photoautotrophs	Chemoautorophs	Heterotrophs
Source of energy	Sunlight	Chemical reactions	Organic compounds
Precursor	CO_2, H_2O, NH^+	NO_3^-/H_2S	Glucose (Sugars)
Biochemical Pathway yielding energy	Photosynthesis	Respiration	Respiration/fermentation
Example	Cyanobacteria	*Thiobacillus* spp.	Green and purple bacteria

Many microorganisms resemble animals and humans, using organic compounds. These are called chemoorganotrophs but when they use inorganic chemicals as energy source, called chemolithotrophs.

D. Staining and Smearing

The colouring agents impart a colour to the colourless microorganisms. Due to this colouration the microorganisms become visible, so as to observe its cell shape, and structures. These stains are composed of a positive and a negative ion, one of which is coloured and is known as chromophore. The basic dyes are in positive ion while acidic dyes in the negative. Since bacteria are towards negatively charged at pH 7.0, thus the coloured positive ion in a basic dye binds to the negatively charged bacterial cell. Some of the dyes are classified as acidic dye and some are basic dye.

1. Negative Staining

In the negative staining, as stated earlier that acidic dye due to negatively charged bacterial surface does not interact with dye's negative ions, the background stains colour. This type of preparation where colourless bacteria are visible against a coloured background is called negative staining. Such dyes are eosin, nigrosine, India ink, etc.

There are three kinds of staining techniques used in microbiological preparations called smear. Smear is a thin film of material (microorganisms) spread over the slide's surface. This fixing procedure kills the microbes. In this procedure, stain is applied and then washed with current of water, slide is blotted with blotting paper so as to absorb excess water and stained microbe is now ready for examination.

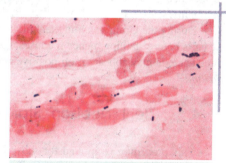

Smear of microorganisms.

2. Simple Staining

The aqueous or alcoholic solution of a single basic dye is called simple staining for example methylene blue, crystal violet, carbol fuschin and safranine. Sometimes a chemical helps to intensify the stain, such chemical is called mordant. The mordant is closely associated with stain, and shows its affinity. In case, stain is applied to the fixed smear, then washed off, slide kept for drying and then it is to be examined under the microscope. Such procedure helps in studying the cellular shape and structures of microorganism.

3.   Differential Staining

The most important differential stain used in bacteriology is Gram's stain which was discovered by Danish Bacteriologist Hans Christian Gram. It is the only procedure based on staining technique which divides bacteria into two large groups, Gram-positive and Gram-negative. The procedure of staining is diagramatically represented below.

Since the crystal violet imparts purple colour to all cells, it is called primary stain. The purple dye after washing off from the smear, is to be covered with iodine (mordant), both type of bacteria appear dark violet or purple in colour. The smear is then washed off with ethanol or ethanol-acetone solution which acts as decolourizing agent, removes the purple colour from the cells of some species. Thus, the ethanol is rinsed off and again stained with safranine, a basic dye, which acts as counter stain, washed again, blotted dry and examined under microscope.

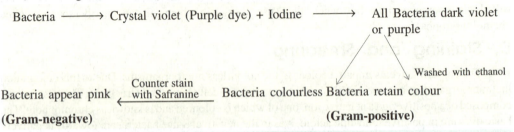

Fig. 3.12 : Procedure of Gram's staining.

Mechanism of Staining : It is believed that the mechanism of staining depends upon the structural differences in their cell walls of Gram-negative and Gram-positive bacteria. This is confirmed on the basis that if cell wall of Gram-positive bacteria is removed and treated with Gram stain the bacteria turned Gram-negative . The cell wall of Gram-positive bacteria consists of a single 20-80 nm homogeneous peptidoglycan or murein layer lying outside the membrane, while the Gram-negative cell wall has a 1-3 nm thick peptidoglycan layer surrounded by a 7-8 nm thick outer membrane.

During the Gram's staining technique the bacterial cell size appear to be about two third than the cells before staining, hence it causes a degree of shrinkage of bacterial cells. On the otherhand, some bacteria may appear Gram-positive and some Gram-negative. Such bacteria are termed *Gram-variable*.

4.   Ziehl Neelson Staining (Acid Fast Staining)

This procedure is applied for those bacteria whose cell wall contains considerable amounts of *mycolic acid* which makes their cells difficult to stain. In Ziehl Neelson staining a heat fixed film is prepared and flooded with strong carbol fuchsin. The slide is heated (for 5 min) so that the carbol fuchsin starts to steam. This helps in the penetration of carbol fuchsin to the cell wall. The smear is washed with filtered (bacteria free) water as tap water may contain *Mycobacterium* which may give false results.

Actually, in the above staining procedure due to the basic nature of carbol fuchsin reaction with mycolic acid in the cell wall produced a yellowish-brown compound that may be leached easily from cells. The leaching process is enhanced by alcohol. Although, the cell wall of mycobacteria protect the dye within their cells when exposed to a solution of 3% HCl in 95% alcohol. When bacterial films (smear) are treated with acid alcohol, only mycobacteria retain the carbol fuchsin and appear red. Methylene blue or malachite green is used as counter stain bacause of their behaviour.

5. Special Stains

The specific part of microbial cell such as endospores and flagella may become visible by imparting colour to them. Such strains also reveal the presence of capsules. Some of the important staining techniques are given below :

(*i*) **Negative Staining for Capsule** : The gelatinous covering over certain microorganisms is called capsule. The capsule is a device to determine the organism's ability to cause a disease i.e. virulence. The capsular material is soluble in water hence easily removed during washing.

For negative staining, the microbial cells are mixed in a solution containing a fine colloidal suspension of coloured particles such as India ink. This provides a dark background and then bacteria can stain with a simple stain such as safranine. The chemical composition of the capsular material generally does not stain with such dyes which make them visible as halos surrounding each stained bacterial cell. In this technique India ink depicts a negative staining technique, hence it makes contrast the capsule in their surroundings.

(*ii*) **Endospore (spore) Staining** : Only six bacterial genera form endospores. Actually these are formed when the bacterial cells undergo unfavourable condition (lack of nutrients, etc.). They protect the microorganisms from adverse environmental condition.The most commonly used endospore stain is the Schaeffer-Fulton endospore stain. In this technique, malachite green acts as primary stain and safranine acts as counter stain.

The smear is treated with malachite green for about 5 minutes till it is heated to steaming. The heat helps the stain to penetrate the endospore wall. The smear is washed with water to remove the malachite green stain from all the cells. Now, safranine is applied to the smear to displace any residual malachite green. The endospore appears green within red or pink cells.

(*iii*) **Flagella Staining** : It is the part of the bacterial cell meant for locomotion. The flagella can not be seen in light microscope. A mordant is used to increase the diameter of flagella until they become visible microscopically when stained with carbol fuchsin.

QUESTIONS

1. Mention the differences between the growth rate and generation time of a microorganism.
2. Describe the methods for measurement of growth. Which method you understand is suitable for the measurement.
3. How does the growth pattern differ when it is measured by total count or by viable count?
4. Discuss the process of appearence of a visible colony on an agar plate. With this view, describe colony on an agar plate. Describe the principle behind the viable cell count.
5. What is batch and continuous cultures? Write about use of chemostat.
6. Explain microbial growth. How will you measure growth by turbidity measurement?
7. Describe how would you dilute a bacterial culture.
8. To which phase of the growth curve do the mathematical analysis of growth is possible ?
9. What are the basic requirements of a microbial cell? Discuss the mode of nutrition in major groups of microorganisms.
10. Write short notes on the following:

(*i*) Stationary phase, (*ii*) Total cell count, (*iii*) Microbial growth,
(*iv*)Chemostat, (*v*) Halophiles, (*vi*) Osmophiles,

REFERENCES

Dubey, R.C. and Maheshwari, D.K. (2002). *Practical Microbiology*. S. Chand & Co. Ltd. Ram Nagar, New Delhi-55.

The Microbial Cells: Morphology and Fine Structure

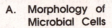

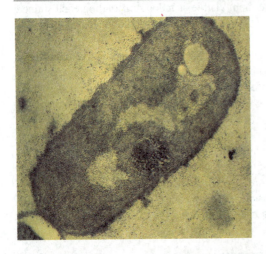

Morphology includes the size, shape and rearrangement of microbial cell. However, these features vary with species of microorganisms. The structure of microbial cells reveals the composition and topography of chemical constituents building the cell wall, the components outside and internal to the cell walls. However, the prokaryotic cells differ from the eukaryotic cells. A comparison of these two types of cells is given in Table. 4.1.

A. Morphology of Microbial Cells

  Size and Shape

Microscopic view of microbial cells.

The size, shape and arrangement of microbial cells vary with species to which they belong. Bacteria are of about 0.1 to 60 × 6 μm in size. However, there is variation in dimension of bacilli (5 x 0.4-0.7 μm), pseudomonads (0.4-0.7 μm diameter, 2-3 μm length) and micrococci (about 0.5μm diameter). Size of some of the bacteria are given in Table 4.2.

It is the rigid cell wall that determines the shape of bacterial cell. Generally, the bacterial cells are spherical (coccus, plural cocci

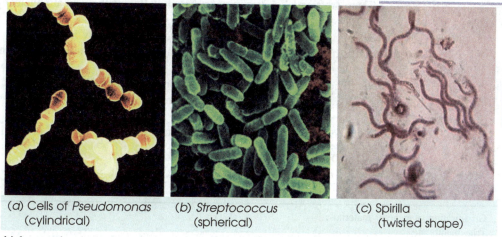

| (a) Cells of *Pseudomonas* (cylindrical) | (b) *Streptococcus* (spherical) | (c) Spirilla (twisted shape) |

which mean berries), elongated rods (bacillus, plural bacilli), helical rods (*Spirillum*, plural spirilli), pear-shaped (*Pasteuria*), lobed spheres (*Sulfolobus*), rods with squared ends (*Bacillus anthracis*), rods with helically sculptured surface (*Seliberia*) and of changing shape (*pleomorphic*), etc (Fig.4.1). The unicellular cyanobacterial cells are usually spherical (*Chroococcus, Scenedesmus, Anacystis*), some are elongated and multicellular.

2. Arrangement

The arrangement of cells are more complex in cocci than bacilli. The arrangement of cells depends upon adherance of cells together after the cell division (Fig.4.1). Different forms of arrangement are given below :

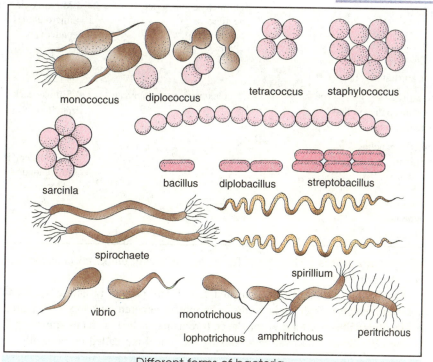

Different forms of bacteria.

(i) **Coccus Forms :** There are several groups of cocci based on the number and arrangement of cells.

(a) **Diplococcus :** Cells divide in one plane and get attached permanently in pairs.

(b) **Streptococcus :** Cells divide in one plane and remain attached to form a linear chain of cells.

(c) **Tetracocci :** Cells divide in two planes and form groups of four cells.

(d) **Straphylococci :** Cells divide in three planes in an irregular pattern producing bunches of cocci.

(e) **Sarcinae :** Cells divide in three planes in regular pattern producing bunches of cocci.

Table 4.1 : Comparison of prokaryotic and eukaryotic cells (based on Prescott, 1996)

Character	*Prokaryotes*	*Eukaryotes*
1. Organisation of genetic material :		
True membrane	—	+
DNA complexed with histone	—	+
Number of Chromosomes	One (besides plasmids)	More than one
Introns in genes	Rare	Present
Nucleolus	—	+
Mitosis	—	+
2. Genetic recombination	Unidirectional transfer of DNA	Meiosis and fusion of gametes
3. Mitochondria	—	+
4. Chloroplast	—	+
5. Plasma membrane with sterol	Absent except mycoplasma and methanotrophs	+
6. Flagella	Sub-microscopic, cone one fibre	Microscopic, 10 microtubules in 9+2 pattern
7. Endoplasmic reticulum	—	+
8. Golgi apparatus	—	+
9. Cell walls except mycoplasma and archaeobacteria	Complex with peptidoglycan	Simple, lack peptidoglycan
10. Simpler organelles		
Ribosomes	70S	80S, except in mitochondria and chloroplast
Lysosomes	—	+
Microtubules	—	+
Cytoskeleton	—	+

(ii) **Forms of Bacillus :** There are a few groups of bacilli unlike cocci as the former divide across their short aries.

(a) **Monobacillus :** The single elongated cells freely present in nature are monobacillus.

(b) **Diplobacillus :** After division the cells remain adhered and appear in paired form.

(c) **Streptobacillus :** After division the cells remain attached in chains appearing like straws.

(d) **Coccobacillus :** The oval cells looking like cocci are called coccobacilli.

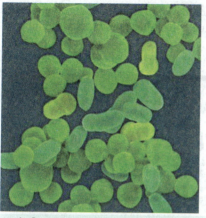

A Coccobacillus prokaryote.

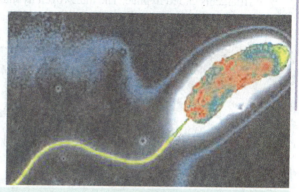

Vibrio cholerae.

There is two meaning of bacillus, one is the form and the second is the genus. For example the bacterium *Bacillus anthracis* causes anthrax disease.

(iii) Forms of Spirilli :

(a) Vibrioid : Bacterial cells having less than one complete twist form vibrioid shape e.g. *Vibrio cholerae.*

(b) Helical : Cells that have more than one twist form a distinct helical shape e.g. *Spirillum* (with flagella).

(iv) Other Forms :

(a) Pleomorphic : Of changing forms e.g. *Rhizobium, Mycoplasma,* etc.

(b) Trichomes : Cells divide in one plane forming a chain which has much larger area of contact between the adjacent cells e.g. *Baggiatoa, Saprospria.*

(c) Palisade : The cells are arranged laterally (side by side) to form a match sticks like structure and at angles to one another e.g. *Corynebacterium diptheriae.*

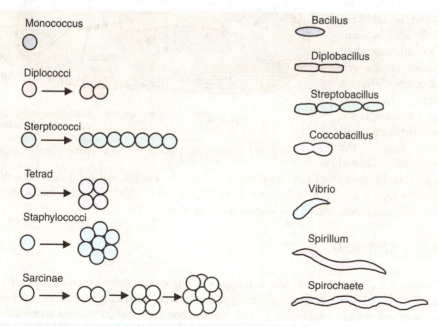

Fig. 4.1: Diagram of cocci, bacilli and spiral bacteria.

(d) **Hyphae :** Some microorganisms form the multicellular, thin-walled, profusely branched filaments called hyphae. The interwoven hyphae are collectively known as mycelium e.g. *Streptomyes, Aspergillus, Penicillium.*

The cyanobacterial cells are comparatively larger than the bacterial cells. In addition, the

| Streptomyces | Aspergillus | Penicillium |

cell size of eukaryotes such as algae, fungi, protozoa, etc. is several times greater than the cells of prokaryotes. The unicellular cyanobacteria are usually spherical or elongated. The fungi are either unicellular (*yeast, candida*) or multicellular hyphal (*Fusarium, Aspergillus*) forms.

B. Structure of Bacterial Cell

Upon observation under microscope there reveal several structural components outside and inside the cell wall. Some of these structures are found only in certain species of bacteria/ fungi. Out of all these only cell wall is common to all microbial cell. However, the cell wall structure differs in different microbial groups; for example the prokaryotic cell wall differs from the eukaryotic cell wall, and

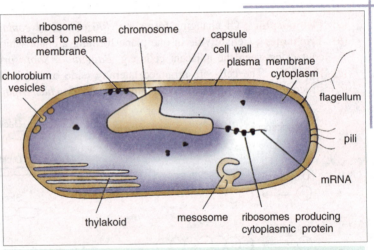

A diagram of structures seen in the prokaryotic (bacterial) cell.

Gram-positive bacteria from the Gram-negative bacteria. Fungal and algal cells also differ. The animal cells lack a rigid cell wall, many have flagella and fimbriae or pili. Inside the cell wall there is cell membrane which encloses cell inclusions.

1.   Capsule

Some of the bacterial cells are surrounded by the extracellular polymeric substances (EPS) which are commonly called capsule glycocalyx (Costerton *et al.*,1981). It forms an envelope around the cell wall and can be observed under light microscope after special staining technique (Fig.4.2). The presence of capsule may be detected by negative staining also such as India ink method.

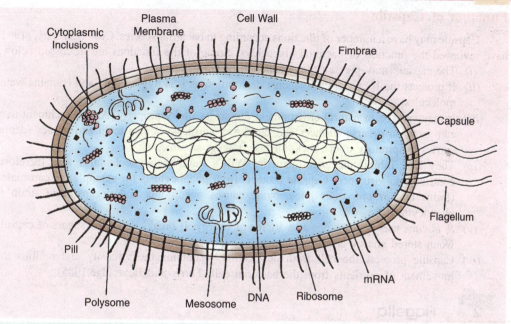

Fig. 4.2 : Structure of a typical bacterial cell (diagrammatic)

Table 4.2 : Cell size of some common bacteria

Bacteria	Disease	length (μm)
Clostridium botulinum	Food poisioning	3.8
C. tetani	Tetanus	2-5
Corynebacterium diphtheriae	Diphtheria	1-8
Mycobacterium tuberculosis	Tuberculosis	0.5-4
Neisseria meningitidis	Meningitis	1
Pasturella pestis	Plague	1-2
Salmonella typhii	Typhoid	0.5-4
Staphylococcus sp.	Boils	0.8
Streptococcus pneumoniae	Pneumonia	1.25
Treponema pallidum	Syphilis	6-14

The capsule is gelatinous polymer made up of either polysaccharide (*Klebsiella pneumoniae*) or polypeptide (*B. anthracis*) or both. The polysaccharides may be of a single type of sugars (homopolysaccharide) or several types of sugars (heteropolysaccharides). The heteropoly- saccharide is synthesized by sugar precursors within the cell. Homopolysaccharide constitutes the capsule of *Acetobacter xylinum*, and heteropolysaccharide (consisting of D-glucose, D-galactose, D-mannose, D-gluconic acid and D-rhamnose) is secreted by *Pseudomonas aeruginosa*. The capsule of pneumococci is made up of hexoses, uronic acids and amino sugars and that of streptococci consists of L-amino acids. The bacterial capsule is species specific and, therefore, can be used for immunological differentiation of related species (Ferris and Beveridge, 1985). Amount of these polymers vary with bacterial species. It is sticky in nature and secreted from the inner side of cell which gets firmly attached to the surface of cell wall. If the substances are unorganised and loosely attached to cell wall, the capsule is called slime layer. The fresh water and marine bacteria form trichomes which are enclosed inside the gelatinous matrix called sheath. Sheath is also found in cyanobacteria and other algae.

Function of Capsule

Capsule may have a number of functions according to bacterial species. Costerton *et al.,* (1981) have reviewed the function of bacterial glycocalyx. Some of the functions are discussed below.

(*i*) The capsule may prevent the attachment of bacteriophages.

(*ii*) It protects the bacterial cells against desiccation as it is hygroscopic and contains water molecules.

(*iii*) It may survive in natural environment due to its sticky property. After attachment they can grow on diverse surfaces e.g. plant root surfaces, human teeth and tissues (dental carries, respiratory tract), rocks in fast flowing streams, etc.

(*iv*) They may inhibit the engulfment by WBCs (antiphagocytic feature) and, therefore, contribute to virulence. Capsule protects from phagocytosis for example the capsulated strains of *Sreptococcus pneumoniae* causes pneumonia and uncapsulated strain is phagocytized.

(*v*) *S. mutans* uses its capsules as a source of energy. It breaks down the sugars of capsule when stored energy is in low amount.

(*iv*) Capsule protects the cell from desiccation, maintains the viscosity and inhibits the movement of nutrients from the bacterial cell (Ferris and Beveridge,1985).

---

2.   Flagella

The motile bacterium may possess a flagellum (plural flagella). The flagellum is hair like, helical and surface appendeges emerging from the cell wall. It is of 20-30 nm in diameter and 15 µm long. It provides various types of motality to the bacterial cell. The flagella of prokaryotes are several time thinner than that of eukaryotes.

In addition, the number and position of flagella vary. The arrangement may be monotrichous (a single polar flagellum e.g. *V. cholerae*), lophotrichous (a clusture of polar flagella e.g. *Spirillum*), amphitrichous (flagella at both the ends either singly or in clusture), cephalotrichous (two or more flagella at one end of bacterial cell e.g. *Pseudomonas*), peritrichous (cell surface evenly surrounded by several lateral flagella e.g. *Proteus vulgaris*) or atrichous (cells devoid of flagella e.g. *Lactobacillus*).

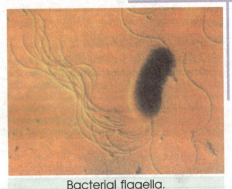

Bacterial flagella.

(*i*) **Structure of Flagella:** The structure and function of bacterial flagella have been described by Simon *et al.* (1978), Doestsch and Sjoblad (1980) and Ferris and Beveridge (1985). A flagellum consists of three basic parts, the basal body, hook and filament (Fig. 4.3).

(*a*) **Basal body:** M. L. De Pamphilis and J. Alder (1971) isolated the basal body of a flagellum of *E. coli* and *B. subtilis* and studied its fine structure and arrangement of rings. The basal body attaches the flagellum to the cell wall and plasma membrane. It is composed of a small central rod inserted into a series of rings.

In Gram-negative bacteria two pairs of rings, the proximal ring and the distal ring, are connected by a central rod. These two pairs of rings i.e. four rings are L-(lipopolysaccharide) ring P-(peptidoglycan) ring, S-(super membrane) ring, and M-(membrane) ring (De Pamphilis and Alder,1971).

The outer pair of rings, L-ring and P-ring, are attached to respective polysaccharide and peptidoglycan layer of cell wall, and the inner pair of rings i.e. S-ring and M-ring are attached with

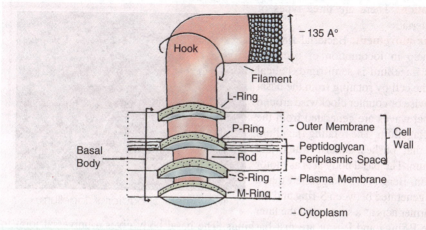

Fig. 4.3 : Different parts of attachment of a flagellum in a Gram-negative bacterium.

cell membrane. The outer rings form a bearing for the rod to pass through it.

In Gram-positive bacteria only the distal (inner) pair of rings is present. The S-ring is attached

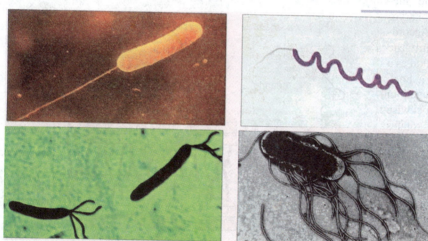

to inside thick layer of peptidoglycan and M-ring is attached to cell membrane (Simon *et al.*, 1978).

(*b*) **Hook :** The hook is present outside the cell wall and connects filament to the basal body. It consists of different proteins. The hook in Gram-positive bacteria is slightly longer than the Gram-negative bacteria.

(*c*) **Filament or shaft :** The outermost long region of the flagellum is called filament or shaft (Fig. 4.3). It has a constant diameter and is made up of globular proteins, the flagellin. the flagellins are arranged in several chains that intertwine and form a helix around a hollow core (Fig. 4.4). The proteins of flagella act to identify certain pathogenic bacteria unlike eukaryotes, the filaments are not covered by a membrane or sheath.

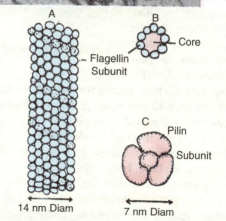

Fig. 4.4 : Molecular structure of bacterial flagellum. A, flagellum under electron microscope; B, end view of the flagellum; C, arrangement of pilin subunits of the pilus.

(ii) **Locomotion:** There are three types of movement in bacteria.

(a) **Flagellar movement:** Bacterial flagella are motile and help in locomotion of bacterial cells. Prokaryotic flagellum is semi rigid, helical rotor that moves the cell by rotating from the basal body either clockwise or counter clockwise around its axis. The helical waves are generated from the base to the tip of flagellum. The rotating flagellum forms a bundle that pushes against water and propel the bacterium. The basal body acts as motor and causes rotation. Berg (1975) suggested that a turning motion is generated between S-ring and M-ring, where the former acts as a starter and the later

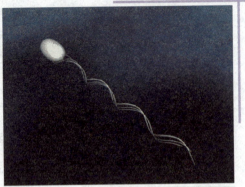

Movement of bacterial flagellum.

acts as rotor. The P-rings and L-ring are just bashings. The basal body gives a universal joint to the cell and allows complete rotation of the hook and shaft both clockwise and counter clock-wise.

Doetsch and Sjoblad (1980) have reviewed the mechanics of prokaryotic flagella. According to them the flagella function as a propeller of a boat. The polar flagellum rotates anticlockwise but the cell rotates clockwise when moving normally. Rotation of flagellum in anticlock-wise direction results in movement of bacterial cell in opposite direction. The peritrichous flagella as a trailing bundle also rotate in anti-clockwise direction

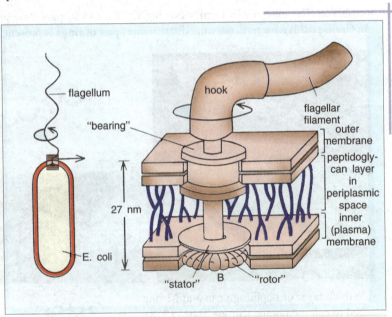

Schematic drawing of the flagellar rotatory 'motor' of *E coli.*

(Fig. 4.5). Cells of spirilla possess a non-helical tuft of polar flagella. The flagella rotate either at one end or both and result in cell movement. There is coordination between the flagella of both ends for rotation and movement of cell.

Energy is required for the movement of flagella. In eukaryotes, energy is generated by ATP. The mechanism of energy generation for the movement of prokaryotic flagella is not known. However, the basal body requires energy to cause motion. Movement of ions between M-ring and S-ring possibly energises the flagellar motor (Doetsch and Sjoblad, 1980).

(b) **Spirochaetial movement :** The spirochaetes show several types of movements such as flexing, spining, free swimming and creeping as they are flexible and helical bacteria lack flagella. Just within the cell envelope they have flagella like structure which are known as periplasmic flagella or axial fibrils or endoflagella. The axial fibrils are present in the space between inner and outer membrane of cell envelope. The mechanism of motility is not known. Berg (1975) postulated that the axial fibrils rotate in periplasmic space and cause the rotation of periplasmic cylinder on the body axis in the opposite direction.

(*c*) **Gliding movement:** Some bacteria such as the species of cyanobacteria (e.g. *Cytophaga*) and mycoplasma show gliding movement when come in contact with a solid surface. However, no organelles are associated with the movement. Except mycoplasma, in others two Gram-negative type cell walls are present. In the members of cytophagales and cyanobacteria, movement helps to find out the substratum e.g. wood, bark, shell, etc for anchorage and reproduction. They secrete slime with the help of which they get attached to the substratum. In *Oscillatoria princeps* fibrils of 5-8 nm thick are present near the cell surface and located helically around the cell. Oscillation is observed in this alga.

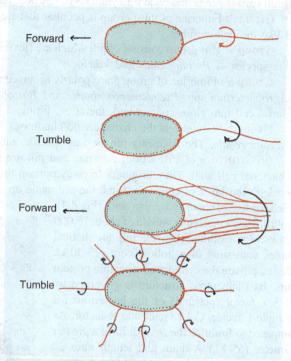

Fig. 4.5 : Flagellar motility in bacteria.

3.   Pili and Fimbriae

Pili and fimbriae are hair like appendages found on surface of cell wall (Fig. 4.2) in Gram-negative bacteria (e.g. Enterobacteriaceae, Pseudomondaceae and Caulobacter). Eukaryotic cells lack pili. The term fimbriae is used for all hair like structure covering the surface of the cell. Pili are genetically governed by plasmids, the number of which varies from 3 to 5. The number of fimbriae is around 1,000. However, a similar structure has also been observed only in *Coryne-bacterium renale*, a Gram-positive bacterium. Pili differ from flagella in being shorter and thinner, straight and less rigid. But they are in large

Oscillatoria princeps.

number. They occur either at the poles of bacterial cell or evenly distributed over the entric surface of the cell. The pili are 0.2-20 µm long with a diameter of about 250 Å.

(*i*) **Classes of Pili:** According to the function pili are of two types :

(*a*) Common pili which act to adhere the cell to surfaces, and (*b*) sex pili which join the other bacterial cell for transfer of genome. Ottow (1975) has classified the pili into the following six groups.

Group 1: Fimbriae of this group act for adherance to a particular surface including the surface of the other cells too. These are about 300 per cell arranged peritrichously. Pili of *Neisseria gonorrhoae*, the causal agent of gonorrhea, help the bacterium to colonize the mucous membranes. In the absence of pili on cell surfaces, mucous colonization and disease development cannot occur.

Group 2: The sex pili of this group have a uniform diameter of about 9 nm and length of about 1-20 µm. There are about 10 pili per cell. They are filamentous and determined by sex factor. The plasmid carries genes that code for synthesis of sex pili. They make contact between two cells.

Group 3: Fimbriae of third group is peculiar that are found in *Agrobacterium*. They are thick and like the hollow tubes.

Group 4: This group consists of pili which are flexible, rod-shaped and polar. These are found in the species of *Pseudomonas* and *Vibrio*.

Group 5 : Fimbriae of group 5 are polarly arranged and contractile in nature. They are found in *Agrobacterium* spp, *Pseudomonas rhodes* and *Rhizobium lupini*. They contract and pull two bacterial cell into close contact and, therefore, promote the conjugation process.

Group 6 : Group 6 is the characteristic bundles of fimbriae found in Gram-positive *Coryne-bacterium renale*. The filaments function as specific antigens.

(ii) Structure of Pili : Both fimbriae and pili are like flagella as both are the appendages on bacterial cell wall. They originate from cytoplasm that protrudes outside after penetrating the peptidoglycan layer of cell wall. Fimbriae are made up of 100% protein called fimbrilin or pilin which consists of about 163 amino acids (Fig.4.4).

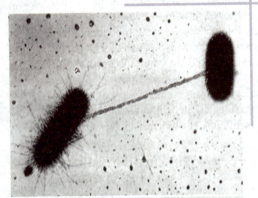

Fimbrilin has a molecular weight of about 16,000 daltons. In addition, the sex pili are helical tubules consisting of a hollow core (25-30Å). The sex pili are the cylinder of repeating protein units. Its filamentous structure is governed by the sex factor (plasmid) of the bacterium for example F factor, Col I factor and R factor. As compared to fimbriae the sex pili have a greater diameter (65-135 Å diam and length upto 20 μm), and a terminal knob of 150-800 Å diameter. There are two types of pili in *E. coli* for example F-pili (determined by F factor) and I-pili (determined by Col I factor. They have several receptor sites on which bacteriophages (e.g. f2, Ms2, M13, etc.) get adsorbed (Ferris and Beveridge,1985).

Structure of pili during conjugation.

(iii) Function of Pili: There are several functions of fimbriae and pili as given below:

(a) Bacteria containing fimbriae are called fimbriate bacteria. Fimbriae have the adhesive properties which attach the organism to the natural substrate or to the other organism. Fimbriae agglutinate the blood cells such as erythrocytes, leucocytes, eptithelial cells, etc.

(b) Fimbriae are equipped with antigenic properties as they act as thermolabile nonspecific agglutinogen.

(c) Fimbriae affect the metabolic activity. The Fim^+ cells (cells containing fimbriae) possess higher rate of metabolic activity than the Fim^- cells (cells devoid of fimbriae). Moreover, they function as aggregation organelles i.e. they can form stellate aggregation on a static liquid medium.

(d) The sex pili make contact between two cells. Since they posses hollow core, they act as conjugation tube. The tip of pilus recognises the female (F^-) cell through which the genetic material of donor (F^+) cell passes to the recipient (female) cell. Only F-pili (not I-pili) contain axial hole (Simon *et al.*, 1978).

4.   Chemotaxis

Chemotaxis is the movement of bacteria towards chemical attraction and away from chemical repellants. Bacteria are attracted towards the nutrients such as sugars and amino acids, and are repelled by harmful substances and bacterial wastes. They also respond to other environemntal fluctuations such as temperature, light, gravity, etc.

(i) Demonstration of Chemotaxis:
Chemotaxis can be demonstrated in laboratory by observing them in chemical gradients. A thin capillary tube is filled with an attractant and put into bacterial suspension by lowering down the capillary tube. From the end of capillary attractant is diffused into the suspension and chemical gradient is established. Bacteria assemble at the end of capillary tube and swim up the tube. The rate of chemotaxis can be measured by the number of bacteria within the capillary after a given time. It has been found that bacteria also respond to even low concentration (10^{-8}M) of some sugars. They increase the responses with increasing

Chemotaxis.

the concentration of sugar. If both the attractants and repellants are present together, bacteria will respond chemicals with the most effective concentration (Adler,1976).

(ii) Chemotactic Behaviour of Bacteria: By using the tracking microscope (a microscope with automatic moving stage that keeps an individual bacterial cell in view) chemotactic behaviour of bacteria can be studied.

E. coli and other bacteria move randomly in absence of chemical radient. Bacteria run in a straight direction or slightly curved line for a few second. Then it stops and tumbles for a short while (Fig. 4.5). The tumble is followed by a run in different directions. If the attractant concentration is higher tumbling behaviour is checked and they run for a long time. The opposite response occurs when a repellent is present. The frequency of tumbling decreases when bacteria moves down the gradient away from the repellent. When a bacterium is provided with attractant gradient they tumble less frequently i.e. has long runs while travelling the gradient. Bacteria compare its current environment with that present a few moment ago (Adler,1976).

(iii) Molecular Mechanism of Chemotaxis: The special proteins called chemoreceptors are supposed to be present in the periplasmic space of plasma membrane that detect the attractants and repellents. These proteins bind to chemicals and transmit signals to the other components of chemosensing system. So far about 20 chemoreceptors for attractants and 10 for repellents have been discovered, a few of them take part in the begining during sugar transport into the cell (Manson, 1992).

Parkinson (1993) has studied signal transduction schemes of bacteria. *E. coli* consists of four different chemoreceptors which are often called 'methyl accepting chemotaxis proteins' (MCPs). These four MCPs are localised in patches often at the end of rod-shaped cells. The MCPs act through a series of proteins. The whole responses triggered within 200 mili second. The mechanism of chemotaxis is shown in Fig. 4.6.

The MCPs are embeded in plasma membrane in such a way that their major parts are exposed on the both sides i.e. periplasmic side and cytoplasmic side. The periplasmic side consists of a binding site for attractants and repellents. The attractants either directly bind to MCPs or first bind to special periplasmic binding proteins and then to MCP. The cytoplasmic portion of MCP molecules usually contains about 4-5 methylation sites containing special glutamic acid residues. Methyls can be added to these glutamic acid carboxyl group by using S- adenosine methionine as the methylating agent. The chemoreceptor proteins associated with chemotactic responses are CheW, CheA, CheB, CheY and CheZ.

The cytoplasmic side of MCP binds two chemoreceptor proteins. First it binds to two molecules of CheW proteins which attach to a CheA protein resulting in formation of a full complex (an MCP

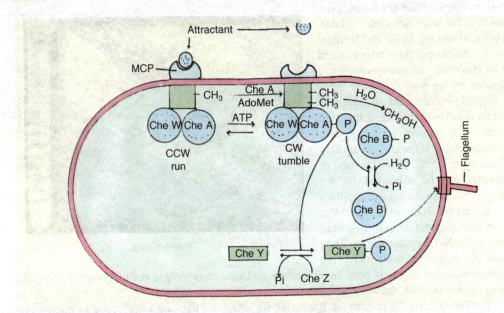

Fig. 4.6 : The mechanism of chemotaxis in *E. coli* (based on Prescott *et al.*, 1996)

dimer, two CheW monomer and a CheA dimer). The chemotatic response arises from a combination of (*i*) the control of CheA phosphorylation by the concentration of attractants/repellent, (*ii*) the clockwise rotation promoted by CheY, and (*iii*) a feed back regulation circuit.

CheA is stimulated by the MCP, when unbound to an attractant, to phosphorylate itself by using ATP through the process of autophosphorylation. Autophosphorylation does not occur when attractant binds to MCP. Phosphorylated CheA provides its phosphate to either CheY of CheB protein receptor. If CheY is phosphorylated it migrates to flagellum, interacts with base protein and results in movement of flagellum clockwise (CW). However, when the concentration of attractant decreases, it results in clockwise rotation and tambling. The Che Z protein removes the phosphate from CheY in about 10 seconds; when the attractant level changes the bacterium cannot remain in tumbling stage for a long time. In the absence of attractant or repellent the system maintains the intermediate concentrations of CheA phosphate and CheY phosphate. Consequently a normal run-tumble swimming state is maintained (Parkinson,1993).

The responses are shown very quickly. This has a short term memory i.e. for a few second of previous attractant/repellent concentration. This adaptation is accomplished by methylation of MCP receptors (Fig. 4.6). Irrespective of the concentration of attractants, methylation reaction is catalysed by the CheR protein. The phosphorylated CheB protein (a methylesterase) hydrolytically removes the methyl group from MCPs. The MCP-attractant complex is a good substrate for CheR protein and a poor substrate for CheB. Due to combining of attractant to MCPs the level of CheY phosphate and CheB phosphate decreases and autophosphorylation of CheA is inhibited. This causes in counter clockwise (CCW) rotation and run, and lowers methylesterase activity resulting in an increase in the process of methylation of MCP. When methylation is increased, it results in alteration in MCP conformation which in turn again maintains the intermediate level of CheA autophosphorylation. Therefore, CheY phosphate and CheB phosphate come back to its normal form and attain the normal behaviour of run tumble. When the attractants are removed, the over methylated MCP sitmulates CheA autophosphorylation. This results in increase in the level of CheY phosphate and CheB phosphate, and in turn tumbling and demethylation of MCP receptor.

5.   The Cell Wall (Outer Membrane)

The cell wall of bacteria is a semirigid and complex structure present beneath capsule and external to the plasma membrane. It is responsible for characteristic shape of the cell. The cell wall protects the plasma membrane and the other cytoplasmic inclusions from adverse environment. It also protects the bacterial cell from bursting when the osmotic pressure of cytoplasm is higher than that of out side of cell wall. It provides support for attachment to the flagella. It rescues the cell from antibodies and harmful chemicals.

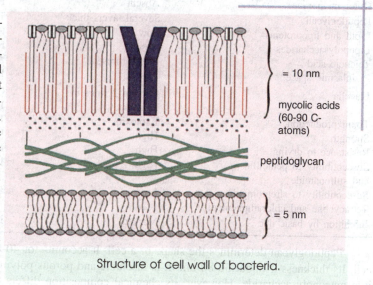

= 10 nm

mycolic acids (60-90 C-atoms)

peptidoglycan

= 5 nm

Structure of cell wall of bacteria.

The cell wall of Gram-negative eubacteria is comparatively thinner than the cell wall of Gram-positive bacteria. Similar is the situation of Gram-negative archaeobacteria. In addition, chemical composition of cell wall of archaeobacteria differs from eubacteria. Also cell walls of eukaryotic microorganisms (e.g. algae, fungi) differ chemically from those of prokaryotes. The cell envelope of Gram-negative bacteria consists of two unit membranes of 75 Å wide, separated by 100Å wide periplasmic space. Peptidoglycan is present in periplasmic space in Gram-negative bacteria. Differences between cell walls of Gram-positive and Gram-negative bacteria are given in Table. 4.3.

Chemical Composition and Wall Characteristics

The cell wall of bacteria is made up of network of peptidoglycan (murein, *murus* means wall). It is present almost on all bacterial cell wall except *Halobacterium* and *Halococcus*. Because these bacteria live in marine water which contains high salt concentration. The osmotic pressure of cytoplasm is more or less similar to outside the cell environment.

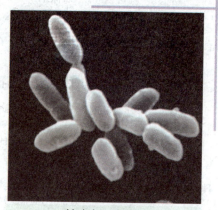

Halobacterium.

Table 4.3 : Differences between cell walls of Gram-positive and Gram-negative bacteria (modified after Tortora *et al.* 1989)

Character	Gram-positive	Gram-negative
Gram's staining	Retain crystal violet and appear dark violet	Pass crystal violet and counter stained by safranine and appear red

Character	Gram-positive	Gram-negative
Outer membrane	Absent	Present
Peptidoglycan	Several layers thick	Thin (single layer)
Lipid and lipoproteins	Low	High
Lipopolysaccharides	Absent	High
Teichoic acid	Mostly present	Absent
Periplasmic space	Absent	Present
Flagella	Contain 2 rings in basal body	Contains 4 rings in Basal body
Toxin production	Exotoxin	Endotoxin
Strength	High	Low
Resistance to drying	High	Low
Susceptibility to penicillin and sulfonamide	High	Low
Susceptibility to streptomycin, Tetracycline and chloramphenicol	Low	High
Inhibition by basic dye	Marked	Much less marked

Peptidoglycan determines the shape of a cell. It accounts for 40-80% of total dry weight of cell. Its thickness is about 30-80 nm. It is insoluble and porous polymer that provides rigidity. It is a mucopolysaccharide. However, its chemical composition differs from species to species. It consists of repeating disaccharides attached to chains of four or five amino acids. The monosaccharides, N-acetylglucosamine (NAG) and N-acetylmuramic acid (NAM) are linked by β-1, 4-glycosidic bond. These are related to glucose attached with amino acid groups. The structural formulae of NAG and NAM are shown in Fig. 4.7.

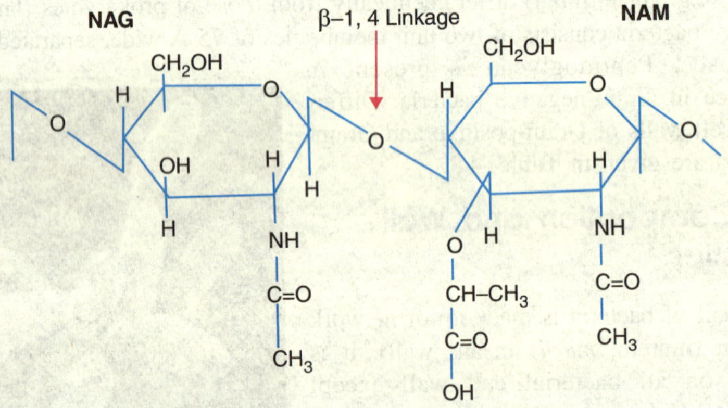

Fig. 4.7 : Chemical structure of N-acetylglucosamine (NAG) and N-acetylmuramic acid (NAM) linked together by β-1,4-linkage.

A tetrapeptide side chain containing four amino acid, (L-alanine, D-glutamate, L-lysine and D-alanine) is attached to each NAM. The third amino acid varies with different bacteria and may be lysine, diaminopametic acid or threonine. For example in *E. coli* instead of L-lysine (the third amino acid) there is mesodiaminopimetic acid. The D and L forms of amino acids alternate to each other. Except peptidoglycan, the amino acids found in protein are L-forms. The parallel tetrapeptide side chains are linked by a pentaglycine peptide cross bridge (PPCB) that contains five amino acid (Fig. 4.8).

The PPCB links L-lysine of one tetrapeptide with D-alanine at the terminal end. Due to extensive cross linking the peptidoglycan becomes a rigid macromolecule of the cell wall (Ferris and Beveridge,1985).

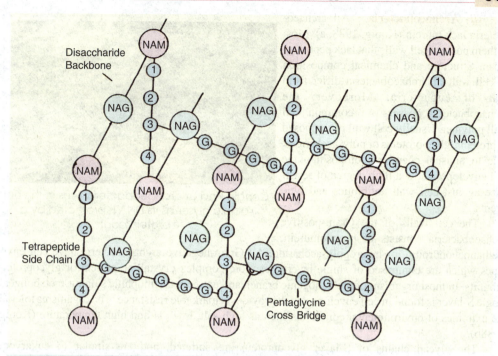

Fig. 4.8 : Organisation of peptidoglycan layer of *Staphylococcus aureus*, 1, alanine, 2, D-glutamate; 3, L-lysine; 4, D-alanine; G, glycine.

(*i*) Gram-positive Bacteria: In most of Gram-positive bacteria, the cell wall contains several layers of peptidoglycan which is inter-connected by side chains and cross bridges (Fig. 4.9B). Peptidoglycan accounts for 40-90% of toal dry weight of cell wall. However, the thickness may vary with types of species from 30 nm to 8 nm. The thickness of peptidoglycan provides rigidity to cell wall. The layers of peptidoglycan are thicker in Gram-positive bacteria than that in Gram-negative bacteria. In most of Gram-positive bacteria peptidoglycan is associated with acidic polymers containing phosphorus called teichoic acid or acidic polysaccharides such as teichuronic acids. Teichoic acids are hydrophilic, flexible and linear molecules. The presence of teichoic acid makes easy to diagnose the bacteria serologically.

Teichoic acids consist of an alcohol (e.g. glycerol or ribitol) and phosphate. Therefore, it is the polymer of glycerol phosphate or ribitol phosphate. Teichoic acids are mainly of three types (*i*) ribitol teichoic acids (found in *S.aureus* and *B.subtilis*), (*ii*) glycerol teichoic acid (e.g. *B. subtilis*), and (*iii*) glucosylglycerol phosphate teichoic acid (*B. licheniformis*). Out of these three only one type is found in a particular bacterium. The acids are linked to layers of peptidoglycan of plasma membrane. The phosphate groups provide negative charge which in turn controls the movement of cations i.e. positive ions across the cells. Teichoic acids possibly play a role in growth of bacterial cell by regulating the activity of an enzyme *autolysin*. The acids prevent the extensive break down and possibly the lysis of cell wall. They also store phosphorus (Rogers *et al.*,1978).

Peptidoglycan of *Staphylococcus aureus* consists of linear carbohydrate backbone, the glycan chain. The glycan chain contains alternate residues of NAG and NAM linked by β-1,4-linkage. The amino acid residues are connected by tetrapeptide and pentaglycine bridge (Fig. 4.8). The wall of most eubacteria contains very low amount of lipid except *Mycobacterium* and *Corynebacterium*. *Mycobacterium* exhibits acid fast staining i.e. the stain from cell wall is not easily decolourized with dilute acid. This is due to the presence of mycolic acid in cell wall. A mycolic acid derivative i.e. trehalose dimycolate plays a role in diseases caused by *M.tuberculosis* and *C.diphtheriae* (Pelczar *et al.*,1986).

(ii) **Archaeobacteria :** All archaeo-bacteria lack murein (Konig, 1988, a). Most of them possess cell walls that lack peptidog-lycan. Structure and chemical composition of cell wall of archaeobacteria differ from that of eubacteria. Moreover, the archaeobacteria possess no common cell wall polymer. Usually cell wall is composed of proteins, glycoproteins or polysaccharides. Due to unusual chemical composition the cell envelopes show a high degree of resist-ance against cell wall antibiotics and lytic agents.

Certain kind of archaeobacteria lives in this hot, acidic glyser as its DNA is protected by a protein coat.

The cell walls of the Gram-positive archaeobacteria consists of pseudomurein, methanochondroitin, or heteropolysaccharide. All Gram-negative archaeobacteria have cell enve-lopes which are composed of single layered or more complex crystalline protein or glycoprotein subunits. In most organism of methanogenic branch and extreme thermophilic sulphur metabolizers, single S-layer is found in cell envelope which provides remarkable resistance. The organisms tolerate the extremes of environmental conditions such as high salt, low pH and high temperature (Konig, 1988b).

The glycan chains of S-layer glycoprotein are sulfated, and are similar to eukaryotic proteoglycans. The S-layer shows a complex spongy three dimensional structure, and in some cases it occupies about 30% of the cell envelope.

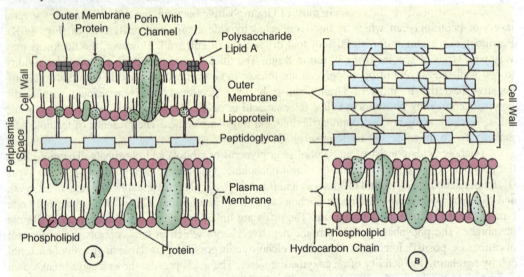

Fig. 4.9 : Structure and chemical composition and cell walls of Gram-negative (A) and Gram-positive (B) bacteria.

(iii) **Gram-negative bacteria :** The Gram-negative bacteria contain peptidoglycan but in very low amount (Fig. 4.9A). They totally lack teichoic acids. Peptidoglycan is situated in periplasmic space and covalently linked to lipoproteins in the outer membrane. The periplasmic space is a space between the outer membrane and plasma membrane which appears like gel and contains a high amount of enzymes and transport proteins. Due to the presence of low amount of peptidoglycan, the cell wall of Gram-negative bacteria can easily be disintegrated (Leive, 1976).

The cell envelope of Gram-negative bacteria is a bilayered structure consisting of mainly lipoproteins, lipopolysaccharides (LPS) and phospholipids. The chemical constituents and arrangement are described in detail as below :

(a) **Lipoproteins:** Lipoproteins occur freely and in bound forms as well. In lipoproteins of outer membrane, protein binds to lipid non-covalently, whereas in lipoproteins of plasma membrane, proteins bind to lipid covalently. Lipoproteins have a molecular weight of about 7000 daltons and consists of about 58 amino acids. Lipoproteins together with matrix proteins form a complex which contains diffusion channels (Di Rienzo *et al.*, 1978). A diffusion channel is enclosed by three molecules of matrix protein leaving a diameter of about 1.5-2 nm.

(b) **Lipopolysaccharides (LPS):** The outer membrane of Gram-negative bacteria is covered by LPS which is made up of polysaccharides covalently linked to lipid A (Fig. 4.10).

Lipid A : It consists of glucosamine, phosphate and fatty acids. The β-1,6-D- glucosamine disaccharide units constitute the carbohydrate components of lipid A.

The hydroxyl and amino groups of this disaccharide are substituted by the constituents such as polysaccharide chain, phosphate or pyrophosphate and fatty acids. The fatty acids provide hydrophilic property to lipid A. There are six fatty acids such as lauric acid (12C), myristic acid (14C) and palmitic acid (16C) (present in the ratio of 1:1:1), three molecules of 3-D-hydromyristic acid (14C) two of which contain amide (–NH_2) linkages with each amino group of two glucosamine. The third is esterified through its hydroxyl (-OH) group to myristic acid (De Rienzo *et al.*, 1978) (Fig. 4.10).

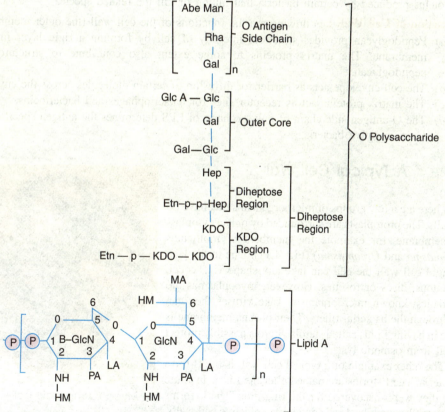

Fig. 4.10 : Chemical organisation of a lipopolysaccharide molecule. Polysaccharide (Abe, abequose; Rtn, ethanolamine; Gal, galactose); Glc, glucose acetyl; GlcN, glucosamine; Hep (L-glycero-D-mannoheptose); KDO (2-Keto-3-deoxyoctonate); Fatty acids-HM, β-hydroxymyristic acid (12C); LA, lauric acid (12C); M, myristic acid (14C); PA, Palmitic acid (16C); P, phosphate.

The esterified group of lipid A is referred to as endotoxin which is toxic when present in blood. Lipid A of LPS is responsible to induce fever and shocks.

(c) **Polysacharide :** The polysacharide portion of LPS of *Salmonella* cell wall is composed of three important components, the inner core, the outer core and the O-antigen side chain. Although the polysaccharide of LPS is also known as O-polysaccharide. The O-antigen side chains in rough strains of Gram-negative bacteria are absent, whereas the smooth strains of *Salmonella* have O-antigen side chains which may extend out the wall surface and attain about 30 nm length. These chains have antigenic property and, therefore, can be distinguished serologically (e.g. species of *Salmonella*). Its role is comparable to that of teichoic acids in Gram-positive bacterial cell walls.

The outer membrane consists of two molecules of glucose sandwitching a molecule of galactose. One glucose subunit is linked to glucose acetyl and the others to galactose. The inner core consists of two regions, the ketodeoxyoctonate (KDO) region and diheptose region. The KDO region comprises of three units of KDO and eight carbon α-keto sugar. The diheptose region consists of two units of L-glycero-D-mannoheptose, the seven carbon heptose sugar (Fig. 4.10).

(d) **Matrix proteins :** The outer membrane of cell envelope does not warrant the entry of all substances since the nutrients are to pass across the membrane. It is impermeable only to macromolecules such as proteins, lipids, etc. The permeability of outer membrane is due to the presence of proteins called porins that form channels (Fig. 4.9). The porins are not specific and allow the small molecules. Certain porins are specific and permit only the specific substances such as vitamin B_{12}, nucleotides, etc. Porins also act as receptor sites for bacteriophages and bacteriocins (the proteins produced by certain bacteria that inhibit or kill the related species).

(ii) **Function of Cell Wall :** Following are the functions of the cell wall (the outer membrane):

(*a*) Peptidoglycan provides structural integrity of cell by forming a rigid layer in outer membrane. The matrix proteins to some extent also contribute to structure with peptidoglycan.

(*b*) The cell envelope acts as barrier for diffusion to certain molecules across the envelope.

(*c*) The matrix proteins act as receptor sites for bacteriophages and bacteriocins.

(*d*) The O-antigen side chain of polysaccharie of LPS determines the antigen specificity of Gram-negative bacteria.

6.   A Typical Cell Wall

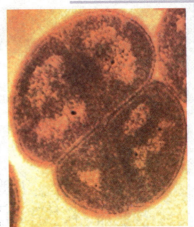
Mycoplasma lacks cell wall.

There are a few groups of microorganisms that have no cell walls. The protoplasts are surrounded by only a cytoplasmic membrane, for example the members of mollicutes (*Mycoplasma* and *Ureoplasma*) (Fig. 4.11). Due to the lack of a rigid cell wall the cell can take any shape viz., cocci, filamentous, discs or rosettes. However, mycoplasmas are the smallest known microorganism. Like viruses they can pass through the bacterial filters. Their plasma membrane is unique in having lipids called sterols. Sterols possibly protect the cell from osmotic lysis.

The other example of a typical cell wall is the L-form of bacteria. The L-forms are named after the Lister Institute where they were discovered for the first time. The L-forms are the small mutant bacteria containing defective cell walls. This form can be produced by inducing the bacteria with certain chemicals and antibiotics. Some L-forms after giving proper nutrition revert to the original forms and the others are stable. The L-forms attain irregular shape. It has been found when the protoplast of *B. subtilis* is placed in a 25% gelatin medium and incubated at 26°C, it reverts back to the normal walled form.

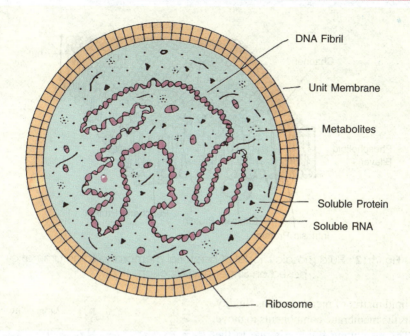

DNA Fibril

Unit Membrane

Metabolites

Soluble Protein

Soluble RNA

Ribosome

Fig. 4.11: A profile of mycoplasma (diagrammatic).

The cell wall of Gram-positive bacteria is degraded when treated with lysozyme. The cellular material is surrounded by plasma membrane. If osmotic lysis does not occur the plasma membrane remains intact. These cells devoid of cell wall are known as protoplast. In contrast the lysozyme destroy the cell wall of Gram-negative bacteria only to some extent. Some of the outer membrane remains intact. Thus the cellular contact, plasma membrane and remaining outer cell wall layer are jointly called *spheroplast*. These are also spherical in shape. Rupturing of protoplast and spheroplast is known as osmotic lysis after placing them in water of dilute solution of salt or sugars. Kawate *et al*. (1961) prepared the protoplast of *B. megaterium* and found that the spherical cells still retained their flagella. This indicates that the flagella are not a part of the cell wall.

7. Plasma Membrane (Cell or Cytoplasmic Membrane)

The plasma memrane also called cell membrane or cytoplasmic membrane is a structure internal to the cell wall.. The term cell membrane was coined by C. Nugeli and C. Cramer in 1855. However, no cell could be alive without a plasma membrane. It is situated just beneath the cell wall and is 7.5 nm thick. It consists of proteins (20-70%), lipids (28-80%), oligosaccharides (1-5%) and water (20%). The plasma membrane consists of a continuous bilayer of phospholipid molecules in which globular proteins are embedded.

(*i*) **The Fluid Mosaic Model :** Singer and Nicolson (1972) have given the fluid mosaic model of plasma membrane to explain its major features (Fig. 4.12). This model is well accepted. According to this model the plasma membrane is quasifluid structure in which lipids and proteins are arranged in a mosaic manner. The globular proteins are of two types: extrinsic (peripheral) proteins and intrinsic (integral) proteins. The extrinsic protein is soluble and, therefore, dissociate from the membrane, while the intrinsic protein is insoluble and could not (or rarely) dissociate. The intrinsic proteins are partially embedded either on outer surface or on inner surface of the bilayer and take part in lateral diffusion in lipid bilayer (Ghosh,1981).

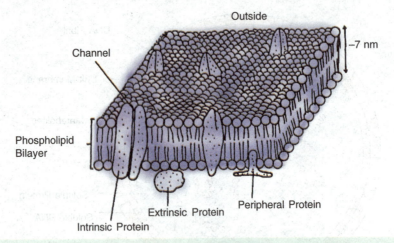

Outside

Channel

Phospholipid
Bilayer

~7 nm

Extrinsic Protein

Peripheral Protein

Intrinsic Protein

Fig. 4.12 : Fluid mosaic model of bacterial membrane (diagrammatic, based on Singer and Nicolson, 1972).

The lipid matrix of membrane has fluidity that permits the membrane components to move laterally. The membrane fluidity is due to the hydrophobic interactions of lipids and proteins. The fluidity is important for a number of membrane functions. Phospholipids and many intrinsic proteins are amphipatic i.e. they possess both hydrophilic and hydrophobic groups.

Phospholipids are the complex lipids which are made up of glycerol, two fatty acids and, in place of a third fatty acid, a phosphate group bounded to one of several organic groups (Fig. 4.13). They have polar (hydrophilic) as well as non-polar (hydrophobic) regions. Polar portion consists of a phosphate group and glycerol, while non-polar portion consists of fatty acids. All non-polar parts of phospholipid make contact only with the non-polar portion of the neighbouring molecules. The polar portion occurs towards outside. This characteristic feature gives the appearance of bilayer. However, between the fatty acid chains proper spacing is maintained by interspersing unsaturated chains throughout the membrane. This type of arrangement maintains the semifluidity of plasma membrane (Ghosh,1981).

The presence of complex lipids becomes a key character of certain microorganisms on the basis of which they can be identified. For example, the cell wall of *Mycobacterium* contains high amount of lipids such as waxes and glycolipids which gives the bacterium a distinctive staining characteristic.

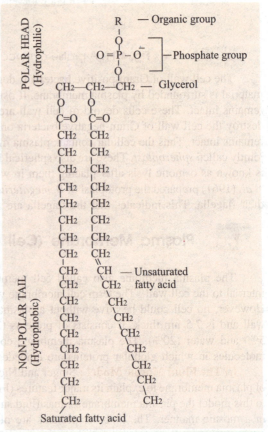

Fig. 4.13 : Chemical composition of a phospholipid molecule.

In some microorganisms such as mycoplasmas and fungi, sterols are found to be associated within the plasma membrane (Fig.4.14). Sterols are structurally different from the lipids. The -OH group in cholesterol makes it a sterol. Sterols are alcohols composed of hydrocarbon rings attached to hydrocarbon chain. The sterols separates the fatty acid chains and check packing which harden the plasma membrane at low temperature. In case of certain bacteria hopanoids are present which have similar role to that of sterols found in certain fungi. These hopanoids donot require oxidation step for their biosynthesis, therefore due to lack of oxygen, anaerobic bacteria also contain hopanoids.

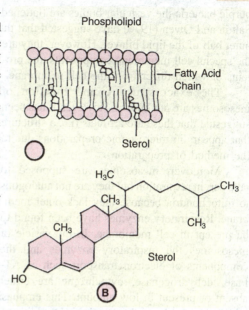

Fig. 4.14 : Presence of steroids in plasma membrane (A), and chemical structure of a steroid (B).

(*ii*) **Function of Plasma Membrane:** The cytoplasmic membrane is the site of many metabolic activities as given below :

(*a*) The organic and inorganic nutrients are transported by permeases through plasma membrane.

(*b*) It consists of enzymes of biosynthetic pathways that synthesize different components of the cell wall such as peptidoglycan, teichoic acids, polysaccharides, lipopolysaccharides and phospholipids.

(*c*) It possesses the attachment sites for bacterial chromosome and plasmid DNA.

(*d*) The inner membrane invaginates to form mesosomes, a site for respiratory activity. The plasma membrane contains about 200 respiratory proteins that have been found to be anchored for the transport of H^+ ions.

(*e*) It provides permeability barrier and thus prevents the escape of cellular materials outside the cell, and facilitate the selective entry of organic and inorganic substances inside. Hence, the plasma membranes shows selective permability.

8. Mesosomes

Mesosomes are the invaginated structures formed by the localized infoldings of the plasma membrane. The invaginated structures comprise of vesicles, tubules of lamellar whorls (Fig.4.15). Generally mesosomes are found in association with nuclear area or near the site of cell division. They are absent in eukaryotes. The lamellae are formed by flat vesicles when arranged parallely. Some of the lamellae are connected to the cell membrane. The lamellar whorl can be observed in *Nitrobacter, Nitromonas* and *Nitrococcus*.

The vesicles are formed probably by invagination and tubular accretion of the plasma membrane. The structure of vesicle becomes interrupted due to constriction at equal distance. The constriction does not cause the complete separation of tubules. Closely packed spherical vesicles are seen in *Chromatium* and *Rhodospirillum rubrum*. In some

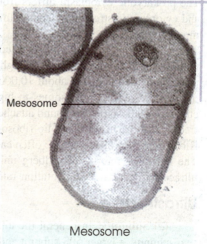

purple bacteria the vesicular bodies are flattened and stacked into the regular plates like thylakoids. Salton and Owen (1976) have suggested that the mesosomes are formed due to vesicularization of outer half of the lipid bilayer. However, they are the special cell membrane components, the proteins of which differ from the cell membrane.

The exact structure and function of mesosomes are not known. However, it has been suggested that these are artifacts (i.e. a structure that appears in microscopic preparations due to the method of preparation).

Moreover, mesosomes are supposed to take part in respiration but they are not analogous to mitochondria because they lack outer membrane. Respiratory enzymes have been found to be present in cell membrane. In the vesicle of mesosomes the respiratory enzymes and the components of electron transport such as ATPase, dehydrogenase, cytochrome are either

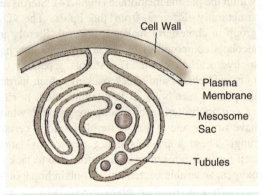

Fig. 4.15 : The bacterial mesosome (diagrammatic)

absent or present in low amount. This emphasizes their inability to carryout transport process in which the membrane is energised. In addition, mesosomes are supposed as a site for synthesis of some of wall membranes (Salton and Owen, 1976).

Mesosomes might play a role in reproduction also. During binary fission a cross wall is formed resulting in formation of two cells. Mesosomes begin the formation of septum and attach bacterial DNA to the cell membrane. It separates the bacterial DNA into each daughter cell. In addition, the infoldings of mesosomes increase the surface area of plasma membrane that in turn increase the absorption of nutrients.

9.   Cytoplasm

Cytoplasm of prokaryotes refers to the internal matrix of cell inside the plasma membrane. Cytoplasm consists of water (80%), proteins, carbohydrates, lipids, inorganic ions and certain low molecular compounds. Cytoplasm is thick and semitransparent. The DNA molecules, ribosomes and the other inclusions are the structure of cytoplasm. In certain cyanobacteria gas vacuoles are found. The prokaryotic cytoplasm differs from the eukaryotic cytoplasm. The former lacks cytoskeleton and cytoplasmic streaming. Compartmentation of organelles are absent in prokaryotes and present in eukaryotes.

(*i*) **Ribosomes :** All living cell contains ribosomes which act as site of protein synthesis. High number of ribosomes represents high rate of protein synthesis and *vice versa*. Cytoplasm of a prokaryotic cell contains about 10,000 ribosomes which account upto 30% of total dry weight of the cell. Presence of ribosomes in high number gives the cytoplasm a granular appearance. The eukaryotic ribosomes are found attached to cell membrane, whereas the prokaryotic mesosomes are free in cytoplasm. Prokaryotic ribosomes are smaller and less dense than eukaryotic ribosomes. Ribosomes of prokaryotes are often called 70S ribosomes and that of eukaryotes as 80S ribosomes. The letter 'S' refers to Svedberg unit which indicates the relative rate of sedimentation during ultracentrifugation. Sedimentation rate depends on size, shape and weight of particles.

Ultrastructure

(*a*) **Subunits:** In general the ultrastructure of ribosomes reveals that these are composed of two subunits, a larger 50S subunit and a smaller 30S subunit. Each subunit is composed of protein and ribosomal RNA (rRNA). Their association and dissociation depend on the concentration of Mg^{++} ions. The structure of a ribosome is very complex. The proteins and RNA are interwined. James A. Lake (1981) presented the structure and function of ribosome. According to him the smaller subunit of ribosome consists of a head, a base and a platform. With the help of a cleft the plateform and head are separated from the base. The larger subunit comprises of a ridge, a central protuberance and a stalk; the former two are separated by a valley (Fig. 4.16).

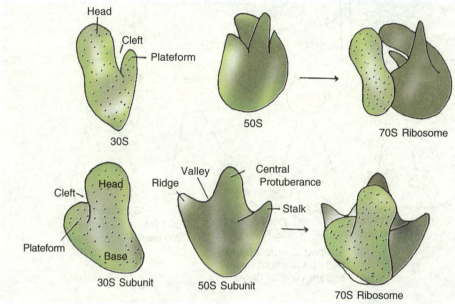

Fig. 4.16 : Three dimensional model of *E.coli* ribosome shown in two different orientation (A and B) (after Lake, 1981)

(*b*) **Chemical composition :** The ribosomes of *E. coli* consist of three types of RNAs, 5S, 16S and 23S, and 53 proteins. The 50S subunit consists of 5S and 23S RNA, and 34 proteins; 30S subunit consists of 16S RNA and 21 proteins. The 5S RNA is 120 nucleotides long, 16S RNA is about 1,600 nucleotides long and 23S RNA is about 3,200 nucleotides long. Base sequence of 5S RNA has been strongly conserved through out the evolution and that of 16S form double stranded hairpin loops. Only 30-35% bases of 16S form single stranded loops. Interactions between rRNA and cellular RNA (mRNA and tRNA) occur (Fig.4.17).

From 30S RNA of *E. coli,* 21 proteins have been isolated that are designated as S1 to S21. Similarly from larger subunit (50S), 31 proteins (L1 to L34) have been isolated. Protein map of ribosome showing their sites on two subunits are presented in Fig.4.18.

The function of ribosome in protein synthesis is a well established fact (see Chapter 10). However, they are not specific in nature. Ribosome of one species can be used for protein synthesis in other species. There are several antibiotics such as streptomycin, neomycin and tetracycline that inhibit protein synthesis on the ribosomes. Even antibodies can kill the prokaryotic microorganisms but not the eukaryoic microorganisms. This is due to differences in prokaryotic and eukaryotic ribosomes.

(*ii*) **Molecular Chaperones :** It was thought for many years that polypeptides after synthesis fold into native stage and this folding is not determined by its amino acids. It is now clear that there are certain helper proteins called molecular chaperones or chaperones which recognise the newly formed polypeptides and fold to its proper shape. Proteins fold rapidly into the secondary structure. This unusually open and flexible conformation is called as molten globule. It is the starting points for slow process which results in correct tertiary structure (Kuwajima,1989).

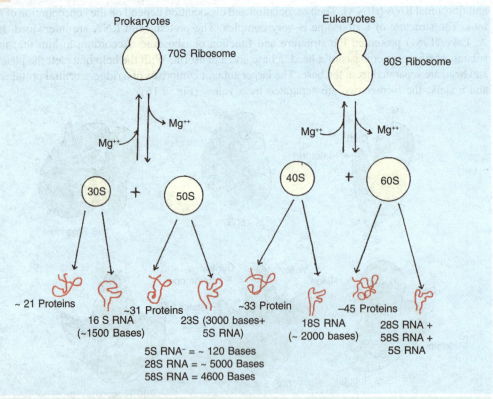

Fig. 4.17 : The rRNA and proteins of prokaryotic (A) and eukaryotic (B) ribosome.

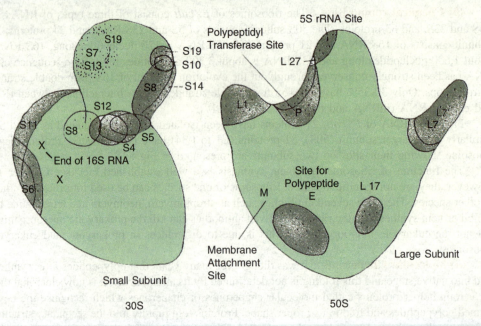

Fig. 4.18 : Protein map showing their sites on small subunit and large subunit of ribosome.

There are several chaperones involved in proper protein folding in bacteria. Chaperones were first identified in *E. coli* mutant that did not allow to replicate phage lambda. In *E. coli* at least four chaperones viz., DnaK, DnaJ, GroEL and GroES, and stress protein GrpE are involved in folding process. They play an important role because after protein synthesis the cytoplasmic matrix is filled with nascent polypeptides and proteins. It is possible that these polypeptides become folded and form a nonfunctional complex. The chaperones check wrong infolding and promote correct folding. The chaperones are found both in the cells of prokaryotes and eukaryotes (Hartl *et al.*,1994).

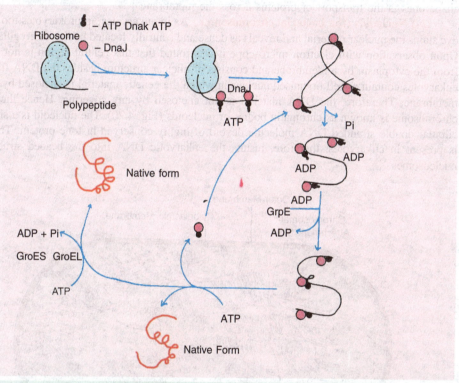

Fig. 4.19 : Mechanism of polypeptide folding by chaperones
(based on Prescott *et al.*, 1996).

After synthesis of a sufficient length of polypeptide from the ribosome, DnaJ binds to unfolded chain (Fig.4.19). DnaK complexed with ATP attaches to polypeptide. These two chaperones check the polypeptide from folding. After binding of DnaK to polypeptide ATP is hydrolysed to ADP which increases the ability of DnaK to bind with unfolded peptide. When polypeptide synthesis is completed, GrpE protein binds with DnaK-polypeptide complex and causes the DnaK to release ADP. Thereafter, ATP binds to DnaK, and DnaK and DnaJ are released from the polypeptide. During these events, polypeptide is folded and reach to its final native conformation. At this stage if polypeptide is partially folded it binds to DnaJ and DnaK, and repeats the same process again. Mostly DnaK and DnaJ transfer the polypeptide to GroEL and GroES where final foldings occur. GroEL is a long hollow barrel shaped complex of 14 subunits which are stacked in two rings, whereas GroES contains four subunits arranged in one ring and can combine to both ends of GroEL. ATP binds to GroEL and changes the ability of the later for polypeptide binding. GroES binds to GroEL and help in binding and release of refolding peptide (Hartl *et al.*,1994).

Heat shock proteins: When *E. coli* cells are exposed to high temperature, metabolic poisons and other stressful condition, the concentrations of chaperones increases. In *E. coli* cultures, at temperature between 30 and 40°C, 20 different chaperones often called heat shock proteins are produced within 5 minutes. These protect the cell from thermal damage and stress, and promote the proper folding of polypeptides. In hypothermophiles (e.g. *Pyrodictum occultum*) that grow at about 110°C, a large amount of chaperones are present (Georgopoulos and Welch, 1991).

Mutant *E. coli* resistant to phage λ produces slightly two changed chaperones like heat shock proteins 60 and 70 (hsp60 and hsp70). The eukaryotic cells have families of hsp60 and hsp70 proteins, and different family members function in different organelles. The mitochondria contain their own hsp60 and hsp70 molecules which are different from those functioning in the cytosol. A special hsp70 helps to fold proteins in the endoplasmic reticulum. The other function of chaperones is the transport of proteins across the membrane.

(*iii*) **Nucleoids (the bacterial chromosome) :** As in eukaryotes, in prokaryotes too the basic dye stains the nuclear material and reveals as dense and centrally located bodies of irregular outline. Upon observation with electron microscope it was found that this central region is not separated from the cytoplasm by a membrane and consists of nuclear structure besides the DNA fibrils. The eukaryotes contain a well organised nucleus in which the genetic material is enclosed by a nuclear membrane. Therefore, the DNA material is not enclosed by any covering. Hence the bacterial chromosome is known as chromatin bodies or nucleoids (Fig. 4.20). The nucleoid is a single long circular double stranded DNA molecule devoid of highly conserved histone protein. The histone is present in eukaryotes, therefore, results the eukaryotic DNA into the beaded structures i.e. nucleosomes.

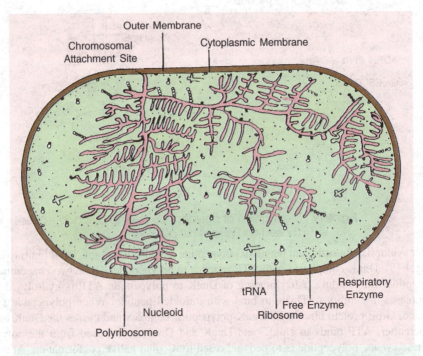

Fig. 4.20 : A cell of *E.coli* showing nucleoid in centre and the other cell components (diagrammatic)

By the turn of 1960s, the nuclear material was studied in detail. In 1963, J. Cairns succeeded in extracting *E. coli* DNA under conditions that minimize its shearing. Autoradiographic studies showed that the DNA was extremely long threads measuring about 1mm in length. A few threads were circular. Hobot *et al.* (1985) found that the fully extended *E. coli* DNA was 1 mm long (4 × 10^6 × 3.4 Å) having molecular weight 3 × 10^9 dalton. Through electron microscope they also observed the compaction of chromosome into irregular shaped nucleoids.

The nucleoid is observed as a coralline (coral like) shaped structure the branches of which spread far into the cytoplasm and over the entire area of the cell. By using serial sections it has been possible to reconstruct the ribosome free area of the nucleoid (Bohrmann *et al.,* 1991).

Two types of nuclear bodies can be observed, an envelope associated nucleoid and an envelope free nucleoid. Associated with the first type a large amount of RNA, proteins, lipids and peptidoglycan are found, whereas the second type contains less amount of it.

Generally, the number of nucleoid per bacterial cell is one, but in some bacteria the number may go even to four or more. The DNA molecule appears to be present in 10 to 80 super coils. Worcel and Burg (1972) proposed the structure of the folded chromosome of *E. coli* and showed as seven loops, each twisting into a super helix. These loops are held together by a core of DNA (Fig.4.21).

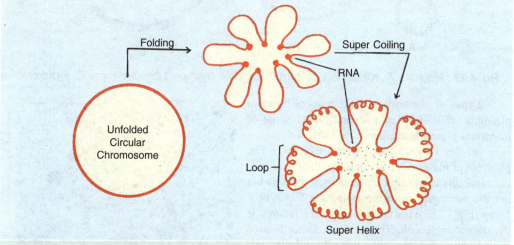

Fig. 4.21 : A model representing the process of folding and super coiling of bacterial chromosome (based on Worcel and Burgi, 1972).

Super coiling may be induced enzymatically. Possibly it may be a factor for the formation of nucleoids. The folded structure was found to be attached to a fragement of cell membrane (Delius and Worcel, 1974). This shows that the bacterial chromosome remains associated to a point of cell membrane. This helps in separation of newly replicated DNA molecules.

(iv) **Plasmids:** During 1950s, working on conjugation process it was found that maleness in bacteria is determined by a transmissible genetic element. When male and female bacteria conjugate, every female is converted into a male. This inherited property of male is called the F (fertility) factor which is transmitted by cell to cell contact. Therefore, F is a separate genetic element. In 1952, J. Lederberg coined the term plasmid as a genetic name for this element. Hence the plasmids may be defined as *a small circular, self replicating and double stranded DNA molecule present in bacterial cell, in addition to bacterial chromosome.* It replicates independently during cell division and inherited by both of daughter cells. Therefore, its function is not governed by the bacterial chromosome.

In 1960, Jacob Schaeffer and Wollman for the first time used the term episome to denote the extrachromosomal genetic element that integrated the bacterial chromosome during replication.

The number of plasmids ranges from one to hundreds or more per bacterial cell. A plasmid contains 5-100 genes that determine several biological functions. Under certain circumstances they provide special characteristics to the bacterial cell and help them in survivability. They may even lose without harming the bacterial cell.

Plasmids are the circular DNA molecule but in resting stage helix twists in right hand direction at every 400-600 base pairs and forms super coils. The twisted form is called covalently closed circular DNA. After cleaving the twists this form is converted into an open circular form of double stranded DNA molecule (Fig.4.22).

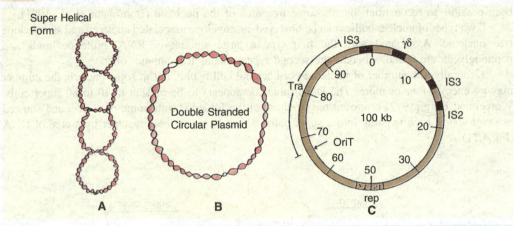

Fig. 4.22 : Plasmid. A, super helical form; B, double stranded circular form; C, F-plasmid.

Types of plasmids: On the basis of function plasmids are divided into several types. Some of important plasmids are given in Table 4.4.

(a) Sex factor or fertility (F) factor : The plasmids of male cells that confer on their host cells the capability to transmit chromosomal markers but not the other properties are called sex factor or F factor (Fig.4.22C). For the first time the F factor was discovered in *E. coli*. The term sex factor is used in two ways : (a) as genetic name for all plasmids which determines host conjugation and their own transfer irrespective of the transfer of bacterial chromosome, and (b) to describe the set of genes on plasmids whose products mediate the conjugation process. Some times the F factor transfers a nonconjugative plasmid also which is present with it in a cell. A new strain was isolated from F⁺ cultures that underwent sexual recombination with F⁻ cells. This new strain had recombinant rate about 10^3 time greater than F⁺ × F⁻ cells. These strains

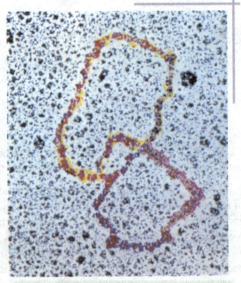

TEM of plasmids in bacterial cytoplasm.

were called high frequency recombination (Hfr) strains. The Hfr strains contained both bacterial DNA and plasmid DNA. For detail see Chapter 8, *Microbial Genetics*.

(b) R (resistance) plasmids : During 1946 several out breaks of dysentery by *Shigella* occurred in Japan and resistance against antibiotic drugs developed. In 1955, in Japan a strain of *S. dysenteriae* causing dysentery was isolated that was resistant to four drugs viz., sulfanilamide, streptomycin, chloramphenicol and tetracycline. By 1964, over 50% all *Shigella* hospital isolates in Japan became multiple drug resistant. Similar strains were also isolated from hospital in London in 1962, and that of *Salmonella* in England in 1965.

K. Ochiai and co-workers in Japan found that the genetic elements governing the drug resistance are transferable to the other bacterial cell through the process of conjugation. Further studies explained that the resistance genes are associated in various combinations as parts of transmissible plasmids known as R factor. Throughout the world R factors spread among populations of enteric bacteria due to rapid use of antibiotics. Different R factors having different sets of resistance genes (ranging from one to eight) are required from different chemicals. Therefore, there are several genes of R- factor which confer resistance against antibodies (Helinsky, 1973).

An R factor consists of two segments of DNA, one the resistance transfer factor (RTF) and the second resistant determinants (r-determinants). The RTF contains genes for replication and transmission of plasmid, and the r-determinants are on the another segment. Some of r-determinants replicate autonomously. The G+C content, molecular weight and buoyant density of these two segments vary.

When two bacteria (one containing R plasmid and the other devoid of R) conjugate the R plasmid is transferred to the later that lacks R plasmid.

Table 4.4 : Some important types of plasmids

Type	Representative	Hosts	Size(kb)/ (Number)	Features
Fertility factor	F factor	*E.coli, Salmonella Citrobacter*	95-100 (1-3)	Sex pilus, Conjugation
R plasmids	RP_4	*Pseudomonas* and other Gram-negative bacteria	54 (1-3)	Sex pilus, conjugation Amp^r, Kan^r, Neo^r, Tet^r
	R100	*E.coli, Shigella Salmonella, Proteus*	90 (1-3)	$Chlo^r$, $Stre^r$, Tet^r, Hg^r
	pSJ23a	*S. aureus*	36	Pen^r, Hg^r, $Gent^r$, Kan^r, Neo^r, Ery^r
Col plasmids	ColE1	*E.coli*	9 (10-30)	Colicin E1 production
	ColE2	*Shigella*	10-15	Colicin E2
Virulence plasmids	Ent (P307)	*E.coli*	83	Enterotoxin production
	Col V-K30	*E.coli*	2	Siderophore for iron uptake
Metabolic plasmids	CAM	*Pseudomonas*	230	Camphor degradation
	SAL	*Pseudomonas*	56	Salicylate degradation
	TOL	*P. putida*	75	Toluene degradation
	P5P4	*Pseudomonas*	—	2,4-dichlorophenoxy acetic acid degradation
		E.coli	—	Lactose degradation

Source : Based on Prescott *et al.* (1996)

(*c*) **Heavy-metal resistance plasmids :** There are several bacterial strains that contain genetic determinant of resistance to heavy metals viz., Hg^{++}, Ag^+, Cd^{++}, Co^{++}, CrO_4, Cu^{++}, Ni^{++}, Pb^{+++}, Zn^{++}, etc (Bopp *et al*, 1983). These determinants for resistance are often found on plasmids and transposons (Summers, 1982).

In the 1970s in Tokyo both heavy metal resistance and antibiotic resistance were found with high frequency in *E. coli* isolated from hospital patients, whereas heavy metal resistance plasmids without antibiotic resistance determinants were found in *E. coli* from an industrially polluted river (Misra *et al.,* 1985). Bacteria that have been found resistant to heavy metals are *E. coli* and *Staphylococcus aureus* (As), *Pseudomonas aeruginosa, P. fluorescens, E. coli* (Cr), *B. subtilis, Alcaligens eutrophus* (Cd), *Shigella* spp, *E. coli* (Hg), *Pseudomonas syringae.*

(*d*) **Col plasmids :** There are many bacterial strains that produce proteinaceous toxins known as bacteriocin which are lethal to other strains of the same genus. Toxins secreted by the strains of *E. coli* are called colicins. It kills the sensitive cells. Synthesis of colicins is specified by the plasmids present in *E. coli* cells but not by bacterial chromosome. These plasmids associated with colicin production are called colicinogeny (Col) factor. There are several Col plasmids such as Col B, Col E, Col I, Col V which produce different types of colicins.

Some Col plasmids carry fertility determinants i.e. a set of genes governing conjugation and transfer of plasmids (e.g. Col B, Col V). These can be called conjugative plasmids. The second type is the non-conjugative plasmid which is non-transmissible by their own means (e.g. col E). However, when a cell contains both the plasmids, it transfers both.

(*e*) **Degradative plasmids:** Much work has been done on degradative plasmid of *Pseudomonas*. The pseudomonads have been found to catalyse a number of unusual complex organic compounds through the special metabolic pathways. Anand Mohan Chakrabarty, an India-borne American scientist, has isolated plasmids from a number of cultures of *Pseudomonas putida* which could utilize a number of complex organic chemicals such as 2,4-D salicylate, 3-chlorobenzene, biphenyls, etc. (Chatterjee *et al.*,1981).

Special genes present on different plasmids confer degradation capacity to the species of *Pseudomonas*. For example, the camphor (CAM) plasmid of *P.putida* encodes enzyme for degradation of camphor, octane (OCT) plasmid degrades octane, XYL plasmid degrades xylene and toluene, NAH plasmid degrades naphthalene, SAL degrades salicylate, etc. These plasmids are transmissible between the strains of species of *P.putida* through conjugation. More interestingly A.M. Chakrabarty succeeded in transferring the four plasmids, OCT, XYL, CAM and NAH present in four differnt strains of *P.putida* into one strain and called it *superbug* (oil eating bug) (Fig. 4.23). The superbug can degrade the above four types of substrate for which four types of plasmids are recognised.

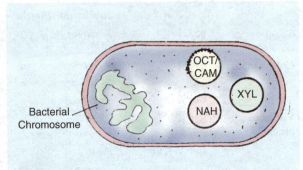

Fig. 4.23 : A strain of *Pseudomonas putida* (superbug) containing four different plasmids (diagrammatic).

(*f*) **Penicillinase plasmid of *Staphylococcus aureus*:** *Staphylococcus aureus* is a Gram-positive bacterial pathogen causing infection of skin and wounds of humans. After treatment with penicillin antibiotic, several penicillin-resistant staphylococci developed by 1950 throughout the world. High level resistance to penicillin was possible due to secretion of an enzyme, penicillinase which degrade penicillin by hydrolysing its β-lactum ring. During 1970s R.P. Novick isolated the genes (Pt) from plasmid encoding penicillinase.

There is a large variety of penicillinase plasmids designated as α, β, γ, etc. on the basis of markers present on them and production of chemically different penicillinases. Penicillinase plasmids also confer resistance against kanamycin, neomycin, tetracycline, streptomycin and chloramphenicol. Molecular weight of Kan^r and Neo^r plasmids has been found to about 15×10^6 daltons and that of Tet^r, Str^r and Chl^r as 3×10^6 Dalton (Lacey,1975).

Penicillinase plasmids do not promote conjugation but these can be transferred from cell to cell by phage transduction at a rate that could be sufficient to balance the spontaneous loss in natural population of staphylococci.

(*g*) **Cryptic plasmids :** During isolation of plasmid DNA from a large number of bacteria, it was found that every bacterium contained a low molecular weight DNA as plasmid. They do not carry any gene, therefore, they are non-functional. These plasmids have been called as cryptic plasmid. It seems that the presence of plasmids is a rule rather than exceptions.

(*h*) **Ti-plasmids** of *Agrobacterium tumifaciens* : *A. tumifaciens* is a Gram-negative soil bacterium that infects over 300 dicots and causes crown gall disease at collar region. It produces tumours, therefore, it is oncogenic. The Ti- refers to tumour inducing plasmids. The size of Ti plasmid ranges from 180-205 Kb. It consists of T-DNA (transfer DNA) of about 20 Kb which comprises of two adjacent independently encoding DNA segments, the right T-DNA (TR-DNA) and the left T-DNA (TL-DNA). T-DNA encodes enzymes for the synthesis of auxin and cytosine which interfere plant metabolism, and develop tumour and enable the infected plant to produce a nitrogenous compound called Opines. Opines are metabolized by the bacterium as a source of carbon and nitrogen.

In addition to T-DNA, the Ti plasmid also consists of several genes such as *Vir* (for virulence), *ori* (for origin of replication), *tra* (for transfer), *noc* (for nopaline catabolism in nopaline plasmid), *arc* (for arginine catabolism), and *occ* (for octopine catabolism in octopine plasmid) genes (Ream, 1989).

A tumor-inducing plasmid from a common bacterium (*Agrobacterium tumifaciens* causes crown gall tumors on willow trees.

The Ti-plasmid are divided into the four groups : octopine Ti-plasmids (e.g. pTiB6, pTiAG), nopaline Ti-plasmid (e.g. pTiT37, pTiC58), leucinopine plasmids and succinamopine plasmids. The first two plasmids octopine and nopaline plasmids have been extensively studied. Diagram of nopaline plasmid is shown in Fig. 4.24A.

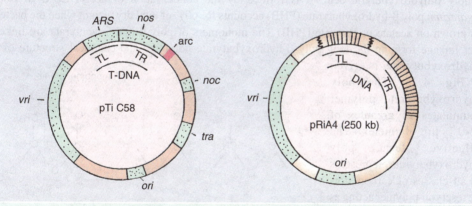

Fig. 4.24 : A nopaline Ti-plasmid of *Agrobacterium tumifaciens* (A), and agropine Ri-plasmid of *A.rhizogenes* (B).

(*i*) **Ri-plasmids:** The Ri (root inducing) plasmids are found in *Agrobacterium rhizogenes* which cause hairy root ditease in plants. The Ri-plasmids are closely related to Ti plasmids (Huffman *et al.,*1984). The size of Ri-plasmids varies from 190-240 Kb; thus they are large sized plasmids. On the basis of opine production in the infected plants the Ri plasmids are put into three groups : manopine Ri-plasmids (e.g. pRiTR7, pRi8196), agropine plasmids (e.g. pRiA4, pRi1855) and cucumopine plasmids (e.g.pRi2659). Diagram of an agropine Ri-plasmids is given in Fig.4.24B.

The mechanism of T-DNA transfer and tumour morphology have been described in detail in *A Text Book of Biotechonology* by R.C. Dubey (1995).

(*v*) **Cytoplasmic Inclusions:** The cytoplasm of prokaryotic and eukaryotic cells contain several reserve deposits which are called inclusions. Some inclusions are common to most of bacteria and some are restricted to certain species only. These inclusions serve as the basis for identification of bacteria. Shiveley (1974) has given an excellent account of inclusion bodies of prokaryotes. The inclusion bodies are of two types (a) free inclusion bodies (e.g. polyphosphate granules and cyanophycean granules), and (b) single-layered non-unit membrane enclosed inclusion bodies (such as poly β-hydroxybutyrate granules, glycogen granules, sulphur granules, carboxysomes and gas vaculoes). The membrane of inclusion bodies is made up of proteins or lipids. Allen (1984) has reviewed the cyanobacterial cell inclusions. Some of the inclusions are discussed here :

(*a*) **Volutin granules:** The volutin granules are also known as polyphosphate granules or metachromatic granules because after staining the bacteria with blue dye (e.g. methylene blue) these granules take stain and appear reddish purple in colour. Polyphosphate is a linear polymer of orthophosphates joined by ester bonds. These granules are found in algae, fungi, protozoa and bacteria. These are present in high amount in *Corynebacterium diphtheriae*, hence it can easily be diagnosed. The volutin granules are composed of polyphosphates i.e. inorganic phosphates which are used in synthesis of ATP. The phosphate is incorporated into nucleic acid. These are generally formed when the cells grow in phosphate rich environment. Growing cells of *Aerobacter aerogenes* contain traces of inorganic phosphate when there is nutrient deficiency, a large amount of phosphate is accumulates when synthesis of nucleic acid gets ceased (Harold,1963).

(*b*) **Polysaccharide Granules:** The polysaccharide granules are found in protozoa, yeasts, fungi and algae. These can be identified by using iodine solution. After reacting with iodine, glycogen turns into reddish brown and starch into blue colour.

(*c*) **Lipid inclusions:** Lipids are found in high amount in several species of *Bacillus, Azotobacter, Mycobacterium, Spirillum*, etc. These are present as storage material and is polymer of poly β-hydroxybutyric acid. It is formed by the condensation of acetyl CoA. In *Bacillus megaterium* poly β-hydroxybutyrate (PHB) accounts for 60% of total dry weight when the bacterium has grown on acetate or butyrate (PHB) The monomers of poly β-hydroxybutyrate are linked by ester linkage forming the long poly β-hydroxybutyrate polymer. The chemical structure of poly β-hydroxybutyrate is given in Fig. 4.25. The poly β-hydroxybutyrate polymer accumulates as granules of 0.2-0.7 μm diameter. The collective term poly β-hydroxybutyrate represents the all classes of carbon storage reservoir polymer acting as a source of energy and biosynthesis. These are found naturally both in Archaea but not in Eukarya. . The presence of lipid inclusions can be demonstrated by using fat soluble dyes, for example sudan dye.

(*d*) **Glycogen:** Glycogen is another storage product of prokaryotes. It is polymer of glucose sub- units looking like

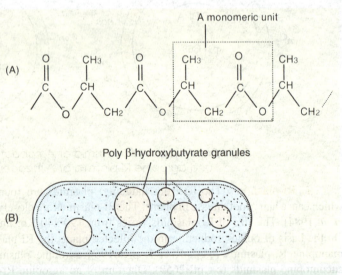

Fig. 4.25 : (*A*) Chemical structure of poly β-hydro-xybutyrate (PHB). (*B*) *Rhodospirillum sodomense* containing granules of poly β-hydroxybutyrate (PHB).

starch. The glucose units are linked by $\alpha(1{\rightarrow}4)$ glycosidic bonds. The branching chains are connected by $\alpha(1{\rightarrow}6)$ glycosidic bonds. Glycogen is also a storage reservoir for carbon and energy.

(e) Carboxysomes (polyhedral bodies): Carboxysomes are found in photosynthetic bacteria, nitrifying bacteria, cyanobacteria and thermobacilli. These are polyhedral or hexagonal inclusions containing ribulose-1, 5-biphosphate carboxylase. This enzyme is required in carbon dioxide fixation during photosynthesis.

(f) Sulphur granules: Sulphur granules are also temporarily stored by some bacteria which are also called a second type of inorganic inclusion body. It is exemplified by photosynthetic purple sulphur bacteria (*e.g. Thiobacillus, Thiospirillum, Thiocapsa, Chromatium*) which are found in anaerobic, sulphide-rich zones of

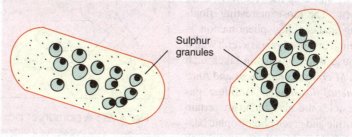

Fig. 4.26 : Cytoplasmic inclusions found in some bacteria. (intracellular sulphur granules found in *Chromatium vinosum*).

lake. They oxidise H_2O to S and internally deposit as sulphur granules within invaginated structures of plasma membrane. Sulphur is oxidised into sulphate as the reduced sulphur source becomes limiting. Fig. 4.26 shows the intracellular sulphur granules found in *Chromatium vinosum*.

(g) Magnetosomes: Magnetosomes are the intracellular chains of 40-50 magnetite (Fe_3O_4) particles of about 40-100 nm diameter found in magnetotactic bacteria (Fig. 4.27). These bacteria are motile, highly microaerophilic spirilla isolated from fresh water habitats such as *Aquaspirillum magnetotacticum*. In addition, some bacteria isolated from sulphide habitats have also been found to possess magnetosomes containing greigite (Fe_3S_4) and pyrite (FeS_2).

Magnetosomes are surrounded by membrane containing phospholipids, proteins and glycoproteins. The membrane proteins of magnetosome precipitate Fe^{3+} as Fe_3O_4 in the magnetosome. The shape of magnetosomes varies from square to rectangular or spike-shaped.

Each iron particle is a tiny magnet. Hence, the bacteria employ their magnetosomes to determine northward and downward directions. With the help of magnetosome they swim to nutrient-rich sediments or locate the optimum depth in fresh water and marine habitats.

Besides, magnetosomes are also found in some eukaryotic algae, heads of birds, dolphins, green turtles and other animals to aid navigation.

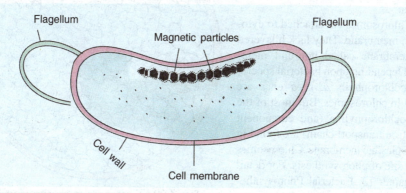

Fig. 4.27 : Magnetotactic bacterium (*Aquaspirillum magnetotacticum*) showing magnetosomes arranged in a chain (diagrammatic).

(h) Gas vesicles: There are many prokaryotic microorganisms found in floating forms in lakes and sea that possess gas vesicles. Gas vesicles are gas-filled structures which contain the same gas in which the organisms are suspended. Gas vesicles provide buoyancy and keep the cells in floating form. The most interesting 'floating and sinking' phenomenon is found in those cyanobacteria which cause water blooms in lakes such as *Microcystis aquaticus* and *Anabaena flos-aquae*. Besides, gas vesicles are also found in certain purple and green phototrophic bacteria and some archaebacteria such as halobacteria e.g. *Thiopedia and Amoebobacter.*

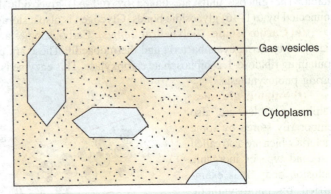

Fig. 4.28 : A portion of bacterial cell showing gas vesicles (diagrammatic).

Gas vesicles look like spindles of varying dimension (300-700x60–110 nm) and numbers (a few to hundreds per cell). They contain a rigid membrane and internally a hollow structure (fig. 4.28). The membrane consists of only proteins called GvpA and GvpC. They are 2 nm thick and impermeable to water and solutes; but they are permeable to most of gases.

Gas vesicles cannot resists high hydrostatic pressure and can be collapsed resulting in loss of buoyancy. The collapsed vesicles cannot resume the normal shape. Gas vesicles attain a density of about 5-20% of the cell. Presence of gas vesicles in photoautotrophic microorganisms provides a mechanism to adjust in a water column to move in a region to rescue from high or low light intensity.

(i) Chlorobium vesicles (Chlorosomes) : In Heliobacteria (strict anaerobic, green coloured phototrophs) bacteriochlorophylls are associated with the cytoplasmic membrane. But in green bacteria photosynthetic apparatus in different. These bacteria consists of a series of cylindrical, flat sheets (lamellae) or ellipsoidal vesicles called chlorosomes or chlorobium vesicles (fig 4.29). Structure of chlorosomes varies in different bacteria.

Chlorosomes are attached to cytoplasmic membrane. They lack bilayered thin membrane called non-unit membrane. Depending upon bacterial species bacteriochlorophylls *a, c, d* or *e* are present in chlorosomes. But most of the bacteriochlorophyll a (and component of electron transport chain) are found in the cytoplasmic membrane. Chlorosomes are the site of photosynthesis. For detail see Chapter 13, Bacterial Photosynthesis.

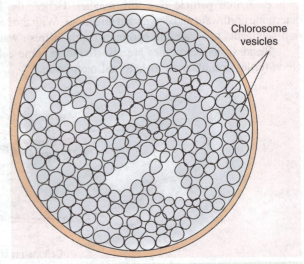

Fig. 4.29 : *Chromatium sp.* showing individual vesicle (diagrammatic).

(vi) Spores and Cysts: Certain species of bacteria produce metabolically dormant structures known as spores. The spores may be endospores (*i.e.* produced inside the cell) or exospores (i.e. produced external to the cell). After return of suitable conditions spores undergo germination and produce vegetative cells.

In addition, some bacteria such as *Azotobacter* produces thickwalled dormant, and resistant spores which are known as cysts. Cysts develop after differentiation of a vegetative cell. To some extent cysts resemble endospores that differ in structure and chemical composition.

(a) **Endospores :** The endospore formation is restricted to six bacterial genera including two Gram-positive bacteria such as *Bacillus* (e.g. *B. megaterium, B. spharericus*) and *Clostridium* (*C. tetani*). They form endospores when there is lack of water or depletion of essential nutrients in the environment. Endospores, therefore, are highly durable and dehydrated resting bodies produced inside plasma membrane of the cells. These have additional layers of thick walls which are made up of peptidoglycan (Fig. 4.30).

Frankein and Bradlay (1957) reported the spores in a majority of species of *Bacillus* and *Clostridium* which differed by surface layer. The surface may be smooth ribbed. Van den Hooff and Aninga (1956) found that the spore coat consists of an outer and an inner layer separated by a space. The outer layer is called exine and the inner layer intine. The central core is separated from the intine by a regular non-osmophilic space or cortex.

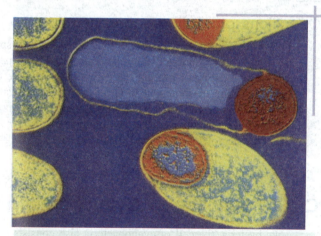

Ribosome
Cortex
Spore Coat
Intine or Core Wall
(Protoplast/core with nucleoid)
Exine or Exosporium

Fig. 4.30 : Structure of an endospore of *Bacillus anthracis* (diagrammatic).

Some of the endospores can remain viable over 100 years. Several species of endospore forming bacteria are the dangerous pathogens. The endospores are of much practical importance in food, industrial and medical microbiology. Several physico-chemical processes are adopted to kill the vegetative cells. If the endospores are produced the process of killing by chemicals is useless. In the processed food the endospores are adapted and produce toxins when suitable growth conditions prevail. In such conditions special methods are adopted. Generally the endospores are difficult to stain for deletion. Therefore, specially pre-

Resistant endospores, here coloured red, formed inside bacteria *Clostridium*.

pared stain *i.e.* the Schaeffer-Fulton endospore stain is commonly used for detection purpose.

Sporulation: The process of endospore formation is called sporulation or sporogenesis. It occurs normally when growth of bacterium ceases due to lack of nutrients. Sporulation is a complex process and occurs in several stages (Fig.4.31). Generally sporulation process requires 10 hours to get accomplished. Formation of endospores begins with the development of ingrowth of plasma membrane known as spore septum which becomes double layered.

The spore septum surrounds the cytoplasm and chromosome. A structure entirely enclosed within cell is formed which is known as forespore. Between the two membrane a thick layer of peptidoglycan is laid down. A thick coat of protein is formed around this structure that provides resistance to endospores. All the endospores contain DNA, small amount of RNA and large amount of organic acid (dipicolinic acid). These cellular components are important for resuming metabolism in later stage.

Spore Germination :
The endospores may be produced terminally, subterminally or centrally depending upon species. After maturity wall layer dissolved and endospores are set free (Fig. 4.31). This process is called spore germination. Germination occurs only in favourable environmental conditions which is accomplished in three stages : activation, germination and outgrowth.

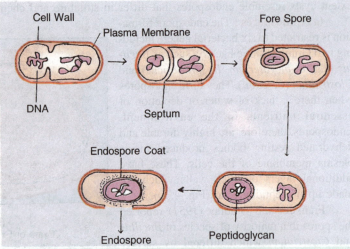

Fig. 4.31 : The process of endospore formation i.e. sporulation in a bacterial cell.

(*b*) **Exospores:** Cells of *Methylosinus* (methane oxidizing bacterium) produce exospores through budding at one end of the cell. The exospores do not contain dipicolinic acid. They can resist desiccation and heat.

Conditions promoting spore germination

Following are some of the conditions that stimulate spore germination :
(*a*) Heat shock at sub-lethal temperature for varying period of time from 1 minute to 1 hour or more (time of heating varies with species, age of spores and conditions of storage).
(*b*) Incubation of spores at 30°C for 5 minutes in water-ethanol mixtures.
(*c*) Addition of yeast extract in active components viz., glucose, a few amino acids with or without nucleotides.
(*d*) Presence of various metal ions such as Ca^{++}, Mg^{++}, etc.
(*e*) Presence of dipicolinic acid.
(*f*) Addition of L-alanine.

C. Cell Structure of Cyanobacteria

Like bacteria, the cell of cyanobacteria also consists of a mucilagenous layer called sheath, the cell wall, plasma membrane and cytoplasm. These are shown in Fig. 4.32 and described below:

1.   Sheath

Usually the cell of cyanobacteria are covered by a hygroscopic mucilagenous sheath which provides protection to cell from unfavourable conditions and keeps the cells moist (Fig. 4.32). Thickness, consistancy and nature of sheath are influenced by the environmental conditions. Sheath consists of pectic substances. It is undulating, electron dense and fibrillar in appearance.

2. Cell Wall

After observing the cyanobacterial cell under electron microscope, it appears multilayered present between the sheath and plasma membrane. The cell wall consists of four layers designated as LI, LII, LIII and LIV (Allen, 1968) (Fig.4.33). The layers LI and LIII are electron transparent, and LII and LIV electron dense. (*i*) LI is the innermost layer of the cell wall present next to the plasma membrane. It is of about 3-10 nm thickness and enclosed by LII. (*ii*) LII is a thin, electron dense layer. It is made up of mucopeptide and muramic acid, glucosamine, alanine, glutamic acid and diaminopamelic acid. The layer LII provides shape and mechanical strength to the cell wall. Thickness of this layer varies from 10 to 1000 nm. (*iii*) LIII is again electron transparent layer of about 3-10 nm thickness. (*iv*) The outermost layer is LIV which is a thin and electron dense layer. It appears wrinkled and is undulating or convoluted.

All the layers are interconnected by plasmadesmata. Numerous pores are present on the cell which act as passage for secretion of mucilage by the cell. Chemically the cell wall of eubacteria and cyanobacteria are much similar.

The chemical constituent of cyanobacteria and Gram-negative bacteria is the presence of mucopolymer which is made up of five chemical substances viz., three amino acids (diaminopamelic acid) and two sugars (glucosamine and muramic acid) in the ratio of 1:1:1:1:2. Similar ratio of these constituents is also found in *E. coli*. However, in some cyanobacteria such as *Anacystis nidulans*, *Phormidium uncinatum* and *Chlorogloea fritschi* the amino acids and sugars are found in different ratios. Moreover, diamino acid is common in all prokaryotes. In addition, lipids and lipopolysaccharides have also been detected in the cells of cyanobacteria.

3. Plasma Membrane

The cell wall is followed by a bilayer membrane called plasma membrane or plasmalemma. It is 70Å thick, selectively permeable and maintain physiological integrity of the cell. Plasma membrane sometimes invaginates locally and fuses with the photosynthetic lamellae (thylakoids) to form a structure called lamellosomes (Fig.4.32). The plasma membrane encloses cytoplasm and the other inclusions.

4. Cytoplasm

Cytoplasm is distinguished into the two regions, the outer peripheral region which is called the chromoplasm and the central colourless region called centroplasm.

(*i*) **Chromoplasm :** The chromoplasm contains the flattened vesicular structures called photosynthetic lamellae or thylakoids (Fig.4.32). Thylakoids may be peripheral, parallel or central. Besides photosynthesis, thylakoids have the capacity of photophosphorylation, Hill reaction and respiration. Depending upon physiological conditions they are arranged accordingly.

Several photosynthetic pigments such as chlorophyll a, chlorophyll c, xanthophylls, carotenoids are present inside the lamellae. On its upper surface *phycobilisomes* (biliproteins) of about 40 nm diameter are anchored by a protein. Phycobilisomes comprises of three pigments: phycocyanin-C, allophycocyanin and phycoerythrin-C. These three pigments harness light in the sequence : Phycoerythrin—Phycocyanin—Allophycocyanin—Chlorophylls.

(*ii*) **Centroplasm:** The centroplasm is colourless and regarded as primitive nucleus devoid of bilayered nuclear membrane and nucleolus. Several grains that can take stain are dispersed in centroplasm. Some people are of the opinion that the centroplasm is the store house of food, and according to the others it is an incipient nucleus (Fig.4.32).

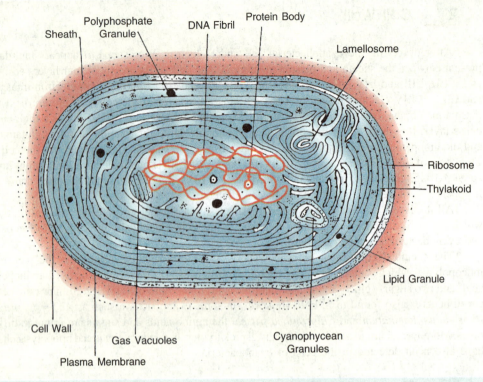

Fig. 4.32 : A typical cell of a cyanobacterium.

5.   Cytoplasmic Inclusions

Several glycogen granules, oil droplets and other inclusions are dispersed in chromoplasm as well as in centroplasm regions.

(*i*) **Cyanophycin:** The cyanobacteria accumulate nitrogenous reserve material called cyanophycin or cyanophycian granules when grown at conditions of surplus nitrogen. These are built with equal molecules of arginine and aspartic acid. These represent as much as 8% of total cellular dry weight. They can be observed under a light microscope as they accept neutral red or carmine.

(*ii*) **Gas Vacuoles:** In many cyanobacteria e.g. *Anabaena, Gloeotrichia, Microcystis, Oscillatoria*, etc. the gas vesicles of viscous pseudovacuoles of different dimensions are found. The cytoplasm

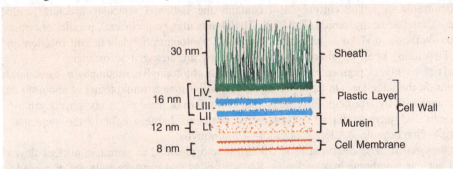

Fig. 4.33 : An enlarged view of different components of wall layer of the cell of a cyanobacterium (diagrammatic).

lacks vacuoles. The vesicles are hollow, rigid and elongated cylinders (75 nm diameter, 200-1000 nm long) covered by a 2 nm thick protein boundary.

The ends of vacuoles are conical (Fig. 4.32). The protein boundary is impermeable to water and freely permeable to gases. Under pressure they get collapsed, and therefore, lose refractivity. The function of gas vesicles is to maintain buoyancy so that the cell can remain at certain depth of water where they can get sufficient light, oxygen and nutrients (Walsby, 1972). Floating and sinking phenomenon is a key feature found in free floating cyanobacteria. Through this mechanism they can escape from harmful effect of bright light.

(*iii*) **Carboxysomes:** Carboxysomes are the polyhedral bodies containing 1,5-ribulose biphosphate carboxylase (Rubisco).

(*iv*) **Phosphate bodies:** See volutin glanules of bacteria.

(*v*) **Phycobilisomes:** Some phototrophic organisms (i.e. cyanobacteria and red algae) contain two accessory pigments such as carotenoids and phycobilins (also called phycobiliproteins), in addition to chlorophyll or bacteriochlorophyll pigments. The carotenoids play a photo-protective role, whereas phycobilins serve as light harvesting pigments.

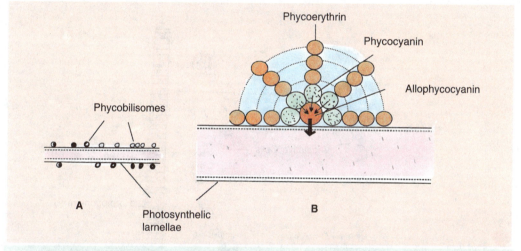

Fig. 4.34 : Phycobilisomes attached to the lamellar membrane (A); arrangement of phycobiliproteins in a phycobilisome attached to lamellar membrane (B); arrows show the direction of transfer of energy (diagrammatic)

Phycobilins are the main light-harvesting pigments of these organisms. Phycobiliproteins are red or blue in colour. These compounds consist of an open-chain tetrapyroles derived biosynthetically from a closed porphyrin ring. The tetrapyroles are coupled to proteins.

Phycobiliproteins are aggregated to form a high molecular weight darkly stained ball-like structure called **Phycobilisomes**. The phycobilisomes are attached to the outer surface of lamellar membrane (Fig. 4.34 A).

Phycobiliproteins includes three different pigments; *(a)* a red pigment *phycoerythrin* which strongly absorbs light at 550 nm, *(b)* a blue pigment *phycocyanin* which absorbs light strongly at 620 nm, and *(c) allophycocyanin* which absorbs light at 650 nm. The pigments in phycobilisomes are arranged in such a way that allophycocyanin is attached to photosynthetic lamellar membrane. Allophycocyanin is surrounded by the molecules of phycocyanin, and the latter by phycoerythrin.

Phycoerythrin and phycocyanin absorb shorter (high energy) wavelength of light and transfer energy to allophycocyanin. Allophycocyanin is closely linked to the reaction centre chlorophyll. Thus energy is transferred from allophycocyanin to chlorophyll *a*. Presence of phycobilisomes makes the cyanobacterial growth possible at the region of lowest light intensities.

(*vi*) **DNA Matrix :** Like other prokaryotes the cyanobacteria also contain naked DNA fibrils dispersed in the centroplasm. DNA material lacks nucleoplasm, and like *E. coli* contains a histone like protein that binds with DNA. The total number of genomes is yet not known but in *Agmenellum* 2 to 3 genomes have been reported. However, base composition of DNA in different cyanobacteria varies, for example in chroococcales (35-71 moles percent G + C), Oscillatoriales (40-67 moles percent G + C), Pleurocapsales (39-47 moles percent G + C) and heterocystous forms (38-47 moles percent G + C). The molecular weight ranges from 2.2×10^9 to 7.4×10^9 daltons.

Like eubacteria, cyanobacteria also have 70S ribosomes. Similar to bacteria cyanobacteria contain covalently closed, non-functional, circular plasmid DNA. These are called cryptic plasmid as the function of cyanobacterial DNA is not known (Kumar and Rai,1986).

6.   Specialized Structures of Cyanobacteria

There are certain specialized structures viz., hormogones, exopores, endospores, nanocysts, heterocysts, exospores, endospores, akinetes, etc. which are produced in cyanobacteria.

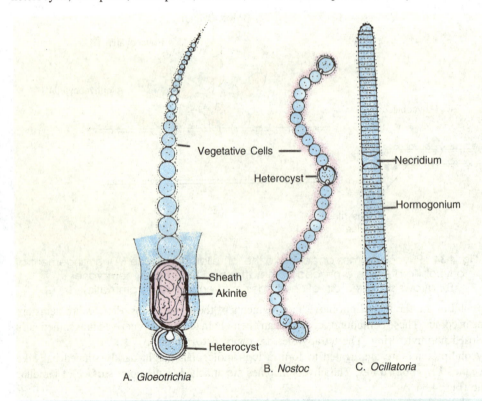

Fig. 4.35 : Diagrams of a *Gloeotrichia* (A), *Nostoc* (B) and *Oscillatoria* (C).

(*i*) **Hormogones and Hormocysts:** Hormogones are the short segments of trichomes produced in all filamentous cyanobacteria. Hormogones are produced by several methods such as fragmentation of trichomes into pieces (e.g. *Oscillatoria*) (Fig. 4.35C), delimination of cells into intercalary groups (*Gloeotrichia*) (Fig. 4.35A), fragmentation and round off the end cells (*Nostoc*) (B), formation of separating disc or necridia and their subsequent degradation (*Oscillatoria, Phormidium*). The hormogones show gliding movement. Each hormogone may develop into a new individual.

Some other cyanobacteria produce hormocysts or hormospores which function similar to hormogones. Hormocysts are produced intercalary or terminal in position. They are highly granulated and cells are covered by a thick mucilagenous sheath. In the cells of hormocysts a large quality of food material is accumulated. During favourable condition hormocysts develop into a new plant.

(*ii*) **Endospores, Exospores and Nanocysts:** The non-filamentous cyanobacteria generally produce endospores, exospores and nanocysts, for example *Chamaesiphon, Dermocapsa* and *Stichosiphon*. The endospores are produced inside the cell. During endospore formation, cytoplasm of the cell is cleaved into several bits which are converted into endospores. After liberation each endospore germinates into a new plant, for example *Dermocapsa*. When the size of endospores is smaller but larger in number, they are called nanospores or nanocysts. Some of the cyanobacteria (e.g. *Chamaesiphon*) reproduce by budding exogenously. The spores produced through this method are called exospores.

(*iii*) **Akinetes:** The members of Stigonemataceae, Rivulariaceae and Nostocaceae are capable to develop the vegatative cells into spherical perennating structures called akinetes or spores such as *Nostoc, Rivularia, Gloeotrichia,* etc (Fig. 4.35A).

During unfavourable conditions, the vegetative cells accumulate much food, enlarge and become thick walled. These are formed singly or in chains. Akinetes possess cyanophycean ganules, hence these appear brown in colour. Under favourable conditions the akinetes germinate into vegetative filaments.

(*iv*) **Heterocysts:** Heterocysts are the modified vegetative cells (Fig.4.35A-B). Depending on nitrogen concentration in the environment heterocyst formation occurs. During differentiation several morphological, physiological, biochemical and genetical modifications take place in heterocyst. They are slightly enlarged cells, pale yellow in colour containing an additional outer investment. They are produced singly or in chains and remain intercalary or terminal in positions. These are found most frequently in Oscillatoriaceae, Rivulariaceae, Nostocaceae and Scytonemataceae.

In heterocysts total amount of thylakoids gets reduced or absent. The photosystem II that generates oxygen, becomes non-functional. The amount of surface proteins that combine with oxygen and create oxygen tense environment is increased. Rearrangement in *nif* gene (nitrogen fixing gene) cluster takes place and expression of nitrogenase and nitrogen fixation are accomplished. In addition, these take part in perennation and reproduction as well.

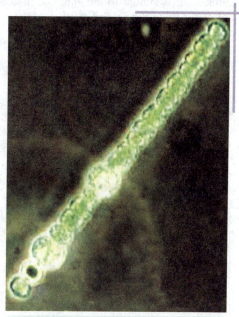

A. *Nostoc* sp. showing heterocysts.

D. Structure of Fungal Cell

Fungi are eukaryotes and, therefore, consist of well developed membrane bound organelles. They occur in both unicellular and multicellular forms. In some groups the hyphae form interwoven mycelia. A large number of hyphal modifications such as rhizomorph, sclerotia, pycnidia, cleistothecia, etc are found in them.

When there is no differentiation of vegetative and reproductive structures such thalli are called holocarpic e.g. *Synchytrium, Olpidium* yeasts, etc. The other type of thalli are called eukarpic where it is differentiated into the vegetative part and the reproductive part. In the eucarpic thallus there are differences in the presence and absence of septa. The hyphae lacking septa are called coenocytic hyphae, for example the member of Oomycetes and Zygomycetes. The coenocytic mycelium remains multinucleate.

The septate mycelium is found in the members of Ascomycetes, Basidiomycetes and Deuteromycetes. If the nuclei are genetically identical the mycelium is said to be homokaryotic but when the mycelium contains nuclei of different genotype, the mycelium is said to heterokaryotic mycelium. In Basidiomycetes each cell of mycelium contains genetically different nuclei which are known as dikaryotic mycelium. Subcellular organization of a fungal cell is discussed below:

1. The Cell Wall

Except slime molds (Myxomycetes), the fungal cell consists of a rigid cell wall and cell organelles. However, composition of cell wall of different fungal groups differs. Chemical analysis of cell wall reveals that it contains 80-90% polysaccharides, and remaining proteins and lipids (Aronson,1965; Barnickie-Garcia,1968). Chitin (a polymer of N-acetylglucosamine), cellulose (a polymer of D-glucose) or other glucans are present in cell walls in the form of fibrils forming layers. In most of the fungi the cell wall lacks cellulose (except Oomycetes) usually chitin and cellulose are found together e.g. *Ceratocystis* and *Rhizidiomyces* contain a form of chitin called fungus cellulose. It is similar to the chitin of insects. Structural formulae of repeating units of cellulose and chitin are given in Fig. 4.36.

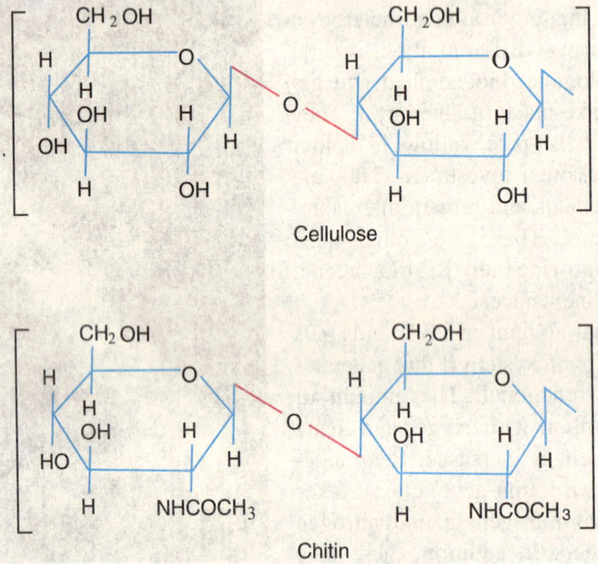

Fig. 4.36 : Structural formula representing the units of cellulose and chitin.

The microfibril layers run parallel to the surface. Several non-fibril materials are also associated with microfibrils. Though chitin is the most usual component yet cellulose is present in cell walls of Oomycetes along with glucans. An amino acid and hydroxyprotein are present in the cell wall of Oomycetes along with cellulose (Novaes-Ledieu *et al*, 1967). Several other substances have also been found to be associated together in cell walls and cell wall components such as proteins enzymes, etc. In *Peronospora* and *Saprolegnia* true cellulose is present, but in *Phytophthora* and *Pythium* cellulose is totally absent and glucans predominates in their walls. In the cell wall of some fungi the presence of chitin has been reported.

The basic constituents of cell walls of Zygomycetes, Ascomycetes and Basidiomycetes is chitin. But in yeasts and some Hemiascomycetidae chitin is absent. Microfibrils of mannans and β-glucan constitute their cell wall. Various chemical substances found in cell walls seem to be correlated with fungal taxonomy (Table 4.5).

2. Plasma Membrane (Plasmalemma, Cell Membrane)

In fungi too the cell wall is followed by plasma membrane that encloses the cytoplasm. It is semipermeable and, in structure and function, it is similar to that of prokaryotes. However, specialized organelles have been reported at the surface of plasma membrane in the region where the fusion of secretory vesicles of cytoplasm occurs. The plasmalemma invaginates and forms a pouch like structure enclosing the granular or vesicular materials.

Table 4.5 : Taxonomy of fungal cell walls (modified after barnicki-garcia, 1968)

Categories	Taxonomic groups	Features
Cellulose-glycogen	Acrasiales	Pseudoplasmodia
Cellulose-glucan	Oomycetes	Biflagellate zoospores
Cellulose-chitin	Hypochytridiomycetes	Anteriorly uniflagellate zoospores
Chitosan-chitin	Zygomycetes	Zygospores
Chitin-glucan	Chytridiomycetes	Posteriorly uniflagallate zoospore
	Ascomycetes	Septate hyphae, ascospores
	Basidiomycetes	Septate hyphae, basidiospores
	Deuteromycetes	Septate hyphae
Mannan-glucan	Saccharomycetaceae	Yeast cells, ascospore
	Cryptococcaceae	Yeast cells
Mannan- chitin	Sporobolomycetaceae	Yeast cells, ballistospores
	Rhodotorulaceae	Yeast (carotenoid pigment)
Polygalacturosamine-galactan	Trichomycetes	Heterogenous group

Moore and McAlear (1962) named it lomasomes. It has been defined as "membranous vesicular material embeded in the wall external to the line of plasmalemma: Lomasomes are formed by plasmalemmasomes which are *"various membrane configurations external to the plasmalemma, often pockets projecting into the cytoplasm and less obviously embedded into wall materials"* (Health and Greenwood, 1970). However, plasmalemmasomes are produced when the balance between wall plasticity and turgor pressur is distributed is such a way that more plasmalemma is produced than is needed to line the cell wall (Health and Greenwood, 1970).

3. Cytoplasm

Cytoplasm is colourless in which sap-filled vacuoles are found. Except chloroplasts many of the familiar organelles and inclusions, characteristic of eukaryotes, are found in fungal cytoplasm. The cytoplasmic inclusions are dead, non-functional and unimportant for fungal survival for example stored food (glycogen and oil drops), pigments and the secretary granules. The cell organelles are endoplasmic reticulum, mitochondria, ribosomes, golgi bodies and vacuoles (Fig. 4.37A). Lomasomes are also present between plasma membrane and cell wall. The organelles are described below.

(i) **Endoplasmic Reticulum :** Presence of endoplasmic reticulum in fungal cytoplasm is observed through electron microscope. It is made up of a system of microtubules beset with small granules. In most of the fungi it is highly vesicular. It is loose and irregular as compared with cells of green plants. In multinucleate hyphae the nuclei may be connected by endoplasmic reticulum.

(*ii*) **Mitochondria :** Numerous small and spherical to elongated bodies known as mitochondria are dispersed in cytoplasm. Mitochondria are covered by an outer double membrane; the inner infoldings form parallel flat plates of irregular tubules called cristae. There is no difference between mitochondria of fungi and green plants. Generally, these are called the power house of the cell. Mitochondria consists of its genetic material (mt-DNA) as a circular double helical molecules devoid of histones and very similar to prokaryotic DNA (molecular weight 1×10^7 dalton). Mitochondria has its own machinery for transcription and translation of organelle specific DNA.

(*iii*) **Golgi Apparatus or Dictyosomes:** Except in Oomycetes (e.g. *Pythium*) and non-fungal eukaryotic cells, Golgi apparatus is of rare occurrence in fungal cells. In Oomycetes and non-fungal eukaryotes, Golgi apparatus consists of stacks of folded membranes functioning in secretion. In the cells of *Saccharomyces* a Golgi apparatus consisting of three flattened sacs can be observed.

(*iv*) **Vacuoles:** Vacuoles are found in the old cells of hyphae. The end of hyphal tip of young hyphae lacks vacuole. With the age, the vacuoles coalesce. Vacuoles are surrounded by a membrane known as tonoplast.

(*v*) **Septum :** Basically three types of septa are found in fungal cells: (a) complete septa which lack a pore and rare in vegetative hyphae (Fig.4.37B), (b) perforated septa containing a pore through which cytoplasmic organelles such as mitochondria and nuclei can pass freely, for example septa found in members of Ascomycetes and Deuteromycotes (C), and (c) dolipore (*dolium* means a large Jar) septa which are found in Basidiomycetes and are rather more complex. The central pore of the septum is surrounded by a curved flange of wall material which is often thickened to form a barrel shaped cylindrical structure (D). These septa are often overlaid by the perforated endoplasmic reticulum (Webster,1980). The central pore cap is known as parthenosome.

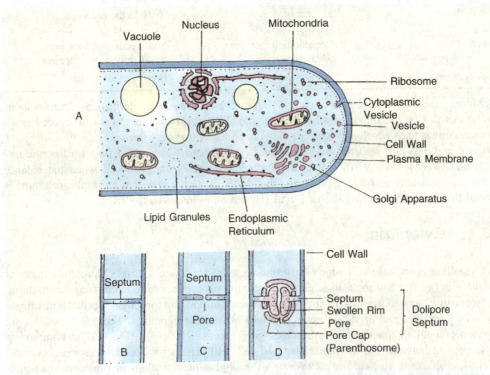

Fig. 4.37 : Different cell organellels of the hyphal apex of *Pythium ultimum (A)*, and different types of septa: complete (B), porus (C) and dolipore (D) septa.

(*vi*) **Cytoplasmic Inclusions :** The cytoplasm consists of various inclusions such as lipid droplets, and glycogen (a typical fungal storage product), the carbohydrate trehalose, proteinaceous material and volutin. The vacuoles contain glycogen. Several metabolites (enzyme, organic acids, etc) are secreted by the cytoplasm. In mature cells, lipids and glycogenes are abundantly present.

(*viii*) **Nucleus :** The cytoplasm contains one, two or more globose or spherical nuclei of about 1-3 µm diameter. A nucleus consists of a bilayered porous nuclear envelope that encloses the chromosomes and nucleolus. The chromosome consists of DNA and a few basic proteins called histones. The DNA material remains in changing stage with cell growth. The nuclear pores permit to interchange the materials betwen the cytoplasm and nucleus.

QUESTIONS

1. Write an essay on size, shape and arrangement of cells in microrganisms.
2. Discuss in brief locomotion in bacteria .
3. Write in detail chemotaxis with emphasis on molecular mechanism.
4. Give an illustrated account of Gram-negative bacteria and its cell wall composition.
5. Give an illustrated account of structure and function of cell membrane.
6. What do you known about plasmids ? Write in detail different types of plasmids.
7. Write an essay on cytoplasmic inclusions.
8. Discuss in detail the cell wall structure and chemical composition of cyanobacteria.
9. Give an illustrated account of fungal cell and its wall composition.
10. Write short notes on the followings :
    (*i*) Capsule, (*ii*) Flagella, (*iii*) Pili, (*iv*) differences between cell wall of Gram-positive and Gram-negative bacteria, (*v*) cell wall composition of archaeobacteria, (*vi*) Gram-positive bacteria, (*vii*) Gram-negative bacteria, (*viii*) Mesosomes, (*ix*) Atypical cell wall, (*x*) Ribosomes, (*xi*) Chaperones, (*xii*) Nucleoids, (*xiii*) Endospores, (*xiv*) Heterocysts, (*xv*) Akinetes, *(xvi)* Chlorosomes, (*xvii*) Synchronous growth; (*xiii*) Magnetosomes, (*ix*) Phycobilisomes.

REFERENCES

Adler, J. 1976. The sensing of chemicals by bacteria. *Sci. Am.* 234:40-47

Allen, M.M. 1968. Ultrastructure of cell wall and cell division of unicellular blue green algae. *J. Bacteriol.* 96:842-852.

Allen, M.M. 1984. Cyanobacterial cell inclusions. *Ann. Rev. Microbiol.* 38:1-25

Aronson, J.M. 1965. The cell wall. In *The fungi : An advanced Tretise I* (Eds. G.C. Ainsworth and A.S. Sussman) pp. 49-76 New York and London : Academic press.

Barnicki-Garcia, S. 1968. Cell wall chemistry, morphology, genesis, and taxonomy of fungi *Ann. Rev. Microbiol.* 22:87-108.

Berg, H.C.. How bacteria swim. *Sci. Am.* 230:36-44

Bohrmann, B.; Villiger,W., Johansen, R.and Kellenberg,E. 1991. Corallin shape of the bacterial nucleoid after Cryofixation. *J. Bacteriol.* 173:3149-3158.

Bopp, L.H.; Chakrabarty, A.M. and Ehrlich,H.L. 1983. Chromate resistance plasmids in *Pseudomonas fluorescens.* J. Bacteriol. 155:1105-1109.

Crains, J (1963). The bacterial chromosomes and its manner of replication as seen by autoradiography. *J. Mol. Bacteriol.* 6:208

Chatterjee, D.K., Kellog,S.T., Hamada,S. and Chakraborthy, A.M. (1981). *J. Bacteriol* 146:639-646

Costerton, J.W., Irwin, R.T. and Cheng,K-J 1981. The bacterial glycocalyx in nature and disease. *Ann. Rev. Microbiol.* 35:299-324.

Delius, H and Worcel, A. 1974. Electron microscopic visualization of the folded chromosome of *Escherichia coli. J. Mol. Biol.* 82:107

De Pamphilis, H.L. and Adler, J. 1971. Fine structure and isolation of the hook-basal body complex of flagella from *Escherichia coli* and *Bacillus subtilis J. Bact.* **105**:384.

DiRienzo, J.M.; Nakamura, K.; Di Rienzo,M. 1978. The outer membrane proteins of Gram-negative bacteria: biosynthesis assembly and function. *Ann. Rev. Biochem.* 47:481-532.

Doetsch, R.N. and Sjoblad,R.D. 1980. Flagella structure and function in Eubacteria. *Ann. Rev. Microbiol.* 34:69-108.

Dubey, R.C. 1995. *A textbook of Biotechnology* 2nd ed. S. Chand & Co. Ltd., New Delhi p 357.

Ferris, F.G. and Beveridge,T.J. 1985. Functions of bacterial cell surface structure. *Bio Sci.* 35:172-177.

Franklin, J.G. and Bradley, D.E. 1957. A further study of spores of species of the genus *Bacillus* in the electron microscope using carbon replicase and some preliminary observations on *Clostridium welchii J. Appl. Bact.* 20:467.

Georgopoulos,C. and Welch,W.J. 1991. Role of the major heat sock proteins as molecular chaperones. *Ann. Rev. Cell. Biol.* 9:601-634.

Ghosh, B.K. ed.(1981) Organisation of prokaryotic cell membrane : CRC Press, Boca Reton, Florida.

Harold, F.M. 1963. Accumulation of inorganic phosphate in *Aerobacter aerogenes*. I. Relationship to growth and nucleic acid synthesis *J. Bact.* 86:216.

Hartl, F.U.; Hlodan, R and Langery,T. 1994. Molecular chaperones in protein foldings: The art of avoiding sticky situations. *Trends Biochem.Sci.* 19:20-25.

Health, I.B. and Greenwood, A.D. 1970. The structure and formation of lomasomes. *J. Gen. Microbiol.* 62:129-137.

Helinsky, D.R. 1973. Plasmids determined resistance to antibiotics : molecular properties of R factors. *Ann. Rev. Microbiol.* 27:437.

Hobot, J.A. *et al.* 1985. Shape and fine structure of nucleoids observed on sections of ultrarapidly frozen and cryosubstituted bacteria *J. Bacteriol.* 162:960-971.

Huffman, G.A., White, F.F. Gordon, M.R. and Nester, E.W. 1984. Hairy-root inducing plasmids : physical map and homology to tumour-inducing plasmids *J. Bacteriol.* 157:269-276.

Kawata, T. *et al.* 1961. Autolytic formation of spheroplasts of *Bacillus megaterium* after ceasation of aeration *J. Bacteriol.* 81:160.

Konig, H. 1988a. Archaeobacterial cell envelops. *Can.J. Microbiol.* 34:395-406

Konig, H. 1988b. Archaeobacteria. In *Biotechnology.* (eds. H.J. Rehm & G. Read), pp.698-728, VCH Verlag; Weinheim.

Kumar, H.D. and Rai,L.C. 1986. *Microbes and Microbial process.* Affilitated East West Press Ltd., New Delhi p. 213.

Kuwajima, K 1989. The molten globular state as a clue for understanding the folding and cooperative of globular protein structure. *Protein* 6:87-103.

Lacey, R.W. 1975. Antibiotic resistance plasmids of *Staphylococcus aureus* and their clinical importance. *Bact. Rev.* 39.

Leive, L. ed. 1976. Bacterial membranes and Walls. Marcel Dekker, Inc. New York pp. 495.

Lake, J.A. 1981. *Sci. Am.* 245(2).

Manson, M.D. 1992. Bacterial motility and chemotaxis. In *Adv. in Microbial Physiol.* Vol. 33 (Ed. A.H. Rose) pp. 277-346. Academic Press, New York.

Misra,T.K. *et al.*, 1985. Mercuric reductase structural genes from plasmid R100 and transposon Tn 501: Functional domains of the enzyme. *Gene* 34:253-262.

Moore, R.T. and McAlear, J.H. 1962. Fine structure of mycota.7. Observations on septa of Ascomycetes and Basidiomycetes *Am. Jr. Bot.* 49:86-94.

Novaes-Ledieu, M. et al., 1967. Chemical composition of hyphal wall of phycomycetes *J. Gen. Micriobiol.* 47:237-245.

Ottow, J.C.G. 1975. Ecology, physiology and genetics of fimbriae and pili. *Ann. Rev. Microbiol.* 29:79-108.

Parkinson, J.S. 1993. Signal transduction schemes of bacteria. *Cell.* 73:857-871.

Pelczar Jr., M.J., Chan,E.C.S. and Krieg,N.R. 1986. Microbiology, Mc Graw-Hill, Inc., New York.

Ream, W. 1989. *Ann. Rev. Phytopathol.* 27:583-618.

Rogers, A.J., Ward, J.B. and Burdett, I.D.S. 1978. Structure and growth of the walls of Gram-positive bacteria. In *Relation Between Structure and Function in the Prokaryotic Cells* (eds. Rystainer, H.J., Rogers and J.B. Ward)pp.139-176, Cambridge University Press, Cambridge.

Salton, M.R.J. and Owen,P. 1976. Bacterial membrane structure. *Ann. Rev. Microbiol.* 30:451-482.

Shiveley, J.M. 1974. Inclusion bodies of prokaryotes. *Ann. Rev. Microbiol.* 28:167.

Simon, M. et.al., 1978. Structure and function of bacterial flagella. In : *Relation Between Structure and Function in the Prokaryotic Cell.* pp. 272-284 (eds. Ry. Stainer,H.J. Rogers and J.B. Ward), 28th symposium of the Society for General Microbiology held at the Univ. of south hampton, April, 1978 Cambridge Univ. Press, Cambridge pp. 369.

Singer, S.J. and Nicolson,G.L. 1972. The fluid mosaic model of the structure of cell membranes. *Science* 175:720-730.

Summers, A.O. 1985. Bacterial resistance to toxic elements. *Trends Biotechnol.* 3:122-125.

Van den Hooff,A. and Aninga,S. 1956. An electron microscope study on the shape of the spores of *Bacillus polymyxa. Anton Van Leeuwenhoek J. Microbiol. Serol.* 22:327.

Walsby, A.E. 1972. Structure and function of gas vacoules. *Bact. Rev.* 36:1

Webster, J. 1980. *Introduction to Fungi.* 2nd. ed. Cambridge University Press, cambridge p. 669.

Worcel, A. and Burgi,E. 1972. On the structure of the folded chromosome of *E. coli J. Mol. Biol.* 71:127.

The Nucleic Acids : DNA and RNA

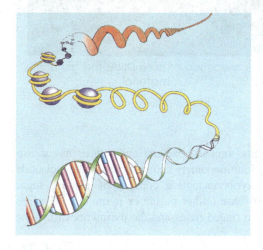

5

The nucleic acids found in viruses and all living organisms (microorganisms, plants and animals) carry the genetic informations. A nucleic acid is the polynucleotide i.e. polymer of nucleotides. Each nucleotide consists of three main constituents: (*i*) a cyclic five carbon sugar, (*ii*) a purine or pyrimidine base, and (*iii*) a phosphate. The sugar is ribose or deoxyribose. Based on the types of sugars, the nucleic acids are of two types, ribose nucleic acid (RNA) and deoxyribose sugar (DNA). Most of the organisms contain DNA but few phages, and plants and animal viruses contains RNA as genetic material.

Generally, nucleic acids are associated with protein to form nucleoprotein. In 1868, for the first time F. Miescher isolated nucleic acids from white blood cells that were acidic in nature to which he called nuclein. Purine and pyrimidines were isolated by Fischer in 1880. In 1881, Zacharis identified nuclein with chromatin. Altaman in 1899 replaced the term nuclein with nucleic acid. Kossel identified the presence of histones and protamines within nucleic acids and was awarded Nobel prize for demonstrating the presence of two purines (adenine and guanine) and two pyrimidines (thymine and cytosine). During 1910s, P.A. Levene discovered the phosphate and pentose sugars called deoxyribose molecules. In 1943, three American microbiologists, Ostwald

James D. Watson

Francis Crick

Avery, Colin MacLeod and Maclyn McCarty, for the first time presented the evidence that DNA is the genetic material and it is made up of genes. In 1953, J.D. Watson (an American biologist) and F.H.C. Crick (a British physicist) presented the double helix model of DNA and they were awarded Nobel prize.

A. Nature of DNA

The DNA is found in all plants, animals, prokaryotes and some viruses. In eukaryotes it is present inside the nucleus, chloroplast and mitochondria, whereas in prokaryotes it is dispersed in cytoplasm. In plants, animals and some viruses the genetic material is double stranded (ds) DNA molecule except some viruses such as $\phi \times 174$. In TMV, influenza virus, poliomyelitic virus and bacteriophages the genetic material is single stranded (ss)RNA molecule (Table 5.1).

A specimen of pure DNA, the genetic material.

1. Chemical Composition

Purified DNA isolated from a variety of plants, animals, bacteria and viruses has shown a complex form of polymeric compounds containing four monomers known as deoxyribonucleotide monomers or deoxyribotids (Fig 5.1). Each deoxyribonucleotide consists of pentose sugar (deoxyribose), a phosphate group and a nitrogenous base (either purine or pyrimidine). Purines bases (adenine and guanine) are heterocyclic and two ringed bases and the pyrimidines (thymine and cytosine) are one ringed bases.

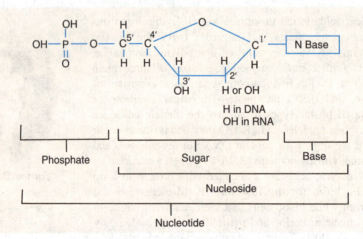

Fig. 5.1 : A typical nucleotide showing its components: base, sugar and phosphate.

The following components of deoxyribonucleotide have been described :

(*i*) **A Five Carbon Ring :** Deoxyribose is a pentose sugar consisting of five carbon atoms. The first and fourth carbon atoms of this sugar combine with one oxygen atom and form a ring. The fifth atom (5') forms –CH_2 group which is present outside this ring. Three -OH groups are attached at position 1', 3' and 5' and the hydrogen atoms combine at position 1', 2', 3' and 4' of carbon atoms (Fig 5.1). In ribonucleotides, the pentose sugar is ribose which is similar to deoxyribose except that there is an –OH group instead of –H at 2' carbon atom. The absence of -OH group in DNA makes it chemically more stable than the RNA.

Table 5.1: Nature of genetic material

DNA/RNA	Examples
Double stranded DNA (dsDNA)	Higher plants, animals, bacteria, animal viruses (polyoma virus, small pox, herpes virus), bacteriophages (T-even)
Single stranded DNA (ssDNA)	Bacteriophages (ϕ ×174 and other bacteriophages), Animal viruses (parvovirus)
Double stranded RNA (dsRNA)	Retrovirus, reovirus, hepatitis-A, animal viruses
Single stranded RNA (ssRNA)	Plant viruses (TMV, tobacco virus), animal viruses (influenza virus, poliomyelitis virus)
	Bacteriophages (F2, ~ R17)

(ii) Nitrogenous Base : There are two nitrogenous bases, purines and pyrimidines. The purines are double ring compounds that consists of 5-membered imidazole ring with nitrogen at 1', 3', 7' and 9' positions. The pyrimidines are single ring compounds, the nitrogen being at position 1' and 3' in 6-membered benzene ring (Fig 5.2). A single base is attached to 1'-carbon atom of pentose sugar

Pyrimidine Ring   Purine Ring

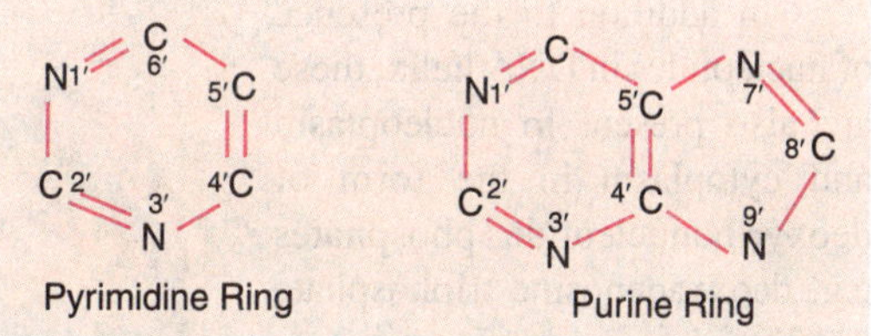

Fig. 5.2 : The pyrimidine and purine rings.

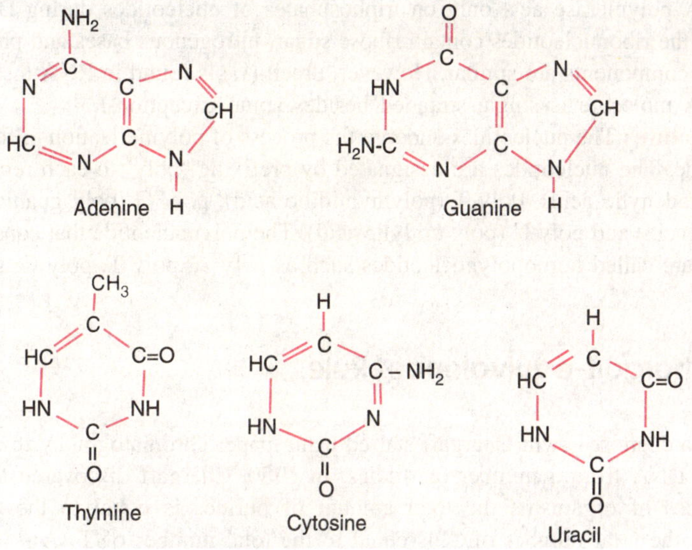

by an *N*-glycosidic bond. Purines are of two types, adenine (A) and guanine (G), and pyrimidines are also of two types, thymine (T) and cytosine (C). Uracil (U) is a third pyrimidine (Fig 5.3). A,G and C are common in both DNA and RNA. U is found only in RNA.

Adenine   Guanine   Thymine   Cytosine   Uracil

Fig. 5.3 : Nitrogenous base of nucleic acids.

(*iii*) **A Phosphate Group:** In DNA a phosphate group (PO_4^{-3}) is attached to the 3'-carbon of deoxyribose sugar and 5'-carbon of another sugar. Therefore, each strand contains 3' end and 5' end arranged in an alternate manner. Strong negative charges of nucleic acid are due to the presence of phosphate groups. A nucleotide is a nucleoside phosphate which contains its bond at 3' and 5' carbon atoms of pentose sugar that is called phosphodiester (Fig 5.4)

2. Nucleosides, Nucleotides and Polynucleotides

The nitrogenous bases combined with pentose sugar are called nucleosides. A nucleoside linked with phosphate forms a nucleotide (Fig 5.1)

Nucleoside = pentose sugar + nitrogenous base

Nucleotide = nucleoside + phosphate

On the basis of different nitrogenous bases the deoxynucleotides are of following types :

(*i*) Adenine (A) = deoxyadenosine-3'/5'-monophosphate (3'/5'-dAMP)

(*ii*) Guanine (G) = deoxyguanosine -5'-monophosphate (5'-dGMP)

(*iii*) Thymine (T) = deoxythymidine -5'-monophosphate (5'-dTMP)

(*iv*) Cytosine (C) = deoxycytidine -5'-monophosphate (5'-dCMP)

In addition to the presence of nucleotides in DNA helix, these are also present in nucleoplasm and cytoplasm in the form of deoxyribonucleotide phosphates e.g. deoxyadenosine triphosphate (dATP), deoxyguanosine triphosphate (dGTP), deoxycytidine triphosphate (dCTP), deoxythymidine triphosphate (dTTP). The advantage of these four deoxyribonucleotide in triphosphate form

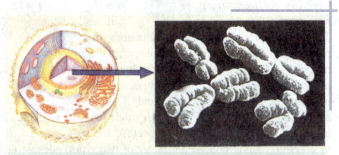

The four kinds of nucleotide subunits of DNA that make chromosomes.

is that the DNA polymerase acts only on triphosphates of nucleotides during DNA replication.

Similarly, the ribonucleotides contain ribose sugar, nitrogenous bases and phosphate. Except sugar, the other components are similar. However, uracil (U) is found in RNA instead of thymine. Generally, RNA molecule is single stranded besides some exceptions.

Polynucleotide : The nucleotides undergo the process of polymerization to form a long chain of polynucleotide. The nucleotides are designated by prefixing 'poly' to each repeating unit such as poly A (polyadenylic acid), Poly T (polythymidilic acid), poly G (poly guanidylic acid), poly C (polycytidilic acid) and poly U (poly uridylic acid). The polynucleotide that consists of the same repeating units are called homopolynucleotides such as poly A, poly T, poly G, poly C and poly U.

3. Chargaff-equivalence Rule

By 1948, a chemist Erwin Chargaff started using paper chromatography to analyse the base composition of DNA from a number of studies. In 1950, Chargaff discovered that in the DNA of different types of organisms the total amount of purines is equal to the total amount of pyrimidines i.e. the total number of A is equal to the total number of T (A − T), and the total

Deoxyguanosine

3'–OH Terminus

Phosphodiester Group

5'–Terminus

Deoxythymidine

Fig. 5.4 : A segment of polynucleotide chain showing dinucleotide linked with phosphodiester bonds.

number of G is equal to the total number of C (G – C). It means that A/T = G/C i.e. A + T/G + C = 1. In the DNA molecules isolated from several organisms regularity exists in the base composition.

The DNA molecule of each species comprises of base composition which is not influenced either by environmental conditions or growth stages or age. The molar ratio i.e. [A] + [T]/[G]+[C] represents a characteristic composition of DNA of each species. However, in higher plants and animals A – T composition was found generally high and G – C content low, whereas the DNA molecules isolated from lower plants and animals, and bacteria and viruses was generally rich in G – C and poor in A – T contents (Table 5.2)

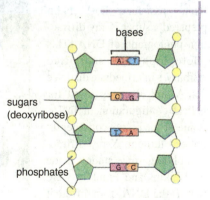

Molecular Structure of DNA.

The two closely related species will have very similar molar % G+C values and vice versa. Thus, the use of base composition has much significance in establishing relationship between two species and in taxonomy and phylogeny of species.

Table 5.2 : Relative amount of nitrogenous bases in DNA isolated from different organisms (modified after Chargaff and Davidson, 1955)

Source	Adenine	Guanine	Thymine	Cytosine	$\dfrac{A+T}{G+C}$
Human sperm	30.9	19.1	31.6	18.4	1.62
Human thymus	30.9	19.9	29.4	19.8	1.52
Sea urchin sperm	32.8	17.7	32.1	18.4	1.85
Wheat germ	26.5	23.5	27.0	23.0	1.19
Yeast	31.3	18.7	32.9	17.1	1.79
Escherichia coli	26.0	24.9	23.9	25.2	1.00
Diplococcus pneumoniae	29.8	20.5	31.6	18.0	1.59
Bacteriophage T_2	32.5	18.2	32.6	16.7	1.86

4. X-ray Crystallographic Studies of DNA

In 1920, the basic chemical composition of nucleic acid was elucidated through the efforts of P.A. Levene. In 1940, for the first time W.T. Astbury gave first three dimensional structure of DNA model studied through X-ray crystallography. He concluded that its flat nucleotides get stacked to form polynuceotide. Each nucleotide was oriented perpendicularly to the axis of molecule present at every 3.4 Å along the stack. M.H.F Wilkins, Rosalind Franklin and coworkers countinued to study the work of Astbury and prepared the highly oriented DNA fibres which facilitated for getting photographs of X-ray diffraction. His female associate, R. Franklin prepared a super X-ray diffraction photograph of DNA. This photograph supported Astbury's intermediate distance 3.4 Å and also suggested for helical configuration of DNA. Some of the points consistently noticed through X-ray diffraction photographic studies made by Wilkins and Franklin were that (*i*) the DNA from different species gave the identical X-ray diffraction besides variation in their base composition,

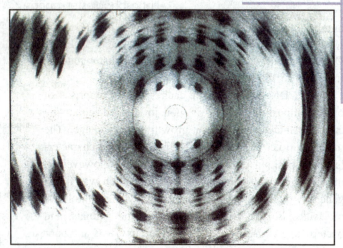

An X-ray diffraction image of DNA.

(*ii*) the DNA molecules are about 20 Å thick and more than 30,000 Å long, and (*iii*) the DNA molecules have repeating structure at every 34 Å.

However, the pertinent experimental facts were achieved through the study of detailed chemical structure of DNA, Chargaff's base pairing rule and Wilkins-Franklin's X-ray diffraction photographic studies. There was an urgent need to get three dimensional structure of DNA molecule. In February 1953, Pauling and R.B. Corey gave a triple helix model of DNA molecule. However, they could not explain the process of DNA replication. It was April 2, 1953 when a young American biologist James D.Watson and an English physicist Francis H. Crick of the Cavendish Laboratory, Cambridge University sent a brief letter to the Journal *Nature* entitled "a structure for deoxyribose nucleic acid" suggesting the three dimensional double helical structure of DNA molecule. For this ground - breaking work, Watson and Crick were honoured with the Nobel Prize in 1962. They shared the prize with Maurice Wilkins of the King's College, London who investigated the X-ray diffraction photographs of the DNA molecule. Unfortunately, Rosalind Franklin who also contributed a lot was deprived, by Cruel death in 1958, of possible sharing of the prize.

5.   Consideration of Watson and Crick in Construction of Double Helix DNA Model

Watson and Crick studied in advance the manuscript of Pauling and Corey in 1953, and X-ray diffraction photographs of DNA taken from Franklin. They concluded that (*i*) polynucleotide chain of DNA has a regular helix, (*ii*) the helix has a diameter of about 20 Å, and (*iii*) the helix

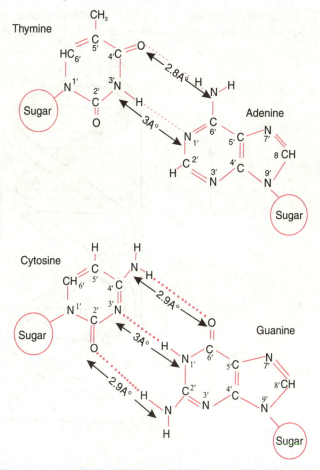

Fig. 5.5 : Pairing between thymine and adenine, and cytosine and guanine.

makes a full turn at every 34 Å along its length and contains a stack of ten nucleotides per turn because the internucleotide distance is 3.4 Å. They considered the density of DNA molecule and concluded that the density of a cylinder having 20 Å diameter and 34 Å length is too low if it contained a single stack of ten nucleotides and the density will be too high if it contains three or more stacks of ten nucleotides. Therefore, the helix must contain two polynucleotide chains or two stacks of ten nucleotides per turn.

They also took into account that if the DNA is to contain heredity information which is inscribed as specific sequence of four bases along the polynucleotide chain, then the molecular structure of DNA must include any arbitrary sequences along its polynucleotide chain. The dimension of purine ring is more than the dimension of pyrimidine ring; therefore, the two chain helix may contain a constant diameter provided a complementary relationship existed between the two nucleotide stacks, one harbouring purine and the other pyrimidine bases. To provide thermodynamic stability they thought over the formation of hydrogen bonds between amino ($-NH_2$) or hydroxyl (-OH) hydrogens and ketone oxygen or amino-nitrogen of the two bases. On the basis of these considerations they got success in constructing a double helix model for the DNA molecule.

6. Watson and Crick's Model for DNA

J.D.Watson and F.H.C. Crick (1953) combined the physical and chemical data, and proposed a double helix model for DNA molecule. This model is widely accepted. According to this model,

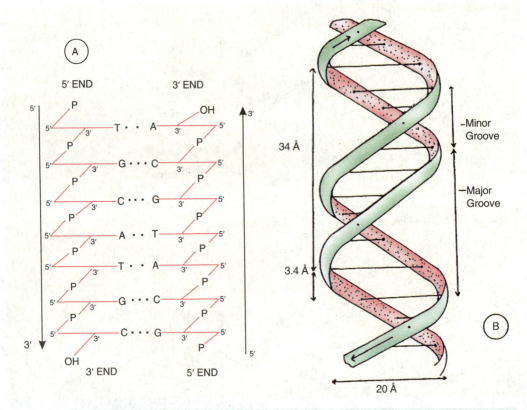

Fig. 5.6 : Antiparallel orientation of the complementary chains (A), and Watson and Crick's model of DNA double helix (B).

the DNA molecule consists of two strands which are connected together by hydrogen bonds and helically twisted. Each step on one strand consists of a nucleotide of purine base which alternates with that of pyrimidine base. Thus, a strand of DNA molecule is a polymer of four nucleotides *i.e.* A, G, T, C. The two strands join together to form a double helix. Bases of two nucleotides form hydrogen bonds *i.e.* A combines with T by two hydrogen bonds (A = T) and G combines with C by three hydrogen bonds (G ≡ C) (Fig. 5.5). However, the sequence of bonding is such that for every ATGC on one strand there would be TACG on the other strand. Therefore, the two chains are complementary to each other *i.e.* sequences of nucleotides on one chain are the photocopy of sequences of nucleotides on the other chain. The two strands of double helix run in antiparallel direction *i.e.* they have opposite polarity. In Fig. 5.6A the left hand strand has 5' → 3' polarity, whereas the right hand has 3' → 5' polarity as compared to the first one. The polarity is due to the direction of phosphodiester linkage.

The hydrogen bonds between the two strands are such that they maintain a distance of 20 Å. The double helix coils in right hand direction *i.e.* clockwise direction and completes a turn at every 34 Å distance (Fig. 5.6 B). The turning of double helix results in the appearance of a deep and wide groove called major groove. The major groove is the site of bonding of specific protein. The distance between two strands forms a minor groove. One turn of double helix at every 34Å distance includes 10 nucleotides *i.e.* each nucleotide is situated at a distance of 3.4Å. Sugar-phosphate (nucleoside) makes the backbone of double helix of DNA molecule (Fig. 5.6B).

The DNA model also suggested a copying mechanism of the genetic material. DNA replication is the fundamental and unique event underlying growth and reproduction in all living organisms ranging from the smallest viruses to the most complex of all creatures including man. DNA replicates by semiconservative mechanism which was experimentally proved by Mathew, Meselson and Frank W. Stahl in 1958. If changes occur in sequence or composition of base pairs of DNA, mutation takes place. Though the presence of adenine, guanine, thymine and cytosine is universal phenomenon, yet unusual bases in DNA molecule are also found. In some bacteriophages 5-hydroxymethylcytosine (HMC) replaces cytosine of the DNA molecule when methylation of adenine, guanine and cytosine occurs. This results in changes in these bases.

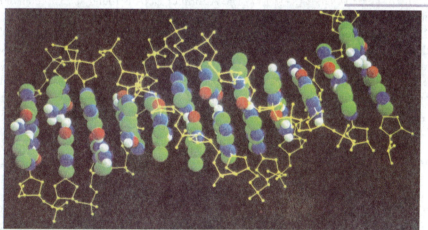

The base pairs (balls) in the centre of the DNA spiral held the two strands together.

7.   Different Forms of DNA

The most common form of DNA which has right handed helix and proposed by Watson and Crick is called B-form of DNA or B-DNA. In addition, the DNA may be able to exist in other forms of double helical structure. These are A and C forms of double helix which vary from B-form in spacing between nucleotides and number of nucleotides per turn, rotation per base pair, vertical rise per base pair and helical diameter (Table 5.3)

(*i*) **The B-Form of DNA (B-DNA):** Structure of B-form of DNA has been proposed by Watson and Crick. It is present in every cell at a very high relative humidity (92%) and low concentration of ions. It has antiparallel double helix, rotating clockwise (right hand) and made up of sugar-phosphate back bone combined with base pairs or purine-pyrimidine. The base pairs are perpendicular to longitudinal axis of the helix. The base pairs tilt to helix by 6.3°. The B-form of DNA is metabolically stable and undergo changes to A, C or D forms depending on sequence of nucleotides and concentration of excess salts.

Table 5.3 : Forms of double stranded DNA helix

Parameters	*Forms of DNA*				
	A	*B*	*C*	*D*	*Z*
Conditions	75% rel. Humidity, Na^+ K^+, Cs^+ ions	92% rel. humidity low ions	66% rel. humidity Li^+ ions	—	low high salt conc.
Base pair per turn	11	10	9.33	8	12 (6 dimers)
Rotation per bp	+32.7°	+36.0°	+38.6°	—	-30.0°
Vertical rise per bp	2.56Å	3.38Å	3.32Å	3.03Å	3.71Å
Helical diameter	23Å	20Å	19Å	—	18Å
Pitch of the helix	28.15Å	34Å	31Å	—	45 Å
Tilt of bp Sugar puckering	20.2Å	6.3°	-7.8°	-16.7°	7°

bp = base pairs; (+), right handed, (–), left handed

(*ii*) **The A-Form of DNA (A-DNA):** The A-form of DNA is found at 75% relative humidity in the presence of Na^+, K^+ or Cs^+ ions. It contains eleven base pairs as compared to ten base pairs of B-DNA (Table 3.3) which tilt from the axis of helix by 20.2°. Due to this displacement the depth of major groove increases and that of minor groove decreases. The A-form is metastable and quickly turns to the D-form.

(*iii*) **The C-Form DNA (C-DNA):** The C-form of DNA is found at 66% relative humidity in the presence of lithium (Lit^+) ions. As compared to A-and B-DNA, in C-DNA the number of base pairs per turn is less i.e. 28/3 or $9^1/_3$. The base pairs show pronounced negative tilt by 7.8°.

(*iv*) **The D-Form of DNA (D-DNA):** The D-form of DNA is found rarely as extreme variants. Total number of base pairs per turn of helix is eight. Therefore, it shows eight-fold symmetry. This form is also called poly (dA-dT) and poly (dG-dC) form. There is pronounced negative tilt of base pairs by 16.7° as compared to C form *i.e.* the base pairs are displaced backwardly with respect to the axis of DNA helix.

(*v*) **The Z-Form of DNA (Z-DNA) or Left Handed DNA:** In 1979, Rich and coworkers at MIT (U.S.A.) obtained Z-DNA by artificially synthesizing d (C-G)3 molecules in the form of crystals. They proposed a left handed (synistral) double helix model with zig-zag sugar-phosphate back bone running in antiparallel direction.

Therefore, this DNA has been termed as Z-DNA. The Z-DNA has been found in a large number of living organisms including mammals, protozoans and several plant species.

There are, several similarities with B-DNA in having (*i*) double helix, (*ii*) two antiparallel strands, and (*iii*) three hydrogen bonds between G-C pairing. In addition, the Z-DNA differs from the B-DNA in the following ways :

(*a*) The Z-DNA has left handed helix, while the B-DNA has right handed helix.

(*b*) The Z-DNA contains zig-zag sugar phosphate back bone as compared to regular back bone of the B-DNA.

(*c*) The repeating unit in Z-DNA is a dinucleotide due to alternating orientation of sugar residues, whereas in B-DNA the repeating unit is a mononucleotide, and sugar molecules do not have the alternating orientation.

(*d*) In the Z-DNA one complete turn contains 12 base pairs of six repeating dinucleotide, while in B-DNA one full turn consists of 10 base pairs i.e. the 10 repeating units.

(*e*) Due to the presence of high number (12) of base pairs in one turn of Z-DNA, the angle of twist per repeating unit *i.e.* dinucleotide is 60° as compared to 36° of B-DNA molecule.

(*f*) In Z-DNA the distance of twist making one turn of 360° is 45Å as against 34Å in B-DNA.

(*g*) The Z-DNA has less diameter (18Å) as compared to the B-DNA (20Å diameter).

(*vi*) **Single Stranded (ss) DNA:** Almost all the organisms contain double stranded DNA except a few viruses such as bacteriophage φ×174 which consists of single stranded circular DNA (Sinsheimer, 1959). It becomes double stranded only at the time of replication. The differences of ssDNA from the dsDNA are as below :

(*a*) The dsDNA absorbs wavelength 2600 Å of ultra violet light constantly from 0 to 80°C, thereafter rise sharply, whereas in ssDNA absorption of UV light increases steadily from 20° to 90°C.

(*b*) The dsDNA resists the action of formaline due to closed reactive site, while the ss DNA does not resist it due to exposed reactive sites.

(*c*) Base pair composition in dsDNA is equal *i.e.* A=T and G=C, in ssDNA the composition of A,T,G,C is in proportion of 1:1.33:0.98:0.75.

(*d*) The dsDNA always remains in linear helical form, while the ssDNA remains in circular form, however, it becomes double stranded only during replication (*i.e.* replicative form).

(*vii*) **Circular and Super Helical DNA:** Almost in all the prokaryotes and a few viruses, the DNA is organised in the form of closed circle. The two ends of the double helix get covalently sealed to form a closed circle. Thus, a closed circle contains two unbroken complementary strands. Some times one or more nicks or breaks may be present on one or both strands, for example DNA of phage PM2 (Fig. 5.7 A).

Besides some exceptions, the covalently closed circles are twisted into super helix or super coils (Bauer *et al.*, 1980) (Fig.5.7 B) and is associated with basic proteins but not with histones found complexed with all eukaryotic DNA.

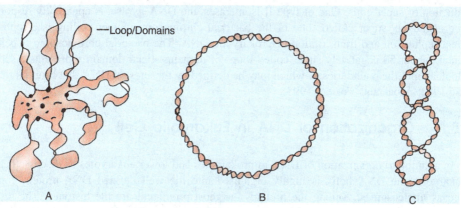

Fig. 5.7 : The forms of DNA. A, Nucleoids of *E.coli*; B, a closed, circular bacterial DNA; C, twisted supercoils of double stranded DNA.

These histone like proteins appear to help the organization of bacterial DNA into a coiled chromatin structure with the result of nucleosome like structure, folding and super coiling of DNA, and association of DNA polymerase with nucleoids. Several histone like DNA binding proteins have been described in bacteria (Table 5.4). These nucleoid-associated proteins include HU proteins, IHF, protein H1, Fir A, H-NS and Fis (Pettijohn, 1988). In archaeobacteria (e.g. *Archaea*) the chromosomal DNA exists in protein-associated form. Histone like proteins have been isolated from nucleoprotein complexes in *Thermoplasma acidophilum* and *Halobacterium salinarum*. Thus, the protein associated DNA and nucleosome like structures are detected in a variety of bacteria (Drlica and Rouviere-Yuniv, 1987). If the helix coils clockwise from the axis the coiling is termed as positive or right handed coiling. In contrast, if the path of coiling is anticlockwise, the coil is called left handed or negative coil.

Table 5.4 : Histone-like proteins of *E.coli* (after Moat and Foster, 1995)

Name	Subunit	Molecular Weight	Function
Hu (hupA)	Hu(α)[HLPIIa]	9,000	Can form nucleosome like structure like H_2B
(hupB)	Hu(β)[HLPIIb] linked as heterogenous dimer	9,000	—
HLP 1	—	17,000	Affects RNA polymerase transcription
H1(hns)	—	15,000	May modulate in vivo transcription
H	—	28,000	Similar to eucaryotic H_2A

The two ends of a linear DNA helix can be joined to form each strand continuously. However, if one of ends rotates at 360° with respect to the other to produce some unwinding of the double helix, the ends are joined resulting in formation of a twisted circle in opposite sense i.e. opposite to unwinding direction. Such twisted circle appears as 8 i.e. it has one node or crossing over point. If it is twisted at 720° before joining, the resulting super helix will contain two nodes (Fig. 5.7B).

The enzyme topoisomerases alter the topological form i.e. super coiling of a circular DNA molecule. Type I topoisomerases (e.g. *E.coli* Top A) relax the negatively super coiled DNA by breaking one of the phosphodiester bonds in dsDNA allowing the 3'-OH end to swivel around the 5'-phosphoryl end, and then resealing the nicked phosphodiester backbone (Moat and Foster, 1995). Type II topoisomerases need energy to unwind the DNA molecules resulting in the introduction of super coils. One of type II isomerases, the DNA gyrase, is apparently responsible for the negatively super coiled state of the bacterial chromosome. Super coiling is essential for efficient replication and transcription of prokaryotic DNA. The bacterial chromosome is believed to contain about 50 negatively super coiled loops or domains. Each domain represents a separate topological unit, the boundaries of which may be defined by the sites on DNA that limit its rotation (Wang,1982; Moat and Foster, 1995).

8. Organization of DNA in Eukaryotic Cell

In addition to organization of DNA in prokaryotes and lower eukaryotes as discussed earlier, in eukaryotes the DNA helix is highly organised into the well defined DNA-protein complex termed as **nucleosomes.** Among the proteins the most prominent are the histones. The histones are small and basic proteins rich in amino acids such as lysine and/or arginine. Almost in all eukaryotic cells there are five types of histones e.g. H_1, H_2A, H_2B, H_3 and H_4. Eight histone molecules (two each of H_2A, H_2B, H_3 and H_4) form an octamer ellipsoidal structure of about 11 nm long and 6.5-7 nm in diameter. DNA coils around the surface of ellipsoidal structure of histones 166 base pairs (about 7/4 turns) before proceeding onto the next and form a complex structure, the nucleosome (Fig. 5.8A-B). Thus a nucleosome is an octamer of four histone proteins complexed with DNA.

A Model of Nucleosome.

The histones play an important role in determining of eukaryotic chromosomes by determining the conformation known as chromatin. The nucleosomes are the repeating units of DNA organization which are often termed as beads. The DNA isolated from chromatin looks like string or beads. The 146 base pairs of DNA lie in the helical path and the histone-DNA assembly is known as the nucleosome core particle. The stretch of DNA between the nucleosomes is known

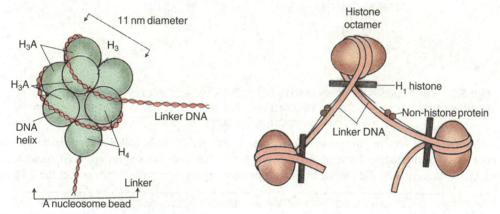

Fig. 5.8 : Internal organisation of nucleosomes. A highly super coiled chromatin fibre; B, a single nucleosome.

as 'linker' which varies in length from 14 to over 100 base pairs. The H_1 is associated with the linker region and helps the folding of DNA into complex structure called chromatin fibres which in turn get coiled to form chromatin. As a result of maximum folding of DNA, chromatin becomes visible as chromosomes during cell division (Kornberg and Llug, 1981).

9.   Palindromic DNA (Palindromes)

In the DNA molecule a variety of base sequences have been observed. Most of them do not have special features. The repeated sequences are of particular interest because they are the site of enzymatic activity and, sometimes give special features to the nucleic acid. For the first time Wilson and Thomas (1974) used the term palindromic DNA. The palindromic DNA or palindromes are the inverted repeats and region of dyad symmetry (Rao, 1972). The length of palindromes may be short by about 3-10 bases or long by about 50-100 base pairs. As compared to prokaryotic DNA, the eukaryotic DNA contains a large palindrome of about several thousand base pairs. Sometimes a spacer separates the two inverted repeats (Fig. 5.9A).

The DNA molecules containing palindromes and inverted repeats may exist in alternative forms. After separation of complementary strands, the intramolecular base pairing may result in a double stranded stretch formed between adjacent complementary sequences. This is known as cruciform structure (Fig. 5.9 B). The cruciform structures have been produced in laboratory but not detected in the DNA isolated from cells.

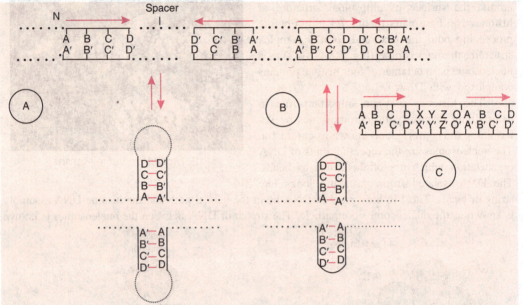

Fig. 5.9 : The possible alternative forms of a DNA molecule containing two inverted repeats and direct repeats. A, inverted repeats separated by a spacer; B, adjacent inverted repeats; C, direct repeats.

Moreover, if repeats are present in the same orientation, it is called direct repeats. They neither exist in alternative form of dsDNA nor have effect on the single advanced molecule.

In addition, if both the palindromes and inverted repeats are present they affect the ssDNA

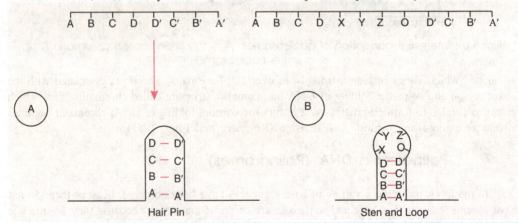

Fig. 5.10 : Formation of hair pin (A), and stem and loop (B) structure in single stranded nucleic acids.

or RNA. They form intra-strand hydrogen bonding between adjacent or nearly complementary sequences. Consquently a palindrome produces an intrastrand double stranded structure which is known as hairpin (Fig 5.10 A). The interrupted inverted repeats result in a structure consisting of a double stranded segment with a terminal single-stranded loop. This structure is called stem and loop (B).

Several possible functions of palindrome have been suggested such as (*i*) its action as recognition sites of DNA for protein and many bacterial restriction enzymes, (*ii*) its action as bacterial restriction enzyme against destruction by foreign DNA, (*iii*) providing structural strength to the transcribed RNA by hydrogen bonding in the hairpin loops, (*iv*) possible involvement of cruciform structure in genetic recombination, and (*v*) its possestion of genes on long palindrome of some lower eukaryotes that code for ribosomal RNA.

B. Structure of Ribonucleic Acid (RNA)

The RNA is usually single stranded except some viruses such as TMV, yellow mosaic virus, influenza virus, foot and mouth disease virus, reovirus, wound tumour virus, etc. which have dsRNA. The single strand of the RNA is folded either at certain regions or entirely to form hairpin-shaped structure. In the hairpin shaped structures the complementary bases are linked by hydrogen bonds which give stability to the molecules. However, no complementary bases are found in the unfolded region. The RNA does not possess equal purine-pyrimidine ratio, as it is found in the DNA.

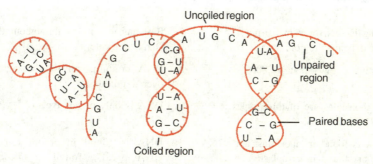

Fig. 5.11 : Molecular structure of ribosomal RNA (rRNA).

Like DNA, the RNA is also the polymer of four nucleotides each one contains D-ribose, phosphoric acid and a nitrogenous base. The bases are two purines (A, G) and two pyrimidines (C, U) (Fig 5.11). Thymine is not found in RNA. Pairing between bases occurs as A-U and G-C. The nucleotides formed by the four bases are adenosine monophosphate (AMP), guanosine monophosphate (GMP), cytosine monophosphate (CMP), and uridine monophosphate (UMP). These are found freely in nucleoplasm but in the form of triphosphates e.g. ATP, GTP, UTP and CTP. As a result of polymerization the ribonucleotides form a polynucleotide chain of RNA (Holley, 1966).

If the RNA is involved in genetic mechanism, it is called **genetic RNA** as found in plant, animal and bacterial viruses. However, DNA acts as genetic material and RNA follows the order of DNA. In such cells the RNA does not have genetic role. Therefore, it is called **non-genetic RNA**. The non-genetic RNA is of three types : (*i*) ribosomal RNA (rRNA), (*ii*) transfer RNA (tRNA) or soluble RNA (sRNA), and (*iii*) messenger RNA (mRNA) or template RNA. These three types of RNA differ from each other in structure, site of synthesis in eukaryotic cell and function.

The mRNA and tRNA are synthesized on DNA template, whereas rRNA is delivered from nucleolar DNA. These RNAs are synthesized during different stages. During cleavage most of the mRNAs are synthesized, whereas tRNA is synthesized at the end of cleavage. Synthesis of rRNA occurs during gastrulation. The total population of rRNA is about 90% of all RNAs.

Like DNA, the RNA is not self – replicating but it has to depend on DNA. Therefore, replication of non-genetic RNA is known as DNA-dependent RNA replication. Moreover, the genetic RNA of viruses is self-replicating i.e. it can form its own several replica copies. For detail see Chapter 15 *Viruses-I*. Differences between the DNA and RNA molecules are given in Table 5.5.

Table 5.5 : Differences between RNA and DNA

RNA	*DNA*
1. RNA is more primitive than DNA	1. DNA originated after RNA
2. RNA is the genetic material of some plant, animal and bacterial viruses.	2. DNA is the genetic material of almost all living organisms
3. Except some viruses (e.g. reovirus) most cellular RNA is single stranded.	3. Except a few (e.g. $\phi \times 174$), most of DNA viruses is double stranded.
4. Pentose sugar is ribose.	4. Pentose sugar is deoxyribose.
5. The bases are adenine, guanine, cytosine and uracil.	5. The bases are adenine, guanine, cytosine and thymine.
6. Base pairing occurs between adenine and uracil (A-U)	6. Base pairs are A-T and G-C.
7. Base pairing is seen only in hairpin structure and helical region.	7. Base pairing occurs throughout the length of DNA molecule.
8. RNA contains a few (about 12,000) nucleotides	8. DNA contains millions of nucleotides e.g. over 4 millions.
9. The RNA molecules are of three types : rRNA, mRNA, tRNA.	9. DNA is only of one type.
10. The mRNA is found in nucleolus, and tRNA and rRNA are found in cytoplasm. They are formed on the DNA.	10. DNA is found in chromosomes. However, DNA is also found in mitochondria and chloroplasts.
11. RNAs translate the message of DNA into proteins.	11. DNA encodes the genetic messages in the form of mRNA transcripts.
12. Genetic RNA uses the enzyme, reverse transcriptase during replication.	12. This enzyme is not required by DNA. DNA after replication forms DNA and after transcription forms RNA.

I. Types of RNA

1.   The Ribosomal RNA (rRNA)

The non-genetic RNAs are synthesized on the DNA template and are present in the nucleolus and cytoplasm. Therefore, the base sequences of rRNA and part of DNA where they are synthesized are complementary. In prokaryotes rRNA is formed on a part of DNA called ribosomal DNA, while in eukaryotes these are formed in nucleolus containing the nuclear DNA. The rRNAs are found in ribosomes and accounts for 40-60% of dry weight. In general, it represents about 80% of total RNA of the cell. The ribosome consists of proteins and RNA. The ribosomes are of different types such as 80S (found in eukaryotes) and 55S (found in mitchondria of vertebrates). The 70S ribosomes of prokaryotes are made up of two subunits, 50S and 30S. The 50S subunit contains 23S and 5S rRNA, whereas the 30S subunit consists of 16S rRNA.

The 80S ribosome consists of 60S and 40S subunit. The rRNA types in both the subunits of plants differ from that of animals (Table 5.6).

Table 5.6 : Types of Ribosomes and rRNA subunits

Source	Ribosomes	Larger subunit		Smaller subunit	
		Size	RNA	Size	RNA
Prokaryotes	70S	50S	23S and 5S	30S	16S
Eukaryotes					
(i) Plants and invertebrates	80S	60S	25S, 5S, 5.8S	40S	16-18S
(ii) Vertebrates	80S	60S	28S, 29S, 5S 5.8S	40S	18S
(iii) Vertebrate mitochondria	55S	40S	16-17S, 5S	30S	12-13S

The rRNA is a single stranded molecule which is twisted at certain points to form helical regions. In the helical region most of the base pairs are complementary and linked by hydrogen bonds. The uncoiled single stranded regions lack the complementary bases. Therefore, in rRNA the ratio of purine : pyrimidine is not equal. The rRNA exists in a living cell for about two generations.

2. The Messenger RNA (mRNA)

The mRNA is transcribed on the DNA template and, therefore, carries the genetic information of DNA. For the first time, Francis Jacob and Jacques Monod (1961) proposed the name mRNA for bearing the transcripts of DNA for protein synthesis on ribosomes. The total population of mRNA in a cell varies from 5 to 10% of the total cellular RNA because the species of mRNA are short lived as these are broken into ribonucleotides by the enzyme ribonuclease. In *E.coli* some of the mRNAs remain alive only for about two minutes. Therefore, the cell does not contain high amount of mRNA at a time. In contrast, the mRNAs of eukaryotes are metabolically stable.

The size of mRNA varies. The smallest protein contains about 50 amino acids (50×3=150 nucleotides needed for monocistronic mRNA molecules). Typically protein has 300-600 amino acids (900–1,800 nucleotide long mRNA). In prokaryotes the polycistronic mRNA is more common than monocistronic mRNA and contains 3000–8000 nucleotides. Polycistronic mRNA contains usually 10 bases long inter-cistronic sequences called spacers.

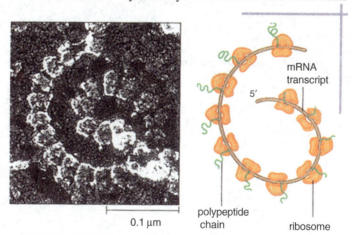

Polysome (many ribosomes simultaneously translating the same mRNA molecule) found in a eukaryotic cell.

The sedimentation coefficient of mRNA is 8S and average molecular wieght ranges from 500,000 to 1,00,000. Since they represent a gene, their length and molecular weight change because a gene contains 100 to 1,500 nucleotides. The mRNAs are transcribed by genes, hence individual mRNA represents a single gene. Therefore, in a cell there will be as much mRNAs as genes, and every mRNA differs from each other. Taylor (1979) has reviewed the isolation of eukaryotic mRNAs. Kozak (1983) has given a comparative account of initiation of protein synthesis in prokaryotes, eukaryotes and the organelles.

Initiation of synthesis of first polypeptide chain of a polycistronic mRNA may begin hundreds of nucleotides from the 5' end. The section of non-translated RNA before coding region is called leader. Untranslated sequences are usually formed at both 5' and 3' ends.

As the mRNAs always remain in single stranded form, it may disrupt the biological activity after being coiled. However, the coils lack complementary bases. Kozak (1991) has discussed the structural features in eukaryotic mRNAs that modulate the initiation of translation. The structure of prokaryotic and eukaryotic mRNA is shown in Fig 5.12 (A-C) and discussed below:

(*i*) **The 5' Cap:** In most of the eukaryotes and animal viruses, 5' end of mRNA contains a cap which is formed after methylation of any of four nucleotides. For example, an mRNA (Fig. 3.12C) contains m7G(5')ppp(5')N where m7G is the methylguanosine and (5')ppp(5') represents a 5-5' triphosphate linked to a base (N) at 5' end. The mRNA binds to ribosome with the help

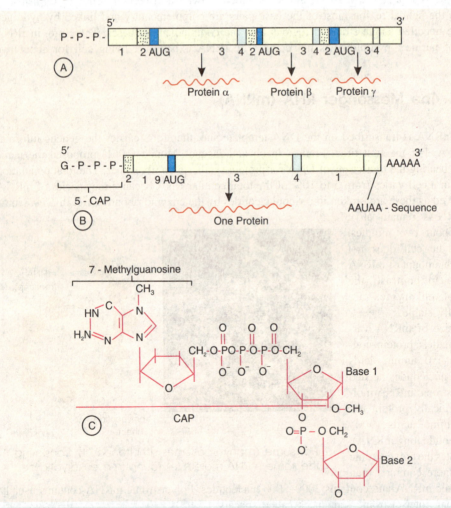

Fig. 5.12 : Structure of prokaryotic (A) and eukaryotic (B) mRNA molecules; C, eukaryotic mRNA showing a detailed structure of 5' cap-7G(5) ppp(5')N; 1, non-coding sequences; 2, ribosome binding sites; 3, coding sequences; 4, stop codons.

of this cap. Therefore, it governs protein synthesis. The bacterial mRNA does not contain 5' cap. Instead they contain a specific ribosome binding site about six nucleotide long which occurs at several places in the mRNA molecules. These are located at 4 nucleotide upstream from AUC. In bacterial mRNA there may be multiple ribosome binding sites called Shine- Dalgarno sequences in the interior of an mRNA chain, each resulting in synthesis of a different protein.

(*ii*) **The Non-coding Regions:** There are two non-coding regions first followed by the cap and the second followed by the termination codon. The non-coding region (NC1) is about 10-100 nucleotides long and rich in A and G residues, whereas the NC2 is 50-150 nucleotide long and contains an AAUAAA residues. Both the non-coding regions do not translate protein.

(*iii*) **The Initiation Codon:** Both in prokaryotes and eukaryotes the initiation codon (AUG) is present which starts protein synthesis. Bacterial ribosomes, unlike the eukaryotic ribosomes, directly bind to start codons in the interior of mRNA to initiate protein synthesis.

(*iv*) **The Coding Region:** It is the most important region of mRNA which is about 1,500 nucleotides long. This region translates a long chain of protein after attaching with several ribosomes. The combination of mRNA strand with several ribosomes is called polyribosome. Therefore, the bacterial mRNAs are commonly called polycistronic mRNA i.e. they encode multiple proteins that are separately translated from the same mRNA molecule. The eukaryotic mRNAs are typically monocistronic i.e. only one species of polypeptide chain is translated per mRNA molecule.

(*v*) **The Termination Codon:** The termination codon is required to give the signal to stop protein synthesis. In eukaryotes the termination codons are UAA, UAG or UGA that terminates the translation process i.e. the process of protein synthesis.

(*vi*) **The Poly (A) Sequence:** The NC_2 is followed by poly (A) sequence in the eukaryotic mRNA. The prokaryotic mRNAs lack poly (A). The polyadenylate or poly (A) sequences of 200-250 nucleotides are present at 3'OH end of mRNA. Poly (A) sequences are added when mRNA is present inside the nucleus.The function of poly (A) sequence in translation is unknown.

3. Differences Between Prokaryotic and Eukaryotic mRNAs.

The mRNAs found in prokaryotes differ from that of eukaryotes in the following ways :

(*i*) The prokaryotic mRNAs are broken down by the ribonuclease into ribonucleotide, hence these are very short lived e.g. 2 minutes in *E.coli*. The short lived nature is due to changes in environmental conditions. In contrast the eukaryotic mRNAs have longer life i.e. they are metabolically more stable than the prokaryotic mRNAs.

(*ii*) In most of the prokaryotic mRNAs translation starts soon when the mRNAs are being transcribed on DNA template, whereas in eukaryotic mRNAs the translation process begins (in cytoplasm) when transcription process is over (in nucleus).

(*iii*) Most of the mRNAs of prokaryotes and bacteriophages are polycistronic *i.e.* contain several sites for initiation and termination of polypeptides due to the presence of several structural genes. Therefore, an mRNA molecule synthesises more than one polypeptide chains. On the otherhand, the mRNAs of eukaryotes are monocistronic i.e. contain only a single initiation and termination codons and synthesize one chain only.

(*iv*) A very little processing in mRNAs of prokaryotes occurs due to short time between transcription and translation. The mRNAs of eukaryotes undergo a considerable processing after being transcribed. The mRNA processing includes : (*a*) polyadenylation (addition of 200-250 poly (A) at 3'OH end), (*b*) capping (formation of a 5' cap at 5' end by condensation of guanosine residues, and (*c*) methylation (addition of methyl *i.e.* –CH_3 group to some of nucleotides) (Perry, 1976).

(*v*) Similarly, the mRNAs of prokaryotes do not contain poly (A) at 3' end, whereas the mRNA of eukaryotes contains 200-250 poly (A) residues at 3' end.

4.   The Transfer RNA (tRNA) or Soluble RNA (sRNA)

Twenty different amino acids required for protein synthesis are present in cytoplasm. Before joining an appropriate amino acid together to form protein, they are activated by attaching to the RNA. Requirement of energy for activation is met from ATP. The RNA which is capable to transfer an amino acid from amino acid pool, possesses capacity to combine with only one amino acid in the presence of an enzyme, *aminoacyl tRNA synthetase*, and recognises the codon of mRNA, is called tRNA or sRNA. For each amino acid there is different tRNA. It is likely that 20 different tRNAs are present in cytoplasm. However in several cases more than one type of tRNA for each amino acid is present. Therefore, there are more tRNAs than the amino acids. For example, about 100 types of tRNAs are found in bacterial cell.

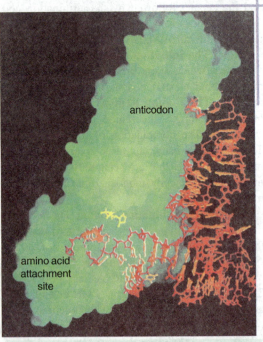

Computer - generated three-dimensional model of one type of tRNA molecule.

The mRNA, contain codes of each of three nucleotides called codon which specifies a single amino acid (see Chapter 7, *The Genetic code*). The tRNA molecules read the coded message on mRNA. Therefore, the tRNA molecules act as interpreter of genetic code.

(*i*) **Structure of tRNA:** The tRNA is dissolved in cytoplasm and is too small to be precipitated even at $1,00,000\,g$. Its molecular weight ranges from 25,000 to 30,000 D and sedimentation coefficient is 3.85. It accounts for 10-20% of the total cytoplasmic RNA. These are synthesized in the nucleus on a DNA template by only 0.025% of total DNA at the end of cleavage stage. As the tRNA synthesis is over on DNA template, a part of ribonucleotide (5'CCA3') is added to the 3' end of each molecule regardless of amino acid affinity, by an enzyme tRNA phosphorylase. It is the special feature of tRNA as compared to mRNA and rRNA. The tRNA molecules contain about 70-93% nucleotides arranged in a single strand at 5→3' end. This sstRNA forms double strands at certain regions with a single stranded loop. The 3' end terminates with –CCA sequence and the 5' end with G or C.

In addition to bases A, G, C and U present in tRNA, certain unusual bases are also found. These unusual bases are absent in other RNAs. The unusual bases are formed by specific chemical modification such as addition of methyl ($-CH_3$) group to form 3-methyl cytosine or l-methylguanosine, deamination of adenosine to inosine, reduction of uracil to dihydrouracil or rerrangement of uracil into pseudouracil. The other unusual bases are methylguanine (Gme), dimethylguanine (Gme2), methylcytosine (Cme), ribothymine (T), pseudouridine (ψ), dihydrouridine (DHU, H_2U, UH_2), inosine (I) and methylinosine (Ime). Most of the bases pair according to Watson and Crick's model but unusual bases do not because of bringing about changes due to substitution or alterations in those positions that take part in hydrogen bonding. The unusual bases protect the tRNA molecules from break down by RNAase. Consequently several non-base paired loops are formed in tRNA.

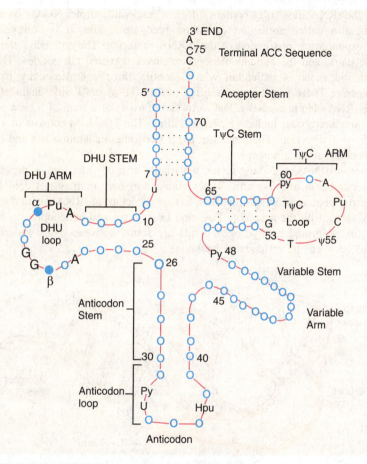

Fig. 5.13 : A clover leaf structure for yeast tRNA. Pu, purine (A, G); Py, pyrimidine (C, U); ψ, pseudouridine; DHU, dihydrouridine; T, ribothymidine. The circle represent the variable bases.

Clover Leaf Model of tRNA : For the first time R. Holley (1968) prepared the clover leaf model for yeast tRNA alanine (tRNA^{ala}) which includes several known functions of tRNA. This model has been well accepted. A typical clover leaf model is shown in Fig. 5.13 which reveals the following features.

(i)   The single polynucleotide chain of all the tRNA molecule is folded upon itself to form five arms e.g. acceptor arm, DHU arm, anticodon arm, variable arm and TψC arm. An arm consists of a stem and a loop. Except the acceptor arm, the other arms consist of their respective stem and loop.

(ii) The acceptor stem consists of 7 base pairs and 4 unpaired bases, the unpaired bases contain a three -CCA bases and a forth variable purine (A or G) at 3' end or polynucleotide chain. The last residue, adenylic acid (A) acts as amino acid attachment site. The 5' end of tRNA contains either (G) or (C).

(iii) The DHU (dihydrouridine) loop constitutes 7-12 unpaired bases, and acts as the site for recognition of amino acid activating enzyme aminoacyl tRNA synthetase. It consists of a total of 15-18 nucleotides (3-4 base pairs and 7-11 unpaired bases) in the loop. It has two variable regions α and β on both sides of guanine residues. These two regions contain 1-3 nucleotides, often pyrimidines.

(*iv*) All the tRNA molecules contain different nucleotide triplet codons on anticodon loop. It is also called anticodon or codon recognition site. It is complementary to the corresponding triplet codon of the mRNA molecule. The anticodon stem consists of 5 base pairs, and the anticodon loop contains 7 unpaired nucleotides. The middle three nucleotides act as anticodon which identify three complementary bases of mRNA molecule. There is a hyper modified purine (HPu) on 3' side chain of anticodon.

(*v*) The tRNA also possesses a TψC arm that consists of a stem of 5 base pairs and a loop of 7 unpaired bases including pseudouridine. The TψC loop consists of a TψC sequence at 5' → 3' direction. The TψC arm has a ribosome recognition site and binds the tRNA molecules to the ribosome.

(*vi*) In some tRNAs with long chain, a variable arm of extra arm is present between the anticodon arm and Tψ arm. The variable arm may or may not contain a stem.

The electron photomicrograph reveals a tertiary structure of tRNA where the different limbs are separately formed by the acceptor, TψC and DHU arms and anticodon arm are visible (Fig. 5.14). These limbs formed by hydrogen bonds are found between bases and ribose-phosphate backbone, and between the residues of backbone.

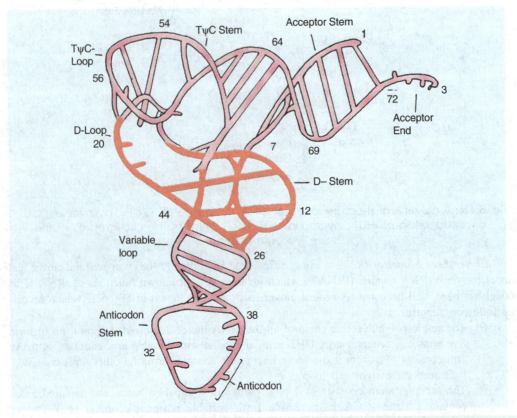

Fig. 5.14 : The three dimensional structure of tRNA.

The tRNA that initates protein synthesis is called initiator tRNA. The initator tRNA of eukaryotes differs from the prokaryotes. The tRNA specifies methionine as the starting amino acid in eukaryotic protein synthesis and *N*-formyl methionine in prokaryotes. Therefore, the two tRNAs specific to these two amino acids are methionyl tRNA (tRNA *f-met*). These two tRNAs differ from each other.

C. Replication of DNA

One of the most important properties of DNA is that it forms its additional identical copies. The process of forming its replica copy is called replication. Replication is the basis of evolution of all morphologically complex forms of life. Howard and Pelc (1953) demonstrated that in eukaryotes replication occurs during interphase between mitotic cycles and also during interphase of meiosis. During interphase of cell division the number of DNA molecules doubles which at anaphase is separated into two daughter cells, and thus equal number of chromosomes is maintained.

However, replication does not occur during entire anaphase but is confined only to synthesis (S) phase. There is a post-mitotic gap (G1) between the telophase and S phase. A second premitotic gap (G2) is between the S phase and prophase. Only S phase involves replication process. The G1 phase is most variable and in many eukaryotic cells it is completed within 3 to 4 hours or even months depending on physiological conditions. Mostly DNA synthesis is accomplished in 7 to 8 hours. In bacteria growing at log phase, DNA synthesis occurs from the time a cell originates to give rise to two daughter cells. It is noteworthy that bacteria divide only through binary fission.

In general, DNA carries out two important functions such as heterocatalytic function and autocatalytic function. The heterocatalytic function is protein synthesis directed by DNA, and autocatalytic function is the synthesis of its own DNA into replica copies. However, replication of DNA in prokaryotes differs from that of eukaryotes.

I.   Watson and Crick's Model for DNA Replication

Each chain of double helix acts as template and is envolved in replication of DNA. Watson and Crick proposed that the hydrogen bonds between the base pairs of two strands are broken and separated from each other. Each purine and pyrimidine base of the strands forms hydrogen bonds with complementary free nucleotides to be involved in polymerization in the cell. The free nucleotides form phosphodiester bonds with deoxyribose residue resulting in formation of a new polynucleotide molecule (Fig. 5.15). This model of Watson and Crick for DNA replication was later on verified experimentally.

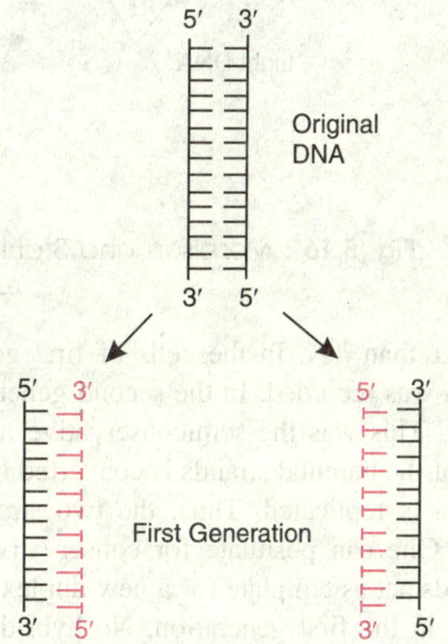

Fig. 5.15 : Semiconservative model for DNA replication as proposed by Watson and Crick.

1.   Experimental Evidence for Watson and Crick's Model for DNA Replication

M. Meselson and F. Stahl (1958) provided the experimental support for Watson and Crick's model for semiconservative nature of DNA replication which is called Meselson-Stahl experiment. They grew *E. coli* cells in medium containing heavy isotopic nitrogen (^{15}N) for several generations. They obtained a population of *E. coli* that contained totally ^{15}N - labelled DNA. Density was measured by density gradient centrifugation in CsCl containing ethidium bromide. Density of ^{15}N-DNA was heavier (1.722 g/cc) than the norrmal DNA (1.708 g/cc). Again the cells of *E. coli* were grown on medium containing less dense isotopic nitrogen (^{14}N) and was allowed to multiply several times. After the first generation, DNA was extracted which was found to be hybrid of ^{15}N-^{14}N (Fig. 5.16). This strand is ^{15}N and the other ^{14}N. It was neither heavier than ^{15}N nor

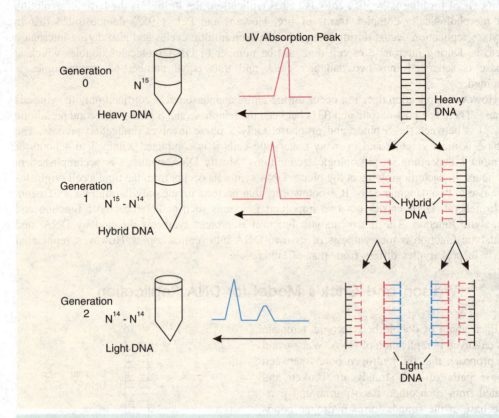

Fig. 5.16 : Meselson and Stahl's experiment demonstrating semiconservative replication of DNA.

lighter than ^{14}N. In the cells of first generation 50% ^{14}N - DNA, and 50% hybrid (^{15}N - ^{14}N) DNA was recorded. In the second generation the ratio of normal and hybrid DNA molecules was 3 : 1. This was the semiconservative nature of DNA replication because in the first generation one of the parental strands is converted into progenies and the other complementary polynucleotide strand is replicated. Thus, the two strands were the hybrid.

One can postulate for conservative mode of DNA replication i.e. both the original DNA strands act as template for a new duplex but is not separated, and results in an old and new double helix in the first generation. No hybrid ^{15}N – ^{14}N-DNA is formed. Therefore, this model could not be supported by Meselson-Stahl experiment.

Also, this experiment does not support for dispersive mode of DNA replication in which model there is no pattern of replication. The parental strands break randomly at several points during replication. Each segment will replicate and rejoin randomly. This results in varying amount of old and new DNA molecules in daughter cells. After first generation instead of a single ^{15}N- ^{14}N hybrid, a wide spectrum of DNA densities are detected. Therefore, the dispersive mode of replication is also ruled out through Meselson and Stahl experiment.

2.   Enzymes Involved in DNA Replication

Both the prokaryotic and eukaryotic cells contain three types of nuclear enzymes that are essential for DNA replication. These enzymes are nucleases, polymerases and ligases.

(*i*) **Nucleases:** The polynucleotide is held together by 3'→5' phosphodiester bonds. The nucleases hydrolyse the polynucleotide chain into the nucleotides. It attacks either at 3' OH end or 5' phosphate end of the chain. The nucleases are of two types (Fig. 5.17-B).

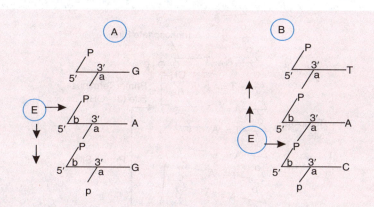

Fig. 5.17 : Exonuclease action on a polynucleotide chain. A, action in 5'→3' direction; B, action in 3'→5' direction; a 3' OH side of phosphodiester linkage; b, 5' side of phosphodiester linkage.

(*a*) **Exonucleases :** The nuclease that attacks on outer free end of the polynucleotide chain is called exonuclease. It breaks phosphodiester bond either in 5'→3' direction (A) or in 3'→5' direction (B). The enzyme moves in either cases stepwise along the chain and removes nucleotides one by one. Thus, the whole chain is digested.

(*b*) **Endonucleases :** The endonucleases attack within the inner portion of one or the double strands. Therefore, a nick is made on double stranded DNA molecule. However, if the polypeptide chain is single stranded (e.g. in DNA viruses), the attack of endonuclease will render the chain into two pieces. On double stranded DNA the nick contains two free ends that in turn act as template for DNA replication. Apart from this, the nicked double helix is distorted due to rotation of free molecules around its intact strand (Goodenough and Levine, 1974).

(*ii*) **DNA Polymerases:** DNA polymerases carryout the process of polymerization of nucleotides and formation of polynucleotide chain. This enzyme is called replicase when it replicates the DNA molecules and inherited by daughter cells. In 1959, for the first time A. Kornberg discovered an enzyme in *E. coli* which polymerized the deoxyribonucleotide triphosphate on a DNA template and produced complementary strand of DNA. This enzyme was called DNA polymerase. Later on it was named as Kornberg polymerase or Kornberg enzyme after the name of discoverer, for demonstrating *in vitro* polymerization of DNA. For the catalysis of polymerization, it requires the four deoxyribonucleotide triphosphates e.g. dATP, dGTP, dTTP and dCTP, a DNA template, a primer for initiation of catalytic activity and Mg^{++} (Fig. 5.18). In prokaryotes, three types of DNA polymerases e.g. polymerase I (Poly-I), polymerase II(Pol II), and DNA polymerase III (Pol III) are found, whereas in eukaryotes three or four polymerases termed as α, β and γ polymerases and mitchondrial (mt) DNA polymerase are present.

The molecular weight of α and γ polymerases are over 100,000 and that of β-polymerase is 30,000-50,000. The α and β polymerases are located in the nucleus. The β-polymerase copies a poly (A) or poly (C) template. The γ-polymerase copies many polyribonucleotides such as poly (A), poly (C), etc. The mtDNA polymerase is like γ-polymerase.

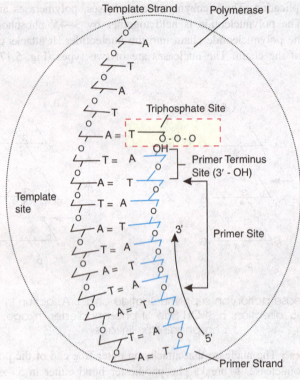

Fig. 5.18 : Diagram of DNA polymerase I of *E. coli.*

(*a*) **Polymerase I (Pol I) :** The Kornberg polymerase is known as Pol I. It is a single peptide chain with a molecular weight of 109,000 D. It is the largest single chain of globular protein known so far. One atom of zinc (Zn) per chain is present, therefore, it is metalloenzyme. In *E. coli,* approximately 400 molecules of Pol I are present.

Early experiments carried out by Kornberg revealed that when artificially synthesized DNA template strands alternating A and T *i.e.* poly d(AT) were incubated with polymerase and four radio-labelled nucleoside triphosphate, radioactive DNA containing alternating A and T was synthesized. Though sufficient amount of dGTP and dCTP was present in the solution but these were not synthesized into DNA because the DNA strand contained only poly dAT. This emphasizes that Pol I synthesizes only complementary copy of the template.

Shape of Pol I has been studied through electron microscope. It is roughly spherical of about 65 Å diameter (Fig. 5.18) which gets attached regularly to the DNA chain. Pol I possesses several attachment sites such as (*i*) a template site for attachment to the DNA template, (*ii*) a primer site of about 100 nucleotides contemporary to a segment of RNA on which the growth of newly synthesized DNA occur, (*iii*) a primer terminus site containing a terminal 3'OH group at the tip, and (*iv*) a triphosphate site for matching the incoming nucleoside triphosphates according to complementary nucleotide of DNA template.

Function: Pol I plays a significant role in polymerization (synthetic) as well as degradation (exonucleolytic) process of nucleotides, Pol I is broken by trypsin into two fragments, a large fragment (MW 75,000) and a small fragment (MW 36,000). The large fragement shows 3'→5' exonuclease activity, and the small fragement shows 5'→3' exonuclease activity. In *E.coli* the following three types of functions of Pol I have been found.

Polymerization: Polymerization is a process of synthesis in 5'→3' direction of short segments of DNA chain from deoxyribonucleoside triphosphate monomers to the 3' –OH end of a DNA strand. It is not the main polymerization enzyme because it cannot synthesize a long chain. It synthesizes only a small segment of DNA. It binds only to a DNA and forms nick in dsDNA. Therefore, it takes part in repair synthesis. In *E.coli* Pol I polymerizes the nucleotides at the rate of 1,000 nucleotides per minute at 37°C. The chief enzyme associated with polymerization is known as polymerase III.

Exonuclease activity

3'→5' exonuclease activity : Pol I catalyses the breaking of one or two DNA strands in 3'→5' direction into the nucleotide components i.e. the nucleotides are set free in 3'→5' direction which is reverse to polymerization direction. Therefore, it is called 3'→5' exonuclease activity. Pol I corrects the errors made during the polymerization, and edits the mismatching nucleotides at the primer terminus before the start of strand synthesis. Therefore, the function of Pol I is termed as repair synthesis.

5'→3' exonuclease activity: Pol I also breaks the polynucleotide chain in 5'→3' direction with the removal of nucleotide residues. Upon exposure of DNA to the ultraviolet light two adjacent pyrimidines such as thymines are covalently linked forming pyrimidine dimers. These dimers block the replication of DNA. Therefore, removal of pyrimidine dimers e.g. thymine dimers (T=T) is necessary. Through 5'→3' exonuclease activity, Pol I removes pyrimidine dimers. Secondly, DNA synthesis occurs on RNA primer in the form Okazaki fragments. Through 5'→3' exonuclease activity Pol I removes RNA primer and seals the gap with deoxyribonucleotides. Its onward movement results in removal of ribonucleotides from the front portion followed by of deoxyribonucleotides behind it.

(*b*) Polymerase II (Pol II): For several years Pol I was considered to be responsible for replicating in *E.coli.* but the work done during 1970s made it clear that Pol I is associated only with repair synthesis and the other enzymes, Pol II and Pol III are involved in polymerization process. Pol II is a single polypeptide chain (MW 90,000) that shows polymerization in 5'→3' direction of a complementary chain. It also shows exonuclease activity in 3'→5' direction but not in 5'→3' direction. The polymerization activity of Pol II is much less than Pol I in *E.coli* cells. About 50 nucleotides per minute are synthesized. *E.coli* cells contain about 40 Pol II molecules. The 3'→5' exonuclease activity of Pol II shows that it also plays a role in repair synthesis or DNA damaged by U.V. light just like Pol I. In the absence of Pol I, it can elongate the Okazaki fragments. Therefore, Pol II is an alternative to Pol I.

(*c*) Polymerase III (Pol III) : DNA polymerase III is several times more active than Pol I and Pol II enzymes. It is the dimer of two polypeptide chains with molecular weight 1,40,000 and 40,000 D respectively. Pol III polymerises deoxyribonucleoside triphosphates in 5'→3' direction very efficiently. Therefore, Pol III is the main polymerization enzyme that can polymerize about 15,000 nucleotides per minutes in *E. coli.* Like Pol II, it cannot polymerize efficiently if the template DNA is too long but can do when ATP and certain protein factors are present. Synthesis of a long template also occurs when an auxillary protein DNA (copolymerase II) is linked with Pol III and produced Pol III-co Pol II complex. In addition Pol III also shows 3'→5' exonuclease activity like Pol II. The 5'→3' exonuclease activity is absent. All the polymerases e.g. Pol I, Pol II and Pol III show 3'→5' exonuclease activity, whereas besides Pol I, the other two polymerases (Pol I and Pol II) lack 5'→3' exonuclease activity. However, some workers have shown both 3'→5' and 5'→3' exonuclease activity in Pol III.

(*iii*) DNA Ligases: The DNA ligases seal single strand nicks in DNA which has 5'→3' termini. It catalyses the formation of phosphodiester bonds between 3'-OH and 5'-PO$_4$ group of a nick, and turns into an intact DNA. There are two types of DNA ligases : *E. coli* DNA ligase and T4 DNA ligase. The *E. coli* DNA ligase requires nicotinamide adenine dinucleotide (NAD⁺) as co-factor, whereas T4 DNA ligase uses ATP as cofactor for joining reaction of the nick (Lehman, 1974) (Fig. 5.19).

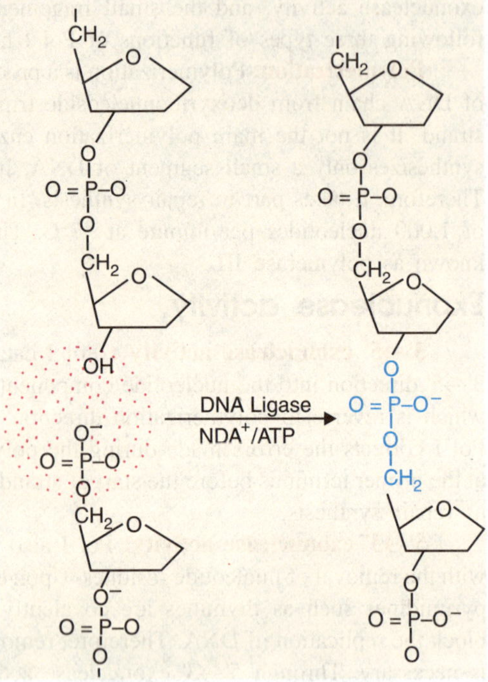

3.   Mechanism of DNA Replication

A. Kornberg (1992) has nicely discussed the DNA replication. In *E. coli* DNA replication has been investigated most extensively. It was thought that in eukaryotes probably similar mechanism operates. However, it has been found that in *E. coli* replication always starts at a very unique site called the origin. In *E. coli* the replicating apparatus contains an enzyme complex at the point where DNA thread is attached. Through this replicating point DNA thread moves and replication is accomplished. In eukaryotes enzyme moves along the DNA thread. It has earlier been described that *E. coli* possesses three types of DNA polymerases, each reads DNA template in 3'→5' direction and catalyses the synthesis of DNA in 5'→3' direction. The polymerases read deoxyribonucleotide triphosphates (dATP, dGTP, dCTP, dTTP) as substrate and a DNA template. To the 3' end of growing point, the nucleotides are added after interaction of 3'-OH end of deoxyribose with alpha (first) phosphate group of substrate releasing pyrophosphate as below.

Fig. 5.19 : Action of DNA ligase in the presence of NAD⁺/ATP.

$$\text{dNTPs (dATP, dGTP, dCTP, dTTP)} \xrightarrow[\text{DNA template}]{\text{DNA polymerase}} \text{DNA} + n\text{PPi}$$

Before the replication begins, DNA double helix must be unwounded to give rise to single strand. Unwinding process occurs very rapidly to form a fork that rotates about 75-100 revolutions per second. The unwinding process is facilitated by helicases. Some important genes and proteins associated with replication are given in Table 5.7. Overall DNA replication is accomplished in the following stages (Fig 5.20)

(i) **Unwinding of Double Helix :** Helicases are responsible for unwinding of double helix. They use energy from ATP to unwind short stretches of helix just ahead the replication fork (Matson *et al.*,1990). After separation of strand it is very necessary to keep them single stranded through single stranded DNA binding proteins (SSB). The SSB is a tetramer with each of four subunits of a molecular weight of 18,500 – 22,000. It may bind as a binding sites of 8-10 nucleotides (Meyer and Lain, 1988) (Fig. 5.20). However there is possibility of leading tension and formation of super coils in helix. The relieving of tension and promotion of unwinding process are done by the enzyme topoisomerases which transiently break one of two strands in such a way that it remains unchanged. It ties or unties a knot in DNA strand. DNA gyrase is one of the *E. coli* topoisomerases that removes super coiling of DNA during replication. Thus there is formation of a ssDNA template.

(ii) **DNA Replication:** DNA replication is accomplished in several steps. The first step is the RNA-primer synthesis on DNA template near origin of replication. Synthesis of RNA primer is very necessary because during DNA replication there is chance of more error in initial laying

down of first few nucleotides to pre-existing DNA template. DNA Pol I and Pol II cannot synthesize DNA without an RNA primer, therefore a special RNA polymerase called primase synthesizes an about 10 nucleotide long short primer. Before priming, preprimer intermediate is formed with the help of six prepriming proteins e.g. dnaB, dnaC, n, n', n'' and i proteins. For the synthesis of primer, primase needs several accessary proteins which combine with primase. The complex of primase-accessory protein is called *primosome*. Therefore, DNA Pol III holoenzyme starts synthesis of DNA in 5'→3' direction at the end of RNA primer (Fig. 5.20B).

Table 5.7 : Some important genes and proteins involved in DNA replication (modified after Moat and Foster, 1995)

Proteins	Genes	Map Location	Mol. wt.	Function
—	oriC	83.5		Origin of replication
Protein I	dnaT	99		Priming
Protein n	priB	95		"
Protein n'	priA	89		"
Protein n''	pri C	10		"
DnaA	dnaA	82	54,000	Initiation, binds oriC
DnaB	dnaB	91	55,000	Mobile promoter, helicase prepriming
DnaC	dnaC	99	25,000	Forms dnaB-dnaC complex
Pol III (a)	dnaE	4	129,000	DNA Pol III holoenzyme, elongation
Primase	dnaG	67	60,000	Priming, RNA primer synthesis
γ-subunit	dnaZx	10	47,500	Synthesis, part of gamma complex
Helix-destabilizing	ssb-1	91	20,000	Single strand binding SSB protein
Helix-unwinding	rep	85	—	Strand separation
E subunit	dnaQ	5	27,000	Proof reading
Dna Pol I	polA	85	109,000	Gap filling, primer degradation
Ligase	lig	52	75,000	Ligation of single stranded nicks in phosphodiester backbone
Dna gyrase (α)	gyrA	48	105,000	Super twisting
(β)	gyrB	82	95,000	Super coil relaxation
Dna Pol II	polB	2	120,000	?
Dna helicase I	—	—	180,000	Unwinding
Dna helicase II	—	—	75,000	Unwinding

The second step is chain elongation. A new DNA strand starts synthesizing by addition of deoxyribonucleoside triphosphates to the 3' end of last nucleotide of RNA primer. DNA synthesis occurs in 5'→3' direction catalysed by the replisome. The replisome has two DNA Pol III holoenzyme complex. It is a very large complex containing DNA Pol III and several proteins. The γ and β-subunits bind the holoenzyme to the DNA template and primer. The α-subunit synthesizes the DNA. One polymerase continuously copies the leading strand (i.e. a strand growing in the direction of replication fork and showing continuous replication). The lagging strand (*i.e.* a strand growing in opposite direction of replication fork and showing discontinuous replication of strand) loops around replisome continuously. There is formation of Y shaped replicating fork at the point where two strands are separated.

On leading strand DNA synthesis occurs continuously because there is always a free 3'-OH at the replication fork to which a new nucleotide is added. But on the opposite strand called lagging strand, DNA synthesis occurs discontinuously because there is no 3' -OH at the replication fork to which a new nucleotide can link. On this strand there is free 3' -OH at the opposite end away from growing point. Therefore, on lagging strand a small (11 bases long) RNA primer must be synthesised by primase to provide free 3' -OH group.

The replication of DNA has two directions, one direction (unidirectional replication) and both the directions (bidirectional replication) from the point of origin. The bidirectional replication is found in most of the bacteria (e.g. *E. coli, Bacillus subtilis, Salmonella typhimurium,* etc.), whereas unidirectional replication occurs in *E. coli* bacteriophages (P$_2$ and 186) and mtDNA of mouse LD cells (Gefter, 1975).

Finally, about 1000-2000 nucleotides long fragment in bacteria and about 100 nucleotides long fragment in eukaryotic cells are synthesized. These fragments are called the Okazaki fragments after the name of a Japanese discoverer, R. Okazaki (Fig. 5.20 B).

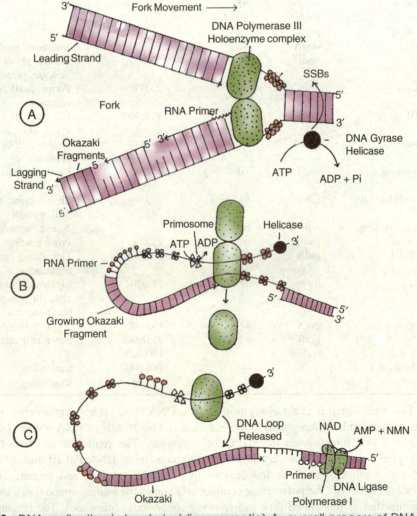

Fig. 5.20 : DNA replication in bacteria (diagrammatic) A, overall process of DNA replication; B, action of replisome, helicases and primosome, and looping of lagging strand around polymerase III; C, completion of Okazaki fragments, release of lagging strand and sealing the gap by DNA ligase (based on Prescott *et al.*, 1996).

(iii) Removal of RNA Primer and Completion of DNA Strand: When the Okazaki fragments are formed, most of lagging strands become duplicated. The RNA primer is removed by DNA Pol I or RNase H. DNA polymerase I synthesizes a short segment of complementary DNA to seal the gap. Possibly Pol I removes the primer nucleotide at a time and replaces it with suitable complementary deoxyribonucleotide (Fig. 5.20)

(iv) Joining of Fragments : At the end, the fragments are joined by DNA ligase that forms a phosphodiester bond between 3'-OH end of growing strand and 5' end of an Okazaki fragment (Fig. 5.20 C). Reaction of DNA ligases is given in Fig 5.19. In mutants defective ligase is produced; therefore, joining of Okazaki fragments is greatly improved.

E. coli DNA ligase derives energy from NAD. It is first adenylated by AMP moiety of NAD releasing the nicotinamide mononucleotide (NMN) (Kornberg,1992). The adenylated ligase reacts with ssDNA having a nick and forms phosphodiester bond. The complete reaction is as below

$$E.\ coli\ \text{ligase} + \text{NAD} \longrightarrow \text{ligase - AMP} + \text{NMN}$$

$$\text{Ligase - AMP} + \text{DNA (with break)} \longrightarrow \text{phosphodiester} + \text{ligase} + \text{AMP}$$

$$\text{AMP} + \text{NMN} \longrightarrow \text{NAD}$$

Obviously, DNA replication is a very complex process. If any error is made during replication, it leads to mutation. *E. coli* makes error about 10^{-6} per gene per generation. The DNA Pol I and Pol III act as proof reader of the newly formed DNA. These move along new DNA synthesized, read mistakes formed due to improper base-pairing and correct those through $3' \rightarrow 5'$ exonuclease activity. Despite all these complexity, replication takes place rapidly in bacteria (750-1,000 base pairs per second) and much slower in eukaryotes (50-100 base pairs per second).

4. Models for DNA Replication

The pattern of DNA replication in prokaryotes differs from that of eukaryotes. These differences are due to the nature of prokaryotic DNA. For example, the circular DNA of *E. coli* replicates at a replication point, the origin. In others the mechanism is different. Some of the models for DNA replication is discussed below.

(i) The Cairns Model for DNA Replication: J.Cairns (1963) was the first to visualize the replicating chromosome of *E. coli* through autoradiographic study. This study revealed that the replicating DNA thread got fixed at a specific site called origin, and was moving in one direction and within a replicating fork where the original strands are synthesized (Fig. 5.21). Further studies have shown that in circular DNA (Fig. 5.22A) the two strands denatured at origin site (B). There is bidirectional DNA synthesis i.e. after initiation there appeared two growing points travelling in opposite directions around the circular DNA molecule (C). Growing points proceed with unwinding of DNA double helix. The process of unwinding creates a torque that is transmitted to the unreplicated part of the DNA molecule resulting in formation of super helix or super twist (D). Super coil prevents its further replication. A temporary

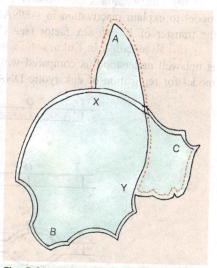

Fig. 5.21 : Diagrammatic presentation of autoradiographic observation of E.coli. K12 DNA replication as observed by Cairns (1963). A, B and C are the three section of chromosome arising at two forks, X and Y.

nick is made on one of the strands by a swivelling protein (w) which contracts this effect. The nick allows the parental strand for their free rotation on each other and finally freed. At the end swivelling protein seals the nick to continue the replication process (E). Replication process goes on and the two growing point converge on the terminus (F).

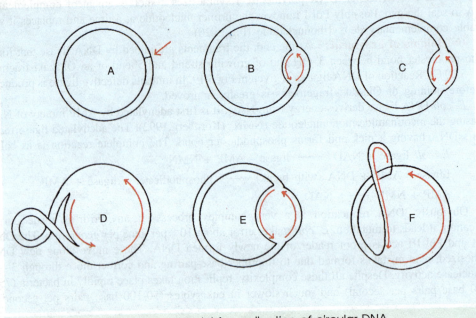

Fig. 5.22 : Cairns model for replication of circular DNA.

(ii) **The Rolling Circle Model :** Gilbert and Dressler (1969) described the rolling circle model to explain reactivation in ssDNA viruses e.g . Ø × 174 (*see* Chapter 15, *Viruses-I*) and the transfer of *E. coli* sex factor (*see* Chapter 8 : *Microbial genetics*).

(iii) **Replication in Eukaryotic Chromosome:** DNA replication in eukaryotic chromosome is not well understood as compared with prokaryotic chromosome. However, the well accepted model for replication of eukaryotic DNA is the bidirectional model (Fig. 5.23). DNA synthesis

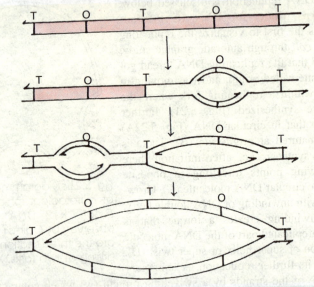

Fig. 5.23 : A bidirectional model for replication of mammalian chromosome. O, origin of replication site; T, termination site.

starts at a mid point of replication unit which is called initiation point (O-origin) (Fig. 5.23 A-B), and progresses in both the directions until reaches the terminal point (T) (C). The replication fork meets at T point (D) on the entire chromosome. There may be thousands of initiation points.

D. Nucleic Acids as Genetic Materials

(i) Evidences from Bacteria: For the first time, an English Health officer, Frederick Griffith (1928) gave an experimental evidence that the DNA was the genetic material. He took two types of a bacterial strain, pneumococci (*Streptococcus penumoniae*) that causes pneumonia in humans and other animals. There were two types of pneumococi, type II and type III. Each type exists in two forms RII, SII, and RIII, SIII forms where 'R' represents the rough, non-capsulated and non-virulent form and 'S' represent the smooth, encapsulated and virulent form.

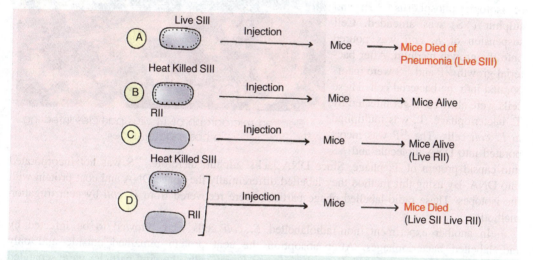

Fig. 5.24 : Griffith's transformtion experiment.

After injection of smooth (virulent) strain, mice were killed (Fig. 5.24) When boiled smooth strain was injected, the mice were not affected and no pneumococci could be recovered from the mice. Upon injection of a mixture of heat killed smooth strain and live rough (non-virulent) strain, the mice were killed and live virulent bacteria could be recovered from blood of dead mice. Griffith called this change of non-virulent strain into virulent strain as transformation, because the virulent strain transformed the non-virulent strain into the virulent strain. This is called Griffith's transformation experiment (Fig. 5.24). The phenomenon of bacterial transformation is called "Griffith effect".

Identification of transforming substance : Griffith could not identify the nature of transforming substance. Oswald T. Avery, C.M. Mac Leod and M.J. Mc Carty (1944), set out experiments to identify the transforming principle. They destroyed cell constituents in extract of virulent pneumococci SIII using enzyme that hydrolyzed DNA, RNA, proteins and polysaccharides. They treated the non-virulent strain separately with bacterial extracts as below :

RII Cells + Purified SIII cell polysaccharide	⟶	R colonies
RII Cells + Purified SIII cell protein	⟶	R colonies
RII Cells + Purified SIII cell RNA	⟶	R colonies
RII Cells + Purified SIII cell DNA	⟶	SIII COLONIES
RII Cells + Purified SIII cell extract + protease	⟶	SIII colonies
RII Cells + Purified SIII cell extract + RNase	⟶	SIII colonies

They concluded that a cell free and highly purified RNA extract of SIII strain could bring about transformation of RII strain into SIII strain. However, this effect was lost when the extract was treated with deoxyribonuclease. Therefore, DNA is the genetic material and carries sufficient information for transformation.

Thereafter, the transforming material (DNA) was also confirmed in several bacteria such as *Bacillus subtilis, Haemophillus influenzae, Shigella paradysenteriae,* etc.

(ii) Evidence from Bacteriophages: Alfred D. Hershey and Martha Chase (1952) carried out several experiments in bacteriophage T_2 that proved the DNA to be the genetic material of T_2. They prepared a culture medium for the growth of *E. coli* that contained phosphorus and sulphur. To this culture medium known amount of isotopic phosphorus (^{32}P) and sulphur (^{35}S) was amended. Cell suspension of *E. coli* was poured onto the growth medium. After bacterial growth ^{32}P and ^{35}S were incorporated into the bacterial cell. These cells were allowed to get infected by T_2 bacteriophage. T_2 was multiplied in *E. coli* cells. The ^{32}P was incorporated into DNA molecule and ^{35}S

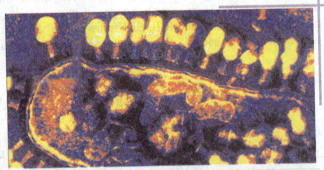

Electron micrograph of T4 virus particles infecting a host bacterial cell.

into capsid protein of T_2 phage. Since DNA lacks sulphur, therefore, ^{35}S was not incorporated into DNA. By using this method they labelled differentially the phage DNA and coat protein with the isotopes. These radio-labelled phage particles were recovered from *E. coli* by centrifugation method.

In another experiment, non-radiolabelled *E. coli* cells were allowed to be infected by radiolabelled bacteriophages. After absoption the coat protein remained outside and only radiolabelled DNA was injected. From the infected *E. coli* cells, the viral particles were separated from the host through agitation. The amount of ^{32}P and ^{35}S in bacteriophage, particles, *E. coli* cells and medium was estimated. The ^{32}P was estimated with the bacterial cells and ^{35}S with protein coat left outside the bacterial wall i.e. the growth medium. This experiment shows that the genetic material of T_2 phage resides in DNA but not in their protein.

For this work A.D. Hershey shared Nobel Prize in medicine for 1969 with M. Delbruck and S.E. Luria. Furthermore, this experiment was repeated by several other workers by taking different bacteriophages. In all the cases, DNA was found to give evidence of genetic material.

(iii) Evidence from Bacterial Conjugation: Conjugation is one of the methods of transfer of DNA from donor bacterium to the recipient bacterium. Lederberg and Tatum (1946) demonstrated when F$^+$ (bacterium containing fertility factor i.e. male) strain of *E. coli* cells are mixed with F$^-$ (cells devoid of fertility factor i.e. female), the F$^-$ cells were converted to F$^+$ cells.

$$F^+ \text{ cells} + F^- \text{ cells} \longrightarrow F^+ \text{ cells}$$

The F$^+$ factor is a segment of plasmid DNA (*see* conjugation, Chapter 8, *Microbial Genetics*). It is evident from this experiment that DNA is the genetic material.

(iv) Evidence from RNA viruses: In 1956, A. Gierer and S. Schramm put example that when healthy tobacco leaves are inoculated by RNA purified from tobacco mosaic virus, lesions developed on healthy leaves which have like those obtained from tobacco mosaic plants.

Like Hershey and Chase, another experiment was doue by H.F. Conract and B. Singler in 1957. They separated RNA from protein coat of TMV and reconstituted them. One group contained parental RNA and protein from mutant TMV, and the other had RNA from the mutant TMV and protein coat from the parental TMV. The reconstituted viruses were allowed to infect healthy tobacco leaves. After infection from the lesions the TMV particles were recovered. In all the cases the progeny TMV particles contained parental RNA type, but not parental protein type.

QUESTIONS

1. What is the chemical nature of DNA? Give an illustrated account of arrangement of different constituents.
2. With the help of suitable diagrams give Watson and Crick's model of double helix of DNA.
3. Discuss in detail the different forms of DNA.
4. Write an essay on organization of DNA in eukaryotes.
5. What is the chemical structure of RNA ? Write in brief different types of RNA found in a cell.
6. Give an illustrated account of differences in prokaryotic and eukaryotic mRNA.
7. Discuss in detail mechanism of DNA replication.
8. Give the experimental evidences for nucleic acids as genetic material.
9. Write in detail the function of different types of DNA polymerases.
10. Write shortnotes on the following :

   (i)   Circular DNA   (ii)   Palindromic DNA
   (iii) Nucleotides,   (iv)   tRNA,
   (v)   mRNA,   (vi)   Single stranded DNA,
   (vii) DNA polymerases   (viii) Nucleases
   (ix) The rolling circle model   (x)   Cairns model for DNA replication.

REFERENCES

Ankenbauer, R.G.1977. Reassessing forty years of genetic doctrine : retrotransfer and conjugation . *Genetics* 145 : 543-549.

Avery O.T., Macleod C.M. and McCarty, M.1944. Studies on the chemical nature of the substance including transformation of pneumococci types. Induction of transformation by a deoxyribonucleic acid fraction isolated from *Pneumococcus* type III.) *Exp.Med.* 79:137.

Bachmann, 1990 Linkage map of *Escherichia coli* K-12 edition 8. *Microbiol Rev.* 54 : 130-197.

Bukhari,A.I.; Shapiro,J.A. and Adhya,S.L. 1977. DNA insertion elements, plasmids and episomes. *Cold spring Harbor Lab.* 782 pp.

Campbell, A *et.al.* 1977. Nomenculature of transposable elements in prokaryotes. In *"DNA Insertion Elements, plasmids and Episomes"* (eds.Bukhari et.al.). *Cold spring Harbor Lab.*, New York.

Clewell,D.1992. Bacterial conjugation. Plenum Press,New York.

Davis B.D.C. 1950 Non-filterability of the agents of genetic recombination in *Escherichia coli J.Bact.* 60:507.

Dressler, D. and Potter, H. 1982. Molecular mechanism in genetic recombination. *Ann.Rev.Biochem.* 51 : 727-761.

Dubnau,D. 1991 Genetic competence in *Bacillus subtilis. Microbiol.Rev.* 55 : 395 - 424.

Ebel, T. Sipis, J.Botstein, *et.al.* 1972 Generalized transduction by phage P22 in *S.typhimurium.* I. Molecular origin of transducing DNA. *J.Mol. Biol.* 71 : 433.

Echols, H. and Court, D. 1971. The role of helper phage in gal transduction. In The *bacteriao phage lambda* (ed. A.D. Hershey),. *Cold spring Harber,* pp 701

Freifelder,D. 1987a. *Moleculer Biology* 2nd ed. Jones are Bartlett pub, Boston.

Freifelder D.1987b. Microbial genetics. Jones & Bartlett Publ. Boston P.601.

Griffith,F.1928. The significance of pneumococcal types *J. Hyg.* 27:113.

Heffrom,F.; McCarthy, B.J.; Ohtsubo,H. and Ohtsubo, E. 1979. DNA sequence analysis of the transposon Tn3: three genes and three sites involved in transportion of Tn3. *Cell* 18 : 1153 - 1163.

Holliday,R. 1974. *Genetics :* 78 : 273.

Jacob, F. and Wollman, E.L. 1961. *Sexuality and the Genetics of Bacteria.* Academic Press : New York.

Kleckner,N.1981.Transposable elements in prokaryotes.*Ann.Rev.Genetics* 15:341-404.

Kowalczykowski,S.C. 1991. Biochemistry of genetic recombination : energetics and mechanism of DNA strand exchange. *Annu Rev.Biophys.Chem.* 20.539-575.

Landy, A.1989.Dynamic, structural and regulatory aspects of lambda site-specific recombination. *Annu-Rev.Biochem*.58 : 913 . 949.

Lederberg J. and Tatum E.L. 1946. Gene recombination in *Escherichia coli. Nature* : 158 : 588.

Lewin,B.1994. *Genes*, Oxford Univ. Press, Oxford p.1257.

Lohman,T.M.1992. *Escherchia coli* DNA helicases : mechanism of DNA unwinding. *Mol. Microbiol*. 6:5-14.

Luria S.E. and Delbruck, M. 1943. Mutation of bacteria from virus sensitivity to virus resistance. *Genetics* 28 : 491.

Mergeay, M. *et.al*.1985. Back transfer : a property of some broad host range plasmids. In *Plasmids in Bacteria* (ed. D.Helinski,S.N.Cofen, D.B. Clewe; D.A.Jackson and A.Hollaender) Plenum press, New york, pp.942.

Mizuuchi,K. 1992a. Transpositional recombination : mechanistic insights from studies of *mu* and other elements. *Annu. Rev Biochem*. 61 : 1011-1051.

Mizuuchi, K.1992 b. *J.Biol Chem*.267 : 21273 -21276.

Moat. A.G. and Foster, J.W. 1995. *Microbial Physiology* : 3rd ed. Wiley-Liss, New York.

Morse, M.L. Lederberg,E.L. and Lederberg.J. 1956. Transduction in *Escherichia Coli.Genetics:* 41:142.

Notani,K. and Setlow,J.K. 1974. Mechanism of bacterial transformation and transfection. *Progr. Nucleic Acid Res. & Mol. Biol*. 14:39.

Prajapati, J.B. and Batish 1988. Trasposable elements-DNA segments that jump : some new horizons. *Everyman's Sci:* 23 (6):198-207.

Radman, M. 1989. Mismatch repair and fidelity of genetic recombination. *Genome* 31:68-73.

Rayssiguier, C.; Thaler,D.S. and Radman,M. 1989. The barrier to recombination between *Escherichia coli* and *Salmonella thyphimurium* is disrupted in mismatch repair mutants. *Nature* 342 (6248):396-401.

Roca,A.I. and Cox, M.M. 1990. The recA protein. *Crit. Rev. Biochem. Mol. Biol*. 25:415-456.

Smith, G.R. 1988. Homologous recombination in prokaryotes. *Microbiol. Rev*. 52:1-28.

Smith,G.R. 1989. Homologous recombination in *E.coli*: multiple pathways for multiple reasons. Cell 58 :807-809.

Shapiro,J.A. 1983. *Mobile Genetic Elements*, Academic Press Inc., Orlands.

Thiry,G.; Mergeary,M. and Faflen,M. 1984. Back mobilization of Tra$^+$ Mob$^+$ plasmids mediated by various IncM, IncN and Inc P1 plasmids. *Arch. Int. physical Bio-chem*. 92 : 6465.

Tope,E. *et. al,* 1992. Determination of the mechanism of retrotransfer by mechanistic mathematical modelling. *J. Bacteriol.* 174: 5953-5960.

Willets, N. and Wilkins,B. 1984. Processing of plasmid DNA during bacterial conjugation. *Microbiol. Rev.* 48:24-41.

Willetts, N. and Skurray, R. 1980. The conjugation system of F$^-$ like plasmids. *Annu Rev. Genet.* 14:41-76.

Whitehouse, H.L.K. 1982. Genetic recombination: understanding the mechanisms. Wiley : New York.

Wollman, E.L. *et.al.* 1956. Conjugation and genetic recombination in *Escherichia coli* K-12. *Cold spring harbor Symp. Quant. Biol.* 21:141.

Zinder, N.D. and Lederberg, J. 1952. Genetic exchange in *Salmonella. J. Bact.* 64:679.

Genes : Concept and Synthesis

6

A. Gene Concept
B. Artificial Synthesis of Gene

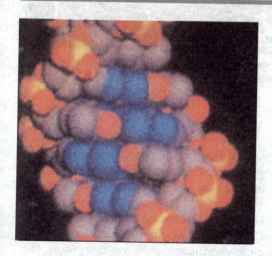

A British biologist, Richard Dawkins in his book *The Selfish Gene* (1989) has stated about the gene "they are in you, and in me; they created us, body and minds and their preservation is the ultimate rationale for our existence, they,.... go by the name of genes, and we are their survival machines". In this chapter the modern concept of gene has been discussed.

A. Gene Concept

Although the role of hereditary units (factors) in transfer of genetic characters over several generations in organisms was advocated by Gregor John Mendel, yet the mystery of the 'hereditary units' was unravelled during early 1900s. In 1909, W. Johanson coined the term 'gene' that acts as hereditary units. However, early work done by several workers proposes various hypotheses to explain the exact nature of genes. In 1906, W. Bateson and R.C. Punnet reported the first case of linkage in sweet pea and proposed the *presence or absence theory*. According to them the dominant character has a determiner, and the recessive character lacks determiner. In 1926, T.H. Morgan discarded all the previous existing theories and put forth the particulate gene theory. He thought that genes are arranged in a linear order on the

T.H. Morgan

chromosome and look like beads on a string. In 1928, Belling proposed that the chromosome that appeared as granules would be the gene. This theory of gene was well accepted by the cytologists. In 1933, Morgan was awarded Nobel prize for advocating the theory of genes.

After the discovery of DNA as carrier of genetic informations, the Morgan's theory was discarded. Therefore, it is necessary to understand both, the classical and modern concepts of gene.

According to the classical concepts a gene is a unit of (*i*) physiological functions, (*ii*) transmission or segregation of characters, and (*iii*) mutation. In 1969, Shapiro and co-workers published the first picture of isolated genes. They purified the *lac* operon of DNA and took photographs through electron microscope.

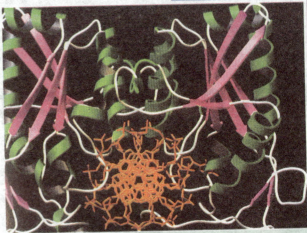

First experimental proof for the role of DNA as a genetic material was given by Avery, MaCleod and Mc Carty.

Avery, MaCleod and Mc Carty (1944) gave the first experimental proof for the role of DNA as genetic material. Therefore, the presence of genes was supposed on DNA. However, in some viruses like TMV, retroviruses, revoviruses, etc. the chemical nature of genes is RNA but not DNA.

In 1908, the British physician Sir E.R. Garrod first proposed one-gene-one product hypothesis. In 1941, G.W. Beadle and E.L. Tatum working at St. Standford university clearly demonstrated one-gene-one enzyme hypothesis based on experiments on *Neurospora crassa*. They made is clear that genes are the functional units and transmitted to progenies over generations; also they undergo mutations. They treated *N. crassa* with X-rays and selected for X-ray induced mutations that would have been lethal. Their selection would have been possible when *N. crassa* was allowed to grow on nutrient medium containing vitamin B_6. This explains that X-rays mutated vitamin B_6 synthesing genes. They concluded that a gene codes for the synthesis of one enzyme. In 1958, Beadle and Tatum with Lederberg received a Nobel prize for their contribution to physiological genetics.

I. Units of a Gene

After much extensive work done by the molecular biologists the nature of gene became clear. A gene can be defined as *a polynucleotide chain that consists of segments each controlling a particular trait*. Now, genes are considered as a unit of function (cistron), a unit of mutation (mutan) and a unit of recombination (recon).

1.   Cistron

One-gene-one enzyme hypothesis of Beadle and Tatum was redefined by several workers in coming years. A single mRNA is transcribed by a single gene. Therefore, one-gene-one mRNA hypothesis was put forth. Exceptionally, a single mRNA is also transcribed by more than one gene and it is said to be polycistronic. Therefore, the concept has been given as one-gene-one protein hypothesis. The proteins are the polypeptide chain of amino acids translated by mRNA. Therefore, it has been correctly used as one-gene-one polypeptide hypothesis.

Moreover, genes are present within the chromosome and their *cis-trans* effect govern the function. Therefore, S. Benzer termed the functional gene as *cistron* (Fig. 6.1 A). Crossing over within the functional genes or cistron is possible. The cis and trans arrangement of alleles may be written as below:

$$\frac{+\ +}{a\ b}\qquad\qquad \frac{a\ +}{+\ b}$$

Cis (wild)   Trans (mutant)

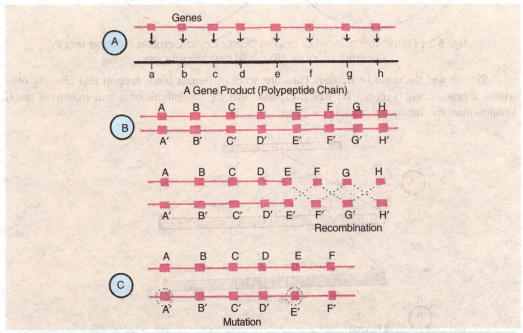

Fig. 6.1 : The genes as a unit of function i.e. cistron (A), recombination *i.e.* recon (B) and mutation i.e. muton (C)

2. Recon

Earlier, it was thought that crossing over occurs between two genes. In 1962, Benzer demonstrated that the crossing over or recombination occurs within a functional gene or cistron. In a cistron the recombinational units may be more than one. Thus, the smallest unit capable of undergoing recombination is known as *recon* (Fig. 6.1B).

Benzer (1955) found that the cultures of T4 bacteriophage formed plaques on agar plates of *Escherichia coli*. Normally T4 formed small plaques of smooth edges, whereas the mutant T4 phage formed the larger plaques of rough edges (Fig. 6.2). The DNA molecule of T4 phage consists of several genes one of which is called rII region. Formation of rough edged plaques was governed by two adjacent genes (cistrons rIIA and rIIB) in mutant bacteriophage (Fig. 6.3 A). Both the regions function independently and consist of 2,500 and 1,500 nucleotides, respectively. In rIIA gene over 500 mutational sites are present where crossing over may occur. Through crossing over exchange of two segments of DNA occurs. If crossing over takes place within the gene by mating two rII mutant of T4 phage, a normal wild type phage can be produced.

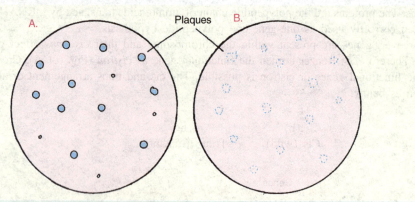

Fig. 6.2 : Formation of normal and smooth edged plaques (A) and rough edged plaques (B) by T4 bacteriophage.

Similar was the result of Benzer. Thus, the work of Benzer lends support that crossing over within a gene occurs (Fig. 6.3 B) which explains that the recombinational unit (recon) is much smaller than the functional unit i.e. cistron.

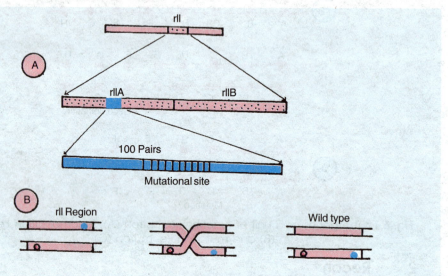

Fig. 6.3 : Diagrammatic presentation of rII region of T4 bacteriophage (A), and demonstration of crossing over within the genes (B).

3. Muton

Benzer (1962) coined the term *muton* to denote the smallest unit of chromosome that under goes mutational changes. Hence, muton may be defined as 'the smallest unit of DNA which may be changed is the nucleotide'. Thus, changes at nucleotide level are possible (Fig. 6.1C). The smallest unit of muton is the nucleotide. Therefore, cistron is

The White-Eye Mutation in *Drosophila*.

the largest unit in size followed by recon and muton. This can be explained that a gene consists of several cistron, a cistron contains many recon, and a recon a number of mutons. However, if the size of a recon is equal to muton, there would be no possibility in recon for consisting of several mutons.

II. Genetic Diversity

There is genetic diversity among the microorganisms. The total number of genes in one microorganism differs, from that in the other. For example, bacteriophage R17 and QB consist of only three genes. SV40 contains 5-10 genes; phage λ consists of about 49 genes, and the other complex viruses possess about 250 genes. *E. coli* contains about 4,000 genes on about 1mm long chromosome.

On the other hand the length of each gene varies in base pair composition. It may vary between 100 and 3,000 nucleotides giving the complexity to proteins.

III. Split Genes

During 1970, in some mammalian viruses (e.g. adenoviruses) it was found that the DNA sequences coding for a polypeptide were not present continuously but were split into several pieces. Therefore, these genes were variously named as split genes or introns (Gilbert, 1978), *interrupted genes* or *intervening sequences* (Lewin, 1980), inserts (Weismann, 1978), *Junk DNA*. For the discovery of split genes in adenoviruses and higher organisms, Richards J. Roberts and Phillip Sharp were awarded Nobel prize in 1993.

As shown in Fig 6.4 a DNA sequence codes for mRNA but the complete corresponding sequence of DNA is not found in mRNA. Certain sequences of DNA are missing in mRNA. The sequences present in DNA but missing in mRNA are called intervening sequences or *introns*, and the sequences of DNA found in RNA are known as *exons*. The *exons* code for mRNA. For the first time W. Gilbert used the term *introns* and *exons*.

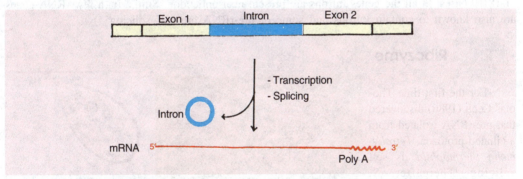

Fig. 6.4 : The split genes have exons separated by introns. Removal of introns through RNA splicing.

After transcription a limited RNA transcript has the intron. Genes coding for rRNA and tRNA may also be intervened. The introns are also found in some eubacteria, cyanobacteria and archaeobacteria.

For some time it was not certain how mRNA is synthesized from a DNA containing introns. Some possible explanations for the mechanism of mRNA synthesis were given: (*i*) DNA rearrangement occurs during transcription with the removal of introns, (*ii*) during transcription RNA polymerase skips the introns and transcribes only exons, (*iii*) individual exon transcribes separately and rejoins to form the complete mRNA, and (*iv*) RNA polymerase may synthesize both introns and exons, and processing of transcripts occurs later on. The transcripts corresponding to introns are removed. Later on it was shown that the fourth mechanism operates in transcription of mRNA.

1.   RNA Splicing

In the initial stage, RNA transcript introns are synshesized which are removed later on by a process called RNA splicing (Fig. 6.4). The junctions of intron-exon have a GU sequences at the intron's 5'-end, and an AG sequence at its 3'OH end. These two sequences are recognised by the special RNA molecules known as small nuclear RNA (snRNA) or snurps (Steitz, 1988).

These together with proteins form small nuclear ribonucleoprotein particles called snRNPs. Some of the snRNPs recognize the splice junctions and splice introns accurately. For example, the UI-snRNP recognizes the 5'-splicing junction, and the U5 snRNP recognizes the 3' splicing junction. Consequently pre-mRNA is spliced in a large complex called a spliceosome (Guthrie, 1991). The spliceosome consists of pre-mRNA, five types of snRNPs and non-snRNP splicing factors (Rosbash and Seraphin, 1991).

Robert and Sharp, the Nobel prize winner in 1993, independently hybridized the mRNA of adenovirus with their progeny or DNA segments of virus. The mRNAs hybridized the ssDNA of virus where the complementary sequences were present. The mRNA-DNA complexes were observed under electron microscope to confirm which part of viral genome had produced the mRNA strand. It was found that mRNA did not hybridize DNA linearly but showed a discontinuous complex pattern. Huge loops of unpaired DNA between the hybridized complexes clearly revealed the large chunk of DNA strand that carried no genetic information and did not take part in protein synthesis. The adenovirus mRNA contained four different regions of the DNA.

The β-globin genes of mice and rabbits, and tRNA genes of yeast tyrosine-tRNA consists of eight genes three of which have been studied in detail. Each gene contains 14 bases (ATTT-AYCAC-TACGA) as intron in the middle. In the same way the pre-tRNA genes contain introns of 18-19 bases. In all the genes introns are present near anticodon. Similarly, a few rRNA genes are also known to contain introns and some of pre-rRNA are self–splicing.

2.   Ribozyme

For the first time Thomas Cech (1986) discovered that pre-rRNA isolated from a ciliated protozoa, *Tetrahymena thermophila,* is self splicing. Thereafter, S. Altman showed that ribonuclease cleaves a fragment of pre-tRNA from one end, and also contains a piece of RNA. This RNA fragment catalyses the splicing reaction i.e. acts as enzyme. Therefore, this RNA segment catalyzing the splicing reaction is called ribozyme. For this discovery

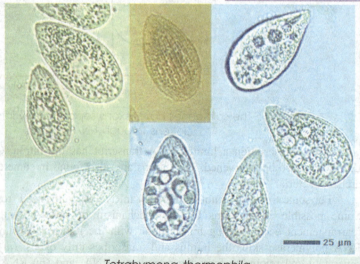

25 µm

Tetrahymena thermophila.

Cech and Altman were awarded the Nobel prize in 1989 in chemistry. The best studied ribozyme activity is the self-splicing of RNA. This process is wide spread and occurs in *T. thermophila* pre-tRNA, mito chondrial rRNA and mRNA of yeast and other fungi, chloroplast tRNA, rRNA and mRNA, and mRNA of bacteriophage.

The rRNA intron of *T. thermophila* is 413 nucleotide long. The self-splicing reaction needs guanosine and is accomplished in three steps: (*i*) the 3'-G attacks the 5' group of introns and cleaves the phosphodiester bond, (*ii*) the new 3'-OH group on the left exon attacks the 5'-phosphate on right exon. Consequently two exons join and remove the intron; and (*iii*) the 3'-OH of intron attacks the phosphate bond of nucleotide 15 residues from its end releasing the terminal fragment and cyclizing the intron (Cech, 1986).

3. Evolution of Split Genes

Before the discovery of split genes in 1977, all the genes analysed in detail were the bacterial genes. Bacteria were considered to resemble with the simpler cell from which eukaryotes must have been evolved. Now, it is supposed that split genes are the ancient condition and bacteria lost their introns only after evolution of most of their proteins. Evidence for the ancient origin of introns has been obtained by the examination of the gene that encodes the ubiquitous enzyme, triose phosphate isomerase (TPI). The TPI is coded by a gene that contains six introns (in vertebrates), five of these are present at the same position as in maize. This shows that five introns were present in the gene before evolution of eukaryotes about 10^9 years ago (Nyberg and Cronhjort, 1992).

The TPI plays a key role in cell metabolism that catalyses the interconversion of glyceraldehyde 3-phosphate and dihydroxy acetone phosphate- a central step in glycolysis and glucogenesis. By comparing this enzyme in various organisms it appears that the TPI evolved before the divergence of prokaryotes and eukaryotes from a common ancestor cell *progenote* (Nyberg and Cronhjort, 1992).

The unicellular organisms under a strong selection pressure minimised the superfluous genome in their cell, whereas there was no such pressure on multicellular organisms. That is why *Aspergillus* has five introns and *Saccharomyces* has none. Precise loss of introns would have occurred by deletion in prokaryotes. The loss of introns requires the exact rejoining of DNA coding

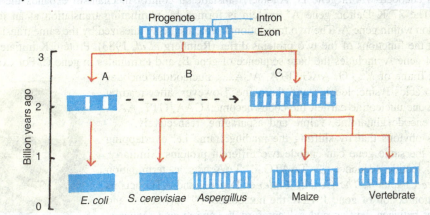

Fig. 6.5 : Outline of evolution of a particular gene. Introns were present in progenote before the evolution of archaeobacteria, eubacteria and eukaryotes. A, eubacteria that formed mitochondria and chloroplasts; B, approximate time of the endosymbiotic events that gave to mitochondria and chloroplast, and C, anaerobic eukaryotes.

sequence. The most likely source of the information is needed for such event in an mRNA transcript of the original gene from which introns are to be removed. The mRNA may be copied back into DNA by reverse transcriptase. The recombination enzymes allow the DNA copies to become paired with the original sequence resulting in intronless form by a gene-conversion type of event. This pathway of intronless has been demonstrated in laboratory in *S. cerevisiae*.

Fig. 6.5 shows an outline how a particular gene evolved? The TPI is thought to be evolved to its final three dimensional structure before eubacteria, archaeobacteria and eukaryotic lineage split off from a common ancestor cell i.e. progenote (Nyberg and Cronhjort, 1992).

IV. Overlapping Genes (genes within genes)

In 1940s, Beadle and Tatum proposed one-gene-one protein hypothesis which explains that one gene encodes for one protein. However, if one gene consists of 1,500 base pairs, a protein of 500 amino acids in length would be synthesized. In addition, if the same sequence read in two different ways, two different amino acids would be synthesized by the same sequence of base pairs. It means, the same DNA sequence can synthesize more than one proteins at different time. It was realized for the first time when the total number of proteins synthesized by ØX174 exceeded from the coding potential of the phage genome. A similar phonemenon is found in the tumour virus SV40 where the total molecular weight of proteins (*i.e.* VP1, VP2 and VP3) synthesized by SV40 genes is much more than the size of the DNA molecule (5,200 base pairs i.e. 1,733 codons). From these observations the concept of overlapping genes has emerged.

For the first time Barrell *et al.* (1970) gave the evidence for the possibility of the above fact based on the overlapping genes found in bacteriophage ØX174. This virus contains an icosahedral capsid with a knob at each vertex enclosing a single stranded circular DNA (see Fig 16.9). Sanger *et al.* (1977) mapped the whole nucleotide sequence of phage ØX174 and phage G4 DNA. Barrell *et al.* (1976) have found the sequences of genes D, E and J, and B, C to overlap in the whole sequence of ØX174.

The ØX174 strand is made up of 5,386 nucleotides of known base sequences. If a single reading frame was used, about 1,795 amino acids would be encoded in the sequence and with an average protein size of about 400 amino acids, only 4-5 proteins could be made. In contrast, ØX174 makes 11 proteins containing a total of more than 2,300 amino acids. The genes A and B have been characterised by Weisbeek *et al.* (1977). The sequence of gene A is now known to contain all of gene B. Gene B is translated in a different reading frame from gene A. Similarly gene E is encoded within gene D. Another translational control mechanism expands the use of gene A. The 37 K Dalton gene A* protein is formed by reinitiating translation at an internal AUG codon within gene A. The two translational proteins are synthesized by the same translational phase but the functions of the two proteins differ (Reinberg *et al*, 1983). Protein K initiated near the end of gene A, includes the base sequence of gene B, and terminates in gene C. For example, a reading frame of .... G, AAG, TTA, ACA...... nucleotides encodes the amino acids lysine, leucine and threonine. However, after reading the frame one nucleotide earlier, the codes become.. GAA, GTT, AAC, A... that encode glutamine, valine and asparagine, respectively.

It is obvious that by shifting the reading frame i.e. overlapping the code, the same gene can encode two different proteins. Similarly, in the nucleotide sequence .... TAATG...., TAA acts as termination codon (*see* Chapter 7, *The Genetic Code*) of D gene, and ATG acts as the initiation codon of gene J. Here the nucleotide 'A' between A and T overlaps between the two codes. Therefore, the amino acid sequence of A* is similar to a segment of protein A. Functions of ØX174 genes are given in Table 6.1. In addition, overlapping genes have also been detected in animal virus SV40, and tryptophan mRNA of *E. coli*.

Har Govind Khorana for the first time chemically synthesized gene in lab.

B. Artificial Synthesis of Gene

For the first time in 1955, Michelson chemically synthesized a dinucleotide in laboratory. Later on in 1970, Har Govind Khorana and K.L. Agarwal for the first time chemically synthesized gene coding for tyrosine tRNA of yeast. For the synthesis of tRNA and rRNA there are specific genes. However, genes of tRNA are the smallest genes containing about 80 nucleotides. In 1965, Robert W. Holley and coworkers worked out first the molecular structure of yeast alanine tRNA. This structure lent support to Khorana in deduction of structure of the gene. A gene is responsible for encoding mRNA, and mRNA for polypeptide chain. If the structure of a polypeptide chain is known, the structure of mRNA from genetic code dictionary and in turn the structure of gene can easily be worked out. There are two approaches for artificial synthesis of the gene, by using chemicals and through mRNAs.

I. Synthesis of a Gene for Yeast alanine tRNA

As mentioned earlier that the molecular structure of yeast alanine tRNA was worked out by R.W. Holley and coworkers in 1965, this information helped Khorana to deduce the structure of alanine tRNA. They worked out that yeast alanine tRNA contains 77 base pairs. It was very difficult to assemble 77 base pairs of nucleotides in ordered form. Therefore, they synthesized chemically the short deoxynucleotide sequences which was joined by hydrogen bonding to form a long complementary strand. By using polynucleotide ligase the double stranded pieces were produced. The complete procedure of synthesising gene for yeast alanine tRNA is discussed in the following steps.

1. Synthesis of Oligonucleotides

In the first approach, fifteen oligonucleotides ranging from pentanucleotide (i.e. oligodeoxynucleotide of five bases) to an icosanucleotide (i.e. oligodeoxynucleotide of twenty bases) were synthesized. The chemical synthesis was brought about through condensation between the –OH group at 3' position of one deoxynucleotide and the –PO$_4$ group at 5' position of the second deoxynucleotide. All other functional groups of deoxyribonucleotides not taking part in condensation processes were protected so that the condensation could be brought about. The deoxynucleotides were protected as below:

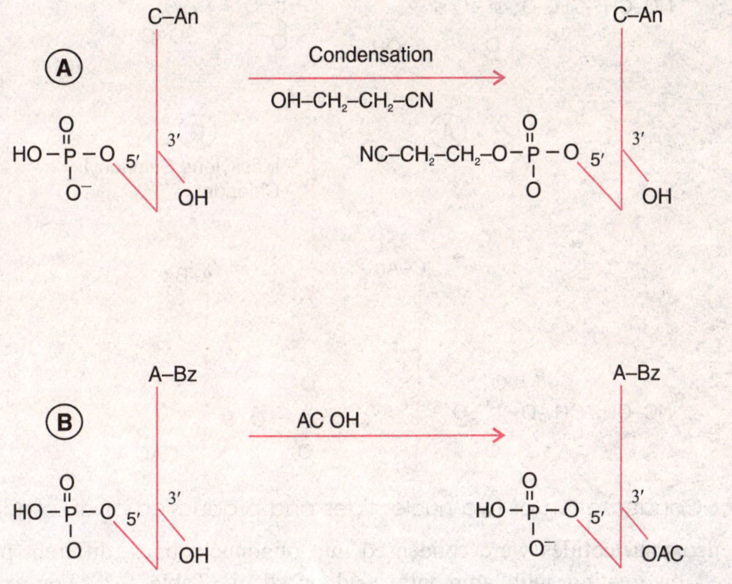

Fig. 6.6 : Protection of 5' –OH group due to condensation of cyanoethyl group (A) and 3'–OH group due to condensation of acetyl group (B) alanine–tRNA.

(*i*) The amino group of deoxyadenosine was protected by benzyl group (BZ), the amino group of deoxycytidine was protected by anisoyl group (An), and the amino group of deoxyguanosine was protected by isobutyl group. These protective groups were removed by treating with ammonia when synthesis was over.

(*ii*) The hydroxyl group (–OH) at the 5' position of receiving deoxynucleotide was protected by cyanoethyl group (HN-CH$_2$-CH$_2$-) (Fig. 6.6A).

(*iii*) The –OH group at the 3' position of second incoming deoxynucleotide was protected by using acetyl group (Ac) (Fig. 6.6B).

Table 6.1 : The φx174 genes and their function (after Moat and Foster, 1995).

Gene	Sequence (kb)	Function
A	56	RF replication, synthesis of viral strand
A*		Shut-off host DNA synthesis
B	13.8	Morphogenesis of capsid
C	9.4	DNA maturation
D	16.8	Capsid morphogenesis
E	9.9	Cell lysis
F	48.3	Major coat protein
G	19.1	Major spike protein
H	35.8	Major spike protein, adsorption
J	4.1	Core protein; DNA condensation
K	—	Unknown.

The different protecting groups used and treatment required to remove the protecting groups are given in Table 6.2. When the groups of deoxynucleotides were protected, the products of Fig. 6.7A and B reacted to form deoxyoligonucleotide as given in Fig. 6.7.

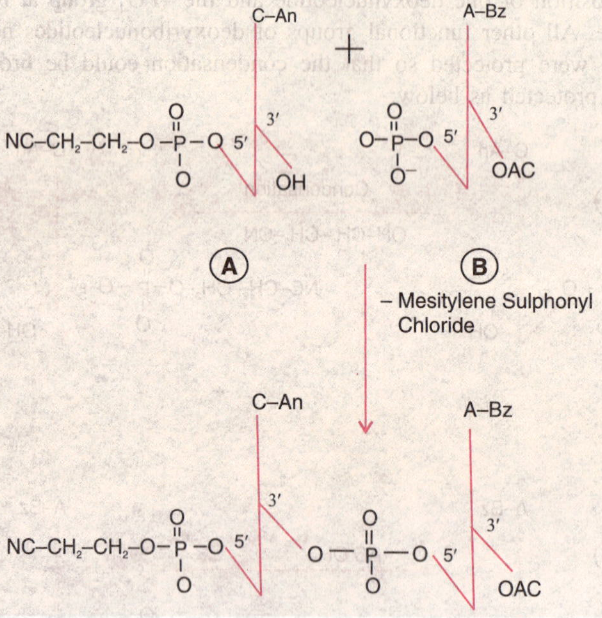

Fig. 6.7 : Condensation of two nucleotides and production of an oligonucleotide.

When deoxynucleotides were condensed into oligonucleotides, different protecting groups were removed by treating with ammonia, acid or alkali (Table 6.2). For example, both the cyanoethyl group at the 5' position and the acetyl group at the 3' position were removed by alkali treatment.

Finally, condensation between the groups of two, three or four nucleotides was brought about. The receiving segment had a free 3'-OH group and a protected 5'-OH group, whereas the incoming segment had a free 5'-OH group and a protected 3'-OH group. After each addition, the protective group at the 3' end had removed so that free 3'-OH group could receive another segment.

2. Synthesis of Three Duplex Fragments of a Gene

By using 15 single stranded oligonucleotides, three large double stranded DNA fragments were synthesized (Fig. 6.7). These three fragments contained (i) segment of A having the first 20 nucleotides with the nucleotides 17-20 as the single stranded, (ii) segment B consisting of nucleotides 17-50 with the nucleotides 17-20 and 46-50 as the single stranded, and (iii) segment C containing the nucleotides 46-77 with the single stranded region 46-50.

3. Synthesis of a Gene from Three Duplex Fragments of DNA

The three segments (A, B, C) synthesized as above were joined by using the enzyme polynucleotide ligase to produce the complete gene for alanine tRNA (Fig. 6.8). The joining of the three fragments was done by any of two following methods:

Table 6.2 : Different protective groups of nueleotides and their removal.

The groups in base or sugar base	Protected by	Protecting groups removed by
A. –NH₂ group (base)		
(i) deoxyadenosine	Benzyl group	Ammonia
(ii) deoxycytidine	Anisoyl group	Ammonia
(iii) deoxyguanosine	Isobutyl group	Ammonia
B. –OH group (sugar)		
(i) –OH group at 5' position of first nueleotide	Monomethoxy trityl group	Acid
(ii) –OH group at 5' position of growing chain	Cyanoethyl	Alkali
(iii) –OH group at 3' position	Acetyl group	Alkali

(i) In one approach, fragment A was joined to B by taking advantage of overlapping in nucleotide residues 17-20. Then, the fragment C was added with the overlap in nucleotides 46-50. Thus, a complete double stranded DNA with 77 base pairs was prepared.

(ii) In the second approach, the fragment B was joined to C. At the end the fragment A was added to nucleotide residues 17-20 to obtain the complete gene for alanine tRNA.

Khorana et al. (1970) prepared this gene in vitro which was used for future work. They

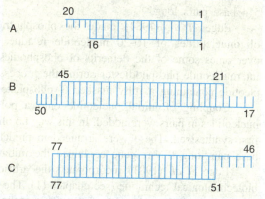

Fig. 6.8: The three duplex DNA fragments (A, B and C) were synthesised for the synthesis of the gene for yeast alanine tRNA.

found that alanine tRNA gene replicated and transcribed into tRNA just like the natural gene. It is not known whether tRNA prepared from artificially synthesized gene had the molecular organisation similar to alanine tRNA or not.

II. Artificial Synthesis of a Gene for Bacterial tyrosine tRNA

In 1975, Khorana and co-workers completed the synthesis of a gene for *E. coli* tyrosine tRNA precursor. *E. coli* tRNA precursors are formed from the larger precursors. The tyrosine tRNA precursor has 126 nucleotides. They synthesized the complete sequence of DNA duplex coding for tyrosine-tRNA precursor of *E. coli*. Though these segments are not the proper structural gene yet are the regions involved in its regulation.

Twenty six small oligonucleotide DNA segment giving rise to tRNA precursor was synthesized which were arranged into six double stranded fragments each containing single stranded ends. These six fragments were joined to give rise to complete gene of 126 base pairs for tyrosine tRNA precursor of *E. coli*.

Khorana (1979) completely synthesized a biologically functional tyrosine tRNA suppressor gene of *E. coli* which was 207 base pairs long and contained (*i*) a 51 base pairs long DNA corresponding to promotor region, (*ii*) a 126 base pair long DNA corresponding to precursor region of tRNA, (*iii*) a 25 base pair long DNA including 16 base pairs contained restriction site for Eco RI. This complete synthetic gene was joined in phage lambda vector which in turn was allowed to transfect *E. coli* cells. After transfection phage containing synthetic gene successfully multiplied in *E. coli*.

Khorana (1979) made the phosphodiester approach for synthesising the oligonucleotides of the biologically active tRNA. The demerits of this approach are: (*i*) the completion of reaction in long time, (*ii*) rapidly decrease in yield with the increase in chain length, and (*iii*) time taking procedure of purification.

III. Artificial Synthesis of a human leukocyte interferon gene

Interferons are proteinaceous in nature produced in human to inhibit viral infection. These are of three types secreted by three genes i.e. (*i*) leukocyte interferon gene (IFN-α gene), (*ii*) fibroblast interferon gene (IFN-β gene) and immune interferon gene (IFN-γ gene). In 1980, Weisman and coworkers published the nucleotide sequence of IFN-α gene. Taking advantage of this information Edge *et al.* (1981) successfully synthesized the total human interferon gene of 514 base pairs long.

Edge *et al.* (1981) made the phosphotriester approach in artificially synthesising 67 oligonucleotides of 10-20 nucleotide residue long segment. The phosphotriester approach overcomes some of the demerits of phosphodiester approach by blocking the function of each internucleotide phosphodiester during the process of synthesis. A completely protected mononucleotide containing a fully masked 3' phosphate triester group is used in this method.

Coupling of initial nucleotide onto a polyacrylamide resin was done to which further nucleotides in pairs were added. In this way 66 oligonucleotides of 14-21 nucleotide residues were first synthesized. These were arranged in predetermined ways and joined chemically. The 514 base pair long IFN-α gene contained the initiation and termination signals.

Edge *et al.* (1981) incorporated the artificially synthesized gene into a plasmid through biotechnological technique (see chapter 11). The recombinant plasmid was transferred into *E.coli* cells which expressed α-interferon. This technique now-a-days is being adopted to produce interferon commercially.

IV. Gene Machine

Recently, fully automated commercial instrument called automated polynucleotide synthesizer or gene machine is available in market which synthesizes predetermined polynucleotide sequence. Therefore, the genes can be synthesized rapidly and in high amount. For example, a gene for tRNA can be synthesized within a few days through gene machine. It automaticaly synthesizes the short segments of single stranded DNA under the control of microprocessor. The working principle of a gene machine includes (*i*) development of insoluble silica–based support in the form of beads which provides support for solid phase synthesis of DNA chain, and (*ii*) development of stable deoxyribonucleoside phosphoramidites as synthons which are stable to oxidation and hydrolysis, and ideal for DNA synthesis.

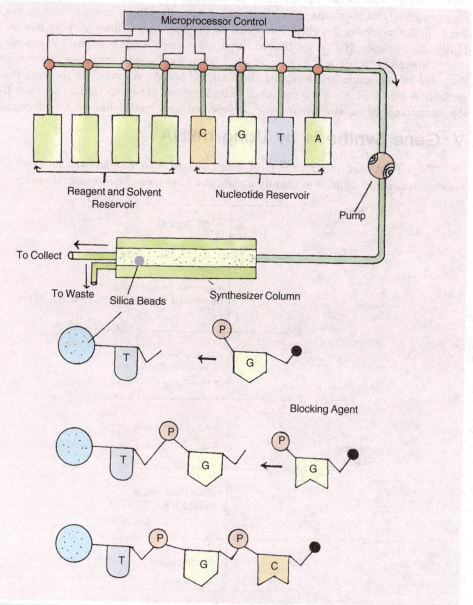

Fig. 6.9 : A gene machine and its working mechanism (diagrammatic)

The mechanism of a gene machine is shown in Fig. 6.9. Four separate reservoirs containing nucleotides (A, T, C and G) are connected with a tube to a cylinder (synthesiser column) packed with small silica beads. These beads provide support for assembly of DNA molecules. Reservoirs for reagent and solvent are also attached. The whole procedure of adding or removing the chemicals from the reagent reservoir in time is controlled by microcomputer control system i.e. microprocessor.

If one desires to synthesise a short polynucleotide with a sequence of nucleotides T, G, C, the cylinder is first filled with beads with a single 'T' attached. Thereafter, it is flooded with 'G' from the reservoir. The right hand side of each G is blocked by using chemicals from the reservoir so that its attachment with any other Gs can be prevented. The remaining Gs which could not join with Ts are flushed from the cylinder. The other chemicals are passed from the reagent and solvent reservoirs so that these can remove the blocks from G which is attached with the T. In the same way this cycle is repeated by flooding with C from reservoir into the cylinder. Finally, the sequence TGC is synthesized on the silica beads which is removed chemically later on.

The desired sequence is entered on a key board and the microprocessor automatically opens the valve of nucleotide reservoir, and chemical and solvent reservoir. In the gene machine the nucleotides are added into a polynucleotide chain at the rate of two nucleotides per hour. By feeding the instructions of human insulin gene in gene machine insulin has been synthesized.

V. Gene Synthesis by Using mRNA

The mRNAs are the transcripts of genes that have to be translated into polypeptides or proteins. It is rather difficult to identify a particular gene on a chromosome. However, it is easier

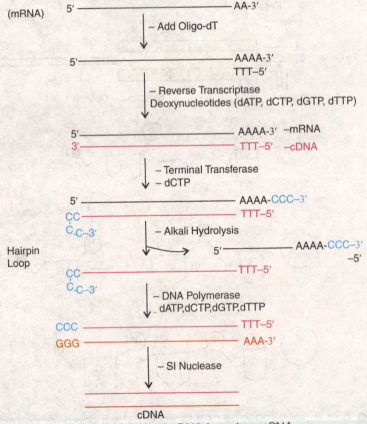

Fig. 6.10: Synthesis of cDNA by using mRNA.

to pick up the mRNAs and synthesize a gene because the total population of mRNA in a cell remains high. The majority of mRNA contains a long poly adenylated tract i.e. about 100 (A) residues at 3' terminus (Kates, 1970). Therefore, the mRNA can be separated from the rest of RNA population of the cell.

The mRNAs are passed through an oligo-dT cellulose affinity column (Fig. 6.10). The poly (A) binds to (T). The oligo-dT segment contains 10-20 nucleotides which hybridizes the poly (A) of mRNA. The oligo-dT provides primer at poly (A) region with a free 3'–OH end. The reverse transcriptase uses the free end and synthesizes a single stranded cDNA in the presence of dCTP, dGTP, dATP and dTTP, and results in mRNA-cDNA hybrid. At the end the enzyme forms a loop by using the last few bases as the template. This results in synthesis of a short 'hairpin' loop at 3'end of the cDNA (Leis and Hurwitz, 1972). The mRNA is degraded from mRNA-cDNA hybrid by using alkali. This phenomenon is known as hydrolysis. Consequently mRNA is separated.

In the next step the single stranded DNA acts as template for the synthesis of double stranded DNA in the presence of polymerase I and all the four deoxynucleotides. The hairpin acts as primer for chain elongation. Finally, a double stranded cDNA is synthesized. SI nuclease is used to cleave hairpin loop and result in double stranded cDNA (Fig. 6.10). For detail see *A Text Book of Biotechnology* by R.C. Dubey (2005).

In 1970, S. Mizutami, H.M. Temin and D. Baltimore discovered the RNA dependent DNA polymerase i.e. reverse transcriptase in retroviruses. This enzyme yields single stranded DNA on RNA template. For the discovery of reverse transcriptase R. Dulbecco, Temin and Baltimore were awarded Nobel prize in 1975.

Land *et al.* (1981) have given an improved method for cDNA synthesis. The single stranded cDNA complementary to mRNA is tailed with oligo-dC tail. This process is facilitated by using the enzyme terminal transferase and adding dCTP nucleotide. The tailing is followed by oligo-dG priming of the second strand synthesis. This checks the formation of hairpin loop. Therefore, SI nuclease is not required.

VI. The Polymerase Chain Reaction (PCR)

The polymerase chain reaction (PCR) provides a simple and ingeneous method for exponentially amplification of specific DNA sequences by *in vitro* DNA synthesis. This technique was developed by Kary Mullis at Cetus Corporation between 1983 and 1985. This technique has made it possible to synthesize large quantities of a DNA fragment without cloning it. The details of PCR techniques and its mechanism are described by Erlich (1989) in his edited book 'PCR Technology'. The PCR technique has now been automated and is carried out by a specially designed machine.

The PCR includes the following three essential steps to amplify a specific DNA sequence (Fig. 6.11):

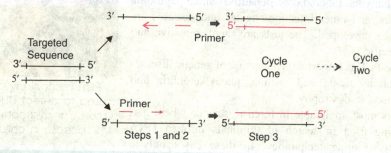

Fig. 6.11 : The working system of PCR. Cycle two follows the steps of cycle one.

(*i*) **Melting of Target DNA:** The target DNA containing sequence (between 100 and 5,000 base) to be amplified is heat denatured (around 94°C for 15 second) to separate its complementary strands (step 1). This process is called melting of target DNA.

(*ii*) **Annealing of Primers:** The second step is the annealing of two oligonucleotide primers to the denatured DNA strands. Primers are added in excess and the temperature lowered to about 68°C for 60 seconds; consequently the primers form hydrogen bonds i.e. anneal to the DNA on both sides of the DNA sequence (step 2).

(*iii*) **Primer Extension:** Finally, deoxynucleotide triphosphates (dATP, dGTP, dCTP, dTTP) and a thermostable DNA polymerase are added to the reaction mixture. The DNA polymerase accelerates the polymerization process of primers and, therefore, extends the primers (at 68°C) resulting in synthesis of copies of target DNA sequence (step 3). Only those DNA polymerases which are thermostable i.e. function at the high temperature are employed in PCR technique. For this purpose the two popular enzymes, Taq polymerase (of a thermophilic bacterium *Thermus aquaticus*) and the vent polymerase (from *Thermococcus litoralis*) are used in PCR technology. These enzymes exhibit relative stability at DNA-melting replenishment after each cycle of synthesis. Also it reduces the cost of PCR and allow automated thermal cycling.

However, after completion of step 3 (of one cycle) the targeted sequences on both strands are copied and four strands are produced. Now, the three step cycle (first cycle) is repeated which yields 8 copies from four strands.

Similarly, the third cycle produces 16 strands. This cycle is repeated about 50 times. Theoretically, 20 cycles (each of three steps) will produce about one million copies of the target DNA sequence, and 30 cycles will produce about one billion copies. In each cycle the newly synthesized DNA strands serve as targets for subsequent DNA synthesis as the three steps are repeated upto 50 times.

For the working of PCR about 10-100 picomoles of primers are required. The concentration of target DNA can be about 10^{-20} to 10^{-15} M (or 1 to 10^5 DNA copies per ml). The PCR machine can carryout 25 cycles and amplify DNA 10^5 times in 75 minutes.

The PCR technology has been improved in recent years. RNA can also be efficiently used in PCR technology. The *rTth* DNA polymerase can also be used instead of the Taq polymerase. The rTth polymerase will transcribe RNA to DNA, thereafter amplify the DNA. Therefore, cellular RNA and RNA viruses may be studied when they are present in small quantities.

Application of the PCR Technology

The PCR technology is extensively applied in the following areas (of molecular biology, medicines and biotechnology)

(*i*) Amplification of DNA and RNA

(*ii*) Determination of orientation and location of restriction fragments relative to one another.

(*iii*) Diagnosis of diseases and causal microorganisms. For example, PCR-based diagnostic tests for AIDS, Chlamydia, tuberculosis, hepatitis, human papilloma virus, and other infectious agents and diseases are being developed. The tests are rapid, sensitive and specific.

(*iv*) The PCR is important in detection of genetic diseases such as sickle cell anemia, phenylketonuria and muscular dystrophy.

(*v*) It is most applicable in forensic science where it is being used in search of criminals through DNA fingerprinting technology. In these cases only a small samples of biological materials are required.

PCR technology can be used to detect Duchenne muscular dystrophy (DMD) disease in children.

(*vi*) It is also applied in diagnosis of plant diseases. A large number of plant pathogens in various hosts or environmental samples are detected by using PCR, for example viroids (associated with hops, apple, pear, grape, citrus, etc.), viruses (such as TMV, cauliflower mosaic virus, bean yellow mosaic virus, plum pox virus, potyviruses), mycoplasmas, bacteria (*Agrobacterium tumifaciens, Pseudomonas solanacearum, Rhizobium leguminosarum, Xanthomonas compestris,* etc.), fungi (e.g. *Colletotrichum gloeosporioides, Glomus* spp.; *Laccaria* spp., *Phytophthora* spp., *Verticillium* spp.), and nematodes (e.g. *Meloidogyne incoginta, M. javanica,* etc.) (Henson and French, 1993).

QUESTIONS

1. What do you mean by a gene? Discuss in detail the different units of a gene.
2. Write an essay on split genes with emphasis on evolution of split genes.
3. Give an illustrated account of overlapping gene.
4. Discuss in detail about different approaches made for artificial synthesis of genes with emphasis on works of H.G. Khorana.
5. How will you synthesize a gene by using gene machine or mRNAs.
6. Write in detail the PCR technology and its applications.
7. Write short notes on the following:
   (*i*) Cistron, (*ii*) Recon, (*iii*) Muton, (*iv*) Split genes, (*v*) Gene machine, (*vi*) Overlapping genes. (*vii*) Har Govind Khorana, (*viii*) cDNA, (*ix*) Yeast alanin-tRNA, (*x*) PCR technology.

REFERENCES

Avery, D.T., Macleod, C.M. and Mc Carty, M. 1944. Studies on Chemical Nature of Substances inducing transformation of pneumococcal types. Induction of transformation by a deoxyribonucleic acid fraction isolated from pneumococcus Type III. *J. Exp* Med 79:137

Barrell, B.G. Air, G.M. and Hutchinson, C.A. 1976 Overlapping genes in the bacteriophage $\phi \times 174$. *Nature* 264 34-41.

Benzer, S. 1955, Fine Structure of a genetic region in bacteriophage. *Proc. Natl Acad, Sci.* 41;344.

Benzer, S. 1962. The fine structure of the gene. *Sci. Am.* January. 217 : 70-84

Cech T.R. 1986. RNA as an enzyme. *Sci : Am.* 255 : 64-75

Dubey, R.C. 1995. *A Text Book of Biotechnology.* S. Chand & Co. Ltd. New Delhi, p. 357.

Edge, M.D., Green, A.R. *et al.* 1981. Total Synthesis of a human leukocyte interferon gene. *Nature.* 292 : 756-762

Erlich, H.A. 1989. PCR *Technology.* Stocton Press, New York (U.S.A.)

Guthrie, C. 1991 Messenger RNA splicing in yeast: Clues to why the spliceosome is a ribonucleoprotein. *Science.* 253 : 157-163

Henson, J.M and French, R. 1993. The polymerase chain reaction and plant disease diagnosis. *Annu. Rev. Phytopath.* 31 : 81-109.

Kates, J. 1970. *Cold Spring Harb Symp. Quant Biol.* 35 : 743-752

Khorana, H.G. 1979. Total Synthesis of a gene. *Science.* 203:614-625

Land, H., Grey, M. Hanser, H et al . 1981 5'-terminal sequences of eukaryotic mRNA can be cloned with a high efticiency. *Nucleic Acids Res* : 9:2251-2266

Leis, J.P. and Hurwitz, J. 1972. *Proc. Natl Acad Sci,* (USA) 69:2331-2395

Moat, A.G. and Foster, S.W. 1995. *Microbial physiology.* Wiley - Liss, New yock P 560.

Nyberg, A.M. and Cronhjort, M.B. 1992. Intron evolution : a statistical comparison of two model. *J. Theor Biol* 157 : 175-190.

Reinberg, D. Zipursky, S.L; Weisbeek, P. Brown, D. and Hurtwitz, J. 1983. Studies on the synthesis of $\phi \times 174$ gene it protein mediated termination of leading strand DNA synthesis. *J. Biol.* chem. 258; 529.

Rosbash, M. and Seraphin, B. 1991. Who's on first ? The U1 and nRNP-5-splice site interaction and splicing. *Trends. Biochem* Sci. 16: 187-190

Sanger, F. Air, G,M. Barrell B.G. Brown, N.L. Coulson, A.R. *et al.* 1977. Nucleotide sequence of bacteriophage $\phi \times 174$ DNA. *Nature.* 265 : 687-695.

Steitz, J.A. 1988. Snurps. *Sci. Am.* 258 : 56-63

Weisbeek P.J. et al. 1977. Bacteriophage $\phi \times 174$: gene A overlaps gene B. *Proc Natl Acad.* Sci 74 :2504.

Genetic Code

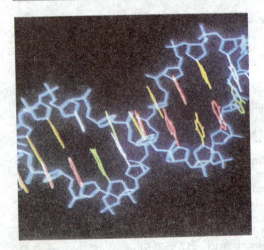

From Chapter 6, it has became obvious that nucleic acids are the genetic material. The nucleic acids being polynucleotide, function to store genetic informations and to replicate. The genetic informations flow from polynucleotide to polypeptide (Albert, 1986; Maizels and Weiner, 1987). It is surprising to note that at the origin of life any polynucleotide that helped to guide the synthesis of a useful polypeptide in its environment would have had a great advantage in the evolutionary struggle for survival (Albert, 1986).

A long chain of a DNA molecule consists of three components, nitrogen bases, deoxyribose sugar and phosphoric acid. Except nitrogen base, the chemical configuration does not change. The nitrogen bases are of four types, adenine, guanine, thymine, cystosine. Therefore, it is likely that the sequence of these bases on a segment of DNA molecule changes. Obviously, the above three components (nucleotides) are involved in restoring the genetic information. Thus, there are four alphabets (A, G, T, C) of DNA. We know that there are 20 different amino acids that constitute a protein as the nitrogen bases constitute nucleotides. Hence, there are 20 alphabets of the language of a protein. From DNA to protein the informations pass through an mRNA. One mRNA carries the

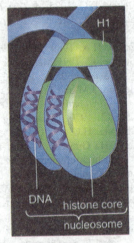

Short stretches of DNA may appear like this when DNA is being replicated.

genetic information of one protein. The synthesis of specific protein under the guidance of mRNA required the evolution of a code by which the polynucleotide sequence specifies the amino acid sequence that makes up the protein. This code is called "genetic code" which is spelled out in a dictionary of words. This code is virtually the same in all living organisms (Maizels and Weiner, 1987).

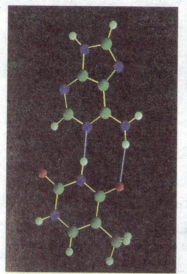 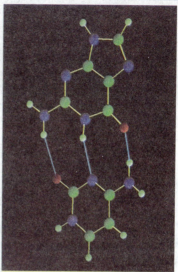

The base pairs of DNA. It shows the four bases that make two pairs – adenine bonds with thymine (A = T) (A), and guanine with cytosine (G –C) (B). Their three hydrogen bonds make GC pairs stronger than AT pairs.

Since the genetic informations flow from 4 alphabet to 20 amino acids of a polypeptide through an mRNA, the problem arose to prepare a dictionary of the codes of mRNA for 20 amino acids. However, it is not possible that one alphabet i.e. single base of RNA may be equivalent to a single alphabet of protein i.e. single amino acid if it is so just four alphabets would be enough for four amino acid. Therefore, a code of more than one alphabet of RNA must be assigned for an amino acid. Any message in coded form is called cryptogram.

A. Cryptogram of DNA

The dictionary of coded language of genetic information is called cryptogram. The simplest and smallest possible code is a single code or singlet code which specifies that one amino acid is synthesized by one nucleotide. Singlet code is not possible for coding 20 different amino acids. Then how many letters would be used as a code for a single amino acid? In addition to the single code, if a code of two alphabets of RNA are used, only $4^2=16$ codes are possible. There are a total of 20 essential amino acids, hence the use of double code gives only 16 possible combinations (code). Therefore, use of a tripple code i.e. a code of three nucleotide bases, can serve the purpose. By using triple codes we get $4^3 = 64$ codons or triplets. Therefore, 64 codes are enough for 20 amino acids. The possible singlet, doublet and triplet codes represented in terms of mRNA language are illustrated in Table 7.1.

Table 7.1 : A Possible Single Code, Double Code and Triple Code.

A	AA AG AC AU
G	GA GG GC GU
C	CA CG CC CU
U	UA UG UC UU

Singlet code (4 words)		Doublet code (16 words)	
AAA	AAG	AAC	AAU
AGA	AGG	AGC	AGU

ACA	ACG	ACC	ACU
AUA	AUG	AUC	AUU
GAA	GAG	GAC	GAU
GGA	GGG	GGC	GGU
GCA	GCG	GCC	GCU
GUA	GUG	GUC	GUU
CAA	CAG	CAC	CAU
CGA	CGG	CGC	CGU
CCA	CCG	CCC	CCU
CUA	CUG	CUC	CUU
UAA	UAG	UAC	UAU
UGA	UGG	UGC	UGU
UCA	UCG	UCC	UCU
UUA	UUG	UUC	UUU

Triplet code (64 words)

Crick *et al.* (1961) provided the first experimental evidence in support to the concept of triplet code of mRNA. When they inserted or deleted single or double base pairs in a particular region of DNA of T4 phage of *E. coli*, the bacteriophage ceased the normal function. In addition, when three base pairs were added or deleted in the T4 DNA, the bacteriophage performed normal function. Based on this experiment they concluded that the genetic code is a triplet code because due to addition or deletion of single or double base pairs the reading sequence was changed, whereas it was returned to normal with addition of a third nucleotide. They also suggested that many amino acids are specified by more than one triplet code i.e. the code is degenerate (Crick, 1966a).

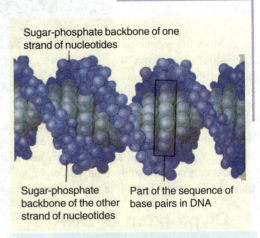

Sugar-phosphate backbone of one strand of nucleotides

Sugar-phosphate backbone of the other strand of nucleotides

Part of the sequence of base pairs in DNA

A gene region of a DNA double helix.

B. Codon Assignments

In 1960s the actual codons were discovered by Marshall Nirenberg, Heinrich Matthaei, Philip Leder and Har Govind Khorana. In 1968, Nirenberg and Khorana shared the Nobel prize with Robert W. Holley, the first person to sequence a nucleic acid (phenylalanyl tRNA). The following *in vitro* and *in vivo* approaches were made to decipher or crack the genetic code.

1.   Codon Assignment with Unknown Sequences

Theoretically it was considered that genetic code should be triplet so that codons must be assigned for 20 amino acids. But it was not known which codon synthesises which amino acid. Therefore, the first approach was made to find out the codons for amino acids by using sequences of mRNA *in vitro*, and secondly to prove the same *in vivo*.

(i) Use of Homopolymers: In the first approach, Nirenberg and Matthaei (1961) synthesized the RNA by using polyuridylic acid (poly-U). This resulted in poly-U along the lenght of mRNA with possible triple UUU codon. When poly-U (RNA) was used for cell free synthesis of polypeptide by using cell free extract of *E. coli* that supplied all the components of protein synthesising machinery, polyphenylalanine was synthesised. It means UUU triplet coded for amino acid phenylalanine. This news about cracking the genetic code was brought out in news papers throughout the world in 1961. Later on they discovered that poly-C induced the synthesis of proline, therefore, CCC triplet was assigned to proline. Poly-A stimulated polylysine synthesis. Hence, the sequence of AAA designates lysine. The experiments with poly-G was unsuccessful because it did not attach to ribosome due to attaining the secondary structures.

(*ii*) **Use of Copolymers:** In the next experiment, Nirenberg and co-workers used artificially synthesised random ribopolynucleotide containing two or three different nucleotides. For example, if we use A and C we will get poly AC. One can calculate the possible combinations in mixed copolymers. From a mixture of A and C, eight possible triple codons (e.g. AAA, AAC, ACA, ACC, CAC, CCC, CCA and CAA) with equal frequency are formed. However, when known amount of A and C is used in synthesis, the proportion of eight codons can be calculated. For example, if A:C = 5:1 (i.e. 5/6 is A and 1/6 is C), the eight possible combinations would be as shown in Table 7.2.

Table 7.2 : Formation of different codons in mRNA by 5:1 ratio of A or C bases.

Base Composition	Codon	Probability	Ratio	
3A	AAA	5/6x5/6x5/6=125/216	125	100
2A1C	AAC	5/6x5/6x1/6=25/216	25	20
	ACA	5/6x1/6x5/6=25/216	25	20
	CAA	1/6x5/6x5/6=25/216	25	20
1A2C	CCA	1/6x1/6x5/6=5/216	5	4
	CAC	1/6x5/6x1/6=5/216	5	4
	ACC	5/6x1/6x1/6=5/216	5	4
3Cz	CCC	1/6x1/6x1/6=1/216	1	0.8

Table 7.3 : Codon assignment based on A:C (5:1) composition in synthesized mRNA.

Base compositon	codons	Amino acids
3A	AAA	Lysine
2A1C	AAC	Asparagine
	ACA	Threonine
	CAA	Glutamine
1A2C	CCA	Proline
	CAC	Histidine
	ACC	Threonine
3C	CCC	Proline

The proportion of amino acids coming out of calculated relative proportion of codons synthesised by poly AC were calculated (Table 7.2). Initially the codons were assigned on the basis of base composition, but not sequences of bases in codons. Hence, by using poly AC six amino acids viz, lysine, asparagine, threonine, glutamine, proline and histidine were found in polypeptide of AC (Table 7.3). However, if the ratio of poly A was more than poly C, the ratio of asparagine to histidine increased accordingly.

2.   Codon Assignment with Known Sequences

The codons assigned to amino acids were found out and a dictionary of codon composition was prepared by using a large number of copolymers of unknown sequences. The next approach was to search out the sequence of codons of known composition.

(i) Codon Assignment Through Fitter Binding Technique: In 1964, Leder and Nirenberg (1964) found when simple trinucleotides of known sequence (i.e. with 5' and 3' ends') are added to ribosomes, it causes the ribosome to attach it to only aminoacyl-tRNA that contains anticodon complementary to trinucleotide (mixed with reaction mixture).

Codon + ribosome + AA- tRNA → Ribosome - codon - AA-tRNA

For example, the trinucleotide GCC is activated to bind alanyl-tRNA but not aminoacyl-tRNA. This shows that GCC is a codon for alanine but not for others.

In another approach it was also found out that ribosome-codon-AA-tRNA complex absorbed on nitrocellulose membrane when passed through it. Nirenberg and coworkers thought when only one amino acid in alternate manner is made radioactive, mixed with rest of 19 amino acids, codons of known sequences, ribosomes and tRNA, and passed through nitrocellulose membrane, then the presence or absence of radioactivity is detected on the basis of adsorption of complex on membrane. The same codon is used at all the time. Radioactivity will be observed on the membrane only when the radioactive amino acid takes part in complex formation. Otherwise there would be no radioactivity. Each amino acid is radiolabelled in each sample with 19 non-radioactive amino acids.

Nirenberg and coworkers incorporated the radioactive amino acids in cell free system for protein synthesis containing known sequence of codon (mRNA). The cell free system for protein synthesis contained ribosomes, enzymes, tRNA, etc. which were separated from the remainder of broken *E. coli* cells by gentle centrifugation. During this process mRNAs are broken, therefore, artificial mRNA of known sequence was incorporated. The artificial mRNA was prepared by using each nitrogen base and an enzyme polynucleotide phosphorylase obtained from *Azotobacter vinelandii* or *Micrococcus lysodeikticus*. Unlike RNA polymerase, the poly nucleotide phospho-rylase does not require DNA as template. In this way Nirenberg constructed polyuridylic acid (UUU...), polycytidylic acid (CCC...), polyadenylic acid (AAA...) and polyguanyl acid (GGG...). The experiment with poly-G was unsuccessful because it did not attach to ribosome.

By this technique the cracked 45 codons for amino acids are arginine, alanine, cysteine glutamine, glycine, isoleucine, leucine, methionine, proline, tryptophane, tyrosine, serine and valine.

(ii) Use of Copolymers of Repetitive Sequences: Almost at the same time Har Govind Khorana and coworkers succeded in determining the exact sequence of nucleotides in many codons discovered previously by Nirenberg or himself. Khorana prepared artificial mRNA with known sequences and used to ascertain the structural isomer of a codon which specified an amino acid. For example, if CU (two bases) are repeatedly present throughout the length, it will give the sequence CUCUCUCU .... In the same way the three bases (ACU) will form the repeatitive sequence ACU ACU ACU ACU ..... Thus, an alternating copolymer e.g. UGU-GUG-UGU-GUG.... was synthesised and used as mRNA *in vitro* in protein synthesising system (Khorana, 1968). Consequently the alternate codons, UGU-GUG-UGU..., synthesised the alternating polypeptide, cysteine-valine-cysteine. ... This confirms that both UGU (cysteine) and UUG (leucine) are different codons though base composition is 2U1G. Similarly, GUG (valine), UGG (tryptophan) and GGU (glycine) are the different codons though all the three have base composition 2G1U. In this way a dictionary of complete genetic code could be prepared (Table 7.4).

3.   *In Vivo* Studies for Codon Assignment

After carrying out experimentation *in vitro* on cracking of genetic code it could not provide evidence, whether the genetic code is for all organisms. Therefore, different molecular biologists

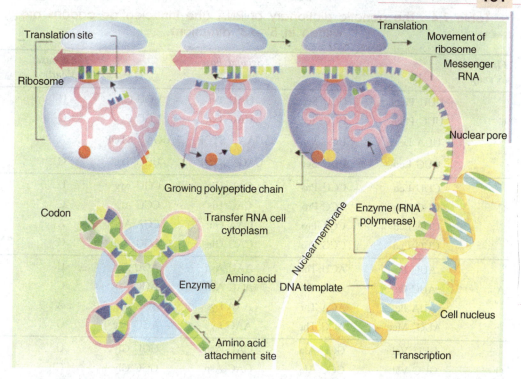

Translation site
Ribosome
Codon
Translation
Movement of ribosome
Messenger RNA
Nuclear pore
Growing polypeptide chain
Transfer RNA cell cytoplasm
Enzyme (RNA polymerase)
Nuclear membrane
Enzyme
Amino acid
DNA template
Amino acid attachment site
Cell nucleus
Transcription

The production of all proteins starts when the genetic message on DNA is transcribed on the messanger RNA. In this only one strand of DNA is transcribed.

by using different techniques proved that the same code is also used *in vivo* as well. It could be possible through (*i*) amino acid replacement studies (tryptophan synthetase synthesis in *E. coli*), (*ii*) framshift mutations on lysozyme of T4 bacteriophage, and (*iii*) comparison of a DNA or mRNA polynucleotide cryptogram with its corresponding polypeptide.

On the basis of *in vitro* and *in vivo* studies, the genetic code could be formulated for all essential amino acids. Each codon is written in genetic dictionary in such a way that appears in mRNA sequence in 5'→3' direction (Table 7.4). In DNA each codon would be complementary and in the reverse order to mRNA on a 5'→3' strand.

C. Patterns to Genetic Code

After going through Table 7.4 a remarkable pattern of genetic code emerged. Following are some of the important features of genetic code:

(*i*) Sixty one codons correspond to amino acids.

(*ii*) Four codons are the signals. There are three stop codons (UAA,UAG,UGA) and one start codon (AUG). Rarely, GUG also acts as start codon.

(*iii*) Amino acids with similar structural property consist of related codons; therefore, the aspartic acid codons (GAU and GAC) are related to glutamic acid codons (GAA and GAG). Similarly, the codons of phenylalanine (UUU,UUC), tyrosine (UAU,UAC) and tryptophan (UGG) start with uracil. This characteristic of codons facitilate to minimize the effect of mistakes arising during translation or mutagenic base substitution.

(*iv*) For many synonym codons specifying the same amino acid the first two bases of the triplet are constant, while the third base varies. For example, all codons starting with CC (e.g.CCU,CCC,CCG) specify proline, and all codons starting with AC (ACU, ACC, ACA, ACG) specify threonine. The flexibility in third codon may be to minimize errors.

Table 7.4 : The genetic dictionary of RNA (the trinucleotide codons are written in 5' → 3' direction).

First base (5'end)	Second base				Third base (3'end)
	U	C	A	G	
U	UUU Phen	UCU Ser	UAU Tyr	UGU Cys	U
	UUC Phen	UCC Ser	UAC Tyr	UGC Cys	C
	UUA Leu	UCA Ser	UAA** STOP	UGA** STOP	A
	UUG Leu	UCG Ser	UAG** STOP	UGG Trp	G
C	CUU Leu	CCU Pro	CAU His	CGU Arg	U
	CUC Leu	CCC Pro	CAC His	CGC Arg	C
	CUA Leu	CCA Pro	CAA Gln	CGA Arg	A
	CUG Leu	CCG Pro	CAG Gln	CGG Arg	G
A	AUU Ile	ACU Thr	AAU Asn	AGU Ser	U
	AUC Ile	ACC Thr	AAC Asn	AGC Ser	C
	AUA Ile	ACA Thr	AAA Lys	AGA Arg	A
	AUG* Met	ACG Thr	AAG Lys	AGG Arg	G
G	GUU Val	GCU Ala	GAU Asp	GGU Gly	U
	GUC Val	GCC Ala	GAC Asp	GGC Gly	C
	GUA Val	GCA Ala	GAA Glu	GGA Gly	A
	GUG* Val	GCG Ala	GAG Glu	GGG Gly	G

* Start codons, ** stop codons.

D. Properties of the Genetic Code

Through the experiments it has been proved that the mRNA codons of the genetic code have the following properties:

(i) The Code is a Triplet: As it has been discussed earlier that singlet and doublets are not adequate to code for 20 amino acid; therefore, the triplet codes that consist of $4^3 = 64$ codons may code for 20 essential amino acids. The triple code of mRNA has been accepted.

(ii) The Code is Degenerate: There are 64 codons in the genetic code for 20 amino acids of which 4 codons are the signals. Therefore, 60 codons are to code for amino acids. It means that more than one codons may be coding for individual amino acid. The number of codons coding for different amino acids are as below:

(a) Tryptophan, methionine .... 1 codon
(b) Phenylalanine, tyrosine, histidine, glutamine, asparagine .... 2 codons
(c) Isoleucine .... 3 codons
(d) Valine, proline, threonine, alanine, glycine .... 4 codons
(e) Leucine, arginine, serine .... 6 codons

The codons that code for more than one amino acid are called degenerate. For example, a codons starting with CC specify proline (CCU, CCC, CCA, CCG) and all codons starting with AC specify threonine (ACU, ACC, ACA, ACG). Unequal distribution of amino acids in protein may be due to this variability in number of codons for amino acids.

(*iii*) **The Code is Non-overlapping:** The genetic code is non-overlapping which means that the same letter does not take part in the formation of more than one codon. The overlapping and non-overlapping codes are given below:

In addition, it has been shown that the bacteriophage Øx174 consists of overlapping genes (Barell *et al.*, 1976). Two genes besides separately coding for its own protein, also take part in coding for the third protein with different amino acid sequences. This is done by a frameshift mechanism (i.e. overlapping code). The entire nucleotide sequence of Øx174 has been given by Sanger *et al.* (1977). For detail see 'overlapping genes' (Chapter 6).

(*iv*) **The Code is Non-ambiguous:** The non-ambiguous means that a particular codon will always code for the same amino acid. It may also be that the same amino acid may be coded by two different codons (degenerate). However, when one codon codes for two amino acids, it is called ambiguous. For example, UUU codon codes for phenylalanine, but in the presence of streptomycin it may code for isoleucine, leucine or serine.

(*v*) **The Code is Commaless:** The genetic code is without comma i.e. no puctuations are required between the two codons. There are no demarkating signals between the two codons. This results in continuous coding of amino acid without interruptions. No codons are left uncoded. The commaless codons may be written as below:

               UUUCUCGUAUCC   - Bases

               Phe-Leu-Val-Ser   - amino acids

The genetic code, however, with comma may be represented as below:

               UUU-CUC-GUA-UCC

Due to deletion of a base (e.g. C), a drastic change in coding of amino acids occurs.

               UUUUCGUAUCC   - Bases

               Phe-Ser-Tyr....   - amino acids

However, if the introns are present, the coding process is interrupted. See introns i.e. split genes in preceeding section.

(*vi*) **The code has polarily:** The code has polarity i.e. it is read between the fixed start and stop codons. The start codon is also known as initiation codon, and stop codon as termination codon. The message of mRNA is read in 5'→'3 direction. The polypeptide chain is synthesized from the amino (-NH$_2$) end to the carboxyl (-COOH) end i.e N→C.

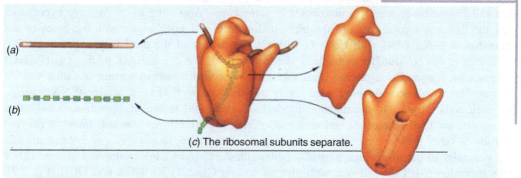

(*c*) The ribosomal subunits separate.

Chain termination stage: (*a*) once a stop codon is reached, the mRNA transcript is released from the ribosome, (*b*) the newly formed polypeptide chain also is formed, (*c*) the ribosomal subunits separate.

However, if the code is read in opposite directions the message will be reversed due to change in base sequence. Therefore, two different proteins will be synthesized. For example, if 5'-AUCGUCUCGUUGACA-3' is read from left to write it will specify: Ile-Val-Ser-Leu-Thr-, and if it is read from right to left it will specify: Thr-Val-Ala-Leu-Leu.

5'-AUCGUCUCGUUGACA-3'	**mRNA codons**
→ Ile-Val-Ser-Leu-Thr→	Amino acids in right direction
←Leu-Leu-Ala-Val-Thr←	Amino acids in left direction

The codon, AUG is the initiation codon. Rarely GUG also acts as initiation codon in bacterial protein synthesis, when AUG is lost by deletion or it is non-functional. In the phage MS2, GUG acts as initiation codon for A protein. However, generally GUG codes for the amino acid valine, and AUG codes for methionine.

Moreover, three of 64 codons are called as non-sense codons because they do not specify any tRNA. These codons are amber (UAG), Ochre (UAA) and Opal or amber (UGA). These also bring about termination of polypeptide chain, therefore, they are also called termination codons.

For the first time the codon UAG was investigated by a graduate student who belonged to the Bernstein family. Bernstein means 'amber' in German. He helped in discovery of a class of mutation. The other two termination codons were also named after the colours just to give the uniformity.

(*vii*) **The Code is Universal :** Though the genetic code has been worked out by using *in vitro* systems of microorganisms, yet there is no doubt of being its universal for all group of microorganisms. In 1967, Nirenberg and his associates demonstrated the universality of the code. They found that aminoacyl tRNAs of *E. coli* (bacterium), *Xenopus laevis* (amphibian) and guinea pig (mammal) use almost the same codon. It has also been shown that when purified mRNAs from rabbit reticulocytes specifying the synthesis of haemoglobin are injected into frog oocytes, these synthesise rabbit haemoglobin by using the translation machinery of frog.

E. The Wobble Hypothesis

Crick (1966b) proposed the 'wobble hypothesis' to explain the degeneracy of the genetic code. Except for tryptophan and methioine, more than one codons direct the synthesis of one amino acid. There are 61 codons that synthesise amino acids, therefore, there must be 61 tRNAs each having different anticodons. But the total number of tRNAs is less than 61. This may be explained that the anticodons of some tRNA read more than one codon.

In addtion, identity of the third codon seems to be unimportant. For example CGU, CGC, CGA and CGG all code for arginine. It appears that CG specifies arginine and the third letter is not important. Conventionally, the codons are written from 5' end to 3' end. Therefore, the first and second bases specify amino acids in some cases. According to the Wobble hypothesis, only the first and second bases of the triple codon on 5'→'3 mRNA pair with the bases of the anticodon of tRNA i.e A with U, or G with C. The pairing of the third base varies according to the base at this position, for example G may pair with U. The conventional pairing (A=U, G≡C) is known as Watson-Crick pairing (Fig. 7.1) and the second abnormal pairing is called wobble pairing. This was observed from the discovery that the anticodon of yeast alanine-tRNA contains the nucleoside inosine (a deamination product of adenosine) in the first position (5'→3') that paired with the third base of the codon (5'→3') (Crick,1966b). Inosine was also found at the first position in other tRNAs e.g. isoleucine and serine. The purine, inosine, is a wobble nucleotide and is similar to guanine which normally pairs with A, U and C. For example a glycine-tRNA with anticodon 5'-ICC-3' will pair with glycine codons GGU, GGC, GGA and GGG (Fig 7.2). Similarly, a seryl-tRNA with anticodon 5'-IGA-3' pairs with serine codons UCC, UCU and UCA

(5'-3'). The U at the wobble position will be able to pair with an adenine or a guanine.

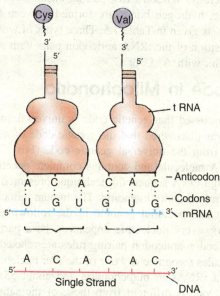

Fig. 7.1 : DNA triplet, mRNA codons and tRNA anitocodons showing Watson-Crick pairing.

According to Wobble hypothesis, allowed base pairings are given in Table 7.5.

Table 7.5 : Wobble base pairings

Third position codon base	First position anticodon base
A	U, I
G	C, U
U	G, I
C	G, I

Due to the Wobble base pairing one tRNA becomes able to recognise more than one codons for an individual amino acid. By direct sequence of several tRNA molecules, the wobble hypothesis is confirmed which explains the pattern of redundancy in genetic code in some anticodons (e.g. the anticodons containing U, I and G in the first position in 5'→3' direction)

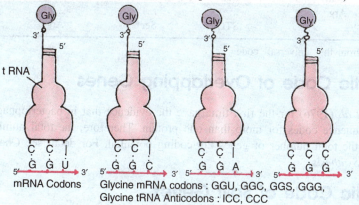

Glycine mRNA codons : GGU, GGC, GGS, GGG,
Glycine tRNA Anticodons : ICC, CCC

Fig.7.2 : Wobble pairing of one glycine tRNA with three codons of mRNA due to Wobble in 5'→3' direction.

The seryl-tRNA anticodon (UCG) 5'-GCU-3' base pairs with two serine codons, 5'-AGC-3' and 5'-AGU-3'. Generally, Watson-Crick pairing occurs between AGC and GCU. However, in AGU and GCU pairing, hydrogen bonds are formed between G and U. Such abnormal pairing called 'Wobble pairing' is given in Table 7.5. Three types of wobble pairings have been proposed: (*i*) U in the wobble position of the tRNA anticodon pairs with A or G of codon, (*ii*) G pairs with U or C, and (*iii*) I pairs with A, U or C.

F. Genetic Code in Mitochondria

Previously we discussed that genetic code is universal and does not undergo any changes. But during 1980s, it was discovered that the genetic code of mitochondria of yeasts, *Drosophila* and mammals differs from the universal genetic code (Fox, 1987). The mitochondrial genome is usually circular DNA molecule and contains complete genetic system. Anderson *et al.* (1981) for the first time through DNA sequencing technique presented the complete sequence of 16,569 nucleotides of human mitochondrial genome. The human mitochondrial genome differs from that of nuclear, chloroplast and bacterial genome in the following respect:

 (*i*) Unlike others, every nucleotide appears to be a part of coding sequence.
 (*ii*) The normal codon-anticodon pairing rules are relaxed in mitochondria, therefore, many tRNA molecules recognize any one of the four nucleotides in the third (wobble) position. There are 22 tRNAs in mitochondria and about 55 tRNAs in universal code.
 (*iii*) The genetic code is different from those of the same codons in other genomes.

However, the genetic code of mitochondria differs from the universal genetic code (Table 7.6). On the other hand the mitochondrial genetic code in different group of organisms also differ. For example, UGA which is a stop codon elsewhere, is read as tryptophan in mitochondria of yeasts, *Drosophila*, mammals and protozoa, but as stop in plant mitochondria. The codon AGG normally codes for arginine, but it acts as stop codon in mitochondria of mammals, and codes for serine in *Drosophila*. Similarly, AUA codon which codes for isoleusine, specifies methionine in yeasts, *Drosophila* and mammal's mitochondria, but not plant mitochondria (Fox, 1987).

Table 7.6 : Some differences between the universal code and mitochondrial genetic codes (after Alberts, 1994)

| Codon | 'Universal' Code | Mitochondrial Codes | | | |
		Mammals	Drosophila	Yeasts	Plants
UGA	STOP	Trp*	Trp*	Trp*	STOP
AUA	Ile	Met*	Met*	Met*	Ile
CUA	Leu	Leu	Leu	Thr*	Leu
AGA	Arg	...	...	...	...
AGG	...	STOP*	Ser*	Arg	Arg

* Codes differ from the 'universal' code.

G. Genetic Code of Overlapping Genes

Barrell *et al.* (1976) for the first time gave the evidence that in bacteriophage ØX174, the same DNA sequence codes for more than one protein. Therefore, the total number of proteins exceeds from the total number of genes i.e. coding potential. For detail see Chapter 6, *Genes: Concept and Synthesis*.

H. Genetic Code of Split Genes

In some animal viruses and eukaryotes the DNA sequence coding for a polypeptide is not continuous but interrupted, and genes were arranged into several pieces. Detailed account of split genes is given in Chapter 6, *Genes: concept and synthesis.*

I. Origin and Evolution of Genetic Code

From the discussion of genetic code some important facts came into light as the degeneracy and universality of the genetic code. It may be supposed that present form of genetic code would have evolved from a more primitive code which must have occurred about three billion years ago. Since the evolution of bacteria, the code has remained fixed. Any mutation that altered the sequence would changed the reading frame of mRNA resulting in changes in specifying amino acids. Wong (1988) has discussed in detail about the evolution of genetic code.

QUESTIONS

1. What are different approaches made for codon assigments?
2. What are the properties of the genetic code?
3. In what ways the Wobble hypothesis explains about the degeneracy of the genetic code? Discuss in brief with suitable illustrations.
4. Write in brief the differences between universal code and mitochondrial code.
5. Discuss in brief the genetic code of overlapping genes.
6. Write short notes on the following:
   (*i*) Cryptogram of DNA,   (*ii*) Universal code,
   (*iii*) Wobble hypothesis,   (*iv*) Genetic code in mitochondria.

REFERENCES

Alberts, B. Bray, B. *et al* 1994. *Molecular Biology of the Cell.* 3rd ed. Gart and Publ Inc. New york

Albert, B.M. 1986. The function of the heriditary materials: biological Catalyses reflect the Cell's evolutionary history. *Am. Zool* 26 : 781-796.

Anderson, S. *et al* 1981. sequence and organization of the human mitoch ondrial genome *Nature* : 290 : 457-465

Barrell, B.G. Air, G.M. and Hutchinson, C.A. 1976 Overlapping genses in the bacteriophage Øx174. *Nature* 264 : 34-41.

Crick. F.H.C. 1966 a. The genetic code. III. *Sci. Am.* 216 : 55-62.

Crick F.H.C. 1966 b. Codon-anticodon pairing : The wobble hypothesis *J.Mol Biol* 19 : 548.

Crick F.H.C. *et al* (1961) General nature of the genetic code for proteins. *Nature* : 192 : 1227.

Fox, T.D. 1987. Natural variation in the gentic code. *Annu Rev. Genetics.* 21 : 67-91.

Khorana, H.G. 1968. Nucleic acid synthesis in the study of the genetic code. In Nobel Letures : Physiology of medicine Vol 4 American Elseving.

Leder, P. and Nirenberg, M. 1964. RNA code words and protein synthesis II Nucleotide sequence of a valine RNA code word. *Proc. Natl. Acad Sci* (USA) 52:420.

Maizels, N and Weiner A.M. 1987. Peptide - Specific ribozymes, genomic tags and the origin of the genetic code. *Cold Spring Harbor Syrup Quant Biol.* 52 : 743-749.

Nirenberg M and Matthaei, J. H. 1961. The dependence of cell free protein synthesis in *E. coli* upon naturally occurring or synthetic polyribonudeotides *Proc. Nat Acad Sci,* 47 : 1588.

Sanger, F. Air, G.M. Barrel, B.G. Brown N.L. Coulson, A.R. *et al* 1977. Nueleotide sequence of bacteriophage Øx174 DNA. *Nature.* 265: 687-695.

Wong. J. T. 1988 Evolution of the genetic code. *Microbial Sci.* 5: 174-181.

Microbial Genetics

8

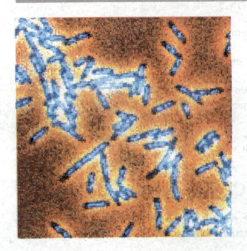

D uring 1860s, an Australian monk Gregor John Mendel under took the first serious study of genetics of pea. For this pioneering work Mendel is called the **father of genetics**. He crossed two varieties of pea by selecting the contrast characters and noted the results i.e. changes in height of plant, and colour, shape and size of seeds. All the work was published in 1865 under the title "Experiments with plant hybridization" in Vol. 4 of *Proceedings of the Natural Science Society*. Based on the results of pea he formulated the laws of heredity which is common among all life forms including humans as well.

During the begining of 20th Century three botanists, Hugo de Vries working on *Oenothera*, Correns working on *Xenia*, pea and maize, and Von Tschermark working on several flowering plants separately derived the results like Mendel. Later on Mendel's work was reconsidered in 1900 and the original work was republished in *Flora* 89:364 (1901).

Obviously, genetics is the science of heredity (inheritance) and the variability of the characters of an organism. Heredity is the process by which exact transmission of genetic information takes place from parents to their progenies. The organisms resemble to each other due to the transmission of hereditary

Greogor John Mendel, father of genetics

informations from a common ancestor. Through many generations, species exist simply because the heredity material remains stable. However, genetic diversity exists within most of the species. Hereditary variability is an important factor for evolution which is governed by cell itself or the environment. The genetic make up adjusts itself according to the environment. Therefore, environment has been constantly influencing the organisms since their origin; that is how evolution took place.

Another break through took place after the identification of chemical nature of hereditary

The garden pea plant (*Pisum sativum*), the focus of Mendel's experiments. A flower has been sectioned to show the location of its stamens and carpel. Sperm-producing pollen grains form in stamens. Eggs develop, fertilization takes place, and seeds mature inside the carpel.

material as nucleic acids i.e. deoxyribonucleic acid (DNA) and ribonucleic acid (RNA) during 1940s (see Chapter 5). Since then much work has been done to unravel the mechanism of inheritance and the factors governing them.

A. Mendel's Laws of Inheritance

Mendel himself did not give any laws. He simply mentioned the conclusions of theoretical and statistical explanations of his experiment. Correns discovered Mendel's work and propounded the whole work into two laws of heredity: (*i*) the law of segregation or law of purity of gametes, and (*ii*) the law of independent assortment.

1. The Law of Segregation

In a heterozygote a character remains in contrast paired form which is known as allele. One of alleles remains in dominant form and the other in recessive form or *vice versa*. Dominant and recessive alleles are adhered together through out the life without getting mixed with each other. During gametogenesis these alleles segregate (separate) from each other with the result that one gamete receives dominant allele and the other recessive allele. After hybridization (in monohybrid cross) in F1 generation all plants look alike, but their gametes after uniting together in F2 generation give the phenotypic (morphological) ratio 3:1.

2. The Law of Independent Assortment

The law of independent assortment states that the alleles segregate quite independently from each other during the formation of gametes i.e. gametogenesis. In F1 generation or dihybrid cross

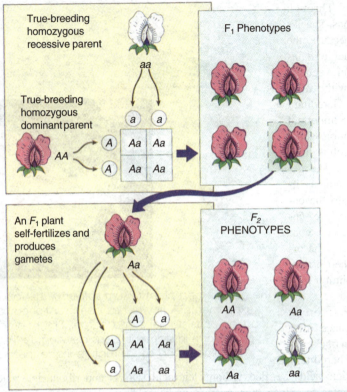

Results from one of Mendel's monohybrid crosses. On the average, the dominant-to-recessive ratio among F₂ plants was 3 : 1.

all the plants would be alike but in F2 generation the ratio of characters of plants was 9:3:3:1. Discussion of Mendel's laws in reference to higher plants is not the aim of this chapter, except brief informations.

B. Microbial Genetics

In prokaryotes the method of cell division differs from that of eukaryotes. No meiosis or mitosis occurs in prokaryotes because of lack of a nucleus. In contrast, in prokaryotes and a few eukaryotic microorganisms most of genetical and molecular works, have been carried out because of easy in handling, fast growth immediate observable characters under any change in conditions and least complex life forms.

Like higher plants and animals, bacteria and the other protists transmit characters to their progenies and in onward generations. Apart from inher-

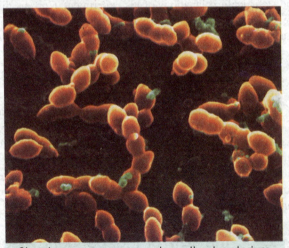

Streptococcus pneumoniae : the bacterium first used to obtain evidence that DNA is the genetic material of organisms.

itance of characters they also show variability in progenies. These changes are referred to as phenotype (i.e. the morphological changes) and genotype (genetical changes). The genotype always remains constant; however, changes occur through mutation which results in morphological changes or phenotypes in progenies.

In addition, the phenotypic expression depends on the environmental conditions as well. For example, phenotypic changes in *Agrobacterium radiobacter* takes place if it is grown on two different growth media. Mucoid colonies are formed on sucrose salt medium and non-mucoid colonies develop on trypticase soy agar medium. Secondly, yeast produces ethanol when grown in oxygen deficient conditions, and it increases its biomass and does not produce ethanol when grown in the presence of suffficient oxygen.

For the first time Luria and Delbruck (1943) gave the experimental evidence for acquisition of drug resistance by sensitive bacteria, and this conceeded to have formed the basis for modern work in the field of bacterial genetics. According to them resistance occurs as a direct adaptation by some bacteria against drug. This was also explained by spontaneous mutation theory.

I. Genetic Notation

If a bacterial cell synthesizes amino acid leucine, it is represented as Leu^+, and if it does not it is denoted as Leu^-. The symbol has capitalized (not italicized) the letters. It denotes that Leu^- has some defective genes which cripple the cell to synthesize leucine. The defective gene is represented as *leu*$^-$ (italicized three letters). However, if more than one gene are needed to synthesize leucine, it is denoted as *leu*A, *leu*B, etc. and the functional gene as *Leu*A$^+$ *Leu*B$^+$, etc.

It should be kept in mind that the bacteria always have a single set of genes i.e. they are haploid. The eukaryotic organisms are haploid as well as diploid but the dominance of these two phase in the life of an organism differs. The diploid cells of organisms contain two sets of genes. The double set of functional genes are represented as *leu*$^+$/*leu*$^-$. They may have normal or defective genes. The functional forms of a gene is called wild type.

However, if a bacterium is resistant to certain antibiotics it is represented by giving the symbols, for example Tet^r (resistant to tetracycline), Amp^r (resistant to ampicillin), Kan^r (resistant to kanamycin), etc. The microorganisms susceptible to the above antibiotics are represented as Tet^s, Amp^s, Kan^s, etc.

Mutations occurring in genotype are also represented by numbers in the order in which they have been isolated. For example, if leucine mutation has occurred on 58 and 79 position it is written as *leu*58 and *leu*79, respectively. More specifically to denote the mutation on a particular gene e.g. A and B, it is written as *leu*A58 and *leu*B79 if mutation has occurred in *leu*A and *leu*B genes, respectively (Friefelder, 1987).

II. Transfer of genetic Material in Prokaryotes

Genetic recombination is the formation of new recombination of genes (i.e. genotypes) through reassortment of genes of two different cells. This occurs between two different chromosomes having identical genes at the corresponding sites. These are called homologous chromosomes derived from two different cells. The chromosomes of progenies differ from those of parental cells. In prokaryotes several mechanisms viz., transformation, conjugation and transduction have been described that mediate the formation of new recombinants.

Transformation refers to transfer of relatively small segment of naked DNA from a donor cell (male) to the recipient (female) cell. Conjugation is the process of gene transfer between cells of opposite mating types that are in physical contact with each other. Transduction involves the transfer of genetic information from a donor cell to a recipient cell through a bacteriophage.

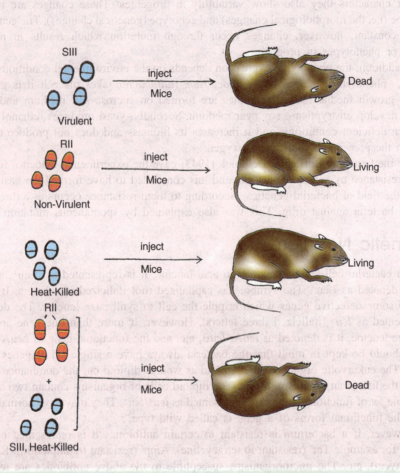

Fig. 8.1 : Griffith's experiment on transformation.

In the above processes not whole but a portion of chromosomes is transferred from a donor to recipient cell. Consequently the female cell becomes partially diploid called merozygote. The ultimate fate of the donated DNA is that whether it becomes incorporated in the recipient genome by recombination process degraded by restriction enzymes of the host or maintained as stable extrachromosomal fragement (Moat and Foster, 1995).

1. Transformation

As mentioned in Chapter 5, bacterial transformation was first discovered by Griffith (1928) between the two strains of *Streptococcus* (*Diplococcus*) *pneumoniae* which causes pneumonia in humans and mice. The results of Griffith are shown in Fig. 8.1. The   RII strain (non-pathogenic, rough colony forming, mutant strain) did not cause death of mice, whereas the SIII strain (virulent and of smooth surface, wild strain) caused death of mice. The heat killed SIII strain also showed results like RII strain. However, when heat killed SIII strain mixed with RII strain was injected in to mice, they died. This induction in change of RII strain was called **transformation**. Griffith thought that the transformation would have been caused by a protein.

Avery, Macleod and Mc Carty (1944) conducted the experiment and demonstrated that heat killed SIII strain transformed RII strain into virulent form which caused death of mice. The transforming ability was not altered by treatment with enzyme or by RNase , but was completely destroyed by DNase. These findings showed that DNA has the ability to carry hereditary information. Major steps of transformation are shown in Fig. 8.2.

Subsequently, transformation has been shown in a number of bacteria such as *Haemophilus, Neisseria, Xanthomonas, Rhizobium, Bacillus, Staphylococcus* and *Salmonella*. Generally transformation does not occur in *E. coli* but it does after the treatment with calcium chloride. Possibly it facilitates the entry of DNA into the recipient cells.

(*i*) **Competence:** Freifelder (1987) has defined competence as a physiological state that permits a cell to take up transforming DNA and be genetically changed by it. On the basis of development of competent state the organisms undergoing transformation can be divided into two groups: organisms always present in competent stage (e.g. *Neisseria*) and the organisms transiently competent in late exponential phase of growth (e.g. *S. pneumoniae*). These differences misrepresent the complex series

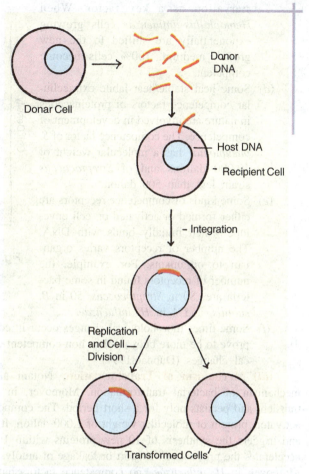

Fig. 8.2 : Steps in bacterial transformation.

of regulatory processes that are required to control this mechanism.

(*ii*) **Factors Affecting Competence:** Competence is not a permanent feature of bacteria but occurs only for a short duration during the life cycle. However, there are several factors that affect competence.

(*a*) Commonly competence is observed towards the end of logarithmic phase of growth i.e. just before the establishment of stationary phase (Fig. 8.3). In *Pneumococcus* duration of competence persists for 7-15 minutes. In *Bacillus subtilis* it persists for several hours.

(*b*) Ability of most of the bacteria to receive DNA efficiently is limited. After proper incubation the cell density increases greatly and yields a population of cells in which uptake of DNA is enhanced by a factor of 10^4 - 10^6. However, the conditions that induce competence and the fraction of competent cells vary from one species to the other. Both the fraction, of culture that become competent and the duration of competent state also vary species to species. For example, 100% *S. pneumoniae* cells become competent while only 20% cells of *B. subtilis* become competent. But the duration of competence in *S. pneumoniae* lasts for a few minutes and in *B. subtilis* for several hours.

(*c*) The growth medium, temperature, degree of aeration, etc. also influence the development of competence in cells. There are three types of competence in *B. subtilis* : nutritional, growth stage-specific and cell type-specific. The best competence develops in glucose minimal medium. Competence develops post exponentially and may involve nitrogen starvation as a key factor. When *Hemophillus influenzae* cells growing exponentially are shifted to the new growth medium, 100% cells become competent.

(*d*) Some heat stable/heat labile extracellular competence factors of proteinaceous in nature are involved in development of competence. The competence factor of *S. pneumoniae* has a molecular weight of 10,000 daltons and of *Streptococcus* strain less than 500 dalton.

(*e*) Some kinds of competence receptors are either formed or activated on cell envelope which initially binds with DNA. The number of receptors varies organism to organisms. For example, the number of receptors found in some bacteria are 80 in *Streptococcus*, 50 in *B. subtilis*, and 4 in *H. influenzae*.

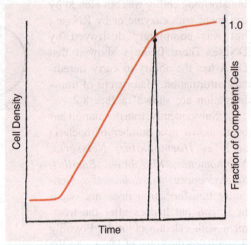

Fig. 8.3 : Development and persistance of competence in *Streptococcus pneumoniae*.

(*f*) Some times morphological changes occur in comptent cells. *B. subtilis* competent cells prove to be more buoyant than non-competent cells which show remarkable physiological changes (Dubnau,1991).

(*iii*) **Mechanism of Transformation:** Notani and Setlow (1974) have described the mechanism of bacterial transformation. Moreover, in *S. pneumoniae* the competent state is transient and persists only for a short period. The competent state is induced by the competence activator protein of molecular weight of 1,000 dalton. It binds the plasma membrane of receptor and triggers the synthesis of 10 new proteins within 10 minutes. The competence factor (CF) accelerates the process of transport or leakage of autolysin molecules into the periplasmic space. Moreover, in *H. influenzae* no competence factors have been reported. Only changes in cell envelope accompany the development of competence state. The cell envelope of competent cells contain increased level of polysaccharide as compared to the cells of log phase. Structural changes in competent cells induce numerous vesicles called transformosomes bud from the surface that contains protein and mediates the uptake of transforming DNA (Goodgal,1982). Transformation is accomplished in the following steps (Fig.8.4).

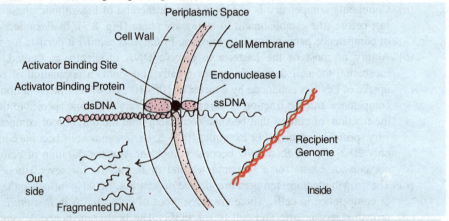

Fig. 8.4 : Diagrammatic presentation of transformaiton in *streptococci* (based on Moat and Foster, 1995).

(*a*) **DNA binding:** As a result of random collision, DNA comes first in the contact of cell surface of competent bacteria (Figs. 8.4 and 8.5 A-B). First the DNA binding is reversible and lasts for about 4-5 seconds. Thereafter, it becomes irreversible permanently. For about 2 minutes it remains in non-transforming state. There after, before 5 minutes it is converted into the transforming state. The period (about 10 minutes) during which no transformation occurs in competent recipient cells is called **eclipse**. Both types of DNA, transforming and non-transforming, bind the cell surface where the receptor sites are located. In *B. subtilis* membrane vesicles in competent cells are found that bind to 20 mg of dsDNA/mg of membrane protein. The competent cells show six fold more DNA binding sites than the non-competent cells.

In *H. influenzae* transformosomes bud forms the surface and contains proteins that mediate DNA uptake. It binds with conserved sequence (5'AAGTGCGGTCA 3') present at 4 kb interval on DNA (Goodgal, 1982). The DNA uptake site contains two proteins of 28 and 52 kilodaltons. After binding, the receptor proteins present the donor DNA to the membrane associated uptake sites (Goodgal,1982).

In *S. pneumoniae* the CF induces the ability to bind DNA molecules.

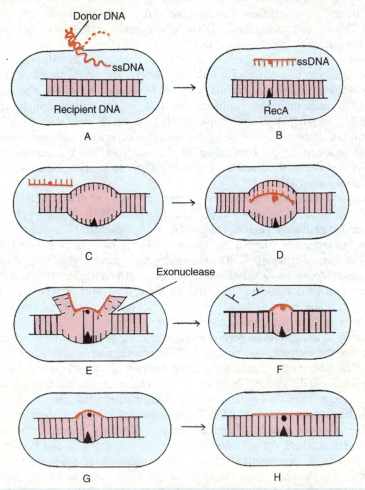

Fig. 8.5 : Mechanism of transformation. A, binding and penetration of donor DNA in competent recipient cell; B, binding of SSB to donor ssDNA and RecA to recipient DNA; C-D, synapsis and assimilation; E, nicking of recipient stand; F, branch migration, trimming and integration; G, sealing of nicks by DNA ligase; H, mismatch repair.

(*b*) **Penetration :** The DNA molecules that bind permanently enter the competent recipient cells. DNA is also resistant to DNase degradation. The nucleolytic enzymes located at the surface of competent recipient cells act upon the donor DNA molecule when it bind, the cell membrane. The endonuclease-1 of the recipient cells which is associated with cell membrane acts as DNA translocase by attacking and degrading one strand of the dsDNA. Consequently only complementary single strand of DNA enters into the recipient cells (Figs. 8.4 and 8.5A).

It has been confirmed by performing the experiments with radiolabelling of donor DNA. The mutant cells of *S. pneumoniae* lack endonuclease - 1, therefore, transformation does not occur. Interestingly in *B. subtilis* degradtion of one strand is being delayed. Hence, both the strands enter the recipient cell. The upper limit of peneterating DNA into the recipient cell is about 750 base pairs.

The size of donor DNA affects transformation. Successful transformation occurs with the donor DNA of molecular weight between 30,00,000 and 8 million dalton. With increasing the concentration of donor DNA the number of competent cells increases. DNA uptake process is the energy requiring mechanism because it can be inhibited by the energy requiring inhibitors. After penetration the donor DNA migrates from periphery of cell to the bacterial DNA. This movement in different bacteria differs. For example in *B. subtilis* this movement occurs for about 16-60 minutes. During this movement, DNA is associated with mesosomes which possibly transport it to the bacterial DNA (Notani and Setlow, 1974).

(*c*) **Synapsis formation:** The single stranded DNA is coated with SSB protein and maintain, the single stranded region in a replication fork (Fig. 8.5B). The single strand of the donor DNA or portion of it is linearly inserted into the recipient DNA (Fig. 8.5 C-D). The bacterial protein like *E. coli* RecA protein, probably facilitates the DNA pairing during recombination. It causes the local unwinding of dsDNA of the recipient cell from the 5' end. How the displaced single strand is cut, still not known ? Base pairing i.e. synapsis occurs between the homologous donor ssDNA and the recipient DNA. Unwinding of the recipient DNA continues at the end of assimilated DNA and allows the fraction of invading DNA to increase base pairs. This processs is called branch migration (F).

(*d*) **Integration :** The endonuclease cuts the unpaired free end of donor DNA or the recipient DNA. This process is called trimming (Fig. 8.5E-F). The nick is sealed by DNA ligase (G). Consequently, a heteroduplex region containing a mismatched base pairs is formed (H). Furthermore, in the progenies whether the donor marker is or is not recovered, it depends on the occurrence of mismatch repair. If the mismatch repair occurs again, it depends whether the unpaired base in the donor or recipient strand is removed (Friefelder, 1987). After replication the heteroduplex forms the homoduplexes, one of these is of normal type and the second is transformed duplex. The normal duplex is from the recipient cell in origin, whereas the transformed duplex is from the donor genome.

The efficiency of integration of genetic markers into the genome of recipient cell varies with different genes that the recipient cell possesses. This genetic trait is called **hex** (high efficiency of integration). The hex system eliminates a large fraction of low efficiency (LE) markers and permits high efficiency (HE) markers to be integrated. Therefore, the hex function is a mismatch-base correction system. The donor genes differing from the recipient genes by a single base pair create a mismatch when integrated initially. The hex mismatch repair system (with LE markers) can correct either of donor strands. Therefore, there is fifty fifty chance for a given marker to be retained. The HF markers correct only the recipient strand (Moat and Foster, 1995).

For the LE markers, hex mismatch repair system unusually removes the mismatched bases of the donor DNA and the cell retains the recipient genotype, whereas for HE markers the same system removes the mismatched bases of recipient DNA and the cell consists of donor genotype. In the later case, after replication of chromosome and cell division the one progeny cell contains the donor genotype and the other has the recipient genotype. These two types of cells can be differentiated through plating method by using the antibiotic markers.

However, for pneumococci it is a general feature that all the strains discriminate between LH and HE markers when transformation has occurred with homologous DNA. The hex⁻ cells (mutant in hex function) fail to discriminate between the two markers and, therefore, integrate all markers with high efficiency, because one of the two daughter cells after cell division contains the genotype.

2.   Transfection

Transfection is the process that involves transformation of bacterial cells with purified bacteriophage DNA resulting in production of the complete virus particles. The organisms not considered naturally transformable (e.g. *E. coli, Salmonella typhimurium*) can be transformed *in vitro*. Transformation can occur by using $CaCl_2$ or electroporation (electric shocks) that brings about alteration in outer membrane.

In 1964, Foldes and Trautner were the first to show the infection of bacterial protoplast with purified nucleic acids and to this phenomenon they gave the term transfection. Since then transfection has been demonstrated in a number of bacteria such as *B. subtilis, E. coli, H. influenzae, S. typhimurium, Staphylococcus, Streptococcus,* etc. When the transfected cells forming colonies on agar plates are lysed, clear zones are visible. These zones are called plaques. Notani and Setlow (1974) have discussed the mechanism of bacterial transformation and transfection. In recent years much work is being done on transfection *in vitro* for genetic engineering purpose. However, it has been an important factor in the success of researches in recombinant DNA technology.

3.   Conjugation

For the first time Joshua Lederberg and Edward L.Tatum (1946) in their brilliant and remarkable experiment presented the evidence for bacterial conjugation i.e. a process of transfer of genetic material by cell-to-cell contact. They procured the two different auxotrophs (the mutant prototrophs lacking ability to synthesize an essential nutrient and, therefore, obtaining it or precurssor from its surroundings) of *E. coli*, mixed them and incubated the two strains for hours in nutrient medium and plated on minimal medium (devoid of biotin, phenylalanine and other amino acids). They used the double or triple auxotrophs to rescue from the chance of reversion. One strain (58-161) required biotin (Bio⁻), phenylalanin (Phe⁻) and cystine (Cys⁻) for their growth, hence designated as Bio⁻ Phe⁻ Cys⁻ Thr⁺ Leu⁺ Thi⁺. The second strain (W677) required threonine (Thr), leucine (Leu⁻) and thiamine (Thi⁻) and designated as Bio⁺ Phe⁺ Cys⁺ Thr Leu⁻ Thi⁻. After incubation the recombinant prototrophic colonies (i.e. microbe requiring the same nutrients as the majority of naturally occurring microbial species) grew on minimal medium. Production of recombinant prototrophs (i.e. Bio⁺ Phe⁺ Cys⁺ Thr Lei⁻ Thi⁻) would have been possible as a result of recombination) between the two auxotrophs (Fig 8.6).The recombinant prototroph had capacity to synthesize all the six growth factors i.e. amino acids.

Lederberg and Tatum (1946) could not give the proof for physical contact of cells required for gene transfer. The evidence for cell-to-cell contact was provided by Bernard Davis (1950) who built a U shaped tube. Two separate pieces of curved glass tubes were prepared and fused at the base to form a U shape with a fritted glass filter between the halves (Fig.8.7). The filter allows the movement of media but not bacteria from both the ends of U tube. Nutrient medium was inoculated with different auxotrophic strain of *E. coli*. When it was inoculated the medium was pumped back and brought forth from the filter to facilitate the exchange of medium present on either side of the filter. The bacterial strain from both the halves of U tube were plated after

4 hour, of incubation onto minimal medium. The bacterial colonies did not appear on medium because cell-to-cell contact could not be established and, therefore, gene transfer did not occur. Hence, no recombiant prototrophs were produced.

(*i*) **Role of Surface Protein in Conjugation :** Conjugation differs from transformation with the fact that in the former physical contact is established between two different strains through a conjugation tube. The genetic material from the donor cell (male) is transferred to the recipient (female) cell. There are special appendages present on bacterial cell surface which are called sex pilus or F pilus which form the conjugation tube. The fertility (F) factor enables the cell to act as donor. The F factor of donor cell includes the informations of sex pili the number of which varies from 1 to 3. The cells containing an autonomous F are referred to as F^+ cells. It replicates independently. There are only 1-3 copies of F factor per cell.

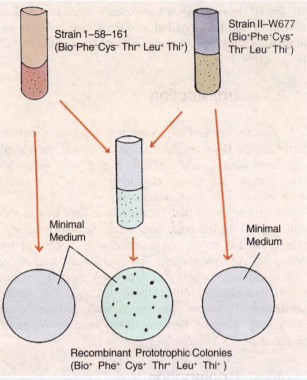

Fig. 8.6 : Demonstration of Lederberg and Tatum's experiment of bacterial conjugation by using triple auxotrophs.

Under certain specific conditions the number of pili per cell goes to five. The number of pili corresponds to the number of copies of F factor. This explains that each F factor synthesizes a single pilus whether it is autonomous replicating conditions (as plasmid) or in integrated conditions (as episome). Moreover, the recipient cells possess receptor sites on cell surfaces which are required for conjugation. Certain bacteriophages e.g. f_2, MS_2, and $Q\beta$ act as donor. The donor *E. coli* cells possess sex pili as well as type I pilus on their cell surfaces. For example phage M12 is adsorbed randomly only on sex pili but not on cell surfaces of recipient bacterial cell (Slewel, 1992).

(*ii*) **The F Factor:** The presence of F factor in a bacterial cell determines its autonomous replication, sex pili formation and conjugal transfer function. Thus, it governs the sexuality and conjugation. Two mating types in *E. coli* K12 have

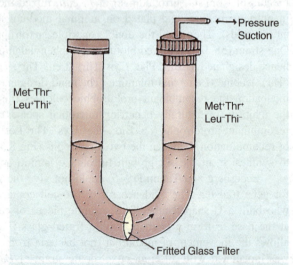

Fig. 8.7 : The U-tube experiment of Bernard Davis to show the need of cell-to-cell contact for conjugation.

been found depending on presence and absence of the F factor. The F factor remains in two stages as plasmid and as episome. The F plasmid replicates independently. However, sometimes it is integrated with the normal chromosome of the bacterium. Therefore, it is referred to as *episome*. The F factors have shown the following significant features :

(a) When F⁺ strain of a bacterium is incubated on a nutrient medium mixed with acridine orange, it is converted into F⁻ strain. Acridine orange is effective only with the growing bacteria as it inhibits the autonomously replicating F factor.

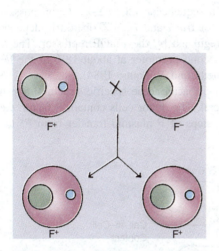

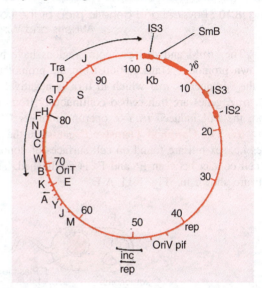

Fig 8.8 : Cross between F⁺ and F⁻ cells.　　　　**Fig. 8.9 :** Genetic map of F plasmid.

(b) A cross between two F⁻ strains does not yield recombinants. It is always sterile as the F⁻ strain cannot undergo conjugation with the other F⁻ strain.

(c) The crosses between F⁺ and F⁺ strain yield F⁺ cells but a very low level. Some of F⁺ cells are converted into F⁻ genotype.

(d) Transfer of F factor from F⁺ to F⁻ in F⁺ x F⁺ crosses occurs at a frequency of about 100% but the production of recombinants occurs at the rate of one per 10^4 to 10^5 cells.

(e) Among F⁺ strains there are certain F⁺ substrains that show about 1000 time more the rate of recombination with F⁻ strains. The substrains are called *high frequency recombination* (Hfr) strains. The Hfr strains are produced when F factor integrates with the bacterial chromosomes.

(iii) Genetic Map of F Plasmid: A genetic map of F plasmid is shown in Fig. 8.9. The F plasmid of *E. coli* is about 100 kb with genes coding for autonomous replication, sex pili formation and conjugal transfer function. The F plasmid contains the transfer (*tra*) region and non-transfer related markers. The non-transfer related markers are the insertion sequences (IS3, δ, γ and IS2), stable DNA degradation (srn B), inhibition of replication by T7, and II phages (*pif*), and a region for replication (*rep*), incompatibility (*inc*) and origin of vegetative replication (*ori*V). Some genes of R plasmid with their functions are given in Table, 8.1.

Willetts and Wilkins (1984) have given the physical and genetic map of transfer region of F plasmid (Fig. 8.10) which is about 32 kb long consisting of about 25 known transfer genes. Twelve genes are involved in F pilus formation (e.g. *tra* A,-L,-E,-K,-B,-V,-W/C,-U,-F,-H,-G). Genes involved in regulation are *fin*P and *tra*J. Stabilization of mating pairs is done by genes *tra*N and *tra*G, conjugative DNA metabolism by *tra*M, *tra*Y, *tra*D, *tra*I and *tra*Z and surface exclusion by *tra*S and *tra*T.

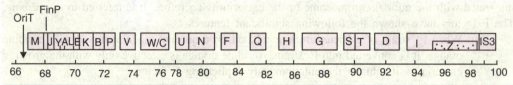

Fig. 8.10 : Physical and genetic map of the transfer (tra) region of F-plasmid (based on Willetts and Wilkins, 1984).

The *tra*M and *tra*J promotor regions have been sequenced and *tra*Y-Z operon possesses its own promotor. Transcription from the promoters for *tra*M and *tra*Y-Z operon is dependent on the product or *tra*J which in turn is negatively regulated by the *Fin*OP repressor. The *tra*I and *tra*Z genes are transcribed continuously from a second promoter at about 18% or the level from the *tra*I induced *tra*Y-Z operon promoter (Willetts and Wilkans, 1984).

(iv) **The Conjugal Transfer Process:** In the Enterobacteriaceae specific structural append- ages i.e. sex pili are found on cell surfaces of donor cell. The donor cells contain F factor. Cell- to-cell contact between F⁺ and F⁻ is established. The steps of F plasmid transfer from F⁺ to F⁻ cell are shown in Fig. 8.11 A-E.

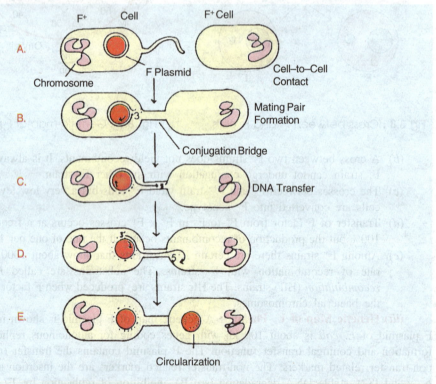

Fig. 8.11 : Steps in F plasmid transfer. A, cell-to-cell contact; B, formation of matching pair; C-D, DNA transfer; E, circularization of transferred DNA.

A pool of preformed subunits is incorporated into mature sex pili. The number of sex pili vary from 1 to 3 per cell. The tip of pilius is involved in the stable mating pair formation (governed by *tra*N and *tra*F genes) when interacts with the *omp*A gene product on the outer surface of the recipient cell. After the initial contact between the tip of pilus and recipient cell (A) the pilus

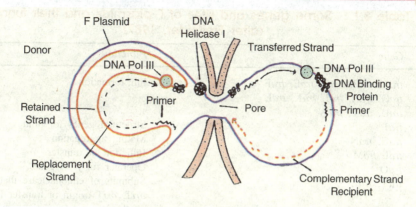

Fig. 8.12 : A model for conjugative transfer of F plasmid (based on Willetts and Wilkins, 1984)

contracts and brings the F⁺ and F⁻ cells into the close proximity (B). This wall to wall contact forms a conjugation bridge involving the fusion of the cell envelopes (Figs. 8.11B and 8.12). At *ori*T site of plasmid a nick is made by *tra*YZ endonuclease yielding in 5'-terminus single strand that invades the recipient cells. The 5'-terminus of DNA binds with a pilot protein and travels gradually through this membrane bridge (probably a pore involving the *tra*D DNA gene product (Fig. 8.12) but not through the pilus itself as it was originally believed. It has been found that the mating mixture of *E. coli* form mating aggregates of 2- 20 cells each rather than only mating pairs. After the formation of mating aggregates transfer of F⁺ DNA starts from *ori*F region as opposed to *ori*V as a plasmid enclosed endonuclease (*tra*I gene product) nicks the F plasmid at *ori*T. The intact strand acts as template and the 5' end strand is transferred to the recipient cell through a rolling circle mechanism of replication (C-D) (*see* Chapter 15).

At the nicked *ori*T site, the *tra*M triggers conjugal DNA synthesis by exposing sufficient *ss*DNA to facilitate the binding of helicase (a *tra*I gene product) or DNA helicase I (Fig.8.12). The helicase I moves on the other strand which is under going transfer for unwinding the plasmid duplex. Helicase I migrates with DNA polymerase III synthesizing the replacement strand of the donor DNA. If the helicase I binds to the membrane complex during conjugation, the concomitant ATP synthesis might provide the motive force to displace the transferred strand into the recipient cell. After entering into the recipient cell, the 5' end strand is attached to the membrane and undergoes replication. Transfer of DNA is associated with synthesis of a replacement strand in the donor cell and of a complementary strand in the recipient cell. Both the processes require *de novo* primer synthesis and the activity of DNA polymerase III holoenzyme. It is assumed that a single strand binding protein coats the DNA and help the conjugal DNA synthesis. This depends upon the nature of the pore. This protein is also transferred from the donor to the recipient cells (Willetts and Wilkins,1984). After the synthesis of complementary strand the F plasmid is circularized (Fig. 8.11 E).

(*v*) **Barriers to Conjugation:** It has been found that some times the cells containing F factor are poor recipient, when conjugative crosses occur. This is due to the presence of surface exclusion. A similar phenomenon (incompatibility) occurs when a F' element is transferred into a recipient cell that already contained F plasmid. The F' element renders F plasmid to become unable for fertility. It has been found out that for surface exclusion two genes (*tra*S and *tra*T) are required with *tra*T protein which is an outer membrane-protein. It is hoped that *tra*T protein may block the stabilization sites of mating pair or inhibits the structural proteins required for stabilization of mating pair.

Table 8.1 :   Some genes and sites of F plasmids and their function
(after Freifelder, 1987)

Gene	Function
traA, traB, traC, traE, traF traG, traH, traJ, traK, traL traU, traV, traW	Pili formation
traJ	Structural genes for pilin
traG, traN	Mating aggregation
traI, traM	Initiation of transfer
traO	Operater for *fin*O gene
traY, traZ	Subunits of endonuclease that nicks *ori*T, *ori*T origin of transfer DNA synthesis.
*ori*V	DNA replication origin.
traS, traT	Surface exclusion (inhibition of mating between F$^-$ containing cells), encodes membrane proteins.
il_z A, il_z B	Lethal zygosis (killing of females by excess Hfr cells)
tp	F replication
inc	Incompatibility of IncF group
*fin*O, *fin*P	Fertility inhibition (found in Col R and F$^-$ like plasmids, but not F)

In addition, in most of the conjugative plasmids e.g. R 100 transfer of DNA is markedly reduced as compared with F. This is because the fertility inhibition system (FinOP) controls the regulatory system of *tra* genes. The *fin*O and *fin*P gene products interact and form a FinOP inhibitor of *tra* gene expression.

(*vi*) **The High Frequency Recombination (Hfr) Strains:** When the chromosome of F$^+$ cell integrates with F plasmid, it is called high frequency recombination (Hfr) cell. The Hfr cells arise from F$^+$ cultures (Fig.8.13A). Even after integration of F into chromosome, the chromosome retains a single, circular DNA molecule. The F acts as it was a part of the chromosome. The F increases the size of chromosome. However F is capable of transferring the whole chromosome from Hfr cells to the F$^-$ culture. The frequency of insertion occurs at about $10^{-5} - 10^7$ per generation i.e. in a bacterial population of 10^7 F$^+$ cells, there is possibility of 1-100 cells in having an integrated F plasmid with chromosome.

Following are some of the differences between F$^+$ cells and Hfr cells:

(*a*) The F factor of Hfr cells is rarely transferred during recombination. In an Hfr x F, the frequency of recombination is high and that of transfer of F factor in low. In contrast in F$^+$ x F$^-$ cross, the frequency of recombination is very low and that of transfer of F factor is high.

(*b*) It takes about 2 minutes for transfer of F, whereas 100 minutes for entire bacterial chromosome transfer. This difference is mainly due to the relative size of F and the integrated chromosome.

(*c*) In a cross between F$^-$ and Hfr cells, F$^-$ cells always remain F$^-$ because of separation of cells before final transfer of ultimate F segment.

(*d*) In a mating between an Hfr leu$^+$ culture and an F$^-$ leu$^-$ culture, F$^-$ leu$^+$ cells arise. The genotype of the donor is not changed because the concurrent replication in the donor replaces the transferred DNA strand.

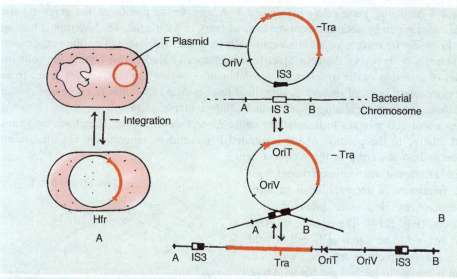

Fig. 8.13 : Integration of F plasmid by a reciprocal exchange between an IS element in F and homologous sequence of bacterial chromosome. A, origin of Hfr cell; B, mechanism of integration.

(*a*) **Mechanism of integration :** The mechanism of integration is shown in Fig. 8.13B. Integration is a reciprocal DNA exchange. From base sequencing study it is clear that an integrated F sequence is flanked by two copies of one of the insertion sequences (IS elements) present in F plasmid. Thus, integration involves homologous recombination between two circular DNA molecules resulting in one circular molecule that contains both the DNAs. Both the IS elements (IS2 and IS3) present on F plasmid and IS element on bacterial chromosome (*E. coli* also contain 5 each of IS2 and IS3 sequences) set as the homologous regions for insertion. The integration of F plasmid depends on *rec*A but rarely independent on *rec*A. The F integration also takes place depending on transposition of IS elements. Thus Hfr cells arise due to homologous recombination between the two indentical IS elements, one is present in chromosome and the other in F plasmid. Secondly, the Hfr cells also arise by forming a cointegrate mediated by an IS element in F plasmid, and by duplicating a target sequence in the chromosome.

There is directionality to conjugal DNA transfer. The direction of transfer from *ori*T is such that the *tra* genes are transferred in the last. The orientation of chromosomal IS element is such that the host gene B is as a proximal marker, whereas gene A is transferred in the last.

After integration, the F DNA replicates along with the host chromosome. When the host *dna*A gene is non-functional, replication of whole chromosome can begin from an integrated F DNA.

(*b*) **The order of chromosome transfer and conjugation mapping:** Wollman *et al.* (1956) determined, by interrupted mating experiments, the order of chromosome transfer from an Hfr donor to an F⁻ recipient cell. The

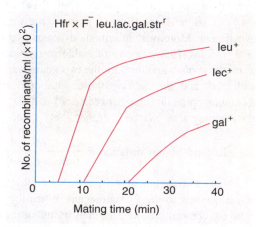

Fig. 8.14 : Gene mapping by Hfr interrupted mating experiment.

time at which a particular gene enters a recipient is related to the portion of the genes on the chromosome. A map can be obtained from the time of entry of each gene. The interrupted mating experiment involves: (*i*) mixing of an Hfr strain with F⁻ strain, (*ii*) interrupting the conjugation at certain intervals by breaking the cells apart in a high speed blender, (*iii*) plating the cells on various types of selective media to select the recombinant cells that had received the genes from Hfr strain before interruption of mating. A simplified linkage map of circular *E. coli* chromosome constructed from interrupted mating experiment is shown in Fig. 8.14.

It takes about 100 minutes to transfer the entire *E. coli* chromosome. Therefore, the genes are mapped relative to the position of the integrated F plasmid by determining the time taken by the gene to be transferred to the recipient cell.

The inerrupted mating experiment also reveals the frequency of recombination of each marker identified by detectable mutation at a particular locus (Fig. 8.14). The genetic markers are *leu*, *lac* and *gal*. The donor Hfr cell is wild type, whereas the recipient is *leu, lac⁻, gal⁻* i.e. the recipient is mutant in *lac* and *gal* but wild type in *leu*. Therefore, the mutant requires lactose or galactose as carbon source. The Hfr is streptomy- cin- sensitive (Str^s) and the recipient is streptomy- cin- resistant (Str^r). After mixing the donor and recipient cells at zero time, the aliquots of mixture are removed at different intervals and mating pairs disrupted by blending. The mixture is plated on minimal media containing (*i*) glucose to select for Leu⁺ recombinant (*ii*) lactose plus leucine to select for Lac⁺ recombinants, and (*iii*) galactose + leucine to select for Gal⁺ recombinants. Colonies growing on these media are the recombinants i.e. the recipient into which the wild type donor gene

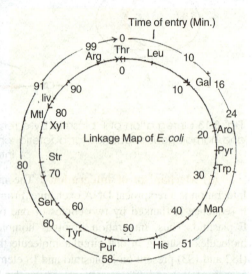

Fig. 8.15 : Simplified linkage map of *E.coli* chromosome constructed from interrupted mating experiment.

was transferred and replaced the mutant gene. Thus, the conjugate transfer of Hfr chromosome is time dependent. Each gene enters the F⁻ cell at a particular time. By measuring different time intervals a graph can be plotted (Fig. 8.14) and linkage map can be constructed (Fig. 8.15).

(c) **High Resolution Mapping:** One gets low resolution mapping by interrupted mating experiment. Moreover, this method is not useful for high resolution mapping within a distance of 2 minutes. Hence, study of stable recombinants rather than the gene transfer is required. For example, if *abc⁺* and *thr⁺* are the two genes transferred, the frequency of colony can be calculated with *thr⁺* and *thr⁻* among those with *abc⁺* genes if *abc⁺* is more frequent than *thr⁺*. The recombinational distance between *abc⁺* and *thr⁺* can be obtained from the proportion of *thr⁻* (*abc⁺ thr⁻*) among total *abc⁺* i.e. [(*abc⁺ thr⁺*) + (*abc⁺ thr⁻*)] as below:

$$\text{Recombination distance} = \frac{(abc^+thr^-)}{(abc^+thr^+)+(abc^+thr^-)}$$

(vii) **Formation of F-Prime (F'):** Integration of F factor is a reversible process. The Hfr cell can revert to the F⁺ state again. When the reversible process occurs the F factor is set free from the chromosome and resumes its autonomously replicating capability. Separation of F factor from the integrated chromosome occurs aberrantly at a low frequency and yields plasmid containing F factor and a small segment of chromosome is called F' cells. The F' is of two types. Type I F' has lost some sequence but carries some host DNA located at one or the other side

of the integrated F. Type II F' contains all of F' plus some host DNA from both sides of the point where F was integrated. In both the condition F' contains a small segment of chromosome.

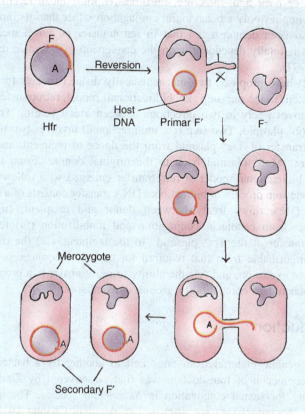

Fig. 8.16 : Diagrammatic representation of F⁻ conjugation and formation of merozygote.

When such primary F' cells are crossed with F⁻ recipients, the F factor is transferred efficiently together in F⁻ converting them into the secondary F' cells. The secondary F' cells are partially diploid hence called as **merodiploid** or **merozygote** because the recipient cells, in addition to its own chromosome, contained a segment of DNA from the donor cell i.e. F' cell (Fig. 8.16). This process of transfer of bacterial DNA from donor cell to recipient cell as a part of sex factor has been called *sexduction* by Jacob and Wollman (1961).

The F' conjugation is very important in the study of microbial genetics to find out whether the allele carried by an F' plasmid in merozygote is dominant or recessive to the chromosomal gene. Study of F' plasmid is also useful in mapping the chromosome since two neighbour genes are picked up by an F factor.

4.   Retrotransfer

Undoubtedly, William Hayes (1952) erected the scientific frame work of bacterial genetics, plasmid biology and horizontal gene transfer (conjugation) mechanism. Later on much work was done. Thomas D. Brock (1990) has delineated two phases of research in bacterial genetics, pre-Hayes and post-Hayes. In his discovery of unidirectional transfer of genetic material, Hayes deduced the inequality between the strains, and identified 58-161 as a donor and W677 as a recipient.

However, during the mid-1980s, the universality of the unidirectional model of conjugation was questioned. IncP1, IncM and IncN plasmids were reported to mediate 'back transfer' or back mobilization from the recipient into the donor (Thiry *et al.,* 1984, Mergeay, *et al.,* 1985). This observation led to the birth of a novel conjugation phenomenon which is known as *retrotransfer,* Retro signifies return or back and, therefore, retrotransfer must mean return transfer or back transfer. Retrotransfer implies only a behavioural conjugation rather than its molecular or genetic mechanism. It has also been demonstrated that in retrotransfer no novel mechanism of DNA transfer is involved. It is totally dependent upon the conversion of a female recipient to a male donor (Ankenbauer,1997).

Top *et al.* (1992) have proposed two mechanistically distinct models for retrotransfer, the one-step and the two-step. In the one-step (i.e. bidirectional) model, retrotransfer is a single event during which DNA moves freely in two directions between a cell bearing Tra$^+$ plasmid and a cell carying a Tra$^+$ Mob$^+$ plasmid. Two step (i.e. unidirectional) involves two transfer events, the first event being the transfer of Tra$^+$ plasmid from the donor to recipient, and the second step the transfer of the Tra$^+$ Mob$^+$ plasmid back to the original donor. From the work done in subsequent years, the bidirectional model of retrotransfer emerged with following characteristics : (i) retrotransfer is a one step process of bidrectional DNA transfer consists of a single conjugative event during which DNA flows freely between donor and recipient, (ii) retrotransfer is mechanistically distinct from canonical conjugation and mobilization (iii) retrotransfer is not dependent upon the transfer of the Tra$^+$ plasmid to the recipient, (iv) the time required for a retrotransfer is indistinguishable from that required for canonical conjugation, (v) retrotransfer is unaffected by surface exclusion, and (vi) the ability to retrotransfer is a property possessed by an exclusive set of plasmid in compatibility groups (Ankenbauer,1997).

5.   Transduction

The transfer of genetic material from one cell to another by a bacteriophage is called transduction. The phenomenon of transduction was first discovered by Zinder and Lederberg (1952) while searching for sexual conjugation in *Salmonella* species. The morphological and chemical structure of bacteriophages are discussed in detail in Chapter 16 *Viruses II.*

The infection by a bacteriophage is accomplished in several stages such as adsorption, penetration, replication, assembly, lysis and release. In brief the virus particles first attaches to specific receptor site on bacterial cell wall surface. The genetic material penetrates the bacterial cell, and replicates independently by using cell machinery of the host. Consequently, the virus DNA is replicated into multiple copies, and synthesises phage proteins. Complete phage particles are assembled and finally cell is lysed resulting in release of virus particles.

Depending on mode of reproduction the bacteriophages are of two types, the virulent phage and the temperate phage. The phages that reproduce by using a lytic cycle are called virulent phages because they destroy the host cell such as T phages, phage lambda (λ), etc. In contrast the temperate phages ordinarily, do not lyse the bacterial cell. The viral genome behaves as episome like F factor and becomes integrated into the bacterial chromosome. The latent form of phage genome that remains within the host without harm and integrates with chromosome is called **prophage**. Bacteria containing prophage are known as lysogenic baceria and the relationship between phage and its host is called **lysogeny.** The lysogenic bacteria can produce phage particles under some conditions, and the phage is able to establish the phenomenon of lysogeny and behaves as temperate phage.

Usually, transduction occurs most readily between the closely related species of same genus of a bacterium i.e. intragenic. This preference is due to requirement for specific cell surface receptor for recognition of the phage. In addition, intergeneric transduction has been shown between the closely related enteric bacteria such as between *E. coli* and *Salmonella* or *Shigella* species. Several genetic traits for example fermentation potential, antigens, chemical resistance are transducible

(Moat and Foster,1995). Transduction is of two types, generalized transduction and specialized transduction.

(*i*) **Generalized Transduction:** Generalized transduction was discovered in 1952 by Norton Zinder and Joshua Lederberg. They were repeating the experiments of Lederberg and Tatum (1946) on conjugation that occurred in *E.coli* K12 taking another bacterium *Salmonella typhimurium*. They selected the following two strains of *S. typhimurium*.

(*a*) **The LA22 Strains :** This strain was unable to synthesize the amino acids, phenyl alanine and tryptophan (Phen⁻ Trp⁻ strain) but could synthesize methionine and histidine (Met⁺ His⁺ strain).

(*b*) **The LA2 strain :** It was unable to synthesize methonine and histidine (Met⁻ His⁻ strain) but could synthesize phenylalanine and tryptophan. It is written as LA2 Met⁻ His⁻ Phen⁺ Trp⁺.

They found that a mixture of two autotrophic strains resulted in prototrophs at the rate of $1/10^5$ cells. The wild type prototrophs could synthesize all the four amino acids (Phe⁺ Trp⁺ Met⁺ His⁺). Though genetic recombination occurred in *S. typhimurium,* yet it was not as a result of conjugation which was confirmed later on.

Each strain was added in an arm of U tube. The two arms of U-tube were separated by a bacteria proof sintered glass filter which allowed free movement of nutrient media but not bacteria (Fig. 8.17). After applying alternate sunction and pressure the culture medium was allowed to pass from one arm to the other. Thereafter, the two auxotrophs present in two separate arms were grown in the same culture medium. The prototrophs were recovered from LA22 culture but not from LA2 culture. From LA2 strain a genetically active filterable agent was produced which formed prototrophs in LA22. It was confirmed that the filterable agent was larger than DNA and resistant to DNase. Later on the filterable agent was confirmed as temperate *Salmonella* phage P22. The P22 was carried in one of the parental strain as prophage. Its presence was ascertained by the fact that it could be destroyed by treating it with P22 antiserum.

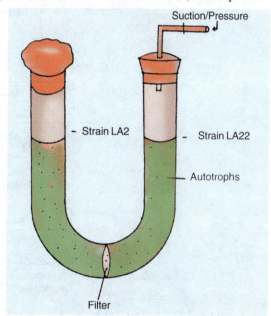

Fig. 8.17 : The Zinder and Lederberg's experiment on transduction.

Generally, P22 exists in lysogenic state in LA22 strain of *S. typhimurium*. Generalized transduction occurs during the lytic cycle of virulent or temperate phages. Outline of generalised transduction is given in Fig. 8.18. After peneteration (A) phage genome multiplies within the cell (B,C), the bacterial chromosome is fragmented into pieces (C). During assembly when virus DNA is packed into protein capsid, by mistake the random fragment of partially degraded bacterial chromosome is also packed (D). Since the upper limit of DNA to be packed in capsid is about 44 kb, some or all viral DNA is left behind. The quantity of bacterial DNA carried by phage DNA depends mainly on the size of capsid. Phage P22 carries about 1% of bacterial chromosome and P1 phage of *E. coli* can carry about 2-25% of bacterial genome. The frequency of such defective phage is about 10^{-5} to 10^{-7} of the total progeny phage produced. This defective phage is also called *generalized transducing particle*. It is the carrier to a other (Ebel Tsipis *et al.,* 1972).

When this defective phage infects another bacterium, the genome is introduced within the host cell (Fig. 8.18 E). The transferred bacterial DNA (exogenote) is integrated into the recipient bacterial chromosome (endogenote) (F). About 70-90% of transferred DNA is not integrated with the recipient chromosome but survives and expresses itself (G).

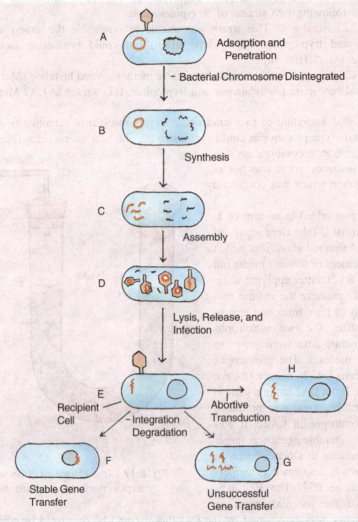

Fig. 8.18 : Outline of generalised transduction by bacteriophages.

Generalised transduction also helps the mapping of bacterial genes because the chromosomal segment which has been transferred by bacteriophage contains hundreds of genes. These are tested by using the genetic markers. For example, two temperate (P1 and 363) transfer markers from one strain of *E. coli* to the other. When coliphage P1 is grown in a wild type *E. coli* cells that utilizes threonine, leucine and sodium azide (Thr+ Leu+ Azi+) and then used to infect a Thr- Leu- Azi- recipient auxotroph, only 3% of threonine prototrophs (Thr+ Leu- Azi-) were produced. A few recombinants were leucine prototrophs (Thr+ Leu+ Azi+). This shows that only a short segment of bacterial DNA is involved and usually only a single marker locus is produced. Rarely, multiple transduction (transduction of several gene loci) occurs. This also shows that Thr+ is closely linked to *leu+* but not *azi+*. Therefore the order of linkage is *thr+ leu+ azi+*.

Abortive transduction : When a segment of bacterial DNA (exogenote) is introduced into another bacterium by a bacteriophage, the exogenote may be integrated into the recipient bacterial chromosome (Fig. 8.18 E-F). This type of transduction is called complete transduction. In contrast, when the exogenote is not integrated into the endogenote and remains free, it is called abortive transduction (H). The recipient bacteria that contain this non-integrated transduced DNA and are partially diploid, are called **abortive transductants**. When the abortive transductant divides, out of two, only two daughter cells contain exogenote in each generation. The other cells do not contain the exogenote.

The cells containing exogenote synthesize the functional enzymes because the transduced DNA specifies the normal complement by enzymes. However, in non–transduced cells the synthesis of functional enzyme gradually decreases with the progress of cell multiplication. This results in formation of slow growing small colonies or microcolonies.

(ii) Specialized (Restricted) Transduction: Certain temperate phages can also transfer only a few restricited genes of the bacterial chromosome to the recipient bacterial cell. This transfer of bacterial genes adjacent to prophage only to the recipient chromosome is called restricted or **specialized transduction** (Fig. 8.19 A-H). For the first time Morse *et al*. (1956) discovered specialized transduction. It is made possible by an error in the lysogenic life cycle of phage. When a phage genome is introduced in the bacterial cell, it becomes integrated with bacterial chromosome as prophage (A). Upon induction the DNA becomes free containing a small segment (about 5 to 10%) of bacterial chromosome (B). It multiplies and disintegrates the bacterial chromosome (C). After assembly of phage DNA plus bacterial chromosome, the bacteriophage is released from the bacterial host (D). Usually this phage is defective and lacks

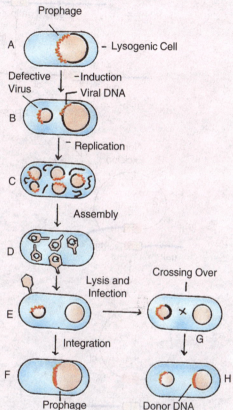

Fig. 8.19 : Specialised transduction by a temperate bacteriophage.

some part of its attachment site. When this defective phage infects a bacterium it introduces its DNA containing a piece of bacterial chromosome (E). The defective phage cannot reproduce without the assistance of helper phage. The genes of the phage can insert with homologous DNA of the infected bacterium (F). Sometimes crossing over occurs between the homologous gene loci of the bacterial chromosome and the donor DNA attached with phage genome (G) resulting in integration of the donor DNA with the recipient DNA (H).

(a) Low frequency transducing (LFT) lysates : The mechanism of specialized transduction and production of LFT are shown in Fig. 8.20. The best studied example of specialized transduction is the phage λ of *E. coli*. The genome of phage λ is inserted into the bacterial chromosome at attachment *(att)* site. The *att* site of both the phage and bacterium are similar but not identical. They can complex with each other. The insertion of phage genome in the bacterial chromosome takes place always between the genes of *E. coli* e.g. *gal* (galactose locus) and *bio* (biotin locus). When prophage is excised out from *E. coli* chromosome, it some times takes with it *gal* or *bio* genes. This occurs due to improper excision of integrated prophage DNA as it occurs in the

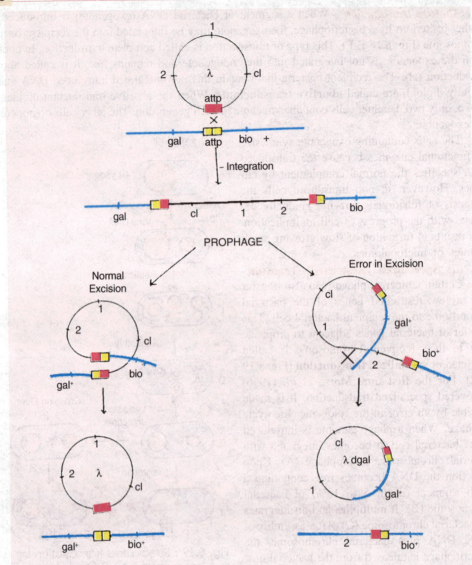

Fig. 8.20 : Mechanism of specialised transduction with phage λ in *E. coli* showing the production of low frequency transducting (LFT) lysates.

formation of F factor. Thus after cell lysis resulting from induction of lysogenized *E. coli*, some normal phages and a few defective transducing particles are released. These particles are called λdgal (λ defective *gal*) because they contain galactose utilizing gene. Since the number of defective particles in cell lysate (product) is low (10^{-5} to 10^{-6}), it is called low frequency transduction (LFT) lysates.

The defecting transducing particles possess a non-functional hybrid integration site which is in some part bacterial in origin and some part phage in origin. They lack some portion of phage genome, but they are capable of transducing and integrating into a new host carrying the original host genes with it. This establishes a merodiploid condition.

(*b*) **High frequency transduction (HFT) lysates :** If there is a normal phage λ genome in the same cell, the defective phage λ carrying the *gal* gene can integrate. The normal phage integrates resulting in two bacterial phage hybrid *att* sites where the defective λdgal can insert phage that acts as helper phage because it helps the integration and reproduction of the defective phage (Fig. 8.21) (Echols and Court,1971). The transductants are unstable because the prophage can be excised by certain stress like UV radiation.

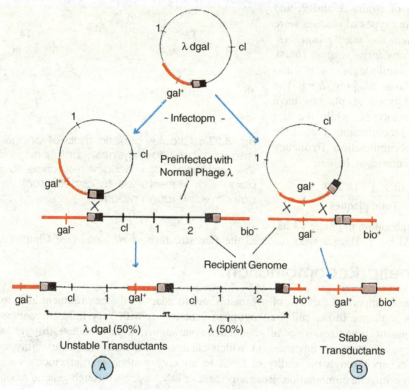

Fig. 8.21 : Mechanism of specialised transduction with phage λ and *E.coli* showing production of high frequency transduction (HFT) lysates.

The UV radiation results in new phages if a normal phage λ has also lysogenized this cell. Hence, induction of this double lysogen (dislysogen) produces high frequency transduction (HFT) lysates containing about 50% λdgal and 50% phage λ. In constrast to LFT lysate, it is shown as HFT lysate because the number of HFT particles is very effective in transduction. The HFT lysates contain transducing particles with a frequency of about 0.1 to 0.5%. However, phage λdgal infects the recipient cells containing *gal* bacterial chromosome (*gal* recipients). Crossing over occurs at the homologous *gal* sites than heterologous *att* sites. This results in production of stable transductants (B).

(iii) **Use of Transduction in Genetic Mapping in Viruses:** Viruses are the very small particles, hence they are considered unsuitable to study inheritance. Because scorable traits are not available with them. However, some of the bacterial characters have been used for inheritance and recombination studies such as plaque morphology (large or small, fuzzy or sharp), host range and virulence. Plaques are the clear transparent area produced on opaque lawn of bacteria grown on the surface of solid medium in Petri plates.

For the first time Alfred D. Hershey attempted to study the inheritance and recombination in viruses by using a cross (*h⁻r⁺ x h⁺ r, where* h = host range, r= rough plaque) in bacteriophage T2. A circular genetic map of phage T2 and T4 is given in Fig. 8.22.

The genetic traits studied in this cross were host range and plaque morphology. The *h⁺* infects strain 1 and *h⁻* infects both strains 1 and 2. The *r⁺* lyses slowly and produce small plaques, whereas *r-* lyses rapidly and produces large plaques. Both the types of phages were used to infect *E. coli* strain 1 to facilitate **mixed infection** or **double infection.** The lysate was spread over a bacterial lawn having the mixture of strains 1 and 2, and analysed. Four types of plaques were recorded: *clear and small plaques (h⁻ r⁺), cloudy and large plaques (h⁺r⁻), cloudy and small plaques (h⁺r⁺), and clear and large plaques (h⁻r⁻).* The former two types of plaques have parental phenotypes, while the last two are the recombinants. Out of the four, the recombination frequency (RF) is calculated as below:

$$RF = \frac{(h^+r^+) + (h^-r^-)}{\text{Total plaques}}$$

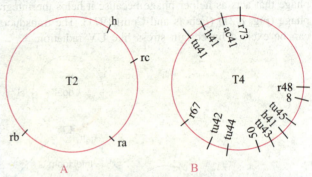

Fig. 8.22 : Circular genetic map of phage T2 (A) (showing only four genes), and phage T4 (B); *h* = host range, *ac* = acridine resistance, *tu* = turbid plaque, *os* = resistance to osmotic sock, *e* = lysis defective, *r* = rough plaque.

Recombination in phage T4 has been studied by S. Benzer who resolved the fine structure of *rII* locus (*see* Chapter 6).

III. Genetic Recombination

The most important features of organisms, are to adapt in the environment and to maintain their DNA sequence in the cells generation to generations with very little alternations. In long term survival of organisms depends on genetic variations, a key feature through which the organism can adapt to an environment which changes with time. This variability among the organisms occurs through the ability of DNA to undergo genetic rearrangements resulting in a little changes in gene combination. Rearrangement of DNA occurs through genetic recombination. Thus, recombination is the process of formation of new recombinant chromosome by combining the genetic material from two organisms. The new recombinants show changes in phenotypic characters (White-house, 1982).

Most of the eukaryotes show a complete sexual life cycle including meiosis, an important event that generates new allelic combinations by recombination. It is made possible through chromosomal exchange resulting from crossing over between the two homologous chromosomes containing indentical gene sequences. Much work was done on eukaryotic genetics until 1945 that laid the foundation of classical genetics. The work on bacterial genetics was done between 1945 and 1965 that advanced the understanding of microbial genetics at molecular level.

1.   Mechanism of Recombination

Basically, there are three theories viz., breakage and reunion, breakage and copying and complete copy choice that explain the mechanism of recombination (Fig.8.23).

(i) Breakage and Reunion: Two homologous duplex of chromosome laying in paired form breaks between the gene loci a and b, and a⁺ and b⁺ (Fig. 8.23A). The broken segments rejoin crosswise and yield recombinants containing a and b⁺ segment, and a⁺ and b segment. This type of recombination does not require the synthesis of new DNA. This concept has been used to explain genetic recombination.

(ii) Breakage and Copying: One helix of paired homologous chromosome (ab and a⁺ b⁺) breaks between a and b (Fig. 8.23B). Segment b is replaced by a newly synthesized segment copied from b⁺ and attached to a section. Thus the recombinants contain ab⁺ and a⁺ b⁺.

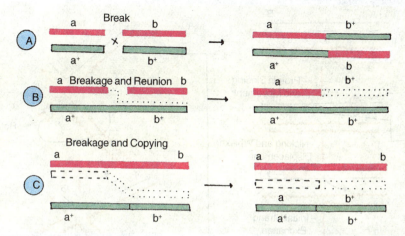

Fig. 8.23 : Three possible mechanisms of recombination.

(*iii*) **Complete Copy Choice:** In, 1931, Belling proposed this theory for recombination of chromosome in higher animals. However, it has been questioned by several workers. Therefore, it has only historical importance. According to this theory a portion of one parental strand of homologous chromosome acts as template for the synthesis of a copy of its DNA molecule. The process of copying shifts to the other parental strand. Thus, the recombinants contain some genetic information of one parental strand and some of the other strand (Fig. 8.23 C).

2. Types of Recombination

Many kinds of recombinations occur in microorganisms. These are classified basically in two groups, (*i*) general recombination, and (*ii*) site specific recombination.

(*i*) **General Recombination :** General recombination occurs only between the complementary strands of two homologous DNA molecules. Smith (1989) reviewed the homologous recombination in prokaryotes. General recombination in *E. coli* is guided by base pairing interactions between the complementary strands of two homologous DNA molecules (Smith, 1989).

Double helix of two DNA molecules breaks and the two broken ends join to their opposite partners to reunite to form double helix. The site of exchange can occur any where in the homologous nucleotide sequence where a strand of one DNA molecule becomes base paired to the second strand to yield heteroduplex just between two double helices (Fig. 8.24). In the heteroduplex no nucleotide sequences are changed at the site of exchange due to cleavage and rejoining events. However, heteroduplex joints can have a small number of mismatched base pairs.

General recombination is also known as homologous recombination as it requires homologous chromosomes. In bacteria and viruses general recombination is carried out by the products of *rec* genes such as RecA protein. The RecA protein is very important for DNA repair, therefore, it is *rec*A dependent recombination.

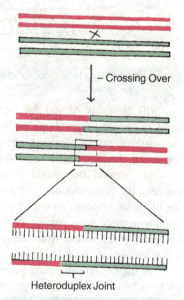

Fig. 8.24 : Formation of heteroduplex joint consisting of base pairs of two different DNA helices.

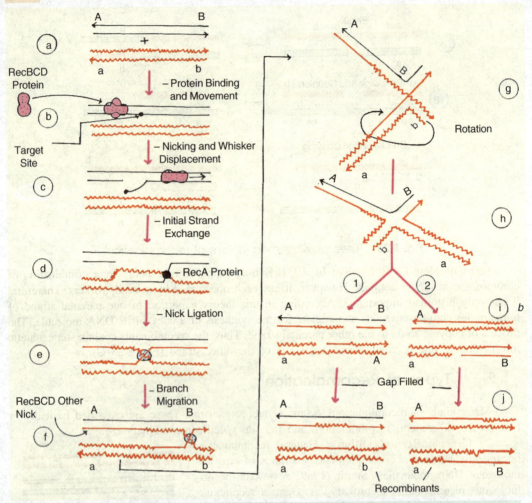

Fig. 8.25 : The Holliday model for reciprocal general recombination (modified after Holliday, 1974).

(ii) **Holliday Model for General Recombination:** Holliday (1974) presented a model to show the general recombination (Fig. 8.25). According to this model recombination occurs in five steps such as strand breakage, strand pairing, strand invasion/assimilation, chiasma (crossing over) formation, breakage and reunion and mismatch repair.

(a) **Strand breakage:** General recombination occurs through crossing over by pairing between the complementary single strand of DNA duplex (*a*). Two homologous regions of DNA double helix undergo an exchange reaction. The homologous region contains a long sequence of complementary base pairing between a strand from one or two original double helices and a complementary strand from the other. However, it is unknown how the homologous region of DNA recognises each other.

A list of recombination genes and their function has been given in Table 8.2. The RecBCD proteins of *rec*BCD or *rec*J genes are required for recombination in *E. coli*. This protein enters the DNA from one end of double helix, and travels along the DNA at double helix, at the rate of about 300 nucleotides per second. It creates a loop of ssDNA along travelling DNA (*b*). It uses energy derived from hydrolysis of ATP molecules. A special recognition site (*a*) sequence of eight nucleotides scattered through out *E. coli* chromosome (*b*) is nicked in the travelling loop of DNA formed by RecBCD protein.

(*b*) **Strand pairing :** The RecBCD proteins act as DNA helicase because these hydrolyse ATP and travel along DNA helix (Lohman,1992). Thus, the RecBCD proteins result in formation of single stranded whisker at the recognition site which is displaced from the helix (c). This initiates a base pairing interaction between the two complementary sequences of DNA double helix.

(*c*) **Strand invasion/assimilation :** A single strand (whisker) generated from one DNA double helix invades the another double helix (*d*). In *E. coli* *rec*A gene produces RecA protein which is important for recombination between the chromosomes like single strand binding (SSB) protein, The RecA protein binds firmly to single stranded DNA to form a nucleoprotein filament. Roca and Cox (1990) have reviewed the structure and function of RecA protein. RecA protein promotes rapid renaturation of complementary ssDNA hydrolysing ATP in the process. RecA protein has several binding sites, therefore, it can bind a ssDNA and subsequently a dsDNA. RecA protein binds first to ssDNA, then search for homology between the donor strand and the recipient molecule. Due to the presence of these sites RecA protein catalyses a multistep reaction (called synapsis) between the homologous region of ssDNA and a DNA double helix. *E. coli* SSB protein helps the Rec protein to carry out these reactions. When a region of homology is identified by an initial base pairing between the complementary sequences, the crucial step in synapsis occurs.

In vivo experiments have shown that several types of complexes are formed between a ssDNA covered with RecA protein and a dsDNA helix. First a non-base paired complex is formed which is converted into a three stranded structure (ssDNA, dsDNA and RecA protein) when a homologous region is found (Lohman,1992). This complex is unstable and spins out a DNA heteroduplex plus a displaced ssDNA from the original helix. Once the homologous regions are encountered and the ssDNA and dsDNA are complexed, a stable D-loop is formed (*d*).

Table 8.2 : Recombination (*rec*) genes and their function
(after Backmann, 1990)

Gene	Map location	Gene Function
*rec*A	58	Complete recombination deficiency and many other phenotype defects including suppression of *tif* DNA dependent ATPase.
*rec*B	61	Structural gene of exonuclease V couples ATP hydrolysis to DNA unwinding.
*rec*C	61	Structural gene of exonuclease V
*rec*D	61	α subunit of exo V.
*rec*E	30	Exonuclease VIII, 5'——>3' dsDNA
*rec*F	83	Recombination deficiency of recB⁻ rec C⁻ sbc B⁻ strains blocks UV induction of λ prophage.
*rec*J	64.6	Recombination deficiency of *rec*B⁻ *rec*C⁻ *sbc*B⁻ strains.
*rec*G	82.6	ATPase, disrupts Holiday structure.
*rec*R	11	Help recA utilize SSB-ssDNA complex as substrate.
*rec*O	56	Promotes renaturation of complementary ssDNA.
*ruv*A	41.6	Complexes with Holiday junction.
*ruv*B	41.6	ATPase, dissociates Holiday junction.
*ruv*C	—	Endonuclease, Holiday junction, resolvase.
*ruv*Q	86.5	DNA helicase.
*ruv*L	83	Recombination deficiency of *rec*B⁻ recC⁻ *sbc*B⁻ strain.
*sbc*A	30	Suppress or of *rec*B⁻ and *rec*C⁻ mutations, controlling gene of *rec*E.
*sbc*B	44	Structural gene for exonuclease I

(*d*) **Branch migration:** The next step is the assimilation of strand and nick ligation (e). The donor strand gradually displaces the recipient strand which is called *branch migration*. After formation of synapsis, the heteroduplex region is enlarged through protein-directed branch migration catalysed by RecA protein. RecA protein directed branch migration proceeds at a uniform rate in one direction due to addition of more RecA protein to one end of RecA protein filament on the ssDNA. Branch migration can take place at any point where two single strands with the sequence make attempts to pair with the same complementary strand. An unpaired region of the other single strand resulting in movement of branch point without changing the total number of DNA base pairs. Special DNA helicases that catalyse protein directed branch migration are involved in recombination.

In contrast, the spontaneous branch migration proceeds in both the directions almost at the same rate. Therefore, it makes a little progress over a long distance.

(*e*) **Chiasma or crossing over formation :** Exchange of a single strand between two double helices is a different step in a general recombination event. After the initial cross strand exchange, further strand exchanges between the two closely opposed helices is thought to proceed rapidly. A nuclease cleaves and partly degrades the D-loop at some points.

At this stage possibly different organisms follow different pathways. However, in most of the cases an important structure called cross-strand exchange (also called Holliday Juncture or chi form or chiasmas, is formed by the two participating DNA helices (g). A chi form of single stranded connections in the cross over region has also been observed under the electron microscope by Dressler and Potter (1982).

The chi form of two homologous helices that initially paired and held together by mutual exchange of two of the four strands where one strand originates from each of the helices (g). The chi form has two important properties, (*i*) the point of exchange can migrate rapidly back and forth along the helices by a double branch migration, and (*ii*) it contains two pairs of strands, one pair of crossing strands and the other pair of non-crossing strands.

(*f*) **Breakage and reunion:** The chi structure can isomerise several rotations (h). This results in alteration

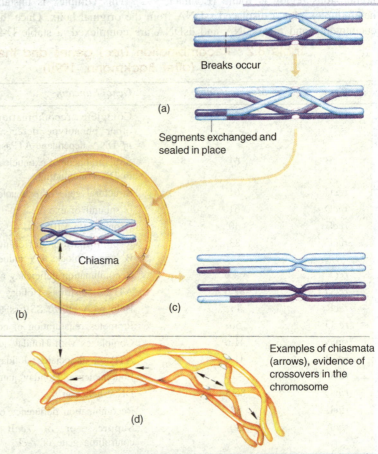

Crossing over and formation of chiasmata. It breaks up old and combination of alleles and puts new ones together in pairs of homologous chromosomes.

of two original non-crossing strands into the crossing strands, and the crossing strands into the non-crossing strands. In order to regenerate two separate DNA helices, breakage and reunion in two crossing strands are required. If breakage and reunion occur before isomerization the two crossing strands would not occur. Therefore, isomerization is required for the breakage and reunion of two homologous DNA double helices resulting from general genetic recombination (Kowalczykowski,1991). Breakage and reunion occur either in the vertical or horizontal plane. If breakage occurs horizontally the recombinants would contain genotype AB/ab with a little change in base sequences at the inner regin (i). However, if breakage occurs vertically the recombinants would contain Ab/aB (J). The RurC protein and RecG protein expressed from *ruv*C and *rec*G genes respectively are thought to be alternative endonucleases specific for Holliday structure.

(*iii*) **Mismatch Repair (Mismatch Proof Reading System):** It is such a repair system which corrects mismatched base pairs of unpaired regions after recombination. This system recognises mismatched function of DNA polymerase. The mechanism involves the excision of one of the other mismatched bases along with about 3,000 nucleotides. This RecFJO is involved in the repair of short mismatch either in the initial stage or at the end of recombination (Radman,1989).

The two proteins MutS and MutL are present in bacteria and eukaryotes. The MutS protein binds to mismatched base pair, whereas MutL scan the DNA for a nick (Fig. 8.26). When a nick is formed MutL triggers the degradation of the nicked strand all the way back through the mismatch, because the nicks are largely confined to the newly replicated strands in eukaryotes, replication errors are selectively removed. In bacteria the mechanism is the same except that an additional protein, MutH, nicks the unmethylated GATC sequences and begins the process.

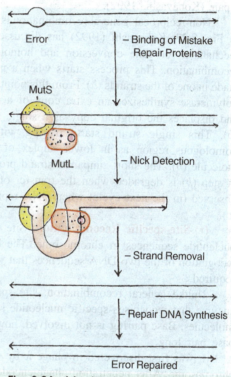

Fig. 8.26 : Mechanism of removal of error in newly made strand by mismatch repair system.

It has been demonstrated in yeast and bacteria that the same mismatch repair system which removes replication errors as in Fig. 8.26 also interrupts the genetic recombination events between imperfectly matched DNA sequences. It is known that homologous genes in two closely related bacteria (*E. coli* and *S.typhimurium*) generally will not recombine, even after having 80% identical nucleotide sequences. However, when mismatch repair system is inactivated by mutation, the frequency of such interspecies recombination increases by 100-fold. This mechanism protects the bacterial genome from sequence changes that would be caused by recombination with foregin DNA molecules entering in the cell (Rayssiguier *et al.* 1989).

(*iv*) **Non-reciprocal Recombination (Gene Conversion):** The fundamental law of genetics is that the two partners contribute the equal amount of genes to the offsprings. It means that the offsprings inherit half complete set of genes from the male and half from the female. One diploid cell undergoes meiosis producing four haploid cells; therefore, the number of genes contributed by male gets halved and so the genes of female. In higher animals like man it is not possible to analyse these genes taking a single cell. However, in certain organisms such as fungi it is possible to recover and analyse all the four daughter cells produced from a single cell through meiosis.

Occasionally, three copies of maternal allele and only one copy of paternal allele is formed by meiosis. This indicates that one of two copies of parental alleles has been altered to the maternal allele. This gene alteration is of non-reciprocal type and is called **gene conversion**. Gene conversion is thought to be an important event in the evolution of certain genes and occurs as a result of the mechanism of general recombination and DNA repair (Kobayashi, 1992).

Non-reciprocal general recombination is given in Fig. 8.27. Kobayashi (1992) has discussed the mechanism for gene conversion and homologous recombination. This process starts when a nick is made in one of the strands (a). From this point DNA polymerase synthesizes an extra copy of a strand and displaces the original copy as a single strand (b). This single strand starts pairing with the homologous region as in lower duplex of DNA molecule (b). The short unpaired strand produced

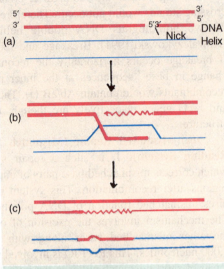

Fig. 8.27 : A model for non-reciprocal general recombination.

in step (b) is degraded when the transfer of nucleotide sequence is completed. The results are observed (in the next cycle) when DNA replication has separated the two non-matching strands (c).

(*v*) **Site-specific Recombination:** Site specific recombination alters the relative position of nucleotide sequences in chromosome. The base pairing reaction depends on protein mediated recognition of the two DNA sequences that will combine. Very long homologous sequence is not required.

Unlike general recombination, site specific recombination is guided by a recombination enzyme that recognises specific nucleotide sequences present on one of both recombining DNA molecules. Base pairing is not involved, however, if occurs the heteroduplex joint is only a few base pair long.

It was first discovered in phage λ by which its genome moves into and out of the *E. coli* chromosome. After penetration phage encoded an enzyme, lambda integrase which catalyses the recombination process (Fig. 8.28). Lambda integrase binds to a specific attachment site of DNA sequence on each chromosome. It makes cuts and breaks a short homologous DNA sequences. The integrase switches the partner strands and rejoins them to form a heteroduplex joint of 7 bp long. The integrase resembles a DNA topoisomerase in rejoining the strands which have previously been broken (Landy,1989). Site specific recombination is of the following two types:

(*a*) **Conservative site-specific recombination:** Production of a very short heteroduplex by requiring some DNA sequence that is the same on the two DNA molecules is known as conservative site-specific recombination. The detail procedure is described in Fig. 8.28.

(*b*) **Transpositional site-specific recombination:** There is another type of recombination system known as transpositional site-specific (TSS) recombination. The TSS recombination does not produce heteroduplex and requires no specific sequences on the largest DNA. There are several mobile DNA sequences including many viruses and transposable elements that encode integrases. The enzyme integrases by involving a mechanism different from phage λ insert its DNA into a chromosome. Each enzyme of integrases recognises a specific DNA sequence like phage λ.

K. Mizuuchi (1992a) reviewed the mechanism of transpositional recombination based on the studies of bacteriophage Mu and the other elements. The enzyme integrase was first purified from Mu.

Similar to integrase of phage λ, the Mu integrase also carries out of its cutting and rejoining reactions without requirement of ATP. Also they do not require a specific DNA sequence in the target chromosome and do not form a joint of heteroduplex.

Different steps of TSS recombinational events are shown in Fig. 8.29. The integrase makes a cut in one strand at each end of the viral DNA sequences, and exposes the 3'-OH group that protrudes out. Therefore, each of these 3'-OH ends directly invades a phosphodiester bond on opposite strands of a randomly selected site on a target chromosome. This facilitates to insert the viral DNA sequence into the target chromosome, leaving two short single stranded gaps on each side of recombinational DNA molecule. These gaps are filled in later on by DNA repair process (i.e. DNA polymerase) to complete the recombination process. This mechanism results in formation of short duplication (short repeats of about 3 to 12 nucleotide long) of the adjacent target DNA sequence. Formation of short repeats is the hall-marks of a TSS recombinatin (Mizuuchi,1992b).

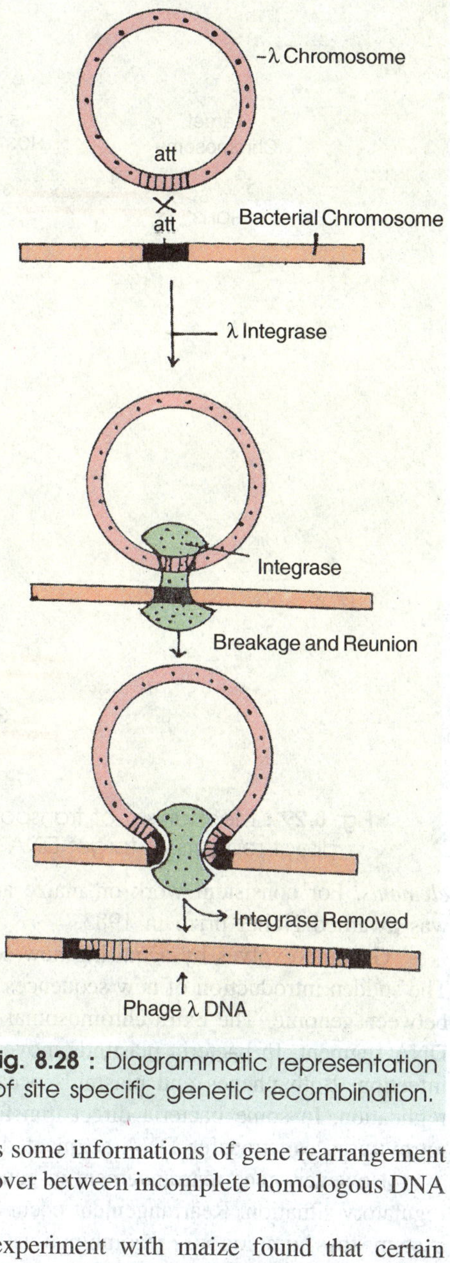

IV. Transposable Elements

For a long time it was assumed that the genes have fixed chromosomal locations. Genetic recombination is a common process of a cell within which exchange between the homologous DNA sequences takes place. Over the past few decades some informations of gene rearrangement have come forth that results from illegitimate crossing over between incomplete homologous DNA sequences.

Fig. 8.28 : Diagrammatic representation of site specific genetic recombination.

In 1940s, McClintock Barbara in her genetic experiment with maize found that certain genetic elements regularly jump to new locations, and thereby affect gene expression. Therefore, kernels of maize ear show variation in colours. Her work was followed 30 years later by the recognition that the bacteria contain mobile DNA sequence. The genetic segments moving from one location to the other were also discovered in bacteria well during 1970s but the frequency was quite low ($10^{-7} - 10^{-2}$ per generation). Several apparent random mutations disrupting gene function in *E.coli* were found to occur due to insertion of a large DNA segments. Many of these sequences had the jumping properties similar to maize elements (Bukhari *et al.*,1977). Later on the properties of such genes of *E. coli* could be known at molecular levels through the method of gene cloning and DNA sequencing (Shapiro,1983). These moving genetic elements have been called with different names such as *jumping genes, movable genes, transposons* and *transposable*

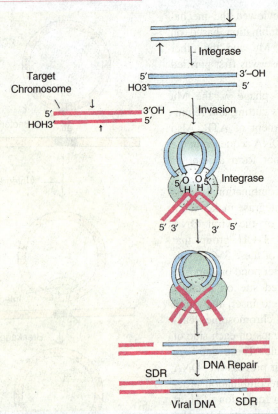

Fig. 8.29 : Mechanism of transpositional site-specific recombination; SDR, short direct repeats of target DNA sequence (modified after Mizuuchi, 1992).

elements. For consistent work on maize and discovery of jumping genes, McClintock Barbara was awarded Nobel prize in 1983.

Genome evolves by acquiring new sequences and by rearrangement of existing sequences. The sudden introduction of new sequences results from the ability of a factor to carry information between genome. The extra chromosomal element provides informations by transferring a small DNA segment. In bacteria plasmids move through conjugation, whereas phages transfer through infection. Both phages and plasmids occassionally transfer host genes along with their own replication. In some bacteria direct transfer of DNA occurs by transformation. In addition, the retroviruses can transfer DNA segment during infection cycle in eukaryotes (Lewin,1994).

However, gene rearrangement in genome may form new sequences by placing them in new regulatory situation. Rearrangement occurs either from recombination under taken by the cellular system for homologous recombination and repair, or from the transportation induced by transposons (Lewin,1994). Movement of transposable element from one chromosome site to another is shown in Fig. 8.30.

Generally, the mobile segments of temperate phages (e.g. λ and Mu phages) which resemble insertion sequences are called transposable elements, and the mobile sequence, of bacterial system are called **transposons**. Thus, the transposable elements have been defined as *DNA sequences which can insert into several sites in a genome.* The phenomenon of moving genetic segments from one location to the other in a genome is known as *transposition.* Transposable elements are recognised by the presence of inverted repeat DNA sequences at their ends. These sequences are necessary for the DNA between them to be transposed by a particular enzyme (transposase) associated with the transposable element.

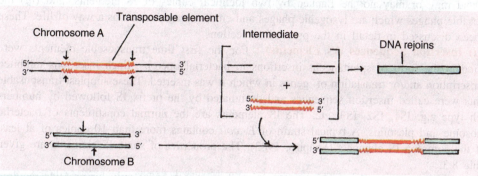

Fig. 8.30 : Movement of a transposable element from one chromosomal site to another. Arrows indicate the site for action of enzyme and cutting of DNA stands.

There are two types of transposition, replicative and conservative transposition. The replicative transposition involves the events of both replication and recombination processes generating the two daughter copies of the original transposable elements, one remaining at the parental site and the other at the target site. In addition, the conservative transposition does not involve replication. Simply the elements move to a new site. When the target site is present within a gene both types of transposition takes place.

1. Nomenclature of Transposable Elements

Campbell *et al.* (1977) have described the nomenclature of transposable elements in prokaryotes. The transposons which were discovered first did not contain any known host genes. Therefore, for historical reasons these were called insertion sequences or IS elements and designated as IS1, IS2, IS3, etc.

Transposons contain easily recognizable bacterial genes especially for antibiotic resistance. The members of this class are designated by the abbreviation, Tn followed by a number such as Tn1, Tn2, Tn3, etc. The number distinguishes different transposons. When it is necessary to refer the genes carried on transposons these are represented by strandard genotypical designations such as Tn1 (*amp*^{r}), where *amp*^{r} refers that the transposon carries the genetic locus for ampicillin resistance.

Moreover, transposon present within a particular gene creates mutation in that gene. For example if mutation occurs at position 135 in *gal*T gene by transposon 4, the notation is designated as *gal*T 135::Tn4.

Transposons have occasionally been designated in non-standard way also, for example γ, δ which is an element present in F plasmid.

However, in eukaryotes no definite pattern of naming of transposable elements is used. For example the transposable elements in *Drosophila* are named Copia, 497 and P, transposable elements of yeast as Ty1 and transposable element of maize as dissociation element and activator (Friefelder,1987).

2. Classes of Transposable Elements

Transposable elements of prokaryotes are generally called transposons. Mainly three classes of transposable elements have been categorized, (*i*) *insertion sequences* or *IS elements* also called simple transposons which contain no genetic information except the DNA sequences necessary for transposition, and also lack host gene, (*ii*) *transposons* which contains antibiotic resistance genes and may or may not be flanked by two identical copies of IS elements, and (*iii*) the transposable phages which are lysogenic phages and employ transposition as a way of life. These have been discussed in detail in the preceeding sections.

(*i*) **Insertion Sequences (IS Elements) :** For the first time transposeble elements were identified in the form of spontaneous insertions in bacterial operons. Such insertions inhibited the transcription and/or translation of genes in which it was inserted. These simplest transposable elements were called insertion sequences and designated by the prefix IS followed by numbers of each type e.g. IS1, IS2, IS3, etc. The IS elements are the normal constituents of bacterial chromosome and plasmids. A typical strain of *E. coli* contains more than 10 copies of at least one of the common IS elements (Shapiro, 1983). The properties of some IS elements are given in Table 8.3.

The IS elements are the autonomous units each of which codes only for proteins required for its own transposition. Each IS element differs in sequence size but resemble in several ways in organization (Table, 8.3). Mostly all of them are of about 1000 bp long ending with inverted terminal repeats of about 10-40 bp long. Structure of an IS elements before and after insertion at a target site is shown in Fig. 8.31.

Table 8.3 : The properties of insertion sequences

IS elements	Occurrence	Length (bp)	Inverted terminal repeat (bp)	Direct repeat (bp)	No. of copies
IS1	*E. coli* chromosome	768	23	9	6-10
IS2	" "	1327	41	5	4-13
IS3	" "	1400	38	3.4	5-6
IS4	*E. coli* K12 chromosome	1428	18	11-12	1-2
IS5	" "	1195	16	4	10-11
IS50 R	Tn5 on PJR67	1534	9	9	—
IS101	pSC101	201	37	—	—
IS903	Tn 903 on R6	1057	18	9	—
IS10R	Tn 10 on R100, pSM14	1329	22	9	—

The inverted terminal repeat has two copies of repeats which are closely related. For example the same sequence of inverted terminal repeat (ITR) of 9 bp (e.g. 987654321 bp long) is encountered from the flanking DNA on either side of it after proceeding towards the elements. The terminal repeats are believed to serve as recognition sequence for transposition enzymes, the transposases in their role of fusing the ends of IS elements with the target DNA. Besides IS1, all the IS elements possess a single long coding sequence that starts just inside at one end of ITR and terminate just before or within inverted repeat at the other end. This codes for the transposase. However, IS1 consists of a more complex organisation with two reading frames. The transposase is produced by making a frame shift translation to allow both reading frames to be used (Lewin, 1994).

After transposition by IS element, a sequence of target site of host DNA at the site of insertion is duplicated. The duplicated segment is known as direct repeat. It is repeated in the same orientation. This duplication of sequence of target site is revealed upon comparision before and after insertion of IS element. Therefore, after transposition one copy of sequence of target i.e. direct repeat is present on both the ends of transposon (Fig. 8.31) (Kleckner, 1981).

However, a transposon results in different sequences of direct repeats, the length of any IS element remains almost constant that is 9 bp. Within a host DNA insertion of most of IS elements occur at a variety of sites. The frequency of transposition varies among different elements. The overall rate of transposition is 10^5-10^4 per element per generation (Lewin, 1994). Different IS

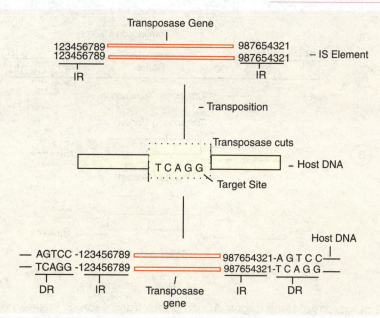

Fig. 8.31 : Diagrammatic presentation of insertion sequence (IS element) and its transposition into host chromosome; IR, inverted repeats; DR, direct repeats of target DNA.

elements contain different number of bases. The elements contain at least two apparent coding sequences initiated by an AUG and termination with an in-phase stop codon. Study of a large number of bacterial transposable elements indicates that each element encodes at least two proteins (Freifelder, 1987).

The insertion sequences can be detected in two ways, (*i*) they interact and inactivates, genes into which they insert, and (*ii*) they may contain promoters which allow RNA polymerase to transcribe and thus turn on adjacent genes. They do not perform any function in bacterial cells except that they may act as natural agent of genetic change by bringing about structure and function. Many structural variations have been observed in IS element regarding the source and host range of IS elements. Two generalization, can be made, (*i*) the same kind of IS element such as IS1 can be found on plasmid, phage genome and chromosome of different bacteria, and (*ii*) several IS elements can function efficiently in bacteria that do not contain the same IS element in their chromosome. For example, IS10 is active in both *E. coli* K12 and *Salmonella typhimurium*, although their chromosome do not contain IS10 sequence. Several copies of IS elements may be found on bacterial genome. For example, 8 copies of IS1 and about 5 copies of IS2 are found in *E. coli* chromosome (Prajapati and Batish, 1988).

(*ii*) **Transposons:** Transposons possess additional genetic proportion for encoding genetic information such as drug resistance unrelated to transposition process, and may or may not be flanked by IS elements. Transposons are of two types, composite transposon and complex transposon.

(*a*) **The composite transposons:** The composite transposons are those which consist of a central region carrying antibiotic resistant genes flanked at both the ends by identical copies of an IS element. Therefore, composite transposons carry drug resistance or other markers in addition to transposition (Fig. 8.32A). These are named with Tn followed by number as described earlier. It is a class of larger transposons. Three frequently studied composite transposons are Tn5, Tn9, and Tn10.

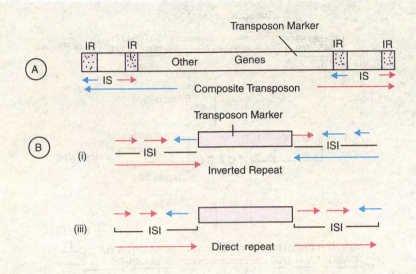

Fig. 8.32 : Diagram of transposons; A, composite transposon; B, inverted (*i*) and direct (*ii*) repeats.

Tn5 element shows kanamycin resistance (kan^r) and consists of 5,400 base pair segment with 1,450 bp inverted repeats at both ends of the segment. It can be transposed from phage λ to the chromosome of *E .coli* and from a locus of *E. coli* chromosome to another locus. If inserted into genes it causes mutations.

Tn9 transposon consists of an R factor derived gene for chloramphenicol resistance (*cam^r*). The enzyme that confers drug resistance consists of 2,638 sequences in the middle of Tn flanked by 768 bp long IS1 element on either side. Tn9 IS1 is present in direct order with small inverted repeat at ends. The *cam^r* segment is translocated from an R-factor to F-episome as well as phage λ through the phage P1. T9 differs from the other transposons in its instability as a high frequency loss of its antibiotic resistance can be encountered.

Tn10 consists of tetracycline resistance genes (*tet^r* genes). It is 9,300 bp long consisting of inverted repeats, on either side of 1,400 base pairs. The IS elements is IS10. Tn10 can be translocated from R222 (a drug resistance plasmid) to phage P22.

Structure of composite transposons **:** In the composite transposons, the IS elements can be in an inverted or direct repeat configuration (Fig. 8.31B). The two ends of IS elements are themselves inverted repeats. The relative orientation (direct or inverted) of the flanking IS elements of a composite transposon does not alter its terminal sequences. Thus a composite transposon with arms of direct repeats has the structure : arm L- central region - arm R. The structure becomes : arm L-central region - arm R, if the arms are inverted repeats. The arrows show the orientation of arms according to the orientation of genetic map of the transposon from left (L) to right (R) (Fig. 8.31B, i-ii). The features of composite transposons are given in Table 8.4.

Moreover, there are some cases where modules of a composite transposon are identical, for example Tn9 (has direct repeat of IS1) or Tn903 (inverted repeats of IS 903 present), where as in certain cases the modules are closely related. Hence, the modules in Tn10 or Tn5 can be distingushed. However, when the modules are identical both can sponsor the movement of transposon as found in Tn9 of Tn903. When the modules differ, they may also differ in functional ability. Therefore, transposition depends entirely on one of the modules for example Tn10 or Tn5.

Thus an IS module when functional, tranposes either itself or the whole transposon. The ability of a single module to transpose the entire composite trnsposon explains the lack of selective pressure for both the modules to remain active. The IS elements code for transposase activity that is responsible both for creating a target site and for reconizing the ends of transposon. Only the ends are needed for a transposon to serve as a substrate for transposition (Lewin,1994).

(*b*) **The complex transposons (TnA transposon family):** The TnA family of transposons includes Tn1,Tn2, and Tn3. TnA consists of quite large elements (about 500 bp). These contain independent units carrying genes for transposition as well as for drug resistance. These are not composite relaying on IS-type transposition modules. The TnA family includes several related transposons of which Tn3 and Tn10 are the best studied ones. TnA family is also known with the name Tn3 family because Tn3 transposon was first discovered in bacterial plasmids in 1974 by Hedges and Jacob. Tn3 is an example of a complex transposon. It takes a modular structure as found in Tn10 and is not based on IS elements. Also, it does not has evolutionary links to IS elements. The basic structure of Tn3 is described for TnA family.

One of the unique features of TnA family is that it limits multiple insertions of the same transposon into a plasmid. Most surprising finding is that the transposition frequency to other plasmids in the same cell is unaffected. Heffron *et al.* (1979) have given the analysis of DNA sequence of transposon Tn3 (Prajapati and Batish,1988).

Table 8.4 : The properties of composite and complex transposons

Transposons	Direction of terminal IS elements	Length (bp)	Target (bp)	Terminal IS elements &size (bp)	Genetic Marker*
Composite					
Tn5	Inverted	5700	9	IS50(1500)	kan^r
Tn9	Direct	2638	9	IS1 (768)	can^r
Tn10	Inverted	9300	9	IS10 (1400)	tet^r
Tn204	Direct	2457	—	IS1(768)	cam^r
Tn903	Inverted	3100	—	IS903(1050)	kan^r
Complex					
Tn1	Inverted	4957	5	38	amp^r
Tn2	Inverted	4957	5	—	amp^r
Tn3	Inverted	4957	5	38	amp^r
Tn4	Inverted	20,000	—	—	amp^r, str^r, sul^r

* Resistance against Kanamycin (kan), Chloramphenicol (Cam), tetracycline (tet), ampicillin (amp), streptomycin (str) and sulfonamide (sul).

Structure of Transposons of TnA family: Transposons of TnA family consist of inverted terminal repeats (of 38 bp long but none are flanked by IS-like elements), an internal *res* site and three known genes e.g. *tnp*A, *tnp*R and amp^r. The gene *tnp*A encodes for transposase and *tnp*R encodes for resolvase. The amp^r gene (5 bp long) is generated at the target site as direct repeat and codes for β-lactamase that confers resistance to ampicillin. The *res* site consists of three subunits: I, II and III (Fig. 8.33)

Mechanism of transposition: The bacterial transposon Tn3 has been extensively studied. Analysis of DNA sequences and its junction with target DNA provides some clue to the mechanism of transposition. Movement of transposons occurs only when the enzyme transposase recognises and cleaves at either 5' or 3' of both ends of transposon, and catalyses at either 5' or 3' of both ends of transposon and catalyses a staggered cut at the target site (Figs. 8.30 and 8.34A). Depending on transposon, a duplication of 3-12 bases of target DNA occurs at the site where insertion is to be done. One copy remains at each end of the transposon sequence.

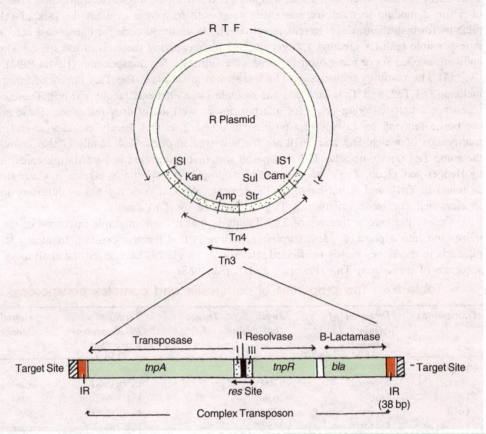

Fig. 8.33 : The structure of R plasmid and transposons. R1 plasmid carries resistance gene for five antibiotics (*i.e.* chloramphenicol-*cam'* streptomycin-*str'*, sulfonamide-*sul'*; ampicillin-*amp'*, and kanamycin-*kan'*). These are found in Tn3 and Tn4 transposons. The resistance transfer factor (RTF) codes for plasmid proteins. The structure of Tn is shown in detail. IR, inverted repeats.

After attachment of both ends of transposon to the target site, two replication forks are immediately formed (Fig. 8.34 B-C). From this stage there starts two path for carrying out onward processes. The first model is the replication path where the transposon replicates and the replicated DNA sealed to flanking sequences generating a cointegrate (D). Cointegrate is resolved by the genetic exchange between the two copies of transposon resulting in a simple insertion and regeneration of donor replicon (E). This model explains the transposition of only TnA family but not explain completely for IS elements of Mu.

The second model (F-G) is the non-replicative path that generates simple insertions without formation of cointegrate. At the prime termini in the target DNA, repair synthesis occurs. The displaced single strand that attaches the transposon to the donor replicon is broken. This forms a simple insertion (G). It is likely that both the pathways can be used but the frequency of simple insertion and cointegrate formation varies.

Thus, for transposition the two enzymes, transposase and resolvase coded by *tnp*A and *tnp*R respectively are required. Transposase recognises the ends of transposon and connects them to the target site. Resolvase provides a site-specific recombination function.

Genetics of Transposition: The genes of transposase and resolvase i.e. *tnp*A and *tnp*R are identified by recessive mutations. The above enzymes accomplish the two stages of TnA mediated transposition. Like IS type elements the transposition stage involves the ends of the elements. A unique feature of TnA family is that a specific internal site is required for resolution.

The mutants of *tnp*A cannot transpose because the enzyme transposase will not be encoded. However, transposase recognises the ends of elements and binds to 25 bp long sequence located within 38 bp of the inverted terminal repeat. Transposase also makes the staggered 5 bp breaks in target DNA where transposon is to be inserted. Resolvase functions in two ways, (*i*) it acts as repressor of gene expressions, and (*ii*) provides the resolvase function. The frequency of transposition gets increased in *tnp*R mutants because tnpR represses the transcription of both *tnp*A and its own gene. Inactivation of tnpR protein allows the increased synthesis of tnpA resulting in the increased transposition frequency. Therefore, the amount of tnpA transposase is a limiting factor in transposition (Lewin,1994).

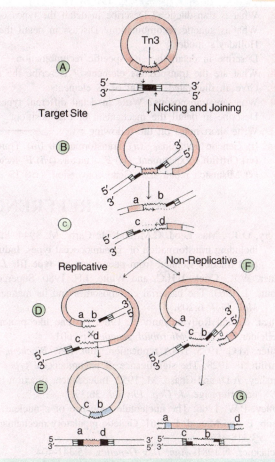

Fig. 8.34 : A model to explain the transposition mechanism of transposon Tn3.

Both the genes, *tnp*A and *tnp*R express divergently from an ATP rich enter-cistronic central region. The effects of *tnp*R are also mediated by its binding in this region. TnpR resolvase gets involved in recombination between direct repeats of Tn3 in a cointegrate structure. But in Tn3, resolution reaction occurs only at a specific site (Heffron, *et al.,* 1979).

The *res* is the site where the recombination carried by *tnp*R resolvase occurs. The *res* site is identified by cis-acting deletions. The deletions block transposition resulting in accumulation of cointegrates. The sites bound by *tnp*R resolvase have been determined by foot printing the DNA-protein complex. It binds independently at each of three sites i.e. I, II and III, each 30-40 bp long. Site I is the region genetically defined as the *res* site. In the absence of site I, resolution reaction does not proceed. However, resolution also involves binding at sites II and III. In the absence of either of II and III sites reaction proceeds poorly. Site I overlaps with the start point for *tnp*A transcription and site II with the start point for *tnp*R transcription (Heffron et al.,1979).

QUESTIONS

1. Give an illustrated account of bacterial transformation with emphasis on Griffith's experiment.
2. What is the mechanism of transformation? Discuss in brief with suitable diagrams.
3. What is conjugation ? Describe the role of surface proteins and F factors in conjugation
4. Write a note on mechanism of integration and Hfr formation.

5. What is transduction? Describe in detail the types of transduction studied by you.
6. What is genetic recombination? Discuss in detail the mechanism of recombination based on Holliday's model.
7. Describe in detail the site specific recombination.
8. What are the transposable elements ? Describe the nomenclature of transposable elements.
9. Give an illustrated account of IS elements.
10. What are transposons? Write in detail different types of transposons.
11. Describe in detail the mechanism of transposition.
12. Write short notes on the following :
    (*i*) Genetic notations, (*ii*) Transformation, (*iii*) Transfection, (*iv*) Conjugation, (*v*) Transduction, (*vi*) Griffith's experiment, (*vi*) F, - factor, (*vii*) F'-factor, (*viii*) Hfr strains, (*ix*) Abortive transduction, (*x*) Mismatch repair, (*xi*) Gene conversion, (*xii*) IS elements, (*xiii*) Transposons, (*xiv*) TnA family.

REFERENCES

Avery, O.T. , Mac Leod, C.M. and Mc Carty, M. 1944. Studies on the chemical nature of the substance including transformation of pneumococcal types. Induction of transformation by a deoxyribonucleic acid fraction isolated from pneumococcus type III. *J. Exp. Med.* 79 : 137.

Bauer, W.R., Crick, F.H.C. and White, J.H. 1980. Supercoiled DNA. *Sci. Am.* 243:118-133.

Crains, J. 1963. The bacterial chromosomes and its manner of replication as seen by autoradiography. *J. Mol. Biol.* 6:208.

Drlica, K. and Rouviereyaniv, J. 1987. Histone like proteins in bacteria. *Microbiol. Rev.* 51:301-319.

Freifelder, D. 1987. *Microbial genetics.* p. 601.

Gefter, M.G. 1975. DNA replication. *Ann. Rev. Biochem.* 44:45-78.

Griffith, F. 1928. The significance of pneumococcal types *J. Hyg.* 27:113.

Hershey, A.D. and Chase, M. 1952. Independent function of viral protein and nucleic acid in growth of bacteriophage. *J. Gen. Physiol.* 36:39.

Holley, P.W. 1966. The nucleotide sequence of a nucleic acid. *Sci. Am.* 214:30-39.

Jacob, F. and Manod, J. 1961. Genetic regulatory mechanism in the synthesis of protein *J. Molecular Biol.* 3:318-356.

Klenckner, N.1981. *Ann. Rev Genetics.* 15:341-404.

Kornberg, A. 1992. *DNA replication* 2nd ed. W.H. Freeman, San Francisco.

Kornberg, R.D. and Klug, A. 1981. The nucleosome. *Sci. Am.* 244 (2):52-64.

Kozak, M. 1983. Comparison of initiation of protein synthesis in prokaryotes, eukaryotes and organelles. *Microbiol. Rev.* 47:1-45.

Kozak, M. 1991. Structural features in eukaryotic mRNA that modulate the initiation of translation. *J. Biol. Chem.* 266:19867-19870.

Kowalczykowsk;, S.C.1991. Biochemistry of genetic recombination : energetics and mechanissm of DNA strand exchange. *Ann. Rev. Biophyso Chem.* 20:539-575.

Lehman, I.R. 1974. DNA ligase : structural mechanism and function. *Science* 186:790-797.

Lewin, B. 1994. *Genes.* Oxford Univ. Press. Oxford. p. 1257.

Matson, S.W. and Kaiser-Roger, K.A. 1990. DNA helicases. *Ann. Rev. Bio-chem* 59:289-329.

Meselson, M. and Stahl, F.W. 1958. The replication of DNA in *Escherichia coli. Proc. Natl. Acad. Sci.* 44:671.

Meyer, R.R. and Lain ,P.S.1990. The single stranded DNA binding protein of *Escherichia coli. Microbiol. Rev.* 54:342-386.

Moat, A.G. and Foster, J.W. 1995. *Microbial Physiology.* 3rd ed. Wiley-Liss, New York, p.580.

Perry, R.P. 1976. Processing of RNA. *Ann. Rev. Biochem.* 45:605-629.

Pettijohn, D. E. 1988. Histone -like proteins and bacterial chromosome structure *J. Biol. Chem.* 263:12793-12796.

Mutation and Mutagenesis

9

A. **Spontaneous Mutation**

B. **Induced Mutations**

C. **Expression of Mutations**

D. **Detection and Isolation of Mutants**

E. **DNA Repair Systems**

F. **Uses of Mutations in Microorganisms**

In 1880s, for the first time Hugo de Vries used the term mutation to describe the phenotypic changes in evening primrose, *Oenothera lamarckiana*. Mutation (Latin *mutare* means to change) refers to any heritable change in nucleotide sequence of a gene of the organisms irrespective of altered phenotypic expression of characters of the organisms. A gene codes for a protein, therefore, the physical and chemical properties of proteins are changed due to alteration in genes. Genes are made up of nucleotide sequences. Hence, changes in a gene refer to changes in nucleotides or nucleotide sequence. For example, if the normal sequence was ATT, the change to AAT may lead to entirely different amino acids into a polypeptide chain. Therefore, mutation is the result of stable and heritable changes in nucleotide sequence of DNA. These changes may include alteration of a single base pair of nucleotides or addition/deletion of one or more nucleotides in the coding region of a gene.

When an amino acid substitution has no detectable effect on phenotype, it is known as silent mutation. On the other hand if a bacterium carries such a mutation in the enzyme that synthesizes an essential amino acid, it can grow very slowly unless the medium is supplied with that substance. This type of mutation is called *leaky mutation*.

Hugo de Vries

An inducible enzyme, β-lactamase is found in *E. coli* which hydrolyses lactose to glucose and galactose, the monosaccharide constituents. When Lac⁺ *E. coli* cells are grown on nutrient medium devoid of lactose, β-galactosidase cannot be detected within the cell and such cultures are phenotypically lactose sensitive (Lac⁻). On the other hand, if lactose is substituted for glucose, *E. coli* cells quickly produce large quantity of the enzyme and permease transports lactose into the cell where it is utilized as carbon source. Here lactose does not change the phenotype of *E. coli* cells but induces the necessary enzyme for its utilization. When lactose is removed from the environment the cells stop producing β-galactosidase and revert to phenotypically Lac⁻ population. Mutation can occur in genotype of Lac⁺ cells.

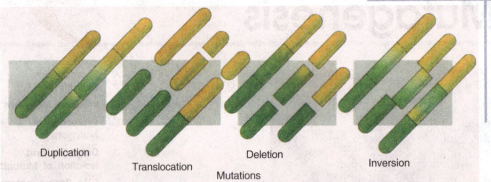

Duplication   Translocation   Deletion   Inversion

Mutations

For an organism to exist with stability, it is necessary that the nucleotide sequence of its gene must not be altered to such an extent that can promote instability. Changes to some extent brings about alteration in phenotypic expression. Though changes in genetic make up is harmful to an organism, but it is necessary to generate variability in organisms and contribute to the process of evolution in nature.

The process of formation of a mutant organism is known as mutagenesis. Mutagenesis occurs in an organism by two mechanisms, (*i*) spontaneously or (*ii*) through physical or chemical agents. The mutation that arises all of sudden without any effort is called spontaneous mutation. The mutation that arises through induction by addition of chemicals (i.e. mutagens) or radiation is called induced mutation.

A.   Spontaneous Mutation

There is no means to know when and which cell will undergo mutation. Any gene of a cell of microorganisms is vulnerable to mutation. It is not sure, however, which gene will mutate. Every gene is subject to mutation, therefore, mutation occurs in a gene spontaneously and there is possibility of mutating the genes in a cell and more probability of occurrence of mutant allele in a given population of a microorganism. In addition, it is not sure that mutation will be beneficial, it may be even detrimental too. Thus, the spontaneous mutation arises randomly in a population of organisms. Spontaneous mutations are rare ranging from 10^{-6} to 10^{-8} per generation depending on the gene and organism.

I.   The Random Nature of Mutation

Before 1940's it was believed that mutation occurs in bacterial population in response to a given selective condition i.e. a medium containing antibiotic substance. But Luria and Delbruck (1943) demonstrated the spontaneous and non-adaptive nature of mutation. These experimentations gave birth to microbial genetics. They investigated the origin of mutation in *E. coli* conferring resistance to phage T1 infection. The number of T1⁻ʳ (r, resistant) mutant cells arising in different cultures of T1^{-s} (s, sensitive) cells was compared with the number found in repeated samples of the same size taken from a single culture. The result was analysed by a statistical test named as fluctuation test (Table. 9.1). Table 9.1 is based on the experiment in which twenty 0.2 ml culture and one 10 ml culture, each containing 10^3 cells/ml of T1^{-s} bacteria were grown to 21 generation. The number of cells increased to 2.8 x 10^9 cells/ml. Each of small culture and ten of 0.2 ml of large culture were plated onto individual plate which were already uniformly spread with 10^{10} particles of T1 phage (probably sufficient for destroying all T1^{-s} cells). After incubation the number of T1^{-r} colonies were counted. The number of bacterial cells inoculated in each plate was the same (5.6 x 10^8) but the number of T1^{-r} colonies depended on whether the cells had grown in small individual culture or in large culture. Out of 20 cultures no T1^{-r} cell was detected in 11 small cultures. In rest of 9 cultures the number of T1^{-r} cells ranged from 1 to 10^7. On the other hand, each of 10 samples of large culture had the same number. This experiment shows that T1^{-r} cells arose by spontaneous mutation at different times in the growth of cultures in the absence of phage T1, therefore the number in different cultures varied greatly. However, if T1^{-r} cells arise in response to phage, these should be about equal numbers in all population of the same size.

Table 9.1 : The number of T1 phage-resistant *E.coli* mutants in small individual cultures and in samples from a large bulk culture (after Luiria and Delbruck, 1943)

Small Individual cultures		Samples from Large culture	
Culture	*T1^{-r}-colonies/plate*	*Culture*	*T1^{-r}-colonies/plate*
1	1	1	14
2	0	2	15
3	3	3	13
4	0	4	21
5	0	5	15
6	5	6	14
7	0	9	20
8	5	10	13
9	0	Variance	15
10	6	Variance/means	0.9
11	107		
12	0		
13	0		
14	0		
15	1		
16	0		
17	0		
18	64		
19	0		
20	35		
Mean	11.4		
Variance	694		
Variance/mean	60.8		

II. Evidence for Spontaneous Mutation

Lederberg and Lederberg (1952) gave the direct evidence for the origin of spontaneous mutation in T1^{-r} cells of *E. coli* without exposing to phage. This was presented by a procedure called **replica plating** (Fig. 9.1). The steps of replica plating involve: (*i*) plating of bacteria on nutrient medium and proper incubation of plates for growth of bacterial colony, (*ii*) preparation

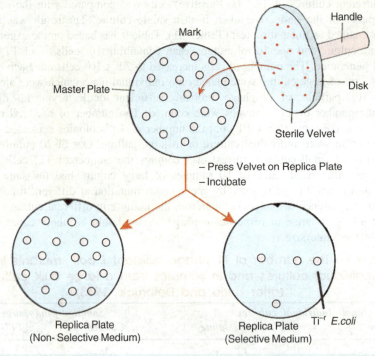

Fig. 9.1 : Replica plating technique for selection of T1⁻ᵣ *E. coli* cells.

of a solid support mounted with a piece of sterile velvet and gentle pressing it onto surface of plate supporting bacterial colonies i.e. the master plate, (*iii*) again gentle press of velvet (onto which some bacteria from each colony had sticked to fibres) on a Pertiplate containing fresh medium that allowed to transfer the bacterial cells on fresh plate i.e. replica plate. Precaution was taken to have the identical position of both the plates and velvet also. The master plate contained 10^7 colonies of bacteria, growing on non-selective medium, whereas the replica plate contained T1 phage particles spread onto medium. After proper incubation a few T1⁻ᵣ cells formed colonies in the same position on each of replica plate as in master plate. The master plate did not contain T1 phage. Therefore, appearance of T1⁻ᵣ cells on replica plate can be explained to occur due to the mutation for resistance that had taken place by chance in the master plate not exposed to T1 phage particles.

III. Mutation Rates

The probability of a gene undergoing mutation in a single generation is known as mutation rate. For the study of population genetics, evolution and analysis of effect of environment mutagens and measurement of mutation rates are important. In bacteria, a mutation can occur at any time during growth in culture. The mutant bacterial cell divides and increases the number at the same rate as the normal cells divide. Therefore, measuring the mutation rates is rather complicated.

Fluctuation test is an important method for estimation of mutation rates in bacteria. As given in column 2 of Table 9.1, a culture may contain many mutants, some mutants or none. If mutation rate per generation is μ, the probability of getting n mutants in a culture of N cells is Po, by the Poisson distribution- $e^{-\mu N}$.

The total number of divisions of individual cell needed to yield N cells from 1 cell is N-1, where N is the large number of bacterial cells.

Thus, $\mu N = \ln Po$

or $\mu = (1/N) \ln Po$ ... (1)

As shown in Table 9.1, 11 of the 20 small cultures did not contain $T1^{-r}$ cells. The average number of N cells per culture was 5.6×10^8.

Thus, the mutation rate can be calculated by putting the values in equation No.1

$$\mu = (1/5.6 \times 10^8) \text{ In } (11/20)$$
$$= 1.1 \times 10^{-9} \text{ per cell per round of replication.}$$

For the first time Benzer (1961) observed that at a particular site mutation occurs with high frequency than the other site within a gene. He mapped several thousand independently isolated rII mutation in T4 phage. Therefore, the favoured sites for mutation at high frequency were called **hot spots**. Coulondre *et al.* (1978) have studied the molecular basis of base substitution hot spots in *E. coli.*

IV. The Origin of Spontaneous Mutations

Mutation involves changes in DNA. Several mechanisms are known that bring about alterations in DNA. These modifications may arise from error in DNA replication, damage to DNA from radiation. Errors occur during replication by substitution of frame shift in DNA sequence.

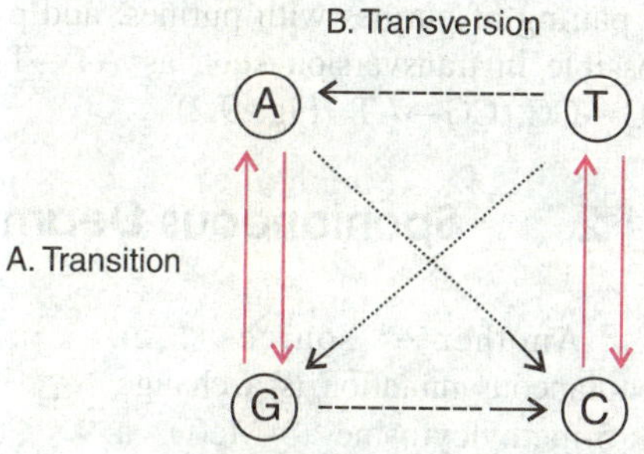

1. Substitution

Substitution of base pairs is the most common mutation. During replication of DNA repair wrong base pairs are incorporated. Base pair substitutions is of two types, transition and transversion.

(i) Transition Mutation (due to Tautomerism):

Transition is the replacement of one purine (e.g. A or G) by another purine (e.g. G or A) or one pyrimidine (e.g. T or C) by another pyrimidine (e.g. C or T). Four types of changes are possible in transition such as AT→GC, GC→AT, TA→CG, CG→TA. (Fig. 9.2). During replication, errors in DNA arise with high frequency. Some of the bases exist in alternative forms having different base pairing properties. Tautomerism is the relationship between two structural isomers that are in chemical equilibrium and readily change into one another. Generally the bases exist in keto form, but at a time they can take on either an imino or enol form (Fig. 9.3). A rare form of adenine can pair with cytosine, and the end form of thymine can pair with guanine. These tautomeric shifts alter the hydrogen bonding characteristics of the bases, and permit for purine substitution or pyrimidine for pyrimidine substitution. These lead to stable modification in nucleotide sequences.

During replication an incorrect base is correctly hydrogen bonded and

B. Transversion

A. Transition

Fig. 9.2 : Transition(A) and Transversion (B) mutations.

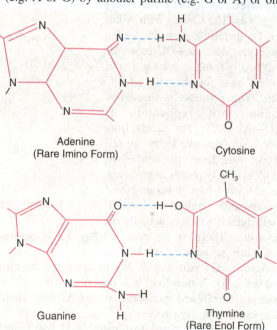

Adenine
(Rare Imino Form)

Cytosine

Guanine

Thymine
(Rare Enol Form)

Fig. 9.3 : Base-pairing between rare imino form of adenine, cytosine, and guanine and enol form of thymine.

incorporated to the template strand. Even the editing system does not recognise it as incorrect. Later on when the base assumes its normal function, the mismatch repair system corrects the mismatching bases at this level. However, if the daughter strand containing the incorrect bases is methylated, the mismatch repair system fails to distinguish between the parental and daughter strands. Therefore, the incorrect bases exist in daughter strand and lead to mutation. Such mutation is known as *transition mutation*. Transition mutations are common, although most of them are repaired by various proof reading function.

In a strand of DNA molecule a pyrimidine base C is substituted by the another pyrimidine base T which is the transition of base pair from one pyrimidine to another. After replication the error is retained in one strand and inherited in the progeny. The other strand remains normal. In the second generation after replication, purine base G is substituted by another purine base giving rise to mutant progeny.

(*ii*) **Transversion mutation:** Transversion mutation involves the substitution of purine by a pyrimidine or a pyrimidine by a purine. This type of mutation is rare due to steric problems of pairing of purines with purines, and pyrimidines with pyrimidines. Eight types of changes are possible in transversion such as AT→TA, AT→GC, GC→CG, GC→TA, AT→TA, TA→GC, CG→GC, CG→AT (Fig. 9.2).

2.   Spontaneous Deamination of 5-Methylcytosine

Another source of spontaneous mutation is a change in 5-methylcytosine (5-MeC), a methyl form of cystoine. Both MeC and C occasionally loose an amino group. The C is converted to U, and MeC to 5-MeU (thymine) (Fig. 9.4). The C pairs with A but not G, therefore, replication of a molecule containing GU base pair will finally result in substitution of an AT pair for the original GC pair. The process in successive round of replication is GU→AU→AT. The uracil from DNA is removed by uracil glycosylase, hence the conversion of C→U rarely results in mutation. In addition 5-MeC loses an amino group and converted into 5-methyluracil which is acutally the thymine. There is no removal mechanism, hence GMeC pair

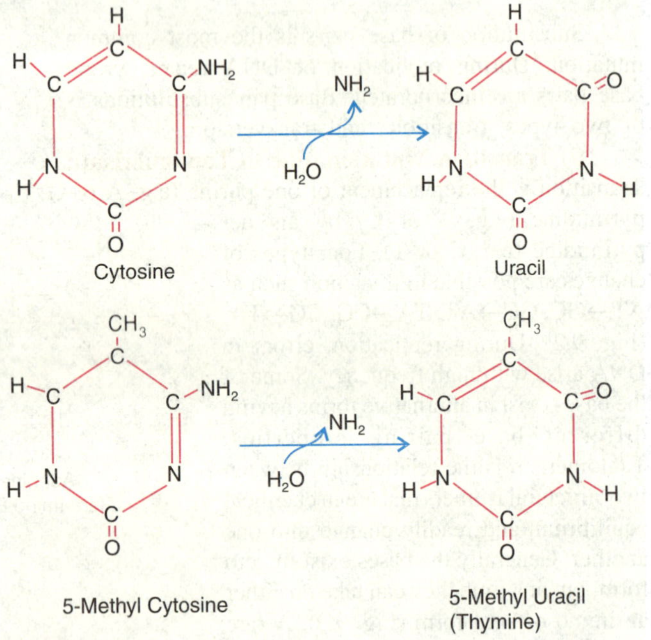

Fig. 9.4 : Deamination of cytosine and 5-methyl cytosine.

becomes a GT pair which may be corrected by mismatch repair system. This system does not recognise MeC, therefore, it is present in a methylated strand. It is converted randomly some times correct GC pairs and some times wrong AT pair. Therefore, MeC acts as mutable site in a gene, and this site is called hot spot.

It should be noted that methylated bases are the normal constituents of DNA in some microorganism but not the mutagens. Many organisms contain both C and MeC. Methylation results in protection of DNA against enzymes synthesized by viruses. The bacteriophage T2 contains 5-hydroxylmethylcytosine instead of cytosine.

3.   Frameshift Mutation

If there is deletion or insertion of one or a few nucleotides in the DNA molecule, this shifts the reading frame of nucleotide sequences resulting in mutation (Streisinger *et al*, 1966). Therefore, such mutation that results from shifting in reading frame backward or forward by one or more nucleotides is called *frameshift mutation*. (Fig. 9.5). Generally this mutation occurs where there is a short repeated nucleotide sequence. Roth (1974) reviewed the frameshift mutation occurring in organisms.

Fig. 9.5. shows an addition of one T due to slippage of the new strand (A), and a deletion (loss of two T base) occurs as a result of slippage of the parental strand (B). Hence, there is shifting in sequence of genetic code (Streisinger *et al*., 1966).

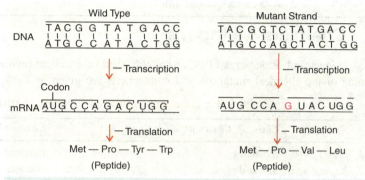

Fig. 9.5 : Frameshift mutation resulting from insertion of a GC base pair (see arrow) and sense strand T.

The same can be explained by an analogy of a code of words. If a gene gives the message: THE DOG CAN EAT JAM, each word of three letters represents a codon. After addition of a letter A after G, the new sentence becomes: THE DOG ACA NEA TJA M, which also is meaningless. Similarly, after deletion of C the new sentences becomes THE DOG ANC ATJ AM, which also is meaningless.

B. Induced Mutations

Any agent that directly causes damage to the DNA alters the base sequence or interferes with repair system will certainly induce mutations in DNA of organisms. The agents causing damage to DNA are called mutagens such as chemicals and radiation. Some common mutagens are listed in Table. 9.2.

Table 9.2 : Type of mutagens (after Friefelder, 1987)

Mutagen	Mode of action	Example	Consequence
Base analogue	Substitutes for a standard base during replication, a new base pair appears in daughter cell.	5-Bromouracil	A:T → G:C and G:C → A:T
		2-aminopurine	A:T → G:C
Chemical mutagen	Alter bases that appear in next generation	Nitrous acid	G:C → A:T and A:T → G:C
		Hydroxylamine	G:C → AT
		Ethyl methane sulfonate (EMS)	G:C → AT, G:C → C:G, G:C → TA
		U V light	All single base pair changes

Mutagen	Mode of action	Example	Consequence
Intercalating agents	Addition or deletion of one or more base pairs	Acridines	frameshifts
Mutator genes	Excessive insertion of incorrect bases or lack of repair of incorrectly inserted bases possible	—	All single base pair change
None	Spontaneous deamination of 5-methylcytosine (MeC)	—	G:MeC → AT

I. Chemical Mutagens

Singer and Kusmierek (1982) have published an excellent review on chemical mutagenesis. Some of the chemical mutagens and mutagenesis are given in Table 9.3, and described below:

Table 9.3 : Different types of chemical mutagens

Class of Chemical	Chemical Mutagens
Acridines	Ethyleneimine (EI)
Mustard	Nitrogen mustard
	Sulphur mustard
Nitrosamines	Diethylnitrosamine (DMN)
	Diethylsulphonate (DES)
	Nitrosomethylurea (NMU)
Epoxide	Ethyleneoxide (EO)
	Diepoxybutane (DEB)
Alkyl sulphonates	Diethylsulphonate (DES)
	Methylmethanesulphonate (MMS)
	Ethylmethanesulphonate (EMS)
Others	Nitrous acid
	Maleic hydrazide
	Hydroxylamine

1.   Base Analogues

A base analogue is a chemical compound similar to one of the four bases of DNA. It can be incorporated into a growing polynucleotide chain when normal process of replication occurs. These compounds have base pairing properties different from the bases. They replace the bases and cause stable mutation. A very common and widely used base analogue is 5-bromouracil (5-BU) which is an analogue of thymine. The 5-BU functions like thymine and pairs with adenine (Fig. 9.6A). The 5-BU undergoes tautomeric shift from keto form to enol form caused by bromine atom. The enol form can exist for a long time for 5-BU than for thymine (Fig. 9.6B). If 5-BU replaces a thymine, it generates a guanine during replication which in turn specifies cytosine causing G:C pair (Fig. 9.6A).

During the replication, keto form of 5-BU substitutes for T and the replication of an initial AT pair becomes an A:BU pair (Fig. 9.7A). The rare enol form of 5-BU that pairs with G is the first mutagenic step of replication. In the next round of replication G pairs with C. Thus, the transition is completed from AT→GC pair.

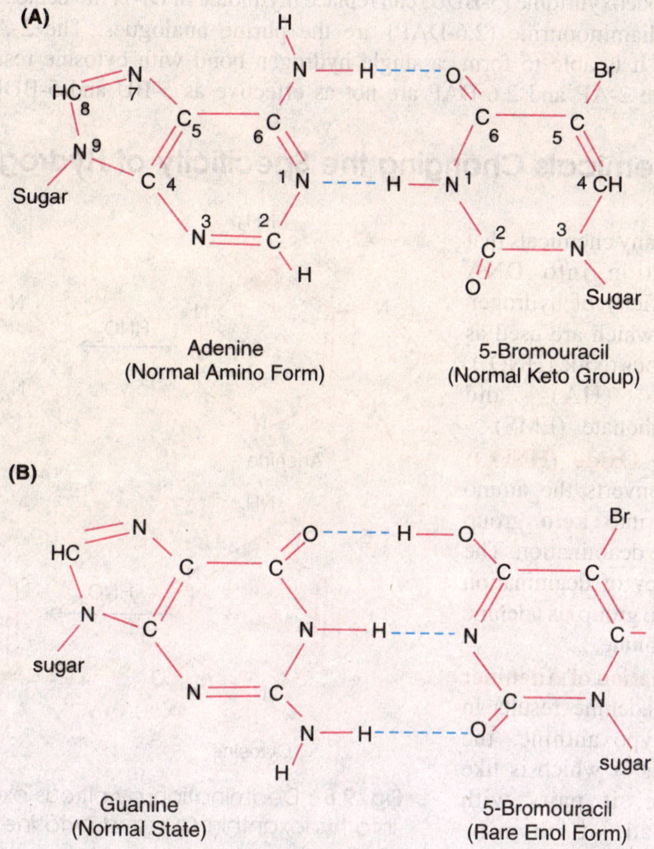

Fig. 9.6 : Mutagenesis by base analogue 5-bromouracil. A, the keto form of 5-BU pairs with adenine; B, 5-BU is tautomarised to enol form and pairs with guanine rather than adenine.

The 5-BU can also induce the conversion of GC to AT. The *enol* form infrequently acts as an analogue of cytosine rather than thymine. Due to error, GC pair is converted into a G:BU pair which in turn becomes an AT pair (Fig. 9.7B). Due to such pairing properties 5-BU is used in chemotherapy of viruses and cancer. Because of pairing with guanine it disturbs the normal replication process in microorganisms.

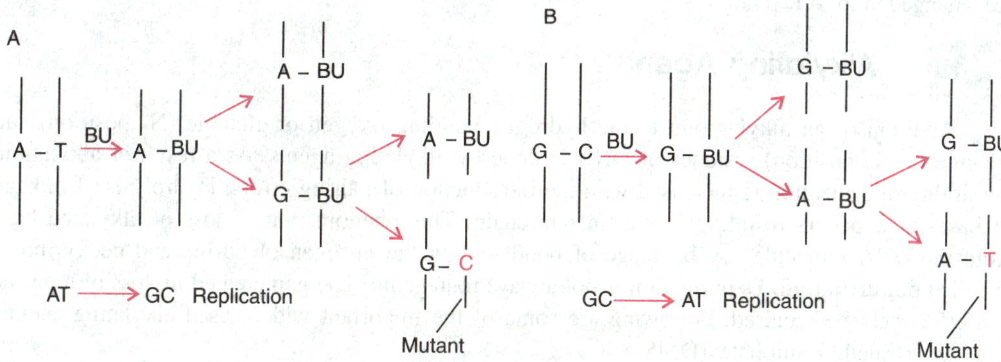

Fig. 9.7 : Mechanism of 5-bromouracil (BU)-induced mutagenesis. A, AT→GC replication; B, GC→AT replication.

The 5-bromodeoxyuridine (5-BDU) can replace thymidine in DNA molecule. The 2-aminopurine (2-AP) and 2,6-diaminopurine (2,6-DAP) are the purine analogues. The 2-AP normally pairs with thymine but it is able to form a single hydrogen bond with cytosine resulting in transition of AT to GC. The 2-AP and 2,6-DAP are not as effective as 5-BU and 5-BDU (Freese, 1959).

2.   Chemicals Changing the Specificity of Hydrogen Bonding

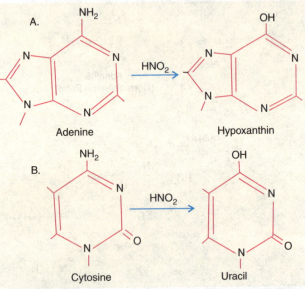

Fig. 9.8 : Deamination by nitrous oxide of adenine into hypoxanthin (A), and cytosine into uracil (B).

There are many chemicals that after incorporation into DNA change the specificity of hydrogen-bonding. Those which are used as mutagens are nitrous oxide (HNO_2), hydroxylamine (HA) and ethylmethanesulphonate (EMS).

(*i*) **Nitrous Oxide (HNO_2):** Nitrous oxide converts the amino group of bases into **keto** group through oxidative deamination. The order of frequency of deamination (removal of amino group) is adenine > cytosine > guanine.

(*ii*) **Deamination of Adenine:** Deamination of adenine results in formation of hypoxanthine, the pairing behaviour of which is like guanine. Hence, it pairs with cytosine instead of thymine replacing AT pairing by GC pairing (Fig. 9.8A).

(*iii*) **Deamination of Cytosine:** Deamination of cytosine results in formation of uracil by replacing -NH_2 group with -OH group. The affinity for hydrogen bonding of uracil is like thymine, therefore, C-G pairing is replaced by U-A pairing (Fig. 9.8B).

(*iv*) **Deamination of Guanine:** Deamination of guanine results in formation of xanthine, the later is not mutagenic. Xanthine behaves like guanine because there is no change in pairing behaviour. Xanthine pairs with cytosine. Therefore, G-C pairing is replaced by X-C pairing.

(*v*) **Hydroxylamine (NH_2OH):** It hydroxylates the C_4 nitrogen of cytosine and converts into a modified base via deamination which causes to base pairs like thymine. Therefore, GC pairs are changed into AT pairs.

3.   Alkylating Agents

Addition of an alkyl group to the hydrogen bonding oxygen of guanine (N_7 position) and adenine (at N_3 position) residues of DNA is done by alkylating agents. As a result of alkylation, possibility of ionization is increased with the introduction of pairing errors. Hydrolysis of linkage of base-sugar occurs resulting in gap in one chain. This phenomenon of loss of alkylated base from the DNA molecule (by breakage of bond joining the nitrogen of purine and deoxyribose) is called **depurination.** Depurination is not always mutagenic. The gap created by loss of a purine can effectively be repaired. Following are some of the important widely used alkylating agents:

(*a*) Dimethyl sulphate (DMS)

(*b*) Ethyl methane sulphonate (EMS) –$CH_3CH_2SO_3CH_3$

(*c*) Ethyl ethane sulphonate (EES) –$CH_3CH_2SO_3CH_2CH_3$

EMS has the specifity to remove guanine and cytosine from the chain and results in gap formation. Any base (A,T,G,C) may be inserted in the gap. During replication chain without gap will result in normal DNA. In the second round of replication gap is filled by suitable base. If the correct base is inserted, normal DNA sequence will be produced. Insertion of incorrect bases result in transversion or transition mutation.

Another example is methyl nitrosoguanidine that adds methyl group to guanine causing it to mispair with thyamine. After subsequent replication, GC is converted into AT transition.

4.   Intercalating Agents

There are certain dyes such as acridine orange, proflavine and acriflavin which are three ringed molecules of similar dimensions as those of purine pyrimidine pairs (Fig. 9.9). In aqueous solution these dyes can insert themselves in DNA (i.e. intercalate the DNA) between the bases in adjacent pairs by a process called **intercalation**. Therefore, the dyes are called intercalating agents. The acridines are planer (flat) molecules which can be intercalated between the base pairs of DNA, distort the DNA and results deletion or insertion after replication of DNA molecule. Due to deletion or insertion of intercalating agents, there occurs frameshift mutations (Fig. 9.10).

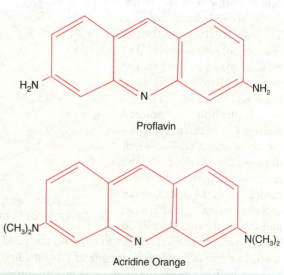

Fig. 9.9 : Chemical structure of two mutagenic acridine derivatives.

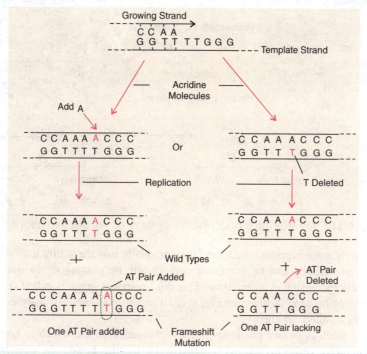

Fig. 9.10 : Mechanism of intercalation of an acridine molecule in the replication fork.

II. Radiations as Mutagens

Radiation is the most important among the physical mutagens. Radiations damaging the DNA molecules fall in the wave length range below 340 nm and photon energy above 1 electrovolt (eV). The destructive radiation consists of ultraviolet (UV) rays, X-rays, γ-rays, alpha (α) rays, beta (β) rays, cosmic rays, neutrons, etc. (Fig. 9.11).

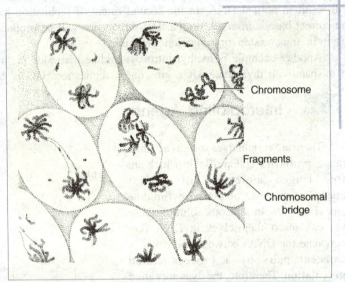

Radiation induced damage can be categorised into the three broad types : lethal damage (killing the organisms), potentially lethal damage (can be lethal under certain ordinary conditions) and sublethal damage (cells do

Radiation-induced chromosomal bridges and fragments in cells of X-rayed anthers of *Trillium*.

not die unless radiation reaches to a certain threshold value). The effect of damage is at molecular level. In a live cell radiation damage to proteins, lipoproteins, DNA, carbohydrates, etc. is caused directly by ionization/excitation, or indirectly through highly reactive free radicals produced by radiolysis of cellular water. DNA stores genetic informations so a damage to it assumes great dimension. It can perpetuate genetic effects and, therefore, the cellular repair system is largely devoted to its welfare (Sah *et al.*, 1987).

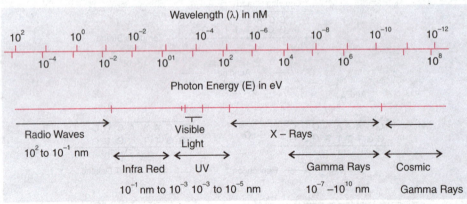

Fig. 9.11 : Wavelengths and photon energy of various radiations.

When the bacteria are exposed to radiation they gradually lose the ability to develop colonies. This gradual loss of viability can be expressed graphically by plotting the surviving colonies against the gradually increasing exposure time. This dose-response graph is called **survival curve**. The survival curve of bacteria is given in Fig. 9.12. The survival curve is analysed by a simple mathematical theory called hit theory.

Hit Theory : Each organism possesses at least one sensitive site which is known as target site. Radiation photons (particles of light) damage or hit the target site and inactivate the organisms. One can derive the equation based on this theory. The equations help to calculate the survival curve for many kinds of populations of N identical organisms exposed to dose D of radiation causing damage. The number dN damaged by a dose dD is proportional to the initial population that has not received radiation; hence dN = KN where,

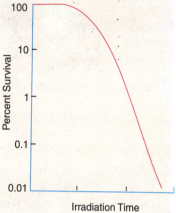

K is the constant which measures the effectiveness of dose.
Integrating this equation from N = No at D=O, we get

$$N = N_o e^{-KD}$$

...(1)

The surviving fraction S = N/No is

$$S = N/NO = e^{-KD}$$

...(2)

A plot of S verus D gives a straight line with a slope of -K (Fig. 9.12). This type of curves are called exponential or **single hit curve**. The exponential curve is obtained when the phages are irradiated with X-rays.

If there is a population of different organisms, and each organism consists of atleast n sites, each site must be hit to inactivate an organism. Therefore, each organism is hit by n times. The probability of one unit being hit by a dose D is, $P = 1-e^{-KD}$), so the probability of Pn will be

$$Pn = (1-e^{-KD})^n$$

The surviving fraction S of the population is 1-pn or

$$S = 1-(1-e^{-KD})n$$

...(3)

Fig. 9.12 : A typical ultra violet light survival curve for a bacterium.

This equation can be expanded as

$$S=1- (1-ne^{-KD} + e^{-nKD})$$

At the large value of D, the higher order terms become negligible as compared to ne^{-KD}. Therefore, at high dose D, $S=ne^{-KD}$ or

$$In\ S = In\ n – KD$$

... (4)

When the equation 3 is plotted for K = 1, various values of n reveals that for small values of D, In S gradually changes (Fig. 9.13). At large value of D, equation 4 dominates and curve becomes linear. Freifelder (1987) has discussed the hit theory in detail.

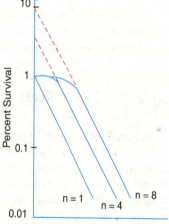

Fig. 9.13 : Survival curve for different values of n (hits' time).

1.   Ultraviolet (UV) Radiation

UV radiation causes damage in the DNA duplex of the bacteria and phages. The UV rays are absorbed and causes excitation of macromolecules. The absorption maxima of nucleic acid = (280 nm) and protein (260 nm) are more or less similar. The DNA molecule is the target molecule for UV rays but not the proteins. However, absorption spectrum of RNA is quite similar to that of DNA. The excited DNA leads to cross-linking, single strand breaks and base damage as minor lesion and generation of nucleotide dimer as a major one. Purines are generally more radio - resistant than the pyrimidine of the latter, thymine is more reactive than cytosine. Hence, the ratio of thymine-thymine (TT), thymine-cytosine (TC), cytosine-cytosine (CC) dimer

(Fig. 9.14) is 10:3:3, respectively. A few dimers of TU and UU also appear. The initial step in pyrimidine dimerization is known to be hydration of their 4 : 5 bonds.

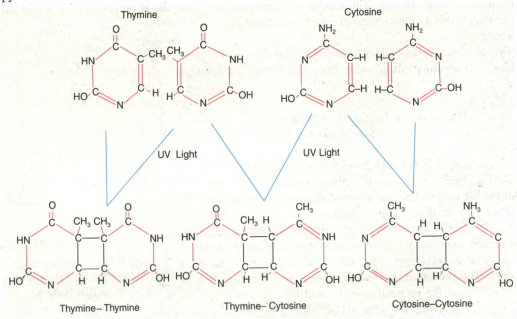

Fig. 9.14 : Formation of pyrimidine dimer induced by UV radiation.

Formation of thymine-thymine (TT) dimer causes distortion of DNA helix because the thymines are pulled towards one another. The distortion results in weakening of hydrogen-bonding to adenines in the opposing strand. This structural distortion inhibits the advance of replication fork.

2.   The X-Rays

The X-rays cause breaking of phosphate ester linkages in the DNA. This breakage occurs at one or more points. Consequently, a large number of bases are deleted or rearranged in the DNA molecule. The X-rays may break the DNA either in one or both strands. If breaks occur in both strands, it become lethal. The DNA segment between the two breaks is removed resulting in deletion. Since both the X-rays and UV rays bring about damage in DNA molecule, they are used in sterilization of bacteria and viruses.

Due to gene mutations individuals of a population show degrees of variations in their traits.

C. Expression of Mutations

The expression of mutation can be observed only when there occurs a detectable and altered phenotype. If a mutation occurs from a most prevalent wild type to a mutant form, it is called forward mutation. In forward mutation, a second mutation may take place resulting in appearance of a wild type organism again. This type of mutation is known as back mutation, reverse mutation or reversion. The reverse mutants are called reverants.

I. Reversion

It is the phenomenon of reverting of a mutant organism to the original phenotype characteristics. Reversion may occur spontaneously by induction or mutagens, for example :

$$\underset{\text{wild type}}{\text{AAA (Lys)}} \xrightarrow{\text{Forward}} \underset{\text{Mutant}}{\text{GAA (Glu)}} \xrightarrow{\text{Reverse}} \underset{\text{Wild type}}{\text{AAA (Lys)}}$$

$$\underset{\text{wild type}}{\text{UCC (Ser)}} \xrightarrow{\text{Forward}} \underset{\text{Mutant}}{\text{UGC (Cys)}} \xrightarrow{\text{Reverse}} \underset{\text{Wild type}}{\text{AGC (Ser)}}$$

The reversion frequency of an organism is a useful criterion for the identification of point mutation. However, if deletion has taken place, reversion will not occur. In bacteria, reversion events can be detected by measuring the ability of formation of bacterial colonies on solid growth medium. For example, if 10^7 Leu$^-$ bacterial cells are plated on solid medium devoid of leucine about 60 colonies are formed. These colonies must be consisting of Leu$^+$ bacteria which would have reverted spontaneously from Lac$^-$ to Lac$^+$ cells. Hence, the reversion frequency can be calculated as below :

No. of cells plated = 10^7

No. of colonies formed = 60

Reversion frequency = $60/10^7 = 6 \times 10^{-6}$

This value of frequency is a feature of the reversion of point mutants. Since, the production of spontaneous reverants occurs randomly, the reversion of a double mutants requires two independent events. Therefore, the reversion frequency would be $(6 \times 10^{-6})^2 = 3 \times 10^{-11}$

I. Second Site Mutation (Suppressor Mutation)

Spontaneous reversion rarely results in a restoration of the wild type base sequences. Only 100 Leu$^-$ mutants found in a population of about 10^9 Leu$^+$ cells and mutation causing mutants would be distributed over 100 different sites. After the growth of these cells, in a second population, there may also be 100 mutants some having mutation at sites present in the first population but most of these having mutation on other sites. In many population there will be about 500 mutation sites. If a reversion occurs at one particular site, it yields wild type base sequence. On average only $100/500 \times 10^9$ would be mutated at that site. Thus, the phenomenon of reversion into wild type phenotype by a second mutation in a different gene is called second site mutation or **suppressor mutation.**

The second site mutation overcomes the effect of first mutation. If the second mutation is within the same gene, the mutation is called a second site reversion or intragenic suppression. Although the reverant phenotypes are wild type, the original sequences of DNA will not be regained.

II. Suppression

If the mutational changes occur in a second gene, it eliminates or suppresses a mutant phenotype which is called suppression or intergenic reversion. This type of suppression has been studied carefully with conditional mutation (described in preceeding section) which develops wild type phenotypes on certain conditions and produces mutant phenotype in other conditions. The major class of this mutation is called suppressor sensitive mutation. It acts like wild type when a suppressor molecule is present. For example, a phage mutant can grow in one strain of bacteria but becomes unable to grow in other strain. Suppressor-sensitive mutations are of two types, nonsense (chain termination) mutation and missense (amino acid substitution) mutations.

1.   Non-sense Mutations

Most of the mutations affect only one base pair in a given location, therefore, these mutations are called point mutation or **gene mutation.** There are several types of point mutations. Non-sense mutation is one type of point mutation. There are 64 codons that code for amino acid out of which three codons (UAA, UAG, UGA) are known as termination codons that do not encode for any amino acid. If any change occurs in any codon, it brings about changes in amino acids which specify an amino acid to termination codon. This process is called **non-sense mutation.** For example, UAC codes for tyrosine. If it undergoes base substitution (C-G), it becomes UAG i.e. a termination codon. This results in synthesis of incomplete polynucleotide chain which remains inactive. Only a fragment of wild type protein is produced which has a little or no biological function unless the mutation is very near to the carboxyl terminus of the wild type protein. The non-sense mutations bring about drastic change in expression of phenotypic characters because in this mutation the structure and function of enzymes are changed.

2.   Missense Mutation

Missense mutation is the second type of point mutation. When one amino acid in a polypeptide chain is replaced by the other amino acid, this type of mutation is known as missense mutation. For example, if a protein valine (non-polar) has been mutated to aspartic acid (polar) due to loss of activity, it can be restored by the wild type phenotype by a missense suppressor that substitutes alanine (non-polar) for asparatic acid.

A missense mutation occurs by insertion, deletion or substitution of a single base into a code, for example the codon GAG specifying glutamic acid could be changed to GUG which codes for valine. Missense mutation that arises from substitution synthesises proteins which differ from the normal protein by a single amino acid. Substitution occurs in three different ways : (i) a mutant tRNA may recognise two codons perhaps by a change in anticodon loop, (ii) a mutant tRNA can be recognised

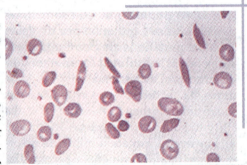

A missense mutation codes for a single amino acid change that causes sickled blood cells and sickle cell anemia.

by a wrong aminoacyl synthetase and be misacylated, and (iii) a mutant synthetase can change a wrong tRNA molecule. However, if a suppressor that substitutes alanine for aspartic acid works with 20% efficiency, every protein to which a cell synthesises atleast one aspartic acid is replaced. In this situation a cell probably cannot survive.

Inspite of substitution of a single amino acid many proteins are still functional. This depends on type and location of amino acid. For example, if a non-polar amino acid in the polypeptide chain is replaced by a polar amino acid, it will drastically change the three dimensional structure of the protein and also change the function. But if the polar amino acid is replaced by the another, there will be little or no effect on protein. Missense mutation plays an important role in providing new variability in organisms and driving the evolution because they are not lethal and remain in the genome.

3.   Silent Mutation

Silent mutation is another type of point mutation which could not be detected until the nucleic acid sequencing is done. Any change in gene does not affect the phenotypic expression because the code is degenerate i.e. more than one code specify an amino acid. For example, if the codon CGU is changed to CGC, still it would code for arginine. Similarly, both AAG and AAA specify alanine. If the codon AAG is changed to AAA, the latter codon will still code for lysine even after change in base sequence of DNA. This mutation is of silent type because even after change in base sequence of DNA, there is no change in the amino acid sequence and expression of phenotype characters.

4.   Frameshift Mutation

As described earlier, frameshift mutation arises from insertion or deletion of one or two pairs of nucleotides within the coding region of the gene. It is also a gene mutation. It is very deleterious and yields mutant phenotypes resulting from the synthesis of non-functional proteins. If frameshift occurs near the end of the gene or there occurs a second frameshift down stream from the first and restores reading frame, the phenotypic effect would not be drastic (Roth,1974).

The other examples of mutation that affect expression of phenotypes are : (i) mutations in regulatory sequences that control gene expression e.g. constitutive lactose operon mutant in *E. coli*, and (ii) mutations in rRNA and tRNA genes (Roth,1974).

D. Detection and Isolation of Mutants

Mutation occurring in microorganisms can be detected and efficiently isolated from the parent organisms of other mutants. While studying mutation one must be aware of wild type characters of an organism so that mutants can easily be detected. Since, mutations are rare about one per 10^7 to 10^{11} cells, it is very important to have a very sensitive detection system so that

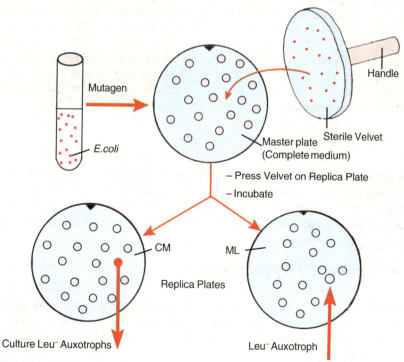

Fig. 9.15 : Replica plating technique for isolation of leucsine auxotrophs (leu⁻); CM, complete medium;, ML, medium devoid of leucine.

the rare mutant may not be missed from detection. Therefore, to rescue from these difficulties the probability of getting alterations and frequency of mutation (one in 10^3 to 10^6 cells) are increased by inducing mutations.

In bacteria and other haploid microorganisms, the detection systems are straight forward because any new allele should be observed immediately. In albino mutation, the deletion is very simple. It requires only change in colour of bacterial colony. The other detection systems are rather complex. Some of the methods of detection have been described as below :

1. Replica Plating Technique

Lederberg and Lederberg (1952) have given replica plating technique. This technique is used to detect auxotrophic mutants which differentiates between mutants and wild type strains on the basis of ability to grow in the absence of an amino acid. Method of replica plating has been described earlier (Fig. 9.1). However, isolation of a leucine auxotroph through replica plating follows the following steps (Fig. 9.15).

(i) Generate the mutants by treating a culture with a mutagen e.g. nitrosoguanidine.

(ii) Inoculate a plate containing complete growth medium and incubate it at proper temperature. Both wild type and mutant survivors will from colonies on the complete medium. This plate containing complete medium is called master plate.

(iii) Prepare a piece of sterile velvet and gently press on upper surface of master plate to pick up bacterial cells from each colony. As pressed on master plate, again gently press the velvet on replica plates containing complete medium in one set and lacking only leucine in the other set. Thus, the bacterial cells are transferred in replica plates in the same position as in master plate.

(iv) Incubate the plates and compare the replica plates with master plate for the bacterial colony not growing on replica plate. The leucine auxotrophs (Leu⁻) will not grow on replica plates devoid of leucine. Isolate and culture Leu⁻ cells growing on complete medium.

Replica plating can also be used to isolate the temperature sensitive mutants. It involves by forming colonies at 30°C and then transferring these colonies on two plates, one incubated at 30°C and the other at 42°C. The colony which grow at 30°C and absent at 42°C certainly consist of temperature sensitive mutation.

2. Resistance Selection Method

It is the other approach for isolation of mutants. Generally the wild type cells are not resistant either to antibiotics or bacteriophages. Therefore, it is possible to grow the bacterium in the presence of the agent (antibiotics or bacteriophage and look for survivors (see Fig. 9.1). This method is applied for isolation of mutants resistant to any chemical compounds that can be amended in agar, phage resistant mutants.

3. Substrate Utilization Method

This method is employed in the selection of bacteria. Several bacteria utilize only a few primary carbon sources. The cultures are plated onto medium containing an alternate carbon sources. Any colony that grows on medium can use the substrate and are possibly mutants. These can be isolated.

Sugar utilization mutants are also isolated by means of colour indicator plates. A popular medium (EMB agar) is used for this purpose. The EMB agar contains two dyes eosin and methylene blue in the medium. Colour of these dyes is sensitive to pH. This medium also contains lactose sugar as carbon source and complete mixture of amino acids. Therefore, both lactose wild type (Lac⁺) and lactose mutant (Lac⁻) cells can grow and form colonies on EMB agar plates. The Lac⁺ cells catabolize lactose and secrete acids, therefore, local pH of the medium decreases. This results in staining of colony to dark purple. On the other hand, Lac⁻ cells are unable to utilize lactose and use some of the amino acids as carbon source. After utilization of amino acid, possibly ammonia is produced that increases the local pH and decolourizes the dye resulting in white colony.

4. Carcinogenicity Test

An understanding has developed to identify the environmental carcionogens that cause mutation and induce cancer in organisms. This method is based on detecting potential of carcinogens and testing for mutagenicity in bacteria.

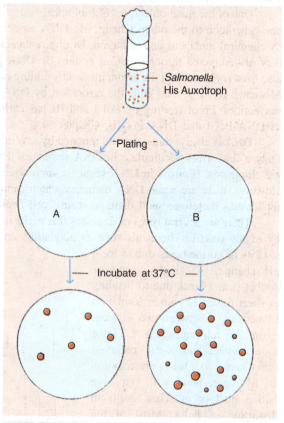

Ames *et al.* (1973) developed a method for deletion of mutagenicity of carcinogens which is commonly known as **Ames test.** It is widely used to detect the carcinogens. The Ames test is a mutational reversion assay in which several special strains of *Salmonella typhimurium* are employed. Each strain contains a different mutation in the operon of histidine biosynthesis. The Ames test follows the following steps (Fig. 9.16).

 (*i*) Prepare the culture of *Salmonella* histidine auxotrophs (His⁻).

 (*ii*) Mix the bacterial cells and test substance (mutagen) in dilute molten top agar with a small amount of histidine in one set, and control with complete medium plus large amount of histidine.

Fig. 9.16 : The Ames test for mutagenicity; A, complete medium containing a large amount of histidine; B, medium containing test mutagen and a small amount of histidine.

 (*iii*) Pour the molten mix on the top of minimal agar plates and incubate at 37°C for 2-3 days. Until histidine is depleted all the His⁻ cells will grow in the presence of test mutagens. When histidine is completely exhausted only the reverants (the mutants which have regained the original wild type characters) will grow on agar plate. The number of spontaneous reverants is low, whereas the number of reverants induced by the test mutagen is quite high. In order to estimate the relative mutagenicity of the mutagenic substance the visible colonies are counted and compared with control. The high number of colonies represents the greater mutagenicity.

A mammalian liver extract is added to the above molten top agar before plating. The extract converts the carcinogens into electrophilic derivatives which will soon react with DNA molecule. In natural way this process occurs in mammalian system when foregin substances are metabolized in the liver. Bacteria do not possess the metabolizing capacity as liver does, therefore, the liver extract is added to this test, just to promote the transformation.

Therefore, several potential carcinogens that generally are not carcinogenic until modified in liver, for example aflatoxins viz., B_1, B_2, G_1, G_2, etc. The Ames test has now been used with thousands of substances and mixtures such as the industrial chemicals, food additives, pesticides, hair dyes and cosmetics.

E. DNA Repair Systems

One of the main objectives of biological system is to maintain base sequences of DNA from one generation to the other. Changes in DNA sequence arise during replication of DNA damage by chemical mutagens and radiation. During replication if incorrect nucleotides have been added, they are corrected through editing system by DNA Pol I and DNA Pol III. The other systems also exist for correcting the errors missed by editing function. It is called **mismatch repair** system. Mismatch repair system edits the errors left by DNA Pol I and DNA III and removes the wrong nucleotides. Proof reading by Pol I and III has earlier been described (see exonuclease activity of DNA Pol I and DNA Pol III, Chapter 9).

DNA is always damaged and mutated by several chemicals and radiation as discussed earlier. Only a few errors accumulate in DNA sequence. The stable errors cause mutation and the rest are eliminated. If errors in DNA sequence are corrected before cell division, no mutation occurs. However, there are some DNA damages which cannot be mutated because the damages are not replicated. Therefore, such damages cause cell death.

There are several types of damages that occur in DNA: *(a)* modification of one or more bases by highly reactive chemicals such as alkylating agents like nitrosoamine and nitrosoguanidine, *(b)* loss of purine bases due to local pH change, (c) single strand or double strand break due to bending or shear forces, (d) dimer formation (dimerisation) between two adjacent pyrimidine molecules (e.g. T-T) due to ultraviolet and X-ray radiation (see Chapter 9). Due to dimerisation no hydrogen bond with opposing purine shall occur. This results in distortion of helix. Most of the spontaneous errors are temporary because they are soon corrected by a process called DNA repair.

1. Mechanisms of DNA Repair

There are four major pathways through which thymine-thymine dimer in DNA is repaired: *light induced repair (photoreactivation)* and *light-independent repair (dark repair)*.

(i) **Photoreactivation:** The UV damages caused in cells are repaired after exposure of cells in visible light. This is called photoreactiva-

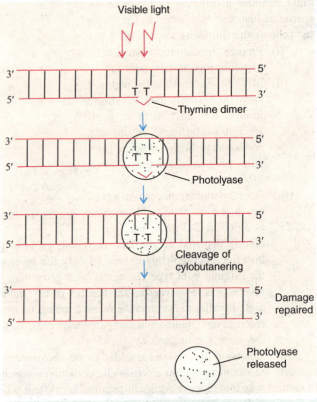

Fig. 9.17 : Photoreactivation for repair of thymine dimer.

tion. In this mechanism an enzyme DNA *photolyase* cleaves T-T dimer and reverse to monomeric stage (Fig. 9.17). This enzyme is activated only when exposed to visible light. The mutant cells lack photolyase. This enzyme absorbs energy, binds of cyclobutane ring to defective sites of DNA and promotes cleavage of covalent bonds formed between T-T. This enzyme is found in several bacteria and placental mammals. Finally, thymine residues are made free and damage is repaired.

Some other photolyases catalyse DNA repair in other ways. The 6-4 photoproduct photolyase repairs the DNA damage i.e. 6-4 photoproduct caused by UV rays. The 6-4 photoproduct is formed due to formation of C4-C6 bond between two adjacent pyrimidines or due to migration of a substituent from C4 position of one pyrimidine to the C6 position of the adjoining pyrimidine. The C4 photoproduct photolyase corrects both the errors.

In *Bacillus subtilis* a spore photoproduct (5-thyminyl-5, 6-dihydrothymine) is produced after UV radiation, but not cyclobutane dimers. In light-independent reaction, photoproduct lyase is formed which repair C-C bond between the two thymines.

(ii) **Excision Repair:** It is an enzymatic process. In this mechanism, the damaged portion is removed and replaced by new DNA. The second DNA strand acts as template for the synthesis of new DNA fragment. Excision repair involves DNA of different lengths such as *(a)* **very short patch repair** *(b)* **short patch repair**, and *(c)* **long patch repair**. The very short patch repair includes the mismatch of a single base, while the latter two deals with mismatches in a long patches of the DNA. The short and long patches of damaged DNA molecules are repaired by *uvr* genes for example *uvr A, B C* and *D* which encode repair endonuclease.

(a) **Base excision repair:** The lesions containing non-helix distortion (e.g. alkylating bases) are repaired by base excision repair. It involves at least six enzymes called DNA glycosylases.

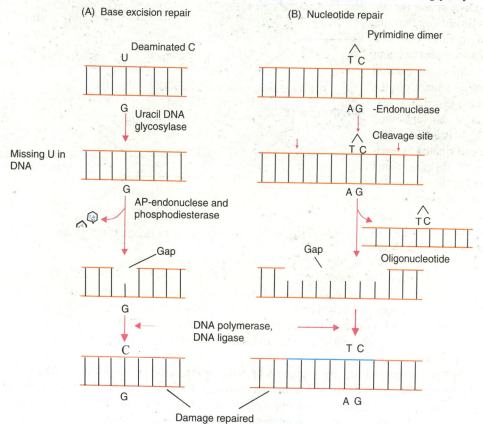

Fig. 9.18 : Excision repair pathways; A, base excision repair, B, nucleotide excision repair.

Each enzyme recognises at least bases and removes from DNA strand. The enzymes remove deaminated cytosine, deaminated adenine, alkylated or oxidised base. Base excision repair pathway starts with a DNA glycosylation. For example, the enzyme uracyl DNA glycosylase removes the uracyl that has wrongly joined with G which is really deaminated cytosine (Fig. 9.18A). Then AP-endonuclease (apurinic or apyriminic site) and phosphodiesterase removes sugar-phosphate. AP- sites arise as a result of loss of a purine or a pyrimidine. A gap of single nucleotide develops on DNA which acts as template-primer for DNA polymerase to synthesise DNA and fill the gap by DNA lygase.

(b) **Nucleotide excision repair:** Any type of damage having a large change in DNA helix causing helical changes in DNA structure is repaired by this pathway. Such damage may arise due to pyrimidine dimers (T-T, T-C and C-C) caused by sun light and covalently joins large hydrocarbon (e.g. the carcinogen benzopyrene). In *E. coli* a repair endonuclease recognises the distortion produced by T-T dimer and makes two cuts in the sugar phosphate backbone on each side of the damage. The enzyme DNA helicases removes oligonucleotide from the double helix containing damage. DNA polymerase III and DNA ligase repair the gap produced in DNA helix (Fig. 9.18B).

(c) **Recombination repair (daughter-strand gap repair):** When excision repair mechanisms fails, this mechanism, is required to repair errors. This mechanisms, operates in the viral chromosome in host cell whose DNA is damaged. This mechanism operates only after replication; therefore, it is also known as *post-replication repair.*

Probably RecA protein in *E. coli* catalyses DNA strand for sister-strand exchange. Thus a single stranded DNA segment without any defect is excised from a strand on the homologous DNA segment at the replication fork. It is inserted into the gap created by excision of thymine dimer (Fig. 9.19). Then the combined action of DNA Pol I and DNA ligase joins the inserted piece. The gap formed in donor DNA molecule is also filled by DNA Pol I and ligase enzymes.

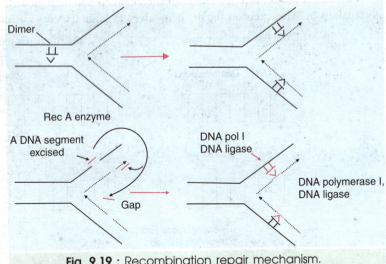

Fig. 9.19 : Recombination repair mechanism.

(d) **Methylation-directed very short patch repair:** Very short patch (VSP) repair is accomplished by involving methylation of bases especially cytosine and adenine. In *E. coli* methylation of adenine and in a sequence of -GATC- is done by the enzyme methylase (a product of *dam* gene) on both strands of DNA. After replication only A of -GATC- of one strand remains methylated, while the other remains unmethylated until methylase accomplishes methylation (Fig. 9.20 A-B).

In *E. coli* repairing activity required four proteins viz., Mutl, MutS, MutU (UvrS) and MutH by the genes *mutL, mutS, mutU,* and *mutH,* respectively. The mut genes are the loci which increase the frequency of spontaneous mutation. MutL recognises the unmethylated -GATC-during transition period. Then MutS binds to mismatches. MutU supports in unwinding the single strand, and single strand DNA binding (SSB) proteins and maintain the structural topography of single strand. MutH cleaves the newly synthesised DNA strand and the protein MutU separates the mismatch strand (A).

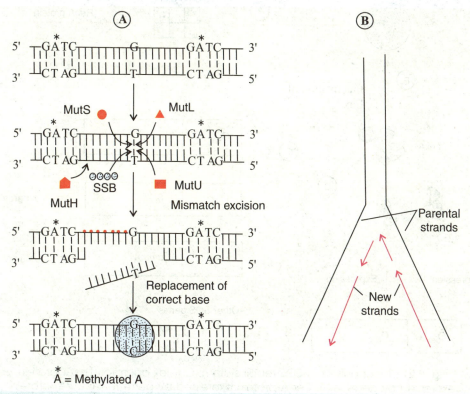

Fig. 9.20 : Mismatch repair. (A) excision of a newly synthesised strand and repair system; (B) arrows shows the region where methylation is not complete and dark region line shows the region where methylation is complete.

However, there is a gradient of methylation along the newly synthesised strand. Least methylation occurs at the replication fork. The parental strand is uniformly methylated. The methylated bases direct the excision mechanisms to the newly synthesised strand containing the incorrect nucleotides (B). During this transition period, the repair system works and distinguishes the old and new strands and repairs only the new strands.

(e) **SOS Repair:** SOS (**S**ave **O**ur **S**oul) repair is a by-pass repair system. It is also called emergency repair. The damage in DNA itself induces the SOS regulatory system which is a complex cellular mechanisms. SOS works where photodimers are formed that lead to cell death. SOS is the last attempt to minimise mutation for survival. It induces a number of DNA repair processes. SOS system works in the absence of a DNA template. Therefore, many errors arise leading to mutation.

Generally genes of SOS system remains in repressed condition caused by a protein LexA. Repression of SOS is inactivated by RecA protease. It is formed after the conversion of RecA protein by DNA damage. DNA damage results in conversion of RecA protein into RecA protease (Fig. 9.21A). RecA protease breaks LexA protein (B). Normally, LexA protein inhibits the activity or *recA* gene (C) and the DNA repair genes (*uvrA* and *umuD*) (D). Finally, DNA repair genes are activated. SOS system does not repair the large amount of damage. When DNA repair is over, RecA protein loses its proteolytic activity. Then LexA protein accumulates and binds to SOS operator and turns off SOS operator (D). However, repression is not complete. Beside, some RecA protein is also produced that inactivates LexA protein.

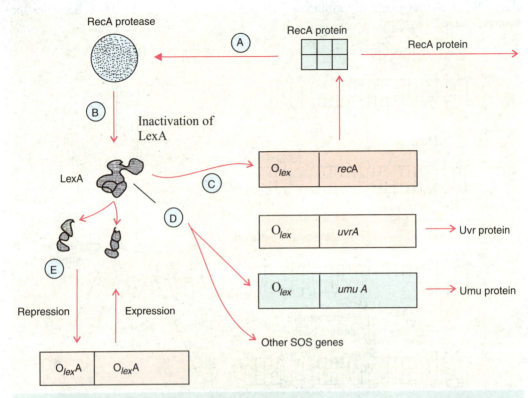

Fig. 9.21 : Mechanisms of SOS repair system. O_{lex}- lex operator, LexA- protein expressed by gene *lexA*; RecA-protein expressed by gene *recA*; UvrA- protein expressed by gene *uvrA*.

F. Uses of Mutations in Microorganisms

The most interesting and important advancement in microbial genetics is the use of mutation for detection of many more special features of biochemical interest, and enhanced production of useful metabolic products. Freifelder (1987) has discussed the uses of mutations in detail in his book 'Microbial Genetics'. Some of example of uses of mutations are given below :

1.   Determination of Function

A mutation defines a function. For example, a wild type *E.coli* cells can uptake lactose from 10^{-5} M solution by a passive diffusion through the cell membrane. But the mutants cannot uptake lactose even at the concentration higher than 10^{-5} M. This shows that the genetically determined process is involved in lactose uptake.

2.   Demonstration of Metabolic Pathways in Microorganisms

It has been demonstrated by isolating the three different classes of gal mutants that galactose is utilized by three distinct genes, *galK*, *galT* and *galE*. The Gal+ cells are grown on medium containing radioactive galactose (^{14}C-Gal). As the ^{14}C-Gal is metabolized, many different radioactive compounds can be found in the growth medium. In the beginning of addition of ^{14}C-

Gal three radiolabelled compounds viz., ^{14}C-galactose 1-phosphate (Gal-1-P), ^{14}C-uridine diphosphogalactose (UDP-Glu) and ^{14}C-uridine diphosphoglucose (UDP-Gal) are detected. If mutation occurs in three different genes, it will block the specific step of metabolic pathway.

For example, the mutant cells containing a galK$^-$ mutation would not be able to metabolize ^{14}C– Gal, therefore ^{14}C galactose remains unutilized because galK gene product converts into the first metabolic product. The galT-mutant cannot convert the first metabolic product into the second product, but accumulates Gal-1-P. This shows that the first step of galactose metabolism is the conversion of galactose by galK gene product to gal-1-P. However, if galE mutant is used the UDP-Gal is found in the culture medium. This shows that galE gene is associated with conversion of UDP-Gal into a product X. On the basis of conversion of these products the biochemical pathway must be as below :

$$ \text{Gal} \xrightarrow{\text{gal K}} \text{Gal-1-P} \xrightarrow{\text{gal T}} \text{UDP-Gal} \xrightarrow{\text{gal E}} \text{X} $$

3. For Understanding The Metabolic Regulation

Several bacterial mutants have been isolated which show changes in amount of a particular protein of its responses to external signals. For example, the enzyme synthesized by galK, galT and galE genes are normally not present in bacteria. These are synthesized only when galactose is supplied in the growth medium. In addition, such mutants have also been isolated in which these enzymes are always present irrespective of presence or absence or galactose in the growth medium. This shows that a regulatory gene must be associated for switching on or switching off the enzyme production.

4. For Matching a Biochemical Entity with a Biological Function

E. coli synthesizes an enzyme, DNA polymerase which polymerizes the DNA. It was thought that DNA polymerase I is also synthesizes the bacterial DNA. In Pol A-mutant of E.coli the activity of polymerase I has been found reduced by 50 time. After biochemical analysis of cell extracts of Pol A- mutants of E.coli two other enzymes, DNA polymerase II and DNA polymerase III were isolated. The purified enzymes synthesized the DNA molecule.

In another study, a temperature sensitive (Ts) mutation in dnaE gene was detected which was not able to synthesize DNA at 42°C , but synthesized the DNA normally at 30°C. From the culture of DnaE$^-$ (Ts) mutant the enzyme DNA polymerase I, II and III were isolated and assayed separately. It was found that DNA polymerase I, II and III were active at 30°C and 42°C, and polymerase III was active only at 30°C but not 42°C.

5. For Locating the Site of Action of External Agents

An antibiotic, rifamycin is known to inhibit RNA synthesis. In the begining it was unknown about the precise activity of rifamycin whether it acts by checking the synthesis of precursor molecule by binding to DNA and in turn by inhibiting the transcription of DNA into RNA, or by binding to RNA polymerase. Two types of rifamycin resistant mutants were isolated. First with altered cell wall in which rifamycin could not enter, and the second with altered RNA polymerase. These findings prove that the antibiotic rifamycin acts by binding with the enzyme RNA polymerase.

6.   For the Production of Useful Products

In addition, mutation in microorganisms for beneficial products was started since the time of Alexander Fleming during the end of 1920s. By using ultraviolet rays penicillin production by a mold *Pencillium chrysogenum* has been increased by about thousand fold greater to what was produced at Fleming's time (*see* Chapter 17, *Industrial Microbiology*).

QUESTIONS

1. What do you know about mutation ? Give evidences of spontaneous   mutation?

2. Write an essay on origin and random nature of spontaneous mutation?

3. What is induced mutation ? discuss in brief about the role of chemicals in induction of mutation?

4. In what ways radiations act as mutagens. Discuss in brief the ultra violet light as mutagen?

5. How does mutation express ? Discuss in brief reversion and suppersion?

6. Give a brief account of methods for detection and isolation of mutants?

7. Write an essay on application of mutations in microorganisms?

8. Give a detailed account of repair systems in microorganisms.

9. Write short notes on the following :

   (*i*) Mutation rates, (*ii*) Transversion mutation, (*iii*) Frame shift mutation, (*iv*) Point mutation, (*v*) Base analogue, (*vi*) Alkylating agents, (*vii*) Intercalating agents, (*viii*) Reversion, (*ix*) Suppression (*x*) Missense mutation, (*xi*) Silent mutation, (*xii*) Replica plating, (*xiii*) Ames test (*xiv*) Spontaneous mutation, (*xv*) Induced mutation, (*xvi*) Excision repair, (*xvii*) SOS repair.

REFERENCES

Ames, B.W. 1979. Identification enviroments chemicals causing mutations and cancer. *Science* **204**: 587.

Ames, B.W *et al.,* 1973. Carcinogenic arte mutagens : a simple test system combining liver homogenates for activation and bacteria for detection. *Proc. Nathl. Acad sci.* **70**: 2381.

Benzer,S. 1961. On the topography of the genetic fine structure of T_4. *Proc. Nathl. Acad.* 47:403.

Coulondre, R. *et al.* 1978. Molecular basis of base substitution hot spots in *E. coli*. Nature : **274** : 775.

Freese, E. 1959. The specific mutagenic effect of base anlogues on phage T_4. *J. Mol.* Biol. 1:87.

Freifelder, D. 1987. *Microbial Genetics. Jones* and Bartlett Publ. Inc. Boston.

Lederberg, J. and Lederberg, E. 1952. Replica plating and indirect selection of bacterial mutants. *J. Bact.* 63: 399.

Luria, S.E. and Dulbruck,M. 1943. Mutations of bacteria from virus senitivity to virus resistance. *Genetics* 28: 491.

Roth, J.R. 1974. Frameshift mutation *Ann. Rev. Genetics* **8**: 319.

Singer, B. and Kusmierek, J.T. 1982. Chemical mutagenesis. *Ann. Rev. Biochem.* 51: 655-693.

Streisinger, G. *et al.* 1966. Frameshift mutation and the genetic code. *Cold Spring Harb. Symp. Quant. Biol.* 31: 77.

Sah, N.K. Khan, M.A. and Husain, S.E. 1987. DNA repair mechanism. *Sci. Reptr.* **24**: 279-284.

Gene Expression and Regulation

10

A. Gene Expression

B. Regulation of Gene Expression

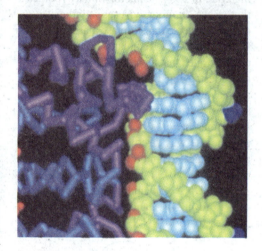

The DNA has two important roles, in the cell, first is replication and the second is expression. DNA replication has been described in detail in Chapter 5, *The Nucleic acids: DNA and RNA.* The second aspect, gene expression and regulation has been described in detail in this chapter.

A. Gene Expression

Gene expression is accomplished by a series of events. The information present in DNA is converted into molecules that determine the metabolism of the cell. During the process of gene expression DNA is first copied into an RNA molecule which determines the amino acid sequence of a molecule of protein. The RNA molecules are synthesized by using a portion of base sequences of single strand of double stranded DNA. This single strand is called **template.** Hence, formation of an RNA transcript is facilitated by an enzyme, RNA polymerase. Therefore, the process of synthesis of an RNA molecule corresponding to a gene is called transcription. By using base sequences and RNA molecule, proteins are synthesized in a definite order. Production of an amino acid sequence from an RNA base sequence is called translation. After completion of translation proteins are synthesized. Therefore, gene expression refers to protein synthesis through two major events, transcription and translation.

Francois Jacob (Left) and Jacques Monod (Right)

Central Dogma

DNA itself cannot directly order for the synthesis of amino acids but forms its transcripts first which is then translated into protein. For the first time in 1958, F. Crick suggested that there is unidirectional flow of informations from DNA to RNA to protein as shown below :

$$\text{DNA} \xrightarrow{\text{Transcription}} \text{RNA} \xrightarrow{\text{Translation}} \text{Protein}$$

This sequencial transfer of information from DNA to protein via RNA is known as *Central Dogma*.

Original Gene

Possible genes after mitosis/ mutation

Mitosis alone
Final Clones

Gene Expression can be controlled by altering the sequence of nucleotides in the DNA itself.

Furthermore, in 1963 H.M. Temin and coworkers reported that Rous sarcoma virus causing cancer contains RNA as genetic material. In 1964, he put forward a hypothesis that RNA tumour viruses synthesise an enzyme, reverse transcriptase that synthesizes DNA from RNA template. It could also be demonstrated that DNA synthesis is prevented after destroying the enzyme RNase. It was also shown that RNA specific DNA was synthesized. However, in lymphocytes of leukemia patients RNA-dependent DNA polymerase was discovered. In 1970, Crick again suggested a modified version of central dogma (Fig. 10.1).

DNA can undergo replication process to form DNA and transcription to form RNA that inturn undergoes translation. RNA also replicates to form RNA. In special cells only (dot lines) RNA synthesizes DNA, and DNA synthesizes proteins. This process is known as RNA directed DNA synthesis (Temin,1972).

I. Transcription (RNA Synthesis)

The tRNA, mRNA and rRNA are involved in the process of transcription. However, an enzyme transcriptase *i.e.* DNA-dependent RNA polymerase is required for the synthesis of RNA by using ribonucleotide triphosphates *i.e* ATP, GTP, CTP and UTP. Structure of a nucleotide has been described in Chapter 5. Structure and function of RNA polymerase are described herewith.

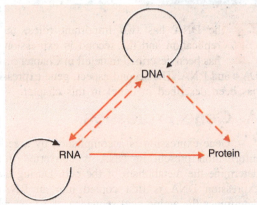

Fig. 10.1 : Modified central dogma i.e. RNA dependent DNA synthesis.

1. RNA Polymerase

On a DNA template elongation of RNA chain at each step is catalysed by RNA polymerase. RNA polymerase is found both in prokaryotes and eukaryotes but the structure and function in these two groups of organisms differ.

In prokaryotes only a single enzyme, RNA polymerase governs the synthesis of all cellular RNAs, whereas in eukaryotes for the synthesis of a cellular RNA several types of RNA polymerases are involved. For example, mRNA, tRNA and rRNA in *E. coli* are synthesized by the same RNA polymerase (Burgess, 1971).

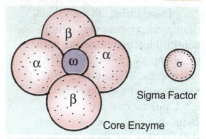

Fig. 10.2 : Structure of RNA polymerase (diagrammatic).

The holoenzyme is a complete RNA polymerase consisting of core enzyme and sigma (σ) factor, hence it is a complex enzyme. The core enzyme of *E. coli* consists of five polypeptide chains, two α subunits, one (β), one β' subunits and one omega (ω) subunit (Fig. 10.2). There are seven polypeptide chains (e.g. β αα δω$_1$, ω$_2$) in core enzyme of *E. coli*. The holoenzyme binds to DNA at specific sites called promoters and transcribes specific length of RNA. Thus σ factor plays a significant role in promoter recognition by RNA polymerase. It can easily be isolated from the holoenzyme. The β subunit consists of a catalytic site for RNA synthesis and binding sites for substrates and products as well. The β subunit plays a role in binding RNA polymerase to the DNA template. The two α subunits assemble the two larger subunits into core enzyme (α2ββ'ω). The function of small subunit (omega, ω) is not known in detail, however, it is supposed that it takes part in unwinding (Lathe, 1976). DNA polymerase subunits, structural genes and their functions are summarized in Table 10.1. The *E. coli* RNA polymerase synthesises RNA at the rate of 40 nucleotides per minutes at 37°C.

(i) **Types of RNA polymerase:** There are three different eukaryotic RNA polymerases that are transcribed by three different sets of genes (Fig. 10.3). They are distinguished by their sensitivity to a fungal toxin, α-amanitin.

(a) **RNA polymerase I (RNA Pol I):** It is located in the nucleolus and synthesises precursors of most rRNAs. It is sensitive to α-amanitin.

(b) **RNA polymerase II (RNA Pol II):** It is located in the nucleoplasm and synthesises mRNA precursors and some small nuclear RNAs. It is very sensitive to α-amanitin.

(c) **RNA polymerase III (RNA Pol III):** It is located in the nucleoplasm. It synthesises the precursors of tRNA, 5S rRNA and other small nuclear and cytoplasmic RNAs. It is moderately sensitive to α-amanitin.

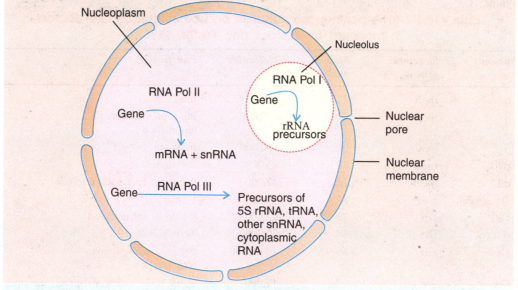

Fig. 10.3 : Function of RNA polymerase I, II and III in eukaryotes (diagrammatic).

In addition, an antibiotic rifampicin commonly inhibits transcription by interfering with β subunit of prokaryotic RNA polymerase. Ishihawa (1992) has presented a current view of RNA polymerase structure and some binding sites for each subunit (Fig. 10.4).

(ii) **The Site of Transcription :** Transcription begins at the promoter site i.e. cistron localised on DNA molecule. Out of two only one strand of dsDNA is transcribed. It has also been demonstrated that in ØX174 for any cistron only one strand of DNA is transcribed into RNA. In addition, in phage λ and phage T4 both strands are transcribed, where one strand serves as template for some genes. The rest of genes are transcribed from the other strand. The DNA strand that acts as template is

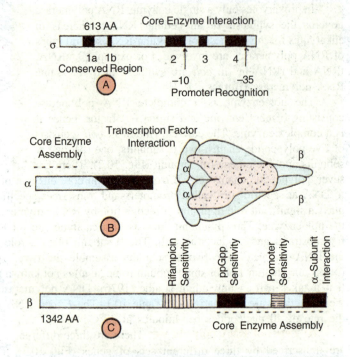

Fig. 10.4 : Functional map of RNA polymerase subunits (based on Ishihawa, 1992).

known as anti-sense strand. The nucleotide sequences of anti-sense strand and mRNA transcripts are complementary. The other strand is called sense strand. The bases of sense strand and that of mRNA are exactly the same.

In some viruses e.g. SP8, ØX174, etc. out of two only one strand is transcribed. In SV40 both the strands are transcribed for different genes. In contrast in T even phages, phage λ, *E. coli* and eukaryotes different regions of both the DNA strand act as template for RNA synthesis.

Table 10.1 : RNA polymerase subunits and other transcription factors (modified after Moat and Foster, 1995).

Subunits	Structural gene	Map position	Molecular weight	Function
α	rpoA	72	36,000	Initiation
β	rpoB	90	150,000	Initiation, elongation, termination
β'	rpoC	90	160,000	Initiation
σ70	rpoD	60	83,000	Initiation
σ32	rpoH	76	32,000	Initiation, heat sock response
Omega (ω)	rpoZ	82	—	Unwinding
CRP	crp	73	22,500	Initiation enhancement
rho (ρ)	rho	84	50,000	Termination
Nus	nus	69-90	20-69,000	Pausing, transcription, termination
Gre	gre	72	17,000	3'→5' RNA hydrolysis, proof reading.

(iii) **Process of Transcription:** The process of transcription is accomplished in the following three main steps : chain initiation, chain elongation and chain termination. The complete process of transcription is outlined in Fig. 10.5.

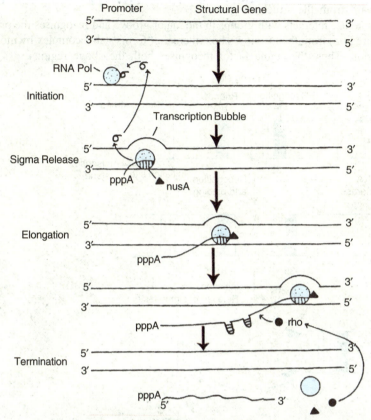

Fig. 10.5 : The process of transcription (diagrammatic).

2. Chain Initiation

During the process of transcription the components required to initiate the RNA chain are the template activated precursors, divalent metal ions (Mg^{++} or Mn^{++}), RNA polymerase and template.

(i) **Promoter Recognition:** The enzyme RNA polymerase plays a key role in recognition and binding of initiation site. Before the formation of complex enzyme, sigma factor interacts with core enzyme at β subunit site. This is required to check transcription of both the strands by core enzyme. The holoenzyme transcribes only one of the two DNA strands. The sigma factor of holoenzyme recognizes the promoter region of the DNA (Fig.10.6).

A specific base sequence (20-200 bases long) is called a promoter. Examination of a large number of promoter sequences of different genes from different bacteria has shown that they have many common features. The promoter sequences are centred at −10 and −35 base pairs from the transcription start point and are implicated in normal promoter function. *E. coli* has the following two distinct hexamer components of promoter :

5'TTGACA 3' .. 16-18 bp.. 5'TATAATPU 3'

3'AACTGT 5' .. 16-18 bp ... 3'ATATTAPY 5'

−35 box   −10 box

The -10 sequence is called Pribnow's box, whereas -35 region is called the recognition site (Pribnow, 1965). The Pribnow's box is found as a part of all prokaryotic promoters. The -35 region is located about 35 base pairs upstream (away from the direction of transcription site). The promoter is upstream from the structural gene.

The RNA polymerase interacts with groups in the major groove and recognises the proper sequence upstream (-35 region) from the Pribnow's box, thereafter forms stable complex by moving laterally to the -10 region. Thus, the sigma factor recognises both the above regions.

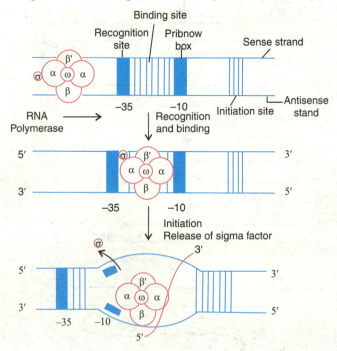

Fig. 10.6 : DNA template showing recognition, binding and initiation sites. (diagrammatic).

(ii) Binding of RNA Polymerase to Promoter: The strength of binding of RNA polymerase to different promoters varies. There are some promoters that have binding sites for proteins rather than RNA polymerase. Therefore, for these promoters the site must be occupied by that protein for RNA polymerase to bind correctly. One of the most common binding sites for this type is one that binds a complex of cyclic AMP receptor protein (CRP). The concentration of AMP governs the activity of these promoters (Bujard, 1980; Freifelder, 1987).

(iii) Unwinding of DNA Double Helix: The super helical nature of the chromosome may play a role in promoter function. Generally, a negatively super coiled chromosome is a better transcriptional template than relaxed chromosome. The torsional stress imposed by super coiling makes the certain area of DNA easier to separate by RNA polymerase. Hence, super coiling affects the expression of some genes more than the others. Binding of omega (ω) factor of RNA polymerase results in unwinding of a DNA helix. Consequently, a short segment of DNA opens.

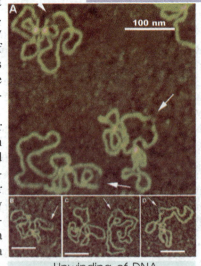

Unwinding of DNA.

The open complex then allows tight binding of the RNA polymerase with subsequent initiation of RNA synthesis.

(*iv*) **Synthesis of First Base of RNA Chain :** The first base of RNA synthesised is always in the form of purine i.e. either triphosphate guanine (pppG) or adenine (pppA). Most of the chains in *E. coli* is started with pppG, whereas in phage T7 and ØX174 it is initiated with pppA. In contrast to DNA replication, initiation of mRNA synthesis does not require primers. Initiation ends after the formation of first inter nucleotide bond (5'pppPupN) (Fig.10.5).

3. Chain Elongation

Chain elongation occurs by the core enzyme that moves along the DNA template. After beginning the elongation, transcription goes on at a rate between 30 and 60 nucleotides per second at 37°C. In general the elongation reaction includes the steps: (*i*) nucleotide triphosphate binding, (*ii*) bond formation between the nucleotide and the 3'–OH of nascent RNA chain, (*iii*) pyrophosphate release, and (*iv*) translocation of polymerase along the DNA template. The activated ribonucleotide triphosphates i.e. ATP, UTP, GTP and CTP are added according to nucleotides of one strand of DNA template. The incoming nucleotide forms hydrogen bonds with a DNA base. Reaction occurs between the nucleotides at 3' end of the RNA molecule and P in the triphosphate group (Hanna and Meares, 1983) leading to removal of phosphates (PPi). The PPi is soon hydrolysed to inorganic phosphate (Pi) (Fig. 10.7A).

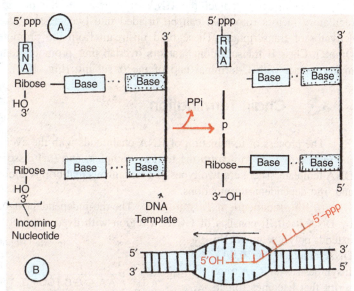

Fig. 10.7 : RNA synthesis; A, steps of polymerization; B, geometry of RNA synthesis; RNA, ribonucleic acid (diagrammatic).

(*i*) **Release of Sigma (σ) Factor :** When 8-9 nucleotide long mRNA transcript is formed, the σ factor is released abruptly from the complex. Consequently this causes a decrease in the affinity of σ from the RNA polymerase-DNA-nascent RNA complex. The release of σ can combine with any of core enzyme and thus it is reused for the initiation of a new chain (Fig.10.5) (Travers and Burgess, 1969).

(*ii*) **Direction of Transcription:** The ribonucleotide triphosphates attach to the first nucleotide of the template and chain growth takes place in 5'→3' direction. The region of template that has been transcribed regains its double helical form behind the bubble and the next region of DNA which is to be transcribed unwinds (Bremer *et al.,* 1965).

The RNA transcript does not elongate uniformly along the template. This is due to the presence of pausing sites at certain regions in the template. It has been found that generally the pausing sequence contains GC rich regions about 16-20 bp, upstream of 3'-OH end of the paused transcript. Also, the region of dyad symmetry present 16-20 bp upstream of 3'-end causes pausing. However, the mechanism of pausing is not clear.

Moreover, it is believed that after release of sigma factor the NusA and NusG proteins become associated with core enzyme and modulate the rate of elongation. At some sites these proteins enhance pausing.

It is interesting to note that the newly formed RNA transcript consists of a triphosphate group at 5' and free –OH group at 3' end (Fig.10.7 B). Moreover, the RNA transcript is larger than the structural gene. Most messages contain a promoter i.e. proximal region of mRNA which is called the leader sequence i.e. the untranslated region before the beginning of translation.

***(iii)* Proof Reading of mRNA:** The RNA polymerase possesses proof reading activity which is analogous to the 3'→5' exonuclease activity of DNA polymerase. The two proteins, GreA and GreB enable the RNA polymerase (which is transcriptionally arrested) to back up and cleave 2-3 nucleotides from 3' end of nascent message. This results in release of RNA polymerase which in turn moves through the arrest point at the 3' end of the gene (Most and Foster,1995).

***(iv)* Protein-Protein Interaction with RNA Polymerase:** There are several regulatory factors that directly contact the RNA polymerase at the promoter region of template. Such regulatory factors (proteins) can be divided into two classes, I and II. Class I proteins act as activator of transcription. These bind upstream from the promoter (e.g. CRP, AraC, Fnr and OmpR). Class II transcriptional factors overlap the promoter region; a few of these regulators make contact with C-terminal end of the α subunit (Fig. 10.4).

3.   Chain Termination

The process of termination of RNA chain ends with the events : (*i*) cessation of elongation, (*ii*) release of transcript from the tertiary complex, and (*iii*) dissociation of polymerase from the DNA template. There are two types of mechanisms that brings about termination: rho–independent and rho-dependent terminations.

***(i)* Rho-independent Termination :** The rho-independent signal for termination is recognised by DNA itself. It consists of GC rich region with dyad symmetry. RNA polymerase reads and extends polyA sequence on DNA template (Fig.10.8A). It synthesises an RNA transcript that becomes folded to form a stem and loop structure of about 20 bases upstream from the 3'-OH terminus with a terminal stretch of 4-8 poly U residues (Fig.10.8B). This RNA stem-loop structure causes RNA polymerase to pause and disrupt the RNA-DNA hybrid at 5' end. Poly U residues are present at 3' end of RNA-DNA hybrid molecule. The U-A hybrid base pairs are relatively unstable; therefore, it causes the 3' end of the hybrid to break and release the mRNA chain.

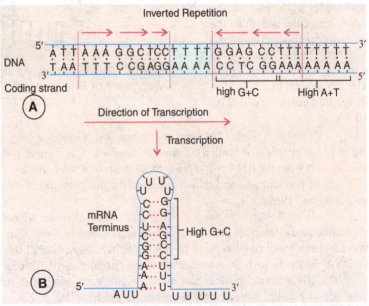

Fig. 10.8 : Base sequence of DNA of *E. coli* where termination occurs (A), and 3' terminus of transcribed mRNA molecule that has formed stem-loop structure (B).

(ii) **Rho-dependent Termination:** Richardson (1983) has given a model for rho-dependent termination of RNA chain (Fig. 10.9). Rho-dependent termination requires the Rho protein (factor) encode by *rho* gene. The rho factor is a hexamer. When a strong pause site is apparently located at specific distance from a promoter region the termination associated with rho-factor occurs. However, it must be clear that a strong pause site present close to the promoter will not act as a site for termination. Rho requires ssRNA as a binding region (Fig. 10.9 A).

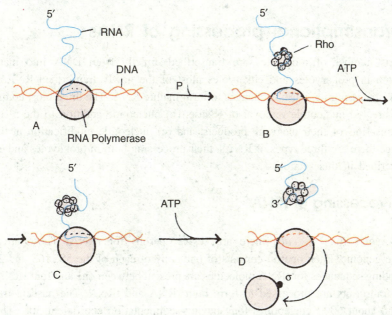

Fig. 10.9 : A model for Rho factor-mediated release of RNA chain (based on Richardson, 1983).

Rho factor moves along RNA from its binding site to the enzyme RNA polymerase. The movement is facilitated by the hydrolysis of ATP that provides energy. Hopefully, the ssRNA winds arround the outside of rho factor in the large binding site. This step precedes the arrival of RNA polymerase at the termination site (B). The RNA polymerase is partially wrapped around the outside of Rho factor. Consequently the Rho protein comes in the contact of core enzyme of polymerase. This contact is facilitated by NusA or NusG proteins. This step is dependent on the hydrolysis of nucleotide triphosphates (C). The process of wrapping goes on continuously and disrupts the non-covalent bonds that hold the newly formed RNA to DNA and RNA polymerase. Activation of an RNA/DNA helicase activity causes the mRNA chain and core RNA polymerase to get dissociated from the DNA template. Thus, the Rho-protein-RNA complex is set free from the RNA polymerase-DNA complex (D). Finally, core RNA polymerase is released and interacts with sigma factor that is used for the initiation of second round of transcription (Richardson,1983). A rho-dependent transcription–termination signal has also been described in bacteriophage F_1 by Mose and Model (1984).

4. RNA Turnover

On the basis of decay rates, the RNA transcribed in this way *in vivo* can be grouped into two : stable and unstable RNA molecules. The stable RNA molecules consist of tRNA and rRNA, whereas mRNAs are unstable. In *E. coli* the total population of tRNA and rRNA is 15-25% and

70-80%, respectively. The mRNA is present only in low amount (3-5%). The reasons for the stability are association of rRNA in ribonucleoprotein complexes (ribosome) and conversion of tRNA and rRNA into the secondary structure. This conversion protects the stable RNAs at 3' terminus from the dissolution of ribonucleases. The cellular exonucleases degrade the mRNA in 3'→5' direction. However, presence of stem-loop structure possibly inhibits the association of endonucleases.

II. Post-transcriptional processing of RNAs

Very less number of molecules are transcribed directly from DNA into mature RNA molecules. Both in prokaryotes and eukaryotes most of the newly transcribed RNA molecules undergo RNA processing i.e. various changes to form mature RNA. The three most common types of alterations are: (a) nucleotide removal of RNase, (b) nucleotide addition to the 5' or 3' ends of primary transcripts or their cleavage products, and (c) nucleotide modification in the base or the sugar. Since there are three types of RNAs, their processing both in prokayotes and eukaryotes is briefly described in this section.

1. Processing of rRNA

In **prokayotes** (e.g. *E. coli*) there are seven operons for rRNA. These are dispersed throughout the genome. Each operon consists of one copy of each of the 5S, 16S and 23S rRNA sequences. Coding sequences for tRNA molecules are present between one and four rRNA operons. The primary transcripts are processed to form both rRNA and tRNA. The nascent transcript is short lived and about 6000 nucleotides long having sedimentation coefficient 30S. The primary RNA transcript folds into many stem-loop structures through base pairing between complementary sequences. This secondary structure of stem – loop allows some of the proteins to bind with it and form **ribonucleoprotein (RNP)** complex. The RNPs are the RNA – protein complexes. Then a methyl group is added to the base adenine by a methylating agent, S-adenosylmethionine. Thereafter, RNase II carries out primary cleavage to from precursors of 5S, 16S and 23S molecules. Sceondary cleavage of precursor is done by RNase M5, M16 and M23, respectively to form mature rRNA molecules.

In **Eukaryotes,** the RNA genes exist in tandem repeated clusters containing 100 or more copies of transcriptional units. They are transcribed in the nucleolus by RNA Pol I to yield a long, single pre-rRNA molecule which contains one copy each of the 18S, 5.8S and 28S sequences. In each organism the precursor has a characteristic size e.g. 7,000 nucleotides in yeast and 13,500 nucleotides in mammals. Many spacer sequences are removed from the long mammalian pre-tRNA molecule (47S) by a series of specific cleavage, first in the **external transcribed spacers** (ETS) 1 and 2 and then in the **internal transcribed spacers** (ITS) to release 20S pre-rRNA from the 32S pre-rRNA molecule (Fig. 10.10). Further both of these precursors must be trimmed and the 5.8S region must base pair to the 28S rRNA before production of the mature molecules. **Methylation** takes place at more than 100 sites to from 2'–O-methylribose. This is carried out by small nuclear RNP (snRNP) particles which are abundant in the nucleolus. Maturing RNA molecules fold and complex with ribosomal proteins. RNA Pol III synthesises the 5S rRNA from unlinked genes which undergoes a little processing.

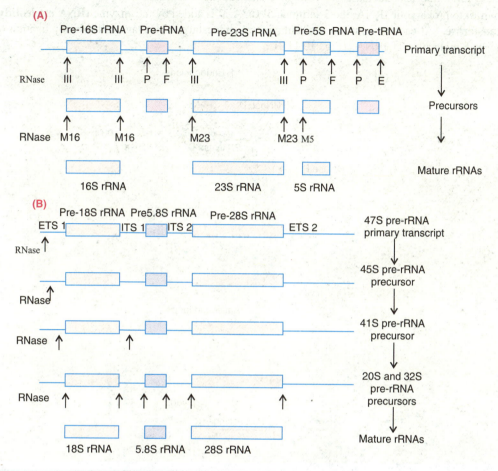

Fig. 10.10 : Processing of rRNA in *E. coli* (A) and mammals (B).

2. Processing of tRNA

In **prokaryotes,** there are many operons (e.g. 7 genes in *E. coli*) which are separated by spacer sequences. Mature tRNA molecules are generated by processing longer pre-tRNA transcripts. Through a specific step, RNases D, E, F and P generate these mature tRNA molecules by exo-and endo-nucleolytic cleavage (Fig. 10.11A). When the primary transcripts have folded and formed stem and loop, an endonuclease (RNase or F) cuts off a flaking sequence at 3' end at the base of stem and generates a precursor with nine extra nucleotides. Then RNase D removes seven of these 3'–nucleotides one at a time. RNase P cleaves to produce the mature 5' end of the tRNA followed by RNase D trimming of remaining 2 nucleotides from 3' end. Lastly, the tRNA undergoes a chain of **base modification** i.e. modification in 20% of bases to form unusual bases such as ribothymidine (rT), pseudouridine (ψ), dihydrouridine (DHU) and inosine (I).

Many **eukaryotic** pre-tRNA molecules are synthesised with a 16 nucleotide 5'–leader, a 14 nucleotide intron and extra 5'– and 3'– nucleotides. However, these are removed during processing. The primary transcript is converted into secondary structure having stems and loops.

These allow the endonucleases to recognise and cut the 5'– leader and two 3'–nucleotides. In contrast to prokaryotic tRNA, the 3'terminal 5' CCA-3' is added by the enzyme **tRNA nucleotidyl transferase.** At each end introns are removed by endocatalytic cleavage followed by ligation of the half molecules of tRNA. (Fig. 10.11B).

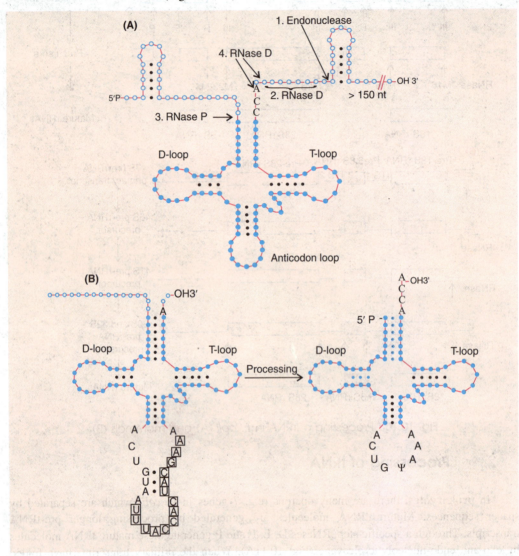

Fig. 10.11 : Pre-tRNA processing in *E. coli* (A) and pre-tRNA^Tyr processing in yeast (B).

3.   Processing of mRNA

In **prokaryotes,** there is a little or no processing of mRNA transcripts. Prokaryotic mRNA is degraded very rapidly from 5' end. Therefore, to rescue from degradation it is translated before being finally transcribed. Ribosomes assemble on unfinished mRNA and first cistron (protein coding region) can be translated very soon. The internal cistrons are partially protected by stem-loop structure formed at 5'– and 3'– ends.

The **eukaryotic** RNA Pol II transcribes different genes from snRNA genes and forms a collection of products which is known as heterogenous nuclear RNA (hnRNA). The pre-mRNA transcripts undergo processing to form mature mRNAs. Processing events are briefly described herewith.

(a) **The hnRNP:** RNA Pol II synthesises the hnRNA, which is mainly pre-mRNA. Soon after synthesis hnRNA is covered by three important hnRNP proteins (e.g. A, B and C proteins) to form **heterogeneous nuclear ribonucleoprotein** (hnRNP) particles. They contain three copies of three tetramers and about 600-700 nucleotides of hnRNA. Possibly the hnRNP proteins keep the hnRNA in a single stranded form and help in various events of RNA processing.

(b) **The snRNP particles:** There are many uracil-rich snRNA molecules denoted by U1, U2, etc. Most of them are transcribed by RNA Pol II which complex with specific proteins and form **snRNPs.** The most abundant are involved in pre-mRNA splicing– U1, U2, U4, U5 and U6. A majority of them is involved in determining the methylation sites of pre-rRNA and located in nucleolus. The major snRNAs containing the sequence 5'–RA(U)n GR-3' bind with eight common proteins in the cytoplasm and become hypermethylated. Thereafter, these are transported back into the nucleus.

(c) **5'-Capping:** Soon after synthesis of about 25 nucleotide-long mRNA chain by RNA pol II, the 5'-end is chemically modified by the addition of a **7-methylguanosine** (m7G) residue (see Fig. 5.12). This 5' modification is called a CAP which is done by addition of a GMP nucleotide to the new transcript. It is added (by an enzyme **mRNA guanyltransferase**) in a reverse polarity (5' to 5' triphosphate bridge) as compared with the normal 3'→5' linkage. This cap acts as a barrier to 5'–exonuclease attack, but it promotes splicing transport and translation processes.

(d) **3' Cleavage and polyadenylation:** The mature 3'–end in most pre-mRNA molecules is generated by cleavage at polyadenylation. For cleavage and polyadenylation reaction specific sequences are present in DNA and its pre-mRNA transcript which consists of 5'-AAUAAA-3' that provides polyadenylation signal. It is followed by a 5'-YA-3' and GU-rich sequence (Fig. 10.12). Collectively these sequences are called **polyadenylation site**. Then about 250 nucleotides long poly (A) tail is added by poly (A) polymerase (PAP) and mature mRNA is generated.

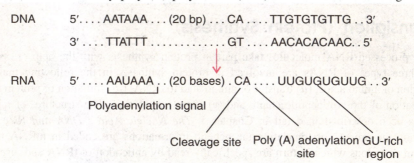

Fig. 10.12 : A typical polyadenylation site.

(e) **Splicing:** The eukaryotic pre-mRNA possesses introns (intervening sequencing) that interrupt the exons (the coding regions). The pre-mRNA is cut, introns are spliced out (removed) and two flanking exons are joined together. This process is called **splicing.** This event takes place in nucleus before transport of mature mRNA in cytoplasm. For splicing, introns require to have a 5'-GU, an AG-3' and a branchpoint sequence. The introns are removed in a two-step reaction as a tailed circular molecule called **lariat** which is degraded later on (Fig. 10.13). The splicing process is catalysed by the snRNPs (U1, U2, U4, U5 and U6) and the other splicing factors. The complex of snRNPs and pre-mRNA holds the upstream and downstream exons close together and the looping out introns is called **splisosome**. Reaction for cleavage of introns and ligation of exons take place inside the splisosome releasing the introns as a lariat.

(f) **Pre-mRNA methylation:** In pre-mRNA (containing the sequence 5'-RRACX-3' where R= purine) a small percentage of A residues becomes methylated at the N6 position.

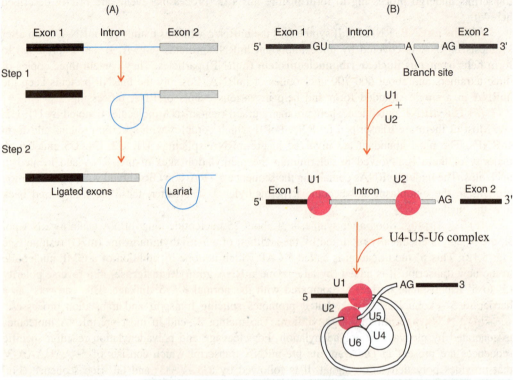

Fig. 10.13 : Splicing of eukaryotic pre-mRNA; A-a two-step reaction, B- involvement of snRNPs in splisosome formation.

III. Translation (Protein Synthesis)

The processed RNA molecules take part in protein synthesis with the help of ribosomes. All the three types of RNAs are involved in protein synthesis in the following main steps : *(i)* activation of amino acids, *(ii)* transfer of amino acid to tRNA, *(iii)* initiation of protein synthesis, *(iv)* elongation of the polypeptide chain, and *(v)* chain termination. The structure of mRNA and tRNA has been described in detail in Chapter 5, *The Nucleic Acid : DNA and RNA.*

During the process of translation the genetic informations are coded in mRNA transcripts in the form of codons which in turn are specifically read by anticodon of tRNA and used to form a polypeptide molecule of defined function. For detail account of codon assignment and its function, see Chapter 7 *The Genetic Code.*

1.   Charging of tRNA

(i) **Activation of Amino Acids:** In protein synthesis only L-amino acids take part. The D-amino acids are screened from the all 20 amino acids. In addition, the other amino acids which are not used in protein synthesis are citrulline, alanine, β-alanine, etc. Each amino acid has a specific aminoacyl tRNA synthetase (charging enzyme) and a specific tRNA. At least 32 tRNAs are required to recognise all the amino acid codons, but some cells used more than 32. However, these may be more than one species of tRNA for a specific amino acid but there is only one charging

enzyme for each amino acid. Its carboxyl group activates the amino acids through being catalysed by its own specific activating enzyme (aminoacyl tRNA synthetase) in the presence of ATP. Consequently aminoacyl AMP synthetase complex is formed which remains in bound form with the activating enzyme.

$$NH_2 \qquad\qquad Mg^{++} \qquad\qquad NH_2\ O$$

R — CH-COOH + ATP $\xrightarrow{\qquad}$ R— CH— C — O — AMP + PPi

Amino acid   Aminoacyl-   Aminoacyl-AMP-synthetase

tRNA-synthetase   (AAS) complex

(ii) Transfer of Amino Acid to tRNA: The process of transfer of activated amino acids to tRNA is called charging of tRNA. The tRNAs are specific to their specific amino acid. Therefore, tRNAs are named according to specific amino acid such as tRNA^{ala} (for alanine), tRNA^{val} (for valine), etc. Therefore, the activated amino acid is transferred to its specific tRNA. The aminoacyl-AMP-synthetase complex formed as above is transferred to tRNA as below :

Aminoacyl-AMP- synthetase complex + tRNA $\longrightarrow$ Aminoacyl – tRNA + AMP + aminoacyl tRNA synthetase

Structure of aminoacyl tRNA is given in Fig. 10.14. The aminoacyl-AMP synthetase reacts with specific tRNA and forms aminoacyl-tRNA complex by releasing the enzyme aminoacyl-tRNA synthetase. This show that the enzyme tRNA synthetase has two specific sites. One site recognises the specific amino acids and the other site recognises the specific tRNA molecule. Thus, the tRNA synthetase brings the specific amino acid and tRNA molecule together. However, these recognition properties are essential for making sure that the specific amino acid is charged on the proper tRNA molecule. In the same way

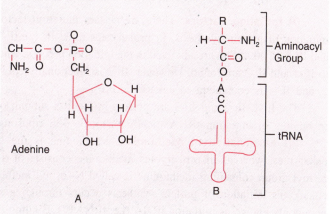

Fig. 10.14 : Structure of aminoacyl adenylate (A), and an aminoacyl -tRNA (B)

the tRNA molecule also consists of two specific sites, one site for recognising its specific aminoacyl-tRNA synthetase and the second (the anticodon) for codon present on mRNA molecule (see tRNA Chapter 5). For the incorporation of an amino acid at proper position in the polypeptide chain, recognition of codon on mRNA by the specific anticodon on tRNA is required (Goodman *et al.,* 1968).

2. Initiation of Polypeptide Synthesis

There are several specific and complex processes (Fig. 10.15) that are involved in the initiation and continuation of the elongation of polypeptide sequence. The essential components required for initiation are : initiation factors, ribosome, mRNA, guanosine triphosphate (GTP) and aminoacyl-tRNase.

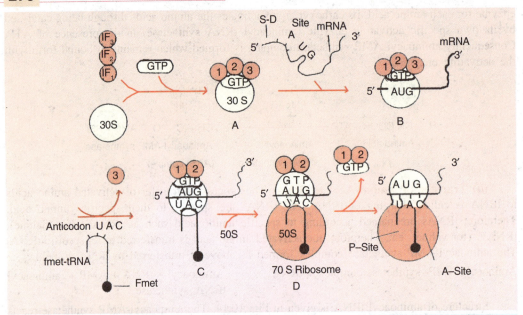

Fig. 10.15 : Initiation of translation (based on Moat and Foster, 1995).

(*i*) **Initiation Factors:** There are certain initiation factors (IF) which are required for the initiation of protein synthesis. In prokaryotes three IF (i.e.IF-1, 9,000 MW; IF-2, 1,15,000 MW and IF-3,22,000 MW) are involved in the initiation process, whereas in eukaryotes no IF equivalent to IF-1 and IF-2 are found. However, IF-2 is functionally equivalent to eukaryotic eIF-2 and eIF-2' and IF-3 is equivalent to eukaryotic eIF-3.

The IF-1, IF-2 and IF-3 are present in the 30S subunit of the ribosome. IF-1 and IF-2 help in binding of initiation tRNA (tRNA^{met}) to the 30S ribosome subunit.

(*ii*) **Formylation of Methionine:** Methionine is the starting N-terminal amino acid in eukaryotes, whereas in prokaryotes methionine consists of a formyl group (-CHO). Therefore, formyl group containing methionine is called N-formyl methionine. In prokaryotes as well as in eukaryotes initiation of protein synthesis occurs through a specific methionyl tRNA which is commonly known as initiation tRNA (i.e. tRNA^{met}). Binding of initiation tRNA with methionine/formylmethionine occurs as below :

In eukaryotes :

Methionine   +   tRNA $\longrightarrow$ Methionine - tRNA (met - tRNA)

In prokaryotes :

Formyl tetrahydrofolate + NH_2-methionyl tRNA $\xrightarrow{\text{Transformylase}}$ *N*-formylmethionyl-tRNA
(*N*-fmet-tRNA)

(*iii*) **Formation of 30S Initiation Complex:** The first step in initiation of protein synthesis is the formation of 30S initiation complex. This complex consists of an mRNA, 30S ribosomal subunit, GTP, IF (1, 2 and 3) and the initiator tRNA i.e. N-fmet-RNA. Formation of 30S initiator occurs in the following steps (the actual order of these steps are not known):

(*a*) The initiation factors (IF-1, IF-2 and IF-3) bind to 30S ribosomal subunit in the presence of GTP to form 30S-IF complex (Fig. 10.15A). However, when the mRNA is absent IF-1 and IF-3 do not form complex neither with 30S subunit nor 50S subunit.

(*b*) The second step involves the association of mRNA and initiator tRNA to the 30S subunit. However, the actual order of these steps vary. The IF-3 can bind to both 30S subunit and to mRNA. The 30S-IF complex binds to mRNA at the site containing initiation codon (in the order of pB reference AUG, GUG, UUG, CUG, AUA or AUU). Each mRNA at its untranslational region consists of a ribosome binding site for every polypeptide in the form of polycistronic message. This ribosome binding site (i.e. 5'-AGGAGGU-3') is known as Shine-Dalgarno sequence which is important in the binding of mRNA to the 30S-IF complex (Fig. 10.15B). The Shine-Dalgarno sequence base pairs to a region present at 3' end of 16S rRNA. This pairing will results in proper position of initiating AUG codon so that it can combine with an initiator anticodon on tRNA.

(*c*) The IF-2 which has combined with GTP, permits the initiator tRNA (i.e.N-fmet-tRNA) to bind to the 30S ribosomal subunit (Fig. 10.15C). Then it allows to the 30S and 50S subunits to get associated. This binding is followed by removal of IF-3 from the 30S-IF complex. Removal of IF-3 is necessary because its presence inhibits the association of two ribosomal subunits. At this stage the initiation complex consists of mRNA associated with the 30S ribosomal subunits, IF-1, IF-2-GTP and fmet-tRNA.

(*iv*) **Formation of the Complete Initiation Complex:** The last step in prokaryotes is the union of the 30S initiation complex with the 50S ribosomal subunit and formation of a complete 70S initiation complex (Fig. 10.15D). This process of union causes the immediate hydrolysis of the bound GTP to GDP + Pi. The process of union is accomplished in the presence of an analogue of GTP (i.e. 5'guanyl methylenediphosphate). Therefore, hydrolysis of GTP and subsequent removal of GDP is essential for the IF-1 dependent release of IF-2 from the ribosome (Fig. 10.15 E) (Moat and Foster,1995). Similarly, in eukaryotes the 40S initiation complex is attached to 60S ribosomal subunit and forms the complete 80S initiation complex.

Structure and binding sites of ribosome have been described in Chapter. 4 (see the cytoplasm). The ribosome has three important binding sites, two are important in protein synthesis. The two binding sites are: the aminoacyl-tRNA binding site (A), and the peptide (P) binding site (Fig. 10.15E). The A site receives all the incoming charged tRNA, whereas the P site possesses the previous tRNA with the new polypeptide (peptidyl tRNA) attached. The fmet-tRNA (initiation tRNA) directly binds with P site, but not A site.

3. Elongation of Polypeptide Chain

As shown in Fig. 10.15 E, at the end of initiation sequence, the 70S ribosome possesses the fmet-tRNA in the P site, whereas the A site is free to receive the next aminoacyl-tRNA according to the codons on mRNA. The addition of amino acids to the growing polypeptide chain as per codon on mRNA is called elongation of chain. The rate of addition of amino acid to the growing polypeptide is about 16 residues per second at 37°C. The 5S rRNA molecule recognises the nucleotide sequence of Tψ loop of tRNA and thus helps in binding of tRNA to the A site. The codons direct the specific aminoacyl-tRNA to form bonds. The bond formation is stimulated by an elongation factor T (EF-T) and GTP. T refers to transferase activity (Clark,1980).

The elongation factor (EF) is a soluble protein which is required for elongation of polypeptide chain. The EF is of two types, EF-T and EF-G. The EF-T is associated with transferase activity, whereas the EF-G is involved in translocation of mRNA. In prokaryotes the EF-T consists of two protein subunits which are called EF-Tu (temperature unstable, MW 44,000) and EF-Ts (temperature stable, MW 30,000). The EF-Tu is most abundant protein in *E.coli* that accounts for 5-10% of the total cellular protein. Both the proteins (EF-Tu and EF-Ts) are needed for binding the aminoacyl-tRNA to the ribosome.

The eukaryotic EF is called EF-1 and EF-2 which has resemblance with the prokaryotic EF-T and EF-G. More specifically the EF-1 is like the EF-Tu in its structure and function. At a time the EF-1 exists in one of the two forms (light form, EF-1L and heavy form, EF-1H). The function of EF-2 is translocation of aminoacyl-tRNA from A site to the P site. The GTP is required to drive the process of chain elongation. Bermerk (1978) has discussed the mechanism of chain elongation on ribosome. Elongation of the polypeptide chain is accomplished in the following two steps (Clark, 1980).

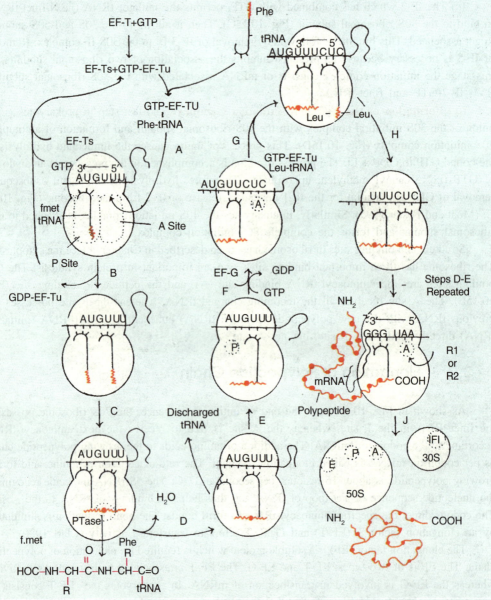

Fig. 10.16 : Events of polypeptide chain formation (diagrammatic, based on Moat and Foster, 1995).

(*i*) **Binding of Aminoacyl-tRNA to the A Site:** The GTP binds to EF-T and splits it into EF-Tu-GTP and EF-Ts. The EF-Tu-GTP can bind to all aminoacyl-tRNA (except the initiator tRNA) and results in formation of GTP-EF-Tu-aminoacyl-tRNA complex (Fig. 10.16A). It is an intermediate complex which is bound to the ribosome. In this step the EF-Ts complex does not play any role. After the aminoacyl-tRNA binds to the A site, GTP is hydrolysed and EF-Tu-GDP complex is released from the ribosome (B). Each aminoacyl-tRNA bound hydrolyses one GDP. The aminoacyl-tRNA may bind to the A site but this binding may not be followed by EF-Tu release from the ribosome. This shows that the purpose of GTP hydrolysis is the release of EF-Tu from the ribosome.

(*ii*) **Peptide-bond Formation:** Soon the enzyme peptidyl transferase (PTas) catalyses the peptide-bond formation. In fact this is catalysed by the 23S rRNA. This process is called peptidyl transfer (Fig. 10.16.C). However, peptide bond formation depends on release of EF-Tu from the ribosome but not on hydrolysis of GTP. The EF-Ts complex recycles the EF-Tu-GDP to EF-Tu-GTP, but does not cause release of EF-Tu from the ribosome as the release of IF-2 depends on IF-1.

When a new aminoacyl-tRNA binds to the A site, peptide bond formation occurs between the starting amino acid (N-fmet-tRNA on prokaryotes and met-tRNA in eukaryotes) and new aminoacyl-tRNA at the P site. The enzyme peptidyl transferase located in 50S ribosomal subunit catalyses the formation of peptide bond between the amino group of new incoming amino acid and the C-terminal of the elongating polypeptide attached to tRNA (Fig. 10.16D). During the process of bond formation, H_2O is eliminated.

4. Translocation

When the peptide bond is formed, the growing peptide chain binds to the tRNA that carries the incoming amino acid and occupies the A site of ribosome. The discharged tRNA after dissociating itself from the peptide chain is released from the P site (Fig. 10.16D-E). It is known so far that ribosome consists of two sites (A and P) but the recent evidences suggest that it consists of three sites: A, P and E. The E site is specific for deacylated tRNA (E).

(*i*) **Mechanism of Translocation :** In the ribosome at site A (aminoacyl-tRNA accepting site) the incoming aminoacyl-tRNA enters where decoding (codon-anticodon recognition) takes place. Thereafter, the ribosome moves along mRNA and, therefore, a change in complex occurs. The movement of ribosome causes the alignment with A site of next codon of mRNA to be translated. Consequently, the peptidyl-tRNA situated at A site is transferred to P site. This event of transfer of peptidyl-tRNA is called translocation (Fig. 10.16E-F). During translocation the events that are accomplished are: (*i*) removal of discharged tRNA from the P site, (*ii*) movement of the peptidyl tRNA from the A site to the P site, and (*iii*) movement of message by one codon.

(*ii*) **Energetics:** The recent model of ribosome shows that (*i*) the incoming charged tRNA binds at the A site, (*ii*) the growing polypeptide attached to tRNA and bound to P site is transferred to the tRNA in the A site, and (*iii*) the newly deacaylated tRNA after translocation is not released immediately but are bound to the E site. Now both the E and P sites are engaged. The other incoming charged tRNA binds to the unoccupied A site. This causes reduction in affinity of the E site for the deacylated tRNA and resulting in release of discharged tRNA from the ribosome. The process of binding of incoming aminoacyl-tRNA to site A continue until termination signal is received (G-I) (Driessen,1992).

In prokaryotes translocation is brought by the EF-G or translocase (MW, 80,000 in which G =GTPase) and GTP hydrolysis are required. EF-G binds to the same site as the EF-Tu. After binding EF-G hydrolyse the ATP to ADP + Pi in the presence of ribosome. It is obvious that during elongation two molecules of GTP are hydrolysed per peptide bond, one is EF-T dependent and the other EF-G-dependent. The EF-G is released from the ribosome after each step of elongation. Since both EF-T and EF-G utilize the same binding site, elongation cannot continue unless EF-G is released (Driessen,1992).

5. Termination of Polypeptide Chain

(i) Receognition of Termination Signal: The polypeptide chain continues the elongating until a termination codon on mRNA reaches to ribosome. The termination codons (UAA-ochre, UAG-amber, UGA-opal or umber) are also called as non-sense codon because no tRNA anticodon pairs with them. It is not necessary that the termination codon is the last codon of mRNA. For example in bacteria and bacteriophages polygenic mRNAs are common and they consist of a number of initiation and termination codons. After translocation brings one of the above termination codons into the A site, the ribosome does not bind with an aminoacyl-tRNA-EF-Tu-GTP complex. Then it receives the signal of termination.

(ii) Release of Polypeptide Chain: When a termination codon is translocated into the A site, the ribosome instead of binding with a complex containing an amino acid, binds with a peptide release factor (RE) (Fig. 10.16 J). In prokaryotes there are three RF proteins (RF-1, MW 44,000; RF-2, MW 47,000, ; RF-3, MW 46,000). The RF-1 is active with UAA and UAG codons and the RF-2 is active with UAA and UGA codons. The RF-3 activates the RF-1 and RF-2; therefore, the RF-3 is called stimulatory (S) factor. In eukaryotes there is only one RF protein (MW 56,500 and 1,15,000) which is active with codons UAA, UAG and UGA. The RF protein exists in two units and both of them remain in active form.

The ribosome binds either with RF-1 or RF-2. However, the RF protein activates peptidyl transferase which hydrolyses the bond joining the peptide to the tRNA at the site P. This results in release of the peptide chain (Fig. 10.16J).

6. Post Translational Processing

After release some of the processing events occur in the polypeptide chain. Such modifications occur both in prokaryotes and eukaryotes as given below:

(i) Removal of fmet from the Polypeptide Chain: The formyl group of the N-terminal fmet is removed by the enzyme methionine deformylase. The enzyme formylmethionine specific peptidase (methionyl aminopeptidase or MAP) hydrolyses the entire formylmethionine residues. All the terminal methionines are not removed because there is involvement of discrimination in channelling of different polypeptide through these two alternative steps. The side-chain penultimate amino acid acts as the discriminating factor. Removal of methionine by MAP depends on the length of side chain. The side chain of the longer length has less possibility for MAP to remove the methionine. The other processings are acetylation (of L12 to give rise L7) or adenylation.

(a) Loss of signal sequences: In some polypeptides, about 15 to 30 amino acid residues are present at N-terminus. These residues act as signal sequence and direct the protein to its ultimate destination. The signal sequences are cleaved by specific peptidases.

(b) Modification of individual amino acid: Some amino acid side chains are specifically modified such as : *(a)* enzymatic phosphorylation by ATP of -OH group of certain amino acids (e.g. serine, threonine, tyrosine), *(b)* binding of Ca^{++} to phosphoresine groups of milk protein, casein, *(c)* addition of carboxyl (-COOH) group to aspartate glutamate residues of some proteins (e.g. blood clotting protein, prothrombin), *(d)* methylation of proteins (e.g. methylation of lysine residues in cytochrome *c*, calmodulin).

(c) **Formation of disulphide cross-links:** Disulphide bridges between cysteine residues of some proteins are formed. Hence, they are covalently cross-linked and attain native from.

(d) **Glycosylation:** Attachment of the carbohydrate side chains during or after protein synthesis is called glycosylation, for example glycoproteins.

(e) **Addition of prosthetic group:** Prosthetic groups get covalently bound to many prokaryotic and eukaryotic proteins. For example, biotin molecule is covalently linked to acetyl-CoA carboxylase.

(ii) **Ribosome Editing:** During the process of translation certain inappropriate aminoacylated - tRNAs enter the A site of ribosome and remain bound to the ribosome. These out number the appropriate amino acylated (aa)-tRNAs. However, the aa-tRNAs remain bound for a long time to the A site for a peptide from P site of ribosome. There are two processes that can reduce the error of surviving polypeptide chain e.g. ribosome editing and preferential degradation of polypeptide chain containing erroneous amino acids (Moat and Foster,1995).

According to the ribosome editing hypothesis the structure of inappropriate peptidyl-tRNA does not correctly complement the structure of mRNA; therefore, it dissociates from the ribosome during protein synthesis. However, the gene *rel*A produces a signal molecule (alarmone), guanosine tetraphosphate (ppGpp) which affects the editing process. The ppGpp interacts with EF-G and results in longer life of peptidyl-tRNA in the A site and enchances the editing process. The gene *pth* synthesises peptidyl-tRNA hydrolase that acts upon the peptidyl-tRNA when it is released. The ribosome free peptidyl-tRNA is hydrolysed by this enzyme. Consequently an intact peptide and an intact tRNA are produced. The defective peptide is degraded by the enzyme (Goldberg and Goff,1986).

(iii) **Protein Folding:** After the synthesis of polypeptide chain, it undergoes spontaneous foldings. The secondary folds are formed between the folded regions. Finally as a result of further folding there develops a tertiary structure of polypeptide chain i.e. protein. Before the terminal amino acids are added in polypeptide chain, the protein reaches to its final shape during the course of chain termination.

According to the recent theories the process of protein folding is complex. Many proteins require assistance to get folded properly. This assistance is provided by the proteins of special kind known as chaperones or chaperonins. These assist polypeptides to self assemble by inhibiting alternative assembling pathway. When chaperones interact with the polypeptide, the chance of incorrect infolding is reduced. In *E. coli* the example of chaperones are GroEl (60 KDa), GroES (10KDa) and DnaK (70KDa). All these are constitutive proteins which increase their concentration when there is stress like heat shock (Ellis and Vander Vries, 1991). For detailed discussion see Chapter 4 -*The Microbial cells : Morphology and Structure.*

7. The Signal Hypothesis (Protein Export)

It is interesting to note that after synthesis of protein, how it is incorporated in membranes or secreted out side by the cell? However, it is believed that the secretory proteins are synthesised by the ribosomes which are attached to the endoplasmic reticulum and released into it. From endoplasmic reticulum they are transported to various cell organelles (in eukaryotes) where from secreted outside the cell through the process of exocytosis. To explain this mechanism Blobel and Dobberstein (1975) proposed a theory known as *signal hypothesis.* Furthermore Blobel (1978) reviewed this hypothesis and postulated that the mRNAs that translate secretary proteins possess on 3′ side of initiation codon (AUG) a group of signal codons. The endoplasmic reticulum consists of ribosome receptor proteins. A polypeptide chain consisting of a special region (signal peptide region) is synthesized by the ribosome. After coming out from the ribosome the signal peptide interacts with ribosome receptor protein and results in formation of a tunnel in the membrane.

However, the membrane tunnel coincides with the ribosomal tunnel. The enzyme signal peptidase breaks the polypeptide chain which is being synthesised. Upon complete synthesis of polypeptide chain, it is released inside the space of endoplasmic reticulum. At the end the ribosomes dissociate from the membrane of endoplasmic reticulum; ribosome receptor proteins get diffused and close the tunnels. This process of entering the proteins into the membranes is also known as *protein export.*

A significant work has been done in recent years on secretory protein translocation system in *E. coli* and the other bacteria as well such as *Bacillus subtilis, Salmonella typhimurium, Pseudomonas fluorescens, Enterobacter aerogenes, Vibrio cholerae, Klebsiella oxycota,* etc.

Most of the secretory proteins are translated first in a form of precursor containing 15-30 amino acids at N-terminus which is called a signal sequence. The signal sequence consists of a hydrophobic region of about 11 amino acid residues and a short stretch of hydrophilic region at N-terminus. The signal sequence is involved to bind the nascent polypeptide to the membrane.

In Gram-negative organisms the pathway for export and secretion of signal sequence-containing proteins is called general secretary pathway. The first step is the *sec* gene product dependent translocation of exported protein outside the cytoplasmic membrane. Brondage *et al.,* (1990) have presented a model of proOmpA (an export protein) translocation across the plasma membrane (Fig. 10.17). The protein SecB (a product of *secB* gene) is a pilot chaperonin. It is associated with the protein that is to be transported i.e. transport protein, for example proOmpA. It is also synthesised on the ribosome. However, in the absence of SecB, the proOmpA aggregates and checks its insertion into the membrane. The protein SecA is basically an ATPase and also forms a part of preprotein translocase in association with integral membrane protein SecY/E. The SecA protein acts as a receptor for proOmpA-SecB complex. Subsequently, the ATP is hydrolysed releasing the proOmpA into the membrane.

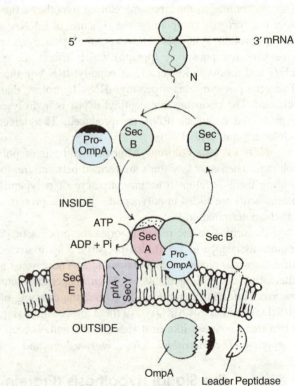

Fig. 10.17 : A model for translocation of proOmpA across the plasma membrane (based on Brondage *et al.,* 1990).

It also drives the over all chaperones and membrane-associated reactions. Once the process of transport has been started at the expense of ATP, further translocation event proceeds through a series of transmembrane intermediates, the energy requirement of which is met by proton motive force rather than ATP hydrolysis. During the process of translocation, the enzyme signal peptidase (LepB) cleaves the signal sequence of the exported protein. It has to enter the periplasm or translocate across the outer membrane. Pugsley (1993) has published the complete general secretary pathway in Gram-negative bacteria where the various branches direct proteins to their final extra cytoplasmic destination (Fig. 10.18).

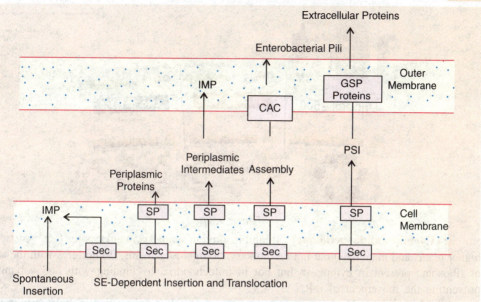

Fig. 10.18 : Main branches of the general secretary pathways (GSP) of Gram-negative bacteria (modified after Pugsley, 1993). IMP, integrated membrane proteins; CAC, chaperone assembly channel; PSI, periplasmic secretion intermediates; SP, signal peptidase.

8. The Inhibitors of Gene Expression

There are several antimicrobial agents that inhibit protein synthesis at two steps, either during transcription or translation. Franklin and Snow (1991) have nicely discussed the biochemistry of the action of antimicrobial antibiotics.

(i) Inhibitors of Transcription

(a) **Rifamycin:** The rifamycin and chemically similar group (streptovarcins) inhibit the initiation of transcription. Rifampin and streptovarcin are the semisynthetic compounds, whereas rifamycins are the naturally occurring antibiotics. They tightly bind with β subunit of RNA polymerase (rpoB) and inhibit initiation of transcription.

(b) **Streptolydigin:** It is similar to rifamycin because it too binds with β subunit of RNA polymerase. In addition, it inhibits both the processes : chain initiation and chain elongation *in vitro*.

(ii) Inhibitors of Translation

(a) **Chloramphenicol :** It inhibits the activity of peptidyl transferase after binding to 50S subunit of bacterial ribosome. Its effect is bacteriostatic i.e. after removal of drug the effect is soon reversed. However, its effect in eukaryotes is the same as in prokaryotes. Bone marrow toxicity results in aplastic anaemia.

(b) **Tetracyclines :** This group of antibiotic shows broad spectrum bacteriostatic activities against Gram-positive and negative bacteria, mycoplasmas, rickettsiae and chlamydiae. These prevent the binding of aminoacyl -tRNA to the A site of 30S ribosome. Moreover, it can bind to several sites of both 30S and 50S subunits.

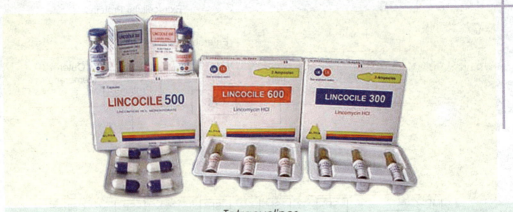

Tetracyclines

(*c*) **Cycloheximide (actidione) :** It inhibits protein synthesis in eukaryotes (e.g. yeasts, fungi, higher plants and mammals) but not the prokaryotic microorganisms. It interferes with the activity of ribosome present in cytoplasm but not in mitochondria, by binding with 80S subunits and preventing the movement of mRNA.

(*d*) **Macrolides :** This is a large group of antimicrobial agents that includes erythromycin, leucomycin, macrocin, carbomycin, chalcomycin, angalomycin, etc. These are active against the Gram-positive bacteria and less active against Gram-negative bacteria but not against the eukaryotes. These interact with 50S subunit of ribosome and inhibit protein synthesis. Also, they stimulate the dissociation of peptidyl-tRNA from the ribosome through abortive translocation step.

(*e*) **Lincomycin :** Lincomycin and the other chemically similar antibiotics inhibit peptidyl transferase through binding to 23S rRNA of the 50S subunit. These affect both Gram-positive and Gram-negative bacteria.

(*f*) **Puromycin :** It binds to a peptide with the C-terminus of growing polypeptide and results in premature termination of polypeptide chain. It interacts with the P site of ribosome but not A site. It works equally well on 70S and 80S ribosomes.

9.   Protein Synthesis in Prokaryotes and Eukaryotes

The mechanism of protein synthesis in prokaryotes and eukaryotes is the same. However, there are some of the following differences :

   (*i*) In prokaryotes 70S ribosomes (30S and 50S subunits) and eukaryotes 80S ribosomes (40S and 60S subunits) carryout protein synthesis.

  (*ii*) The bacterial mRNAs are polycistronic (i.e. possess many initiation and termination sites), whereas the eukaryotic mRNAs are mainly monocistronic (i.e. contain only one functional site for protein synthesis).

 (*iii*) Mostly in bacterial mRNA translation starts while transcription goes on DNA. But in eukaryotes after transcription, mRNAs migrate to cytoplasm through nuclear pore. Translation occurs on ribosomes in cytoplasm. Therefore, during translation mRNA is not associated with DNA.

  (*iv*) In prokaryotes N-formyl methionine is the first amino acid which starts translation, whereas in eukaryotes the first amino acid that starts translation is methionine.

   (*v*) The initiation factors of polypeptide chain (in prokaryotes) are IF-1, IF-2 and IF-3, whereas in eukaryotes there are several IF factors.

  (*vi*) The elongation factors in prokaryotes are EF-Tu, FF-Ts and EF-G, whereas in eukaryotes the elongation factors are only EF-1 and EF-2. The EF-1 is similar to EF-T (i.e. EF-Tu + EF-Ts).

(*vii*) In eukaryotes the release factor (RF) recognizes the three termination codons (UAA, UAG and UGA), whereas in prokaryotes there are three release factors (RF-1, RF-2, RF-3). RF-1 recognises the termination codons UAA, and UAG, RF-2 recognises UAA and UGA, and RF-3 is associated with stimulatory activity.

B. Regulation of Gene Expression

The DNA of a microbial cell consists of genes, a few to thousands, which do not express at the same time. At a particular time only a few genes express and synthesize the desired protein. The other genes remain silent at this moment and express when required. Requirement of gene expression is governed by the environment in which they grow. This shows that the genes have a property to switch on and switch off .

It is clear from the Chapter 7, *The Genetic Code* that 20 different amino acids constitute different protein. All are synthesised by codons. Therefore, synthesis of all the amino acids requires energy which is useless because all the amino acids constituting proteins are not needed at a time. Hence, there is need to control the synthesis of those amino acids (proteins) which are not required. By doing this the energy of a living cell is conserved and cells become more competent. Therefore, a control system is operative which is known as *gene regulation*.

There are certain substrates called inducers that induce the enzyme synthesis. For example, if yeast cells are grown in medium containing lactose, an enzyme lactase, is formed. Lactase hydrolyses the lactose into glucose and galactose. In the absence of lactase, lactose synthesis does not occur. This shows that lactose induces the enzyme lactase. Therefore, lactase is known as inducible enzyme. In addition, sometimes the end product of metabolism has inhibitory effect on the synthesis of enzyme. This phenomenon is called feed back or *end product inhibition*.

From the out going discussion it appears that a cell has autocontrol mediated by the gene itself. For the first time Francois Jacob and Jacques Monod (1961) at the Pasteur Institute (Paris) put forward a hypothesis to explain the induction and repression of enzyme synthesis. They investigated the regulation of activities of genes which controls lactose fermentation in *E. coli* through synthesis of an enzyme, β-galactosidase. For this significant contribution in the field of biochemistry they were awarded Nobel Prize in Medicine in 1965.

Gene expression of prokaryotes is controlled basically at two levels i.e. transcription and translation stages. In addition, mRNA degradation and protein modification also play a role in regulation. Most of the prokaryotic genes that are regulated are controlled at transcriptional stage. Other control measures operating at different levels are given in Table. 10.2.

Table 10.2 : Types of gene regulation in prokaryotes operating at different levels

Levels of control	Means of control	Examples
1. Transcriptional control	DNA binding proteins	Helix-turn-helix in phage λ
		Zinc fingers in *Xenopus*
	The *lac* operon	*E.coli*
2. Catabolic control		Cyclic AMP in *E.coli*
	The *gal* operon	
	The arabinose operon	*E.coli*
	The *trp* operon	*E.coli*
3. Translation control	The *arg* regulon	*E.coli*
4. The membrane- mediated regulation	The *put* system	*Salmonella typhimurium*
5. Osmotic control	Turgor	*E.coli*, etc.
6. Through Electron transport	Stringent control	Response to amino acid starvation

I. Transcriptional Control

It is a general strategy in a living organism that chemical changes occur by a metabolic pathway through a chain of reactions. Each step is determined by the enzymes. Again synthesis of an enzyme comes under the control of genetic material i.e. DNA in living organisms. Enzymes (proteins) are synthesised via two steps : transcription and translation. Transcription refers to synthesis of mRNA. Structural components of mRNA are given in Fig. 5.12. Transcription is regulated at or around promoter region of a gene. By controlling the ability of RNA polymerase to the promoter the cell can modulate the amount of message being transcribed through the structural gene. However, if RNA polymerase has bound, again it can modulate transcription. By doing so the amount of gene product synthesized is also modulated. The coding region is also called structural gene. Adjacent to it are regulatory regions that control the structural genes. The regulatory regions are composed of promoter (for the initiation of transcription) and an operator (where a diffusible regulatory protein binds) regions.

The molecular mechanisms for each of regulatory patterns vary widely but usually fall in one of two major groups: negative regulation and positive regulation. In negative regulation an inhibitor is present in the cell and prevents transcription. This inhibitor is called as repressor. An inducer i.e. antagonist repressor is required to permit the initiation of transcription. In a positive regulated system an effector molecule (*i.e.* a protein, molecule or molecular complex) activates a promoter. The repressor proteins produce negative control, whereas the activator proteins produce positive control. Since the transcription process is accomplished in three steps (RNA polymerase binding, isomerization of a few nucleotides and release of RNA polymerase from promoter region), the negative regulators usually block the binding, whereas the activators interact with RNA polymerase making one or more steps. Fig. 10.19 shows the negative and positive regulation mechanism of the genes. In negative regulation (A) an inhibitor is bound to the DNA molecule. It must be removed for efficient transcription. In positive regulation (B) an effector molecule must bind to DNA for transcription.

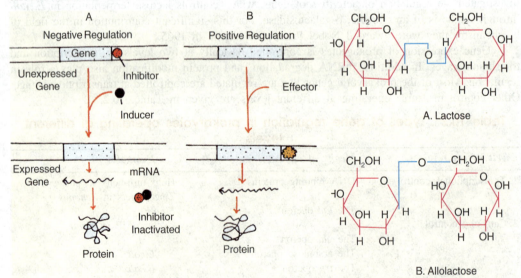

Fig. **10.19** : The negative (A) and positive (B) regulation genes.   Fig. **10.20** : Chemical structure of lactose (4-D glucose B-D-galactopyranoiside) (A), and allolactose (B).

1. The *Lac* Operon

For the first time Jacob and Monod (1961) gave the concept of operon model to explain the regulation of gene action. An operon is defined as several distinct genes situated in tandem, all controlled by a common regulatory region. Commonly an operon consists of repressor, promoter, operator and structural genes. The message produced by an operon is polycistronic because the information of all the structural genes resides on a single molecule of mRNA.

The regulatory mechanism of operon responsible for utilization of lactose as a carbon source is called the *lac* operon. It was extensively studied for the first time by Jacob and Monod (1961). Lactose is a disaccharide which is composed of glucose and galactose (Fig. 10.20).

The lactose utilizing system consists of two types of components; the structural genes (*lac*Z, *lac*Y and *lac*A) the products of which are required for transport and metabolism of lactose and the regulatory genes (the *lac*I, the *lac*O and the *lac*P). These two components together comprises of the *lac* operon (Fig. 10.21a). One of the most key features is that operon provides a mechanism for the coordinate expression of structural genes controlled by regulatory genes. Secondly, operon shows polarity i.e. the genes Z,Y and A synthesise equal quantities of three enzymes β-galactosidase (by *lac*Z), permease (by *lac*Y) and acetylase (by *lac*A). These are synthesized in an order i.e. β-galactosidase first and acetylase in the last.

(i) **The Structural Genes:** The structural genes form one long polycistronic mRNA molecule. The number of structural gene corresponds to the number of proteins. Each structural gene is controlled independently, and transcribes mRNA molecules separately. This depends on substrates to be utilized. For example, in *lac* operon three structural genes (Z,Y and A) are

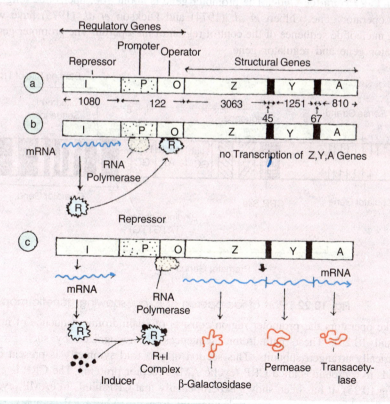

Fig. 10.21 : The *lac* operon; (*a*) genetic map (the numbers show the number of base pairs that comprise each gene); (*b*) repressed state; (*c*) induced state.

associated with lactose utilization (Fig. 10.21A). β-galactose is the product of *lac*Z that cleaves β-1,4 linkage of lactose and releases the free monosaccharides. This enzyme is a tetramer of four identical subunits each with molecular weight of 1,16,400. The enzyme permease (a product of *lac*Y) facilitates the lactose to enter inside the bacterium. Permease has molecular weight of 46,500. It is hydrophobic. The cells mutant in *lac*Z and *lac*Y are designated as Lac⁻ i.e. the bacteria cannot grow in lactose-free medium. The enzyme transacetylase (30,000 MW) is a product of *lac*A whose no definite role has been assigned.

The *lac* operon consists of a promoter (P) and an operator (O) together with the structural genes. The initiation codon of *lac*Z is TAC that corresponds to AUG of mRNA. It is situated 10 bp away from the end of operator gene. However, the *lac* operon cannot function in the presence of sugars other than lactose.

(ii) The Operator Gene: The operator gene is about 28 bp in length present adjacent to *lac*Z gene. The base pairs in the operator region are palindrome *i.e.* show two fold symmetry from a point (Fig. 10.22) (Gilbert and Maxam,1973). The operator overlaps the promoter region. The *lac* repressor proteins (a tetramer of four subunits) bind to the *lac* operator *in vitro* and protect part of the *lac* operator *in vitro* and protect part of the promoter region from the digestion of DNase. The repressor proteins bind to the operator and form an operator-repressor complex which in turn physically blocks the transcription of Z,Y and A genes by preventing the release of RNA polymerase to begin transcription (Fig. 10.21b) (Maniatis and Ptashne,1976).

In bacteriophage λ there are two operators the OL and OR (Fig. 16.13B) which have different base sequences. Lambda repressor (gpcI) is rapidly synthesized, binds to OL and OR and inhibits the synthesis of mRNA and production of proteins gpcII and gpcIII (Fig.16.13C).

(iii) The Promoter Gene: The promoter gene is about 100, nucleotide long and continuous with the operator gene. Gilbert *et al.* (1974) and Dickson *et al.* (1975) have worked out the complete nucleotide sequence of the control region of *lac* operon. The promoter gene lies between the operator gene and regulator gene.

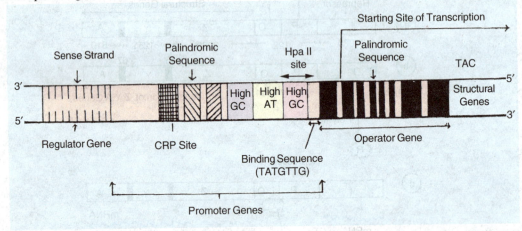

Fig. 10.22 : Part of *lac* operon of *E. coli* showing genetic map.

Like operators the promoter region consists of palindromic sequence of nucleotides (Figs. 10.22 and 10.23). These palindromic sequences are recognized by such proteins that have symmetrically arranged subunits. This section of two fold symmetry is present on the CRP site that binds to a protein called CRP (cyclic AMP receptor protein). The CRP is encoded by CRP gene (Fig.10.25). It has been shown experimentally that CRP binds to cAMP (cyclic AMP found in *E. coli* and other organisms) molecule and form a cAMP-CRP complex. This complex is required for transcription because it binds to promoter and enhances the attachment of RNA polymerase to the promoter. Therefore, it increases transcription and translation processes. Thus, cAMP-CRP is a positive regulator in contrast with the repressor, and the *lac* operon is controlled by both positively and negatively.

According to a model proposed by Pribnow (1975) the promoter region consists of three important components which are present at a fixed position to each other. These components are (*i*) the recognition sequence, (*ii*) the binding sequence, and (*iii*) an mRNA initiation site.

The recognition sequence is situated outside the polymerase binding site that is why it is protected from DNase. Firstly, RNA polymerase binds to DNA and forms a complex with the recognition sequence. The binding site is 7 bp long (5'TATGTTG) and present at such region that is protected from DNase. In other organisms the base pairs do not differ from more than two bases. Hence, it can be written as 5' TATPuATG. The mRNA initiation site is present near the binding site on one of the two bases. The initiation site is also protected from DNase. However, there is overlapping of promoter and operator in *lac* operon. Moreover, there is a sequence 5'CCGG, 20 bp left to mRNA initiation site.This is known as *Hpa*II site (5'CCGG) because of being cleaved at this site by the restriction enzyme *Hpa*II (*see* Chapter 11, *Gene Cloning in Microorganisms*).

(*iv*) **The Repressor (Regulator) Gene:** Repressor gene determines the transcription of structural gene. It is of two types : active and inactive repressors. It codes for amino acid of a defined repressor protein. After synthesis the repressor molecules are diffused from the ribosome and bind to the operator in the absence of an inducer. Finally, the path of RNA polymerase is blocked and mRNA is not transcribed. Consequently, no protein synthesis occurs. This type of mechanism occurs in the inducible system of active repressor.

Moreover, when an inducer (e.g. lactose) is present, it binds to repressor proteins and forms an inducer-repressor complex. This complex cannot bind to the operator. Due to formation of complex the repressor undergoes changes in conformation of shape and becomes inactive. Consequently, the structural genes can synthesise the polycistronic mRNAs and the later synthesizes enzymes (proteins).

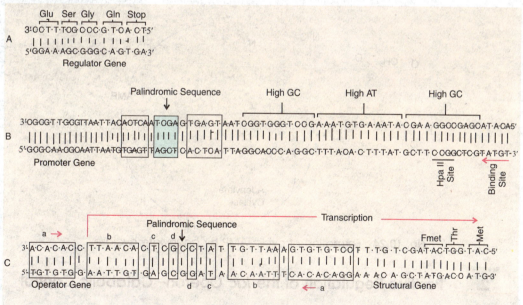

Fig. 10.23 : The nucleotide sequence of *E.coli lac* operon showing regulator gene (A), promoter gene (B) and operator gene and part of structural genes (C). The vertical arrows in B and C make the axis of two fold symmetry of palindrome.

In contrast, in the reversible system the regulator gene synthesises repressor protein that is inactive and, therefore, fails to bind to operator. Consequently, proteins are synthesised by the structural genes. However, the repressor proteins can be activated in the presence of a co-repressor. The co-repressor together with repressor proteins forms the repressor-co-repressor complex. This complex binds to operator gene and blocks protein synthesis.

Jacob and Monad (1961) could not identify the repressor protein. Gilbert and Muller - Hill (1966) succeeded in isolating the *lac* repressor from the Lac mutant cells of *E. coli* inside which the *lac* repressor was about ten times greater than the normal cells. The *lac* repressor proteins have been crystallized. It has a molecular weight of about 1,50,000. It consists of four subunits each has 347 amino acid residues and molecular weight of about 40,000 daltons. The repressor proteins have strong affinity for a segment of 12-15 base pairs of operator gene. This binding of repressor blocks the synthesis of mRNA transcript by RNA polymerase.

The *lac* operon is induced when *E. coli* cells are kept in medium containing lactose. The lactose is taken up inside the cell where it undergoes glycosylation i.e. molecular rearrangement from lactose to allolactose. The galactosyl residue is present on 6 rather than 4 position of glucose (Fig. 10.20). Glycosylation is done by β-galactosidase that is constitutively present in the cell before induction. Allolactose is the real inducer molecule. The *lac* repressor protein is an allosteric molecule with specific binding sites for DNA and inducer. Allolacctose binds to *lac* repressor to form an inducer-repressor complex. Binding of inducer to repressor allosterically changes the repressor lowering its affinity for *lac*O DNA. Consequently repressor is released from *lac*O due to changes in three dimensional conformation. This is called allosteric effect. After being free *lac*O allows the RNA polymerase to form mRNA transcript. Here, allolactose acts as the effector molecule and checks the regulatory protein from binding to *lac*O (operator) gene.

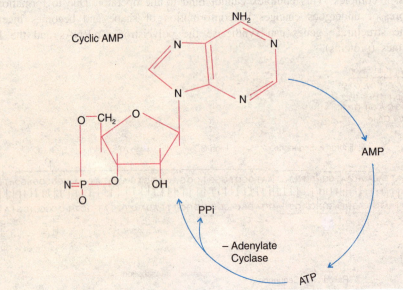

Fig. 10.24 : Formation of cyclic AMP by adenylate cyclase.

---

2.   Positive Regulation of the *lac* Operon– Catabolic Control

Cyclic AMP (cAMP) is the small molecule which is distributed in animal tissues, and controls the action of many hormones. It is also present in *E. coli* and the other bacteria. The cAMP is synthesized by the enzyme adenyl cyclase. (Fig. 10.24). Its concentration is directly regulated by glucose metabolism.

The *Lac* operon has an additional positive regulatory control mechanism to avoid the wastage of energy during the synthesis of lactose-utilizing proteins while there is adequate supply of glucose. When *E. coli* grows in a medium containing glucose the cAMP concentration in the cells falls down. This mechanism is poorly understood. However, the note worthy point is that cAMP regulates the activity of *lac* operon (and other operons also).

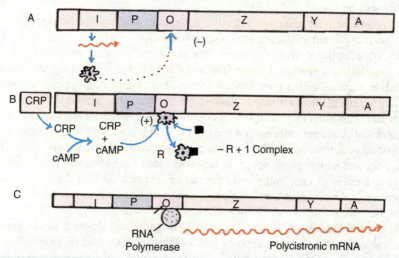

Fig. 10.25 : Regulation of *lac* operon. A, inhibition of transcription; B, catabolic control of cAMP by binding of complex to *lac* O; C, synthesis of mRNA after removal of repressor (R) protein.

In contrast when *E. coli* cells are fed with alternate carbon source e.g. succinate, cAMP level increases. The *cya* locus expresses the enzyme adenylate cyclase that converts the ATP to cAMP. How does cAMP increase the process of transcription, is not known clearly. It has been shown experimently that cAMP binds to the proteins expressed by *crp* locus which is known as cAMP receptor protein (CRP) or catabolic activator protein (CAP) (Fig.10.25). Therefore, CRP-cAMP complex binds to the CAP-binding site present on *lac* promoter. The CRP-cAMP bound complex promotes the helix destabilization downstream, and facilitates RNA polymerase binding. This results in efficient open promoter formation and in turn transcription (Spassky *et al.*, 1984).

3. Regulation of Gene Expression in Eukaryotes

There is much variation and complexity in regulation of genes in eukaryotes. Because in eukaryotes different genes are expressed at different developmental stages of cells or different tissues under the influence of different types of stimuli imposed by external environment. Eukaryotic DNA undergoes several changes such as double stranded, linear thread, nucleosome, fibres, chromatid and chromosomes. Gene expression and regulation take place only when DNA is in double stranded linear form. Moreover, if the promoter or regulator region of any gene is organized into chromosome, initiation of transcription does not take place. Therefore, changes in state of chromatin occur by chromatin remodelling which results in gene activation. Thus packaging of DNA influences gene expression. In majority of cases regulation of gene expression takes place at transcription level. Regulation of expression at processing or translation level may also occur in eukaryotes.

Gene expression can be regulated at several steps in the pathway from DNA to RNA to protein in a cell as described below.

- **Transcriptional control:** Controlling the gene expression during transcription
- **RNA processing control:** Control of processing of primary RNA transcripts to from mature mRNA
- **RNA transport control:** Control of transport of mature mRNA from nucleus to cytoplasm
- **Translational control:** Selection of mRNAs in cytoplasm to be translated by ribosome.
- **mRNA degradation control:** Selective degradation of certain mRNA molecules in the cytoplasm, or
- **Protein activity control:** Selective activation, inactivation or compartmentalization of specific protein molecule after their synthesis.

Only transcriptional control ensures that no superfluous intermediates are synthesized.

(i) **Regulation Through Transcriptional Factors:** Unlike prokaryotes, there are multiple DNA binding proteins called transcription factors that control transcription in eukaryotes. These proteins are grouped into two major classes: the general transcriptional factors (GTFs) and the regulatory transcriptional factors (RTFs) As discussed earlier the eukaryotic RNA polymerase fails to recognize the promoter directly. Therefore, the GTFs bind first the promoter directly (TATA sequence of all prokaryotes). RNA polymerase starts transcription at promoter site. The RTFs bind the regulatory site of the genes which is far away from the promoter. The RTFs bind to all the regulatory sequences of gene and control the rate of assembly of GTFs at the promoter. The RTFs either increase or decrease the transcription. When transcription is increased, this property is called activator. The decreasing level of transcription is called repression.

(ii) **Britten-Davidson Model for Gene Regulation:** Regulation at transcription level involves both activation and repression of genes. Because genes may be switched on in some cases and switched off in others. Various models have been proposed for regulation of gene expression in eukaryotes. In 1969, Britten and Davidson proposed a model called gene battery model or Britten-Davidson model which is very popular. This model was further elaborated in 1973.

According to this model, there are four class of sequences: *(i) producer genes* (which are comparable to structural genes of prokaryotes), *(ii) receptor site* (comparable to operator gene in bacterial operon), *(iii) integrator gene* (comparable to regulator gene synthesizing an activator RNA which may or may not synthesize protein before it activates the receptor

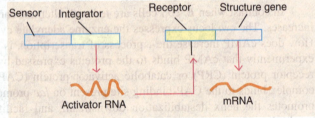

Fig. 10.26: Britten-Davidson model for gene regulation.

site), and *(iv) sensor site* (regulates the activity of integrator gene which can be transcribed only after activation of sensor site). The four classes of sequences are interrelated (Fig. 10.26).

In this model producer gene and integrator gene are involved in transcription, whereas the receptor and sensor sequences help in recognition without participating in RNA synthesis. It has been proposed that receptor site and integrator gene are repeated several times so that the activity of a large number of genes may be controlled in the same cell, same activator may recognize all the repeats, and several enzymes of one pathway may be synthesized simultaneously.

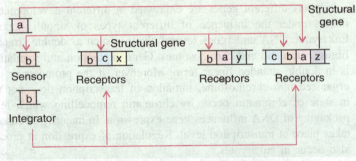

Fig. 10.27: Britten-Davidson model showing redundancy of receptor sites.

Transcription of the same gene is done in different developmental stages. This is achieved by several receptor sites and integrator genes. Each producer gene possesses many receptors sites, each site responds to one activator (Fig. 10.27) so that several genes can be recognized by a single activator. But at different time the same gene may be activated by different activators.

A set of structural genes controlled by one sensor site is called **gene battery.** Several sets of genes may be activated when major changes are required. If one sensor site gets associated with them, transcription of all integrators may be caused at the same time. Thus, transcription of several producer genes is caused through receptor sites.

QUESTIONS

1. Write an extended note on transcription with the help of suitable diagrams.

2. Write in detail the process of chain initiation in transcription.

3. Give a detailed account of translation of mRNA.

4. What is translocation ? How does it occur during protein synthesis ?

5. Write in detail the post-translational processing of polypeptide chain.

6. Give an illustrated account of signal hypothesis?

7. Write an essay on inhibitors of transcription and translation.

8. What do you mean by lactose operon ? How does it function ?

9. Give a detailed account of post-transcriptional processing of RNAs.

10. Discuss in brief about the post translational modification of polypeptide.

11. Write in detail on regulation of gene expression in eukaryotes.

12. Write short notes on

    (i) Splisosome,   (ii) 5'- Capping, (iii) rRNA processing, (iv) mRNA processing, (iv) Britten-Davidson model of gene regulation

13. Write short notes on the following :

    (i) Central dogma, (ii) RNA polymerase, (iii) Transcription, (iv) Sigma proteins (v) Rho, (vi) Ribosomes, (vii) Ribosome editing, (viii) Protein synthesis in prokaryotes and eukaryotes, (ix) lac operon, (x) Catabolic control.

REFERENCES

Bermek,E. 1978. Mechanism in polypeptide chain elongation on ribosomes. *Prog. Nucl. Acad. Res. Mol. Biol.* **2**: 63.

Blobel, G. 1978.Mechanism for intracellular compartment of newly synthesized proteins. In FEBS, 11th meeting, Copenhagen, 1977, Vol. 43. Symposium A2,pp. 99-108. Pergamon Press, Oxford.

Bremer, H. *et.al.* 1965, Direction of chain growth in enzymatic RNA-synthesis. *J. Mol. Biol.* **13**:540.

Brondage, L. *et al.* 1990. *Cell* **62**: 649-657.

Bujard, H. 1980. The interaction of *E.coli* RNA polymerase with promoters. *Trend. Biochem.* Sci. 5:274.

Burgess, R.R. 1971. RNA polymerase. *Ann. Rev. Biochem.* **40**:711-740.

Clark, B 1980. Elongation step of protein synthesis. *Trends Biochem.* 5:205.

Dickson, R.C.; Albelson, J. *et al.* 1975. Genetic regulation- the *lac* control region. *Science* **181**:27.

Driessen, A.J.M. 1992. Bacterial protein translocation : kinetics and thermodynamics role of ATP and the proton motive force. *Trend. Biochem.* Sci. 17:219-223.

Ellis, R.J. and Van der Vries,S.M. 1991. Molecular chaperones. *Ann. Rev. Biochem.* **60**:321-347.

Franklin, T.J. and Snow, G.A. 1991. Biochemistry of Antimicrobial Action. Chapman & Hall, New York.

Gilbert, W. and Muller-Hill,B. 1966. Isolation of *lac* repressor. *Proc. Nat. Acad. Sci.* **58**:2415.

Gilbert, W. and Maxam,A. 1973. The nucleotide sequence of the *lac* operator. *Proc. Natl. Acad. Sci.* 70:3581.

Gilbert, W.; Maziel,N. and Maxam, A. 1974. Sequences of controlling regions of the lactose operon. Cold Spring Harbor Symp. 38:854.

Goodman, H.M. *et al.*, 1968. Translational initiation in prokaryotes. *Ann. Rev. Microbiol.* 35:365.

Goldberg, A.L. and Goff,S.A. 1986. The selective degradation of abnormal proteins in bacteria. In *Maximising Gene Expression*. eds. W. Reznikoff & L. Gold), Butterworth, Stoneham,M.A.

Hanna, M.M. and Meares,C.F. 1983. Topography of transcriptional: path of the leading end of nascent RNA complex *Proc. Natl. Acad. Sci.* 80:4238.

Ishihama, A. 1992. *Mol. Microbiol.* 6:3283.

Jacob, F. and Monod, J. 1961. Genetic regulatory mechanism in the synthesis of protein. *J. Mol. Biol.* 3: 318-356.

Lathe, R. 1976. RNA polymerase of *E.coli. Curr. Topics Microbiol. Immunol.* 83:37.

Maniatis, T. and Ptashne,M. 1976. A DNA operator. *Sci. Am. Jan.*, p-64.

Moat, A.G. and Foster,J.W. 1995. *Microbiol Physiology*. Wiley-Liss, Inc., New York pp-590

Mosses, D.B. and Model,P. 1984. A Rho-dependent transcription termination signal in bacteriophage f1. *J. Mol.* Biol. 172:1

Pribnow, D. 1975. Nucleotide sequence of an RNA polymerase binding site at an early T7 promoter. *Proc. Natl. Acad. Sci.* 72:784.

Pugsley, A.P. 1993. The complete general secretory pathway in Gram-negative bacteria. *Microbiol. Rev.* 57:50-108.

Richardson, J.P. 1983. Involvement of a multistep interaction between rho protein and RNA in transcription termination. In *Microbiology*-1983 (eds. D. Schlessinger), American Society for Microbiology, Washigton, D.C.

Spassky, A; Busby,S. and Buc,H. 1984. On the action of the cyclic-AMP receptor protein complex at the E. coli lactose and galactose promoter regions. *EMBOJ.* 3:43-50.

Temin, H.M. 1972. RNA directed DNA synthesis. *Sci. Amer.* 226,24.

Travers, A.A. and Burgess,R.P. 1969. Cyclic reuse of RNA polymerase sigma factor. *Nature* 222:537.

Gene Cloning in Microorganisms
(Genetic Engineering)

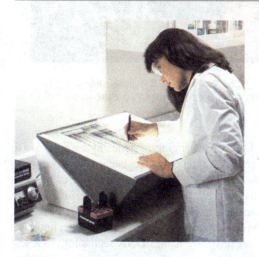

hapter 7 describes the physical and chemical nature of genetic material. In chapter 8 concepts and synthesis of artificial genes have been discussed. However, chapter 8 deals with the natural phenomena of genetic recombination that occur during normal sexual reproduction involving the breakage and rejoining of DNA molecules of the chromosomes. Genetic recombination is of fundamental importance to the living organisms for reassortment of genetic material.

Transgenic E.coli with luciferase genes

1. Concepts of Gene Cloning

Moreover, for centuries human beings have been altering the genetic make up of organisms by selective breeding of plant and animals. The deliberate modification in genetic material of an organism by changing the nucleic acid directly is called *genetic engineering or gene cloning or gene manipulation* and is accomplished by several methods which are collectively known as *recombinant DNA (rDNA) technology*. Recombinant DNA technology begins a new area of research and applied aspects of biology. Therefore, it is a part of *biotechnology* which is gaining momentum and much boost in recent years.

However, in breeding programmes much work has been done on alteration of nucleotides by several parasexual or conjugation methods in different group of organisms. Now-a-day, a large number of mutagenic agents are available that mutate the genes (see chapter 9). It is likely that the changed genes may be beneficial, neutral or lethal. Moreover, the conventional breeding programmes are time taking for making sure that the genes have been altered. In contrast, the rDNA technology has solved several problems which hardly or never are possible through the conventional methods.

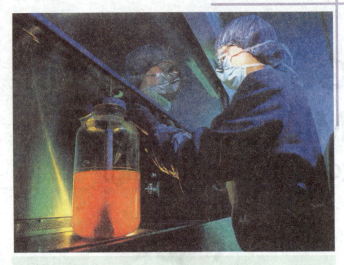

Genetically engineered cells in this flask are used for the production of a useful protein to cure disease.

Gene cloning or genetic engineering can be defined as "changing of genes by using *in vitro* processes" (Dubey, 2005). A unified definition of genetic engineering has been given by Smith (1996) as "*the formation of new combinations of heritable material by the insertion of nucleic acid molecules produced by whatever means outside the cell, into any virus, bacterial plasmid or other vector system so as to allow their incorporation into a host organism in which they do not naturally occur but in which they are capable continued propagation*. In brief, gene technology is the modification of the genetic properties of an organism by using rDNA technology. Genes are like the biological software filled with programmes that govern the growth, development and function of organism. By changing in programme of the software it is possible to bring about alteration in the characters of a given organism (Smith,1996).

A gene of known function can be isolated from its normal location by biochemical methods *in vitro*. Moreover, a gene can be synthesized by using gene machine (see chapter 6 section *artificial synthesis of gene*). The isolated genes can be transferred into the microbial cells (that of course do not contain) via a suitable vector. The transferred gene replicates normally and is handed over to the next progeny over generations. After confirmation for its presence through biochemical procedures clone of the same cell is produced.

2.   History of Recombinant Technology

The first break through of rDNA technology occurred with the discovery of restriction endonucleases (restriction enzyme) during the late 1960s by Werner, Arber and Hamilton Smith. The restriction enzymes were discovered in microorganisms. These enzymes protect the host cell from the bacteriophage DNA. The restriction enzymes are described in the preceding section. In 1969, Herbert Boyer isolated restriction enzyme *Eco*RI from *E. coli* that cleaves the DNA between G and A in the base sequence GAATTC as below :

$$5'\text{-}GAATTC\text{-}3' \xrightarrow{Eco\text{RI}} 5'\text{-}G \quad + \quad AATTC\text{-}3'$$
$$3'\text{-}CTTAAG\text{-}5' \qquad\qquad 3'\text{-}CTTAA\text{-}5' \qquad\qquad G\text{-}5'$$

In 1970 *Howard Temin and Davin Baltimore* independently discovered the enzyme reverse transcriptase from retroviruses. Later on this enzyme was used to construct a DNA called complementary DNA (cDNA) from any RNA (see Fig. 6.11).

In,1972 David Jackson, Robert Symons and Paul Berg successfully generated rDNA molecules. They allowed the stickly ends of complementary DNA by using an enzyme DNA ligase. In 1973 for the first time S.Cohen and H. Boyer developed a recombinant plasmid (p^{SC101}) which after using as vector replicated well within a bacterial host. In, 1975, Edwin M.Southern developed a method for detection of specific DNA fragements for isolation of a gene from complex mixture of DNA. This method is known as the *Southern blotting technique*. Some milestones of recombinant DNA technology have been summarized as below :

1976 - First prenatal diagnosis by using gene specific probe.

1977 - Methods for rapid DNA sequencing, discovery of split genes and somatostanin by rDNA.

1979 - Insulin synthesized by using rDNA; first human viral antigen.

1981 - Foot and mouth disease viral antigen cloned.

1982 - Commercial production of *E. coli* of genetically engineered human insulin. Isolation, cloning and characterization of human cancer gene.

1983 - Engineered Ti-plasmid used to transform plants.

1985 - Insertion of cloned gene from *Salmonella* into tobacco plant to make resistant to herbicide glyphosphate ; Development of PCR technique.

1986 - Development of gene gun.

1989 - First field test of genetically engineered virus (baculovirus) that kills cabbage looper caterpillars.

1990 - Production of first transformed corn.

1991 - Production of first transgenic pigs and goats, manufacture of human haemoglobin, first test of gene therapy on human cancer patients.

1994 - The Flavr Savr tomato introduced; the first genetically engineered whole food approved for sale. Fully human monoclonal antibiodies produced in genetically engineered mice.

1997 - World's first mammalian clone (Dolly) developed from a non-reproductive cell of an adult animal through cloning by nuclear transplantation.

3. Prospects of Genetic Engineering

Genetic engineering holds the potential to extend the range and power of every aspect of biotechnology. The techniques will be used widely to improve the existing microbial processes through improving the existing cultures and discarding the unwanted bye-products. Of course, within this decade the rDNA technology will establish fully the basis of new microorganisms with new metabolic properties. In this way the branches of industrial microbiology, environmental microbiology, agriculture microbiology will certainly gain much from the techniques of genetic engineering. Certainly the genetically engineered microorganisms such as bacteria and fungi have influence the traditional processes of baking, cheese making and texture.

Enzymes produced in genetically engineered mircroorganisms are used in beer making, baking, cheese making and detergent making.

However, some marvellous new discoveries have revolutionized the above area either by improving the efficiency of product or bringing about qualitative changes. Much impact has been seen in production of life saving drugs and removing non-functional genes and delivering the functional genes in animals. Several vaccines have been prepared and tested for its applicability.

In general, the gene cloning has been described in two parts : strategies and applications.

A. Strategies of Genetic Engineering

According to the requirement there are different routes of genetic engineering and a person prefers as per need and choice. Fig. 11.1 gives a general outline of DNA cloning. The overall strategies involve the major four steps: (*i*) formation of DNA fragments, (*ii*) splicing of DNA into vectors, (*iii*) introduction of vectors into host cells, and (*iv*) selection of newly acquired DNA.

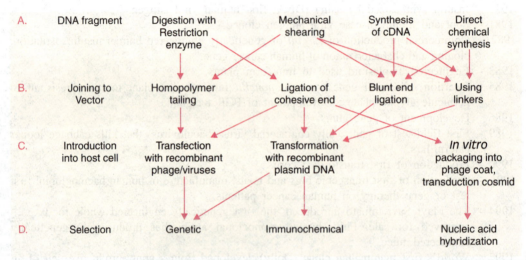

Fig. 11.1 : A generalized outline for DNA cloning. Arrows show some preferred routes (after Old and Primose, 1981)

I. Formation of DNA Fragments

The DNA fragments to be cloned are called foreign DNA or passenger DNA. The DNA fragment of known function is selected and identified. It is isolated from the organism by several *in vitro* biochemical methods. In addition, the DNA fragments can be constructed chemically by using mRNA of gene machine (see chapter 6, Section : *Artificial synthesis of genes*).

By using mRNA, the complementary DNA (cDNA) molecules are produced. All these processes are possible only due to enzymes, the nucleases (Fig. 5.17A-B), DNA polymerases, DNA ligases (see Chapter 5, Fig. 5.19) and restriction endonucleases or restriction enzymes.

1. Restriction Enzymes

Restriction enzymes occur in bacteria as a chemical weapon against the invading viruses and cut both the strands of DNA when certain foreign nucleotides are introduced in a bacterial cell. Now a days many restriction enzymes are known. The first restriction enzyme was isolated in 1970 from *Haemophilus influenzae*. However, different restriction enzymes present in different bacteria recognize different nucleotide sequences.

The restriction enzymes cleave a DNA to generate a nick with a 5' phosphoryl and 3' OH termini. The broken nucleotides form a DNA duplex and exhibit two fold symmetry from a point. In some cases cleavage in two strands are staggered to produce single strand short projections opposite to each other with blunt ends of mutually cohesive stickly ends which are identical and complementary sequences are called palindrome sequences or palindromes. Therefore, when read from 5'→3', both strands have the same sequence (Glover, 1984). Now-a-days a large number of restriction enzymes are commercially available. Some of the commonly used restriction enzymes are given in Table 11.1.

Table 11.1 : Source of restriction enzyme, cleavage sites and products of cleavage (after Dubey, 2005)

Microorganisms	Restriction enzymes	Cleavage sites		Cleavage product
Bacillus amyloliquefaciens H	BamHI	↓ 5'-GGATCC-3' 3'-CCTAGG-5' ↑	5'-G 3'-CCTAG	GATCC-3' G-5'
B. globigii	BglII	↓ 5'-AGATCT-3' 3'-TCTAGA-5' ↑	5'-A 3'-TCTAG	GATCT-3' A-5'
Escherichia coli RY13	EcoRI	↓ 5'-GAATTC-3' 3'-CTTAAG-5' ↑	5'-G 3'-CTTAA	AATTC-3' G-5'
Haemophilus influenzae Rd	HindIII	↓ 5'-AAGCTT-3' 3'-TTCGAA-5' ↑	5'-A 3'-TTCGA	AGCTT-3' A-5'
H. parainfluenzae	HpaI	↓ 5'-GTTAAC-3' 3'-CAATTG-5' ↑	5'-GTT 3'-CAA	AAC-3' TTG-5'
Klebsiella pneumoniae	KpnI	↓ 5'-GGTACC-5' 3'-CCATGG-3' ↑	5'-GGTAC 3'-C	C-3' CATGG-5'
Streptomyces albus G	SalI	↓ 5'-GTCGAC-3' 3'-CAGCTG-5' ↑	5'-G 3'-CAGCT	TCGAC-3' G-5'
Staphylococcus aureus 3AI	Sau3AI	↓ 5'-GATC-3' 3'-CTAG-5' ↑	5'- 3'-CTAG	GATC-3' -5'

Arrows indicate the recognition sites.

(i) Action of Restriction Enzymes

The restriction enzymes recognise the cleavage sites and result in cohesive ends as shown in Table 11.1 and given below.

(a) **Action of EcoRI**

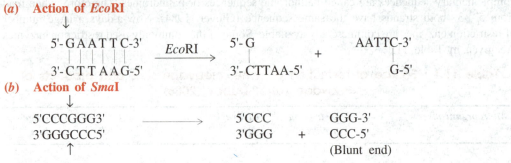

(b) **Action of SmaI**

5'CCCGGG3' ⟶ 5'CCC   GGG-3'
3'GGGCCC5'   3'GGG + CCC-5'
                              (Blunt end)

---

2.   Use of Linkers, Adaptors and Homopolymer Tails

Some times after use of restriction enzymes there is no generation of stickly ends. DNA molecule is fragemented that contains blunt ends as above. In such situation, the recombination frequency lowers due to non-availability of suitable sites on the DNA to be manipulated. Hence, cleavage sites can be added as linkers, adaptor molecules or homopolymer tails.

Linkers are the chemically synthesized dsDNA oligonucleotides containing on it one or more cleavage sites for restriction enzymes. Linkers are ligated to the blunt ends of DNA to be cloned by using T4 DNA ligase. Then these are cut with specific restriction enzyme to generate DNA fragments with sticky ends (Fig. 11.2A).

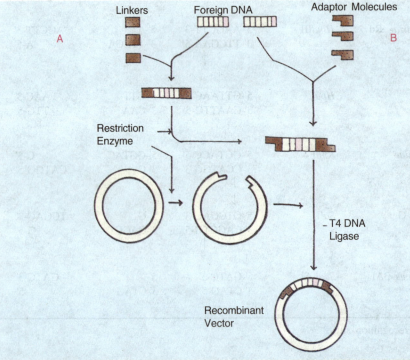

Fig. 11.2 : Use of linker (A) and adaptor (B) molecules in the formation of recombinant plasmid.

The adaptor molecules are the chemically synthesized DNA molecules with preformed cohesive ends. The adaptors are used when the target cleavage site for the restriction enzyme used is located within the DNA sequence (i.e. foreign DNA) to be cloned (Fig.11.2B).

The homopolymers are oligo dA (AAA........A) sequences and oligo dT (TTT....T) sequences. These sequences are added to the foreign DNA fragments to be cloned and vector used by terminal transferase enzyme. The homopolymers are used in such a way that if foreign DNA is to add oligo dA, the vector will be added with oligo dT. Consequently, there develops oligo dA and oligo dT tails on foreign DNA and vector, respectively. When mixed together they anneal to form the covalently closed circular rDNA molecules (Fig. 11.3).

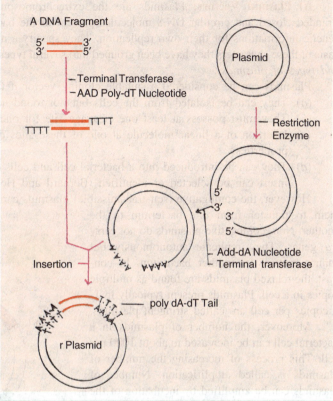

Fig. 11.3 : Homopolymer tails.

II. Splicing of DNA into Vectors

The small sequences of DNA can be spliced or joined into the vector DNA by an enzyme, DNA ligase resulting into the creation of artificial DNA vector i.e. recombinant vectors.

1. DNA Ligases

DNA ligases seal the cut ends of two DNA molecules. It has been described in detail in Chapter 5 : *The Nucleic Acids* (section C : *Replication of DNA*, Fig. 5.19).

2. The Cloning Vectors

Vectors are those DNA molecules that can carry a foreign DNA fragment when inserted into vectors. These are also called as vehicle DNA because they act as carrier of genes to be cloned into a recipient cell. There are many types of vectors used in genetic engineering experiments such as plasmids, bacteriophages, cosmids, phasmids and shuttle vectors.

Moreover, vectors can be introduced into the host cell, thereafter they replicate independently by using the host enzymes. The cloned (inserted) foreign DNA segment can be amplified along the vector inside the host cell.

(*i*) **Plasmid Vectors:** Plasmids are the extrachromosomal, self replicating and double stranded closed and circular DNA molecules present in the bacterial cell. They contain several genetic informations for their own replication. They specify a number of host properties. On the basis of these characters they have been grouped into several types (see Chapter 2, section- *plasmids and types of plasmids*).

Plasmids can be considered a suitable cloning vector if they bear the following features:

(*a*) They can be isolated from the cells but not found naturally.

(*b*) They must possess at least one cleavage site for one or more restriction enzymes.

(*c*) Insertion of a linear molecule at one of these sites does not alter its replication properties.

(*d*) They can be introduced into a bacterial cell and cells carrying plasmid with or without insert can be selected or identified (Bernard and Helinski,1980).

However, the conjugative (self-transmissible) plasmids carry transfer (*tra*) genes that enable them to transfer from one bacterium to the another. Non- conjugative plasmids do not carry *tra* genes. They replicate autonomously but cannot transfer to another bacterium. In contrast, the relaxed plasmids are found as multiple copies in a cell. Plasmids present typically in 1-2 copies per cell are called stringent plasmids.

Moreover, the number of plasmids in a bacterial cell can be increased to about 1000 per cell. This process of increasing the number of plasmids is called amplification. Number of plasmids can be amplified by incubation of the host cells with the antibiotic chloramphenicol. It inhibits the proteins required for replication of chromosome but does not inhibit plasmid replication.

Some of the important plasmids are given in Table 11.2. A physical map of plasmid pBR322 is shown in Fig. 11.4. Plasmid pBR322 is one of the most widely used multicopy cloning vector of *E. coli* and is a hybrid vector of COI

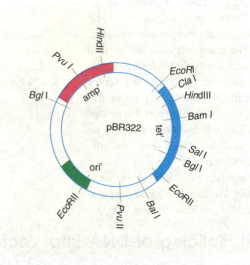

Fig. 11.4 : Plasmid pBR322.

EL and genes coding for resistance against tetracycline and ampicillin. Plasmid pBR322 consists of genes for origin of replication (*ori*) for resistance to tetracycline (*tet^r*) and ampicillin (*amp^r*), unique recognition for 20 restriction enzymes.

Different types of plasmids of *E. coli* and *Bacillus* are listed in Table 11.2A-C. Different species of *Pseudomonas, Streptococcus, Salmonella*, etc. possess a variety of plasmids of different size and markers.

(*ii*) **Bacteriophage Vectors:** The filamentous bacteriophages are M13, Fd and F1 that contain single stranded circular DNA molecule. They are used for cloning of DNA fragments. Examples of phage vectors are given in Table. 11.3. In addition, bacteriophage λ is a phage of *E. coli* and contains double stranded DNA molecule of about 49.5 kilobase-pairs (kb). The ssDNA exists as a linear molecule that consists of 12 nucleotides long single stranded 5' ends. However, there is possibility to introduce the foreign DNA up to 25 kb in length into the lambda genome without making them incapable to infect *E. coli* and replicate properly.

Table 11.2 : Some of the plasmid cloning vectors.

Vectors	Genetic markers	Cloning site(s)
A. General purpose plasmid vectors in *E. coli* (after Thomson,1982)		
p^MB9	Colicin immunity	*Eco*RI, *Sma*I, *Hpa*I, *Hind*III, *Bam*HI,
p^BR322	Resistance for tetracycline (*tet^r*) and ampicillin (*amp^r*)	*Eco*RI, *Bal*I, *Pvu*III, *Bam*HI, *Sal*I, *Sph*I, *Ava*I, *Pvr*I
p^BR325	*tet^r*, *amp^r*, *cm^r*	*Eco*RI, *Pst*I, *Prv*I, *Hin*III, *Bam*HI, *Sal*I, *Ava*I
p^ACYC184	*tet^r*, *cm^r*	*Eco*RI, *Hin*III, *Bam*HI, *Sal*I
p^AC105	colicin immunity	*Eco*RI
B. Special purpose plasmid vectors in *E. coli* (after Thompson, 1982)		
p^MF3	*Amp^r*	*Eco*RI, *Hind*III, *Bam*HI
p^RK2501	*tet^r*, *Kan^r*	*Sal*, *Hind*III, *xho*I, *Eco*RI, *Bgl*I
p^BRH1	*Amp^r*	Promoters having *Eco*RI can express *tet^r* on cloning
p^KY2289	*Amp^r*	DNA insertion at EcoRI or *Xma*I site allows cloning to grow on plates containing mitomycin C
C. Cloning plasmids of *Bacillus* (after Subbaiah *et al.* 1985)		
p^UB110	*Kan^r*	*Eco*RI,
p^HV33	*Amp^r*, *tet^r*, *Cm^r*	*Eco*RI, *Hind*I, *Bam*HI, *Sal*I, *Pvu*I, *Ara*I
p^BD6	*Kan^r*, *Str^r*	*Bam*HI
p^BC16	*Tet^r*	*Eco*RI
p^BC16-1	*Tet^r*	*Eco*RI, *Hind*III
p^GY31	*Tet^r*	*Eco*RI

Bacteriophage λ consists of several non-essential regions. Therefore, the non-essential regions can be replaced with almost equal length of foreign DNA molecule to be cloned. However, there are certain advantages of phage cloning vectors over the plasmids.

(a) DNA can be packed into the phage particles and transduced into *E. coli* with high efficiency.

(b) The upper size limit of insertion of foreign DNA into the phage is about 35 kb in length.

(c) Screening and storage of recombinant DNA is easier in phage λ than plasmids.

Two types of cloning vectors have been constructed : insertion vector and replacement vector. *Insertion vectors* are those in which relatively a fragment of DNA of short length has been inserted without removal of any portion of phage DNA. The *replacement vectors* are those in which case first the non-essential region is cut and removed and almost equal length of foreign DNA is inserted. Thus, the foreign DNA replaces the non-essential region of phage λ. For detail see Dubey, R.C., *A Textbook of Biotechnology*.

Phage λ DNA has been widely used as cloning vector because of ease in handling and screening of a large number of recombinant DNA containing phages. That is why it is used as a biological tool in building the gene bank, for example rat, yeast, mouse and human gene bank (Dahl *et al.,* 1981).

(iii) Cosmid Vectors: Based on the properties of phage DNA and Col El plasmid DNA, a group of Japanese workers (Fukumaki *et al.*, 1976) showed that the presence of a small segment of phage λ containing cohesive end (*cos* site) on the plasmid DNA molecule is a sufficient prerequisite for *in vitro* packaging of this DNA into the infection particles. Therefore, the cosmids are the hybrid vectors derived from plasmids and phage λ (only *cos* site). Cosmid was constructed for the first time by Collins and Hohn (1978). Examples of cosmid vectors are given in Table11.3.

Table 11.3 : Examples of some cosmids and phage vectors

Vectors	Genetic/Phenotypic Marker(s)	Cloning site
A. Cosmid Vectors		
p^{JC74}	Amp^r, El imm.	—
p^{JC720}	El imm. Rif^r	
p^{HC79}	Amp^r, tet^r	
B. Phage Vectors (after Bibb et al., 1980)		
M13^{mp2}	Blue plagues	EcoRI (white plaques)
M13^{mp5}	Blue plague	HindIII (white plaques)
fd^{101}	Amp^r, Kam^r	PstI, HindIII, SmaI
fd^{107}	Amp^r	PstI, SalI, HindIII, EcoRI
fd^{tet}	tet^r	EcoRI, HindIII, Ava

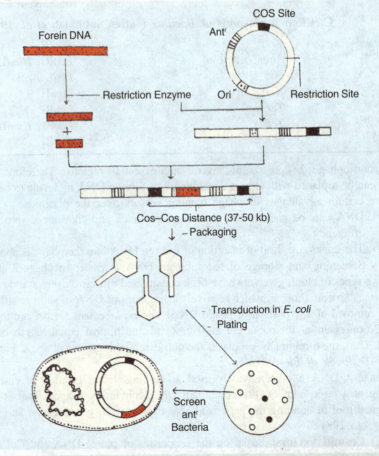

Fig. 11.5 : Cloning of a cosmid vector.

Cosmids lack genes coding for viral proteins, therefore, neither viral particles are formed nor cell lysis occurs. The special features of cosmids similar to plasmids are the presence of origin of replication, markers (conferring resistance against antibiotics), a special cleavage site for the insertion of foreign DNA, and their small size.

Phage λ consists of 49.5 kb long DNA in which *cos* site has 12 base pairs. The *cos* site is ligated to 5 kb pairs long plasmid containing markers. Therfore, it is possible to insert about 45 kb long foreign DNA, which is larger than it would be possible to be done in phage λ or plasmid vector. (Dahl *et al.,* 1981).

Procedure of DNA cloning by using cosmid vector is shown in Fig.11.5. After infection, the DNA of cosmid vector cyclizes at *cos* site and replicates as plasmid and expresses the drug resistance markers. In recent years, a number of cosmid vectors has been constructed from *E. coli,* yeast and mammalian cells and gene bank has been built up (Dahl, *et al.,* 1981).

(iv) Phagemids Cloning Vectors : Like cosmids, phagemids are also the *hybrids of a plasmid and a phage.* The plasmids are restricted to intracellular state while the phage particles can exist extracellularly as the infection particles. Kahn and Helinski (1978) reconstructed the plasmid of Col El artificially and allowed to get packed in *vitro* into bacteriophage particles. The phage particles containing plasmid DNA were allowed to infect bacterium. Thus the hybrid vector was termed as phagemids. This insertion of plasmid into phage genome is reversible and called as 'lifting' the plasmid. It generates a phage genome containing *att* site and one or more plasmid site(s). These new genetic recombinations called as phagemids (Brenner *et al.* 1982). The phagemids contains functional *ori* genes of plagemid and of phage λ. Moreover, they may be allowed to propagate as plasmid of phage in appropriate *E. coli* strains. After reversal of lifting process plasmids are released.

(v) Shuttle Vector: The shuttle vectors are the plasmids that are designated to replicate in different host systems i.e. in prokaryotes and eukaryotes. A shuttle vector is constructed by using bacterial origin of replication in a yeast plasmid. Thus the origins of replication of different host systems such as *E. coli* or yeast are combined in one plasmid. This is why any gene inserted into shuttle vector can be expressed either in bacterium *E. coli* or *Bacillus* or yeast cells. Therefore, shuttle vectors transform *E. coli* cells with greater efficiency than the original organisms. For example, a prokaryotic gene for β-lactamase is expressed in *E. coli.* Genetic map of a shuttle vector constructed by *E. coli* plasmid and yeast is shown in Fig. 11.6. The important gene loci are *ars* (autono-

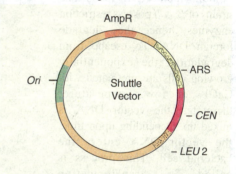

Fig. 11.6 : Genetic map of a shuttle vector constructed by yeast and *E.coli*: *ARS*, autonomously replicating sequence; *CEN*, centromere of Yeast; *LEU2* compliments of defective gene encoding for leucine; *ori*, origin of replication of prokaryotes; *amp* ᵣ, amplicillin resistance gene.

mously replicating sequence), cen (centromere of yeast), *leu-2* (complements of a defective gene encoding for leucin, *ori* (origin for replication in prokaryotes) and *amp*ʳ (ampicillin resistance).

3.   Insertion of the Foreign DNA into a Vector

The foreign DNA of desired function to be cloned is either synthesized artificially following chemical methods or gene machine or synthesized by using mRNA of known function (see Chapter 6, section *Artificial Synthesis of the Gene*). The cDNA can also be procured from cDNA bank also. The vectors are used as desired. Both the vector and foreign DNA are treated with restriction enzyme to generate cohesive (stickly) ends of identical base pairs (Fig. 11.7A-B). Both the DNA molecules are mixed together with ATP and T4 DNA ligase which forms phosphodiester bonds between the stickly ends and thus permanently seal them (C). Thus a hybrid of foreign DNA molecule and vector is obtained which is called recombinant DNA or rDNA. The rDNA is now circularized. In order to get efficient rDNA molecules addition of cohesive ends on both termini of foreign DNA and vector DNA is necessary. There are three ways of generating efficient rDNA molecules as described earlier (see linkers, adaptors and homopolymer tails).

III. Introduction of Recombinant Vector into the Recipient Cells (transformation)

After preparing the rDNA of a vector, it is allowed to enter into a suitable host cell for expression of foreign DNA. Originally, the procedure of *transformation of bacterial cells by rDNA of vector was developed by Mandell and Higa (1979).* The strains of E. coli possess restriction enzymes, hence they degrade foreign DNA. To escape from degradation, the exponentially growing cells are pretreated with $CaCl_2$ at low temperature; thereafter, the vector DNA is mixed up. Depending upon time, the vector DNA enters the bacterial cell. This process is called transformation (Fig.11.7C-D).

For the first time phage λ was used to transfer the foreign DNA into E. coli cells. Thus, the process of transfer of DNA by a phage virus is called transfection similar to *transduction* (see Chapter 8, *Microbial Genetics*, section *Transformation*).

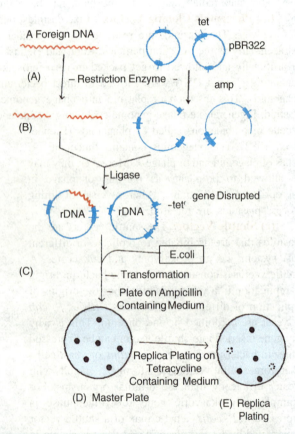

Fig. 11.7 : Steps of gene cloning (diagrammatic).

| 1. | **Expression of Cloned Genes (in _E. coli_)** |

There are many factors that influence the expression of a cloned gene in *E. coli*. Some of them are discussed below :

(i) Commonly vectors are constructed such that the DNA insert is flanked by promoters for two different viral RNA polymerases of phages T3, T7 or SP6. Thus efficiency of promoters influences the expression of cloned gene.

(ii) The foreign gene of interest should be cloned into a high copy number plasmid such as pBR322. Replication of pBR322 is controlled by an RNA molecule called RNA I.

It binds to a second RNA molecule called RNA II and, therefore, inhibits replication of chromosomal DNA.

However, sometimes copy number is controlled by promoters whose activity is regulated by changing the temperature. Such vectors are called runaway plasmid vectors. After increasing the temperature from 30°C to 38°C control of plasmid over replication is lost. This results in increase in copy number of plasmid.

(*iii*) Upstream of inhibition codon (AUG) the sequences are known to influence the translational efficiency. This is called Shine Dalgarno (SD) sequence (5' UAAGGAGGU3') which acts as a ribosome binding site. While constructing a plasmid, presence of a SD sequence should be kept in mind.

IV. Selection of Clones

Not all, but some of the transformed bacterial cells contain foreign gene where it functions properly. Such cells are selected from the mass of cells by using antibiotics. For example the plasmids of rDNA so obtained (Fig. 11.7 C) are introduced in *E. coli* cells. The transformed cells are plated onto nutrient medium containing ampicillin. The antibiotic marker gene (*amp*^{r}) present on pBR322 determines resistance to ampicillin. Therefore, such cells will grow on medium and form colonies. This is called as master plate.

To ascertain the presence or absence of *tet*^{r} gene in the cloned plasmid, a replica plating from the master plate is done as described earlier (see chapter 8, *Microbial Genetics*).

Replica plates are incubated for the growth of bacterial colonies. The appearance of colonies on replica plates is compared with the master plate and those colonies that fail to grow on replica plate are supposed to have plasmids where *tet*^{r} gene is destroyed while those growing on replica plate show that both genes (*amp*^{r} and *tet*^{r}) are present in plasmid.

Moreover, the foreign DNA cloned in plasmid vector is now functioning or not is also to be tested and selected from thousands of cells. There are several approaches, which ascertain about the presence and functioning of foreign DNA into transformed cells. Foreign DNA of known function transferred through vector in *E. coli* cells can be observed by *colony hybridization (nucleic acid hybridization) method, in vitro translation method and immunological tests.*

1. Colony Hybridization (Nucleic acid Hybridization) Technique

The colony hybridization technique has been given by *Grustein and Hogness* (1975). It is based on the *availability of radioactively labelled DNA probe*. A probe is a radioactively labelled (P^{32}) oligonucleotide (20-10 nucleotides long) with a sequence complementary to at least one part of the desired DNA. The probe may be even partially pure mRNA of related gene that defects corresponding rDNA. DNA probes have commercial significance in disease diagnosis, finger printing, microbiological tests and research as well.

This technique follows replica plating of bacterial colonies on nitrocellulose filter disc (Fig. 11.8A) which is then placed on gelled nutrient medium and both master plate and disc are incubated to grow the bacterial colonies. Colonies growing on nitrocellulose filter paper derive nutrients from gelled medium through pores (B). The filter paper is removed and placed on blotting paper soaked with 0.5 N NaOH solution. The alkali lyses bacteria and denatures their DNA. Then the disc is neutralized by tris (hydroxymethyl) amino methane-HCl buffer by maintaining high concentrations of the salt. Thus, the DNA binds to the disc in the same pattern as the bacterial colonies did. The filter disc is baked at 80°C so that DNA may fix properly (C).

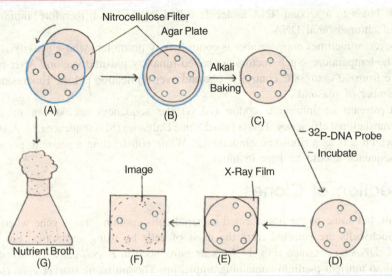

Fig. 11.8 : Colony hybridization technique.

The disc is incubated with a solution containing radioactively labelled probe (P^{32}). The probe hybridizes any bound DNA that contains sequences complementary to the probe (D). Disc is washed thoroughly to remove unhybridized probes and it is passed through X-rays (E-F). Colonies that develop positive X-ray image (G) are compared with master plate and picked up and multiplied in nutrient medium.

2. Plaque Hybridization Technique

Plaque hybridization technique is analogous to the colony hybridization technique. It has been given for phages (Benton and Davis,1977). Similarly this technique follows (*i*) cleavage of cellular DNA and bacteriophage DNA separately, (*ii*) ligation of foreign DNA fragments randomly with phage DNA resulting in recombinant DNA, (*iii*) allowing the rDNA for packaging and producing hybrid phages artificially (*iv*) plating of recombinant phages with host bacteria (phage clones are produced on bacterial lawns), (*v*) isolation of phage clones and establishment of genomic library (each colony of the bacterial lawn is a recombinant clone carrying different DNA fragments), (*vi*) second stage plating of phage clones to be analysed with host bacteria, (*vii*) replica plating on nitrocellulose filter paper and incubation of filter paper on gelled medium, (*viii*) lysis of bacteria with alkali and neutralizing with HCl buffer, (*ix*) adding radiolabelled probe and passing of filter paper through X-ray, (*x*) detection of clone from master plate and selection of desired radiolabelled phage.

3. The Southern Blotting Technique

In 1975, Edwin M. Southern (a molecular biologist) published a procedure *for detecting specific DNA fragments* so that a particular gene could be isolated from a complex DNA mixture. The procedure starts with digestion of DNA population by one or more restriction enzymes (Fig. 11.9). Consequently *DNA fragments* of unequal length are produced. This preparation is passed through agarose gel electrophoresis that results in separation of DNA molecules according to their respective size. The DNA fragments present in gel are denatured by alkali treatment. Then gel is put on top of buffer saturated with filter paper. Upper surface of gel is then covered with nitrocellulose filter and overlaid with dry filter paper. The dry filter paper draws the buffer through gel. Buffer contains single stranded DNA. Nitrocellulose filter paper binds DNA fragments strongly when comes in contact of it. After baking at 80°C, DNA fragments are permanently fixed to the nitrocellulose filter. Then the filter is placed in a solution containing radio-labelled DNA probe of known sequences for about 20 minutes. These are complementary in sequence to the blot transferred DNA.

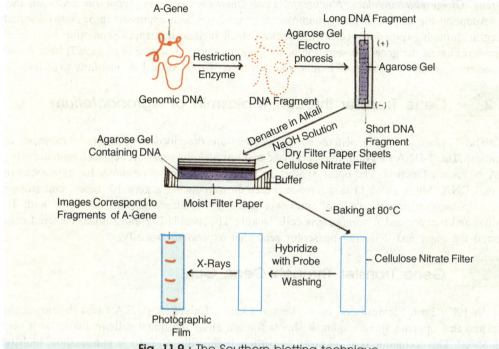

Fig. 11.9 : The Southern blotting technique.

The radiolabelled nucleic acid probe hybridizes the complementary DNA on nitro cellulose in about 20 minutes. The filter is thoroughly washed to remove the probe. The hybridized regions are detected autoradiographically by placing the nitrocellulose filter in contact with a photographic film. The images show the hybridized DNA molecules. Thus, the sequences of DNA are recognised following the sequences of nucleic acid probe.

B. Gene Cloning in Eukaryotic Microorganisms

Most of the studies of gene cloning have been done in prokaryotes (particularly bacteria). In addition, many difficulties are associated with them in gene cloning experiments : (*i*) correction of introns of eukaryotic mRNAs, (*ii*) failure of transfer of equal number of plasmids into daughter cells during the cell division and yield of two types of cells, one with plasmid and the second without plasmid, post translational modifications, (*ii*) threat for hazardous effect, and (*iv*) expression to ensure for the presence of a stable plasmid in the bacterium when applied on large scale for commercial applications. Due to these problems the attraction of the use of eukaryotic cells is obvious. Much work on construction of vectors and expression of insert genes in eukaryotic cells has been done. Some procedures applied for gene cloning in eukaryotes are described in this context.

1. Transformation in Filamentous Fungi

Like bacteria, transformation in filamentous fungi have also been demonstrated. Hinnen *et al.,* (1978) first described transformation of protoplasts of *Saccharomyces cerevisiae* by *E. coli* plasmid. Later on transformation was demonstrated in *Neurospora crassa*. The situation of filamentous fungi is different from the others as there is no convincing evidence for autonomously replicating plasmids, but sequences having functions similar to ARS of shuttle vector have been isolated. There are several examples where transformation of fungal protoplasts has been carried out such as *Aspergillus nidulans, A. niger, A. oryzae, Cephalosporium acremonium, Coprinus cinereus, Glomerella cingulata, Mucor sp., Penicillium chrysogenum, Sepotoria nodorum*, etc.

Adopting this approach many mammalian genes have been expressed in *S. cerevisiae* and aspergilli through expression vector (i.e. vector which include sequences promoting high level of transcription of the gene and secretion of the gene products). Gwynne *et al.* (1987) have found the secretion of about 1g/litre of human interferon from transformed *A. nidulans* mycelia.

2. Gene Transfer through Ti-plasmid of *Agrobacterium*

The Ti-plasmids of *Agrobacterium tumifaciens* are described in Chapter 4 (see *types of plasmids*). The T-DNA of Ti-plasmid is excised and *nos* and *ops* genes are replaced with a foreign DNA of known function. The opine synthetase is retained to act as promoter for expression of inserted DNA. Murai *et al.* (1983) have succeded in introducing a gene for bean plant storage protein (phaseolin) into the cells of sunflower plant. They integrated *phaseolin* gene with Ti-plasmid and transformed *A. tumifaciens* cells lacking Ti-plasmid. The transformed bacterial cells infected the plant and delivered phaseolin gene that expressed normally.

3. Gene Transfer Through Gene Gun

In 1987, Prof. Stanford and co-workers at Cornell University (U.S.A.) first developed the gene gun that operates like shotgun. It shoots foreign DNA into plant cells or tissues at a very high speed. This technique is also known as particle bombardment, particle gun method, biolistic process, microprojectile bombardment or particle acceleration. This technique is applicable for those plants which hardly regenerate and do not show sufficient response to gene transfer through *Agrobacterium* for example, rice, wheat, corn, sorghum, chickpea and pigeon-pea. Successs has been acheived in shooting missing DNA into the chloroplast of *Chlamydomonas* and mitochondria of yeast (Bonynton *et al.* 1988; Johnson *et al.* 1988).

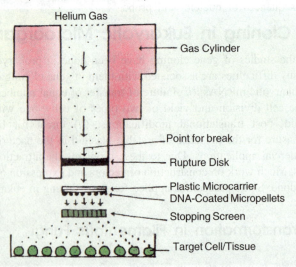

Fig. 11.10 : Working system of particle bombardment gun.

This apparatus consists of chamber connected to an outlet to create vacuum (Fig. 11.10). A 0.22 caliber blank shell shoots a spray of DNA coated metallic microprojectiles (micropellets) into the cells. This device uses high pressure gas to propel the DNA coated projectiles. When pressure of cylinder exceeds the bursting point of plastic disc, it gets ruptured. Helium shock waves propel the plastic microcarrier containing DNA coated micropellets. The transformed cells are regenerated onto nutrient medium and thereafter analysed for expression of foreign DNA.

4. Gene Bank or Genomic Library

Gene bank or genomic library is a complete collection of cloned DNA fragments which comprises of the entire genome of an organism (Dahl *et al.*, 1981). Gene bank is constructed by a shotgun experiment method where whole genome of a cell is cloned in the form of random and unidentified clones. The steps of constructing gene bank of an organism have been previously described in this chapter, section-*plaque hybridization technique* (steps i-iv).

Table 11.4 : Some human peptides and proteins synthesized by gene technology (based on Prescott *et al.*, 1996)

Peptides/ Proteins	*Potential Use*
α_1-antitrypsin	Treatment of emphysema
α, β, γ- interferons	Antiviral, antitumour agent
Blood clotting factor VIII	Treatment of haemophilia
Erythropoetin	Treatment of anaemia
Growth hormone	Growth promotion
Insulin	Treatment of diabetes
Interleukins-1,2,3	Treatment of immune disorder and tumours
Relaxin	Aid to childbirth
Serum albumin	Plasma supplement
Somatostatin	Treatment of acromegaly
Streptokinase	Anticoagulant
Tissue plasminogen activator	Anticoagulant
Tumour necrosis factor	Cancer treatment

C. Application of Gene Technology (Genetic Engineering for Human Welfare)

The cloned genes are utilized commercially in various fields such as pharmaceuticals, industry, agriculture, pollution control, medical science, etc. Some of the examples have been discussed here.

I. Production of pharmaceuticals

The production of medically useful human peptides and proteins (e.g. human growth hormones, insulin, somatostatin and interferon) are of much importance (Table 11.4).

Genetic engineering has made it possible to extract and store the DNA of whole organisms indefinitely.

1. Recombinant Human Growth Hormone (hGH)

The pituitary gland of humans produces growth hormones that regulate the growth and development. However, in children stunted growth occurs due to deficiency of the hormone which is called pituitary dwarfism. Such childern are regularly treated with growth hormone which is procured from the pituitary glands of deceased persons. The injections of hGH has been found effective in children.

Now, the hGH is available as recombinant protein. The hGH-coding DNA sequence is linked with the bacterial signal sequence of *E. coli*. The hGH is secreted into the periplasmic space of bacterial cell by the signal peptides where from the protein is purified. The hGH lacks terminal methionine, hence it is called *met-less hGH*.

In the USA, the recombinant growth hormone is extensively used for farm animals for increased milk production and leaner meat. But for the safety of food produced such methods may lead ethical problems.

2. Recombinant Insulin

Insulin is a peptide hormone secreted by the sets of Langerhans of pancreas. It catabolizes glucose in blood. Insulin is a boon for the diabetics whose normal function for sugar metabolism generally fails. However, diabetes affect a significant percentage of world population. The diabetics take daily injection of insulin for its control. Previously insulin for injection had been isolated from the pancreas of cows, pigs, etc. It was quite effective for diabetics but some patients developed antibodies against insulin as it was antigen because insulin of human and animals has antigenic differences.

Insulin consists of two short polypeptide chains : A (21 amino acid long) and B (30 amino acid long). These two chains are linked by two sulfide bridges. These two peptides are connected by a third peptide chain-C (35 amino acid long). The precursor of insulin is pre-

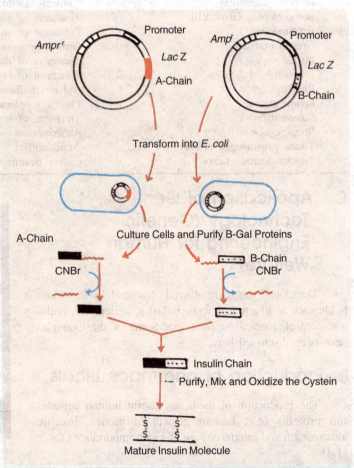

Fig. 11.11 : Production of recombinant insulin in *E.coli.*

proinsulin which is about 109 amino acids long. The structure of pre-proinsulin is as below :

$$NH_2 - (peptide)\ \beta\text{-chain-}(peptide\ C)\text{-A chain-COOH}$$

For the production of mature insulin molecule, post translational proteolysis of pre hormone is essential. Itakura *et al.* (1977) chemically synthesized DNA sequences for two chains (A and B) of insulin and separately inserted into the plasmid pBR322 by the side of β-galactosidase gene of *E. coli*. The recombinant plasmids were separately transferred into *E. coli* cells which secreted fused β-galactosidase-A chain, and β-galactosidase-B-chain separately. These two chains were isolated by detaching from β-galactosidase through CNBr (cyanogen bromide) . It was obtained in pure form to about 10 mg/24 g of healthy and transformed cells (Sasson, 1984). Production of recombinant insulin by *E. coli* is given in Fig. 11.11. However, after addition of extra methionine codon to the N-terminus of each gene A and B, detachment of proinsulin could be possible. The chains A and B are joined together *in vitro* to constitute a native insulin by sulphonating the two peptides with sodium dissulphonate and sodium sulphite.

The human insulin (humulin) is the first therapeutic product by means of recombinant DNA technology by Eli Lilly & Co. (USA).

2.   Recombinant Vaccines

For the production of recombinant vaccines, genes for desired antigens are identified and cloned into suitable vectors. The vectors are introduced into suitable hosts for expression. Production of recombinant vaccine through this method has several advantages. However, the major problem associated with them is the low level of immunogenicity (of recombinant proteins). Some of the recombinant vaccines are described in this section.

(*i*) Vaccine for Hepatitis B virus : The characteristics of Hepatitis B virus (HBV) are described in Chapter 15 (*Virus-I*), and the disease in Chapter 22 (*Medical Microbiology-II*). After infection, HBV fails to grow and even in cultured cells it does not grow. This property has been explained to be due to inhibition of its molecular expression and development of vaccines. Plasma of human contained varying amount of antigens. Three types of viral proteins are recognised to be antigenic : (*i*) viral surface antigen (HBsAg), (*ii*) viral core antigen (HBcAg), and (*iii*) the e-antigen (HBeAg).

Recombinant vaccine for HBV was produced by cloning HBsAg gene of the virus in yeast cells. The yeast system has its complex membrane and ability of secreting glycosylate protein. This has made it possible to build an autonomously replicating plasmid containing HBsAg gene near the yeast alcohol dehydrogenase (ADH) I promoter (Fig. 11.12). The HbsAg gene contains 6 bp long sequence

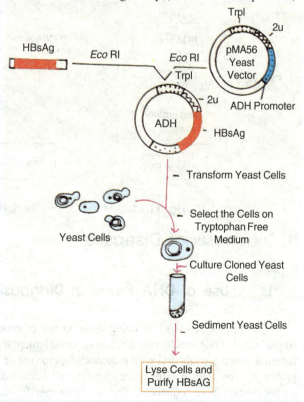

Fig. 11.12 : Expression of HBsAg gene in yeast.

preceding the AUG that synthesises N-terminal methionine. This is joined to ADH promoter cloned in the yeast vector PMA-56. The recombinant plasmid is inserted into yeast cells. The transformed yeast cells are multiplied in tryptophan-free medium. The transformed cells are selected. The cloned yeast cells are cultured for expression of HBsAg gene. This inserted gene sequence expresses and produces particles similar to the 22 µm particle of HBV as these particles are produced in serum of HBV patients. The expressed HBsAg particles have similarity in structure and immunogenicity with those isolated from HBV-infected cells of patients. Its high immunogenicity has made it possible to market the recombinant product as vaccine against HBV infection.

Indigenous Hepatitis-B Vaccine : India's first genetically engineered vaccine (Guni) against HBV developed by a Hyderabad based laboratory (Shantha Biotechnics Pvt. Ltd.) was launched on August 18,1997. India is the fourth country (after the U.S.A., France and Belgium) to develop this highly advanced vaccine. The indigenous yeast-desired HBV-vaccine is one third the cost of the imported vaccine. This new vaccine had undergone human clinical trials at Nizam's Institute of Medical Sciences, Hyderabad and K.E.M. Hospital, Mumbai. The clinical trials clearly proved that the seroprotection is about 98%. It was found more effective than the imported vaccine. The Drug Controller General of India has permitted it for commercial manufacture.

(ii) **Vaccine for Foot and Mouth Disease (FMD) Virus:** FMD is a very serious disease of animals caused by an RNA virus belonging to the picorna virus group. It consists of ssRNA molecule of 8,000 nucleotides surrounded by a capsid. The capsid is made up of 60 copies of four proteins: VP1, VP2, VP3 and VP4. Only VP1 has a little immunogenic activity. The gene coding for VP1 has been identified and cloned on pBR322. The recombinant plasmid was introduced in *E. coli*. About 1,000 molecules of VP1 per bacterial cell were synthesized (Kupper *et al.,* 1981). An outline of making vaccine for FMD virus is as below :

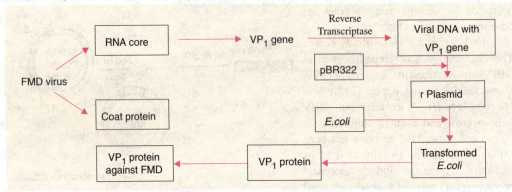

Fig. 11.13 : Steps for production of FMD vaccine.

II. Diagnosis of Diseases

1. Use of DNA Probe in Diagnosis

Recently, much work is being done on use of probe in disease diagnosis. DNA probes are single stranded oligonucleotide sequences, complementary to a particular DNA sequence of desired function which are labelled with radionuclide, enzyme or fluorescent molecule. Very specific DNA probes are constructed. The specificity lies in such a way that the other related species or strains do not contain those sequences. The unrelated specific sequences of known parasite are recognised by using DNA hybridization technique. Then a DNA sequence, not found in any species is identified, cleaved by restriction enzyme and inserted into a cloning vector (plasmid). The bacterial cells are transformed by the recombinant vector. The transformants are multipled. Finally, the foreign DNA fragment is retrieved from the host cells. The DNA sequences of the parasite, thus obtained are labelled with radioisotope and used as a probe. The probe can also be chemically synthesized. Following are the steps for diagnosis of a particular disease :

(a) Isolate the parasite from the infected tissue of the patient. Extract the DNA from the parasite and purify it.

(b) Break the DNA by using restriction enzyme.

(c) Electrophorise the DNA solution containing DNA fragment of different lengths by using agarose gel to get a smear of DNA.

(d) Attach DNA to more firmer support by Southern blotting technique. Thus, the filter paper carries the exact replica of the DNA adhered to it.

(e) Hybridize the immobilized DNA on filter paper by incubating it with radiolabelled probe. Probe DNA complementary to certain DNA sequence of parasite DNA sticks to it and forms the hybrid.

(f) Wash filter paper to remove unbound probe; pass filter paper through X-ray film. The hybridized specific sequences appear as dark bands. Thus a parasitic diease is diagonised positive. If dark bands do not appear the parasite is reported to be absent.

This diagnosis system is very effective for viruses, bacteria and protozoa. Tuberculosis caused by *Mycobacterium tuberculosis* is diagnosed by this method. Genprobe Inc, California has marketed a complete testing system of tuberculosis. Similar effort has also been made for diagnosis of leprosy, *Kala Azar, malaria,* etc.

2. Use of PCR in Disease Diagnosis

The functioning of the PCR is described in Chapter 6 : (section *artificial synthesis of the gene*). The PCR can detect even a single organism that has infected the humans which is present even in low number. From the suspected patients, sample of sera is taken and the region from DNA samples is amplified as described in Chapter 6. The PCR as diagnostic tool may be used in some diseases as given in Table 11.5.

Table 11.5 : Use of the PCR as diagnostic tool (based on Balasubramanian *et al.,* 1996).

Types of Disease	Name of Disease	Causal agents
Bacterial infections	(i) Tuberculosis	Mycobacterium tuberculosis
	(ii) Lyme disease	Borrelia burgdoferi
	(iii) Listeriosis	Listeria monocytogenes
Viral infections	(i) Hepatitis	Hepatitis B virus
	(ii) AIDS	Human Immunodeficiency virus (HIV)
Animal parasites	(i) Malaria	Plasmodium vivax
	(ii) Filariasis	Wuchereria bancrofti

III. Insect Pest Control

Mosquitoes are a menace and vector for transmission of several human diseases such as malaria, filaria, encephalitis, dengu, plague, etc. The increase in their population is augmented by environmental pollution particularly polluted water outlets from houses, huts, water in coolers of any site where stagnant water is found. In addition, they cause serious diseases to crop plants with great loss in yield.

To kill these insect pests, insecticides are being produced worldwide that in turn pose environmental pollution leading to health hazards in animals and humans. Who can forget *en masse* killing of more than 4,000 people, many animals, birds and plants, when methylisocyanate (MIC) gas leaked out in night of 2/3 December, 1984 from the underground reservoir of Union Carbide Factory of Bhopal (M.P.). Many gas affected people are suffering even today. Small incidences of gas leakage and human death are many and occurring day by day. In this grim scenario, total ban on insecticide chemicals, and formulation and production of microbial bio pesticides are urgently required.

In recent years, attempts have been made to produce microbial insecticides. i.e. biopesticides Biopesticides are the preparations of chemicals/microbial cells basically from bacteria, fungi and viruses for killing of insect pests. The examples are baculoviruses, iridovirus, entomopox virus, *Bacillus thuringiensis, B.popilliae, B. sphaericus, B. moritai, and species of Aspergillus, Coelomomycos, Entomophthora, Fusarium, Paecilomyces* (Aizawa,1982).

1.   Bacterial Biopesticides (Bioinsecticides)

Bacillus thuringiensis is a wide spread, spore forming bacterium which is found in soil, litter and dead insects. It produces toxins viz., α-, β- and δ-exotoxins, and δ-endotoxin which can be obtained in crystalline form. β-endotoxin is composed of a glycoprotein subunit (Bulla *et al.* 1977,1978). These toxins have insecticidal properties. *B. thuringiensis* has been found as a strong antagonist against larvae of lapidoptera. After ingestion of spores larvae are damaged as the rod shaped bacterial cell secretes at the opposite end a single large crystal in the cell. This toxic crystal is proteinaceous in nature. It gets dissolved in alkaline juice of caterpillar's digestive cavity (Riviere,1977).

This toxin crystal is secreted by a plasmid present in bacterial cell. This plasmid has been transferred in cells of *B.subtilis* and *E. coli* where it successfully expressed the toxin (Schneff and Whiteley,1981).

Microbial biopesticides have been produced by many companies by using genetically engineered microbial cells of preparation of *B. thuringiensis*. In the U.S.A. commercial formulation of β-endotoxin from *B.thuringiensis* (Bt) has been banned. In other countries like France, formulations of Bt in the form of wettable powder and water suspension have been recommended for use.

2.   Production of Transgenic Plants

Now, scientists have started producing transgenic plants instead of producing Bt prepara-tions. Transgenic plants are those plants in which a gene of foreign origin i.e. other organisms has been introduced. Recently, the scientists at the US Multinational Monsanto Co., Australia, Cotton Seed distributors and the Council of Scientific and Industrial Research Organisation (CSIRO) of Australia, have produced genetically engineered species of cotton known as *Killer cotton*. It kills the predators especially the bullworm (*Heliothis sp.*). The leaves of transgenic plants, produced through cell culture, secrete lethal toxins. When the bullworm eats upon cotton leaves, toxins is taken up by them. Toxin activates in their guts and results in death of bullworm. Toxin does not harm the spiders, humans and other mammals.

Bt Cotton, a transgenic plant.

B. thuringiensis (Bt) β-endotoxin have been cloned in *E. coli*. Such transformed *E. coli* produced larvicidal proteins in large amount which accumulate in cells to form large crystals (Kumar and Kumar,1992). Similarly, transgenic tomato plants have been produced through tissue culture/cell culture. Introduction of Bt gene in a plant is shown in Fig. 11.14.

Today biopesticide sales have attained 3% of the global chemical pesticide market and expected to reach 10% in ten years time. The bacteria based biopesticides have extremely popular in view of their efficiency and from the knowledge of genetic basis of their inseticidal activity (Balasubramanian *et al.*, 1996). *B. thuringiensis* var. *israelensis* and *B. sphaericus* emerged as the two major microbes for mosquitoes and black flies. When these ingest the bacilli which dissolve in alkaline pH of the gut, release the endotoxin contained in the inclusion bodies. The multiple toxic peptides bind specifically to the larval gut cells and dissolve the membrane by causing leakage of Ca⁺⁺ resulting in paralysis of gut and death of larvae.

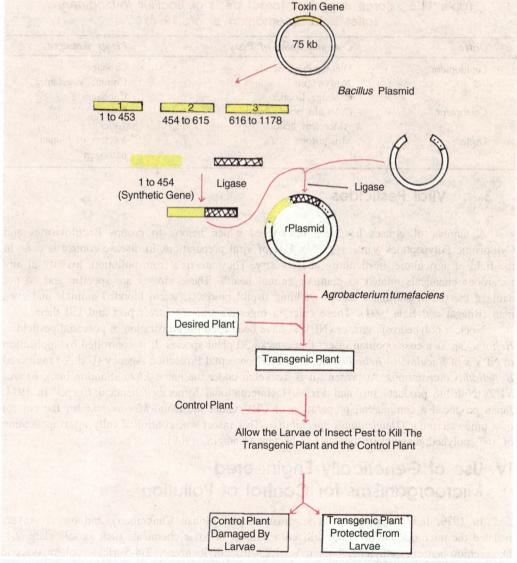

Fig. 11.14 : Construction of a transgenic plant and expression of Bt gene against insect larvae.

The biopesticides are preferred as biological alternative to environmentally toxic chemicals. The biopesticides offer many advantages such as (*i*) their nontoxic effects on non-target pathogen/ microorganisms/pests (unlike chemicals), (*ii*) they are biodegradable, and (*iii*) development of resistance by the insects to a low extent against the bipyramidal crystal shaped proteins (Balasubramanian *et al.*, 1996). The target pests of *B. thuringiensis* are given in Table 11.6.

Additional types of transgenic plants have been produced in the last few years by using modern molecular biology. A synthetic toxin gene was created in which natural A-T-rich gene regions were replaced with G-C content sequences. This was done to minimize degradation of mRNA in plant. After successful insertion of modified artificial gene into plant, over 0.2% of transgenic plants have been produced. Attempts are being made to commercialize the crops. The research has been carried out with soyabean, oilseed rape, sugar beet and sunflower, and also in monocots such as maize, rice and wheat (Gasser and Fraley,1992).

Table 11.6 : Some important target pests of *Bacillus thuringinensis* (after Balasubramanian *et al.*, 1996)

Order	Common name of Pests	Plants damaged
Lepidoptera	Diamondback moth	Cabbage
	Armyworms	Cotton, Vegetables
	Cabbage Looper	Cabbage
Coleoptera	Colorado potato beetle	Potato
	Alder leaf beetle	Alder
Diptera	Mosquitoes	Vectors of human pathogen

3. Viral Pesticides

A number of viruses has been discovered which belong to groups Baculoviruses and Cytoplsmic Polypeptides Viruses (CPV). Use of viral preparations in disease control is done in the field of agriculture, horticulture and forestry. They are free from pollution, toxicity of any hazardous chemicals related to plant or animal health. These viruses are specific and do not damage the useful pollinator insects yielding useful products, warm blooded animals and even man (Bhagat and Bala,1991). These enter in digestive tract of insect pest and kill them.

Nuclear polyhedrosis viruses (NPV's) have been used for preparation of potential pesticides. *Heliothis* sp. is a cosmopolitan insect that attacks 30 plant species. It is controlled by application of NPV's of *Baculovirus heliothis*. In 1975, Environmental Protection Agency (U.S.A.) registered *B. heliothis* preparations. At present, it is marketed under the name *Elcar* (Sandoz Inc.), *Biotrol* VHZ (Nutrilite products Int) and *Virom*/H (International Minerals Chemical Corp.). In 1974, Japan produced a commercial preparation of CPV under the name *Matsukemin* for the control of a pine caterpillar (*Dendrolimus spectabilis*). This insect was controlled fully after application of 10^{11} polyhedral inclusion bodies per hectare (Katagiri, 1969).

IV. Use of Genetically Engineered Microorganisms for Control of Pollution

In 1979, India-born American scientist, Anand Mohan Chakrabarty and his coworkers isolated the microbial cultures that utilized a number of toxic chemicals such as salicylate, 2.4-D, 3 chlorobenzene, ethylene, biphenyls, 1,2,4-trimethylbenzene, 2-4-5-trichlorophenoxyacetic acid, etc (Chatterjee *et al.*, 1981, Kellog *et al.*, 1981).

Genes responsible for degradation of environmental pollutants, for example toluene, chlorobenzene, acids and other pesticides and toxic wastes have been isolated. For the degradation of every toxic compound one gene is required. However, it is not like that one plasmid can degrade all the toxic compounds of different groups. A.M.Chakrabarty categorised the plasmids into four groups as below :

   (*i*) OCT plasmid that degrades octane, hexane and decane.
  (*ii*) XYL plasmid that degrades xylene and toluene.
 (*iii*) CAM plasmid that decomposes camphor and,
 (*iv*) NAH plasmid which degrades naphthalene.

Chakrabarty produced a new genetically engineered microbial strain called superbug (oil eating bug) after introducing all the four plasmids from different strains into one single cell of *Pseudomonas putida*. This superbug is such that which degrades all the four types of substrates for which four separate plasmids were required (Fig. 11.15).

In addition, attempts have been made to render plants resistant to environmental stresses, and have shown much success. The genes for detoxification of glyophosphate herbicides were isolated from *Salmonella*. These genes were successfully cloned and intered into tobacco cells through Ti-plasmid of *A. tumifaciens*. The plantlets regenerated from transformed cells showed resistance against the herbicide. Thus, herbicide-resistant tobacco and corn plants have been developed. It is now known that many plants suffer from stress when treated with herbicides. Therefore, such type of plants seems much importance. The herbicide resistant plants will not suffer from stress when a herbicide is used to control weeds in a crop field (Prescott *et al.*, 1996).

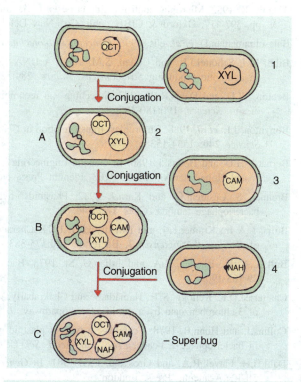

Fig. 11.15 : Construction of superbug. 1, 2 and 3 are the bacterial cells containing different plasmids; A, B and C are the conjugants.

QUESTIONS

1. Describe in brief gene cloning. What are the various biological tools required for it ?
2. Discuss in detail the restriction enzymes and their mode of action ?
3. What is a cloning vector? Describe in detail a cloning vector studied by you ?
4. What are the restriction enzyme linkers, adaptors and homopolymer tails ? Discuss in detail their significance in gene cloning.
5. Write an essay on colony hybridization technique.
6. Write an essay on genetic engineering in medical science.
7. Write an essay on gene cloning for human welfare.
8. Write short notes on the following :
      (*i*) Restriction enzyme,   (*ii*) Plasmids,   (*iii*) Cosmids,

(*iv*) Phagemid,   (*v*) Bacteriophage lambda,   (*vi*) pBR322,
(*vii*) cDNA,   (*viii*) Shuttle vector,   (*ix*) Particle bombardment gun,
(*x*) Transgenic plants,   (*xi*) Microbial bioinsecticides, (*xi*) Viral pesticides,
(*xii*) Recombinant insulin, (*xiii*) Recombinant vaccines,   (*xiv*) DNA probe,
(*xv*) Superbug,   (*xvi*) *Agrobacterium tumifaciens*, (*xvii*) *Bacillus thuringiensis*
(*xviii*) *Escherichia coli.*

REFERENCES

Aizawa, K. 1982. Microbial control of insect pests. In *Advances in Agric. Microbiol.* (ed. N.S. Subba Rao). pp. 397-417, Oxford & CIBH publi. Co., New Delhi

Balasubramanian,D. *et al.* eds. 1996. *Concepts in biotechnology.* Universities press (India) Ltd., Hyderabad.

Bibb, M.J. Schottel,J.L. and Cohen, S.M. 1980. A DNA cloning system for interspecies gene transfer in an antibiotic producing *Streptomyces. Nature.* 284:526-531.

Benton, W.D. and Davis,R.W 1977. Screening λgt recombinant clones by hybridization to single plaques in situ. *Science.* 196:180-182.

Bonynton, J.E. *et al.* 1988. Chloroplast transformation in *Chlamydomonas* with high velocity microprojectile. *Science* **240**: 1534-1538.

Bernard, H.U. and Helinski 1980. In Genetic Engineering : Principles and Methods. Vol. II (eds. Setlow), J. and Hollaender,A.), pp 133-168, Plenum Press, New York.

Brenner, S. Cesareni,G. and Karn,J. 1982. Phasmids: hybrids between Col EL plasmids and *E. coli* bacteriophage lambda. *Gene* **17**:27-44.

Bulla, L.A Jr., Kramer,K.J. and Davidson,L.I. 1977. Characterisation of the entomocidal parasporal crystal of *Bacillus thuringiensis.* J. Bacteriol. **130**:375-383.

Bulla,L.A. Jr., Rhodes, R.A. and Julian,G. St. 1978. Bacteria as insect pathogens. *Ann. Rev. Microbiol.* **29**:163-190.

Chatterjee,D.K. Kellogg, S.T., Hamada,S. and Chakrabarty,A.M. 1981,. Plasmid specifying total degradation of 3-chlorobenzoate by a modified orthopathway. *J. Bact* **146** : 639-646.

Collins,J. and Hohn,B. 1978. Cosmids : a type of plasmid gene cloning vector that is packageable *in vitro* in bacteriophage λ heads. *Proc. Nat. Acad. Sci.* USA **75**:4242-4246.

Dahl,H.H. Flavell,R.A. and Grosveld, F.G. 1981. In *Genetic Engineering,* Vol 2 (eds.Williamson,R.) pp-50-127.Academic Press, London.

Dubey,R.C. 2005. *A text Book of Biotechnology.* 4th ed. S. Chand & Co. Ltd., Ram Nagar New Delhi.

Fukumani, Y. Shimada, K. and Takagi, Y. 1976. *Proc. Nat. Acad.Sci.* (USA). **73**:3238-3243.

Gasser, C.S. and Fraley, 1992. Transgenic crops. *Sci. Am.* **271**(6):76-82.

Glover, D.M. 1984. *Gene Cloning- The mechanism of DNA Manipulation.* Chapman & Hall, London p.222.

Grunstein, M. and Hogness, D.S. 1975. Colony hybridization: a method for the isolation of cloned DNAs that contain a specific gene. *Proc. Nat. Acad. Sci. USA.* **72**:3961-3965.

Gwynee, D.M. *et al.* 1987. *Biotech.* **5**:713-719.

Hinnen,A. Hicks, J.B. and Frink,G.R. 1978. Transformation of yeast. *Proc. Nat. Acad. Sci.* USA **75**:1929-1933.

Itakura,K. *et al.* 1976.Expression in *E.coli* of a chemically synthesized gene for the hormone somatostatin. *Science.* **198**:1056-1063.

Johanston,S.A. *et al.* 1988. Mitochondrial transformation in Yeast by bombardment with microprojectiles. Science **210**:1538-1541.

Kahn, M. and Helinski, D.R. 1978. Construction of a novel plasmid-phage hybrid : use of the hybrid to demonstrate ColEl DNA replication *in vitro* in the absence of a ColEl- specified protein. *Proc. Nat.Acad. Sci.* (USA) **75**. 2200-2204.

Katagiri, K. 1969. A review on microbial control of insect pests in forest in Japan. *Entomophaga.* **14**:203-214.

Kellogg,S.T., Chatterjee,D.K., and Chakrabarty,A.M. 1981. Plasmids assisted molecular breeding new technique for enhanced biodegradation of persistant toxic chemicals. *Science.* **214**:1135-1135.

Kumar,V.A. and Kumar, A. *Everyman's Science.* **27**:101-107.

Mandell,M.and Higa,A. 1979. Calcium dependent bacteriophage DNA infection. *J. Mol. Biol.* **53**:159-162.

Old,R.W. and Primrose,S.B. 1981. *Principles of Gene manipulation : An introduction to Genetic Engineering.* 2nd ed. Blackwell Oxford.

Riviere, J. 1977. *Industrial Application of Microbiology* (eds. M.O. Moss and J.E. Smith), Surrey University Press, London.

Sasson, A. 1984. *Biotechnology: Challenges and Promises* UNESCO, Paris.

Schnepf, H.E. and Whiteley,H.R. 1981. Cloning and expression of the *Bacillus thuringiensis* crystal protein gene in *E. coli Proc. Nat.Acad. Sci.* (USA) **78**:2893-2897.

Smith, J.E. 1996. *Biotechnology,.* 3rd ed. Cambridge Univ. Press, Cambridge, p.236

Southern,E.M. 1975. Detection of specific sequences among DNA fragments separated by gel electrophoresis. *J. Mol. Biol.* **98**:503-517.

Subbaiah, T.V., Shailubhai,K. and Pandey,N.K. 1985. Vectors for cloning in various hosts. In *Review in Biotechnology*, Series I pp.95-116. Biotech Board, DST., New Delhi.

Thompson,R. 1982. Plasmid and phage M13 cloning vectors. In *Genetic Engineering,* Vol III (ed. Williamson), pp.95-116, Academic Press, Orlando.

Microbial Metabolism

12

A. Metabolism

B. Metabolic pathways

C. Enzymes

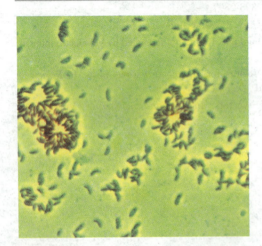

For every biological process, there are two basic requirements: (*i*) energy and, (*ii*) nutrients. The basic cell building require ments are supplied by the nutrition. Nutrition not only provides energy but also acts as precursor for growth of microorganism. The nutritional requirement of an organism depends on the biochemical composition of the cell.

For growth apart from energy the organisms also require a source of carbon, nitrogen, sulfur, phosphorus, oxygen and other metal ions. Microorganisms get these from the environment. On the basis of source of energy, there are two types of organisms: phototrophs and chemotrophs. Phototrophs utilize the energy trapped in the solar radiations (solar energy) for their general metabolic requirement and convert it into chemical energy, while chemotrophs use the energy trapped in the chemical bonds (chemical energy) for their metabolism.

Carbon is the chief requirement of any organism. According to the source of carbon, organisms can be either autotrophs or hetrotrophs. Autotrophs or self-feeders use carbon dioxide of the atmosphere as their source of carbon. Thus, these organisms fix the atmospheric carbon (present in the form of carbon dioxide) into carbohydrates required for growth and metabolism. These organisms are independent to other organisms as far as carbon source is

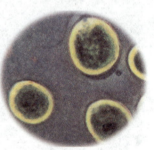

Phototrophs such as the green algae utilize the solar energy for their metabolic requirement.

concerned. Heterotrophs (feeders on others) depend upon other organisms for carbon i.e. they depend upon ready made source of carbon already fixed by autotrophs or present in other heterotrophs.

Lithotrophs and organotrophs are the two types of microorganisms based upon the source of electron. Lithotrophs utilize inorganic compounds as electron donors.

Thus, depending upon nutritional requirements (taking carbon energy source into consideration), the organisms may be: (*i*) photoautotrophs, (*ii*) photoheterotrophs, (*iii*) chemoautotrophs (*iv*) chemohetrotrophs.

1. Phototrophs

Photoautotrophs include the photosynthetic bacteria, algae and green plants. These organisms fix carbon utilizing the energy of solar radiations through a process called photosynthesis. They are also called the primary producers of simple procedures because such organisms are capable of self production of organic material from simple inorganic sources (sunlight for energy and carbon dioxide for carbon). Photoautotrophs form the first trophic (food) level in any food chain. These organisms have a very simple food requirements and can grow on very simple defined media (where chemical composition is known).

Photoautotrophs are the organisms which require solar radiations for their energy but depend upon ready made carbon source in the form of organic compounds. These include green non-sulfur bacteria and purple non-sulfur bacteria (for details see Chapter 13).

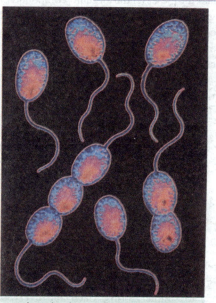

An electron micrograph of individual cells of purple bacteria.

2. Chemotrophs

Chemoautotrophs utilize the energy present in chemical compounds and carbon in the form of carbon dioxide. These include organisms like *Beggiatoa* (utilizing hydrogen sulfide as energy source), *Nitrosomonas* (utilizing ammonia as energy source), *Thiobacillus* (utilizing iron), *Nitrobacter* (utilizing *nitrites*) etc. These organisms also have simpler nutritional requirements and can be grown on defined media. Generally, same organic compound can act both as a source of energy as well as carbon. These organisms are the chief concern to man because all the parasites or pathogens of man, animals and plants are chemoheterotrophs. Apart from pathogens all the saprophytes are also chemoheterotrophic.

A. Metabolism

The metabolism of any organism can be divided into two major categories, (*i*) energy generating or degrading pathways i.e. catabolism, and (*ii*) energy consuming or biosynthetic pathways i.e. anabolism. Two biomolecules, adenosine triphosphate (ATP) and nicotinamide adenine dinucleotide (NAD) or nicotinamide adenine dinucleotide phosphate (NADP) are the main link between the two processes of anabolism and catabolism.

As stated before, metabolism in any organism includes two processes anabolism and catabolism. These two processes include all the biochemical reactions of living organisms.

1. Anabolism

It is a process in which essential biomolecules required for growth are generated by the utilization of energy. The chief biomolecules required are carbon like glucose, ribose, glycerol, pyruvate, etc. Some of the biomolecules act as the central metabolic intermediates for all types of carbon and nitrogen compounds required for growth. Some microorganisms can themselves make all the essential organic compounds required for growth as in the case of autotrophs. Such organisms can be grown on simple and chemically defined media. On the other hand, some of the microorganisms which are unable to make most of the organic compounds from atmosphere, are called fastidious organisms. These can only be grown on complex media with different growth factors.

For the biosynthesis (anabolism) of different essential biomolecules, following anabolic processes take place in organisms:

(a) Synthesis of carbohydrates like glucose, sucrose, cellulose, etc.
(b) Synthesis of lipids, glycolipids, phospholipids, etc.
(c) Synthesis of amino acids and protein.
(d) Synthesis of nucleic acids.
(e) Synthesis of other growth requirements like vitamins, hormones, etc.

2. Catabolism

All the processes in which the nutrients taken in the form of biomolecules as food are broken down (digested) to release energy are called catabolism. Catabolic processes also convert complex organic compounds, stored in the cells of microorganisms (like glycogen granules, polyphosphate, etc.) to simpler forms. Each and every organic compound whether it is carbohydrates, protein or fat, can be catabolized according to the requirement of organism. Similar to anabolism certain biomolecules act as link between catabolic and anabolic processes. Apart from these, ATP and NAD(P)H also act as a link between the two types of pathways. The chief catabolic processes involved in cell metabolism are : (a) glycolysis, (b) pentosephosphate pathway (PPP), (c) Entner doudoroff pathway (EDP), (d) tricarboxylic acid cycle (TCA), (e) fermentation, (f) glyoxylate (g) lipid hydrolysis, (h) protein hydrolysis

3. Adenosine Triphosphate (ATP)

ATP is the chief energy carrier molecule in all type of organisms. ATP is required for the anabolic (generative) process of an organism by which organic macromolecules required for the growth are synthesized. ATP is formed by the catabolic (degradative) process in which macromolecules are broken down and energy is generated.

The structure of ATP was first deduced by Lohman in 1930 and confirmed by Alexander Todd et al., in 1948. ATP consists of an adenosine (adenine plus ribose sugar) and three phosphate groups. These three phosphate groups are bounded by high energy bonds to adenosine unit and whenever one or two of these phosphate groups are removed from the ATP, large amount of energy is released (Fig. 12.1).

ATP $\longrightarrow$ ADP + Pi + energy

Adenosine triphosphate   Adenosine diphosphate   Inorganic phosphate

ADP $\longrightarrow$ AMP + Pi + energy

Adenosine monophosphate

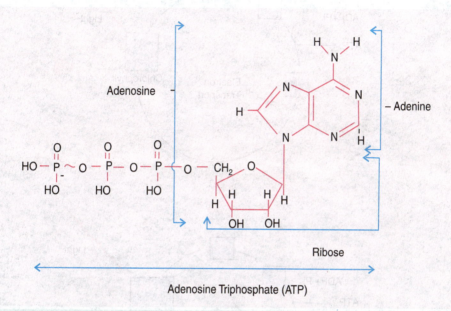

Fig. 12.1 : Chemical structure of an ATP molecule.

This energy is the instant energy utilized in the various anabolic processes of the cell. When the ATP is utilized in anabolic process, it is broken down to ADP or AMP releasing phosphate and energy. The ATP utilized in anabolic process is replenished by catabolic process by reversal reactions. The trapping of chemical energy, released by the oxidative reactions of the cell, in the form of ATP, is called phosphorylation. There are three type of phosphorylation.

(*i*) **Photophosphorylation :** It occurs in the presence of light by photosynthetic organisms mainly in the photoautotrophs and photoheterotrophs.

In the photosynthetic cells, green pigment (chlorophyll) present in the chloroplast or bacteriochlorophylls present in the cell membrane of thylakoids, trap energy from solar radiations which become activated, releasing electrons. These electrons then pass through a series of carriers. The energy thus released is trapped in the form of phosphate bonds in ATPs. Photophosphorylation is of two types. For details see in the Chapter 13.

(*a*) **Cyclic photophosphorylation:** In this type of photophosphorylation electrons released from chlorophyll molecules (due to excitement by light energy) return back to the same chlorophyll molecules by the same route using electron carriers. Such pathway is cyclic in nature as the electrons pass, energy is released which is trapped in the form of ATP.

(*b*) **Non-cyclic photophosphorylation:** In this type of photophosphorylation electrons released from chlorophyll molecules do not return back to the same chlorophyll molecules, but instead are received by nicotinamide adenine dinucleotide phosphate (NADP). From NADP electrons enter the electron transport chain whereby oxidative phosphorylation occurs and ATP is formed. The electrons released from chlorophyll molecules are replenished in the chlorophyll molecules by photolysis of water.

$$2 H_2O \longrightarrow 4 H^+ + 4e^- + O_2$$

Thus, we can see that oxygen is released in this type of photophosphorylation. Oxygen is toxic for anaerobes and, therefore, this type of phosphorylation is not found among anaerobic microorganisms (Fig.12.2).

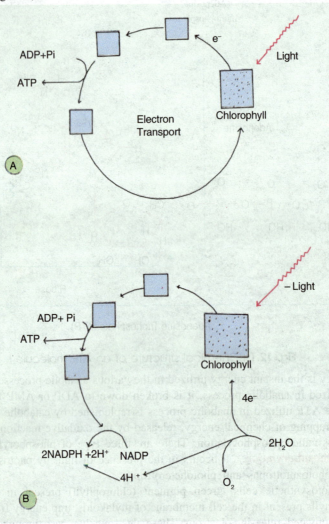

Fig. 12.2 : Photophosphorylation in aerobic microorganisms.

(ii) Oxidative Phosphorylation : In this phosphorylation, the electrons collected by certain electron carriers like NAD, NADP and FAD from various sources, are passed into electron transport chain. Finally, electrons reach oxygen or some other inorganic molecule (like iron, nitrate, etc.), which act as final electron acceptors. The transfer of electrons from one carrier to another released energy, which is used to generate ATP from ADP.

Oxidative phosphorylation occurs in the inner membrane of mitochondria in eukaryotes and in plasma membrane of prokaryotes. One molecule of NADPH generates three molecules of ATP, when it enters the electron transport chain for oxidative phosphorylation, however, one molecule of FADH generates only two molecules of ATP. This is due to the fact that FADH enters the ETC later than NADPH.

(*iii*) **Substrate Level Phosphorylation :** In substrate level phosphorylation, ATP is generated by the transfer of high energy phosphate bond from any other metabolic compound to ADP, for example :

$$GTP \quad + \quad ADP \quad \longrightarrow \quad ATP \quad + \quad GDP$$

Guanosine   Guanosine
triphosphate   diphosphate

(*iv*) **Nicotinamide Adenine Dinucleotide Phosphate (NADP):** Nicotinamide adenine dinucleotide phosphate (NADP) and Nicotinamide adenine dinucleotide (NAD) are the carriers of electrons (protons) in the cells. Therefore, NADP and NAD serve as the reducing power of the cells in the form of NADP/NADPH$_2$ or an NAD\NADH$_2$.

Reduced substrate + NADP $\longrightarrow$ Oxidised substrate + NADPH$_2$

Oxidised substrate + NADPH$_2$ $\longrightarrow$ Reduced substrate + NAD

NADP or NAD functions as coenzymes of a large number of oxidoreductase enzyme. They act as electron acceptors during enzymatic removal of hydrogen atoms from specific substrate molecules. Finally, reduced NADP or NAD i.e. NADPH$_2$ or NADH$_2$ release energy, ATP is generated.

Structurally, NAD consists of a nicotinamide moiety (the protein which undergoes reversible reduction), two adenine nucleotide. In NADP, a phosphate group is esterfied to the second adenine nucleotide at 2-hydroxyl group (Fig.12.3).

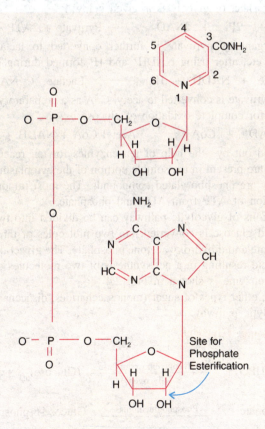

Fig. 12.3 : Chemical structure of nicotinamide adenine dinucleotide (NAD).

B. Metabolic Pathways

1.   Glycolysis

Glycolysis is the most important type of mechanism by which organisms obtain energy from organic compounds in absence of molecular oxygen. As it occurs in the absence of oxygen, therefore, it is also called anaerobic fermentation.

Since living organisms arose in the environment lacking oxygen, anaerobic fermentation was the only method to obtain energy. However, glycolysis or anaerobic fermentation is present in both aerobic and anaerobic organisms. Most higher organisms have retained the glycolytic pathway of degradation i.e. glucose to pyruvic acid as a preparatory pathway for complete aerobic catabolism of glucose. Glycolysis also serves as an emergency mechanism in anaerobic organisms to produce energy in the absence of oxygen.

The complete pathway of glycolysis from glucose to pyruvate (Fig.12.4) were elucidated by Gustav Embden (who gave the manner of cleavage of fructose 1, 6-diphosphate and pattern of subsequent steps) and Otto Meyerhof (who confirmed Embden's work and studied the energetics of glycolysis), in late 1953s. Therefore the sequence reaction from glucose to pyruvate is also called Embden–Meyerhof pathway of glycolysis (EMP).

The overall balance sheet of glycolysis is given below :

$$\text{Glucose} + 2ADP + 2Pi + 2NAD \longrightarrow \text{Pyruvate} + 2ATP + 2NADH + 2H^+$$

In anaerobic organisms pyruvate is further converted to lactate or other organic compounds like alcohol, etc, after using $NADH^+$ and H^+ formed during glycolysis :

$$\text{Pyruvate} + NADH + H^+ \longleftrightarrow \text{Lactate} + NAD$$

In aerobes the pyruvate is converted to acetyl CoA as a preparatory step for entrance into tricarboxylic acid cycle, for complete oxidation of glucose.

$$\text{Pyruvate} + NAD^+ + CoA \longrightarrow \text{Acetyl CoA} + NADH + H^+ + CO_2$$

Glycolysis is carried out by the help of ten enzymes for ten reactions of the glycolytic pathway. These enzymes are present in the soluble portion of the cytoplasm. All the intermediates of the glycolytic pathway are phosphorylated compounds. The most important use of phosphate groups is in the production of ATP from ADP and phosphate.

The complete reactions of glycolytic pathway can be divided into two stages. In the first stage, ATP is utilized and glucose is converted into two molecules of three carbon compounds, glyceraldehyde 3-phosphate and dihydroxy acetone phosphate. The glyceraldehyde 3-phosphate is converted into pyruvic acid resulting in a net synthesis of two molecules of ATP. The complete reaction with respective enzyme is shown in Fig.12.4

Apart from glucose, other types of sugar (monosaccharides, disaccharides, polysaccharides) can also enter the glycolytic pathway.

(a) Polysaccharides e.g. Glycogen

$$(\text{Glucose})n + HPO_4^{2-} \xrightarrow[\text{Glycogen}]{\text{Glycogenphosphorylase}} (\text{Glucose})_{n-1} + \text{Glucose-1– phosphate}$$

$$\text{Glucose-1-phosphate} \xrightarrow{\text{Phosphoglucomutase}} \text{Glucose-6-phosphate}$$

Glucose 6 - phosphate can enter as an intermediate of glycolysis.

(b) Disaccharides e.g. Sucrose

$$\text{Sucrose} + H_2O \xrightarrow{\text{Fructuranosidase}} \text{Glucose} + \text{Fructose}$$

$$\text{Maltose} \quad + \quad H_2O \quad \xrightarrow{\text{Glucosidase}} \quad 2 \text{ Glucose}$$

$$\text{Lactose} \quad + \quad H_2O \quad \xrightarrow{\text{Galactosidase}} \quad \text{Glucose} \quad + \quad \text{Galactose}$$

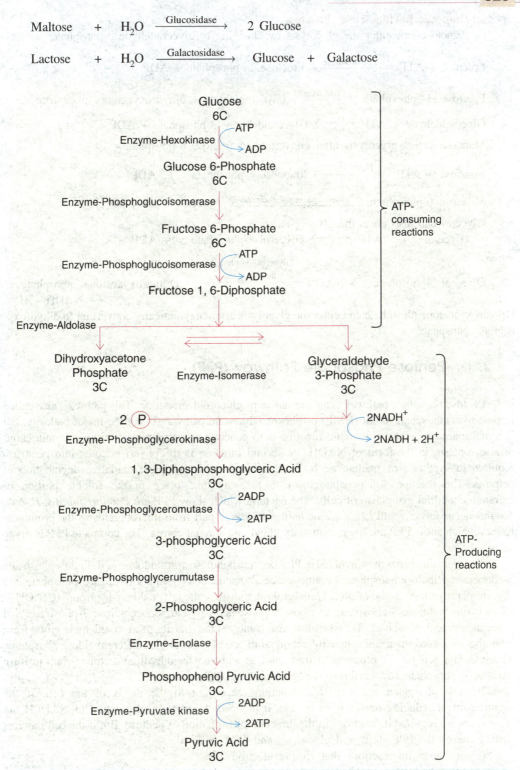

Fig. 12.4 : Glycolytic pathway showing synthesis of two ATP molecules and pyruvic acid (glycolysis).

(*c*) **Homosaccharides : e.g. Fructose**

Fructose can enter the glycolysis by changing to glyceradehyde 3-phosphate.

$$\text{Fructose + ATP} \xrightarrow{\text{Fructokinase}} \text{Fructose 1- phosphate + ADP}$$

$$\text{Fructose -1- phosphate} \xrightarrow{\text{Aldolase}} \text{Glyceraldehyde + dihydroxyacetone phosphate}$$

$$\text{Glyceraldehyde + ATP} \longrightarrow \text{Glyceraldehyde- 3-phosphate + ADP}$$

Mannose enters glycolysis after converting to fructose 6-phosphate

$$\text{Mannose + ATP} \xrightarrow{\text{Hexokinase}} \text{Mannose 6-phosphate + ADP}$$

$$\text{Mannose 6-phosphate} \xrightarrow{\text{Mannose phosphate isomerase}} \text{Fructose 6-phosphate}$$

Glycerol can also enter the glycolysis.

$$\text{Glycerol + ATP} \longrightarrow \text{Glycerol 3-phosphate + ADP}$$

$$\text{Glycerol 3-phosphate + NAD}^+ \xrightarrow[\text{dehydrogenase}]{\text{Glycerol - 3 - phosphate}} \text{Dihydroxyacetone phosphate}$$
$$+ \text{NADH + H}^+$$

Dihydroxyacetone phosphate can enter the glycolysis after enzymatically converting to dihydroxy-acetone phosphate.

2. Pentose Phosphate Pathway (PPP)

Pentose phosphate pathway is an alternative of glucose degradation. This pathway, also called hexose monophosphate shunt (HMP) or phosphogluconate pathway is not the major pathway, but is a multipurpose pathway. Its main function is to generate reducing power in the extramitochon-drial cytoplasm in the form of NADH. Its second function is to convert hexoses into pentoses, required in synthesis of nucleic acids. Its third function is complete oxidative degradation of pentose. The reactions of phosphogluconate pathway take place in the soluble portion of extramitochondrial cytoplasm of cells. The bacteria which show PPP are *Bacillus subtilis, E. coli, Streptococcus faecalis* and *Leuconostoc mesenteroides*. Apart from microorganisms the prominent tissues which show PPP are liver, mammary gland and adrenal cortex. The complete PPP is given in Fig 12.5.

There are three enzymes involved in PPP i.e., transketolase, transaldolase and ribulosephosphate 3-epimerase. Ribulose phosphate 3-epimerase catalyzes the conversion of ribulose 5-phosphate into the epimer xylulose 5-phosphate. Transketolase transfers the glycoaldehyde group (CH_2OH—CO—) from xylulose 5-phosphate to ribose 5-phosphate to yield sedoheptulose 7-phosphate and glyceraldehyde-3-phosphate. Transketolase also catalyzes the transfer of glycoaldehyde group from a number of 2-keto sugar phosphate to carbon atom one of a number of different aldose phosphate. Transaldolase acts on the products of transketolase and transfer dihydroxyacetone group to form fructose 6-phosphate and erythrose 4-phosphate (Fig.12.5).

Pentose phosphate pathway thus functions according to the needs of the cell. If the requirement of reducing power is more then it proceeds towards the formation of NADPH but if pentoses are required it functions in the direction of formation of pentose. But if the cell requires instant energy the PPP stops and glycolysis and TCA proceed.

(*a*) To anabolic reactions that require electron donors:

(*b*) To Calvin-Benson Cycle (dark reactions of photosynthesis)

(*c*) To synthesis of nucleotides and nucleic acids

(*d*) To step e of glycolysis

(e) To glucose 6-phosphate which can enter the pentose phosphate pathway or glycolysis
(f) To synthesis of several amino acids.

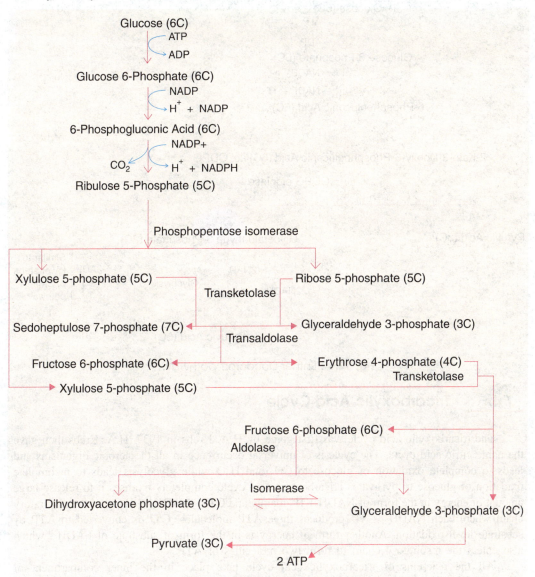

Fig. 12.5 : The Pentose phosphate pathway (see text for clarification of *a-f* steps of the pathway).

3.   Entner-Doudoroff Pathway

Apart from glycolysis, Entner-Doudoroff pathway is another pathway for oxidation of glucose to pyruvic acid. This pathway is found in some Gram-negative bacteria like *Rhizobium*, *Agrobacterium* and *Pseudomonas*. In this pathway each molecule of glucose, forms two molecules of NADPH and one molecule of ATP. The complete pathway is shown in the Fig. 12.6.

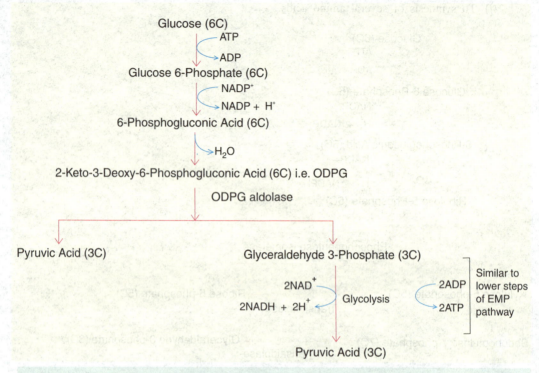

Fig. 12.6 : Entner-Doudoroff pathway.

4.   Tricarboxylic Acid Cycle

The tricarboxylic acid cycle was first given by H.A. Krebs in 1937. H.A. Krebs then gave the name citric acid cycle. The cycle is of universal occurrence in all the aerobic organisms and leads to complete oxidation of glucose to CO_2 and H_2O while glycolysis leads to incomplete oxidation of glucose to pyruvate. Tricarboxylic acid cycle completely oxidises it to release large amount of energy in the form of NADH + H^+ mainly and GTP. $NADH^+$ + H^+ enter the respiratory chain where each $NADH^+$ + H^+ produces three ATP molecules. GTP is converted to ATP by substrate level oxidation. Another form of energy is in the form of substrate of $FADH_2$, which also enters the respiratory chain to form two molecules of ATP.

All the reactions of tricarboxylic acid cycle take place in the inner compartment of mitochondrion. Some of these enzymes occur in the matrix of inner compartment, while rest of them occur on the inner mitochondrial membrane. For the start of the cycle, the pyruvate formed in the glycolysis is first converted to acetyl Co-A by preparatory reaction.

$$Pyruvate + NAD^+ + CoA \longrightarrow acetyl\ CoA + NADH + H^+ + CO_2$$

The reaction is irreversible and is not itself a part of the tricarboxylic acid cycle. It is carried out with the help of the enzyme pyruvate dehydrogenase. Acetyl CoA then enters the cycle after combining with oxaloacetate to form citrate, after which a cycle of reactions occurs (Fig. 12.7) leading to the formation of six CO_2, eight NADH + H^+, one $FADH_2$ and one molecule of glucose.

There are few key steps in the tricarboxylic acid cycle which control the cycle as per need of the cell. The first of these controls is the preparatory reaction. The activity of pyruvate dehydrogenase is reduced in the presence of excess ATP and again increases in the absence of

ATP. There are two more steps which can control the cycle. These are the isocitrate dehydrogenase reaction (which requires ADP as positive regulation), and succinate dehydrogenase reaction (promoted by succinate, phosphate and ATP). However, the key control of the cycle is the reaction carried out by citrate synthase. This is the primary control of the cycle.

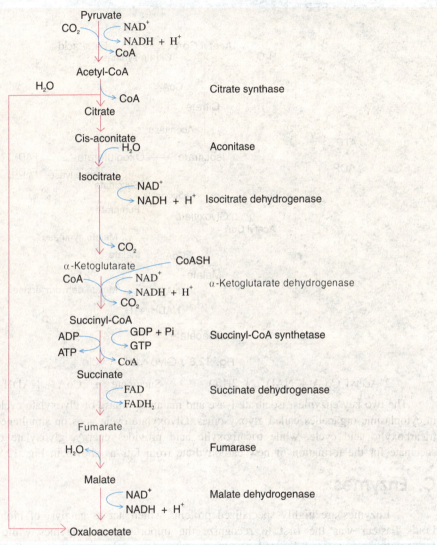

Fig. 12.7 : Tricarboxylic acid cycle.

5. Glyoxylate Cycle

It is anaplerotic reaction in which oxaloacetate is taken from TCA cycle to meet out the demand of carbon requirement for amino acid biosynthesis. Hence, these intermediates have to be replenished via an alternate route, called **anaplerotic pathway** i.e. glyoxylate pathway. This cycle operates for **gluconeogenesis.** Glyoxylate cycle was given first by Krebs and H.R. Kornberg. This cycle is a modified form of tricarboxylic acid cycle found in plants and those microorganisms which utilize fatty acids as the source of energy in the form of acetyl CoA. In this cycle the CO_2 evolving steps of tricarboxylic acid cycle were by-passed and instead a second molecule of acetyl CoA is utilized (which condenses with glyoxylate to form malate). Succinate is a by product, used for biosynthesis, particularly in gluconeogenesis. The overall reaction of glyoxylate cycle is given below.

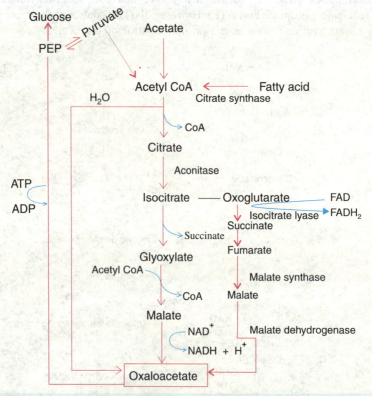

Fig. 12.8 : Glyoxylate cycle.

$$2 \text{ Acetyl Co-A} + NAD^+ + 2H_2O \longrightarrow \text{Succinate} + 2CoA + NADH + H^+$$

The two key enzymes, isocitrate lyase and malate synthase of glyoxylate cycle are localised in cytoplasmic organelles called glyoxysomes. Glyoxylate cycle goes on simultaneously with the tricarboxylic acid cycle, while tricarboxylic acid provides energy, glyoxylate cycle provides succinate for the formation of new carbohydrate from fats as shown in Fig. 12.8.

C. Enzymes

Enzymes are highly specialized proteins which act as catalyst of biological system. Louis Pasteur was the first to recognize the importance of enzymes while studying the fermentation process and denoted it as "ferment"–an integral part of living cells. It was Edward Buchner who in 1897 extracted the enzyme from yeast cells, responsible for fermentation of sugar to alcohol. In 1926, James B. Sumner isolated and crystallized urease and also postulated that all enzymes are proteins. Today we know, this is true but with exception of Ribozymes which is a catalytic RNA.

Enzymes are the large globular proteins with molecular weight ranging from 13,000 to millions Dalton. The catalytic efficiency of an enzyme depends upon it's three dimensional conformation. Moreover, these biocatalysts are highly specific for the reaction as well as other conditions e.g. each enzyme has it's own optimum pH, temperature, etc. for maximum performance. Some enzymes require an additional moiety known as cofactors for their biological activity.

1. Mechanism of Enzyme Action

Like any other catalyst, enzymes also act by lowering down the activation energy of any reaction. Activation energy is required to attain the 'transition state' which is a state of event comprising various molecular changes like breakage, formation or rearrangement of bonds and development of changes. The 'transition state' is a high energy state and falls between the process of conversion of substrate into products. This state is attained at the top of energy barrier which has to be overcome so as to complete the reaction. A biocatalyst facilitates the process by virtue of active sites-which are pockets for accommodating the substrate molecules on enzyme. Active sites are lined by amino acids, which bind to substrate by chemical groups of side chains and catalyze chemical transformation. This also explains the high specificity of enzymes for their substrate accordingly which can be compared to 'lock and key' (Fig. 12.9). The set of events occurring in whole process are as follows:

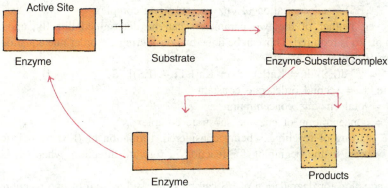

Fig. 12.9 : Lock and key model for mechanism of enzyme action given by Fischer.

(*i*) Substrate with its particular corresponding location binds to active site to form enzyme-substrate complex.

(*ii*) At the active site, energy is derived because of enzyme-substrate interaction (i.e. binding energy) viz., the major source of free energy used by enzymes to lower the activation energies of reactions. It leads to rearrangement of molecules, breaking and formation of bonds-resulting in the product.

(*iii*) Products are released and enzyme recovered to catalyze further unlimited times the same reaction. Similar to any other catalyst, enzymes do not affect reaction equilibria. However, reaction reaches the equilibrium relatively faster.

2. Enzyme Kinetics

A number of approaches are now available to study the mechanism of enzyme action including knowledge of complete 3-D structure, site directed mutagenesis and protein engineering, still central approach is to determine the rate of reaction and how it is affected by different experimental parameters-more precisely-'The enzyme kinetics'.

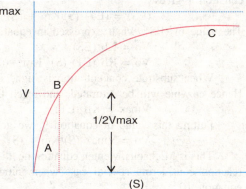

Fig. 12.10 : Kinetics of enzyme – substrate relationship (Michaelis-Menten hypothesis).

Enzyme kinetics follows the principles of general chemical reaction kinetics, however, show a distinctive feature of saturating (Fig. 12.10). At lower substrate concentration, the initial reaction velocity is proportional to substrate concentration (1st order reaction). Further increase in substrate concentration does not affect the reaction rate and the latter became constant (zero order reaction).

In hypothetical one substrate reaction Michaelis-Menten theory, enzymes first combine with substrate to from enzyme-substrate (ES) complex. This ES complex then breaks in second step to release free enzyme and product as-

$$E+S \quad \underset{\xleftarrow{\hspace{1cm}}}{\overset{K a}{\xrightarrow{\hspace{1cm}}}} \quad ES \qquad (1)$$

$$ES \quad \underset{\xleftarrow{\hspace{1cm}}}{\overset{K b}{\xrightarrow{\hspace{1cm}}}} \quad E+P \qquad (2)$$

According to above equations, initial velocity of complete reaction equals the breakdown of enzyme substrate complex. Hence,

$$Vo = Kb (ES) \qquad (3)$$

Where, Vo = initial velocity,

(ES) = concentration of enzyme substrate complex

However, neither of the two parameters can be determined directly, and an alternative expression of Vo is required. This can be done by considering 2nd order rate equation for formation of ES from E and S.

$$d(ES)/dt = Ka(E) (S) = Ka[ET] - (ES)] (S) \qquad (4)$$

Where, Ka= second order rate constant

(ET) = total enzyme concentration

(ES) = concentration of enzyme substrate complex

Since, starting of reaction is being considered, formation of (ES) by reaction (2) may be neglected because in the beginning for reaction in forward direction, when (S) is high and (P) is zero.

Rate equation for degradation of ES can be expressed as sum of two reactions-first reaction yielding the product and second reaction yielding E + S. Then,

$$-d (ES) /dt = Ka (ES) + Kb (ES) \qquad (5)$$

However, in the steady state, when rate of formation of ES is equal to its breakdown, then equation 4 = 5

Thus,

$$(ES)/dt = -d(ES)/dt \qquad (6)$$

Or,   Ka [ET]-[ES] (S) = Ka (ES) + Kb (ES)

Or,   [S] [(ET)-(ES)]/(ES) =Ka + Kb/Ka = Km   (7)

(Ka+Kb/Ka), i.e. expressed as Km is known as Michaelis-Menten constant. Rearranging the equation 7 gives,

$$(ES) = (ET) (S)/Km + (S) \qquad (8)$$

The value of (ES), when expressed in equation

$$Vo = Kb (ES)$$

$$Vo = Kb (ET) (S)/ Km + (S) \qquad (9)$$

When substrate concentration is high, all enzymes in system are present as ES complex. Hence enzyme will be saturated, and reach the maximum velocity (V max), given as

$$Vmax = Kb (ET)$$

Putting this value in equation 9, we get

$$Vo = Vmax (S)/Km + (S) \qquad (10)$$

This is Michaelis-Menten equation i.e. the rate equation for a one substrate enzyme catalyzed reaction considering special case when initial reaction rate is exactly half of Vmax, then by equation 10

$$Vmax/2 = Vmax (S)/Km + (s)$$

Or,   Km + (S) = 2 (S)

Or,   Km = (S)

Thus Michaelis-Menten constant is equal to substrate concentration at which initial reaction velocity is half of maximum velocity.

Km varies according to substrate, pH, and temperature and is not a fixed value. Reciprocating the equation 10, we get

$$1/Vo = Km + (S)/ Vmax (S)$$

Or, $1/Vo = Km/ Vmax (S) + 1 / Vmax$ (11)

It is Lineweaver-Burk equation, which gives a straight line when 1/Vo is plotted against 1/S (Fig. 12.11)

Effect of pH and tempera-ture on enzymatic action: Enzymes have an optimum pH and temperature (or a range) at which they are maximally active. The side chains in amino acid play the central role. The side chains in the active site may act as weak acid or base and their alignment depends on the state of ionization. Hence ionization state is important to decide the conformation of enzyme molecule and hence the activity. Dissimilarly,

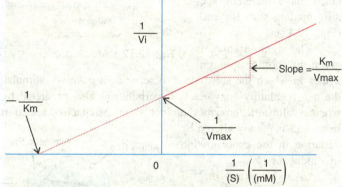

Fig. 12.11 : Enzyme-substrate relationship (Lineweaver-Burk Plot).

higher temperatures result in the rearrangement of bonds. Because of this the conformation of active sites get altered and activity is affected. Enzymes with more cysteine residues are thermally more stable because of the formation of –S-S bridges between the peptide chains.

3.   Allosteric Enzymes

These are regulatory enzymes, their catalytic activity depends upon non-covalent binding of a specific metabolite at a site other then active site. In many multi-enzyme system, the end product inhibits the pathway by inhibiting an enzyme of initial reaction. This phenomenon is is generally known as *feed back inhibition*. The inhibited enzymes are allosteric enzymes, which accommodate the final product on allosteric sites. Interaction of modulator (end product) with allosteric site alters the three dimensional structure of catalytic site and thus affects it's activity. Modulators may inhibit, or induce the target allosteric enzyme, hence called as negative and positive modulators respectively.

(i) **Properties of Allosteric Enzymes:** Allosteric or Regulatory enzymes have multiple subunits (Quaternary Structure) and multiple active sites. Allosteric enzymes have active and inactive shapes differing in 3D structure. Allosteric enzymes often have multiple inhibitor or activator binding sites involved in switching between active and inactive shapes. Allosteric enzymes have characteristic "S"-shaped curve for reaction rate vs. substrate concentration. Why? Because the substrate binding is "Cooperative." And the binding of first substrate at first active site stimulates active shapes, and promotes binding of second substrate.

A modulator is a metabolite, when bound to the allosteric site of an enzyme, alters its kinetic characteristics. The modulators for allosteric enzyme may be either stimulatory or inhibitory. A stimulator is often the substrate itself. The regulatory enzymes for which substrate and modulator are identical are called **homotropic**. When the modulator has a structure different then the substrate, the enzyme is called **heterotropic.** Some enzymes have more then one modulators. The allosteric enzymes also have one or more regulatory or allosteric sites for binding the modulator. Enzymes with several modulators generally have different specific binding sites for each (Fig. 12.12).

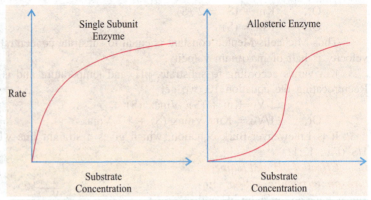

Fig. 12.12 : An allosteric enzymes activity by its substrate.

The sigmoid curve is given by homotropic enzymes in which the substrate also serve as a positive (stimulator) modulator (12.13). Curve for the non-regulatory enzymes is hyperbolic, as also predicted by the Michaelis-Menten equation, whereas allosteric enzymes do not show Michaelis-Menten relationship because their kinetic behaviour is greatly altered by variation in the concentration of modulators.

(ii) Mechanism of Action of Allosteric Enzymes: Two general models for the interconversion of inactive and active forms of allosteric enzymes have been proposed:

(a) Simple sequential model : This model was proposed by Koshland Jr. in the year 1966. According to this theory, the allosteric enzyme can exist in only two conformational changes individually.

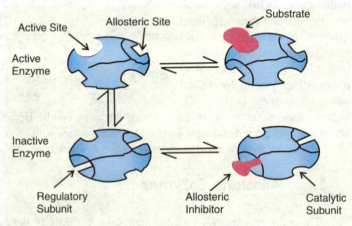

Fig. 12.13 : Mechanisms of allosteric effect.

Consider an allosteric enzyme consisting of two identical subunits, each containing an active site (Fig. 12. 14A). The T (tense) form has low affinity and the R (relaxed) form has high affinity for substrate (12.4). In this model, the binding of substrate to one of the subunits induces a T → R transition in that subunit but not in the other subunits.

(ii) Concerted or Symmetry Model : This model was proposed by Jacques Monod and his colleagues in 1965. According to them, an allosteric enzyme can exist in still two conformations, active and relaxed or inactive form. All subunits are either in the active form or all are in inactive form. Every substrate molecule that binds with enzyme increases the probability of transition from the inactive to the active site. The effect of allosteric activators and inhibitors can be explained quite easily by this model. An allosteric inhibitor binds preferably to the T form whereas an allosteric activator binds to the R form (Fig. 12.14B). An allosteric inhibitor shifts the R →T conformational equilibrium towards T. Whereas an allosteric activator shifts it toward R. The result is that an allosteric activator increases the binding to substrate of the enzyme, whereas an allosteric inhibitor decreases substrate binding (Fig. 12.15). Symmetry is conserved in this model but not in the sequential model.

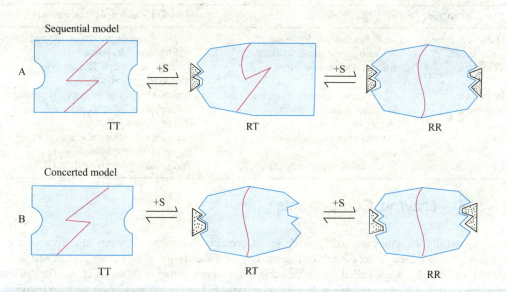

Fig. 12.14 : Kinetic models of allosteric enzymes (A) ; Simple sequential model (B) Concerted or symmetry model.

4. Classification of Enzymes

Generally, the name of enzymes ends with the suffix *ase*. Until recently enzymes were named arbitrarily after the name of the discoverer. However, this is not a practical approach, because some of the enzyme have completely irrelevant names. Therefore, now a new system has been proposed to combat this difficulty and to include all the new enzymes being discovered. This system has been recommended by International Committee for Nomenclature of Enzymes in 1973. This system includes six major classes including other subclasses according to the type of reaction catalyzed (Table 12.1). Now-a-days, enzymes are generally assigned two names: the recommended name and a systematic name. Recommended name is the common name. It is small and easy to use. Systematic name is given according to the reaction in the six classes of the enzymes.

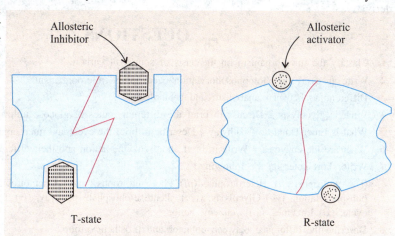

Fig. 12.15 : Effect of activator and inhibitor on substrate binding.

Table 12.1: Classification of enzymes

Class of enzyme	Example	Reaction carried by the enzyme
Oxidoreductase	Dehydrogenases (Pyruvate dehydrogenase etc.), oxidase	Redox reactions in which oxygen or hydrogen are transfered
Transferase	Kinase, Deaminase etc.	Transfer of groups like acetyl, phosphate, amino etc.
Hydrolase	Glucosidase, Lipase, Fumerase, etc.	Addition of water or removal of water
Isomerase	Triosephosphate isomerase etc.	Isomerization of molecules i.e. rearrangement of atoms.
Lyase	Decarboxylases, Aldolases etc.	Carry out addition or removal of group of atoms without water
Ligase	Synthetases, Synthases etc.	Formation of bonds with ATP.

5.   Enzyme Components

Generally, enzymes are solely made up of proteins. But many enzymes also contain a non protein portion which is necessary for the activity of the enzyme. The protein part of enzyme in these types of enzymes is called *apoenzyme* and the nonprotein part is called *cofactor*. The cofactor may be a metal ion or an organic molecule (coenzyme). The complete enzyme with both cofactor and apoenzyme is called *holoenzyme*.

In the enzyme with metal ion as cofactor are ions such as Mg^{+2}, Fe^{+2}, Fe^{+3}, K^+, etc.. These may serve as primary catalytic centres or a bridging group to bind substrate and enzyme together. Such enzyme with a metal ions as cofactor are sometimes called metalloenzymes e.g. phosphotrans ferases (which have Mg^{+2} as metal cofactor), cytochromes and catalase (which have Fe^{+2} and Fe^{+3} metal ion).

The coenzyme, which is a complex organic molecule, may be any of the vitamins (trace organic molecule that are vital to the function of all cells and required in the diet). When the coenzyme is very tightly bound to the enzyme molecules, it is usually called prosthetic group. Some of the important coenzymes are nicotinamide derivatives, flavin adenine dinucleotide (FAD) derivatives.

QUESTIONS

1. Classify the microorganism on the basis of source of carbon.
2. Write an essay on photophosphorylation occurring in microorganisms.
3. Differentiate between anabolism and catabolism.
4. What is glycolysis ? Discuss in brief the different routes of glucose utilization.
5. What is Enter-Doudoroff pathway ? Describe in brief the cycle and microorganisms in which it occurs.
6. What are the enzymes ? Write in brief about enzyme action and their different structural components.
7. Write short notes on the following :

    (*i*) Autotrophs, (*ii*) Heterotrophs, (*iii*) Photoautotrophs, (*iv*) Chemotrophs, (*v*) ATP, (*vi*) Photo-phosphorylation, (*vii*) Glycolysis, (*viii*) Pentose-phosphate pathway, (*ix*) TCA cycle, (*x*) Glyoxylate cycle, (*xi*) Enzymes, (*xii*) enzyme action.

8. Describe the difference between cofactor and prosthetic group.
9. Define enzymes. Discuss the kinetic relationships with substrate.
10. Write in brief on the following:

    (*a*) Line-weaver Burk plot   (*b*) Michaelis-Menten hypothesis

    (*c*) Rate of reaction   (*d*) Km value

Bacterial Photosynthesis

13

All the living organisms require energy to carryout their different activities of life. For this energy is needed which comes by the oxidation of carbohydrates, proteins or fats. Photosynthesis is an anabolic process in which electrons are released from excited photo-pigments by radiation, travel through a series acceptors converting radiant energy to electronic energy and then later to chemical energy. Similar to the green plants, there are certain chlorophyll (pigment) containing bacteria which synthesise foods from simple substances like carbon dioxide and water.

Photosynthesis, in bacteria, is defined as "the synthesis of carbohydrates by the chlorophyll in the presence of sunlight, CO_2 and reductants taken from air and oxygen do not evolve as by product, except in cyanobacteria."

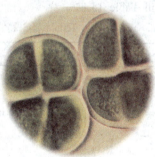

Purple non-sulphur bacterium

$$2H_2A + CO_2 \xrightarrow{\text{Light and bacteriochlorophyll}} (CH_2O)x + 2\ A + H_2O$$

A. History of Photosynthetic Prokaryotes

Before ninteenth century it was considered that the photosynthetic machinery is present in purple bacteria because these bacteria showed movement towards light (phototactic) and growth is also induced by light (W. Engelmann, 1885). On the otherhand, S. Winogradsky, a German botanist observed that some purple bacteria can utilize hydrogen disulphide to sulphate with intracellular deposition of sulphur. C.B. Van Niel (1930) defined various metabolic versions of anoxygenic photosynthesis and demonstrated that it is the characteristic mode of energy yielding metabolism in both purple and green bacteria.

The photosynthetic purple bacteria use a variety of hydrogen donors in place of water (e.g. H_2S or various organic compounds). In some anaerobic photosynthetic bacteria using hydrogen donors other than hydrogen or water (e.g. succinate) not only CO_2 is reduced to $NADPH_2$ but also atmospheric nitrogen is reduced to ammonia. Such nitrogen fixation occurs at the expense of photic energy. This fact is of considerable importance in view of the significance of nitrogen fixation in the economy of nature, therefore, it appeared that they lack photosystem II (PS-II), which among other things, in green plants is involved in O_2 production (from OH). The photosynthetic bacteria show no enhancement (Blinks and Van Niel, 1963).

Parson and Cogdell (1975) isolated functional complexes from photosynthetic bacteria. The reaction centre from the purple non-sulphur bacterium, *Rhodopseudomonas sphaeroides*, contains four molecules of chlorophyll and two molecules of bacteriopheophytin (like *b* chlorophyll but the Mg replaced by two H^+), one or two molecules of ubiquinone and one atom of ferrous iron together with three polypeptides of apparent molecular weight in the region of 28, 24 and 21 KD (Cogdell, 1983).

All the photosynthetic bacteria are divided into 35 groups. The group 10 contains anoxygenic phototrophic bacteria, while group 11 belongs to oxygenic phototrophic bacteria. The anoxygenic group (no evolution of oxygen) has purple and green bacteria, while oxygen evolving group has only cyanobacteria. Another type of oxygenic bacteria has recently been discovered and placed under prochlorophyta. Prochlorophyta acts as a bridge between cyanophyta and chlorophyta (or green algae).

B. Classification of Photosynthetic Bacteria

The photosynthetic bacteria are divided into two broad groups, Anoxygenic photosynthetic bacteria and oxygenic photosynthetic bacteria. Classification of photosynthetic bacteria is given in Table 13.1

Table 13.1 : Characteristics of photosynthetic bacteria.

Category	Type	Pigment	C Source	Electron donor
Anoxygenic	Purple bacteria	BChl *a* and *b*	Organic C and/ or CO_2	H_2, H_2S, S
"	Green bacteria	BChl *c, d, e*, small amount of Chl *a*	CO_2	H_2, H_2S, S
Oxygenic	Cyanobacteria	Chl *a*, phycobilins	CO_2	H_2O
"	Prochlorophytes	Chl *a* and *b*	CO_2	H_2O

1.   Anoxygenic Photosynthetic Bacteria

The anoxygenic photosynthesis depends on e^- donors such as reduced sulphur compounds, molecular hydrogen or organic compounds. The ammonium salts are generally used as nitrogen source. Nitrogen fixation has been reported in some bacterial species. Some can grow chemoautotrophically under aerobic/microaerobic condition. Fatty acids, ethanol and organic acids serve as carbon sources. They are found in fresh water, brackish water, marine and hypersaline water. They may be classified into seven groups:

Sub-group 1: Globules of sulphur are found inside the cell e.g. *Amoebobacter, Chromatium, Lamprobacter, Thiocapsa, Lamprocystis, Thiocystis, Thiodictyon, Thiopedia, Thiospirillum.*

Sub-group 2: Globules of sulphur appears outside the cell e.g. *Ectothiorhodospira.*

Sub-group 3: Globules of sulphur may appear outside the cell. Most genera depend upon growth factor. Cells of *Rhodobacter, Rhodocyclus, Rhodomicrobium, Rhodopila, Rhodopseudomonas, Rhodospirillum* grow by photoassimilation of simple organic substrates.

Sub-group 4: Internal membrane systems or chlorosomes are absent. Spiral shaped cell wall has no lipopolysaccharide. Cells contain bacteriocholophyll g and carotenoids. Reduced sulphur compounds are not utilized. They are photoheterotrophic e.g. *Heliobacillus* (glider).

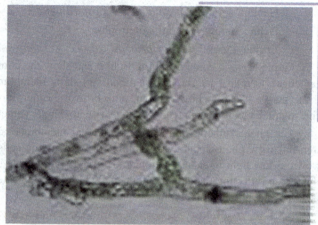

Cells of *Chloronema* are arranged in multicellular filaments

Sub-group 5: Globules of sulphur appears outside but never inside the cell. Bacteriochlorophylls are located in chlorosomes. Simple organic substances are photo-assimilated only in presence of sulphide and bicarbonates e.g. *Ancalochloris, Chlorobium, Pelodictyon, Prosthecochloris.*

Sub-group 6: Cells are arranged in multicellular filaments that show gliding motility and utilize organic substances e.g. *Chloroflexus, Chloronema, Heliothrix, Oscillochloris.*

Sub-group 7: Cells grow chemoheterotrophically under aerobic conditions; no growth occurs under anaerobic conditions. They contain bacteriochlorophyll *a* and carotenoids e.g. *Erythrobacter.*

Anoxygenic photosynthetic bacteria have been divided into three groups on the basis of pigmentation; purple bacteria, green bacteria and heliobacteria (Table 13.2).

Table 13.2 : Characteristics and classification of anoxygenic bacteria.

Category	Name of organism	Family	Characteristics
Purple-Sulphur	*Chromatium, Ectothiorhodospira Thicapsa, Thiospirillum*	Chromatiaceae	Photolithotrophs
Purple non-Sulphur	*Rhodomicrobium Rhodopseudomonas Rhodospirillum*	Rhodospirillaceae	Photoorganotrophs
Green-Sulphur	*Chlorobium, Pelodictyon,*	Chlorobiaceae	Photolithotrophs
Green non-Sulphur	*Chloroflexus, Heliothrix, Oscillochloris*	Chloroflexaceae	Chemoorganotrophs and phototrophs

(i) Purple Bacteria (Proteobacteria): The anoxygenic phototrophs grow under anaerobic conditions in the presence of light and do not use water as *e*-donor as higher plants. The pigment synthesis is repressed by O_2. They grow autotrophically with CO_2 (C source) and hydrogen or reduced sulphur compounds act as e^- donor. Photoheterotrophy (i.e. light as energy source and an organic compound as a carbon source) also supports growth. Some purple bacteria also show chemoorganotrophy i.e. can grow in dark under similar conditions. The purple bacteria comprise of five sub-groups as given below :

- Alpha group: *Rhodospirillum, Rhodopseudomonas, Rhodobacter, Rhodomicrobium.*
     Rhodopila, Rhodovulvum
- Beta group: *Rhodocyclus, Rhodoferax, Rubrivivax*
- Gamma group: *Chromatium, Thiospirillum*
- Delta group: Nil non-phototrophs
- Epslon group: *Bdellovibrio, Myxococcus, Campylobacter, Helicobacter*

Purple bacteria contain Bchl *a*, Bchl *b* and show the photosynthetic membranes in flat sheets (lamellae). Certain bacteria (*Chromatium* sp.) show membrane as individual vesicle. The colour of the purple bacteria shows brown, pink brown-red, purple-violet based on carotenoid contents. The internal membrane extends to give rise to photosynthetic pigments.

The photosynthetic pigments and internal membrane are influenced by light intensity. At high intensity, photo-apparatus is inhibited, whereas the cells get packed with membranes when grown at low light intensity. Carotenoids give rise to purple colour; mutants lack carotenoids are blue green reflecting the actual colour of BChl *a*. Purple bacteria are of two types; purple-sulphur bacteria and purple non-sulphur bacteria.

(a) **Purple sulphur bacteria (family; Chromatiaceae):** They are Gram-negative bacteria which contain BChl *a* and BChl *b* and grow chemolithotrophically in dark with thiosulphate as e^- donor. They are also chemoorganotrophs, utilize acetate, pyruvate and few other compounds. The mole % of G+C varies from 46-70. The cells of purple-sulphur bacteria are large than green bacteria and packed with intracellular sulfide deposition. They are found in anoxic zone of lakes and sulphur springs (obligate anaerobe). They contain vesicles that are enclosed within a thin membrane that is not directly associated with the cell membrane called *vesicular thylakoid.* They are photolithotrophs and motile in nature e.g. *Ectothiorhodospira, Chromatium, Thiocapsa, Thiospirillum, Thiodictyon, Thiopedia* etc.

(b) **Purple non-sulphur bacteria (family: Ectothiorhodospiraceae old name Rhodospirillaceae):** They also contain BChl *a* and *b* and use low concentration of sulphide. The concentration of sulphide utilized by purple-sulphur bacteria proved toxic to this category of bacteria. Earlier, scientists thought that these bacteria are unable to use sulphide as an e^- donor for the reduction of CO_2 to cell material, hence named them non-sulphur. They deposit sulphur extracellularly. Some non-sulphur bacteria grow anaerobically in the dark using fermentative metabolism, while the others can grow anaerobically in dark by respiration in which e^- donor may be an organic compound/inorganic compound as H_2. This group is most versatile energetically due to broad requirements and are photoorganotrophs *i.e.* use organic acids, amino acids, benzoate and ethanol. They also grow as chemoorganotrophs and require vitamins. They are heterogenous group due to the presence of both polar and peritrichous flagella. Some can utilize methanol for phototrophic growth when grown anaerobically. The following reaction occurs inside their cell.

$$2CH_3OH + CO_2 \longrightarrow 3(CH_2O) + H_2O$$

The DNA base composition is 61-73 mole % (G+C) and the sulphur granules are formed outside the cell. Examples of purple non-sulphur bacteria are *Rhodomicrobium, Rhodopseudomonas, Rhodospirillum, Rhodocyclus,* etc.

(ii) **Green Bacteria:** Instead of green in colour, these are brown due to the presence of carotenoids components. They are Gram-negative. Hence, colour is not a suitable basis for these bacteria. They contain BChl *c*, BChl *d* or BChl *e* plus small amount of Bchl *a*. The photosynthetic

apparatus is *chlorosomes* which consist of a series of cylindrical structures underlying and/or attached to cytoplasmic membrane and are quite different with lamellae. These vesicles are enclosed within a thin membrane devoid of bilayer but consist of transporter proteins located in the cytoplasmic membrane. They do not require vitamins for their growth.

$$CO_2 + 2H_2S \xrightarrow{\text{Light}} CH_2O + H_2O + 2S$$

$$CO_2 + 2H_2 \xrightarrow{\text{Light}} CH_2O + H_2$$

$$2CO_2 + Na_2S_2O_3 + H_2O \xrightarrow{\text{Light}} 2NaHSO_4$$

Green bacteria are of two types: green sulphur bacteria and green non-sulphur bacteria.

(a) **Green sulphur bacteria (family : Chlorobiaceae):** They are non-motile, rods, spiral and cocci. Some have appendages i.e. *prosthecae.* Chlorosomes are present in the cell. They do not possess gas vesicles (*Chlorobium*) except in *Pelodictyon.* They are strictly anaerobic and obligate phototroph. Most of them assimilate simple oxygenic substances for photosynthetic growth if sulphur source is present. They deposit sulphur extracellularly. Some *(Chloroflexus)* grow chemoorganotrophically, hence they are non-sulphur green type. The mol % G+C is 45-58. Examples of these bacteria are *Chlorobium, Prostheochloris, Pelodictyon, Chloroherpeton.*

(b) **Green non-sulphur bacteria (family: Chloroflexaceae):** The green non-sulphur bacteria are filamentous, gliding bacteria, thermophilic in nature. The pigments are Bchl *a,* Bchl *c,* β- and γ- carotenes. The chlorosomes are present when grown anaerobically. They are photoheterotrophic and photoautotrophic and show gliding movement. They do not deposit sulphur. The mol % G+C contents vary 53-55. Example is *Chloroflexus.*

(iii) **Heliobacteria:** Based on 16S rRNA sequencing and other morphological and biochemical characters, helicobacter are quite different with other anoxygenic photosynthetic bacteria. They are Gram-positive, rod shaped, motile either by gliding or by means of flagella. The mol % G+C is between 50 and 55, and at present comprises of three genera and five species such as *Heliobacterium, Helophilum* and *Heliobacillus.* Most of them produce endospores and grow up to 42°C. The heliobacteria are green in colour. The bacteriochlorophyll is associated with the cytoplasmic membrane, hence lamellae and chlorosomes are absent. The endospores contain a dipicolinic acid similar to *Bacillus* and *Clostridium.* Most of the heliobacteria are found in tropical soils of paddy fields. They contain BChl *g* having vinyl ($H_2C = CH_2$) group on ring I of the tetrapyrrol molecules similar to bacteriochlorophyll *a* but difference lies in ring II of the tetrapyrrole.

2.   Oxygenic Photosynthetic Bacteria

The oxygenic photosynthetic bacteria are unicellular or multicellular and possess bacteriochlorophyll *a* and carry out oxygenic photosynthesis. They contain phycobilins. One group, prochlorophytes lack phycobilins, but contain both bacteriochlorophyll *a* and *b.* They are mostly represented by Gram-negative cyanobacteria having only membrane. Many possess extracellular sheath called glycocalyx or capsule or merely mucilage or slime. The flagella are not present but they show gliding movement. The light harvesting pigments are phycobilin proteins, phycoerythrin, phycocyanin, bacteriochlorophyll *a* and carotenoids but sheath capsule may contain yellow pigment called scytonemin or red-blue pigment gloeocapsin which may mask cellular pigmentation. Phycobilins form phyobilisomes on both surfaces on double unit internal membrane called thylakoids. Bacteriochlorophyll *a* and carotenoids are part of it. Photosynthesis is oxygenic and autotrophic but chemoautotrophy also occurs. Photosynthates get accumulated in the form of glycogen, polyphosphate granules. Carboxysomes and gas vesicles are present. Heterocysts have modified thylakoids containing low contents of photosynthetic pigments and lack of photosystem II. Akinetes are thick walled resistant storage cells. Some non symbiotic bacteria lack typical cynobacterial wall is termed *cyanelles*. They are put into the following five sub-groups:

Sub-group I: They are unicellular or non-filamentous aggregates of cells held together by outer walls or a gel-like matrix. Binary fission occurs on one, two or three planes symmetric or asymmetrically or by budding. Examples of the members are *Gloeothece, Synechococcus, Gloeocapsa, Gloeobacter.*

Sub-group II: In the members of this group, reproduction takes place by internal multiple fission with production of daughter cells smaller than ½ the parent or by multiple and binary fission e.g. *Dermocorspa, Xenococcus, Pleurocapsa.*

Sub-group III: Binary fission occurs in one plane only. Trichomes are composed of cells which do not differentiate into heterocysts or akinetes e.g. *Spirulina, Arthospira, Oscillatropia, Phormidium, Lyngbya.*

Sub-group IV: One of few cells of each trichome differentiate into heterocysts, at least when concentration of external combined nitrogen is low. Some genera also produce akinetes e.g. *Cylindrospermum, Anabaena, Nodularia, Calothrix, Nostoc.*

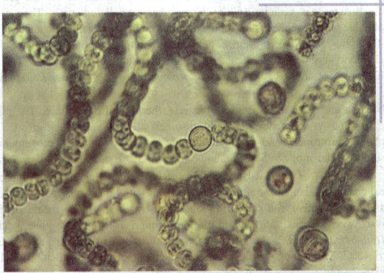

Nostoc with heterocysts (× 550).

Sub-group V: Binary fission occurs periodically or commonly in more than one plane giving rise to mulitseriate trichomes or trichomes with true branches or both e.g. *Stigonema, Cyanobotrytis. Westicella, Loriella, Nostichopsis.*

3.   Members of Prochlorophyta

On the otherhand, group II in section 19 of *Prochloron* was first discovered as an extracellular symbiont growing either on the surface or within the cloacal cavity of marine colonial ascidian invertebrates. These bacteria are unicellular, spherical and 8.30 μm in diameter. The mol. % G+C is 31 to 41. In addition, *Prochlorothrix* is free living, consists of cylindrical cells that form filaments. It has been found in Dutch lakes. The DNA has a high mol. G+C content (53%).

These bacteria are unicellular or filamentous branched/unbranched. They have bacteriochlorophylls *a* and *b* and lack accessory red or blue bilin pigment e.g. *Prochloron, Prochlorothrix.* The presence of 5S and 16S rRNA sequences shows affinity with the cyanobacteria.

The prochlorophytes bear the following characters:

(*i*) They possess the characters of both prokaryotes and eukaryotes.

(*ii*) The prokaryotes contain thylakoid which are spread towards periphery but the prochlorophytes contain paired thylakoids. This shows evolution of thylakoid from single to paired form.

(*iii*) Prochlorophytes consist of both chlorophyll a and chlorophyll b, whereas cyanobacteria contains only chlorophyll a. The chlorophyta consists of both chlorophyll a and chlorophyll b.

(*iv*) Phycobilisomes are grass-green in colour because they lack phycobilin pigments.

(*v*) Their 5S and 16S rRNA show affinities with cyanobacteria.

(*vi*) Possibly a common ancestor has given rise to prochlorophytes, cyanophyta and plant chloroplasts.

4.　　Unclassified Bacteria

There are certain other photosynthetic bacteria isolated recently but did not find any place in Bergey's manual due to their discovery in recent years. *Porphyrobacter neustonensis*, an aerobic bacteriochlorophyll synthesizing budding bacterium from fresh water and *Roseobacter denitrificans* was discovered by Fuert and others in the year 1993. The genus *Erythromicrobium sibericus* was isolated by Yukov and his Research team in the year 1991, while it is interesting to note the discovery of one photosynthetic as well as nitrogen fixing *Rhizobium* B T Ai by Evans and others in 1990. *Pseudomonas radiora* is again an interesting photosynthetic bacteria.

C. Photosynthetic Pigments

Because of the presence of carotenoids in all photosynthetic tissues their role is anticipated in photosynthesis. The tissue/cells rich in carotenoids devoid of chlorophyll do not photosynthe-size. Light energy absorbed by carotenoids appears to be transferred to chlorophyll a or bacterio chlorophyll a and utilized in the photosynthesis. It was observed if light is absorbed directly by chlorophyll a, generally found to be less efficient in photosynthesis than that of light absorbed by accessory pigments such as phycocyanin or phycoerythrin. All the photosynthetic bacteria contain chromatophores which have β-carotene, xanthophyll (carotenoids), and phycobilisomes (phycocyanin and phycoerythrin). Most of the carotenes are present in photosystem I while phycobilisomes are present in photosystem II.

1.　　Bacteriochlorophylls

Photosynthetic bacteria appear in different colour suspension. They are green, purple, violet, red, blue-green, brown coloured due to the presence of photosynthetic pigments in their photosynthetic apparatus. The pigments can be measured by their absorption spectra. Generally chlorophyll absorbs maximally at < 450 nm, carotenoid at 400-550 nm and phycobiliproteins of cyanobacteria at 550-650 nm. Sometimes chlorophyll pigments absorb maximum in infra red region of 650-1000 nm. There are several types of chlorophyll molecules present in bacteria called bacteriochlorophyll (BChl). They differ from each other mainly by the presence or absence of a double bond between carbon atoms 3 and 4 and by the substituents on the porphyrin ring. The main absorption maxima of chlorophyll *a* in cyanobacteria is between 680 and 685 nm, whereas bacteriochlorophylls *c, d* and *e* present in green sulphur bacteria show absorption maxima between 715-755 nm. Most of the purple bacteria contains BChl a of 850-890 nm. BChl *b* shows absorption maxima at 1020 to 1035 nm (Table 13.3). They are found in *Rhodopseudomonas viridis, Ectothiorhodospira halochloris* and *Thiocapsa pfeningii*.

Table 13.3 : Absorption maxima (long wavelength) of bacteriochlorophyll molecules.

Pigment	Adsorption maxima (nm)	In methanol
Bacteriochlorophyll *a*	830-890	771
Bacteriochlorophyll *b*	835-850	794
	1020-1040	
Bacteriochlorophyll *c*	744-755	660-66
Bacteriochlorophyll *cs*	750	667
Bacteriochlorophyll *d*	705-740	654
Bacteriochlorophyll *e*	719-726	646
Bacteriochlorophyll *g*	670,788	765

Sometimes, type of binding and position of the BChl molecule in the pigment protein complex in photosynthetic apparatus given rise to different spectral forms to the individual chlorophyll molecule. For instance, BChl *a* in purple bacteria has four spectral froms: BChl 800, BChl 820, BChl 850 and BChl 870-890 (in *Chloroflexus*), however, BChl *a* has two distinct absorption maxima at 808 and 868 nm (Table 13.3).

(i) **Carotenoids :** These accessory pigments (400-500 nm, absorption spectrum) are found phototrophs. They are yellow, green, red-brown in colour and absorb light in blue region. They are insoluble in water and embedded in membrane. The presence of an oxo or aldehyde group can give them a deep colouration. They have long hydrocarbon chains of C_{40} compounds (tetraterpenoids) with tertiary hydroxyl or methoxy groups with alternating C-C and C=C bonds (conjugate bonds system).

Fig. 13.1 : Structure of β- carotene, a carotenoid.

(Fig. 13.1). The carotenoids do not function directly in ATP synthesis however, transfer energy to reaction center. They are photo-protective and quench singlet oxygen. The members of Chromatiaceae bear okenone, while iso-renceratene is present in Chlorobiaceae. Carotenoids are also reported to be present in certain airborne pigmented bacteria and in halophiles (*Halobacter* sp.). They have a defensive role in such bacteria.

(ii) **Bacteriorhodopsin:** These are present in halophiles in which light-mediated synthesis of ATP not involves BChl but low aeration insert a proton into their memberane. Rhodopsin in retinal carotenoid like protein absorb light and transfer H^+ across the membrane. They are purple in colour but change occurs in high aeration to O_2 limiting condition. They absorb light in green region (570 nm) and convert *trans* form to *cis* form during absorption of light hence molecules relax and return to more stable form in the dark following uptake of proton from cytoplasm. A separate light driven pump is present which is called halorhodopsin to pump Cl^- in to the cell as anion for K^+. For detailed description see Chapter 26, *Extremophiles.*

(iii) **Phycobilins:** Phycobilins are water soluble proteins which contain covalently bound linear tertra pyrroles (bilins) as chromophores present inside the granules. They occurs on outer surface of thylakoid membrane of cyanobacteria. Phycobiliproteins are red or blue in colour, contains tetrapyrroles coupled to proteins. Some of the phycobiliproteins are phycocyanin and allophycocyanins. On aggregation, these proteins form a structure called *phycobilisomes* which are attached to photosynthetic membranes.

2. Metabolism in Photosynthetic Bacteria

The green bacteria are strictly aerobic organisms that are obligately photosynthetic. They utilize H_2S, H_2 or thiosulphate as an electron donor (in place of H_2O as in cyanobacteria and algae) and CO_2 as the carbon source.

$$CO_2 + 2H_2S \xrightarrow{\text{Light}} (CH_2O) + H_2O + 2S$$

$$2CO_2 + NA_2S_2O_3 \longrightarrow 3H_2O \xrightarrow{\text{Light}} 2NAHSO_4 + 2 (CH_2O)$$

$$CO_2 + 2H_2 \xrightarrow{\text{Light}} (CH_2O) + H_2O$$

The purple bacteria contain two groups: the purple sulphur bacteria (Chromatiaceae), which utilize H_2S as an electron donor, and the purple nonsulphur bacteria (Ecto-thiorhodaceae), which depend on organic compounds for their metabolism. Lipids containing short chain fatty acids are suitable substrate.

$$CO_2 + 2\ CH_3CHOH\ CH_3 \xrightarrow{\text{Light}} (CH_2O) + H_2O + 2CH_3COCH_3$$

Poly-β hydroxybutyrate $(C_4 H_6 O_2)n$ is the major storage reserve material in such organisms. The photosynthetic bacteria found in deeper water are called meromictic where conditions are anaerobic but light is available. They are helpful in early evolutionary forms of life due to their independence without oxygen.

The bacteriochlorophylls shows absorption spectrum in an acetone-methanol mixture near ultra-red spectrum at 770 nm. Most of the photosynthetic bacteria contain bacteriochlorophyll as given in Table 13.1.

Table 13.4 : Photosynthetic bacteria and their pigments.

Name of the Photosynthetic bacteria	Pigment System
Cyanobacteria	B Chl.a
Green bacteria	B Chl. c or B Chl.d and small amount of B Chl. a
Purple bacteria	B Chl. a or B Chl. b

3. Photosynthetic Electron Transport System (ETS)

In the phototrophic bacteria, cytochrome a and other type of cytochrome oxidase are not present because photosynthesis takes place under anaerobic conditions and there is no need of interaction with molecular O_2. The electron transport system i.e. the mechanism of reduction of NADP to NADPH + H+ called electron transport system in photosynthesis, while the mechanism of production of ATP from ADP and pi with the help of light energy is called photophosphorylation.

The ETS consists of an intermediate electron acceptor (I), a primary electron acceptor (X), secondary electron acceptor (Y) which is generally ubiquinone and b and c types of cytochromes. The nature of I is unknown. All the electron transport carriers are asymmetrically located in the cell membrane, just to set up the hydrogen ion gradient (Fig. 13.2).

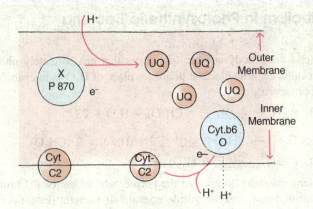

Fig. 13.2 : Asymmetrical location of transport carriers on the outer and inner membrane of photosynthetic bacteria.

(i) Purple Bacteria: Both purple and sulphur bacteria have anoxygenic photosynthesis (i.e. no O_2 evolution occurs during photosynthesis). Recently, the similar phenomena has also been discovered in heliobacteria. The electron transport from NAD or NADP to $NADH_2$ or $NADPH_2$ comprises of electron transport system (ETS). On the other hand production of ATP is called phosphorylation. It occurs in the presence of light, hence it is called photophosphorylation. Both ATP and reducing power are required to reduce CO_2 for carbohydrate synthesis.

In addition, in purple bacteria light energy is trapped in the reaction centre by their surrounding antennas which provide a large surface for capturing the light. The transfer of light energy from antennas to reaction centre (P870) take place in *excitons*. A special BChl a pigment accepts the electrons which later on moves via different electron carrier molecules, bacteriopheophytin (BPh), quinone $_A$, quinone $_B$ and quinone pool. After reduction of these quinone molecules, electron transport occurs slowly to Cyt bc_1, Cyt c_2 and finally to reaction centre. The whole electron transport is cyclic during which proton motive force develops to yield ATP formation. In this process no consumption of electron takes place as found in ATP formation during respiration. The electron comes from reduced sulphur compounds such as H_2S, S^o or thiosulphate, H_2 in case of photolithotrophs and of succinate, malate or butyrate in photoorganotrophs as given below:

$$6CO_2 + 12 H_2S \longrightarrow C_6H_{12}O_6 + 6H_2O + 12S^o$$

The electrons are transferred from reduced carrier to $NADP^+$ so as to give rise to NADPH, involved direct transfer or from more-electro positive quinone to $NADP^+$. In later case, electrons from the quinone pool are forced backward against the electropotential gradient to reduce $NADP^+$ to NADPH. This process is called *reversed electron transport*. In this process, membrane potential is required to utilize the electron donor of high reduction potential such as quinones. Most of the chemolithotrophic bacteria have this phenomenon (Fig. 13.3A)

(ii) Green Bacteria : Among green bacteria electron flow occurs after accepting the electron by P840 of high electropotential (0.5 V). Since it is too strong negative, sometimes primary electron acceptor directly reduce ferredoxin and pyridine nucleotide (NADP). No reverse electron flow is required similar to photosystem I of cyanobacteria. Electron transfer also may occur via iron sulphur protein complex to quinone, Cyt bc_1, Cyt C_{553}. Electrons are finally accepted by reaction centre. The reaction centre is then capable to absorb energy leading to ATP production in cyclic reaction because electrons repeatedly move in a closed circle. Electron flow in green bacteria is given in Fig. 13.3 B.

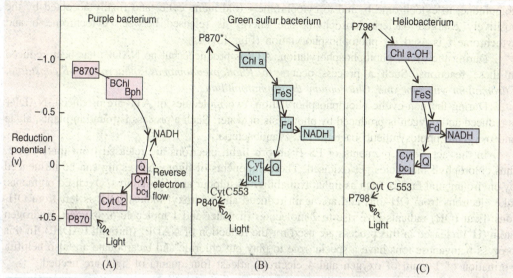

Fig. 13.3 : Electron flow in purple bacteria (A), green bacteria (B) and heliobacteria (C).

(iii) **Heliobacteria:** The reaction centre P798 absorbs the light energy and photosynthetic electron flow occurs via modified form of chlorophyll *a* called hydroxychlorophyll *a* –Fe-S-Q-bc_1 Cyt - Cyt C_{553} to reaction centre which is slightly different from green sulphur bacteria. In both the bacteria NADH production is light-mediated. The primary electron acceptor in such bacteria has reduction potential of -0.5 V. If it is reduced, it is able to reduce NAD^+ directly, hence reverse electron flow does not require for reducing NAD^+ as shown in Fig. 13.3C.

4. Mechanism of Photosynthesis: Cyclic and Noncyclic Photophosphorylation

As stated earlier, the absorption of light is by a pigment molecule which after absorption delivers the energy to electron carriers that can transduce the energy into chemical form. The function of chlorophyll molecule is absorption of photons. The energy contained in the excited pigments is channelled through light-harvesting pigments (also called-antenna molecules) into the reaction centre in which photochemical reactions take place. From this reaction centre, high energy electrons are released. The reaction centre consists of 4 bacteriochlorophyll molecules. Out of four, two bacteriochlorophyll molecules participate in absorption of light and transfer it to an electron. Such molecules are called special pair. The other two bacteriochlorophyll molecules appear to be inactive. After excitation, electrons are released and thus bacteriochlorophyll becomes positively charged.

The electron is now transferred to an iron-containing heme protein called ferredoxin, and here electron moves to

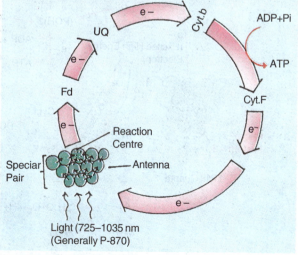

Fig. 13.4 : Cyclic phosphorylation in photosynthetic bacteria.

ubiquinone, to cytochrome b, and to cytochrome f (old name cyt.c) and finally accepted by the positively charged bacteriochlorophyll. The energy is released between cytochrome b and cytochrome f is used for photophosphorylation (Fig. 13.4).

During the cyclic photophosphorylation, ATP is produced but no $NADP^+$ has been reduced in these reactions. Such a process occurs in *Rhodopseudomonas sphaeroides, R.capsulata, Chromatium vinosum* and *Chlorobium thiosuphatophilum.*

During the non-cyclic photophosphorylation, two molecules of ATP are produced, NADP+ is reduced and oxygen is produced by photolysis of water. Such a process is found in plants, algae and oxygenic photosynthetic bacteria i.e. cyanobacteria.

In this act, when pigments of PS II absorb light, electrons are released from this system, thus chlorophyll molecules get oxidised. The components of light - harvesting can be influenced by environmental factors such as light, availability of sulphur and phosphorus. Oxidised pigments, take electrons from OH-ions and are again reduced and hydroxyl radical (OH) is left from OH-ions. Four (OH) radicals give rise to 2 molecules of water and 1 molecule of oxygen. Hydrogen atom (H^+) released in this process are used for the reduction of $NADP^+$ (through $FADH_2$). In this process manganese ions have a special role to play but chloride and bicarbonate are also helpful. Formation of 1 atom of oxygen and 4 electrons atleast four quanta of light are needed.

$$4H_2O \longrightarrow 4H^+ + 4OH^-$$
$$4OH^- \longrightarrow 4(OH) + 4e^-$$
$$4(OH) \longrightarrow 2H_2O + O_2$$
$$\overline{2H_2O \longrightarrow O_2 + 4H^+ + 4e^-}$$

Electrons getting from bacteriochlorophyll molecule of PS II, reduce a substance Q the quencher of PS II whose chemical nature is not known. From reduced 'Q', high energy electrons are transferred to another unknown substance 'B', then to plastoquinone. From reduced plastoquinone (PQH), electrons moved to cytochrome-f which gets reduced. From reduced cytochrome-f, electrons are taken by plastocyanin (PC), which is thus reduced. From reduced plastocyanin, electrons are accepted by oxidised P870. In this way, plastocyanin works as a plug between photoact II and photoact I. Electrons from water are passed on to NADPH.H (Fig. 13.5)

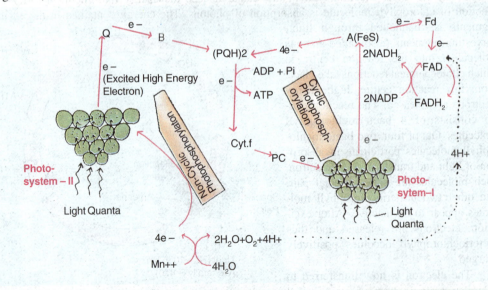

Fig. 13.5 : Non-cyclic and cyclic photophosphorylation in bacterial photosynthesis

In the non-cyclic photophosphorylation, high energy electrons are accepted by PS I, from where electrons are accepted by A (Fe S), these electrons thereafter moved to ferredoxin (Fd), the Fd is oxidised from reduced Fd, the electrons are taken by FAD which gets reduced to $FADH_2$

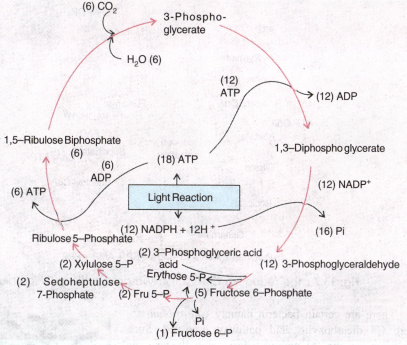

Fig. 13.6 : The reductive pentose-phosphate (Calvin) pathway.

(here hydrogen comes from photolysis of water), electrons and protons (Hydrogen) are accepted by NADP and its get reduced to $NADPH_2$ as shown in Fig. 13.6.

The hydrogen attached to $NADPH_2$ is used for CO_2 reduction in dark reaction (calvin cycle). The $NADPH_2$ is called reducing power of the cell.

It appears that PS II alone can carry out the entire process of non-cyclic photophosphorylation. Thus, the noncyclic reduction of Fd need not require PS I.

5. Carbon Dioxide Assimilation or Dark Reaction (Calvin Benson Cycle)

Similar to chemoautotrophs, photoautotrophs fix carbon dioxide via either the reductive pentose phosphate cycle (calvin cycle) or the reductive C_4 dicarboxylic acid pathway. In the calvin-cycle, the first stable product is phosphoglyceric acid (PGA) while in C_4 cycle, the first stable product is oxaloacetic acid and malate. Both are the primary products of photosynthesis.

In the bacterial photosynthesis, reduction of 1 mol of CO_2 to the oxidation level of carbohydrate involves the oxidation of 2 mol of NADPH and the hydrolysis of 3 mol of ATP. Only two of the enzymes, phosphoribulokinase and ribulosebisphosphate carboxylase are specific to photosynthetic organisms. Six turns of the cycle results in the conversion of 6 mol of CO_2 into 1 molecule hexose (Fig. 13.6).

$$6CO_2 + 6H_2O + 18\ ATP + 12\ NADPH + 12\ H^+ \longrightarrow F\text{-}6\text{-}P + 18\ ADP + 12\ NADP^+ + 17\ Pi$$

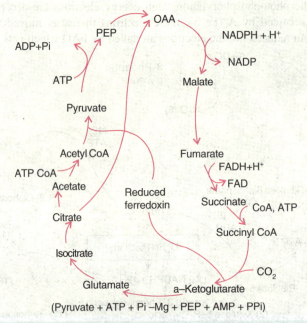

(Pyruvate + ATP + Pi –Mg + PEP + AMP + PPi)

Fig. 13.7 : The reductive C_4 - Carboxylic acid cycle in *Chlorobium* spp.

There are certain bacteria namely *Chlorobium* sp. where C4 dicarboxylic acid pathway involved. Such organisms use the C4 pathway possessing the enzyme pyruvate orthophosphate dikinase which synthesizes phosphoenol pyruvate (PEP).

It is interesting to note that certain bacteria such as *Chlorobium thiosulphatophilum* (green sulphur bacteria) *Chromatium* (purple bacteria), and *Rhodospirillum rubrum* (purple non sulphur bacteria) require Pi in addition to Mg^{2+} and ATP for the formation of PEP from pyruvate indicated that photosynthetic assimilation of CO_2 in bacteria does not require PEP synthase to form PEP from pyruvates as observed in case of green plants. Hence, they require orthophosphate dikinase (Fig. 13.7) in the photosynthetic assimilation of CO_2. The carboxylic acid cycle is essentially a reverse of the TCA cycle.

$$\text{Pyruvate} + \text{ATP} + \text{Pi} \longrightarrow \text{PEP} + \text{AMP} + \text{PPi}$$

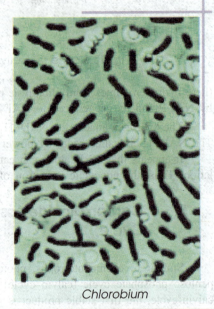

Chlorobium

QUESTIONS

1. What are the major differences between photosynthesis by bacteria and algae ?

2. Explain why photosynthesis is an oxidation-reduction process?

3. Write short answers on the following

   (a) Calvin cycle

   (b) Photophosphorylation

   (c) Bacteriochlorophylls

   (d) Sources of Oxygen

4. Write in brief about the characters of different groups of photosynthetic bacteria.

5. Give a historical account on the bacterial photosynthesis.

6. Write an account on the following

   (*a*) Photosynthetic electron transport system.

   (*b*) Purple sulphur bacteria.

   (*c*) Prochlorophyta.

7. Are there any differences in the mechanism of CO_2 fixation between C_4 bacteria and C_3 bacteria?

8. What is the significance of light in the photosynthetic process of green and purple bacteria? Compare their photosynthetic processes.

9. How the reducing power or ATP is generated in the purple bacterium?

10. What are the major function of Carotenoids and Phycobilins in phototrophic bacteria?

11. Discuss the Electron transport system of any group of photosynthetic microorganisms.

REFERENCES

Bassham, J.A. and B.B. Buchanan 1982. Carbon dioxide fixation pathways in plants and bacteria. In *Photosynthesis*. Vol. II Govindjee (Ed.) Academic, Ny, PP 141-189.

Giovannoni, S .J.S. Turner, G.J. Olsen, S. Barns, D.J. Lane and N.R. Pace 1988. Evolutionary relationships among cyanobacteria and green chloroplasts. *J.Baetcriol* 170 : 51–64.

Jones, C.W. 1982. Bacterial Respiration and photosynthesis,pp.3584-3592 American Society for Microbiology, Washington, D.C.

Kiley, P.J. and S. Kaplan 1988. Molecular genetics of photosynthetic membrane biosynthesis in *Rhodobacter sphaeroides*. *Microbiol Rev.* 52 : 50-69.

Youvan, . D.C. and B.L. Marrs. 1987 Molecular mechanisms of photosynthesis. *Sci Am.* 256 : 42-48.

Zuber, H. 1986 Structure of light - harvesting antenna complexes of photosynthetic bacteria, cyanobacteria and red algae. *Trends Biochem*. Sci 11:414-419.

Nitrogen Fixation

14

A. Symbiotic Nitrogen Fixing System

B. Genetics of N_2 Fixing Microorganisms

C. Nitrogen Fixation Mechanisms

The value of crop rotation in improving the crop field was well known to Greeks who observed that if a legume crop is followed by corn, it gave better yield. It was the work of Lawes and Gilbest in 1891 who showed that legumes had the inherent ability to add nitrogen to the soil. In 1888, Hellriegel and Wilfarth demonstrated that nitrogen gains in peas (*Pisum sativum*) took place only in the presence of soil microorganisms and that the root nodules were necessary to the process. Beijerinck isolated the N_2 fixing bacteria from the root nodules in the same year and named it *Bacillus radicicola*. The leguminosae, third largest family of plants, comprises of about 750 genera containing 16000-19000 species. About 15-16% of the total legumes have been studied for bacteria-root interaction processes. The symbiotic association between legumes and rhizobia is by far the most important contributor to the world's supply of biologically fixed N_2 to agriculture. In 1893, a German botanist, Winogradsky, discovered nitrogen fixation in a free-living heterotrophic bacterium, *Clostridium pasteurianum*.

Winogradsky

The associations formed by leguminous plants in which soil bacteria of the genera *Rhizobium* and *Bradyrhizobuim* colonize roots are the established genera for the Nitrogen fixation process. They are traditionally more popular due to their symbiotic association with agricultural crops. In recent years, however, new symbionts have been discovered and some important genera are given in Table 14.1.

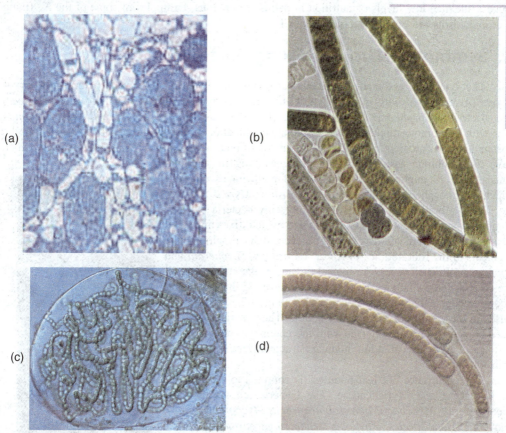

(a)

(b)

(c)

(d)

Photographs of some of the symbiotic and free living micro-organisms. A, cells of root nodule forming species of *Rhizobium*; B, a filament of *Scytonema* sp., C, *Nostoc* sp. (Courtesy: Dr. S.K. Goyal, I.A.R.I., New Delhi); D, growth of *Aulosira* (Courtesy: CST, U.P. (DBT Centre) Lucknow).

In virtually all nodulated legumes the nodules are on the roots, but in the tropical weed, *Sesbania* they occur on the stem as well as on root. The nodulated bacteria in this case is of the genus *Azorhizobium caulinodans*.

The element nitrogen is an essential constituent of all-living things, the proteins and nucleic acids are their major nitrogenous constituents but some other also contain nitrogen atoms, and the gas which composes 80% of our atmospheric N_2, is correctly termed 'dinitrogen'. This dinitrogen is fixed ($N_2 \rightarrow NH_3$) by bacteria; hence these are called "diazotrophs". More important diazotrophs are those which fix N_2 in association with a plant because fixed N is supplied in the plant where it is required.

Nitrogenous fertilizer is made from atmospheric dinitrogen by following reaction which is called 'Haber Process'.

$$N_2 + 3\ H_2 \longrightarrow 2\ NH_3$$

The hydrogen is generated from natural gas and the reaction required high atmospheric pressure (4000 atm.) and relatively high temperature (500°C). Besides, lighting and ultraviolet radiation cause the formation of nitrogen oxides, these gases reached in soil through rain or dew and add total N input in soil. It is clear now to accept that about 60% of the input of world's soil and water nitrogen is today supplied by nitrogen fixing microorganisms. Nitrogen fixation plays a major role in 'global biological productivity'. Hence the possibilities of exploiting this biological process alleviate low levels of nutrition in plants are obvious. Examples of some of the N_2 fixing microorganisms are given in photoplate 14.1.

A. Symbiotic Nitrogen Fixing Systems

The symbiotic association of cyanobacteria with fungi (lichen), cyanobacteria with bryophytes (*Anthoceros*), with pteridophytes, (*Azolla*) with gymnospermes (coralloid root of *Cycas*) and bacteria (*Rhizobium, Bradyrhizobium, Azorhizobium, Sinorhizobium* and *Mesorhizobium etc.*) with leguminous plants are under mutual beneficial relationship (symbiosis) in which both the host and bacteria are benefitted. There are number of other angiosperms (excluding legumes) which have the symbiotic association with nitrogen-fixing microorganisms. About 15 angiospermic plants are of nonlegumes category which fix atmospheric nitrogen, for example, *Alnus, Myrica, Purshia,* etc. The symbiotic microorganisms are not only bacteria but also comprises actinomycetes such as *Frankia* which fixes nitrogen. There are some indications of the existence of haemoglobin like pigments in the root nodules of *Alnus, Elaeganus, Shepherdia* and *Hippophae*. In such cases, the hyphal threads of the endophyte fill the cortical cells which increase in volume resulting into a primary nodule recognizable on the root as a 'swelling'. The lateral roots arises in the vicinity of the primary nodule. Their meristem undergoes branching and get infected with the endophyte results in the formation of a typical adult nodular structure referred to as a 'rhizothamion'.

Frankia nodules

The occurrence of "leaf nodules" is confined to the families of Rubiaceae and Myristicaceae. The bacteria are isolated and identified as *Mycobacterium rubiacearum, Mycoplana rubra, Flavobacterium species, Phyllobacterium rubiacearum* and *Klebsiella rubiacearum*. It is interesting to note that these bacterial isolates do not fix nitrogen.

Several species of *Podocarpus* possess numerous small nodules on the root system. The most common endophyte is a non-septate fungus resembling the fungal component of endotrophic mycorrhizae. This nodulated root system demonstrates that it performs the process of nitrogen fixation very slowly.

1.   Root Nodulating Symbiotic Bacteria

Rhizobium forms nodules and participate in the symbiotic acquisition of nitrogen. The rod shaped bacteria, 0.5 - 0.9 × 1.2 - 3.0 µm long, motile, Gram - negative, non-spore forming, utilize organic acid salts as carbon sources without gas formation; while the cellulose and starch are not utilised. The growth is optimum at 27°C (pH 6.8) and colonies appeared as circular, convex, semitranslucent, raised and mucilagenous, usually 2 - 4 mm in diameter. Production of an acid reaction occurs in mineral salt medium. Some strains of rhizobia and agrobacteria show a close relationship in DNA base composition. All species (except *Agrobacterium radiobacter,* syn. *Rhizobium radiobacter*), incite hypertrophies on plant roots. Nodules are incited by strains of rhizobia on root of leguminous plants and leaves of certain plants in the families Myristicaceae and Rubiaceae by strains of *Phyllobacteria*. The strains of *Rhizobium* are fast - growing, where generation time lasts about 6 h besides showing some other differences with rest of the members of family - Rhizobiaceae.

Some plants bear stem nodules (*Sesbania* species) by *Azorhizobium caulinodans*. The strains bear flagella, hence cells are motile (peritrichous flagella on solid medium but one lateral flagellum in liquid medium). They also fix nitrogen. These are oxidase and catalase positive and can not oxidise mannitol.

The *Bradyrhizobium* strains are slow growers where generation time is about 12 h or more. The motility occurs by one polar or subpolar flagellum. The growth on carbohydrate medium is accompanied by exopolysaccharide (EPS) slime. Some strains can grow chemolithotrophically (utilize inorganic salts) in the presence of H_2, CO_2 and low level of O_2. The bacteroids in root nodules are slightly swollen rods with rare branching or coccus forms. Their main symbiotic partner is soybean, while other bradyrhizobia produce nodules in the plants such as *Lotus, Vigna, Lupinus, Ornithopus, Cicer, Leucaena, Mimosa, Lablab, Acacia* and *Dalbergia*. Now the strains of bradyrhizobia are designated as name of the host plant in parentheses e.g. *Bradyrhizobium (Lotus)* species.

Recently, it has been observed that some rhizobial strains which are fast growers nodulate soybean, (generally, bradyrhizobia nodulate soybean). These fast growers are identified as a separate genus *Sinorhizobium* which are rod shaped, usually contain poly-β-hydroxybutyric acid (PBHA) granules, non-spore forming, Gram-negative, motile, aerobic. Most strains grows at 35°C (pH 6-8). Recently several new species have been added (Table 14.2).

Most of the rhizobia are discovered only in last decade. Hence, it is not surprising if more and more host which bear bacterial nodules, may not contain the traditional strains of *Rhizobium* or *Bradyrhizobium*. *Mesorhizobium*, a new genus of the family Rhizobiaceae has been named on the basis of whole sequence studies of 16s rRNA. Some of the species of *Rhizobium* namely, *R. loti, R. huakii R. ciceri, R. mediterraneum* and *R. tianshanense* now known as *Mesorhizobium*.

Table: 14.1 : Some symbiotic nitrogen fixing bacteria

Recognised genera	Recognised species
Rhizobium	*Rhizobium etli, Rhizobium galegae, Rhizobium gallicum, Rhizobium giardinii, Rhizobium hainanense, Rhizobium huautlense, Rhizobium indigoferae, Rhizobium leguminosarum, Rhizobium loessense, Rhizobium lupini, Rhizobium mongolense, Rhizobium sullae, Rhizobium tropici, Rhizobium undicola, Rhizobium yanglingense.*
Mesorhizobium	*Mesorhizobium amorphae, Mesorhizobium chacoense, Mesorhizobium ciceri, Mesorhizobium huakuii, Mesorhizobium loti, Mesorhizobium mediterraneum, Mesorhizobium plurifarium, Mesorhizobium tianshanense.*
Sinorhizobium	*Sinorhizobium americanus, Sinorhizobium arboris, Sinorhizobium fredii, Sinorhizobium kostiense, Sinorhizobium kummerowiae, Sinorhizobium abri Sinorhizobium medicae, Sinorhizobium indiaense, Sinorhizobium meliloti, Sinorhizobium morelense, Sinorhizobium saheli, Sinorhizobium terangae, Sinorhizobium xinjiangense.*
Bradyrhizobium	*Bradyrhizobium elkanii, Bradyrhizobium japonicum, Bradyrhizobium liaoningense, Bradyrhizobium yuanmingense*
Azorhizobium	*Azorhizobium caulinodans*
Other rhizobia	*Methylobacterium nodulans, Burkholderia tuberum, Burkholderia phymatum, Ralstonia taiwanensis, Devosia neptuniae*

Table: 14.2 : Some free-living nitrogen fixing bacteria

Obligate aerobic	Facultatively anaerobic	Anaerobic
Azotobacter chroococcum	Klebsiella pneumoniae	Clostridium pasteurianum
Azotobacter vinelandii	Bacillus polymyxa	Chlorobium limicola
Azomonas agilis		Chromatium okenii
Achromobacter sp.	Pseudomonas species	Desulfovibrio desulfuricans
Arthrobacter globiformis		Desulftomaculum nigrificans
Azospirillum lipoferum		Methanobacterium formicicum
Beijerinckia indica		Methanosarcina barkeri
Derxia gummosa		Rhodomicrobium vannielii
		Rhodopseudomonas palustris
		Rhodospirillum rubrum
		Rhodobacter capsulants
		Heliobacterium chlorum
		Methylomonas communis

All the rhizobia live freely in soil in the root region of both leguminous and non-leguminous plants. Generally, they can enter into symbiosis only with legumes. If legumes are bigger partner in this process and rhizobia are samller, such a relationship develops which is named as microsymbiosis and strains are called microsymbionts. The nodules become senescent after a period.

Rhizobium root nodules on a bean plant.

Breznak *et al.* (1973) and French *et al.* (1976) demonstrated nitrogen fixing activity in the termite gut. It could be accomplished with the aid of nitrogen fixing bacteria e.g. *Enterobacter* sp. and *Desulfovibrio* species (Kuhnigh *et al.,* 1996). With culture independent methods using oligonucleotide probes specific for nitrogenase (Noda *et al.* 1999), the presence of nitrogen fixing bacteria from different systematic positions has also been demonstrated. Recently, Koustiane *et* al. (2001) have isolated bacterial strains related to *Rhizobium* (Kuhnigk *et al* 1994) which were found to be related to *Rhizobium fredii and R. meliloti.*

2.   Process of Root Nodule Formation

The 'rhizobia' live freely in soil and as soon as they come in contact with suitable host, starts the process of infection. There is an initial contact between the bacteria and host which depends upon recognition. Recent evidences suggests that polysaccharides on the surface of invasive bacteria are involved in binding of these cells to constituents (lectins) on the surface of the roots. The factors or proteins located in the nodules are called nodulins while on bacterial surfaces, named as bacteriociins which help in nodulation. Generally, nodulation starts from the following processes:

(i) **Curling and Deformation of Root Hairs:** Invasion of rhizobia occurs through root hairs. Fine studies of infected root hairs showed the continuation of the wall of the infection thread with the cell wall of the root hair which lend support to the invagination hypothesis. The physiological events leading to infection can be summarised below:

Normal root hair
↓
Exudation of organic substances by roots
↓
Accumulation of 'rhizobia' in the rhizosphere
↓
Conversion of tryptophan to IAA
↓
Root hair curling and deformation

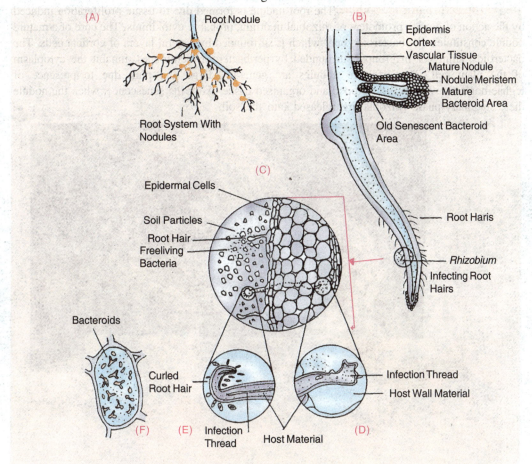

Fig. 14.1: Development of root nodule in a leguminous plant by a species of *Rhizobium*; A, root with nodules; B, a nodule showing different regions C, T.S. of a root showing epidermal hairs and cortical cells with bacteroids; D, a cell showing intracellular infection thread; E, curled root hair infected with bacteria; F, a cortical cell filled with pleomorphic bacteroid.

(ii) Formation of Infection - thread and Formation of Nodule: It is interesting to note that such bindings occurs between compatible (bacteria - host) partners. The tip of curled root hair bends and the bacteria (rhizobial polysaccharide and DNA) penetrate and grow in the form of an infection tube. Meanwhile, the polysaccharides react with a component of root hair cell to form an 'organizer'. The 'organizer' induces the production of polygalacturonase (PG) followed by depolymerisation of cell wall pectin. In such process, incorporation of rhizobia into cell wall occurs which participate in 'intussusception' i.e. taking in of rhizobia by root hair and its conversion into organic tissues. The infection tube or thread branches into the central portions of the nodule, and the bacteria released into their symbiont's cytoplasm multiply. The nucleus of the root hair cell guides the rhizobia.

(***iii***) **Development of Nodule:** Immediately, at the time of release of rhizobia into cytoplasm of the host cortical cells, rapid cell division (called hyperplasia) takes place in the cortical cells. Inside these cells, the bacteria alter their morphology into larger forms called bacteroids. The root cells are stimulated due to this infection to form a tumor like nodule of bacteroid-packed cells (Fig. 14.1). The host cells chromosome number of the area become double. The doubling of the chromosome number occurs in the nodules of polyploids as well as diploid legumes.

(***a***) **Structure of root nodule:** The root nodule is formed due to tissue proliferation induced by the action of growth promoters of rhizobial in origin, probably cytokininis. The core of a mature nodule constitutes the 'bacteroid zone' which is surrounded by several layers of cortical cells. The bacteroids, singly or in groups, surrounded by peribacteroid membranes, inhabit the cytoplasm of the plant cell. The effective nodules are generally large and pink due to presence of leghaemoglobin with well developed and organised tissue. After the senescence, when the nodule dies, stationary-phase rhizobia are released into the soil.

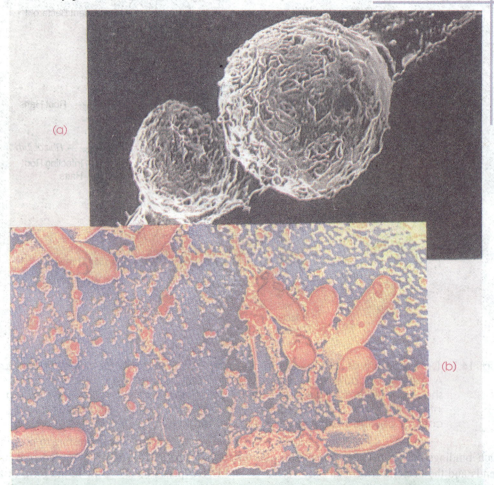

Scanning electron micrograph of root nodules on pea caused by *Rhizobium leguminosarum* (a) ; The orange cylinders are individual *Rhizobium* bacteria on the root hairs of pea (b).

(b) **Function of the nodule 'bacteroids':** The present evidences cited the fact that bacteroids are the sites of nitrogen fixation. The isotopic (^{15}N) studies indicated that bacteroids are the primary sites of nitrogen fixation. Further, in contrast to the free-living rhizobia, the bacteroids are unable to utilise sugar, and secrete ammonium ions which are apparently incorporated into organic compounds by glutamine synthetase present in the surrounding plant cell. This process indicates the involvement of true mutual symbiosis in which role played by leghaemoglobin in the fixation of nitrogen is much more significant. The formation of leghaemoglobin is a specific effect of the symbiosis. The relative capacity of the plant - bacterial association, once established, to assimilate molecular nitrogen is called *'effectiveness'*.

Leghaemoglobin: A red pigment similar to blood haemoglobin is found in the nodules between bacteroids and the membrane envelopes surrounding them. Leghaemoglobin, the prefix 'leg' indicating its presence in legume root nodules, is a haemoprotein having a haeme moiety synthesised by the bacteria attached to a peptide chain which represents the globin part of the molecule, is encoded by a plant gene. The molecular weight of leghaemoglobin is about 16,000-17,000 daltons while the blood haemoglobin has around 66,000 daltons mol wt. The prosthetic group protohaem is synthesized by the bacteroids, while the synthesis of the protein part involves the plant cell. The induction of leghaemoglobin enhances the transport of oxygen at a low partial pressure to the nodules and maintains a steady supply of oxygen at low concentration of the nodule. It is not analysed in cyanobacterial symbiotic system or in other higher plants such as *Frankia* and *Parasponia* which fix nitrogen without leghaemoglobin. The presence of the leghaemoglobin seems to provide full protection against oxygen damage to the N_2 fixing enzymes (Table 14.3).

Table 14.3 : Some nitrogen fixing actinorrhizae and cyanobacteria

A. Actinomycetes :
  (i) *Frankia* and *Alnus, Myrica,*
      Casuarina, Ceanothus, Elaeganus, Hippophae
      Purshia, Dryas, Coriaria, Shepherdia,
      Cercocarpus, Discaria, Arctostaphylos

B. Cyanobacteria:
   (i) *Anabaena cylindrica, A. inaequalis*
  (ii) *Plectonema boryanum*
 (iii) *Gloeocapsa*
  (iv) *Nostoc muscorum*
   (v) *Lyngbya oestuarii*
  (vi) *Gloeothece alpicola*

(c) **Free-living nitrogen fixing bacteria:** There are 54 genera comprising in 22 families as well as 3 thermophilic archaeobacteria which fix nitrogen in free-living form. Winogradsky in 1893 demonstrated biological nitrogen fixation in *Clostridium pasteurianum,* and in 1901 Beijerinck detected in *Azotobacter*. About 30 genera of cyanobacteria representing 10 families have been shown to fix atmospheric nitrogen. Cyanobacteria are heterocystous, filamentous, nonheterocystous filamentous, unicellular, reproducing by binary fission or budding, and by multiple fission.

Other N_2 fixing symbionts

   (i) Lichens (Cyanobacteria - fungus association)
  (ii) Liverworts (*Anabaena - Anthoceros*)
 (iii) Pteridophytes (*Anabaena azollae - Azolla*)
  (iv) Gymnosperms (*Nostoc - Cycads*)
   (v) Angiosperms (*Nostoc - Gunnera*)
  (vi) Termites (with Citrobacters)
 (vii) Human intestine (Klebsiellae)

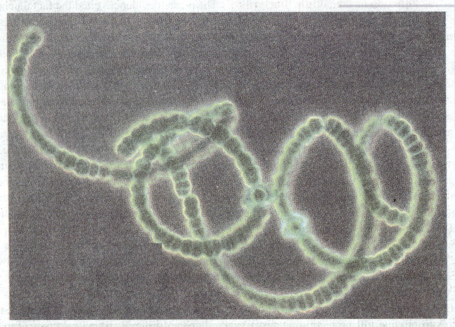

The filamentus cyanobacterium *Anabaena*.

5. Metabolism of Nitrogen Fixation (The Oxygen Problem)

Nitrogenase enzyme, which is chiefly associated with fixing of nitrogen is inactivated by oxygen. This inactivation may be reversible or irreversible, the latter clearly being a more serious problem. How it arose is an open question? One suggestion is that oxygen sensitivity results from evolution of nitrogenase in days before oxygen was present in atmosphere. An alternative proposal is that in order to reduce such a stable substrate as nitrogen gas, nitrogenase is invariably able to pass electrons to stronger oxidizing agents of comparable size, such as oxygen. Thus nitrogenase-oxygen complexes would be expected and if stable, these would prevent the reduction of nitrogen. Reversible inactivation may be a way of regulating its activity, oxygen may also regulate nitrogenase synthesis. Each and every type of nitrogen fixing organism, therefore, has some means to counter the effects of oxygen.

(*i*) **Anaerobic Microorganisms:** There are two types of anaerobic nitrogen fixers: photosynthetic and nonphotosynthetic. In non-photosynthetic, ATP for nitrogen fixation is provided by substrate level phosphorylation. Certain species like *K. pneumoniae* fix nitrogen better when grown microaerophilically. This is because it can use glucose. Low oxygen coupled with low combined nitrogen may stimulate nitrogenase synthesis.

In *Rhodospirillum rubrum* photosynthesis does not involve oxygen evolution but generates ATP. Nitrogenase activity is switched off in dark and switched on again in light. This is due to reversible modification of dinitrogenase reductase enzyme. Inactive form of dinitrogenase reductase has an ADP-ribosyl group specifically attached to it. In the presence of light this group is removed by an enzyme, DRAG (dinitrogenase reductase activating glycohydrolase), and in dark it is replaced using NAD and Mg-ADP using enzyme DRAT (dinitrogenase reductase ADP-ribosyl transferase). DRAT is controlled by ammonium and glutamine, hence it ensures that nitrogenase is switched off when combined nitrogen is available. This phenomenon also occurs in *Azotobacter*,

Rhizobium, K. pneumoniae and *Methylococcus capsulatus*. Therefore, photosynthetic bacteria do not have problem of internally generated oxygen and they may have to only guard against low levels of oxygen from outside. Reversible inhibition of nitrogenase by oxygen can also be influenced in *R. rubrum* by DRAG/DRAT system.

(*ii*) **Cyanobacteria:** These microorganisms carry out higher level of photosynthesis i.e. generate oxygen in light. They have also to face the problem of externally supplied oxygen. These organisms solve the problem of oxygen inactivation of nitrogenase by separating photosynthesis and nitrogen fixation in space or time.

Heterocystous cyanobacteria carry out nitrogen fixation in heterocysts and photosynthesis in vegetative cells. These organisms can fix nitrogen in light as well as dark. In dark it requires oxygen and ATP for nitrogen fixation generated by respiratory processes, and in light ATP is generated in heterocysts which contains only part of photosynthetic apparatus. Heterocysts lack both O_2 evolution and CO_2 fixation but retain the ability to generate ATP and reductant. Heterocysts are connected to vegetative cells by large pores which permit a photosynthate disaccharide to heterocysts and glutamine out of heterocysts. Sugars passing in, help to generate NADPH by pentose phosphate pathway or scavange oxygen via respiratory processes. Heterocyst wall allows sufficient nitrogen in but restrict oxygen diffusion to a level which can be removed by respiration.

Non-heterocystous cyonobacteria separate photosynthesis and nitrogen fixation in time (day/night respectively). Aerobic organisms can fix nitrogen in light and dark, but when grown in alternating light/dark cycles, fix 95% nitrogen during dark. Thus, it separates photosynthesis from nitrogen fixation. Atmospheric oxygen is used to generate ATP by respiration. In certain cyanobacteria control of nitrogenase is at transcriptional level acting like an endogenous rhythm.

(*iii*) **Free Living Aerobic Microorganisms:** *Azotobacter* is the most studied and best example of free living aerobic nitrogen fixers. It produces another FeS protein which complexes with two nitrogenase proteins to form a three membered oxygen stable and inactive complex. This is a short term response to increased pO_2. The overall system represents a series of modifications of respiration e.g. cytochrome O increases as pO_2 decreases and coupled with an increase in phosphorylating efficiency. There is, therefore, a branched pathway of respiration for different conditions (Fig. 14.2). Under low pO_2 conditions, electron flow from NADPH via cytochrome *d* to oxygen, by passing phosphorylation. Under high pO_2, the pathway from malate and NADPH is followed. Thus at high pO_2, oxygen is taken up rapidly and inactivated.

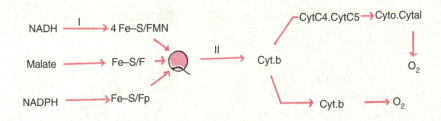

Fig. 14.2 : Electron transport pathways in *Azotobacter* (after Haddock and Jones, 1977).

Under low pO_2, oxygen may limit phosphorylation. Carbon compounds and reducing equivalents then tends to accumulate. This leads that acetyl CoA, instead of entering tricarboxylic acid cycle is used with NADPH for synthesis of poly β hydroxybutyrate, a major reserve product.

Colony morphology is also altered similarly to changes in pO_2 e.g. *Azospirillum* when grows without combined nitrogen on semisolid agar can move to that part of culture which has a suitable pO_2 for nitrogenase activity.

Association with an oxygen consuming organism is a natural way of finding a niche with required level of oxygen. *Azospirillum* sp. do this by colonizing with a wide variety of plant roots. Therefore, cyanobacteria show a number of associations with wide variety of organisms e.g. *R. capsulata* fixes nitrogen under aerobic conditions with *Bacillus megaterium*. *Chloropseudomonas ethylicum* now appears to be a mixture of atleast two species, one being aerobic and non-nitrogen fixer and other nonaerobic but nitrogen fixer.

(iv) **Symbiotic Microorganisms :** These microorganisms fix nitrogen by formation of nodules in the roots of the symbiotic plant. Nodules have a variable diffusion resistance to oxygen, coupled with leghaemoglobin (a haemoglobin like pigment found in nodules) provide a very good system for tailoring oxygen demand to supply rhizobia; best example of these type of organisms can also develop branched respiratory pathways and multiple forms of cytochromes linked to varying oxygen sensitivity.

Control of oxygen diffusion is additionally associated with intercellular space system which may vary with species and/or environment and through the development of special walls in nitrogenase containing vesicles.

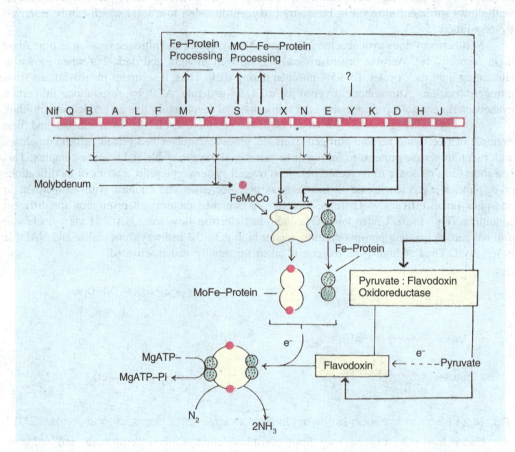

Fig. 14.3 : Regulation N_2 fixation in *Klebsiella pneumoniae* by *nif* gene products (based on Gallon and Chaplin, 1987).

B. Genetics of N₂ Fixing Microorganisms

1. Bacterial Nodulation Genes and Regulation of *nod* Gene Expression

In *Rhizobium* species, nodulation genes together with other symbiotic genes, are located on large-plasmids (Sym plasmids). Sym plasmids vary from 50 to over 600 kb in *R. leguminosarum* bv. *trifoli* to 1200 to 1500 kb in *R. meliloti*. Nodulation genes, *nod* and *nol* genes, are classified as regulatory, common and host specific. Regulation of *nod* genes is controlled by the *nod*D gene, of which all rhizobia tested so far contain one or more copies. In conjunction with plant flavonoids or other phenolic compounds, *nod*D proteins act as transcriptional activators of inducible *nod* genes. Different *nod*D proteins respond to specific plant signal molecules and, therefore, contribute to host-specificity of nodulation. Most *nod*D genes are constitutively expressed, other are autoregulated.

The common *nod*ABC genes are structurally and functionally conserved among *Rhizobuim, Bradyrhizobuim* and *Azorhizobium* strains. The inactivation of these genes completely abolishes root-hair infection and nodule formation. Putative function of nodulation gene is given in Table 14.4.

Table 14.4 : Possible functions of rhizobial Nod proteins (Based on Michiels and Vanderleyden, 1995)

Protein	Homologies
NodC	Chitin and cellulose synthases
NodD	LysR family of DNA-binding proteins
NodE	β- ketoacyl synthases
NodF	Acyl carrier proteins
NodH	Sulphotransferases
NodL	Acetyltransferases
NodM	D-Glucosamine synthase
NodP	ATP sulphurylase
NodQ	APS kinase

The *nif* gene products of *K. pneumoniae* and their function are given in Table 14.5

Table 14.5 : *Nif* gene products and their function in *Klebsiella pneumoniae*

Genes	Function
Q	Not known
B	Synthesis or processing of FeMoCo
A	Regulatory-*nif* A product activates the other operons
L	Regulatory
F	A flavoprotein involved in electron transfer in nitrogenase
M	Processing of Fe-protein
V	Influences specificity of MoFe-protein
S	Not known
U	Not known
X	Not defined genetically; presence deduced from physical map and cloning of *nif* DNA
N	Like B
E	Like B
Y	Discovered in the same way as X
K	Codes for B-subunit of MoFe-protein
D	Codes for subunit of MoFe-protein
H	Codes for subunit of Fe-protein
J	May be involved in electron transfer to nitrogenase

2. Nif Genes and their Regulation

(i) *K. pneumoniae:* The N_2 fixation (*nif*) genes are organized into a regulon of 17 genes, consisting of seven or eight operons each of which is transcribed into a single, usually polycistronic mRNA. Although only five of the gene products have been purified and properly characterized, functions have been assigned to all of the genes except for *nif*X and *nif*Y (Fig. 14.3).

Regulation of *nif* gene expression has two elements, an external system designated *ntr* and an internal system mediated by *nif* A and *nif* L. The *ntr* system responds to conditions of nitrogen starvation by activating genes that enable the organism to utilize 'unusual' nitrogen sources such as arginine, proline, and histidine as well as N_2 itself, in the last case by switching on the *nif* genes. The inter-relationships between external and internal regulation of the *nif* genes in *K. pneumoniae* and the conditions under which nitrogenase synthesis occurs, are summarised in Fig. 14.4.

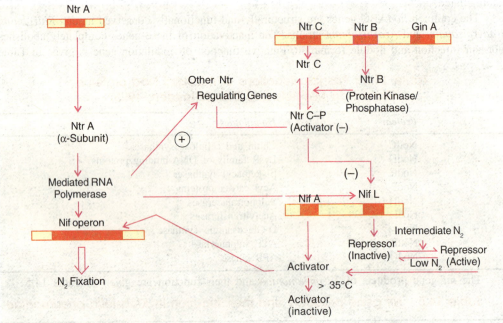

Fig. 14.4 : Regulation of *nif* gene expression in *Klebsiella pneumoniae* (based on Heselkorn, 1986)

Actually, the *ntr*A gene product (NtrA) is a-factor of RNA polymerase which recognizes the *nif* and, other *ntr* - regulated genes. These promoters have a structure different from that of typical bacterial promoters. *Ntr*A allows RNA polymerase to bind at the *nif* promoters and to initiate transcription there. The *ntr*B gene product (NtrB) is an enzyme that functions both as a protein kinase and as a phosphatase, the substrate of which is NtrC (the *ntr*C gene product). Whether kinase or phosphatase activity predominates depends upon the nitrogen status of the bacterium, and the consequence of this is that, under condition of starvation, NtrC-P acts as an activator of, among other operons, *nif* L and *nif* A. The *nif*A product is an activator of transcription of other *nif* genes, whilst the *nif* L product, in the presence of either intermediate concentrations of fixed nitrogen or O_2, inactivate the *nif*A product, thereby preventing transcription of other *nif* genes.

(ii) In Cyanobacteria: In heterocystous cyanobacteria, the acquisition of nitrogenase activity in response to nitrogen deficiency is accompanied by the differentiation of vegetative cells into a specialised structure called 'heterocysts'. This process has been studied by Haselkorn (1986).

All non-heterocystous cyanobacteria possess the genes *nif* H, *nif* D, *nif* K as a cluster (Kallas *et al.* 1985) which is similar to *K.pneumoniae*. In the DNA of vegetative cells of heterocystous cyanobacteria the gene *nif* K is separated from the genes *nif* D and *nif* H as observed by Haselkorn *et al.* (1986). During the differentiation the intervening DNA of about 11000 base pairs (11 kb) is excised as a circle resulting in a clustered *nif* HDK operon as studied in *Anabaena* PCC 7120. This excision is catalysed by the product of a gene, *xis*A, located within the excised 11 kb region as shown in Fig. 14.5.

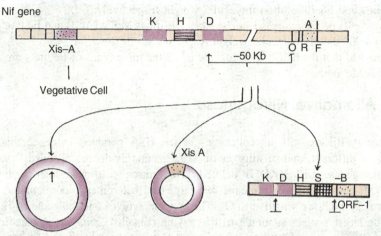

Fig. 14.5 : Rearrangement of *nif* genes in *Anabaena* PCC 7120 during heterocyst differentiation (based on Heselkorn, 1986 and Heselkorn *et al.*, 1986)

A second rearrangement has been observed during heterocyst differentiation by Heselkorn *et al.* (1986). This occurs, in the region of *nif* S, a gene involved in *K. pneumoniae*. In this arrangement, a segment of DNA of approximately 50 kb is excised, again as a circle, and after this rearrangement an operon with the structure *nif* B: ORF -1: *nif* S: ORF-2 is formed as shown in Fig. 14.5. The excision occurs between ORF-1 and *nif* S, which is not catalysed by the *xis* A product. It is interesting to note that no expression of the *nif* H gene from *Anabaena variabilis* was found when it was inserted with its promoter into *K. pneumoniae* which gave the evidence that *nif* gene promoters in *Anabaena* PCC 7120 differ in structure from both *E. coli* promoters and the *nif* promoters of *K. pneumoniae*.

C. Nitrogen Fixation Mechanisms

1. Nitrogenase Types, Structure and Function

The biological conversion of atmospheric nitrogen to ammonia taken place with the help of an enzyme called Nitrogenase. This enzyme is anaerobic in nature and when it comes in contact with oxygen or air, it becomes inert. The proteinaceous enzyme is made up of 2 subunits mainly called larger MoFe protein (2,20,000 dalton mol. wt) and another smaller Fe protein (55000 dalton mol. wt.). Postgate, a British microbiologist named them as Kp_1 and Kp_2. As stated it contains iron and molbydenum atoms. They need Mg^{2+} ions for activation and can convert ATP to ADP during functioning. It is inhibited by ADP and also reduces several other substrates with triple bonded molecules (similar to N=N). The enzyme can reduce hydrogen ions to gaseous hydrogen, even when N_2 is present, and also have an ability to reduce acetylene to ethylene.

The nitrogenase purified from three species of bacteria showed its following nature as given below. This indicates that nature of enzyme varies in different nitrogen fixing organisms as far as the size of proteins (in both MoFe and Fe components) is concerned.

(*i*) **The Cp Type:** According to Postgate, the properties of the nitrogenase are the Cp type (*Clostridium pasteurianum*): It has MoFe (Cp_1) and Fe (Cp_2) proteins which have 2,20000 dalton and 55,000 dalton molecular weight, respectively. The half-life of enzyme is quite short.

(*ii*) **The Kp Type (*Klebsiella pneumoniae*):** It has MoFe (Kp_1) and Fe (Kp_2) which have 2,18000 daltons and 66,700 dalton molecular weight respectively.

(*iii*) **The Ac Type (*Azotobacter chroococcum*):** It has MoFe (Ac_1) and Fe (Ac_2) which have 2,27000 and 64,000 dalton molecular weight respectively.

It is observed that the half-life of enzyme of all the three types of Fe units are much shorter than that of FeMo units.

2.   Alternative Nitrogenase

Professor P. Bishop and his colleagues in the USA obtained evidence that *Azotobacter vinelandii* has a different kind of nitrogenase. The genetic evidences revealed that their normal genes for nitrogenase (*nif* YKDH) deleted. This nitrogenase was later isolated from *A. chroococcum*. It consists of two proteins, one large and heteromeric, one smaller very like the regular Fe-protein; both are sensitive to O_2. The enzyme evolves one molecule of H_2 and reduces acetylene. The larger protein subunit part of enzyme contains *vanadium* in place of Mo ion of the conventional system in this 'new' nitrogenase. V-nitrogenase repressed by Mo suggests that it provides *Azotobacter* with a physiological 'back-up' nitrogen-fixing system for use in case of lack of Mo.

3.   Substrates for Nitrogenase

For enzyme activity a suitable substrate is required so as to bind all the active sites of enzyme to get a product. The overall reaction in the enzymic reduction of atmospheric nitrogen to ammonia could be postulated as follow:

$$N_2 \ (N{\equiv}N) \quad \rightarrow \quad HN{=}NH \quad \rightarrow \quad H_2N{-}NH_2 \quad \rightarrow \quad H_3N + NH_3$$

Nitrogen   Di-imide   Hydrazine   Ammonia

It is interesting to note that cell free extract of *Azotobacter* and *Clostridium* converted nitrogen in the same way as the free-living bacterial cells. This finally led to the initial isolation and purification of the enzyme from *C. pasteurianum* and *A.chroococcum*. Further, the enzyme was responsible for the adsorption and reduction of N_2 gas.

Although there are several substrates of nitrogenase but except H_2 most of the substrates are nonphysiological substrates because the inhibitors are actually reduced by nitrogenase. Besides this, acetylene (HC=CH) is an important and one of the reliable substrates for measuring the enzyme activity by 'acetylene-reduction' test. It is also important to note that most of the substrates have triple bond in their molecules similar to nitrogen (N≡N).

Following are the substrates reactive to nitrogenase:

Substrate	Products(s)
N_2	NH_3
N_2O	H_2O, N_2
N_3	NH_3, N_2
C_2H_2	C_2H_4
HCN	CH_4, NH_3, CH_3, NH_2
CH_3CN	C_2H_6, NH_3

In the reduction of acetylene, the product formed is ethylene i.e. $HC \equiv CH \rightarrow H_2C\text{-}CH_2$ which requires two electrons, whereas reduction of nitrogen to ammonia requires six electrons. Nitrogen is readily reduced in comparison to rest of the substrates.

If there is N_2 fixation, simultaneously H_2 is also produced by some of the nitrogen fixing-microorganisms. N_2 fixation is correlated with each other. Nitrogenase reduced the H^+ ion and formed H-D in the presence of deuterium. The H-D reaction suggests the involvement of a bound di-imide intermediate in N_2 fixation.

In view of H_2 evolution during nitrogen fixation, the following reaction has been suggested:

$$N_2 + 8\ H^+ + 8e^- \longrightarrow 2NH_3 + H_2 + 16\ ADP + 16\ Pi$$

The energy for nitrogenase reaction comes from the cellular metabolic cycles in the form of ATP. This is met out by photophosphorylation, oxidative phosphorylation or phosphoroclastic dissimilation. In later process, keto-acid is dissimilated to acetyl phosphate, CO_2 and H_2.

$$
\begin{array}{c}
CH_3 \\
| \\
C = O \\
| \\
COOH
\end{array}
+ \ CoA \xrightarrow[\text{dehydrogenase}]{\text{Pyurvate}} \underset{\text{Acetyl phosphate}}{CH_3CO - SCoA} + CO_2 + H_2
$$

Pyruvate functions both as electron donor and as energy source. In the phosphoroclastic reactions, pyruvate forms acetyl phosphate which in the presence of ADP gives rise to ATP. The reductants are the naturally occurring electron carrier proteins called *ferredoxin* and *flavodoxin*. Dithionite ($Na_2S_2O_4$) and certain dyes such as methyl viologen and benzyl viologen can also serve as artificial extracellular sources of electron donors. This enzyme system catalyzes the transfer of electrons from pyruvate or hydrogen to ferredoxin or flavodoxin.

(*a*) **Ferredoxins or dinitrogenase:** Ferredoxins are electron carrier, discovered by Mortenson and Caruahan in the year 1962 from *C. pasteurianum*. It is naturally occurring e^- carrier iron-sulphur (Fe-S) protein (reversible). It has now been isolated from number of cyanobacteria, photosynthetic bacteria and even from higher plants. The ferredoxins are involved in various physiological processes such as photosynthesis in plants and pyruvate metabolism in anaerobic bacteria. The ferredoxin involved in N_2 fixation contains one cluster of four iron and four sulphur atoms in the molecule. Many similar iron-sulphur clusters are the part of iron atoms of nitrogenase. The whole cluster behaves as oxido-reductive unit. The electron paramagnetic resonance indicates that the ferredoxins of aerobic nitrogen fixing bacteria such as *Azotobacter* behave slightly differently from those of anaerobes such as *Clostridium pasteurianum*. In *C. pasteurianum*, ferredoxin is the actual protein which reacts with nitrogenase and provides the reducing power for the conversion of N_2 to NH_3.

(*b*) **Flavodoxins or dinitrogen reductase:** The bacteria grow under limited iron supply i.e. nutritional stress condition and produce flavodoxin. It was also isolated in the beginning from *C.pasteurianum*. It is interesting that it was found to replace ferredoxin as an electron carrier in a large number of reactions. An electron carrier named 'azotoflavin' has been isolated from *Azotobacter vinelandii* possessing biological activity similar to ferredoxins. In *K.pneumoniae* and *A.chroococcum*, flavodoxins are the primary reductant to nitrogenase. *A. chroococcum* flavodoxin is blue when half-reduced and colourless when fully reduced. The flavodoxin reduced form is quinone and semiquinone form. This form is unusually stable to oxidation by air, which may be why this protein rather than a ferredoxin is particularly suitable to *Azotobacter's* aerobic way of life. Flavodoxins do not contain iron-atoms; their oxido-reducible centre is yellow, fluorescent molecule called a 'flavin'.

Role of Pyruvate and Ferredoxin-Nitrogenase Reaction. The Fig.14.6 shows the active site of the enzyme for substrate reduction. This enzyme is believed to be composed of a Mo-Fe dinuclear site bridged by sulphur having the proper size and electron characteristics to provide

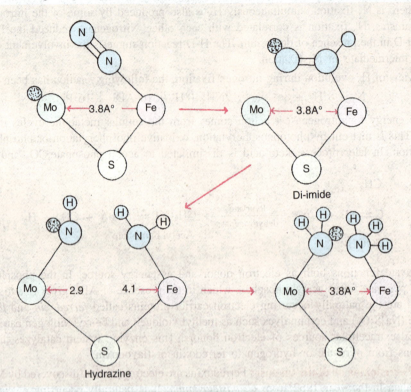

Fig. 14.6 : Proposed intermediates and dinuclear active sites for N_2 reduction by nitrogenase.

Mo-Fe distance of 3.8 Å. This distance is specific so as to accomodate various nitrogenase substrates including nitrogen and to exclude others. The first reaction in nitrogen reduction is the formation of a linear complex of nitrogen with the Fe atom of nitrogenase. This is followed by transfer of electrons from Mo which is the end point of the electron activating system resulting in the formation of di-imide which is stabilized by hydrogen bonding from the protein as well as the metal nitrogen bonds. Successive addition of electrons produces hydrazine followed by cleavage of N—N bond to fill two molecules of NH_3. The increase in the N—N bond length during reduction is accompanied by compensating changes in the MNN angles so that Mo—Fe distance remains constant.

3.   Electron Proteins

MoFe (kp_1) proteins plays a key role in nitrogen fixation (substrate binding and reduction) with Fe protein (kp_2) that assists in transfer of electron from flavodoxin to the bigger subunits and ATP consumption. Neither of the two proteins subunits can function independently (Fig. 14.7). Fe proteins has four Fe centres and equal number of sulphur, whereas the number of Fe centres in MoFe protein varies. It is 22-24 in *C.pasturianum*. An equivalent number of inorganic sulphur is also present. In contrast to Fe protein, MoFe protein possesses two additional Mo atoms. Mo is suggested to play a vital role in H_2 evolution which accompanies N_2 fixation.

N_2 reduction process starts with the transfer of single electron (2 ATP molecules) from flavodoxin to smaller Fe protein subunit Kp_2. At this stage two ATP molecules combine with two Mg^{2+} ions to form complex, Mg ATP which attaches to Fe protein and energizes it to transfer electron from the iron atoms in Fe protein to MoFe protein, prior reaching the bound substrate. The MoFe protein becomes unstable and to counter balance this electron a H^+ ion formed by dissociation of H_2O comes to attach at Mo atom of MoFe protein. The electron flow from flavodoxin to MoFe protein via Fe protein is a repeated process. Each time an electron is transferred, there is a consumption of two molecules of ATP. The second electron is again balanced by another H^+ ion. When the third electron is transferred to MoFe protein, the H^+ ions are displaced by nitrogen leading to the evolution of one molecule of hydrogen. This third electron again balances by attachment of one H^+ ion to nitrogen i.e. HN=NH.

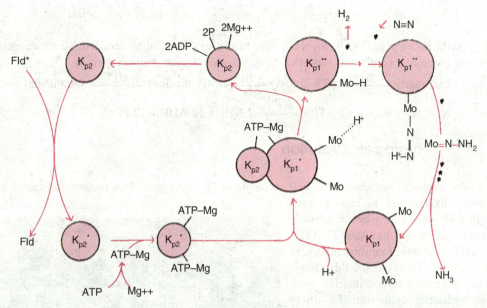

Fig. 14.7 : Current scheme for nitrogen fixation as shown by John Postgate (1982) in *Klebsiella pneumoniae*; Fld, flavodoxin; Kp_1, bigger protein; Kp_2, smaller protein; ●, electrons.

This process of electron transfer is continued till net 8 electrons transfer thereby reducing N_2 to NH_3. The series of reactions involved by electron transfer is depicted in Fig. 14.7.

H^+ ——————— Counter balance 1st e^- (electron)

H^+ ——————— Counter balance 2nd e^- (electron)

$H^+ + H^+$ ———→ H_2 evolved

H^+ .... N=N .... H^+ ←——— N≡N
(3rd e^-)   (4th e^-)

H^+ ...... HN. —— .NH ...... H^+
(5th e^-)   (6th e^-)

H^+ ...... H_2N. —— .NH_2 .... H^+
(7th e^-)   (8th e^-)

$NH_3 + NH_3$

i.e. $2NH_3$

Thus, net 8 e⁻ and 8H⁺ are involved in reducing N_2 to two molecules of ammonia. One electron utilizes two moles of ATP. Hence, net 16 moles of ATP are used. The H_2 evolution is mediated by transfer of net two electrons, thereby leading to the loss of four ATP molecules. It is obvious that the process of nitrogen reduction cousumes 12 ATP and the complete process of biological nitrogen fixation is still more expensive.

The enzyme reaction should be written formally as:

$$N_2 + 8H^+ + 16 \text{ ATP} + 8 \text{ e}^- \longrightarrow 2 \text{ NH}_3 + H_2 + 16 \text{ ADP}$$

Nitrogen fixation is a reductive process where N_2 is reduced to give NH_3, an inorganic product. Such studies have been confirmed by autoradiography (use of $^{15}N_2$). On the otherhand, the first organic product formed is glutamic acid. On the basis of oxidation number, following scale is proposed:

$$\underset{+5}{NO_3} \longrightarrow \underset{+3}{NO_2} \longrightarrow \underset{+1}{N_2O_2} \longrightarrow \underset{\text{Zero}}{N_2} \longrightarrow \underset{-1}{NH_2OH} \longrightarrow \underset{-3}{NH_3}$$

On the above scale, NO_3 is highly oxidized and ammonia is highly reduced. Therefore, during nitrogen fixation, a continuous reduction ranges from oxidation number zero to - 3.

$$\text{Dinitrogen} \xrightarrow{\frac{2H^+}{2e^-}} \text{(Di-imide)} \xrightarrow{\frac{2H^+}{2e^-}} \text{(Hydrazine)} \xrightarrow{\frac{2H^+}{2e^-}} \text{(Ammonia)}$$

$$\text{or, } N_2 + 6e^- + 6H^+ + 12 \text{ ATP} \xrightarrow{Mg^{2+}} 2 \text{ NH}_3 + 12 \text{ ADP} + 12 \text{ Pi}$$

4. Hydrogen Evolution

For every N_2 molecule fixed, one molecule of H_2 is evolved. This process is expensive to nitrogen fixation due to involvement of 16 ATP molecules, whereas similar fixation can be had by using 12 ATP molecules. It means four molecules are just wasting. To overcome this process, the enzyme responsible for normal hydrogen evolution (i.e. that formed in the normal metabolism of these bacteria, not via nitrogenase) is called hydrogenase, and it can cataylse both the uptake and evolution of hydrogen. The hydrogen is trapped by hydrogenase in *Azotobacter* and recycled. The use of recycled hydrogen to generate some extra ATP may contribute to the unusually efficient ATP economy of *Azotobacter*. Some rhizobia also show similar reaction besides cyanobacteria and photosynthetic bacteria which show photoevolution of hydrogen. Non-H_2-evolving called 'tight' symbiosis fixes more nitrogen

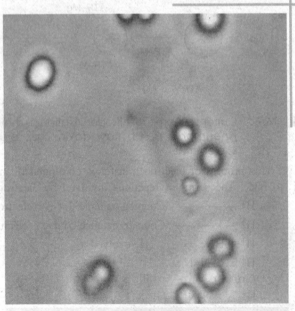

Azotobacter.

per unit of solar energy than 'loose' (H_2-evolving) one. The non-leguminous associations have been "tighter" than in most of the leguminous associations.

ֿ QUESTIONS

1. Write an historical account about nitrogen fixation. Mention in brief about the different group of microorgamisns fixing nitrogen?

2. Define nitrogen fixation and its process in non-heterocystous cyanobacteria?

3. What are the properties of nitrogenase? Write in detail about the types, structure and its function.

4. Give an illustrated account of process of infection, nodule formation in symbiotic system.

5. Write in detail about the mechanism of nitrogen fixation.

6. What are carrier mediated protein? Write in detail about role played by them in nitrogen fixation.

7. Discuss in detail about metabolism of nitrogen fixation in different group of microorganisms.

8. Write an account on *nod* and *nif* genes. Give a brief note about genetics of nitrogen fixation in diazotrophs.

9. Write critically the differences between genetics of nitrogen fixation in diazotrophs with that of cyanobacteria.

10. Write short notes on the following

   (*a*) Microsymbiont   (*b*) Stem nodulating Rhizobia   (*c*) Leghaemoglobin
   (*d*) Ferredoxin   (*e*) Flavodoxin   (*f*) nif operon
   (*g*) *Hup* gene   (*h*) *nod* gene   (*i*) Nitrogenase
   (*j*) Free-living N$_2$ fixer   (*k*) Hydrogen evolution   (*l*) Bacteroids.

REFERENCES

Breznak, J.A., Brill, W.J., Mertins, J.W. and Coppel, H.C. (1973). Nitrogen fixation in termites. *Nature* 244, 577-580

French, J.R.J., Turner, G.L. & Bradbury, J.F. (1976). Nitrogen fixation by bacteria from the hindgut of termites. *J. Gen. Microbiol.* 95, 202-206.

Gallon, J.R. and Chaplin, A.E. (1987). *An introduction to nitrogen fixation.* Cassell, Eastboume, Sussex.

Kallas, T., Coursni, T,. and Rippka, R. (1985). *Plant Mol.* Biol 5,321

Kuhnigk, T., Brake, J., Krekeler, D., Cypionka, H. and Konig. H. (1996). A feasible role of sulphate-reducing bacteria in the termit gut. *Syst. Appl. Microbiol.* 11:139-149.

Haselkorn, R. (1986). *Ann. Rev. Microbiol.* 40,525

Haselkorn, R. Golden, J.W., Lammers, P.J., and Mulhgan, M.E. (1986). *Trends Genet.* 2,255.

Michiels, J. and J. Vanderleyden (1995). Molecular basis of the establishment and functioning of a N$_2$. fixing root nodule. *World J. Microbiol. Biotech.* 10,612-630.

Noda, S., Ohkuma, M., Usmani, R., Horikoshi, K. and Kudo, T. (1999). Culture independent characterization of a gene responsible for nitrogen fixation in the symbiotic microbial community in the gut of the termit *Neotermus koshunensis. Appl. Env. Microbiol.* 65: 4935-4962.

Viruses-I
(Plant and Animal Viruses, Viroids and Prions)

15

A. General Concept
B. Plant Viruses
C. Animal Viruses
D. Viroids
E. Virusoids
F. Prions

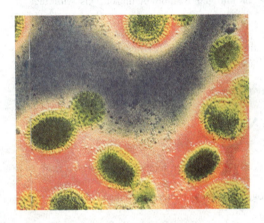

In 1884, Charls Chamberland developed a porcelain filter that allowed the passing of fluid but not bacteria. The filter was useful in sterilization of liquids. A. Mayer (1886) working in Holland demonstrated that the sap of mosaic leaves of tobacco plant developed the mosaic symptom when injected into the healthy plants. However, the infectivity of the sap was destroyed when it was boiled. He thought that the causal agent was the bacteria. However, after inoculation with a large number of bacteria, he failed to develop a mosaic symptom.

In 1892, for the first time a Russian botanist Dmitri Iwanowski filtered the sap of diseased tobacco plant through the porcelain (Chamberland filter) designed to obtain bacteria but the infectious agent was filtered through the pores of porcelain. After injections of filtered sap he found the development of mosaic symptom. He concluded that the infectious agent was the bacterium but smaller than it. The infectious agent was called filtrable virus i.e. poisonous fluid.

M.V. Beijerinck (1898), a Dutch microbiologist and Professor of Microbiology at Technical University of Delft (Holland), put a concept *contagium vivum fluidum* (living infectious fluid). This idea was considered revolutionary. Thereafter, several infectious filtrable

Dmitri Iwanowski

fluids were characterized from various sources. Beijerinck observed that the behaviour of filtrable virus was different from the bacteria.

By the end of third decade of 20th century, two major technical break throughs occurred: (*i*) development of technique of ultracentrifugation, and (*ii*) electron microscopy.

An Americal chemist, Wendell M. Stanley (1935) crystalized the virus causing mosaic on tobacco plant. He found that the crystals were also infectious when inoculated on healthy tobacco plants. He concluded that the viruses were not like a typical cell of the living organism. For this work Stanley was awarded the Nobel Prize in 1946.

The two British biochemists, F.C. Bawden and N.W. Pirie (1938) analysed the crystallized particles and demonstrated that these were made of protein and ribonucleic acid (RNA), for which Pirie was awarded the Nobel Prize. Gierrer and Schramm (1956) proved the real nature of nucleic acid as infectious agent and genetic material. Inside the infected cells virus uses the nitrogenous and the other compounds and replicates its genome. Later on it was confirmed by Fraenkel-Conrat (1956) that the genetic material of tobacco mosaic virus (TMV) is the RNA.

One of the interesting historical phenomena related to the virus was the use of broken tulip. During 16th and 17th century a variety of tulip (an ornamental plant cultivated for its flower) was very popular due to the presence of striations and different shades on petals. Such tulip plants were called broken tulip. In Holland and France, people were very crazy for broken tulip which was much prized and thought as status symbol. The development of striations and shades was due to the infection of a virus.

Frederick W. Twort (1915) in England and Felix d'Herelle (1916) in Pasteur Research Institute (Paris) observed independently that in cultures the bacterial colonies were lysed by some agent and this effect could be transmitted from colony to colony. The highly diluted material from a lysed colony, even after passing through a bacterial filter, transmit the lytic effect. After heating the filtrate lytic property was destroyed. Twort suggested that the lysis would have been due to a virus. d'Herelle (1917) rediscovered the phenomenon (commonly known as Twort - d'Herelle phenomenon) and coined a term bacteriophage i.e. bacteria eater. Hershey and Chase (1952) studied in detail T_2 bacteriophage and demonstrated that the phage DNA carries the genetic information, and infection occurs upon penetration of phage DNA into living cells.

Moreover, in 1894 a Japanese worker Hashimoto discovered the relationships between an insect, *Nephotettix apicalis* var *cinticeps* and virus associated with rice dwarf disease. Later on the three American workers, Ball, Adams and Shaw (1907) proved the relationship between the curly top of sugarbeet and a leaf hopper, *Eutettix tenella*.

The other development includes the discovery of mycoviruses in the cultivated button mushroom by Hollings (1962), discovery of cyanophages (viruses eating upon blue-green algae) by Safferman and Morris (1963), satellite virus by Kassanis (1966), viroids by Diener and Raymer (1967), prion by Prusiner (1982), and HIV by LuC Montagnier (1983).

During the last few decade much informations have been gathered on isolation and culture of viruses, replication processes, preparation of maps, immunization processes, genetic engineering, molecular biology, vaccine development, etc.

A. General Concepts

I. What is a Virus ?

Virus is a parasite in all types of organisms. They infect animals, plants, bacteria, algae, insects, etc. So far the exact nature of viruses is ambiguous whether they are living or non-living organisms. If we look into life, it is a complex set of processes taking place through the action of proteins governed by the nucleic acid. The nucleic acid of the living organism is functional in all time. Outside the living cell, viruses remain inactive. Therefore, they cannot be said as living organism. In addition, if we consider the diseases caused by them they act as pathogen against bacteria, fungi, protozoa, etc. So from this angle viruses may be regarded as exceptionally simple living organism or as an exceptionally complex aggregation or non-living chemicals. Then how may a virus be defined. They are specially small, filterable and obligate intracellular parasite requiring a living host for its multiplication.

Lwoff (1957) defined that *"Viruses are infectious, potentially pathogenic nucleoprotein with only one type of nucleic acid which reproduce from their genetic material are unable to grow and divide, and devoid of enzymes.*

Luria and Darnell (1968) defined that *"Viruses are entities, whole genome of which are element of nucleic acid that replicate inside the living cells using the cellular synthetic machinery and causing the synthesis of specialized elements that can transfer the viral genome to other cell.*

While defining the microorganisms viruses are set apart on the basis of certain characters as given below by Lwoff and Tournier (1971) :

(*i*)   They are all potentially infectious,

(*ii*)   Presence of a single nucleic acid,

(*iii*)   Incapability to grow the genetic material only,

(*iv*)   Reproduction from the genetic material only,

(*v*)   Absence of enzymes for energy metabolism (Lipman system),

(*vi*)   Absence of ribosomes,

(*vii*)   Absence of information for the production of enzymes in the energy cycle,

(*viii*)   Absence of information for the synthesis of ribosomal proteins, and

(*ix*)   Absence of information for the synthesis of ribosomal RNA and soluble tRNA.

II. Occurrence

Viruses occur on a very wide hosts such as plants (angiosperms, gymnosperms, ferns, algae, fungi) and animals (protozoans, insects, fish, amphibians, birds, mammals, humans). They cause very serious diseases in crop plants, ornamental plants and forest trees resulting in decreased growth, yield and mortality.

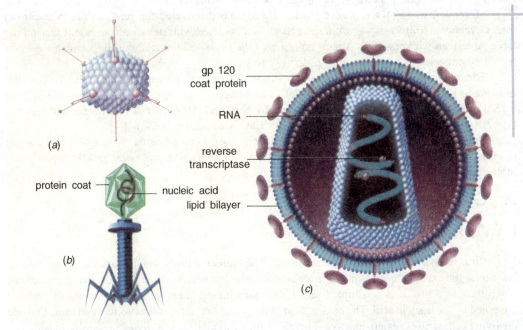

Virus diversity : (*a*) Adenovirus,   (*b*) T- even bacteriophage,   (*c*) HIV

In animals like dogs, cats, and humans several serious diseases are known since the time immemorial. Important viral diseases of humans are discussed in Chapter-22 *Medical Microbiology-II*. Total number of viruses in different host taxa according to their genome types are given in Table 15.1

Table 15.1 : Number of viruses in different host taxa according to genome type (after Hollings, 1982).

Phylum	Virus Genome			
	ssDNA	*dsDNA*	*ssRN*	*dsRNA*
Vertebrates	23	202	440	86
Invertebrates	6	143	30	12
Flowering plants	9	07	371	12
Bacteria	47	168	27	1
Fungi	?	?,2	1	25

ss, Single stranded, and ds, Double stranded genome

III. Morphology of Viruses

1. Shape

Viruses are of different shapes such as spheroid or cuboid (adenoviruses), elongated (potato viruses), flexuous or coiled (beet yellow), bullet shaped (rabies virus), filamentous (bacteriophage M13), pleomorphic (alfalfa mosaic), etc.

2. Size

Viruses are of variable sizes. Initially their sizes were estimated by passing them through membranes of known pore diameter. In recent years, their size is determined by ultracentrifugation and electron microscopy. Size vary from 20 nm to 300 nm in diameter. They are smaller than bacteria; some are slightly larger than protein and nucleic acid molecules and some are about of the same size (small pox virus) as the smallest bacterium and some virus (virus of lymphogranuloma, 300-400 um) are slightly larger than the smallest bacterium.

3. Viral Structure

The complete assembly of the infectious particle is known as virion. A virion consists of a nucleic acid core surrounded by a protein coat or capsid. The complete set of virion is known as nucleocapsid. In turn the nucleocapsid may be naked or enveloped by a loose covering. The capsid is composed of a large number of subunits known as capsomers.

Chemically the envelope is made up of proteins and glycoproteins. Due to the presence of lipid the envelope seems flexible and loose. Envelope is composed of both the host and viral components i.e. protein (virus specific) and carbohydrates (host specific).

There are certain projections on the envelope known as spikes which are arranged into distinct units. The morphological types of virus observed through electron microscopy and crystallography have been categorised into the following three groups :

(i) **Helical (cylindrical) Viruses:**

　　(a) Naked capsids e.g. TMV and the bacteriophage M13, etc.

(*b*) Enveloped capsid e.g. influenza virus, etc.

(*ii*) **Polyhedral (icosahedral) Viruses :**

(*a*) Naked capsid e.g. adenovirus, polio viruses, etc.

(*b*) Enveloped capsids e.g. herpes simplex viruses, etc.

(*iii*) **Complex Viruses :**

(*a*) Capsids not clearly identified e.g. vaccinia virus, etc.

(*b*) Capsids to which some other structures are attached e.g. some bacteriophages,etc.

(*i*) **Helical Viruses :** The helical viruses are elongated, rod-shaped, rigid or flexible. Their capsid is a hollow cylinder with a helical structure. Capsid consists of monomers arranged helically in a rotational axis. The helical capsids may be naked (e.g. TMV) or enveloped (e.g. influenza virus).

(*a*) **Naked viruses :** One of the examples of naked viruses is the TMV. For the first time Stanley (1935) isolated TMV in the crystalline form from leaf sap of the infected tobacco plants. Thereafter, a lot of work was done on TMV. This virus is rod shaped measuring about $280 \times 150\text{-}180$ µm. It consists of a protein tube with a lumen of 20Å which encloses a single stranded (ss) helix of coiled RNA. Protein coat of the virus contains a number of identical subunits (monomers) which are arrranged in helical manner. A capsid consists of several capsomers each composed of a few monomers. Forty nine monomers (each with molecular weight 12,000) take three turns of the helix and give a total of 2,130 subunits of the rod. Each subunit is made up of 150 amino acid residues forming a single polypeptide chain. Genetic material i.e. ssRNA has the molecular weight of 2.06×10^6 dalton. The RNA consists of 6,500 nucleotides in total and coiled to form a helix with a radius of 80Å which is enclosed in a helix of 85Å radius of similar pitch (23Å) formed by monomers of protein. Each turn of the RNA helix consists of at least 49 nucleotides. There is a cylindrical hole of 20Å radius. The capsid protects the RNA molecule.

(*b*) **Enveloped viruses :** When the helical viruses are enclosed within an envelope they are known as enveloped helical viruses, for example influenza virus. The envelope is composed of a viral protein and the host cell components i.e. lipid and carbohydrates. The envelope consists of numerous spikes. The helical capsid exists in folded form inside the envelope and some times may show pleomorphic appearance.

(*ii*) **Polyhedral (Icosahedral) Viruses :** There are several animal, plant and bacterial viruses which have either naked or enveloped icosahedral shape. Polyhedral structure has the three possible symmetries such as tetrahedral, octahedral and icosahedral. The viruses are more-or-less spherical. Therefore, icosahedral symmetry is the best one for packaging and bonding of subunits. Several inter molecular bonds of low free energy is formed. An icosahedron is a regular polyhedron with 20 triangular faces and 12 corners. The capsomers of each face form an equatorial triangles and 12 intersepting points or corners (Fig. 15.1 A-C).

Basically, there are two types of capsomers, the pentamers and hexamers. The pentamer is a clusture of 5 monomers and the hexamer is a clusture of 6 monomers (C). The monomers are linked together by bonds. Thus, the capsomers are also linked together by bonds which are weaker than those of monomers due to their breaking into the capsomers during the purification of viral particles. However, certain capsomers only in certain number may be present in icosahedral capsid, theoretically the minimum number may be 12. Therefore, in different viruses the number of capsomers differ, for example 12 capsomers in ∅X174, 32 in poliovirus, 72 in polyomavirus, 92 in reovirus, 162 in herpesvirus, 252 in adenovirus and 812 in tipula iridescent virus.

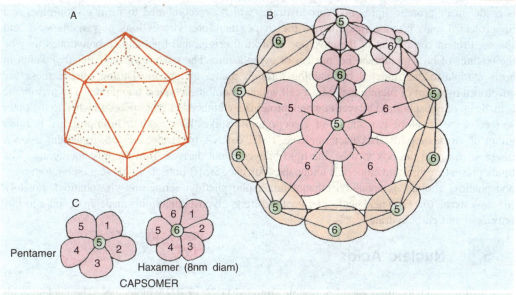

Fig. 15.1 : Icosahedral symmetry (A), architecture (B) and subunits of capsomers (C).

(*a*) **Naked icosahedral viruses:** Naked icosahedral viruses are turnip yellow mosaic virus (TYMV), poliovirus, adenovirus and bacteriophage ØX174, QB, etc. Adenoviruses are large (80 nm diameter), icosahedral containing dsDNA. The capsid has the ring like capsomers each containing pentamers and hexamers. A total of 32 hexamers and 12 pentamers are found in the capsid. The hexamers are polygonal discs of 8 nm diameter with a hole of 2.5 nm diameter in the centre.

At the surface of capsid these spike like structures form 12 points of the five fold symmetry. The capsomers are assembled to construct the capsid in a specific geometrical pattern where the pentamers form the corners of the icosahedron and the hexamers occupy the internal space (Casper and King, 1962).

(*b*) **Enveloped icosahedral viruses :** There are some of the enveloped icosahedral viruses, for example herpes virus where the capsid is enclosed inside an envelope of 30nm thickness that is made up of a glycoprotein-lipid complex. The envelope consists of spikes on its surface. The capsid is spherical and of about 100 nm diameter enclosing a dense core of dsRNA molecule. The capsid contains 162 capsomers, (12 pentamer capsomers at apices and 150 hexamer capsomers at the faces).

(*iii*) **Complex Viruses:** The viruses which have the unidentifiable capsids or have the capsids with additional structures are called complex viruses, for example, vaccinia virus and T-even bacteriophages. In addition, the other variations in the structure of complex viruses are also found such as (i) a definite capsid absent (e.g. vaccinia virus), (ii) capsid present and consists of a tail. The second group consists of different types, for example, tadpole shaped viruses (with head and tail, e.g. T-even phage), viruses with tail less head (phage λ, T_1, T_5), virus with brick shaped and devoid of flattened cylinder (pox virus), bullet shaped capsid viruses (e.g. nuclear polyhedrosis or cytoplasmic polyhedrosis viruses).

4. Envelope

There are certain plant and animal viruses and bacteriophages, both icosahedral and helical, which are surrounded by a thin membranous envelope. This envelope is about 10-15 μm thick. It is made up of protein, lipids and carbodydrates which are combined to form glycoprotein and lipoprotein. Lipids provide flexibility to the shape, therefore, viruses look of variable sizes and shapes. Protein component of the envelope is of viral origin, and lipid and carbohydrates may be the features of host membrane i.e. nuclear or cytoplasmic. The ssRNA viruses after replication in hosts cytoplasm are released by budding through plama membrane. During release these are enveloped by a part of membrane of host cell which resembles a typical membrane. The membrane is made up of phospholipid bilayer in which proteins are embedded. The spikes attached to the outer surface of the envelope is made up of glycoproteins. Spikes have agglutination proteins. In other group of viruses e.g. dsDNA bacteriophage, PM2, dsRNA bacteriophage Ø6, iridescent dsDNA insect viruses and the pox viruses, the lipid bilayer is not derived from the host membrane. The lipids present in viral envelope fall under the four classes, (*i*) phospholipids (e.g. sphingomyelin, phosphatidyl choline, phophatidyl ethanolamine, phosphatidyl serine and phosphatidyl inositol), (*ii*) cholesterol, (*iii*) fatty acid, and (*iv*) glycolipids (e.g. glycosphingolipids made up of sphingosine, fatty acid and carbohydrate).

5. Nucleic Acids

Viruses contain either single or double stranded DNA or RNA molecules. The nucleic acids may be in linear or circular form, and have plus or minus polarity.

Nucleic acids of several plant viruses occur in their respective particles in one pieces. However, the reovirus is known to contain the nucleic acid in 10 segments (Millward and Graham, 1970). The segmented nucleic acid is also found in wound tumour virus and influenza virus.

6. Proteins

Proteins found in viruses may be grouped into the four categories : (*i*) envelope protein, (*ii*) nucleocapsid (structural) protein, (*iii*) core protein, and (*iv*) viral enzymes.

(*i*) **Envelope Protein :** Envelope of the viruses consists of proteins specified by both virus and host cell. Membranes of all class of enveloped viruses contain glycoprotein. In influenza virus the main protein is the carbohydrate free protein which comprises of 50% of envelope protein and 35-40% of virion protein as a whole. Herpes viruses, poxviruses and leukoviruses do not contain protein in the envelope specified by them. It has been found that the envelope proteins are enclosed by the genome of arboviruses, rhabdoviruses and myxoviruses. In addition, the glycoproteins differ virus to virus. For example, one glycoprotein in rhabdoviruses, two glycoprotein in paramyxoviruses and four glycoprotein in orthomyxoviruses have been found.

(*ii*) **Nucleocapsid Protein:** The viral capsids are made up totally of proteins of identical subunits (promoters). The helical capsids contain single type of protein and icosahedral capsid contains several types of protein. For example, TMV contains single protein types, adenovirus contain 14 protein types, T4 bacterophage contains 30 protein types, etc.

(*iii*) **Core Protein:** Protein found in nucleic acid is known as core protein, for example nucleoproteins of influenzavirus, and proteins V and VI of adenoviruses.

(*iv*) **Viral Enzymes :** In animal viruses especially in the enveloped viruses, many virion specific enzymes have been characterized, for example RNase and reverse transcriptase in retrovirus, protein kinase in herpes and adenoviruses, DNA dependent RNA polymerase in poxvirus.

7.   Carbohydrates

A substantial amount of carbohydrate specified by either host cell (e.g. arbovirus) or viral genome (e.g. vaccinia virus) is found in viral envelope. For example galactose, mannose, glucose, fucose, glucosamine, galactosamine are found in influenza virus, parainfluenza virus, SV5 and sindbis virus. The carbohydrates are hexoses and hexamines which are present in the form of glycoprotein and/or glycolipids.

IV. Classification

In 1927, Johanson was the first to attempt for the classification of plant viruses. Traditionally, viruses have been named according to the diseases caused by them by adding a suffix virus e.g. poliovirus, influenzavirus, etc. In addition, the bacteriophages were named after the laboratory codes e.g. QB, Øx174, M13, etc. The cyanophages were named after the host they lysed and the serological differences among them e.g. LPP1, LPP2, etc. The bacteriophages and cyanophages are described in Chapter 16, *Viruses-II*.

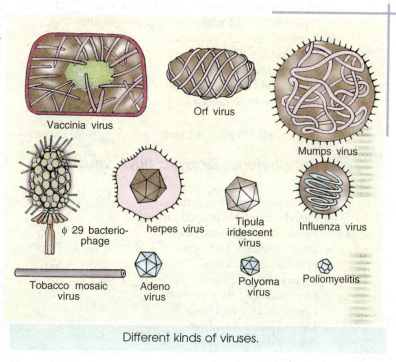

Different kinds of viruses.

Holmes (1948) followed the Linnean system of bionomial nomenclature and put the viruses into an order virales and three suborders :

(*i*) *Phaginae* : Viruses attacking on bacteria.

(*ii*) *Phytophagine* : Viruses attacking on plants.

(*iii*) *Zoophaginae* : Viruses attacking on animals.

In 1962, A. Lwoff, R. Horne and P. Tournier proposed a system of classification of viruses which is commonly referred to as LHT system of classification. It was adopted by the provisional Committee on Nomenclature of Viruses (PNCV) formed by the International Association of Microbiological society. The LHT system of classification is based on (*i*) the nature of nucleic acid, (DNA or RNA), (*ii*) symmetry or viral particle (helical, icosaherdal, cubic, cubic-tailed), (*iii*) presence or absence of envelope, (*iv*) diameter of capsid, and (*v*) number of capsomers forming the capsid. The LHT system of classification is as below :

Phylum : Vira

1.　Subphylum : Deoxyvira (DNA Viruses)

Class　　: Deoxy helica (helical symmetry)
Order　　: Chitovirales (enveloped)
Family　 : Pox viridae (poxviruses)
Class　　: Deoxycubica (cubical symmetry)
Order　　: Haplovirales (no envelope)
Family　 : Microviridae- 12 capsomers e.g. ØX174
"　　　　: Parvoviridae- 32 capsomers e.g. a rat virus
"　　　　: Papiloviridae- 72 capsomers e.g. papovaviruses.
"　　　　: Adenoviridae- 252 capsomers e.g. adenoviruses.
"　　　　: Iridoviridae - 812 capsomers e.g. insect viruses
Order　　: Peplovirales (mantle viruses)
Family　 : Herpes viridae- 162 capsomers (herpes viruses)
Class　　: Deoxy binala (viruses with head and tail)
Order　　: Urovirales
Family　 : Phagoviridae (bacteriophages)

2.　Subphylum ; Ribovira (RNA viruses)

Class　　　: Ribohelica (helical symmetry)
Order　　　: Rhabdovirales (rodshaped viruses)
Suborder : Rigidovirales (plant viruses)
Family　　: Dolichoviridae- 12-13 nm
"　　　　　: Protoviridae - 15 nm
"　　　　　: Pachyviridae - 20 nm
Suborder : Flexiviridales (plant viruses)
Family　　: Leptoviridae - 10-11 nm
"　　　　　: Mesoviridae - 12-13 nm
"　　　　　: Adroviridae - 15 nm
Order　　　: Sagovirales
Family　　: Myxoviridae - 9 nm
"　　　　　: Paramyxoviridae - 18 nm
"　　　　　: Stomato viridae
Class　　　: Ribocubica (cuboidal symmetry)
Order　　　: Gymovirales - 32 capsomers
Family　　: Napoviridae (plant viruses, picorna viruses)
"　　　　　: Reoviridae - 92 capsomers (reoviruses)
Order　　　: Toga virales
Family　　: Arboviridae (arboviruses)
"　　　　　: Encephaloviridae

The LHT system of classification is neither a natural classification nor shows any evolutionary phylogenetic relationship. However, it has been widely criticized, besides evoking the considerable interest among the virologists. Now a days, this system of classification is getting much attention.

Bellet (1967) proposed a system of classification. This system of classification is based mainly on two criteria of the viral particles (*i*) molecular weight, and (*ii*) percentage of guanine + cytosine of the nucleic acid. Serological, antigenic and phenotypic properties were also considered.

Gibbs (1969) proposed a system of classification for plant viruses which is known as Gibbs system of classification. The criteria to classify the viruses are: (*i*) shape of capsid, (*ii*) mode of transmission, (*iii*) type of vector, (*iv*) symptoms on host after infection, and (*v*) the nature of accessory particles. He put 135 known viruses into the 6 broad groups.

Casjens and King (1975) classified the viruses on the basis of nucleic acid types, symmetry, presence or absence of the envelope and site of assembly of the envelope i.e. nuclear and cytoplasmic. The classification of Casjens and King (1975) is given below:

1. ssRNA viruses

Helical

(*i*) Rigid rods (plants) : TMV, tobacco rattle virus, barley stripe mosaic virus.

(*ii*) Flexous rods (plants) : Potato X and Y viruses, clover yellow mosaic virus.

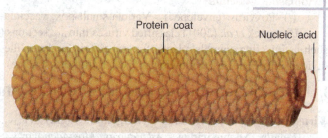

Molecular organization of the tobacco mosaic virus (TMV) (after De Robertis and De Robertis, Jr., 1987).

Icosaherdal

(*i*) Spherical plant viruses.

(*a*) With 180 identical capsomers (T=3) e.g. cowpea chlorotic mosaic virus, cucumber mosaic virus, turnip yellow.

(*b*) With 60 subunits of two structural proteins (T=1) e.g. cowpea mosaic virus.

(*ii*) Bacteriophages : e.g. R17, Fr, F2, QB, MS2

(*iii*) Picorna Viruses (animal viruses) :

(*a*) Human entroviruses : poliovirus

(*b*) Rodent cardioviruses : Encephalomyocarditis virus, mengovirus.

(*c*) Rhinoviruses : Human respiratory infection viruses.

(*d*) Foot and mouth disease virus.

Envelope

(*i*) Spherical : Togavirus, yellow fever virus.

(*ii*) Bullet shaped : Rhabdovirus e.g. rabies

(*iii*) Spherical : Paramyxovirus e.g. measles, or filamentous Myxovirus e.g. influenza virus.

(*iv*) Spherical   : Corona virus e.g. acute upper respiratory tract infection virus, severe acute respiratory syndrome (SARS) virus

: Arena virus e.g. lymphocylic chloromeningitis

: Oncoviruses e.g. leukemia, sacroma

2. dsRNA viruses

Segmented genome

(*i*) Animal viruses : Reovirus, blue tongue virus of sheep, cytoplasmic polyhedrosis virus of silk worm.

(*ii*) Plant viruses : Wound tumour virus of plants, Maize rough dwarf virus, Rice dwarf virus.

Enveloped : Bacteriophage Ø6

3. ssDNA viruses

Icosaherdal :

(*i*) Bacteriophages : Øx174, S13

(*ii*) Parvoviruses : Animal and insect viruses

Helical

Bacteriophages : fd, F1, M13

4. dsDNA viruses

Icosaherdal complex (tailed)

(*i*) *E. coli* phages : T4, P2, T3, T5, T7

(*ii*) *S. typhimurium* phage : P22

(*iii*) *B. subtilis* phage : ∅29, cyanophages

Enveloped : Bacteriophage PM2

Nuclear Assembly

(*i*) Papovavirus : Polyomavirus, SV40, human wart virus.

(*ii*) Adenovirus : Respiratory disease in birds and mammals (icosaherdal)

(*iii*) Herpsvirus (enveloped) : Cold sores, shingles, infections mononucleosis cervical sarcoma of uterus, Burkitt's lymphoma.

Cytoplasmic Assembly

(*i*) Poxvirus (enveloped) : Variola-small pox, vaccinia-immunity to small pox.

Dimmock *et al.* (2001) classified viruses into six sections (on the basis of their host preference) such as infecting animals, plants, fungi, bacteria, and satellite viruses and viroids. Each section is divided into 7 classes (following revised Baltimore scheme), and each class into families.

I.   Virus of Vertebrates and Invertebrate Animals

Viruses with double stranded DNA genomes (class 1)

Family: **Adenoviridae** - *Adenovirus, Mastadenovirus, Aviadenovirus, Atendovirus,*

Family: **Ascoviridae** - *Spodoptera frugiperda* ascovirus 1 a

Family: **Asfraviridae** - swine fever virus.

Family: **Baculoviridae** - *Nuclepolyhedrosisvirus, Granulovirus*

Family: **Herpesviridae** - *Simplexvirus* (herpes simples virus), *Varicellovirus*

Family: **Iridoviridae** - *Iridovirus, Chloriridovirus, Ranavirus* (virus of frog)

Family: **Papilomaviridae** - Over 60 papilomaviruses are known, oncogenic.

Family: **Polydnaviridae** - *Ichnovirus* (infects wasps), *Bracovirus.*

Family: **Polyomaviridae** - SV40, murine polyomavirus, oncogenic.

Family : **Poxviridae** - *Orthopoxvirus, Avipoxvirus, Parapox virus, Entomopoxvirus*

Viruses with ssDNA genomes (class 2)

Family: **Circoviridae** - *Circovirus* (chicken anaemia virus).

Family: **Parvoviridae** - *Parvovirus, Dependovirus, Densovirus.*

Viruses with dsRNA genomes and virion-associated RNA-dependent RNA polymerase (class 3)

Family: **Birnaviridae** - *Aquabirnavirus, Avibirnavirus, Entomobirnavirus*

Family: **Reoviridae** - *Orthroreovirus, Orbivirus, Rotovirus, Coltivirus, Cypovirus*

Viruses with positive sense ssRNA genomes (class 4)

Family: **Arteriviridae** - *Artivirus* (equine arteritis virus)

Family: **Astroviridae** Astrorivirus of human, bovine and duck.

Family: **Caliciviridae** - Human *Calicivirus,* Hepatitis E virus.

Family: **Coronaviridae** - Coronavirus, Torovirus (berne virus of house)

Family: **Flaviviridae** - *Flavivirus, yellow fever virus,* Hepatitic C *virus, Pestivirus Hepacivirus* (hepatitis C virus of human).

Family: **Nodaviridae** - *Alphanodavirus, Betanodavirus*

Family: **Picornaviridae** - *Exterovirus* (polio virus), *Hepatovirus, Aphthovirus* (human Hepatitis A virus), *Cardiovirus, Rhinovirus* (common cold virus).

Family: **Tetraviridae** - Naudaurelia beta virus.

Family: **Togaviridae** - *Alphavirus* (Arbovirus), *Rubivirus* (rubella virus of humans).

Viruses with negative-sense/ambisense ssRNA genomes and virion-associated RNA-dependent RNA-polymerase (class 5)

Family: **Arenaviridae** - Arenavirus, Lassavirus.

Family: **Bornaviridae** - Borna disease virus.

Family: **Bunyaviridae** - *Bunyavirus, Hantavirus, Tospovirus*

Family: **Filoviridae** - Filovirus (Marburg and Ebova virus).

Family: **Orthomyxoviridae** - *Influenzavirus* A, B abd C, *Thogotovirus.*

Family: **Paramyxoviridae** - Paramyxovirus, *Rubulavirus, Respirovirus, Pneumovirus*

Family: **Rhabdoviridae** - *Vesiculovirus, Lyssavirus*

Viruses with RNA genomes that replicate through a DNA intermediate (class 6)

Family: **Retroviridae** - Murine leukemia virus, *Spumavirus, Lentivirus* (human immunodeficiency virus), *Visna virus, Deltaretrovirus, Alpharetrovirus.*

Viruses with a DNA genome that replicates through an RNA intermediate (class 7)

Family: **Hepadnaviridae** - *Orthohepadnavirus* (Hepatitis B virus,) *Avihepadnavirus.*

II. Viruses that Multiply in Plants

Viruses with ssDNA genomes (class 2)

Family: **Circoviridae** - *Nanovirus*

Family: **Geminivirus** - *Curtovirus, Mastrevirus, Begomovirus*

Viruses with dsRNA genomes and virion-associated RNA-dependent RNA Polymerase (class 3)

Family: **Partitiviridae** - *Alphacryptovirus, Betacryptovirus*

Family: **Reoviridae** - *Fijivirus, Oryzavirus, Phytoreorvirus*

Viruses with (+) ssRNA genomes (class 4)

Isometric Virions

Family: **Comovirus** - *Comovirus* (cowpea mosaic virus), *Fabavirus* (broad bean wilt virus-1), *Nepovirus* (tobacco ring spot virus)

Family: **Luteovirus** - *Luteovirus* (barely yellow dwarf virus), *Polerovirus* (potato leaf roll virus), *Enamovirus* (bean enation mosaic virus)

Family: **Sequiviridae** - *Sequivirus* (parsnip yellow fleck virus), *Waikavirus* (rice tongro spherical virus)

Family: **Tombusviridae** - *Avenavirus* (oat chlorotic stunt virus), *Carmovirus* (cernation mottle virus), *Dianthovirus* (carnation ring spot virus), *Necrovirus* (tobacco necrosis virus A), *Tombusvirus* (tomato bushy stunt virus).

Isometric virions and virions that are short rods

Family: **Bromovirus** - *Alfamovirus* (alfalfa mosaic virus), *Bromovirus* (brone mosaic virus), *Cucumovirus* (cucumber mosaic virus)

Virions that are rigid rod

The following genera have not been assigned to a family

Benyvirus (beet necrotic yellow vein virus), *Furovirus* (soilborne wheat mosaic virus), *Hordeviridae* (barely stripe mosaic virus), (*Pecluvirus* (Peanut clump virus), *Pomovirus* (potato mop top virus), *Tobamovirus* (tobacco mosaic virus), *Tobravirus* (tobacco rattle virus)

Virions that are flexuous rods

Family: **Closteroviridae** - *Closterovirus, Crinivirus*

Family: **Potyviridae** - *Potyvirus* (potato virus Y) *Ipomavirus* (sweep potato mild mosaic virus), *Macluravirus, Rymovirus* (ryegrass mosaic virus), *Tritimovirus* (wheat streak mosaic virus), *Bymovirus* (barley yellow mosaic virus)

Viruses with (-) / ambisense ssRNA genomes and a virion-associated RNA-dependent RNA polymerase (class 5)

Family: **Bunyaviridae** - *Tospovirus* (tomato spotted wilt virus)

Family: **Rhabdoviridae-** *Cytorhabdovirus, Nucleorhabdovirus*

Viruses with DNA Genomes that Replicate through an RNA intermediate (class 7)

Family: **Caulimoviridae** - *Caulimovirus* (cauliflower mosaic virus), *Badnavirus* (commelina yellow mottle virus).

III. Viruses Multiplying in Algae, Fungi and Protozoa

Viruses with dsDNA Genomes (class 1)

Family: **Adenoviridae** - fungal virus *Rhizidiovirus* (*Rhizidiomyces* virus),

Family: **Phycodnaviridae** - viruses infecting *Chlorella, Parmaecium, Hydra.*

Viruses with dsDNA Genomes and a virion-associated RNA-dependent RNA Polymerase (class 3)

Family: **Hypoviridae** - *Hypovirus,*. Cryphonectria virus 1

Family: **Partitiviridae** - *Partitivirus* (Gaeumannomyces graminis virus 019/6A), *Chrysovirus* (*Penicillum chrysogenum* virus),

Family: **Totiviridae** - *Totivirus* (infects *Saccharomyces cerevisiae*), *Giardiavirus* (infects *Giardia lamblia,*) *Leishmaniavirus* (infects *Leishmania* spp.)

Viruses with (+) ssRNA genomes (class 4)

Family: **Barnavirus** - Mushroom bacilliform virus

Family: **Narnaviridae** - *Narnavirus* (*Saccharomyces cerevisiae* 20S narnavirus), *Mitovirus* (*Cryphonectria parasitica* mitovirus 1 NB631)

IV. Viruses (Phages) Multiplying in Archaea, Bacteria, *Mycoplasma* and *Spiroplasma*

Viruses with dsDNA gnomes (class 1)

Viruses that have head-tail structure

Family: **Sipoviridae** - Bacteriophages T1, phage λ, chi (ψ) and phage phai (φ) 80

Family: **Myoviridae** - 'T-even' coliphages T2, T4, T6, Mu, P1, P2, PBSI, SP8 SP50

Family: **Podoviridae** - Coliphages T3, T7, and P22, φ 29

Viruses that do not have a head-tail structure

Family: **Fuselloviridae** - *Sulfolobus* virus 1

Family: **Tectiviridae** - icosahedral particle, similar to adenovirus

Family: **Cortiviridae** - *Alteromonas* phage PM2 (infects *Pseudomonas*)

Family: **Plasmaviridae** - *Acholeplasma* phage L2 (infects *Mycoplasma*)

Family: **Rudiviridae** - *Sulfolobus* viruses SIRV-1 (infects thermophilic *Archaea*)

Family: **Lipothrixviridae** - *Thermoproteus* virus 1 (infects *Archaea*)

Viruses with dsDNA Genomes (class 2)

Family: **Inoviridae** - *Inovirus* (e.g. Phage M13, phage fd), Plectivirus (*Acholeplasma* phage MV-L51)

Family: **Microviridae** - *Microvirus* (φX174), *Bdellovibriovirus, Spriomicrovirus*

Viruses with dsRNA genomes and a virion-associated RNA-dependent RNA polymerase (class 3)

Family: **Cystoviridae** - *Pseudomonas* φ 6

Viruses with (+) ssRNA Genomes (class 4)

Family: **Leviviridae** - Bacteriophages R17, MS2, and Qβ

V. Satellite Viruses and Satellite Nucleic acids of Animals, Plants, Fungi and Bacteria

Satellite nucleic acids with dsDNA genomes (class 1): *Mycoviridae* helper bacteriophage P2, (enterobacteria P4 satellite)

Satellite nucleic acids with ssDNA genomes (class 2): *Parvovirinae, Dependovirus*
Satellite nucleic acids with ssDNA genomes (class 2): Geminiviruses
Satellite nucleic acids with dsRNA genomes (class 3): Satellite of Saccharomyces cerevisiae M virus
Sattelite viruses with (+) ssRNA genomes (class 4): Tobacco necrosis virus satellite virus, chronic bee-paralysis virus-associated satellite virus
Satellite nucleic acids with (+) ssRNA genomes (class 4): A non-structural protein e.g.arabis mosaic large satellite RNA
Satellite nucleic acids with (-) ssRNA genome (class 5): *Deltavirus* - Hepatitis delta virus infects humans
Satellite nucleic acids with ssRNA genomes (unclassified as makes no mRNA): Velvet tobacco mottle virus satellite RNA

VI. Viriods (Unclassified as Made no mRNA)

Viroids are small, circular, infectious ssRNAs (246-370 nucleotide) with are never encapsulated and have no helper virus and are serious plant pathogens.
Family: **Pospiviroidae** - potato spindle tuber viroid (PSTV), coconut cadang cadang viroid
Family: **Avsunviroidae** - avocado sunblotch viroid

V. Isolation and Cultivation of Viruses

Viruses cannot multiply outside a living host cell, however, their isolation, enumeration and identification become a difficult task. Instead of a chemical medium, they require a living host plant, animal or bacterial cells.

1. Cultivation of Bacteriophages

The bacteriophages infect the specific bacterial host, hence a defined bacterium is required for the cultivation of bacteriophages. A sample of bacteriophage is mixed with known host bacteria in molten nutrient agar. The nutrient agar containing bacteriophage and its hosts cell is poured into steile Petri dishes. The mixture of virus and bacteria solidifies into a thin top layer. The Petri plates are incubated at optimum temperature in an incubator. A single virus infects a bacterial cell, multiplies and releases hundreds of new viral particles. The released viruses in turn infect the healthy bacteria in the surroundings. As a result of bacterial lysis, the area where bacterial colonies were growing turns into clearings or plaques. These plaques are visible against the bacterial lawn. During the course of time the uninfected bacteria multiply rapidly and produce turbid growth in the Petri plate. Theoretically, each plaque represents for the presence of at least a single virus. Hence the total number of viruses present in suspension is expressed as plaque forming units (pfu). Like bacteriophages, the cyanophages are also cultivated on Petri plates by using the cyanobacteria as the host.

2. Cultivation of Animal Viruses

Some of the viruses can be cultured in the living animals such as mice, rabbits, guinea pigs, etc. After incubation, animals are carefully examined for the development of signs or symptoms. The animals may be killed.

(*i*) **Use of Embryonated Eggs :** However, some viruses can be cultivated in the embryonated eggs. If they grow on eggs it will be a cheaper and convenient method. A hole is made in the shell of embryonated egg, and a viral suspension or tissue containing viral particles is introduced into the fluid of egg. If virus grows in egg, it gives the signal of death of embryo or damage of embryonic cells, formation of lesions on membrane, etc.

(ii) Use of Animal Cell Culture : In recent years, use of animal cell culture *in vitro* as the growth medium for the viruses has replaced the embryonated egg. Cells are cultured artificially in laboratory. The cultured animal cells act as bacterial cell. To start the cell culture, small slice of animal tissue is treated with enzymes which separates each cell. The cells are suspended in a solution of proper osmotic pressure, nutrients and growth hormones that are required for the cell growth. The cells get attached to the surface of glass and form a layer. Viruses inoculated in the cell suspension destroy the monolayer. The phenomenon is called cytopathic effect which is detected as the plaques formed by bacteriophages.

The grown cells obtained from the sliced tissue are known as primary cell lines. Primary cell lines start degenerating after a few generations. Therefore, the diploid cell lines can be established from human embryos. The diploid cell lines can be maintained for about 100 generation. However, for the growth and continuous work on viruses, one requires the continous cell lines. Only the transformed cells are used as continuous cell lines because they can be maintained for indefinite generations. Therefore, the transformed cells are known as immortal cell lines. HeLa cell line is one of the transformed cell lines which was obtained from the cervix malignant tumours of Henrietta Lacks of 31, who died in 1951 after the spread of cancer through out her whole body even after intensive irradiation. After several generations many cell lines lost all the original characters except their support for the viral propagation. Maintenance of cell culture line needs the trained persons, otherwise it gets contaminated and destroyed.

3.   Virus Identification

It is very difficult to identify a virus isolate because they could be observed only under electron microscope and tested serologically by using the human antibodies. They are isolated, purified, crystallized and characterized (particularly the base-sequence of nucleic acid).

In 1966, Gibbs and his associates suggested for the following eight approved characters for the identification of a virus :

    (i)   The nucleic acid types.
    (ii) The number of strands in the nucleic acid.
    (iii) The molecular weight of nucleic acid.
    (iv) Percentage of nucleic acid in a virion.
    (v)   The forms of particles.
    (vi) The forms of nucleocapsid.
    (vii) The host and
   (viii) The vector

This proposal was accepted with some modification by the International Committee on Nomenclature of Viruses (ICNV). These parameters are designed in abbreviations while describing a virus. They named the formula as cryptogram (Table 15.2).

4.   Viral Multiplication

A virus needs a living host cell to multiply. It invades the host cell and takes over metabolic machinery of the host. Consequently, depending on virus types, cell death occurs releasing thousands of similar viral particles. Multiplication of plant viruses and animal viruses are discussed in the preceding sections. Bacteriophages are described in Chapter 16: *Virus-II*

B. The Plant Viruses

In 1892, Iwanowski demonstrated that the TMV, even after passing through bacteria-proof filter, transmitted the mosaic disease through sap. In 1935, Stanley crystallized TMV and demonstrated that it was made up of nucleoprotein. Later on through electron microscopic studies, crystallography, ELISA test and the other modern techniques much informations were gathered.

However, most of the plant viruses contain RNA as genetic material. Majority of

(Testing for Aids by ELISA Test).

them contain rod shaped ssRNA. In addition, the other viruses are bacilli form or bullet-shaped or pleomorphic. Plant viruses or reoviridae contain dsRNA as genetic material e.g. the wound tumour virus. Cauliflower mosaic virus and dahlia mosaic virus contain dsRNA as their genetic material. No other ssDNA containing plant virus is known so far except Geminivirus. The families of plant viruses are given in Figs 15.2 and 15.3.

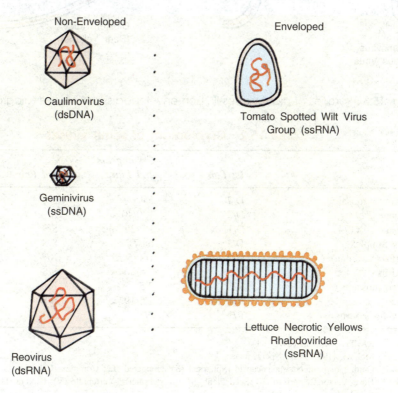

Non-Enveloped

Caulimovirus
(dsDNA)

Geminivirus
(ssDNA)

Reovirus
(dsRNA)

Enveloped

Tomato Spotted Wilt Virus
Group (ssRNA)

Lettuce Necrotic Yellows
Rhabdoviridae
(ssRNA)

Fig. 15.2 : Families of plant viruses-I: ssDNA, dsDNA and ssRNA, dsRNA (enveloped) viruses (diagrammatic).

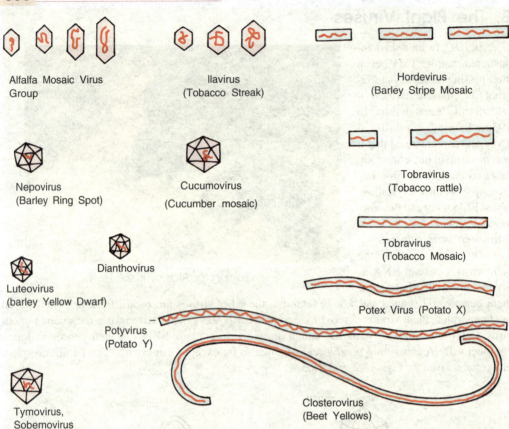

Alfalfa Mosaic Virus
Group

Ilavirus
(Tobacco Streak)

Hordevirus
(Barley Stripe Mosaic

Nepovirus
(Barley Ring Spot)

Cucumovirus
(Cucumber mosaic)

Tobravirus
(Tobacco rattle)

Tobravirus
(Tobacco Mosaic)

Dianthovirus

Luteovirus
(barley Yellow Dwarf)

Potex Virus (Potato X)

Potyvirus
(Potato Y)

Closterovirus
(Beet Yellows)

Tymovirus,
Sobemovirus
Tombus Virus

Fig. 15.3 : Families of plant viruses-II : non-enveloped ssRNA viruses (diagrammatic).

Table 15.2 : Cryptogram of some viruses

Virus	Cryptogram			
	1st Pair	2nd Pair	3rd Pair	4th Pair
Cauliflower mosaic	D/2	5/15	S/S	S/Ap
Coliphage T4	D/2	130/40	X/X	B/O
Coliphage ØX174	D/1	1.7/25	S/S	B/O
Herpes Simplex	D/2	68/7	S/S	V/O
Poxvirus (Vaccinia)	D/2	160/5.75	X/*	V/O
Poliovirus	R/1	2.5/30	S/S	V/O
Tobacco mosaic	R/1	2/5	E/E	S/O
Tobacco necrosis	R/1	1.5/19	S/S	S/fu
Yellow fever virus	R/*	*/*	S/*	V,I/Di

1st Pair : NA Type/ Strand (D/R, 2/1)
2nd Pair : Mol wt. (106) or NA/ percentage
3rd Pair : Outline of particle/ nucleocapsid [Spherical (S); elongated (E); Complex (C)]
4th Pair : Host/ Vector Bacterium (B). Seed Plant (S), Invertebrate(I), Vertebrate (V), Diptera (Di), Coleoptera (Cl).,
Aphid (Ap)., Fungus (fu).

* Unknown

Vectors play a key role in transmission of plant viruses. The vectors belong mostly to the order Hemiptera (e.g. bugs, aphids), and some to the orders Orthoptera (e.g. grasshopper and locusts) and Coleoptera (e.g. beetles) of Arthropoda.

In 1966, a plant virus committee was established by the executive committee of the International Committee of Nomeclature of Viruses. This committee recomended a classification of viruses. This classification includes the viruses which have been studied well. These groups are as below :

1. Tobravirus group.	2. Tobamovirus group.
3. Potexvirus group.	4. Carlavirus group.
5. Potyvirus group.	6. Cucumovirus group.
7. Tymovirus group.	8. Comovirus group.
9. Nepovirus group.	10. Bromovirus group.
11. Tombusvirus group.	12. Caulimovirus group
13. Alfalfa mosaic virus group	14. Pea enation mosaic virus
15. Tobacco necrosis virus	16. Tomato spotted wilt virus

Some of the important plant viruses are described in this section.

I. Tobamovirus Group

1. Tobacco Mosaic Virus (TMV)

(Tomato plants showing symptoms of TMV infection.

TMV is the most serious pathogen causing mosaic on tobacco leaves. It is transmitted by artificial inoculation but not by insect vectors. TMV is the most resistant virus known so far of which the thermal death point is 90°C for 10 minutes. This is the first virus that was crystalized in 1935 by W.M.Stanley in the U.S.A.

(*i*) **Symptoms :** TMV damages the solanaceous plants. However, it can infect the other plants too. Several strains of TMV has also been reported. After infection, it develops symptoms of lightening of leaf colour along the veins in early stages. Thereafter, it turns into light and dark green mosaic symptoms. Along the veins green colour turns into dark green and the internal region turns into chlorotic. Some times dark green blisters appear in the leaf blade. If the plants are infected early in season they become stunted. However, symptoms vary with varieties of tobacco. The virus reduces the yield as well as quality of the products i.e. the nicotine content is decreased by 20-30 %.

(*ii*) **Virus Structure:** Franklin *et al.* (1957) have described the structure of TMV. It is rod-shaped helical virus measuring about 280×150 μm with a molecular wieght of 39×10^6 daltons. The virion is made up of 2,130 protein subunits of identical size. The protein subunits are arranged around a central hole of 4 nm (40Å) (Fig.15.4 A-B). Each protein subunit is made up of a single polypeptide chain which possesses 158 amino acids, the molecular wieght of which is 17,500 daltons. Inside the protein capsid there is a single stranded RNA molecule which is also spirally coiled to form helix. Virus RNA consists of 6,500 nucleotides. In one turn the RNA contains 49 nucleotides. Total number of protein subunits counting in three turns is 49 i.e. 49/3 subunits per turn. Therefore a single protein subunit is linked with 3 nucleotides of RNA. Arrangement of capsomers on RNA is shown in Fig.15.4 B.

(*iii*) **Protein Synthesis :** Takeba (1975) demonstrated the direct entry of TMV into the isolated protoplast from mesophyll cells of tobacco. After making entry, RNA rapidly starts uncoating by removing the subunits from the capsid by using host enzymes. The parental RNA is localized in nucleus but not in cytoplasm. It performs two important functions: (*i*) it acts as mRNA and directs and the synthesis of protein, and (*ii*) functions as template for synthesis of complementary strand.

The virus RNA utilizes the amino acids, ribosomes and tRNA of the host and synthesizes the complementary strand and proteins i.e. coat proteins of 17,500 daltons and two other polypeptides (of molecular weight 160,000 and 140,000 daltons). The ratio of nucleic acid and protein differs with each virus. Nucleic acid is about 5-40% of the virus and protein 60-70%. Each protein subunit of TMV consists of 158 amino acid making a total number to about 17,531.

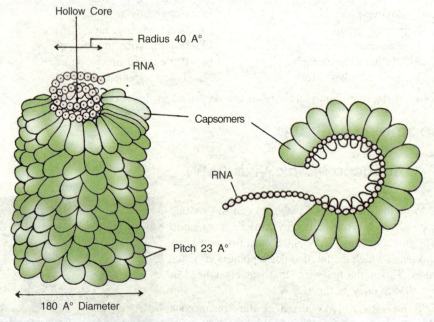

Fig. 15.4 : Structure of tobacco mosaic virus. A, helical model of TMV showing structure of RNA containing capsomers; B, arrangement or capsomers on RNA in one turn.

(*iv*) **Transmission :** TMV is transmitted through the cell sap of host and enters a new host through wound incision. Wound is caused in plant due to various cultural operations such as clipping or topping the shoot. It is not seed transmitted but acts as seed contaminant. It is also transmitted by wind and water.

Various control methods of the disease are regular roguing diseased plants and weeds, sanitation and use of resistant varieties.

II. Potex Virus Group

1. Potato Virus X (PVX)

PVX is a ssRNA virus which is distributed throughout the world. Development of symptoms on different varieties of potato differs. There are several strain of PVX. The infected tubers transmit the viral particles. It can remain alive for about 5 months but can be inactivated at 74°C. A large number of solanaceous plants such as tobacco, datura, *Solanum nigrum*, egg plants, pepper, tomato, etc. can also be infected by PVX.

(*i*) **Symptoms** : PVX is also called by the other names such as *Solanum* virus, potato latent virus mottle virus. The infected potato shows a wild mosaic between the veins on foliage. Some times it is not visible clearly. In many cells PVX do not impart any symptoms. The infected plants become dwarf and deformity in foliage occurs. Symptoms produced on different hosts differ considerably. In the varieties *Epicure, crest, Key Edward* the virus produces top necrosis.

(*ii*) **Structure of PVX** : Unlike TMV, PVX consists of a flexous rod in helical symmetry of the dimension of about 515 x 11.2 nm (see Fig. 15.3). Capsid is made up of identical protein subunits forming a helix of 3.3 nm pitch with a hole of 3 nm diameter. Possibly, a single subunit is associated with 3 or 4 nucleotides.

Strains of PVX exist in nature which can be detected serologically or on the basis of symptoms in tobacco and *Capsicum* plants, thermal inactivation point (TIP 68-67°C), reaction in potato cultivars determining top necrotic response.

In India research on viruses and viral diseases was started in 1940 when import of seeds from Europe was banned and available seed stocks degenerated within the short period. The Indian Council of Agricultural Research started a scheme for production of disease-free seed potato which came into operation at Kufri (now the Central Potato Research Institute) at Shimla in 1940. PVX was first isolated in 1945 by Vasudeva and Lal. Thereafter, several group of this virus was reported.

(*iii*) **Epidemiology :** PVX spreads through rubbing, contact of plants and tubers, seed cutting knife, farm implements, clothing and animal fur. In stores it can spread by sprout contact also. Root clubbing may also transmit PVX to a limited extent. In nature a wide range of PVX-infected plant species including weeds *(Dahlia, Solanum chacoense)* serve as the source of virus infection.

(*iv*) **Transmission** : Besides a fungus *Synchytrium endobioticum* causing potato wart also transmit through oospores. Like TMV, it is also sap transmitted and no vector has been reported so far. It spreads through contact between the healthy and diseased plant, core grafting and dodder (*Cuscuta compestris*). However, it perennates in the diseased seed stocks. Average tuber infection varies from 13 to 23%.

(*v*) **Yield Loss:** PVX causes 10-30% yield loss. But the combined effect of PVX and PVY or PVA causes severe disease and reduce yield loss drastically. PVX-infected plants turn more or less susceptible to both early blight *(Alternaria solani)* and late blight *(Phytophthora infestans)* due to secretion of antifungal substances in leaves. Extracts and washings from PVX-infected plant leaves suppressed the germination of *A. solani* conidia due to the presence of pathogenesis-related proteins (PRP). A comparative assessment of yield loss due to different potato viruses and viroids is given in Table 15.3.

Table 15.3 : Yield loss due to potato viruses/viroids (based on Khurana and Singh, 1996).

Potato viruses	Yield loss (%)
Potato virus X	10-30%
Potato virus Y	80%; in autumn 14-20%, in spring 44-57%
Potato virus S	3-20%
Potato virus M	10-30%
Potato Virus A	40%

(*vi*) **Control:** There are different methods which can minimise the yield loss and control the viral infection such as plant roguing, seed certification, use of resistant varieties, etc. Recently, transgenic PVX-resistant cultivars have also been produced by incorporating the viral coat protein gene in the potato plants through tissue culture methods (Khurana and Singh, 1996).

III. Potyvirus Group

2.   Potato Virus Y (PVY)

PVY is also known as *Solanum* virus 2, potato virus 20 and potato leaf drop streak virus. It is distributed throughout the world on a wide range of hosts in Solanaceae, for example tomato, potato etc. Several strains of PVY have also been reported which differ in physical features. The thermal death point is 52°C for 10 minutes and longevity 24-48 hours at the laboratory conditions.

(i) **Symptoms :** Symptoms caused by PVY vary according to environment and host varieties. It ranges from weak mosaic to foliage necrosis. The common symptoms are leaf drops streak or acropetal necrosis. Secondary infected plants show less necrosis but are dwarfed with brittle, crinkled and branched leaves. In combination with PVX, it causes rugose mosaic of potato. This disease causes a severe damage to the plants. It shows variable symptoms in different varieties of potato. The foilage is mottled which in turn becomes wrinkled, punctured and remarkably reduced in size too. The margin of leaflets gets rolled downwardly and plant is stunted. Abnormal hairiness of leaves and dwarfing of whole plant occurs. When the plants are severely infected they die before producing tubers. Host range of PVY is mainly Solanaceae but the virus also causes disease in Chenopodiaceae and Leguminosae.

(ii) **Structure of PVY:** Potato virus Y (Potyvirus) is a non-enveloped flexuous filamentous rods with helical symmetry containing liner ss(+)RNA (molecular weight $3\text{-}3.5\times10^6$ Daltons). It is cytoplasmic virus but some have nuclear inclusions. It is not persistent in aphid vectors. It is the major virus of potatoes because it spreads easily and reduces the yield. There are three types of strains of PVY viz., PVY° (common strains), PVN^n (tobacco vein necrosis strains), PVY^C (stipple streak strains). These strains are identified according to severity of symptoms in tobacco, potato, etc. PVY is divided into three subgroups: *(i)* Potyvirus (aphid-transmitted virus), *(ii)* Bymovirus (fungus-transmitted virus), and *(iii)* Rymovirus (mite-transmitted virus).

(iii) **Transmission :** The small sized tubers are responsible for transmission of PVY under field conditions. An insect vector, *Myzus persicae* effectively transmit PVY on the diseased plants.

(iv) **Epidemiology:** PVX spreads in nature mainly through both infected tubers and aphids. Infected plants show a source of inoculum. It replicates in cytoplasm of infected cells and involves a membrane-bound replicase system.

(v) **Yield Loss:** PYV is the most important virus causing about 80% loss according to variety, environment and virus strains. PVY can cause complete failure of the crop. In combination with PVS or PVA, PVY induces synergistic symptoms called rugose and crinckle mosaic, respectively PYV infected plants are more susceptible to early and late blights.

(vi) **Control:** A large number of varieties having high degree of resistance to PVY have been bred using *Solanum tuberosum* and *S. stoloniferum*. Six resistant genes have been identified in *S. tuberosum* (Khurana and Singh, 1996)

IV. Tymovirus Group

1.   Cucumber Mosaic Virus (CMV)

CMV occurs in several countries including India. It infects about 100 plants belonging to about 20 families. It occurs most frequently on the members of Cucurbitaceae, tomato, potato, carrrot, spinach, etc.

(i) **Symptoms:** Leaf mottling, crinkling and curling of edges occur on cucumber. Petioles and internodes are shortened resulting in stunted and compact appearance of the plant. If the plants are

infected early in season, flowers do not develop, and if infected late only a few small sized grains are formed. The fruits too are mottled with light yellow and white patches. The older leaves eventually dry. On the other plants symptoms differ.

(*ii*) **Structure of CMV :** CMV is icosahedral weighing to about 5.5×10^6 daltons. It contains RNA of four classes, the three larger are required for infection.

(*iii*) **Transmission :** CMV is transmitted through sap and by insect vectors too such as *Aphis gossypi, Myzus persicae, Macrosiphum euphorbiae* etc. It does not persist in vectors.

IV. Tomato Spotted Wilt Virus (TSWV)

Tomato spotted wilt virus is a serious viral diseasese which is found throughout the world. It is caused by a complex of four or five strains of TSWV. It has a very wide host range infecting about 200 plant species, failling under several species. It infects both dicotyledon and monocotyledon plants.

(*i*) **Symptoms :** The first sign of infection is the appearance of dark stained granular area in the cytoplasm. In the infected plants slighty bunch appears at the growing points, thereafter, the tender leaves curl. In older leaves margin becomes dark brown coloured resulting in development of irregular dark patches on lower leaves. Gradually the symptoms spread to entire plant with stunting growth. In diseased plant fruits are not formed or if formed they show colour variation and mottling symptoms.

(*ii*) **Structure of TSWV :** TSWV is a pleomorphic myxovirus. The diameter of viral particles in the infected tissue is about 70 nm which is enclosed by a membrane of 50Å. It contains ssRNA as the genetic material. Exact size of the virus is not known so far.

(*iii*) **Transmission :** TSWV is sap transmissible. An insect vector, *Thrips tabaci* transmits the virus only at stage of larva.

V. Cauliflower Mosaic Virus (CaMV)

The cauliflower mosaic is caused by CaMV. The CaMV consists of open circular dsDNA as genetic material with single strand discontinuity like hepadna virus. The DNA is linear *in situ* but gets circularized after extraction.

CaMV shows icosahedral symmetry of the capsid with a 50 nm diameter. It consists of more than one protein shell. In the cytoplasm of infected cauliflower leaves, CaMV forms characteristic X bodies which are rounded structure (Shephered and Wakeman, 1968).

(*i*) **Symptoms :** From the cell walls of infected leaves, there arise finger like processes. It has also been observed that mitochondria and nuclei become abnormal in the infected cells of host leaves.

(*ii*) **Transmission :** CaMV is transmitted by aphids.

VI. Potato Leaf Roll Virus (Polerovirus)

Leaf curl of tomato is caused by Polerovirus (potato leaf roll virus). Polerovirus falls in the family **Luteovirus.** Leaf curl of tomato has been reported from many part of the world. It is most common is winter season in India. It has non-enveloped particles with isometric symmetry. It possesses one linear (+) ssRNA as genetic material.

(*i*) **Symptoms:** The symptoms that develops in the infected plants are dwarfing,

Tomato plant showing symptoms of yellowlleaf curl

puckering, twisting and curling towards the dorsal side of leaves, mottling, vein clearing, excessive branching, shortening of whole plant, and partial or complete sterility in the plants.

(ii) **Transmission:** This virus spreads through the contaminated seed of freshly infected fruits. Besides, it is also spread by vector. The chief vector is *Bremisia tabaci.*

(iii) **Control:** So far no definite control measures have given fruitful result. However, developing resistant varieties, spraying with Ekatox (0.02%) and Regor (0.05%) at 10 days interval reduce the infection of plants.

VII. Rice Tungro Virus

Tungro (yellow–orange) is the most important virus disease of rice in the South and South East Asian countries. It is a major constraint in stable rice production in this region causing annual loss of about US $ 1.5 billion.

(i) **Causal Agent:** Tungro is caused by joined infection of two distinct viruses, *rice tungro spherical waikavirus* (RTSV) and *rice tungro bacilliform badnavirus* (RTBV). RTSV has non-enveloped isometric particles with one (+) ssRNA. RTBV is rod shaped, non-enveloped particle of 130×30 nm. It consists of circular dsDNA genome of 8 kb size. Replication is nuclear and its genome does not integrate with host genome.

(ii) **Symptoms:** The diseased plants are much stunted and show mottling and yellow-orange discoloration of leaves. The infected plants produce poor panicles with empty glumes that impart dark brown colour. Infected plants bear reduced tillering. Basically, RTBV causes typical tungro symptoms, whereas RTSV alone causes no symptom except mild stunting. But the typical tungro symptoms are intensified by RTSV. Thus tungro symptom develops in a synergistic manner.

(iii) **Transmission:** Both RTSV and RTBV are transmitted by aphids and green leafhopper (*Nephotettix virescens*) in the semi-persistent way. RTBV is transmitted by leafhopper from the host and cause disease only when RTSV is also present in the infected plants.

(iv) **Control:** Management of this disease by development and use of conventional tungro resistant rice cultivars has been totally unsuccessful due to breakdown of resistance. Because this was primarily against the vectors and not against the virus. Earlier approaches for tungro resistance was based on the development of viral symptoms without assessing the presence or absence of tungro virus. The control measures include use of tungro-resistant varieties, roguing vector control, destruction of volunteer plants, etc. Plant oils (e.g. 2% coconut and mustard oil) was found good to control vector of rice tungro virus.

VIII. Mosaic Disease of Sugarcane

Sugarcane is one of the important crops of India which has been sown since the dawn of human civilisation. Besides the attack of several pathogens and pests, it is infected by many viruses which cause different types of diseases such as Fiji disease, chlorotic streak, mosaic, ratoon stunting, Australian dwarf, etc. Mosaic disease of sugarcane has been described in this section.

In 1921, this disease was noticed first. But it is widespread in all sugarcane growing areas of India. It becomes a potential threat through development of pathogenic strains. More than 20% loss is caused by this diseases alone where juice quality remains unimpaired.

(i) **Causal Agent:** Mosaic of sugarcane is caused by sugarcane mosaic virus (SCMV) or *Saccharum* virus 1 or sugarcane virus 1. It is a member of family Potyviridae. There are six different strains of SCMV that cause mosaic disease.

(ii) **Symptoms:** For the development of symptoms, SCMV must be inside the living tissues especially in phloem established bundles. Six weeks after planting there develop pale patches or blotches in the green tissues of the leaves. Moreover, the diseased leaves show mottling, chlorotic

or light coloured, irregularly spread stripes or streaks. An area of proteolysis develops in one part of the cytoplasm in the cells of affected plants. This areas shows a vacuolated mass having X bodies. These can be heavily stained and observed under microscope.

(iii) **Transmission:** The SCMV particles are transmitted through diseased cane setts which are used as seeds. Perennial grasses also act as host and a source of reservoir of inoculum of SCMV.

(iv) **Control:** There are several methods which have been recommended for the control of mosaic disease of sugarcane such as : (a) use of healthy setts as seeding material, (b) systemic roguing of infected cane when infection is not severe, (c) elimination of perennial grasses acting as weeds and host, (d) avoiding ratooning practices if plants are heavily infected, (e) using resistant varieties, (f) operating of good quarantine, etc.

IX. Transmission of Plant Viruses

Plant viruses are the obligate parasites which need cellular machinery of living host for their multiplication, propagation and survival. Moreover, wind or water cannot help in their transmission. The knowledge for transmission of plant virus is necessary for control of viral diseases of plants. Some of the ways of plant virus transmission have been described in this section.

1.   Mechanical Transmission

In nature plant viruses are mechanically transmitted from diseased to health plants by rubbing leaves together, injecting plant extract, by action of animals, etc. Viral particles remain adhered to plant surfaces, epidermis or hairs. During rubbing the cells are broken and viral particles are liberated in the damaged cells. Transmission through this mechanism occur in such plants which are closely planted.

Similarly viral particles attached on surface of animal body are transmitted when they rub their body first on infected plants and then on healthy plants. Viral particles enter through the injuries made by animals. Similarly birds also transmit viruses by this method, for example TMV.

Potex virus survives on human clothing and agricultural tools such as cutting knives, sickles, etc. These tools are frequently used in agriculture and horticulture. Hence their use first on infected plants then healthy plants facilitates viral transmission.

Indicator Plants: The test plants that react promptly and characteristically to particular viruses are called 'indicator plants' or differential hosts'. Only susceptible plants act as indicator plant but not the virus resistant plants. Usually the cotyledons of cucumber and primary leaves of cowpea and beans are inoculated by viruses, because older plants are less susceptible. Virus resistant plants consist of some special inhibitory substance that prevent the transmission of viruses. Table 15.4 shows some of the indicator plants commonly used for inoculation purpose.

Table 15.4 : Some common test plants.

Plant species	Plant age	Number of leaves
Beta vulgaris	5 weeks	5
Chenopodium amaranticolor	2 months	6
Cucumis sativus	10 days	cotyledons
Datura stramonium	5-8 weeks	1 or 2
Lycopersicon esculentum	3-4 weeks	2 or 3
Nicotiana tabacum	5 weeks	3 or 4
Phaseolus vulgaris	10 days	2 primary
Solanum tuberosum	2 weeks	4
Trifolium incarnatum	5 weeks	3 or 4
Vigna sinensis	10 days	2 primary

2.   Vegetative and Graft Transmission

Generally viral particles are present almost in all parts of systematically infected plants. Virus will certainly be transmitted to the progeny, if any part of such mother plants in used for vegetative propagation through rhizomes, bulbs, corns, tubers, cuttings, etc. Therefore, one must not use the infected vegetative part of vegetatively propagating plants such as dahlia, chrysanthemums, carnations, potatoes, etc.

Grafting technique (placing the cut end of one plant onto immediate contact of tissues of other plants to establish a union product in one plant) has been well practised in India since time immemorial. There is wide variety of grafting techniques such as stem grafting or wedge grafting, tuber grafting (in potatoes), root grafting, etc. Grafting is widely used commercially for propagation of plants. For example, more than 4,000 varieties of mango have been possible due to grafting. Systemic virus can be transferred from infected portion of a plant to the healthy portion of other one e.g. colour breaking of tulip, apple mosaic. *Nicotiana glutinosa* plants die when tomato or tobacco plants systematically infected with TMV are grafted with it because TMV produces necrosis of leaves and buds.

3.   Pollen Transmission

When pollens consisting for viruses fall on stigma of female plants, they germinate and eventually facilitate the virus to infect the ovules of plants. Such viruses are called pollen-borne viruses. Example of pollen-transmitted viruses are: barely stripe virus, tobacco ring spot virus, bean common mosaic virus, fruit ring spot virus. *Dhatura* mosaic virus is transmitted through pollen to seeds in 79% offspring.

4.   Seed Transmission

Seed transmission of viruses is very rare but many viruses are known to be seed transmitted. However, a very low level (0.1%) of seed transmission had epidemiologically been found out. Some example of seed transmitted viruses are bean mosaic virus, tomato ring spot virus, tobacco ring spot virus, cowpea mosaic virus, cucumber mosaic virus, mung and urd bean mosaic viruses, etc. Seed transmitted viruses are present in embryo, endosperm or seed coat. After seed germination, virus-infected seedlings are produced.

A substantial yield loss is caused by viruses. For example, soyabean mosaic virus results in 10-20% decline in seed weight with seed discolouration. Since seeds are carried for a long distance within a country or across the country, viruses are transmitted over long distances. Seed-borne viruses get good chance of survival between two seasons and during adverse conditions. Besides, nematodes also transmit the seed-borne viruses.

5.   Nematode Transmission

There are some plant pathogenic nematodes that feed roots of plants. Such nematodes also act as vector for some viral pathogens. **Vectors** are the organism that assist in transmission of viruses. Examples of some plant parasitic nematodes are: *Longidorus, Paratrichodorus, Trichodorus* and

Xiphinema. These nematodes transmit viruses for about 10 months but transmission does not involve viral replication inside the vector.

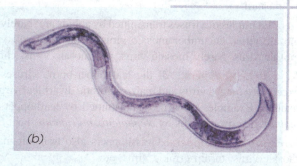

Xiphinema (a) and Trichodorus (b) are virus transmitting nematodes.

There is virus-vector specificity for transmission by nematodes. *Nepoviruses* (polyhedral) are transmitted by *Xiphenema* and *Lingidorus,* and *Tobraviruses* (tubular) are transmitted by *Paratrichodorus* and *Trichodorus.* In soil two types of virus transmission can be observed: *(a)* some of the viruses remain active in debris of roots are leaves. Healthy plants that come in contact of debris are infected, *e.g.* TMV, *(b)* virus is transmitted to healthy plants by nematodes that inhabit the soil *e.g.* potato rattle virus.

6.   Fungal Transmission

There are many soil-inhabiting fungi which transmit several viruses, for example *Olpidium brassicae, Polymyxa graminis, Synchytrium endobioticum* and *Spongospora subterranea.* These fungi are obligate endoparasite of higher plants. Their zoospores infect the roots of new hosts, introduced viruses and produce virus specific symptoms. These fungi acquire virus from virus-infected plants which persists in soil for several months or years. *O. brassicae* transmit tobacco necrosis virus, *P. graminis* transmit wheat mosaic virus and *S. subterranea* transmit potato mop top virus.

7.   Insect Vector Transmission

Vectors are highly mobile and play an important role in natural ecology of viruses. In many cases relationship between virus and vector is an intimate biological association, besides mere mechanical transfer. Most of the viruses are transmitted by vectors. They have slender, needle-like mouthparts to which they pierce the cells and suck cell sap of host plants. Virus is transmitted to healthy plants when the viruliferous insects feed the plant tissue.

Viruses are divided into three groups, on the basis of length of the period and relationship with insect: *(a)* **non-persistent** [where acquisition period i.e. feeding period on diseased plant is for a short period (10-60 second); thereafter rate of transmission decreases because viruses survives for a short period], *(b)* **semi-persistent** (acquisition period is of 12-24 hours and transmission occurs immediately after acquisition for 1-3 hour. Hence, virus survives for a few hours in the vectors), and *(c)* **persistent** (viruses survives for a week or month in the vectors; hence, transmission occurs only where the inoculation feeding lasts at least for some hours). Some of the important insect vectors and viruses transmitted by them are described in this section.

(i) **Aphids:** Aphids are the most notorious and important groups of plant vector. They are found in large numbers during spring and winter seasons. They show preference towards feeding

of the hosts, for example *Aphis craccivora* and *A. fabae* prefers beans; *A gossypii* prefers cotton, cucurbits, chilli and brinjal; *Myzus persicae* prefers tobacco; *Liaphis erysimi* infects crucifers such as mustard.

Examples of viruses transmitted by aphids are: barely yellow dwarf virus, potato virus S, alfalfa mosaic virus, cucumber mosaic virus, potato leaf roll virus, lettuce necrotic yellow virus, bean mosaic virus, barely mosaic virus, pea mosaic virus, chilli mosaic virus, etc.

(ii) **Leafhoppers:** All the leafhopper-borne viruses are transmitted in a persistent manner, except rice tungro virus and maize chlorotic dwarf leaf virus which are transmitted in semi-persistent manner. Examples of viruses transmitted by leafhoppers are: maize streak virus by *Cicardulina mobila,* rice dwarf virus by *Nephotettix nigripictus,* beat curly top virus by *Circulifer tennelus,* etc.

(iii) **Grasshoppers:** Grasshoppers also transmit many viruses such as TMV, turnip yellow mosaic virus, turnip crinkle virus, etc.

(iv) **Beetles:** More than seventy four spécies of beetle are know to transmit virus. Examples of beetle-transmitted viruses are cow pea mosaic virus, turnip yellow mosaic virus, squash mosaic virus, southern bean mosaic virus.

8.   Dodder Transmission

Dodders are the trailer or climber parasitic plants which grow forming bridge between two plants. *Cuscuta reflexa* is the most famous dodder plant that lacks leaves. They belong to the family Convolvulaceae. Dodders wind around the host and penetrate its haustoria into host tissue sending up to vascular tissue. Haustoria acquire virus from the infected plants that are eventually transmit to the new hosts. Dodder transmitted viruses are sugar beet curly top virus, tomato bushy stunt virus, tobacco rattle virus, etc.

X. Effect of Viruses on Plants (symptoms of virus infection)

Since viruses are the obligate parasites, they impart several external and internal symptoms on host plants. Some of the important symptoms are described in this section.

1.   External Symptoms

There is a variety of external symptoms that appear on infected plants. Some of the most common symptoms are described below:

(i) **Mosaic:** The most common effect of virus infection is the development of light (chlorosis) and dark-green areas on leaves giving a mosaic symptom. There are over 80 viruses that produced mosaic symptoms. The most common is the tobacco mosaic, bhindi mosaic, cotton mosaic, cowpea mosaic, etc. Chlorotic and dark-green colours are due to disturbances of chloroplast and decrease in chlorophyll content.

(ii) **Chlorosis, Vein Clearing and Vein-Banding:** Chlorosis refers to loss of chlorophyll and breakdown of chloroplast caused by viruses. Vein clearing symptoms

Chlorosis in grapes is causes by virus infection.

develop adjacent to veins before mottling or chlorosis of tissue, while broader bands of green tissue in chlorosis or necrosis is called-vein-banding, for example potato vein banding.

(iii) **Ringspots:** Ringspot is characterised by formation of concentric rings or broken rings of infected dead cells. The ringspot may be chlorotic rings rather than necrotic rings. When necrosis occurs in rings alternating with normal green area, such spots are called necrotic ringspots. One of the common examples is the ringspots caused by tobacco ringpot virus.

(iv) **Necrosis:** Besides localised cell death in necrotic local lesions or ringspots, the other parts of plants also suffer from necrosis in certain areas, organs (e.g. leaves, fruits, seeds, tubers) or entire plant. Affected leaves show scattered necrotic patches of dead tissues. Commonly potato virus X and Y impart necrotic lesions.

(v) **Leaf Abnormalities:** Due to virus infection leaves show abnormal growth like leaf curling, leaf rolling (upwardly or downwardly), crinkling, puckering (depression), etc. The other abnormalities may also develop in leaves such as smaller, blistered and thickened leaves. Examples are leaf curl of papaya, leaf roll of potato, etc.

Grapes leaves showing necrosis.

(vi) **Stunting:** Virus-affected parts of plants (e.g. leaf, flower and fruits) are reduced. The petioles and internodes are shortened in bean infected by bean yellow mosaic virus. In some of banana virus infected banana cultivars, the fruit emerges from the side of the pseudo-stem instead of coming from the top.

(vii) **Flower Symptoms:** Colour-breaking symptom is very common in the flowers of many virus-infected plants. The colour breaking means streaks or sector of tissues with such colour that are different from the normal one. This happens due to loss or increase of anthocyanin pigment in petals. In addition to colour-breaking, flowers are of reduced size and deformed shape. One of the most popular example is the tulip flower breaking and wall flower *(Cherianthus cheri)* breaking.

2. Internal Symptoms

(i) **Histological Abnormalities:** Plant parts showing external symptoms also displays anatomical and histological abnormalities within the plants. These abnormalities include necrosis, hypo-and hyper-plasia. In leaves showing mosaic symptom, mesophyll cells are smaller and less differentiated. Some of the abnormalities are discussed in this section. The xylem elements in grap vine infected by grape vine fan leaf virus contain characteristic lignified strands commonly known as **endocellular cordons**. Tyloses are formed in xylem elements of barley leaves infected by barely leaf dwarf virus and other leaf yellowing viruses. The phloem cells degenerate or die and callose deposition occurs on phloem sieve plates. Starch accumulation in chloroplasts occurs in TMV-infected plants.

(ii) **Cytological Abnormalities:** Many cytological abnormalities are seen when virus-infected cell are studied cytologically. Moreover, some viruses affect specific organelles only. Tymovirus induces formation of marginal vesicles in chloroplasts. Tobacco rattle virus modifies mitochondria and aggregates to form inclusion bodies. TMV particles are found in cytoplasm, whereas TMV of

strain U5 are round in chloroplast and nuclei. Rhabdovirus particles are found in spaces between outer and inner nuclear membrane. Nucleocapsid of virus is found inside the nucleus. Particles of tomato spotted wilt virus are found in membranous bags, those of beet yellow virus occurs in bundles and particles of tobacco rattle virus are packed around mitochondria. In geminivirus-infected plants, viral particles are present inside nuclei. Moreover, nuclei may be granular with reduced chromatin. Nucleoli are hypertrophied or disintegrated.

Chloroplasts may be rounded, swollen and clumped together in the cells. Chloroplasts may be fragmented and colour may turn to colourles.

The major cytological effect of viruses is the development of **inclusion bodies** *i.e.* intracellular structures produced as a result of viral infection. They contain virus particles, or any materials related to virus or cell materials. They may be nuclear or cytoplasmic. Some of them are large sized which can be seen by light microscope. Inclusion bodies may be crystalline, amorphous or pinwheel shaped. TMV produces crystalline inclusion bodies and potyviruses produce pinwheel inclusion bodies (Fig. 15.5). Amorphous inclusion bodies look like certain microbes and are called *X-bodies*. Basically X-bodies are granular, highly infectious having vacuoles and located adjacent of nucleus. These are inducted by viruses where virus components are synthesised. Some time inclusion bodies help in diagnosis and classification of viruses.

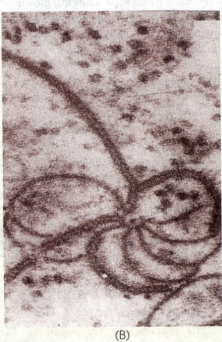

(A)   (B)

Fig. 15.5 : Crystalline inclusion bodies produced by TMV in tobacco (A); pinwheel inclusion bodies produced by potyviruses (B).

XI. Serological Tests for Diagnosis of Plant Viruses

Like other viruses, plant viruses also act as antigens. Their protein coat possesses antigenic properties. However, specific antigens of each virus differ from the other as well as host to which they infect. All serological tests depend upon surface properties of coat proteins. Each antigen will react with its antiserum (antibodies) which may be demonstrated *in vitro* by mixing viral preparation (antigen) with blood serum (antibody). Study of these sensitive and specific reactions of antigen and antibody is called **serology.** Serological studies help in rapid diagnosis of viruses (present in seeds and seed stocks and diseased plants), strain differentiation and virus identification.

(i) **Preparation of Viral Antigens:** The crude antigen is prepared by using fresh leaf tissues/ diseased seeds/seedstocks in coating buffer. Tissues are macerated well so that viral particles could be released from damaged cells to out side. Extract is centrifuged and supernatant containing viral particles is used to detect the presence of viruses.

The experimental animals (e.g. rabbit) are injected by purified viral preparations/ viral coat protein/ proteins from disrupted virus/proteins expressed from cDNA. Animal is repeatedly injected at a few days interval. Ten days after the last dose, blood is collected from the experimental animal and allowed to clot for 2 hours at room temperature. After keeping in the refrigerator overnight the serum is procured through centrifugation. The serum is stored in a refrigerator or lyophilized. Antiserum has been prepared against a large number of plant viruses such as TMV, PVX, etc. Some of the methods (e.g. precipitation test, ring test, agglutination test, complementation test) have been described in Chapter 19. Some other serological methods are described below.

(ii) **Histochemical Test by Immunofluorescence:** This method is suitable to detect viral particles present in protoplasts of plant cells. There are many fluorescent dyes such as rhodamine B (emits red light), Texas red (emits red light), CY3 (emits orange light) and fluorescein (emits green light). Thin sections of virus-infected seeds/seeds stocks are cut using a microtome. A fluorescent dye is mixed with γ–globulin extracted from the antiserum. Sections containing viral particles are stained with fluorescent stain. Fluorescent dye-tagged immunoglobulin enters into the infected tissues. The antigen-antibody reaction takes place. When specimen is exposed to UV light, the cells emit fluorescence of varying colours (based on type of dyes) inside which immunoglobulin-tagged dye has bound with antigen. This gives a positive result. The fluorescence can be detected through a fluorescent filter fitted in a fluorescent microscope.

(iii) **Agar Gel Double Diffusion Test:** This test is performed to determine the antigenic relationship of the viruses. Petri plates containing gelled agar plates are used to prepare wells. Wells are prepared in such a way that one well must be in centre and the others to its periphery. The peripheral wells are filled with purified viral suspension and central well with similar amount of known antiserum. If antigens react positively with any antiserum, fusion of precipitin line (spur) is observed. This shows that the two isolated strains are identical. Since both the reactants diffuse into the well, it is called double diffusion test.

(iv) **Direct Antigen-coated ELISA (DAC-ELISA):** This technique was developed by Hobbs *et al.* (1987). In this method an assay is performed on a polystyrene microtitre plate. The crude antigen is prepared by using fresh leaf tissues of infected parts of the plant in coating buffer and extract is prepared through maceration of tissues. Pure extract is obtained after centrifugation. The extract is placed into wells and incubated at 37°C for 1 hour. Then antigens are washed.

The plates are blocked with milk powder solution for 1 hours. Polyclonal rabbit antibody and anti-rabbit immunoglobulin alkaline phosphatase conjugates are used at 1:1000 and 1:30,000 dilution, respectively. The reaction is read at 405 nm after 1 hour of adding substrate (*p*-nitrophenyl phosphate) by an ELISA- reader.

(v) **Double Antibody Sandwich ELISA (DAS-ELISA):** This technique was developed by Permar *et al.* (1990). DAS-ELISA techniques has been used for detection and identification of strains of viruses. Polyclonal antiserum (PAb) detects the presence of virus up to 10^{-3} dilution. Differentiation of mild and severe strains is done using monoclonal antibodies (MAb). Strain differentiation achieved by MAb depends on its structural epitope.

Small amount of infected plant tissues is transferred in phosphate buffer and macerated. Wells of ELISA plate are coated with γ-immunoglobulin (IgG). Plate is incubated for 1-4 hours and washed with water. Enzyme-tagged with IgG diluted in conjugated buffer is added to each well and incubated for 1-4 hours. It is washed and substrate solution is added. After 15 minutes optical density is measured at 405 mn in an ELISA reader.

(vi) **Indirect Double Antibody Sandwich ELISA (I-DAS-ELISA):** Nikolaeva *et al.* (1998) found that MAb (e.g. MCA-13) cannot distinguish between two isolates among the severe strains

of virus. To overcome this problem, they developed a set of primary/secondary combination of trapping and detecting antibodies in an I-DAS-ELISA.

Nikolaeva put forth the information on epitope-specificity of a range of viral specific antibodies which helped to develop a serological system to distinguish between viral isolates. In this techniques a polyclonal antibody (e.g. Ch-110) is used for coating the plate and a secondary antibody R-109 was used before adding the conjugate.

The tissues of test sample are macerated and left for 6 days at 4°C. The sample is processed as described for DAS-ELISA. A dilution series of two antibodies is run to standardize the dilution combination CH110/R 109. Based on optimum combination dilution for testing strain was chosen as 1:1,00,000/ 1:15,000. The other details are as described for DAS- ELISA. Early colour development is considered to be non-specific.

(vii) **Detection of Virus by Nucleic Acid Spot Hybridisation (NASH) using DNA Probe:** This technique is used for detection of viral DNA in symptom-less host. Small samples (e.g. leaves, seeds, etc.) are collected from each plant after 21 days of inoculation. Total DNA from these samples is extracted and stored at −70°C for NASH. Prepared DNA sample is heat-denatured to get a concentration of 2.5 and 0.25 microgram of total DNA. DNA is dotted on nitrocellulose membrane. The membrane is dried and baked at 80°C for 2 hours in vacuum.

Virus-specific DNA probe is prepared from the clones of virus present in a plasmid (e.g. pUC18) after incorporating (α-^{32}P)-dCTP by random priming method. The probes of different size are generated using restriction enzymes that separates the viral DNA insert from the plasmid. Further restriction of the viral DNA by different restriction enzymes yields the viral DNA fragments of different size. These fragments are tagged with (α-^{32}P)-dCTP.

The radio-labelled probes are added to pre-hybridisation solution. The baked nitrocellulose membrane is transferred in pre-hybridisation solution. Pre-hybridisation and hybridisation are done at 65°C for 4 hours and 18 hours, respectively. The hybridised blots are washed in 2X saline buffer thrice for 15 minutes at 65°C. The membrane is exposed to X-ray film for 24-48 hours at −70°C and auto-radiographs are prepared. The hybridised viral DNA will form blots and indicates the presence of suspected virus in the sample.

(viii) **Electron Microscopy:** Electron microscopy helped in identifying the presence of viruses in plant tissues. For example, TMV studies were done by Stanley (1935). TEM studies are made in the leaf dip preparation which were infected by potyviruses. TEM studies revealed the presence of flexuous rods (750×12 nm). Ultra-thin sections of infected chilli leaves revealed the presence of cytoplasmic inclusion called as 'pinwheels' inclusion bodies which are typical of potyvirus group.

(ix) **Dot Immunobinding Assay (DIBA) or Dot ELISA:** In this method nitrocellulose membrane is used as solid substrate for ELISA. Because nitrocellulose membrane can easily be transported to detect a large number of samples in field. Using small quantities of antiserum virus present even in small quantities can be detected. Plant extract from infected tissues is prepared and blotted onto nitrocellulose membrane or enzyme-linked γ-immunoglobulin is coated onto it. A substrate is added to develop final colour. The substrate (nitroblue tetraxolium, 5-bromo-4-chloro-3-iodylphosphate *p*-toludine salt and formamide) is hydrolysed and produces bright stain which can be observed by eyes or a reflectance densitometer.

C. The Animal Viruses

In human's the viral diseases are known since the ancient times in India and China, for example small pox, influenza and common cold. Viral diseases like small pox were linked with super natural causes and people used to perform offerings to Goddess *Shitala* throughout the country assuming that *Shitala* had incarnated in the sufferers. Therefore, the sufferers had been given due attention by regular cleaning of cloths and beds, and maintaining the sanctity from all round. However, during the spread of epidemics the villagers got assembled and poured the aqueous preparation of extracts of some medicinal plants all around the village (at least one turn) just to prevent the entry of disease from the neighbouring villages having the sufferers. This had been practised even up to 1970. Still no body can forget the Bihar epidemic which occurred during 1973-74. It is also linked with unavailability of vaccination and other control measures.

In 1798, Edward Jenner discovered vaccination against small pox but during that period it was not associated with any virus or any other living organisms. So far more than 750 vertebrate viruses are known which cause serious diseases together, their antigenic serotypes increase the number to about 1000. Though animal viruses have been classified according to symptoms/organ affected/animals by the early men, yet scientific endeavour has also been made after the discovery of instruments.

I. Classification

On the basis of types of nucleic acid the classification of viruses is given in Table 15.5.

Baltimore (1971) classified the animal viruses in the following six groups according to the relationships between virion, nucleic acid and mRNA transcription. The RNA within the virion is known as plus (+) or sense strand because it acts as mRNA, whereas the newly synthesized RNA which is complementary in base-sequence to the original infectious strand is called minus (-) or antisense strand. It acts as template to produce additional (+) strand which may act as mRNA.

Class 1. dsDNA viruses

The mRNA is synthesized on a dsDNA genome template ($\pm$ dsDNA $\rightarrow$ (+) mRNA) which usually occurs in a cell. Following are the example of some viruses :

Papovaviruses : Polyomavirus, SV40
Poxviruses : Vaccinia virus
Adenoviruses : Human adenovirus
Herpesviruses : Herpes simplex virus type I and type II, Epstein-Barr virus.

Table 15.5 : The animals viruses (after Watson *et al.*, 1987)

Family	Virions	Genome (kb)	Example
1. dsDNA viruses			
Adenoviruses	Naked, icosahedral	35-40	Human and Animal adenoviruses
Herpes virus	Enveloped, icosahedral	120-200	Herpes simplex, chicken-pox, Epstein-Bar virus
Papovavirus	Naked, icosahedral	5-8	SV40, polyoma, papilomaviruses
Poxvirus	Enveloped, complex	120-300	Small pox, vaccinia
2. ssDNA viruses			
Parvovirus	Naked, icosahedral	4-5	Adeno-associated virus, Human parvovirus
3. (+) ssRNA viruses			
Coronavirus	Enveloped, helical	16-21	Human common cold like diseases, mouse hepatitis virus
Picorna virus	Naked, icosahedral	7	Polio, common cold virus, foot and mouth disease virus

Togavirus	Enveloped, icosahedral	17	Encephalitis virus
4. (-) ssRNA virus			
Paramyxoviruses	Enveloped, helical	15	Measles, mumps
Rhabdovirus	Enveloped, helical (Bullet shaped)	12-15	Rabies virus
Orthomyxovirus	Enveloped, helical	14	Influenza
5. dsRNA viruses			
Reovirus	Naked, icosahedral	18-30	Reovirus, human and animal diarrhoea virus
6. RNA-DNA viruses			
Retrovirus	Enveloped, icosahedral (diploid)	7-10	Rous sarcoma, Avian mouse, Mammary tumour virus
7. DNA-RNA viruses			
Hepadnavirus	Enveloped, icosahedral	13	Hepatitis-B virus, humans, rodents and birds

1 kb = 1,000 base pairs

Class 2. ssDNA viruses

In such viruses an intermediate DNA is synthesized before the synthesis of mRNA transcript (+ ssDNA → + mRNA). The mRNA has the same polarity as the DNA.
Parvoviruses : Adeno-associated viruses, mouse minute virus.

Class 3. (+) ssRNA viruses

The RNA has similar polarity as the mRNA. Viruses of this class have been grouped into the following two classes :
Subclass 3a : Individual mRNA encodes a polyprotein which is broken later on to form viral proteins.
Picornaviruses : e.g. polio virus.
Subclass 3b : From (+) ss RNA two types of mRNA molecules are transcribed, one is of same length as virion RNA and the other is a fragement of virion RNA.
Togaviruses : Alpha viruses (group A), sindbis virus, semliki forest virus, Haviviruses (group B) e.g. dengu virus, yellow fever, St. Louis encephalitis virus are the important examples.

Class 4. (–) ssRNA viruses

The virion RNA is complementary to mRNA. Following two types of viruses are found in this class :
Subclass 4a : The ssRNA genome encodes a series of monocistronic mRNA.
Rhabdoviruses : e.g. Mumps virus, measles virus, sendai virus.
Subclass 4b : Each segment molecule of the genome acts as template for the synthesis of mRNA which are monocistronic or encodes polyprotein.
Orthomyxoviruses : e.g. Human influenza virus
Bunya viruses : e.g. Bunyawera virus
Arenaviruses : e.g. Lassa virus

Class 5. dsRNA viruses

All the viruses of this class have segmented genome. Each chromosome encodes a single polypeptide. The dsRNA acts as template and asymetrically synthesize (+) mRNA.
Reoviruses : e.g. reovirus of humans.

Class 6. RNA-DNA viruses

In these viruses (+) ssRNA directs the synthesis of (-) DNA which in turn acts as template for the transcription of mRNA (RNA→(-) DNA→ + RNA). Virion RNA and mRNA are of the same polarity.

Retroviruses : e.g. Rous sarcoma virus, mouse leukemia virus.

Class 7. DNA-RNA viruses

This group consists of DNA containing hepatitis B Viruses.

II. Multiplication of Animal Viruses

Animal viruses differ from phages in mechanism of entering the host cell. This is due to differences in host cell i.e. one is prokaryotic and the other eukaryotic in nature. It is accomplished into the following stages :

(*i*) **Attachment :** Animal viruses like bacteriophages posses the attachment sites with the help of which it attaches to the receptor sites present on host cell surface. The receptor sites are the proteins or glycoproteins present on membrane surface of the host cell. The attachment sites of one group of viruses differ from the others. Distribution of these proteins plays a key role in tissue and host specificity of animal viruses. For example, poliovirus receptors are found only in human nasopharynx, gut and cells of spinal cord. While receptors of measles virus are present in most tissues. Differences in nature of polio and measles can be explained through the dissimilarities in the distribution of receptor proteins of host cells to which viruses get adsorbed. In some naked viruses (e.g. adenoviruses) the attachment sites are small fibres at the corners of icosahedron. In enveloped viruses (e.g. myxoviruses) the attachment sites are the spikes present on the surface of envelope.

For example, influenza virus has two types of spikes: H (haemagglutinin) spikes and N (neuraminidase) spikes. The H spikes attach to the host cell receptor site and recognise siatic acid (N-acetyl neuraminic acid). Influenza neuraminidase helps the virus in penetrating the nasal and respiratory tract secretions by degrading mucosal polysaccharides. However, the receptor sites also vary from person to person.

(*ii*) **Penetration:** After the attachment, virus penetrates the host cell. In enveloped animal viruses, penetration occurs by endocytosis, a process of bringing nutrients into the cell. If a virion attaches to a small outfolding i.e. microvillus on plasma membrane of a host cell, it will unfold the virion into a fold or plasma membrane forming a vesicle. When the virion is enclosed within a vesicle, its envelope is disintegrated and the capsid is digested resulting in release of nucleic acid in cytoplasm.

The detailed mechanism of entry of virus in not clear. However, the following three mode of entry of viruses occurs in them:

(*a*) **Direct penetration:** Some naked viruses (e.g. poliovirus) undergo a major change in capsid structure after adsorption to plasma membrane. This change facilitates the release of nucleic acids into cytoplasm (Fig. 15.6A)

(*b*) **Fusion with plasma membrane:** The envelop of enveloped viruses (e.g. paramyxoviruses) fuses directly with host plasma membrane and nucleocapsid is deposited in cytoplasmic matrix where uncoating is done (Fg. 15.6B). When virus is within the capsid a virus polymerase attached with nucleocapsid transcribes the virus RNA.

(*c*) **Endocytosis:** Most enveloped viruses enter the host cell through engulfment by receptor-mediated endocytosis and form coated vesicles. Virions attached to coated pits with the protein clathrin. Lysosomes help in uncoating of virion inside the cytoplasm. (Fig.15.6C).

(A) Direct penetration by naked viruses

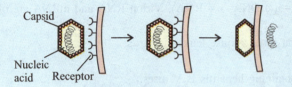

(B) Enveloped virus fusing with plasma membrane

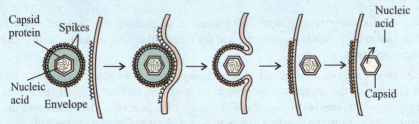

(C) Entry of enveloped virus by endocytosis

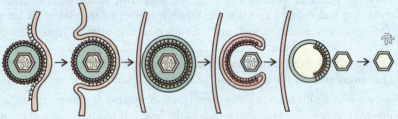

Fig. 15.6 : Mode of entry of animal viruses. A, direct penetration; B, fusion with plasma membrane; C, endocytosis.

(*iii*) **Uncoating :** It is a process of separation of viral nucleic acid from the protein coat. This process is not fully understood. In some viruses uncoating is done by lysosomal enzymes of the host cell which degrade protein coat and make the nucleic acid free in cytoplasm. In poxviruses, the viral DNA synthesizes a specific protein after infection. Thus, it varies with virus groups.

(*iv*) **Replication :** The replication process of DNA viruses differs from that of RNA viruses. However, in some DNA viruses multiplication occurs in cytoplasm (e.g. poxviruses) and in some others replication occurs in the nucleus of host (e.g. parvoviruses, papovaviruses, adenoviruses, herpes viruses). Multiplication of RNA viruses is more or less the same as in DNA viruses except the mechanism of formation of mRNA among the different group (Fig.15.7). Replication of DNA and RNA viruses has been discussed with their respective groups in the preceding sections.

(*v*) **Assembly and Release :** After replication of genetic material and synthesis of viral proteins assembly of viral particles occurs inside the host cell. Thereafter, they are released from the host cell.

III. Examples of Animal Viruses DNA Containing Viruses

Some of the DNA containing animal viruses are shown in Fig.15.8.

1. Papova Viruses

Papova viruses are one of the four important dsDNA viruses (e.g. papovaviruses, adenoviruses, herpes viruses and pox viruses) which produce tumour in many animals. The term *papova* is derived from the first two letters of the three prototypes, rabbit papilloma virus, polyoma virus and simian vacuolating virus-40 (SV40). The other important viruses of this group are JC virus (associated with neurological degeneration), BX virus (which suppresses immune system of humans), K virus of mice, etc.

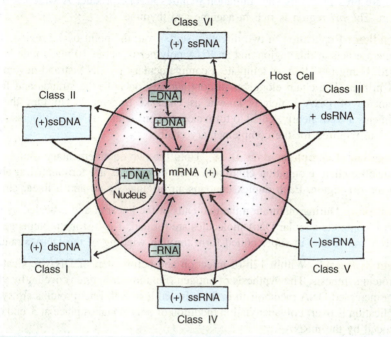

Fig. 15.7 : Formation of mRNA transcripts by nucleic acids of different types of animal virus inside the host cell. Class I-adenovirus, herpesvirus, papovavirus; Class II-parvoviruses; Class III-revoviruses; Class IV-picornaviruses, togaviruses; Class V- arenaviruses, rhabdo viruses, orthomyxo viruses, bunyaviruses; Class VI- retroviruses.

Capsid is of 45-55 nm, naked, icosahedral; virion consists of dsDNA and protein. Capsid is made up of 72 capsomers which are built by 420 subunits (Fig.15.8A). Capsid contains one major polypeptide (VP1) and two identical minor polypeptide (VP2 and VP3). Virus enters the cell and migrates to the nucleus where it replicates. The dsDNA encodes the early proteins and capsid proteins.

2. SV40

SV40 is an oncogenic virus. It is naked and icosahedral in morphology with a diameter of 45 nm. Capsid consists of 72 capsomers. SV40 is similar to polyoma virus in size and structure. Polyoma is associated with tumour in mice.

The dsDNA in its native form is supercoiled (i.e. covalently closed circle) helix having the sedimentation coefficient of 21S. Total G+C content of nucleic acid is 41%. After breaking the phosphodiester bond, single stranded DNA helix is converted into a relaxed circular form. This form has the sedimentation coefficient of 16S. A linear form (of 14S) is formed after double stranded break in the supercoil.

(i) **Replication :** Virus enters the cell and directly migrates to the nucleus. Replication of the viral RNA takes place inside the nucleus. Before the replication begins, early proteins are synthesized

in the nucleus of the infected cells. The mechanism of DNA replication can be divided into the following four stages :

(*a*) **Initiation :** DNA replication begins at a site known as origin of replication as the *ori* genes are present at this site. Initiation requires a gene product A which is a globular protein. The *ori* region is rich in adenine and thymine.

(*b*) **Elongation :** Replication in two direction starts from the point of *ori* region. The RNA polymerase acts at this region and an RNA polymer of about 10 nucleotide in length is formed. Using (+) DNA as template a complementary (-) DNA strand develops on the RNA primer. The chain elongates discontinuously on both the strands and form short fragments of DNA which is known as Okazaki fragments. In turn the Okazaki fragments are covalently sealed to form a continuous strand. DNA polymerase and DNA ligase are required for the complementary chain.

(*c*) **Segregation of complementary DNA :** Until the two complementary strands reach the termination, chain elongation continues. Both the strands are terminated at about 180° from the *ori* region. Each duplex contains an original strand and a linear strand.

(*d*) **Maturation :** During maturation the two ends of the linear strand is sealed by the ligase and two complete circular DNA molecules are formed. The histone proteins get attached to DNA and results in super coiled form through winding of the DNA strands.

(*ii*) **Protein Synthesis :** Within 12h of infection and before start of DNA replication, there begins early protein synthesis. The synthesis of antigen (i.e. tumour antigen) occurs by viral DNA which results in increased DNA metabolism in the infected host cell. Late proteins are synthesized when DNA replication is over. Polyadenylation (addition of poly A) takes place at 3' end of mRNA which is not coded by the mRNAs.

3.   Adenoviruses

Adenoviruses are icosahedral in morphology and contain naked capsid of 80 nm which is composed of about 252 capsomers. Capsomers contain 240 hexons and 12 pentons. The hexons are antigenically different from proteins. On the surface of capsid spikes are present from 12 points (Fig.15.8 B).

There is a linear dsDNA molecule which may be converted into a circular structure. The dsDNA consists 12-14% by weight of the particle. The genome is about 35 kb long. At the ends of the chromosome there are 100-140 nucleotide long inverted repeats, which are identical but with opposite polarity. Replication of adenoviruses takes place similar to herpes virus (Fig.15.8 C).

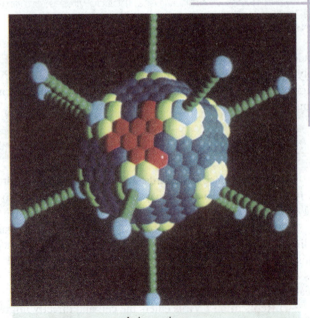

Adenovirus.

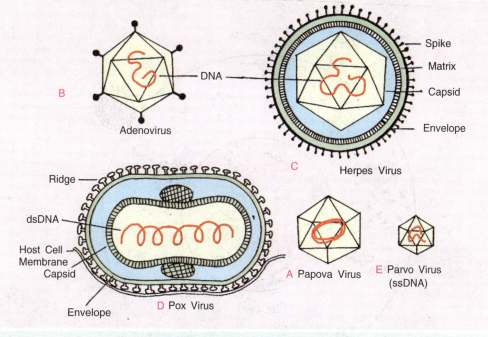

Fig 15.8 : DNA containing animal viruses.

4. Herpes Viruses

Herpes viruses are associated with several diseases in humans and the other animals such as cold sores, genital herpes, shingles, cervical and nasopharyngeal carcinoma, and abortion in horses. All of them are morphologically identical. They are large with a varying diameter from 140-170 nm. The nucleocapsid is spherical icosaherdal and of about 100 nm diameter which is enclosed in a 30 nm thick glycoprotein lipid envelope. Thickness of the envelope indicates for the presence of a matrix. The nucleocapsid contains the spikes (Fig.15.8C).

The capsid measures about 100 nm which is made up of hollow columnar capsomers (162) forming an icosahedral symmetry. There are a total of 150 hexamers and 12 pentamers.

The nucleic acid is a linear dsDNA which is converted into a circular DNA molecule upon a slight digestion either 3' or 5' end. It has also been observed that the DNA consists of two fragments (i.e. one long and the other short DNA molecule) which are covalently linked at the identical region, but they replicate independently or may be the recombination product of different orientation.

Virus attaches to the host cell, penetrates cell wall and outer envelope and released in the cytoplasm (Fig.15.9). The nucleocapsid migrates to the nucleus. Transcription of a portion of the viral DNA (i.e. early genes) occurs. The mRNA is synthesized which migrates into the cytoplasm. Translation of mRNA occurs where early proteins are synthesized.

Replication of viral DNA and also transcription and translation of viral gene occur inside the nucleus. This results in synthesis of later proteins and viral DNA molecules. Consequently capsid proteins are synthesized in cytoplasm, which migrate to nucleus of the host cell. Maturation of viral DNA takes place in nucleus. The capsid proteins assemble and form the complete viral particles which are released from the host cell.

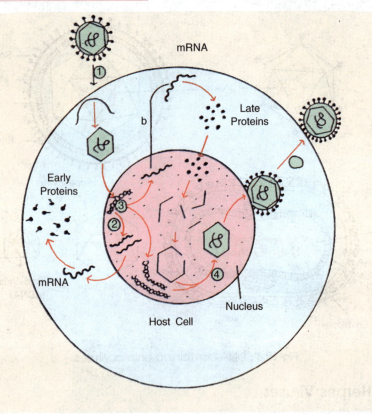

mRNA

Late Proteins

b

Early Proteins

mRNA

Nucleus

Host Cell

Fig. 15.9 : Multiplication of herpesvirus. 1, penetration; 2, transcription of early genes; 3, replication (a, DNA replication, b, late transcription); 4, assembly; 5, release of virion.

5.   Pox Viruses

Pox viruses infect the birds, insects and vertebrates including mammals. They are the largest and complex viruses that unlike papovaviruses, adenoviruses and herpes viruses replicate in cytoplasm of the infected cell (Fig.15.8D). Some important pox viruses are orthopox virus, variola major, variola minor, vaccinia, rabbit pox virus, entomopox virus, etc. Except parapox virus (e.g. the orf virus), virion of all the pox viruses are oval or birck shaped. Complex structure of pox viruses are the biconcave proteinaceous core, ellipsoidal lateral bodies and envelope. It measures about 160 × 200 nm.

Pox viruses contain dsDNA which consists of about 400 kb nucleotides. In vaccinia virus, the DNA is in circular form. The DNA encodes more than 75 proteins. Multiplication of pox viruses has been described by Moss (1974).

RNA Containing Viruses

There are different groups of viruses that differ each other due to the morphology and capsid size as well as the number and characteristics of the genetic material (Fig.15.10). They multiply inside the host cell. The major differences among the multiplication of process of these viruses lie how mRNA and the viral RNAs are produced. Some of the important groups are discussed below:

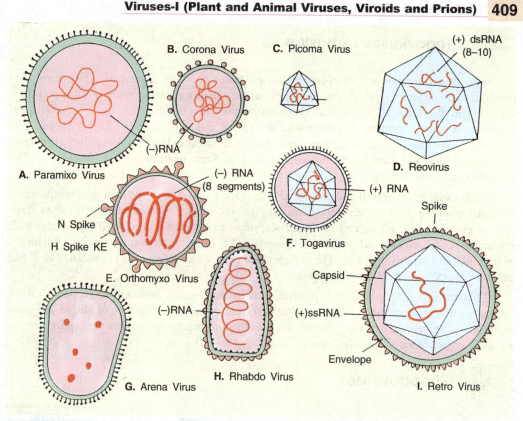

Fig. 15.10 : RNA containing animal viruses.

1. Picornavirus

As the term denotes (*pico*=small, *rna*=RNA) picorna viruses are the smallest in size (18-30 nm). They are icosaherdal and contain a (+) ssRNA because it acts as mRNA. There are four groups of picorna viruses : (*i*) human enterovirus which are found in alimentary canal e.g. poliovirus, ECHO (enteric cytoplasmic human orphan) virus causing paralysis, diarrhoea, (*ii*) cardioviruses of rodent e.g. encephalomyocarditis virus, (*iii*) rhinovirus which causes respiratory infection like common cold, bronchitis and foot and mouth disease virus e.g. FMD virus in cattles.

All the subgroups are structurally identical. They have icosahedral capsid of 18-30 nm which is composed of 32 capsomers (Fig. 15.10G). Virus consists of capsid proteins such as VP1,VP2,VP3,VP4 and several other viral proteins. Virion is a (+) ssRNA at one end of which is 75 nucleotides of adenosine (poly A) sequence. After attachment, penetration and uncoating of the capsid, the (+) ssRNA is translated into the two main proteins: (*i*) the first protein that inhibits the synthesis of RNA and host protein, and (*ii*) the second protein that acts as RNA dependent RNA polymerase. The second enzyme catalyses the synthesis of (–) ssRNA strand which is complementary in base sequence of the original strand. The (–) ssRNA strand acts as template to synthesize (+) ssRNA strands. The (+) strand may act as mRNA for the translation of capsid proteins, become incorporated into capsid protein to form a new virus, or may serve as a template for continued RNA multiplication (Fig.15.7, Class IV).

Two types of proteins, capsid proteins (VP_1, VPs, VP_3, VP_4) and several non-capsid virus-specific proteins (NcVP1-NcVP10) are synthesized. NcVP1 undergoes cleavage and produces NcVP1a and NcVP1b (Tortora *et.al.*, 1989).

2.   Togaviruses (Arboviruses)

Togaviruses are the arthropod-borne viruses, hence they are named as arboviruses. The arthopods carry different types of viruses, therefore, the use of term arboviruses is not appropriate. However, on the basis of serological types togaviruses are divided into the two serological subgroups, A and B. Serotype A is known as alpha virus and serotype B is called flavivirus. The alpha viruses are insect borne which cause encephalitis in humans. The flaviviruses include dengue virus, yellow fever virus and St. Louis encephalitis virus. Some of them are also insect-borne. The non-arthropod borne togaviruses are rubella virus and lactic dehydrogenase elevating virus.

Togaviruses are spherical and enveloped (Fig. 15.10F). The envelope is of host origin which is derived from the plasma membrane consisting of lipid bilayer. The envelope consists of three types of glycoproteins: E1,E2 and E3. Nucleocapsid is icosahedral and of 50-70 nm diameter. It consists of core protein and membrane protein. The later is different from rhabdoviruses, myxoviruses and paramyxoviruses. The nucleocapsid encloses an infectious (+) ssRNA. The 3' end contains poly A and the 5' end is capped.

Togaviruses multiply similar to picorna virus in the infected host cell. From the (+) RNA genome (−) RNA is transcribed which translate two types of mRNAs. One type of mRNA is a short strand. It synthesizes the envelope protein, whereas the long strand synthesizes the capsid proteins and itself is incorporated into the capsid (Fig. 15.7, Class IV).

3.   Rhabdoviruses

Rhabdoviruses are bullet-shaped with one flattened end. Rhabdo is a Greek word which means a rod, but the term rod is not accurate to its morphology (Fig.15.10 H, Table 15.4). Rhabdo viruses are found in vertebrates, invertebrates and plants. The examples of rhabdoviruses are rabies virus, vesicular stromatitis virus (VSV) and potato yellow dwarf virus.

VSV is the most extensively studied rhab-dovirus. It is a mild pathogen of cattle. VSV contains a nucleocapsid (70-180) which is en-closed by an envelope on its outer surface, consists

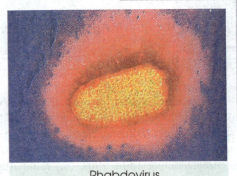

Rhabdovirus.

of numerous spikes of glycoprotein on inner surface and the membrane (M) protein. Inside the nucleocapsid, there is a (−) ssRNA molecule which is cylindrically coiled to make the core hollow.

After infection VSV starts both transcription and replication phenomenon in the host cell. The RNA of the virus is a (-) strand which acts as a template for mRNA synthesis, but does not act directly as mRNA. It lacks poly A at 3' end and not capped at 5' end. Therefore, its RNA is not infectious. It also contains an RNA dependent RNA polymerase in the virion. The other proteins present in virion are the major nucleocapsid (N) protein and minor nucleocapsid (L) protein.

The RNA polymerase uses (-) RNA strand as the template and produces (+) RNA strand. The (+) RNA strand acts as mRNA and as a template for the synthesis of viral RNA (Fig.15.5, Class V).

Rabies virus is discussed in detail in Chapter 22, *Medical Microbiology-II*.

4.   Orthomyxoviruses

The most important representative of orthomyxoviruses is the influenza virus (Fig.15.8 E) which causes influenza (i.e. flu) in humans. This disease is of medical importance has been discussed in detail in Chapter-22, *Medical Microbiology-II*.

Orthomyxovirus particles are pleomorphic and of 80-120 nm diameter. They are enveloped helical viruses having a capsid of 6-9 nm diameter. The nucleocapsid is ribonuclease sensitive which is a special character different from the other non-segmented viruses. The envelope contains outer lipid and inner protein layers. The surface of the protein layer is covered with numerous spikes. The spikes are of two types. Haemagglutinin (H) spikes and Neuramidase (N) spikes. The number of spikes is about 500. With the help of spikes the virus recognises and attaches the cell of hosts, before infection. Haemagglutinins refer to agglutination of RBCs when mixed with the viral particles. The number of N spikes per virion is about 100 (Fig. 15.11). It differs from H spike both in number and function. The N spikes help the virus to separate from the infection to RBCs, and also stimulate the formation of antibodies. The spikes provide the antigenic properties.

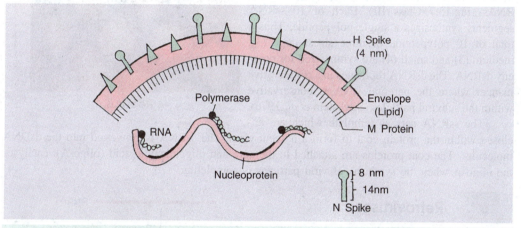

Fig. 15.11 : Diagram of a part of influenza virus showing envelope, spikes and (–) ssRNA.

On the basis of differences in H and N antigens, several antigens have been identified. After the removal of spikes by treating the viruses with pectolytic enzymes, their infectivity is lost. On the inner side of envelope there is M protein which provides rigid frame work to the particle and thus stabilizes the lipid bilayer.

The envelope encloses eight (–) ssRNA strands of different molecular wieght. Similar to diploma virus, each gene is found on an individual RNA segment. There is a constant shifting in gene segment of animal and human influenza viruses. This gene shifting results in changes in viral characters.

5.   Reoviruses

The term *reo* is an acronym for respirtory, enteric and orphan as the reoviruses are found in the respiratory and enteric (digestive) system of humans. They either are non-pathogenic or of low virulence; therefore, they are known as orphan viruses.

Reovirus particle is naked, icosahedral and enclosing the segmented dsRNA molecules (Fig. 15.10D). The viral particle is made up of about 8 proteins e.g. two mu ($\mu 1$, $\mu 2$), three sigma ($\sigma 1$, $\sigma 2$, $\sigma 3$) and three lamda ($\lambda 1$, $\lambda 2$, $\lambda 3$) proteins. The $\mu 2$, $\sigma 1$, and $\sigma 2$ form the outer protein shells of the capsid, σ3 constitutes spikes. The proteins (λ1, λ2, λ3, μ1 and σ3) form the core of virus particles. The viral particles possesses the activity of enzymes also such as ssRNA→dsRNA polymerase, oligouridenylic acid synthetase and RNA transcriptase.

The genome consists of 10 segments of (+) dsRNA the size of which is large (3), medium (3) and small (4). One of the unique features of the (+) strand is that on the 5' end it consists of the structure : 7 methyl GPPP-G-2-0-methyl-PC. Similarly, the (-) strand contains ppGPuPyP at 5' end but they are not methylated and capped. The viral RNA and mRNA do not have poly A at 3' end which is an exception among the mammalian RNA.

Reovirus enters the host cell through the vacuoles of phagocytes which fuses with lysosome within 20 minute and form a phagolysosome. Viral protein hydrolyses the lysosome in 2-3 hours and release the sub-viral particles. The RNA polymerase transcribes the 10 segments of dsRNA into 10 segments of (+) ssRNA. The ssRNA mol-

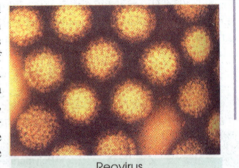

ecules are converted into mRNA by capping and methylation at 5' end. The mRNA takes part in transcription of polypeptide chain and also acts as template for the replication of polypeptide chain of RNAs (Fig.15.7 Class III). Each of 10 ssRNA segments synthesizes a single polypeptide. Thus a total of 10 polypeptide of three sizes: large (3), medium (3) and small (4) are synthesized for progeny dsRNA. The dsRNA replicates in a conservative manner where the parental RNA is conservative within the subviral particles (Silvestein *et al.* 1976).

Reovirus.

The ssRNA acts as template which are enclosed within the protein coat to form a nascent virus. The ssRNA is converted into the dsDNA molecules. The coat proteins are attached to the core and oligoadenylic acid (oligo-A), catalyses the oligo-A when the synthesis of viral particles is completing.

6.   Retroviruses

Retroviruses are also known as oncornaviruses because they are oncogenic (cancer cuasing). Most of the retroviruses infect vertebrates including fish, reptiles, birds, mammals, etc. They are known to cause leukemia and sarcoma in chickens and lymphoma and mammary carcinoma in mice. So far it is not clear whether they cause cancer in humans or not. They are known as retroviruses because their genome synthesised DNA molecule which is a reverse phenomenon. Of the most studied retroviruses is rous sarcoma virus (RSV), leukemia viruses of chicken and mice, and human immuno deficiency viruse (HIV1, HIV2) that causes AIDS (acquired immuno deficiency syndrome).

The retrovirus particles are enveloped, icosahedral measuring about 100-120 nm. Virions consist of a nucleocapsid which is derived from the plasma membrane of host cell. The envelope encloses the virion. On the outer surface of virion several spikes

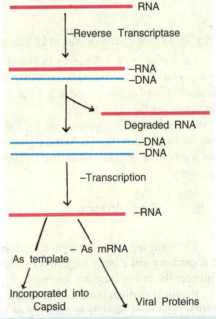

Fig. 15.12 : Nucleic acid replication and protein synthesis in retroviruses.

are present which contain glycoproteins. The glycoprotein recognises the specific receptor site present on host cell. The core capsid is composed of three major internal proteins (gp30) inside which are reverse transcriptase and RNA. Molecular weight of glycoprotein varies according to virus particle. Hence, the nomenclature of structural protein is done accordingly. However, the major internal protein is known as gp30 (mol. weight 30,000) and the glycoprotein as gp70 (mol. weight 70,000). The structural proteins act as antigens and thus they induce antibodies (Varmus, 1988). The retroviruses has been reviewed by Bishop (1978).

The viral genome is unique. It becomes diploid and contains two linear 35S ssRNA strands of 7-10 kb. The RNA molecules are linked at their 5' end with a small 5S/7S RNA segment. The small RNA segment consists of 4S tRNA which is of host origin. The 4S tRNA acts as primer for reverse transcriptase (see Chapter.20, *Human Cancer*).

Replication of retroviruses is very interesting. They carry their own enzyme reverse transcriptase. This enzyme was discovered by Temin and coworkers. Temin (1977) has given an excellent account of this enzyme. Varmus (1987) has discussed the action of reverse transcriptase and its role in retroviruses. Reverse transcriptase uses the viral RNA as template and synthesizes a DNA strand and forms RNA-DNA complex. In turn the ssDNA replicates to forms a dsDNA which is transcribed back into RNA. The RNA acts as mRNA for viral protein synthesis (Fig.15.12) and incorporated into the new virions. Before transcription, the viral DNA must be integrated into the host chromosome. This integrated DNA is called provirus that never comes out of the chromosome (Temin,1976).

AIDS virus, which is one of the very important retroviruses, is described in detail.

HIV (AIDS Virus)

An international committee of the scientists has proposed the name human immunodefiency virus (HIV) for such retrovirus that causes acquired immuno–deficiency syndrome (AIDS) in

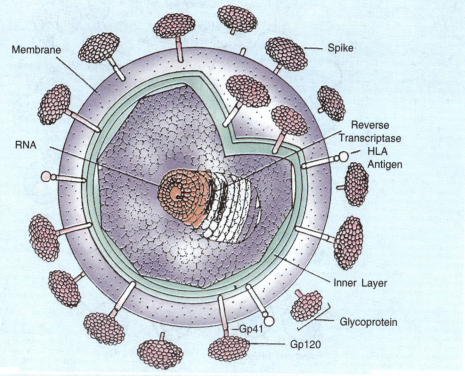

Fig. 15.13 : A model of AIDS virus (i.e. HIV) (diagrammatic).

humans. HIV has antigenic variants of HIV-1 and HIV-2. It destroys the immune system and, therefore, in the victims the immunity to combat the pathogenic disease is lost and patients die afterwords. HIV is of 100-120 nm diameter containing an protein envelope to which spicules of glycoprotein are attached. The envelope encloses an icosahedral capsid core that possesses identical macromolecules of RNA as genetic material (Arunachalam, 1987).

Three dimensional structure of the viral envelope appears like a sphere made up from an assembly of 12 pentamers and 20 hexamers (Fig. 15.13). In the centre at each corners of these hexamers a glycoprotein (gp120) is found which plays a key role in binding the HIV to its host cells. The gp120 (glycoprotein of mol. weight 1,20,000 dalton) is attached to another protein called gp41 which crosses through the viral membrane. The gp120 is connected with gp41 through the chemical connections and constitute the target of neutralizing the antibodies which are capable of blocking the virus and checking its capacity for infection. Immediately under the envelope the proteins of the core surrounding the central mass also form a polygonal structure. The central mass contains two helix of RNA molecules in the folded form to which gets attached the reverse transcriptase. Reverse transcriptase controls the transformation of RNA into DNA (Planta and Wain-Hobson, 1988).

(*i*) **Working of immune System in the Presence of HIV :** Infection is carried out through the spicules present on envelope of HIV which attach to the target cell. The gp120 has affinity for a specific molecule, the T4 or CD4 molecule exposed on the surface of certain cells i.e. lymphocytes. The lymphocytes secrete T4 molecules and play a major role in the immune system, and AIDS symptoms are expressed after their destruction. Function of immune system is a complicated process (Fig.15.14).

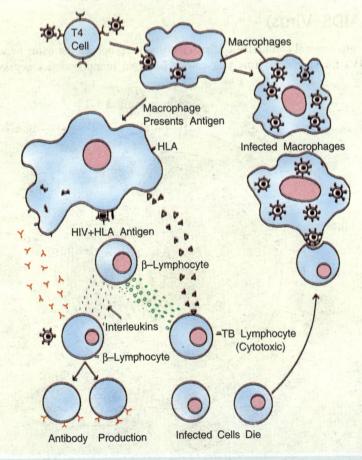

Fig. 15.14 : Working of immune system in the presence of HIV.

The infected host cells rapidly produce antibodies against gp120 which is recognised as a foregin molecule or an antigen. Some of the antibodies attach to the envelope and block the connection of virus to T4 molecules of the target cell. Even the large quantity of neutralizer antibodies is not capable of stopping the dissemination of HIV. Thus the infected individual causes infection to his/her sexual partner of those receiving blood (Barre-Sinoussi *et al.*, 1983). The provirus may remain silent for a long period without showing any sign of presence.

In response to introduction of an infected cell into the organism the macrophages are the first to intervene. They have the capacity to catch the viral particles, then cut them into pieces and prevent as antigen to the T4 lymphocytes. The T4 lymphocytes become incapable to secrete interleukins, a soluble substance which can communicate to other cells of immune system for the presence of antigens. The T4 lymphocyte is known to induce the proliferation and maturation of B-lymophocytes which then produce antigen specific antibodies. As the disease envolves, it loses the capacity to produce antibodies. The infected macrophages serve as the reservoir of viruses and disseminate HIV to all the tissues. They become weakened and their function becomes diminished (Asjo *et al.*, 1987).

The action of T4 lymphocytes takes place on another category of lymphocytes (T8 lymphocyte) which can either kill the infected cell (T8 cytotoxic effect) or inhibit the immune response (T8 suppressors).

The T4 markers are not present on T8 lymphocytes. They are capable of killing the cells through direct contact and, therefore, they are called cytotoxic lymphocytes. T lymphocytes have the long life of several months to years and represent to about 60-80% of total lymphocyte population. T8 lymphocytes are resistant to HIV. It carries on its surface a protein of an antigen against which the T8 lymphocyte has been sensitized. It also carries on its membrane surface the antigens of histocompatibility of self markers called the HLA antigen (human lymphocyte system A). But their role of antiviral protectors turns them into formidable demolition agents. Through the elimination of infected cells, they clean up the organisms or its infected centres, while doing this they eliminate the cells which carry out immune functions. This is how T4 lymphocytes, B-lym-

Electron Micrograph of a human white blood cell infected with the AIDS virus.

phocytes and infected macrophages can be rapidly destroyed when they are recognised as the targets by the cytotoxic lymphocytes. Inflammatory reactions are brought about by the release of toxic products which attack the surrounding tissues such as pulmonary inflammation of immune origin among the HIV carriers (Venet *et al.*, 1985; Young *et al.*, 1985). Unfortunately, in case of AIDS, T4 cells are first attacked which are destroyed later on. The other cells such as B-lymphocytes and macrophages are also destroyed. Thus, in overall struggle the immune system is destroyed and HIV gets upper hand.

(ii) **Replication of HIV in Target Cell :** In the infected cell, HIV replicates after coming out of capsid core (Fig.15.13). The RNA molecules act as mRNA. It possesses the enzyme reverse transcriptase that synthesizes ssDNA molecules with the help of tRNA. Consequently, a ssDNA molecule is synthesized which later on synthesizes the dsDNA molecule.

Reverse transcriptase has a problem of copying a ssRNA into dsDNA capable of integrating with the cellular DNA. Therefore, a complex process occurs for efficient integration with cellular DNA. This complex process is known as the formation *of long terminal repeats* (LTRs) at both the ends of provirus. The provirus is transported to the nucleus and it becomes circular. The circle is opened at specific site and the provirus is integrated at the specific site of the cellular DNA. Therefore, transcription of mRNA and translation of viral proteins occur.

The provirus contains the genetic informations required for viral protein synthesis. It has a length of 9600 nucleotides. Like all retrovirus genome, HIV genome contains the three basic genes, the *gag* gene, *pol* gene and *env* gene. The *gag* gene encodes for the protein of internal structure, the *pol* gene encodes reverse transcriptase and the *env* gene encodes for envelop glycoprotein. There are five other genes such as Q, R, *tat*, *art/trs* and F. The function of *tat* and art/trs genes are known and they are connected with replication and expression of the virus (Fig.15.15). When synthesis of core proteins envelope protein and reverse transcriptase is over in the infected cell, packaging of HIV particles takes place. Later on the viral particles are released from the infected cells by budding process (Planta and Wain-Hobson,1988).

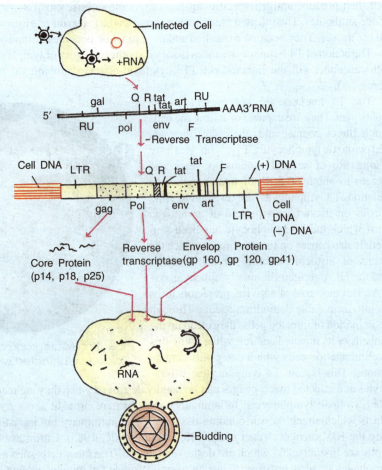

Fig. 15.15 : Replication of HIV in an infected cell.

D. Viroids

Until 1970s, viruses were considered as the smallest infectious agent. The discovery of viroids has proved that the infectious entities smaller than virus exist in nature. For the first time T.O. Diener and W.B. Raymer (1967) discovered potato spindle tuber viroids (PSTV) which caused a disease in potatoes. This disease resulted in loss of millions of dollors. Moreover, Diener (1971) advanced the concept of viroids on the basis of newly established properties of the infectious agent responsible for potato spindle tuber disease. Their properties differ basically from those of conventional viruses in atleast following five important features :

   (*i*) The pathogen exists *in vivo* as an encapsulated in the RNA,

   (*ii*) Virion like particles are not detected in the infected tissues,

   (*iii*) The infectious RNA is of low molecular weight,

   (*iv*) Despite its small size, the infectious RNA replicates autonomously in susceptible cells i.e. no helper virus is required, and

   (*v*) The infectious RNA consists of one molecular species only.

   (*vi*) **Host Range :** The other plants susceptible to viroids are potato, citrus, cucumber and chrysanthemum, hops, stunt, tomato banchy top, etc.

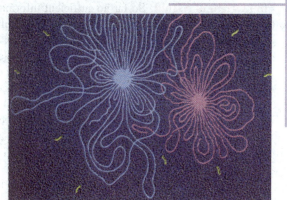

Electron Micrograph of Viroid particles that cause potato spindle tuber disease.

The host range of PSTV is the members of Solanaceae and Compositae. Recently, mild strains of PSTV were observed in cultivars Kufri, Chandramukhi, Kufri Jyoti of potato and wild solanums in Himachal Pradesh in India but these are not economically important in India as compared to the USA, Canada, USSR and China. Symptoms vary according to the cultivars, strain and age of infection. Stem and petiole are more acute than the other parts. Diseased tubers get elongated i.e. with pointed ends having numerous eyes and heavy brows. The viroids are contangious and spread mainly through mechanical injury/contact but also through pollen and true seeds from the infected plants. The control measures are the use of diseased free seeds, early roguing and avoiding cutting of potato tubers (Khurana and Singh, 1997).

   (*iii*) **Viroid Genome :** Viroids are low molecular weight nucleic acid (1.1-1.3×10^5). They are the only known pathogen that do not code for any protein. They differ from viruses in lacking protein coat. The PSTV has been found to be present in nucleus of the infected cells but not the other subcellular organelles of potato. About 200 to 10,000 copies of PSTV are found in each cell. These are just a small fragment of RNA molecule which are commonly circularized, and remain as naked RNA strand consisting of about 250-370 nucelotides. The genes lack the initiation codon (AUG) for protein synthesis. Mostly the nucleotides are paired resulting in dsRNA molecule due to the presence of intra molecular complementary regions. Hence it appears as rods. The dsRNA has closed folded, three dimensional structure (Fig.15.16). The closed single stranded circle has extensive intrastrand base pairing and interspersed unpaired loops. Viroids have five domains. Most changes in viroids. Pathogenicity seems to arise from variation in the pathogenicity domain (P) and left terminal domain (TL). The other domains are the central conserved region (CCR), variable domains (V), and rigid terminal domain (TR). The folded structure probably protects it form the attack by cellular enzymes. RNA does not code for any protein just like introns (Diener, 1993).

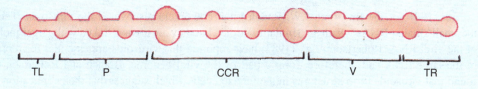

Fig. 15.16 : Diagram of structural organisation of viroid.

PSTV is restricted only to plants. However, no conclusive evidence is available for their presence in the animals. Moreover, a few animal diseases were suspected to be caused by viroids but no specific immunety occurred (Diener, 1971).

(*iv*) **Origin of Viroid:** So far no sufficient information is available that can lend support for the origin of viroids. Following are some of the speculations regarding the origin of viroids.

(*a*) Viroids are supposed to be the primitive viruses and must have originated from cellular RNAs. This view has been emphasized by Watson *et al.* (1987). In most of the healthy plants, RNA synthesis on RNA template must occur. Viroids would have originated from this RNA as they did not induce the biosynthetic machinery of their host from their own replication.

(*b*) Except tRNA and 5S RNA, several low molecular weight RNAs have been found to be associated with several virus infections such as tobacco leaves infected by TMV, in *E. coli* infected by QB phage, in oncogenic RNA viruses, etc. It is supposed that viroids would have been originated from virus induced low molecular weight RNAs which later on adapted as autonomously replicating infectious entities. Therefore viroids provide the evidence that they are the degenerated virus entities (Riesner and Gross,1985).

(*c*) With the discovery of spilt genes and RNA splicing in eukaryotes it has been suggested that viroids might have originated through circularization of spliced out introns. If such excised sequences would persuit the extensive intramolecular base-pairing (as viroids do) and if they are circularized they might become established and escape degradation. If such introns would compromise the appropriate recognition sequence they might be transcribed by host enzymes capable of functioning as an RNA polymerase and thus escape from the control mechanism of host cell (Diener, 1979).

(*v*) **Replication of Viroids :** There is no convincing evidence for the replication of viroid genome. It is likely that nucleic acid codes for an enzymes replicase which is essential for its replications. However, there are two possibilities for genome replications, RNA dependent replication and DNA dependent replication :

(*a*) **RNA directed replication :** According to this scheme, it appears that RNA directed RNA polymerase are present to a limited extent in the normal cell of plant and *E. coli* which may synthesize the RNA molecules directed by the RNA.

(*b*) **DNA directed replication :** The viroids are transcribed from a cellular DNA of the host cell complementary to viroid RNA. In the infected cell new DNA may be produced with the infecting viroid RNA which serves as template. This makes the assumption for the presence of reverse transcriptase i.e. RNA directed DNA polymerase. From this the viroid RNAs are synthesized.

$$\text{Viroid RNA} \xrightarrow[\text{Reverse Transcriptase}]{\text{Infection}} \text{DNA} \longrightarrow \text{Viroid RNA}$$

Branch and Robertson (1984) have analysed the viroid specific nucleic acids on tomato plants infected by PSTV. They conclude that (*i*) viroids replicate by direct RNA to RNA copying, (*ii*) the host cells possibly contain the machinery needed for replication of viroid RNA.

E. Virusoids

In 1981 for the first time J.W. Randles and coworkers discovered the virusoids. These are the small circular RNAs which are similar to viroids but they are always linked with larger molecules of the viral RNA. Robertson *et al.* (1983) have reported that virusoids are essential for replication of the large RNA and, therefore, form a part of the viral genome. Branch and Robertson (1984) found that virusoids form structures in the infected cells which suggest that their replication cycles resemble with those of the PSTV and several other viroids (e.g. velvet tobacco mosaic virus). The other virusoids like a small satellite (extra RNA molecule associated with virus) is capable to replicate in virus infected host cell.

F. Prions

An infectiours agent different from both viruses and viroids can cause disease in animal and humans. This agent is called prions (i.e. proteinaceous infectious particles). In 1957 during the visit to New Guinea, Carleton Gajdusek was informed by a local physician about half of deaths of the tribals in the remote places due the a misterious disease. Later on this disease was named as *Kuru* i.e. shaking or shivering by the local inhabitants. Mostly women and childern exhibit involuntary tremors at the early stage of disease. Thereafter, the patients pass through stages of increased debilitation and dementia leading to death. This disease spreads due to a kind of ritual cannibolism involving the eating of parts of brains of the deceased as a mark of homage to the later. The patients dying of Kuru did not show any of symptoms of infection of central nervous system. No virus toxin or chemical abnormalities could be isolated in the laboratories of Australia.

In 1957, William Hadlow, veterinary pathologist who was working on scrapie, a neurological disorder of sheeps and goats, came across the neuropathological specimens prepared by Gajdusek (a person who died of *Kuru*). He observed remarkable similarities between the abnormalities found in brains of *Kuru* victim and the sheeps and goats dying of scrapie. The disease which was known for about 200 years was called *scrapie* because the affected animals rub their skin and scrape much of their wool due to loss of coordination of movement. Gajdusek *et al.* (1966) proved the infectious nature of *Kuru* by injecting the brain suspension of *Kuru* victims to chimpanzees. The clinical symptoms of kuru developed after 18 to 21 months of intercereberal inoculations. Now the incidence of *Kuru* has decreased due to the elimination of cannibalism.

In 1970s Stanely B. Prusiner (a biochemist at the university of California, San Francisco, U.S.A.) with his coworkers made attempts to isolate and identify this infectious agent responsible for scrapie. After exhaustive research for about 10 years Prusiner (1982) concluded that (*i*) the causal agent of scrapie is smaller than any known virus, (*ii*) most unexpectedly it is composed of a protein having a molecular weight of 27,000 to 30,000 daltons. No nucleic acid from this infectious agent could be isolated (Prusiner,1982). He christened these highly unusual pathogens as *Prions* i.e. the proteinaceous infectious particles. Besides scrapie, prions are now known to be causal agents for neurological disorders such as bovine spongiform encephalopathy (mad cow disease) in cattle and Creutzfeldt Jakob disease (CJD), Gerstmann-Straussler-Scheinker (GSS) disease, Kuru and fatal familiar insomnia in humans. For the outstanding work on prions, Prusiner was awarded Nobel Prize in Medicine in 1997.

The mad cow disease has reached epidemic proportion in Great Britain because the cattles are fed with bone meal made from cattle. However, Kuru has been found only in New Guinea. Prions are about 100 time smaller than a virus. It has been estimated about 27 nm by cellulose nitrate membrane filtration and 7 nm by X-ray target size determination. These lack nucleic acid at all. This leads to a number of questions. What is the nature of its genome ? Do they violate the basic rules and regulations of central dogma ? How can a protein enter a host cell and directs the replication? It consists of only hydrophobic protein of 33 to 35 Kilo daltons which is often called Prp (prion protein). The *prp* gene is present in many normal vertebrates. It is supposed that a changed Prp is at least responsible for the disease. Electron microscopic studies revealed that a large number of Prp molecules (about 1,000 in number) aggregate to form prion rods. These rods are typically 10-20 nm in diameter and 100 to 200 nm in length (Prusiner,1984, Banerjee, 1987).

(i) **Spread of Prion**

The mechanism of spread of the disease is not fully understood. Prusiner (1994) has reviewed the biology and genetics of prion disease. There are several hypotheses for the spread of this disease.

(*a*) Some workers are of the opinion that disease is transmitted by the *prp* alone. The infectious pathogen is Prp which has been chemically modified. When abnormal *prp* enters the normal brain cells, possibly it binds with normal *prp* and induces the enzymes that modify its structure into the abnormal confirmation. This newly formed abnormal Prp in turn attacks the other normal Prp molecules. But the other researchers feel that this hypothesis is not adequate.

(*b*) According to the opinion of the other group the genetic informations cannot be transmitted by the protein between the hosts. Proteins are now known to carry the genetic information, therefore, possibly the infectious agent is a *Virio*, a tiny scrapie-specific nucleic acid coated with Prp protein. The nucleic acid may not be transmitted but interacts host cells to cause disease. Many strains of scrapie agent have been isolated from the infected cells. This hypothesis seems to be consistent with the findings (Prusiner,1994).

(*c*) Others believe that prion diseases are caused by unknown viruses with usual properties (Prusiner, 1994).

(*ii*) Artificial Prions: Some significant studies related to molecular biology a gene manipulation have been carried out in recent years. It has been found that the entire open reading frame (ORF) of all known mammalion and avial Prp gene eliminates the possibilities that Prp (scrapie core protein) arises from RNA splicing. The two exons of the syrian hamster *Prp* gene are separated by a 10 kb intron,exon-1 encodes a portion of the 5' untranslated leader sequence while exon -2 encodes the ORF and 3' untranslated region. The mouse and sheep *Prp* genes contain three exons. The promoter region contains multiple copies of G-C rich repeats and are devoid of TATA box.

Transgenic mice expressing chimeric *prp* genes were constructed by Scott *et al.* (1992). One hamster/mouse Prp gene classified MH2M Prp contains five amino acid substitutions encoded by hamster Prp while another construct designated MH2M Prp has two substitutions. Transgenic MH2M Prp mice were susceptible to both hamster and mouse prion, whereas three lines expressing MHM2 Prp were resistant to hamster prions. The brain of transgenic (MH2M Prp) mice dying of scrapie contained chimeric Prp^{sc} and prions with an artificial host rage favouring propagation in mice that expressed the corresponding chimeric Prp.

The prions were also transmissible at reduced efficiency to non-transgenic mice and hamster. These findings provide genetic evidence for homophilic interactions between Prp in the inoculum and Prp synthesized by the host.

QUESTIONS

1. Write the brief history of development of virology.
2. What is a virus ? Write in brief its general feature and occurrence.
3. Give an illustrated account of morphology and chemical structure of different types of viruses ?
4. Write the criteria of virus classification with special references to Casjens and King (1975) .
5. Write an essay on isolation and cultivation of animal viruses and bacteriophages
6. Describe the structure of TMV and symptoms developed on tobacco plant.
7. Give an illustrated account of families of animal viruses.
8. Write an essay on multiplication of animal viruses.
9. What do you know about papova viruses with special reference to SV40.
10. Write in detail about HIV, working of immune system and AIDS in humans.
11. Write short notes on the following :
    (*i*) Icosaherdal symmetry, (*ii*) Enveloped viruses, (*iii*) Herpes virus, (*iv*) Togaviruse, (*v*) Rabies virus, (*vi*) Influenza viruses, (*vii*) Reovirus, (*viii*) Retroviruses, (*ix*) Viroids, (*x*) Prions.
12. Write in detail about transmission of plant viruses.
13. Give a detailed account of serological tests for diagnosis of plant viruses.
14. Write short notes on the following:
    (*i*) Mosaic disease of sugarcane, (*ii*) Pinwheel bodies (*iii*) Indicator plants, (*v*) Rice tungro virus

REFERENCES

Arunachalam,S. 1987. AIDS : An update. *Sci. Reporter* : 24:459-463.

Aiken, J.M. and Marsh R.F. 1990. The search for scrapie agent nucleic acid. *Microbial. Rev.* 54(3):242-246.

Asjo,B. *et al.,* 1987. *Virology* **157**:359.

Baltimore, D. 1971. Expression of animal virus genome. *Bacterol. Review.* **35**:236.

Banerjee, S.N. 1987. Do prions carry genetic material ?. *Sci. Rept.* **24**(5):291.

Barre-Sinoussi, F. *et al.* 1983. *Science* :220 :868.

Bawden, F.C. and N.W. Pirie 1938. A plant virus preparation in fully crystalline state. *Nature* 141:513-514.

Bishop, J.M. 1978. Retroviruses. *Ann. Rev. Biochem.* **47**:35-88.

Beijerinck,M.W. 1898. Over een contagium vivum fluidum als oorzaak van Vlekziekte tabaksbladen. *Verhandel Kohinl. Akad. Welenschap, Afdel Wis Natuurk,* 7:229-235

Bellet, A.J.D. 1967. Classification of viruses. *Virology* **37**:117-123.

Branch,A.D. and Robertson,H.D. 1984. A replication cycle for viroids and other small infectious RNase. *Science* **223**:450-455.

Casjen,S. and Kings. 1975. Virus assembly. *Ann. Rev. Biochem* **44**:555-611.

Casper D.L.D and Klug,A. 1962. Physical principles of regular viruses. *Cold Sring. Harb. Symp. for Quant. Biol.* 27:1

Diener,T.O. 1971. *Virology* 43:75-89.

Diener,T.O. 1979. Viroids and viroid Diseases. Wiley, New York.

Diener,T.O. 1993. The viroid : Big punch in a small package. *Trends Microbiol.* **1**:289-294.

Diener,T.O. and Raymer,W.B. 1967. Potato spindle with properties of a nucleic acid *Science* 158:378-381.

Fraenkel-Conrat, H. 1956. The role of nucleic acid in the recognition of TMV *J. Amer-Chem. Soc.* **78**:882.

Franklin, R. ; Klug,A. and Holmes, K.C. 1957. Structure of tobacco mosaic viruses. *Ciba Found Symp.* 29.

Gajdusek,D.C.; Gibbs,C.J. Jr. and Alpers,M. 1966. Experimental transmission of kuru like syndrome to chimpanzees. *Nature* **209**:794-794.

Gibbs, A.J. 1969. Plant virus classification. *Adv. Virus Res.* 121:263-328.

Gierer, A. and Schramm, G. 1956. Infectivity of ribonucleic acid from tobacco mosaic virus. Nature:**177**:702-703.

Harshey, A.D. and Chase,M. 1952. Independent function of viral protein and nucleic acid in growth of bacteriophage. *J. Gen. Physiol.* 36:39

D'Herelle, F. 1917. Sur un microbe invisible antagonisite antagoniste des bacilles dysenteriques. *C.R. Acad. Sci. Paris* **165**:373.

Hobbs, H.A. *et al.* 1987. Use of direct coating and protein A coating ELISA procedure for detection of three peanut viruses. *Plant Disease* **71:**747-749.

Hollings, M. 1962. Viruses associated with a die back disease of cultivated mushroom. *Nature.* 146:962:963.

Holmes, F.O. 1948. *Bergey's Manual of Determinative Bacteriology,* 6th ed. (Abstract, in : Phytopath., Williams and Wilkins Co. Baltimore **38**:314

Kassanis, B. 1966. Properties and behaviour of sattelite virus. In *Viruses of plants* (eds. A.B.R. Beemester and D. Dijkstra). North Holland Armsterdam pp. 177-187.

Khurana, S.M.P. and Singh,R.A. 1997. Virus, viroid and mycoplasma disease of potato in India and their control. In *"Himalayan Microbial Diversity"* (eds. S.C. Sati, J. Saxena and R.C. Dubey) Today's and Tommorow Printers and Publishers, New Delhi, 219-247.

Luria, S.E. and Darnell, S.E. 1968. *General virology* Ind. ed, Wiley, New York.

Lwoff, A., Horne,R. and Tournier, P. 1962. The classification of viruses. *Ann. Rev. Microbiol.* 20:45-74.

Nikolaeva, O.A. *et al.* 1998. Serological differentiation of the citrus tristeza virus isolates causing stem pitting in sweet orange. *Plant Disease* **82:**1276-1280.

Mayer, A. 1886. *Landwirtsch Vers. Sta.* 32:451-467.

Moss, B. 1974. Reproduction of poxviruses. *Comprehensive Virol.* 3:405

Millward, S. and Graham,A.F. 1970 Structural studies on reovirus: discontinuities in the genome. *Prot. Natl. Acad. Sci.* (U.S.A) 65:422-429.

Permar, T.A. *et al.* 1990. A monoclonal antibody that discriminates strains of citrus tristeza virus. *Phytopath.* **80:** 244-228.

Plata, F. and Wan-Hobson,S. 1988 AIDS: Immunity and vaccination. *The World Scientist* pp. 27:14.

Prusiner, S.B. 1982. Novel proteinaceous infectious particles cause scrapie. *Science* 216:136-144.

Prusiner, S.B. 1984. Prions. *Sci. Amer.* (Oct.) 15:45-57.

Prusiner, S.B. 1994. Biology and genetics of prion diseases. *Ann. Rev. Microbiol* 48:655-686.

Riesner,D. and Gross,H.J. 1985. Viroids. *Ann. Rev. Biochem.* 54:53-564.

Robertson,H.D.; Howell,S.H.; Zaitlin,M. and Malmberg,R.L. (eds.) (1983). *Plant infectious agents: viruses, viroids, virusoids and satellites.* Cold Spring harbour Lab.. Cold spring harbour, New York

Safferman, R.S. and Morris, M.E. 1963. Algal viruses : isolation, *Science* 140:679:680

Scott, M.R. *et al.* 1992. Chimeric prion protein expression in cultured cells and transgenic mice. *Protein sci..* 1:986-997

Shepherd, R.S. and Watesman, R.J. 1968. DNA in cauliflower mosaic virus. *Virology* 36:150-152.

Silverstein, S.C.; Christman,J.K. and Acs, G. 1976. The reovirus replicative cycle. *Ann. Rev. Biochem* **45:**375-408.

Stanley,W.M. 1935. Isolation of a crystalline protein possessing the properties of the tobacco mosaic virus. *Science* **81:**644-645.

Takeba,I. 1975. The use of protoplast in plant virology. *Ann. Rev. Phytopath.* **13:**105-25.

Temin, H.M. 1972. RNA synthesis. *Sci. Am.* 226:24-33.

Temin, H.M. 1976. The DNA provirus hypothesis. *Science.* 192:1075-1080.

Twort, F.W. 1915. An investigation on the nature of the ultramicroscopic viruses. *lancet* **189**:1241.

Tortora, G.L.; Funke, B.R. and Case,C.L. 1989. *Microbiology: An Introduction* 3rd ed. The Benjamin Cummings publ. Co. Inc. California.

Varmus, H. 1987. Reverse transcription. *Sci. Amm.* **257**:56-64.

Varmus, H. 1988. Retroviruses. *Science* **240**:1427-1435

Venet, A. *et al.* 1985. *Clin. Res. Physiol.* **21**:535.

Watson, J.D. *et al.* 1987 *Molecular Biology of the Gene* 4th ed. Benzamin/ Cummings, California.

Young, K.R. *et al.* 1985. *Ann. Int.* Med. **103**:522.

Viruses-II
(Bacteriophages, Cyanophages and Mycoviruses)

16

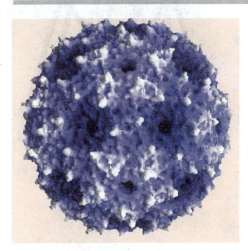

I n 1915, for the first time Frederick W. Twort in England observed that bacterial colonies sometimes were lysed. The lytic effect was transmitted from one colony to the other. The transmission of lytic effect continued even after dilution and passing them through bacterial filter. However, after heating the solution, the lytic effect was destroyed. He thought that the lytic effect would have been done due to the virus. In 1917, D'Herelle at Pasteur Institute (Paris) rediscovered this phenomenon i.e. 'Twort-D'Herelle phenomenon' and coined the term bacteriophage (bacteria eater). Since then a large number of bacteriophages has been discovered. They contain a single stranded (ss) DNA, double stranded (ds) DNA, ssRNA or dsRNA as genetic material.

A. Bacteriophages

I. Classification

On the basis of presence of single or double strands of genetic material, the bacteriophages are of the following types:

Frederick W. Twort

1.   The ssDNA Bacteriophages

(*i*) Icosahedral phages   = ØX174, ØR, St-1, 6SR, BR2, U3 and G series e.g. G4,G6,G13,G16. All are like ØX174.

(*ii*) Helical (filamentous)

(*a*) The Ft group: They are F specific phages and absorb to the tip of F type sex pilus e.g. *E. coli* phages (fd, fl, M_{13}).

(*b*) If group: They are absorbed to I-type sex pilus specified by R factors e.g. If_1, IF_2, etc.

(*c*) The third group is specific to strains carrying RF_1 sex factor.

2.   The dsDNA phages

Following are the examples of dsDNA phages:

(*i*) T-even phages of *E. coli* e.g. T2, T4, T6.

(*ii*) T-odd phages of *E. coli* e.g. T1, T3, T5, T7.

(*iii*) The other *E. coli* phages e.g. P1, P2, Mu, , Ø80

(*iv*) The phages of *Bacillus subtilis* e.g. PBSX, PBSI, PBS_2, SPO_1, SPO_2.

(*v*) The phages of *Salmonella* e.g. P1, P22.

(*vi*) The phage of *Shigellsa* a e.g. P2.

(*vii*) The phage of *Haemophilus* e.g. HP1.

(*viii*) The phage of *Pseudomonas* e.g. PM2.

Structure of Bacteriophage T4.

3.   The ssRNA phages

Examples of the ssRNA bacteriophages are as below.

(*i*) **Group I:** *E. coli.* phages such as f2, MS2, M12, R17, fr, etc.

(*ii*) **Group II:** The QB phages.

4.   The dsRNA phages

Example: The Ø6 bacteriophage.

II. Morphological groups of bacteriophages

The structural characteristics of bacteriophages have been determined through electron microscopy. All the phages contain a nucleic acid enclosed in a capsid which is made up of capsomers. The capsomers contains protomers i.e. protein subunits. Diagram of families of bacteriophages are given in Fig. 16.1. Bradley (1967) has described the following six morphological types of bacteriophages.

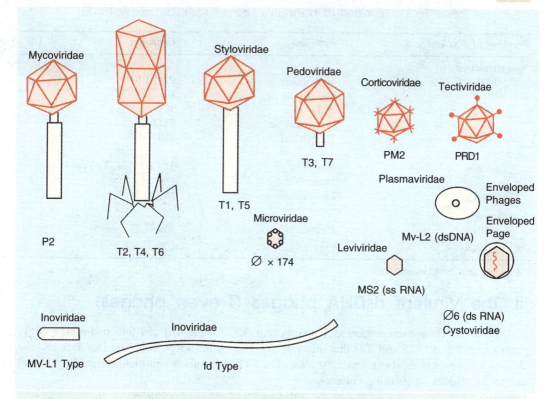

Fig. 16.1 : Families of bacteriophages (diagrammatic).

Type A: This type of virus has hexagonal head, a rigid tail with contractile sheath and tail fibers e.g. dsDNA T-even (T2, T4, T6) phages.

Type B: This type of phage contains a hexagonal head but lacks contractile sheath. Its tail is flexible and may or may not have tail fiber, for example dsDNA phages e.g. T1, T5 phages.

Type C: Type C characterized by a hexagonal head and a tail shorter than head. Tail lacks contractile sheath and may or may not have tail fiber, for example dsDNA phages e.g. T3, T7.

Type D: Type D contains a head which is made up of capsomers but lacks tail, for example ssDNA phages (e.g : ∅X174)

Type E: This type consists of a head made up of small capsomers but contains no tail, for example ssRNA phages (e.g. F2, MS2).

Type F: Type F is a filamentous phage, for example ssDNA phages (e.g. fd, f1).

Further a group G is recently discovered which has a lipid containing envelope and has no detectable capsid, for example a dsDNA phage, MV-L2. Types A, B, C are unique bacteriophages. Types D and E are also found in plants and animals, and Type F is found in some plant viruses too.

Nomenclature of viruses was done according to laboratory codes which do not follow any particular guidelines. It is a haphazard way of nomenclature. The bacterial virus subcommittee of the International Committee on Taxonomy of Viruses (ICTV) has recommended the names for families of bacteriophages which end with viridae (Mathews, 1982) (Table 16.1).

Table 16.1 : Families of bacterial viruses (after Bradley, 1982)

Presence of Nucleic acid	Families	Examples
Non-enveloped		
dsDNA	Myoviridae	P2*, T2 **
	Styloviridae	λ
	Podoviridae	T7
	Corticoviridae	PM2
	Tectiviridae	PRD1
ssDNA	Inoviridae	MV-L1 type, fd Type
	Microviridae	ØX174
ssRNA	Leviviridae	MS2
Enveloped		
ssDNA	Plasmaviridae	MV -L2
dsRNA	Cytoviridae	Ø6

* Isometric-head, ** Elongated head

III. The Virulent dsDNA phages (T-even phages)

There are three even-numbered T-phages (e.g. T2, T4 and T6) and four odd-numbered T-phages (e.g. T1, T3, T5 and T7) that infect *E.coil*. Out of the seven coliphages of T-series, the virulent T4 phages is most extensively studied. The T4 phage infects non-motile strain B of *E. coli* and forms plaques on growing colonies.

1.   Morphology

The T4 phage is tadpole shaped and consists of the five important sub-structures such as the head, head-tail connector, tail base plate and fibers (Fig.16.2). The viral particle is naked icosahedral and tailed. The head is an elongated, bi-pyramidal, hexagonal like prism consisting of two 10-faced equatorial bands. It consists of a pyramidal vertex at either ends also. Size of head is 95 × 65 nm which consists of about 2,000 identical protein subunits (capsomers).

The phage consists of a long helical tail which is connected to the head with a connector having a collar with attached whiskers. Around the tail, fibres remain folded and held at midpoint by the whiskers. Size of the tail is 80 × 18 nm. It consists of an inner tubular core (having a hole of 25 Å) which is surrounded by a contractile sheath. Protein subunits (144) arranged in 24 rings each containing 6 subunits, constitute the sheath. The sheath connects the head at one end and base plate at the other end (Favre *et al*, 1965).

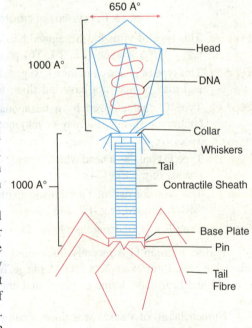

Fig. 16.2 : Structure of bacteriophage T2.

From the head at distal end, there is a hexagonal base plate attached to an end of tail. The base plate contains six spikes or tail fibers at its six corners. The spikes are 130 × 2 nm in size.

The spikes have two parts, the proximal half fibre, and the distal half fibre. The former is attached to the base plate and the later helps in recognition of specific receptor sites present on cell surface of the bacterial cell wall.

A dsDNA molecule of about 50 µm is tightly packed inside the head. The DNA is about 1,000 times longer than the phage itself. It is circular and terminally redundant. Unlike other DNA, all the T-even phages contain 5-hydroxymethylcytosine instead of cytosine, so that synthesis of phage can occur easily.

2. Multiplication

The multiplication processes of T-even phages are a little complex. It is accomplished in certain stages viz, adsorption, penetration, synthesis, replication, maturation (assembly) and release.

(i) **Adsorption:** T4 phage particles interact with cell wall of *E.coli*. The phage tail makes contact between the two, and tail fibres recognise the specific recognition sites present on cell surfaces (Fig. 16.3A). However, any part of the bacterial cell may act as specific receptor sites e.g. flagella, pili, carbohydrates and proteins presents in the membrane or cell wall itself. Specific receptor in the bacterium is the lipopolysaccharide (LPS) or lipoprotein or both. The LPS of tail fibres act as receptor in phages T3, T4 and T7, whereas lipoprotein acts as receptor in T2 and T6. In T5 phage both LPS and lipoprotein acts as receptors. When these receptors are absent, the bacteria become resistant to phage. Therefore, it is not necessary that every interacting phage is adsorbed on bacterial cell. However, if phages are added artificially in excess, multiple adsorption of upto 200-300 phages per bacterium may occur (Schlegel, 1986).

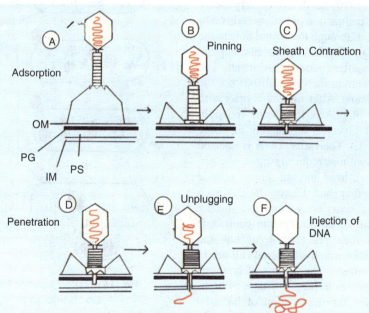

Fig. 16.3 : Process of T4 infection. OM, outer membrane; IM, inner membrane; PG, peptidoglycan; PS, periplasmic space; A; adsorption; B, pinning; C, tail sheath contraction; D, penetration; E, unplugging; F, injection of T4 DNA.

When the contact is made between tail fibres and bacterium, it becomes unfolded from the tail. Unfolding and release of tail fibres from whiskers is governed by certain cofactors, for example tryptophan is needed (1 mg/ml) (by Benzer strain T4B), and in some T phages the absence of tryptophan (e.g. in Doerman strain T4D) and indole (e.g. in Hershey strain T2H) are not required.

High concentration (0.2-0.6 M) of glucose is inhibitory for adsorption of T4 to bacterial cell. The whole process of interaction of phage to recognition of specific receptor site is known as *landing* (Fig. 16.3A). After landing there starts the second process of adsorption which is known as *pinning* (B). Before pinning the phage can move attached with tips of tail fibres to cell surface until it finds the site for pinning of spikes. Pinning is the irreversible process. All the activities before pinning are reversible. Possibly it occurs at the point where both plasma membrane and cell wall are attached (Favre *et al*, 1965).

(ii) Penetration : The process of peneteration occurs by mechanical and also enzymatic digestion of cell wall. At the recognition site phage either digests certain cell wall structure by viral enzymes e.g. lysozyme or activates the degradative enzymes of the host. But the real penetration of phage into the host is mechanical.

After pinning the tail sheath contracts and, therefore, appears shorter and thicker (Fig.16.3C, Fig.16.4A). Sheath contracts by rearrangement of discs from 24 to 12. The phage uses its energy in tail contraction as ATP, and calcium has been detected in purified T-even phages. The activity of phage contraction is comparable to that of microsyringe. Kollenberger (1970) has described very nicely the process of contraction of hollow stylus into the bacterial cell through the central hole in the base plate (D). The base plate through the centre enlarges after contraction of sheath. The stylus can penetrate the cell wall but not the plasma membrane. After making contact with a component of plasma membrane (possibly phosphatidyll glycerol) unplugging of phage DNA begins (E). Thereafter, DNA is injected into the cell without requiring metabolic energy (F). Phage head and tail remain outside the cell (Hershey and Chase, 1952). Such empty protein coats are known as 'ghosts'. Phage T1 and T5 do not contain contractile sheath, even then peneteration of tail and injection of DNA into the bacterial cell occur.

(iii) Synthesis: The process of synthesis of macromolecules leading to phage multiplication involves the degradation of bacterial chromosomes, protein synthesis, and DNA replication.

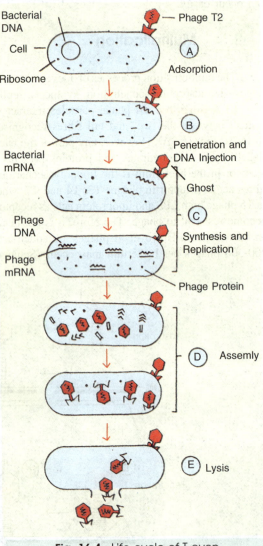

Fig. 16.4 : Life cycle of T-even bacteriophage.

(Figure labels: Bacterial DNA — Phage T2; Cell; A Adsorption; Ribosome; B Penetration and DNA Injection; Bacterial mRNA; Ghost; C Synthesis and Replication; Phage DNA; Phage mRNA; Phage Protein; D Assemly; E Lysis)

(a) Degradation of bacterial chromosome: After infection by T-even phage, the bacterial chromosome is unfolded due to relax of superhelical twisting. The nucleoid, which is present in the centre, is disrupted and DNA helix is attached with plasma membrane at different points. This process is facilitated by nucleoid disrupting gene (T4 *dna* gene) products. The RNA polymerase-I binds to membrane simultaneously (Fig. 16.4C). After 3 minutes of infection, protein synthesis by bacterial host stops, and partial break down of bacterial chromosome occurs. However, before 3 minutes of infection replicative synthesis of bacterial cell takes place. However, upto 10 minutes non-conservative DNA synthesis occurs just to repair the damage in the bacterial DNA caused by nucleases encoded by phage. The RNA polymerase and ribosomes are modified which initiate protein synthesis of phage but not of bacterial cell.

(b) **Protein synthesis:** The phage DNA contains early and late genes, therefore, after a few minutes of introduction of genome bacterial mRNA and protein synthesis stop. The early and late genes transcribe (at different intervals) the early mRNA and late mRNA respectively. After 5 minutes of infection the early mRNA of phage have been detected in bacterial cell. Late mRNA synthesis starts after 10 minutes of infection. After 20 minutes both the early and late mRNAs are found in the host cell. The left strand of T4 DNA transcribes the early genes and the right strand of DNA transcribes the late genes. The early genes are of two types, immediate early and delayed early genes. The immediate early genes are transcribed after 1.25 or 2.5 minutes of infection by using the bacterial RNA polymerase. These genes also code enzymes that degrade bacterial chromosome. The delayed early genes transcribe the mRNA from 2.5 to 3.75 minutes. The delayed early genes synthesize the phage enzymes which produce 5-hydroxy-methylcytosine (5HMC) by using cytosine of the bacterial DNA. At this stage the bacterial restriction enzymes could not degrade the phage genome. Use of chloramphenicol can warrant the transcription of the delayed early mRNA.

The late phage gene products are the lysozyme and the structural components of new phage particles such as the heads, tails and fibres.

(iii) **Replication of T4 DNA:** During the first 10 minutes after infection of the phage DNA, no phage could be recovered from the infected bacterium. This time interval is called as eclipse period (Fig.16.7). Moreover, replication of T4 DNA is a complex process. T4 DNA differs from the typical DNA molecules as the former contains no cytosine, but consists of modified base and 5-HMC which pairs with guanine. Serveral glucose like sugars are covalently linked to the 5-HMC. There are five important events of T4 DNA replication:

(a) Degradation of host DNA,

(b) Synthesis of 5-HMC,

(c) Prevention of incorporation of cytosine into the T4 DNA,

(d) Glucosylation of T4 DNA, and

(e) Enzymology of DNA replication.

Generally, the bacterial DNA is degraded by phage coded exonuclease to deoxynucleotide monophosphate (dNMP). Then the dNMP builts up dATP, dTTP, dGTP and dCTP by the bacterial enzymes which provide sufficient dNTP to synthesize 30 T4 DNA molecules. Therefore, replacement of cytosine to 5-HMC is done which becomes glucosylated. This modified nucleotide is protected from degradation by the bacterial restriction enzyme. Timing of life cycle in minutes at 37°C is outlined in Table 16.2.

Table: 16.2 : Events of life cycle of T4 bacteriophage in bacterial cell after penetration.

Time (min) after penetration	Events
0	Adsorption on bacterial cell wall and penetration of T4 DNA
1	Inhibition in synthesis of host DNA, RNA and Proteins
2	Synthesis of first mRNA.
3	Degradation of bacterial DNA
5	Initiation of synthesis of T4 DNA
9	Synthesis of late mRNA
12	Completion of synthesis of head and tail.
15	Appearance of first phage particle
22	Bacterial lysis and release of about 300 progeny phages.

After 6 minutes of infection, replication of T4 DNA begins, then the rate of replication reaches maximum after 10-12 minutes. DNA synthesis is not done by T4 DNA polymerase, but a short RNA primer is synthesized which presumably starts each step of DNA synthesis. On RNA primer T4, DNA polymerase extends the DNA strand by using one of the strands as template. DNA polymerase possesses 3' exonuclease activity but lacks 5' exonuclease activity.

Gene 32 encodes the T4 DNA binding proteins which is required for replication and recombination. It maintains the DNA in single stranded state. The genetic map of T4 phage is given by King (1968) (Fig.16.6). It has been found that the different regions in phage genome regulate the specific operations like promotion of synthesis of viral DNA and of RNA molecules (King, 1968). T4 DNA replicates by rolling circle mechanism also. After viral DNA synthesis, the late proteins are formed.

(iv) Assembly (Maturation) : When the structural proteins and phage DNA are synthesized, assembly of these components begins to give rise the mature phages (Fig. 16.4D and Fig.16.5). About 135 T4 DNA molecules are identified which account for 90% of the DNA. About 20% genes are to be identified and 50 genes take part in morphogenesis of this phage. The process of assembling the phage particles is known as maturation. Assembly of phage components begins in a well defined order, such as the head, tail, and tail-fibres synthesized independently, are combined to yield phage particles (Fig. 16.5). After 20 minutes of infection about 300 new phages have been assembled. Some steps in assembly of coliphage T4 have been described by Edger and Lielausis (1968).

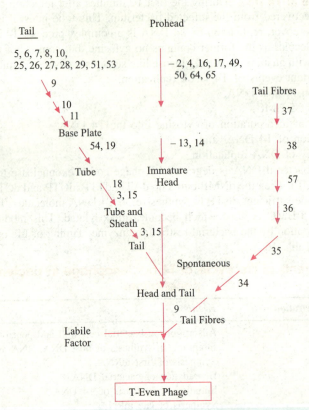

Fig. 16.5 : Different morphogenetic pathways synthesising the T-even phage. The number represents phage genes whose products are involved in carry out each step.

J.D. Watson and F.C. Crick (1972) postulated that the assembly of preformed parts into a virion is a process of condensation, and the shape and design of a viral capsid is determined by the specific bonding properties of its identical capsomers. However, it appears that the maturation is a sequential event in which each step requires the previous ones.

(v) **Lysis and Release :** After eclipse period maturation of phage particles starts where they accumulate inside the bacterial cell. The progenies are released by the sudden bursting i.e. lysis of host cell wall (Fig 16.4E). Lysozyme is attached to the tip of the tail of phages. Time taken from infection to lysis is known as latent period. Yield of phage per bacterium is known as burst size. Starting with infection by a phage of the bacterial cell to release the newly synthesized viral particles with respect to the time is known as one step growth curve (Fig 16.7). This growth curve was first introduced by Ellis and Delbruck (1939).

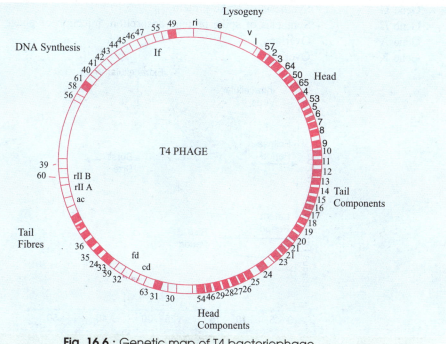

Fig. 16.6 : Genetic map of T4 bacteriophage.

IV. The ssDNA phage: ØX174

1. Morphology

The bacteriophage ØX174 was discovered by R.L. Sinsheimer at California Institute of Technology. It is one of the ssDNA phages of *E.coli* which has been most extensively studied. The phage particles are naked and icosahedral having a diameter (without spike) of 24-29 nm. The weight of a virus particles is 6.2×10^6 daltons. The capsid is made up of capsomers, each consisting of five structural units. Morphologically, the capsomers are probably angular, hollow and pentagonal from the centre of which projects a single spike situated at one apex of the icosahedron (Fig 16.8). Hence, there are 12 spikes in one phage particle. Individual spike is constituted by H protein (encoded by one gene) and G protein (encoded by five genes). These interact with F gene protein. The H gene protein assists the adsorption of phage to the bacterial cell, in addition to functioning as pilot protein and helping the injection of DNA into the bacterial cell.

The capsid encloses a single and cicular (+) ssDNA (molecular weight 1.7×10^6 daltons) consisting of eleven genes arranged in an order *A, B, C, D, E, J, F, G, H, A** and *K*. Gene *A** and *K* overlap to each other. The gene *B* lies within gene *A*, and gene *E* is present within gene *D*. The gene *C* overlaps with genes *A* and *D*. Therefore, from two genes four proteins are encoded instead of two. Sanger *et al* (1977) have shown that the phage contains 5386 nucleotides and mapped the nucleotide sequence of ØX174 (Fig. 16.9). They discussed the function of these genes as below:

Gene *A*	:	Replication of phage by introducing single stranded break in the DNA.
Genes *B,C,D*	:	Synthesis and packaging of ssDNA of progenies.
Gene *E*	:	Bacterial cell lysis.
Gene *J*	:	Internal protein.
Gene *F*	:	Major capsid protein
Gene *G*	:	Synthesis of spike protein
Gene *H*	:	Synthesis of spike protein, adsorption, injection of phage DNA.
Gene *A**	:	Overlaps
Gene *K*	:	Overlaps

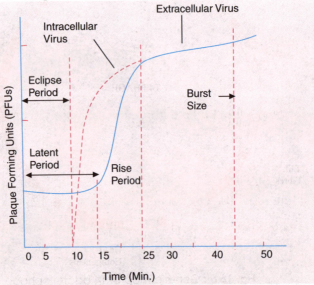

Fig. 16.7 : Growth curve of T-even phage.

2.   Multiplication

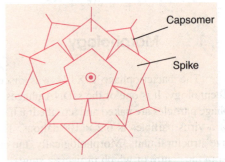

Fig. 16.8 : Structure of bacteriophage ØX174 (diagrammatic).

(*i*) **Adsorption :** The cell wall surface of *E.coli* contains specific receptor sites. For example, the receptor site of ØX174 present on outer membrane of cell wall of *Salmonella typhimurium* is a lipopolysaccharide. The phage gets adsorbed on bacterial cell surface through any one of 12 icosahedral vertices. The spikes differ each other in specificity for attachment. For the attachment to host cell wall the phage needs Ca^{++} or Mg^{++} ions. Therefore, to find out the specific receptor site, the adsorption may be reversible.

(*ii*) **Injection:** There starts eclipse period. Viral DNA is injected into the bacterial cell. Phage gene H encodes proteins which act as pilot protein, conveys the DNA into the bacterial cell.

(*iii*) **Synthesis:** After the introduction of phage DNA into the bacterial cell, there starts synthesis of viral DNA and protein. However, these processes are rather very complex in ØX174. Mitra (1980) has reviewed the DNA replication in viruses. The phage genome is in circular form consisting of an infectious hairpin duplex. Phage genome acts as template (+ strand). It replicates immediately after infection which is accomplished in three different stages:

    (*a*) Synthesis of a replicative from (RF) of viral DNA,

    (*b*) Replication of parental RF to progeny RF DNA, and

    (*c*) Conversion of RF molecules into rolling circle molecules.

    (*a*) **Formation of replicative form (RF) of DNA :** Denhard and Hours (1978) have described the DNA replication of ØX174. Soon after peneteration, the ssDNA of phage synthesises a complementary RF DNA and results in formation of the double stranded DNA. The newly synthesized RF strand is known as (–) strand (Fig. 16.10A-F). Synthesis of RF of DNA is completed in certain stages. Soon after entry of ssDNA, the DNA unwinding proteins extend ssDNA leaving the hairpin duplex which is a promoter region. This step requires for the presence of ssDNA, dnaB protein, dnaC protein, unwinding protein and two protein factors (the X, Y, and Z, and ATP). These proteins form an intermediate substrate for the synthesis of dnaG protein (molecular weight 60,000 dalton) (Weiner, McMacken and Kornberg, 1976).

    The unwinding proteins of which about 800 molecules are present, binds to ssDNA but not dsDNA. This protein establishes the ssDNA in a state which can act as a template for synthesis of its complementary strand. DNA polymerase is used since this process occurs before transcription of mRNA. It allows the initiation of DNA synthesis at the point of origin of replication present in dsDNA. The presence of RNA primer in ssDNA is essential for the synthesis of DNA. However, after pre-priming with dnaB and dnaC proteins, dnaG protein catalyses the RNA priming in ØX174. The RNA polymerase is not required for the synthesis of RNA primer. On RNA primer a complementary (-) strand is extended by using the (+) ssDNA as the template. Chain elongation of primed DNA takes place by DNA polymerase III holoenzyme (i.e. DNA polymerase III plus DNA copolymerase III) in Okazaki fragments (the newly synthesized DNA in the form of short segments which later on are covalently joined to yield a continuous (–) strand) (Ogawa and Okazaki, 1980). This structure is known as RF II which is converted to RF when gap is sealed. DNA polymerase through 5'→3' exonuclease activity removes the RNA primer from (–) DNA strand. This gap is filled up by the activity of DNA polymerase I. The two ends of the DNA molecule are ligated by the enzyme DNA ligase which results in a circular DNA molecule. Thereafter, the cellular poly-

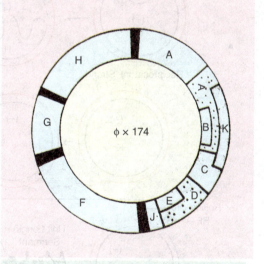

Fig. 16.9 : Genetic map of bacteriophage ØX174. Inner arrows show the initiation of site for mRNA of genes A, B and D. Dark spaces show the intergeneric spaces. The inner number indicates the untranslated nucleotides (based on data of Sanger *et al.*, 1977).

merase forms a pool of progeny RF molecules (RF → RF). The ssDNA strands are also synthesized.

    (*b*) **Replication of RF (RF → RF) :** The RF molecules produced in this way replicate to form a pool of progeny RF molecules (Fig. 16.10 G-M). For the replication of parental RF molecule, the A gene protein is needed that makes a nick on the viral strand of RF molecule. Possibly the protein acts as hairpin in (–) DNA superhelix existing at palindromic sequences. For the replication of RF molecule the host cell machinery is required that includes proteins encoded by *rep* gene and

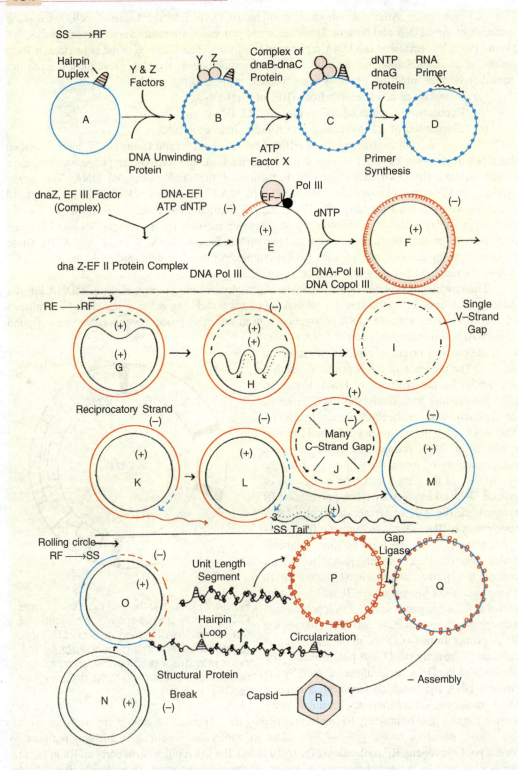

Fig. 16.10 : Replication of ØX174.

*dna*B, *dna*C, *dna*E, *dna*G, *dna*H and *dna*Z genes. Probably the origin of replication is present in gene A. Two models, reciprocatory strands model and rolling circle model, have been suggested for the replication of RF DNA duplex.

Reciprocatory Strand model : After the formation of RNA primers synthesis of a new strand begins at the origin of replication present on gene A (Fig. 16.10 G-J). In ∅X174 the RNA priming is catalyzed by dnaG protein when a short sequence of DNA is formed. Positive superhelical twists accumulate which impose the strain. The viral strand is synthesized around the genome continuously in unidirectional and in clockwise directions. The growing point moves for its origin to terminus around the circular genome.

Moreover, the complementary strand primed by RNA is synthesized in discontinuous manner. Synthesis of complementary strand is lagged behind the viral strand. After synthesis of the nascent complementary strand, there occurs the repeated exchange reactions between the nascent strand and bacterial strand. The two parental strands are completely unwound. A strand exchange reaction releases two separate, circular but incomplete duplexes. One duplex has a single gap in the nascent viral strand (V-strand gap), and the other duplex has several gaps that separate nascent complementary strand of DNA segments (many C-strand gaps.).

Rolling circle model : Dressler and Wolfson (1970) suggested that the rolling circle mechanism results in replication of RF → RF as well as RF → ss. Replication of RF → RF takes place by semi-conservative method (Fig. 16.10 KM.). A nick is made on the outer strand, and the 5' end tail serves as template for synthesis of a small DNA segment. The growing point moves from its joint by the enzyme DNA ligase to produce a dsDNA molecule. In turn, each strand of dsDNA molecules acts as a template for the synthesis of complementary DNA molecule.

(c) Synthesis of ssDNA (RF → ss) : For the first time Gilbert and Dressler (1968) suggested the rolling circle mechanism for ssDNA synthesis of ∅X174 (Fig. 16.10 N-R). They demonstrated that during the first round of replication RF molecules are produced. The rolling circle mechanism operates during the first stage of replication and produces ssDNA molecules.

However, no complementary strand is synthesized on the tail during RF → ss stage of synthesis.

To begin the replication a virus coded enzyme makes a specific single stranded nick in the plus strand of the RF. This generates the 3'-OH and 5'-phosphate ends on the strand. The deoxynucleotides (dATP, dCTP, dGTP, dTTP) are added to free 3'-OH end. The open single strand rolls off the circle as a free tail with the progress of synthesis. By using the minus strand as a template and deoxynucleotide precursor a new plus strand is synthesized. With the rolling of the strand, structural proteins of phage bind to the elongating tail. Nuclease acts within the hairpin loops present on DNA and release the plus strand which becomes circularized by binding the cut ends and forming the complete hairpin loop. The enzyme ligase fills up the gap.

(iv) Assembly and Release : The phage proteins are synthesized in the bacterial cell. Soon after synthesis, the progeny ssDNA molecules are packed into phage particles (Fig. 16.10R). (Denhard and Hours, 1978). The events of phage assembly and release from the host cell are similar as described for T2 phage.

In 1967, success has been achieved in artificial synthesis of ∅X174 by A. Kornberg and coworkers at Stanford University (Goullan *et al*, 1967) and by R.L. Sinsheimer at California Institute of Technology.

V. Bacteriophage Typing

The virulent phages destroy the bacterial hosts after their maturation. If the phages destroy the pathogenic bacteria, they are of much use in medical science. So far there is no evidence of the use of bacteriophage therapeutically, because the phages do not persist in body though kill bacteria. Therefore, the bacteriophages are primarily used in identification of bacterial strains. The laboratory procedure is called bacteriophage typing and is used in identification of some strains of bacterial pathogens such as the staphylococci and typhoid bacilli. Lawn of a single strain of a bacterial species is prepared on specific growth medium. Virulent phage is allowed to infect the bacterial cells and produce plaques. The strains of bacteria are characterized by their resistance or susceptibility to lysis by specific phage.

VI. Phage Lambda (λ) : Temperate dsDNA Phage

Phage lambda is a virus of *E. coli* K12 which after entering inside host cell normally does not kill it inspite of being capable of destroying. Therefore, it leads its life cycle in two different ways, one as virulent virus and the second as non-virulent. The virulent phase is called lytic cycle and the non-virulent as temperate or lysogenic one, and the respective viruses as virulent phage and temperate phage, respectively. The other temperate lamboid phages are 21, Ø80, Ø81, 424, 434, etc.

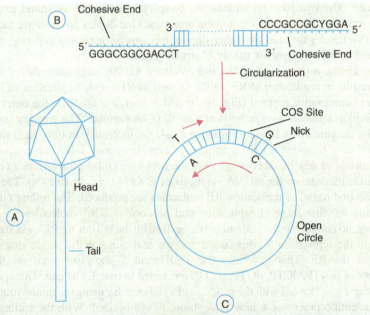

Fig. 16.11 : Structure of bacteriophage lambda (λ). (A), single standard cohesive ends of 12 bases (B), and circularization of complementary DNA to form *cos* site (C).

1. Morphology

Morphological structure of phage λ is given in Fig. 16.11A. The phage λ contains double stranded (ds) circular DNA of about 17 μm in length packed in protein head of capsid. The head is icosahedral, 55 nm in diameter consisting of 300-600 capsomers (subunits) of 37,500 daltons. The capsomers are arranged in clusters of 5 and 6 subunits i.e. pentamers and hexamers. The head is joined to a non-contractile 180 μm long tail by a connector. There is a hole in capsid through which passes this narrow neck portion expanding into a knob like structure inside. The tail possesses a thin tail fibre (25 nm long) at its end which recognises the hosts. Also the tail consists of about 35 stacked discs or annuli. Unlike T-even phage, it is a simple structure devoid of the tail seath.

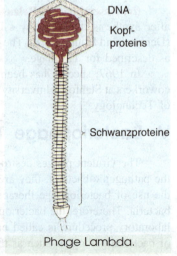

DNA

Kopf-
proteins

Schwanzproteine

Phage Lambda.

2.   DNA and Gene Organization

Lambda DNA is a linear and double stranded duplex of about 17 µm in length. It consists of 48, 514 base pairs of known sequence. Both the ends of 5' terminus consists of 12 bases which extend beyond the 3' terminus nucleotide. This results in single stranded complementary region commonly called cohesive ends. The cohesive ends form base-pairs and can easily circularize. Consequently a circular DNA with two single strand breaks are formed. The double stranded region formed after base pairing of complementary nucleotides is designated as COS. The 12 nucleotides of cohesive ends and process of circularization are shown in Fig. 16.11B-C. The events of circularization occurs after injection of phage DNA into *E.coli* cell where the bacterial enzyme, *E.coli* DNA ligase converts the molecule to a covalently sealed circle.

3.   Life Cycles

The phage λ leads two life cycles, the lytic cycle and the lysogenic cycle after injecting its DNA into *E.coli* cell. In the lytic cycle, phage genes are expressed and DNA is replicated resulting in production of several phage particles. The lytic cycle ends with lysis of *E.coli* cells and liberation of phage particles. This lytic cycle is a virulent or Sintemperate where phage multiplies into several particle.

In addition, the lysogenic cycle results in integration of phage DNA with bacterial chromosome and becomes a part of host DNA. It replicates along with bacterial chromosome and is inherited into progenies. The phage DNA integrated with bacterial chromosome is called prophage. The prophage is nonvirulent and termed as temperate phage. The bacteria containing prophage are called lysogenic bacteria, and the prophage stage of viruses as lysogenic viruses. After treatment of lysogenic bacteria with UV light, X-rays or mitomycin, the prophage can be separated from bacterial chromosomes and enter the lytic cycle. This process is known as induction.

4.   Genetic Map of Lambda

The genetic map of phage λ is given in Fig 16.12. The remarkable characteristics of the map is the clustering of genes according to their functions. For example the head and tail synthesis, replication and recombination genes are arranged in four distinct clusters. These genes can also be grouped into three major operons viz. right operon, left operon and immunity operon. The right operon is involved in the vegetative function of the phage e.g. head synthesis, tail synthesis and DNA replication leading lytic cycle. The left operon is associated with integration and recombination events of lysognic cycle. The immunity operon products interact with DNA and decide whether the phage will initiate lytic cycle or lysogenic cycle. Singer *et al* (1977) have given the nucleotide sequence of ØX174. Genetic map of bacteriophage has been given by Echols and Murialdo (1978). Function of some important genes is summarized in Table 16.3 and briefly described below.

(*i*) **Head Synthesis Genes:** At the left end of phage genome the head genes viz, *A, W, B, C, D, E* are located which are associated with phage DNA maturation and head proteins.

(*ii*) **Tail Synthesis Genes:** The genes *F, Z, U, V, G, H, M, L, K, I, J* are clustered just right to head genes and code for tail proteins.

(*iii*) **Excision and Integration Genes:** The gene *xis* codes protein that excises the phage DNA from the bacterial chromosomes, and *int* coded protein is involved in integration of phage DNA into the bacterial chromosome.

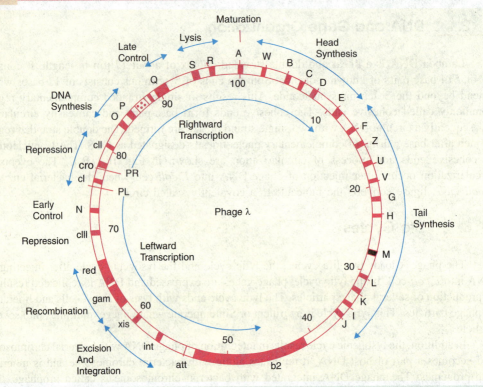

Fig. 16.12 : Genetic map of phage λ showing some important genes and their function; PL, leftward promoter; PR, righward promotor.

(iv) **Recombination Genes:** The two genes *int* and *xis* codes at *att* P for site-specific recombination. The three *red* genes code for three proteins at normal frequency for general recombination. The *red*L codes for exonuclease, *red* B for beta-protein and *red* V for gamma protein. The gamma protein inhibits exonuclease V.

Table 16.3 : Some important phage genes and their function.

	Genes	*Functions*
1.	*A, W, B, C, D, E*	Code for DNA maturation and phage head proteins
2.	*E, Z, U, V, G, T, M, L, K, I, J*	Synthesis of phage tail protein
3.	*b2*	Deletion for non-essential region of phage DNA
4.	*att*	Site for integration of phage DNA into bacterial chromosome.
5.	*int, xis*	Code for integrase and excisionase required for integration and excision of phage DNA into and from bacterial chromosome, respectively
6.	*gam, red*	Codes for gamma proteins that inhibits exonuclease V, Kills host
7.	*cIII*	Activates *cro*, codes for *cI* and *int* gene products.
8.	*N*	Product extends transcription from PL and PR, antiterminator for tL or tR

	Genes	Functions
9.	PL	Leftward promoter that promotes transcription initiation from N through *int*
10.	cI	Codes for immunity repressor
11.	PR	Rightward promoter that initiates transcription from *cro* through O region
12.	cro	Depresses transcription from PL and PR, directly inhibits repressor synthesis
13.	cII	Helps in synthesis of *cI* and *int* gene products.
14.	O,P	Control replication of phage DNA
15.	Q	Positive regulation for transcription of phage late genes *S* to *J* from PR Promoter.
16.	S	Product shuts off synthesis of phage DNA and affects hosts membrane
17.	R	Codes for phage endolysin endopeptidase.
18.	A	Phage maturation.

(*v*) **Positive Regulation Gene:** The genes *N* and *R* are the positive regulation genes. The proteins coded by these genes increase the rate of transcription of other genes. Protein coded by *N* gene induces the transcription of *cII, Q, P, A, red, gam, xis* and *int*, whereas the protein coded by *Q* gene stimulates the transcription of head, tail and lysis genes. The *N* and *Q* genes are also required in plaque formation, in the absence of which the number of phage particles would be less but not zero.

(*vi*) **Negative Regulation Genes:** The *cI* gene acts as a repressor and its product maintains the prophage in the lysogenic form in bacterial host. Moreover, the *cII* and *cIII* assist the *cI* gene in lysogeny. The proteins encoded by *cro* binds to *PL* and *PR* and reduce the expression of *cI, N, red* and *xis* genes. The interactions between *Q* proteins encoded by *cro* and phage repressor occurs in host cell and the result decides the operation of lytic or lysogenic cycle. The choice between lysogeny and lysis has been discussed in the preceeding section.

(*vii*) **DNA Synthesis Genes:** The two genes *O* and *P* are involved in synthesis of phage DNA. The origin of DNA replication lies within the coding seqence for gene *Q* which encodes a protein for initiation of DNA replication, and the gene that generates the cohesive ends is located adjacent to one of the ends. The function of gene *N* is required in transcriptional process of these genes.

(*viii*) **Lysis Genes:** The *S* and *R* genes control the lysis of bacterial cell envelope which occurs at the end of lytic cycle.

5. Choice Between Lytic and Lysogenic Cycles

Soon after circularization of genome and start of transcription, the gpcII and gpcIII accumulate. The gpcII binds to PRE (promoter for repressor establishment) and stimulates binding of RNA polymerase (Fig 16.13A). The gpcIII protects gpcII from degradation by host nucleases. Lambda repressor (gpcI) is rapidly synthesized (B), binds to OL and OR, and inhibits the synthesis of mRNA and production of gpcII and gpcIII (proteins) (C). The repressor activates the promoter for repressor maintenance (PRM) which induces *cI* gene to be transcribed continuously at a low rate. This process goes on continuously and ensures the stable lysogeny when it is established (C).

During the course of time the gpcro also accumulates. It binds to OL and OR, turns of the transcription repressor gene *cI* and repress PRM function (D). The repressor (gpcI) can block *cro* transcription. Therefore, there is a race between the production of gpcI and gpcro proteins. The detail of this competition is not yet clear but the environmental factors influence the result of this race for choice of the two cycles. If the repressor wins the competitions, the circular DNA is inserted into the *E. coli* genome. The amount of gpcro and the outcome of competition with gpcI decides the establishment of lysogenic or lytic pathways.

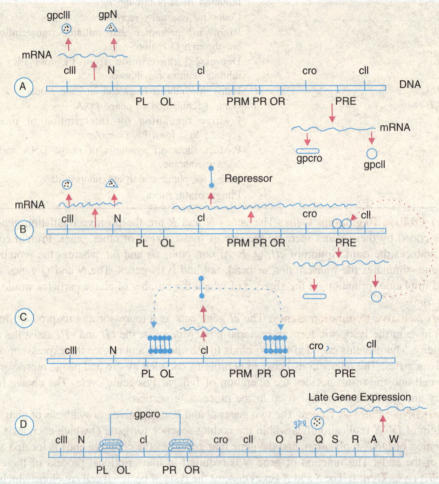

Fig. 16.13 : Choice between lysis and lysogeny (diagrammatic, modified after Prescott *et al.*, 1996)

6.   The Lytic Cycle

As described earlier, lambda consists of two types of life cycles, the lytic and lysogenic. The lysogenic life cycle would be described later on. The events of lytic cycle, starting with adsorption, at 37°C occurs as below:

Time (t) = 0 minutes : Phage adsorbed and DNA is injected
        = 3 minutes : First early mRNA is synthesized
        = 5 minutes : Two classes of early mRNA are synthesized
        = 6 minutes : DNA replication begins
        = 9 minutes : Synthesis of late mRNA begins
        = 10 minutes : Synthesis of structural protein occurs
        = 22 minutes : First phage particle is completed
        = 45 minutes : Lysis of bacterial cell envelope and release of progeny phage

Life cycle of phage λ is 45 minutes long, as compared to 22-25 minutes long life cycle of T4 phage. In general the life cycle of most phages at 37°C varies between 22 and 60 minutes. After injection, the linear phage DNA is circularized (Fig. 16.11B). The *cos* site results in a circular form of DNA which starts the two events: (*i*) transcription and translation of DNA, and (*ii*) replication of DNA. As a result of replication numerous DNA molecules are formed in the form of concatemers. Structural proteins for head and tail are synthesized through translation of mRNA. During maturation assembly of phage particles occurs. The lytic cycle ends with lysis of host cell envelope and releases lambda particle. The events of lytic cycle are discussed in detail as below:

(*i*) **Circularization of Phage DNA:** After injection the linear phage DNA is delivered into bacterial cell. The cohesive ends form hydrogen bonds. The *E.coli* DNA seals the breaks and the DNA is converted into closed circle within three minutes after infection.

(*ii*) **Transcription:** After circularization of phage DNA, transcription begins by the modified RNA polymerase of host. The modification enables the polymerase to ignore certain termination sites. Fig. 16.14 shows a detailed version of genetic map of phage. There are three regulatory genes *cro*, *N* and *Q*, three promoters, left promoter (PL), right promoter (PR1 and PR2). The DNA replication genes *O* and *P*, and five termination sites (tL1, tR1, tR2, tR3 and tR4). The L and R series are transcribed leftward and rightward, respectively from the complementary DNA strands.

On the basis of transcription the genes are grouped into three classes, immediate early genes (*N* and *cro*), delayed early genes (located left to *N* e.g. *cIII*, *gam*, *red*, *xis* and *int*, and right to *cro* e.g. *cII*, *O*, *P* and *Q*) and late genes. Consequently three classes of mRNAs are transcribed e.g. immediate early, delayed early and late mRNAs. Early mRNAs are transcribed from both left and right strands, whereas late mRNAs are transcribed from left to right, and the left transcribes from right to left.

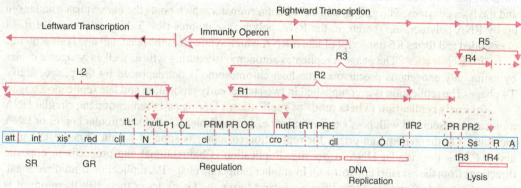

Fig. 16.14 : Genetic map of phage λ showing genes arranged in different groups and their association with leftward and rightward transcription. Description of genes and their products are given in the text. SR, specialised recombination; GR, generalised recombination.

First of all *O* and *P* genes are transcribed whose gene products are necessary for DNA synthesis. This is followed by transcription of gene coding for structural proteins, packaging system and finally lytic proteins. Before transcription of *O* and *P* genes, two immediate early mRNA transcripts are formed that codes for regulatory proteins responsible for turning on and off to the leftward transcription and rightward transcription when ever required. The phage has two promoters, PL and PR which initiates the synthesis of RNA molecules L1 and L2. Initially transcription terminates at the sites tL1 and tR1. L1 encodes gene product N (gpN) which is the delayed early gene product and a major regulatory factor that controls certain regions of DNA when it is transcribed. After synthesis, gpN binds to *nut*L and *nut*R sites present at left and right side of the promoters. When RNA polymerase moves along with the DNA, it picks up the gpN. The gpN enables the polymerase to ignore the termination sites (tL1 and tR1) resulting in longer transcripts L1 and

L2. Therefore, the gpN acts as antiterminator and neutralizes the effect of tL1 and tR2. Hence, the gpN controls the expression of most vital function (Epp and Pearson, 1976).

After inhibition of termination, transcription occurs in leftward direction and extends upto b2 region, and the rightward transcription extends upto Q gene. The rightward transcription permits the synthesis of O and P gene products. The leftward transcript consists of a *red* region that codes for two proteins required for genetic recombination. The O, P and *red* proteins have catalytic property, therefore, they are not made continuously. When sufficient amount of gpcro (gene product of *cro*) encoded in R1 is available, it binds to leftward operator (OL), and the repressor activity of *cro* turns off the synthesis of all leftward mRNA (Hu and Szybalski, 1979).

The tR3 is the termination site of mRNA even after modification by gpN. Therefore, the rightward transcription terminates at tR3. However, during the time of early transcription sufficient amount of mRNA is produced by rightward transcription. Thus the O and P proteins become sufficient for DNA replication. After binding of gpcro to OL the concentration of gpcro increases, and also binds to OR and turns off the rightward mRNA synthesis. Therefore, wasteful synthesis of O and P proteins does not occur, and aberrant and deleterious DNA synthesis are also checked (Hu and Szyblaski, 1979)

The gene product of *cII* (gpcII) is encoded in the same transcript containing O and P during the early transcription. The sufficient amount of gpcII acts as late promoter and delays late mRNA synthesis. After turning of rightward transcription by gpcro, the gpcII is not synthesized. This relieves the inhibition of late mRNA synthesis. At this time the gpQ (a positive regulator) is required that begins the late mRNA synthesis and neutralizes the third right terminator (tR3) and allow to proceed transcription through vegetative genes to J gene and into the b2 region of DNA molecule. The gpQ turns on late mRNA synthesis which translates structural and assembly proteins, maturation system and the lysis enzymes. The gpQ is also called antiterminator which binds the *qut* sequence and taken up by RNA polymerase. Therefore, the RNA polymerase ignores tR4. Taking advantage of it R4 is extended and forms R5 transcript of late mRNA which synthesizes the head, tail and lysis proteins.

(iii) Replication: The phage λ replicates autonomously during lytic as well as lysogenic cycles by using only exogenous precursors. The host chromosome is not degraded by the phage, unlike T4 phage. The replication is accomplished in two stages, early replication and late replication stages.

(a) Early replication (Theta mode of replication): During early replication the circular DNA molecule is associated with host's cell membrane, and replicates to produce circular copies of DNA molecule. On the genome an origin for replication (*ori* site) is situated within O gene. The gpO and gpP nick the circular DNA at *ori* site. Replication is bidirectional and proceeds in opposite directions from the *ori* site (Fig 16.15A). In another temperate phage P2, replication is unidirectional. The replication fork moves around the circle and forms the Greek letter theta (θ); therefore, it is called theta mode of replication.

The two branch points are called replication forks at which the non-replicated original duplex joins the two daughter chromosomes. At the end of replication two identical copies of circular DNA molecule are formed. Thus the theta mode increases the number of templates for transcription and futher replication. For the first time Cairns (1963) reported the theta mode of replication in *E. coli*. For detail see Chapter 5, *Nucleic Acids: DNA and RNA*.

(b) Late replication (rolling circle mode of replication): After the synthesis of circular copies of DNA the progenies dissociate from the cell membrane and switch over from theta to rolling circle model of replication. By the time heads and tails have been synthesized and the sequence-specific cutting system called the terminase (Ter) system (Ter proteins are the components of an empty head) becomes active resulting in predominance of rolling circles. A nick is made at a point on outer strand of duplex (Fig. 16.15 B). Circle rolls and a new strand is synthesized at 3' end. The 5' end single stranded DNA is displaced. Finally, the displaced strand contains a long ssDNA molecule of one parental strand and other newly synthesized strand. The rolling circle has the two types of *cos* sites, one in the circle and the second in the linear branch. In the linear branch there are several *cos* sites. Such branch is called concatemeric branch. During replication the *cos* site of rolling circle does not open because in the open strand replication cannot occur. If two *cos* sites are present in circular DNA, one is cut by Ter system resulting in free end in concatemer and removal of phage DNA. Thus, Ter-cutting requires two *cos* sites or one cos site and a free cohesive end on a single DNA molecule. The Ter system was first identified-by genetic analysis of tandem dilysogens i.e. cells having two adjacent prophage.

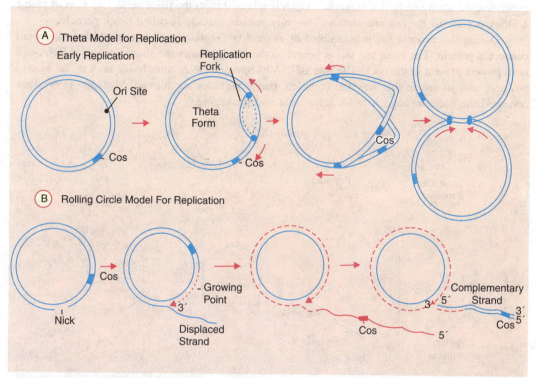

Fig. 16.15 : Replication of bacteriophage λ DNA. A, theta (θ) mode of replication; B rolling circle mode of replication.

The cut dsDNA of phage thus liberated from the concatemeric branch contains 12 nucleotide long ssDNA that acts as cohesive ends. The unit length genome synthesized by this mechanism contains exactly the phage genome. The gpgam is required to inhibit recBC endonuclease V of the host. Otherwise the concatemers would be broken down.

The two process cutting at *cos* sites and packaging of phage genome are coupled.

(iv) Assembly : Hohn and Katsura (1977) have described about the structure and assembly of phage. During the process of maturation, the phage particles (head and tail) are independently assembled. The genes encoding for DNA maturation and phage head proteins are: nul, A, W, B, C, nu3, D, E and F. The genAes that code for phage tail are: Z, U, V, G, H, M, K, L, I and J. There are bacterial genes (*groES* and *groEL*) that also help in assembly of phage particles.

Different steps in DNA packaging in phage λ head assembly have been described by Kaiser *et al* (1975). The process of assembly begins with aggregation of several copies of four head proteins which built up a scaffolded prohead (Fig.16.16). The prohead is only a sphere supported with an internal supporting system. Therefore, it looks like a wheel. Many phages form this type of scaffolded prohead. In the second stage, gene product of *groES* and *groEL* genes i.e. GroEs and GroEL proteins interact and form GroES-GroEL complex. This complex binds to scaffold prohead. The scaffolding is removed by bacterial protease. In some phages, scaffolding falls away and is reused. Gene products of *nul* and *A* (gpnul and gpA) that contain Ter system interact a short base-sequence near one *cos* site with a point on the head. Later on, it becomes the region for head-tail attachment. The phage DNA folds into the head. A change in conformation in E protein occurs after a small amount of DNA enters into the head. The gpE and the changed E protein causes the formation of icosahedral head. The gpF1 plays a role in expansion of spherical particles to icosahedron. A small amount of gpD enters into head and the head is filled with λ DNA by the unknown mechanism.

When the next *cos* site reaches the head during the process of filling, it is cut by Ter system generating the sticky ends. The unpacked DNA is released from the filled head. Insertion of phage λ DNA occurs till the *cos* site comes. The fully packed particle is called black particle.

During this process tail is assembled by several tail proteins, and terminated by a head-tail connector protein. The complete tail is bound to the head through the short piece of ssDNA and head protein present in the neck. The free ssDNA of released DNA binds to the neck of the second prohead, and so on. In this way the complete phage particles are formed. Murialdo (1991) has reviewed the bacteriophage λ DNA maturation and packaging.

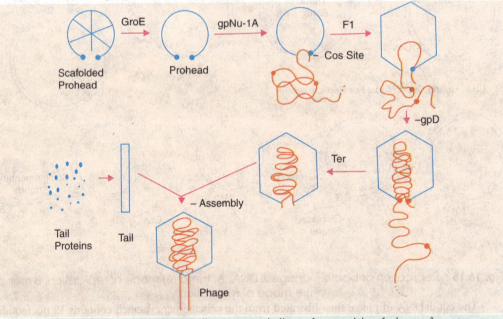

Fig. 16.16 : Diagrammmatic representation of assembly of phage λ.

(*v*) **Lysis :** Inside the bacterial cell about 100 particles are assembled within an hour. The two genes S and R of λ take part in bacterial lysis. The gpS stops metabolism of bacterial cell, and gpR lyses the cell wall. Finally progenies are released from the destroyed cell.

7.   Lysogenic Cycle

The alternative cycle of phage λ where progenies are not produced is called lysogenic pathway. The phage genome is integrated into bacterial chromosomes. The host cell survives for indefinite time. The host cells that contain integrated phage DNA, i.e. prophage, is called the lysogen. The prophage multiplies for several generations. The prophage is excised from the bacterial chromosomes due to stimulation by UV irradiation or mitomycin C. After excision of DNA, the phage leads lytic cycle and the host cell is killed.

The lysogenic bacteria bear the two key features, immunity to superinfection by other phage λ and induction under certain environmental conditions to enter into lytic cycle. The immunity to superinfection and establishment of lysogeny in lysogens is conferred in the presence of λ repressor coded by *cI* gene. In a lysogen, repressor is always synthesized to bind operators OL and OP resulting in blocking of RNA polymerase activity. Thus the repressor prevents the transcription of all prophage genes except its own. As a result of blocking of PL, transcription of *N* gene does not occur. Similarly blocking of PR prevents the early transcriptional genes i.e. *O, P* and *O* genes. The gpcII activates the specific promotor site (PRE) for transcription of *cI* gene and synthesis of the repressor (gpcI). Therefore, the maintenance of lysogenic state requires that the synthesis of repressor must be continued. Establishment of lysogenic state occurs as described below:

(i) **Integration (Insertion)** : The process of firmly joining of phage DNA with bacterial chromosomes is called integration or insertion (Fig. 16.17). The significant features of insertion mechanism have been given by Allen Campbell in 1962 who obtained a λ prophage map by three factor crosses with two *l* genes and a *gal* (galactose utilizing) gene marker of bacterial chromosomes. The two important features of the Campbell model are: (*i*) the formation of a circular DNA from the linear DNA, and (*ii*) at specific loci in a phage and bacterial DNA, a single reciprocal recombination results in the insertion of phage DNA into the bacterial chromosome (Campbell, 1976).

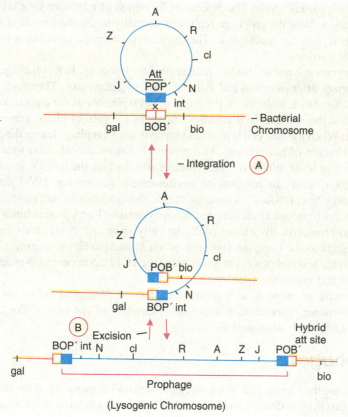

Fig. 16.17 : Campbell model of integration (A) of phage λ DNA into the bacterial chromosome and its excision (B).

(a) **The Attachment sites:** The specific loci are called the attachment sites (Fig.16.17). The attachment site of phage is designed as *att*P which consists of two halves, P.P. Similarly, the attachment site of bacterial chromosomes designated as *att*B, and its two halves as B.B. The dots (.) between the two *att* sites are the points where crossing over occurs. This point has a common base pairs and designated as O. Thus the complete *att* sites are designated as POP' and BOB'. The

phage *att* site is located between the *int* and *J* genes, and bacterial *att* site is situated between *gal* (galactose) and *bio* (biotin) genes.

(*b*) **Mechanism of Integration:** The essence of mechanism which is called Campbell model is the circularization of DNA followed by physical breakage and rejoining of phage and bacterial DNA between the POP' and BOB'. The Ter endonuclease makes staggered nicks in λ DNA. The gpcII stimulates transcription of the *int* gene at the same time as that of *cI* gene. The *int* gene codes for synthesis of an integrase enzyme which becomes in plentiful before 1 repressor turns off transcription. The *att*PoP' and *att*BOB' match each other. The integrase with the help of a special bacterial protein catalyses the physical exchange of viral and bacterial DNA strand. The circular DNA is integrated into the *E.coli* chromosomes as a linear DNA between *gal* and *bio* genes, and is called prophage (Fig 16.17). The phage DNA joins with bacterial chromosomes by covalent bonds. The process by which the λ DNA is inserted into the bacterial chromosome is called site-specific recombination. They pint recognises the *att* sites and brings about the reciprocal crossing over between the POP' and BOB' sites.

(*ii*) **Replication:** The bacterium containing a complete set of phage genes is called lysogen and the life cycle as lysogenic cycle. The process of formation of a lysogen by a temperate phage is called lysogenization. Now the prophage replicates normally under the control of the bacterium by normal bacterial replication mechanism. The replication prophage contributes to viral growth and produces phage particles.

However, integration is not an absolute requirement for lysogeny. In *E.coli* phage, P1 is similar to λ which circularizes after infection and starts synthesizing repressor. Therefore, it remains as an independent circular DNA molecule in the lysogen and replicates as the chromosome. After the bacterial cell divides, the daughter cell contains one or two copies of phage genome.

(*iii*) **Excision:** When the host cell is unable to survive, the λ prophage leaves the *E.coli* genome and begins the production of new phages. This process is known as induction which is triggered by a drop in λ repressor level. Whenever, the repressor will decline, the lytic cycle will commence. In addition, induction occurs in response to environmental factors e.g. UV light or chemical mutagens that damage host DNA. This damage causes the synthesis of recA protein, which acts as protease and cleaves repressor chain between the two domains. RecA protein binds to λ repressor and stimulates it to proteolytically cleave itself. An early gene (*xis* gene) codes for synthesis of excision are that binds to the integrase (*int* gene product) and enables it to excise the prophage. Thus, the excision is the reversal process of integration (Fig 16.17). After excision phage is converted to its circular form and enters the lytic cycle.

During the course of excision, as a result of mistake, the bacterial *gal* or *bio* gene remains included in phage genome. This mistake occurs at a frequency of one in a million. For detail see 'Transduction' in Chapter 8, *Microbial Genetics*.

B. Cyanophages

Cyanophages are the viruses that attack on cyanobacteria i.e. members of the blue-green algae in general. For the first time Safferman and Morris (1963) isolated a virus from the waste stabilization pond of Indiana University (U.S.A) that attacked and destroyed the three genera: *Lyngbya*, *Plectonema* and *Phormidium*. Therefore, they named the virus by using the first letter of the three genera as LPP. Thereafter, several serological strains of LPP were isolated and named as LPP-1, LPP-2, LPP-3, LPP-4 and LPP-5. The viruses are commonly called as blue-green algal viruses or cyanophages. They screened 78 host organisms and found the *cyanophages* only in 11 filamentous cyanobacteria.

After the discovery of LPP-1, a large number of cyanophages was discovered by the other workers including R.N. Singh and coworkers from Banaras Hindu University, India (Table 16.4). Padan and Shilo (1973) reviewed different types of cyanophages.

1.   Morphology

Morphology of LPP-1 has been studied in detail as compared to the other cyanophages. The cyanophages differ morphologically (Fig. 16.18) as well as in physico-chemical properties. The salient features of a few important cyanophages are summarized in Table 16.5.

The LPP-1 group of cyanophages has an icosahedral head and a tail and are similar to T3 and T7 bacteriophages, whereas the N-1 group resembles with T2 and T4 phages (Fig. 16.18). Like T-even phages the tail may be contractile or non-contractile. In some groups the tail is absent. The AS-1 group has the largest cyanophages. The group G-III and D-1 are serologically related but do not show any relationship with T-phages.

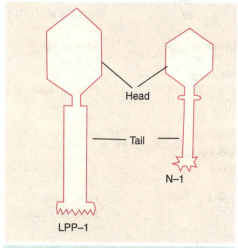

Fig. 16.18 : Diagram of cyanophages.

2.   Growth Cycle

Like bacteriophages, the cyanophages too follow the same one-step growth curve. The growth cycle resembles with that of T4 phages, however, the latent and rise period in different cyanophages differ (Table 16.5).

Replication of genetic material of cyanophages has been reviewed by Padan and Shilo (1973) and Sherman and Brown (1978). The growth cycle of LPP-1 has been studied in greater detail.

LPP-1 is adsorbed on host surface and the DNA is injected into the host cell leaving the protein coat outside the cell wall. How does the DNA is injected, its mechanism is not known. However, soon after injection of the genome the rate of protein synthesis is reduced and gradually blocked at the end of 5 hour of injection. The phage multiplies in the invaginated photosynthetic lamellae or in virogenic stroma. After injection the following three types of proteins are formed.

(*i*) The earliest proteins soon after injection upto 4th hour,

(*ii*) Earlier proteins after two hours of injection to completion of lysis,

(*iii*) Late proteins or structural proteins after 4th hour to host lysis.

Table 16.4 : Some important cyanophages.

Cyanophages	Sources	Host
LPP Group		
LPP-1	Waste stabilization ponds Indiana, U.S.A.	*Lyngbya, Plectonema* and *Phormidium*
LPP-2	do	Same but serologically different
LPP-3	Russia	do
LPP-4	Banaras Hindu University by R.N. Singh and H.N. Singh (1962)	do
LPP-5	do	do

Cyanophages	Sources	Host
G-III Group		
Long tailed	Polluted water, B.H.U.,	*Plectonema borryanum* India by R.N. Singh(1967)
Unnamed	Stream, Japan	*Oscillatoria princeps*
D-1	Scotland	Same as LPP-1
N-Group		
C-1	Polluted water, B.H.U.,	*Cylindrospermum* sp.
AR-1	do	*Anabaenopsis circulans, Raphidiopsis indica*
N-1	do	*Nostoc muscorum*
SM-Group		
SM-1	Waste stabilization ponds Indiana, U.S.A.	*Synecococcus elongatus* and *Microcystis aeruginosa*
SM-2	Fresh water	do
AS-1	Polluted water	*Anacystis nidulans* and *Synechococcus cedrorum*
AS-1 M	do	do, also *M. aeruginosa*

Table 16.5 : Physico-chemical and morphological characteristics of some cyanophages.

	Characters	LPP1	LPP2	N1	SM1	AS1
1.	Morphology	Icosahedral	Icosahedral	Icosahedral	Icosahedral	Hexagonal
2.	Tail, size(nm)	Short 20×15	Short, 20×15	Long 110×10	Absent —	Long, 243×22
3.	Nature of tail	Non-Contractile	Non-Contractile	Contractile	Absent	Contractile
4.	Head diam (Å)	586	573	550	880	900
5.	Class	C	C	A	C	A
6.	Relationship with coliphage	T3-T7	T7	T2,T4	T7	P1,P2
7.	DNA mol wt.(Daltons)	27×10^6	—	38×10^6	56-62×10^6	—
8.	G+C Content(%)	53	–	37	66-67	53-54
9.	Sedimentation coefficient (S)	526-550S	490S	539S	820S	254S
10.	Buoyant density in CsCl2 (g/cm3)	1.71 at 25°C	1.48	1.498	1.72 at 25°C	1.49
11.	Mg requirement	+(1mM)	+(1mM)	+	–	–
12.	Stability					
	(i) pH range	5-11	5-11	—	5-11	4-10
	(ii) Temperature	4-40°C	4-40°C	—	4-40°C	—
	(iii) Temp. for inactivation	35°C	55°C	—	55°C	55°C
13.	Growth					
	(i) Latent period (h)	7	6	7	32 min.	8
	(ii) lytic period (h)	14	14	14	48	12
14.	Burst Size (pfu/cell)	200-350	200-300	100	100	50

After three hours of infection, degradation of the host DNA begins and by the end of 7th hour it is converted into acid soluble material. However, complete degradation of the host DNA does not occur. Sufficient amount of dergraded DNA material is used up in building of viral DNA. It has also been demonstrated that in virogenic stroma synthesis of viral DNA takes place. The latent period differs in different viruses, for example 7 hours in LPP-1 and N-1. Thereafter, there starts the rise period which also varies with the viral types. At the end, after maturation and assembly, the progeny cyanophages are released almost from each cell leaving aside the lysed cell. The burst size for different cyanophages follows the following sequence: 350 plaque forming units (pfu) in LPP-1, 100 pfu in N-1, 50 pfu in AS-1, 100 pfu in SM-1. The plaques can be observed either on algal lawn growing on nutrient agar or in broth cultures.

After infections, several physiological processes are disturbed such as respiration, photosynthesis, host DNA metabolism and nitrogen fixation.

Waste stablization ponds, eutrophic lakes and polluted water support the luxurient growth of cyanobacteria. These can be obnoxious bloom in water reservoirs like lakes and result in fish mortality. Therefore, the cyanophages can play a significant role in control of blooms. So far the problems with them are that they are specific to genus and difficult to isolate.

Cyanphages can control the growth of cyanobacteria found in polluted water.

C. Mycoviruses (Mycophages)

The viruses associated with fungi are called mycoviruses, and also mycophages. They are often typically latent but some induce symptoms. They are wide spread in all taxonomic groups of fungi. Much less is known about the mycoviruses of the lower fungi.

During 1950s, several disorders in fungi were described and some authors suspected for the involvement of viruses. For the first time Hollings (1962) gave the conclusive evidence of viruses that infected the cultivated mushrooms, *Agaricus bisporus* causing die back disease. The most characteristics and consistent features of mushroom virus diseases are the loss of crop and the degeneration of mycelium in the compost. Subsequently mycoviruses from different taxonomic group of fungi were described. So far at least 5,000 fungal species are known to contain mycoviruses. It is interesting to note that most of species of *Penicillium* and *Aspergillus* have been found to be infected with viruses, whereas in other genera the viruses have not been found so frequent.

Ecologically the mycoviruses are exceptional, possibly unique among the viruses, For their existence they appear to be intracellular, a life style for which they are well fitted.

1. Types of Mycoviruses

So far very few mycoviruses have been fully characterized, and most are only the 'virus like particles' (VLPs) in electron micrograph of partially purified extracts from the fungus or some times from thin section studies. Several different morphological types of VLPs have been observed, some corresponding fairly close with well known viruses of the other host taxa. Some of the examples of mycoviruses are given in Table 16.6.

Table 16.6 : Virus particles reported in fungi (modified after Hollings and Stone, 1971)

Fungal Species	Virion size (nm)
Agaricus bisporus	25-50,19-50 (rods)
A.campestris	25-50,19x50 (rods)
Alternaria tenuis	30-40
Aspergillus foetidus	40-42
A.glaucus	25
A.niger	40-42
Endothia parasitica	300, club shaped
Gaeumannomyces tritici	29
Laccaria laccata	28
Pencillium brevicompactum	40
P.chrysogenum	35
P.funiculosum	25-30
P.notatum	25
P. stoloniferum	40-45
Peziza ostracoderma	17 × 350 (rods)
Stemphilium botryosum	25, VLPs
Saccharomyces cerevisiae	40

Some of the isometric particles (105-110 nm diameter) i.e. viruses containing capsid roughly spherical polyhedron, superficially resemble iridoviruses, and some others (about 50 nm diameter) resemble caulimoviruses. Tailed DNA phage particles and paired 20 nm isometric particles of possibly geminivirus type have also been reported. Most of the virus particles recorded in fungi have been isometric with a genome of several species (Hollings, 1982).

The mycoviruses have a heterogeneous properties with a diameter ranging from 25-50 nm and particle weight from $6-13 \times 10^6$ dalton. They possess 1-8 segments of dsRNA with a total molecular weight of $2-8.5 \times 10^6$ dalton. All the examined samples had only a single capsid protein but of varying molecular weight from 25 to 130×10^3 dalton in different viruses. However, the mycroviruses with more than one capsid polypeptide have also been reported in the purified preparations of many potyviruses, the other viruses of higher plants and in *Aspergillus foetidus* virus-S (Afv-s) (Buck, 1980)

2.   Taxonomic Affinity

The mycoviruses poses a frustrating taxonomic problem. However, no serological relationship have been detected between any mycovirus and morphologically similar viruses in higher plants. Similarly, no mycoviruses could be demonstrated to infect higher plants. In addition, the isometric ds RNA mycoviruses have no clear affinity with dsRNA viruses of the other host taxa or taxonomic unity among the viruses with dsRNA genome (Table 16.7).

It could be noted that the dsRNA mycoviruses have evolved on more than one occasion and that the dsRNA genome reflects adaptation to the conditions within the fungal cell. Therefore, the members of the International Committee for Virus Taxonomy resisted to the temptation to set up a large taxonomic group to accomodate all dsRNA mycoviruses. Two such groups now has been designated.

(*i*) The *Penicillium chrysogenum* virus group, and

(*ii*) The *P. stoloniferum* virus - S (PsV-s) group.

The member viruses within each of these groups are serologically related (Hollings, 1979)

Table 16.7 : Taxonomic affinity of mycoviruses with other viruses of plants or animals.

Mycovirus particles	Affinity with other viruses	Example
Rod-shaped particles	(*i*) Tobamovirus type (*ii*) Uncertain affintiy	*Peziza ostracoderma* *Lentinus edodes* *Mycogone perinisa* *Armillaria mellea*
Filament particles	Potexvirus Type	*Boletus edulis*
Isometric particles	Herpes virus Type	*Thraustochytrium sp* *Phytophthora infestans*
Bacilliform particles	Alfalfa mosaic Type	*Agaricus bisporus*

3.   Replication of Mycoviruses

Buck (1979, 1980) has reviewed the replication of mycoviruses inside the fungal cell. However, the major difficuilty to understand the replication strategies of the dsRNA mycoviruses is the lack of methods to get *in vitro* infection of a healthy fungal cell by free virus preparation (Lecoq *et al*, 1979). Hopefully, the dsRNA seems unable to act as mRNA in the fungal cell and may some times inhibits protein synthesis. Buck (1980) has reported the host cell enzymes capable of transcribing the ssRNA and dsRNA *in vitro* and probably dsRNA *in vivo*. Highly specific virus coded RNA polymerases are necessary for effective *in vivo* transcription and replication of dsRNA. Such polymerase has been demonstrated in a number of dsRNA mycoviruses. Probably, the polymerase remain confined within the virus particle during the replicative cycle of mycoviruses (Buck, 1980).

Within undivided genome of virus L or *Saccharomyces cervisiae*, two types of RNA polymerase activity has been detected: ds→ssRNA (transcriptase) and ss→dsRNA. These activity permits viral L dsRNA to replicate synchronously as occurs in reovirus with several differences.

In AfV-S the transcriptase activity has been noted and possibly replication occurs through semiconservative strand displacement (Buck 1979, 1980). This differs fully from semi-conservative reovirus system. Also, PsV-S shows replicase activity *in vitro* giving rise to dsRNA progeny molecules that remain encapsulated. This type of semiconservative replication and strand displacement as found in adenovirus DNA would suggest a synchronous DNA. *In vivo* studies suggest that replication of PsV-S dsRNA may infact be asynchronous.

4.   Example of Mycoviruses

(*i*) **Mycoviruses of Mushrooms :** At least six viruses and VLP have been reported from the cultivated mushrooms, *A.bisporus* nearly from all countries where it is grown widely. The mycoviruses occur in a mixture of cells and are extremely hard to separate. So far it is unknown about what effects the individual mycovirus causes? In some laboratory it could be demonstrated that the presence of viruses in sporophores of mushrooms resulted in reduction in crop yield, and decreased in mycelial growth on malt agar of cultures taken from the sporophores, whereas in some other it was disputed (Hollings and Stone, 1971).

In recent years, there has been many reports of normal yields and mycelial growth from virus infected mushroom crops with the suggestions that the mycoviruses are not pathogenic (Hollings, 1982). The possible reasons for the conflicting reports are:

(a) The presence of morphologically indistinguishable viruses of differing virion size in a single fungal cell,

(b) Existence of variants within one specific virus that contain additional (or fewer) dsRNA segments associated with pathogenicity. The number and size of dsRNA segments vary in different mushroom viruses. Totally, intracellular life of mycoviruses encourage the persistance of such variants. These combined presence resulted in disease, and

(c) The development of tolerance against infection in present day genotypes than the former races of mushroom. This would have been through the suppression of virus replication (Kaltin and Levine, 1979) and through the production of a mycotoxin (patulin) in several species of *Penicillium* and *Aspergillus* (Petroy and Worden, 1979).

Under certain conditions one or more mycoviruses may exist as DNA proviruses that can initiate rapid virus replication and the associated disease.

(ii) **Mycoviruses in Plant Pathogenic Fungi:** Due to the presence of mycoviruses in pathogenic fungi, the virulence of pathogens gradually declines resulting in even death of fungi. Fungus isolates of take all of cereals (caused by *Gaeumannomyces graminis*) containing only one kind of VLPs were mostly more pathogenic than virus free isolate, whereas the isolates with both kinds of virus particles tended to be less pathogenic than either of the other two classes.

In addition, a highly pathogenic isolate of *G. graminis* from wheat roots gradually lost the virulence over a period of 17 months in culture. In virulent isolates no viruses could be detected. After a few months, 35 nm virions and later on 26 nm virions were observed in increasing quantities resulting in gradual loss in pathogenicity of the fungus (Hollings, 1982).

In *Helminthosporium victoriae*, the cause of Victoria blight of oat, two serologically unrelated mycoviruses designated as 190S and 145S from their sedimentation value occurred in hypovirulent cultures. Hypovirulence could be transferred to normal culture by hyphal anastomosis (Ghabrial, 1980). From more than 40 plant pathogenic fungi, mycoviruses have been reported but no consistent correlation with hypovirulence could be established.

Dodds (1980) has given a strong circumstantial evidence that the cytoplasmically transmissible hypovirulence in chestnut blight fungus (*Endothia parasitica*) is correlated with the presence of dsRNA segments, and in at least one strain, with usual clubshaped particles upto 300 nm long.

5.   Mycoviruses and Interferon

Until 1960s, an extensive study was done on the induction of interferon (antiviral chemical compound) by culture filtrates and cell extracts of some isolates of *P.stoloniferum* and *P.funiculosum* in mice (Banks *et al*, 1968). Viral dsRNA was the inducer of interferon when viral particles were injected into the animals. The dsRNA prepared from the virus induced interferon production in mice as did extracts from the culture filterate containing both the viral particles and RNA. Free RNA was apparently more potential stimulant of interferon production than the virus particles (Hollings and Stone, 1971). Its effects were more rapid though less lasting than whole virus.

This aroused the prospects of clinical and veterinary sciences. A number of preparative processes for industrial production of mycovirus extracts were patented. Some success was also achieved in suppressing the virus infection such as common cold. Soon it was reported that it has several serious side-effects. However, it is still unclear that these effects are derived from toxicity of breakdown products of the injected materials or are an integral part of the interferon response. Cosequently, funding of projects got ceased. This aspect has been well reviewed by Kleinschmidt (1979) and Dubey *et al* (1985).

6.   The Killer Phenomenon

In recent years, much emphasis has been laid on the production of 'killer toxin' and mycovirus dsRNA segment. Some strains of *S.cerevisiae* and *Ustilago maydis* (corn smut fungus) secrete extracellular toxins that either kill or suppress the growth of same or related fungal species but each killer strain is immune to its own toxin.

The most sensitive race of *S.cerevisiae* contains the viral particles of 40 nm diameter with a single dsRNA designated as L. The Killer strains also contain a dsRNA of molecular weight of about $1.1 - 1.4 \times 10^6$ designated as M. The coat proteins of both are indistinguishable serologically or in electrophoresis. The M dsRNA encodes the labile glycoprotein killer strain (confirmed by *in vivo* translation), and is believed to determine the immunity factor. The M is found only in cells containing the L dsRNA. The L is supposed to encode both RNA polymerase and the coat protein for L and M RNAs (Hollings, 1982).

Maize plant infected with *Ustilago maydis*.

D. Rhizobiophages

Rhizobiophages reduce rhizobial population in soil and negatively affect the nitrogen fixing abilities of these bacteria with the host legume plant. Rhizobiophages can be used to distinguish between rhizobial strains through "phage typing". Furthermore, these phages are potential biocontrol agents useful for reducing the number of susceptible rhizobial cells in soils, thus decreasing nodule occupancy by the undesirable indigenous bacterial strain and thereby increasing the nodule occupancy by superior strains used as inoculants. The use of specific bacterial viruses as biocontrol agents requires the identification of symbiotically competent, rhizobiophage-resistant rhizobia that have an ability to promote the growth and yield of their specific legume host.

QUESTIONS

1. What do you know about the bacteriophages? Discuss in detail the morphology of bacteriophages.
2. Give a detailed account of T4 phage.
3. Discuss in detail the replication cycles of ØX174.
4. Describe the morphology, DNA and gene organisation in phage lambda.
5. Write an essay on lytic cycle of phage lambda.
6. Give an illustrated account of phage lambda.
7. Write short notes on the followings:

   (*i*) T4 phage,   (*ii*) Gene map of ØX174,   (*iii*) Choice between lytic and lysogenic cycles,
   (*iv*) Cyanophage,   (*v*) Mycoviruses,   (*vi*) Rhizobiophages

REFERENCES

Banks G.T. *et al.* 1968. Viruses in fungi and interferon stimulation. *Nature.* 218-542: 545

Bradley, D.E. (1967) Ultrastructures of bacteriophages and bacteriocins. *Bacteriol. Rev.* 31: 230-314

Buck , K.W. 1980. Viruses and killer factors of fungi. In *The Eukaryotic cell.* (eds. G.W.Gooday) D. Lloyd and APJ Trinci), PP. 329-375. Cambridge univ. Press: London

Cairns J. 1963. The Chromosome of *E.coli. Cold spring Harb. Symp Quant. Biol.* 28:43.

Campbell, A. 1976. How Viruses insert their DNA into the DNA of a host cell. Sci: *Amer.* Dec. P. 102.

Denhard, D.T. and Hours, C. 1978. The present status of Øx174 DNA replication. PP. 696-704. Symposium on DNA synthesis: present and Future.

Detroy, R.W. and Worden, K.A. 1979. Interactions of fungul viruses and secondary metabolites, In *Fungal Viruses.* eds. H.P. Molitoris, M. Hollings

& H.A. Wood. PP. 94-107. Springer-Verlag, Heldelberg.

Dodds, S.A. 1980. Association of Type-1 virus-like dsRNA with club-shaped particles in hypovirulent strains of *Endothia parasitica. Virology.* 107: 1-12

Dubey R.C., Mishra, R.C. and Dwivedi, R.S. 1985. Mycoviruses and interferon. *Science Reporter* (Oct.). 22: 427. .

Edger, R.S. and Lielausis, T. 1968. Some steps in assembly of coliphage T4. *J. Mol. Biol.* 32: 261

Ellis, E.L. and Delbruck, M. 1939. The growth curve of bacteriophages. *J.gen. Physiol.* 22: 365,384.

Epp, C. and Pearson, M.L. 1976. Association of bacteriophage λN gene protein with *E.coli* RNA polymerase. In *"RNA polymerase* (R.Losick and M.Chamberlin, eds.) Cold Spring Harbor.

Echols, H. and Murialdo, H. 1978. Genetic map of bacteriophage lambda. *Microbiol.* Rev. 42:577.

Favre, R. *et al.* (1965). Studies on the morphogenesis of the head of phage T-even. I. Morphological, immunological, and genetic characterization of polyhead. *J. Ultrastructure* Res. 13: 318.

Ghabrial S.A. 1980. Effects of fungal viruses on their hosts. *Annu. Rev. phytopath.* 18: 441-461

Gilbert W. and Dressler , D. 1968 DNA replication : The rolling circle model. *Cold spring Harbor symp on Quantitative Biology* 23:473

Goullan R, Kornberg, A and Sinsheimer, R.L.1967 Synthesis of ØX174 using the phage DNA as template. *Proc. Natl. Acad. Sci.* (USA) 58:2321

Harshey A.D. and Chase M 1952 Independent function of viral protein and nucleic acid in growth of bacteriophasge *J.Gen Physiol.* 36:39.

Hohn, T. and Katsura 1977 Structure and assembly of bacteriophage λ. *Curr Topics Microbiol. Immunol.* 78:69.

Hollings M. 1962 Viruses associated with a dic back disease of cultivated *mushroom.Nature* 196:962-963

Hollings M. 1979 Taxonomy of fungal viruses. In *Fungal viruses* ed. H.P. Molitoris, M Hollings & H.A. wood pp 165-175 Springer verlag, Heideiberg.

Hollings, M. and Stone, O.M. 1971. Viruses that infect fungi. *Annu Rev. Phytopath* 9:93-118.

Hollings, M. 1982 Mycoviruses and plant pathology. *Plant Dis.* 66:1106-1112

Hu, S.L. and Szybalski, W. 1979. Control of rightward transcription in coliphage lambda by the regulatory function of phage genes *N* and *Cro. Virol* .98:424

Kaiser, A.D. ; Syvanen, M. and Masuda, T. 1975. DNA packaging steps in bacteriophage lambda head assembly. *J. Mol Biol.* 91: 175.

King, S. 1968. Gentic map of coliphage T4. *J.Mol. Biol* 39:261.

Kleinschmidt, W.S. 1979. Biochemistry of interferon and its inducers. *Annu Rev. Biochem* .41:517-547

Koltin, Y. and Levine, R. 1979. Fungal viruses and Killer factors - *Ustilago maydis* Killer proteins. In *Fungal viruses* (eds H.P. Molitors *et al.*), p. 120-129, Springer verlag, Heidelberg.

Lecog, H. *et al.* 1979 Infectivity and transmission of fungal viruses. In *Fungal virses* (eds H.P. Molitoris *et al*) pp 34-47, Springer Verlag, Heidelberg.

Mathews, R.E. 1982. Classification and nomenclature of viruses. *Inter Virology* 17:1-199.

Mitra, S. 1980. DNA replication in viruses. *Ann Rev. Genetics* 14: 347-397.

Murialdo, H. 1991. Bacteriophage lambda DNA maturation and packaging. *Ann. Rev Biochem.* 60 : 125-153.

Ogawa, T and Okazaki, T. 1980. Discontinuous DNA replication. *Ann Rev. Biochem.* 49 : 421-457.

Padan, E and Shilo, M. 1973. Cyanophages : Viruses attacking blue-green algae. *Bact Rev.* 37: 343-370.

Safferman, R.S. and Morris, M.E. 1963. Algal virus : Isolation. *Science* 140:679-680.

Sanger, F. *et. al.* 1977. Nucleotide sequence of ØX174 DNA .*Natur*e 265 : 687-695.

Sherman, L.A. and Brown, R.M. 1978. Cyanophages and viruse of eukaryotic algae - In *"Comprehensive virology* (eds. Fraenkel-conrat, H. and Wanger, R. R.) pp. 145-233 Plenum press, New York.

Schlegel, H.G. 1986. *General Microbiology*. Cambridge University Press, London p. 587.

Watson, J.D. and Crick, F.C. 1972. Assembly of viruses. *Ann. Rev. Microbiol* 26:

Industrial Microbiology

17

A. Fermenters and Fermentative Microbes

I. History and Design of Fermenters

De Becze and Liebmann (1944) used the first large scale (above 20 litre capacity) fermentor for the production of yeast. But it was during the first world war, a British scientist named Chain Weizmann (1914-1918) developed a fermenter for the production of acetone. Since importance of aseptic conditions was recognized, hence steps were taken to design and construct piping, joints and valves in which sterile conditions could be achieved and manufactured when required.

For the first time, large scale aerobic fermenters were used in central Europe in the year 1930's for the production of compressed yeast (de Becze and Liebmann, 1944). The fermenter consisted of a large cylindrical tank with air introduced at the base via network of perforated pipes. In later modifications, mechanical impellers were used to increase the rate of mixing and to break up and disperse the air bubbles (Fig. 17.1). This process led to the compressed air requirements. Baffles on the walls of the vessels prevented a vortex forming

Bacteria or yeasts are cultured in large fermenters.

in the liquid (Fig. 17.2). In the year 1934, Strauch and Schmidt patented a system in which the aeration tubes were introduced with water and steam for cleaning and sterilization.

The decision to use submerged culture technique for penicillin production, where aseptic conditions, good aeration and agitation were essential, was probably a very important factor in forcing the development of carefully designed and purpose-built fermentation vessels. In 1943, when the British Govt. decided that surface culture was inadequate, none of the fermentation plants were immediately suitable for deep fermentation. The first pilot fermenter was erected in India at Hindustan Antibiotic Ltd., Pimpri, Pune in the year 1950.

1. Basic Functions of Fermenters

The main function of a fermenter is to provide a controlled environment for growth of a microorganism, or a defined mixture of microorganism, to obtain a desired product while bioreactors refer to production units of mammalian cell culture. The following criteria are used in designing and constructing a fermentor.

The vessel should be capable of being operated aseptically for a number of days and should be reliable for long term operation. The adequate aeration and agitation should be provided to meet the metabolic requirements of the microbes. However, the mixing should not damage the microorganism. The power consumption should be low and temperature and pH control system should be provided. The evaporation losses from the fermentor should not be excessive. The vessel should be designed to require the minimal use of labour in operation, harvesting, cleansing and maintenance. It should have proper sampling facility. The vessel should be constructed to infuse instead of flange joints. The cheapest and best material should be used and there should be adequate service provisions for individual plants.

2. Types of Fermenter

A fermenter with a vessel and computerised controller.

Fermenters (bioreactors) can be classified into two main types based on shape, (*i*) tabular and (*ii*) stirred tank. Cooling coils are provided to maintain constant temperature inside the bioreactor. It can be operated asepticaly for many days and simple in construction. The disadvantages are high power requirement, shearing on the organisms caused by vigorous agitation and inhibition exercised by the product.

(*i*) **Fluidized Bed Bioreactor.** It is more popular in chemical industry rather new to biochemical industries. These are mostly used in conjuction with immobilized cells or enzyme system and are operated continuously.

(*ii*) **Loop or Air Lift Bioreactor.** In the conventional bioreactor, oxygen is supplied by vigorous agitation of the bioreactor content. The heat is generated which is a problem in conventional type. In air lift fermentor, cooling becomes simpler due to the position of inner or outer loop.

(*iii*) **Membrane Bioreactor.** These consist of a semipermeable membrane made up of cellulose acetate or other polymeric materials. The primary purpose of the membrane is to retain the cells within the bioreactor, thus increasing their density, while at the same time allowing metabolic products to pass through the membrane.

(*iv*) **Pulsed Column Bioreactor.** The essential component of a pulsed column bioreactor is a column bioreactor generator connected to the bottom of the column. A pulsed column bioreactor can be utilized as an aerobic bioreactor, enzyme bioreactor or as a separation unit since its original successful application was in the extraction of uranium.

(*v*) **Bubble Column Bioreactor.** Multistage bubble coloumn bioreactor are suitable in the equivalent batch process. It is necessary to vary the environmental conditions over the course of the reaction. In this bioreactor, it is possible to provide different environmental conditions in various stage. The system may not be suitable for fungal fermentation due to high oxygen demanding system.

(*vi*) **Photo Bioreactor.** For the growth and production of photosynthetic organisms, a light source is required. In photobioreactor, there is an important 'reactant', the photons which must be absorbed in order to react and produce products. So the design of the light source is critical in the performance of this type of bioreactor. One of the most interesting photochemical reactors is the annular reactor. In this source of radiation is a cylinder with an annular section, which enclose the lamp completely. The nutrient passing from the product is removed from the top. This is used for *Spirulina* (SCP) and other algal protein production.

(*vii*) **Packed Tower Bioreactor.** It consists of cylindrical column packed with inert material like wood shavings, twigs, cake, polyethylene or sand. Initially, both medium and cells are fed into the top of the packed bed. Once the cells adhered to the support and were growing well as a thin film, fresh medium is added at the top of the packed bed and the fermented medium removed from the bottom of the column. This is used for vinegar production, sewage effluent treatment and enzymatic conversion of penicillin to 6-amino penicillanic acid.

The design of fermenter involves the co-operation between experts in microbiology, biochemistry, mechanical engineering and economics.

3.   Construction of Fermenters

The criteria considered before selecting materials for constructions of a fermenter are: (*a*) the materials that have no effect of sterilization, and (*b*) its smooth internal finish - discouraging lodging of contamination. The internal surface should be corrosion-resistant.

There are two types of such materials: (*i*) stainless steel, and (*ii*) glass which are used in fermenter. According to American Iron and Steel Institute (AISI), if a steel contains 4% chromium, it is called stainless. The long and continuous use of stainless steel sometimes shows pitting. It is also important to consider the materials used for aseptic seal. Sometimes it is made between glass and glass, glass and metal and metal joints between a vessel and detachable top or base plate. On pilot scale, any material to be used will have to be assessed on their ability to with stand pressure, sterilisation, corrosion and their potential toxicity and cost.

(*i*) **Control of Temperature.** Since heat is produced by microbial activity and mechanical agitation, then it is sometimes necessary to remove it. On the otherhand, in certain processes extra heat is produced by using thermostatically controlled water bath or by using internal heating coil or jacket meant for water circulation.

(*ii*) **Aeration and Agitation.** The main purpose of aeration and agitation is to provide oxygen required to the metabolism of microorganisms. The agitation should ensure a uniform suspension of microbial cells suspended in nutrient medium. There are following necessary requirements for this purpose: (*a*) the agitator (impeller) for mixing; (*b*) stirrer glands and bearings meant for aseptic sealing; (*c*) baffles for checking the vortex resulting into foaming; (*d*) the sparger (aeration) meant for introducing air into the liquid (Fig. 17.3).

(*a*) **The agitator (Impeller) :** The size and position of the impeller in the vessel depends upon the size of the fermenter. In tall vessels, more than one impeller is needed if adequate aeration agitation is to be obtained. Ideally, the impeller should be 1/3 or 1/2 of the vessel diameter (D) above the base of the vessel. The number of impeller may vary from size to size to the vessel (Fig. 17.1 and 17.2).

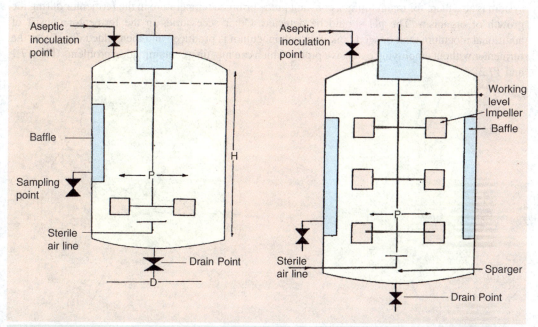

Fig. 17.1 : Different parts of fermenter (diagrammatic) .   Fig. 17.2 : Diagram of a fermenter with multi-bladed impeller.

(*b*) **Stirrer gland and bearing:** Four basic types of seals assembly have been used: the packed gland seal, the simple bush seal, the mechanical seal and the magnetic drive.

(*c*) **Baffles:** The baffles are normally incorporated into agitated vessel of all sizes to prevent a vortex and to improve aeration efficiency. They are metal strips roughly one-tenth of the vessel diameter and attached radially to the walls.

(*d*) **Sparger:** A sparger may be defined as a device for introducing air into the liquid in a fermenter. It is important to know whether sparger is to be used on its own or with mechanical agitation as it can influence equipment design to determine initial bubble size. Three basic types of sparger have been used and may be described as the porous sparger, the orifice sparger and the nozzle sparger.

4. Design and Operation

These are designed to provide support to the best possible growth and biosynthesis for industrially important cultures, and to allow ease of mainipulation for all operations associated with the use of the fermenters. These vessels must be strong enough to resist the pressure of large volume of agitating medium. The product should not corrode the material nor contribute toxicity to the growth medium. This involves a meticulous design of every aspect of the vessel parts and other openings, accessories in contact, etc.

In fermentations, provisions should be made for the control of contaminating organisms, for rapid incorporation of sterile air into the medium in such a way that the oxygen of air is dissolved in the medium and therefore, readily available to the microorganisms and CO_2 produced from microbial metabolism is flushed from the medium. Some stirring devices should be available for mixing the organism through the medium so as to avail the nutrients and oxygen. The fermentor has a possibility for the intermittent addition of antifoam agent. Some form of temperature control efficient heat transfer system is also there for maintaining a constant predetermined temperature in the fermentor during the growth of organism. The pH should be detected. Other accessories in the fermentor consist of additional inoculum tank or seed tank in which inoculum is produced and then added directly to the fermenter without employing extensive piping which can magnify contamination problems (Fig. 17.1 and 17.2).

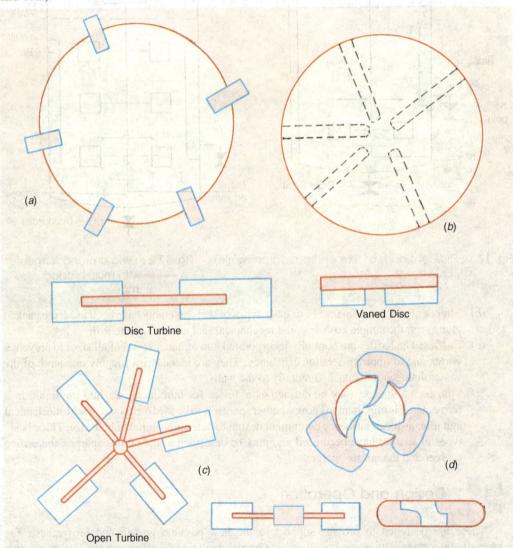

Fig. 17.3 : Types of impellers. (*a*) disk turbine; (*b*) vaned disk type; (*c*) open turbine type; (*d*) margin propeller type.

Use of Computer in Fermenter

Computer technology has produced a remarkable impact in fermentation work in recent years. Integration of computers into fermentation systems is based on the computers capacity for process monitoring, data acquisition, data storage, and error-detection. Some typical on-line data analysis functions include the acquisition measurements, verification of data, filtering, unit conversion, calculations of indirect measurements, differential integration calculation of estimated variables, data reduction, tabulation of results, graphical presentation of results, process stimulation and storage of data.

A large bank of outdoor industrial scale fermenter.

5. The Achievement and Maintenance of Aseptic Conditions

It is necessary to be able to sterilize and keep sterile fermenter and its contents throughout the complete growth. The following operations may have to be performed to achieve and maintain aseptic conditions during fermentation : (*i*) sterilization of the fermentor, (*ii*) sterilization of the air supply, (*iii*) aeration and agitation, (*iv*) the addition of inoculum, nutrients and other supplements, (*v*) sampling, (*vi*) foam control, (*vii*) monitoring and control of various parameters.

(*i*) **Sterilization:** The fermenter can be sterilized by steam under pressure. The medium may be sterilized in the vessel or separately, and subsequently added aseptically. As every point of entry and exit from the fermenter is a potential source of contamination, steam should be introduced through all the entry and exit points except the air outlet from which steam should be allowed to leave. Sterile air will be required in very large volumes in many aerobic fermentation processes. Although, there is a number of ways of sterilizing air, only three have found permanent application. These are heat filteration through fibrous material and filteration through granular material. Heat is generally too costly for full scale operation. The choosen filter must remove microbes to a high level of efficiency, to be relatively cheap, robust and have a low pressure drop. Air filter must also be sterilized in association with the fermenter. An alternate approach is to use a steam packed air filter. At the beginning of sterilisation cycle the valve A should be (*a*) closed (*b*) steam passed through valves B and C and bled out D, (*c*) steam passed through and out of G, (*d*) when steam is issuing freely from valve D, valve E may be opened and steam circulated.

Computer control room for a large fermentation plant.

Once the sterile cycle is complete, valves B and E are closed and A is allowed to open, air to pass through the heated filter. Valve D is closed, valve E opened thus introducing air into the fermenter to achieve positive pressure in the vessel.

(*ii*) **Aeration and Agitation:** Already described earlier.

6. Aseptic Operation and Containment

Containment involves prevention or escape of viable cells from a fermentor or down stream equipment and is much more recent in origin. Containment guidelines was issued in 1970s (East *et al.,* 1984; Flickinger and Sansone, 1984).

Criteria to assess risk was explained by Collins (1992) in the following manner: (*a*) the known pathogenicity of microorganism, (*b*) the virulence or level of pathogenicity of the microbes are the diseases it causes mild or serious, (*c*) the number of organisms required to initiate an infection, (*d*) the route of infection, (*e*) the known incidence of infection in the community and the existence locally of vectors and potential reserve, (*f*) the amounts or volume of organism in the fermentation process, (*g*) the technique or process used, (*h*) ease of prophylaxis and treatment.

7. Batch fermentation

Batch fermentation, is a process in which a large volume of nutrient medium is inoculated to proceed for the harvest and recovery of the product. This ends the batch fermentation as the vessel is cleansed and resterilized for the subsequent batches. This is a close system in which all the nutrients are initially added to the vessel and inoculated. Subsequent treatments include maintaining adequate aeration by providing stirrers and pH control by the addition of an acid or alkali. In aerated systems, antifoam agents such as palm oil or soybean oil are added. Large scale growth of microorganisms tends to generate heat in the system. Temperature control is maintained by providing water circulation system around the vessel for heat exchange.

In large fermenter, the fed-batch method is employed to control the concentration of critical nutrients. For example, penicillin production is controlled by maintaining low glucose level in the medium because its accumulation in the medium results into catabolite repression of secondary metabolite. In batch fermentation, the growth of microorganism follows the characteristics growth curve with a lag phase followed by a log phase, finally reaching the stationary phase due to limitation of nutrients and other factors. When complex nutrient solutions are used, 2 lag phases frequently occurs separated by a second log phase. This phenomenon is called diauxy and arises due to one of the substrates utilized preferentially. The presence of one substrate represses the breakdown of other substrate.

8. Fed-batch Fermentation

In such (fed-batch) cases, substrate is added in increments as the fermentation progresses. The formation of many secondary metabolites is subject to catabolite repression by high concentration of glucose, other carbohydrates, or nitrogen compounds. In such situations, in the fed-batch method the critical elements of the nutrient solution are added in small concentration in the beginning of the fermentation and these substances (substrates) continue to be added in small doses during the production phase.

It is difficult to measure the substrate concentration directly and continuously during the fermentation. Indirect parameters, which are correlated with the metabolism of the critical substrates, have to be measured in order to control the feeding process. For example, in the process of organic acids the pH is to be used to determine the rate of glucose feeding. Sometimes dissolved O_2 or the CO_2 content in the exhaust air are monitored.

9. Continuous Fermentation

Fermentation may be carried out as batch, fed-batch or continuously in a vessel or fermenter. During continuous fermentation, some part of the components (include media and inoculum) of upstream process are withdrawn intermittently and replacement or with drawn substances are made by adding the fresh medium or nutrients. The withdrawn part from the fermenter is used for recovery of the products. During continuous fermentation, the equipment is always in use and secondly inoculum is not required in subsequent addition of the nutrients. It is carried out in three different process as given below:

(i) **Single stage:** In which a single fermenter is inoculated then kept in continuous fermentation. **"Recycle" fermentation** in which, a portion of the withdrawn culture or the residual unused substrate including with drawn culture, is recycled in fermenter. During this process, the substrate is further utilized for product formation and inoculum can also be recycled.

(ii) **Multiple-stage Fermentation:** It involves two or more stages with the use of two or more fermenters in sequence. In such instances, the microbial growth occurs in first stage fermenter followed by a synthetic stage in the next fermenter. This happens in case those metabolities that are not related with growth. The microbial activity in continuous fermentation can be controlled either of the following.

(iii) **Turbidostat:** In this the total cell population is held constant by employing a device that measures the culture turbidity so that it regulates both the nutrient feed rate and the culture withdrawn rate from the fermenter. Sometimes, the inoculum added is multiplied so quickly that its level increased too much, hence medium is added further so as to dilute the inoculum. However, by adding the nutrient, the microbial growth should be maintained in the log phase. During this process unused nutrient is lost from the withdrawl/harvested culture. On the other hand, in the chemostat the nutrient and harvest culture withdrawl rate at constant values. This growth rate can be controlled by toxic product formation during the fermentation, pH and even the change of temperature. Actually, it is necessary to maintain a constant cell population in the fermenter. The flow rate also related to the growth of organism. If it is low than it can allow the culture to go into maximum stationary phase. Too high flow rate can adversely affect a number of microorganisms namely, *Streptomyces, Chlorella, Aerobacter, Azotobacter, Bacillus, Brucella, Clostridium, Salmonella, Penicillium, Saccharomyces, Torula* etc. which are being used in continuous process of fermentation. Some of the products such as beer, has been commercialized. The sewage treatment by activated sludge system has been considered as commercialized continuous fermentation, a process in which mixed microbial population acts on a heterogeneous substrate.

There are certain requirements for continuous fermentation. To know the microbial growth, its behaviour is requisite. To evaluate the contamination and mutation, prolonged incubation may cause contamination except *Torula* yeast on sulphate waste liquor. To check the contamination problems, antibiotics or other chemicals are added to continuous fermentation to hold the level of contaminant growth. The multiple continuous fermentation is advantageous for checking mutation. It is therefore, necessary to reduce their rate of occurrence so that these cells do not multiply. Since everytime, fresh medium is being added hence it is observed quite often that some part of it remain utilized. In certain cases, it is possible to separate the residual nutrient substrate from the harvested culture so that it is recycled through the fermenter.

Fermentation media sometimes require strong mixing if the media is viscous. Adequate mixing is a problem. To overcome the problem, mathematical assumptions are necessary while processing for continuous fermentation. It is also necessary to evaluate the microbial cells do not adhere/attach to the surface of the fermenter. Sometimes, filamentous fungi grow in piping and valve. This is immobilized state and cultures do not expose to nutrient added. This condition makes difficult a proper regulation of inflow and outflow for the fermentation. It is also observed that conditions for the microbial growth and product formation differ. Hence, the optimum conditions for microbial growth do not align with the optimum conditions required for product formation. In a single stage fermentation, a compromise must be laid in the nutrient and physical conditions during the fermentation, growth in the first stage and product formation in the second or succeeding stages are required. The multistage chemostat is useful in the utilization of multiple C source in the production of secondary metabolites. This system is complex hence in industries limited applications are those as in continuous brewing.

10. Scale-up of Fermentations

The determination of the proper incubation conditions to be employed with large scale production tanks as based on information obtained with various sized smaller tanks is called "scale-up". This process allows to carry out laboratory procedure at industrial scale. It is the best way to obtain fermentation information for production tanks directly in large tanks. However, this is not practical for (*a*) new fermentation, (*b*) variation studies on a fermentation already in production, and (*c*) valid experiment cannot be carried out with only a single tank; one or more tanks are required as experimental controls. Aside from these considerations, costs, media also affect the scale-up.

Use of Erlenmeyer Flasks: The conventional methods provide a poor production because of the poor aeration characteristics associated with this vessel.

Use of Baffle Flasks: In this, flask with glass baffles projecting into the medium from bottoms or sides provide better aeration than Erlenmeyer flask.

Use of Small Laboratory Fermenters: One to ten to twelve litre size fermenters are most ideal for this type of studies, since their aeration and agitation conditions can be varied and since the overall fermentation conditions of these tanks more closely resemble with those of the larger production tanks. These tanks allow fermentation studies on a scale that has meaning in relation to production tank but without too great an expense for media, labour, power input, *etc*.

Experience with particular fermentation equipment and previous fermentations is the only real guide in translating scale up.

(*i*) Sterilization of Gases and Nutrient Solutions: In virtually all fermentation processes, it is mandatory for a cost effective operation to have contamination free seed culture at all stages from the preliminary culture to the production fermenter. A fermenter or bioreactor can be sterilized by destroying the organism with some lethal agents, or by removing the viable organisms by a physical process such as filteration.

Nutrient media as initially prepared contain a variety of different cells and spares derived from the constituents of the culture medium, the water and the vessel. These must be eliminated by a suitable means before inoculation. A number of means are recommended but in practice for large scale sterilisation, heat is the main mechanism used.

(*ii*) Stock Cultures: It is extremely important to maintain microorganisms for extended periods in viable conditions, and in situation which did not alter their desired product formation capacity. This condition is also true for strains used in biological assays. Thus, microbial species procured from various culture collection centres are maintained in viable conditions and known as stock-culture collection. The stock culture generally retain all the characteristics initially described. Stock culture collection centres have been established through out the world to help microbiologists in obtaining cultures for various studies. These centres also help in classifying a newly isolated organism. There are two types of stock cultures: working stocks and primary stocks.

(*a*) Working Stocks: These stocks are used frequently and they must be maintained in vigorous and uncontaminated conditions on agar slants, agar stabs, spore preparations, or broth culture, and one held under refrigeration. They must be checked constantly for possible changes in growth characteristics, nutrition, productive capacity and contamination.

(*b*) **Primary Stocks:** The cultures that are held in reserve for presently practical or new fermentation for comparative purposes, for biological assays, or for possible later screening programmes. These are not maintained in a state of physiological activity. Transfers from these cultures are made only when a new working stock culture is required, or when the primary stock culture is subcultured to avoid death of the cells. Thus, primary stock cultures are stored in such a manner as to require the least possible numbers of transfers over a period of time. Further, these are stored at room temperature, are maintained in sterile soil, or in agar or broth overlayed with sterile mineral oil.

Agar and broth culture without mineral oil also are refrigerated. The culture in milk of agar are maintained frozen at low temperature. Finally, primary stock cultures are lyophilized or frozen-dried, and stored at low temperature. The culture of *Blakeslea trispora* used in β - carotene production, cannot be stored at refrigeration temperature because they die relatively quickly. However, at room temperature transfers are being made to fresh medium when the cultures become nearly dried out.

II. Culture Preservation

1. Methods of Preservations

A number of methods are used for maintaining organisms in a viable condition over a long period of time. Different microbes behave differentially using a specific condition of growth. Therefore, a method useful for one species may not be applicable to another. Some of the methods are as given below:

(*a*) **Agar Slant culture:** Agar slants are prepared *in vitro*. After inoculation slants are incubated for a period of 24h and then stored in a refrigerator. These cultures require periodic transfer after six months.

(*b*) **Agar slant culture covered with oil:** The agar slants are incubated after inoculation until profuse growth appears. These are then covered with sterile mineral oil to a depth of 1 cm above the tip of the slanted surface. Transfers are made by removing a loopful of growth, touching the loop to the glass surface to drain off excess oil in the medium and then preserving the initial block culture.

(*c*) **Saline suspension:** High concentration of sodium chloride is used as inhibitor of bacterial growth. Bacteria are suspended in 1% salt solution in screw cap tubes to prevent evaporation. The tubes are stored at room temperature and transfers are made on agar slants.

(*d*) **Preservation at very low temperature:** The organisms are suspended in a nutrient broth containing 15% glycerol, or in skimmed milk containing 7.5% glucose. The suspensions are frozen and stored at -15°C to -30°C. The ready availability of liquid nitrogen (-196°C) has provided another means of preservation of stock cultures. In this procedure, the cultures are frozen with a protective agent (glycerol or dimethyl sulfoxide) in sealed ampules. The frozen cultures are kept in liquid nitrogen flask.

(*e*) **Preservation by drying in vacuum:** The organisms are dried over $CaCl_2$ in a vacuum, then stored in the refrigerator. The organism survives longer than when air dried.

(*f*) **Lyophilization or freeze drying:** The microbial suspension is placed in small vials. A thin film is frozen over the inside surface of the vial by rotating it in a mixture of dry ice or alcohol or acetone at a temperature of –78°C. The vials are connected to a high vacuum line. This dries the organism while still frozen. Finally, the ampules are sealed off in a vacuum with a small flame. These cultures can then be stored for several years at 4°C. To revive microbial cultures, it is merely necessary to break up the vial aseptically to which suitable sterile medium is added. After incubation, growth appears which allows them for further transfer. The process permits the maintenance of a large number of cultures without variations in the characteristics of the cultures which generally reduces the danger of contamination.

2. Stock Culture Collection Centres

(a) American Type Culture Collection, 12301, Parklawn Drive, Rockville, Maryland, USA.

(b) Indian Collection of Industrial Microorganisms, National Chemical Laboratory, Pune.

(c) Institute of Pasteur, Paris (France).

(d) Institute of Microbial Technology, Sector 39-A, Chandigarh.

(e) National Collection of Type cultures, Central Public Health Laboratory, Colinadate Avenue, London.

(f) Microbiological Type Culture Collection, 4-54, Juso-Nishinocho, Higashiyodogowa-ku, Osaka, Japan.

An anaerobic transfer chamber with an air lock for introducing equipment.

(g) Commonwealth Mycological Institute, Ferry Lane, Kew, Furrey, England.

(h) Centre de collections de types Microbeins, 19, Rue Cesar-Roux, Lausane, Switzerland.

(i) Central Bureau Voor Schimmel cultures, Javaloan, 20, Baarn, Neetherlands.

(j) National collection of Industrial Bacteria, Department of Scientific and Industrial Research, Torry Research Station, PO BOX 31, 135, Abbey Road, Aberdeen, Scotland.

III. Criteria used for the Selection of Microorganisms for fermentation

The selection of microorganisms used in fermentation processes and the methods used for the maintenance of these organisms are among the most important decisions that have to be made in designing an industrial fermentation process. The microbes should have following attributes:

(i) The strain must be genetically stable

(ii) The strain should be readily maintained for reasonably long period of time

(iii) The strain must readily produce many vegetative cells, spores or other structures

(iv) The strains should grow vigorously and rapidly after inoculation into inoculum vessel in the fermentation unit

(v) The strain should be in pure culture, free from other microorganisms including bacteriophages

(vi) The strains should be amenable to change by certain mutagens or a group of mutagenic agents

(vii) The strains should be able to protect themselves from contamination.

1.   Methods of Culture Maintenance

There are three methods for culture maintenance which seem to be generally used in the fermentation industries: (*a*) drying organisms on soil or some other solid, (*b*) storing organisms on agar slants, and (*c*) removing the water from the cells or spores by lyophilization and storage of dried product.

(*i*) **Preservation of culture by drying.** These are different methods of drying of cultures given below:

(*a*) **Dried on Silica Gel:** The higher survival rate was noticed at 4°C in comparison to storage at room temperature.

(*b*) **Dried on soil:** About 50% of the total cells remain viable after 20 years of storage. It is observed that 92-96% cells remain viable after 4 years. Iijima and Sakene (1973) developed a method of drying bacterial cultures and bacteriophage under a vacuum at 2 to 5°C where a cotton plug acts as a dessicant to remove the water from the cells and the cells are more gently treated than in the lyophilization process. A some what similar method found useful with yeast after adding $CaCO_3$ to the suspension and allow it to dry on the powder.

2.   Maintenance of Cultures by Storage with Limited Metabolic Activity.

(*a*) **Storage on Agar Slants:** Recent studies have suggested that storage under oil for 10 months did not change carbohydrate assimilation pattern for *Mucor racemosus, Cunnighamella echinuata, Penicillium cyclopium* and *Aspergillus niger*. Elliot (1975) found 95% viability when above microbes stored under oil for 1 year and when stored for 2 years, viability was 79%.

(*b*) **Storage of spores in water:** Long term viability has been noted when spores of various fungi suspended in sterile distilled water and stored in a refrigerator. Similar success has been obtained with the bioassay organisms such as *Saccharomyces cerevisiae* and *Sarcinia lutea* suspended in weak buffer and stored in a refrigerator for more than a year.

(*c*) **Storage at frozen temperature:** Yamasato *et al* (1973) studied the viability of 259 strains belonging to 32 genera suspended in 10% glycerol and stored at -53°C for 16 months. About 10% of the Gram - positive bacteria and 3% of the Gram-negative bacteria lost viability quickly. Honey was suggested as a better adjuvent for frozen storage than glycerol.

Preservation by storage of cells or spore suspension in liquid nitrogen has been widely used since the initial advantages described by Sokolski *et. al* (1984). Daily and Higgens (1973) reported the inclusion of 10% glycerol with 5% of either lactose, maltose, or raffinose in the suspending solutions increased the viability of spores, vegetative cells and Streptomycete mycelial fragments. Moore *et. al* (1975) observed the survival of plant pathogenic bacteria, was enhanced by suspending the cells in 10% skim milk prior to freezing and storage. The brewing yeasts have been difficult to maintain by lyophilization. They were successful when the cell suspension was mixed with 10% glycerol and frozen at 1°C/min stored at -196°C. Similar success was noted in case of *Thiobacillus ferrooxidans*. On the otherhand, *Lactobacillus bulgaricus* concentrated cell suspensions lost viability after storage in liquid nitrogen and the addition of known cryoprotective agents to cell suspensions of the labile strains before freezing provided little or no protection. They also found that supplementing the growth medium with Tween 80 improved the storage stability of all strains. Fatty acid composition has also a direct relation with the survival of organisms.

(*iii*) **Preservation by Lyophilization:** As stated earlier, in addition to that the sterile glass ampoules are suspended in a carrier or protective agent such as sterile bovine serum or skim milk, rapidly frozen at low temperature, and dried in a high vacuum, the ampules are then sealed and stored in a refrigerator. If properly prepared and stored, most lyophilized cultures will remain viable for more than 10 years. In experiments with mycophages and bacteriophages, lyophilisation of suspensions in 5% sodium glutamate and 5% gelatin solution resulted in long-term survival. *E. coli* phage T4 showed damaged head coats after lyophilization while the tail assembly did not damage and this may have the result of both osmotic shock and drying phase. In fact, the losses in titre of myco-bacteriophages during the lyophilisation can perhaps be related to the particle morphology, size, chloroform sensitivity, nucleic acid contents and osmotic sensitivity. Ashwood Smith and Grant (1976) examined the incidence of mutants in *E.coli* and observed that single stranded DNA breaks either during or immediately after lyophilisation. It was concluded that addition of freeze-drying protective agents does not significantly affect the number of mutant cells, survival of *E.coli* and oxygen has no role in mutation induction.

B. Production of Microbial Products

Wild strains of microorganism produce low amount of commercially important products, therefore investigation needs to be made to increase the productivity of selected organisms. To achieve the increased yield optimization, the culture medium and growth conditions are the pre-requisites but the potential productivity of the microorganisms is controlled or regulated by its genome, therefore, genome must be modified so as to enhance the productivity. The modified microorganisms have different nutritional requirements with different culture conditions such as temperature, pH etc. These factors provide benefits to the process of strain improvement involving genome modification of the microorganisms followed by reappraisals of its cultural requirements.

I. Strain Improvement of Microorganism

1. Methods of Strain Improvement

A mutant requiring oleic acid for neomycin formation by *Streptomyces fradiae* showed a decrease in the intracellular level of neomycin precursors in the mutant. On the other hand, supersensitive mutants of β-lactam antibiotics are another example. Recent approaches towards strain improvement are given below:

(i) **Role of Plasmid:** Plasmid genes are involved in antibiotic production in *Streptomyces* spp. Although, plasmids are involved in genetic characteristics on curing experiments. Involvement of plasmids in biosynthesis of aureothricin and kasugamycin in *Str. kasuaensis* was demonstrated more than decades ago by Okanishi *et al* (1970). The genetic study using *Str. venezuelae* ISP 5230, a chloramphenicol (CM) producer, contains most of the structural genes for the CM biosynthetic steps treated between *met* and *ilu* on the chromosome and the plasmid played role in increasing CM production. A linear plasmid like DNA (pSLA2) of 11.2×10^6 dalton molecular weight from *Streptomyces* sp. produced antibiotics.

(ii) **Protoplast Fusion:** Protoplast fusion is one of the useful techniques for obtaining hybrids or recombinants of different microorganism strains. Various studies have been carried out by using protoplast fusion in *Streptomyces, Saccharomyces,* and fungi. Protoplast formation in *Sterptomyces* was first reported by Okanishi and his team in the year 1966. Further, they have worked on formation, stabilization and regeneration of protoplast of *Str. griseus* and *Str. venezuelae.* Fusion of yeast protoplasts has been reported with *Sacchromyces cerevisiae.* Technique for protoplast fusion in *Brevibacterium flavum,* has been used for strain improvement.

(iii) **Mutation:** Screening after major subjection of a parent strain to physical or chemical mutagen greatly increased the probability of finding improved strain.

(a) **Major mutations:** It involves the selection of mutants with a pronounced change in a biochemical character of practical interest. Such variants are commonly used in genetic studies and are generally 'low mutants'. They are isolated routinely from population surviving after prolonged exposure to a mutagen, for example, selection of non-pigmented *Penicillium chrysogenum* strains with high penicillin production. The initial strain of *Sterptomyces griseus* (a streptomycin producing organism) synthesized the small amount of streptomycin but its variant was isolated which produced greater amount of streptomycin. For further improvement it is also necessary to study the biosynthetic pathways which contribute to the identification of precursors as in case of a modified tetracycline synthesized by a mutant strain of *Str. aureofacies*. The molecule got changed at the C-5 position and was almost devoid of antibiotic activity. Another mutant strain S-604 synthesized 6-dimethyl tetracycline, a new antibiotic, not elaborated by the parent strains, proved to have several advantages. Today it is one of leading commercial forms of tetracycline.

(b) **Minor mutations:** It plays a dominant role in strains improvement. By definition, such mutation affects only the amount of product synthesized. Such variants are usually phenotypically similar to the parent, with rapid and abundant mycelial and conidial development. A 10 to 15% increase in conidial population exposed to moderate doses of a mutagen, obtained after repeated isolation of minor (positive) variants and using each succeeding strain for further mutation and selection. Such increases have also been obtained by repeated selection without the introduction of mutagen. In this case, the population to be tested must be large and assay for the desired product also must be accurate and specific. This technique fetched importance in improving *P. chrysogenum.* For example, Wisconsin series were the famous Q-176 culture with significantly improved antibiotic titres, and strains BL3-D10, which does not produce the characteristic and trouble some chrysogenin pigment. All further mutant selections over the next decade were derived from Q-176.

2. Mutation Concept for Strain Development

Strains selected as obvious variants after exposure to mutagen are usually inferior in their capacity for accumulation of antibiotic. Improvements are extremely few and their selection and evaluation is extremely important.

Mutagen dose is important. Mutants sought for major mutation rates are best isolated from populations surviving prolonged doses of mutagen, whereas variants for increased productivity are generally isolated from population surviving intermediates dose level.

Strain with enhanced altered morphology, etc. may be inherently better producers but may require considerable fermentation development. Step wise selection implies small increment in productivity, the productivity increases, and the probability of getting hyper producing strains decreases.

Variant strains may require special propogation and preservation procedures and actual production gains depend also on stability and reliability of performance. Though, strains may prove better in their productivity at laboratory scale, there is no guarantee that enhanced productivity will occur in production fermenters. The long term pilot plant studies are often necessary before any enhanced strains potential can be realized in actual production.

3. Isolation of Mutant Classes and Their Use in Microbial Processes

(i) **Localized Mutagenesis and Computation:** Localized mutagenesis affecting the small selected regions of the chromosomes, offers a promising new approach. Mutation programmes can be directed to maximize mutations in any marked area on the chromosome, specially the areas known to

affect the formation of end products. Isolation of strains in unknown loci linked to the revertant site can be done by a heterokaryon method or by the use of temperature sensitive mutants.

(ii) **Sexual and Parasexual Processes:** In fungi, the vegetative mycelium is haploid and can be propogated almost indefinitely by serial transfer of hyphal fragments and can also be propogated by asexual spores/conidia. Two strains of opposite mating types (A or B) are required to initiate the sexual cycle and allow to mate by mixing the conidia of mating type A with mycelia on appropriate media. After a period of nuclear division and migration, fusion between A and B nuclei takes place. Each fused nucleus (diploid) undergoes meiosis to form four haploids, which divide mitotically into eight nuclei contained in ascus.

Few of the industrially important fungi form heterokaryon in which rare diploid nuclei result from the fusion of two haploid nuclei. This process is called **Parasexuality**. Although, recombination is fewer fragments in the parasexual cycle compared to the meiotic process it can occur by mitotic crossing over or by other mechanism. The importance of mitotic crossing over or recombination is that it makes possible genetic analysis and controlled breeding in organisms with no sexual cycle. Strain improvement through parasexual cycle has been reported in *P. chrysogenum* and in one study, a homozygous diploid representing parent was an efficient producer of penicillin V.

II. Alcohol Production

Simple organic compounds act as feed stock for chemical industry. Microbial production of one of the organic feed stocks from plant substances such as molasses is presently used for ethanol production. This alcohol was produced by fermentation in the early days but for many years by chemical means through the catalytic hydration of ethylene. In modern era, attention has been paid to the production of ethanol for chemical and fuel purposes by microbial fermentation. Ethanol is now-a-days produced by using sugar beet, potatoes, corn, cassava, and sugar cane (Fig. 17.4).

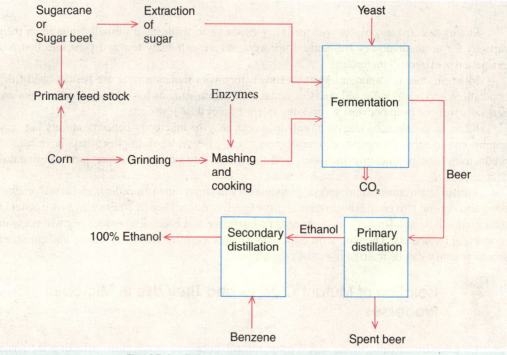

Fig. 17.4 : Ethanol production from molasses.

Both yeasts (*Saccharomyces cerevisiae, S. uvarum S. carlsbergensis, Candida brassicae, C. utilis, Kluyveromyces fragilis, K. lactis*) and bacteria (*Zymomonas mobilis*) have been employed for ethanol production in industries. The commercial production is carried out with *Saccharomyces cerevisiae*. On the otherhand, *S. uvarum* has also largely been used. The *Candida utilis* is used for the fermentation of waste sulphite liquor since it also ferments pentoses. Recently, experimentation with *Schizosaccharomyces* has shown promising results. When whey from milk is used, strain of *K. fragilis* is recommended for the production of ethanol. It is also found that *Fusarium, Bacillus* and *Pachysolen tannophilus* (yeast) can transform pentose sugars to ethanol.

Theoretically, it is interesting to note that fermentation process retains most of the energy of the sugar in the form of ethanol. The heat of combustion of solid sucrose is 5.647 MJ mol-1, the heat of combustion of glucose is 2.816 MJ mol^{-1} but the heat release is 1.371 MJ mol-1. The equations are given below:

1st equation: Because $C_{12}H_{24}O_{11} \longrightarrow 4C_2H_5OH + 4CO_2$
                                    (sucrose)   (ethanol)

          Hence,   5.647 $\longrightarrow 4 \times 1.371 = 5.184$

                                                            i.e. 97% conversion

2nd equation: In this case,

                        $C_6H_{12}O_6 \longrightarrow 2C_2H_5OH + 2CO_2$
                          (glucose)   (ethanol)

Hence,   2.816 $\longrightarrow 2 \times 1.371 = 2.742$

                                                            i.e. 97% conversion

Thus, the above reactions show that 97% sugar transforms into ethanol. But in practice, the fermentation yield of ethanol from sugar is about 46% or one hundred grams of pure glucose will yield 48.4 grams of ethanol, 46.6 g of CO_2 3.3 grams of glycerol and 1.2 g of yeast. The biosynthesis of ethanol is given in Fig. 17.4.

It is noteworthy that the ethanol at high concentration inhibits the yeast. Hence, the concentration of ethanol reduces the yeast growth rate which affect the biosynthesis of ethanol. It can produce about 10-12 % ethanol but the demerit of yeast is that it has limitation of converting whole biomass derived by their ability to convert xylulose into ethanol. The *Zymomonas* has a merit over yeast that it has osmotic tolerance to higher sugar concentration. It is relatively having high tolerance to ethanol and have more specific growth rate.

1. Preparation of Medium

Three types of substrates are used for ethanol production: (*a*) starch containing substrate, (*b*) juice from sugarcane or molasses or sugar beet, (*c*) waste products from wood or processed wood. Production of ethanol from whey is not viable. If yeast strains are to be used, the starch must be hydrolysed as yeast does not contain amylases. After hydrolysis, it is supplemented with cellulases of microbial origin so as to obtain reducing sugars. About 1 ton of starch required 1 litre of amylases and 3.5 litre of glucoamylases. Following steps are involved in conversion of starch into ethanol (Fig. 17.5).

On the otherhand, if molasses are used for ethanol production, the bagasse can also give ethanol after fermentation. Several other non-conventional sources of energy such as aquatic plant biomass, wood after hydrolysis with cellulases gives ethanol.

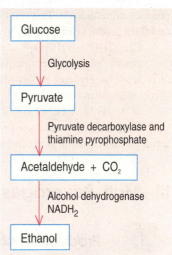

Fig. 17.5 : Biosynthesis of ethanol.

Sulphite waste-liquor, a waste left after production of paper, also contains hexose as well as pentose sugar. The former can be microbially easily converted.

2.   Fermentation

Ethanol is produced by continuous fermentation. Hence, large fermenters are used for continuous manufacturing of ethanol. The process varies from one country to another. India, Brazil, Germany, Denmark have their own technology for ethanol production.

The fermentation conditions are almost similar (pH 5, temperature 35°C) but the cultures and culture conditions are different. The fermentation is normally carried out for several days but within 12h, starts production. After the fermentation is over, the cells are separated to get biomass of yeast cells which are used as single cell protein (SCP) for animal's feed. The culture medium or supernatant is processed for recovery of ethanol (Fig. 17.4). Ethanol is also produced by batch fermentation as no significant difference is found both in batch and continuous fermentation. Although as stated earlier within 12h *Saccharomyces cerevisiae* starts producing ethanol at the rate of 10% (v/v) with 10-20g cells dry weight/lit. The reduction in fermentation time is accomplished use of cell recycling continuously in fermentation.

3.   Recovery

Ethanol can be recovered upto 95 percent by successive distillations. To obtain 100 percent, it requires to form an azeotropic mixture containing 5 percent water. Thus 5 percent water is removed from azeotropic mixture of ethanol, water and benzene after distillation. In this procedure, benzene-water ethanol and then ethanol-benzene azeotropic mixture are removed so that absolute alcohol is obtained.

Neuberg's Fermentation

Yeasts utilize pyruvate during fermentation resulting in the formation of an intermediary product, acetaldehyde. This trapped by hydrogen sulfite to yield the acetaldehyde in precipitated form and fluid product formation is glycerol as shown below:

$$CH_3CHO + NaHSO_3 \longrightarrow CH_2\text{-}CHOH\text{-}SO_3Na$$

Now in place of acetaldehyde, dihydroxyacetone phosphate acts as a hydrogen acceptor which is reduced to glycerol-3-phosphate. After removal of phosphate i.e. dephosphorylation, it gives glycerol as given below:

$$C_6H_{12}O_6 + H_2SO_3 \longrightarrow CH_2\text{-}CHOH\text{-}SO_3Na + Glycerol + CO_2$$

Neuberg's fermentation process is categorized as reward and third fermentation. The first fermentation equation is given below:

$$2Glucose + H_2O \longrightarrow C_2H_5OH + acetate + glycerol + 2CO_2$$

III. Malt Beverages

1.   Production of Beer

For more than 6000 years man has been using microorganisms to produce alcohol (beer) from starch of barley grains, or wine from sugar in grapes.

Brewing, the production of malt beverages, though have its origin in Mesopotamia where it is said that about 40% of total cereal production was used for this purpose. It is likely that beer was a later discovery than wine. Although, most of the beer is produced by using barley and other cereals are occasionally used. Wheat beers such as Berliner Weisee and the gueuze lambic beers of Belgium are notable exceptions. One reason for barley's pre-eminence is that the grain retains the husk which affords protection during storage and transport and also acts as an aid to filteration during wort separation.

(*i*) **Malting Process:** The barley kernels (*Hordeum vulgare*) separated from the stalk and chaff, are allowed to germinate under controlled temperature and humidity to generate enzyme systems which partially degrade endospermic starch and protein. For this, the barley grains are soaked in water for 1-2 days and after sprouting kernels stopped by drying in hot air. During this process, some

Fermentation occurs in large deep tank during beer making.

flavour and colour components are also formed. The dried malt (source of amylase and proteinase) crushed before it is used.

(*ii*) **Mashing Process:** The crushed malt and adjuvent such as corn grits or rice are each made into a mash with brewing water, and combined for further enzymatic degradation and subsequent extraction of sugars and proteins. The process of making malt as soluble as possible by using enzymes, adjuncts etc. is called *mashing*. This process allows the malt amylases and proteinase to degrade starch into maltose, and protein into peptone and peptides respectively. The enzymatic hydrolysis is influenced by temperature and pH. Therefore, both these factors are adjusted to get partially and totally degraded enzymatic products. The remaining extract (liquid) is called *wort*. The wort is boiled with flower of hops (*Humulus lupulus*) which imparts longer durability, preservation during beer processing. The wort now is used for fermentation. The boiled wort is cooked, inoculated or "pitched" with a pure yeast culture (*Saccharomyces carlsbergensis* : bottom type) and fermented. Each beer producing industry or brewery maintains its own strain. The fermentable sugars are converted to ethyl alcohol and trace quantities of flavour ingredients while the non-fermentable carbohydrates are also left in the beer. The fermentation process is completed within 8-14 days. During the later stages, the bottom, yeast 'breaks' flocculate and settle. The bacterial growth is not desired during the fermentation. The yeast generated during the fermentation is separated from the beer by centrifugation or natural setting. After storage for several weeks, recarbonation and two or more filteration are necessary to remove the yeast. The finished product is kept for aging or maturing for several months to years. The "green" beer is stored or "laged" in vats at 0°C for several weeks, the proteins alongwith remaining yeast and other substances precipitate. The flavour is added due to ester formation and aroma develops. The body changes from harsh to smooth (Fig. 17.6 and 17.7).

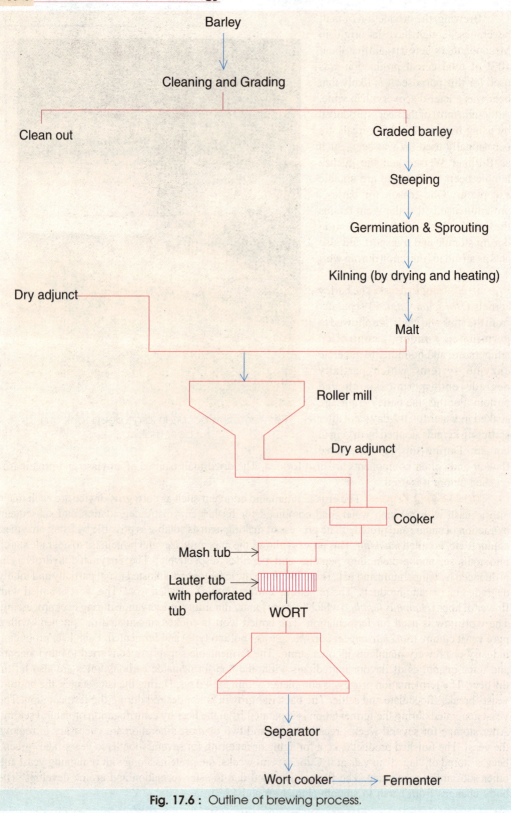

Barley

Cleaning and Grading

Clean out

Graded barley

Steeping

Germination & Sprouting

Kilning (by drying and heating)

Dry adjunct

Malt

Roller mill

Dry adjunct

Cooker

Mash tub

Lauter tub
with perforated
tub

WORT

Separator

Wort cooker ⟶ Fermenter

Fig. 17.6 : Outline of brewing process.

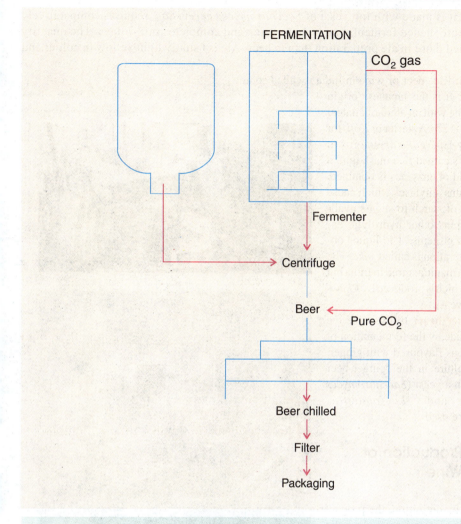

Fig. 17.7 : Different steps of brewing process.

(*iii*) **Finishing:** The beer is later on carbonated to CO_2 content (45-52%). The beer is now cooled, clarified or filtered and packed in bottles, consor barrels, such process is called "*finishing*".

2. Other Malt Products

(*a*) **Bock Beer:** It is prepared from roasted germinated barley seeds, hence the beer is dark in colour with a high alcohol content. Brewing involves the high concentrations of malt and hops. It also requires longer aging.

(*b*) **Pilsener:** It is lager type beer (fermentation at low temperature is carried out for the product formation). The beer is light in colour which contains little fermentable carbohydrate.

(*c*) **Lager beer:** The fermented product or beer is allowed for aging process in the cold. The bottom fermenting yeast is used for the preparation of lager beer.

(*d*) **Ale:** It is made with top yeast of *Saccharomyces cerevisiae*, requires comparatively high temperature, hence fermentation is more quick and completes in 5-7 days. The quantity of hops are used more in ale preparation than in beer. Ale is usually light yellow in colour and tart in taste.

(*e*) **Sonti:** Rice beer or wine in India is called sonti.

(*f*) **Sake:** It is the Japanese origin rice beer or wine with an alcohol content varying from 4-17%. A starter or *Koji*, for sake is made by *Aspergillus oryzae* grown on soaked and steamed rice mash until a maximum yield of enzymes is obtained. The koji contains amylases which cause the hydrolysis of starch to sugars available to yeast plus other hydrolytic enzymes such as proteins. The liquor obtained after filteration is called *sake*. The rice beer is also manufactured in India but called 'sonti' which is produced by *Rhizopus sonti* and yeast.

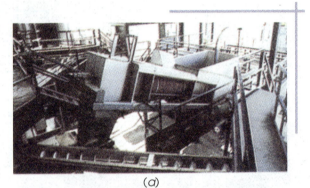

(*a*)

(*g*) **Ginger beer:** In this case the beverage is made by the fermentation of a sugar solutions flavoured with ginger. The starter culture in the 'ginger-beer plant' in which a yeast (*Saccharomyces pyriformis*) and *Lactobacillus vermiformis* are used.

3. Production of Wine

Wine is by definition, the product of grapes obtained after normal alcoholic fermentation by yeast.

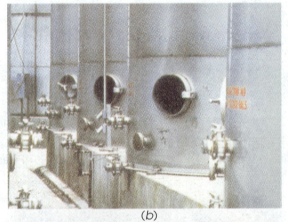

(*b*)

(*i*) **Microbial Process:** Wine is basically the transformation of sugars of grapes by yeast under anaerobic conditions into ethanol, carbon dioxide and small amount of byproducts, D-glucose and D-fructose, the two principal sugars of grape juice yield ethanol and carbon-dioxide. The natural yeast flora of crushed grapes and fermentation equipment are less commonly used. In such fermentation, a succession of yeast and bacterial population is found in the fermenting grape. Wine can be produced by the fermentation of the fruit juices, berries, honey, etc.

(*c*)

Commercial wine making.
(*a*) Equipment for transporting grapes for crushing.
(*b*) Large tanks for wine fermentation.
(*c*) Barrels where aging process takes place.

(*ii*) **Wine from Grapes:** Two kinds of grapes are cultivated and used for wine production. The red grapes (where the skin of grape is red) gives rise red wine, while the white grape juice after fermentation gives white wine. Sometimes, the red pigments may add in the juice additionally.

When the grapes have the derived sugar content (14-20 percent) should be harvested for the extraction of juice. For this, the grapes must be stemmed, cleaned and crushed and then sodium or potassium meta-bisulphite should be added to check the undesirable organisms. The crushed grape is known as *must*. Sometimes, it is difficult to control the natural microbial population present in the must. It is, therefore, recommended to use the pure culture of wine yeast of proven quality. Such strains are usually used in conjunction with proper concentration of sulphur dioxide to which the wine yeast is resistant and wild yeast and bacteria are sensitive.

(*iii*) **Fermentation:** This process is carried out by adding 2-5% of wine yeast, namely *Sacharomyces cerevisiae* var. *ellipsoideus* in the *must*. The whole contents are mixed twice a day by punching the "cap" of floating grapes so as to allow profuge growth of yeast due to increase in aeration in the beginning. This process aids in the extraction of colour (white or red) depending upon the quality of grapes used. Now, the mixing should be stopped and anaerobic fermentation is allowed to carry out. The temperature is maintained at 24-27°C for red wine for 3-5 days and 10-21°C for white wine for 7-12 days. The heat generated during the fermentation may allow to increase the temperature. Sometimes lactobacilli may grow at such a temperature which contaminate the product. Hence, artificial cooling is required.

(*iv*) **Recovery:** The fermented juice is drawn off from the residue (Pomace) and stored under atmosphere of carbon dioxide for carrying out further fermentation for about 7-11 days at 21-29°C. If *dry wine* is to be obtained then whole residual sugar should be fermented to yield ethanol.

The wine may be pasteurized before aging. During pasteurization, protein is precipitated and removed. It is cooled, filtered and transferred in the wooden tanks made up of white oak or red wood or plastic concrete tank for aging. Aging imparts flavour, aroma, sanctity and colour to the wine. After aging, the wine is clarified, barreled or bottled. The final alcohol content in wine varies from 6 to 9 percent by weight or 8 to 13 percent by volume.

4. Types of Wine

On the basis of production of carbon dioxide, wines are of two categories.

(*a*) **Still wine:** Those wine in which no carbon dioxide is produced during the fermentation is called still wine.

(*b*) **Carbonated wine:** Those wine in which considerable quantity of carbon di oxide is produced is called carbonated wine.

Again, on the basis of presence of sugars, these are called *dry wine,* which have no unfermented sugar, while *sweet wines* contain sugar. On distillation of wine '*brandy*' is obtained. The alcohol percentage increased upto 21 percent in brandy. *Table wines* have low alcohol and devoid of sugar.

(*c*) **Sherry:** It is more popular in France. Hence, called French dry sherry. It is made from ripe and dried grapes with high sugar contents. Sometimes, the grapes infected with *Botrytis cinerea* are selected for sherry production.

5. Microbial Deterioration and Spoilage of Wine

Defects in wine are mainly due to metal and other reactants which are used for wine processing. White wines may turned brown and red wine may have precipitation by peroxidases or oxidizing enzyme of certain molds. Various microorganisms such as *Acetobacter, Lactobacillus, Leuconostoc,* wild yeasts, molds etc. spoil wine. The species of *Acetobacter aceti* or *Gluconobacter oxydans* impart an undesirable process called *acetification*. In this process, glucose may oxidise and may give a `mousy' or sweet sour taste to wine. Spoilage may also occur due to abnormal fermentation by natural microflora that result in low alcohol content, high volatile acidity, undesirable flavour and cloudiness in wine. Some common defects in wine are either due to bacterio-fermentative process by lactobacilli that result in acid production from sugars. Such spoilage is called *tourne*. On the otherhand, bitter taste in wine also occurs due to the fermentation of glycerol. Such process is called *amerture*. The liberation of carbon dioxide or gassiness results due to the action of heterofermentative lactis, is called *pousse*.

6. Distilled Beverages or Liquors

(*a*) **Rum:** It is obtained by distillation from alcoholically fermented sugar-cane juice, syrup, or molasses.

(*b*) **Whiskey:** It is obtained after distillation from saccharified and fermented grain mashes. Whiskeys are prepared by using wheat, corn, rye *etc.* Whiskeys are made by novel strains of *Saccharomyces cerevisiae* var. *ellipsoideus*.

(*c*) **Brandy:** It is obtained from grape wine after distillation. The brandy can also be obtained by using apple, peach and apricot.

Peaches and apriocots are also used for making Brandy.

C. Microbial Production of Organic Acids

1. Vinegar Production

The vinegar is in fact derived from the french "*vinaigre*" means sour wine. The vinegar mainly depends upon the type of local alcohol production. e.g. rice vinegar in Japan or sugarcane vinegar in India.

There are mainly two steps in vinegar production. In the first stage yeast converts sugar into ethanol anaerobically while in the second step ethanol is oxidized to acetic acid aerobically by *Acetobacter* and *Gluconobacter*. This process is called acetification.

Both the bacteria are Gram-negative, catalase positive, oxidase negative and strictly aerobic. *Acetobacter* species are the better acid producers and are commonly used in industries.

Vinegar is an alcoholic fermentation product and if enough acetic acid is present, it is a legal vinegar. The *Saccharomyces cerevisae* var. *ellipsoideus* carried out the conversion of sugar into ethanol in the first step, while the species of *Acetobacter* or *Gluconobacter* convert ethanol into acetic acid.

(*i*) **Substrate:** Vinegar is produced by using fruit juices, starchy vegetables, malted cereals such as barely, rice, wheat, corn, sugar cane syrups, molasses, honey, alcohol *etc.*

From the fruit juice such as apple juice, the vinegar produced is known as cider vinegar.

If it is produced by using Ale then it is called Alegar. While malted grain made malt vinegar and ethanol produced sprit vinegar.

(*ii*) **Method:** Mainly there are two methods: (*a*) slow or "let home" process; (*b*) quick or "French orleans" process.

(*a*) **Slow process:** In the slow process, the alcoholic liquid (substrate) is not allowed to move during acetification and fermented juice or malt liquors are used for acetic acid production.

Apple juice is specially used for alcoholic production by batch fermentation process. The barrel is partially filled with the fermented juice and it is allowed to undergo the process of acetification. The vinegar bacteria "food" is supplemented from the previous batch of vinegar. A film of vinegar bacteria called "mother of vinegar" should grow on the surface of the liquid which indicates that ethanol is oxidised into acetic acid and formed vinegar.

Demerits of this process are the absence of productive strain, and poor yield of ethanol gives inferior quality of vinegar.

(*b*) **Quick process:** In this process about 1/4 barrel is filled with raw vinegar from previous run supplemented with active vinegar bacteria. This is to be acetified so as to check competing microbes. Now the barrel is filled by alcoholic hydrolysate up to half filling and keep watch, examine if bacteria are growing in a film on the top of the liquid. This process generally takes weeks to months at 21 to 29°C. The recovery is made in part by withdrawing and replacing by equal quantity of alcoholic liquor. This is a continuous fermentation.

Major difficulty in this process is dropping of the gelatinous film of vinegar bacteria which results in retardation of the acetification. To avoid this, a raft or floating frame work is provided to support the filter. Another important process is called quick generator process.

Generator process: This involves the movement of ethanol liquid during acetification. This ethanol liquid is trickled over the film of vinegar bacteria. A simple generator or cylindrical tank made up of wood is used (Fig. 17.8).

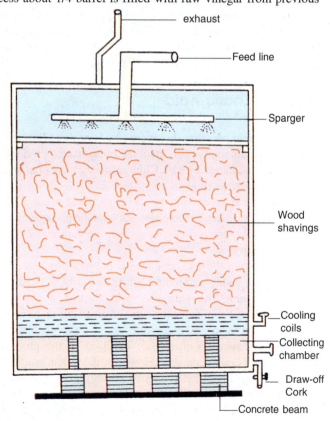

Fig. 17.8 : Diagram of a generator for vinegar production.

In starting a new generator, the slime of vinegar bacteria must be established before vinegar is produced. Hence, in the beginning the middle section of the tank is filled with raw vinegar, that contained active vinegar bacteria. To inoculate the shaving with the desired bacteria an alcoholic liquid acetified with vinegar is slowly trickled through the generator to build up bacterial growth on the shaving and then is recirculated.

The fringe generator is also recommended for vinegar production which contains a large cylindrical airtight tank equipped with a sprinkler at the top. The cooling coils are fitted in the lower part of the middle section. The recirculation of the vinegar is done from the bottom collection chamber. The generator gives high yield of acetic acid and leaves little residue *i.e.* alcohol.

(*d*) **Makin's process:** A fine mist of a mixture of vinegar bacteria and nutrient alcoholic solution is sprayed through jet nozzles into a chamber. The mist is kept in circulation by filtered air for a while and is allowed to fall to the bottom for circulation, collection, *etc.*

(*e*) **Submerged process:** A stirred medium containing 8-12% of ethanol is inoculated with *Acetobacter acetigenum* and is held at 24-29°C. The bacteria grow in a suspension of fine air bubbles fermenting liquid.

(*f*) **Finishing:** It is the slow process, the vinegar produced is less harsh while in quick process improved vinegar is produced as far as its taste, body and flavor are concerned. The unit of vinegar is in grams. While it is commercially and internationally defined in terms of grain which is 10 times that of gram.

Defects: If it is (storage vessel) of steel or iron the ferric ion is oxidised into ferrous ion. The tannin, phosphate or proteins are formed which make the vinegar hazy in appearance. If the barrel is of tin or copper, cloudiness develops which results into darkening of vinegar.

The animal pests such as mites, fruit flies, vinegar eel (*Anguillula aceti*) attack the film of acetic acid bacteria which cause it to sink and deteriorate vinegar. Some times *Leuconostoc, Lactobacillus* contaminate the vinegar imparting off flavour.

2. Lactic Acid

Lactic acid was first discovered by Scheele (1789) from sour milk. Later on, Pasteur (1857) identified the microorganism involved in lactic acid production. In the year 1881, first commercial production was started by M/S Clinton Processing Company, Clinton, Iowa (USA). This was based on the fermentation process.

Lactic acid production by using chemical process was not economical and recovery and purification were also not upto the mark, hence continuous efforts were made to improve the process. Moreover, the requirement in plastic industry was of very high purity.

$$COOH$$
$$|$$
$$H\!-\!C\!-\!H$$
$$|$$
$$CH_3$$

Lactic acid

(*i*) **Fermentation :** Lactic acid is produced by several microorganisms which differ in their ability to produce either D (–) lactic acid, L (+) lactic acid or the racemic mixture. The particular acid formed seems to be characteristic of the individual microorganism. The racemic mixture is formed due to the production of an enzyme called `recemase'. The lactic acid recovered is optically active but becomes inactive due to the action of the enzyme.

Various microorganisms are involved in the production of lactic acid. *Rhizopus oryzae* produces only L(+) lactic acid. However, the production is quite slow and yield is also low. There are mainly two important processes based on end product formation.

(*a*) **Homofermentative Process:** This process involves certain bacteria namely, *Lactobacillus delbruckii, L. bulgaricus, L. pentosus, L. leichmanii, L. casei, Streptococcus lactis, etc.* These bacteria utilize the EMP pathway to produce pyruvic acid which is then reduced by the lactate dehydrogenase to lactic acid. All the microbes are considered to be anaerobic, although they can withstand some oxygen. The end product is lactic acid with traces of others.

(*b*) **Heterofermentative Process:** This process involves the action of *Leuconostoc mesenteroides* which produces lactic acid, carbon dioxide, ethanol, acetic acid, water and few other products.

(*ii*) **Medium and Manufacturing Process:** The culture medium contains semirefined sugar, (molasses or whey contains semirefined sugar), molasses or whey starch, maltose, lactose, sucrose, calcium carbonate with ammonium hydrogen phosphate. The malt sprouts are mixed and pH is kept between 5.5 and 6.5.

Lactic acid is quite corrosive, hence metals are avoided, consequently wooden fermenters are used. The thermophilic clostridia results in the production of some butanol and butyric acid which are the major contaminants in lactic acid production.

The colonies of *L. delbruckii* are transferred into large culture vessel kept at 45-55°C. Each stage of culture building requires 16-18*h*. A slight excess of calcium carbonate is present in each stage. The inoculum volume is usually 5% and the fermentation is carried out for 5-10 days. The sugar be reduced to 0.11% or less during the fermentation because residual sugar makes the recovery of better quality of lactic acid difficult. Aeration and agitation are required (Fig. 17.9).

(*iii*) **Recovery:** To the fermentation medium, $CaCO_3$ is added, pH adjusted to 10, broth is heated and filtered. Lactic acid is converted to calcium lactate. It decomposes residual sugar which kills bacteria. The H_2SO_4 is added to remove Ca as $CaSO_4$. Lactic acid is recrystallized as calcium lactate. The activated charcoal is added to remove impurities and lactic acid is recovered.

In the solvent extraction procedure free lactic acid is extracted with isopropyl ether directly. It is washed with cold water in a centrifuge to get lactic acid. In another process, methyl ester of free lactic acid is prepared. The fermentation broth is distilled to get ester. It is boiled with water (hydrolysis) which allows the ester to decompose. The

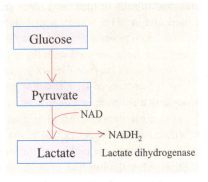

Fig. 17.9: Lactic acid production.

product is distilled to get aqueous solution of lactic acid and distilled product is methanol. The lactic acid can also be recovered as tertiary alkaline salt. Hence in the fermentation broth, organic solvent is added to get salts of lactic acid which is decomposed and free acid is released.

(*iv*) **Uses:** Since it is a weak acid with good solvent properties, it polymerises readily for the production of polymers. It provides acidity in foods and beverages and served as a preservative in food stuff. The delining of hides in leather industries is also carried out by its utilization. Certain other industries such as textile and laundry use lactic acid in fabric treatment. Calcium lactate is employed in baking powder and as a source of calcium in pharmaceutical industries. The pure form of lactic acid is used in plastic industry.

3. Citric Acid

This acid was first produced commercially by John and Edmund Sturage company in UK in the year 1826. Scheele (1789) reported the isolation and crystallization of the four constituents of lemon juice. Grimoux and Adams (1880) synthesized citric acid from glycerol. Wehmner (1893) observed the occurrence of citric acid as a microbial product by using *Penicillium* and *Citromyces*. It was Millard (1922) who recorded accumulation of citric acid in culture of *Aspergillus niger* under condition of nutrition deficiency. Meanwhile, Currie (1917) reported

$$CH_2 \!-\! COOH$$
$$HOC \!-\! COOH$$
$$CH_2 \!-\! COOH$$

Citric acid (2-hydroxy-1,2, 3 propane tricarboxylic acid)

better yield while using *A. niger*. In 1923, Pfizer began operating fermentation based process in USA.

(*i*) **Fermentation:** *Aspergillus niger* has been the choice for the production of this primary metabolite citric acid for several decades. A large number of other microorganisms (fungi and yeast) such as *Aspergillus clavatus, A. wentii, Penicillium luteum, P. citrinum, Mucor pyriforms, Candida lipolytica, C. oleophila, C. guillermondis, Hensenula spp. Torulopsis spp., Pichia spp., Debaromyces daussenii* etc. have also been used for citric acid production in industries. The advantages of using yeast, rather than *A. niger* are the possibility of using very high initial sugar concenteration together with a much faster fermentations. This combination gives a high productivity run to which must be added the reported insensitivity of the fermentation to variations in the heavy metal content of the crude carbohydrates.

From 1965 onwards, yeasts are used for citric acid production using carbohydrate and n-alkanes. In all the processes, a variety of carbohydrates such as beet molasses, cane molasses, sucrose, commercial glucose, starch hydrolysate etc. used in fermentation medium. The starchy raw material is diluted to obtain 20-25% sugar concentration and mixed with a nitrogen source (ammonium salts or urea) and other salts. The pH of the medium is kept around 5 when molasses is used and at pH 3 when sucrose used. The fermentation is carried out by any of the processes:

(*a*) *Koji process or solid state fermentation.* It is a Japanese process in which special strains of *Aspergillus niger* are used with the solid substrate such as sweet potato starch.

(*b*) *Liquid surface culture process.* In this case, *A. niger* floats on the surface of a solution.

(*c*) *Submerged fermentation process*: It is the process in which the fungal mycelium grows throughout a solution in a deep tank.

(*a*) **Koji process:** Mold is used in the preparation called Koji to which wheat bran was substituted in the sweet potato material. The pH of the bran is adjusted between 4 and 5, and additional moisture is picked up during steaming so as to get the water content of the mash around 70-80%. After cooling the bran to 30-60°C, the mass is inoculated with a koji which was made by a special strain of *A. niger* which is probably not as possible to the presence of ions of iron as the culture strains used in other process. Since bran contains starch which on saccharification by the amylase enzyme of *A. niger* induces citric acid production. The bran after inoculation, is spread in trays to a depth of 3-5 cm and kept for incubation at 25-30°C. After 5-8 days, the koji is harvested and citric acid is extracted with water.

(*b*) **Liquid surface culture process:** In this case aluminium or stainless steel shallow pans (5-20 cms deep) or trays are used. The sterilized medium usually contains molasses and salts. The fermentation is carried out by blowing the spores of *A.niger* over the surface of the solution for 5-6 days, after which dry air is used. Spore germination occurs within 24 hours and a white mycelium grows over the surface of the solution, Eight or ten days after inoculation, the initial sugar concentration (20-25%) reduced to the range of 1-3%. The liquid can be drained off and any portion of mycelial mat left becomes submerged and inactivated. The small quantity of citric acid is produced during the growth phase. This is called primary metabolite. The mycelium can also be reused.

During the preparation of fermentable sugar from molasses, sucrose is the main carbohydrate along with some glucose as well as protein, peptide, amino acids, and inorganic ions. This is to be subjected to heat; so it contains saccharic acids and related compounds in traces. The initial sugar concentration is about 20-25%. The removal of metallic ions or reduction in quantity of undesirable ions in sucrose syrup by adsorption with a combination of $CaCO_3$, colloidal silica, tricalcium phosphate and starch are other important steps. The iron is also precipitated by addition of calcium ferrocyanide.

Initially, the pH remains in the range of 5-6, but on spore germination, pH approaches the range of 1.5-2 as ammonium ions are removed from the solution. It is important to mention that at initial pH of 3-5 some oxalic acid is also produced. The presence of iron also favours oxalic acid production, and of yellow or yellow green pigments in the mycelium sometimes secreted into the culture solution and are difficult to remove during product recovery and purification (Fig. 17.10)

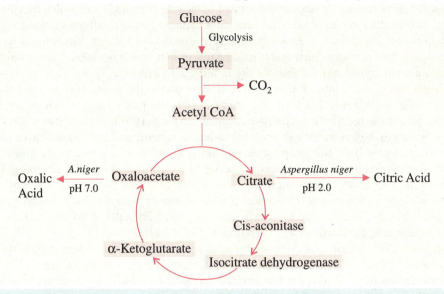

Fig. 17.10 : Biochemical pathway of oxalic acid and citric acid.

(c) **Submerged culture process:** This process is quite economical. In this case, the organism (*Aspergillus japonicus*) which is a black *Aspergillus* is slowly bubbled in a steam of air through a culture solution of 15 cm depth. Since the organism shows subsurface growth and produces citric acid in the culture solution, the yields are inferior in comparison to liquid surface culture fermentation.

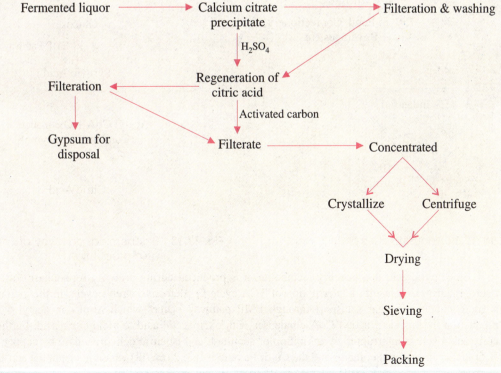

Fig. 17.11 : Process of recovery of citric acid.

The earlier workers used shaking culture and extracted Mollard's phosphate deficiency concept to submerged fermentation. But they could not realize the role of metallic ions which are commonly occurring as impurities in phosphate salts. Aeration is required for the continuous fermentation. The addition of copper ion is must to ensure that the new growth is of the right biochemical type. The antifoam agents are necessary. Such agents must be free of iron, cobalt or nickel. Continuous culture techniques are not considered suitable for use in citric acid product.

(*ii*) **Recovery:** The culture filtrate used to be hazy due to the presence of residual antifoam agents, mycelia and oxalate. The $Ca(OH)_2$) slurry is added to precipitate calcium citrate.

After filtrations the filtrate is transferred and treated with H_2SO_4 to precipitate Ca as $CaSO_4$. This is subjected to the treatment with activated carbon. It is demineralized by successive passages through ion exchange beds and the purified solution is evaporated in a circulating granulator or in a circulating crystallizers. The crystals are removed by centrifugation. The remaining mother liquor is returned to the recovery stream. The solvent extraction can also be performed by adding 100 parts tri-*n*-butyl phosphate and 5-30 parts *n*-butyl acetate or methyl isobutyl ketone which are to be mixed with the filtrate. The solvent is then extracted with water at 70-90°C. Citric acid is further concentrated, decolourized and crystallized (Fig. 17.11).

(*iii*) **Uses:** Citric acid is sold in the market as an anhydrous crystalline chemical, as crystalline monohydrate, or as a crystalline sodium salt. Citric acid is used in the soft drinks, jams, jelly, wines, candies and frozen fruits. It also used in artificial flavours. In medical applications, it is used in blood transfusion and as effervescent product. In cosmetic industries, various astringent lotions have citric acid where it is used to adjust pH and acts as sequestrant and in hair dressing and added in hair setting fluids. It also acts as synergists with antioxidants for oils and delay browning in sliced peaches (Fig. 17.12).

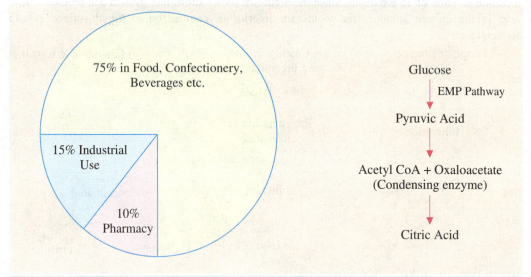

Fig. 17.12 : Industrial uses of citric acid.   **Fig. 17.13 :** Biochemical pathway of citric acid production.

Since citric acid is a primary metabolite, it is produced during growth (trophophase) of the microorganism as a result of interruption of TCA cycle *i.e.* defective Kreb's cycle. In this process a good amount of sugar transport through EMP pathway occurs which results in acetyl CoA production. In this case, acetyl CoA condenses with oxaloacetic acid to yield citric acid. Further, citric acid cycle is interrupted by inhibition of aconitase and isocitric dehydrogenase by copper or H_2O_2. Fe is an essential cofactor and the Cu or Fe ratio is 0.3: 2 (mg/lit) which interrupt the activity of these enzymes to pH 2.0 (Lockwood, 1979) as shown in Fig. 17.13.

D. Production of Antibiotics

Antibiotics (Fr. *anti* = against; *bios* = life) are chemical substances secreted by some microorganisms which inhibit the growth and development of other microbes. Most of them are produced by actinomycetes, specially the genus *Streptomyces,* and filamentous fungi.

The study of antibiotics began by the discovery of penicillin in 1929, when Alexander Fleming proved that the filtrate of a broth culture of *Penicillium notatum* has antibacterial properties in relation to Gram-positive bacteria. In 1940, E.Chain and H.Florey obtained a relatively stable preparation of penicillin. Clutterbuck (1932) studied the chemistry of penicillin and observed that it can be extracted from organic acid solvents from aqueous solution of low pH, and it was heat labile. But its activity was lost during evaporation of solution to dryness. By keeping low temperature during extraction, it was used to demonstrate the curative properties (Abraham *et al.* 1941). Florey and Heatley (1941) shifted from (U.K.) to USA during world war aid form the American Govt. By the time American troops entered in France in 1944, sufficient amount of penicillin was available for saving the wounded soldiers.

Strain improvement for Secondary metabolites production

The most wide spread method or increasing product yield in antibiotic producing microorganisms is still random mutation by treatment with mutagenic agents. Morphological differentiation has been studied in relation to cephalosporin synthesis in *Cephalosporium acremonium* by causing mutation for elucidation of the biosynthesis pathways for antibiotics cephalosporin synthesis.

In streptomycete, the ability to produce aerial mycelium is sometimes associated with production. Another approach is by raising blocked mutants for product cosynthesis. An end product is synthesized by two such blocked mutants (secretor and converter) when they are propagated together i.e. cosynthesis. These studies successfully elucidated the biosynthetic pathway for butirosins using blocked mutants of *Bacillus circulans*. Similarly, for new antibiotic production, biosynthetic blocked mutants for secondary metabolites have been evolved. This is called mutational synthesis e.g. macrolide antibiotics production.

Some times, mutants' resistance to antibiotics or precursors are effective for strain improvement. Corynesin (Chloramphenicol analog) is produced by *Corynebacterium hydrocarboclastus*. Another example is the growth inhibition of *Streptomyces hygroscopicus* by valine, which produced more maridomycin than the parent strain. A mutant requiring oleic acid for neomycin formation by *Streptomyces fradiac* showed a decrease in the intracellular level of neomycin precursors in the mutant. On the other hand, a supersensitive mutant to β-lactam antibiotics is another example.

1. Penicillins

Penicillins, a group of several penicillin, differ from one another in the side chain attached to its amino group. Most of these penicillins are 6-aminopenicillanic acid derivatives and all have β-lactam ring which is responsible for the antibiotic activity (Fig. 17.14). Pencillin acts against Gram-positive bacteria and inhibit their cell wall synthesis. These are non-toxic to mammals except for certain allergic reactions. More than 100 penicillins have been synthesized so far.

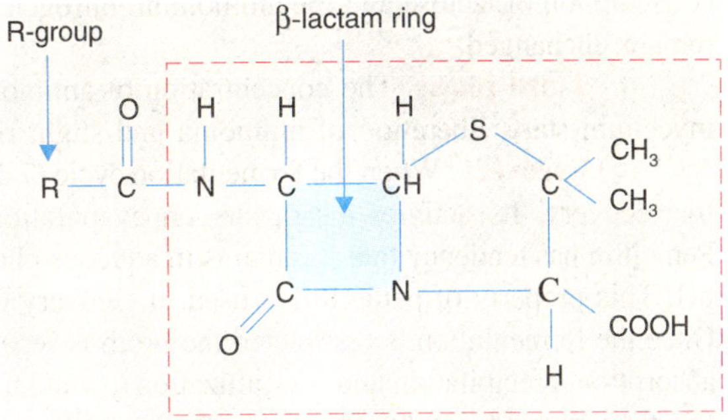

Fig. 17.14 : Structure of β-lactam nucleus. The positions where chemicals are substituted are indicated by *R*.

Penicillium species required the production medium which contains lactose (1%), calcium carbonate (1%), cornsteep liquor 8.5%, glucose (1%), sodium hydrogen phosphate (0.4%) and phenyl acetic acid (0.5g). The pH is kept between 5 and 6 and temperature for incubation is 23-25°C. Aeration and agitation are necessary.

(*i*) **Fermentation:** Penicillin is produced by *Penicillum chrysogenum* Q-176, a fungus that can be grown in stirred fermenters. The inoculum under aerobic condition (seed) can be produced when there is glucose in sufficient amount in the medium. If a particular penicillin is produced, specific precursor (substance added prior or simultaneously with the fermentation which are incorporated without any major change in the molecules) is added in the medium for e.g. phenyl acetic acid or its derivatives such as ethanol amide to get penicillin G (Fig. 17.15). The antifoam agents such as vegetable oil (corn or soybean oil) is added to the medium before sterilization.

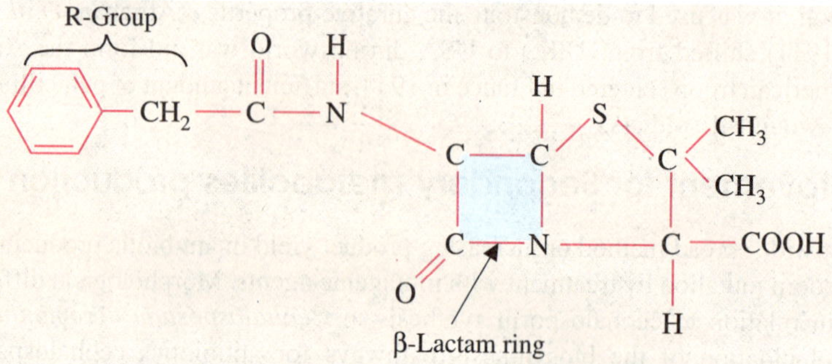

Fig. 17.15 : Penicillin G, a natural product.

The spore suspension is inoculated in flasks, each containing 15 g barley seeds. These flasks are vacuum dried, to which sterilized quartz is added.

The preparation of inoculum takes place on barley seeds. The flask containing 15 g barley seeds are to be mixed with mother culture, and incubated at 25°C for 7 days. The spores developed on barley seeds are suspended in distilled water to make spore suspension. After testing the antibiotic activity, the seeds containing flasks are ready for seeding in fermenter. Three phases of growth can be differentiated during cultivation of *Penicillium chrysogenum*.

(*a*) **First phase.** In this phase, growth of mycelium occurs, yield of antibiotic is quite low. Lactic acid present in corn steep liquor is utilized at a maximum rate by the microorganism. Lactose is used slowly. Ammonia is liberated into the medium resulting into rise in pH.

(*b*) **Second phase.** There was intense synthesis of penicillin in this phase, due to rapid consumption of lactose and the ammonium nitrogen (NH_3N). The mycelial mass increases; the pH remain unchanged.

(*c*) **Third phase.** The concentration of antibiotic decreases in the medium. The autolysis of mycelium starts liberation of ammonia and slight rise in pH.

(*ii*) **Recovery:** When the fermentation cycle (7 days) is completed, the whole batch is harvested for recovery. Its activity disappears on evaporation to dryness, hydrolysed to penicilloic acid. Penicillin has tendency that it remains in aqueous phase at normal pH and in solvent phase at acidic pH. This property of penicillin is used in recovery of potassium penicillin from natural solutions. Once the fermentation is completed the broth is separated from fungal mycelium and processed by absorption, precipitation and crystallization to yield the final product (Fig. 17.16). This basic product can then be modified by chemical procedures to yield a variety of semisynthetic penicillins such as ampicillin, amoxycillin, etc.

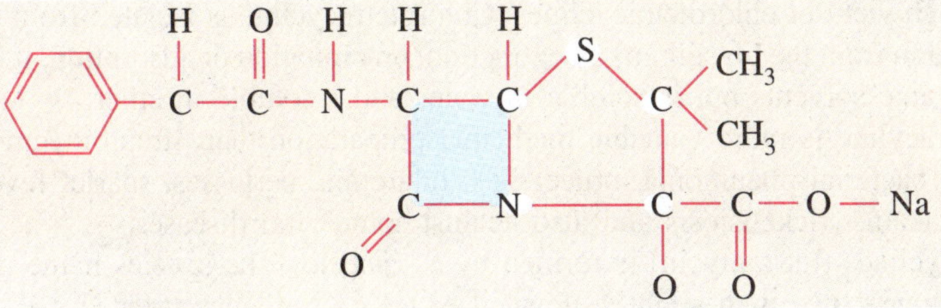

Fig. 17.16 : Structure of sodium penicillin G.

(*iii*) **Units of Antibiotics:** The potency of antibiotics is normally expressed in terms of units/ml of solution or in 1 mg (U/mg). In most of the cases, the antibiotic vial shows the quantity expressed in terms of weight as well as activity for example 1 mg of the penicillin G (benzyl penicillin) is equivalent to 1665 units.

2. Tetracyclines and Chloramphenicol

Due to the broad antibiotic activity of the tetracyclines (against Gram-positive and Gram-negative bacteria, rickettsia, and some large viruses), these compounds are widely used in medicine. The tetracyclines are also valuable for relatively low toxicity.

There are three important actinomycetes namely, *Streptomyces aureofaciens, S. ramosus* and *Nocardia sulphurea* which form tetracyclines. Several semisynthetic preparations are obtained from the other tetracyclines such as 6-dimethyl tetracycline, 7-chlorotetracycline, 7-chloro-6-demethyl chlorotetracycline, 7-bromotetracycline, 5-oxytetracycline including deoxycycline and metacycline by chemical modification of the oxytetracyclines. (Fig. 17.17)

Streptomyces aureofaciens produces aureomycin as observed by Duggar (1948, 1949). The name of the antibiotic was corresponded to the name of the species *aureofaciens* and secondly, due to the golden colour of its crystals. Now this antibiotic is named as chlorotetracyclines due to its chemical nature. (Fig. 17.18)

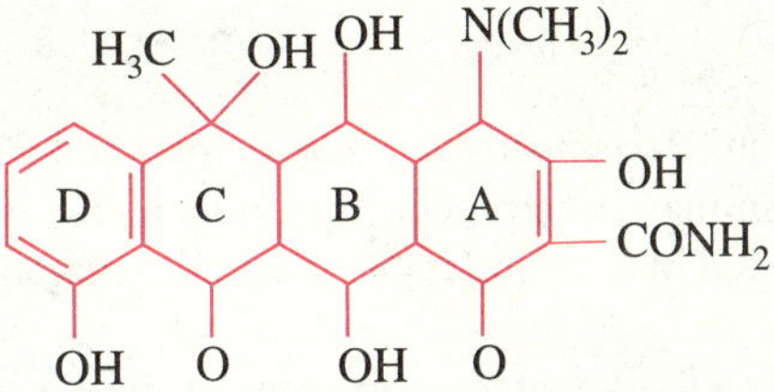

Fig. 17.17 : Structure of oxytetracycline.

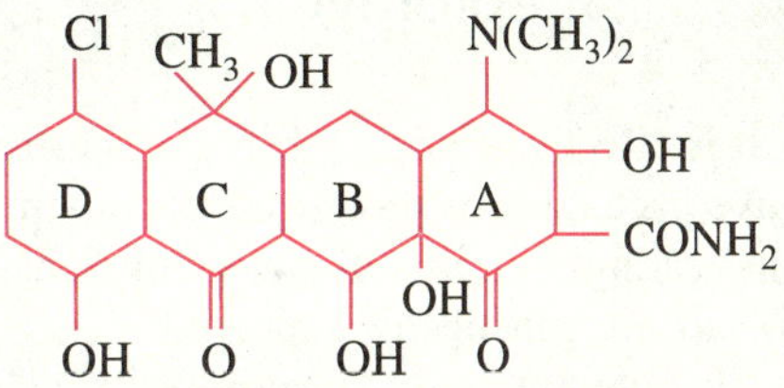

Fig. 17.18 : Structure of chlorotetracycline.

(*i*) **Media Composition:** Following medium is normally used for the production of chlorotetracyclins.

Sugar	3%
Corn steep liquor	1%
$CaCO_3$	1%
$(NH_4)_2SO_4$	0.2%
NH_4Cl	0.1%

The pH is kept at 6-7. The conditions of the medium for the growth and biosynthesis of chlorotetracyclines are important. The yield of antibiotic depends upon pH, age of the inoculum, and composition of the medium. Aeration of the culture is also very important for the biosynthesis of antibiotic. When the culture is grown in submerged conditions continuous aeration is required to ensure the high yield of chlorotetracyclines. Chlorotetracycline is isolated from the culture fluid (after its separation from the mycelium) by extraction, precipitation or adsorption. It is poorly soluble in common organic solvents but is soluble in water and insoluble in ether.

Chlorotetracyline is more valuable medicinal preparation than streptomycin or penicillin. It is used to treat bacterial pneumonia, brucellosis, tularemia, pertusses, scarlet fever, anthrax, etc. It is also used against rickettsiosis and also against some viral diseases.

Oxytetracyclines (tetrramycin) is formed by *S. rimosus*. The species name is indicating due to fissure like appearance with slightly elevated edges on the agar-surface.

S. griseoflavus, S. armilatus, S. aureofaciens var. *oxytetracyclini* also produce oxytetracyline. It requires following contents for its growth medium.

Corn steep liquor	0.5%
Starch	3.0%
Ammonium sulphate	0.4%
Sodium chloride	0.5%
Calcium carbonate	0.5%

In 24-28 hours of cultivation, submerged spores of streptomycete develop and then the secondary mycelium grows. This process requires aeration and there is liberation of protein N into the mycelium. The ammonium salt and nitrates favour the biosynthesis of oxytetracylines, starch, glucose, maltose, galactose, glycerol etc. are also used as carbon source. Lactose and saccharose are not at all used by *Streptomyces* species. The consumption of carbohydrates increases, the pH of the medium decreases, and the biosynthesis of the antibiotic slows down.

When tetracyclines are taken in low concentration, these act bacteriostatically and only become bacteriocidal when the concentration increases. The chlorotetracyclines and oxytetracyclines inhibit phosphorylation processes by preventing the inclusion of phosphorus into nucleic acids. The synthesis of proteins is essentially stopped while that of nucleic acid continues and sometimes is stimulated as well. Chloramphenicol in low concentration inhibits the protein synthesis. The antibiotic prevents assimilation of amino acids but inhibits the formation of polypeptide chain. It also inhibits selectively the growth of many bacteria but fails to produce an appreciable effect on the growth of yeasts, fungi, protozoa, or animal cells in the same concentration.

3. Streptomycin

It is effective against tuberculosis causing organism, *Mycobacterium tuberculosis* and Gram-negative bacteria. The prolonged use of streptomycin in mass, can produce neurotoxic effects and loss in hearing. Since its discovery by Schatz, Burgie and Waksman (1944) most of the strains of *S. griseus* are genetically improved.

(*i*) **Structure of Streptomycin:** The commercial available streptomycin is basically hydrochloride of streptomycin ($C_{21}H_{39}N_7O_{12}.3HCl$) with calcium chloride. During the production of streptomycin, mannosidostreptomycin or hydroxystreptomycin is also produced in the early fermentation. This salt is not economical and is easily converted to streptomycin by the action of *S. griseus* (Fig. 17.19). No precursor in reported to increase the yield.

(*ii*) **Media Composition:** Following is the composition of medium (lit^{-1}).

Soybean meal	10g
Glucose	10g
Peptone	5g
Meat extract	5g
Sodium chloride	5g

The pH is kept at 7.6-8.0 before sterilization and after inoculation the culture is incubated at 28°C.

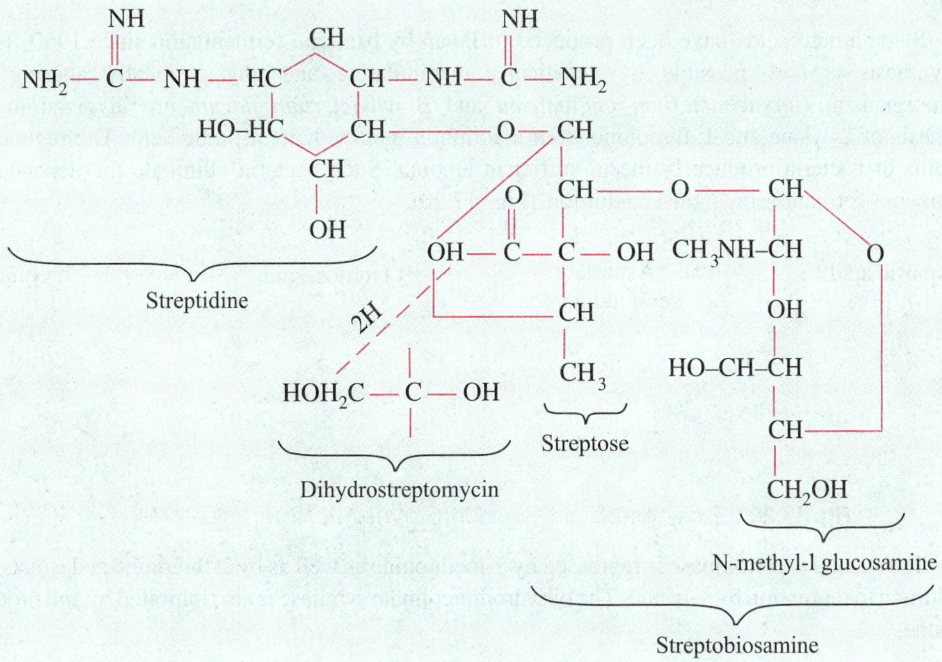

Fig. 17.19 : Structure of streptomycin.

(iii) **Production:** Streptomycin produced by *S. griseus*, an actinomycete that can be grown in stirred fermentation due to strong requirement of high aeration and agitation. The spores can be produced on medium which provide enough sporulated growth to initiate liquid culture of mycelium. The optimum fermentation temperature is approximately 28°C and the whole process completes within 5-7 days. There are three main steps in the production process of streptomycin.

(a) **First phase:** Growth of mycelium occurs, the proteolytic activity of *S.griseus* releases ammonia from the soybean meal, the carbon from soybean meal utilizes and induce growth but glucose is utilized at a minimum rate. The yield of streptomycin produced is low. The pH rises.

(b) **Second phase:** The streptomycin synthesized at a rapid rate in this phase, due to rapid utilization of ammonia and glucose. The total incubation period lasts from 24 h to 6 - 7 days. No mycelial growth occurs in this phase. The pH remains from 7.6 to 8.

(c) **Third phase:** The sugar depletes from the medium resulting into cease of streptomycin production. The cells lyse, releasing ammonia resulting into raised pH. Before lysis, fermentative material is harvested for recovery of streptomycin.

(iv) **Recovery:** After filteration, the broth is treated with activated carbon and then streptomycin is eluted with dilute acid. The eluted streptomycin is then precipitated by solvents, filtered, and dried before further purification. In another process, culture filtrate is acidified, filtered and neutralized. It is passed through cation exchange column to absorb streptomycin. Later it is dissolved in methanol, after filteration acetone is added to yield streptomycin. About 1200 µg/ml yield is obtained. The yield is not affected by contamination except due to actinophages as found in penicillin production.

(v) **Unit of Antibiotic:** One unit of streptomycin activity, however, is equivalent to the microgram of free base.

E. Production of Amino Acids

Some amino acids have been produced in Japan by bacterial fermentation since 1950. Their biosynthesis is now possible by genetically manipulating and physiologically altering the microorganism. *Corynebacterium glutamicum* and *Brevibacterium flavum* are involved in the synthesis of L-lysine and L-threonine from a common intermediate, aspartic acid. The regulatory mutants of bacteria produce lysine in sufficient amount. Such bacteria eliminate the dependence of enzyme formation on inducer addition (Fig. 17.20).

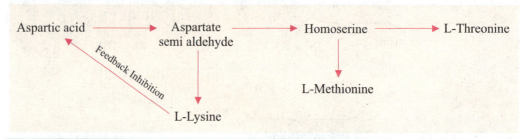

Fig. 17.20 : Biosynthesis of L-lysine in *Corynebacterium glutamicum*.

The second aspartokinase is repressed by 1-methionine as well as by 1-threonine and isoleucine in addition to repression by 1-lysine. The dihydrodipicolinate synthase is also inhibited by end product 1-lysine.

1. L-Lysine production

Generally, *Escherichia coli* and *Enterobacter aerogenes* are used for the formation of diaminopalmelic acid and for the decarboxylation of the diaminopalmelic acid by an enzyme DAP decarboxylase (*E. aerogenes* origin) respectively. *E. aerogenes* is an auxotroph requires L-homoserine or a mixture of L-threonine and L-methionine. The lysine-histidine, double auxotrophic mutant of *E.coli* (ATCC 13002) produces diaminopalmelic acid on a molasses medium with a yield of 19-24 g/litre. The entire fermentation solution including the cell material is subsequently incubated with *E. aerogenes* (ATCC 12409) at 35°C. After 20h, the DAP is quantitatively decarboxylated to L-lysine ($C_6H_{14}N_2O_2$).

(i) Fermentation: For industrial production of lysine, the seed culture is to be prepared by using glucose (20g) peptone (10g), meat extract (5g), sod. chloride (2.5g) in 1 litre of tap water. Culture obtained is reinoculated for second seed culture in media containing sugarcane molasses (200g), soyprotein hydrolysate (18g) mixed in tap water (1 litre), inoculated by *Brevibacterium flavum*, double auxotrophic mutant. The acetate is used as C source. After preparation of first, second and main culture, the production medium is used. The fermentation media consist of glycerol, corn steep liquor, ammonium sulphate. In addition, calcium carbonate is employed in the production medium. The pH is left neutral and incubation is carried out for 72h at 28°C with high aeration. The yield of lysine is as high as 75g/litre Lysine, earlier was produced by two stage processes using two different organisms. But now a days single stage process using mutants of *Conynebacterium*, *Brevibacterium etc.* are grown on a synthetic medium containing glucose, an inorganic nitrogen source and a small concentration of either homoserine, methionine, etc. in addition to a small concentration of biotin.

(ii) Recovery: Although, lysine is bound in the cell but due to mutations in the producing strains, it is secreted out and recovered.

(iii) Strain Improvement for Lysine Production: The biosynthesis of lysine is depicted in Fig. 17.20. Certain analogs of end products act as false feed back effectors and inhibit microbial growth.

To resist the feed back inhibition of aspartate kinase by lysine and threonine, mutants resistant to 5- (2- aminoethyl)-L- cystine (AEC) a lysine analog, been reported in *Brevibacterium flavum*. In such cases, lysine production may be increased to the extent of 57mg/ml. The strain does not accumulate threonine in the medium because normal feed back inhibition of homoserine dehydrogenase still operates. Some other amino acids, arginine, histidine, proline, valine, leucine, citrulline etc. are also improved by isolating analog-resistant mutants (regulatory mutants).

Quorn, a mycoprotein is produced by the growth of microorganisms.

2. L-Glutamic Acid

Kinosita *et. al* (1957) observed that L-glutamic acid ($C_5H_9O_4N$) is produced by using bacterial isolate, *Micrococcus glutamicus* (syn: *Corynebacterium glutamicum*). Although, this organism is not an auxotroph but requires biotin in the medium for growth. This amino acid is present both intracellularly as well as leaked out in the medium subjected to optimum biotin level available for fermentative production of L-glutamic acid. If excess of biotin is present in the medium, there is heavy cell growth, but lactic acid production starts.

Some microorganisms such as *Corynebacterium herculis*, *C. lilium*, *Arthrobacter globiformis*, *Microbacterium salicinovorum*, *Brevibacterium divaricatum*, *B. aminogenes*, *B. flavum*, *Bacillus megaterium* are other glutamic acid producing species.

Modifications of the permeability of a microorganism is provided by the glutamic acid fermentation. The permeability of *Corynebacterium glutamicum* may be controlled by the composition of the culture medium (Fig. 17.21)

(*i*) **Fermentation:** The medium contains glucose (121g), ammonium

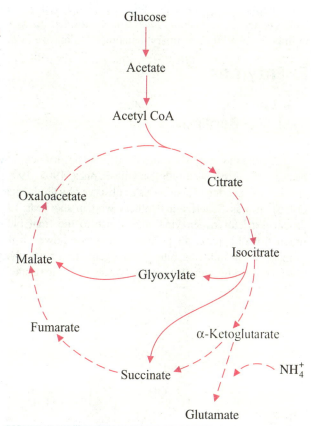

Fig. 17.21 : Formation of glutamate (the dotted line indicates the route of glucose conversion into glutamate while the complete line indicates the regeneration of oxaloacetate via glyoxylate cycle).

acetate (5g), molasses (6g), potassium hydrogen phosphate (1.2 g), potassium sulphate (6 ppm), manganese sulphate (6 ppm), antifoam agent (0.1 ml) in one litre distilled water. The size of the inoculum remain 6 percent. The fermentative organism is *Brevibacterium divaricatum* (NRRL B-231). The incubation was carried out for 16 hours at 35°C. At the beginning of the fermentations 0.65 ml per litre of olive oil is added. The pH is set at 8.5 with ammonia and is automatically maintained at 7.8 during fermentation. After growth of the culture (about 14 h), the temperature is increased from 32-33°C to 38°C. The glucose feeding is done until the fermentation is completed.

(*ii*) **Recovery:** The glutamic acid content is analysed hourly. The fermentation is stopped after 30-35 hours with a yield of 100 g per litre. If molasses from starch saccharification is substituted for glucose, the glutamic acid yield is 94 g per litre after 36 hours.

(*iii*) **Strain Improvement for Glutamic Acid Production:** Some times elimination of the permeability barrier in the membrane proved effective or improved. A natural biotin auxotroph that has altered membrane permeability found suitable for the hyperproduction of glutamic acid. The oleic acid auxotroph of *Br. thiogenitalis* showed good amount of glutamic acid production. Oleic acid auxotrophy is not applicable for production of glutamic acid from n-paraffin because the biosynthetic pathway of fatty acids from n-paraffin is different from that of glucose. Mutants with altered permeability in the membrane to glutamic acid led to the success in obtaining a glycerol auxotroph from *Corynebacterium alkanolyticum*. Protease production by kabacidin resistant strains of *Fusarium* sp is another example of participation of permeability barrier in production. Some times conditional mutants that behave as lethals under one set of conditions termed restricted and as wild type under, permissive conditions help in improvement of the primary metabolites as in case of *Br. lactofermentun*.

F. Enzymes

1. Pectinases

During the survey of crops for disease occurrence, it was observed that certain soft-rot causing bacteria produced pectolytic enzymes (Jones, 1905, 1909).The wilt disease induced by *Fusarium* species also produced pectinases as observed by Gothoskar *et.al* (1955) and Waggoner and Dimond (1955). It was Scheffer and others who in the year 1950 contributed that *Verticilluim* species produces pectolytic enzymes that cause to the final plugging of xylem of infected plants. The production of appropriate pectolytic enzymes, however, may not be the only characteristics needed to make a wilt inducing fungus pathogenic to a particular species of plant. The pectic substances, maker of middle lamellae, are polymers of galacturonic acid (Fig. 17.22 and 17.23)

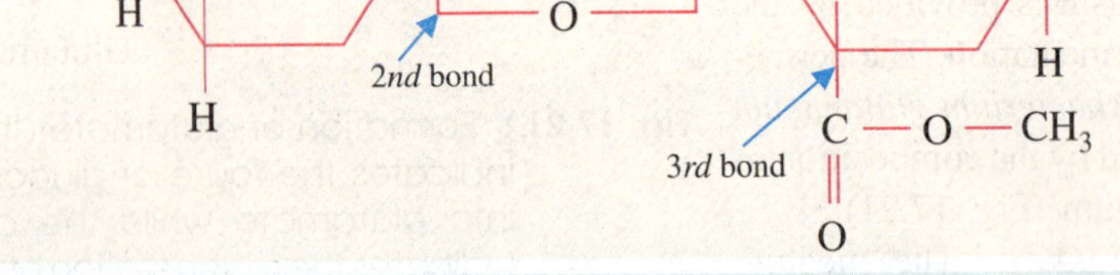

Fig. 17.22 : Structure of pectin.

Two groups of pectolytic enzymes are produced by plant pathogens. One is *pectinesterase* or *pectin methyl esterase* which breaks the ester bond *i.e.* 1st bond and removes CH$_3$ group. The resulting products are pectic acid and methyl alcohol. The other enzyme is *transeliminases* which breaks 1, 4 glycosidic bond *i.e.* 2nd bond and removes H from bond 3rd, thereby unsaturating the ring between 4th and 5th C atoms.

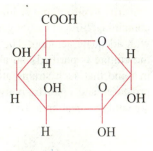

Galacturonic Acid

Fig. 17.23 : Structure of galacturonic acid.

(*i*) **Production:** A number of commercial firms produces fungal pectinases using *Aspergillus niger* and *Aspergillus wentii*. Sometimes *Rhizopus* and *Penicillium* species are also used for pectinases production.

The mycelium is developed on a medium containing pectin or a pectin like compound, a nitrogen source, such as yeast or malt extract, ammonia or peptone etc. and mineral salts. The fermentation with *A. niger* runs for 60-80 h in fed-batch cultures at pH 3-4 at 37°C using 2% sucrose and 2% pectin. The purification is simple. Since the pectinase is present both in the cells (intracellular) as well as excreted to the medium (extracellular), the enzyme is recovered from both sources.

(*ii*) **Harvest and Recovery:** At the time of harvest, the mycelium is dried and ground and its pectinase is extracted with water. It is then precipitated from this aqueous solution and from the culture broth by procedure as described for amylases extraction.

(*iii*) **Uses:** Pectinases are used to clarify fruit juices and grapes *must*, for the maceration of vegetables and fruits besides extraction of olive oil. By treatment with pectinases for a period of 1-2 h at 54°C or for a period of 6-8 h at 18°C, the yield of fruit juice during pressing is considerably increased. The commercial enzyme preparation contains at least two types of pectinases differing from each other by the extent to which they degrade pectin.

2. Invertase – (Saccharase or Sucrase)

This enzyme splits sucrose into glucose and fructose. It is widely distributed in nature. *Saccharomyces fragilis, S. cerevisiae* and some other *Saccharomyces* sp. are the richest source of enzyme invertase. Invertase is produced in industries from baker's or brewers yeast.

(*i*) **Production:** Industrially, enzyme invertase is produced by special strains of yeast which grow on bottom. The medium contains sucrose, an ammonium salt, phosphate buffer and other minerals. The pH is adjusted to 4.5. Fermentation is carried out for about 8 h at 28-30°C.

(*ii*) **Recovery:** The yeast cells are filtered off, compressed, plasmolysed and autolyzed. The invertase extracted may be dried or held in a sucrose syrup. The enzyme can also be purified by dialysis.

(*iii*) **Uses:** Invertase is used in confectionery to make invert sugar for the preparation of ice creams, etc. Chocolate-coated candies are also prepared. This enzyme imparts decrystallization of sugar in syrups on standing. Now-a-days, it has also been used in the manufacture of artificial honey.

3. Proteases

This group of enzymes catalyze the hydrolysis of the protein molecule. The enzyme is actually a mixture of proteinases and peptidases. These hydrolyze polypeptide fragments as amino acids.

(*i*) **Microorganisms involved:** Various bacteria, such as *Bacillus, Pseudomonas, Clostridium, Proteus, Serratia,* species and fungi, namely *Aspergillus niger, A. oryzae, A. flavus,* and *Penicillium roquefortii* are the sources of proteolytic enzymes.

(*ii*) **Production:** A high yielding strain is selected and inoculated in special culture media containing 2-6% of carbohydrate, protein and mineral salt. It is incubated for 3-5 days at about 37°C with adequate aeration. The filtrate is concentrated, and the enzymes are used in this form from the culture is purified and absorbed onto some inert material such as saw dust. It is always kept in mind that such strains are suitable for enzyme production which can give rise high yield of proteases and comparatively low yield of other enzymes. Many different media such as those containing wheat bran, soybean cake, alfalfa meal, are proved to be better for protease production.

Cheese manufacture.

(*iii*) **Uses:** Bacterial proteases help in digestion of fish liver to release fish oil, to the tenderization of meat. Such enzymes are also used in beverage industries in clarification and maturing of malt beverages. On the otherhand, fungal proteases are active in the production of soy sauce and other continental foods. They also remove protein haze from the beer and to hydrolyse the gelatinous protein material in fish waste. Cheese manufacturing is a batch process and current practice is to use a milk-clotting enzyme in a soluble form. Immobilization of proteases for milk coagulation has received renewed interest and potential applications have recently been reported. Use of immobilized proteases would permit renneting of milk as described by Garg and Johri (1993).

4. Lipases (Glycerol Ester Hydrolase)

A large number of microorganisms are capable of using natural oils and fats as a carbon source for their growth. The enzyme responsible for the breakdown of the oils and fats prior to their digestion by microorganisms are extracellular *lipases* which catalyse the hydrolysis of triglycerides to free fatty acids, partial glycerides and glycerol.

(*i*) **Microorganisms used in Fermentation:** *Candida cylindracae, Candida rugosa, Aspergillus niger, Penicillium cyclopium, Humicola langinosa, Mucor javanicus, Rhizopus arrhizus, Geotrichum candidum* have been reported for extracellular lipase production. Extraction of extracellular lipase from *Pseudomonas aeruginosa* has been studied by Jalger *et al.* (1992).

(*ii*) **Production:** Olive oil is being used as the substrate for the lipase enzyme. Stable emulsion of olive oil was prepared by vigorous mixing of the oil with a solution of an emulsion stabilizer Triton X-100 in 50 mM potassium phosphate buffer (Isobe *et.al*, 1988). The fermentation should be carried out at 29°C for 96 h supplied with 1% carbon source. The type of oil used proved to be important both lipase production and fungal growth.

(*iii*) **Recovery:** The mycelia were filtered and the culture filtrate was subjected to lipase assay. Lipase determination can be done either by stirring method or by standing method.

(iv) **Uses.** Miyoshi Oil and Fat Co. Japan is using *Candida cylindracae* lipase to hydrolyse oils for the production of soaps. Microbial liapases can also be used as catalysts for interesterification reactions (Matsuo *et al.*, 1980). The application of lipolytic enzymes to flavour development in dairy products has been revived by Arnold *et al* (1975). Some Italian cheese are manufactured using lipases. Microbial lipases are also used in the formulation of clothes, washing products etc.

5. Cellulases

The cellulases are synthesized by a large number of microorganisms, including fungi, bacteria, actinomycetes etc. Many bacteria utilize cellulose by cell bound enzymes, but fungi secrete cellulase into the growth medium. The extracellular cellulases secreted by *Trichoderma ressei, T. koningii, Penicillum pinophiluns, Fusarium solani, Phanerochaete chrysosporium,* etc. are among the best examples of cellulases. Examples of aerobic cellulolytic bacteria are *Cellulomonas* sp., *Cellubrio* sp., *Microbispora bispora,* and *Thermomonospora* sp. Some of the anaerobic cellulolytic bacteria are *Acetovibrio cellulolyticcus, Bacteroids cellulosovens, Clostridium thermocelluns, Ruminococcus albus.*

(i) **Production :** These bacteria can be grown on basal medium given by Mendels and Sternberg (1976). This medium consists of all requirements for growth. The ingredients lit^{-1} are; potassium dihydrogen phosphate (2g), ammonium sulphate (1.4g), magnesium sulphate (0.3g), calcium chloride (0.3g), urea (0.3g), peptone (1.0g), trace elements, ferrous sulphate (0.005g), manganese sulphate (0.0016g), zinc sulphate (0.0014g), cobalt chloride (0.002g) and pH 5.5.

The medium containing the enzyme-producing microorganisms (*Trichoderma reesei*) was then incubated at 35±1°C for 12 days. The mycelial mass is removed and the filtrate thus obtained acts as enzyme (crude). Cellulase is a multienzyme complex, consisting of endo-β-glucanase, exo-β-glucanase and β-glucosidase. The endo-β-glucanase or carboxymethyl cellulase or Cx is assayed by determining the amount of reducing sugar released from the specific substrate carboxy methyl cellulose (CMC). The mixture is boiled for 5 minutes at 100°C in a boiling waterbath. It is cooled and optical density is recorded at 540 nm in a spectrophotometer. Exo-β-glucanase is measured by using Whatman No. 1 filter paper strip (50 mg) as substrate. The strip is dipped in 1.0 ml of citrate phosphate buffer (pH 5.5) and 0.5 ml culture filtrate. This mixture was incubated at 50°C for 60 min. After incubation, 3.0 ml of DNS reagent is added and mixture is boiled for 5 min, cooled and diluted with distilled water. The optical density was recorded at 550 nm is spectrophotometer. The above enzymes give the approximate values of cellulases produced by any organism of industrial importance.

(ii) **Recovery:** Cellulases can be recovered by following processes:
* *Process A:* By recycling of the insoluble residue as a source of cellulolytic activity.
* *Process B:* By adsorption of the soluble remaining celluloytic activity onto fresh substrate.
* *Process C:* By combining both process A and process B.

In the process A, hydrolysate is centrifuged to get wet solid (bottom) phase. After washing, hydrolysis is carried on fresh substrate. The net production of sugar is determined which is further used to measure FPA/g (filter paper activity per gram) of substrate. In the process B, recovery of soluble activity of the liquid supernatant phase of hydrolysate 1 is contacted at room temp. After centrifugation, the bottom phase is used to start hydrolysis 2. The net production of sugar is obtained which can be calculated in terms of FPU/g. On the otherhand, maximum recovery may be obtained by combining both process A and B.

(iii) **Uses:** Cellulases are being used in clarifying juices and in improvement of beer production. These are helpful in essential oils extraction from the plant tissues besides increasing the digestibility and nutritive value of plant products. Cellulases are also exploited in bioconversion of lignocellulosic waste into useful products such as glucose-fructose syrup, single cell protein cultivation and as fermentable sugars to alcohol.

G. Vitamins

1. Vitamin B12 (Cyanocobalamine)

G.R. Minot and W.B. Murphy (1926) studied that liver extracts cure pernicious anaemia in human beings. In 1936, both of them shared a Nobel prize in medicine for this discovery. After this, E.L. Ricke and L. Smith (1948) isolated and crystallized vitamin B12 from liver extracts. This vitamin is present in a very small amount in every animal tissue including human blood 2×10^{14} µg/ml) but it is synthesized exclusively by microorganisms such as *Butyribacterium rettgeri, Bacillus megaterium, Streptomyces olivaceus, Micromonospora* sp., *Klebsiella pneumoniae etc*. High yields have been obtained from *Propionibacterium freudenreichii, P. shermanii* and *Pseudomonas denitrificans*. The most important industries manufacturing this vitamin are: Farmitalia S.P.A. (Italy), Glaxo Lab., Ltd. (England); Merck & Co., Inc (USA). Rhome Poulenc S.A. (France); Roussel UCLAF (France); G. Richter Pharmaceutical Co. and Chinoin in Hungary.

Vitamin B12 consists of a base structure corrin and a tetrapyrrole ring which differs from the porphyrin ring system in that the methane bridge between rings A and D is missing.

The molecular biology and genetic engineering techniques have improved strains of *Rhodopseudomonas protamicus* (hybrid between *Propionibacter ruber* and *Rhodopseudomonas spheroides*) produces 135 mg/litre vitamin B12.

Thus, vitamin B12 is not a single compound, but a group of closely chemically related cobamides. These cobamides are also called pseudo B12 group. They consist of cobalt porphyrin nucleus to which is attached ribose and phosphate. Various cobamide differ in the purine, benzimidazole, or other base, found in the nucleotide like portion of the molecule. Vitamin B12 analogue having other heterocyclic bases (purines or substituted benzimidazoles) are either spontaneously produced by microorganisms or are produced after the addition of these substances in the culture medium. The chemical process requires about 70 reactions, hence not viable process (Table 17.1).

(i) Fermentation: The nutritionally rich crude medium with glucose as a major carbon source is used in a two-stage process with added cobalt chloride (10-100 mg lit⁻¹). In a preliminary anaerobic phase (2-4 days), 5-deoxyadenosylcobinamide is mainly produced; in a second phase,

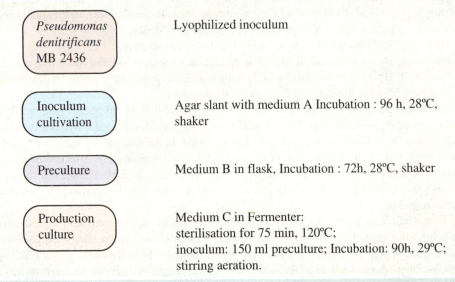

Fig. 17.24 : Steps of vitamin B12 production by using *Pseudomonas denitrificans* MB 2436.

which is aerobic (3-4 days), the biosynthesis of 5, 6-dimethyl benzimidazole takes place, so that 5-deoxyadenosyl cobalamine (B12) can be produced. This compound is completely intracellular and bound to the cell which after heat treatment released in solution form after 10 min at 80-120°C at pH 6.5-8.5. This process is applicable in case of *Propionibacterium freudenreichii* ATCC 6207 and *P. shermanii*, ATCC3673.

While using *Pseudomonas denitrificans*, there is one-stage process that occurs during the entire fermentation. Similar to the previous process, cobalt is added but here we also add 5,6-dimethylbenzimidazole as supplement. It has also been observed that addition of betaine induce the yield. In place of glucose, other cheap C sources such as hydrocarbon and higher alcohol are also found promising. Methanol proved better carbon source (Fig. 17.24).

Table 17.1 : Composition of media *A*, *B* and *C* for *Pseudomonas denitrificans* MB 2436

Media	Composition in g/litre		pH
A	Sugarbeet molasses:	60	
	yeast extract	1	
	N-Z-Amine	1	
	$(NH_4)_2HPO_4$	2	7.4
	$MgSO_4.7H_2O$	1	
	$MnSO_4.H_2O$	0.2	
	$ZnSO_4.7H_2O$	0.02	
	$Na_2MoO_4.2H_2O$	0.005	
	Agar	25	
B	Medium A but without agar		
C	Sugarbeet molasses	100	
	yeast extract	2	
	$(NH_4)_2 HPO_4$	5	
	$MgSO_4.7H_2O$	3	
	$MnSO_4.H_2O$	0.2	7.4
	$CO(NO_3)_2.6H_2O$	0.188	
	5-6, Dimethyl benzimidazole	0.025	
	$ZnSO_4.7H_2O$	0.02	
	$Na_2MoO_4.2H_2O$	0.005	

(*ii*) **Uses:** Vitamin B12 is produced by intestinal microorganisms. However, humans obtain vitamin B12 from food, since the B12 synthesized by microorganisms in the larger intestinal tract cannot be assimilated. For scoine and poultry feeds, 10-15 mg vitamin B12 is added per ton of feed, since animal protein can be replaced with less expensive vegetable protein if the vegetable protein is fortified with *viz.* B12. It has also a role to play in biological nitrogen fixation.

2.   Riboflavin (Vitamin B2)

Kuhn, Gyorgy and Wagner Jauregg in 1933 isolated riboflavin (also called lactoflavin) from whey of milk where it is present in free riboflavin form. It is also present in other foods (liver, heart, kidney, or eggs) as flavoproteins which contains the prosthetic group FMN (flavin mononucleotide) or FAD (flavin adenine dinucleotide). Several microorganisms namely, *Clostridium acetobutylicum, Mycobacterium smegmatis, Mycocandida riboflavina, Candida flareri, Eremothecium ashbyii* and

Foods containing riboflavin (Vitamin B2)

Ashbya gossypii are used in commercial production *A. gossypii* has got an ability to resist against riboflavin accumulation. Riboflavin is produced also by chemical synthesis but biotransformation of glucose to D-ribose by mutants of *Bacillus pumilus* and subsequent chemical conversion to riboflavin produced 50% of world wide production.

It is an alloxazine derivative which consists of a pteridine ring condensed to a benzene ring. The side chain consists of a C_5-polyhydroxy group, a derivative of ribitol (Fig. 17.25). Riboflavin is produced in the following steps :

(a) Media preparation and biosynthesis of Riboflavin. For Riboflavin production, basic medium consists of corn steep liquor 2.25%, commercial peptone 3.5%, soybean oil 4.5% but it can be supplemented further by addition of different peptones, glycine, distiller's solubles, or yeast extract. The glucose and inositol increase the production of riboflavin. The medium should be kept at 26-28°C at pH 6.8 for 4-5 days incubation.

Fig. 17.25 : Structure of riboflavin.

After inoculation the submerged growth of *Ashbya gossypi* is supported by insufficient air supply. The excess air inhibits mycelial production and reduces the riboflavin yield. The fermentation progresses through three phases.

(b) First phase: In this phase, rapid growth occurs with small quantity of riboflavin production. The utilization of glucose occurs resulting into decrease in pH due to accumulation of pyruvate. By the end of this phase, the glucose is exhausted and growth ceases.

(c) Second phase: Sporulation occurs in this phase. The pyruvate decreases in concentration. Ammonia accumulates because of an increase in deaminase activity. The pH reaches towards alkalinity.

(d) Third phase: There is a rapid synthesis of cell-bound riboflavin (FMN and FAD). This phase is accompanied by rapid increase in catalase activity subsequently cytochromes disappear.

As the fermentation completes, the autolysis takes place which releases free riboflavin into the medium as well as retained in the nucleotide form. It is also observed that certain purines also stimulate riboflavin production without simultaneous growth stimulation.

The riboflavin is present both in solution and bound to the mycelium in the fermentation broth. The bound vitamin is released from the cells by heat treatment (1h, 120°C) and the mycelium is separated and discarded. The riboflavin is then further purified. The crystalline riboflavin preparation of high purity have been produced using *Saccharomyces* fermentation with acetate as sole C source.

(*ii*) **Uses:** It is essential for the growth and reproduction of both humans and animals and, thus, it often is recommended as a feed additives for the animal nutrition. The riboflavin deficiency in rats causes stunted growth, dermatitis, and eye damage. Ariboflavinosis is a disease in humans caused by riboflavin deficiency.

H. Steroids Biotransformation

These are complex organic compounds, as shown in Fig. 17.26. Since the several chemical reactions involved in transformation of basic 4-membered ring into various other products, hence biological transformation is important due to the reason that only few steps are required in getting products. For example 32 reactions or steps were required to get cortisone from deoxycholic acid, by Sasset (1946). Hence, it proved to be uneconomical. Now-a-days, soybean for stigmasterol, diosgenin from the *Dioscorea* plant were found to be easy and economical sources. Microbial preparations of many steroids by their enzyme action at the specific site led to the synthesis of novel varieties of steroids. Such processes are more viable and specific. Steroid transformation is different with that of microbiological process due to the fact that in later that organic acids, solvents, antibiotics are synthesized from the ingredients in

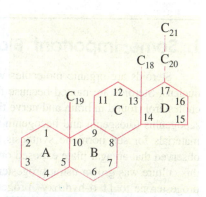

Fig. 17.26 : Structure of steriod nucleus showing numbering system.

the medium which also serve as substrate for growth and reproduction of microorganisms.

(*i*) **Microorganisms :** Several fungi, such as *Rhizopus nigricans, Curvularia lunata, Aspergillus* sp., *Penicillium* sp., *Gliocladium* sp., *Fusarium solani* and yeasts are some of the important organisms of steroid biotransformation. Besides, bacteria namely *Corynebacterium simplex,* and actinomycete *Nocardia restrictus* are known for biotransformation.

(*ii*) **Production:** The production of steroids can be carried out by the following steps.

(*a*) **Cultivation of microbe:** The appropriate microbe is grown in a way to get maximum growth in a short period of time. Generally, glucose or sucrose are recommended as carbon source and corn steep liquor as nitrogen source.

(*b*) **Incorporation of steroids and inhibitors into the medium:** A suitable amount of steroid (0.25 to 1.0 g lit^{-1}) to be transformed is first dissolved with the desired solvent (solvent should not inhibit with the microbe) and then is to add in the medium after the growth of the organism. The desired inhibitors are added to inhibit the undesired enzyme activities.

(*iii*) **Transformation of Steroids:** Microbes transform the steroid in a few hours to several days.

(*a*) **Separation and purification:** For this, regular sample is to be analysed chromatographically using TLC. The spots are eluted and their quantity can be measured spectrophotometrically. For this, the transformation product is extracted with a suitable solvent and then purified by using column chromatography or crystallization.

The structure is determined by using classical organic chemistry methods. Some important transformations are shown as below (Table 17.2).

Table 17.2 : Some important steroid transformations.

Microbe involved	Substrate into product	Reaction
Corynebacterium simplex	Cortisol → Prednisolone	Nuclear double
	Cortisone → Prednisone	bonds in ring A.
Rhizopus nigricans	Progesterone → 11 α	Hydroxylation
	Hydroxy progesterone →	at 11 α
Yeast	Androstenedione	Hydroxylation
	Testosterone	at 17 β

1. Some Important Biotransformations

Steroids are organic molecules which have in common *a per* hydrocyclopentaphenanthrene nucleus. Steroids are named because they are related to sterols which are abundant in nature e.g. cholesterol (found in brain and nerve tissue), stigmasterol (in vegetable oils), ergosterols (in yeast), sapogenins (diosgenin and hecogenin in plants) etc. The sapogenins are extremely useful starting materials for sex hormone synthesis and later on for corticoides and contraceptive drugs. It is observed that an agar plate exposed on the window yielded a culture of the genus *Rhizopus*. When this culture was grown using progesterone as a substrate, they found that it unexpectedly converts progesterone to 11 α-hydroxy progesterone in higher yield (Fig. 17.27).

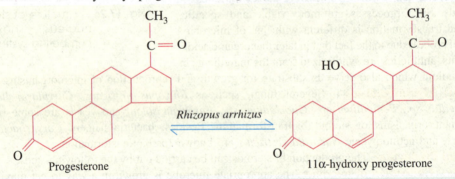

Progesterone   11α-hydroxy progesterone

Rhizopus arrhizus

Fig. 17.27 : Steroid transformation by chemical means I: progesterone into 11 α-hydroxyprogesterone.

The chemical synthesis of steroid hormones is quite difficult due to the reason that it requires several steps. The process is quite costly and may provide low yields because of certain rather difficult chemical steps in the process. For example the total chemical transformation of the cattle bile steroid deoxycholic acid to cortisone requires 37 chemical steps (Fig. 17.28).

These reactions often proceed with apparent base when mediated by microorganisms. Certain microorganisms can introduce hydroxyl groups at any of several of the carbon atoms of the steroid molecule.

Reckstein's compounds
or Cortexolone ——————————→ Hydrocortisone
Curvularia lunata
or
Cunnighamella blakesleana

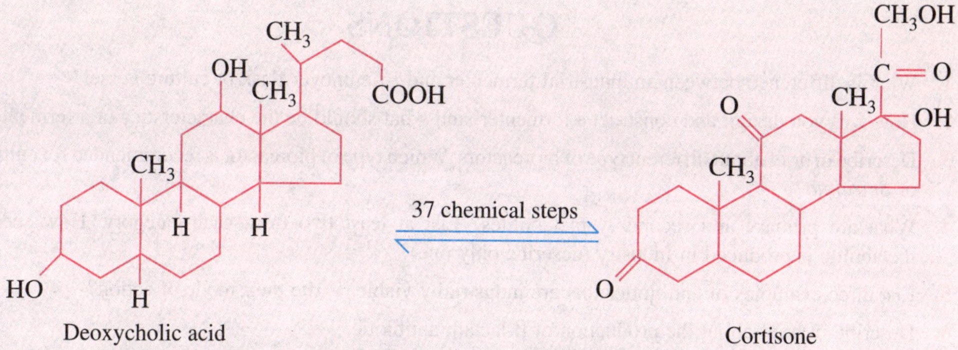

Fig. 17.28 : Steroid biotransformation. II: deoxycholic acid into cortisone.

Actuallly, in this process the hydroxylation occurs at 11P position. The reaction is given below (Fig. 17.29).

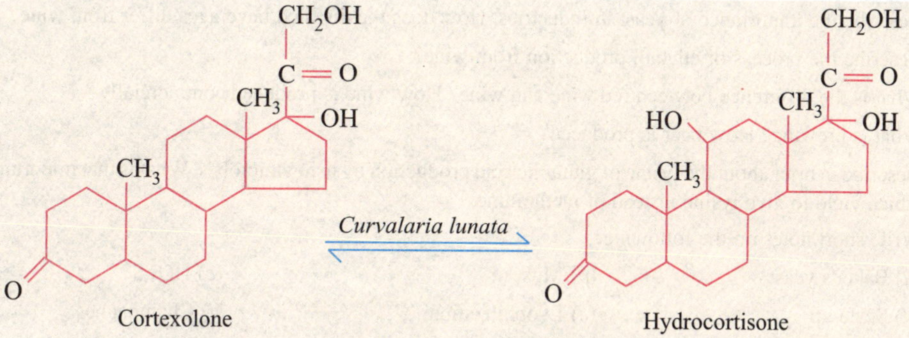

Fig. 17.29 : Steroid transformation. III: cortexolone into hydrocortisone.

In this case, carbon 11 of the steroid nucleus is of particular interest, because an oxygen atom at this point required for biological activity such as that required for the treatment of rheumatoid arthritis, and introduction of oxygen atom at this point is difficult by chemical means.

Table 17. 3 : Steroid biotransformation by certain microorganism

Microorganism	Reaction	Substrate to product
Corynebacterium simplex	Introduction of 1, 2 position	Hydrocortisone to Prednisolone
Rhizopns nigricans	Hydroxylation at 11 α position	Progesterone to 11- α hydroxy-progesterone

These days steroid biotransformation is carried on in submerged cultures in large fermenters. These transformations differ from the conventional process in that the products are not synthesized from the medium ingredients. Steroid precursors are added into the culture towards the end of growth phase. In recent years, purified enzymes have also been tried in place of organisms in steroid transformation.

QUESTIONS

1. What is difference between an industrial fermenter and Erlenmeyer flask or culture vessel?

2. How will you design and construct a fermenter, and what should be the characterstics of a fermenter?

3. Describe in brief about different type of bioreactors. Which type of bioreactor is recommended for cultivation of *Spirulina*?

4. What are primary and secondary metabolites? List at least two from each category. How secondary metabolite is produced in Industry (describe only one)?

5. List three examples of antibiotics that are industrially viable. Write their mode of action?

6. Describe in brief about the production of β-lactam antibiotic.

7. Name the requirements and its use in improving production of vitamin B12.

8. Define microbial bioconversion. Explain the importance of microbial reaction over chemical reactions.

9. List different types of enzymes produced commercially. Describe the process of manufacturing of any one enzyme.

10. Describe the importance of yeast in Industries. How does the distilled beverages differ from wine?

11. Describe the process of ethanol production from wine?

12. What is the difference between red wine and wine? How wine is produced commercially.

13. What is brewing? How beer is produced?

14. Describe in brief about the lysine or glutamic acid production by using microbes. What is that important step which yield to give lysine instead of methoinine.

15. Write short notes on the following:

    (*a*) Baker's yeast   (*b*) Must   (*c*) HOPS

    (*d*) Scale up   (*e*) Lyophilization   (*f*) Champagnes

    (*g*) Liquid nitrogen   (*h*) Idiophase and trophophase   (*i*) Koji

16. Write different mode of preservation of microorganisms.

17. How the microorganisms are maintained and elaborate the process of culture maintenance?

18. How citric acid is produced in Industries?

19. What are steroids? Name few microorganisms and their transformation products.

20. Draw and label the structure of vitamin B12 and riboflavin.

21. Write a detailed account on microbial synthesis of any one vitamin.

22. What is fermentation? Describe different type of fermentation processes carried out in industry.

23. Write short noes on the following:

    (a) Batch fermentation   (b) Fed-batch   (c) Continuous fermentation

24. How the strains are improved for the following product formation:

    (a) Lysine   (b) Glutamic acid   (c) Antibiotics

25. Write the process of cellulase production in industry.

REFERENCES

Aharonowitz, Y., and Cohen, G. 1981. The microbiological production of pharmaceuticals. *Sci. Am.* 245.140 152.

American Association for the Advancement of Science 1991. *Frontiers in biotechnology. Science.* 252 : 1585-1756.

Buchta, K. 1983a. Lactic acid, pp. 409 - 417. In Rehm, H.J. and G. Reed (eds.) *Biotechnology* 3, Verlag Chemic, Weinheim.

Buchta, K. 1983b. Organic acid of minor importance, pp. 467-478. In Rehm, H.J and G. Reed (eds) *Biotechnology* 3, Verlag Chemic, Weinheim.

Cruegar, W., and Cruegar, A. 1990. Biotechnology : *A Textbook of Industrial Microbiology*. Inded, edited by Thomas D. Brock, Sunderland, Mass., Sinauer Associated.

Demain, A.L. and Solomon, N.A. 1986. Industrial Microbiology. *Sci. Am.* 245 : 66 - 75

Florent, J. 1986. Vitamins, PP. 115-158. In Relm, H.J. and G. Reed, *Biotechnology* 4, Verlag chemic.

Fox, J.L. 1992. Contemplating large-scale use of engineered microbes. ASM NEWS 58: 191-196.

Garg. S.K. and Johri, B.N. 1993. Immobilization of milk clotting proteases. *World J. Microbiol and Biotech*. 9, 139-144.

Glazer, A.N and Nikaido, H. 1995. *Microbial Biotechnology. Fundamentals of Applied* Microbiology W.H. Freeman, New York.

Godfrey, T. and West, S. (eds). 1996. *Industrial Enzymology-The application of enzymes in Industry*. 2nd eds. Stockhom, New York.

Hong, M.,Y. Nomura, and M. Iwahara. 1986. Novel method of lactic acid production by electrodialysis fermentation. *Appl. Env. Microbiol.* 52 : 314 - 319.

Hopwood, D.A. 1992. The genetic programming of industrial microorganisms. *Sci. Am.* 245 : 90 - 102.

Kamikubo, T., M. Hayashi, N.Nishio and S. Nagai. 1978. Utilization of non-sugar sources for vitamin B12 production. *Appl. Env. Microbiol.* 35 : 971-973.

Lockwood, L.B. 1979. Production of organic acids by fermentation, pp. 355-387. In Peppler, H.J. and D. Perlman (eds.). *Microbial Technology*. vol. 1 Academic Press, New York.

Mazumdar, T.K., N. Nishio, M. Hayashi and S. Nagi 1986. Production of corrinoids including vitamin B12 by *Methanosarcina barkeri Biotechnol. Letters* 8 : 843 - 848.

Meharia, M.A. and M. Cheryan 1986. Lactic acid from acid whey permeate in a meunbrane recycle bioreactor. *Enzyme Microbiol. Technol* 8: 289 - 292.

Perlman, D. 1978. Vitamins, pp. 303 - 326. In Rose, A.H (ed.) *Economic Microbiology*, Vol-2, Primary products of metabolism. Academic press, London.

Phaff. H.J. 1981. *Industrial Microorganisms. Sci. Am.* 245 : 76 - 89

Primrose, S.B. 1991. *Molecular Biotechnology*. 2nd ed. Oxford : Blackwell Scientific publishers.

Sikyata, B. 1995. *Techniques in Applied Microbiology - Progress in Industrial Microbiology*. Vol. 31. Elsevir science, Amsterdam.

Vining, L.C. and Stuttard, C. (eds.) 1994. *Genetics and biochemistry of Antibiotic production*. Butterworth. Heinmann, Stoneham, MA.

Microbiology of Milk, Dairy and Food

18

A. Microbiology of Milk and Dairy Industries

B. Microbial Contamination and Spoilage of Poultry, Fish and Sea Food

C. Oriental Foods

The first milk drawn from animals always contains microorganisms. Most of bacteria come from dairy utensil, and milk-contract surfaces, milking machines, milk handlers and other similar sources. The bacteria include lactic streptococci, coliform bacteria, psychrotrophic Gram-negative bacteria. The negative rods, are thermoduric which survive pasteurization e.g. enterococci, and bacilli. Disease free dairy personal and utilization of sanitary equipment help in reducing the number of bacterial contaminants from external sources.

For the production of different dairy products, it is required to have appropriate culture of microorganisms. The pure culture or mother culture or stock culture are available in lyophilized or freeze-dried form. Stock cultures of desired organisms may be maintained in the dairy cultures of lactic streptococci, *Leuconostoc*, and *Lactobacillus* are used. On the otherhand, starter culture is produced by mixing 2% pure culture in 600 ml milk. This is heated at 88°C for 30 minutes and then cooled to 21°C and incubated to give starter culture.

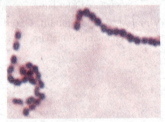

Lactic Streptococci

A. Microbiology of Milk and Dairy Industries

Milk is secreted by mammals for the nourishment of their young ones. It is in liquid form without having any colostrum. The milk contains water, fat, protein and lactose. About 80-85% of the proteins is casein protein.

1. Microbiology of Milk

Due to moderate pH (6.4-6.6), good quantity of nutrients, high water content etc. make milk an excellent nutrient for the microbial growth. It is mainly the udder interior, teats surrounding environment and manual milking process, make the source of contamination. These organisms mainly are micrococci, streptococci, and the diphtheroid *Corynebacterium boris*. An inflammatory disease of mammary tissues called mastitis is due to many microorganisms such as **Staphylococcus aureus**, *Escherichia coli, Streptococcus agalactiae, S. uberis, Pseudomonas aeruginosa* and *Corynebacterium pyogenes*. This disease causes economic loss in the dairy industries. Severely contaminated teats give upto 10^5 colony forming unit (CFU) in 1 ml in the milk. Sometimes the animal feed and manure can be a source of infection in human beings. Such pathogens are *E. coli*, *Salmonella* and *Bacillus* species. *Clostridium butyricum* can get into milk from silage fed to cows. This causes the problem called late blowing in some cheese.

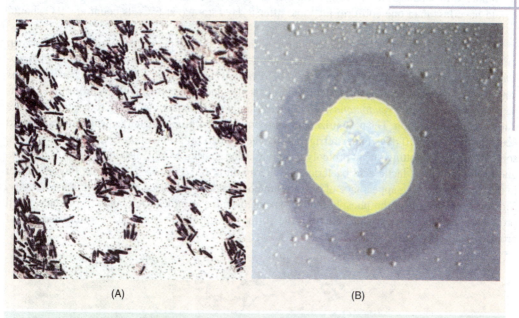

(A)   (B)

E. coli, a bacillus causing mastitis (A) , causes economic loss in the dairy industries (B).

Measures to minimize the contamination: The dairy industry needs enough cleaning, preventing muddy area wherever possible not to allow to leave urine and faeces, shaving udders and trimming tails. There should be a regular washing of teats with warm water containing some disinfectants. Such precautions are to be taken to minimize milk contamination so as to improve the microbiological quality of milk. During the storage of milk at low temperature in refrigerated holding tanks until its use temperature remains below 70°C and most of the organisms grow at such a temperature. In raw milk, Gram-negative rods of the genera *Pseudomonas, Acinetobacter, Alcaligenes, Flavobacterium, Aerobacter* spp. and Gram-positive *Bacillus* spp. are dominant microorganisms. The presence of psychrotrophs is a traditional test for the microbiological quality of milk based on the reduction of a redox dye such as methylene blue. Psychrotrophs reduce the dye poorly.

Heat treatment is based on thermal destruction of microorganisms. To protect against milk-borne diseases in human beings, heat treatment and particularly pasteurization was found most suitable. The following types of heat treatment applied in the milk are given below in the Table 18.1.

Table 18.1 : Heat treatment of milk

Condition	Temperature and holding time
Sterilized	> 100°C for 20-40 min.
High temperature short time (HTST)	71.7°C for 15 seconds
Low temperature holding (LTH)	62.8°C for 30 min.

A simple phosphatase test is recommended to determine whether milk has been properly pasteurized or not. Milk has the alkaline phosphatase inactivated by the time / temperature combinations applied during pasteurization. To determine about the complete pasteurization of milk, if it is free from microorganisms contaminating raw milk, a chromogenic substrate is added. The alkaline phosphatase present in the milk will hydrolyse the substrate producing a colour which can be compared to standards to determine whether the milk is acceptable or not. The pasteurized milk should have less than one coliform ml^{-1} and after 5 days storage at 6°C, its count at 21°C should be less than 10^5 cfu /ml.

Sometimes, raw milk may also contain a number of organisms called thermodurics that can survive mild pasteurization treatments. These are generally Gram-positive bacteria namely *Micro-bacterium, Micrococcus, Enterococcus,* and *Lactobacillus,* but Gram-negative, *Alcaligenes tolerans* may also survive. The spoilage of pasteurized milk is also due to the growth of psychrotrophic Gram-negative rods such as *Pseudomonas, Alcaligenes* and *Psychrobacter* which may contaminate the milk after pasteurization. The pasteurized milk produced under proper manufacturing process should keep for more than 10 days under refrigeration. The spoiled milk imparts off odours and flavours, sometimes shows clotting due to proteolytic activity. The souring of milk also indicate milk spoilage due to the growth of lactic acid bacteria. The thermoduric *Bacillus cereus* causes a flavour defect in the milk by showing appearance of bitty cream phenomenon produced by the lecithinase activity. This enzyme hydrolyse the phospholipids associated with the milk fat globules to produce small proteinaceous fat particles which float on the surface of hot drinks and adhere to the surfaces of crockery and glasses.

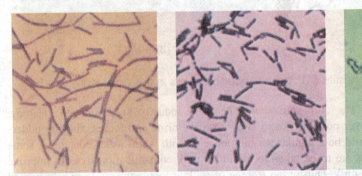

Lactobacillus lactis : Indispensable to food and dairy industry.

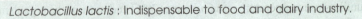

2.   Milk Products

A number of other milk products viz. cheese, yoghurt, milk powder, sweetened-condensed milk, cream, butter etc. are produced from milk. Some of products are shown in Fig. 18.1.

(i) Yoghurt : Yoghurt is derived from a turkish word 'Jugurt' which is the most popular fermented milk in the world now-a-days. It is made from milk, skimmed milk or flavoured milk. For the preparation of yoghurt, the milk should be free from contamination. The solid content (not fat) should be between 11-15% which can be obtained by adding skim or whole milk powder in fresh milk that normally contains 8% solids. The product can be further improved by adding small amount of modified gums which bind water and impart thickening to the product. At this stage, the size of the fat particles in the milk should be around 2μm because this improve the milk's viscosity, product's stability and milk appear form. The milk is then heated at 80-90°C for 30 min., starter culture is added to it. Heating improves the milk by inactivating immunoglobulins, remove excessive oxygen to produce microaerophilic environment which supports the growth of starter culture. Besides, heating also induce the interactions between whey or serum proteins and casein which increase yoghurt viscosity. The milk is now cooled to 40-43°C so as to allow fermentation using starter organisms such as *S. salivarius* sub sp. *thermophilus* and *Lb. delbruckii* sub sp. *bulgaricus* together at a level of 2% by volume (10^6-10^7 cfu/ ml). It is to be carried out for about 4h during which lactose is converted into lactic acid, pH decreases to a level of 6.3-6.5 to 4.6-4.7. The flavour in yoghurt is due to acetaldehyde which should be present at 23-41 mg/kg. Both the organisms are lacking in alcohol dehydrogenase, and produce acetaldehyde from glucose portion of lactose via pyruvate and due to the action of threonine aldolase. Diacetyl, an important flavour giving agent is sometimes present in yoghurt. Finally, after completion of fermentation, the yoghurt is cooled to 15-20°C before adding fruity flavoured agents. It is then cooled to 5°C which helps in keeping upto 3 weeks.

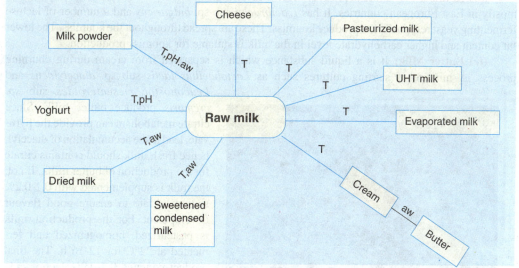

Fig. 18.1 : Milk and milk products. T, elevated temperature; pH, reduced pH; aw, reduced water pressure.

Microbial pathogens or contaminants generally do not occur in yoghurt due to high acidity and low pH. But sometimes, it is spoiled by acidoduric microbe such as molds and yeasts. Yeasts, particularly lactose fermenters *Kluyveromyces fragilis* and *Saccharomyces cerevisae* are particularly important but the yeast-like fungus, *Geotricum* and molds such as *Mucor, Rhizopus, Aspergillus, Alternaria* and *Penicillium* also spoil yoghurt. A good yoghurt should contain not more than 10 yeast g^{-1} with almost complete absence of coliform and molds.

(*ii*) **Kefir:** Kefir is infact, fermented milk, produced by a mixed lactic acid bacteria and alcoholic yeast. The microflora responsible is not spread uniformly throughout the milk but is supplemented as discrete kefir 'grains'. The kefiran, i.e. large layers of polysaccharide material folds upon to produce a cauliflower like florets produce kefir. The outer smooth layer contains lactobacilli while the inner, rough side contains yeast and lactic acid bacteria. The capsular homofermentative *Lactobacillus kefiranolaciens* produces kefiran. *Lactobacillus kefir* contributes the required effervescence in the product. Several yeasts such as *Candida kefir*, *Saccharomyces cerevisiae* and *S.exiques* have been observed, the latter utilize

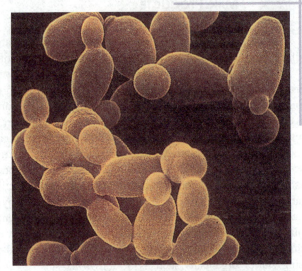

(*Saccharomyces cerevisiae*, an yeast involved in the production of fermented milk).

galactose preferentially glucose. For commercial production of kefir, milk is homogenized and heated at 85-95°C for 3-10 minutes. It is then cooled to 22°C before addition of kefir grains at a size of 5%. The fermentation should be carried out for 8-12 h. The product contains 0.8% acidity, 0.01% to 1.0% ethanol, carbon dioxide, acetaldehyde, and diacetyl etc.

(*iii*) **Koumiss :** Koumiss is produced from mare's milk which is greyish white drink, produced mostly in East European countries. It has *L.delbruckii* sub sp. *bulgaricus* and a number of lactose fermenting yeast responsible to produce koumiss. These are spread throughout the koumiss. The lower fat content and higher carbohydrate level in the milk is suitable for koumiss production.

(*iv*) **Butter Milk:** It is a liquid substance which is separated from cream during churning process. A mixture of starting cultures such as *Lactobacillus lactis* sub sp. *diacetylactis* and *Leuconostoc mesenteroides* sub *sp. cremoris* produce butter milk. Since citrate metabolism can provide the pyruvate, lead to the accumulation of diacetyl, hence fresh milk should contains citrate for the production of butter millk. If not, the milk is supplemented with 0.1-0.2% sodium citrate to ensure good flavour development. For the production, milk is pasteurized, homogenized and fermented at 22°C for 12-16 h. The final product contains 0.7-0.9% lactic acid.

(*v*) **Butter:** Lactic acid producing bacteria are responsible for subsequent separation of butter fat in the churning process. These organisms (*Leuconostoc citrovorum, Streptococcus cremoris* or *S. lactis*) produce a small amount of

(*Streptocccus*, lactic acid producing bacterium, is responsible for separation of butter fat).

acetoin which is spontaneously oxidized to diacetyl. This substance gives butter and similar products with their buttery flavour and aroma. When the pH reaches about 4.3, it ceases growth, but its enzymes attack the citrates in the milk and produce diacetyl. But neither *S. ceremoris* nor *Leuconostoc* alone can produce the desired result in commercial practice.

(*vi*) **Cheese :** There are about 2000 varieties of cheese made from mammalian milk. Cheese is thought to have originated in south western Asia some 8000 years ago. The Romans encouraged technical improvements and stimulated the development of new varieties during their invasions in Europe between 60 B.C. and A.D. 300. The cheese name is derived from Latin name *caseus*.

There are two groups of cheese: fresh cheese and ripened cheese. The fresh cheese are made up of milk coagulated by acid or high heat e.g cottage cheese, while ripened cheese are made through lactic acid bacterial fermentation and coagulated by an enzyme preparation. The curd is removed and salted and whey is separated. The salted curd is held in controlled environment. During this process, various physical and chemical changes occur to give a characteristic flavour and texture. Just as the variety of grape influences the flavour and bouquet of wine, so the mammalian origin of milk influences the flavour and aroma of a natural ripened cheese. There are mainly three categories of cheeses: (*a*) soft cheese, (*b*) hard cheese, and (*c*) semisoft cheese.

The soft cheeses are ripened by the enzymes from yeast and other fungi that grow on the surface, hard cheese ripened by lactic acid bacteria which grows throughout the cheese, die, autolyze and release hydrolytic enzymes. The semisoft cheeses are ripened by proteolytic and lipolytic organisms which soften the curd and give it flavour.

Microbiology of cheese : The large number of microorganisms play a role in the ripening process. On the first day of cheese making process, the microbial number in the starting material ranges from one to two billion gm^{-1}. Therefore, the production declines because of insufficient oxygen, high acidity and the presence of inhibitory compounds that are produced as the cheese ripens. It is mainly the action of their cellular enzymes on lactose, fat and proteins that creates the ripened cheese flavour. The gas forming culture of *Propionibacterium shermanii* is essential for giving swiss cheese its eye, or holes and flavour.

The specificity of cheese depends upon the varieties of microorganisms used. The process of cheese making, involves nine steps: (*a*) preparing the milk, (*b*) forming a curd, (*c*) cutting, (*d*) cooking, (*e*) separating the whey, (*f*) salting the residue, (*g*) applying microbes, (*h*) pressing the curd, and (*i*) ripening the young cheese.

Mostly, a ripened cheese is made from raw or under pasteurized milk which must be held for atleast 60 days. During that time the salt, the acidity, the metabolic compounds of ripening and the absence of oxygen usually destroy food-poisoning organism. During preparation of the milk, some colouring agents can be added which includes β-carotene and extracts of plants e.g. *Bixa orellana* and *Capsium* spp. Milk is transformed into smooth, solid curd more commonly known as chymosin which converts milk curd at 32°C in 30 min. Rennet can be extracted from *Mucor miehei*, *M. pusillus* and *Endothica parasiticus*. The rennin attacks casein and forms lattice or curd. The protein in this chymosin curd is called paracasein, because it is bound largely with calcium, it appears initially as dicalcium paracasein. In the third step, the wire knives or cutting bars are used to reduce the large bed of curd into small cubes of 1.5 cm. This step increases the surface area. During the cooking period, small cubes contracts and expel whey. For cheddar and related cheeses the optimum temperature is 37°C while Emmentaler and Gruyere curds are cooked at about 54°C. The cooking continues for a period ranging from 1 hour to an hour and half. Next process is salting where cheese maker applies dry salt to the curd. The immature cheese in saturated brine is immersed for about 2 to 72 hours. In the pressing stage, sometimes external pressure is applied on the wet, warm curd which is confined in wood, plastic or metal form or a cloth bag. Pressing is the end of preparatory phase of making a ripened cheese (Table 18.2).

Table 18.2 : Some common cheeses in International market

Type	Cheese	Method of Ripening
Hard Cheese	Cheddar	Cured at 36-50°F for eight weeks to 12 months.
	Emmentaler (swiss)	Formation of eyes in three to four weeks at 72°F and 80-85% relative humidity. Ripened at 40°F and higher temperature for 2 to 10 months.
	Gruyere	Formation of eyes at 60°F in 4 weeks. Cured for 80 days or more at 50-60°F.
Semi-soft cheese	Roquefort	Mold-ripened by *P. roqueforti* cured for 3 months at 48°F and 95% relative humidity. Wrapped in foil and stored in a cool room for 2 to 3 months.
Soft	Carrembert	Ripened by *P. candidum* on frames or shelves at 55°F and about 95% relative humidity for 12 days.

(vii) **Rennin (milk-coagulating enzymes) :** The milk coagulation enzyme, rennet, has its importance in cheese production. Certain commonly used cultures which have an ability to convert lactose to lactic acid and lower the pH, are *Streptococcus lactis, S. cremoris, S. thermophilus, Lactobacillus helventicus* and *L. bulgaricus*. These organisms bring the desired curd structure and flavour with minimum gas production.

Rennet is an extract from the fourth stomach of 3 to 4 week old calves which have been fed on milk. The enzyme so purified, called rennin, chymase or chymosin. Rennin production of microbial origin is now widely recommended due to good coagulation of casein without hydrolysis, good odour and structure of cheese, nontoxic, low protein denaturation in order to prevent the development of bitter taste during ripenning process and low lipase activity, to check the ranicidity development.

(Aspergillus sp., a fungus reported to produce rennin).

There are several genera of fungi reported to produce rennin. These are species of *Aspergillus, Candida, Coriolus, Rhizopus, Mucor, Penicillium, Torulopsis* etc. *Mucor pusillus, M. miehei* and *Endothica parasiticus* are widely used.The enzyme is an acid protease, which is stable at pH 4.0-5.5 with a molecular weight of 34000-37500 dalton at 50°C temperature.

Fermentation: A culture consists of soy meal (3%), glucose (1%), skim milk (1%), $NaNO_3$ (0.3%), K_2HPO_4 (0.05%), $MgSO_4.7H_2O$ (0.025%) at pH 6 litre^{-1} is mixed, autoclaved and fermentation takes 48 hours at 28°C. The extracellular enzyme is concentrated after separating the mycelium. Recently cDNA from calf protein has been cloned in *E. coli* resulting into the microbial production of calf renin enzyme.

3.   Biotechnology of Dairy products

In the dairy, fermented milks are produced by inoculating pasteurized milk with a known culture of microorganisms, sometimes referred to as a starter culture, which can be relied onto produce the desired fermentation, thus assuring a uniformly good product. *Streptococcus thermophilus* and *Lactobacillus bulgaricus*, are used as starter cultures in the preparation of yoghurt. Some important milk products and their associated microorganisms are given in Table 18.3.

Table 18.3 : Microorganisms associated with fermented milk

Fermented milk	Principal microbe involved
Cultured buttermilk	*Streptococcus lacti, S. cremoris* with aroma producing bacteria, *Leuconostoc* spp.
Acidophilous milk	*Lactobacillus acidophilus*
Yoghurt	*Streptococcus thermophilus, L. bulgaricus*
Kefir	*S. lactis*
	L. bulgaricus, Lactose-fermenting yeast

Milk is also a starting material for preparation of butter, cheese, curd, condensed milk with other ingredients to prepare ice–cream mixes, pudding etc. Transformation of milk constituents into a variety of fermented milk products carried out by lactic acid bacteria which act as starter culture.

Dairy or lactic streptococci, defined as strains of *S. lactis, S. cremoris,* etc which have been utilized for centuries in the production of fermented milk products. Their main role is to produce lactic acid from lactose (milk-sugar). The quality of such fermented products is also dependent upon the ability of starter organisms to produce flavour or aroma compounds.

Similarly lactobacilli are of commercial importance as they are widely used in the dairy fermentations, alcoholic beverages, pickling etc. Pediococci, the lesser studied organisms of lactic acid bacteria (LAB), are equally important as they are responsible for a great variety of fermenting reactions. In other words lactic acid bacteria are the back bone of dairy industry. Their economic value, of course, depends upon the (*a*) efficient fermentation of lactose into lactic acid, (*b*) ability to hydrolyse milk proteins (proteolytic activity), and (*c*) to utilize citrate to produce aroma compounds and their drug and bacteriophage resistance to minimize variations in the desirable characteristics of these organisms.

(i) Plasmid Concept in Starter Culture : In the 1930s proteolytic activity and citrate utilization by the various strains of group streptococci were reported to be unstable. They caused problems in milk fermentation due to variation in their behaviour. In the subsequent years, many strains were found to be susceptible to viral attack, to produce

(Milk is a starting material for preparation of cheese).

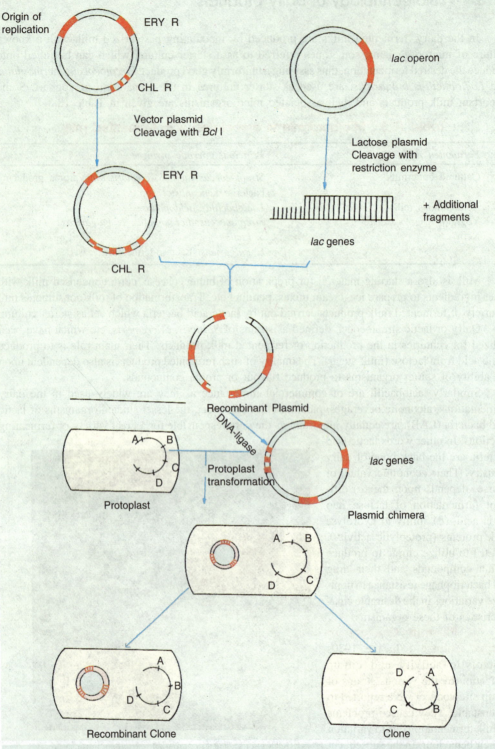

Fig. 18.2 : Gene manipulation to construct stable DNA recombinants of dairy starter culture.

bitter taste and other undesirable flavours and exhibit sensitivity towards slightly high temperature, inorganic ions, antibacterial drugs, etc.

Mekay (1972) stated that lactose metabolism in lactic sreptococci was governed not by chromosomal genes but by certain elements called plasmids. Soon a thorough search for possible linkage of plasmid DNA with other properties in lactic acid bacteria began attracting scientists from all over the world. Between 1972 and 1985 research has proved that lactic acid bacteria possess a wide complement of plasmid DNA which codes for their diverse properties.

(ii) **Gene Manipulation :** As the number of suitable strains for use in food/dairy fermentations on a large scale is limited, a need exists to obtain new and improved strains that can perform efficiently for long durations. Plasmid DNA studies of the existing lactic acid bacteria have opened up possibilities for using rDNA technology. They could be used more judiciously by dairy industry for manufacture of high quality fermented products. With the help of gene manipulation technology it has now become possible to construct stable recombinants of these organisms.

Kondo and Mckay (1982) reported successful transfer of plasmid DNA into *S. lactis*. By using rDNA technology, these workers were able to transfer Lac+ character from one strain of *S. lactis* to another strain of *S. lactis*. Much effort in this area is now focussed on plasmid mapping to determine the position of genes of interest. When genes are localized on a plasmid, the nucleotide sequences of such genes can be determined. After knowing their nucleotide seqnences not only the regulation of such genes but also their expression in the industrially important microorganisms, where they express at a much faster rate, can be understood.

Unfortunately, the potential of genetic engineering has not been exploited in the dairy industry because (*a*) dairy organisms possess complex nutritional requirements, which pose difficulty in culturing them, (*b*) genetics of lactic acid bacteria is not yet fully understood and only a few genetic markers are known which cause problems in the selection technique, (*c*) lactic acid bacteria produce products which are held within cells which further limits the utility of this technique. Dairy starter genetics showed progress rapidly there advances coupled with new methods of gene manipulation (Fig. 18.2).

B. Microbial Contamination and Spoilage of Poultry, Fish and Sea Food

Poultry includes chicken meat, goose, duck, squab, etc. The average number of bacteria associated with skin of bird is 1,500 per centimeter. Contamination of the skin and lining of the body cavity occurs during washing, plucking and evisceration. *Enterobacter, Alcaligenes, Escherichia, Bacillus, Flavobacterium, Micrococcus, Proteus, Pseudomonas, Staphylococcus, Corynebacterium* and *Salmonella* are those bacteria which have been isolated from poultry and poultry products. Various yeasts have also been isolated. The incidence of salmonellae on poultry carcasses has been reported that 20% birds are *Salmonella* positive.

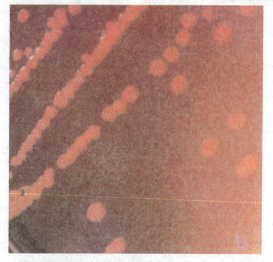

(*Salmonella*, isolated from poultry, is one of the chief bacteria involved in spoilage).

Bacteria are the chief source of spoilage. Most of these grow on the skin, the lining of the body cavity, and on the cut surfaces. The decomposition products diffuse slowly into the meat. Due to severe bacterial infections, sometimes skin imparts a bad odour. Eviscerated poultry held at 10°C or below is spoiled by *Pseudomonas* and to a lesser degree by yeasts. Above this temperature micrococci usually predominate followed by *Alcaligenes* and *Flavobacterium*.

The cut up poultry develops a slime that is accompanied by an odour described as tained and sour. This is due to the infection of various species of *Pseudomonas* and *Alcaligenes*. *Pseudomonas* and *Achromobacter* spoil poultry in polythylene bags.

<table>
<tr><td>**1.**</td><td>## Microbial Contamination of Meats</td></tr>
</table>

The important contamination comes from external source during bleeding, handling and processing. During bleeding, skinning, and cutting, the main sources of microorganisms are the body parts of the animal and the intestinal tract. The contaminating bacteria on the knife soon will be found in meat in various parts of the carcass, carried by blood and lymph. The exterior of the animal harbors large numbers of microorganisms from soil, water, feed and manure, as well as its natural surface flora and the intestinal contents. Knives, cloths, air and hands, clothing of the workers can serve as an intermediate sources of contaminants. During handling contamination may come from carts, boxes or other containers, from other contaminated meat from air and from personnel. Sometimes, it comes during refrigeration. The psychrotrophic bacteria may also contaminate meat.

The various equipments, grinders, sausage stuffers, casing, and ingredients in special products e.g. fillers may add organisms on surfaces touching the meats.

Molds of many genera may reach the surfaces of meats and grow there. The species of *Cladosporium*, *Sporotrichum*, *Geotrichum*, *Thamnidium*, *Mucor*, *Penicillium, Alternaria*, and *Monilia* often contaminate meat. Some of the important bacteria are *Pseudomonas*, *Alcaligenes*, *Micrococcus*,

An extensive *Penicillium* growth involved in spoilage.

Streptococcus, Sarcina, Leuconostoc, Lactobacillus, Proteus, Flavobacterium, Bacillus, Clostridium, Escherichia, Salmonella and *Streptomyces* involved in meat spoilage. Many of these can grow at chilling temperature. Sometimes, human pathogens can also act on meat contamination. In the retail market and in the home additional contamination usually takes place. The refrigerater containers serve as sources of spoilage.

(*i*) **Microbiology of Meat-curing Brines (Saline water) :** The brines usually contain lactic acid bacteria except at the surface, where micrococci and yeasts may develop. The lactis are chiefly lactobacilli and pediococci. Presence of micrococci imparts red colour in meat. Besides micrococci, brines contain a special mixture of cocci and Gram-positive rods that form tiny colonies on agar medium. They are halotolerant to halophilic and reduce nitrates to nitrites. Some beef curing brines have been found to contain micrococci, lactobacilli, streptococci, *Achromobacter*, vibrios, and perhaps pediococci in some small numbers.

(*ii*) **General Types of Meat Spoilage :** On the basis of conditions (aerobic or anaerobic) whether they are caused by bacteria, yeasts or molds spoilage is divided into two categories

(*iii*) **Spoilage of meats under aerobic conditions**

(*a*) **Surface slime:** There are different categories of spoilage. Some are given below: It is caused by species of *Pseudomonas, Alcaligenes, Streptococcus, Leuconostoc, Bacillus* and *Micrococcus*. Some species of *Lactobacillus* can produce slime.

(*b*) **Changes in colour of meat pigments :** The red colour of meat called its "bloom", may be changed to shades of green, brown or grey as the result of the production of oxidizing compounds e.g. peroxides or of hydrogen sulfide by bacteria. *Lactobacillus* and *Leuconostoc* are reported to cause the greening of sausage.

(*c*) **Changes in fats :** The oxidation of unsaturated fats in meats takes place chemically in air and may be catalysed by light and copper. Lipolytic bacteria may cause some lipolysis and also may accelerate the oxidation of the fats. Butter fat becomes tallow on oxidation and rancid on hydrolysis; but most animal fats develop oxidative rancidity when oxidized, with off-odour due to aldehydes and acids. Rancidity may be caused by lipolytic species of *Pseudomonas* and *Achromobacter* or by yeasts.

(*d*) **Phosphorescence :** Phosphorescent or luminous bacteria grow on the surface of the meat e.g. *Photobacterium* sp.

(*e*) **Change in surface colour :** The surface of the meat appears red due to *Serratia marcescens,* blue due to *Pseudomonas syncyanea* and yellow because of *Micrococcus* spp. and *Flavobacterium* sp. It becomes greenish blue to brownish black due to *Chromobacterium* species.

(*f*) **Off-odours and off tastes:** "Taints" or undesirable odours and tastes, appear in the meat due to growth of bacteria. "Souring" cold storage flavour or taints is an indefinite term for a stale flavour. Actinomycetes may be responsible for a musty or earthy flavour.

Aerobic growth of molds may cause meat to become sticky, whisker, black spot, white spot, green patches, decomposition of fats with off-odours and off taste.

(*iv*) **Spoilage under Anaerobic Conditions:** Facultative and anaerobic bacteria are able to grow within the meat under anaerobic conditions and cause spoilage. Following are the major spoilage:

(*a*) **Souring :** This could be caused by formic, acetic, butyric, propionic and higher fatty acids, or other organic acids such as acetic or succinic acid. It is due to action of meat's own enzyme during aging or ripening. The anaerobic production of fatty acids or lactic acid by bacterial action, and proteolysis without putrefaction may cause souring.

(*b*) **Putrefaction :** Anaerobic decomposition of protein results with the production of foul smelling compounds. *Pseudomonas* and *Alcaligenes* putrify meat. Hydrogen and carbon dioxide formation by clostridia make meat putrify.

(*c*) **Taint :** Usually, it imparts bad odour and bad taste. Besides air, temperature has important influences on the spoilage of meat. Many bacteria and molds produce slimes, discolouration, and spots of growth on the surface and may cause souring such as *Pseudomonas, Lactobacillus, Alacaligenes, Leuconostoc, Streptococcus* and *Flavobacterium* species.

2.   Spoilage of Different Kind of Meat

(*i*) **Fresh meat :** Pork is the only meat which spoils more readily than other meats because of its high content of Vitamin B. Lactic acid bacteria chiefly *Lactobacillus, Leuconostoc, Streptococcus,* and *Pediococcus* are present in either meats and can grow even at refrigerated temperature. Lactic acid bacteria cause slime production especially in the presence of sucrose give green colouration and souring.

(ii) Fresh Beef : Fresh beef undergoes the changes in the haemoglobin and myoglobin, the red pigment in the blood muscles, respectively, so as to cause loss of bloom and the production of reddish brown methaemoglobin and metmyoglobin (Fig. 18.3) and the green grey brown oxidation pigments by action of oxygen and microorganisms action. *Pseudomonas* and *Micrococcus* grow in beef, held at 10°C or lower.

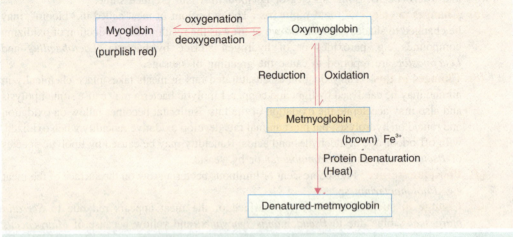

Fig. 18.3 : Different steps of meat spoilage.

(iii) Hamburger : At room temperature, it usually putrifies. Among the genera reported are *Bacillus, Clostridium, Escherichia, Enterobacter, Proteus, Pseudomonas, Alcaligenes, Lactobacillus, Leuconostoc, Streptococcus, Micrococcus* and *Sarcina. Penicillium* and *Mucor* grow on hamburger.

(iv) Fresh Pork Sausage : It is made up of ground fresh pork to which salt and spices have been added. It must be kept in refrigerators. *Alternaria* has been found to cause small dark spots refrigerated links. Souring, the most common type of spoilage at 0°C to 11°C has been attributed to growth and acid production by *Lactobacillus* and *Leuconostoc*.

(v) Cured meat : Curing salts make meats more favourable to growth of Gram-positive bacteria, yeasts and molds than to Gram-negative bacteria which usually spoil meats. The load of microorganisms on the piece of meat to be cured may influence the deterioration and will affect the curing operation.

(vi) Dried Beef or Beef Hams : Beef hams are made spongy by species of *Bacillus*, sour by a variety of bacteria, red by *Halobacterium salinarium* or due to *Bacillus* species. Gas in jars of chipped dried beef has been attributed to a denitrifying aerobic organism that resembles *Pseudomonas fluorescens*. The gases are oxides of nitrogen. *Bacillus* species produce CO_2.

(vii) Sausage : In the presence of enough moisture, micrococci and yeasts can form a slimy layer. With less moisture, molds may produce fuzziness and discolouration.

Various bacteria are responsible for the greening due to the production of peroxides. Surface slimyness often accompanies the greening. Bacon and Ham are the other meats which become putrified by various bacteria.

3.   Growth of Microorganisms in Meat

Due to high moisture, rich in nitrogenous food, plentified supply of minerals and accessory growth factors, usually has some fermentable carbohydrate (glycogen) and a favourable pH allows most of the microbes to grow. The following factors influence the growth of microorganism: (*a*) the kind and amount of contamination with microorganisms and the spread of these organisms in the meat, (*b*) physical properties of meat, (*c*) chemical properties of meat, (*d*) availability of oxygen, and (*e*) temperature

Fish and other Sea Food: Various bacteria involved in the spoilage are part of the natural flora of the external slime of fishes and their intestinal contents. At higher temperature *Micrococcus* and *Bacillus* species involved in fish spoilage while at ordinary temperature, species of *Escherichia, Proteus, Serratia, Sarcina* and *Clostridium* are found. Seafood includes fresh, frozen, dried, pickled and salted fish, as well as various shellfish. Fresh water fishes also get contamination from the surrounding microflora present in the water in which they live. The main genera that covers outer surfaces are: *Pseudomonas, Alcaligenes, Micrococcus, Flavobacterium, Corynebacterium, Sarcina, Serratia, Vibrio, Salmonella* and *Bacillus species.*

The presence of bacteria in the water content depends upon the temperature of water. Both psychrophiles and mesophiles are present in the water. Fresh water fish carry fresh water bacteria. Boats, boxes, bins, fish ponds and fish houses including fisherman soon become heavily contaminated with these bacteria and transfer them during cleaning. Most of the fishes pass large amount of water through their bodies pick up soil and water microorganism in this way, including pathogens if they are present.

The numbers of microorganisms on the skin of fish can be influenced by the method of catching. The other sources of contamination includes potatoes, spices and flavours used in fish cake. The microbial contents of fresh fish vary in the microbial fish products.

(*i*) **Microbiology of Fish Brines :** The temperature of brine and salt concentration influences the bacterial activity. The kind and amount of bacteria also vary which result to fish spoilage. Contamination comes from the fish, which ordinarily introduces species of *Pseudomonas, Alcaligenes* and *Flavobacterium.* The continuous use of fish brine may contribute additional pathogens from successive lots of fish and because of salt tolerant bacteria (micrococci, etc.).

(*ii*) **Spoilage :** Autolysis, oxidation or bacterial activity spoil fish like meat. The fish flesh is autolysed more quickly due to the presence of fish enzymes and because of the less acid reaction of fish flesh that favours microbial growth. Many of the fish oils seem to be more susceptible to oxidative deterioration than most animal fats. The lower the pH of fish flesh, the slower in general bacterial decomposi-

Spoilage of cottage cheese and cream cheese : No longer fit for consumption.

tion. Lowering the pH of fish results from the conversion of muscle glycogen to lactic acid.

(*iii*) **Factors Influencing Kind and Rate of Spoilage**

(*a*) **The kind of fish :** The various kinds of fish differ considerably in their perishability. Certain fatty fish deteriorate rapidly because of oxidation of unsaturated fats of their oils.

(*b*) **The condition of fish when caught :** Fish full of food when caught, are more readily perishable than those with an empty intestinal tract.

(*c*) **The kind and extent of contamination of the fish flesh with bacteria:** These may come from mud, water handlers, exterior slime, the intestinal content of the fish and are supposed to enter the gill of the fish, from which they pass through the vascular system and thus invade the flesh, or to penetrate the intestinal tract and thus enter the body cavity. In general, the greater the load of bacteria, the more rapid the spoilage. The contamination may take place in the net, in the fishing boat, on the docks, or later, in the plants. This process is accelerated by the digestive enzymes attacking and perforating the gut wall and viscera, which in themselves have a high rate of autolysis.

(*d*) **Temperature:** Bacterial growth is delayed at lower temperature. Cooling or chilling should be as rapid as possible 0 to -1°C. The high temperature reduce the life of fish. The prompt and rapid freezing of the fish is more effective in preservation.

(*e*) **Use of an antibiotic ice or dip:** Some antibiotics are recommended to avoid spoilage.

(*iv*) **Characteristics of Spoilage:** The bright colours of fish fade and dirty, yellow-brown discolouration appear. The slime on the skin of fish increases, especially on the flaps and gills. The eyes gradually sink, and on shrinkage the pupil becomes cloudy and the cornea opaque. Flesh becomes soft and juice exudes when squeezed. The discolouration takes place towards tail due to oxidation of haemoglobin and the different types of odours evolved during spoilage and cooking will bring out these odours more strongly.

(*v*) **Control of Spoiling Microoganisms :** Keeping microorganisms away from meats is called *Asepsis*. This process begins with avoidance from contamination. Before slaughtering, the animal should be carefully washed with water so as to remove dust from hair and hoof. The knife may also introduce microorganism in the circulating blood. During evisceration contamination may come from the internal body parts like intestine, the air, the water, cloths, brushes used on the carcass. Some organism may come from the surface soil and from workers.

Once meat is contaminated with microbes their removal is difficult. The use of hot water or sanitizer sprays under pressure is a procedure of decreasing the bacterial number. Moldy or spoiled portion of large piece of meat may be trimmed off.

Meats have been reported to have a shorter storage life in films with less permeable to water. Cured meats are packed in an oxygen- tight film with evacuation. It checks the growth of aerobes especially molds, reduces the rate of growth of staphylococci.

(*a*) **Use of Heat :** Canning of meat differs from product to product to be preserved. Most meat products are low-acid foods and are good culture media. Rates of heat penetration range from fairly effective in meat soups to very slow in tightly packed meats or in paste. Various additives such as spices, chemical salts and flavour also affect the heat processing. The process becomes more effective. On the basis of. heat processing, canned meat can be divided into two groups: (*a*) meat that are processed in one attempt (shelf-stable canned meat), and (*b*) meat that are heated enough to kill part of the spoilage organisms but must be kept refrigerated to prevent spoilage. This type of meat is known as non-shelf stable. The shelf-stable meat is processed at 98°C and the size of container is usually less than a kilogram. Cured meat temperature for processing should be 65°C and the container used during packing is of up to 22 kg.

Hot water treatment is also a method to remove the microbes from meat surfaces. But this may lessen nutrients and can damage colour. Heat applied during the smoking of meat and meat products helps in reducing microbes. The cooking of meats for direct consumption greatly reduces the microbial content and hence lengthen the keeping time. Precooked frozen meat should contain few viable microbes. More meat is preserved by the use of low temperature either by chilling or freezing. Chilling is more common.

Meat can be preserved promptly and rapidly to temperature near freezing and chilling at only slightly above the freezing point. Meat may be held in chilling storage for a limited time with little change from their original constitution. Enzymatic and microbial changes in the foods are not prevented but are slowed down considerably. Cooler temperature prevents growth but slow metabolic activity may continue. The storage time can be prolonged in an atmosphere containing added CO_2 and O_3. Ships equipped for storage of meat in a controlled atmosphere of CO_2 have been employed successfully. Increasing amounts of CO_2 inhibit microbial growth but also enhance the formation of metmyoglobin resulting into loss of colour. Storage time can be increased by the pressure of 2.5 to 3 ppm of ozone in the atmosphere. Ozone is an active oxidizing agent, that may give an oxidized or tallowy flavour to fats. Few bacteria, molds and yeasts can grow in meats at low temperature are known as psychrotrophic bacteria (*Staphylococcus, Alcaligenes, Micrococcus, Leuconostoc, Flavobacterium* and *Proteus*).

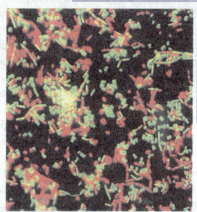

Micrococcus : A psychrotrophic bacterium that can grow in meats at low temperature.

(*b*) **Freezing or frozen storage:** Under the usual conditions of storage of frozen foods microbial growth is prevented entirely and action of food enzymes is greatly retarded. The lower the storage temperature the slower will be any chemical or enzymatic reaction. The preservation of frozen meats is increasingly effective as the storage temperature drops from -12.2°C to -28.9°C.

The freezing process kills the bacteria. The rate of freezing of meat and other food items depends upon a number of factors, such as the temperature, circulation of air, or refrigerant, size and shape of package, kind of food etc. Sharp freezing referes to freezing in air with only natural air circulation or at best with electric fans. The freezing temperature is usually -23.3°C or lower but may vary from -15 to -29°C, and may take from 3 to 72 hours. This is called slow freezing. Quick freezing is accomplished by one of the three methods: (*a*) direct immersion in a refrigerant, (*b*) indirect contact where the meat is in contact with the passage through which the refrigerant at -17.8 to -45.6°C flows, and (*c*) air-blast freezing where frigid air at -17.8 to -34.4°C is blown across the materials being frozen. Certain items now are being frozen into liquid nitrogen.

The advantages of quick freezing are: (*a*) smaller ice crystals formation and less destruction of intact cells of the food, (*b*) a shorter period of sodification and therefore, less time for diffusion of soluble materials and for separation of ice, (*c*) more prompt preservation of microbial growth, and (*d*) more rapid enzyme action.

(*c*) **Irradiation:** The ultraviolet rays serve to reduce number of microorganisms in the air and to inhibit or kill them on the surfaces of the meat reached directly by the rays. Irradiation also is used in the rapid aging of meats at higher than the usual chilling temperature to reduce the growth of microorganisms, especially molds, on the surface. Some oxidation, favoured by UV rays, and hydrolysis of fats may take place during aging.

(*d*) **By drying:** Some types of sausages are preserved primarily by their dryness. In dried beef, made mostly from cured, smoked beef hams, growth of microorganism may take place before processing and may develop in the "pickle" during curing, but numbers of organism are reduced by the smoking and drying process. Organisms may contaminate the dried ham during storage and the slices during cutting and packing. Salting and smoking are usually employed during meat drying.

Another method of drying pork involves a short addition of lecithin as an antioxidant and stabilizer. Drying may be in vaccum in trays, or by other methods. The meat for drying should be of good bacteriological quality.

5.　Food Preservation Methods

There are several methods used for preservation of food. Some of the methods are described below :

(*i*) **Physical Preservation Method**

(*a*) **Asepsis:** Asepsis is the process of prevention of growth of microorganisms and their contaminants or both. The covering or wrapping, prevents primarily contamination during handling, protect the process foods from microbial contamination.

(b) **High temperature:** The killing of microorganisms at high temperature is supposed to be caused by the denaturation of the proteins and especially by the inactivation of enzymes required for metabolism. The heat treatment also kills microorganisms or their spores. The heat treatment selected will depend on the kinds of organisms to be killed.

Certain factors are known to affect the heat resistance of cells of spores and must be kept in mind when microorganisms for the destruction of an organisms considered. These are:-

The temperature-time relationship, initial concentration of spores (or cells), previous history of the vegetative cells or spores, composition of substrate in which cells or spores are heated. Moisture (moist heat is a much more effective killing agent than dry heat), and Hydrogen-ion concentration (pH) are the other factors which play role in killing of microbes. At neutral pH many spores or cells are more heat resistant. An increase in acidity or alkalinity hastens killing by heat, but change towards the acid side is more effective than a corresponding increase in alkalinity.

The heat resistance of microorganisms usually is expressed in terms of their thermal death time (TDT) which is defined as the time it takes at a certain temperature to kill a stated number of cells or spores under specific conditions. This sometimes referred to the absolute thermal death time. Whereas, the thermal death point is the temperature necessary to kill all the organisms in 10 min.

Pasteurization is another method in which microorganisms are killed by heat treatment usually involves the application of temperature below 100° C. The heating may be generated by hot water, dry heat, or electric currents and products are cooled promptly after the heat treatment. The high-temperature for a short time (HTST), whereas the low-temperature-long-time (LTH) are given in pasteurization. For e.g. the minimal heat of market milk is at 62.8°C for 30 min in the holding method; at 71.7°C for at least 15 sec. in the HTST methods; and at 137.8°C for at least 2 sec in the unpasteurized methods.

(c) **Canning:** Canning is defined as the preservation of foods in sealed containers and usually implies heat treatment as the major factor in the preservation of spoilage, Spallanzani (1765) preserved food by heating in a sealed container. Nicholas Appert, who has been called the "father of canning", performed heating of foods in sealed containers and also published direction of canning.

Heat Process for Canning: The heat processes necessary for the preservation of canned foods depend on the factors that influence the heat resistance of the most resistant spoilage organism and those which affect heat penetration. With higher retort temperature the times would be shortened, and processes would vary with the varieties of foods canned, the sources used, the can size and shape, the initial temperature of the food and other factors. HTST

A canning operation.

heat processes, now used for some fluid foods, require special equipment for sterilizing the food in bulk, sterilizing the containers and lids and filling and sealing the sterile containers, under aseptic conditions. The dole process is an example of the HCF, or heat-cool-fill method. In the Martin HTST

Sterile Canning System, mixed liquid and solid pieces are heated directly by contact with high-temperature steam before aseptic canning. When a particular heat resistant spoilage organism is feared, an HTST treatment may be given to a liquid food before canning, followed by a milder heat treatment of the food in the can. In the SC, or sterilizing and canning process, sterilization of food is accomplished before the can is sealed. In the PFC, or pressure-filter-cooker method, the food is sterilized by high pressure steam and filled in to the can, then the can is sealed and the heat processing is continued as long as necessary before cooling. Heat is also being combined with other preservative agencies e.g. antibiotic, irradiation, or chemicals, e.g., hydrogen peroxide.

The Cooling Process: Following the application of heat, the containers of food are cooled as rapidly as is practicable. The can may be cooled in the retort or in tank by immersion in cold water or by a spray of water. Glass containers and large cans are cooled more gradually to avoid undue strain or even breakage. This tempering process involves the use of warm water (or spray). The temperature of water is lowered as cooling processors. Final cooling of containers usually is by means of air currents. Heating process carries out by Canning. In this case, the food is sterilized before filling and sealing the sterile containers. Heat is also combined with other preservative agencies such as antibiotics, irradiation or chemicals e.g. hydrogen peroxide. On the other hand, following the application of heat, the containers of food are cooled rapidly as possible by immersing in cold water. Final cooling of containers usually is by means of air currents. Low temperatures slow down the growth and metabolic activities of microorganism.

Desiccation: Foods contain moisture, hence drying of the food in necessary by removal of water. Various methods of drying such as sun-drying, artificially produced heat etc. are used for desiccation.

Anaerobiosis: It is a process in which the replacement of air by CO_2 or by an inert gas may bring anaerobic conditions. For example spores of some of the aerobic spore forming organisms are specially resistant to heat and may survive in canned food but are unable to germinate due to lack of oxygen.

Food preservation by canning.

(ii) Chemical Preservation Methods: Food additives are specially added to prevent the deterioration or decomposition of food, has been referred to as chemical preservatives. The food deterioration may be due to microorganisms, by food enzymes, or by purely chemical reactions. The inhibition of the growth and activity of microorganisms may inhibit microbes by interfering with their cell membrane, enzyme activity, or their genetic mechanism. Lactic, acetic, propionic and citric acids or their salts are used as preservatives. Citric acid used in syrup, drinks, jams, and jellies as a substitute for fruit flavours and for preservation. Lactic and acetic acids are added to drinks of various kinds, green olives etc. sodium or calcium propionate is used extensively in the prevention of mold growth and rope development in baked foods, cheese foods etc. These are effective against molds, yeast and bacteria. Propionic acid is a short chain fatty acid ($CH_3 CH_2$-COOH) and like some other fatty acids, perhaps affects the cell membrane permeability. The sodium salt of benzoic acid has been used extensively as an antimicrobial agent in foods such as jam, jellies, and margarine, carbonate beverages, fruit salads, pickles, fruit juices etc. Sorbic acid is used as a direct antimicrobial additive in foods and as a spray, dip or coating on packaging materials. It is widely used in cheese, cheese products, baked goods, beverages, syrups, fruit juices, jellies, fruit cocktails, dried fruits, pickles and margarine. It is most effective against yeast and molds but are less effective against bacteria. Monochloroacetic acid, peracetic acid, dihydroacetic acid and sodium diacetate, have been recommended as preservatives but not all are approved. Dihydroacetic acid has been used to impregnate wrappers for cheese to inhibit the growth of molds and as a temporary preservative for squash. Acetic acid in the form of vinegar is used in pickles, pickled sausages, and pig's feet. Acetic acid is most effective against yeast and bacteria. Combination of various salts of nitrites and nitrates used in curing solutions and mixtures for meats. Nitrates decompose nitric acid, which forms nitrosomyoglobin when it reacts with the heme pigment in meats and thereby forms a stable red colour. Sulphur dioxide and sulfite are used in the wine industries to sanitize equipment and to reduce the normal flora of the grape must.

The fumes of burning sulfur are used to treat most light-coloured dehydrated fruits; while dehydrated vegetables are exposed to spray of natural sulfites before drying. Ethylene and propylene oxide are used as sterilants. Ethylene oxide is most effective than propylene oxide. It kills all type of microorganisms. They are thought to act as strong alkalyting agents attacking labile hydrogen. The primary use have been as sterilants for packaging materials fumigation of ware-house, and cold sterilization of numerous plastics, chemicals, pharmaceuticals, syringes, and hospitals supplies, fully dried fruits, dried eggs, gelatin, cereals, dried yeast. Sugar and salts lower the moisture content and thus, have an adverse effect on microorganism. Sodium chloride is used in brines that prevents or inhibits the growth of microorganisms. Sugars such as glucose or sucrose, owe their effectiveness as preservatives to their ability to make water unavailable to organisms and to their osmotic effect. Examples of foods preserved by higher sugar concentrations are sweetened condensed milk, fruits in syrups, jellies, and candies.

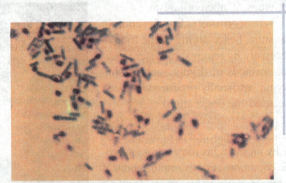

All spores of the bacterium can be inactivated by botulinum.

(iii) **Wood Smoke:** Smoking of foods usually adds desired flavours and acts as preservatives besides, improvement in the colour of the inside of meat. Wood smoke contains a large number of volatile compounds that may have bacteriostatic or bactericidal effect. Formaldehyde is considered as most effective of these compounds, with phenols and cresols next in importance. Other compounds in smoke are aliphatic acids from formic through caproic acid. Primary and secondary alcohols, ketones, and acetaldehydes and the aldehydes, waxes, resins and catechol, methyl catechol and pyrogallol and methyl ester. Wood smoke is more effective against vegetative cell than against spores. The woods oat, apple, maple, beech, birch, walnut, mahogany and hardwood such as chickory are used.

Many antibiotics have been tested on the raw foods chiefly proteinaceous like meats, fish and poultry. Aureomycin (chlorotetracycline) has been found superior to other antibiotics tested because of its broad-spectrum activity. Terramycin (oxytetracycline) is almost as good for lengthening the time of preservation of foods. Antibiotics have been combined with heat in attempts to reduce the thermal treatment necessary for the preservation of low and medium acid canned foods. It has been suggested that a botulinum cooking is enough to inactive all spores of *Clostridium botulinum.*

6.   Use of Preservation

Curing : Curing of meats is for the purpose of preserving by salts without refrigeration, but most cured meats of the present day have other ingredients added and are refrigerated, and may also be smoked and hence dried to some extent. Sodium chloride, sugar, sodium nitrite and vinegar are the common ingredients used in curing. Sodium chloride is used as a preservative and flavouring agent. Its primary purpose is to lower the a.w. (available water). Sugar adds flavour and act as an energy source for nitrate reducing bacteria in the curing solution or pickle. Sodium nitrite is bacteriostatic in acid solution, especially against anaerobe. It has also served as a reservoir from which nitrate can be formed by bacterial reduction during the long cure. Sodium nitrite is the source of nitrite oxide which has some bacteriostatic effect in acid solution.

C. Oriental Foods

Traditionally, oriental foods are used in many countries including India. Some of the oriental foods are described in this section.

1. Mycoprotein

Product such as tempeh and koji contain a significant amount of mould biomass (mycelium) which is a source of food. Mycoproteins, marketed in some foreign countries under the name Quorn, is essentially the mycelium of *Fusarium graminerarum* grown in continuous culture in a medium containing glucose, ammonium salts and a few growth factors. It is essential to reduce the level of RNA which is about 10% on mycelial dry weight to below the levels likely to lead to kidney-stone formation or gout. This can be achieved by mild heating prior to filtration which activates the RNases and leads to RNA reduction. Such products contains 44% of protein and is high in 'fiber' contents.

(*i*) **Tempeh :** This is the fermented food of Indonesians. The most popular type is produced from soybeans and also called as tempeh kedele. Whole clean soybeans are soaked overnight in water to hydrate the beans. A bacterial fermentation occurs decreasing pH to 4.5-5.3. The beans are dehulled and the moist cotyledons cooked result-ing pasteurization of the substrate. It destroys the trypsin inhibitor and lectins contained in the beans. After cooking, the beans are drained and spread on to bamboo trays for cooling.

The fermentation is carried out by mixed cul-ture of molds, yeasts and bacteria but most important component appears to be *Rhizopus oligosporus*. After 48h incubation at 33-35°C, the mycelium develops, pH rises to around 7, fungal protease increase the free acid content of the product and lipases hydrolyse over to neutral fat present to free fatty acids. Fresh tempeh has a nutty odour and flavour which can be consumed after frying in oil.

Tempeh, a fermented food.

Except thiamine, other vitamins increase to varying degree during fermentations. Tempeh is the important source of vit. B 12 for people subsisting on a largely vegetarian diets.

Tempeh has been stopped from 1988 due to food poisoning by *Pseudomonas* growing in the product and elaborating the toxious acid production.

(*ii*) **Soy Sauce :** This is a representative of product which have mould activity in two stage fermentation procces.

(*a*) **Koji stage:** The production of the soy sauce involved two stage fermentation process. The molder starter used is often called koji. In case aerobic conditions allow molds to grow on the substrate producing a range of hydrolytic enzymes. The soybeans are mixed with roasted wheat in equal proportions and inoculated with seed koji. i.e. a mixture of substrate and strains of *Aspergillus oryzae*. The moulds are grown about 5 cm deep for 2-3 days at 25-30°C (Fig. 18.4).

 (b) **Moromi stage :** Moromi stage is a mash process in which conditions are made so as not to allow moulds to grow. In soy sauce production this is obtained by mixing koji with an equal volume of brine to give a final salt concentration of 17-20%. The yeasts and lactobacilli dominate the microflora produced a number of flavoured compounds and convert half of the soluble sugars to lactic acid (2-3%) and ethanol (1%). The halophiles lactic acid bacterium *Tetragenococcus halophilus* (formely *Pediococcus halophillus*) and the yeasts *Zygosaccharomyces rouxi* and *Torulophilus,* etc. have been identified as being important organisms. This stage lasts up to a year

or more at the end of which the mash is filtered to remove the solid residues which may then be mixed with brine to undergo a second fermentation and produce a lower grade product. After pasteurization, filtration and maturation the product is bottled.

 (iii) **Miso :** Molds are involved in the preparation of most of the oriental foods. Miso is prepared by the starter termed koji (by japanese) and chou (Chinese), molds serve the source of enzymes. In this case *Aspergillus oryzae* is used to ferment steamed polished rice in shallow trays at 35°C. The koji is mixed with a mash of crushed and steamed soybeans to which salt is added. The fermentation is now allowed for about 6-7 days at 28°C and 60 days at 35°C. Finally, it is ground to form a paste which is used with other foods.

Miso : an oriental food.

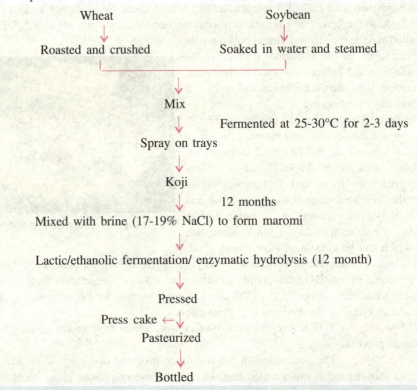

Fig. 18.4 : Flow diagram showing steps of soy sauce production.

(*iv*) **Idli** : It is a fermented food of southern India and is delicacy of the dinning table of Northern Indians. It is a product made from rice and black gram (*Vigna mungo*) in equal parts. After soaking overnight in water, both the ingredients are washed, mixed and ground, fermented overnight at room temperature. When the batter has risen enough, it is cooked by steaming. *Leuconostoc mesenteroides* grows first in batter, leaving it, and is followed by *Streptococcus faecalis*.

Table 18.4 : Some fermented foods

Food	Ingredients	Geographical Distribution
Beer	Barley	Wide spread
Cheese	Milk	Wide spread
Chicha	Maize and others	S. America
Idli/Dosa	Rice and black gram	India
Sushi	Fish	Japan
Kefir	Milk	Hungary,Poland,Romania, Bulgaria
Kimichi	Vegetables	Korea
Leavened bread	Wheat	Europe, N. America
Yoghurt	Milk	Wide spread
Nam	Meat	Thailand
Poi	Taro	Hawaii
Puto	Rice	Phillipines
Salami	Meat	Wide spread
Soy sauce, miso	Soybeans	S.E. Africa, Japan
Tempeh, Tofu	Soybeans	Indonesia, Japan
Tibi	Fruit	Mexico

(*v*) **Natto** : It is a soybean made product. The soybeans are wrapped in rice straw and fermented for 1-2 days. The package becomes slimy on the outsides. The microbe involved *Bacillus natto*, grows in natto, release trypsin like enzymes responsible for ripening process.

(*vi*) **Minchin** : In this case starch free but gluten rich wheat is used. The moist raw glutens placed in a closed jar and allowed to ferment for 2 to 3 weeks, after which it is salted. A typical specimen was found to contain several species of molds, some of bacteria and a few of yeast, the final product is boiled, baked or fried.

(*vii*) **Poi** : The bulbs and the corns of the taro plant are steamed for 2-3 hours, cooled, trimmed, scrapped and then finally ground. It is mixed with water as per desire and then can be consumed. On the other hand, if fermentation is carried out at room temperature for atleast a day during the first 6 hours of fermentation, the poi swells or puffs slightly with a change in the colour. After this, certain microbes in particular *Pseudomonas* spp., chromogenic bacteria, coliform group of bacteria become prominent. *Lactobacillus pasteurians*, and *L. delbruckii, L. brevis, Streptococcus lectis* sub sp. *lactis* and *S. kefir* consist of predominant flora. The fruity odour and pleasing taste impart in poi by the cultures of *Geotrichum candium*. The abnormal fermentations are likely to be of the butyric acid type.

2.   Food—Feed Source

Yeast such as *Saccharomyces cerevisiae* (sak-q-ro- mises se-ri-visei) or *Candida utilis* can be grown in large quantities, but they lead to adverse gastrointestinal reaction in humans. Also their high nucleic acid content exceeds the human body's capacity to metabolize them and may precipitate ailments such as gout. Yeasts are, therefore, most likely to be used as an animal food supplement. Microbes, which can usually double their weight in few hours and often in less than an hour, may solve this problem. When used as a food source, microbe are referred to as single cell proteins (SCP).

Yeasts are cultured in large cylindrial sterile fermenters.

(i) **Baker's Yeast:** Those strains which give good yield of cells, efficient sugar fermentation, easy mixing in water, etc. are used in the mash or medium chosen for their cultivation, should be stable, should remain viable in the cake or dried form for a reasonably long period, and should produce CO_2 rapidly in the bread dough when used for leavening.

The starter culture should be built from the original mother culture through several intermediate cultures and the commercial production of baker's yeast is on cane or beet molasses, or mixture of both mineral salts mash that contains molasses, nitrogen in the form of ammonium salts, urea, malt sprouts, etc., inorganic salts as phosphates and other mineral salts, and accessory growth substance, grain or yeast, or small quantities of vitamin precursors or vitamins such as biotin, pentothenic acid and inositol. The pH is adjusted to about 4.3 to 4.5, and the incubation temperature is around 30°C. During the growth of the yeast the medium is altered at a rapid rate, and molasses is added gradually to maintain the sugar level at about 0.5-1.5%. In the beginning, the culture is aerated but after 8-12 hours its rate is decreased. Addition of sugar and ammonia is checked. After 4-5 budding cycles, the yeast is centrifuged out in the form of a cream which is put through a filter press to remove excess liquid. The mass of yeast is made into cakes of different sizes after incorporation of small amounts of vegetable oils (Fig.18.5).

Active dry yeast now is made by drying the yeast cells to less than 8% moisture. Cells so dried are grown especially for the purpose and are dried carefully at low temperature. So that most of the cells will survive and will retain for some month their ability to actively leaven dough. Baker's yeast can be prepared from grain mashes, waste sulfite liquor from paper mills, wood hydrolysate, and other materials. The commercial production of baker's yeast involves the development of an inoculum through a large numbers of aerobic stages.

Although, the production stages of the process may not be operated under strictly aseptic conditions, a pure culture is used for the initial inoculum, thereby keeping contamination to a minimum in the early stages of growth. Reed and Nagodawithana (1991) discussed the development of inoculum for the production of baker's yeast and noted a process involving 8 stages the first 3 being aseptic while the remaining stages were carried out in open vessels.

Mushroom, an edible delicacy and dietary supplement.

(ii) **Mushroom Nutriceuticals:** According to a UNESCO sponsored symposium held in Manila 1993, nutriceuticals are a new class of compounds extractable from either the mycelium

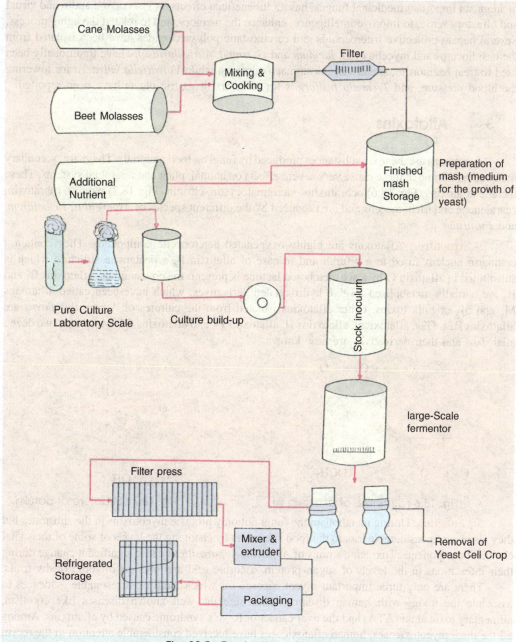

Fig. 18.5 : Production of baker's yeast.

or fruiting body of mushrooms. The term *neutriceutical* comprises of both the nutritional and pharmaceutical (i.e. medicinal) features. These are consumed in the form of capsules or tablets as a dietary supplements and have potential therapeutic applications. A regular intake enhances the immune response of the human body' thereby increases the resistance to disease. In some cases, it causes regression of a disease state.

Application: Preparations of *Lentinus edodes* and *Schizophyllum commune* are the examples. Lantinan (a purified polysaccharides used in the treatment of certain cancers and infectious diseases) and *Schizophyllum* (extract used to treat cervical cancer) are pharmaceutical products. *Ganoderma* *lucidum*, an important medicinal fungus, has multibeneficial effects on viscera, the audio and visual and olfactory sense, to improve intelligence, enhance the memory, and to related the aging process. Several hepato-protective triterpenoids and carcinostatic polysaccharides have been isolated from the basidiocarps and mycelia of *G. lucidum* and *G. tsugul. Auricularia spp.* have traditionally been used to treat haemorrhoids and various stomach ailments, while *Volvariella volvacea* for lowering the blood pressure, and *Tremella fusiformis* for sufferers of gastric ulcers have been reported.

3.   Aflatoxins

Mycotoxins are the toxic substances produced by fungi on food materials. These are 'secondary metabolites'. Some of these cause very severe effects on animal, plant and microbial systems. These include aflatoxins (Fig. 18.6) ochratoxins, sterigmatocystin, citrinin (Fig. 18.7) patulin, rubratoxin, zearalenone and trichothecens and are produced by the different species of *Aspergillus, Penicillium,* and *Fusarium*.

(*i*) **Structure :** Aflatoxins are highly oxygenated heterocyclic compounds. They contain a coumarin nucleus fused to a bifuran and in case of alfatoxin B, a pentanone structure which is substituted in aflatoxin G by a six membered lactone is present. In some animals, aflatoxins B_1 and B_2 are partially metabolised to give hydroxylated derivatives, which have been called aflatoxins M_1 and M_2 or milk toxins. Other aflatoxins isolated from the cultures of *Aspergillus flavus* are aflatoxins B2a, G2a, aflatoxicol, aflatoxins H, aflatoxins P1 and aflatoxins Q1. More than two dozen aflatoxins and their derivatives are now known.

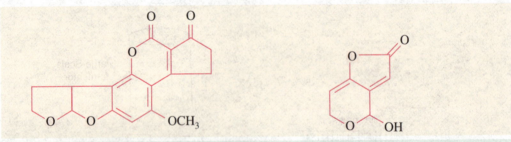

Fig. 18.6 : Structure of aflatoxin B1.   Fig. 18.7 : Structure of patulin.

(*ii*) **Function:** During metabolism the fungi not only produce mycotoxins in the substrates but they also cause considerable loss to the food substrates by changing the levels of some of their vital chemical components. Toxigenic strains of *A. flavus, A. parasiticus* cause significant change during their infestations in the levels of sugar, protein, ascorbic acid and phenols of some fleshy fruits.

There are only three important mycotoxicoses for which there exist resonable evidences to associate the toxins with human diseases. These are the well known diseases like ergotism, alimentary toxic aluki (ATA) and the liver cancer or Reye's syndrome caused by aflatoxins. Among all the three mycotoxicoses, human aflatoxicoses have received considerable attention in the recent years.

Because of the carcinogenicity of the aflatoxins in various laboratory and farm animals, the possible ability of these toxins to produce liver tumours in human is, of course, an area of great interest. However, acute and chronic effects of aflatoxins in human beings can only be investigating the diet associated with such cases.

(*iii*) **Aflatoxin - Producing Potential of Fungi :** Toxin producing potentials vary with the nature of the fungal organism and it may remain confined to specific or even at strain level. More than 150 species of fungi are known to produce different types of mycotoxins in laboratory as well as under natural conditions.

The concern about mycotoxins producing potentials of moulds has increased since 1960 i.e. after the discovery of 'Turkey-K-disease' which was attributed to aflatoxin elaborated by *Aspergillus flavus*, since then several hundred strains belonging to this group have been isolated from various food and feed materials.

All the strains of *A. flavus*, however, do not possess the capacity of elaborating aflatoxins. Screening of *A. flavus* isolates for aflatoxin production also received considerable importance in India. Aflatoxins elaboration by *A. flavus* on various food substrates largely depends upon the nature of the substrates and the environmental conditions. Moisture and temperature are the two most important factors which exert decisive influence on mycotoxin elaboration.

Besides environmental factors, aflatoxin elaboration also varies with the nature. A particular danger from consumption of aflatoxin contaminated food and feed lies in the possibility that some may damage the hereditary material of man. If this is so, what is the magnitude of the effect, are regrettably unknown. Surely, one of the greatest responsibilities of our generation is our temporary custody of the genetic heritage received from our ancestors. We must make every responsible effort to ensure that this heritage is passed on to future generation undamaged. Socially responsible scientists should be concerned about the potential hazards of such mycotoxin-induced mutations for two more reasons i.e. the genetic impairments, and somatic hazards

The former may enhance our existing burden of disease and premature death because it is a well realised fact that about 40% of our health problem is of genetic origin.

Consumption of aflatoxin contaminated food may damage the hereditary material of humans.

Somatic hazards include the manifestations of the action of mycotoxin on DNA and other cellular constituents. Needless to say that our existing genetic load is a summation of these three categories of genetic damage (replication, translation and transcription) as well as the availability of surface areas of the substrates.

Among all the mycotoxins, aflatoxins occupy key position with regards to carcinogenic effects on animals and human systems. These are one of the potent hepatocarcinogens known so far and can induce carcinoma (Table 18.5).

Table 18.5 : Mycotoxins present in different commodities and associated mycotoxicoses

Mycotoxins	Mycotoxicoses	Commodities
(A) Aspergillus toxin		
(i) Alfatoxin B1, B2, G1, G2	Liver cancer	Corn, peanuts, milk
(ii) Sterigmatocystin	Carcinogenesis	Green coffee, grains
(iii) Ochratoxin	Renal tumour	Corn, coffee, wheat flour bread
(B) Fusarium toxins		
(i) Moniliformin	Onlyai disease	Rice
(ii) Fumornisins	Leukoencephalomalacia promote cancer	Corn, wheat flower Corn flakes
(iii) Trichothecenes (T-2 toxin,Deoxyninalend),	Dermatitis,oesophageal digestive disorder	Corn, wheat, wine, cancer, Commercial cattle feed,
(iv) Zearelenone	Cervical cancer abortion	Corn meal, corn flakes, walnut

Mycotoxins	Mycotoxicoses	Commodities
(C) Penicillium toxins		
(i) Citroviridin	Cardiac bacteria	Mouldy peanut,
(ii) Citrinin	Kidney damage	dry fruit, rice, corn
(iii) Cyclopiazonic acid	Kodua poisoning	Cheese crust, corn
(iv) Patulin	Capillary damage in	Cider, apple juice,
	vital organs	jam, scented beetle nut
(v) Penicillinic acid	Edema carcinogenesis	Corn, bean, apple,
(vi) Penitrem A	Bloody diarrhoea death	Mouldy cream cheese
(vii) Rubratoxin	Liver disease	Mouldy grains
(D) Other mycotoxins		
(i) Agaricus toxins	Kwashiorkor	*Agaricus bisporus,*
		frozen mushrooms
(ii) Amanitins	Mushroom poisoning	Mushrooms
(E) Ergot alkaloides		
(i) Ergosine, Ergometrin,	Ergotism	Flour of rye, wheat,
Ergocristine		triticale, baby cereals

(*iv*) **Control of Aflatoxins :** Elimination of these compounds from the food and animal feed is necessary. Three basic approaches i.e. prevention, inactivation and detoxification have been proposed to control the mycotoxins.

A large number of chemical fungicides and fumigants like propionic acid, acetic acid, ethylene bromide, sulfur dioxide, luprosil etc. have been found to be very effective in preventing mould growth and aflatoxin elaboration. Propionic acid and crystal violet were found to exhibit greatest antifungal activity in liquid medium against toxigenic strains of *A. parasiticus*. These compounds also significantly prevented the aflatoxin elaboration by *A. parasiticus*. Prevention of alfatoxin production by the use of plant product have achieved alfatoxin inhibition.

Several attempts have been made to detoxify aflatoxin contaminated food materials by physical, chemical and biological means. However, physical separation procedures have been found to be most successful. An alternative method would be to use UV light to detect potentially contaminated lots and segregate them out but use of X-rays and UV rays have been found to be ineffective in minimizing alfatoxin level in food stuffs. Detoxification of alfatoxin have been achieved through heat treatments. Destruction of alfatoxin in food materials has also been suggested through roasting, frying in oils, spray drying, baking and autoclaving. Sunlight reduces alfatoxin level in peanut oil.

QUESTIONS

1. Name the important steps to produce cheese. Describe in detail.
2. What is whey? Which fungal genus has been used in the production of cheese?
3. Alfatoxins are produced by fungi, can you name it? What do they harm human beings?
4. Describe in general how food spoilage occurs. What factors influence the nature of the spoilage organism responsible?
5. List some antimicrobial substance found in milk and other food products. What is their significance?
6. Describe the microbial flora of milk and how to minimize contamination in milk.
7. Write short notes on the following:
   (*a*) Yoghurt (*b*) Kefir and koumiss (*c*) Butter milk (*d*) Butter (*e*) Curing
8. Describe asepsis? What are those different measures helpful in keeping microbes away?

9. Write an essay on microbial contamination, preservation and spoilage of poultry.
10. Write on the following :

    (*i*) Microbiology of meat curing brines, (*ii*) Symptom of spoilage of meat,

    (*iii*) Sausage, (*iv*) Mycoprotein (*v*) Baker's yeast, (*vi*) Mushroom neutraceuticals
11. Describe the physical and chemical methods of food preservations.
12. Write short notes on the following:

    (*a*) TDT (*b*) Canning (*c*) Wood smoke

REFERENCES

Atkinson, G.F. 1961. *Mushrooms*, 2nd ed. Hofner Publishing Co Newyork.

Atkins, F.C. 1983. *Mushroom J.* 125 : 168-171.

Banwart, G.J. 1989. *Basic food microbiology*, 2nd ed. Van Nostrand Reinhold Co., New york.

Board, R.G. 1983. *A Modern Introduction to Food Microbiology*. Blackwell Scientific Publications, London

Cliver, D.O. 1990. *Food-borne diseases*. Academic Press, San Diego.

Doyle, M.P. 1989. *Food-borne Bacterial Pathogens*. Marcel Dekker, Inc., NewYork

Elliker, P.R. 1949. *Practical Dairy Bacteriology*. McGraw-Hill Book company, New york

Frazer, W.C. and Westhoff, D.C. 1988. *Food Microbiology*. 4th ed. McGraw-Hill,   New york

Forss, D.A. 1964. Fishy flavour in dairy products. *J. Dairy* Sci. 47: 245-250

International dairy federation 1980. Factors influencing the bacteriological quality of raw milk. Doc. No. 120, IDF, Brussels, Belgium.

Jay, J.M. 1991. *Modern Food Microbiology*, 4th ed. Van Nostrand Reinhold Co., New york.

Keogh, B.P. 1971. Reviews of the progress of dairy science. Bacteriology. The survival of pathogens in cheese and milk powder. *J. dairy* Res. 38:91-111.

Kosikowski, F.V. 1985. Cheese: *Sci. Am.* 252; 88-99.

Moreau, C. 1979. *Moulds,toxins and foods*, John wile and Sons, New york

Rose, A.H. 1981. The Microbiological production of food and drink. Sci. Am. 245:   126-139

Smith, J.E. and Moes, M.O. 1987. *Mycotoxins, formation, analysis and   significance*. John wiley and Sons, New york.

Wolcott, M.J. 1991. DNA-based rapid methods for the detection of food borne pathogens. *J. Food prot.* 54: 387-401.

Immunology, Defense Mechanism, Immunity and Diagnosis

19

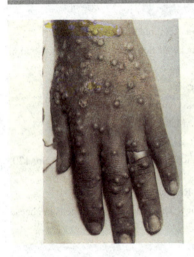

Edward Jenner

A large number of pathogenic microorganisms are present in the environment that continuously interact to our body. But our body's defense system prevents them firstly from entering into the skin, and secondly inactivating them if, however, entered inside. Thus, the defense mechanism involves at two levels, one if they have not infected, and the second after infection in body. Ability of our body toward off disease through these mechanisms is known as *resistance,* and lack of resistance is known as *susceptibility.* Therefore, the body's resistance can be divided into two broad groups: the non-specific resistance, and the specific resistance or immunity (against a particular microorganism).

A. Historical Perspectives

The phenomenon of immunity (Latin *immunis* means 'exempt') can be traced back to Thucydides, the great historian of Peloponnesian War. In 430 BC he wrote that a plague spread in Athens, only those could nurse the sick who recovered from the plague and could not contract the disease second time. In the 15th century the first recorded attempts were made by Chinese and Turks to induce immunity deliberately through a technique called **variolation.** In this method dried crusts from smallpox pustules were either inhaled or inserted into small cuts in the skin. In 1718, Lady Mary Wortley Montague (wife of a British Ambassador) performed variolation on her own children and observed its positive effect. In 1798, Edward Jenner, the English physician significantly improved the method. He introduced the fluid from a cowpox pustule into the milk maids and others having smallpox and protected them. He intentionally infected a 8-year old child with fluid from a cowpox pustule and then smallpox. The child did not develop smallpox.

A major advancement was made in the area of immunology by Louis Pasteur. He cultured pathogenic bacteria causing cholera and demonstrated that chickens injected with cultured bacteria developed cholera. Surprisingly, when he injected the chickens with old culture of cholera, the chickens became ill; thereafter they got recovered. When such chickens were injected with fresh culture of cholera, the chickens were completely protected from the disease. Pasteur put forth a hypothesis that *the ageing has weakened the virulence of the pathogen.* Hence such **attenuated** or weakened strain might be administered to protect the disease. He called such attenuated strain of pathogens as **vaccine** (Latin *vacca* means cow) in the honour or Jenner's work with cowpox inoculation.

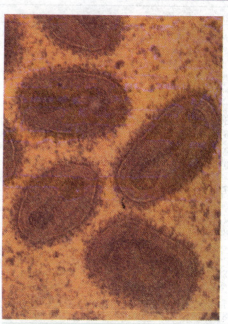

Electron micrograph of the smallpox virus

Pasteur extended his findings to other diseases by attenuating the pathogenic strains. In 1881, Pasteur vaccinated one group of sheep with heat-attenuated anthrax bacillus (*Bacillus anthracis*). Then he challenged the vaccinated sheep and some unvaccinated sheep with a virulent culture of the *Bacillus*. He noted that all the vaccinated sheep survived. This experiment became a milestone in the area of immunology. In 1885, Pasteur administered his first vaccine to a young boy bitten by a rabid dog who survived and later became the custodian at the Pasteur Institute.

In 1890, Emil von Behring and S. Kitasato first gave the mechanism of immunity. In 1901, von Behring was awarded with Nobel prize in medicine. He demonstrated that the **serum** (non-cellular component of coagulated blood) from animals, previously, immunised to diphtheria, transferred the immune state to unimmunised animals. The active state of immune serum could neutralise the toxin of bacteria, hence it was also named as *antitoxin*. Developments

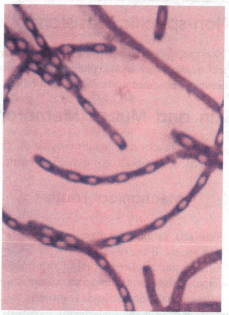

Bacillus anthracis, the heat attenuated strain of which was used by Pasteur for Vaccinating sheep

made during 20th century is most significant. The names of Nobel prize winners in the area of immunology are listed in Table 19.1

Table 19.1 : Nobel prize winners in the area of immunology.

Year	Recipient	Country	Research work done
1901	Emil von Behring	Germany	Serum antitoxin
1905	Robert Koch	Germany	Cellular immunity to tuberculosis
1908	E. Metchnickoff	Russia	Role of phagocytosis in immunity
	P. Ehrlich	Germany	Role of antitoxin in immunity
1913	C. Richet	France	Anaphylaxis
1919	J. Border	Belgium	Complement-mediated bacteriolysis
1930	K. Landsteiner	USA	Discovery of human blood group
1951	M. Theiler	South Africa	Development of yellow fever vaccine
1957	D. Bovet	Switzerland	Antihistamine
1960	F. Macfarlane Barnet,	Australia	Discovery of acquired immunological
	P. Medawar	Britain	tolerance
1972	R.R. Porter	Britain	Chemical structure of antibodies
	G.M. Edelnan		
1977	R.R. Yalow	USA	Development of immunoassay
1980	George Snell,	USA	Major histocompatibility complex
	J. Daussct,	France	
	B. Benacerraf	USA	
1984	C. Milstein,	Britain	Monoclonal antibodies
	G.E. Kohler	Germany	
	N.K. Jerne	Denmark	
1987	S. Tonegawa	Japan	Gene rearrangement in antibody production
1991	E.D. Thomas	USA	Transplantation immunology
	J. Murray	USA	
1996	P.C. Dohert	Australia	Role of major histocompatibility complex in
	R.M. Zinkernagel	Switzerland	antigen recognition by T cells

B. Non-specific Resistance

Non specific resistance is the defense of our body from any kinds of the pathogens. It includes skin and mucous membrane, phagocytosis, fever, inflammatory response, and the production of antimicrobial substances other than antibodies (Tortora, 1989).

I. Skin and Mucus Membranes

Skin and mucus membranes provide the first step of defense to the body against invasion of the pathogen. It acts both as mechanical barrier as well as chemical factors.

1. Mechanical Factors

Skin acts as an outer barrier of keratinized epithelium to microorganisms, chemicals and nonliving agents. It consists of over 15% of dry weight of the body. It contains two portions, the dermis (inner and thicker portion of skin) and epidermis (the outer thinner portion influenced by the external environment). Epidermis comprises of tightly packed layers of epithelial cells. The upper layer of epithelial cells is dead. It protects the inner tissues. As a result of cuts, burns, wounds, etc. infection of skin and underlaying tissues frequently occurs. When the skin frequently remains moist, the chances for skin infection by fungal pathogens get increased (Mc Nabb and Tomasi, 1981).

Mucus membranes lack the thickened layer but have the other features that provide defense. They line the gastrointestinal, respiratory, urinary and reproductive tract. The epithelial layer of mucus membrane secrets mucus which is a free moving liquid produced by globlet cells. It consists of inorganic salts, many organic molecules, loose epithelial cells and leucocytes. Mucous prevents the tract from dessication. Some pathogens *e.g. Treponema pallidum, Mycobacterium tuberculosis, Streptococcus pneumoniae*, etc. attached to mucus (if are in sufficient number) can penetrate the membrane. Mucus offers less protection than the skin (Mc Nabb and Tomasi, 1981)

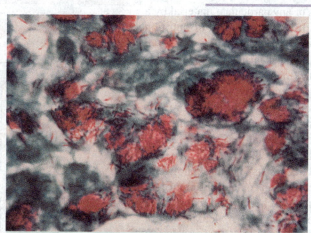

(*Mycobacterium*, that can peneterate the mucus membrane)

(*i*) Lachrymal Apparatus. Lachrymal apparatus is found in eyes and also associated with defense against eye infection. It forms and drains away the tears. Lachrymal gland is present towards the upper and outermost part of both the eye socket. This gland produces tears which is spread over the surface of eye ball through blinking. Continuous washing action protects the eyes from setting on eye surface. Whenever microorganisms come in contact of surface of eyes the lacrimal glands start secreting tears heavily and either dilute or wash away the microorganisms or irritating substances on eye surface.

(*ii*) The Other Glands. Salivary glands produce saliva that also wash microorganisms from teeth and mucus membrane of mouth. Similarly cleansing of urethra and vagina by the flow of urine and vaginal scretion, respectively also wash microorganisms from the respective sites and provide some sorts of defense.

The mucus membrane of nose possesses mucus coated hairs that filter the air after inhaling, and trap microorganisms, dust, etc. However, the cells of mucus membrane of the lower respiratory tract are covered with dust and microorganisms which have been trapped towards the throat. This so called ciliary escalator keeps the mucus blanket moving towards the throat at the rate of 1-3 cm/h. After coughing or sneezing the escalators speed up (Tortora, 1989).

2. Chemical Factors

There are certain chemical factors of skin and mucus membrane that play roles in providing defense such as gastric juice, enzymes, sebum, etc. Sebaceous (oil) glands of the skin produce oily substance which is known as sebum. Sebum prevents the hair desiccation and becoming brittle, and form a protective film over skin. Sebum contains unsaturated fatty acids and to some extent acetic acid. Sebum inhibits the growth of microorganisms. This secretion lowers down the pH between 3 and 5, and arrest the growth of many microorganisms.

Skin also possesses sweat glands that produce perspiration. Perspiration removes the wastes and wash microorganisms from skin surface and maintains body temperature. Perspiration contains the enzyme lysozyme that dissolves cell wall of Gram–positive and a few Gram–negative bacteria. The other sources of lysozyme are saliva, mucus, tears, nasal secretion and tissue fluids.

Gastric juice is secreted by the glands of stomach. It contains HCl, digestive enzymes and a little amount of mucus. Very low pH (1.2) *i.e.* high acidity of gastric juice of stomach kills the bacteria and bacterial toxins. However, the enteric pathogens are protected by the food particles and, therefore, enter the intestine through the gastrointestinal tract.

II. Phagocytosis

Phagocytosis (means *eat* and *cell*) refers to ingestion of microorganisms or any particulate material by a cell. It is also a method of nutrition of some protozoa such as *Amoeba*, but the mechanism discussed here is related to the defense mechanism of body provided by white blood cells through phagocytosis. Before discussing the mechanism of phagocytosis we should learn the components of our blood.

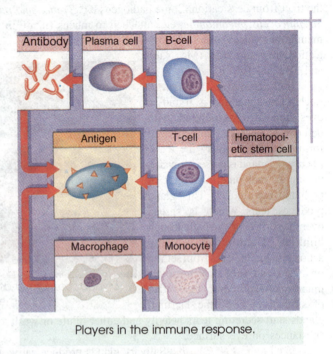

Players in the immune response.

Blood consists of fluid known as plasma which contains different constituents such as erythrocytes or red blood cells (RBC), leukocytes or white blood cells (WBC) and thrombocytes or platelets (Table 19.2). The leukocytes can be divided, on the basis of granules in their cytoplasm, into granulocytes and agranulocytes. Granulocytes contain three types of blood cells (e.g. neutrophils, basophils and eosinophils) and agranulocytes contain two types of cells (lymphocytes and monocytes) (Fig. 19.1).

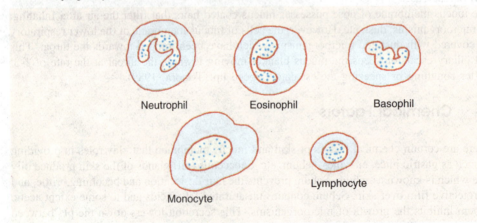

Fig. 19.1 : Different types of leukocytes; A, granulocytes; B, agranulocytes.

The granules of cytoplasm can be observed under the microscope. After staining these granules take different stains with a mixture of acidic (eosin) and basic (methylene blue) dyes the granules of neutrophils take red and blue stains respectively. Granules or basophils stain blue with methylene blue, and that of eosinophils stain red with eosin.

Table 19.2 : Classification of blood cells.

Types of cells	Number of cells/ml (million)	Function
Erythrocytes (red blood corpscells)	4.8 – 5.5	Transport of O_2 and CO_2
Leukocytes (white blood cells)	5000 – 9000	
A. Granulocytes		
(*i*) Neutrophils (Polymorphs)	60-70% of leucocytes	Phagocytosis
(*ii*) Basophils	0.5 – 1% of leucocytes	Production of heparin and histamine
(*iii*) Eosinophils	2 – 4% of leucocytes	Phagocytosis
B. Agranulocytes		
(*i*) Lymphocytes	20–25% of leucocytes	Antibody production
(*a*) B-lymphocytes	do	Cell mediated immunity
(*b*) T-lymphocytes	do	Phagocytosis
(*ii*) Monocytes	3 – 8% of leucocytes	
Thrombocytes (Platelets)	250,000 – 400,000	Blood clotting

Neutrophils can enter an infected tissue and kill microorganisms and foreign particles. The basophils can release substances like *heparin* (an anticoagulant) and histamine (in inflammation and allergic responses).

Eosinophils are the phagocytes. After microbial infection or hypersensitivity their number increases. In agranulocytes the granules are absent. These are of two types, lymphocytes and monocytes.

The lymphocytes are of two types, B-lymphocytes and T–lymphocytes. The B-lymphocytes derived its name from its site of maturation in the *bursa* of *fabricius* in birds. The name turned out to be apt for its major site of maturation in mammales in bone marrow. The B-lymphocytes depend on the activity of *bursa* tissues, whereas the T lymphocytes (derive its name from thymus) depend on the thymus for their activity. Thymus contains T cells but not B cells; similarly bone marrow consists of only B cells but not T-cells. Both the lymphocytes occur in lymphoid tissues (e.g. tonsils, lymph nodes, spleen, thymus gland, thoracic duct, bone marrow, appendix, lymph nodes in respiratory-gastrointestinal and reproductive tracts). The T- and B- cells cooperate in the presence of a third cell, Mechnikov macrophage. These provide immunity.

Monocytes mature into macrophages and act as phagocytes. The leukocytes are derived from stem cells in bone marrow and enter the lymph system (lymph node, spleen, thymus, etc.).

1.   Types of Phagocytes

The phagocytes are of two types, granulocytes (microphages) and monocytes (macrophages). When a microbe infects granulocytes (neutrophils), monocytes move to the infected area. During migration, monocytes enlarge in size and called macrophages. Since these macrophages are migratory, they are also termed as wandering macrophages. Some macrophages remain at a fixed position e.g. in the liver (Kupffer's cells), lungs (aleolar macrophages), nervous system (microglial cells), branchial tissue, bone-marrow, spleen, and lymph nodes and peritoneal cavity surrounding abdominal organs. These macrophages are called fixed macrophages which constitute the mononuclear phagocytic system.

2. Mechanism of Phagocytosis

After infection the number of WBC increases in blood during the initial phase of infection.

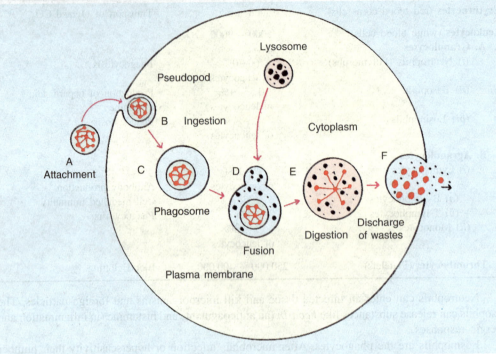

Fig. 19.2 : Mechanism of phagocytosis. A, attachment; B, ingestion; C, phagosome; D, fusion; E; digestion; F, discharge of wastes.

At this stage they are phagocytic in nature. As the infection progresses the number of monocytes increases. They phagocitize the remaining dead or living microbial cells. When blood and lymph containing microbial cells pass through the organs with fixed macrophages, the cells of mononuclear phagocytic system kill them through phagocytosis. The mechanism of phagocytosis can be divided into the following four steps (Fig. 19.2) (Wood, 1951).

(i) **Chemotaxis.** It is a phenomenon of chemical attraction of phagocytes to microorganisms. The chemotactic chemicals which attract the phagocytes are the components of WBC, and damaged cells, peptides derived from complements and microbial products.

(ii) **Attachment.** The plasma membrane of phagocyte gets attached to the surface of a microbe or foreign material (Fig. 19.2 A). When there is a large capsule M protein attachment can be hampered. For example M protein of *Streptococcus pyogenes* inhibits the attachment of a phagocyte to their site. Similarly, *Klebsiella pneumoniae* and *Streptococcus* possess a large capsule and get escaped. However, the large sized microorganisms or foreign material is trapped in blood clots, blood vessels or fibres of connective tissues. If the cell wall of microorganisms is coated with certain plasma protein promoting the attachement of microbe to phagocytes, only then they can be phagocytized. The coat proteins are called *opsonins* and the process of coating of plasma protein is known as *opsonization*.

(iii) **Ingestion.** After attachment the plasma membrane of phagocyte extends short projections known as pseudopods which engulf the microorganisms or foreign materials. This process is known as ingestion (Fig. 19.2B). The extension of pseudopods continues until they contact and fuse, and surround the microorganism inside a sac which is known as phagocytic vacuole or phagosome (C).

(*iv*) **Digestion.** After engulfment phagosome comes in the contact of lysosome that contains digestive enzymes and bactericidal chemicals (C). After making contact the membrane of phagosome and lysosome gets fused (D) and a single layered large structure is formed which is called phagolysosome (E). Within 10-30 minutes the contents of phagolysosomes degrade the microorganisms or foreign materials. Lysosomes also contain lysozyme that breaks the peptidoglycan of bacterial cell wall. Lysozyme is more active at pH 4 which is an optimum pH of phagolysosomes because of production of lactic acid by phagocytes. In addition, lysozyme also contains myeloperoxidase which binds with chloride ions to viruses and bacteria and finally kills them. After complete digestion of the foreign material the phagolysosome migrates towards the boundary of membrane and discharges the wastes (F) (Wood, 1951).

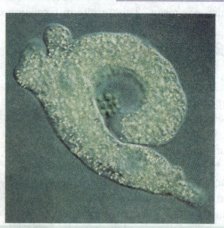

Amoeba, a protozoan with pseudopods which contact and fuse surrounding it inside a sac known as phagocytic vacuole.

Interestingly, toxin producing streptococci can kill the phagocytes and *Mycobacterium tuberculosis* can multiply within the phagolysosome itself and desintegrate the phagocytes. Also, the causal organism of brucellosis can remain dormant for several months or years inside the phagocytes. At this situation the role of immunity becomes vital (Wood, 1951).

III. Inflammation

As a result of damage of body tissues, inflammation in the surrounding areas occurs. However, there are several causes of tissue damage such as physical agents (i.e. heat, radiant energy, electricity or sharp objects) infection of pathogens, chemicals (acids, bases, gases), etc. Therefore, mainly four symptoms characterize inflammation viz., swelling, pain, redness and heat. For our system the inflammatory responses are beneficial and have the following functions:

(*i*) Inflammation possibly destroys the harmful agents and removes them or their by-products from the infected site.

(*ii*) If the harmful agents are not destroyed, it wards off the injurious agents and its bye products.

(*iii*) It repairs or replaces the tissues damaged by the injurious agents or their bye-products.

Inflammation process occurs in the three stages, vasodilation and increased permeability of blood vessels, phagocyte migration, and repair (Collier, 1962).

1. Vasodilation and Increased Permeability of Blood Vessels

After the damage of tissue, blood vessel is dilated where damage has occurred. Permeability of blood vessels also increases. As a result of vasodilation (*i.e.* increase in diameter of blood vessels) flow of blood to damaged area is increased. This is the reason why damaged area turns into red, and inflammation is induced due to heat.

Vasodilation is caused by histamine, a chemical released from the damaged tissue due to injury. In blood plasma another group of chemical (kinin) is present which too causes vasodilation. Collier (1962) has discussed the role of chemical mediators (kinin) in inflammation. Kinin after being activated attracts neutrophils to injured area. From the damaged cells a substance, prostaglandin is secreted which is also associated with vasodilation. Due to increase in permeability of blood vessels, the clotting factors are delivered to the injured area where blood clots prevent the growth of microorganisms. This results in formation of pus in a localised spot (Collier, 1962).

2.   Phagocyte Migration

Bretscher (1987) has given a comprehensive account of movement of phagocytes. Within an hour of inflammation, phagocytes (neutrophils and monocytes) appear and begin to stick on the inner surface of lining (endothelium) of blood vessel as the flow of blood gradually starts decreasing. The process of sticking of phagocytes is known as *margination.* Thereafter, the second phenomenon, diapedesis occurs within two minutes. Diapedesis is a process of sneezing of phagocytes between the endothelial cells of blood vessels and reaching to damaged area.

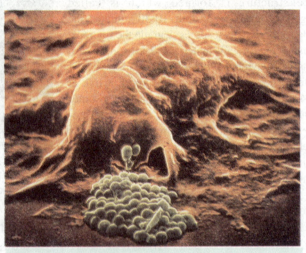

(Phagocytosis by a macrophage : This scanning electron micrograph (×3,000) shows a macrophage devouring a colony of bacteria).

Attraction of neutrophils occurs through chemotactic substances such as kinins, the components of complement system and secondary metabolites of microorganisms.

Production and release of granulocytes from bone marrow ensures the steady stream of neutrophils. Monocytes follow the granulocytes into the infected area as inflammation continues. In the early stage of infection granulocytes predominate but they are short lived. When monocytes are produced in tissue they undergo changes and become wandering macrophages which predominate during later stages. They are several times larger and potential enough to phagocytize the damaged tissue, destroyed granulocytes and infectious microorganisms. After phagocytosis, the granulocytes or macrophages themselves die. After a few days the area contains dead phagocytes, damaged tissue and fluid which collectively are known as pus. Pus formation subsides later on, and gradually destroyed after a few days (Bretscher, 1987).

3.   Repair

Repair is a process through which the tissue replaces the dead cells at the end of inflammation. During the active phase of inflammation repair starts but completes after removal of dead or damaged cell, the ability of which depends on the tissues involved. For example, skin has a high capacity for regeneration, whereas nervous tissues in the brain and spinal cord do not regenerate at all. When the stroma (supporting connective tissue) or parenchyma (functioning part of tissues) produces new cells, the damaged tissue is repaired (Tortora *et al*, 1989).

IV. Fever

An abnormally high body temperature is known as fever which is caused by bacterial or viral infection or bacterial toxins. It is obvious that hypothalamus (a part of brain) controls the body temperature and, therefore, some times it is called the body's thermostat as it sets the temperature normally at 37°C (98.6°F). When antigens affect hypothalamus, body's temperature goes up. For example, when phagocytes ingest the Gram–negative bacteria, the lipopolysaccharide of bacterial cell wall i.e. endotoxin is released that induces the phagocyte also to release interleukin-1 (endogenous pyrogens). Interleukin-1 helps the production of T-lymphocytes. In turn, interleukin induces-hypothalamus to produce prostaglandins that result the hypothalamus to a higher temperature that causes fever.

Fever perists for a long duration until bacterial endotoxin or interleukin-1 is released. At high temperature the body responds with constriction of blood vessels, increased rate of metabolism and shivering (chilling). Chilling disappears after body's temperature has reached the setting of thermostat. Until the endotoxins are not completely removed, body's temperature remains high. Thereafter, it is maintained normally at 37°C.

In addition, fever inhibits the growth of some microorganisms in body. At high temperature body's tissue repairs quickly, and the effect of interferon is intensified.

V. Antimicrobial Substances

Besides certain chemical factors described earlier, some antimicrobial substances (e.g. proteins of the complement, and properdin systems and interferon) are also produced by the body after microbial infection.

1. Complements and Properdin

In classical antigen-antibody complex, certain blood proteins also get associated and complement the immune response. These serum proteins are known as complements. Similarly three serum proteins (e.g. properdin itself, factor B and factor D) which are commonly known as properdin, play a role in alternate pathway. Both types of proteins are related to the defense system. Properdin system is composed of the above three serum proteins which altogether constitute a high proportion of serum protein.

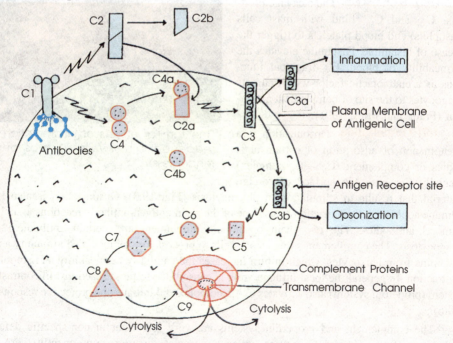

Fig. 19.3 : Pathway of complement and properdin activation.

Therefore, about 20% different types of proteins (*i.e.* complements) which are found in normal blood serum. These are designated as C_1, C_2, C_3, etc. Complements are very important to both non-specific and specific defense against the microbial infection. Proteins of complement and properdin systems act in ordered sequence or *cascade*. The classical pathway initiates when the antibodies bind with antigens (bacteria or other microbial cells). After a pair of antibodies recognise and bind to antigens, the C_1 protein (which consists of 3 protein subunits) binds to antibodies and activated (Fig. 19.3). In turn C_1 acts as an enzyme, activates C_2 and C_4, and splits C_2 and C_4 proteins (C_2 into C_{2a} and C_{2b}, and C_4 into C_{4a} and C_{4b}. C_{2a} and C_{2b} combines to form another enzyme that splits C_3 into C_{3a} and C_{3b} (Ross, 1986).

The antibodies are not involved in the initiation of the alternate pathway, but interactions between protein properdin system and certain polysaccharides initiate this pathway. The polysaccharides are found on the cell wall of most of the bacteria, fungi and foreign RBCs of mammals. Properdin pathway interacts with Gram–negative bacteria, the cell wall of which contains lipopolysaccharide that releases lipid A (an endotoxin) and trigger the alternate pathway (Ross, 1986).

C_3 is cleaved both by classical and alternate pathways into C_{3a} and C_{3b}; C_{3a} is an active fragment. These fragments induce the three processes, cytolysis, inflammation and opsonization.

(*i*) **Cytolysis.** It is a process of leaking of cellular contents of foreign cells through breaking their plasma membrane by the complements. C_{3b} initiates a series of reactions involving C_5, C_6, C_7, C_8 and C_9 which is collectively known as the membrane attack complex (MAC) (Fig. 19.3). The activated proteins attack the microbial cell membrane and form a circular transememembrane chanels (lesions). Through these lesions loss of ions and cytolysis occur. Use of the complements in this process is known as complement fixation which laid a basis for clinical test.

(*ii*) **Inflammation.** The cleavage products, C_{3b} and C_{5b}, bind with mast cells (basophils) and blood platelets to trigger the release of histamine. Histamine elevates the permeability of blood. C_{5a} fragment functions as a chemotactic factor which attracts phagocytes to the site of complement activation (Frank and Fries, 1991).

(*iii*) **Opsonization.** Opsonization is a phenomenon of adsorption of certain antibodies or complement ($C_3 - C_5$ complex) specially C_{3b} onto the surface of foreign

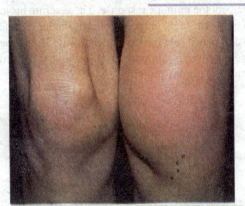

(The familiar effects of inflammation : reddened skin, swollen tissues and tenderness).

material that results in stimulation in phagocytosis (Fig. 19.3). Opsonization is also known as immune adherance. Opsonization is also one of the main antigen-antibody reactions associated with humoral antibodies. The two main opsonins (complement and certain antibodies) stimulate phagocytosis. The complement stimulates T-cells to process for cell mediated immunity and release histamine from leukocytes, which in turn increases the capillarly permeability and smooth muscle contraction. In general the local inflammation is caused due to these reactions. In contrast, another system (properdin system) also activates C_3-C_5 complex and initiates the protective responses (Ross, 1986).,

The complement and properdin systems are very important in non-specific defense. The deficiency of C_1, C_2 and C_4 causes collagen vascular disorder, consequently there develops hypersensitivity. C_3 deficiency increases susceptibility to bacterial invasion and C_5 deficiency (through C_9) causes susceptibility to infection of *Neisseria meningitidis* and *N. gonorrhoeae*.

VI. Interferon

Viruses totally depend on their host cells for multiplication. However, during the course of multiplication the host cells may or may not be damaged. It is very difficult to check the virus multiplication without affecting the host cells. Interferons (IFN) are such class of antiviral proteins produced by certain animal cells after stimulation. Now-a-days interferons are used in causing immunity. Interferons are host specific but not virus-specific. It means that interferons produced by human cells will show antiviral activity only in humans but not in another mammals. In contrast 4 interferons produced against a virus will also act against a number of other viruses. Even in humans different types of cells produce different interferons. Human interferons are of the following three types:

(*i*) Alfa interferon (α-IFN or leucocyte IFN)

(*ii*) Beta interferon (β-IFN or fibroblast IFN), and

(*iii*) Gamma interferon (γ-IFN or immune IFN)

With each principal group there are various subtypes of interferons. In humans interferon is produced by fibroblast in connective tissues, lymphocytes and other leukocytes. The virus infected cells produce interferons in very low quantity which is diffused towards uninfected neighbouring cells. It reacts with plasma or nuclear membrane receptor and induce healthy cells to produce mRNA for the synthesis of the antiviral proteins. These proteins act as enzyme and disrupt translation of viral mRNA, polypeptide chain elongation, etc. Since interferon is in low quantity, it does not badly affect the host cells. Its effect remains only for a very short duration. Interferons do not have any effect on viral multiplication in cells already infected (Friedman and Finters, 1984).

Owing to its importance much emphasis is being laid on artificial production of interferon. For the first time clinical trial of interferon was done in 1981 to determine its anticancer effects. In recent years several companies have applied the recombinant DNA technology to produce interferon in certain bacteria.

C. Immunity

Immunity is the ability of body to specifically counteract with foreign organisms or substances. A person may develop or acquire immunity after the birth. The acquired immunity is not inherited but it is specific resistance to infection developed during the life of the individual. However, it results from the production of antibodies and sensitized lymphocytes.

I. Types of Immunity

Immunity is of two types, naturally acquired immunity and artificially acquired immunity.

1. Naturally Acquired Immunity

Naturally acquired immunity (NAI) is of two types, naturally acquired active immunity and naturally acquired passive immunity.

(*i*) **Naturally Acquired Active Immunity:** Naturally acquired active immunity is obtained when a person is exposed to antigens in the course of daily life and the immune system responds by producing antibodies and specialized lymphocytes. For some diseases immunity is life long for example, measles, chicken pox and yellow fever.

(*ii*) **Naturally Acquired Passive Immunity:** Naturally acquired passive immunity involves the normal transfer of antibodies from a mother to her infants. An expectant mother is able to pass some of her antibodies to her foetus across the placenta. This mechanism is called placental transfer. If the mother is immune to such diseases as diphtheria, rubella or polio, the newly borne infant will be immune to these disease. Certain antibodies also pass to her nursing infant to breast milk, especially in first secretion called colostrum. In the infant, generally the immunity costs only as long as the transmitted antibodies are active usually a few weeks or months.

2.   Artificially Acquired Immunity

Artificially acquired immunity is of two types, artificially acquired active immunity and artificially acquired passive immunity.

(*i*) **Artificially Acquired Active Immunity:** Previously prepared antigens are injected into the susceptible individual who produces antibodies and specialized lymphocytes. This process is known as *vaccination* or *immunization*.

(*ii*) **Artificially Acquired Passive Immunity:** It involves injection of immune serum in the susceptible individuals. A person bitten by snake might be injected with antibodies from a horse that is immune to snake venom.

II. Types of Immune System

There are two types of immune systems, (*i*) the humoral immune system, and (*ii*) the cell mediated immune system.

1.   The Humoral Immune System

The humoral imune system involves the antibodies that get dissolved in extracellular fluid such as blood plasma, lymph and mucus secretion. These were formally known as humors. The humoral immunity is conferred through B-cells that develop from stem cells of bone-marrow in adults and the liver in embryos. However, RBCs, neutrophils, macrophages and other types of WBCs are produced from the same stem cells. Origin of B-cells and T-cells is described earlier.

The specialized lymphocytes *i.e.* B-cells of this system responds when exposed to antigens. B-cells secrete antibodies in correspondence to antigens. This system responds mostly against bacteria (and their toxins) and viruses.

(*i*) **The Antigens:** The antigen (Ag) or immunogen is a large organic molecule capable of stimulating the production of specific antibody with which it may chemically combine. Usually this response involves the formation of antibodies or highly specialized T-cells. The ability of the antigens to induce antibody formation is known as *antigenicity*.

The nature of antigens: The majority of antigens are proteins, nucleoproteins (nucleic acid + proteins), lipoproteins (lipid + protein), glycoprotein (carbohydrate + protein) or large poly-saccharides. The above compounds are also the components of invading microorganisms: the capsules, cell walls, flagella, pili, and toxins of bacteria, bacterial coat and cell surfaces of many organisms.

The whole surface of antigen cannot be recognised by the antibodies because these can identify and counteract with specific regions of the antigen surface which is called antigen determinants. This counteraction depends on size and shape of antigenic determinant and the chemical nature of antibodies as well (Fig. 19.4 A).

There are several antigens containing different determinants attached to their surfaces. Therefore, different types of antibodies are required to recognise these determinants. Consequently,different types of antibodies are produced by our immune system. Size and shape of antigen determinants and the chemical nature of antibodies determines the nature of interactions.

There are many antigens which possess different types of determinants on their surface. The different determinants are identified by different antibodies, however, it is the immune system that may produce several antibodies against a single antigen. Most of antigens have molecular weight of 10,000 or more. The low molecular weight antigen is called hapten. It is too small to stimulate antibody formation. The haptens are not functional unless attached to a carrier molecule, usually a serum protein (Fig. 19.4B). Both function as antigen and provoke the immune response. If an antibody is formed against the hapten, the later reacts with antibody independently even in the absence of a carrier. For example, penicillin acts as α-hapten which is not antigen itself even then some people develop allergy against it. There are several antigens that contain more than one antigenic determinant sites. There may be different kinds of sites on one antigen. At least two binding sites may be present on a single antibody.

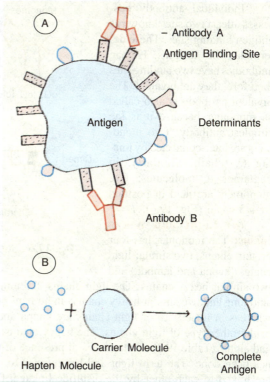

Fig. 19.4 : Antigens containing determinants as antigen binding site for antibodies (A), and formation of antigen (B).

(*ii*) **The Antibodies:** The antibodies belong to such a class of proteins that is called immunoglobulins (Ig). The Ig can identify and bind the antigens resulting in complete break down.

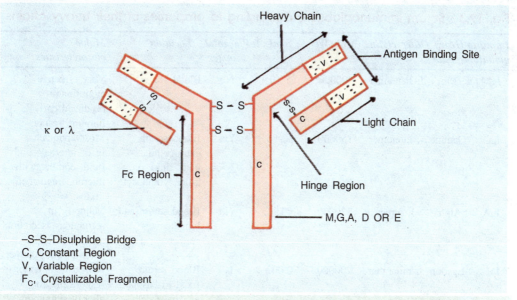

–S–S–Disulphide Bridge
C, Constant Region
V, Variable Region
F_C, Crystallizable Fragment

Fig. 19.5 : Primary structure of a typical antibody molecule (Y-shaped molecule).

A typical antibody is made up of four polypeptide chains, two light chains and two heavy chains (Fig. 19.5).

Individual antibody possesses atleast two sites known as antigen binding sites. These determines the antigens. Human antibodies have two binding sites therefore, they are bivalent. The bivalent antibody is also called as monomer because it is the simplest antibody. Two monomers are inter-connected by joining (J) chain. Similarly, in pentamer Ig molecules, five monomers are held in position by a J-chain (Fig. 19.6).

(*a*) **Light and heavy chains:** The monomer has four protein chains, two similar light chains (kappa and lambda) and

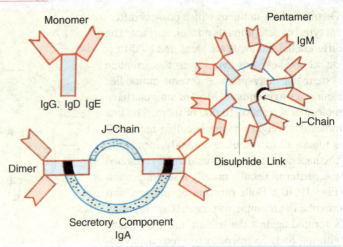

Fig. 19.6 : Structure of immunoglobulins: monomers, dimer and pentamer.

two similar heavy chains. The light chains 220 amino acid long and the heavy chains 440 amino acid long and two similar heavy chains. In humans 60% light chains are kappa and 40% are lambda, whereas in mice 95% of light chains are kappa and 5% are lambda. Each antibody contains one or the other type of light chains. The heavy chains are of 5 types viz., mu, delta, gamma, epsilon and alpha. (Table 19.3). On the basis of presence of these heavy chains, nomenclature of antibodies has been done. The term light and heavy refer to their relative molecular weights. These chains are linked to each other by the disulfide linkage. Due to linkage of light and heavy chains with disulfide and other bonds the monomer looks a flexible and Y-shaped structure. Heavy and light chains are folded into domains, each containing 110 amino acid residues and intra-chain disulfide bond that form a loop. It attains the three dimensional structure. The domains form discrete structural regions.

Fig. 19.3 : Serum immunoglobulins according to properties of their heavy chains

Immuno globulins	Heavy chain	Other chains	Mol. Wt	% total Ig	Placental transfer	Location	Major characteristics
IgM	Mu	J, Pentamer	900000	5-10	No	Blood serum	Very effective agglutinator, produced early in immune response
IgG	Gamma	-, monomer	150000	80-85	Yes	Blood serum, extra-cellular body fluid, on killer lymphocyte	Most abundant Ig of internal body fluid, combat with microbes and their toxins.
IgA	Alpha	J, dimer	320000	15	No	Blood serum and lymphocytes surfaces	Major Ig in seromucus, secretion, and on defends external body helper
IgE	Epsilon	-, monomer	200000	0.002	No	Blood serum fixed to basophils and most cells	Raised in infection responsible for allergy symptoms.
IgD	Delta	-, monomer	185000	0.2	No	Blood serum and on lymphocytes in new borns	Transient present on lymphocyte surface.

(*b*) **Variable regions:** The terminal ends (100-110 amino acids) of both the heavy and light chains i.e. Y arms are of variable (V) in nature because of changes in amino acid sequence at the ends. Therefore, these ends are called V-regions. Thus, two V-regions (one of heavy chain and the other of light chain) form one antigen binding site. On each antibody two such sites are located. Hence a single antibody molecule contains only one type of light chain and one type of heavy chain inspite of being made up of more than one monomer. This shows that one antibody possesses only one type of antigen binding site (Alzari *et.al*, 1989).

However, more than one antibody can combine with an antigen. The antigens can be aggregated into clumps, if two antigen binding sites on an antibody combines with antigenic determinants on two different antigens. These clumps can be an important factor in diagnosis of a few diseases. (Alzari *et al*, 1988)

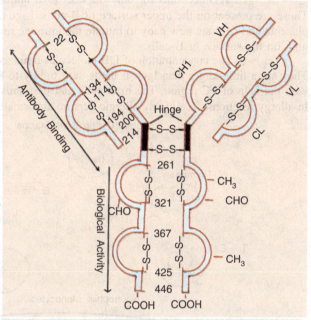

(*c*) **Constant region:** The lower part of Y-shaped monomer antibody is called constant (C) region. The term C refers to relatively invariable feature of amino acid sequence of both the heavy and light chains. The stem of Y-shaped monomer is known as FC (crystallizable fragment) region. An antibody molecule attaches to its host at FC region. Based on sequence of C regions, five different types of sequences of heavy chain and two sequences of light chain have been categorised. A different class of immu-

Fig. 19.7 : Secondary structure of immunoglobulin derived from amino acid sequencing.

noglobulin is determined by each heavy chain sequence. The length of C region is about 330 amino acids for α, γ and δ, and 440 amino acids for μ and ε (epsilon) (Fig. 19.7).

(*iii*) **Immunoglobulin (Ig) Isotypes:** There are five classes of Ig molecules such as IgG (gamma), IgM(mu), IgA (alpha), IgD (delta) and IgE (epsilon) (Fig. 19.6). The heavy chains have been named with greek notations. Each class of Ig molecules play a different immune response. Special features of different Ig molecules are given in Table 19.2 and briefly described herewith.

(*a*) **IgG:** The antibodies account for 50-80% of total antibodies present in serum. The maternal molecule can pass the placenta and provide passive immunity to the foetus. These can also pass the walls of blood vessels and enter in tissue fluid. These can bind to bacteria and viruses, and also can neutralize the toxins secreted by them. After binding with antigens, these enhance the effectiveness of phagocytic cells to engulf and ingest them.

(*b*) **IgM:** About 5-10% IgM molecules are found in serum. It has a pentamer structure of the antibody (Fig. 19.6). When exposed to antigens, it is the IgM molecule which appears first. In the beginning the concentrations of IgM molecules in blood declines and that of Ig increases. After a second exposure to antigens the concentration of IgG molecules is increased in blood serum.

IgM molecules are especially effective at cross linking particulate antigens and causing their aggregation because of its numerous antigen-binding site. It can increase the digestion of target cells by the phagocytic cells as IgG. IgM molecules predominate and is involved in ABO blood group antigens on the surface of RBCs.

After infection of the pathogens, IgM appears first. Due to short life it is valuable in disease diagnosis. The high concentration of IgM against a pathogen in blood of a sick person denotes that the disease is really caused by the pathogen.

(c) **IgA:** The concentration of IgA in blood serum remains about 15% of total antibodies. They are found in body secretions e.g. saliva, sweat and secretion from gastrointestinal tract and colostrum as well. During their transport from blood to secretary tissues, Ig A gets attached to a protein caused secretary component, which protects IgA from enzymatic degradation, and facilitates its entry into secretary tissues. It checks the attachment of pathogens to mucosal surfaces, and protects the gastrointestinal tracts of infants from infection.

(d) **IgD:** IgD accounts for only 0.2% of total antibodies of serum. It resembles with IgG. These are present on the upper surface of B-cells. Like others, they also cannot pass across the placenta and help the new baby to initiate the immune response as their population remains very high on the surface of B-cells.

(e) **IgE:** The concentration of IgE molecules remains around 0.002% of the total antibodies. They are a little larger than IgG. It binds very tightly to the receptors (mast cells and basophils) with the help of FC region. The mast cells and basophils are the specialized cells that take part in allergy reactions. In a highly allergic person, abnormally a high concentration of IgE is found.

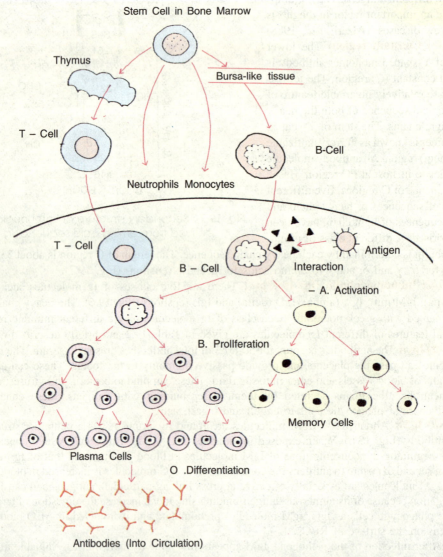

Fig. 19.8 : Differentiation of B-cells and T-cells from stems of bone marrow or foetal liver, interaction of antigen to B-cells and production of antibodies by B-cells.

(*iv*) **Mechanism of Humoral Immunity:** The mechanism of humoral immunity is accomplished after interaction of antigen with B-cells, antibody production and antigen-antibody binding.

(*a*) **Interaction of antigen with B-cells:** The B-cells after production from the stem cells (Fig. 19.8) migrate to lymphoid organs. When antigen interacts and contacts the receptor, the later gets stimulated. The B-cells differentiate into antibody producing cells through three steps: activation, proliferation and differentiation (Fig. 19.8 A-C). Moreover, from the beginning it passes through several stages such as lymphoid stem cell, progenitor B-cell, pre-B-cell, immature B-cell, mature B-cell, activated B-cell, plasma cell and memory cell. The plasma cells divide to produce a large number of genetically identical cells known as clones. Some of the progeny cells are differentiated into the antibody producing cells i.e. plasma cells. The plasma cells secrete antibodies specific to antigens. Some of the other B-cells act as memory cells (Fig. 19.8B). Each plasma cell is capable to produce about 2,000 antibodies per second but they remain alive for a few days only.

(*b*) **Production of antibodies:** Antibody production may or may not depends on T-antigens. **Antibodies production against T-dependent antigens:** Antibody production by B-cells is also cooperated by the other cells such as T-cells and macrophages. For example, B-cells produce antibodies in cooperation with T-cells against a type of antigen known as T-dependent antigen (Fig. 19.9). The T-dependent antigen consists of bacteria, some of proteins, RBCs, etc. The macrophages or dentric cells (highly branched cells) play a significant role. The B-cells need the help of antigen presenting cells (APC) (i.e. the macrophages and dentric cells that attack the T-dependent antigens and partically digest them) and specialized helper T-cells to produce antibodies against T-dependent antigens. This process is accomplished in the following steps:

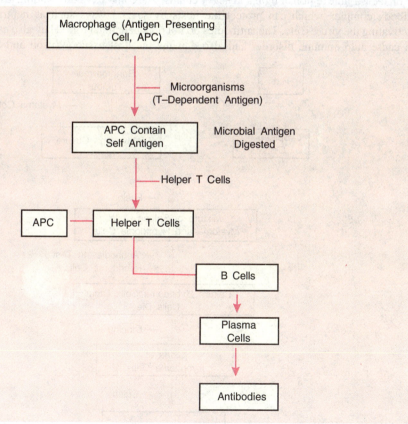

Fig. 19.9 : Activation of B-cells by the helper T-cells to secrete antibodies against T-dependent antigens.

The APC attaches the T-dependent antigens and partially digests them. The APC usually takes up the polypeptide fragments of the antigen and keep them on the cell surface.

The specific T-cells which are known as helper T-cells interact with the APC. The antigen fragments and certain self' antigens are recognised by helper T-cells and carry them on the surface of APC. The self antigens are actually the cell surface proteins that are the components of the major histocompatibility complex. They serve as the self antigens, the signals by which the immune system distinguishes the self from non-self (Hood *et al,* 1984).

The foreign antigen is recognised by the helper T-cells only when it gets combined with a self antigen on the surface of APC. This results in stimulation of suitable B-cells. For that the possession of self antigens by B-cells is very necessary. Helper T-cells are required by all immune responses that induce IgG, Ig and IgE. However, a single complex of the antigen and the self antigen present on the cell surface of B-cells are recognised by T-cell receptors (Hood *et.al,* 1984).

Antibody production against T-independent antigen: The T-independent antigens, without the interference of T-cells, can also induce a response from B-cells. The T-independent antigens contain the repeating subunits of polysaccharide or protein, for example bacterial flagella (composed of proteins) and lipopolysaccharide layer of Gram-negative bacteria. Multiple bonds are formed with B-cells by these antigens. But the immune response by these antigens is weaker as compared to T-dependent antigens. They bind to B-cells which produce antibodies only of the class IgM.

(c) **Antigen-antibody binding:** Lerner and Tramontano (1988) have nicely described how antigen binding sites or antibodies react with antigens. Specific antibody recognises the specific antigen. This depends on the ability of antibody. However, each antibody consists of antigen binding site on the variable region of light and heavy chains. The complex of this binding is known as antigen-antibody complex which can protect the hosts by several ways such as neutralizing the toxin, inactivating the viruses, etc. The antibodies do not play beneficial role. They also can initiate allergy, can cause auto-immune disorder and also damage the host tissue (Wilson and Stanfield, 1993).

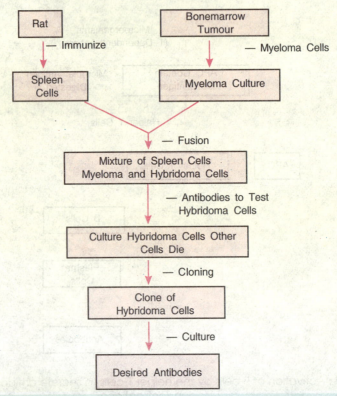

Fig. 19.10 : Outline of production of monoclonal antibodies.

(*v*) **Monoclonal Antibodies:** So far the production of a large number of antibodies in response to antigens has been described. All the antibodies are mixed in serum which are the product of a large number of clones of B-cells. Therefore, they are termed as polyclonal antibodies. When a single clone of B-cells produces antibodies, all the antibodies are alike and, therefore, called monoclonal antibodies. Monoclonal antibodies are produced by B-cells. But the B-cells do not proliferate on artificial medium (Milstein, 1980).

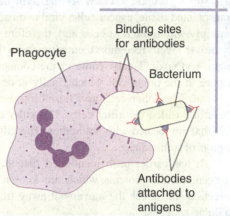

For the first time Kohler and Milstein (1975) got success in producing monoclonal antibodies from the hybrid B-cells. The hybrid B-cells were obtained after fusing the myeloma cells (cancerous cells) with B-cells of spleen obtained after injecting the antigen in mouse (Fig. 19.10). Myeloma cells have the feature of cell division and B-cells have the characters of antibody production. Therefore, the hybrids could successfully proliferate on the artificial media, and produce antibodies. After a succes-

Pathogens such as bacteria are covered by antigens which stimulate B-cells to produce monoclonal antibodies)

sive subculturing of hybrids and clone selection, pure line hybrids was obtained. This new technology of Kohler and Milstein (1975) for production of hybrid cell line is known as *hybridoma technology*. In recent years, through the hybridoma technology antibodies of desired properties can be produced. The monoclonal antibodies are very useful because (*i*) they are uniform, (*ii*) they are highly specific, and (*iii*) they can be quickly produced in adequate amount. For a detailed account readers may consult the book *Biotechnology* by R.C. Dubey (2005).

2. The Cell Mediated Immune System

The cell mediated immune system directly involves the specialized lymphocytes known as T-cells. After differentiation, the T-cells migrate to lymphoid organs. They do not secrete antibodies but they contain antibodies like molecules called antigen receptors which is attached to their surfaces. Several kinds of T-cells are found. This system of immunity is most effective against bacteria or viruses when present within the phagocyte or even infected host cells, or in infected host cells against protozoa, fungi and helminths, transplanted tissues and cancer (Tortora *et al*, 1989). The receptors help the T-cells to interact with a variety of antigens. After recognizing the antigens, T-cells differentiate in a variety of effector T-cells. Only the effector cells recognise the antigen and regulate the immune system.

(*i*) **Types of Effector T-cells:** The effector T-cells are of three main types such as (*i*) helper T-cells, (*ii*) suppressor T-cells, and (*iii*) cytotoxic T-cells.

(*a*) **The helper T-cells:** The helper T-cells play a variety of important roles. They contain a surface antigen which provide a vigorous immune response. Some helper T cells provide T dependent antigens to B-cells, and some others (e.g. delayed hypersensitivity T-cells) are associated with certain allergic reactions and rejection of transplanted tissues. When the antigen interacts, the delayed hypersensitivity T-cells secrete *lymphokines* that recruit defense cells like macrophages. The delayed hypersensitivity T-cells also defend the body against the development of cancer.

(*b*) **Suppressor T-cells:** Not much information is available on suppressor T-cells but the general concept is that they check the conversion of B-cells into plasma cells as well as the T-cells. Hence, they check the immune response, and help the body to develop tolerance. Therefore, both the helper cells and suppressor T-cells are known as regulatory T-cells as they regulate the immune response of the body against the antigens.

(*c*) **Cytotoxic T-cells:** As the term denotes, cytotoxic T-cells destroy the target cells e.g. transplanted tissue, cancer cells, viral and bacterial infections, etc. Viruses and some of the bacteria multiply within the host cells and, therefore, they rescue from the attack of antibodies. Hence, the cytotoxic T-cells recognise their antigens on the surface of host cells that produce viruses and destroy them. Cytotoxic T-cells come in the contact of host cell, release a protein (perforin) that makes a pore in the target cell which finally is destroyed.

(*d*) **The killer cells and lymphokines:** Some of the effector cells pose cytotoxic effect and therefore, called the killer cells. The killer cells are not very specific. They can invade any such cell that are located with antibodies. The killer cells contain the receptors which combine with FC region of antibodies.

It has earlier been mentioned that the delayed hypersensitive T-cells when stimulated by antigen, secrete proteins which are known as lymphokines. It attracts the macrophages to the infection site, check the movement away from the infection site and activates them to destroy the cellular antigens.

(ii) Mechanism of Cell Mediated Immunity

The T cells are very specific (like B-cells) to only a specific antigen. But unlike B-cells, the T-cells do not respond to antigens. It responds to the antigens present on cell surfaces. However, before showing the responses, the antigens on cell surfaces need to be processed by the antigen presenting cell (APC) (Fig. 19.11) as described earlier. Thus, the APC possesses antigens on its

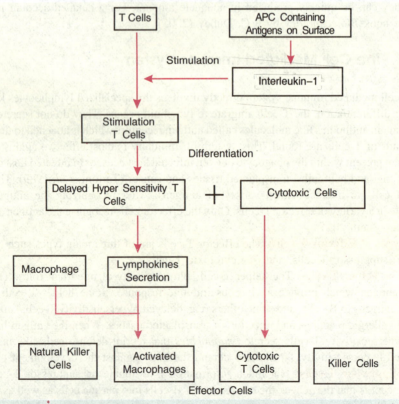

Fig. 19.11 : Outline of cell mediated immunity.

surfaces. The T-cells also respond against the MHC. The T-cells display the associative recognition i.e. the T-cells recognise the antigen only when it is in close association of an MHC antigen. An APC cell secretes interleukine-1 when stimulated by an antigen. Interleukin-1 is a monokine, a biologically active substance secreted by macrophages that activates the T-cells. The activated T-cells in turn secrete interleukin-2 and surface receptors for interleukin-2 which binds the former. After the surface receptor binds to the interleukin-2, the T-cells start proliferating and differentiating into different types of effector cells, cytotoxic T-cells, killer cells, natural killer cells and activated macrophages (Fig. 19.11).

3. Genetics of Antibody (Organisation of Immunoglobulin Genes)

From the previous discussion, it has become clear that the B-lymphocytes of the immune system produce antibodies in the presence of antigen.

The antibodies are formed from the assembly of some protein chains. In mammals, enormous amount of antibodies are produced. For this, there must be millions of genes for each antibody. But how can it be possible because a mammalian genome does not contain more than about a million of genes, out of which only a fraction of genome directs the synthesis of antibodies. This clearly shows that neither the germ cells nor embryonic cells contain a complete set of all genes but have the basic genes which are shuffled during developmental stages of B-lymphocytes.

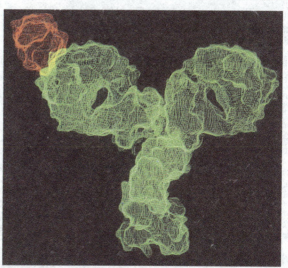

A computer generated 3-D image of an antibody molecule (green) bound to an antigen (red).

In germ line DNA, multiple gene segments encode a single immunoglobulin heavy or light chain. These gene segments are carried in the germ cells that cannot be transcribed and translated into heavy and light chains until these are arranged into functional genes. During the differentiation of B-cells in bone marrow these gene segments are randomly shuffled about 10^8 specificities by a dynamic genetic system. This is maintained by germ line theory. Differentiation of B-cells from a progenitor B-cell to a mature cell involves an ordered progress in rearrangement of immunoglobin genes. When the process of B-cell division is over, a mature immunocompetent B-cell contains a single functional variable region DNA sequence for its heavy chain and a single functional variable region DNA sequence for its light chain.

The somatic variation theories maintained that the genome contains relatively a small number of Ig genes which generate a large number of antibody specificity in the somatic cells through the mechanism of either recombination or mutation. Still these two theories could not explain how the stability be maintained in the C-region while some diversifying mechanism is involved to degenerate the V-region (Kuby, 1994).

This has been confirmed through the recent evidences that a single variable region sequence specific for a particular antigen can be associated with the multiple sequences of C-region of heavy chains. It means that the different isotypes of antibodies (*i.e.* IgG, IgM) can be expressed having identical sequences of V-region.

(*i*) Dryer and Bennett's two Gene Model: In 1965, W.Dryer and J. Bennett in their classical theoretical paper suggested for encoding of immunoglobulin chains. The two separate genes encode two different chains, one the light chain and the other heavy chain. They hypothesized that the two genes must come together and form a complete set of genes that can transcribe and translate the full message and can yield a single heavy or light protein chain. This hypothesis predicted the well established theory of one-gene-one polypeptide hypothesis of Beadle and Tatum for that they were awarded Nobel prize in 1958 with J. Lederberg. However, they could not provide the experimental data, it was merely a theoretical framework. After the development of technologies this hypothesis lent support to disclose the mystery of gene organization at molecular level.

(*a*) Experimental evidence of gene rearrangement: For the first time, Hozumi and Tonegawa (1976) provided the experimental evidence for the rearrangement of two separate genes encoding the V and C-regions of immunoglobulin during the course of differentiation of B-lymphocytes, and produce millions of antibodies. For this novel work, Tonegawa was awarded Nobel prize in 1987 in medicine and physiology.

They used the newly developed Southern Blotting Technique (see Fig. 11.9). They took myeloma cells because they are like Plasma cells and produce large amount of single antibodies, and prepared radiolabelled RNA *i.e.* ^{32}P-mRNA for κ-light and heavy chains, and also for constant chain. ^{32}P-mRNA was used as probe to test two kinds of cells, embryonic cells (that do not produce antibodies) and B-cells (produces antibodies) (Fig. 19.12)

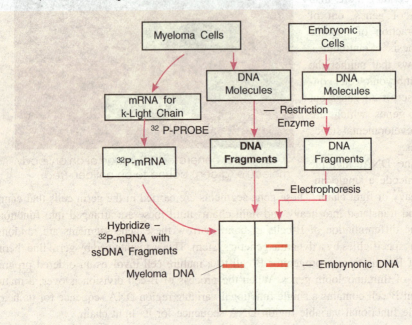

Fig. 19.12 : Experimental demonstration for rearrangement of genes encoding k-light chain.

The DNA of both myeloma cells and embryonic cells was treated with restriction enzymes and subjected to gel electrophoresis. The gel was then sliced, DNA fragments were eluted from the slice, denatured into single stranded DNA and finally incubated with ^{32}P-mRNA encoding κ-light chain. The ^{32}P-mRNA probe hybridised with two bands from the germ line embryonic DNA, but with only a single band from the differentiated myeloma DNA. This clearly reveals that in the fully differentiated plasma cells, which is represented by the myeloma cells, the genes for V and C regions had gone rearrangement. Now they are present together on a single restriction DNA fragment, that is why the ^{32}P-mRNA probe hybridized with a single band only.

(*ii*) **Multigene Organization of Immunoglobulin Gene:** The result of Hozumi and Tonegawa (1976) are analogous to the theoretical two gene model of Dryer and Bennett (1965). This provides evidence for organization of multigene family into the immunoglobulin gene. In the embryonic cells the DNA encoding C-regions is far away from the DNA that encodes for V-region. In plasma cells (*i.e.* cells producing antibodies, also B-cells) and C and V-regions are together (Fig. 19.13). The κ and λ light chains and the heavy chains are encoded by separate multigene families situated on different chromosome, that contain a series of coding sequences

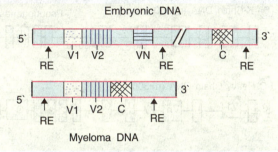

Fig. 19.13 : Arrangement of gene segments of λ-light chain in embryonic and myeloma DNA molecules. RE, restriction enzyme (based on the results of Hozumi and Tonegawa, 1976).

which are known as gene segments. The κ and λ light chain families contain L, V, J and C gene segments, whereas the heavy chain family contains L, V, D, S and C gene segments (Chen and Alt, 1993).

During the process of differentiation of B-cells, a long segment of DNA is deleted resulting in close rearrangement of V (*i.e.* V_2 segment to J-segment). The RNA transcript of immunoglobulin gene that contains intron (non-coding segment within the gene) is processed and correct transcript of mRNA is formed which is translated into a polypeptide light chain (Chen and Alt, 1993).

The rearranged VJ-gene segments encode the V-region of the light chain, whereas VDJ-gene segments encode the V-region of the heavy chain. The C-segments encode the C-region of the light or heavy chain of the gene segment encodes a short signal sequence. The signal sequence guides the light or heavy chain through endoplasmic reticulum but is broken before assembly of the immunoglobulin molecule. Therefore, the amino acids that correspond to L-gene segment do not appear in light or heavy chains (Chen and Alt, 1993).

(*a*) **Multigene families of λ-chain, κ-chain and heavy chain:** For the first time Tonegawa (1983) gave the evidence that V-region of light chain is encoded by two gene segments (V_1 and V_2) by closing the germ-line gene encoding V-region of mouse λ light chain. The complete sequence of nucleotide was determined. When it was compared with known sequences of the λ-chain V-region, a discrepancy was found. In mouse λ-chain multigene family contains two V gene segment (V_1 and V_2), four J-gene segment (J_1, J_2, J_3 and J_4) and four C gene segment (C_1, C_2, C_3, C_4). The arrangement of the gene segments is shown in Fig. 19.14A.

The gene segments, J_4 and C_4, are defective, therefore, called pseudogenes. A functional V-region of λ-chain gene consists of two coding segments *i.e.* exons (a V-gene segment and a J-segment) which are separated by a non-coding sequence (*i.e.* intron) in unrearranged germ line DNA. In humans, there are an estimated 100 V-gene segments, 6 J-segment and 6 C-segments.

In mouse, the κ-chain gene family consists of about 300 V gene segments, five J-segment (one segment is pseudogene) and a single C-gene segment. Unlike λ chain there are no sub classes of κ-light chain as only one C-gene segment is found (Fig. 19.14 B). In humans, the k-chain gene family consists of about 100 V-gene segments, 5J segments and a single C-segment.

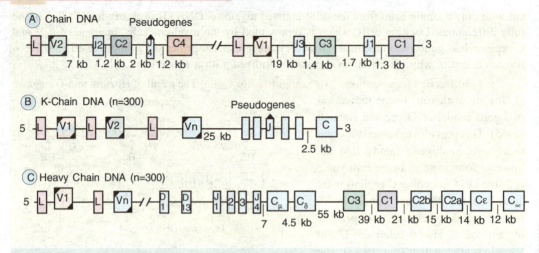

Fig. 19.14 : Germ line organization of λ-light chain (A), κ-light chain (B), and heavy chain (C) gene segments in the mouse (modified after Kuby, 1994).

Number of Antibodies possible through the Combinatorial Joining of Mouse Germ Line Genes

l light chains	V regions = 2
	J regions = 3
	Combinations = 2 × 3 = 6
κ light chains	V_κ regions = 250 – 350
	Combinations = 250 × 4 = 1,000
	= 350 × 4 = 1,400
Heavy chains	V_H = 250 – 1,000
	D = 10 – 30
	Combinations = 250 × 10 × 4 = 10.000
	= 1,000 × 30 × 4 = 120,000
Diversity of antibodies	k-containing : 1,000 × 10,000 = 10^7
	1,400 × 120,000 = 2 × 10^8
	l-containing 6 × 10,000 = 6 × 10^4
	6 × 120,000 = 7 × 10^5

In mouse the heavy chain multigene family of immunoglobulin is like λ and κ chain but a little complex. It is located on chromosome 12. The heavy chain multigene family consists of about 200-1000 V-gene segments, 13 D-gene segments, four J-gene segments and a series of C-gene segments (Fig. 19.14C). Each V-gene segment has a leader sequence a short distance up stream from it. The J-gene segments are located down stream from the D-gene segment. Each C-gene segment encodes the C-region of an immunoglobulin heavy chain isotype. Similarly each C-gene segment encodes separate domain of the heavy chain C-region. In mouse C-gene segments are arranged in the order Cμ - Cδ - Cγ3 - Cγ1 – Cγ2b - Cγ2a - Cε - Cα (Kuby, 1994).

4.   Gene Rearrangements in Light - and Heavy Chain DNAs

Rearranged κ and λ genes of light chain contain gene segments in the order from 5' to 3' end- a short leader (L) gene segment, an intron (noncoding sequence), a joined VJ-gene segment, a second intron, and a C-gene segment (Fig. 19.15). The rearranged light chain sequence is transcribed by RNA polymerase and yields a light chain primary RNA transcript. Intron is removed by RNA processing enzyme. The mRNA upon translation produces the light chain protein.

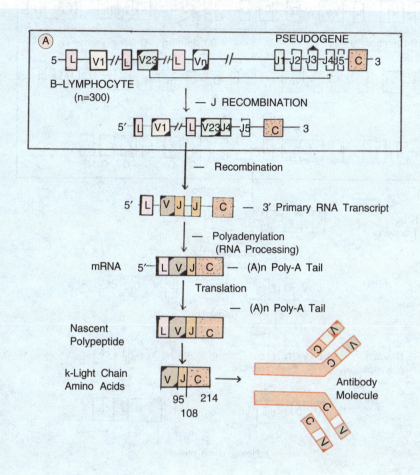

Fig. 19.15 : Steps leading to rearrangement of κ-light chain and processing events necessary to form finished κ-light chain protein. A, germline κ-light chain DNA (in undifferentiated cell), and B, rearranged κ-light chain DNA (in differentiated B lymphocyte). L, leader sequence; V, gene for variable region; J, joining sequence between V and C gene; C, gene for constant region.

Two separate rearrangement events occur within the variable region before the generation of a functional immunoglobulin heavy chain gene. A D-gene segment first joins to a J-segment and form a DJ-segment. Therefore, the DJ-segment moves and joins a V-segment. Thus, a VDJ unit is formed which encodes complete variable region (Fig. 19.16). Thus, the rearranged heavy chain contains the gene subunits in the following sequence starting from 5` end -a short L segment, an intron, VDJ-segment, second intron, and a series of C-gene segments. A short distance upstream from each heavy chain, a leader sequence at promoter gene is located. After gene rearrangement RNA polymerase binds to promotor sequence and transcribe the entire heavy chain gene (Chen and Alt, 1993).

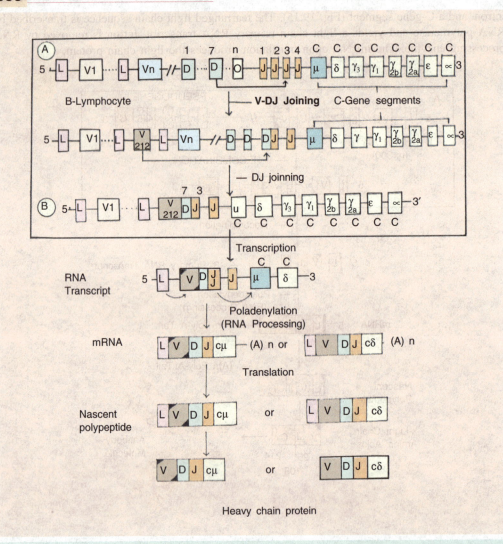

Fig. 19.16 :   The heavy chain gene rearrangement and RNA processing events for the production of finished heavy chain protein. Several products of IgM and IgD chains are formed as a result of differentiated RNA processing. A, germline heavy chain DNA; B, rearranged heavy chain DNA.

By differential processing immature B-cells express IgM alone, and mature cells co-express IgM and IgD. When an antigen stimulates the mature B-cells, additional rearrangement of their heavy chain C-gene segment can occur resulting in expression of different isotypes of antibodies (IgG, IgA and IgE) by some antigens (Chen and Alt, 1993).

(iii) **Antibody Diversity:** Tonegawa (1983) published an excellent paper on somatic generation of antibody diversity. Various mechanisms are involved in generating antibody diversity. These mechanisms can generate to about 10^8 possible combinations of antibodies. This results from random joining of multiple V, J and D germline gene segments and random association of a given heavy chain and light chain in a given cell. For example, any of the 300 - 1,000 V-gene segments of heavy chain can combine with any of the 13 D segments and any of 4 J segments. The possibility

of generating the enormous amount of diversity with these combinations would be as 1.6×10^4 ($300 \times 13 \times 4 = 1.6 \times 10^4$) (Table 19.4). Similarly 300 V-gene segments of κ-light chain randomly can combine with 4 J segments of the same, and can generate 1.2×10^3 possible combinations ($300 \times 4 = 1.2 \times 10^3$). The combinatiorial diversity of the δ-light chain DNA is much less ($2 \times 3 = 6$) as compared to κ-light chain DNA (Kuby, 1994).

Table 19.4 : Antibody diversity in mouse (Modified after Kuby, 1994)

Mechanism of diversity	Heavy Chain	Light Chain	
		Kappa (κ)	Lambda (λ)
A. Estimated number of Segments			
Multiple germline gene segments			
V	300–1000	300	2
D	13	0	0
J	4	4	3
B. Possible number of combinations*			
Combinatorial V-J and VDJ joining	$300 \times 13 \times 4 = 1.6 \times 10^4$	$300 \times 4 = 1.2 \times 10^3$	$2 \times 3 = 6$
Combinatorial association of heavy and light chains	$> 1.6 \ 10^4 \times (> 1.2 \ 10^3 + > 6) = \gg 1.9 \times 10^7$		

* Antibody diversity in mouse is similar to that in humans.

In the beginning both Cμ and Cδ gene segments are transcribed. Polyadenylation and RNA splicing delete the introns, and process the primary RNA transcript into mRNA which encodes either Cμ or Cδ. The two molecules of mRNA (one for Cμ and other for Cδ) are translated to yield leader polypeptide. Later on leader polypeptide of the nascent polypeptide is cleared and the complete μ and δ chains are produced (Tonegawa, 1983).

Leader (1982) has postulated the following stages in antibody formation: (*i*) firstly, synthesis of λ heavy chain in the pre-B-cells, thereafter its joining with specific variable region, (*ii*) association of light chains and δ-chains, and production of complete antibodies of the IgM and IgD classes, (*iii*) concurrent appearance of both the IgM and IgD antibodies on the surface of B-cell, (*iv*) attachment of the antigen to a receptor on the surface of B-cells, (*v*) cell proliferation (with the selected Ig molecules) and production of a clone of B-cells producing specific antibodies, and (*vi*) disappearance of the IgD and IgM from the cell surface and the secretion by the cell of IgM, IgG, IgE or IgA molecules.

Differential RNA processing of Ig heavy chain primary transcript produces either membrane bound antibody or secreted antibody.

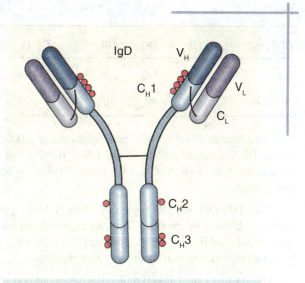

Immunoglobulin D : The structure shows disulfide bonds linking protein chain (in black), carbohydrate side chains are in red.

D. Major Histocompatibility Complex (MHC)

Major histocompatibility complex (MHC) is a tightly linked cluster of genes present on chromosome 6 in humans (and chromosome 17 in mice) which encodes the MHC proteins. The MHC proteins are present on plasma membrane of almost all human tissue/cells. The MHC proteins participate in intercellular recognition and antigen presentation to T lymphocytes. Generally, a group of linked MHC genes is inherited as a unit from parents. These linked groups are called **haplotypes**. MHC genes are polymorphic (*i.e.* there are a large number of alleles for each gene). Also they are polygenic (i.e. there are a number of different MHC genes). Human MHC molecules are also called **human leucocyte antigens** (HLA).

In the mid 1930s Peter Gorer (England) established the concept of rejection of foreign tissue due to an immune response to cell surface molecules. This gave the birth to the study of histocompatibility antigens. He identified four types of genes (I to IV) which encode blood cell antigens. During 1950 George Snell (U.S.A.) pioneered the concept that antigens encoded by the genes took part in the rejection of transplanted tumours. He called these genes as **histocompatibility genes**. For this work Snell was awarded the Nobel prize in 1980.

1. Classes of MHC Molecules

The MHC genes are organized into three classes I, II and III which express three classes of molecules Classes I, II and III, respectively (Table 19.5). Classes I MHC genes consists of A, B and C gene loci. They secrete glycoproteins which are referred to as Class I MHC molecule. Glycoproteins are expressed on the surface of about all nucleated cells. Class I MHC molecules present the peptide antigens to T_C cells.

Table 19.5 : Organisation of major histocompatibility complex (MHC) genes in human in chromosome 6.

Class	Class II			Class III			Class I		
Regions	DP	DQ	DR	C_4, C_2	and DF	B	C	A	
Gene	DP	DQ	DR	C Proteins	TNF-α	HLA-	HLA-	HLA-	
Product	αβ	αβ	αβ		TNF-β	B	C	C	

The human Class I MHC gene spans about 2,000 kb (about 20 genes) at the telomeric end of the HLA complex, whereas the Class II MHC genes (about 1,000 kb) are located at the centromeric end of HLA. Class III genes (flanked by about 10,000 kb long) located between the two genes.

The DP, DQ and DR region of Class II MHC genes in humans encode the Class II MHC molecules called glycoproteins. They are expressed on antigen presenting cells such as macrophages, dendric cells and B cells, and present the processed antigenic peptides to T_H cells. Class II molecules have specialised function in the immune response.

Both Class I and Class II molecules have common structural features. They have role in antigen processing. In addition, the Class III MHC gene is flanked by Class I and Class II regions and encodes molecules critical to immune function. Class III MHC molecules consist of complement components C4, B2, BF, inflammatory cytokines, including tumour necrosis factor (TNF) and heat shock proteins.

2. Structure of MHC Molecules

The Class I molecule is a transmembrane glycoprotein consisting of two chains: α-chain or heavy chain (of 42 KD molecular weight) non-covalently associated with a light chain called β_2–microglobulin (molecular weight 12 KD). The α–chain is organized into three extracellular domains (α, α_2, α_3) and a 'transmembrane segment' (hydrophobic) followed by a short stretch of hydrophilic 'cytoplasmic tail' (Fig. 19.17A). These are encoded by A, B and C regions of HLA complex and expressed on the surface of plasma membrane of almost all cells except erythrocytes. β_2–microglobulin molecule is expressed by different chromosomes. Association of the α– chain with β_2–microglobulin is must for expression of Class I molecules on cell membrane. The α_1 and α_2 form the antigenic-binding cleft located on top of surfaces of molecule.

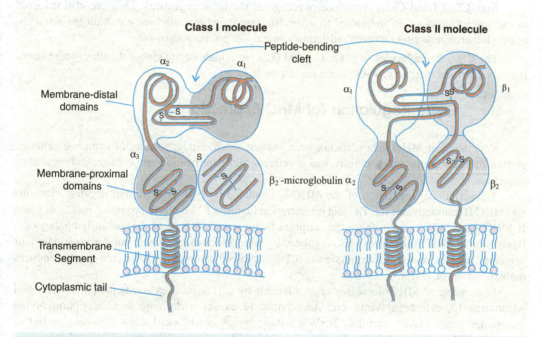

Fig. 19.17 : Structure of MHC molecules; A, Class I molecule; B-Class II molecule.

Class II MHC molecules are also trans-membrane glycoprotein encoded by separate MHC genes. They contain two different α and β chains of 33 and 28 KD, respectively. These two chains are associated non-covalently (Fig. 19.17B). Further, both chains fold to give two domains (β_1 and β_2 domains in other domain), one is membrane proximal domain and the second is membrane-distal domain. Like Class I MHC molecules, the class II molecules also contain trans-segment and a cytoplasmic anchor segment. Each chain of Class II molecule contains two external domains (α_1 and α_2 in one chain) and β_1 and β_2 domains in other chain.

3. Function of MHC Molecules

MHC provides both cell mediated and humoral immune responses, while antibodies react only with antigens, and most of the T cells recognise antigen only when it gets combined with an MHC molecule. Hence, MHC molecules act as antigen-presenting structure. The MHC partly determines the response of an individual to antigens of infectious microorganisms. Therefore, it is implicated in susceptibility to disease and in the development of autoimmunity. Recently, it has been explained that the natural killer cells express receptors for MHC Class I antigens. The receptor-MHC interaction result in inhibition/activation.

Both Class I and Class II MHC molecules present the processed endogenous antigen to CD8 T cells. Class II molecules present the processed exogenous antigen to CD4 T cells. Class I molecules identifies mostly all the cells of the body as 'shelf'. Also they induce the production of antibodies which introduced into host with different Class I molecules. This is the basis for MHC typing when a patient is to undergo for antigen transplantation. Class II molecules comprise of the D group of MHC. They stimulate the production of antibodies. But they are required for T cell communication with macrophage and B cells. Part of T cells receptor recognises Class II molecules on the adjacent cell before cytokine secretion by T cells. This is necessary for immune response.

Both Class I and Class II molecules recognise the microorganisms. They are also involved in the susceptibility of an individual to a specific non-infectious diseases e.g. multiple sclerosis, acute glomerulonephritis, tuberculoid leprosy, paralytic poliomyelitis, etc.

The Class III molecules (e.g. C_2, C_{4a} and C_{4b}) participate in the classical pathway and factor B in the alternate pathway of the immune responses.

4. Gene Regulation (of MHC Expression)

Regulation of MHC genes has not been studied much. Understanding of complete genomic map of the MHC complex hopefully will accelerate the identification and coding, and regulatory sequences.

Transcriptional regulation of the MHC is mediated by both positive and negative elements e.g. MHC II transactivator (cll TA) and transcription factor (RFX) binds to promoter region of Class II MHC gene. Any error in these transcription factor causes a type of disease in lymphocytes. Expression of MHC molecules is also regulated by many kinds of cytokines. Interferons and tumour necrosis factor increases the expression of Class I molecules on cells. Interferon-gamma induces the expression of cIITA.

Expression of MHC decreases after infection by certain viruses e.g. hepatitis B virus, and adenovirus 12, cytomegalovirus, etc. Adenovirus 12 causes a decrease in transcription of the transporter genes (TAP1 and TAP2). When these genes are blocked, class I molecules fail to assemble with β_2–microglobulin. Decreased level of Class I molecules promotes viral infection. Expression of Class II molecules by B cells is down-regulated by INF-gamma. Corticosteroids and rostaglandins decrease the expression of Class II molecules.

E. Antigen-Antibody Reactions (Diagnostic Immunology)

About a century ago, Robert Koch tried to develop a vaccine against the pathogen, *Mycobacterium tuberculosis*. He observed that after injection of cells of *M. tuberculosis* in guinea pigs, the site of injection became red. After 2-3 days, the injected site seemed a little bit swollen. Even today we can observe a similar symptom when a person is injected for the treatment of the pathogen *M. tuberculosis*. This study gave the birth to a new area of immunology to which we call as *serology*. On the basis of serological tests, several abnormalities in blood can be diagnosed in recent years. The antigen-antibody reaction depends on non-covalent interactions including hydrogen bonds, ionic bonds, hydrophobic interactions and Van der Waals interactions. The strength of this interaction depends on the number of these weak noncovalent interactions between antigen and antibodies (Kuby, 1994). Some of the antigen - antibody reactions are described in the preceding sections.

I. Precipitation Reactions

The reaction of soluble antigens with IgG or IgM antibodies to form a large interlocking aggregates (lattices) is called precipitation reaction. The precipitates formed by antibodies are known as precipitins. The precipitation reactions occur in two stages: (*i*) rapid interactions within a second between antigen and antibodies and formation of complex, (*ii*) slow rate of reaction completing even within a few minutes or hours and forming lattices from antigen-antibody complexes. When the antibodies and antigens are in proper ratio, precipitation reactions normally occur. When there is excess amount of either of two, no visible precipitate is formed. One can produce the optimal ratio of these two by putting antigens and antibody adjacent to each other and waiting for their diffusion together. In precipitation test, a precipitation ring appears which display the creation of optimal ratio. This zone is known as the zone of equivalence (Fig. 19.18).

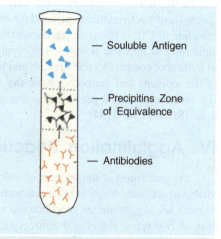

Fig. 19.18 : The precipitation ring test.

II. Immunodiffusion test (IDT)

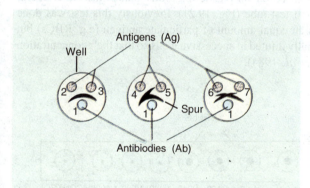

Fig. 19.19 : The Ouchterlony immunodiffusion test. 1, antiserum filled in wells, 2-7, different types of antigens; A, line of identity (diffusion of antigens and antibodies does not occur, therefore, they react and precipitate to form the dark line); B, lines of partial identity (antigens are not identical but they share many antigenic determinant sites and related to each other); C, line of non-identity (antiserum contains antibodies against antigens in wells). The antigens are not related because they diffuse across each others' zone of precipitation.

Immunodiffusion tests are performed in a gelled agar medium. One of the IDTs is *Ouchterlony* test (Fig. 19.19). In Ouchterlony test wells are cut, into which a purified antiserum (a serum containing antibodies) is added, and to each surrounding well, soluble form of test antigens are added. Thereafter, a line of visible precipitate is formed between the wells where after diffusion optimal ratio of antigen-antibody is formed. Through the Ouchterlony test, the presence of antibodies in the serum against more than one antigens at a time can be demonstrated. Through this test, identical, partially identical and different types of antigens can also be found out (Stites *et al*, 1984).

III. Counter Current Immunoelectrophoresis Test (Counter Immunoelectrophoresis (cie))

CIE not only depends entirely on diffusion of antigen and antibody in a gel, but also uses electrophoresis for their rapid movement (Fig. 19.20). By using this method protein can be separated within an hour. CIE is useful for the diagnosis of bacterial meningitis and the other diseases. The principle of CIE is based on the movement of antigens and antibodies to opposite poles after applying electric current in buffers of correct electric strength and pH, because some of the antigens and antibodies have the opposite charges. If a reaction occurs, a precipitation line appears within an hour (Stites *et al*, 1984).

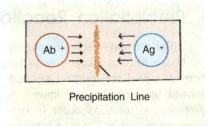

Precipitation Line

Fig. 19.20 : Counter immunoele-ctro-phoresis (CIE). Ag, antigen; Ab, antibodies

IV. Agglutination Reactions

Agglutination is the process of linking together of antigens by antibodies and formation of visible aggregates. Agglutination reactions involve particulate antigens i.e. soluble antigens adhering to particles. Agglutination reactions are very sensitive, readable and available in several varieties. It is of two types, direct and indirect agglutination tests.

1. Direct Agglutination Test

Direct agglutination test diagnoses antibodies against a large number of cellular antigens such as RBCs, bacteria and fungi. This test is carried out in plastic microtiter plates that have several small shallow wells. Each well acts as small test tube (Fig. 19.21). Previously this test was done in test tubes. However, each well contains an equal amount of particulate antigen (e.g. RBCs) but the amount of antibodies in the serum is serially diluted in successive wells so that their concentration may be half of the previous well (Stites *et al*, 1984).

If one starts with more antibodies, more dilutions will be required to lower the amount to a point at which agglutination does not occur. This is the measure of titer or concentration of serum antibody (Tortora, *et al*, 1989). In a positive reaction, agglutination occurs, and sufficient antibodies are present in the serum to link the antigen together. This results in formation of antibody-antigen mat which sinks to bottom of well (A). However, in the negative reaction, agglutination does not occur and insufficient antibodies are present to cause the linking of antigens. The particulate antigens roll down the sloping sides of the well, and

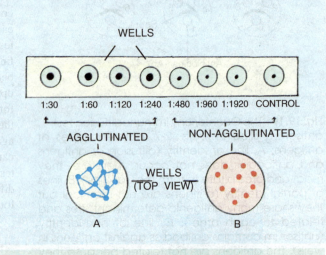

WELLS

1:30   1:60   1:120   1:240   1:480 1:960 1:1920   CONTROL

AGGLUTINATED   NON-AGGLUTINATED

WELLS (TOP VIEW)

A   B

Fig. 19.21 : Direct agglutination test.

form a pellet at the bottom. In this example the antigen titer is only 80 since the well with a 1 : 80 concentration is the most dilute concentration that gives a positive reaction. It may also be demonstrated that before illness, blood of persons does not have any antibody, whereas titer develops significantly with the progress of disease. This change in titer is called servotiter (Stites *et al*, 1984).

2. Indirect (Passive) Agglutination Tests

This type of diagnostic tests are very rapid particularly for the detection of streptococci. If the antigens are adsorbed onto particles (e.g. RBCs, latex beads, bentomile clay), soluble antigens can respond to agglutination test. Antibody reacts with the soluble antigen adhering to the particles. Therefore, the particles agglutinate with each other as these do in the direct agglutination tests.

3. Haemagglutination

Haemagglutination is the phenomenon of clumping of RBCs. When the RBCs are agglutinated by certain viruses such as those causing mumps, measles, influenza, etc. it is called viral haemagglutination. In the serum of a person, certain antibodies act against the antigens (of these viruses), the antibodies neutralize them after reaction. The haemagglutination test is widely used for the diagnosis of a number of viruses including those as above.

V. Opsonization

Opsonization has been described earlier under `Antimicrobial substances'.

VI. Complement Fixation Reactions

A group of 20 or more serum protein is collectively known as complement. During reaction, the complement binds to antigen-antibody complex and is used up or fixed. This process of complement fixation may be used to measure even very small amount of antibody that does not produce a visible reaction such as precipitation or agglutination. Therefore, it is necessary to use indicator system. This method is used in diagnosis of diseases such as leptospirosis, mycoplasmal pneumonia, Q fever, polio, rubella, histoplasmosis, coccidiodomycosis and streptococcal infections. The test requires patient's serum, test antigen, complement from guinea pig and antibodies of sheep RBCs to determine whether sheep RBCs may be lysed by guinea pig complement. The test is accomplished in the following two stages:

Stage 1: The patient's serum is heated at 56°C for 30 minutes so that the complement should be inactivated. The heated serum is diluted and then added to known amount of specific antigen and complement (Fig. 19.21). The test antigen may correspond to the diseases. For example, if a patient is suffering from a disease caused by streptococci the test antigen would be the streptococcal antigen. If the patient's serum contains antibodies against streptococci, the test antigen will form complement sequence. This mixture is again incubated for about 30 minutes. At this point, no antigen-antibody reaction occurs (Stites *et al*, 1984).

Stage 2: In stage 2, the complement fixed by antigen-antibody reaction is detected by an indication system. This system consists of sheep RBCs containing specific antibodies attached to their surfaces. When these are added to complement, haemolysis of RBCs occurs that impart changes in colour of the mixture. This shows that the complements have not been fixed during the first stage, therefore, these become available to cause haemolysis (Fig. 19.22). This indicates that the patient has no streptococcal pneumonia. However, if the guinea pig complements are destroyed, they will not be able to cause the lysis of RBCs. On the other hand, if the complements are fixed (by antigen-antibody reaction) during the first stage, these will not be available to cause haemolysis during the second stage. This indicates that the patient has the infection of streptococci.

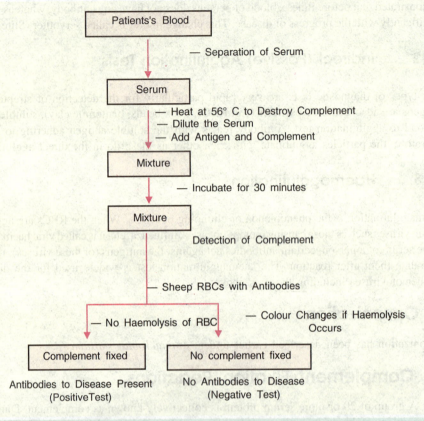

Fig. 19.22 : Procedure for complement fixation test.

VII. Neutralization Reactions

The neutralization reactions are the reactions of antigen- antibody that involve the elimination of harmful effects of bacterial exotoxins or a virus by specific antibodies. These neutralizing substances i.e. antibodies are known as antitoxins. This specific antibody is produced by a host cell in response to a bacterial exotoxin or corresponding toxoid (inactivated toxin). The antitoxin reacts with exotoxin and neutralizes it. These antitoxins can be artificially induced in animals such as horses. Thus, the antitoxin of animal sources in turn can be injected into human which provides a passive immunity against a toxin present in human body produced by the pathogens causing diphtheria, tetanus, etc.

1. Diagnosis of Viral Infections

Neutralization test is very useful in diagnosis of viral infections in humans. After introduction of a virus, antibodies are produced in response and bind to receptor sites present on the viral surfaces. After binding of antibodies, viral particles fail to reach to the cells. Thereafter, the virus is destroyed. Artificially, the virus is capable of destroying their cell-damaging effect in cell culture or embryonic eggs can be used to determine the presence of antibodies against them. However, when serum contains antibodies against a particular virus, the antibodies will not allow the virus to infect the cell in cell culture, consequently the cells will not be damaged.

2. Schick Test

Schick test measures the level of immune system of a person to the infection of diphtheria. When testing the status of immunity, a small amount of diphtheria exotoxin is inoculated in the skin of a person. Depending on ability and quantity of antitoxin, positive or negative responses develop. If serum antitoxin in body would be in sufficient amount to neutralize the exotoxin, no visible reaction will occur. In control, when serum antitoxin is in insufficient amount the exotoxin will damage the tissues at the site where incision was made, and will produce a swollen and reddish area which is converted into brown within 4 or 5 days. This shows that the immune response is not present to a satisfactory level (Stites *et al,* 1984).

VIII. Radioimmunoassay (RIA)

It is such a technique which is highly sensitive and can measure even the less concentration (i.e. 0.001 µg/ml) of antigen or antibody. In 1960, for the first time this technique was developed by S.A. Berson and R. Yalow when they were engaged in determining the concentration of insulin and antinsulin complexes in diabetics. Thereafter, Berson died, and significance of this technique was realised. In 1977, Yalow was awarded a Nobel prize.

There are two methods of measuring RIA: the liquid phase and the solid phase RIAs.

1. Liquid Phase RIA

The liquid phase RIA is based on competitive binding of radiolabelled antigen and unlabelled antigen, to a high affinity antibody. The antigen labelled with ^{125}I is mixed with such a concentration of antibody that can just saturate the antibody. Therefore, the increasing amount of antigen (unlabelled) of unknown concentration is added. The two types of antigens now compete for available sites of the antibody. The antibody does not differentiate the labelled antigen from the unlabelled one. Upon gradually increasing concentration of unlabelled antigen, the labelled antigen could be displaced from the binding sites available on antibody. The labelled antigens are made free in the solution. The amount of labelled antigen in solution is measured, and the concentration of unlabelled antigen can be determined.

2. Solid Phase RIA

In solid phase RIA, either antigen or antibody is immobilized on a solid phase matrix. It is simple and easy in handling as compared to liquid phase RIA.

IX. Enzyme-linked Immunosorbent Assay (ELISA)

The principle of ELISA is similar to RIA, but differs slightly. In RIA radiolabelled antigen is used, whereas in ELISA enzyme is used that reacts with a colourless substrate and

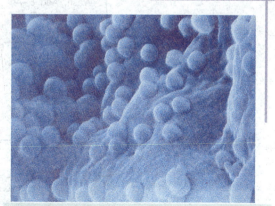

ELISA kits being developed to test for AIDS antibodies, the disease caused by HIV. The viral particles bud out of an infected lymphocyte.

develops a coloured reaction product. There is a large number of enzymes such as alkaline phosphatase, horse radish peroxidase, and p-nitrophenyl phosphatase which are employed in ELISA. As compared to RIA, this assay is both cheaper and safer. On the basis of known concentration of antigen or antibody a standard curve is prepared from which the unknown concentration of sample is measured. A microtiter plate with numerous shallow wells is used in this method. It is very useful in testing for AIDS antibodies. However, now-a days a number of ELISA kits have been developed and are in current use.

1.   Indirect ELISA

It is used to measure antibody. Known antigen is coated on the plastic lining of the wells of microtiter plate which is made up of polystyrene latex. To test for the presence of antibodies against this antigen in the patient, his blood serum is added to the wells (Fig. 19.23A). If the patient's serum contains antibody specific to antigen, the antibody will bind to the absorbed antigen otherwise not. After incubation the wells are washed and the enzyme, labelled with antihuman gamma globulin (anti-Hgg), is added to the wells. Anti-Hgg can react with antigen antibody complex. The mixture of wells is washed to remove the excess of unbound labelled anti- Hgg. Finally the correct substrate for the enzyme is added which is hydrolysed by the enzyme and develops a colour. Varying concentrations of antibody in serum shows changes in the intensity of colour. This method is very useful in detection of antibodies to HIV, *Salmonella*, *Yersinia*, *Brucella*, *Treponema* and streptococci.

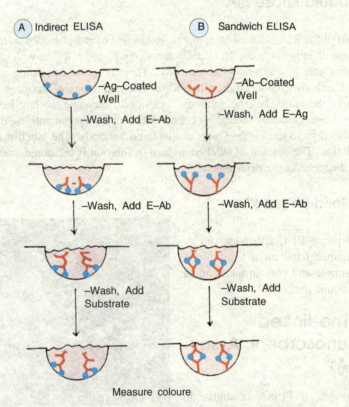

Fig. 19.23 : Enzyme-linked immunosorbent assay (ELISA) technique for detection of antibody (A) and antigen (B).

2. Double Antibody Sandwich ELISA

This method detects antigen. In this case antibody (antiserum) is immobilised on the surface of wells of microtiter plate (Fig. 19.22B). A test antigen is added to each well and allowed to react with the bound antibody. It is incubated during this period. If antigen combines specifically with antibody absorbed to wells, the antigen will be retained even after washing and unbound antigen would be made free. Thereafter, a second enzyme-linked antibody (e.g. alkaline phosphatase tagged to antibody) is added to react with bound antigen. It is again incubated for a few seconds, the enzyme labelled antibody reacts with the antigen-antibody complex already formed in the wells and results in the development of a `sandwich'. The mixture in wells is washed again to remove the excess of labelled enzyme. A chromogenic substrate e.g. nitrophenyl phosphate is added which reacts with enzyme and develop yellow colour. The reaction can be stopped simply by changing the pH or denaturing the enzyme. The change in colour is measured visually or spectrophotometrically. Change in colour shows the presence of desired antigen in the sample. This technique is useful in detection of toxins of *Vibrio cholerae, E. coli, Staphylococcus enterotoxin*-A and antigens of rotavirus.

X. Fluorescent Antibody (FA) Technique

The FA technique is used to detect the microorganisms present in clinical specimens, and specific antibodies present in serum. If the antibodies bind to cell or tissues, it can be observed by tagging the antibody with a fluorescent dye such as fluorescein isothiocyanate and rhodamine. Both the dyes can conjugate the FC region of antibody without affecting the specificity and make the antibody fluorescent when exposed to UV light. Fluorescein absorbs blue light (490 nm) and emits yellow green fluorescence (517 nm). Similarly, rhodamine absorbs the yellow green light (515 nm) and emits deep red fluorescence (546 nm). The FA technique is very useful in testing for rabies within a few hours with 100% accuracy.

There are two methods of FA test, direct FA test and indirect FA test. Direct FA test is used to identify the microorganisms present in clinical specimen. The specimen containing antigen is fixed onto a slide and, thereafter, fluorescein-labelled antibodies are added on the specimen. It is incubated for a few minutes. The slide is washed to remove unbound antibody and observed under the UV microscope for yellow-green fluorescence.

The indirect FA test is useful for the detection of specific antibodies in serum formed by a microorganism. This method follows the following steps: (a) fix a known antigen onto a slide, (b) add a test serum (microorganism-specific antibody reacts with antigen and forms a bound complex, (c) add fluorescein-labelled anti-Hgg to the slide, (d) incubate and wash the slide, (e) examine the slide under fluorescence microscope. The development of fluorescence confirms the presence of antibody specific to antigen fixed on slide.

QUESTIONS

1. What do you know about defense system of body? Describe in detail the chemical means of defense.
2. What is the composition of blood? Discuss the mechanism of phagocytosis.
3. What is the role of complement in immune system? Write different processes involved in complement pathways.
4. Discuss in detail different types of immunity and role of antibodies in providing immune responses.
5. Give an illustrated account of chemical and physical organisation of antibodies.
6. Write an essay on mechanism of humoral immunity.
7. What is the cell-mediated immune system? Write the mechanism involved in this system.

8. Discuss in detail about the organisation of immunoglobulin genes with special reference to multigene organisation.

9. Give an illustrated account of multigene families of λ-chain, κ-chain and heavy chain.

10. Write an essay on gene rearrangement in light and heavy-chain DNA.

11. Give an illustrated account of differentiation of B-cells and T-cells from stem in bone marrow.

12. Write short notes on the following.

(*i*) Phagocytosis, (*ii*) Inflammation, (*iii*) Antibody, (*iv*) Immunoglobulins, (*v*) Interferon, (*vi*) Opsonization, (*vii*) Antigen, (*viii*) Monoclonal antibodies, (*ix*) Precipitation, (*x*) Agglutination, (*xi*) Ouchterlony test, (*xii*) Counter immunoelectrophoresis, (*xiii*) B-cells, (*xiv*) T-cells, (*xv*) Antibody diversity, (*xvi*) Haemagglutination, (*xvii*) Complement-fixation reaction, (*xviii*) Neutralization reaction, (*xix*) Radioimmunoassay, (*xx*) ELISA test, (*xxi*) Fluorescent antibody technique.

13. What is major histocompatibility factor? Discuss in detail the major classes, types and function of MHC molecules.

REFERENCES

Alzari, P.M. *et al*. 1988. Three dimensional structure of antibodies. *Ann. Rev. Immunol* **6** : 555.

Bird, R.E. *et al*. 1988. Single chain antigen binding proteins. *Sci*. **242**: 423 - 426.

Bretscher, M.S. 1987. How animal cells move. *Sci. Am*. **257** : 72 - 90.

Chen, J. and Alt, F.W. 1993. Gene rearrangement and B-cell development. *Curr. Opinion. Immunol*. **5** : 194

Collier, H.O.J. 1962. Kinins. *Sci. Am*. **207** : 111-118.

Frank, M.M. and Fries, L.F. 1991. The role of complement in inflammation and phagocytosis. *Immunol Today* **12** : 322.

Friedman, R.M. and Finter, N.B. 1984. Interferon: *Mechanism of Production and Action. ed*. vol. 3. Elsevier, New York.

Hood, L.E. *et al*. 1984. *Immunology*. 2nd ed. Menlopark, Benjamin/Cumming, California.

Hozumi, N. and Tonegawa, S. 1976. Evidence for somatic rearrangement of immunoglobulin genes coding for variable and constant regions. *Proc. Natl. Acad. Sci*. (U.S.A). **73** : 3628.

Kohler, G and Milstein, C. 1975. Continuous cultures of fused cells secreting antibody of predefined specificity. *Nature* **256** : 495.

Kuby, J. 1994. *Immunology*. 2nd ed. W.H. Freeman & Co., New York, p.660.

McNabb, P.C. and Tomasi, T.B. 1981. Host defense mechanism at mucosal surfaces. *Ann. Rev. Microbial* **33** : 477 - 496.

Milstein, C. 1980 Monoclonal antibodies *Sci. Am* **243** : 66.

Ross, G. E. (ed). 1986. *Immunology of the Complement System. An Introduction for Reaserch and Clinical Medicine*. Academic Press, New York.

Stites, D.P., Stobo, J.D, Fundenberg, H.H. and Wells, J.V. 1984. *Basic and Clinical Immunology* 5th ed, Loss Altos, Langes Medical Publ., California.

Tonegwa, S. 1983. Somatic generation of antibody diversity. *Nature,* **302** : 575.

Tortora, G.J.; Funke, B.R. and Case, C.L. 1989. *Microbiology: An Introduction*, 3rd ed., The Benzamin/ Cummings Publ. Co., Inc. California.

Wilson, I.A. and Stanfield, R.L. 1993. Antibody - antigen interactions. *Curr. Opinion. Stru. Biol* **3** : 113.

Wood, W.D. 1951. White blood cells versus bacteria. *Sci. Am*. **184** : 48-52.

The Infectious Diseases: Epidemiology, Parasitism and Chemotherapy

20

A. Epidemiology
B. Parasitism
C. Antimicrobial Chemotherapy

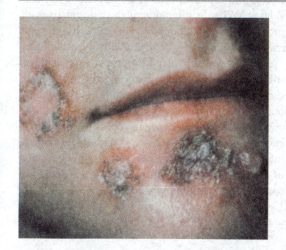

A disease which is caused by viruses, bacteria, fungi, protozoa and helminths after being transmitted from one host or reservoir to the other host, is known as infectious diseases. These may be mild, severe or deadly to the host. Study of cycle of infectious diseases is very important as far as disease control is concerned.

However, a look at the outbreaks of infectious diseases in the past few years gives an idea of the seriousness and instability of the current situation, for example (i) cholera in South America representing the 7th pandemic, and cholera in South Asia representing the 8th pandemic, (ii) yellow fever in Kenya and Liberia, (iii) plague (in Surat) in 1994, and dengue haemorrhagic fever (in Delhi) in 1996 in India, (iv) ebola haemorrhagic fever in Zaire and most recently in Gabon, (v) diphtheria in Eastern Europe and countries of the former Soviet Union, and (vi) the relentless geographic spread of HIV/AIDS throughout the world. This chapter deals with epidemiology, parasitism and antimicrobial chemotherapy of the infectious diseases.

Paul Ehrlich
Ehrlich was a pioneer in the development of chemotherapy for infectious diseases.

A. Epidemiology

The classical study of infectious diseases was conducted by a

British physician, John Snow, between 1849 and 1854. During this period a series of outbreaks of cholera occurred in London. He found that cholera was spread by drinking water from a pump that was contaminated with raw sewage containing the pathogenic microorganisms. Thereafter, the pump was removed. Consequently the number of cholera cases remarkably declined. The second example is the outbreak of "Typhoid Marry" between 1896 and 1906 is New York City (see Chapter 22, section - *Typhoid fever*).

1. Terminologies

An *infectious disease* is any change in a state of normal health in which part or the whole body of an individual does not function properly due to the presence of an infectious agent or its products. The phenomenon of growth, multiplication and establishment of an infectious agent in host tissues or within the cells in known as *infection*. When a microorganism or agent lives on expense of the host *i.e.* derives nutrients for its growth, it is called *parasite*. However, if a parasite or its products cause(s) disease, the former is known as the *pathogen*. The ability of a pathogen to cause disease is termed as *pathogenicity*. The series of events involved for proper establishment of a pathogen in host cells/tissue (*i.e.* the process of disease development) is known as *pathogenesis*.

Epidemiology (*epi* means upon, *demos* means population, *logy* means study) is the science that deals with occurrence, determination, distribution and control of a disease. An individual who studies the epidemiology is called *epidemiologist*. When a disease occurs occasionally and at irregular intervals in a human population, it is known as *sporadic disease* e.g. typhoid. A disease maintaining a steady, low level frequency at a regular interval is called *endemic disease* e.g. common cold. However, a sudden increase in occurrence of a disease beyond a limit is called *epidemic* (upon the people). If the occurrence of a disease increases within a large population over a wide region, it is called *pandemic* (*pan* means all) (Mandell *et al.*, 1995).

2. Frequency of a Disease

Frequency of a disease refers to its repeated occurrence as fractions in a given population. To measure the frequency, the epidemiologists use statistics and find out the rate of increase over the pre-existing cases. It is measured as an increase over per 100 or per 1000 individuals. By measuring frequency one can speculate how severe a disease is? However, it is also related to *morbidity* or *mortality*. Morbidity is the number of individuals becoming ill by a specific disease within a susceptible population during a defined period. It is measured as below:

$$\text{Morbidity rate} = \frac{\text{Number of new cases of a disease during a specific period in a population}}{\text{Total number of individuals in the population}}$$

Similarly, mortality rate refers to death of individuals due to a specific disease with respect to size of population of sufferers with the same disease. It can be measured by using the following formula:

$$\text{Mortality rate} = \frac{\text{Number of deaths due to a given disease}}{\text{Size of total population of sufferers with same size}}$$

3. Characteristics of Infectious Diseases

The infectious diseases have characteristic signs and symptoms. Signs are objective changes in body, for example fever. On the basis of fever a disease can be recognised. **Symptoms** are the subjective changes for example pain, loss of appetite, etc. which are felt by the patients. In a broad sense symptom is used for sign as well. In addition, a *disease syndrome* includes a set of signs and symptoms due to a particular disease; for example an AIDS patient experiences disease syndrome.

Moreover, the characteristic symptoms of a disease develop during certain phases. The knowledge of the phases helps in recognition of a disease. For example, *incubation period* which refers to time required after infection to the appearance of signs/symptoms. Incubation period varies organism to organism. Second is the *prodromal stage i.e.* the period during which there is onset of signs and symptoms of a disease. It cannot be clearly found out. Third, the *period of illness* which is a phase during which the disease gets fully established and becomes most severe with characteristic signs and symptoms. The immune system is triggered. The last characteristic phase is the period of decline when signs and symptoms disappear and the disease is recovered gradually. This stage is known as convalescence (Mandell *et al.*, 1995).

4.   Herd Immunity

Development of immunity in a large percentage of a population resisting infection and spread of pathogen is called herd immunity. A large proportion of susceptible population is immunized at a time by the Public Health Officials just to maintain high level of herd immunity. The increase in susceptible individual is constantly monitored. Because new individuals are added in a given population due to birth and migration. The continuous monitoring helps in prevention of spread of infection agents and its survival in the patients.

Generally the pathogens do not change their nature because they are continuously transferred from one individual to the other and perpetuate in susceptible individuals. In addition, there are some pathogens that go on continuous changes and cause new epidemics, for example AIDS, influenza virus and *Legionella* bacteria. This feature of pathogens is called *antigenic shift* which is genetically determined major character of the pathogens. Due to antigenic shift, they are not recognised by the immune system of the host. For example, antigenic shift in influenza virus occurs due to hybridization between two antigenic types, or two different influenza viruses (*i.e. serovars*) or animal virus and human virus. Sometimes smaller antigenic changes also occur in pathogens just to escape from immune system of the host. These smaller changes occurring in pathogen time to time are called *antigenic drift*. When resistance in a given population is so high (herd immunity), the pathogen cannot infect humans. In such situation, these infect animals.

Moreover, due to antigenic shift or drift again the population of susceptible individuals increases. In this situation the Public Health Officials have to make sure that about 70% of individuals must be immunized so that the herd immunity could be maintained (Mandell *et al*, 1995).

5.   Disease Cycle

The infectious disease cycle is the chain of events that include epidemiological story of the infectious agents. Since infectious organisms perpetuate in hosts and are transmitted through vectors, knowledge of disease cycle helps in control of the disease. The following aspects are linked with disease cycle.

(*i*) **Sources of Disease:** A source is the site or location from where the pathogens spread to a new host either through environmental factors (e.g. soil, water, air, food) or indirectly through animate (animals or humans).

(*ii*) **Reservoirs:** Reservoirs are the natural environment (soil water or air) or susceptible animal hosts where pathogens survive. A list of pathogens surviving in animals is given in Table 20.1.

(*iii*) **Carriers:** Carriers are the individuals already infected with pathogens. Humans are the most important carriers of certain pathogens. Four types of carriers have been recognised: *active carriers* (who have an overt clinical case of disease), *convalescent carriers* (who have recovered from the infectious disease but contain pathogens in sufficient number), *healthy carriers* (who harbour pathogens without being affected), and *incubatory carriers* (who incubate pathogen into a large number without falling ill).

6.   Transmission of Pathogens

For perpetuation of disease and survival of the pathogen, transmission from one host to other occurs by any of four main routes: air borne, contact, vehicle and vector-borne.

(*i*) **Air-borne Transmission:** The pathogens remain suspended in air and are transmitted through droplet nuclei *i.e.* small particles (1-4 µm diameter) left from evaporation of large particles (10 µm diameter). The droplet nuclei remain in air for hours or days and carried to individuals because the pathogens cannot grow in air. Examples of some air borne diseases are chickenpox, flu, measles, mumps, viral pneumonia, diphtheria, pneumonia, tuberculosis, meningitis, etc.

(*ii*) **Contact Transmission:** Some of the pathogens spread when contact of host is done with the reservoir of pathogen. In

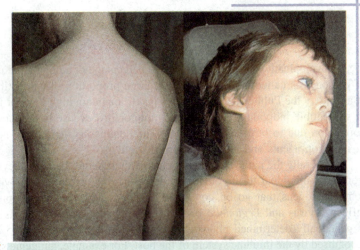

(a) Skin rashes seen during measles.   (b) Typical glandular swelling associated with mumps.

other words contact refers to person-to-person contact through touching, kissing, or sexual contact. The diseases that spread through contact are herpes and boils (through contact of oral secretions or body lesions), infection of staphylococci (by nursing mothers), and AIDS and syphilis (through placenta or blood to blood contact).

Table 20.1: Non-human reservoirs of infectious agents (based on Youmans *et al*, 1985).

Disease	Etiological agent	Nonhuman hosts	Transmission to humans
Anthrax	*Bacillus anthracis*	Cattle, horse, sheep, goat, cats dogs, birds	Direct contact ingestion/ inhalation of spores.
Encephalitis	Arbovirus	Birds, rats	Mosquito
Giardiasis	*Giardia lamblia*	Rodents, cats, dogs, cattle	Contaminated water
Glanders	*Pseudomonas mallei*	Horses	Skin contact, inhalation

Disease	Etiological agent	Nonhuman hosts	Transmission to humans
Plague	*Yersinia pestis*	Domestic rats,	Flea bite
Rabies	Rabies virus (Rhabdovirus group)	Dogs, bats, cats, cattle	Bite of rabid animals
Q Fever	*Coxiella burnetii*	Cattle, sheep, goat	Inhalation of infected soil or dust.
Tuberculosis	*Mycobacterium bovis*	Dogs, cats, cattle	Milk, direct contact

(*iii*) **Vehicle Transmission:** Vehicle refers to inanimate materials such as utensils, towels, beddings, surgical materials, needles, food, water, etc. Bacteria spreading through food and causing food poisoning are *Staphylococcus, Bacillus cereus, E.coli, Vibrio cholerae, Salmonella typhi, Clostridium difficile*, etc.

(*iv*) **Vector-borne Transmission:** A living organism that transmit a pathogen is known as *vector* such as vertebrates (e.g. dogs, cats, bats, goats, sheep, etc) or arthropods (e.g. fleas, mites, insects, ticks, etc.). For example, flies carry *Shigella* on their feet from faeces to food materials.

Mosquito is the vector of pathogen *Plasmodium vivax* causing malaria.

Moreover, when pathogen does not undergo morphological and physiological changes within the vector it is called harborage transmission e.g. the plague pathogen, *Yersinia pestis*. When the pathogen undergo morphological changes within the vector, it is called biological transmission e.g. *Plasmodium vivax*.

7. Control of Infectious Diseases

Since the infectious diseases are spread by several agents and cause epidemics, it can be controlled by one or several measures. This involves the breaking of the links of disease cycle, eliminating the reservoirs of disease and making the individuals resistant (*i.e.* immunization).

(*i*) **Breaking the Links of Disease Cycle:** As discussed earlier the pathogens survive on animate or inanimate for sometime where from transmitted to suitable host. Therefore, if links between two stages of disease cycle are broken, further spread of the pathogen does not occur. This includes general sanitation methods: (*a*) pasteurization of milk, (*b*) destruction of vectors by spraying insecticides (*e.g.* thiodon, malathion, etc), (*c*) chlorination of water supply, (*d*) inspection of food and individuals handling it, etc.

(*ii*) **Elimination of Source of Infection:** The source of infection can be eliminated by (*a*) adopting quarantine (legal prohibition of entry of goods, animals, etc from one country to other or one state to other within a country) and isolating the carriers, (*b*) destruction of animal reservoir (*e.g.* the cattle infected with foot and mouth disease virus are killed in other countries), (*c*) treatment of sewage (to check water borne transmission of pathogens), and (*d*) use of chemicals by individuals to eliminate the pathogens.

(*iii*) **Immunization of Individuals:** For increasing the level of herd immunity, mass immunization programmes were launched. For example during 1960's in India mass immunization programme of children against chicken pox was done. Secondly, Pulse Polio Immunization was launched in December 1995. Thereafter, this programme is conducted twice a year in India.

However, at International level, several programmes have been launched by the World Health Organisation (WHO), for example AIDS, etc.

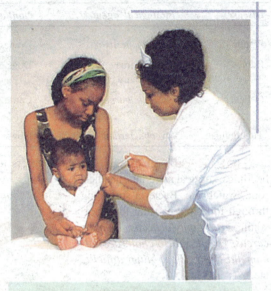

Immunising the child against diseases.

B. Parasitism

Relationships of the parasites with their respective hosts are called parasitism. There are two important processes that play a significant role in parasitism, infections and intoxications.

1.   Infection

Infection refers to entry of a pathogen into host tissues after transmission, and its growth and reproduction resulting in disease. The events of infection are as below.

(*i*) **Attachment and Colonization by the Pathogen:** The host contains its own normal microbial flora (see chapter 24, *Microbial Ecology*). A pathogen enters into host and compete normal microbial flora. Pathogens are equipped with certain specialized structures (e.g. haemagglutinin, fimbriae, capsule, lectin, mucus gel, pili, receptors, etc) that provides high degree of specificity to a particular tissue and help in colonization. These specialized molecules or structures are also known as adherance factors or adhesins. The adhesins bind to complementary receptor sites present on cell surface of the host.

(*ii*) **Entry of the Pathogen.** After attaching to the surface of epithelium, pathogen enters deeper into the epithelium through (*a*) producing lytic substances and dissolving host tissues, (*b*) disrupting the cell surface, or (*c*) degrading carbohydrate-protein complex between the cells. Moreover, entry of the pathogen inside body is also facilitated by already made breaks, lesions or ulcers in mucus membrane, wounds, burns or abrasions on skin, wounds created by arthropods, etc.

After entering inside host tissues the pathogen disseminates throughout the whole body. This is accomplished by specific products and/ or enzymes that help spreading as given below:

(*a*) *Coagulase* by *Staphylococcus aureus* that protects the pathogen from phagocytosis and keeps it isolated from other defense mechanism of the host,

(*b*) *Collagenase* by *Clostridium* sp. that breaks collagen and allow to spread the pathogen,

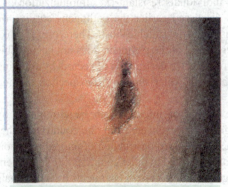

Pathogens enter the body through wounds.

(c) *elastase and alkaline protease* by *Pseudomonas aeruginosa* that breaks laminin associated with basement membrane,

(d) *haemolysins* (by staphylococci, *E. coli, Clostridium perfringens*) that break down erythrocytes causing anaemia and make iron available to the microorganism for growth,

(e) *immunoglobulin* A *protease* by *Streptococcus pneumoniae* that cleaves immunoglobulin A into Fab and Fc fragments,

(f) *lecithinase* by *Clostridium* spp. that breaks down lecithin of plasma membrane and helps the pathogen to spread,

(g) *porins* by *Salmonella typhimurium* that inhibits leukocyte phagocytosis through activation of the adenylate cyclase system, etc.

(iii) **Growth and Multiplication of the Pathogen:** After different mechanisms ultimately the pathogen reaches to terminal lymphatic capillaries surrounding the epithelial cells, which merge into large lymphatic vessels. The lymphatic vessels connect the circulatory system and discharge the pathogen therein. Finally, pathogen reaches to all the body organs. The pathogen gets proper nutrients, optimum pH and temperature in certain specific areas. This results in growth and multiplication of the pathogen. Moreover, some pathogens have evolved specificity to multiply either inside host cells, in tissues, in blood plasma, and also can generate its own nutrient-gathering mechanism. Sometimes the metabolic products of blood become toxic and results in a condition called *septicemia*.

2. Intoxications

Intoxications are the diseases caused due to entry of toxins (Latin *toxicum* means poison) into the host systems or body organs. Intoxications may occur in the presence or absence of the pathogen. The conditions resulted from a toxin is called *toxemia*. Toxins are of two main types: exotoxins and endotoxins.

(i) **Exotoxins:** Exotoxins are heat-labile, soluble proteins secreted by the pathogen that after release is circulated to other part of body tissues or target cells. However, after treatment with formaldehyde the toxicity of exotoxins is lost but their antigenic properties are retained. They have ability to induce the production of antibodies *i.e.* antitoxins which react with toxins and neutralize them. In this inactivated form these exotoxins are known as toxoid. Now-a-days toxoids are used as vaccines to immunize against several diseases e.g., tetanus, diphtheria, botulism, etc. The general characteristics of exotoxins are that these are (a) synthesized by exotoxin genes present on plasmid of the pathogen, (b) heat-labile protein which becomes inactivated between 60 and 80°C, (c) toxic in low doses *i.e.* in µg/kg of substrate, (c) specific in their action, (d) easily inactivated by iodine, formaldehyde, etc to form toxoids (which show immunogenic properties), (e) highly immunogenic and induce the production of neutralising antibodies (antitoxins), (f) named after the name of disease *e.g.* botulinum toxin, diphtheria toxin, etc., and (g) of different types based on their mode of action such as neurotoxins (affecting nerve tissues), cytotoxins (affecting general tissues), enterotoxins (affecting intestinal mucosa) (Gill, 1982) (Table 20.2).

Table 20.2 : Example of exotoxin producing bacteria (based on Prescott *et al.,* 1996).

Exotoxins	*Bacterial species secreting toxins*
Cytotoxins	*Clostridium difficile, Shigella* spp; *Staphylococcus aureus, Vibrio parahaemolyticus*
Enterotoxins	*Clostridium perfringens, E.coli, Klebsiella* spp., *Salmonella* spp., *Vibrio cholerae.*
Neurotoxins	*Bacillus cereus, Clostridium botulinum, C. tetani, S. aureus*

Mechanism of action: The mechanism of transport of exotoxin can be explained with its structural subunits (A and B) of polypeptide chain that constitute it. Enzymatic subunit A has toxic effect on cell after entering inside, whereas the subunit B binds to target cells. Subunit B does not exert toxic effect on the cell. It interacts with its specific receptor sites present on target cell. Thereafter, subunit A that lacks binding site is delivered inside the cell. This mechanism is called AB model for transport of exotoxins.

The mechanism of action of exotoxins is quite complex. It affects by several ways such as: inhibiting the protein synthesis, disrupting the membrane transport, causing damage to plasma membrane, etc.

(ii) **Endotoxins:** The outer membrane of cell wall of most of the Gram-negative bacteria consists of lipopolysaccharide (LPS) (see Chapter 4, lipopolysaccharide - Fig. 4.10). LPS is toxic in certain conditions and, therefore, it is called *endotoxin* because it is tightly bound to cell wall and released only after lysis of microorganisms. Lipid A portion is the toxic component of LPS. It is a complex array of lipid residues. Lipid A component is associated with all the properties of the microorganism. Moreover, endotoxins are (*a*) heat-labile, (*b*) toxic only at high doses (i.e. mg/kg amount of substrate), (*c*) capable of producing systemic effects such as fever, shocks, blood coagulation, weakness, diarrhoea, inflammation, and fibrinolysis i.e. enzymatic breakdown of fibrin (which is a major component of blood clots) (Gill, 1982; Rietschet and Brade, 1992).

Mechanism of action. Both *in vivo* and *in vitro* endotoxins initially activate Hageman Factor (i.e. blood clotting factor VII) which in turn exerts the following four systemic effects.

(*a*) they start the blood clotting cascade that leads to blood coagulation, thrombosis, and acute disseminated intravascular coagulation which in turn depletes platelets and various clotting factors and results in clinical bleeding;

(*b*) they can activate the complement system which leads to inflammation;

(*c*) they can activate fibrinolysis; and

(*d*) they can trigger a series of enzymatic reactions that lead to the release of bradykinins (a potent vasoactive peptide causing), and other vasoactive peptides which cause hypotension.

For comparison a summary of characteristics of exotoxins and endotoxins is given in Table 20.3.

Table 20.3 : Differences between exotoxins and endotoxins.

Features	Exotoxins	Endotoxins
Source	Secreted by microorganisms	Secreted from cell walls of Gram-negative bacteria.
Chemical Nature	Protein	Lipopolysaccharide
Lethal dose	Small amount	High amount
Heat tolerance	Heat labile *i.e.* easily inactivated by boiling	Heat-stable *i.e.* tolerate autoclaving
Immunology	Can be converted to toxoid and readily neutralised by antitoxin	Cannot form toxoid; neutralization with antitoxin is not possible or possible with much difficulty.
Pharmacology	Highly specific in mode of action	Action is governed by blood change shock and pyrogenicity

C. Antimicrobial Chemotherapy

1. History of Development of Antibiotics

There are several ways of killing the microorganisms such as sterilization, pasteurization by using high temperature, UV light, high pressure, steam, chemicals, etc. However, when the pathogenic microbe is inside the human body none of these methods are feasible except using antimicrobial chemotherapeutic agents that include *antibiotics* (of microbial origin) and artificially synthesised compounds (drugs).

Beginning of chemotherapy dates back to the work of a German physician, Paul Ehrlich (1854-1915). He propounded that a chemical which kills pathogens but not human cells can be used for the control of a disease. However, he hoped that a dye, tryphan blue, which was active against the trypanosome and showed toxicity probably as "magic bullet" (binding around the pathogen and destroying them), could be used therapeutically. Subsequently he tested a large number of arsenicals against syphilis–infected rabbits and found them active against syphilis. In 1927, a German chemical industry, I.G. Farbenindustrie, started producing a large number of chemicals and other dyes under the guidance of Gerhard Domagk. Domagk found that Prontosil Red (a new dye for staining leather) was nontoxic against animals and toxic against streptococci and staphylococci. He published this result in 1935. In 1935 a French scientist, Therese Trefouel found that Prontosil Red was converted in the body to sulfanilamide which is the active compound. For the discovery of this sulfadrugs, Domagk was awarded Nobel Prize in 1939.

Alexander Fleming discovered the antibacterial properties of Penicillium.

In 1896, penicillin was discovered by a 21 year old French medical student named Ernest Duchesne but his work was forgotten. It was rediscovered by a Scottish physician Alexander Fleming. During the First World War Fleming was interested to findout such compounds that would kill pathogens causing wound infection. Surprisingly, one day on 28th September 1928, he found that *Penicillium notatum* (a fungus) colonized the Petri plates inoculated with staphylococci and inhibited the growth of the later. From this observation a new era of medical microbiology was borne. Fleming observed that in the area where secondary metabolites were secreted, the bacterial cells of staphylococci were destroyed and in that area there developed a clear inhibition zone. He tried to isolate the inhibitory active compound to which he called penicillin but he was not successful. He published several papers upto 1931, and left the research work later on. Fleming's work was much appreciated throughout the world. In 1939, H. Florey who was Professor of pathology at Oxford University and busy in testing antibacterial activity of sulfonamides, lysozymes and others, thoroughly studied the paper's of Fleming. Thereafter, one of the associates of Florey, named Ernst Chain obtained *P. notatum* from Fleming, and started culturing the fungus and purifying penicillin. Florey and Chain jointly contributed a lot. They were much helped by a biochemist, Norman Heatley. Heatley developed the technique for assay and purification of crude penicillin for further experimentation. In 1940, Florey and Chain found that mice suffering from staphylococci or streptococci survived well when injected with penicillin. Thereafter, successful human trials were also reported by them. In 1945, Fleming, Florey, and Chain were awarded Nobel Prize for the discovery and production of penicillin (Hare, 1970).

In 1944, Selman Waksman discovered a new antibiotic called streptomycin from *Streptomyces griseus* which is a member of actinomycetes. He was awarded Nobel prize in 1952 for the discovery of streptomycin. Streptomycin helped the patients to screen about 10,000 strains of soil bacteria and fungi. Later on chloramphenicol, neomycin, tetracycline and terramycin were discovered during 1950s.

2. General Features of Antimicrobial Drugs

Some of the features of antimicrobial drugs are given below:

(*a*) Antibiotics are biologically active against a large number of organisms even in extremely low concentration. It is interesting to observe that 0.000001 g/ml of penicillin G has a pronounced effect on bacteria sensitive to this antibiotic. Therefore, the minimal inhibitory concentration (MIC) of an antibiotics effective against different microorganisms, or MIC of different microorganisms for a given microbe differs. MIC refers to the lowest concentration of drugs required to kill the microbial pathogen. It is also called minimal lethal concentration.

(*b*) These are very selective in posing toxicity *i.e.* they must kill or inhibit the pathogenic microbe without having harmful effect, or having least harm to the host. This is called selective toxicity. For example penicillin acts against Gram-positive bacteria (*e.g.* streptococci), while streptomycin acts against Gram-negative bacteria *e.g. E. coli*.

(*c*) The degree of selective toxicity is expressed in terms of therapeutic index which is the ratio of toxic dose to the therapeutic dose. The therapeutic dose is the drug level required for treatment of a particular infection, while the toxic dose is amount of drug at which it becomes too toxic to the host.

(*d*) The range of effectiveness of antimicrobial drugs varies. These may be *narrow-spectrum drugs* (effective only against a limited number of pathogens), or *broad-spectrum drugs* (effective against a large number of pathogenic microbes). However, on the basis of groups of microorganisms, the antimicrobial drugs are also classified as antibacterial (effective against bacteria), antifungal (against fungi), antiprotozoan (against protozoan) and antiviral (against viruses).

(*e*) Antibiotics are secondary metabolities *i.e.* they are not required during the growth by microorganisms. Therefore, these accumulate either inside the cell or secreted outside the cell. Some of the antibiotics are volatile as well. In recent years, the natural molecule of the antibiotics are being chemically or biologically modified, and new ones being produced. Such modified antibiotics are known as semisynthetic antibiotics *e.g.* ampicillin, methicillin, carbenicillin, etc. The semisynthetic antibiotics are more important as compared to natural antibiotics. Thus, the semisynthetic antibiotics are natural antibiotics that have been chemically modified by the addition of extra chemical group so that it could not be inactivated by pathogens.

(*f*) The potency of antibiotics is normally expressed in terms of *units/ml* or solution or in one milligram [U (units)/mg]. After the discovery of antibiotics it was found that 1 mg of streptomycin is equivalent to 1000 units. Now-a-days the antibiotic vials' label shows the quantity expressed in terms of weight as well as activity. For example, 1 mg of the penicillin G (benzylpenicillin) is equivalent to 1667 units.

3. Mechanism of Action of Antimicrobial Drugs

After administration of antimicrobial drugs against a particular pathogen, a patient feels relief. This is due to killing of infectious microorganisms involved with a specific disease. It is interesting to know how does it act? The mechanism of action of each antimicrobial drug differs. However, on the basis of their mode of action, it can be grouped as below:

(*i*) **Inhibition in Synthesis of Bacterial Cell Wall:** Penicillins stop the synthesis of cell wall of Gram-positive bacteria that are synthesizing new peptidoglycan. It inhibits transpeptide enzymes which are responsible for cross-linking of the polypeptide chain of the bacterial cell wall peptidoglycan (see Figs. 4.8 and 4.9). It also activates cell wall lytic enzymes. The other antibiotics that inhibit cell wall of bacteria are ampicillin, carbenicillin, methicillin, bacitracin, vancomycin, cephalosporin, etc.

(*ii*) **Disruption of Cell Wall:** Polymyxin B binds to plasma membrane and disrupts its structure and properties of permeability.

(*iii*) **Alteration in Membrane Function:** Gramicidin can alter the function of cell membrane. The disorganisation of the state and function have been well noticed. It brings about changes in permeability of cell membrane and causes a quick loss of amino acids, minerals, phosphorus and nucleotides. Also it inhibits the energy metabolism bringing about alteration in permeability e.g. polymyxin, nystatin, etc.

(*iv*) **Inhibition in Protein Synthesis:** Protein synthesis is the universal phenomenon of a living cell. However, some antibiotics inhibit protein synthesis of pathogenic microorganisms and act differently. Streptomycin binds to 30S subunit of the bacterial ribosome and cause misreading of mRNA and, therefore, inhibits protein synthesis at the final stage (i.e. during transport of aminoacyl tRNA to ribosomes. However, it does not affect the initial stage i.e. activation of amino acids (see Chapter 10). Similar is the effect of gentamycin and tetracycline. Chloramphenicol and erythromycin bind to 50S ribosomal subunit and inhibits peptide chain formation and elongation respectively by inhibiting peptidyl transferase.

(*v*) **Inhibition in Synthesis of Purines and Pyrimidines:** Mitamycin C inhibits the growth of many bacteria, protozoa, etc and arrests the growth of tumour cells. The chromatophore of actinomycin is incorporated between DNA base pairs, whereas mitamycin blocks the synthesis of DNA without affecting RNA and protein synthesis. Rifampicin inhibits DNA-dependent RNA polymerase and, therefore, blocks protein synthesis.

(*v*) **Inhibition in Respiration:** Antimycins inhibit the growth of some fungi. This group of antibiotics inhibit oxidation of succinate. Valinomycin is active against Gram-positive bacteria and inhibits oxidative phosphorylation. Gramicidins are inhibitors of phosphorylation.

(*vii*) **Antagonism of Metabolic Pathways:** There are some of the antibiotics that block the functioning of metabolic pathways through competitive inhibition of key enzymes. Such valuable drugs are known as antimetabolites. For example sulfonamides (*e.g.* sulfonilamide, sulfamethoxazole and sulfacetamide) and other drugs inhibits folic acid metabolism through competing with *p*-aminobenzoic acid. Trimethoprim blocks tetrahydrofolate synthesis through inhibition of the enzyme dihydrofolate reductase. Isoniazid disrupts either pyridoxal or NAD metabolism and functioning.

4. Drug Resistance

Drug resistance is a great problem for successful treatment of an infectious bacterial disease. Bacteria acquire drug resistance and spread within its population. However, the mechanism of drug resistance in two different bacteria differs for a single class of drug. In addition, resistance may arise spontaneously and selected. Moreover, mutants are not formed by direct exposure to a drug. When pathogens check the entry of a drug inside their cell envelope, they become resistant to those drugs. For example penicillin G cannot penetrate the outer membrane of Gram-negative bacteria; therefore, these are resistant to penicillin G (Clowes, 1973).

Secondly, bacteria pump the drugs out side their cell envelope after drugs have entered the cell. For example, some pathogens have plasma membrane translocases that expel drugs outside the envelope. These transport proteins are called multi-drug resistance pumps. In addition, many bacteria resist the drugs by chemical modification. For example, β-lactam ring of many penicillins is hydrolysed by the enzyme penicillinase.

Resistance in bacteria may also arise through using alternate pathway to bypass the sequence inhibited by the agent, or increase the production of the target metabolites.

Table 20.4 : Antimicrobial drugs used for certain diseases/pathogens

Pathogens	Disease	Drugs
A. Bacterial Diseases and Antibacterial Drugs		
Corynebacterium diphtheriae	Diphtheria	Erythromycin, Penicillin G
Streptococcus pneumoniae	Pneumonia	Penicillin G, erythromycin or cephalosporin
Escherichia coli	Urinary tract infection	Ampicillin, cephalosporin, sulfonamide
Staphylococcus	Wound infection, pneumonia, boils	Penicillin G or V Cephalosporin or vancomycin
Neisseria gonorrhoea	Gonorrhea	Amoxicillin, ciftriaxone
Salmonella typhi	Typhoid fever	Ampicillin, chloramphenicol, ceftriaxone
Shigella dysenteriae	Dysentery	Ciprofloxacin, quinolones, trimethoprim-sul famethoxazole
Mycobacterium tuberculosis	Tuberculosis	Isoniazid plus rifampin with/without pyrazinamide
Treponema pallidum	Syphilis	Penicillin G, tetracycline, ceftriaxone
B. Fungal Diseases and Antifungal Drugs		
Candida albicans	Candidiasis in different body organs	Nystatin, Fluconazole
Trichophyton, Microsporum and Epidermophyton	Cutaneous mycosis	Griseofulvin (orally), Miconazole, ketoconazole and imidazole (as cream or solution)
Blastomyces dermatitidis, Coccidioides immitis Histoplasma capsulatum	Blastomycosis Coccidioidomycosis Histoplasmosis	Amphotericin B, Fluconazole 5-flucytosine

D. Origin of Drug-resistance

Drug-resistance in bacteria arises through the drug-resistance genes present on bacterial chromosome and plasmids. The plasmids conferring drug resistance are called resistance plasmids (R plasmids). (For detail see Chapter 4, section *types of plasmid*). Plasmid associated genes are known to confer resistance to chloramphenicol, penicillins, cephalosporins, aminoglycosides, erythromycin, tetracyclines, sulfonamides, etc. The plasmid resistance genes are transferred to progeny cells through genetic recombination *i.e.* conjugation, transformation and transduction (Davies, 1994) (see chapter 8, *Microbial Genetics*, section *transfer of genetic materials in prokaryotes*).

Efforts can be made to discourage the emergence of drug resistance. Drugs can be given in high amount to kill susceptible, and mutant bacteria. Two drugs can be given simultaneously in a hope that any of two will destroy the pathogen. Thirdly, new antibiotics must be used time to time. Example of some of antimicrobial drugs is given in Table 20.4.

See chapter 17, *Industrial Microbiology* for details of antibiotic structure and production.

QUESTIONS

1. What do you mean by epidemiology? Write in brief the significance of study of epidemiology.
2. What is the disease cycle? Write in brief the different components of disease cycle.
3. Write an extended note on transmission of pathogens.
4. What is parasitism? Write in brief the processes of parasistism.
5. Write an essay on general features of antimicrobial drugs.
6. Give a brief account of action of antimicrobial drugs.
7. What is drug resistance? Write in brief the origin of drug resistance.
8. Write an essay on history of development of antibiotics.
9. Write short notes on the following:

    (*i*) Herd immunity, (*ii*) Disease cycle, (*iii*) Transmission of pathogens, (*iv*) Control of infectious diseases, (*v*) Parasitism, (*vi*) Infection, (*vii*) Intoxication, (*viii*) Exotoxins, (*ix*) Endotoxins, (*xi*) Antimicrobial drug resistance, (*x*) Antimicrobial drugs.

REFERENCES

Clowes, R.C. 1973. The molecule of infectious drug resistance. *Sci. Am* 228: 19-27.

Davies, J. 1994. Inactivation of antibiotics and the dissemination of resistance genes. *Science* **264:** 375-382.

Gill, D.M. 1982. Bacterial toxins: A lable of lethal amounts. *Microbiol. Rev* **46** : 86-88.

Hare, R. 1970. *The Birth of Penicillin*. Atlantic Highlands, N.J. Allen and Unwin.

Mandell, G.L., Benett, S.E. and Dolin, R. 1995. *Principles and Practice of Infectious Diseases*. 4th ed. New York: Churchill Livingstone.

Prescott, L.M., Harley, J.P. and Klein, D.A. 1996. *Microbiology* 3rd ed., WmC. Brown Publ., Chicago.

Rietshcel, E.T. and Brade, H. 1992. *Bacterial endotoxins*. *Sci. Am*. **267** (2) : 54-61.

Youmans, G. *et al.* 1985. *The Biological and Clinical Basis of Infectious Diseases*., W.B. Saunders, Philadelphia.

Medical Microbiology I : Protozoan and Fungal Diseases

21

A. Protozoan Diseases
B. Fungal Diseases

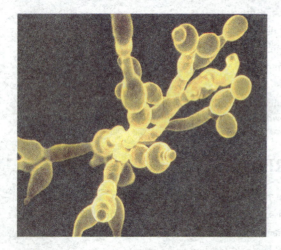

A large number of protozoa and fungi occur in nature. Some of them cause serious diseases in human and other animals. Some of them are fatal also. This chapter covers some important protozoan and fungal diseases.

A. Protozoan Diseases

There is a large number of animal parasites that cause diseases in humans such as arthropods (lice, fleas, etc), worms (liver flukes, tapeworm, pinworm, etc), and protozoa. In this section some human diseases caused by only protozoa are described.

I. *Toxoplasma gondii*

Toxoplasma gondii is an intracellular parasitic sporozoan (Fig.21.1). It causes a disease known as toxoplasmosis when transmitted either from soil or other animals (e.g. domestic cats). The exact nature and life cycle of *T. gondii* are not known (Fig.21.2). The oocytes of the parasite released from faeces (of infected cats, horses, sheep, goats, swines) are inhaled or engulfed by human or other animals.The sporozoites emerge as trophozoites which can reproduce in tissues of a new host.

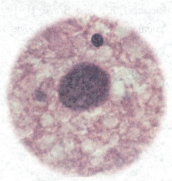

Toxoplasma gondii

After infection in adults symptoms may or may not develop. However, the minor symptoms resemble with that of viral influenza or mononucleosis. The most remarkable symptom develops in pregnant women where the parasite crosses the placenta within three months of pregnancy, and established inside the foetus resulting in infection of uterus. Finally, it results in congenital abnormality or stillbirth of a child. However, if the women is infected at later stages of pregnancy, symptoms occur. During pregnancy about 1% women get infected and about 25% infants after birth show symptoms. Therefore, at this stage the pregnant women should be careful about the cats. Oocytes in their faecal material must be checked.

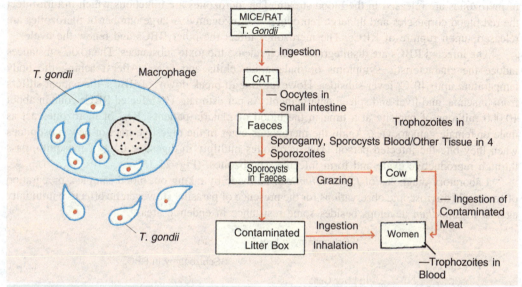

Fig. 21.1 : Structure of *Toxoplasma gondii*.   Fig. 21.2 : Life cycle of *Toxoplasma gondii*.

In addition, when the pathogen is outside the foetus, it migrates to intracellular macrophages, gets established as intracellular parasite, becomes centered in the central nervous system and causes encephalitis.

Examination of tissues and *T. gondii* is useful for diagnosis of taxoplasmosis. Antibodies may be detected by ELISA and direct fluorescence tests (see Chapter 19, *Immunology*). To control the infection antiprotozoan drugs e.g. pyrimethamine should be given in combination with sulfonamides.

II. *Plasmodium*

Plasmodium is a common spore-forming sporo-zoan parasite on human which causes malaria. Until 1935, malaria was a common disease in humans creating a serious problem and causing death *en masse*. But with the development of awareness and research programmes, antimalarial drugs were formulated. Still in African and Asian countries including India, malaria has not been fully eradicated despite of efforts from Government organizations. As per recent estimates about 300 million people suffer from malaria and about 2-4 million die each year.

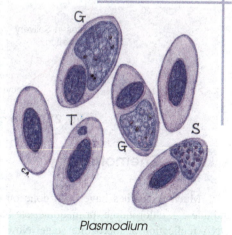

Plasmodium

Four pathogenic species of *Plasmodium* viz., *P. vivax*, *P. falciparum*, *P. malariae* and *P.ovale* have been recognised. *P. falciparum* is the most dangerous species followed by *P. vivax*.

1.   Life Cycle

Malaria is associated with a mosquito vector, *Anopheles*.The life cycle of *Plasmodium* is very complex.The mosquito carries the inocula of *P. vivax* where it multiplies and produces the sporozoites. After mosquito bite, the sporozoites present in saliva are delivered in the blood of human. Thirty minutes after bite, the sporozoites enter in liver cells and undergo reproductive schizogamy through a series of intermediate stages. At the last stage of schizogamy, a large number of merozoites are released in the blood stream. The merozoites are infectious which in turn infect the red blood corpuscles and undergo reproductive schizogamy. A large number of merozoites are released upon rupture of RBCs. The merozoites infect the other RBCs and renew the cycle.

The infected RBCs are disintegrated which release the toxic substances. The toxic substances induce the characteristic symptoms of malaria i.e. chills and fever. After reaching the body temperature upto 40°C, fever subsides. However, due to break down of RBCs, the patient suffers from anaemia, and liver and spleen enlargement. As per estimate 1% infected RBCs contain about 10,000 millions of parasite at a time in the blood of malaria patients. Rest of merozoites act as male or female gametocytes. Again the merozoites enter in the digestive tract of new mosquitoes when they bite the infected person. The merozoites multiply in digestive tract of mosquito, pass through reproductive cycle and form infective sporozoites. (Fig. 21.3)

Laboratory diagnosis of *Plasmodium* can be made by taking out blood samples, preparation of smear and microscopic observations for the presence of parasite. However, no effective immunity against *Plasmodium* develops besides some resistance in endemic areas.

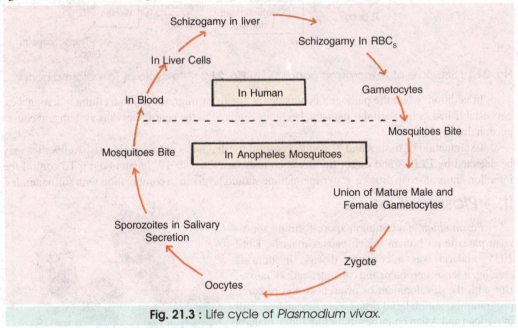

Fig. 21.3 : Life cycle of *Plasmodium vivax*.

2.   Chemotherapy

Much researches have been done on chemotherapy of malaria. Quinine and derivatives of quinine e.g. chloroquine, primaquine, etc have been used for several years. Choloroquine inhibits DNA replication at merozoite stage. Now-a-days, resistance to these drugs have been reported. Therefore, in recent years drug combinations have been recommended e.g. Fansidar which is a combination of pyrimethamine and sulfadoxine. The combinations of drug act synergistically. However, effective control of malaria is not known except chemotherapy.

3. Malarial Vaccines

In recent years, much effort has been made for the preparation of effective vaccine through gene mapping, preparation of monoclonal antibodies, and protein synthesis. In 1975, Rath and Victor at New York University, Medical Care Centre, identified a surface antigen of the sporozoites known as circumsporozoite (CS). The CS antigen was used to produce the monoclonal antibodies against *Plasmodium*. In 1983, the CS gene was cloned to produce the CS protein in large quantities and to analyse and prepare antibodies (Miller *et al.*, 1986). In 1987, for the first time human test for vaccine was done at the university of Maryland, School of Medical Science (U.S.A.) and the effectiveness of vaccine was determined in 1989. But much research work in needed on the preparation of vaccine for its use at several stages of the parasite.

4. Malaria Eradication Movement

Several times malaria eradication movement was launched by the government organizations to eradicate mosquitoes by spraying DDT, but the mosquitoes developed resistance. However, no significant control measure has been reported so far.

III. *Balantidium coli*

Balantidium coli causes balantidiasis or balantidial dysentery. It is the only largest and ciliated pathogenic protozoan of human intestine (Fig. 21.4). *B. coli* enters in the intestine when its cysts present in contaminated food or water are ingested by humans. After ingestion the parasite reaches to the colon, wall of cysts dissolved and trophozoites are released. The trophozoites feed upon bacteria, faecal debris and tissue of colon. When the faeces get dehydrated and pass through colon, the parasite becomes encysted, discharged through faeces and contaminate the water and food. In colon, the parasite remains in two forms i.e. vegetative form (trophozoite) and cyst form. It lives in large intestine and rarely invades the epithelial lining and cause abdominal cramping, vomiting, weight loss. Due to invasion of epithelium, there develops ulcer in large and small intestines with total dysentery. The cysts can be observed upon microscopic observation of faeces.

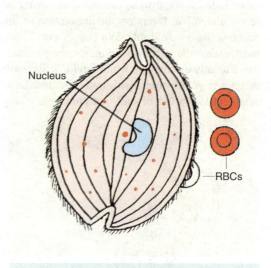

Fig. 21.4 : Structure of *Balantidium coli*.

IV. *Trichomonas vaginalis*

Trichomonas vaginalis is an inhabitant of vagina and urethra (Fig.21.6). It grows and increases its number over the normal microbial community, when acidity of vagina is disturbed from normal pH (3.8-4.4). The change in pH of vaginal fluid may be due to a loss of normal acid-producing

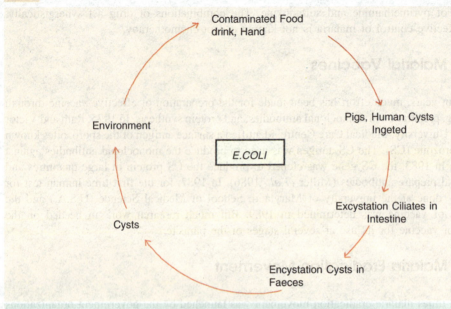

Fig. 21.5 : Life cycle of *Balantidium coli.*

bacterial flora or use of oral contraceptives. In such situation, it causes trichomoniasis or *'trich'* in female and rarely in males. *T. vaginalis* is also transmitted by sexual intercourse or contaminated examination tools, towels, catheters, etc. After infection, it causes a mild inflammation of vagina, cervix and volva. It renders the dissolution of infected surface tissues. Consequently, a yellow or cream-white ooze of foul odour comes out. This type of disease is also caused by *Candida albicans* and *Gardnerella vaginalis*. In males, the prostate, seminal vesicle and urethra may be infected and consequently discharge a white liquid. *T. vaginalis* is observed under the microscope by taking the samples from the discharged fluid. The parasite is also found in semen or urine in males. Effective treatment is the use of oral metronidazole (flagyl), and the other antiprotozoan drugs.

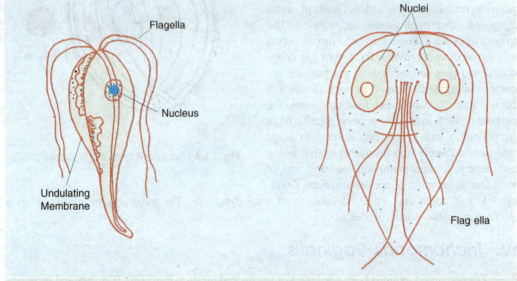

Fig. 21.6 : Structure of *Trichomonas vaginalis.*
(trophozoite)

Fig. 21.7 : Structure of *Giardia lamblia*

V. *Giardia Lamblia*

Giardia lamblia is a flagellated (8 flagella) protozoan parasite of human intestine (Fig.21.7) that causes the prolonged diarrhea in humans called *giardiasis*. It is a water-borne disease occurring throughout the world. The parasite is found in human faeces. The flagella of *G. lamblia* help to attach firmly at the intestinal wall of human. It does not infect the intestinal wall but increases its number in the lumen and interferes the food absorption. A large number of cysts are formed which are released with faeces. This disease persists for about a

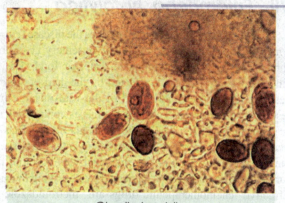

Giardia Lamblia.

week. It is characterised by malaise, nausea, weakness, weight loss and internal cramps with cronic greasy diarrhoea. The greasy consistency of faeces is the key feature of giardiasis. About 7% of total population is carriers of this disease. In addition, wild mammals especially beavers also shed the cysts into faeces which in turn contaminate water. The encysted protozoa is resistant to chlorine. Therefore, filtration of water supply becomes very important for elimination of cysts from water. Effective chemotherapeutic agent for giardiasis is the use of metronidazole or quinacrine hydrochloride.

VI. *Trypanosoma*

Trypanosoma is known as haemoflagellate because it infects the blood. Disease caused by the parasite is known as trypanosomiasis or sleeping sickness. *T. gambiense* and *T. rhodosiense* cause African sleeping sickness, whereas *T. cruzi* is responsible for chagas disease i.e. American trypanosomiasis. Wild animals such as rodents, opossums and armadillos are the reservoir for *T. cruzi*. The arthropod vectors are the kissing bug and the bed bug (*Cimex lectularis*). *T. gambiense* and *T. rhodosiense* are transmitted by *Glossina*, the tse-tse fly. The kissing bug bites the person near the lips. The trypanosomes grow inside the gut of bugs. When the bugs defecate during feeding, the bitten human often rubs the faeces into the wound of the other abrasions of skin or eyes. The parasite passes through the several stages of life cycle at the site of inoculation. At the inoculation site i.e. commonly near the eyes a swollen lesion develops. The pathogen infects the lymph and is carried by the blood to the other part of body organs such as spleen, liver and heart.

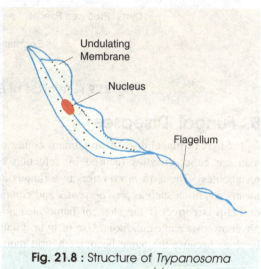

Fig. 21.8 : Structure of *Trypanosoma gambiense.*

In response to the parasite when inflammation occurs, it damages the macrophages and central nervous system. The loss of tissue progresses that leads to weakness, loss of appetite and apathy. Patients usually fall into coma and finally die. At this stage control of chaga's disease is very difficult. Laboratory diagnosis can be done by observing the parasite in blood (Fig. 21.8).

VII. *Entamoeba histolytica*

The name of pathogen itself defines the disease (*ent*=inside, *amoeba*=varied shape, *histo*=tissue, *lytica*-bursting). It causes amoebiatic dysentery i.e. amoebiasis throughout the world. The parasite spreads through the contaminated water or food. The normal level of chlorine does not kill the amoeba. The cysts are not affected by the acid of stomach i.e. HCl. Only the vegetative cells are destroyed. On the epithelial cell of the wall of large intestine the vegetative cells multiply. This results in severe dysentery where the faeces contain *mucus* and blood. The life cycle of *E. histolytica* is given in Fig. 21.9. The pathogen feeds upon RBCs and damages the tissue of gastrointestinal tract. During severe cases, amoebae may enter the tissues of vital organs and cause abscess in liver, lungs, intestinal wall, etc. The intestinal wall becomes perforated. If abscess is formed it is treated surgically. The pathogen is identified by observing the amoebae in faeces and RBCs in trophozoites. Effective chemotherapeutic drugs are metronidazole plus iodoquinol.

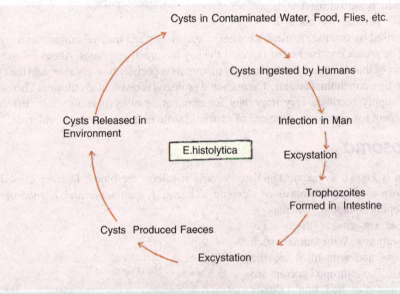

Cysts in Contaminated Water, Food, Flies, etc.

Cysts Ingested by Humans

Cysts Released in Environment

E.histolytica

Infection in Man

Excystation

Cysts Produced Faeces

Trophozoites Formed in Intestine

Excystation

Fig. 21.9 : Life cycle of *Entamoeba histolytica.*

B. Fungal Diseases

Study of fungal diseases in humans comes under **Medical Mycology**. The fungal diseases of man are either *mycoses* (caused by infection of fungi) or *toxicoses* (caused by toxic fungal metabolites). The term *myco* refers to a fungus and *osis* (a Greek suffix) or *iasis* (a Latin suffix) means condition such as *phycomycosis* and *candidiasis,* respectively. Using suffix of either Greek or Latin is correct if the stem of name also is derived from Greek root or Latin root such as phycomycosis and candidiasis. Use of hybrid names with Greek stem and Latin suffix should not be done because the name becomes hybrid names, for example a parasitic disease, ascariasis.

1.   Mycoses

In 1835, Bassi observed that a disease of the silk worm (*muscardine*) was caused by a fungus. Soon the etiology of favus, a type of ringworm and thrush in man were recognised. The attention of many investigators was attracted since the early discovery in medical mycology and parasitic reaction of many fungi and animal were paid due attention. Sabouraud (1910) published his classic

'Les Teignes' after a study of 20 years. During this period several fungi such as *Actinomyces bovis, Candida albicans, Crytococcus neoformans, Coccidioides immitis, Blastomyces dermatitidis, Sporothrix schenckii* and *Histoplasma capsulatum* were isolated, cultured, identified and described. Later on much impetus to study on infection in man was given. Upto 1925, diverse group of disease studies provided a broad foundation of medical mycology for its later development. Dermatologists and pathologists made most of the early studies without the collaboration of a professional mycologist. However, during the past four decades (upto, 1940) the mycologists paid due attention on the study of fungi of medical importance. Moreover, the study of fungal disease was coupled with that of immunity associated with mycoses. In addition to

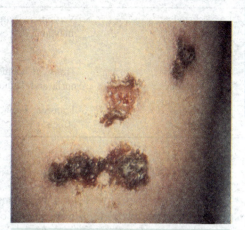

This photograph shows cutaneous ulcers formed due to mycoses.

defense mechanism and immunity as described in Chapter 19, *Immunology*, the other factors providing resistance to disease are food, general nutrition and vigorous good health.

2. Mycotoxicoses

Mycotoxicosis refers to poisoning by ingestion to fungal toxins i.e. mycotoxins in food which have been deteriorated by the growth of toxin-producing fungi. One of the classical examples is the ergot production by *Claviceps purpurea* in rye seeds. The other examples of mycotoxins are aflatoxins (by *Aspergillus flavus*), patulin (by *Pencillium claviforme*), ochratoxin (by *A. ochraceus*), zeralenone (by *Fusarium graminearum*). Mycotoxin production has been described separately in Chapter 18, *Microbiology of Milk, Dairy and Foods*.

Mycetismus (mushroom poisoning) : Use of edible mushroom has been in practice throughout the world since centuries B.C. Early naturalists collected the mushrooms from the field and tested their edibility or poisoning. Ingestion of toxic species due to erroneous identification resulted in acute discomfort, severe illness or even death. Moreover, a series of papers on mycetismus has been published by Ford (1923) which report about the cause of nausea, vomiting and diarrohea and some times death within 24 hours. The mushroom poisoning has its effects on gastrointestinal and nervous system (Table.21.1). Different types of mushroom toxins, their chemical constituents, affected organs and toxin producers have been listed in Table 21.2.

Table 21.1 : Classification of mycetismus (after Ford, 1923)

Diseases	Symptoms	Poisonous mushroom
Mycetismus gastrointestinalis	Mild gastrointestinal disorder	Russula emetica Boletus satanus, Lactarius torminosis Entoloma lividum
Mycetismus choleriformes	Severe gastrointestinal disorder	Lepiota morgani Amanita phalloides

Diseases	Symptoms	Poisonous mushroom
Mycetismus nervosus	Nervous system, gastro-intestinal distress	Amanita muscaria A. patherina Inocybe infelix Clitocybe illudens
Mycetismus sanguinareus	Transient haemoglob-nuria and Jaundice	Helvella esculenta
Mycetismus cerebralis	Transient hallucinogenic effects	Psilocybe sp. Panaeolus sp.

I. The Phycomycoses

The phycomycoses have several synonyms such as *mucormycoses, entomophthoromycoses, oomycoses,* etc. The *phycomycoses* are caused by such fungi that had been previously grouped in the class Phycomycetes. Recently, the fungi classified under Phycomycetes have been put under two sub-divisions Mastigomycotina (with two classes Oomycetes and Chytridiomycetes) and Zygomycotina (with two classes Zygomycetes and Trichomycetes).

Clark (1968) proposed the name mucormycoses for those mycoses caused by the members of the order Mucorales, and entomophthoromycoses for those mycoses caused by the fungi of the order Entomophthorales.

1. Epidemiology

Phycomycosis is sporadic in occurrence which is found in both the sexes and all races. It occurs mostly in patients suffering from *diabetes mellitus*. The fungi are thermophilic but growing between 35°C and 40°C. These also grow on plant leaves, decomposing debris, humus, fruits and other sugary substrates, intestinal tract of frogs and lizards. From these surfaces the conidia are dispersed in the environment and cause mycoses in men and animals.

2. Clinical Types

The phycomycoses are the fatal disease of man and animals in whom the vital organs are infected by the species of *Absidia, Mucor* and *Rhizopus*. Mycoses caused by *Absidia, Mucor, Rhizopus* and *Cunninghamella* (Table 21.3) occur usually in diabetic patients. In rabbits which have been made diabetic with alloxan, cerebral and pulmonary phycomycoses have been produced with *Rhizopus oryzae* and *R. arrhizus*. The characteristic features of mycoses are the invasion and growth in nasal mucosa and the underlying tissue followed by rapid destruction of tissue. In severe cases the hyphae extend to brain and invade the blood vessels. Some of the following clinical types are disscused below :

(i) Phycomycosis of Orbital and Central Nervous System: The pathogens infect the paranasal sinuses and spread in nasal mucosa and underlying tissues. It is followed by rapid disintegration of tissue resulting in severe cellulitis to the area of orbital. In severe cases even patients dies between 2 and 10 days. If the mycelia continue to grow at the same speed these may penetrate the lungs. Predisposing factor for establishment of mucormycosis may be the diabetic acidosis.

(ii) Disseminated Mucormycosis: In addition to the above infection sites, pulmonary lesions, gastrointestinal ulcers and other example of disssemination may also take place. *Cunninghamella* *elegans* is the etiologic agent of pulmonary infection. Some members of mucorales may cause abortion in cattles as well.

Table 21.2 : Mushroom toxins (after, Wieland, 1968)

Mushroom toxins	Chemical substances	Organ affected	Examples
1. Protoplasmic poisons			
(i) Amanita toxins (e.g. phalloidin, phalloin, phallisin, phallicidin, phallin B)	Five phallotoxins intestine blood	Liver cells	Amanita phalloides A. verna, A. tennifolia A. bisporigena
	Five amotoxins (e.g. cyclic octopeptides, gamma amanitin, amanin, amanullin		
(ii) Helvella toxins	Helvellic acid	Liver, nervous system	Helvella esculenta, H. gigas, H. underwoodii
2. Compounds with neurologic effects	Muscarine, Pilzatropine α-psilocybin, psilocin	Nervous system heart, lung (mascarine) spp., Psilocybe mexicana	Amanita muscaria Boletus spp, Russula
3. Gastrointestinal irritants	Resin like substances	Intestine, Liver	Lactarius torminosa Tricholoma pardinum

(iii) Skin Ulcer : *Mortierella* spp. are associated with dermal ulcers. In deep subcutaneous areas a few fragments of coenocytic hyphae can be observed.

(iv)   Entomophthoromycosis   : Entomophthoromycosis is of two types on the basis of causal organism. *Basidiobolus ranarum* causes the disease, *entomophthoromycosis basidiobolae*, whereas the mycosis caused by *Conidiobolus coronatus* is known as *entomophthoromycosis conidiobolae.*

Lie-Kian-Joe *et al.* (1956) reported three cases of mycotic infection by *B. ranarum* from Indonesia. *B. ranarum* infects sub-cutaneous tissues and muscle fascia. It spreads quickly over neck, arms and upper chest. It could be soon healed up without any treatments. However, some other cases were reported involving

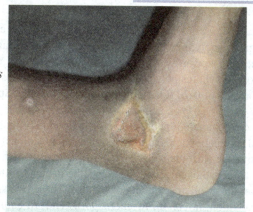

Skin Ulcer.

additional areas where no natural healing could be observed. The lesions developed seem firm, lens shaped and sharply demarcated. The demarcated lesions could be lifted from muscle fascia after insertion of fingers in a fold of skin under the edge of lesion. Moreover, the skin becomes hyperpigmented or discoloured. Infection by the pathogen occurs either through insect bites or piercing of contaminated thorns into the epidermal layer (Emmons *et al.*, 1977).

Entomphthoromycosis conidiobolae (=phycomycosis entomophthorae) is caused by *E.coronatus* in man and horses. The mycosis could be observed as nasal polyps and large granulomas of nasal cavity. Lesions begin first in the inferior turbinate thereafter, spread in submucosa, sutures and paranasal sinuses. Finally, the subcutaneous tissues of the face is covered. Moreover, the lesions spread beneath the facial muscles and widely covers the face. The patients feel discomfort from swollen tissue on the face but not pain. Clark (1968) studied this disease in Nigerians and found that the ratio of male and female was 10: 1.

Table 21.3 : Some examples of phycomycosis, pathogens and diseases.

Mycoses	Fungal pathogens	Disease
Mucormycosis	*Absidia corymbifera*	In bovine walls and lumen of blood vessels, Vascular thrombosis
	Mortierella wolfii	Bovine mycotic abortion
	Mucor pusillus	Pneumonitis in lungs
	Cunninghamella elegans	Pulmonary mycosis, lymphosarcoma, occlusion of pulmonary blood vessels.
	Rhizopus oryzae	Paranasal sinuses, cerebral infection in diabetics
Subcutaneous Phycomycoses		
(*i*) Entomophthoromycosis Basidiobolae	*Basidiobolus haptosporus*	In man, sub-cutaneous tissues, thorax vessels and lymph nodes, and upper arms
(*ii*) Entomophthoromycosis Conidiobolae	*Conidiobolus coronatus*	Nasal polyps in man and horse, massive subcutaneous invasion face
Saprolegniosis	*Saprolegnia ferax*	Common fish mycosis
Hyphomycosis	*Hyphomycoses destruens*	Mycosis of horses and mules

3.   Culture Characteristics

(*i*) *Absidia corymbifera*: It grows rapidly and covers the Petri plates within a few days. Colony is grey, wooly; mycelium coenocytic with numerous sporangiophores of a few millimeter in length emerging upwardly from the internodal portions on the stolon (Fig.21.6) and freely branched. Sporangiophores possess rhizoids. Sporangiophores branch to form corymbs and terminate to form sporangia of 0-70 μm diameter with a columella. Spores are spherical to oval, 2-3x3-4 μm in size.

(*ii*) *Rhizopus* : Species of *Rhizopus* produce simple but branched sporangiophores in some species. Sporangiophores develop individually or in groups from rhizoids. The adjacent rhizoids and sporangiophores are interconnected by stolons. Sporangiophores arise at the nodes i.e. directly above the rhizoids (Fig. 21.10). *R. oryzae*, which is used in making a fermented food (Koji) in Japan is pathogenic also. It infects the patients suffering from diabetes mellitus. *R. arrhizus* arises from the swollen segments of hyphae. These two species are the most frequent pathogens in men. In addition, *R. cohnii, R. microsporus* and *R. equinus* have been isolated from mycotic abortions in animals.

(*iii*) *Mucor* : *Mucor* produces grey mycelium on the nutrient medium. *Mucor* possesses simple or branched sporangiophores that come out from the surface of substratum. Neither stolons nor rhizoids are formed by this fungus. Sporangia are produced on sporangiophores which are large sized (Fig. 21.11). Sporangium consists of a columella in the centre. Spores are spherical and smooth walled. The pathogenic species in men are *M. ramosissimus, M. pusillus, M. racemosus*, etc.

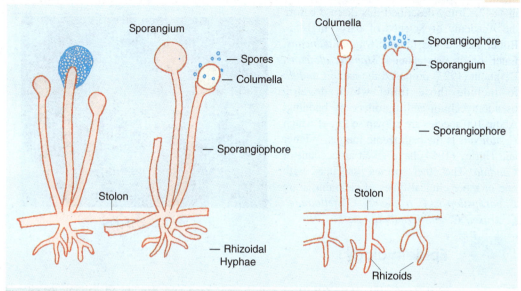

Fig. 21.10 : Thallus of *Rhizopus* showing sporangiophores, sporangia, rhizoids and stolon.

Fig. 21.11 : Thallus of *Mucor* sp.

(iv) Basidiobolus haptosporus : It causes entomophthoromycosis basidiobolae in Indonesia, tropical Africa and Southern Asia. The etiologic agent was identified as *B. ranaum* in some text. Srinivasan and Thirumulachar (1967) concluded its correct name as *B. haptosporus*. On the Sabouraud's agar it forms a grey to yellowish, thin, flat, glabrous and radially folded colony. Hyphae are 8 to 20 µm in diameter with occasional septa which become very numerous with the production of spores. Surface of the colony becomes covered within white bloom composed of very short aerial hyphae, sporangia, chlamydospores and zygospores. Sporangiospores are 30-50 µm in diameter (Emmons *et al.*, 1977).

(v) Candidiobollus coronatus (syn. Entomophthora coronata) : It grows rapidly on all common media. At first colonies are globose, abundant, short aerial. Conidia are globose and multinucleate. Conidia are capable of replicating. The hyphae are produced as the colony matures. In *C. coronatus* no sporangia are formed but the conidia (10-20 µm diameter) are produced on hyphae which are forcibly discharged from the tip of conidiophores. Hyphae produce a short conidiophore at the tip of which a secondary conidium is borne. This process of replication continues until the nutrients are exhausted.

4. Therapy

Diagnosis of mucormycosis on sinuses, orbits and mininges could not be done before its termination into fatal disease. However, the risk of mucormycosis could be reduced by controlling diabetes and using amphotericin B. So far no specific and successful chemotherapy is known. Martinson (1971) in one case reported a good response by sulfamethoxazol and trimethoprim.

II. Candidiasis

Mycosis caused by *Candida albicans* is called candidiasis. It is an acute or chronic superficial or disseminated mycosis. It is cosmopolitan in distribution. *C. albicans* grows prolifically, in the sufficiency of diets high in fruits and carbohydrates. The frequency of its occurrence in the intestinal tract of man in different climatic areas and socio-economic environment differs. For the first time in 1842, Gruby described this disease before the Academy of Science of Paris. In 1853, Robin named the fungus *Oidium albicans*. Later on it was named *Monilia albicans*. Berkhout (1923) proposed the name *Candida* to include those fungi which develop pseudomycelium and reproduce by budding. About 100 species are known so far, of which *C. albicans* is the pathogenic fungus. Winner and Hurley (1964) have given an account of *Candida*. The other species associated with one or more clinical types of candidiasis are *C. parapsilopsis, C. tropicalis, C. stellotoidea C. krusei, C. guilliermondii,* etc.

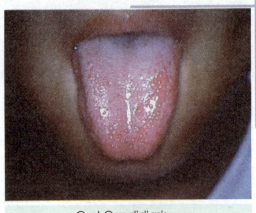

Oral Candidiasis.

1. Epidemiology

Mostly oral candidiasis is observed in new borns, aged and debilitated patients (rendered susceptible by another disease) of both the sex. However, due to long and continued therapy with broad spectrum antibiotics its association becomes possible. In diabetics, house wives or fruit canners, the intertriginous candidiasis becomes severe. It also occurs in such person who frequently have prolonged immersion of hands in water. Infection by *C. albicans* in endogenous, therefore, it is found in mouth and intestinal tract of a large number of humans. From these spots it spreads and causes skin/nail infection.

2. Clinical Types

It is difficult to evaluate the presence of *C. albicans* in cultures procured from various clinical materials. Some of the following types are described below:

(*i*) **Oral Candidiasis :** Discrete or confluent patches of cream white to grey pseudomembrane composed of hyphae and yeast of *C. albicans* cover the buccal mucosa and the other oral surfaces. A red oozing surface is seen after removing the mass of fungus and debris. The deep oral candidiasis may be painful. Severe oral thrush occurs with breathing. Oral lesions may be associated with hypertrophy of the papillae of the tongue, and *C. albicans* grows freely in this environment. Oral candidiasis is a disease mainly of a new borns and of debilitated or aged patients and in persons treated for weeks or months with broad spectrum antibacterial antibiotics. Probably infection occurs through the birth canal.

(*ii*) **Bronchocandidiasis (= Bronchomoniliasis) :** Bronchocandidiasis is characterized by cough, varying amount of sputum and appearance of peribronchial thickening of hazy linear fibrosis. It is found in any chronic abnormal condition of the respiratory tract.

(*iii*) **Pulmonary Candidiasis :** The characteristic features of pulmonary candidiasis are low grade fever cough wih mucoid/blood-streaked sputum, pleurosy and effusion. Patchy bronchopneumonic lesions convert into globular pneumonia when cases become severe. The pathogen can be isolated frequently and with care from sputum of any bronchopulmonary disease. *C. albicans* also colonizes the mucosal surface of respiratory tract damaged by the other microorganisms.

(*iv*) **Endocarditis :** Candida endocarditis is characterised by fever, heart murmur, congestive heart failure and anaemia. These features resemble with bacterial endocarditis but differ from bacterial disease in high frequency of large vegetation on the valves, and emboli to spleenic, renal and iliac arteries. These cases are found in drug addicts. In such persons, the source of infection is contamination through blood in an oxygenator. The etiological agents are *C. parapsilosis, C. guilliermondii, C. krusei* or *C. tropicalis.*

(*v*) **Meningitis** : *C. albicans* causes most frequently the meningeal candidiasis. However, causes of this disease are gradually increasing. Predisposing factors for meningitis in patients are previous administrations of antibacterial or corticosteroid therapy of other debilitating conditions.

3. Culture Characteristics

C. albicans is a small, oval budding yeast like fungus, $2.5 \times 4 \times 6$ μm in dimension. Cell elongates and develops pseudomycelium. In sputum both budding cells (i.e. blastospores or yeasts) and pseudomycelia can be seen (Fig. 21.12). The presence of typical chlamydospores are found in *C. albicans* but not in others. It produces spherical clustures of blastospores or yeasts on three media. It produces short filaments within one or two hours on egg albumin or human serum.

It grows rapidly (24-48 h) at room temperature on Sabouraud's glucose agar medium. The size of colonies are moder-

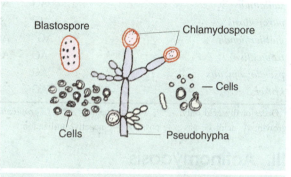

Fig. 21.12 : Structure of *Candida albicans* showing yeast cells, pseudohypha, chlamydospores, and blastospore.

ate, smooth and pasty, and have yeast odour. Older colonies have honey comb like appearance in centre and develop radial furrows. Glucose and maltose after fermentation are converted into acid and gas. Lactose is not fermented.

Different species can be identified by (*i*) types of colony developed on blood agar plates at 37°C in 10 days, (*ii*) types of growth in Subouraud's glucose broth at 37°C (48h), (*iii*) morphology and development of blastospores, chlamydospores, pseudomycelia on corn meal agar at room temperature, and (*iv*) the fermentation reactions after 10 days at 37°C in glucose, maltose, surcose and lactose. Culture characteristics of different species are given in Table 21.4. Development of toxemine in mice by intravenous infection of soluble substances from *C. albicans, C. robusta* and *C. reukaufi* or with suspension of killed *C. albicans* has been found. *C. albicans* possesses a capsular form of polysaccharide that shows pyrogenic activity. Mannans have been extracted from the cell wall of *C. albicans.*

4. Therapy

Some of the candidiasis can improve even without therapy. For example, vulvovaginal candidiasis gets improved after delivery. However, it can be controlled by treatment of diabetes mellitus by diet and drugs, and discontinuance of antibacterial treatment as soon as possible.

Nystatin in the form of ointment or Candid cream is recommended as topical therapy of *Candida* infection forming skin lesions, whereas powder, suspension or tablets for oral or vaginal uses. Use of sodium carprylate, sodium or calcium propionate or gentian violet are effective topical therapy. However, amphotericin-B, and 5-fluorocytosine are patent antifungal agents which are being used for the treatment of *Candida* infection (Herrell, 1971). In systemic candidiasis amphotericin B is given intravenously.

Table 21.4 : Characteristic feature of some common species of *Candida*

Species of Candida	Growth on surface of Sabouraud's agar	Chlamydospore on corn meal agar	Fermentation				(Utilization of carbon sources)			
			Glucose	*Maltose*	*Saccharose*	*Lactose*	*Glucose*	*Maltose*	*Saccharose*	*Lactose*
albicans	–	+	AG	AG	A	–	+	+	–	–
stellatoidea	–	–	AG	AG	–	–	+	+	–	–
parapsilosis	–	–	AG*	A	A	–	+	+	+	–
guilliermondii	–	–	AG**	AG	AG**	–	+	+	+	–
tropicalis	+	***	AG	AG	AG	–	+	+	+	–
pseudotropicalis	–	–	AG	–	AG	AG	+	–	+	+
krusei	+++	–	AG	–	–	–	+	–	–	–

*,Occasionally acid only;   **, weak fermenter;   ***, *C. tropicalis* produce chlamydospores under certain condition of growth; A, acid; G, gas ; +, positive ; –, negative utilization.

III. Actinomycosis

Actinomycosis is a chronic actinomycetous disease. It spreads to contiguous tissues without limitation by anatomical barriers. Sinus tracts are formed which drain suppurative lesions. In pus firm, lobulated or angular colonies of the etiologic agent are present. These colonies assume a diameter of 1-2 mm and are white to yellow. Being granular in size these colonies are called sulphur granules (Emmons *et al.*, 1977).

For the first time Lebert (1857) described actinomycosis in man. Harz (1877) named the organims as *Actinomyces bovis*. Until 1891, *A. bovis* could not be isolated successfully on culture medium. Lord (1910) observed *A. israelii* in the normal mouth growing in and about carious teeth and crypts of the tonsils. Later on endogenous origin of infection by this agent was confirmed by various workers.

1. Epidemiology

Actinomyces israelii is the etiologic agent in man and *A. bovis* in cattle or swine. In addition, *Bifidobacterium eriksonii* has been isolated from pleural fluid and lung absscess. All the three species are anaerobic. Usually the lesions of actinomycosis contain bacteria that help infection. These bacteria are called 'associates'. In cervicofacial and thoracic actinomycosis fusiform bacteria and anaerobic streptococci may be associated with *A. israelii*. In abdominal actinomycosis, *E. coli* and several Gram-negative associates are also found.

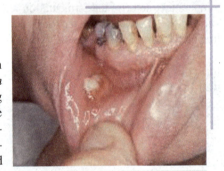

Actinomycosis.

Actinomycosis is sporadic in occurrence but not contagious. However, the infants or children acquire *A. israelii* (as a part of normal oral flora) from other infected persons.

Mainly there are three important sites of primary lesions of actinomycosis e.g. facial, thoracic and abdominal. Skin lesions occur rarely. However, actinomycosis resembles several chronic inflammatory diseases. Therefore, it must be identified very carefully by the pathologists because on this basis specific treatment is given. It produces sulphur granules in tissues which may be sparse.

2.   Culture Characteristics

The two species differ to each other in formation of firm and discrete colonies in broth without turbidity by most strains of *A. israelii*. Typical strains of *A. bovis* produces turbidity with dense sediments (Lord, 1933). *A. israelii* forms long but fragile hyphae with branches arising just below the septa. The older cells grow and add in length. The hyphae also grow terminally and add in length. This occurs continuously. Georg *et al.* (1965) have given differential characteristics of species of *Actinomyces* (Table.21.5).

Table 21.5 : Different characteristics of *Actinomyces* spp.

Parameters	A. bovis	A. israelii	A. naeslundii	Bifidobacterium eriksonii
Aerobic growth	2+	1+	2+	0
Anaerobic (95% N_2 +5% CO_2)	4+	4+	4+	4+
Nitrate reduction	0	80% +	90%+	0
Starch Hydrolysis	4+	≠ or 0	(or 0	4+
Acid from mannitol	0	80% +	0	+
Acid from mannose	≠ or 0	+	+	+

(*i*) *A. israelii* (syn. *Nocardia actinomyces*) : *A. israelii* is a saprophyte of oral cavity. It looks in amorphous form when grown on surface of teeth as in carious cavity and resembles with diphtheroids. It occurs on tonsilar crypts always in association with bacteria as white to yellowish granules of cheesy consistency. In many actinomycoses, it forms in tissues the sulphur granules which are white to yellowish in colour and rounded or lobulated in shape. In the centre of granules hyphae of *Actinomyces* and cellular debris are found. Towards the periphery of granules hyphae grow radially and sometimes extend beyond the surface of granules. The host tissues deposit a sheath of eosinophilic material on the individual hyphal tip and the emptive granule. The crushed granules are stained; they show Gram-positive diphtheroid hyphal fragments of 0.5-1.0 µm diameter (Negroni and Bonfiglioli, 1937).

If *A. israelii* is incubated anaerobically on agar surface, the colony consists of a radiating system of hyphae that looks as spidery. Later on colony becomes white subspherically with lobed surfaces (Fig. 21.13). It penetrates the agar and causes a dimpling of the surface.

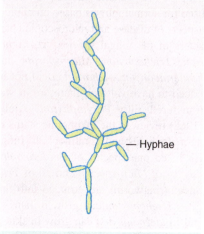

— Hyphae

Fig. 21.13 : Fragile hyphae of *Actinomyces israelii* (diagrammatic).

The entire colony may be removed; after crushing it gives a cheesy consistency (Slack *et al.*, 1969; Emmons *et al.*, 1977.)

A. israelii produces acid but no gas in the presence of glucose, galactose, lactose, maltose, raffinose, saccharose and xylose. Acid is not produced from mannitol and inulin.

(*ii*) *A. bovis*: *A. bovis* is isolated from cattle, and swine and man. It differs greatly from *A. israelii* in fragility of hyphae and production of turbidity in broth cultures. It hydrolyses starch, whereas most strains of *A. israelii* do not. Usually, *A. bovis* does not reduce nitrate, while *A. israelii* does. Similarly *A. bovis* does not utilize xylose or mannose but not the strains of *A. israelii* (Pine *et al.* 1960; Emmons *et al.*, 1977).

3.   Therapy

Traditionally, solution of potassium iodide (KI) has been used but only partial success has been achieved with KI. However, KI has been found effective in actinobacillosis in cattle. During the early phase of treatment intravenous or intramuscular benzylpenicillin is commonly employed (Nichols and Herrell, 1948). Thereafter, oral benzylpenicillin or phenoxymethyl penicillin is given for about six months. In addition, chloramphenicol is also recommended. Observation for side effects should also be taken into account. Moreover, for the cure of actinomycosis, surgical drainage of abcess is necessary. The affected tissues must extensively excised out.

IV. Dermatophytosis

Dermatophytosis is also known as dermatomycosis, but the term dermato-phytosis (skin plant) contrasts the later as the superficial infection on the kerati-nized area of the body such as skin, hairs and nails. Dermatophytes are parasites of man and animals. Sabouraud (1910) adopted four generic names which was used for dermatophytes based on clinical aspects of disease and microscopic char-acters of the fungal species viz., *Acho-rion, Trichophyton, Microsporum* and *Epidermophyton*. *Achorion* could not be traced to have validity; therefore, it was transfered into *Trichophyton*. Ajello (1968) reviewed the taxonomic status of dermatophytes and keratinophilic fungi.

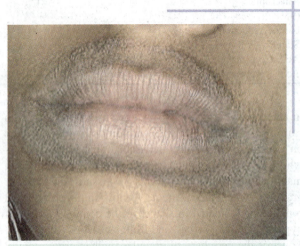

Dermatophytosis, a fungal infection on the facial skin.

He listed one species of *Epidermophy-ton*, 14 of *Microsporum* and 20 of *Trichophyton*. Therefore, the fungi responsible for infection of tineas (ringworm) are grouped into the above three generea.

Trichophyton can grow in hair, skin and nail; *Microsporum* can grow in hair and skin only and *Epidermophyton* can grow in skin and occasionally in nails. The scaly annular skin lesions are generally called as tinea. Tinea are further classified according to the affected parts such as (*i*) *Tinea pedis* (athletes foot), (*ii*) *Tinea capitis* (ringworm of scalp), and (*iii*) *Tinea carporis* (ringworm of non-hairy skin of body).

These fungi are keratinophilic because they are specialised to invade the keratin of the body. By the layman such infections are called athletes foot, Jockey itch and ringworm.

Microsporum audouinii causes extensive epidemic of ringworm of scalp among shcool children in certain area, whereas *Trichophyton tonsurans* causes epidemics in the adult and adolescent. Based on natural occurrence, dermatophytes can be grouped into the following three divisions :

   (*i*)   *Anthropophilic* - Fungi infecting only men.
   (*ii*)   *Zoophilic* - Fungi infect men and animals.
   (*iii*)   *Geophilic* - Fungi widespread in nature and found in soil. Except a few, they are saprophyte. Disease causing soil fungi are known as related species.

1.   Epidemiology

There are little differences in intensity of dermatophytosis with sex, race or occupation. In tinea pedis wearing of shoes favours the growth of the fungus and the symptoms become more severe in winter than summer, because patients wear sandles in summer but not in winter. Therefore, occurrence of Tinea pedis, Tinea corporis and Tinea capitis is more in temperate climate rather than tropical region.

Geographically *Trichophyton rubrum* and *Microsporum oudouinii* are more frequent in the U.S.A., *T. tonsurans* in Mexico, and *M. ferrugineum* in Japan.

Transmission of *M.oudouinii* occurs from the fallen and infected hairs and desquamated epithelium. Use of contaminated comb, hair brush, cap is also helpful in transmission. Similarly, *T. tonsurans* is also transmitted.

E. floccosum and *T. violaceun* are transmitted from person to person through contaminated towels or cloths. *Microsporum canis* is transmitted by infected dogs and cats, *T. equinum* and *T. verrucosum* by dogs, mouse, horse, guinea pigs and other animals. *M. gypseum* and *M. fulvum* is found in soil growing on keratinous debris. *M. gypseun* has been isolated from soil by baiting technique i.e. by burying wool in soil (White *et al.*, 1950). However, occurrence of dermatophytes as saprophyte in soil growing on keratinous debris of human or animal origin is an established fact.

2.   Clinical Types

(*i*) Tinea Pedis (Ringworm of Foot) : Lesions begin in the web between the fourth and fifth toes. The fissures are bordered by narrow zone of peeling epidermis. The lesion extends to involve the web which thereafter is covered by dead white macerated epidermis. After removal of this dead epidermis, red coloured new epidermis is seen. The red coloured epidermis is attacked by the fungus. When lesions become

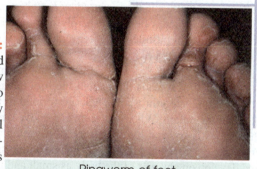

Ringworm of foot.

chronic the infection covers the sole arc, heel and whole dorsal surface of the foot. Similar type of symptom developing on hand is called tinea mannum.

The causal agent of these two types of tinea are *T. rubrum* and *T. mentagrophytes,* and *Epidermophyton floccosum.*

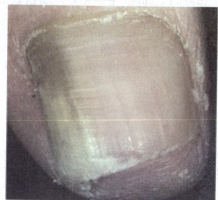

Onychomycosis, ringworm of nail.

(*ii*) Tinea unguium (Onychomycosis, Ringworm of Nail) : When *Cinea unguium* is chronic, the nails are severely infected. This involves the nails that are deformed by shoe and appressed to the infected skin of an adjacent toe. *Onychomycosis* of the nails of finger is associated with fungal hyperkeratolic desquamation of the hand and fingers. The infected nails of toes and hands have similar symptoms. Lesions have a chalky crumbling consistency, pitted or striated surface, thickened and discoloured and hypertrophied nail beds. This results in a raised and thickened nail which overlies a spongy mass of keratinized cells and debris. The fungi causing onychomycosis are *T. rubrum* and *T. mentagrophytes* (Sinki, 1974).

(*iii*) **Tinea corposis (Ringworm of Glabrous Skin)** : Ringworm of glabrous skin is characterised by a circular lesion with inflammations of various degree. The circular lesion is due to uniform radial growth of the pathogen on skin. On marginal zone skin turns into red colour possibly due to severe host reactions. This is found mostly in children but in adults too. Primary lesions develop on the face, shoulders, arms of the other exposed parts of the body and also exposure to the infected pets or cattles e.g. cats, dogs, horses, mouse etc.

Secondary infection may be caused by autoinoculation or exposure to these infected animals. Children or adults who are fond of these animals receiving primary inoculum through the above infected pets. The etiologic agents in cats and dogs are *Microsporum canis. T. mentagrophytes* which are often associated with dogs. These are transmitted to the animals. *T. rubrum* in tinea corporis is lacking. These develop often due to tightly covered clothing such as on waist (Sinki, 1974).

(*iv*) **Tinea barbae (Sycosis, Ringworm of the Beard, Barber's itch)** : The symptoms appear as a small pustular follicles on beard and the other areas of neck and face. However, due to severe infection hairs are shed and the patients make a spontaneous recovery. The infection is mycotic and a large number of arthrospores are found. It occurs mostly in farmers. This disease is caused by *T. verrucosum* and acquired from cattle. *T. mentagrophytes* is acquired from horse or dog.

(*v*) **Tinea cruris (Eczema Marginatum, Ringworm of Groin)** : It is an acute, chronic and severe pruritic dermatophytosis of groin, and perineal areas. It is transmitted through careless exchange of towels or clothing, especially on shipboard and locker rooms. The etiologic agents are *T. rubrum* and *E. floccosum*. Lesions caused by *T. rubrum* develop widely over the body especially buttocks and waist, and those caused by *E. floccosum* rarely extend beyond groin, and perineal areas. The lesions are red, often dry, sharply marginated with the epidermal scales. Sometimes, lesions are in the form of pustules and after removal of top of lesions with pointed scalpel, the causal agent can be diagnosed (Sinki, 1974).

(*vi*) **Tinea favosa (Favus)** : It is the most chronic ringworm acquired during infancy. If favus is not treated, it spreads throughout the life. Favus is characterized by formation of a dense mass of mycelium and arthrospores that are formed in hair follicle and take the form of inverted cone, the favus scutulum. When scutulum is removed, a red moist of oozing base is observed. Much of the lesions is covered by thick crust of mycelium and epidermal debris. After proper therapy the crusts are destroyed and lesions get healed with extensive scarring. Lesions may have a mousy smell and develop on any part of the body as small and isolated lesions. There are three etiologic agents, *T. schoenleinii, T. violaceum* and *M. gypseum.*

(*viii*) **Tinea capitis (Ringworm of Scalp):** *Tinea capitis* has dry diffuse and scaly lesions of the scalp on which are found black dots characterised by broken off hairs at follicle. When the black dot ringworm develops, dermatophytes grow into the hair follicle, peneterate the hair shaft and grow downwardly inside the hair (Emmons *et al.*, 1977). Due to the decomposition of keratin, hairs become fragile and get broken below the surface. In the adjacent epidermis some growth of hyphae also occurs but the hyphal growth in follicle is confined only inside the hair shaft. This type of infection is known as enothrix infection which is caused by *T. tonsurans, T. violaceum* and *T. schoeleinii. Tinea capitis* is caused largely by the species of *Microsporum* viz., *M. audouinii* and *M. canis.*

3. Culture Characteristics

(*i*) *Epidermophyton* : *Epidermophyton* consists of a single species, *E. floccosum* that attacks skin and nails but not hairs. It is characterized by numerous oval to widely clavate macroconidia (6-10 $\times$ 8-15 μ) with rounded distal ends. Conidia are smooth walled and produced in a clustur directly from the sides of hyphae. Microconidia are absent (Fig.21.14A).

(*ii*) *Microsporum* : The species of *Micro-sporum* invade the hairs and skin but rarely nails. It is characterised by obovate to fusiform, echinulate, multiseptate macroconidia having thick walls (upto 4 μ in *M. canis*) (Fig. 21.14B). The outer surface of wall may be pitted or spring at least near the distal end. Macroconidia are 7-20 × 35-125 μm and usually have 4-15 septa. In some species microconidia of 2.5-3.5 × 4.7 μ are also found. The important species are *M. audounii, M. canis, M. gypseum, M. fulvum, M. gallinae, M. distortum* and *M. cookei*.

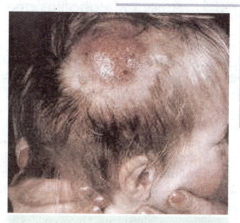

Ringworm of scalp.

(*iii*) *Trichophyton*: The species of *Tricho-phyton* attack the hair, skin and nails. Due to lack of spores in some species, it becomes difficult to characterize it. They can be grouped artificially on the basis of colony characteristics. However, better growth and sporulation can be induced when grown on suitable medium containing accessory growth substances. This genus is characterized by clavate or pyriform macroconidia (4-8 × 8-50 μ) with smooth walls having 0-4 septa. The microconidia are formed in grape like clusture. The microconidia are clavate (2-3 × 3-4 μ) (Fig. 21.14C). This genus is characterized by the production of knobby protuberances near the ends of cells of peridial hyphae. Some important species are *T. mentagrophytes, T. rubrum, T. concentricum, T. violaceum, T. verrucosum, T. simii, T. tonsurans,* etc.

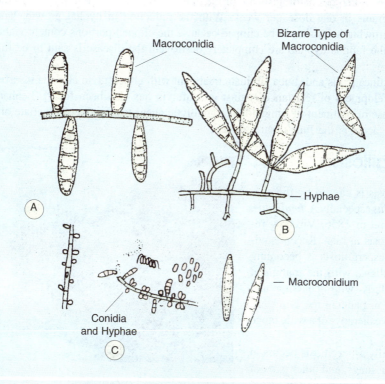

Fig. 21.14 : Conidial structure of dermatophytes. A, *Epidermophyton*; B, *Microsporum*; C, *Trichophyton*.

4. Fungal Metabolites and Immunity

Dermatophytes produce neither exotoxins nor endotoxins. However, its 2-3 months old culture filtrate or saline extracts of dried and powdered culture have shown allergies i.e. eliciting positive skin test in sensitized individuals. Such skin testing material is called as trichophytin. The dermatophytes contain group-specific antigens. From both of *T. mentagrophytes* and *T. rubrum* a protein fraction has been found to exhibit positive precipitin reactions in the sera of rabbits immunized by the homologous fungus. However, in heterologous sera no cross reactions occur with these antigens. Some antigens common to *T. rubrum* and *T. metagrophytes*, and *Penicillum* spp have been demonstrated by gel diffusion and haemagglutination test. After intracutaneous injection of trichophytin, development of cellular antibodies has been demonstrated in animals. The immunity may lost for a short period to several years depending on host types. Cellular antibodies start developing between 24 and 48 hours and may lost for 7 days (Grappel *et al.*, 1974).

Immunologically, active glycopeptides and polysaccharides from several dermatophytes have been isolated and purified by various methods (Bishop *et al.*, 1966). The polysaccharides have no trichophytin activity, while glycopeptides are active. Chemically, these antigens from different species are alike but in serological reaction they differ a little.

5. Therapy

Oral administration of griseofulvin has been found to warrant the penetrations of epidermis by dermatophytes. Some dermatologists are of the opinion to give one gram per day in four doses, and other 3-4 grams in one dose per week. With the reduction in hyphal growth, hair growth increases. The growing hairs should be clipped because the clipped portions contain viable fungus hyphae. When the fungus hyphae are completely destroyed, the disease is said to be fully cured (Blank, 1960).

In case of tinea pedis and tinea circinata treatment with griseofulvin alone is insufficient and some times development of resistance against griseofulvin has also been found (Lenhart, 1970). Therefore, the use of ointment of fungicidal keratinolytic agent is effective. Tolnaftate or a salt of one of the fatty acids is the fungicidal ointment.

V. Aspergillosis

Aspergillosis is caused by the species of *Aspergillus* especially *A. fumigatus*. For the first time in 1856, Virchow reported aspergillosis in man. Recent studies reveal that aspergillosis is becoming important in patients with the malignant disease particularly with leukemia and lymphoma and in patients receiving immunosuppressive therapy for a wide range of illness.

Occasionally aspergillosis occurs in sinuses, bronchi, lungs, and other parts of body as well. *A. fumigatus* is the recognised pathogen of birds, animals and man

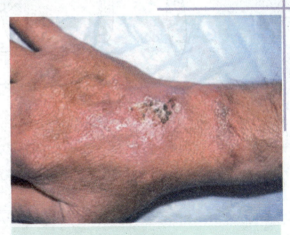

Aspergillosis.

as well. It is generally an occupational disease occurring among those who handle and feed squabs, and also furs and hairs. *A. fumigatus* generally causes pulmonary aspergillosis. The other species of *Aspergillus* viz., *A. niger, A. clavatus, A. terreus* and *A. phialiseptus* are also known to be associated with pulmonary aspergillosis.

1. Epidemiology

Aspergillosis is sporadic and world wide in distribution. It is not contagious but the sources of infection are exogenous. It does not differ with age, sex or race. The contaminated moldy cereal grains are also one of the sources. *A. fumigatus* is also found during early stages of decomposition of vegetables. On trimmed portion of trees in city streets, the wood chips are over grown by *A. fumigatus*. While chips are stored in piles it becomes full of conidia of this fungus and a source for lung infection in millions of human. *A. fumigatus* being a saprophyte luxuriantly grows on decaying organic materials. Conidia that act as propagating inoculum spread in certain cases in air, and enter through inhalation into the lungs of human and get established inside the lung tissues.

2. Clinical Types

Many clinical types of aspergillosis have been reported that differ with species of *Aspergillus*.

(*i*) **Pulmonary Aspergillosis** : *A. fumigatus* is the most virulent and versatile among the aspergilli. It causes pulmonary aspergillosis. It is characterized by granulomatous lesions in lung parenchyma, which may spread through hyphae to kidney and the other organs. The course of disease is marked by mild fever, cough, weight loss and toxemia. In the other type, without, invading the lung parenchyma, "a fungus ball" may be produced by *A. fumigatus* and *A. niger*. Fungus ball is a compact mass of mycelium and cell debris. Often *A. niger* grows in mix culture with *A. fumigatus* (Segretani,1962, Ikenoto *et al.*, 1971).

(*ii*) **Myocarditis** : Myocardial aspergillosis can be fatal in the patients which have primary and predisposing disease. The patients debilitated by cytotoxic or immunosuppressive drugs may also suffer from mycocarditis. This is caused often by *A. fumigatus* and rarely by *A. flavus* (Williams, 1974).

(*iii*) **Otomycosis** : Otomycosis is caused by several fungal species but more specifically by *A. niger*. It grows on cerumen, epithelial scales and deep in external canal. Otomycosis has been described in detail in the preceding section.

(*iv*) **Bronchopulmonary Aspergillosis** : This type of aspergillosis is characterized by frequent pyrexial attacks associated with severe cough purulent occasionally blood-tinged sputum with flecks of whitish or brownish material containing mycelium and a high eosinophilia.

3. Culture Characteristics

On Sabouraud's glucose agar medium, *A. fumigatus* grows rapidly and forms white cottony colony which later on turns like velvet. After production of spores, it becomes dark green and powdery. It produces a typical and smooth walled conidiophore (2-8 × 300-500 μ) containing a vesicle. The upper half of vesicle is covered by phialides which produce dark-green and spherical conidia in chains. The tip of conidiophore widens gradually into a dome-shaped vesicle of 20-30 μ diameter. Phialides are borne on the upper half of two third part of the vesicle and are of 5-10 × 2-3 μ dimension. There are no seconday phialides. Conidia are directly produced by the phialides (Fig. 21.15).

A. fumigatus and *A.niger* produce endotoxins (Henrici, 1939). However, neurotoxin and other toxins are also isolated which possibly are involved in haemorrhagic lesions produced in birds and animals. Toxins of this fungus may cause abortions in cattle and sheep.

Endotoxin is a polysaccharide of low toxicity for rabbits which causes monocytosis and finally death. From the culture filtrate of *A. oryzae*, a chemical substance (aspergillin O) has been extracted. It has shown proteolytic, fibrinolytic and anticoagulant properties. In addition, the polysaccharides of *A. fumigatus* have been found antigenic. Precipitins could be demonstrated in the blood sera of previously inoculated rabbits. Immune serum of rabbit has prevented the effect of toxins in experimental animals.

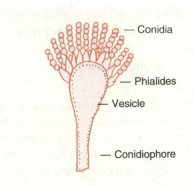

Fig. 21.15 : Conidiophore of *Aspergillus fumigatus* (diagrammatic).

4. Therapy

Surgery is suggested in certain pulmonary aspergillosis. Treatment of aspergillosis with sodium iodide (NaI) intravenously and potassium iodide orally has cured the disease effectively. The only known useful antibiotic for aspergillosis is amphotericin B.

VI. Otomycosis (Mycotic Otitis)

Otomycosis is the mycosis of ear which is characterized by inflammation, scaling, pruritis and pain. It develops as a result of superficial chronic or subacute infection on the outer ear canal. Otomycosis is caused by fungi especially *Aspergillus niger* and *A. fumigatus*. *A. terreus* is the most important agent in Japan. In addition, the species of *Scopulariopsis*, *Polypaecilum*, *Mucor*, *Rhizopus*, *Candida* and dermatophytes are also occasional etiological agents. Merely by isolation of any fungus from the ear, pathogenicity cannot be claimed (Yamashita, 1963).

1. Epidemiology

No body is immune to otomycosis at any age group. The etiologic agent is found abundantly in the environment. This disease occurs throughout the world more specifically in humid and warm conditions. In contrast, otitis esterna is of bacterial origin and occurs in swimmers.

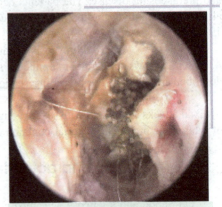

Photograph of ear canal showing Otomycosis.

2. Clinical Types

If the bacterial infection has not complicated the cases, simple otomycosis is characterized by inflammation, pruritus and exfoliation of epithelium of the ear. When the ear canal is occluded by mycelial plug, cerumen and epithelial debris, it results in partial deafness. Inflammation of skin above the cartilagenous tissue and scaling may extend to inner end of the ear canal which causes pain. Perforation of tympanic membrane is of rare occurrence.

Otomycosis differs from bacterial otitis, the later is exudative with a foul odour. In otitis *Corynebacterium*, *Escherichia*, *Pseudomonas*, *Proteus*, *Micrococcus* or *Streptococcus* may be associated.

A clinical diagnosis should be confirmed by microscopic observation of the mycelial plug and scales in order to be sure for treatment through antifungal or antibacterial drugs. When there is *A.niger* otitis, hyphae with globose sporangiophores and black coloured spores are demonstrated in the sample.

3. Therapy

Usually otomycosis is chronic and may occur even after cure. Therefore, ear canal should be cleaned carefully by using 5% aluminum acetate solution. This will reduce edema and remove cerumen and epithelial debris present in ear canal which later on is colonised by bacteria. Treatment of lesions with 70% alcohol, aqueous solution of 0.02-0.1% of phenyl mercuric acetate, 1% thymol in metacresyl acetate is recommended. One should avoid swimming until the infection is fully cured. Neomycin, bacitracin, chloromycetin or other antibacterial ointment should be used in case of primary or mixed bacterial otitis externa.

VII. Penicillinosis

The species of *Penicillium* is ubiquitous in distribution. Consequently it contaminates the eczematoid lesions, open ulcers, urine, sputum, etc. through air- borne conidia and grow rapidly in culture from pathological specimens. Therefore,during its isolation from the samples precaution should be taken to avoid contamination. It has been accepted that pulmonary mycosis is caused by *Penicillium* (Emmons *et al.,* 1977).

Huang and Harris (1963) have reported that *P. commune* caused disseminated pulmonary and cerebral penicillinosis in a patient with acute leukemia and gastrointestinal candidiasis. This fungus grew profusely in the lungs and brain resulting in vascular invasion, thrombosis and infection of lungs. In addition from some cases of penicillinosis,other species of *Penicillium* viz., *P.bertai, P.glaucum, P.bicolor* or *P.spinulosum* were isolated but from some cases no *Penicillium* could be isolated. This creates uncertainity about the role of species of *Penicillium* as etiologic agent of penicillinosis (Fig. 21.16).

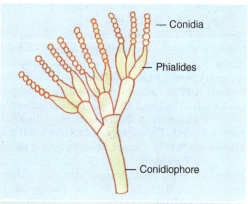

Fig. 21.16 : Conidiophore of *Penicillium* sp.

P. marneffei is the etiologic agent of an unusual mycosis in the bamboo rat (*Rhizomys sinensis*) which is found in high number in the tissues within polymorphonuclear leukocytes, reticulocytes and giant cells (Emmons *et al.,* 1977). Mycelium is grey and floccose cells are elongated, hyphae septate reaching a length of 15 to 20 μ. It produces conidiophores that form phialides (9-11 x 2.5 μ). Phialides produce spherical to subspherical smooth walled conidia measuring 2.3 μ in diameter.

QUESTIONS

1. Give a detailed account of *Plasmodium* with special reference to *P. vivax.*
2. Discuss in detail phycomycosis, epidemiology, clinical types and culture characteristics.
3. What do you know about candidiasis ? Write in brief clinical types.
4. Write an essay an actinomycosis with emphasis on epidemiology, clinical types and therapy.
5. What is dermatophytosis ? Discuss in detail the culture characteristics of etiologic agents.

6. Give a detailed account of clinical types and therapy of dermatophytosis .
7. Write short notes on the following :
   (i) *Taxoplasma*, (ii) *Trichomonas*, (iii) *Giardia*, (vi) *Trypanosoma*,   (v) Mycoses,
   (vi) Mycotoxicoses, (vii) Mycetismus, (viii) *Candida albicans*, (ix) *Trichophyton*,
   (x) Aspergillosis, (ix) Otomycosis, (xii) Penicillinosis

REFERENCES

Ajellow, L. 1968. A Taxonomic review of the dematophytes and releated species. *Sabouradia* 6 : 147-159.

Berknout C.M. 1923. Les generes *Monilia oidium* oospora et *Torula*. Thesis, Univ. Utrecht.

Bishop, C.T.; Perry, M.B. and Blank, F.1966. The water soluble polysaccharides of dermatophytes. *Can.J.Chem*.44:2291-2217.

Blank H. *Ed.* (1960) Griseoflavin in dermatophytes. An international symposium sponsored by Univ. of Miami, *Arch*. Dermat. 81 : 649-882.

Clark, B.M. (1968) Epidemiology of phycomycosis. In *Systemic Mycosis* (eds. Wolstenholme, G.E.W. and porter, R.). J and Achurchill London.

Emmons C.H., Binford C.H. Utz J.P. and kwon-Chung, K.J. 1977. *Medical Mycology*. Lea and Febiger,Philadephia pp. 592.

Ford, WW. 1923. A new classification of mycetismus (mushroom poisoning). *Trans. Ass. Amer.Phys*. 38: 225-229.

George, L.K.;Robertstand, G.W.; Brinkman, S.A. and Hicklin,M.D. 1965.A new pathogenic anaerobic *Actinomyces* species. *J. Infect.Dis*. 115: 88-89.

Grappel, S.F.; Bishop, C.T. and Blank, F. 1974. Immunology of dermatophytes and dermatophytosis. Bact. *Rev*. 38: 222-250.

Henrici, A.T. 1939. An endotoxin from *Aspergillus fumigatus*. *J. Immunol*. 36: 319-338.

Herrel, W.E. 1971. The antifungal activity of 5-fluorocytosine. *Clin.Med*. 78: 11.

Huang, S.N. and Harris, A.S. 1963. Acute disseminated penicillosis. *Am.J.Clin.Pathol*.39: 167-174.

Ikenoto,H.;Watanabe,K. and Mori, T. 1971. Pulmonary aspergillosis. *Sabouraudia*. 9:30.

Lenhart, K.1970. Griseofulvin resistance in dermatophytes. *Experientia*. 26: 109-110.

Lie-Kian-Joe,Njo-Injo,T.E. et al. 1956. *Basidiobolus ranarum* as a cause of subcutaneous physomycosis in Indonesia. *A.M.A. Arch.Dermt*.74: 378-383.

Lord. F.T. 1933. The etiology, pathogenesis and diagnosis of actinomycosis. *Med. Clin.N.Amer*.16: 829-844.

Martinson,F.D.1971. Chronic phycomycosis of the upper respiratory tract, rhinophycomycosis entomophthorae. *Am.J.Trop.Med*.9: 143-148.

Miller,L.H. *et al*. 1986. Research toward a malaria vaccine. *Science*.234: 1349-1355.

Negroni,P. and Bonfiglioli. 1937. Morphology and biology of *Actinomyces israelii*. *J. Trop*.Med.Hyg.40:226-232. and 240-246.

Nichols,D.R. and Herrell, W.E. 1948. Penicillin in the treatment of actinomycosis. *J. Lab.Clin. Med*. 33: 521-5525.

Pine, L. et al. 1960. Studies of the morphological, physiological and biochemical characters of *Actimomyces bovis*. *J. Gen. Microbiol*. 23: 403.

Sabouraud, R. 1910. Les Teignes. Paris, Masson et Cie.

Segretain, G. 1962. Pulmonary aspergillosis. *Lab. Investigatin*. 11: 1046-1052.

Sinski, J.T. 1974. Dermatophytes in human skin, hair and nails. Springfield, Charles; C. Thommas.

Slack, L.M. et al. 1969. Morphological, biochemical and serological studies on 64 strains of *Actinomyces israelii. J. Bact*.97: 873.

Srinivasn, M.C. and Thirmulachar, M.J. 1967. Studies on *Basidiobolus* species from India. *Mycopathomycol. Appl*. 33: 54-64.

Tyler, V.E. 1963. Poisonous mushrooms. *Prgr. Chem. Toxicol*. 1: 339-384.

White, W.L. 1950. Fungi in relation to the degradation of woolen fabrics. *Mycologia*. 42: 199-223.

Wieland, T. 1968. Poisonous principles of mushrooms of the genus *Amanita*. *Science*. 159: 949-952.

Williams, A,H. 1974. Aspergillus myocarditis. *Am.J.Cl. Pathol*. 61: 247-256.

Winner, H.L. and Hurley, R. 1964. *Candida albicans*. Boston, Little, Brown.

Yamashita, K. 1963. Fungal flora in the ear, nose, throat and mouth of man. *Japanese J. Med. Mycol*. 4: 136-149.

Medical Microbiology II: Bacterial and Viral Diseases

22

A.   Bacterial Diseases
B.   Viral Diseases

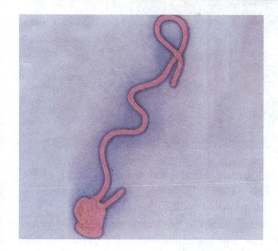

Since the time immemorial, man has been constantly infected by microorganisms. Time to time catastrophic effects had been noticed from almost each country. In India, *en masse* death of hundreds of human population has been reported to be caused by either plague or measles as well as tuberculosis or dysentery. In the villages of North India small pox had been found very severe. The sufferers were given due attention from all shorts of sanctity and cleanpness assuming that Goddess, *Mata Shitala*, has blessed the victim. All these had been taken with religious view rather than microbiological. Gradually, people became aware of the causal agent of small pox. However during 1960s mass movement for eradication of small pox in India was launched with vaccination programme.

Robert Koch

A. Bacterial Diseases

Though the number of bacterial species is very high, yet only a few bacteria are associated with human disease. In this section bacterial disease are described according to source of infection.

I. Air-borne Diseases

Air has been described as *Pran* (means life) in ancient litera-ture. However, it is a source of spread of pathogens too. The air-borne bacterial pathogens cause diseases in respiratory tract and skin such as tuberculosis, whooping cough, pneumonia, diphtheria, etc. Generally these diseases spread in a densely crowded population and polluted environment.

1.   Tuberculosis

For the first time Robert Koch identified *Mycobacterium tuberculosis* as the causal agent of TB (tuberculosis). During that period TB caused 1/7 of death in Europe. At present TB is the global health problem. About 20% of the world's population is suffering from TB and about 8 million people are victimised each year. Approximate annual death is 3 million.

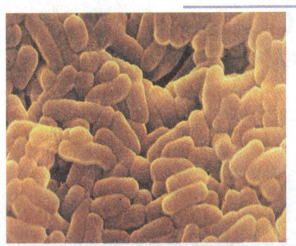

Commonly tuberculosis occurs among the homeless, malnourished persons or alcoholic drinkers. This bacterium spreads through droplet nuclei and the respiratory route. Bovine tuberculosis is caused by *M. bovis* in cows and cattles. It is equally dangerous. *M. bovis* spreads to humans through contaminated milk.

Electron micrograph of *Mycobacterium tuberculosis*, the rod like bacteria that is the causative agent of tuberculosis in humans.

Mycobacterium is rod-shaped and acid fast (stained with acid fast technique by Ziehl-Neelsen Carbolfuchsin stain). Therefore, the cells take red stain. The bacterium infects the respiratory tract and established in lung tissues. After being phagocytosed by macrophages the bacteria are enclosed into the small and hard tubercles which are the characteristic feature of the disease. Hence it gives the name of disease as tuberculosis. In X-rays tubercles can be observed. The symptoms include cough, pain in chest, fever and secretion called sputum. The sputum appears red or rust coloured if mixed with blood in lung cavity. Bacteria remain alive in macrophages. Sometimes the tubercle lesions liquefy and form air-filled cavities where from bacteria can spread new foci of infections throughout the body. This spreading is called miliary tuberculosis due to development of millet seed like many tubercles (Bloom, 1994).

(*i*) **Immunity:** Against infection by *M. tuberculosis* patients develop a cell mediated immunity which involves the sensitized T cells. It is the basis of **tuberculin skin test.** In this test a purified protein derivative of *M. tuberculosis* is injected into the patients (**Mantoux test).** If the pathogen is present in body of patient, sensitized T cells react with these proteins. Thereafter, a hypersensitivity reaction occurs with in 48 hours. Consequently around the injected site there appears hardening and reddening area.

(*ii*) **Diagnosis of Tuberculosis :** Laboratory diagnosis of tuberculosis includes isolation of the acid-fast bacterium, chest X-ray by commercially available DNA probe, HPLC test, the Mantoux or tuberculin skin test.

Chemotherapy is done by adminstering isonizid plus rifampin, ethambutol and pyrazinamide. These are administered simultaneously for 12 to 24 months. BCG (Bacillus-Calmette-Guerine) vaccine is used for treatment of tuberculosis.

2.   Diphtheria

Diphtheria (Greek *diphthera* means membrane, and *ia* means conditions) is a serious air-borne contagious disease. It is caused by *Corynebacterium diphtheriae* which is a Gram-positive bacterium. Like tuberculosis, diphtheria also occurs in poor people living in crowded condition.

C.diphtheriae is club-shaped and contains many metachromatic granules in cytoplasm. It is associated with leathery membrane. In India, it has a major problem. It is inhaled through droplets and reaches to respiratory tract and infects it. Typical symptoms of diphtheria include a thick mucopurulent (containing mucus and pus) nasal discharge, vomiting, headache, fever, cough and stiffness in the neck and back.

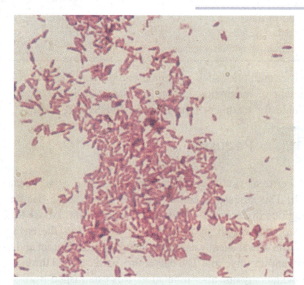

The bacteria produces diphtheria toxin which is an exotoxin that causes an inflammation and greyish pseudomembrane on respiratory mucosa. The exotoxin is absorbed by the body into the circulatory system and transported through out the body.

Corynebacterium diphtheriae, a Gram-positive bacterium causes diphtheria.

Hence, it may destroy kidney, nervous tissues and even heart by producing toxic proteins which is called diphtheria toxin.

Diphtheria is diagonised by culturing the bacterium and observation of pseudomembrane in throat. To neutralize the effect of endotoxin, generally antitoxin is given. Penicillin and erythromycin are prescribed for the treatment of infection.

Sometimes cutaneous diphtheria also develops when *C. diphtheriae* infects skin, skin lesions or wounds causing slow healing ulceration. This disease occurs when people of over 30 years age have a weakened immunity, particularly living in tropical regions. Worldwide immunization programme is being launched with the DPT (diphtheria-pertussis-tetanus) vaccine.

3.   Meningitis

Meningitis (Greek *meninx* means membrane and *itis* means inflammation) is an inflammation of meninges (membranes) of brain or spinal cord. This disease is caused by bacteria, fungi or viruses and, therefore, divided into two: bacterial meningitis or septic meningitis and the aseptic meningitis syndrome (Table 22.1).

Due to a large number of causes of meningitis, accurate identification of causitive agents must be done before the treatment of disease. The respiratory secretion of carriers acts as a source of meningitis. The bacteria colonize the nasopharynx and cross the mucosal barrier. Thereafter, these enter the blood stream and cerebrospinal fluid. Consequently, they cause inflammation of the meninges.

Table 22.1 : The causal agents of meningitis

Types of menigitis	Causal agents
Bacterial meningitis	Streptococcus pneumoniae, Haemophilus influenzae type B, Gram- negative bacilli, Mycobacterium tuberculosis, Nocardia asteroides, Staphylococcus aureus.
Aseptic meningitis syndrome (agents requiring antimicrobials)	Fungi, amoebae, syphilis, mycoplasmas, viruses

Specific antibiotics such as penicillin, chloramphenicol, cefotaxime, ofloxine, etc are used for the treatment of meningitis. Moreover, vaccines against *S. pneumoniae* and *N. meningitis* are also available for treatment (Quagliarello and Scheld, 1992).

4.   Pertussis

Pertussis (Latin *per* means intensive, and *tussis* means cough) is also known as whooping cough. This disease is caused by a Gram-negative bacterium, *Bordetella pertussis*. It is highly contagious disease and affects the children first. About 95% population of the world is affected by whooping cough, and about 5,00,000 patients die each year.

The causal agent was first isolated by Jules Bordet and O. Gengau in 1906. It is a small, fragile, and Gram-negative rod. The bacterium is transmitted through droplets of infected persons and enters into the healthy one after inhalation. It is established from 7 to 14 days of infection. The bacterium is attached to the ciliated epithelial cells of upper respiratory tract. It produces an adherance factor called filamentous haemagglutinin that recognises complementary molecule of the host cell. After being attached to the epithelium, the bacterium produces several toxins. These toxins develop the symptoms. The most important toxin is pertussis toxin. In its presence the tissues become susceptible to histamine and serotonin, and response of lympocyte is increased.

In addition, *B. pertussis* also produces an extracellular enzyme (adenylate cyclase) and a tracheal cytotoxin and a dermonecrotic toxin. These destroy the tissue of epithelium. Due to secretion of thick mucus the ciliary action is impeded and the ciliated epithelial cells die.

Laboratory diagnosis of this bacterium includes: culture of the bacterium, fluorescent antibody, staining of smears from nasopharyngeal swabs and serological tests. The chemotherapeutic agents are chloramphenicol, tetracycline or erythromycin. The children of 2-3 years age are recommended for vaccination with DPT vaccine.

5.   Streptococcal Pneumonia

Streptococcal pneumonia is caused by the bacterium, *Streptococcus pneumoniae* (syn. *Diplococcus pneumoniae*) which is an inhabitant of normal microflora of nose. Therefore, streptococcal pneumonia is considered an endogenous infection. *S. pneumoniae* is a Gram-positive bacterium containing a capsule of polysaccharide. The capsule provides virulence to

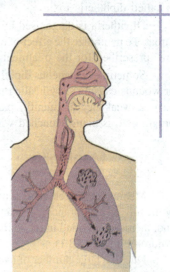

Development of streptococcal pneumonia : Growth of streptococci occurs on damaged ciliated epithelium.

the bacterium because of inability of binding of antibodies to the bacterium. Pneumonia is caused by the bacterium present in respiratory tract. Moveover, about 60 to 80% of all respiratory diseases known so far is caused by this bacterium.

Pneumonia generally develops in those persons who already have viral infection of the respiratory tract, physical injury of tract, diabetes or alcoholism.

S. pneumoniae multiplies rapidly in alveolar spaces of lung. The alveoli are filled with blood and fluid which finally become inflammed. That is why after coughing red-coloured sputum comes out from the lung. When entire lobes of lung are infected, it is known as lobar pneumonia. When both sides are infected, it is called double pneumonia. However, if the branchial tree is infected showing scattered patches, it is called branchopneumonia. The symptoms of pneumonia are chilling, breathing with difficulty and chest pain. Pneumonia is diagnosed by X-ray, culture of the bacterium and biochemical test.

Presence of capsule around the cell wall provides virulence to the pathogen. This also inhibits the binding of antibodies to the cell; thus it inhibits phagocytosis. The antibiotics recommended for the treatment of pneumonia are penicillin G, cephotaxime, ofloxacin and ceftriaxone. The persons, sensitive to penicillin, can take erythromycin or tetracycline. For debilitated persons a pneumococcal vaccine (Pneumovax) is also available.

II. Food-Borne and Water-borne Diseases

Food and water are the essential requirements of our body. There are many bacteria that contaminate food and water, and through these they enter in our intestime. Contaminated food causes stomach and intestinal disorder that is known as food-poisoning. The food-poisoning is caused due to secretion of exotoxin by the bacteria that contaminate food. After ingestion bacteria colonize the grastrointestinal tract, infect the tissue and secrete exotoxin. This condition is known as *food intoxication* because only the toxin is ingested and living bacteria are not required. The enterotoxin poisoning produces the symptoms: nausea, vomiting and diarrhoea. Worldwide death by diarrhea is second to respiratory disease. Diarrhoea occurs maximum in children causing 5 million death in Asia alone. A summary of food borne diseases is given in Table 22.2. Some of the food- and water-borne diseases are discussed in this section.

Table 22.2 : Some food-borne Infectious Diseases

Food	*Disease*	*Organisms*	*Incubation and characteristics*
Meats, fish, eggs Poultry, dairy products	Salmonellopsis	*S. typhimurium,* *S.enteritidis*	8-48h, enterotoxin and colitis
Milk, Pork, poultry product, water	Campylobacteriosis	*Campylobacter jejuni*	2-10 days heat-labile toxin
Raw milk, undercooked ground beef	Colitis and *E.coli* diarrhea	*E.coli*	24-72 h, haemorrhagic colitis
Egg products, puddings	Shigellosis	*Shigella sonnei* *S. flexneri*	24-72 h
Tofu, milk, meat product	Yersiniosis	*Yersinia enterocolitica*	16-48 h
Sea food Shellfish	*Vibrio* *Gastroenteritis*	*V.parahaemo lyticus*	16-48 h

1. Cholera

Since the time immemorial cholera (latin *Chole* means bile) has caused pandemics in Asia, and Africa. It is caused by a Gram-negative, slightly curved bacterium, *Vibrio cholerae,* that contains a single polar flagellum (Fig. 22.1). It was cultured for the first time in 1883 by Robert Koch. It has several serological groups such as 01, 02, 0139 and two bio-types *V. cholerae* and EL Tor. The mortality rate without treatment is 50%, and with care and treatment 1%.

Contaminated food causes stomach and intestinal disorder known as food poisoning.

This bacterium is acquired after taking food and water contaminated by faecal material from patients or carriers. In 1961 the El Tor biotype of *V.cholerae* 01 strain was the cause of cholera pandemic and in 1996 strain *V.cholerae* 0139 emerged in Calcutta in India. The incubation period of the bacteria is 24-72 hours. It adheres to the mucosa of small intestine and secretes a cholera toxin, choleragen. Choleratoxin is a protein. It possesses two functional units : an enzymatic subunit A and an intestinal receptor-binding subunit B. The subunit A acts as diphtheria toxin. It enters into the

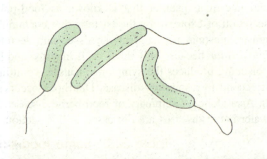

Fig. 22.1 : Structure of *Vibrio cholerae.*

epithelial cells of intestine, binds to ADP - ribosyl group and activates the enzyme adenylate cyclase. Thus, *choleragen* induces the secretion of water and chloride ions and inhibits absorption of sodium ions. The patients loose water and electrolyte. This results in cramping of abdominal muscles, vomiting, fever and watery diarrohea. During infection a person can lose 10-15 litres of fluid. Finally patients die due to loss of water and electrolytes and increased amount of blood proteins that leads shocks and collapse of circulatory system (Blake *et al,* 1980; Kaper *et al,* 1995).

Laboratory diagnosis of the bacterium is by culture from faeces, and agglutination reaction with specific antisera. The patients are immediately rehydrated with NaCl plus sucrose solution. The antibiotics recommended to patients are tetracycline, trimethoprime, sulfamethoxazole or ciprofloxacin. Proper sanitation of water supplies must be done regularly.

2. Botulism

Botulism (Latin *botulus* means sausage) is a type of food poisoning caused by *Clostridium botulinum.* It is an obligately anaerobic, endospore forming, Gram-positive, rod-shaped bacterium which is commonly found in soil as well as in aquatic deposits. The home canned food, not sufficiently heated to kill contaminated *C. botulinum* endospores, acts as source of infection. The endospores are ingested. They germinate and produce botulinum toxin during vegetative growth.

The toxin has effects on nervous systems, hence it is a neurotoxin. It binds to synapses of motor neurons and breaks the synaptic vesicle membrane protein, synaptobrevin. Consequently it inhibits the exocytosis and release of neurotransmitter acetylcholine. This causes failure of muscle contraction in response to activity of motor neurons leading to paralysis.

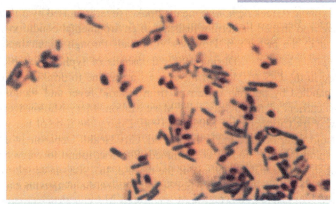

Clostridium botulinum, a causal agent of botulism.

Symptoms occur within 18-24 hours of ingestion of toxin. Symptoms include blurred vision, difficulty in swallowing and speaking, muscle weakness, nausea and vomiting. About one-third patients die without treatment due to failure of respiratory or cardiac system. In the U.S.A. about 100 cases of infant botulism are reported each year. Therefore, infant botulism is most common.

Laboratory diagnosis is by inoculation of mice with patient's blood serum, stools or vomitus to prove toxigenicity. Haemagglutination test is also done. Prevention and control of botulism can be done by (*i*) strictly adopting food processing practices by food industry, (*ii*) educating the people for safe preserving of home made food materials, (*iii*) discouraging the infants of below one year age from feeding honey, and (*iv*) giving large doses of botulism antitoxin to patients.

3.   Shigellosis (or Bacillary Dysentery)

Shigellosis is a diarrhoeal illness. It results from inflammatory reaction of the intestinal tract caused by *Shigella*. Four species of *Shigella* are known to be associated with shigellosis. *Shigella* is a Gram-negative, non motile, facultative and rod-shaped bacterium which is transmitted by the four 'Fs' i.e. food, fingers, faeces and flies. It is most prevalent among children of 1 to 4 years age. World wide deaths by bacillary dysentery is around 500,000 patients per annum. In 1984 a severe outbreak of shigellosis was noticed from West Bengal.

After acquiring the bacterium, its proper establishment in intracellular space of epithelial cells of colon occurs within 1-3 days. The mucosal cells phagocytose the bacteria. Bacteria disrupt the phagosome membrane and reproduce within it. It also secretes endotoxin and exotoxin that do not spread beyond the epithelium of colon. Consequently, a watery fluid containing blood, mucus and pus starts coming out at intervals. When the case becomes serious the colon is ulcerated.

The disease lasts an average of 4 to 7 days in adults and becomes self-limiting, whereas in infants and children it is fatal because of malnutrition, neurological disorders and failure of kidney. Preventive and control measures are: good personal hygiene and clean water supply, and treatment with trimethoprim-sulfamethaxazole or fluoroquinolones. Development of antibiotic-resistant strains of *Shigella* has also been reported.

Typhoid Fever

Typhoid (Greek *typhoides* means smoke) fever is caused by *Salmonella typhi* which is a Gram-negative, rod-shaped bacterium resistant to environmental conditions. Fresh water and food act as reservoir of the bacterium. However, it spreads through contaminated water.

Typhoid Mary : During 1900s thousands of typhoid fever cases and a few deaths were reported in the U.S.A. Most of these cases arose due to drinking of contaminated water of eating foods handled by persons suffering from typhoid fever and shedding *S. typhi*. One of the most famous carriers of this disease was Mary Mallon. Mary Mallon worked as cook in seven houses in New York City between 1896 and 1906. During the time of her work in these homes, 28 cases of typhoid fever occurred. The New York City Health Department arrested Mary and admitted to the hospital. Mary's stool was examined. She was found to carry typhoid fever bacteria but she did not show external symptoms of the disease. In 1908, an article was published in the *J. Amer. Med. Ass.* as *"Typhoid Mary"*. After being released she pledged not to act as cook. But she changed her name and began to work as cook again. For five years she spread typhoid by shedding bacteria. She was again arrested and held in custody for 23 years until she died in 1938. During her life time, she was linked with 10 out breaks of typhoid fever, 53 cases and 3 deaths. Thus, Mary was one of the most famous typhoid carriers.

S. typhi, after ingestion/drinking of water, colonize the small intestine, penetrate the epithelium and spread to lymphoid tissues, blood, liver and gall bladder. Symptoms include fever, headache, abdominal pain and malaise. This remains as such for several weeks. After 3 months most of the patients do not shed bacteria, whereas a few of them do for prolonged time without external symptoms. In these patients bacteria grow in gall bladder, and reach to intestine.

Laboratory diagnosis is made by demonstrating the bacteria in stool, urine and blood, and also through serological test (widal test). The preventive and control measures include (*i*) purification of drinking water, milk pasteurization, and prevention of handling of food by carriers, (*ii*) complete isolation of the carriers, (*iii*) use of antibiotics such as ceftizoxane, trimethoprim-sulfamethoxazole or ampicillin, and (*iv*) use of vaccine for high risk individuals.

III. Soil-Borne Diseases

Tetanus

Tetanus (Greek *tetanos* means to stretch) is caused by *Clostridium tetani. C. tetani* is a Gram-positive, anaerobic and spore form-ing bacterium, the endospores of which are found in soil, dust and faeces of many farm animals and humans. However, bacterium can exists in air, water and human in-testine. Moreover, it can survive in dead anaerobic matter because the bacterium is basically a saprobe rather than a parasite. In India incidence of tetanus is high.

Transmission of the bacte-rium takes place through skin wound. The endospores enter in

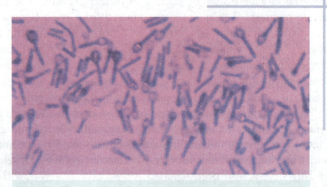

Spores of tetanus causing, Gram-positive bacterium *Clostridium tetani.*

wounds/breaks caused in skin. These germinate if oxygen tension is low. However, the bacteria living in intestine do not cause disease. These are discharged through faeces.

In the wound the neurotoxin, *tetanospasmin*, is released after the death and break down of bacteria. Tetanospasmin is an endopeptidase that breaks the synaptic vesicle membrane protein, synaptobrevin. This in turn inhibits exocytosis and release inhibitory neurotransmitters at synapses within the motor nerves of spinal cord. This results in stimulation of skeletal muscles that lose its control. Tetanospasmin causes tension and twisting in skeletal muscles present around the wound. The jaw muscles become tight. Consequently, there develops trismus ("lock jaw") which is an inability to open the mouth due to spasm or contraction of master muscles. In severe contraction of muscles opisthotonos takes place. It is a situation when back bows backwardly, and back and heels approaches each other to form an arch. Due to contraction of diaphragm and intercostal respiratory muscles deaths occur.

Immunization and control measures: Tetanus is prevented by using tetanus toxoid. The tetanus toxoid is a formaldehyde-treated toxin that is precipitated in the form of salt of aluminum to increase its immunizing potency. The tetanus toxiod is given with DPT programmes. After a few months of birth, initial dose is given. The second dose should be given after 4-6 months of the first dose. Finally, the reinforcing dose should be given 6 to 12 months after the second dose. Between the age of 4 to 6 years, a final booster is given. A single booster dose can provide protection for 10 to 20 years.

One to too many doses over a period of years result in hypersensitivity reactions. Therefore, when an individual has a wound infection, he should be given booster doses.

The tetanus can not be controlled because of survival of bacterium in soil for a long duration. Treatment of tetanus is not very effective. About 30-90% cases become fatal. Hence, preventive measures must be adopted by active immunization with toxoid, prophylactic use of antitoxin, and using penicillin.

2. Anthrax

Anthrax (Greek *anthrax* means coal) is caused by Gram-positive, spore-forming *Bacillus anthracis*. It is highly infectious disease that spread through contact of infected animals (goat, sheep, cattles) with humans having abrasion or cut on skin. Consequently, the endospores come in the contact of cuts, infect skin and develop (between 1-15 days) cutaneous anthrax. In addition, when endospores are inhaled, they result in pulmonary anthrax; similarly, when endospores are ingested, they cause gastrointestinal anthrax. However, endospores can remain viable in soil for a long duration.

It infects skin, wounds and ulcerate the skin. Pulmonary anthrax is similar to influenza. The disease becomes fatal when the bacteria infect blood stream. The bacteria secrete anthrax toxin (a complex exotoxin system composed of three proteins). The toxin develops symptoms that include headache, fever, and nausea.

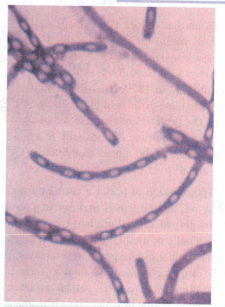

Bacillus anthracis, a Gram-positive spore forming bacterium causing anthrax.

The disease is diagnosed directly through identifying the bacteria, culture of bacteria and serologically. The control meaures are vaccination of cattles, animals, and humans handling them. The antibiotics recommended for patients are penicillin G, or penicillin G plus streptomycin. The other soil-borne bacterial diseases are given in Table 22.3.

Table 22.3 : Some important soil-borne bacterial diseases of humans.

Bacteria	Disease	Affected parts	Characters	Treatment	Vaccines
Bacillus anthracis	Anthrax	Blood, Skin, intestine, lung	*G$^+$, spore-forming rod	PenicillinG	For animals
Clostridium tetani	Tetanus	Nerves at synapses	G$^+$, spore-forming,aerobe	Antitoxin, Penicillin	Toxoid in DPT
C. perfringens	Gas gangrene	Muscles, nerves blood cells	do	Penicillin	-
Leptospira interrogans	Leptospirosis	Kidney, liver,spleen	Spirochete	Penicillin	-
Listeria monoc-ytogenes	Listeriosis	Blood, Nervous system	G$^+$, rod	Tetracycline	-

*G$^+$, Gram-positive

IV. Sexually Transmitted and Contact Diseases

There are several bacterial diseases in human which are transmitted sexually or through contact (Table 22.4). A few of them are discussed in detail.

1.   Gonorrhea

Gonorrhea (Greek *gono* means seed, and *rhein* means to flow) is caused by *Neisseria gonorrhoeae* which is named after Albert L.S. Neisser who cultured the bacterium first in 1879. It is small Gram-negative, oxidase-positive, diplococcus bacterium. The bacteria are also referred to as gonococci. Gonorrhea is a sexually trans-mitted disease of the mucus membrane of genito-urinary tract, eyes, rectum and throat.

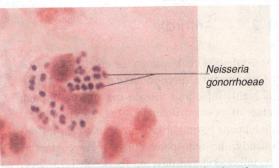

Neisseria gonorrhoeae

Gram stain of male urethal exudate showing *Neisseria gonorrhoeae* (diplococci).

Transmission of bacteria occurs during intercourse. Bacteria adhere to microvili of mucosal cells by pili and protein II that function as adhesion. Due to attachment bacteria are not washed away by normal vaginal discharge or by strong flow of urine. In female, the bacterium infects the epithelial cells of urethra and cervix. It takes about 7-21 days to develop symptoms. These occur in some vaginal discharge. In severe cases, passage of fallopian tube is blocked by pus and scar tissues resulting in sterility. Female feels pain and burning sensation while urinating. Normal menstruation cycle in the patient is interrupted. Gonococci desseminate most often during menstruation, a time in which there is increased concentration of free irons available to the bacteria. In both the sexes gonococcal infection may occur leading to gonorrheal arthritis, gonorrheal endocarditis, or gonorreal pharyngitis. As the newborns pass through an infected birth canal, gonorrheal eye infection occurs in them. This disease is called ophthalmia neonatorum or conjunctivitis of the newborns'. However, about 50% of infected females transmit this disease unknowingly.

Table 22.4 : Bacterial diseases of man transmitted sexually or through contact.

Bacteria	Disease	Characteristics	Affected organs	Symptoms
A. Sexually Transmitted Diseases				
Calymmatobacterium granulomatis	Granuloma inguinale	Mycoplasma	Urethra, fallopian tube, epididymis	Pain on urination, dischargeetc
Haemophilus ducreyi	Chancroid	G⁻, rod	External genital organs	Soft chancres
Gardnerella vaginal	Vaginits	G⁻, rod	Vagina	Foul smelling discharge
Mycoplasma hominis	Mycoplasmal urethritis	Mycoplasma	Urethra, fallopian tube, epididymis	Pain on urination discharge
Neisseria gonorrhoeae	Gonorrhea	G⁻, Diplococcus	Urethra, cervix fallopian tube, eyes, pharynx	Pain on urination, discharge salpingits
Treponema pallidum	Syphilis	Spirochete	Skin, nervous system, Cardio-vascular	Chancre Gumm
B. Contact Bacterial Disease				
Chlamydia rachomatis	Trachoma	Chlamydia	Eyes	Pain in eyes
Haemophilus influenzae biotype II	Bacterial conjunctivitis	G⁻ rod	Eyes	Eye reddening, pain
Mycobacterium leprae	Leprosy	Acid-fast rod	Skin,bones, peripheral nerves	Lesions with deformity in organs
Staphylococcus aureus	Staphyloccal skin diseases	G⁺ staphylococci	Skin	Lesions
Treponema pertenue	Yaws	Spirochate	Skin	-

G⁻, Gram-negative; G⁺, Gram-positive bacteria.

In males incubation period of the bacteria is 2 to 8 days. Primary infection occurs in urethra. The symptoms include a thin and watery discharge followed by white to creamy pus from penis. The males fell frequent painful urination with burning sensation. The flow of sperms is blocked when epididymis is infected. This leads to sterility. Symptoms in male are more serious than that in female. The bacteria also infect eyes after being transmitted through fingertip, towel, etc and cause corneal infection which is known as keratitis.

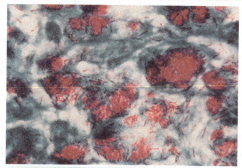

Mycobacterium leprae, a causal bacterium of leprosy.

Diagnosis of *N. gonorrheae* in laboratory is done based on oxidase-reaction, Gram-stain reaction and colony and cell morphology. Performance of confirmation test is also necessary.

Effective methods for control of this sexually transmitted disease are public education, diagnosis and treatment of asymptomatic patients, use of condom and quick treatment of infected individuals. About 60% of this disease occurs in individuals of 15-24 years age.

The recommended antibiotics for treatment of gonorrhea are (*i*) penicillin G plus probenecid, (*ii*) ampicillin plus probenecid, (*iii*) ceftriaxone or ofloxacin plus doxycycline for 7 days, or (*iv*) streptomycin. To prevent conjunctivitis of the newborns, tetracycline, erythromycin, povidone-iodine, or siliver nitrate in dilute solution is placed in the eyes of newborns. Penicilin-resistant strains in 1980, and tetracyline-resistant strains in 1980 have been reported.

2. Syphilis

Venereal syphilis (Greek *syn* means together, and *philein* means to love) is a contagious sexually transmitted disease caused by a spirochete, *Treponema pallidum*. Congenital syphilis is a disease which is acquired in uterus from the mother.

During the end of 15th century, syphilis was recognized in Europe. According to one hypothesis syphilis is of New World Origin. Christopher Columbus (1451-1506) and his crew acquired it in West Indies and introduced into Spain when returned from their historic voyage. The other hypothesis holds that syphilis has been endemic for centuries in Africa where from it was transported to Europe through the migrants during 1500.

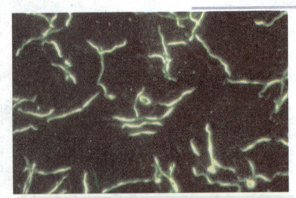

Treponema pallidum, a spirochete that causes syphilis.

In 1530 Girolamo Fracastoro, a poet and physician of Italy, has written about syphilis as a disease transmission of "seeds" which took place through sexual contact. In 18th century venereal transmission of syphilis was shown. The term venereal is derived from venus, the Roman Goddess of love. In 1838, P. Ricord demonstrated the various stages of syphilis. In 1905 Fritz Schaudinn and Erich Hoffmann discovered the causitive bacterium. In 1906 August Von Wassermann introduced the diagnostic test that bears his name. In 1909, P. Ehrich applied the therapy of syphilis by using arsenic derivative, arsephenamine or salvarsan.

T. pallidum moves by axial filament and enters the body through mucus membrane or abrasion or wound or hair follicle on the skin. Sexual intercourse is the most common way of transmission and contact of the bacteria. However, there is 10% chance of acquiring bacteria from single expose to an infected sex partner. In untreated adults the disease occurs in the following three stages:

(*i*) **Primary Syphilis:** In the primary stage incubation period is of about 10 days to 3 weeks. This stage is characterized by a small painless and reddened ulcer or *chancre* (French *canker* a descrutive sore). Chancre has a hard ridge that looks as the infection site and consists of spirochetes. Disease is transmitted if contact with chancre is made during sexual intercourse. About in 30% cases chancre disappears and disease does not spread further. In others the spirochetes are distributed throughout the body via blood streams.

(ii) **Secondary Syphilis :** The second stage is established within 2 to 10 weeks after the primary lesions. This stage is characterized by a skin rash. At this stage about 100% cases are serologically positive. At this stage symptoms include loss of patches of hairs, malaise and fever. There are flat wart like lesions which are filled with spirochetes.

(iii) **Tertiary Syphilis :** After several weaks the disease becomes latent. During latent period disease is not infectious except transmission from mother to foetus. This is called congenital syphilis. After several years (more than a decade) a tertiary stage develops in approximately 40% of cases when secondry syphilis remains untreated. The symptoms of tertiary syphilis include formation of degenerative lesions (gummas) in the skin, bone, cardiovascular and nervous systems. At this stage number of spirochetes is drastically reduced. This results in mental retardness, blindness, a "shuffle" walk leading to paralysis. Several famous persons suchas Henery VIII, Adolf Hitler, Oscar Wilde, etc. have been sufferer of syphilis.

Syphilis is diagonised through clinical history, thorough physical examination, immuno-fluorescence examination of fluid from lesions to observe typical mortality. The serological tests include: *(i)* test for nontreponemal antigen by VDRL (Venereal Disease Laboratories test), RPR (Rapid Plasma Reagin test), Complement fixation test or the Wassermann test, and *(ii)* test for treponemal antibodies (FTA-ABS, fluorescent treponemal antibody-adsorption test; TPI, *T. pallidum* immunization; *T. pallidum* complement fixation; TPHA, *T. pallidum* haemagglutination). Preventive and control measures of syphilis are: *(i)* public education, *(ii)* prompt treatment of new cases, *(iii)* treatment of source of infection and contact, *(iv)* maintaining sexual hygiene, and *(v)* prophylaxis (condoms). The highest incidence of syphilis is in the age group between 20 and 39 years.

3. Leprosy

Leprosy (Greek *Leprosei* means scaly, scabby, rough) or Hansen's disease is caused by *Mycobacterium leprae*. This bacterium was first observed in 1874 by Gerhard Hansen, a Norwegian physician. So far the bacterium could not be cultured on artificial medium. *M. leprae* is acid-fast rod, slow grower and heat-sensitive bacterium.

Leprosy is a contact disease that degenerates the tissues and deforms the body organs. Humans are the only reservoir of the disease. According to WHO report, there are about 10 millions of cases especially in South Asia, Africa and America.

Disease is transmitted to children when prolonged exposures to infected persons are made. For family contact nasal secretions are the infectious materials. It takes about 3 to 5 years to establish properly. The bacterium infects peripheral nerves and skin cells and enters intracellularly. The earliest symptoms of leprosy are slightly pigmented eruptions of skin which are several centimeter in diameter. In early stage about 75% individuals heal the lesions due to development of cell-mediated immunity. In 25% leprosy develops due to weak immune response. There are two forms of leprosy.

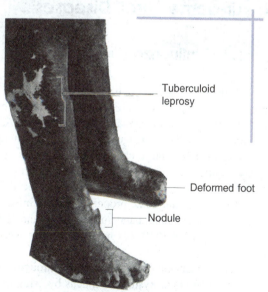

In tuberculoid leprosy, the skin within the nodule is completely without sensation. The deformed foot is associated with lepromatous leprosy.

Tuberculoid leprosy

Deformed foot

Nodule

(*i*) **Tuberculoid (neural) Leprosy:** Tuberculoid leprosy is mild, nonprogressive and associated with delayed type hypersensitivity reaction to antigens on the surface of *M. leprae*. The nerves are damaged and regions of the skin that have lost sensation are surrounded by a border of nodules. This results in atropy of muscles and disfiguring of skin and bones, and curling of fingers. The outer ear and nasal cartilage degenerate.

(*ii*) **Lepromatous (progressive) Leprosy:** In the infected individuals, who do not develop hypersensitivity, there occurs progressive form of disease which is known as lepromatous leprosy. A large number of *M. leprae* develops in the cells of skin. This type of leprosy is characterized by lepromas (tumour like growths) on the skin and along respiratory tract. The bacteria kill skin tissue resulting in progressive loss of facial feature, fingers, toes, etc. Disfiguring nodules form all over the body. The nerves are less damaged as compared to tuberculoid leprosy.

Since the bacterium cannot be cultured *in vitro*, it can be diagnosed by acid-fast staining, and serologically such as fluorescent leprosy antibody adsorption test, DNA amplification, ELISA test. Long-term treatment with sulfone drug, dapson, and rifampicin with or without clofazimine has been recommended. *Mycobacterium* W vaccine is also being used.

B. Viral Diseases

There are more than 400 different viruses that cause disease in humans. Chapter 15 deals with classification of animal viruses and their morphology, and a brief description of genetic material and diseases caused by them. The viral diseases are generally grouped into the four types on the basis of typical symptoms produced on body organs: (*i*) pneumotropic diseases (respiratory tract infected by influenza, etc.), (*ii*) dermotropic diseases (skin and subcutaneous tissues affected by chicken pox, herpes simplex, measles, etc), (*iii*) viscerotropic diseases (blood and visceral organs affected by yellow fever, dengue fever, etc.), and (*iv*) neurotropic diseases (central nervous system affected by rabies, polio, encephalitis, etc.) (Table 22.5). However, on the basis of their source of infection they may be grouped as airborne diseases, food and water-borne diseases, arthropod-borne disease, direct contact diseases, and other viral diseases.

I. Air-borne Viral Diseases

1.   Influenza (Flu)

Influenza (Italian, to influence) or flu is a viral diseased caused by orthomyxovirus in respiratory system. Fig. 15.8E shows the structure of influenza virus. Based on antigenic properties of N and H spikes influenza virus is classified into A, B and C groups. However, it frequently changes its antigenicity. These changes are referred to as antigenic variations. The small variations are called antigenic drift, and large changes are known as antigenic shift (see Chapter 20, section *Epidemiology*). Antigenic variations occur yearly in group A, less frequently in group B and unknown in group C.

Changes in antigenic properties lead to infection in humans. This is why it causes worldwide pandemic; for example, pandemics of 1918 and 1957 were severe, whereas that of 1977 was mild. Pandemic of 1918-1919 was most destructive where 550,000 deaths were recorded in the U.S.A. alone.

Virus enters the respiratory system through inhalation or ingestion of contaminated respiratory secretion. After hydrolysing the mucus by using neuramidase present in spikes, the virus adhers to epithelial cell with the help of haemagglutinin spike protein. Consequently, some part of plasma membrane of the cell bulges inwardly and form a vesicle. The vesicle encloses the virus. When pH decreases, haemagglutinin molecules undergo conformational changes. The hydrophobic ends of haemagglutinin spring outward and extend outward the vesicle. Consequently, nucleocapsid is released into the cytoplasm.

Symptoms of influenza are chills, fever, headache, malaise and muscular pain. These symptoms appear when epithelial cells die due to attack of activated T cells. The flu is recovered in 3 to 7 days. Influenza alone is not fatal. Death is caused probaly due to secondary bacterial invasion by *Staphylococcus aureus*, *Streptococcus pneumoniae* and *Haemophilus influenzae*. The recommended antiviral drugs are amantadine and

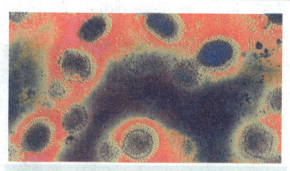

The protein "spikes" projecting from the influenza virus coats attach to plasma membranes of cells in the human respiratory system, helping the virus gain entry into the cells.

rimantidine. Salicylic acid must be avoided in youngs of below 14 years age. One should avoid the contact and droplets of the patients.

Table 22.5 : Example of some viral diseases of humans based on tissues affected.

Disease	Affected Organ	Transmission	Vaccine
A. Pneumotropic diseases			
Influenza	Respiratory tract	Droplets	Available at high risk
Adenovirus infection	Lungs	Droplets	Not available
Rhinovirus infection	Upper respiratory tract	Droplet and Contact	Not available
B. Dermotropic diseases			
Chicken pox (Varicella)	Skin and nervous system	Droplets, Contact	Not available
Herpes simplex	Skin, pharynx Genital organs	Contact	Not available
Measles (Rubeola)	Skin, respiratory tract	Contact, droplets	Attenuated viruses
Mumps	Salivary gland	Droplets	Attenuated viruses
Small pox (variola)	Blood	Contact droplets	Cowpox virus
German measles (Rubella)	Skin	Contact droplets	Attenuated viruses
C. Viscerotropic diseases			
Yellow fever	Liver	*Aedes aegypti*	Inactivated virus (Mosquito)
Dengue fever	Blood	*A. aegypti*	Not available
Hepatitis A	Liver	Food, water, Contact	Not availabe
Hepatitis B	Liver	Contact with body fluid	HgsAg fragments
AIDS	T-lymphocytes	Contact with body fluid	Not available
D. Neurotropic diseases			
Rabies	Brain	Contact with body fluid	Inactivated viruses
Polio	Intestine, brain spinal cord	Food, water, contact	Inactivated / attennuated virus
Arborial	Brain	Arthropod	Not available
Slow vius disease	Brain	Not Known	Not available

2. Measles (Rubeola)

Measles (Latin *rubeus* means red) is highly contagious disease caused by Morbillivirus of the family *Paramyxoviridae*. It is endemic throughout the world. Virus is a helical RNA virion and closely related to mumps or Rs virus. Its size is 125-250 nm with 18 nm diameter of nucleocapsid. The genetic material is single stranded RNA of molecular weight of $5\text{-}6 \times 10^6$ daltons. Capsid contains spikes consisting of haemagglutinin property (see Chapter 15, Fig. 15.8A). The virus enters through the respiratory tract or conjunctiva of the eyes. The membrane co-factor protein (i.e. the complement regulator CD46) acts as the receptor for measles virus.

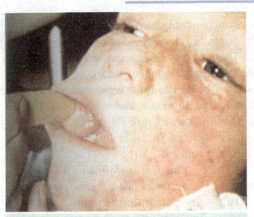

Measles (rubeola).

The incubation period is 10 to 21 days. The first symptom is a nasal discharge on about 10th day followed by cough, fever, headache and conjunctivitis. Faintly pink coloured muscles like skin eruptions occur within 3 to 5 days that last about 5 to 10 days. Lesions of oral cavity contain diagnostically important bright-red *koplik's* spots with a bluish-white region in the centre of each one. Thereafter, infrequent progressive degeneration of the central nervous system occurs. This is called subacute sclerosing panencephalitis.

There is no specific treatment of measles. The use of attenuated measle vaccine (Attenuvax) alone or in combination (MMR vaccine i.e. measles, mumps, rubella) is recommended for children. However, it has been found that measles infects 44 million people and kills 1.5 million per annum throughout the world. Since the beginning of public health immunization programme, about 98% cases have decreased.

3. Mumps

Mumps is an acute disease that occurs mainly in school age children. It is caused by paramyxovirus (see Chapter 15; Fig. 15.8A). The virus is transmitted in saliva and respiratory droplets. It enters the respiratory tract of non-immune individuals through contact. The symptoms are swelling and tenderness of the salivary (parotid) glands after 16 to 18 days of infection. Therefore, this disease is also known as *epidemic parotitis*. Swelling persists for about 1-2 week with low fever. In the postpubescent males certain complications e.g. meningitis and inflammation of epididymis and testes associated with the disease occur. In adults mumps is dangerous for reproductive system. Some males may develop *orchitis* i.e. swelling of testicles to 3 or 4 times of normal size. This results in decline in sperm counts. In contrast, in females, there may develop oophoritis i.e. lower back pain and enlargement of ovaries.

Mumps: A child with diffuse swelling of the parotid glands due to mumps virus.

Prevention of mumps is keeping the children with mumps free from school or other activities for 2 weeks. Therapy of mumps is the use of a live attenuated mumps virus vaccine. It is given as part of the tripple MMR vaccine.

4. Rubella (German Measles)

In the 1800s, Rubella (Latin *rubellus* means reddish) was first described in Germany, hence it is called German measles. It is caused by rubella virus which is a member of the family *Togaviridae* and contain ssRNA. It causes skin disease on children of 5 to 9 years of age. Rubella is distributed throughout the world, and occurs during winters and spring seasons. Virus spreads through droplets and sheds through respiratory secretions of the patients. After entering into body, it takes about 12 to 23 days to get established. Small red spots develop on body that last in about three days. Rash of small red-spots is followed by a light fever. The rash is immunologically mediated, and not caused by the virus. This results in appearance of rash and diappearance of virus from the blood.

Rubella is a mild disease, however, it can be disastrous in the first three months of pregnancy. This leads to fatal death, premature delivery or several congenial defects in heart, eyes, ears of the growing baby. This disease is called congenital rubella syndrome. All children and women of child bearing age can be infected by rubella. Therefore, the live attennated rubella vaccine is recommended to them.

5. Small Pox (Variola)

Small pox (variola) was very dangerous and fatal of all viral diseases. Smallpox has played a major role in reducing Indian resistance to the European colonization of North America. In Mexico Indian poplulation declined by 90% within 100 years of contact with Spanish. This has been due to small pox and other diseases. Before contact with Europeans, 10-12 million Indians lived in north of Rio Grade. In addition, in 1600 A.D. about 72,000 lived in New England, but by 1674 A.D. around 8,600 remained in New England. Such catastrophe is considered with the

The picture shows a child suffering from *small pox*. The last recorded case of this viral disease was in 1977, in Somalia.

fact of European contact with native Americans. Europeans have already suffered from small pox during 16th century. They only carried over the inoculum. Moreover, many American cities e.g. Boston, Philadelpha and Plymouth, grew upon cities of previous Indian villages.

Virus is brick-shaped and consists of dsDNA of largest size, 270 nm which is packed in nucleocapsid. This is surrounded by a swirling series of fibres. For detail see Chapter 15 (pox virux, Fig. 15.6D).

The early symptoms are chills, fever and general prostration. When temperature falls, pinkish red spots (macules) develop first on scalp and forehead and later on neck and extremities. In the last macules develop on trunk. Thereafter, these spots are converted into pink pumples or papules.

These turn into fluid filled vesicles. The term *variola* is derived from Latin *varus* which means pimples. In the advanced stage the vesicles got coalesced to form pustules that cover the whole surface of body; when the pustules burst, pus comes out. The pustules are deep in skin as compared to chicken pox that appears on surface. Sometime the pustules form soft crust. After falling the crusts, pitted scars or pocks are formed. Thus the name small pox differs from the pock developed from the larger lesions of syphilis or chicken pox.

Small pox is transmitted by contact with skin or body fluid such as urine, blood or droplets. In 1798, Edward Jenner (an English physician) found that milk maids and other, who were working with dairy cattle, contracted the closely related mild form of pox called cowpox or vaccinia (Latin *vacca* means cow). But those, who had contracted cowpox and had recovered from it, did not contract small pox. Therefore, Jenner procured some material from cowpox for small pox, and established the process of vaccination. This method of vaccination became very famous in 1806. Napoleon ordered his army to be vaccinated by the above method. This process of vaccination is based on the principle that cowpox virus after entering the blood increases the production of antibodies. Since both cowpox and small pox viruses are similar, the antibodies are equally effective on both of them.

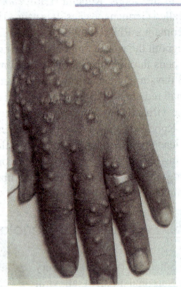

Small pox : Back of hand showing single crop of small pox vesicles.

In 1967, the World Health Organisation (WHO) received funds and launched a Global eradication programme. Through proper surveillance small pox cases were identified and vaccinated. The last case of small pox had been in 1974 in Bangladesh, in 1976 in Ethiopia and in 1977 in Somalia. In 1979, WHO has announced that small pox has been eradicated from the earth, except that which has been kept in four research laboratories of the U.S.A., Russia, U.K. and Japan with the permission of WHO. The WHO is now trying to decide whether or not to destroy the remaining stock of small pox virus.

II. Insect-borne Diseases

The arthropod-borne viruses (arboviruses) are transmitted by arthropods that suck blood of one man, and subsequently the others. The arthropods act as vector. Viruses multiply in arthropods without causing any disease (Calisher, 1994). Six major arbovirus diseases are given in Table 22.6.

Table 22.6 : Occurrence of six major arbovirus diseases in man.

Disease	Vectors	Mortality
1. California encephalitis	*Aedes* sp.	Rare
2. Colorado tick fever	Ticks	Rare
3. Eastern equine encephalitis	*Aedes* sp.	50-70%
4. St. Louis encephalitis	*Culex* sp.	10-30%
5. Venezuelan equine encephalitis (VEE)	*Aedes* sp., *Culex* sp.	20-30%
6. Western equine encephalitis (WEE)	*Culex* sp.	3-7%

1. Yellow Fever

Yellow fever is the first human viral disease caused by yellow fever virus which was discovered by Walter Read in 1901. Carlos Huan Finley demonstrated experimentally that it is transmitted by a mosquito, *Aedes aegypti* from one human to other human in urban areas. However, in sylvatic (sylvatic means in the woods or affecting wild animals) it is transmitted by mosquitoes to monkeys and from monkeys to humans.

Yellow fever is caused by a flavivirus that is endemic in many tropical areas, such as Mexico, South America and Afria. After entering in humans, it spreads to local lymph node and multiplies therein. Later on it moves to liver, spleen, kidneys and heart. The symptoms are fever, chills, headache and backache along with vomiting. In severe cases, the affected organs show lesions with haemorrhage.

No treatment of yellow fever is available so far. It is diagnosed serologically. Vaccine contains attenuated yellow fever 17 D strain or the Dakar strain virus. Therefore, adopting vaccination and control of insect vectors are the steps for control of this disease.

2. Dengue Fever

Dengue fever is caused by an RNA containing virus of icosahedral symmetry. The virus capsid is enveloped. This disease has been found severe in areas of southeast pacific and in Southeast Asia. However, a pandemic effect spread in Delhi and adjoining areas in India in 1996.

Symptoms involve severe fever and prostration followed by acute pain in limbs and muscles. Patients experience bone-breaking pain; therefore, it is also called break-bone fever.

Dengue fever is transmitted by a mosquito, *Aedes* that lays eggs in stagnant fresh water such as in ditches, coolers, etc. The virus is transmitted when the mosquitoes inject their saliva in human body. It is interesting to note that only female bites frequently for getting human blood for egg production. After 9 to 10 days of consuming blood the mosquitoes become infectious, thereafter, dies whether transmit dengue or not.

After infection, human develops dengue haemorrhagic fever. Thus, there are four types of virus of dengue. It was found in Delhi that all the virus were not equally effective. Only one out of four was dangerous.

III. Food-borne and Water-borne Diseases

Since the beginning of recorded history, food and water have been a potential carriers of viral diseases such as hepatitis A, hepatitis E, poliomyelitis, etc. Poliomyelitis has been discussed in this section.

1. Polio (Poliomyelitis)

Poliomyelitis (Greek *polios* means gray, and *myelos* means marrow or spinal cord), polio or

Ancient Egyptian with polio : Note the withred leg.

infantile paralysis is caused by the poliovirus which is a member of family *Picornaviridae*. Polio is of very ancient origin. Various Egyptian hieroglyphics of about 2000 B.C. show persons with withered legs and arms. In 1840, Jacob von Heine, a German orthopaedic described the clinical features of poliomyelitis and found that spinal cord was the area where individuals suffered. In 1890, Oscar Median (a Swedish pediatrician) recognised that a system phase (of fever development) occurred early, and complicated by paralysis. In 1908, Karl Landsteiner and Willian Popper successfully transmitted the disease to monkeys. In 1949, J. Enders, T. Weller and F. Robbins discovered that the poliovirus could be propagated *in vitro* in cultures of human embryonic tissues of non-neural origin. This led to a milestone for the development of vaccine. In 1952, D. Bordian recognised three serotypes of poliovirus. In 1953, J. Salk successfully immunized humans with formalin-inactivated poliovirus and in 1955, this vaccine (IPV) was licensed. Salk vaccine is administered in a series of three intramuscular injections. In 1962, the live attenuated polioviral vaccines (oral polio vaccines, OPV) was developed by Albert Sabin. The vaccines prepared by Salk (1954) and Sabin (1962) are known as Salk vaccine and Sabin vaccine, respectively.

Poliovirus is a (+) ssRNA virus of molecular weight 2.5×10^6 daltons (7.7 kb) with an icosahedral capsid. It has a diameter of about 27-30 nm (see Fig.15.8C). There are three types of polio: Type I (*Brunhidle* strain) causing moderate cases of paralysis, Type II (*Lansing* strain) causing greatest number of paralysis and Type III (*Leon* strain) rarely causing paralysis. It enters the body through contaminated water/food, reaches to intestine and survives therein. It multiplies in mucosa of tonsils and or gastrointestinal tract. The initial symptoms are fever, headache, nausea, vomiting and intestinal cramps resulting in loss of appetite. Some times the virus enters blood stream and causes viremia. In 99% cases, viremia is transient, and in 1% cases, the viremia persists, and the virus enters the central nervous system and causes polio. Poliovirus has high affinity for anterior horn motor nerve cells of the spinal cord. After infecting these cells, it multiplies and destroys them resulting in motor and muscle paralysis.

In developing countries 4 of every 1,000 children born annually have polio. It can be controlled only through vaccination. Government of India launched two major vaccination programmes in 1996 and in subsequent years with the name *Pulse Polio* Immunization. It is hoped that polio control by vaccination on global basis would be possible by the year 2010. The vaccines which are available at present are effective for all the three types of strains of polio virus. A child below 5 years age must be vaccinated at two years intervals.

IV. Direct Contact Disease

<table><tr><td>1.</td><td>

Viral Hepatitis

</td></tr></table>

Hepatitis (plural hepatides) (Greek *hepaticus* means liver) refers to inflammation of liver resulted through any infection. At present there are nine viruses known that cause hepatitis. Out of nine, two are herpes viruses and Epstein-Barr virus, and seven hepatotropic viruses. However, of the seven, only five are well characterised (Table 22.7).

Hepatitis B (serum hepatitis): Hepatitis B is caused by the hepatitis B virus (HBV). It is classified as an Orthohepadnavirus within the family Hepadnaviridae. The HBV consists of a circular dsDNA in a capsid that is enveloped. The HBV has complex structure (Fig. 22.2) It consists of three antigenic particles: (*i*) a spherical particle (22 nm diameter), (*ii*) a spherical particle (42 nm diameter) containing DNA and DNA polymerase called the **Dane particle,** and (*iii*) tubular or filamentous particle of variable length. The Dane particles in this form are ineffective because they are in unassembled form. The unassembled particles contain hepatitis B surface antigen (HBsAg). If these are present in blood, they represent that the person is infected (Tiollais and Buendia). This laid down a basis for the production of first recombinant DNA vaccine for humans developed by recombinant DNA technology (see Chapter 11, *Gene cloning in Microorganisms*, Section *recombinant Hepatitis B Vaccine*).

Table 22.7 : Characteristics of hepatitis viruses.

Disease/Virus	Classification	Genome	Transmission	Prevention
Hepatitis A	Picornaviridae, Hepatovirus	RNA	Food, Water	Killed HAV
Hepatitis B	Hepadnaviridae, Orthohepadnavirus	DNA	Blood, needle, body secretion, placenta	Recombinant HBV vaccine
Hepatitis C	Flaviviridae, Pestivirus or Flavivirus (?)	RNA	Blood	Routine blood screening
Hepatitis D	Unclassified	RNA	Blood	HBVvaccine
Hepatitis E	Caliciviridae (?)	RNA	Faecal/oral	Maintain Hygeine

HAV (Havrix Vaccine), HBV (hepatitis B virus); *, Hepatitis F and G are the newly discovered viruses and not well characterized so far.

Normally, the HBV is transmitted through blood transfusion, contaminated equipment, drug users' unsterile needles, or any body secretion (i.e. saliva, sweat, semen, breast milk, urine, faeces, etc).

However, HBV can pass from blood of infected mother to the foetus through placenta.. About 200 million people are victimized each year. Hepatitis B is more dangerous than the cancer. In India, the number of hepatitis B patients is increasing gradually.

The symptoms of hepatitis B vary. It includes fever, loss of appetite, nausea, abdominal abnormality and fatigue. After 1-3 months of incubation, HBV infects liver, destroys hepatic cells and releases the enzyme transaminase into blood stream. It is followed by jaundice. Consequently a product of haemoglobin (bilirubin) is accumulated in the skin and other tissues. This results in yellow appearance of skin, eyes, and urine as well. Chronic hepatitis B infection also causes

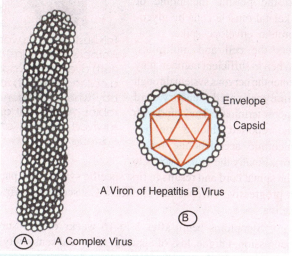

A Viron of Hepatitis B Virus

Envelope
Capsid

Ⓑ

Ⓐ A Complex Virus

Fig. 22.2 : Hepatitis B virus. A, a complex structure consisting of several viruses; B, structure of a single viral particle.

the primary liver cancer (hepatocellular carcinoma). Hepatitis B is known as a human carcinogen only to tobacco smokers.

The preventive and control measures are (*i*) avoiding contact with HBV-infected blood and secretions and immunizing needles, sticks, (*ii*) intramuscular injection of hepatitis B immuno globulin within 7 days, (*iii*) use of recombinant vaccine such as Engerix-B, Recombivax HB, etc, (*iii*) vaccination of other individuals such as carriers of HBV (household members, sex partners, blood recipients, homosexual males, international travellers).

2.   Rabies

Rabies (Latin *rabere* means madness) is caused by a number of different strains of neurotropic viruses. Most of them belong to the genus Lysavirus of the family Rhabdoviridae. Virion is bullet-shaped and enveloped that consists of (-) ssRNA (see Chapter 15; Fig. 15.8H).

When symptoms of rabies are fully materialised, mortality rate of humans increases. Most of the wild animals (e.g. foxes, coyotes and wolves, and street dogs, bats, cats, etc) transmit the disease. Canine rabies is endemic. Virus multiplies in the salivary gland of these animals.

When these animals bite humans or other healthy animals, viral particles are spread through saliva. Virus attaches to the plasma membrane of skeletal muscle cells by glycoprotein envelope spike. It enters the cell and multiplies. When in sufficient number, they enter the nervous system through unmycelinated sensory and motor terminals. They bind with the reported binding site (i.e. nicotinic acetylcholine receptor). Finally the virus reaches to

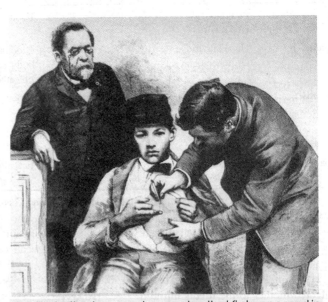

This illustration is a wood engraving that first appeared in *Harper's Weekly* in 1885. During the year, Louis Pasteur (left) oversaw the administration of a rabies vaccine by a colleague (right) to Joseph Meister (Center), a young boy who had been bitten repeatedly by a dog with rabies. As a result of the vaccination, Joseph Meister survived the bite and became a custodian in Louis Pasteurs laboratory-now called the Pasteur institute.

the spinal cord and results in first specific symptoms (pain at wound site). As the virus disseminates, a progressive encephalitis develops. Consequently, virus spreads. throughout the body including saliva.

Symptoms begin after 2 to 16 weeks of exposure of virus. There are anxiety, irritability, depression, fatigue, loss of appetite, fever and sensitivity to light and sound. Consequently, it results in paralysis. In about 50% patients painful spasm of throat and chest muscles occur when the patients swallow liquid. Therefore, patients do not take liquid. This stage has been called **hydrophobia** (fear of water). When the region that regulates breathing is destroyed, it results into death of patients (Kaplan and Koprowski, 1980).

Rabies can be diagionised by observing the *Negri bodies* (mass of viruses or subunits of unassembled viruses) under light microscope. However, today diagnosis is done by direct immunofluorescent-antibody (DIFA) of brain tissue, isolation of viral particles, testing the Negri bodies, and rapid enzyme-mediated immunodiagnosis test.

Prevention and control of rabies involve pre-exposure vaccination of dogs and cats and post exposure vaccination of humans. The effective vaccines are available in market.

3. Cold Sores

Cold sores or fever blisters (herpes labialis) are caused by herpes simplex virus type 1 (HSV-I). Similar to other herpes, it also contain dsDNA present in enveloped icosahedral capsid (see Fig. 15.6C).

Transmission occurs through direct contact of epithelial tissue with virus. At inoculation sites, blisters develop due to destruction of virus mediated host tissue. In severe cases, blisters occur on the surface epidermis of lips, mouth and gums. Blisters may heal within a week. After primary infection the virus travels to nerve ganglion, where it remains in latent stage for the life time of infected person. Certain stresses such as sunlight, fever, emotions may reactivate the virus. After being reactivated it moves from nerve ganglion to down a peripheral nerve to the border of lips or other part of the face, and produce the second fever blisters. Recurring infection may occur in eye causing eye inflammation. This causes a major blindness to patients (Stevens, 1989).

The antigen of virus can be detected by several enzyme immunoassay kits such as VIDAS, HERPCHECK, etc.

4. AIDS (Acquired Immuno-Deficiency Syndrome)

It is the first great pandemic of second half of 20th century. AIDS is caused by the *human immunodeficiency virus* (HIV) which is a lentivirus within the family *Retroviridae*. Schematic

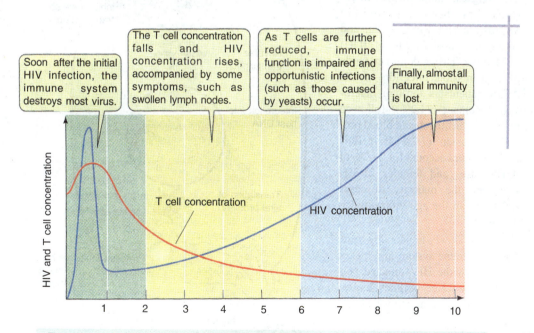

The case of a HIV infection : HIV infection may be carried, unsuspected, for many years before the onset of symptoms.

diagram of the HIV virion is given in Fig. 15.11. For detailed description of virus, and working of immune system see Chapter 15 (section *Retroviruses*).

According to WHO, 19 million people are infected worldwide and over 40 million people will have to be infected by the year 2000. Upto June 1995, about 476, 899 cases and deaths have been reported in the USA.

(*i*) **Transmission:** It is a transmissible disease. The groups, most at risk acquiring AIDS are (in descending order of risk) homosexual/bisexual men, intravenous drug users, prostitutes, heterosexual (who have intercourse with drug users, prostitutes and bisexuals), transfusion patients or haemophilics (who receive donated blood), users of contaminated needles and syringe, and children born of infected mothers.

About 90% of children from HIV positive mothers were infected during pregnancy, at birth or during breast-feeding. The mortality rate is about 92% of the diagnosed persons. Worldwide about 5.3 million people were infected in 1996 and some 5.8 million were infected in 1997. However, 2.3 million people died of AIDS in 1997 i.e. a 50% increase over 1996.

(*ii*) **AIDS in India:** According to a report (November, 1997) of United Nations, India has the largest number of HIV-infected persons. However, the first case of AIDS in India was registered in 1986. Since then, HIV prevalence has been reported in all states and Union territories. By a report of India Government, upto October 31, 1997, out of 3.20 million individuals practising risk behaviours and suspected AIDS cases (who were screened for HIV infection), 67,311 were found seropositive and 5002 cases of AIDS have been reported.

Highest number of HIV infections has been reported in Maharastra and Tamil Nadu, and among injectable drug users (IDU) of Manipur. The predominant mode of transmission infection of the AIDS patients is through heterosexual contact (74.2%), followed by blood transfusion and blood produced infusion (7.0%) and IDU (7.2%). Males account for 78.7% of AIDS cases and females 21.3%. However, the majority of cases (89%) are in the age group of 15-44 years.

HIV slowly breaks down the body's immune system and makes it easier for the HIV infected persons to get a variety of illness known as "opportunistic infections" (Table 22.8).

Soon after the record of first case of AIDS in 1986, a National AIDS committee was constituted. Thereafter, the Government launched the National AIDS con-

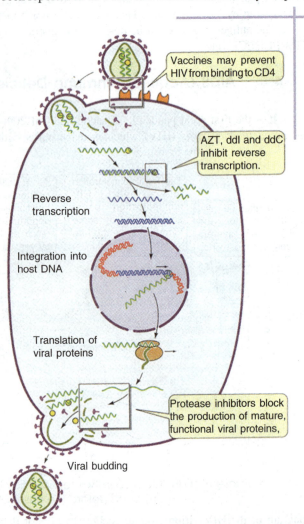

Vaccines may prevent HIV from binding to CD4

AZT, ddI and ddC inhibit reverse transcription.

Reverse transcription

Integration into host DNA

Translation of viral proteins

Protease inhibitors block the production of mature, functional viral proteins,

Viral budding

Strategies to combat HIV reproduction : These widely used drugs block specific steps in the HIV life cycle.

trol programme which focussed attention on increasing awareness of HIV/AIDS, screening of blood for HIV and testing of individual's practising risk behaviours. In 1992, the National AIDS Control Organisation (NACO) which works under the Ministry of Health was established to prepare the programme for making aware the youths. NACO is functioning in the following ways:

(*a*) It makes multimedia strategy for intersectoral collaboration with the Ministry of Information and Broadcasting, mass media, advocacy, involving NGOs, and implementing training and research.

(*b*) It has implemented a school education programme in 18 states, and developed the curriculum and training. To create awareness each year December 1 is celebrated as World AIDS Day.

(*c*) Counselling, training and provision of counselling services.

(*d*) Social marketing to increase the use of affordable quality condoms.

All over the country serosurveillance centres and sentinel sites have been established for monitoring HIV trends overtime in certain sections of the general population. In Manipur a pilot continuum of care programme has been implemented. The Bombay High Court has given the judgement for financial compensation by a company to an employee who was fired on being diagnosed HIV-positive. The Govt. of India has made active effort to involve NGOs (Non Government Organisations) participation. The process of funding NGOs has been decentralised.

(*iii*) **Prevention and Control of AIDS:** Prevention and control of AIDS involve: (*a*) the screening of blood and heat treatment of blood product to destroy the virus, (*b*) education to people about AIDS, (*c*) education about protected sexual behavior and practices including the use of condoms, (*d*) education of intravenous drug users to avoid sharing needles and syringe, (*e*) regular testing for HIV, and (*f*) taking helps from NGOs.

Attempts are being made to develop a vaccine that can (*a*) stimulate the production of neutralizing antibodies which can bind to the virus envelope and prevent from entering the cell, and (*b*) promote the destruction of those cells which are already infected by the virus. So far no vaccine has been produced possibly due to continuous changing in antigenic properties of the virus.

Table 22.8 : Disease processes associated with AIDS (based on MMWR (RR17), 1993

1. Candidiasis of bronchi, trachea, or lungs
2. Candidiasis, oaesophageal
3. Cervical cancer, invasive
4. Coccidioidomycosis, disseminated or extra pulmonary
5. Cryptosporidiosis, chronic intestinal *Cyclospora*, diarroheal disease
6. Cytomegalovirus disease (other than liver, spleen or lymph nodes)
7. Cytomegalovirus retinitris (with loss of vision)
8. Encephalopathy, HIV-related
9. Herpes simplex: Chronic ulcers or bronchitis, pneumonitis or oesophagitis
10. Histoplasmosis, disseminated or extrapulmonary
11. Isosporiasis, chronic intestinal
12. Kaposi's sarcoma
13. Lymphoma, Burkitt's or equivalent term) immunoblastic, primary, of brian
14. *Mycobacterium avium, M. tuberculosis*, etc.
15. *Pneumocystis carinii*, pneumonia
16. Pneumonia, recurrent
17. Progressive multifocal leucoencephalopathy
18. *Salmonella septicemia*, recurrent
19. Taxoplasmosis of brain
20. Wasting syndrome due to AIDS

IV.   The Other Diseases

See Kuru and scrapie in Chapter 15 (section prions).

QUESTIONS

1. Discuss in brief the air-borne bacterial diseases of human.

2. What are the food diseases? Give a detailed account of cholera, shigellosis and typhoid.

3. Give a detailed account of tetanus and anthrax.

4. Discuss in brief the sexually transmitted diseases with emphasis on gonorrhea, syphilis,   etc.

5. What is leprosy? Discuss in brief about the causal organism, types of leprosy and control measures.

6. Write an essay on air-borne viral diseases

7. What is small pox? Write a note on present situation of small pox in India.

8. Write in detail about dengue fever with emphasis on causal organism, vector and control measures.

9. What is polio? Discuss in detail the hazards of polio with emphasis on survival of disease and control measures.

10. What is Hepatides? Write in brief Hepatitis B virus, its spread and control.

11. What do you mean by AIDS? Write in detail about HIV, working of immune system and control measures.

12. Write short notes on the following

    (*i*) Tuberculosis, (*ii*) Diphtheria, (*iii*) Meningitis, (*iv*) Pneumonia, (*v*) (*vi*) Cholera,   (*vi*) Typhoid (*vii*) Botulism, (*viii*) Tetanus, (*ix*) Shigellosis, (*x*) Anthrax, (*xi*) Syphilis, (*xii*) Leprosy, (*xiii*) Flu, (*xiv*) Measles, (*xv*) Mumps, (*xvi*) Small pox, (*xvii*) Typhoid,   (*xviii*) Rubella, (*xix*) Yellow fever, (*xx*) Dengue fever, (*xxi*) Polio, (*xxii*) Hepatitis B, (*xxiii*) Rabies, (*xxiv*) HIV, and (*xxiv*) AIDS.

REFERENCES

Blake, P.A.; Weaver, R.E. and Hollis, D.G. 1980. Diseases of humans cuased by vibrios. *Ann. Rev. Microbiol.* 34: 341-367

Bloom, B. 1994. *Tuberculosis: Pathogenesis, protection and control*: Am. Soc. Microbiol,

Calisher, C.H. 1994. Medically important arboviruses of the United States and Canada. *Clin. Microbiol. Rev.* 7:89-116

Kaper, J.B., *et al.* 1995. Cholera.. *Clin. Microbiol. Rev.*" *Clin. Microbiol. Rev.* 8(1): 48-86.

Kaplan, M.M. and Koprowski, H. 1980. Rabies. *Sci. Am.* 242: 120-134

MMWR, 1993. Revised classification system for HIV infection and expanded surveillance case: definition for AIDS among adolescents and adults. *No. RR17*.

Stevens, J.G. 1989. Human herpesviruses: a consideration of the latent state. *Microbiol Rev.* 53: 318-332

Tiollais, P. and Buendia, M.A. 1991. Hepatitis B virus. *Sci. Am.* 264: 116-124

Quagliarell, O.V. and Scheld, W. 1992. Bacterial meningitis: pathogenesis, physiology and progress. *N. Engl. J. Med.* 327 (12): 864-871.

Human Cancer

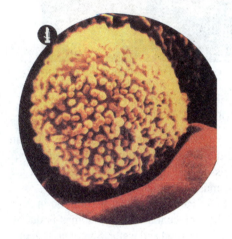

Cells are the fundamental unit of the life. They constitute tissue and organs in animals. In the human body about 10^{13} cells of different kinds are found which form different organs such as heart, liver, lungs, blood, etc. Cells require nutrients to divide and grow resulting in growth of body organs. As far as cell division is concerned, some cells rarely divide or never divide such as nerve cells, whereas the others like skin and progenitors of blood cells divide throughout the life to replace the others which die every day. Therefore, it appears that a cell divides in a most controlled manner, if not, it loses its control and deviates from the normal rule and thus causes chaos in the respective area. This loss of control over cell division and unnecessary production of the contiguous mass of cells is called tumour.

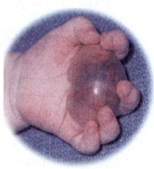

Photograph of extra growth in the form of tumour.

Tumours are of two types. If a cell divides when it should not, but none the less, stays within its normal location, its growth produces mass of cells which is called *benign tumour*. It is not a cancer. It can be surgically removed when becomes large. On the other hand some cells begin to divide and also acquire the ability to invade surrounding tissues and to move to alien sites within the body. For example, if abnormal cell division occurs in a single cell of lungs, a mass of cells are formed. Some of these tumorous lung cells break off from the mass and migrate to the other organs and initiate the cancerous growth of a new tumour which is called *metastasis*. Metastatic lung cells multiply within the invaded organ and first destroy the architecture followed by the function of the invaded organ. The tumours that are highly invasive of surrounding tissues and that can metastasize are called malignant tumour or cancer. It spreads to vital organs and make them deadly (Watson *et al, 1987*).

There are over 200 different types of cancer which are named after cell types and origin. For example, hepatomas (cancer of liver cells), melanomas (cancer of pigment-producing melanocytes), etc. They are named according to the tissue types. *Carcinomas* are the

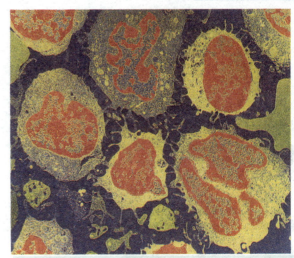

Hairy-Cell Leukemia : Transmission electron micrograph (× 3,100) of abnormal B-lymphocytes. The lymphocytes are caused with characteriotic hair like membrane derived protrusions.

cancers of epithelial cells (covering body organs e.g. skin, gut). It accounts more than 90% human cancers of which cancers of lungs, breast and colon are the most common and fatal. *Lymphomas* or *leukemias* are the cancers of blood cells or vascular tissue and immune system. *Sarcomas* are the cancers of supporting tissues such as blood vessels, muscles, bone, fibroblast, etc.

Carcinomas account for more than half of deaths. Liver cancer is the most common in Asia and Africa, while rest are in Western countries. The highest percentage of carcinomas is due to contact of external epithelia with the environment particularly air, food and water. Sarcomas together with lymphomas account for only 8% of human cancers.

A. Causes of Cancer

Researches done for the half of 20th century reveal several causes that induce cancer in experimental animals. The causal agents are certain chemicals, radiation and viruses that behave as mutagens by acting at the level of DNA. However, it has also been proved that cancer is a genetic disease caused by multiple mutations within the DNA of a cell. Consequently, the cell loses its control over the growth. The causes of cancer are discussed in detail.

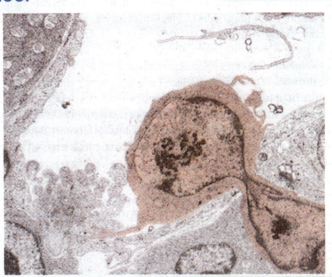

A cancer cell squeezes into a small blood vessel through the cell wall.

1. Hereditary Changes

Like a normal cell, a cancer cell divides and cuts off two daughter cells which, inturn, divide to result four cells, and so on. Thus, the very character of cancer cell (*i.e.* loss of control over growth) is passed on regularly generation to generations. Such features have been observed in tumours developing in animals as well as in tissue culture studies. This reveals that genetic changes within chromosomal DNA may underline the cancer phenotype. The work of Klein *et al.* (1971) lends support that the cancer is caused due to the genetic changes on DNA molecules. They fused cancer cells with the normal cells and obtained hybrids of different ploidy levels. The tetraploid cells (4n) commonly had a normal phenotype. However, these cells did not develop tumour when injected into genetically identical hosts. But most of tetraploid cells were not stable, and divided to produce cells containing lower number of chromosomes.

The non-cancerous hybrids were obtained by fusing normal and cancerous cells. However, from non-cancerous hybrids subtetraploid cells were obtained which were cancerous. This reveals that loss of one or more chromosomes resulted in re-expression of cancerous genes. Therefore, cancer phenotype is a recessive genetic trait.

2. Somatic and Multiple Mutations

Chromosomal abberation occurring in somatic cells, non-reproductive cells, results somatic mutation. The cancerous cells contain abnormal chromosomes. It is presumed that in somatic cells, abnormal chromosome arises due to breaking and rejoining of two other chromosomes. The abnormal chromosome helps the somatic cells to get transformed into cancer. Although, a few cancers are associated with childhood, yet with age the other types may increase sharply. This is because of accumulation of several mutations within a cell that increase with age. For example, a tumour called retinoblastoma results from multiple mutations in the persons who have inherited one of the mutations that can cause cancer. Further at least two and possibly 6 or 7 specific mutation are required to produce a cancer cell.

3. Chromosomal Aberrations

Chromosomes are observed best at the metaphase stage of cell division because the chromosomes are most condensed at this stage. Chromosomes are stained either with Giemsa stain or others which develop bands. These bands can be identified easily. In human, chromosomal abnormalities of tumours include both number and structure. During anaphase of mitosis, one/many chromosome fail to migrate at distant places resulting in production of daughter cells which may be *trisomic* (containing both of the chromosomes of a pair *i.e.* 2n+1) or monosomic (devoid of a pair of chromosome *i.e.* 2n-1).

The structural abnormalities include *deletion* (loss of a portion of chromosome), *inversion* (reunion of intercalary segment of a chromosome in reverse order due to two breaks in a chromosome), *reciprocal translocation* (mutual exchange of chromosomal segments between non-homologous chromosomes) and *amplification* (multiple copies of specific DNA sequences within a chromosome). Yunis (1986) in a human chromosome map showed 400 bands defined by Giemsa staining and observed several abnormalities involved in specific chromosome rearrangements.

4. Environmental Factors

The environmental factors that are the probable cause of cancer include radiations, diets, personal habits, occupation, smoking, etc. All the agents that cause cancer are known as carcinogens.

Ultraviolet (UV) light and ionizing forms of radiation (IFR) are the potent carcinogens since a long time. UV light and IFR break chromosomes and delete the genetic material resulting in changes in genes. Occassionally, the damaged cells are also killed. This results in cancer. Examples of cancer caused by radiation are thyroid cancer (by *X*-rays), skin cancer (by UV light *e.g.* *Xeroderma pigmentosum*). Cancer caused by the other factors are lung cancers which are caused by cigarette smoking (80-90% of all lung cancers), mesotheliomas by asbestos, liver cancer (by hepatitis *B* virus which in some unknown cases cause 80-90% of liver cancers) (Ames *et al.*, 1973).

(A common form of lung cancer due to cigarette smoking where we can see several irregular-shaped cancer cells invading the healthy cells which line a bronchial tube)

5. Chemical Carcinogens

Organic and inorganic chemicals inducing tumour in animals were long been known. But conversion of chemical carcinogens into strong mutagens by enzymes in body could be known upto 1960s. Miller and Miller (1966) discovered the interaction of chemicals with macromolecules of cells, and the mechanism of chemical carcinogenesis. They found that chemicals such as benzpyrene, benzanthracene and aflatoxin B1 (found in mouldy peanuts) had no specific affinity for DNA, but their carcinogenicity is due to their binding with certain proteins resulting in inactivation of proteins. However, the chemical carcinogens enter into the cell and metabolize into derivatives. The derivatives are several times more powerful mutagenic and carcinogenic than their precursors.

Benzpyrene itself does not cause genetic damage in body cells. It is converted by the carcinogen metabolizing liver enzymes (*e.g.* cytochrome P-450, a family of hydrolases) into epoxide derivative which is carcinogenic, and chemically reacts with DNA (Fig. 23.1). Generally, these enzymes act on hydrocarbons and foreign compounds, and produce harmless metabolite (Miller and Miller, 1981).

Mutagenicity Test: Activation of chemical carcinogens into mutagenic derivatives by liver enzymes could be demonstrated experimentally. Ames's test is the most widely used method (Ames *et al*, 1973) (for detail see Chapter 9).

There are certain compounds which itself are not carcinogenic but synergistically promote tumour with carcinogens. For example, when a mouse skin was painted with a carcinogen, benzpyrene followed by multiple application of phorbol ester (12-

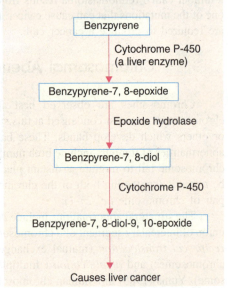

Benzpyrene

↓ Cytochrome P-450 (a liver enzyme)

Benzypyrene-7, 8-epoxide

↓ Epoxide hydrolase

Benzpyrene-7, 8-diol

↓ Cytochrome P-450

Benzpyrene-7, 8-diol-9, 10-epoxide

↓

Causes liver cancer

Fig. 23.1 : Conversion of benzpyrene into carcinogenic derivatives by liver enzymes.

o-tetradecanoyl phorbol 13-acetate, TPA) a few tumours developed. Benzpyrene acts as initiator and TPA as promoter (Hecker *et al.*, 1981) (Fig. 23.1).

6.   Viruses

There are large number of viruses which spread horizontaly due to being highly contagious agent in a wide spread human population, but cause cancer only in a small percentage of individuals. In most of the cases, the viral genome after infection, is integrated with host's chromosome and alter the expression of latter, and fails to produce viral particles. Therefore, in the infected host cells, presence of virus could not be detected. In many rodents it has been found that their body possesses a unique resistance system imposed by the antigens. Similarly, in human, surface antigens play a significant role in protection against cancer. Specialized white blood cells called

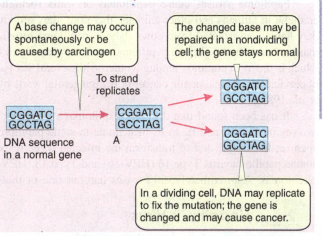

Proto-oncogene products stimulate cell division : Mutations can affect any of the several ways proto-oncogenes stimulate cell division, thus causing cancer.

lymphocytes carryout immune response to body lymphocytes. They are of two types: B cells and T cells. B cells syntheise antibodies, whereas T cells interact with antigen through special membrane bound surface protein called T cell receptor. In contrast, some viruses induce cancer in humans as well as in other mammals. Until 1950s, it could be demonstrated by using new-borne animals that viruses possess cancer-inducing potential.

There is a large number of viruses which transform cultured cells *in vitro*. They can be grouped into two: DNA viruses and RNA viruses. These two groups of viruses are discussed in detail in Chapter 15.

Oncogenes and proto-oncogenes: Although the environmental factors and chemicals induce cancer by bringing changes in DNA, yet no specific cancer causing gene responsible for transformation of normal cells to cancerous cells has been investigated so far. However, it is well known that viruses that cause cancer in experimental animals can also transform cultured cells in laboratory test, either through encoding the viral gene products that alter cellular growth or through integration of viral genome to cellular genome that alter expression of cellular gene. Thus, viral or cellular genes, the proteins of which can change in cell behaviour and induce cancer are called oncogenes. In contrast, by chance when the normal cellular genes, after transduction onto viral genome, have got potential to become oncogenes, due to alteration in their expression, are called proto-oncogenes or cellular oncogenes (*c-oncs*) (Bishop, 1985).

B. DNA Tumour Viruses

The viruses that contain DNA as their genetic material and induce tumour are called DNA tumour viruses; for example, *papova viruses* (SV40, polyoma- and papilloma-viruses), *adenoviruses*, herpes viruses, poxviruses and hepatitis B virus (Table 23.3). SV40, polyoma and adenoviruses are the most widely studied viruses.

In most of the cases viral origin of human cancer could not be proved, however, in several cases, viral cause of human cancer seems clear. Some of the examples are given below:

1. Papilloma Viruses

Papilloma viruses cause papillomas or warts (benign tumour) of cutaneous and mucosal epithelia in humans and other animals, therefore, they are known as papilloma viruses. Moreover, they have been isolated from cows (bovine papilloma virus), rabbits (shope papilloma virus), and many other animals. Though papilloma viruses are known to be associated with cervical carcinoma which is the most common among women with several sex partners, yet no direct evidence of cause of cervical carcinoma, penile cancer or benign genital warts by papilloma viruses is available (Durst *et al*. 1983).

It has been found that different papillomaviruses have similar organization of genome. The two regions i.e. E_6 and E_2 to E_5 of genome in some papilloma viruses have been found to encode open reading frame and to transform the infected cells, respectively. In addition, two isolates i.e. human papillomavirus Type 16 (HPV-16) and Type 18 (HPV-18) are the most common in cervical carcinomas. Approximately, 80% cases harbour one of these viruses (Schwarz, 1985).

2. Hepatitis B Virus

Chronic Hepatitis B virus (HBV) infection leads to liver cancer in humans throughout the world. The virus is least frequent in Canada and most frequent in China and Africa. In India and Russia, they are moderately distributed. In China and some part of African countries 10-150 cases of liver cancer (per 100,000 population) per year were recorded, whereas in India and some of the African and American countries 3-10 cases of liver cancer (per 100,000 population per year) have been recorded. Types and structure of HBV are discussed in Chapter 22.

Persistant infection of HBV also cause cirrhosis. Thus, HBV infection results in 80-90% of liver cancer and, in turn, 500,000 deaths annually throughout the world.

As far as mechanism of development of liver cancer in humans is concerned, it is not fully known. However, no direct evidence for the presence of transforming gene in HBV is known so far. Moreover, based on the experimental findings it is hoped that HBV integrates into cellular DNA and activate a cellular proto-oncogene. In some cultured tumour cells, deletion of chromosome at the site of viral integration has been observed. Therefore, HBV may also acts as mutagen (Melnick, 1983). First Indian HBV vaccine has been developed in Hyderabad (see Chapter 11).

3. Epstein-Barr Virus

Epstein-Barr virus (EBV) belongs to family herpesviruses. EBV virions are enveloped. Capsid is icosahedral that encloses viral genome. It causes a tumour of mature B cells called Burkitt's lymphoma. This disease is named after the name of discoverer, Burkitt. EBV is found throughout the world but most frequently in children in certain parts of East Africa. Burkitt's lymphoma is found in 90% children of 6 years age in some tropical areas like East Africa, whereas children of the same age group in the USA have 30-40% of disease. EBV is also associated with nasopharyngeal carcinoma and mononucleosis. The later is a disease of excessive lymphoid cell proliferation spread through kissing. EBV infects the cells of oropharyngeal cells and readily replicates within the cells. Nasopharyngeal carcinoma (NPC) is a tumour of adults in China caused by EBV. It has been found that this disease is contributed by smoke, chemicals and other environmental factors.

So far it is not known that how EBV transforms cancer cells in humans but viral genes kill B cells of blood. Burkitt's lymphoma possesses specific chromosome translocations on which the immunoglobulin gene loci and the *c-myc* proto-oncogene are present. The *myc* oncogene is the gene of several defective genes of avian retroviruses that induce myelocytomas (tumours of myeloid cells), sarcomas and carcinomas which certainly play role in cell division and expressed in normal growing cells and tumour cells. In humans, *c-myc* gene is located on the long arm of chromosome 8 at band q 24 and spans about 5 kb. In *t* (8:14) translocations immunoglobulin heavy constant chain is fused head to head with *c-myc*. The fusion alters the *c-myc* gene expression resulting in the development of tumour.

Tumour cells generally secrete antibodies that possess highly characteristic chromosome translocations. The translocations involve in chromosome 8, and one of the three chromosomes that carry antibody light or heavy-chain genes (*i*) chromosome 14 (heavy chain genes), (*ii*) chromosome 2 (gamma light-chain genes) or (*iii*) chromosome 22 (kappa light-chain genes). Translocations which involve chromosome 8 and 14 are designated as *t* (8:14). The *t* (8:14) is the most common in Burkitt's lymphoma.

4. SV40 and Polyoma Viruses

During 1950s, SV40 was isolated as the contaminant in the cell cultures of rhesus monkey which was used to prepare Salk and Sabin polio vaccines. However, it kills the cultured cells of African green monkeys, and induces tumours when injected into new-borne hamsters. Polyoma (i.e. many tumours) virus induces tumours occasionally in ovary, mammary gland, liver, adrenal gland, thymus, etc. in rats, mice, rabbits, hamsters, etc. Induction of cancer by these two viruses in human is not clearly known so far.

Both SV40 and polyoma viruses belong to family papovavirus. SV40 is very similar to papilloma virus. SV40 virions are naked, icosahedral structure composed of three viral coded proteins i.e. VP1, VP2 and VP3. It consists of small, double-stranded circular DNA of 5243 base pairs. Its genome possesses three regions, early region and late region, and a third intermediate regulatory region. The early gene products of SV40/polyoma are involved in replication of virus by acting on viral DNA and cellular genes required for viral DNA replication. The role of different early genes in cell transformation has been studied by synthesising cDNA and introducing into cells. It was found that polyoma large T induces indefinite growth in primary cells, whereas a middle T induces the remainder of the transformed phenotype. SV40 large T protein interacts with a cellular protein (p53) present in the nucleus, the stability and amount of which increase in SV40 transformed cells. Role of SV40/polyoma small T protein is still not clear.

C. RNA Tumour Viruses (Retroviruses)

The viruses that contain RNA as genetic material and cause tumours in humans and other animals are called RNA tumour viruses. The RNA viruses that have capacity to cause cancers are all morphologically similar, and probably are descended from a single ancestral virus. Viruses containing single/double stranded RNA genome are given in Table 15.4 and Fig. 15.10. All the RNA tumour viruses are put into retrovirus group. The retrovirus family is classified into the following three sub-families (Watson *et al*, 1987):

(*i*) *Oncoviruses*: Oncoviruses are the oncogenic and closely related non-oncogenic retroviruses of reptiles, fish, rodents, cats, primates, etc. For example, Rous sarcoma virus, other retroviruses carry cell derived oncogenes.

(*ii*) *Lentiviruses*: They are the slow viruses that induce neurological impairment and chronic pneumonia; for example, caprine arthritis encephalitis virus (CAEV) of goats and HIV/AIDS virus.

(*iii*) *Spumaviruses.* Spumaviruses are the viruses that induce persistent infection without evidence of pathogenesis in a number of mammals; they are also known as foamy viruses.

Retroviruses were first isolated as cancer causing agents in chickens. For the first time P. Rous (1911) isolated rous sarcoma virus (RSV) which is still studied as a model cancer-causing agent. Recently, several human cancer-causing retroviruses have been found out.

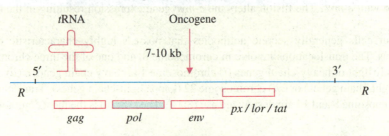

Fig. 23.2 : A retrovirus genome (diagrammatic). The boxes denote the open reading frame. Most of the naturally occurring retroviruses encode only *gag, pol* and *env* genes. The mRNA for *env, src, sor* or *orf* are generated by RNA splicing.

1. Retroviruses

(*i*) **Virion and Genome Structure:** Morphology and structure of retroviruses have been discussed in Chapter 15. Although there are variations in constitution of genome in retroviruses, yet typical retroviruses have three protein coding genes: *gag, pol* and *env* (Fig. 23.2) which translate glycoproteins. The *gag* gene (encoding a group specific antigen) codes a precursor polyprotein that is cleaved to yield the capsid protein; *pol* gene is cleaved to yield reverse transcriptase and an enzyme involved in proviral integration, and *env* gene encodes the precursor to the envelop glycoprotein (analogous to the G protein of VSV or influenza HA protein). Recently, a fourth type of gene called *px*, *lor* or *tat* has been found at the 3' end of HTLV-I and II genomes which encodes a transcriptional activator to the E1A protein of adinovirus increasing in expression of provirus. Besides, a few retroviruses possess an additional gene called *onc* gene which gives ability to induce certain types of cancer at a rapid rate (Varmus, 1982).

(*ii*) **Replication of Retroviruses :** Retroviruses replicate in several stages.

(*a*) **Adsorption and insertion:** Virus gets absorbed at receptor site on cell surface (Fig. 23.3A) and infection occurs. Soon after infection (+) RNA strand is introduced into body cell which acts as mRNA. Single stranded viral RNA genome is copied into a linear molecule or double stranded DNA directed by reverse transcriptase. Reverse transcriptase

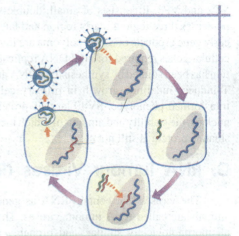

Life cycle of the Rous sarcoma virus: (1) The genome of RSV is composed of RNA, that of animal cells is composed of DNA (2) on infecting an animal cell, the RNA *RanA* is added to the animal cell (3) there, a DNA copy of it is made (4) Some or all of this DNA copy can become incorporated into the animal cell's DNA (provirus)

begins by adding deoxynucleotides to the tRNA primer. It adds 100-200 nucleotides to 'R' regions and U5 ends. At 3' and 5' ends, retrovirus genome contains a long untranslated region called as

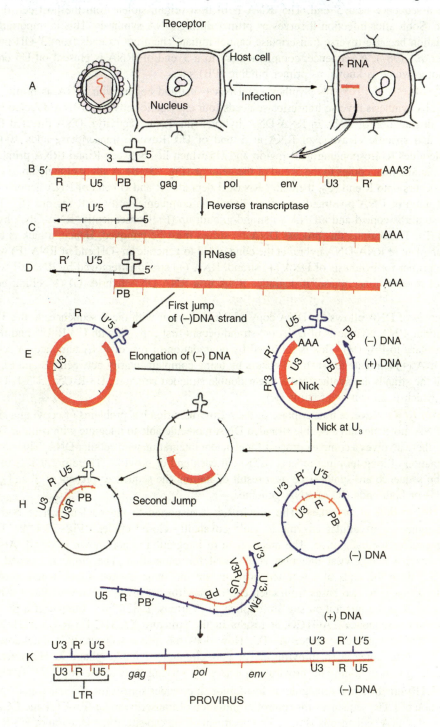

Fig. 23.3 : Replication of retrovirus in a cell : formation of provirus.

'R' region. R region includes signals needed for replication and translation. 'R' is a short and directly repeated sequence present at each end of genome as well as tRNA molecules which is held by base-pairing to a sequence near 5' end. The tRNA is of host origin, stolen from the host cell during replication. Soon after infection it serves as primer during DNA synthesis. This is important for chain initiation because reverse transcriptase cannot initiate chain but extends from 3'-OH primer (Watson *et al.*, 1987). A sequence, complementary to a 3' end of tRNA is present on U5 end or RNA genome which is known as primer binding (PB) site.

(*b*) **Synthesis of Provirus:** Synthesis of DNA (–) strand begins with tRNA as primer. For proviral DNA synthesis reverse transcriptase needs four enzymatic activities : RNA directed DNA synthesis, degradation of RNA in RNA-DNA hybrids (the RNase activity), DNA directed DNA synthesis, and specific cleavage of RNA at 5' end of U3. Reverse transcriptase adds 100-200 deoxynucleotides (corresponding to R region and U5 which lies between R and tRNA primer) to tRNA primer and then comes to the 5' end of the genome (Fig. 23.3B). A slight elongation of DNA (-) strand occurs upto R end (C), thereafter RNaseH degrades R and U5 end of RNA genome (D). Then first jump of DNA (-) strand occurs to 3' end via complementarity of R regions (E). Thus, (+) RNA strand is copied and a DNA (-) strand is built up (F). Consequently, RNA-DNA hybrid is formed (F). To continue this process DNA (-) strand must be removed. Hence, a nick is made at RNA template of RNA-DNA hybrid at the edge of U3 to generate an -OH end of RNA (F) which can act as primer for synthesis of DNA (+) strand. RNA (+) strand is removed by RNaseH which is a part of reverse transcriptase. It degrades RNA in a RNA-DNA hybrid (G) (Moelling *et al.*, 1971).

Removal of DNA allows the DNA copy of R to base pair with the R sequences at the 3' end of the genomic RNA. Synthesis of DNA (+) strand occurs first copying with U'3 R' U'5 and tRNA (complementary to PB) (H). This is followed by degradation of rest of RNA template. A second jump of DNA (+) template to (-) strand occurs by using complementarity between PB and PB'. (I) when both the strands are synthesized a linear double stranded structure, U3-R-U5.....U3-R-U5, is produced which is known as provirus (K).

Role of LTRs: Reverse transcriptase of the retrovirus virion has problems of copying a single stranded RNA molecule into a double stranded DNA molecule able to integrate with cellular DNA. Therefore, there involves a complex process for efficient integration with cellular DNA. This process is the formation of long terminal repeats (LTRs) at each end of provirus. Therefore, LTRs of 300 to 1,000 bp sequence are synthesized as a result of uniting the sequences present at 3' end (U3), 5′ end (U5) or both ends (R) of RNA genome.

(*c*) **Integration of provirus:** The provirus is transported to the nucleus where both the ends get joined by a cellular enzyme to yield covalently closed circles. (Fig. 23.4 A). These serve as precursor for integration. The mechanism of integration is not known in detail. At least one viral enzyme called viral integrase is essential for integration. The circle is opened at a specific site, since the ends of integrated provirus are the same. At the joints, where provirus and cellular DNA join, two bases from each end of provirus are always removed, and 4-6 bases of cellular DNA at the integration site are duplicated (Varmus, 1982). For example, if a provirus integrates with a sequence TAGTCG, it results in the structure TAGTC-U3-R-U5-AGTCG as a result of duplicating the sequence AGTC. How does this process occur, it is not known in detail so far. After integration, provirus never gets circularized unlike phage λ. However, the infected cells go on dividing and provirus replicates by serving as template for RNA synthesis.

The LTRs are the site for integration. In addition, it provides signals for efficient transcription. At U3 portion of LTR, sequences for control of initiation of transcription i.e. TAATA and CCAAT are present. TAATA sequence is about 25 bp upstream of the cap site at the junction of U3-R, and CCAAT is a 80 bp upstream adjacent to it (C). However, the direct repeated sequence (DRS) (70-100 bp long) acts as a transcriptional enhancer. Short inverted repeated sequence (IRS) are present bordering LTRs. Polyadenylation signals and site of poly A addition are present on one side of LTR.

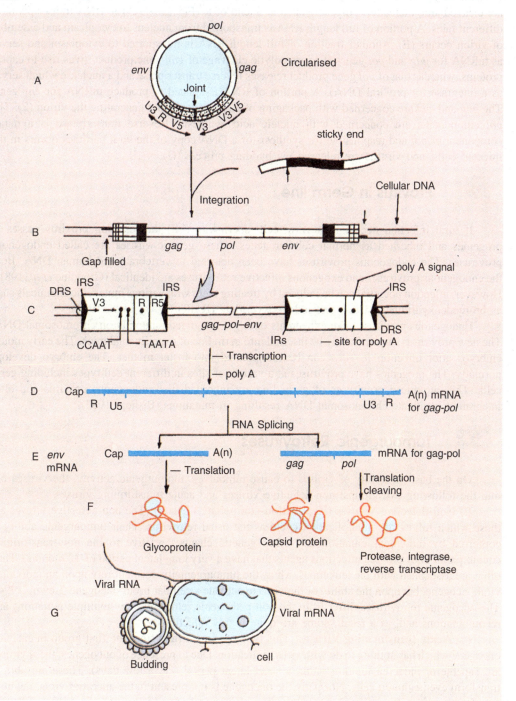

Fig. 23.4 : Integration of provirus; transcription, translation and virion formation by mRNA.

(*d*) **Translation, assembly and release:** The cell machinery processes the RNA transcripts synthesized by RNA polymerase II from the integrated retroviral provirus. The RNA transcripts are cleaved and, thereafter, polyadenylated at 3' end of R (Fig. 23.4D). The RNA transcripts have different fates. A portion of full length RNA is transported from nucleus to cytoplasm and assembly of virion occurs (E). Another fraction of full length RNA is transported to cytoplasm and serves as mRNA for *gag* and *pol* genes (E) (proteolytic cleavage of *gag* gene product gives rise to capsid proteins, whereas that of *pol* gene product releases reverse transcriptase and a nuclease which serves as a integrase for proviral DNA). A portion of RNA is spliced to produce mRNA for *env* gene. The *gag* and *env* are concerned with packaging of RNA genome and generating the virion (F). The *pol* components are concerned with nucleic acid synthesis. Reverse transcriptase is a major component of *pol* and responsible for synthesis of a DNA copy of the viral particles occurs in the infected cells, and viruses are released via budding process (G).

2. Provirus in Germ line

If viral infection occurs in a germ cell (sperm/egg cell) the resulting provirus passes to progenies and inherited as normal cellular genes. These genetic viruses are called endogenous proviruses. The endogenous proviruses have been detected in vertebrate and human DNA. Both the endogenous proviruses and exogenous infective retroviruses are identical (Groudine *et al*, 1981). However, infection potential can be induced by treating them with various mutagenic chemicals such as bromodeoxyuridine (BUdR) and iododeoxyuridine (IUdR) (Lowy *et al*. 1971).

Endogenous proviruses like exogenous ones get integrated randomly on chromosomal DNA. The new proviruses can artificially be inserted into germ line of experimental mice. The early mouse embryos after infection *in vitro* can be reimplanted into foster mother. The embryo develops normally. The progenies have proviruses at a variety of sites in different cell types including germ cells. The introduced proviruses of germ line are inherited like endogenous provirus and after integration inactivate chromosomal DNA resulting in mutations (Bishop, 1983).

3. Tumourgenic Retroviruses

On the basis of ability of viruses to cause tumour *i.e.* tumourgenic activity, the viruses fall into the following two groups: non-defective viruses and acute transforming viruses.

(*i*) **Non-defective Viruses (Replication-competent Viruses):** The non-defective viruses are those which follow the multiplication cycle as the usual retrovirus. Their tumourgenic ability is conferred by individual mutation and other genetic changes relative to the non-transforming counterparts. They provide infectious agents that have a very long latent period (3-12 months). They often are associated with the leukemias. Thus, the tumourgenicity does not rely upon an individual viral oncogene, but upon the ability of the virus to activate a cellular proto-oncogene (Lewin, 1994). The oncogenic retroviruses differ from the non-pathogenic retroviruses by multiple mutations and recombinations and, as a result, some are quite highly infectious.

(*ii*) **Acute-transforming Viruses:** This type of virus differs from the first group in carrying oncogene which has nothing to do with virus replication. Due to presence of oncogenes, these viruses are capable of inducing tumour within a short latent period (*i.e.* a few days). These are able to transform even cultured cells. Normally the oncogene is not present in the ancestral virus, i.e. non-transforming virus. It originated as a cellular gene that was taken up by the virus through transduction during infection cycle in a normal cell where it may exchange part of its own sequence for a cellular sequence. Though this type of event is rare, yet generates a transducing virus. Oncogenes of some acute-transforming viruses are given in Table 23.1. So far 50 acute-transforming retroviruses are known.

Except Rous sarcoma virus (RSV), all the acute transforming retroviruses are defective due to loss of replication, properly by the exchange with cellular sequences. Therefore, they are replication-defective, and cannot replicate by itself. Hence, they need a wild type `helper' virus that provides the function for replication which was lost during the recombination.

4. Cellular and Viral Oncogenes

Except avian erythroblastosis virus (AEV), usually the oncogenic activity is found in a single gene of retroviruses. Normally the cellular sequence is not oncogenic. In contrast, the DNA tumour viruses need 2 or 3 genes to cause cancer. The viral oncogenes and their cellular counter parts are described by using prefix v for viral and c for cellular oncogenes. Therefore, the oncogene carried by RSV is called v-*src*, and the proto-oncogene related to its cellular genome is called c-*src*. During the infection of a retrovirus, the v-oncogene is taken in the form of RNA. Upon comparison it seems that the organisation of viral gene corresponds to the mRNA or the c-oncogene rather than its genomic organisation. Uninterrupted coding sequences are found in v-*onc*, whereas alternating exons and introns (missing sequences) are found in c-*onc*.

So far about 30 c-*onc* genes have been identified through their presence in retroviruses and their number is expected to reach to about 100 (Watson *et al.*, 1987). The viral genes are functionally indistinguishable from the cellular genes, but are oncogenic because they are expressed in much greater amount or in inappropriate cell types. Moreover, c-*onc* genes intrinsically lack oncogenic properties, but may be converted into oncogenes whose devastating effects reflect the acquisition of new properties (Lewin, 1994).

When the chicken c-*src* gene was cloned, it was found that it contained 13 exons and spanning some 8 kb. Like most of the normal cell genes, almost all the proto-oncogenes have intervening sequences. However, v-*oncs* never have whole introns. v-*oncs* are transmitted as a part of a viral RNA. Any intron, initially derived from its c-*onc* parent, would be removed during normal processing resulting in intron-free mRNA that becomes the viral oncogene and template for subsequent DNA synthesis and integration of provirus (Takeya and Hanafusa, 1983).

Table 23.1: Oncogenes of some acute transforming retroviruses causing disease

Oncogene	V- oncoprotein	Retroviruses/species	Tumour
src	$p60^{src}$	Rous sarcoma virus (RSV)/chicken	Sarcoma
mos	$p37^{env-mos}$	Moloney murine sarcoma virus (Mo-MSV)/mouse	Sarcoma
fos	$p55^{fos}$	Finkel-Biskis-Jinkins (FBJ)-MSV/mouse	Osteosarcoma
sis	$p28^{env-sis}$	Simian sarcoma virus (SSV)/monkey	Sarcoma
myc	$p100^{gag-myc}$	Avian myelocytomatosis (MC29)/ chicken	Carcinoma, Sarcoma, myelocytoma
myb	$p45^{myb}$	Avian myeloblastosis virus (AMV)/chicken	Myeloblastosis
erbA,	$p75^{gag-erbA}$	Avian erythroblastosis virus (AEV)/chicken	Fibrosarcoma, Erythroblastosis
erbB	$p65^{erbB}$	do	do
abi	$p90-160^{gag-abl}$	Abelson murine leukemia virus (MLV)/ mouse	B-cell lymphoma
rel	$p64^{rel}$	Reticuloendotheliosis virus (REV)/turkey	Reticuloendotheliosis
H-ras	$p21^{ras}$	Harvey murine sarcoma virus (H-MSV)/rat	sarcoma, Erythroleukemia
K-ras	$p21^{ras}$	Kisten murine sarcoma virus (K-MSV)/rat	do

It has been found that many of the v-*oncs* are altered relative to their cellular counterparts. Transduction of c-*onc* involves the loss of part of the cellular gene, and fusion of the remainder to a viral gene frequently to *gag*. Provirus is integrated in the vicinity of a c-*onc* gene. Thereafter, deletion in intron between c-*oncs* occurs that brings the provirus near to c-*onc* sequence. RNA transcripts of this DNA carries the retrovirus packaging sequence at its 5' end and finally it is packed. Reverse transcriptase makes copy of both the genomes by copy choice type mechanism so as to transfer viral 3' end sequence which is required for replication to the novel *onc* gene carrying genome (Watson *et al.*, 1987).

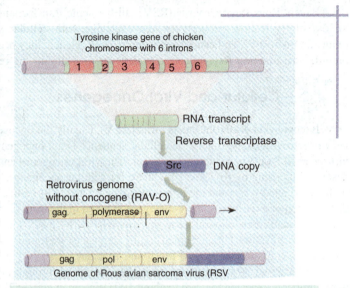

Structure of the Rous avian sarcoma virus (RSV) : the virus contains only a few virus genes, encoding the virus envelope proteins (called the env and gag genes) and reverse transcriptase which produces a DNA copy of its RNA genome (called pol for polymerase). It also contains the gene src (for sarcoma)

5.   Relationship Between Viral and Animal Oncogenes

To assess the relationships, the oncogenes were screened from a large number of retroviruses, cancerous human tissues and their counterparts from normal tissues. The nucleotide sequence of all the genes were determined. Such study was done for the first time in Kirsten sarcoma virus (KSU) causing sarcoma in rats. Like KSV, an oncogene *i.e.* ras onc (rat sarcoma) was also detected in Harrey sarcoma virus. The sequence of *ras* onc was the same (except for a single base change) as the sequences found in human bladder carcinoma, and human lung and colon carcinomas. A similar sequence but with different base change was found in a human neuroblastoma and fibrosarcoma. This variant was called N-*ras*. Studies of a large number of viruses and tumours (human cancer of colon, lung, pancreas, skin, breast, brain and white cells) have led to identification of more than 20 different oncogenes.

Table 23.2 : Some oncogenes and their sources

Oncogenes	Sources
abi	Mouse
B-*lym*	Chicken and human lymphoma cells
H-*ras*	Rats sarcoma, human and rat carcinoma
k-*ras*	Rat sarcoma, human carcinoma, sarcoma
mos	Mouse sarcoma, mouse leukemia cells
myb	Chicken leukemia and human leukemia cells
myc	Chicken leukemia, human lymphoma cells
N-*ras*	Human leukemia and carcinoma cells

6.   Oncogene Families

About 30 oncogenes have been discovered in the genome of retroviruses out of 50 different transforming retroviruses. On the basis of similarity in structure and function of amino acid

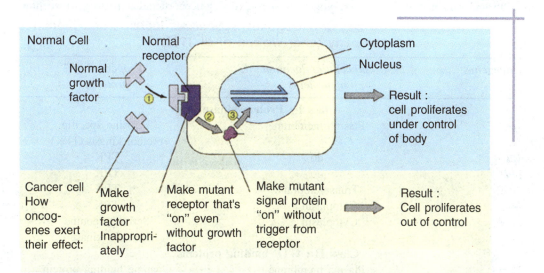

Growth control and oncogenes : (1) Growth in normal cells is controlled by growth factors, proteins made by one cell (2) the growth factor diffuses to the target cell and binds to a receptor on the cells's surface (3) this binding triggers a change in a signal protein which causes the nucleus to divide and the cell to proliferate under control. In contrast to normal cells, the cancer cells have mutations in genes that encode the growth factor. These mutations of the normal gene are called oncogenes and they cause the cell to proliferate out of control.

sequences, the oncogenes can be grouped into the families of the related genes. Table 23.3 shows that some oncogenes resemble pp60^{src} (phosphoprotein of 60 kilo dalton encoded by *src* gene) in the sense that they share homology of amino acid sequences with one another, and possess protein kinase activity. In some cases homology of amino acid is so close that one can expect that different oncogenes have evolved from a common ancestor, in spite of differences in nucleotide sequences. For example *src* and *yes* products have 82% homology; but nucleotide sequences of the genes have 62% homology. Also *src* and *abi* are the related oncogenes, but human c-*src* is more closely related to *Drosophila* c-*src* than c-*abi*. It can be presumed that c-*src* and c-*abi* would have evolved from the common ancestor well before the separation of vertebrates and insects (Watson *et al;* 1987). Most of the oncogenes showing protein kinase activity are tyrosine specific, and two are serine/ threonine residue specific (Table 23.4). Until the normal functions of oncogenes expressed, proteins are well understood, the significance of highly conserved family will probably be not clear. Oncogenes are divided into the five classes.

Class I: It consists of protein kinase encoding genes which have homology of amino acid sequence to *src* gene products. Moreover, most or these proteins are linearly present at the surface of plasmamembrane like pp60^{src}. In contrast, two oncogenes (*erb*B and *fms*) encode transmembrane proteins which are derived from cellular genes. It encodes receptors for polypeptide growth factor.

Class II: The members of the *ras* family encode 21 K dalton proteins, the amino acids of which have similarity with that located at the inner surface of plasma membrane. It binds with guanine nuclecotide and have GTP are activity.

Class III: Only *sis* oncogene is known that encode B-chain of platelet-derived growth factors (PDGF).

Class IV: The oncogenes (*myc, fos* and *myb*) of this class encode proteins that are located in the nucleus. It alters the expression of other cellular genes.

Class V: Only one oncogene, *erb*A, is known to this class which encodes a protein with weak amino acid sequences which are homologous to steroid (e.g., glucocorticoid and estrogen) receptors.

Table 23.3 : Classes of oncogenes, subcellular location and protein property (after Watson *et al*, 1987)

Oncogenes	Subcellular location of proteins	Properties of proteins
Class I : Protein kinase		
src	Plasma membrane	Tyrosine specific protein kinase (TPK)
yes	do	TPK
abl	do	TPK
*erb*B	Transmembrane	TPK
fes	Cytoplasm	TPK
mos	Cytoplasm	Serine/threonine protein kinase
Class II : GTP binding proteins		
H-*ras*	Plasma membrane	Guanine binding protein with GTPase activity
K-*ras*	do	do
Class III : Growth factors		
sis	secreted	Derived from a gene encoding PDGF
Class IV : Nuclear proteins		
myc	Nucleus	—
myb	do	—
fos	do	—
ski	do	—
Class V : Hormone receptor		
*erb*A	Cytoplasm	Thyroid hormone receptor
Unclassified		
rel	—	—
ets	—	—

There are several other oncogenes as well the structure and function of which are not known in detail. Therefore, they have not been classified so far.

Table 23.4 : Classes of oncoproteins.

Oncoproteins	Oncogenes
Tyrosine kinase	*abl, fes, fgr, fps, ros, src, yes*
Other protein kinases	*erb*B, *fms, mil, mos, raf*
Growth factor	*sis*
Nuclear protein	B-*lym, fos, myb, myc, ski*
GTP binding proteins	H-*ras*, K-*ras*, N-*ras*

7. Oncogene Proteins (the Oncoproteins)

There are several proteins which are encoded by oncogenes and proto-oncogenes the activity of which must be involved in several functions related to cell growth. Oncogenes are divided into 5 groups according to the location of oncoproteins in the transformed cell (Fig. 23.5).

The transformed cells lose control on its regulation. These may acquire new properties e.g ability to metastasize. Several phenotypic properties get changed as compared to a normal cell. Some of the oncoproteins are discussed below:

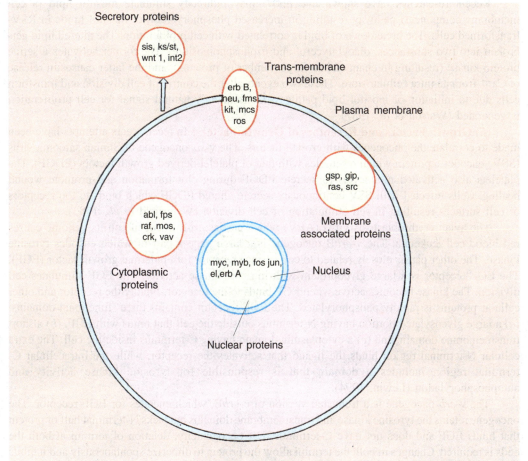

Secretory proteins

sis, ks/st,
wnt 1, int2

Trans-membrane
proteins

erb B,
neu, fms,
kit, mcs
ros

Plasma membrane

gsp, gip,
ras, src

abl, fps
raf, mos,
crk, vav

Membrane
associated proteins

Cytoplasmic
proteins

myc, myb, fos jun,
el,erb A

Nucleus

Nuclear proteins

Fig. 23.5 : Oncogenes are divided into five groups according to the location of oncoproteins in the transformed cell.

(i) **Protein Kinases:** There are several oncogenes which encode protein kinases, especially those proteins that phosphorylate tyrosine or threonine and lead in transformation of cells. The first protein of this kind studied in detail was p36 which is an abundant protein found on inner surface of the plasma membrane. Its highest concentration is found in epithelial and endothelial cells particularly on columnar epithelium of intestine. The v-*src* gene of RSV encodes a phosphoprotein of 60 K dalton (p60^{src}), in RSV transformed cells. The amino terminus of p60^{src} has covalently attached fatty acid which provides oncogenic potential to it. The p60^{c-src} differs from p60^{v-src} in a few amino acids mostly at *c*-terminus and like v-*src*, it is localised at the inner surface of the plasma membrane.

The purified p60$^{v\text{-}src}$ was found to have protein kinase activity. The p60^{src} catalyses the transfer of terminal phosphoryl group of ATP to the -OH group of a tyrosine residues in its substrate resulting in 10 fold higher level of transformed cells (Hunter and Sefton, 1980)

There is another protein, vinculin, whose tyrosines are phosphorylated. In the normal cells, vinculin is present in small patches (*i.e.* adhesion plaques) on the cell surfaces. These patches help to adhere to surfaces and to one another, and serve an attachment sites for microfilaments. Vinculin is also present on these patches and serves as connecting anchor between actin cables and the membranes. After photophosphorylation vinculin is diffused throughout the cell and fails to organise the morphology of transformed cells (Sefton *et al.,* 1981).

Recent researches have shown that p60^{src} may indirectly stimulate inositol lipid (a cell membrane component) pathway resulting in increased phosphorylation of inositol lipids in RSV-transformed cells. The breakdown of lipid is correlated with cell proliferation. The altered lipid gets broken into two substances, diacylglycerol and triphosphoinositide. The former activates a serine protein kinase (resulting in changed in a large number of proteins), and the latter causes in release of Ca^{++} from an intra cellular store. These two events alter the control of cell division and transform cells due to initiation of inositol lipid pathway. Consequently natural signal for cell proliferation is generated (Watson *et at,* 1987).

(*ii*) **Growth Factors and Receptors of Growth Factors:** In recent years attempts have been made to correlate the oncogenes with growth factors. The v-*sis* oncogene of Simian sarcoma virus (SSV) encodes a protein which resembles with that of platelet-derived growth factors (PDGF). The platelets also activated monocytes to secrete PDGF during clot formation and promote wound healing. Cells infected with SSV continuously secrete altered PDGF which binds to the receptors of cell surfaces resulting in self stimulation of cell division (Waterfield *et al.,* 1983).

The avian erythroblastosis virus (AEV) which carries *erb*B oncogene on its genome causes red blood cell leukemia. The v-*erb*B oncogene is a large gene a part of which encodes tyrosine kinase. The other part is closely related to cell surface receptor for epidermal growth factor (EGF). The EGF receptor is a large glycoprotein with intrinsic kinase activity. The EGF stimulates cell division. The kinase becomes active when EGF binds to the receptor. Finally the receptor and other cellular proteins is rapidly phosphorylated. The EGF receptor contains three functional domains: (*a*) a large glycosylated portion having N-terminus outside the cell that binds with EGF, (*b*) a short transmembrane domain, and (*c*) a cytoplasmic domain having C-terminus inside the cell. The extra cellular N-terminal region binds the ligand that activates the receptor, while the intracellular C-terminal region includes a domain that is responsible for tyrosine kinase activity and autophosphorylation (Lewin, 1994).

The v-*erb* oncogene is a truncated version of c-*erb*B, which encodes for EGF receptor. The oncogene retains the tyrosine kinase and transmembrane domains but lacks N-terminal half of protein that binds EGF and does not have C-terminus. For oncogenicity, deletion of termini at both the ends is required. Changes in both the termini allow the protein to dimerize spontaneously and inhibits transformation activity. Therefore, the altered activity may be responsible for oncogenicity applied to the growth factor receptor (Lewin, 1994). The altered receptor produced by v-*erb*B disrupts the regulatory mechanism in which the receptor is a crucial link when EGF binds to the altered receptor, the tyrosine kinase is activated.

(*iii*) **The *ras* Protein, (p21 KDal).** Now-a-days *ras* oncogenes have received attention of researchers due to its possible involvement in many human tumours. The vertebrate *ras* genes constitute a small multigene family with three functional members: (*a*) H-*ras* (or C-H-*ras*, the cellular oncogenes, homologous to V-H-*ras*, the oncogenes present in Harvey sarcoma virus), (*b*) K-*ras* (or C-K-*ras*, present in Kirsten sarcoma virus as V-K-*ras*), and (*c*) N-*ras* (not found on any retrovirus genome, but detected when a transforming gene from a human neuroblastoma was cloned).

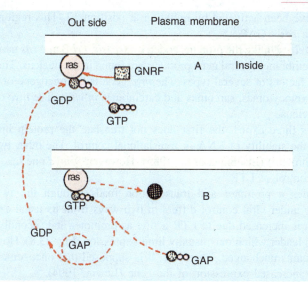

Fig. 23.6 : Diagrammatic representation of *ras* gene controlled by two proteins. A, inactive state when *ras* binds to GDP; B, active state when *ras* binds to GTP. Activated *ras* acts on target protein (modified after Lewin, 1994).

All the *ras* genes encode oncoprotein of 21 kilo dalton (p21 *ras* Kdal). In general, the v- and c-*ras* oncoproteins bind GTP and other guanosine nucleotides. However, c-*ras* oncoprotein has GTP activity as well. G protein, a cellular protein binds guanine nucleotide and show GTPase activity. These proteins become activated after combining with GTP and inactivated when GTP connects the GTP to GDP and then *ras* binds to GDP (Watson *et al.*, 1987). The GNRF (guanine nucleotide release factor) stimulates the *ras* for exchange of GDP with GTP and, therefore, activates *ras*. GAP stimulates the hydrolysis of GTP and thus inactivates *ras* (Fig. 23.6). Several different GAPs have been found with specificity of different GTP binding proteins. Thus, *ras* genes can be activated by activating GNRF or by inactivating GAP. The oncogenic *ras* remains constitutively in GTP bound form and the activated *ras* acts on target protein (Lewin, 1994).

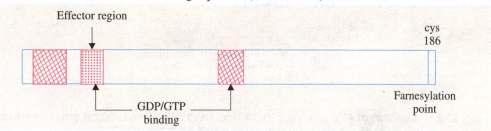

Fig. 23.7 : Discrete domains of *ras* proteins responsible for guanine nucleotide binding effector function and membrane attachment (after Lewin, 1994).

Lewin (1994) has discussed the following three groups of regions (Fig. 23.7) responsible for the characteristic activities of *ras* genes:

(*a*) The region between 5-22 and 109-120 are specific for guanine nucleotide binding by their homology with other c-binding proteins. Some nucleotides that activate the oncogenic potential of *ras* lie within these regions.

(*b*) The *ras* oncogene is attached to the cytoplasmic face of the membrane by farnesylation, close to the C-terminus. For *ras* function membrane attachment is essential.

(*c*) The effector domain is the region that reacts with target molecule when *ras* has been activated. The activity of the region between residues 30-40 is required for the oncogenic activity of *ras* proteins that have been activated by mutation at position 12. This region is also required for the interaction with *ras* - GAP.

(iv) **Nuclear Protein:** Unlike the proteins coded by *src, sis, erb*B and *ras* genes inside, outside or within the plasma membrane, several gene products are found in the nucleus. The nuclear proteins include transcriptional factors of several types. The *myc* gene is the oncogene of several defective retroviruses causing myelocytomas, sarcomas and carcinomas *in vitro*. The *myc* onc coded protein is found inside the nucleus.

The *myc* onc has three exons, the first does not translate the protein but regulate c-*myc* expression perhaps in the stability of RNA or translational control. The other two encode for the c-*myc* proteins (HLH protein) (Eisenman *et al*, 1985). However, v-*myc* one has lost its first exon and fallen under the control of LTR.

The LTR provides a promoter and transcription reads through the two coding exons. Transcription of c-*myc* under viral control differs in two ways from its usual expression. Firstly, the level of expression is increased due to LTR acting as promoter and secondly, transcript lacks its usual non-translated leader which may usually limit expression (Fig. 23.8). However, the coding sequence of c-*myc* remains unchanged. Therefore, it is supposed that oncogenicity in due to loss of normal control and increased expression of the gene (Lewin, 1994).

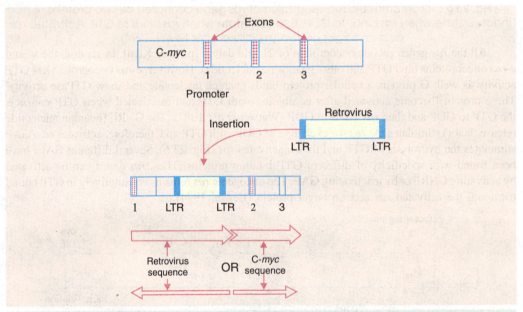

Fig. 23.8 :   Insertion of ALV at c-*myc* locus occurring at various position and activating genes in different ways (modified after Lewin, 1994).

8.   Tumour Suppressors

Cancers are caused due to changes in cellular genes resulting in alteration in gene products. It is also initiated when tumour suppressor genes in cellular DNA are lost or mutated. A mutation that activates a single allele is tumourgenic. This has been discussed with two examples.

(i) **Retinoblastoma (RB) Genes:** Retinoblastoma is a human childhood disease which induces a tumour on retina. It occurs both as a heritable trait and by somatic mutation. In somatic mutation, the parental chromosomes are normal and both RB alleles are lost by somatic events. It is associated with deletion of a band 14 (13 q14) of human chromosome 13 (Fig. 23.9). This disease arises when

both copies of the RB genes are inactivated, which means that the proteins coded by RB genes are absent from retinoblastoma cells. Therefore, the cause of this disease is the loss of protein function. Loss of RB genes also causes the other cancers such as osteosarcoma and small cell lung cancer (Lewin, 1994).

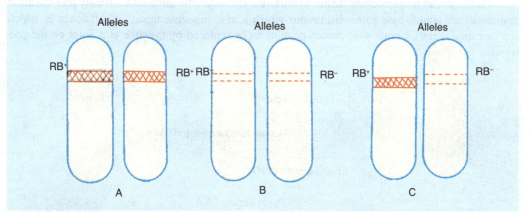

Fig. 23.9 : Diagram of RB gene present on q14 of chromosome 13 of germ cell. A, normal allele. Retinoblastoma is caused when RB (absent) is on both the alleles (B). Loss of one allele (RB⁻) in germ cells results in carrier (C) with wild phenotype.

The RB gene product is a nuclear phosphoprotein that influences cell cycle. The RB is phosphorylated during the cell cycle, but probably at the end of G (not in resting cell i.e. Go/G1), RB is dephosphorylated during mytosis. However, NPRB (non-phosphorylated form of RB) prevents cell proliferation. This activity must be transiently suppressed in order to pass through the cell cycle accomplished by the cyclic phosphorylation. It may also be suppressed when a tumour antigen sequesters the NPRB and thus inhibits its function. For example SV40 T antigen (oncoprotein of tumour virus) binds only to the NPRB. The influence of RB on cell cycle is not known in detail, but it seems plausible that its basic function may lies in some interactions that control progression through the cycle (Lewin, 1994).

(ii) **Protein p53:** P⁵³ is a nuclear phosphoprotein of molecular weight 53 kilo dalton which acts as tumour suppressor due to inhibition of cells in culture by various oncogenes. For the first time p53 was discovered in SV40 transformed cells associated with T antigen. A large increase in the amount of p53 protein was found in many transformed cells or cell line derived from tumours. The p53 acts as transcriptional factor which binds with other genes and control their expression. In addition to transcription, it also plays role in cell cycle control (i.e. cell growth) in other metabolic functions and as tumour suppressor. It halts abnormal growth in cell and thus prevents cancer.

An International conference on cancer was held in India in October, 1994 where several theories on p53 function were discussed. In 1983, it was found that the gene which gives out instructions for coding p53 is actually an oncogene because it immortalized the cells when closed p53 gene was introduced in cell line. However, the transforming form of p53 was found to be a mutant form of protein. But in 1986, this theory was toppled by the other which concluded that p53 is an antioncogene. It becomes oncogenic when becomes mutated and is not able to cope with the malignant process. This theory was proved in 1990 by a study which showed blocked the proliferation of cells. A normal cell has capacity to grow in an unrestrained manner that usually is inhibited by p53. The loss of p53 gene causes a restrained growth.

The p53 protein is made up of 393 amino acids and any change in any one of 393 amino acids cripples the protein. The crippled protein can no long carry out its surveillance duty on cell proliferation and cancer can thrive unchecked.

Mutations in p53 gene are quite common. More than 5% of all human cancer types bear such mutations. Above 90% of mutations in p53 had led to a change in the identity of an amino acid which alters the conformation of protein and affects its function. In humans the cancers which are caused due to mutations are of brain, breast, lung, liver, oesophagus, stomach, colon, pancreas, skin and thyroid. About 50% lung cancers and 40% of breast cancer are due to p53 gene mutation.

Different carcinogens cause distinctly different characteristic mutations. Usually p53 mutation in tumours are simple base pair substitutions which lead to missense mutation. Aflatoxin B, which is a common dietary carcinogen, causes guanine to be replaced by thymine at a point on the p53 protein.

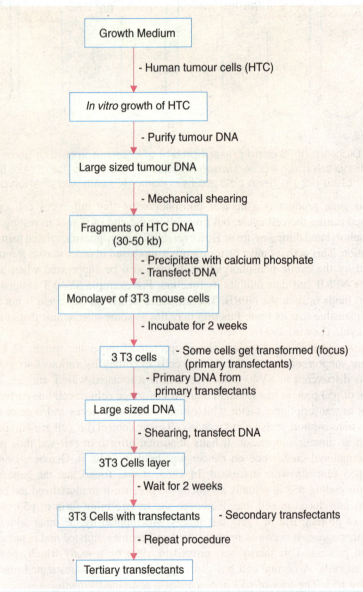

Growth Medium

- Human tumour cells (HTC)

In vitro growth of HTC

- Purify tumour DNA

Large sized tumour DNA

- Mechanical shearing

Fragments of HTC DNA
(30-50 kb)

- Precipitate with calcium phosphate
- Transfect DNA

Monolayer of 3T3 mouse cells

- Incubate for 2 weeks

3 T3 cells   - Some cells get transformed (focus)
                (primary transfectants)
- Primary DNA from
  primary transfectants

Large sized DNA

- Shearing, transfect DNA

3T3 Cells layer

- Wait for 2 weeks

3T3 Cells with transfectants   - Secondary transfectants

- Repeat procedure

Tertiary transfectants

Fig. 23.10 : Detection of cellular oncogene by transfection technique through transforming 3T3 mouse cells.

In general p53 is the guard of the genome but when the p53 protein is mutated it becomes doubly dangerous. Cells with defective p53 protein have a variety of functions. This pleiotropy makes it difficult to determine which of this effects is directly connected to the tumour suppressor function stem from affecting the ability of p53 to interact with other proteins or to bind to DNA (Lewin, 1994).

9. Detection of Human Cancer DNA

During 1980s, a major goal of molecular biologists has been to identify the cellular genes of human tumours *i.e.* human oncogenes. Successful line of research to achieve this goal was the development of molecular cloning technologies, analysis of oncogenes, analysis of specific chromosomal abnormalities in humans and the ability to assay for single copy cellular genes using DNA transfection technique (Watson *et al*, 1987).

(*i*) **Transfection:** The transfection technique as given by Graham and Vander Eb (1973) follows: (*a*) growth of human cancer cells *in vitro*, (b) purification of DNA from human tumour cell lines, (c) mechanical shearing of large sized DNA to break into pieces of about 30 to 50 kb, (*a*) dissolution of DNA fragments in phosphate buffer and precipitation by adding calcium chloride, (*e*) pouring of solution onto a monolayer of mouse 3T3 cells (*i.e.* an established cell lines of mouse embryo cell used frequently in molecular biological experiments), and (*f*) allowing the fine precipitate to settle down onto the cells (Fig. 23.10). The 3T3 cells take up DNA efficiently. Therefore, 3T3 cells are selected as recipients. After two weeks, DNA from human tumour cell lines can induce transformed foci in monolayer of mouse 3T3 cells, whereas DNA from normal cells does not induce foci under the same experimental conditions in control. These transformants could induce tumours in animals.

This shows that human cancer cells indeed sustained genetic alterations that can be assayed by DNA transfection. The number of foci increases linearly with the amount of DNA added. The foci of 3T3 cells transformed by human tumour DNA are called primary transfectants. When DNA from primary transfectants is purified and, following the above procedure, fresh 3T3 cells are allowed to be transfected by DNA molecules, foci of transformed cells occur. These foci are called secondary transfectants. Like control, a DNA of normal 3T3 cells cannot induce transfection. The procedure can be repeated serially to get tertiary transfectants and so on. This reveals that a single gene that acts in a dominant fashion must be responsible for transformation of the tumour DNA by using DNA transfection technique. Several human *ras* oncogenes have been detected from established cell lines derived from tumour from different body organs such as bladder, breast, lung, etc. (Table 23.5).

(*ii*) **Cloning of Transforming Genes:** For molecular cloning of transforming genes from human tumours, DNA of transformed cells (*i.e.* from secondary or tertiary transfectants of 3T3 cells) are isolated and digested with specific restriction enzymes. The recognition sites for each specific restriction enzyme differ. Steps for gene cloning are described in Chapter 11. For hybridization purpose, the *Alu* repeat is used as a hybridization probe because about 3,00,000 copies of *Alu* family sequence are scattered throughout the genome. However, there is possibility of such a sequence to be near any particular human cancer gene. In Southern blot specific human oncogene from secondary or tertiary transfectant 3T3 cells images after hybridization with complementary *Alu* probe. The Southern blot of DNA is isolated and sequence is analysed. For the first time, Shih and Weinberg (1982) got success in isolating a transforming sequence from a human bladder carcinoma cell line (*i.e.* ES cell line) by using mouse 3T3 transfectants. By constructing a λ library of chromosomal DNA isolated from secondary/tertiary 3T3 transfectants now it is possible to clone many human carcinogenes.

Table 23.5 : Human *ras* oncogenes detected by DNA transfection/ transformation assay of mouse 3T3 cells.

Human Cancer	ras Oncogenes*		
	H	K	N
Carcinomas			
Bladder	3	1	–
Breast	1	–	–
Colon	–	2	1
Liver	–	–	2
Lung	1	5	1
Ovary	–	1	–
Pancreas	–	2	–
Sarcomas			
Fibrosarcoma	–	–	1
Rhabdo mycosarcoma	–	1	1

* About 10–20% cell lines are positive for transfectable *ras* oncogenes. Number varies depending on tumour types.

Most of the human oncogenes detected by DNA transfection of 3T3 cells was found to be activated human *ras* genes which is homologous to the genes first identified as oncogenes of the Harvey and Kirsten mouse sarcoma retroviruses. Three functional human *ras* genes were detected e.g. H-*ras*-1, K-*ras*-2 and N-*ras*. Due to mutation any one of these can behave as oncogene. The normal cellular gene can be converted into oncogene due to a single-base change that alters the 12 gene and, in turn, 12th *ras* genes encode proteins of 21 k dalton located at the inner surface of plasma membrane that possess GTPase activity. Furthermore, a large number of cellular oncogene has been cloned so far.

D. Cancers in India

India is a highly populated country next to China, in the world and so is the position of incidence of cancers. India has the fourth highest incidence of breast and cervical cancers in the world, whereas within the country Delhi and Chennai account the highest incidence respectively. The age adjusted rate (43.5 per 100,000) is the next to that of Cali and Columbia which has the highest incidence in the world.

The annual age-adjusted incidence of cancer in India is only about 115-120 per 10,00,000 persons. It is one of the lowest in Asia and less than half of those of Western countries. About one out of 13 to 16 males (and 10 to 13 females) in India get cancer during their life time (0 to 64

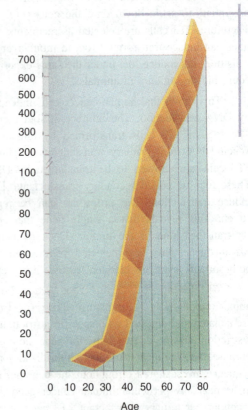

The annual death rate from cancer is a function of age : A plot of age versus death rate is linear, suggesting that several independent events are required to give rise to cancer.

years) as against one in 4 in the USA. Shivaramkrishnan (1994) has summarised the geographical distribution of cancers in India (Table 23.6).

1.   Diversity of Site-Specific Cancer

The cancer has more generalised distribution in males than in females. Site specific distribution of cancer throughout the world (also in India) in men and women is the same. Cancers at site specific to women (*i.e.* cervix, breast and ovary) constitute to about half of the total cancers, whereas the cancers of the prostate, testes and penis together hardly account to 15% of all cancer among males. However, the cancers at the other sites such as oesophagus, stomach, lung, etc. are the leading ones among males. In addition, there are exceptions too; for example, pharyngeal cancer at Hubli and Dibrugarh. The leading cancer in India in males varies from place to place; such a diversity is not observed in females. The cancer of cervix uteri and breast are the predominant ones in females at all the places with little variation in relative proportions.

Table 23.6 : Geographical distribution of cancer in males in India
(after Sivaramkrishnan, 1994)

Region	*Cities*	*Cancer sites*
North		Kashmir, *Kangri* cancer
	Chandigarh	Oesophagus and lung cancer
	Delhi	Larynx, tongue and lung cancer
West	Ahmedabad	Lung, tongue
East	Kolkata	Oropharynx, laryngopharynx, larynx
	Guwahati	Hypopharynx
	Dibrugarh(Assam)	Pharynx, oesophagus
	Kohima(Nagaland)	Nasopharynx
Central	Gwalior (M.P)	Larynx, tongue
	Bhopal	Lung, tongue
	Nagpur (Maharashtra)	Oesophagus, larynx, tongue
	Mumbai	Lung, oesophagus
	Pune	Oesophagus, lung, larynx
	Vishakhapatnam (A.P.)	Hard palate (*Chulta* cancer)
South	Hubli (Karnataka)	Pharynx, oesophagus
	Bangalore	Stomach, lung, oesophagus
	Chennai	—do—
	Tiruvanantpuram (Kerala)	Buccal cavity, lung

(*i*) **Males:** The leading cancers in India in males are of lung, oesophagus, stomach, oral cavity, pharynx, larynx, prostate and rectum. Lung cancer is leading in Bombay followed by Delhi, Ahmedabad, Bhopal, Chennai and Bangalore. Oesophagus cancer is common both in men and women. But it is high in Maharastra (leading in Pune, Nagpur and Bombay), Karnataka (in Hubli) and Assam (in Dibrugarh). Stomach cancer is common both in men and women, and leading among men in Chennai and Bangalore.

Oral cancer is typical in Indian sub-continent which differs from rest of the world (e.g. Canada, Czechoslovakia, Australia, etc.) where lip cancer is most prevalent. In India, the pre-dominant oral cancers are of lung, buccal mucosa, lower alveolus, and palate. Tongue cancer is high in the region between Delhi and Gwalior (in North), and Ahmedabad and Nagpur (in West) with Bhopal in middle.

However, cancer of buccal mucosa is common throughout the India. Cancer of pharynx is leading in males in Hubli (North Karnataka) and Assam, while that of oropharynx and laropharynx are very common in Bengal in Calcutta (W.Bengal). Nasopharynx cancer is seen among Nagas and in Manipur. Cancer of larynx is leading among men in Gwalior, Maharashtra and Gujarat. Cancer of prostate is leading among men in Jaipur.

(*ii*) **In Females:** Among the women, the most common cancers are of the cervix, uteri, breast, ovary, mouth, oesophagus, stomach, pharynx, lungs, larynx as well as ovary which constitute the major cancer burden at present. Cancer of the cervix uteri is the leading among women except in Bombay, Ahmedabad and Tiruvanantpuram where the incidence of brest cancer is slightly high. The relative frequency of cervical cancer is exceptionally low among Assamese women (15% only) and breast cancer about 12%. The incidence of cervical cancer varies between 25 and 50%, and breast cancer 10 and 30% of all the cancers among women in different places. Together they account for about 40-50% of all cancers among female except in Assam.

2. Factors Affecting the Incidence of Cancer

The local habits, life style, age, sex and religion are known to influence the incidence of cancer in India. For example, Bombay is a cosmopolitan city that harbours the four major religion groups i.e. Hindu, Muslims, Christians and Parsees. According to the Bombay Cancer Registry (1994), the incidence of cancer of the cervix uteri is the highest among Hindu and the least among the Parsees. The breast cancer is the highest among Parsees and lowest among the Hindu. The Muslims and Christians have intermediate incidence in both the cases.

Due to peculiar habit of the local inhabitants, *Kangri* cancer (in Kashmir) and *Chulta* cancer (in and around Vishakhapatnam) occur. Chulta cancer occurs in hard palate due to reverse smoking. Many people wrap cured tobacco in dried tobacco leaves and make it into cigar which is locally called *Chulta*. It is often smoked with the lighted end inside the mouth. A malignant transformation of the hard palate ensures, particularly in an area close to the burning end of cigar. Reverse smoking is also practised in Goa mainly by Christian women. The Christian women smoke *dhumti*, a homemade cigar, which is made by wrapping tobacco inside a jack tree leaf.

In Karnataka, the estimates show that every year 35,000 new cases were

Lung cancer cells : These cells are from a tumor located in the alveolus of a lung.

added with one third of them occurring in sites related to use of tobacco. Late school and college-going students become victims to tobacco habit; and in villages the incidence rate is gradually increasing due to lack of awareness. In Bangalore, about 4,000 new cases of cancer are being reported every year, and two people die every year due to tobacco related diseases.

According to Kidwai Memorial Institute of Oncology, Bangalore, 18% women suffer from lung cancer due to the use of tobacco. In addition, emission of pollutants from vehicles also are contributory factors. Moreover, 60% of state Road Transport Corporation bus drivers and conductors are addicted to smoking. So far, much attention has been paid by the Government and non-government organisation to create awareness about tobacco related diseases. Unfortunately, the anti-cancer compaign has failed to create effective impact in the minds of people. Therefore, much is needed in future to check the spread of this fatal disease.

Photograph of lung cancer in an adult human : the bottom half of the lung is normal. The top half has been taken over completely by a cancerous growth.

QUESTIONS

1.   What is a cancer? Discuss in brief the causes of cancer.
2.   Give a detailed account of viruses as the cause of human cancer.
3.   What are the DNA tumour viruses? How are they associated with cell transformation?
4.   What are the retroviruses? Give in brief the replication of a retrovirus genome in eukaryotic cell.
5.   In what way a retrovirus genome replicates in eukaryotic cells.
6.   Give a brief account of acute transforming viruses.
7.   Write an eassy on cancers in India.
8.   Write short notes on the following
     (*i*) Oncogenes and proto-oncogenes, (*ii*) v-*onc* and c-*onc*, (*iii*) DNA tumour viruses, (*iv*) SV40, (*v*) Hepatitis B virus, (*vi*) Retroviruses, (*vii*) LTRs, (*viii*) Provirus, (*ix*) Acute transforming virus, (*x*) Oncogene proteins, (*xi*) Oncogene family, (*xii*) Tumour suppressors, (*xiii*) p53, (*xiv*) Transfection.

REFERENCES

Ames, B.N. *et al.* 1973. Carcinogens are mutagens: a simplest test system combining homogenates for activation and bacteria for detection. *Proc. Natl. Acad. Sci (USA),* **70:** 2281-2285.

Bishop, J.M. 1983. Cellular oncogenes and retroviruses. *Ann. Rev. Biochem* **52.** 301-354

Bishop, J.M. 1985. Viral oncogenes. **Cell** 42 : 23-38

Durst, M.G. *et al.* 1983. *Proc. Natl. Acad. sci.* (USA) 80:3812-3816

Eisenman, R.N. *et al.* 1985. v-myc and c-myc encoded proteins are associated with the nuclear matrix. *Mol. Cell Biol* 5: 114-126

Groudine, M. *et al.* 1981. *Nature,* **292** : 311 - 317.

Hecker, E. *et al.* 1981. *Carcinogenesis : A comprehensive Survey.* Vol. 7 cocarcinogenesis and Biological effects of tumour promotors. Raven Press, New York.

Hunter, T and Sefton, B.M. 1980. Transforming gene product of Rous sarcoma virus phosphorylates tyrosine. *Proc. Natl. Acad. Sci.* **77** : 1311 - 1315.

Klein, G. et al. 1971. The analysis of malignancy by cell fusion *J. Cell Sci.* 8: 659-672.

Lewin, B. 1994. *Genes.* V. Oxford Univ. Press, Oxford, p.1257.

Lowy, D.R. *et al.* 1971. Murine leukemia virus: high frequency activation *in vitro* by 5-iododeoxyuridine and 5-bromodeoxyuridine. *Science.* 174 : 155.

Melnick, J.L. 1983. Hepatitis B virus and liver cancer. In *Viruses Associated with Human Cancer* (ed. L.A. Phillip) pp. 337-367, Dekker, New York.

Miller, E.C. and Miller, J.A. 1981. Search for ultimate carcinogens and their reactions with cellular macromolecules. *Cancer.* **47:** 2327-2345.

Moelling, K.O. *et al.* 1971. Association of viral reverse transcriptase with an enzyme degrading the RNA moety of RNA-DNA hybrids. *Nature New Biot* 234: 240-243.

Rouse, P.1911. A sarcoma of the fowl transmissible by an agent separable from the tumour cells. *J.Exp. Med.* **13:** 397-411.

Schwarz, E. *et al.* 1985. Structure and transcription of human papilloma virus sequences in cervical carcinoma cells. *Nature*, **314** : 111-115

Sefton, B.M. 1981. Vinculin: a cytoskeletal target of the transforming protein of Rous sarcoma virus. **Cell.** **24:** 165-174.

Shih, C. and Weinberg, R.A. 1982. Isolation of a transforming sequence from a human bladder carcinoma cell line. **Cell:** 161-169.

Sivamkrishnan, V.M. 1994. Cancer sites its victims. The *Hindustan Times.* 70 (276) : 16.

Takeya, T and Hanafusa, H 1983. Structure and sequence of cellular gene homologous to the RSV *src* gene and the mechanism for generating the transforming virus. *Cell*, **32:** 881-890.

Varmus, H.E. 1982. Form and function of retroviral proviruses. *Science*, **216:** 812-821

Waterfield, M.D. *et al.* 1983. Platelet-derived growth factor is structurally related to the putative transforming protein p28[sis] of simian sarcoma virus. *Nature*, 304 : 35-39.

Watson, J.D. *et al.* 1987. *Molecular Biology of the Gene.* 4th ed. The Benjamin/Cummings Publ. Co. Inc; California, p.1163.

Yunis, J.J. 1986. Chromosomal rearrangements, genes and fragile sites in cancer: clinical and biological implications. In *Important Advances in Oncology* (eds. V Devita, S; Hellman; Sand Rosenberg, s.), Philadelphia, Lippincott.

Cellular Microbiology

24

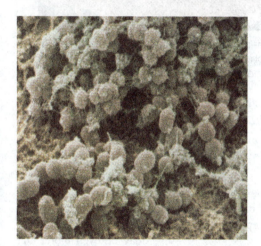

A revolution was made in medicine and clinical science after the introduction of antibiotics in the 1950s and 1960s. During this period a little work was done on the mechanism involved i.e. how bacteria infect multicellular host and develop disease. In the 1990s it became clear that in spite of use of antibiotics to kill bacteria, each year 3 million people die due to tuberculosis, 3-4 million due to diarrhoea, 1-2 million due to malaria, hepatitis and measles and millions by other diseases throughout the world. Some new bacterial diseases were also reported; for example *Helicobacter pylori,* the causal agent of gastric ulcer, and *E. coli* 0157 causing diarrhoea.

Much attention has been paid on bacterial infection stimulated by the growing resistance of bacteria to antibiotics. This has happened at such a time when advances made in microbiology, molecular biology and eukaryotic cell biology are able to give answer of possible questions of infection process.

Electron micrograph of Helicobacter pylori, the causal agent of gastric ulcer.

A. Introduction

Cossart *et al.* (1996) coined the term **cellular microbiology** by synthesising the three related disciplines (*cell biology, molecular biology* and *microbiology*). However, individually cell biology, molecular biology and microbiology are being studied in detail as separate subjects. But cellular microbiology bridges the gap between these two disciplines and provides a current synthesis of the relevant science.

Study of microbiology has advanced the knowledge of immunology that how do bacteria trigger the immune system and T lymphocytes recognise antigen, and how do bacterial exotoxins affect eukaryotic cells. Advances in molecular biology fed back into microbiology. It is obvious how bacteria produce intercellular signalling molecules to determine cell density? Much advances have been made in the development of techniques in molecular biology, for example 16S rRNA sequencing, *in situ* expression

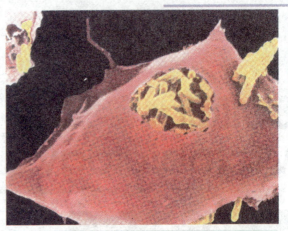

The scanning electron micrograph depicts a macrophage (red) phagocytosing a bacterium, *Mycobacterium tuberculosis* (yellow).

technology, signature *h*-tagged mutagenesis, differential display PCR (DD-PCR), yeast two hybrid analysis and phage display (Henderson *et al.*, 1999).

1. Bacterial Diseases

Some important bacterial diseases have been described in Chapter 22 and immune responses triggered by bacteria in Chapter 19. A large number of bacteria, both normal and pathogens, occur on our body. Over 1,000 species of bacteria live on epithelial surface of human body. The number may be even more. Out of these only a few of them cause disease in humans. In spite of less number they have vanished human population in several regions of many countries. Between 1917 and 1921, typhus alone infected 20 million Russians out of which 3 million people died. About 22-23 million people die each year only due to tuberculosis, diarrhoea, malaria, acute respiratory problem, AIDS, hepatitis and measles (Henderson *et al.*, 1999).

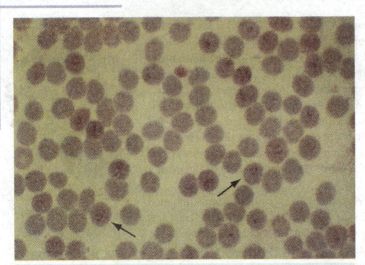

Plasmodium, the causal agent of malaria.

Most of the pathogenic bacteria causing disease were discovered during 19th century (Table 24.1). But it could not be fully understood how do bacteria cause disease ?

Table 24.1: Discovery of major bacterial diseases (based on Henderson *et al.*, 1999).

Year	Disease	Causal agent
1877	Anthrax	*Bacillus anthracis*
1878	Pus formation	Streptococci, staphyllococci
1879	Gonorrhoea	*Neisseria gonorrhoea*
1880	Typhoid fever	*Salmonella typhae*
1882	Tuberculosis	*Mycobacterium tuberculosis*
1883	Cholera	*Vibrio cholerae*
1883	Diphtheria	*Corynebacterium diphtherae*
1884	Tetanus	*Clostridium tetani*
1885	Diarrhoea	*Escherichia coli*
1886	Pneumonia	*Streptococcus pneumoniae*
1887	Meningitis	*Neisseria meningitidis*
1888	Food poisoning	*Salmonella enteritidis*
1892	Gas gangrene	*Clostridium perfringens*
1894	Plague	*Yersinia pestis*
1896	Botulism	*Clostridium botulinum*
1898	Dysentery	*Shigella dysenteriae*
1900	Paratyphoid	*Salmonella paratyphi*
1903	Syphilis	*Treponema pallidium*
1906	Whooping cough	*Bordetella pertussis*

After transmission, the pathogenic bacterium comes in the contact of host cell and involved in several stages of cell-to-cell interaction until the symptoms develop. The stages include: *adhesion* of bacteria to host's epithelium, *invasion* of epithelium, *multiplication* of bacteria inside the host tissue invoking host response, and *evasion* of protective host response for survival of pathogen (small proteins called cytokines provide immune response in host and bacteria attack such cytokines), *tissue destruction* and *release of bacteria* from host tissues (Fig. 24.1).

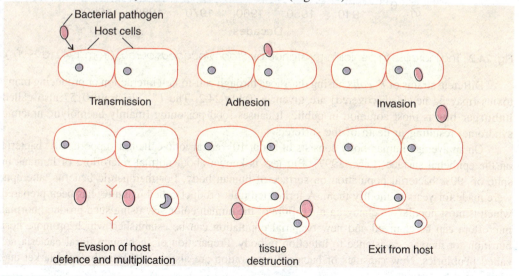

Fig. 24.1 : Interaction between bacterial cell and host cell, and various stages of interaction.

The virulence factors (ability to produce a variety of molecules) govern the bacteria to adhere, invade and evade host defences and cause tissue damage. The virulence factors have been grouped into *adhesins* (responsible for adhesion to host cells), *invasins* (responsible for invasion), *impedins* (to overcome host defence mechanisms), and *aggressins* (to damage host cells). The another class of bacterial virulence factor has been proposed as **modulins** which induce the synthesis of cytokines.

2. Emergence of Cellular Microbiology

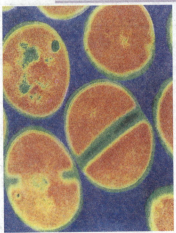

By 1980s it became clear that bacterial infection may result in human death. Development of antibiotic resistance in bacteria continued resulting in the emergence of new bacterial diseases. In 1940s, *Staphylococcus aureus* was discovered as penicillin-resistant commensal bacterium. Now this bacterium is resistant to all forms of penicillin hence known as methicillin (some multiple)- resistant *S. aureus* (Fig. 24.2).

Staphylococcus aureus, a penicillin resistant commensal bacterium.

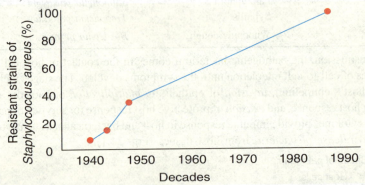

Fig. 24.2 : The increase in penicillin-resistance by *Staphylococcus aureus* (from 1940 to 1990s).

Different strains of *E. coli* causing diarrhoea through cell-to-cell interaction or producing many toxins (may be newly discovered) are given in Table 24.2. The *E. coli* strain 0157 also called **huburger bug** is most common in public. It causes food poisoning (mainly haemolytic uraemic syndrome) resulting in death of the sufferers.

On an average human body consists of about 10^{13} eukaryotic cells but supports 10^{14} bacteria on the epithelial surface (Pace, 1997). But less is known about normal microflora of humans in spite of 90% bacterial population on surface of human body. For therapeutic benefits, attempts were made for years to modify them. A fermented milk product called Yakult® has been prepared which is most popular in changing microflora of the human colon. By using such product, normal microflora can be changed and new bacterial population can be established which promote host nutrition or increase resistance to infection in body. Preparation of such microfloral bacteria are called **probiotics**. New capsules of bacterial preparation are also available in Indian Market and easily used by patients.

Table 24.2 : Different strains of *E. coli* causing diarrhoea (based on Henderson *et al.*, 1998).

Strains	Chemical synthesis	Mechanism
Enteropathogenic	Watery diarrhoea	Pili, Type III secretion
Enterohaemorrhagic (*E. coli* 0157)	Bloody diarrhoea	Shiga like toxin
Enteroinvasive	Dysentery	Cellular invasion and cell-to-cell spread
Enterotoxigenic	Watery diarrhoea	Colonisation factor
Enteroaggregation	Watery diarrhoea, persistent disease	Heat stable toxins
Diffusely adherent	Watery diarrhoea, persistent disease	Fimbriae, heat stable toxins
Cytolethal toxin producing	Diarrhoea	Cytolethal distending toxins

During the 1980s, due to rapid development, the two branches, molecular biology and cell biology were much advanced. During 1990s, application of these methodologies to microbiology enabled the emergence of **cellular microbiology**. You know that bacteria are regarded as complex and beautifully adaptable organisms which respond to varied environmental conditions, whereas multicellular organisms depend on intercellular communication. Earlier, presence of such communication was not thought to operate in bacteria because bacteriologists use to grow them in axenic culture on artificial substrates lacking any signals.

Now it is established fact that bacteria respond through cell-to-cell signalling mechanism called **quorum sensing** (*see* preceding section for detail). This ability determines the maintenance of cell density and also switch on or switch off particular genes. Quorum sensing was earlier discovered in marine bacterium, *Vibrio fischeri* establishing symbiotic association in light organs of certain marine fishes. But its operation in other bacteria is also known. It provides a virulence mechanism in bacteria and permits them to increase their number without inducing the virulence genes.

(i) **Signal Molecules:** Bacteria respond to signals coming from eukaryotic host cells. Example of the mammalian signal molecules are serotonin, catecholamines, insulin and cytokines (e.g. interleukin 1 and 2, tumour necrosis factor and transforming growth factor alpha). Kapreylants and Kell (1996) suggested a term 'microendocrinology' to denote prokaryotic-eukaryotic communication. Besides, bacteria produce cytokines (i.e. local cell-to-cell regulatory molecules) which regulate the production of local hormone. Cytokines also control immune responses in vertebrates.

(ii) **Small Peptides:** Antibacterial small peptides are also important in the interaction of bacteria with host cells. These peptides bind the bacterial cell wall, form pore like structure and results in loss of intracellular fluids. Hence, these kill both Gram-positive and Gram-negative bacteria and provide a defence to us. Hundreds of such peptides have been discovered from organisms including bacteria also.

(iii) **Adhesins:** The intimacy of contacts between bacteria and the host cells has been emphasised by cellular microbiology. The first step of infection is the adherence of bacterial cell to host cell. Bacteria produce adhesins which selectively bind them to suitable host cell. Many surface molecules (e.g. lipids, glycolipids, carbohydrates and proteins) are present on host cell which form ligands with bacterial adhesins. Adhesins trigger cytokine produc-

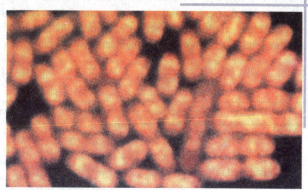

Salmonella, produces *adhesins* which selectively bind them to suitable host cell.

tion which causes inflammation to check the entry of pathogens. As a result of adhesion bacteria release a number of specific proteins into cytoplasm of target cell. This influences the behaviour of eukaryotic cells. *Shigella flexneri* and *Salmonella typhimurium* use this mechanism to steal the cytoskeletal machinery of the host and facilitate the bacterium to enter the host cell.

The protein kinases and phosphatases are injected into the host cell. Bacteria participate 'in biomolecular dialogue' as the host cells. Ability of bacteria to bind and enter the cell and biomolecular dialogue between bacterium and host cell are described in Sections C and D.

(iv) **Exotoxins:** Exotoxins are the major chemical weapon of the infectious bacteria. A large number of exotoxins have been discovered. They act on host cells in various ways. Cell biologists are using exotoxins as probes of cell function. Using exotoxins in understanding of cell signalling has now been determined. Besides, exotoxins are being used as therapeutic agents.

With the development of new molecular techniques, understanding of bacteria-host cell interaction has become clear. Following these techniques you can identify and analyse the genes which get induced both in bacteria and host cell during infection. Some of these new techniques are listed in Table 24.3.

Table 24.3 : Techniques of molecular biology which determine genes involved in bacterial infection (based on Henderson *et al.*, 1999).

Techniques of molecular biology	Applications
Allelic exchange	Gene replacement
DD-PCR	
RNA arbitrary-primer PCR	Identification of genes expressed under different conditions
Substrate hybridisation	
Genome mapping	Identification of homologous genes
Homologous recombination	Production of animals which lack specific gene products
In vitro gene expression technology	Identification of genes involved in infection
Mutagenesis (site-directed)	Identification of gene function by gene disruption
Phage display	Identification of interacting gene products
PhoA fusion cloning	Cloning of exported proteins
Yeast two-hybrid analysis	Identification of protein-protein interactions within cells

Some of the techniques of gene manipulation has been discussed in Chapter 11. A new branch called genomics has emerged which deals with complete DNA sequencing of bacteria. Genomes of some bacteria have completely been sequenced (Table 24.4). On the other hand, a question has always been asked what genes are switched on or switched off when a pathogen comes in contact of host cell. To answer this question several techniques have been developed. One of such techniques is the *in vitro* expression technology which can identify genes that are switched on by the bacterium in the host cell. Differential display polymerase chain reaction (DD-PCR) identifies genes expressed under different conditions i.e. cultured conditions and within the cells. Signature-tag mutagenesis technique can detect the genes which are involved in infection process.

Table 24.4 : Genomes of some bacterial pathogens which have been completely sequenced.

Genome size (mega base, Mb)	Pathogen
0.58	*Mycoplasma genitalium*
0.81	*Mycoplasma pneumoniae*
1.14	*Treponema pallidum*
1.40	*Mycobacterium tuberculosis*
1.44	*Borrelia burgdorferi*
1.66	*Helicobacter pylori*
1.83	*Haemophilus influenzae*

3. The Cellular Biology Underlying Prokaryotic - Eukaryotic Interactions

This unit is important to furnish certain basic information to understand the onward sections. This section provides a brief overview of key points in microbilogy and cell biology which are needed to grasp the matter.

(i) **Bacterial Ultra-structure:** Bacteria are the smallest organisms living independently. Their size, shape and dimension vary. These have been discussed in detail in Chapter 4, *The Microbial Cells: Morphology and Fine structure*. Mobile genetic elements (transposons) are discussed in Chapter 8, *Microbial Genetics*.

(ii) **Gene Expression:** Gene expression (gene transcription) and regulation are described in Chapter 10, *Gene expression and regulation*.

(iii) **Pathogenicity Islands:** Bacterial pathogenicity is governed by inter-species transfer of virulence factors contributed by plasmid and transposons. Pathogenicity is becoming apparent also. The interspersed elements (within the genomes of many pathogenic bacteria) are large distinct that encode virulence-associated genes. These chromosomal loci have been termed as pathogenicity islands (PAIs). This is the other important mechanism which contribute to evolution of microorganisms. Pathogenicity island has been nicely reviewed by Lee (1996) and Groisman and Ochman (1997).

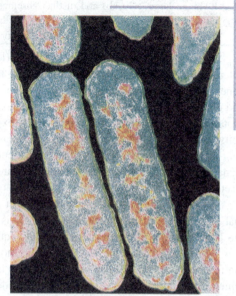

Haemophilus influenzae, a bacterial pathogen whose genome has been completely sequenced.

Uropathogenic *E. coli* is the the first bacterium in which PAIs were described. The α-haemolysin genes were described to be present on a large region of chromosome. They get lost spontaneously resulting in development of an avirulent strain. Common features of PAIs are given in Table 24.5. The *E. coli* haemosin-associated PAIs (PAI-I) are large (PAI-I is 70 kb and PAI-II is 190 kb) and carries genes encoding adherence-mediated P fimbrae (haemolysin and P fimbrae, PAI-II). They are found in pathogenic isolates but not in non-pathogenic isolates. PAI-I and PAI-II contain short direct repeats of 16 kb and 18 kb, respectively. These chromosomal elements have a lower G+C content (41%) as compared to host bacterial chromosome (51%).

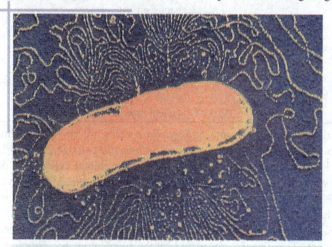

E. coli. bacterium shown burst open so that the DNA within has spilled out.

Table 24.5 : Pathogenicity island showing common features.

1. They carry virulence-associated genes
2. PAIs are pathogen-specific
3. PAIs are large (>30 kb) and distinct chromosomal units often flanked by direct repeats
4. G+C content of PAIs differs from those of host bacterial chromosome
5. They get associated with tRNA genes and/or insertion sequences

A second class of PAIs is the LEE (locus of enterocyte effacement; PAI-III) found in enteropathogenic *E. coli* strains. LEE is a 35 kb long chromosomal element that contain multiple genes providing a property of complex virulence. This property of LEE is the induction of attaching and effacing lesions on enterocytes. LEE encodes a Type III secretion system and several secreted proteins. Type III secretion system which exports proteins directly to cell surface and directly delivers bacterial virulence proteins in cytoplasm of host cells. It is triggered by contact with host cell; therefore, referred to as 'contact-dependent secretion'. LEE encoded molecules induce a chain signal transduction within enterocytes rearrangement of cytoskeletal and formation of pedestal. Bacterium sits on the pedestal.

LEE is stable and not flanked by direct repeats. It has distinct G+C (39%) as compared to that of *E. coli* chromosomes (51%). If LEE is introduced into a non-pathogenic strain of *E. coli*, the latter gets the property of single step acquisition for the attachment and effacement.

In addition to *E. coli* PAIs, certain chromosomal loci of other pathogenic bacteria demonstrated the features like PAIs. These features include the presence of large distinct chromosomal elements which differ in G+C contents, expression of defined virulence functions, unstability in some loci, flanking of direct repeats (Table 24.6).

Table 24.6 : Pathogenicity islands of some bacteria (based on Handerson et al., 1999).

Pathogens	PAIs 'PAI/host)	Size (kb)	G+C	Direct repeats (kb)	Phenotype
Escherichia coli	PAI-1	70	40/51	16	Haemolysin production
	PAI- II	190	40/51	18	Haemolysin, P fimbrae production
	PAI-III (LEE)	35	39/51	-	Induction of attaching and effacing lesions on enterocytes
Salmonella typhimurium	SP-1	40	42/52	-	Invasion of macrophages
	SP-2	40	45/52	-	Survival in macrophages
	SP-3	17	7	-	Survival in macrophages
Vibrio cholerae	VPI	39.5	35/46	13	Colonisation, expression of CTXϕ receptor
Yersinia pestis	HPI	102	46-50/ -50	IS100	Haemin storage, iron uptake

Many of virulence-associated genes of pathogenic *Salmonella* species are encoded within pathogenicity islands. Four of such genes are known so far such as *Salmonella* pathogenicity (SPI-1, SPI-2, SPI-3 and SPI-4). SPI-1 gene products induce apoptosis (programmed cell death) of *Salmonella*-infected macrophages, whereas SPI-2 has 45% G+C content and is required for survival of *Salmonella* macrophages. SPI-3 is essential for the growth of bacteria at low concentration of Mg^{++} and survival within macrophages.

The virulence of *Vibrio cholerae* is associated with the synthesis and secretion of **cholera toxin** encoded by *cixA* and *cixB* genes. These genes are carried by the filamentous bacteriophage CTXϕ. The toxin co-regulated pilus (TCP) is the bacterial receptor for phage infection. It acts as adherence determinant for the bacteria. The TCP gene is present on a 39.5 kb pathogenicity island which is termed as VPI. Acquisition of VPI gene of *V. cholerae* allows it to colonise the intestine of humans and animals.

A cluster of virulence-associated genes encoded by 10 kb long chromosomal element is present in *Listeria* species pathogenic to humans and animals. These genes are absent in non-virulent pathogenic species. This locus lacks direct repeats.

Stable PAIs of *Salmonella* have been analysed and found that they were acquired early in the evolution of the genus *Salmonella*. The donor organism become neither extant nor recognised as the donor. Recent PAIs acquisition events and emerging pathogenic strains of

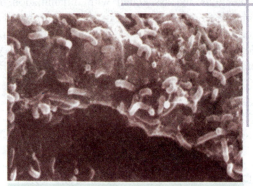

Vibrio cholerae : Numerous cells are visible here on the surface of the small intestine of a mouse.

both donor and recipient bacteria can be identified by molecular analysis combined with genome sequencing. Large horizontally acquired chromosomal elements that encodes discrete functions are wide spread among bacteria. The horizontal transmission of a large chromosomal elements (500 kb) among soil bacteria have recently been described. The detailed mechanism of transfer, integration, retension, and regulation of PAIs are provided by the bacteria that receive the PAIs. For example, nodulation and nitrogen fixation genes are encoded by chromosomal elements which establish symbiotic association with roots of leguminous plants. In non-symbiotic mesorhizobia this 'symbiotic island' is transmissible through laboratory matings. The element flanked by 17 bp direct repeats integrated into a tRNA gene. It encodes a bacteriophage-like integrase within its left end. Also it offers a marvellous opportunity to witness the evolution.

(iv) **Eukaryotic Cell Structure:** Description of cell structure of eukaryotes is not required assuming that the readers have studied biology in their previous classes.

B. Prokaryotic and Eukaryotic Signalling Mechanism

In human body, about 10^{13} cells are organised to constitute many specialised tissues. Activities of all these cells are integrated to keep the organisms healthy. The microorganisms are capable of manipulating the message of host cells for their own need. There are a wide range of biomolecules which establish cell-to-cell signalling. In eukaryotes one of these signals are the proteins, and one class of proteins is the cytokines that control immune response. Signalling mechanisms, both in eukaryotes and prokaryotes, are briefly discussed in this section.

1. Eukaryotic Cell-to-Cell Signalling

The integrative nature of biological systems could be understood after the pioneering work of Claude Bernard (1813-1878) of France. He gave the concept of the *miliew interieur* and suggested the system of ductless gland (i.e. endocrine glands) for integrating function and maintaining homeostasis. In 1902, Bayliss and Staling demonstrated a marked flow of pancreatic juice in dogs after injecting an acid extract of duodenum. Starling coined the term 'hormone' (Greek, *I excite*) for such intercellular messenger molecules. An American physiologist Water Cannon coined the term 'homeostasis' (a condition that may vary but is relatively constant).

There are three types of signalling systems in multicellular organisms like mammals: *neuronal, endocrine* and *cytokine* signalling (Fig. 24.3). Neuronal signalling occurs over very long distance i.e. brain to toe. Synaptic junctions communicate rapidly. Many chemicals are involved in signalling at junctions and associated with inflammation. Endocrine signalling involves the release of a hormone from its gland and its transport to blood to a limited number of cells in the target tissue. It occurs at a long distance and limited by the rate of blood flow and diffusion from blood to tissues. Most of the intracellular signalling is that of the cytokines. Much of this signalling occurs through paracrine signalling (over short distance cell to nearby cell) or by auto-signalling (stimulation of the cell producing the cytokines).

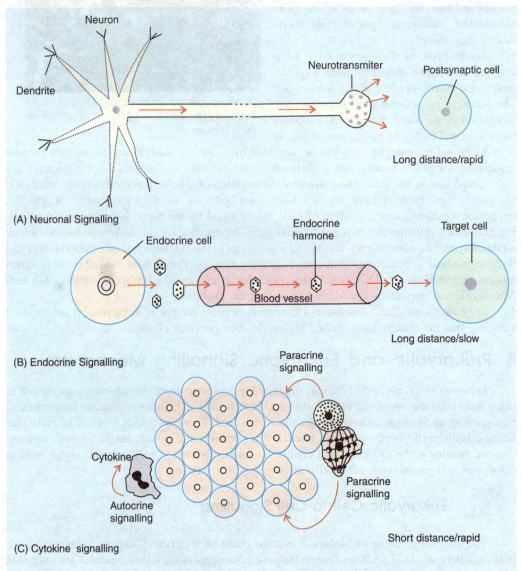

Fig. 24.3 : Three major forms of intracellular communication in mammals. A- neuronal signalling; B- endocrine signalling; C- cytokine signalling.

Certain bacterial endotoxins target neuronal signalling, hence these are called neurotoxins as produced by *Clostridium tetani* and *Cl. botulinum*. These have metalloproteinase activity and cleave specific intracellular proteins. Thus they prevent the neurotransmitters. A recent discovery also points to the interaction of a bacterial toxin with neuroendocrine signalling and synapsis of cytokines.

(i) **Endocrine Hormone Signalling:** Endocrine hormones are mostly produced by specific glands (such as pituitary, hypothalamus, and parathyroid glands) and glandular tissues (such as pancreas and intestine). There are three major groups of hormones: *peptide hormones* (produced especially by the intestine which has neurotransmitter-like activity), *steroid hormones* (produced by adrenal cortex, gonads and skin), and *thyrosine derivatives* (e.g. thyroid hormones T3, T4, etc. and catecholamines, noradrenaline, adrenaline and dopamine having neurotransmitter activity).

After secretion from glands, endocrine hormones are circulated as free hormone or bound to carrier proteins, for example a serum protein, *albumin*. It binds to several circulating hormones and exerts action (only in bound form) to specific cell receptors in target tissues. The peptide hormones bind to specific membrane receptors which result in specific intracellular signalling pathway. On the other hand the steroid and sterol hormones enter into cells and bind to cytoplasmic receptor proteins and then move to nucleus and act as a factor for transcription.

Endocrine hormones cotrol the energy metabolism through insulin and glycogen, adrenaline and noradrenaline involving production and breakdown of carbohydrate stored as glycogen in the liver and muscle. The lack of control of this system is visible in diabetes. Bacterial infection results in hormonal imbalance in body. Certain bacteria and viruses affect neural tissue. *Mycobacterium leprae* and *Treponema pallidum* have a tropism for nervous tissue.

The tissue of the gastric and intestinal mucosae are highly regulated. They respond and produce several endocrine signals including gastro-intestinal hormones e.g. gastrin, secretin, cholecystokinin and guanylin. *E. coli* alters fluid imbalance in the intestine and causes diarrhoea. Now seven different strains of *E. coli* have been reported which induce different pathological symptoms. The enterotoxigenic *E. coli* strains produce heat-labile toxin (LT) and heat-stable toxin (ST). ST is the first bacterial analogue of an endocrine hormone (guanyl) that activates guanyl cyclase and control fluid release from intestinal cells so that mucin layer could be kept wet (Fig. 24.4). There are other heat-stable guanyl-like toxin of other strains of *E. coli* and other bacteria.

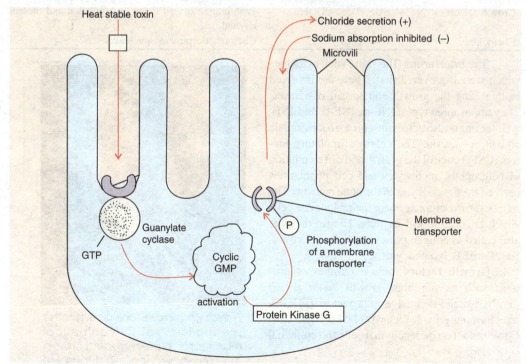

Fig. 24.4 : Mechanism of action of the *E. coli* heat-stable toxin on release of chloride ions from intestinal mucosae (diagrammatic).

(ii) **Cytokines:** Cytokines are a large group of over 1000 proteins which are involved in cell-to-cell signalling and control the inflammatory response to bacterial infection. These are polypeptide hormones secreted by a cell that affects growth and metabolism of the same cell (autocrine signalling) or another cell (paracrine signalling). Their over production causes disease. These are found at the site of infection by the agents. These induce lipid mediators (prostaglandins, leukotrienes, lipoxins, platelet-activating factor and the mediators from mast cells (e.g. histamine and enzymes such as tryptase).

(a) **Nomenclature of cytokines:** Cytokines are divided into six sub-families (Table 24.7) on the basis of several criteria such as historical types, sequence homology, localisation of chromosomes and biochemical actions. In 1979, the term **interleukin** (*inter*: between, *leukin*: leukocytes) was coined to denote the proteinaceous factors which modulate the function of the other leukocytes. At present there are over 20 interleukins (IL-1 to IL-18). The endotoxin-injected mice expressed a tumour necrosis factor (TNF) grouped under **cytotoxic cytokines**. The TNF kill certain tumour cell lines via induction of apoptosis and are potent pro-inflammatory molecules. The TNF receptor family is itself membrane-bounded proteins (e.g. CD27, CD30 and CD40).

Table 24.7 : Cytokines: nomenclature and sub-families (based on Henderson *et al.*,1999).

Family	Examples	Biological activities
Interleukins	IL-1 to IL-18	Manly lymphoid-lineage growth factors
Cytotoxic cytokines	TNFα, TNFβ, CD40L	Pro-inflammatory molecules with cytotoxic/apoptotic potential
Interferons	INF-α, INF-β, INF-γ	Antiviral action with immunological effect
Colony-stimulating factors	IL-3, M-CSF, G-CSF	Myeloid growth and differentiation factor
Growth factors	EGF, TGF, PDGF	Proliferation of cell types e.g. epithelial and mesenchymal cells
Chemokines	IL-8, MCP	Chemotactic proteins for leukocytes

The **interferons** (IFNs) are such cytokines which were discovered first. These are involved in inhibiting the growth and spread of viruses. They are of three types: INF-α, INF-β, and INF-γ. Interferons also act against protozoa, rickettsia and mycobacteria. The colony-stimulating factors (CSFs) control the growth and differentiation of neutrophils, monocytes and cell populations derived from monocytes in the bone marrow. The monocytes/macrophages are the phagocytic cells which engulf and kill bacteria. Hence, they are also called as antigen-presenting cells and stimulate T and B lymphocytes.

Growth factors include families of proteins such as fibroblast growth factor (FGF) family, platelet-derived growth factors (PDGF), transforming growth factor-β (TGFβ). The FGF cytokines act on mesenchymal cells and epithelial cells also.

The peptide chemotactic factors are called **chemokines** which is a large sub-group of

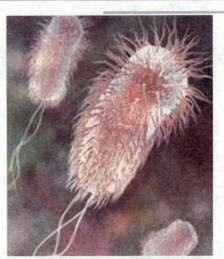

A micrograph of *E. coli* (used for the production of hypothetical mammalian protein product protein X) which shows localized deposits of interleukin-2 (bubble-like areas) within the cells (x 2000).

cytokines. Chemokines have molecular mass of 8-10 kDalton, with 20-50% sequence homology at protein level and cysteine as conserved residues which form disulphide bonds within the molecules. On the basis of chromosomal location of genes and protein structure, chemokines are divided into two families: α-chemokine and β-chemokine families. A third family of chemokines discovered in 1994 currently has one member called lympholactin which is a strong attractant of T cells.

(b) **Receptors of cytokines:** Cytokine receptors have high affinity for their ligand. The number of individual receptor present on target cell is low. On the basis of sequence homology and structural motifs cytokine receptors are grouped into a small number of families. At present there are nine receptors for CC chemokines (CCR), five receptor for CXC chemokines and CXCR1, one receptor for fractalkine. The cytokine receptors are shed from cell via proteolytic cleavage. Cell surface metalloproteinases (sheddases) help the release of cytokine receptors. The released receptors bind the soluble cytokines and inhibit their activity or stimulate the cytokine–receptor lacking cells.

(c) **Biological action of cytokines:** Cytokines play a role in physiological development. They are found at all developmental stages in mammals. On the other hand, cytokine receptors present on cell membrane also play a physiological role. They act as portals for vital entry into cells. For example HIV enters through binding to cytokine receptors. Similarly, herpes simplex virus enters through binding the TNF receptor family.

After binding receptors induce selective intracellular signalling resulting in switching on or switching off of particular genes and production of cyclooxygenase II, and nitric oxide (NO) is synthesised after induction of nitric oxide synthetase.

Aspirin and ibuprofen are the non-steroid anti-inflammatory drugs which block cyclooxygenase activity. These drugs reduce pain and fever as the prostaglandins and prostacyclin lower threshold in pain nerve resulting in a relief of pain and fever. Various molecules are produced after binding cytokines to cytokine-receptor which produce pathology [prostaglandins, NO, tissue plasminogen activator (tPA) and plasminogen activator inhibitor and collaginases]. Tissue damage is directly induced by collagenase and tPA.

Besides, cytokines also induce the synthesis of their own and other cytokines which result in a complex network of interactions. Cytokines can also modify the behaviour of cells in many ways. Various action of cytokines on cells are shown in Fig. 24.5.

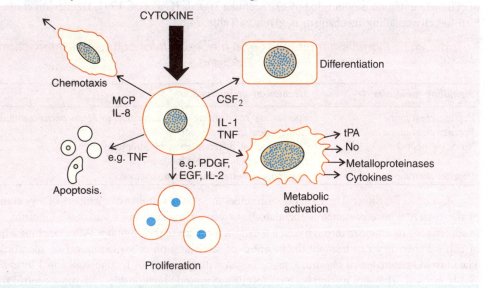

Fig. 24.5 : Different types of action of cytokines on cells.

2.   Prokaryotic Cell-to-Cell Signalling: Quorum Sensing and Bacterial Pheromones

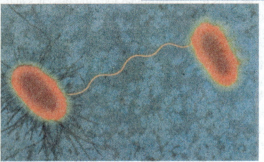

Bacterial conjugation : the transfer of DNA occurs through a special large, hollow sex pilus, shown here connecting a pair of *E. coli*. One bacterium (at top right) acts as a donor, transferring DNA to the recipient.

Until the 1980s, no attention was paid that bacteria could talk to one another. Thereafter, examples were put forth for cell-to-cell signalling in bacteria. Conjugation is one of the methods of DNA transfer between two bacteria. To establish conjugation, both the bacteria must establish cell-to-cell contact. *Enterococcus faecalis* is a Gram-positive mammalian pathogen. Its aggregation in controlled by the secretion of small peptide pheromones. Pheromones induce adhesin production; consequently bacteria form cell clumps which facilitate conjugation. Several pheromones have been isolated which are hepta- or octa-peptides found in low concentration (5×10^{-11}M).

Endospores of *Clostridium tetani* are regarded as resting forms of bacteria and a part of virulence mechanism. In contrast some bacteria such as a myxobacterium under adverse environmental conditions undergo complex morphological changes. *Polyangium vitellinum* forms cyst-like structure consisting of an outer covering of polysaccharide to resist from dehydration.

Myxococcus xanthus forms myxospores (fruiting body) and alternate with vegetative cells. This programme is triggered by starvation which causes morphological changes within 4 hours. A dense mound-shaped structure is formed when a cell density of bacteria has reached to about 10^5. After 20 hours of starvation the cells inside this mound differentiate into myxospores. Myxospores are heat- and starvation-resistant dormant cells. They germinate during favourable conditions and produce vegetative cells. Again myxospores are formed when conditions are unfavourable. This type of cell differentiation is controlled by extracellular signals (Hancock, 1997; Henderson *et al.*, 1999). Cell-to-cell signalling mechanism is given in Table 24.8.

Table 24.8 : Signalling molecules and mechanism of cell-to-cell interaction in bacteria.

Signalling molecules	Mechanism of interaction
Peptide pheromone	Sporulation and fruiting body formation in *Myxococcus xanthus*
Peptide pheromone	Conjugation in *Enterococcus faecalis*
Modified peptides	Morphological differentiation in Streptomycetes
	Antibiotic produced by *Streptomyces*
Peptide pheromone	Autoinducer behaviour in many bacteria

Quorum Sensing: The term quorum refers to 'a fixed number of members of any committee of the society whose presence is mandatory for proper transaction of business'. Quorum sensing in bacteria is a mechanism through which they take a census of their number. After reaching a quorum of cell number they can transact the business of switching on or switching off of specific genes. The current knowledge of quorum sensing began with the study of luminescence in *Vibrio fischeri* and *V. harveyi*. They are marine bacteria forming symbiotic relationship with monocentrid fish and with bobtail squids (e.g. *Euprymna scolopes*). The bobtail squid consists of very high concentration of *V. fischeri*. The light organ is supposed to be part of a counter illumination the details of which are not clear.

The newly hatched squids develop symbiotic association with only certain strains of *V. fischeri*. Within hours after hatching, light organ is colonised by *V. fischeri*. The light organ positively selects only certain strains of *V. fischeri* and negatively selects the others to exclude colonisation of other bacteria present in sea water. It is not known how this selection is made. One of the possible mechanisms may be the expression of specific adhesin for *V. fischeri* by epithelium of light organ. The epithelium is exposed to trypsin which directly triggers a specific morphogenetic response in the squid. This results in formation of the complex.

(a) **Mechanism of quorum sensing:** It is the feedback control system. Bacteria continuously produce a small amount of signal called *autoinducer*. Most of the Gram-positive bacteria produce autoinducer which are acylhomoserine lactones (AHLs). *Staphylococcus aureus* and other bacteria produce peptide autoinducers. *E. coli* and *S. typhimurium* produce a quorum sensing molecule of 1 kDalton. These extracellular inducers are diffused out.

Besides, bacteria also recognise the presence of autoinducer. The bacterial membrane protein does this function. It acts both as receptor of autoinducer and activator of gene transcription. *V. fischeri* produces luminescence. *V. fischeri*

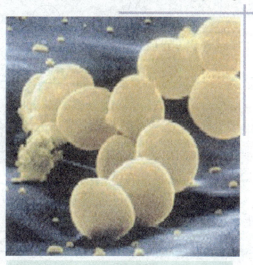

Staphylococcus aureus, a bacterium which produces peptide autoinducers.

system is the best studied quorum sensing system. Luminescence is associated with *lux* operon system which consists of two main regulatory genes *luxI* and *luxR* (Fig. 24.6) and other genes (*luxCDABEG*) which synthesise chemicals to produce light. *LuxI* encodes a protein which catalyses the synthesis of a wide range of AHLα. Autoinducer of *V. fischeri* is *N*-(3-oxo-hexanoyl)-L-homoserine lactone. *LuxR* encodes a protein which acts both as a receptor for AHL and as a transducer of the signal that activates the other genes of *lux* operon. The *luxCDABEG* genes are expressed after binding AHL to the luxR protein (Fig. 24.6). The *luxA* and *luxB* genes synthesise the α- and β- subunits of bacterial luciferase. The other genes encode polypeptides which facilitate the synthesis of the substrate and produces light.

(b) **Quorum Sensing as a virulence mechanism:** In addition to *V. fischeri*, there is a large number of Gram-negative bacteria which produce AHLs to quorum sense. These are medically important bacteria, for example *Pseudomonas aeruginosa, Proteus mirabilis, Serratia liquefaciens* and *Yersinia enterocolitica*. In these bacteria LuxI/LuxA homologues are involved in quorum sensing system. *Ps. aeruginosa* utilises two quorum systems, the *las* and *rhl*. The *las* operon expresses LasR protein which is similar to LuxR and acts as transcriptional activator in the presence of PAI of *Pseudomonas*. The LasI (the LuxI homologue) produces AHL. The autoinducer of *P. aeruginosa* at a threshold concentration swich on a group of virulence gene including *lasB, lasA, apv* and *toxA* (Hartman, 1998).

The *rhl* system is the second quorum sensing system which involves RhIR (the transcriptional activator protein along with the autoinducer (*N*-butyryl-*L*-homoserine lactone) synthesised by RhIR. This quorum sensing system results in production of extra virulence factor e.g. elastase which cleaves and inhibits the interleukin-2 (the key host defence cytokines). The *las* system is dominant which is activated before the *rhl* system. Many Gram-positive bacteria use oligopeptide as signalling molecules. For example, two different peptides are secreted by *Bacillus subtilis*. These are necessary

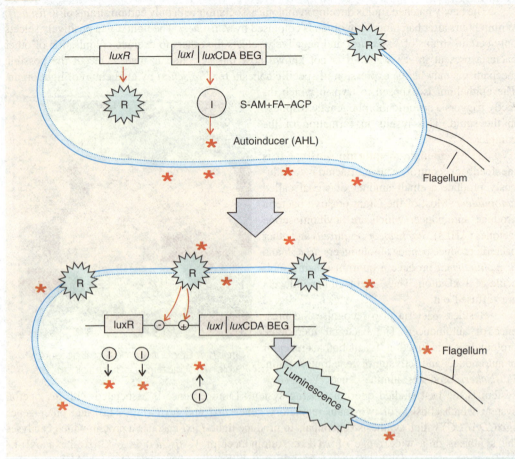

Fig. 24.6 : Operation of *luxI/luxR* system and luminescence in *Vibrio fischeri.*

for competence (ability for DNA uptake) and sporulation. In *Staphylococcus aureus*, a locus *agr* controls the expression of many virulence factors, namely exotoxins, capsular polysaccharide type 8 and V8 protease. An octapeptide quorum sensing autoinducer is encoded by the *agr* lucus which induces the *agr* locus. The quorum sensing autoinducer interacts with host defence system and inhibits the albeit at high concentration (Fig. 24.7).

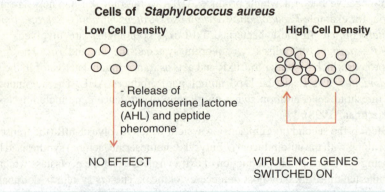

Fig. 24.7 : Action of AHLs on human immune defence system.

3. Intracellular Signalling (Signal Transduction)

Eukaryotic cells and bacteria release a large number of signals and establish communication. The method of action is binding the signals with the protein receptors present on surface of large cell and triggering a series of intracellular reactions called *intracellular signalling* or *signal transduction*. Besides, many signals (steroids and bacterial autoinducer) enter the cell, interact with signalling system and establish signal transduction. For establishing intracellular signalling, one must fully understand the operation of any cell from their origin to death.

The first signalling molecule (cyclic adenosine monophosphate or cAMP) was known during early 1960s. However, importance of intracellular signalling could be realised after the discovery of changes made by mutagenesis in signalling pathway which results in cellular transformation which is now called as *cancer*. One could understand their function by mutagenising their cellular function. It is now known that bacteria can change the eukaryotic cell signalling and invade the cells. The bacterial toxins can hijack the control of host cells. Similarly complexity of bacterial signalling is also known. Now an enormous amount of information is available on cell signalling and signal transduction pathway both in prokaryotes and eukaryotes.

(i) **The Signalling System:** Signalling system is very complex which may be compared to electronic circuits. You know that electronic system is such that can integrate, modulate and amplify inputs and generate output signals when switched on or switched off after getting suitable signals.

The signalling systems include few basic type of modules. There are four main processes, but the signalling system uses the one or more processes. The types of signalling modules used in intracellular signalling are: *(a)* receptor kinases (e.g. tyrosine kinase, serine kinase, histidine kinase), *(b)* receptor non-kinases

A tumor of lung cancer visible as a large pale mass. Smoking results in changes made by mutagenesis in signalling pathway which results in cancer.

(e.g. serpentine, cytokine, His-Asp phosphorelay), *(c)* protein kinase (intracellular enzymes e.g. cycline families, Asp kinase), *(d)* lipid modifying intracellular enzymes (e.g. p13K, p15K, PLC), *(e)* cyclic nucleotides (e.g. cAMP, cGMP), and *(f)* metal ions (e.g. Ca^{++}).

The four main processes are shown in Fig. 24.8. These are *(a)* protein phosphorylation by kinases, *(b)* small molecule and protein interaction, often involving phosphates, *(c)* protein-protein interaction, often mediated by common motifs (specific protein sequences) which frequently results in membrane recruitment when one component is tethered to a membrane, and (d) protein and DNA interaction that promotes gene expression or gene inhibition.

When many signals are interconnected in a series or parallel, the intracellular signalling becomes complex (Henderson *et al.*, 1999). General principles of these modules, the way of processing and onward transmission of signals in prokaryotes and eukaryotes are the main topics of discussion.

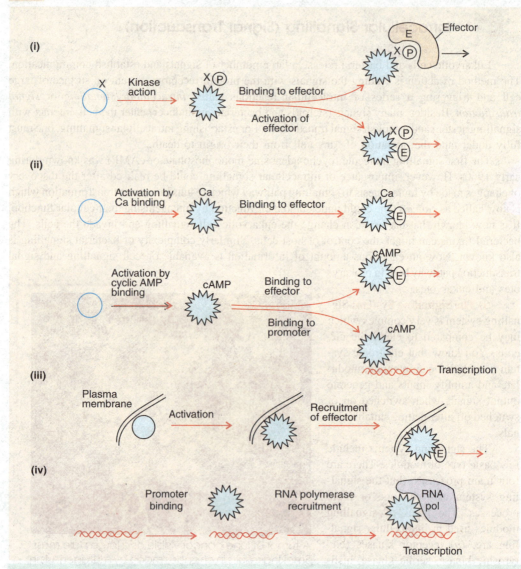

Fig. 24.8 : Signalling molecules that use different types of interactions. *(i)* protein phosphorylation, where X=Tyr, Ser, Thr, His or Asp, *(ii)* interaction between small molecules and proteins e.g. Ca++, or cAMP, *(iii)* interaction between protein and protein, and *(iv)* interaction between protein and DNA which regulates transcription.

(ii) The Basic Building Blocks used in Signalling

(a) **Protein phosphorylation:** Protein phosphorylation is closely linked to cellular signalling. It exists in all signalling modules. The terminal γ-phosphate is directly transferred from ARP (in some cases) to an acceptor protein by a protein kinase. The activity of the acceptor is modified for example mitogen-activated protein (MAP) kinases in eukaryotes and histidyl-aspartyl phosphorelay in bacteria. In some cases, indirect phosphorylation of protein also occurs (e.g. in G protein when binding of GTP activates their function, while GDP binding inactivates). There are *secondary messengers* which are used in intracellular signalling such as phosphorylated inositols or cyclic nucleotides (cAMP, cGMP).

Kinases are regulated by any of a number of mechanisms: threonine and/or tyrosine phosphorylation, legend occupancy resulting in autophosphorylation or interaction with small molecules (e.g. cAMP or Ca^{++}).

• *Histidine kinases:* These are found in bacteria, lower eukaryotes and plants as transmembrane protein. They are stimulated to undergo self-phosphorylation by ligand occupancy.

• *Protein phosphatases:* Proteins which remove phosphate groups from proteins are called protein phosphatases. Protein kinases add phosphate group to proteins and play a key role in activation of signals. Specific phosphatases e.g. dephosphorylate phosphotyrosine and phosphoserine/phosphothreonine play a key role in control of proliferation, differentiation and cell cycle. Phosphoproteins take part in signalling. They moderate the phosphorylation status by regulating the balance of phosphatases and kinases.

(b) **Nucleotide-binding proteins:** The three nucleotides (GTP, cGMP and cAMP) play a major role in the intracellular signalling. Structure of cAMP is given in Chapter 10, *Gene Expression and Gene Regulation.*

• *GTP-binding proteins:* There is a set of eukaryotic proteins (G proteins) that show GTPase activity. They bind to GTP and remove the terminal phosphate of GTP and produce GDP bound to G protein. This cycle operates similar to ATP and ADP cycles. When GDP dissociates from the G protein and GTP binds again, the cycle is repeated (Fig. 24. 9). G proteins are of two types: the *heterotrimeric G proteins* (the dominating proteins), and the *small G proteins* or membrane of Ras superfamily (the intermediate member of the signalling pathways).

The heterotrimeric G proteins consist of three different subunits- α, β and γ subunits. The α-subunit has GTP-binding domain; hence Gβ has a role in signal transduction. The βγ subunits transmit signals by non-covalent interaction with effector molecules. Activation of G proteins and dissociation of α-subunit from βγ subunits are given in Fig. 24.9. GTPase activity results in after binding the Gα subunit with GDP and subsequent reassociation with Gβγ and down regulation.

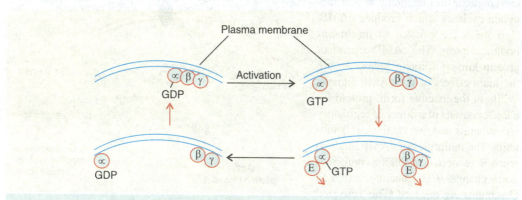

Fig. 24.9 : The function of membrane-bound heterotrimeric G proteins having α, β and γ subunits.

The small G proteins (Ras superfamily or p21 family) play a key role in many cellular functions such as proliferation and differentiation (Ras family), cytoskeletal organisation (Rho) and nuclear membrane transport (Ran). The activity of small G proteins is modulated after interaction with several classes of proteins (Fig. 24.10). GDP-dissociation inhibitors (GDI) inhibit the loss of bound GDP and keep the G proteins in an inactive form to attenuate signalling from the activated G proteins. GTPase activity is stimulated by GTPase-activating proteins (GAP). The removal of the bound GDP is helped by guanine nucleotide exchange factors (GEF) which enable the GTP to bind and activate G proteins. Some of these factors have shown to be proto-oncogenes.

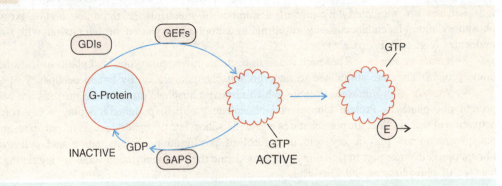

Fig. 24.10 : Functioning of small G proteins.

• *Cycline Nucleotide-binding proteins:* In 1950s, cAMP was identified as the first intracellular signalling molecules. It mediates hormone action and acts as the second messenger (i.e. molecules transmitting the primary signal that has been received at the cell membrane). The cAMP mediates the response to chemo-attractants. The adenylate cyclase and guanylate cyclase regulate the concentration of cAMP and cGMP, respectively.

The soluble bacterial adenylate cyclases produce cAMP which binds to cAMP receptor protein (CRP) and activate them. CRP is a transcription factor. The cAMP influences the expression of many of genes. Consequently bacteria become able to express metabolic enzymes which are required during growth. The cAMP also regulates the expression of the other genes which can cause pathogenesis.

In eukaryotes heterotrimeric G proteins regulate the membrane-bound ade-
nylate cyclases which produce cAMP. G proteins are coupled to transmem-
brane receptors. The cAMP-dependent protein kinases (protein kinase A) are the main effects of the cAMP signals. While in the inactive form, protein kinase A consists of a dimer of regulatory (A) subunits and two catalytic (C) sub-
units. The molecules of cAMP binds to reach R subunits and induce conforma-
tional changes. Consequently activated C subunits are released. This activated protein kinase A phosphorylates many substrates on serine or threonine (Fig. 24.11).

Both the cycles work in eukary-
otes by direct binding to proteins which form cation channels. Binding events result in opening of the channel.

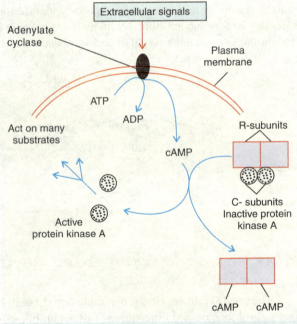

Fig. 24.11 : Production of cAMP and activation of protein kinase A.

The G-protein-linked cell surface receptor generates small intracellular mediators through cAMP pathways (Fig. 24.12).

(c) **Role of intracellular concentration of Ca++ in cell signalling:** Calcium is found in cytoplasm and maintained in a very low concentration (10-100 nM). But its concentration varies

with cell cycle, exogenous source or release from the stores. It gets complexed in membrane bound vesicles acting as stores. A highly specific protein **calmodulin** (CaM) binds to Ca^{++} and transmit the signal. Ca-binding to CaM brings about changes in conformation of CaM. Consequently CaM interacts with many effectors including CaM-modulated kinase. The most extensively studied CaM is the phosphatase calcineurin which is associated with several cellular activities such as NO synthesis, apoptosis, induction of T lymphocytes. In eukaryotic cells Ca^{++} acts as a second messenger. Fig. 24.12 shows the two major pathways by which G-protein-linked cell surface receptors generate small intracellular mediators.

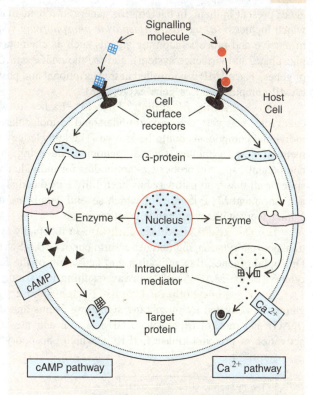

Fig. 24.12 : Generation of small intracellular mediators by G-protein-linked cell.

(d) **Role of phosphorylated lipids in cell signalling:** In eukaryotes lipids are involved in signalling process. Cellular phospholipases attack the lipid moieties of the membrane to produce different types of signalling molecules. For example, phosphatidylinositol lipids play a role in cellular stimulation. They have inositol as head, the six-membered carbon ring with a -OH group on each carbon. On the basis of phosphorylation status of inositol head group, several phosphatidylinositols are found in the cells. The activity of three enzymes triggers their signalling role. These are: phosphoinositide 5'-kinase (PI5K), phosphoinositide 3'-kinase (PI3K), and phospholipase C (PLC). Extracellular signals regulate all these enzymes.

(e) **Regulation of transcription:** Both types of cells are able to respond to any signal by changing their gene expression. In a signalling pathway the end point acts as signal. Regulators causing changes in expression of many genes in bacteria are called 'global regulators'. In prokaryotes post–transcriptional events regulate expression of many of the transcriptional factors for example cAMP-mediated CRP-DNA interactions. In prokaryotes, phosphorylation or protein-protein interactions regulate the control of transcriptional factors and also select the other factors to the promotors. Besides, some other factors also get translocated from cytoplasm to the nucleus and regulate transcription.

(f) **Role of cell membrane in signalling:** Cell membrane acts as boundary of the cell through which extracellular signal has to enter. In bacteria histidine kinases act as receptor and directs signals across the membrane. Besides, there are many signal molecules which are associated with cell membrane because the end effect is membrane-associated. The components can be well organised in three-dimensional way in cell membrane. The signalling components recruit the other molecules to the membrane where they interact with other factors. For example, GTP-bound Ras activates Raf kinase to recruit Raf to the membrane where the membrane-bound kinase activates it through phosphorylation.

(iii) **Prokaryotic Signalling Mechanisms:** Intracellular signalling is very complex like electronic circuit. Genome size of different bacteria varies and those organisms work according to

genes present in them. In bacteria the generic mechanism of regulation is called signalling systems which includes: *(a)* the *histidyl-aspartae phosphorelay systems* (the main module of bacteria used to receive and process incoming signals such as chemotaxis, response to osmolarity, oxygen and phosphate, and virulence system), and *(b)* the *cAMP and CRP* (involved in regulation of hundreds of genes. The cAMP is controlled at transcriptional and post-transcriptional levels. Binding of CRP-cAMP complex induces gene expression).

(iv) **Eukayotic Signalling Pathway:** Earlier it was thought that signalling process in eukaryotes was very complex to understand in molecular terms. Fragmented understanding about individual components could be known. The knowledge of signalling expanded with the development of new techniques such as genome sequencing, increasing number of reagents (isolated components, specific probes e.g. antibodies for individual components and selective inhibitors). In spite of all these, no pathway has been fully elucidated. The best characterised pathway is the Ras activation and MAP kinases of which several details are unclarified. They are interconnected and cannot work without reference to others.

(a) **The Ras/MAP kinase pathway:** In this pathway the ligand occupancy leads to receptor transphosphorylation, membrane recruitment of GrB2, Sos (son-of-sevenless) where sevenless is the Drosophila (gene), Ras activation and activation of the Ras kinase, and stimulation by phosphorylation of the MAP kinase pathway resulting in various sequelae in the cells.

(b) **The phospholipase C/inositol triphosphate pathway:** The phospholipase C, beta or gamma is activated by membrane signalling events and cleaves PIP2 to produce diacylyglycerol (DAG) and inositol triphosphate (IP_3). These activates the release of Ca^{++} ions and results in activation of proteion kinase C (PKC), which phosphorylates many additional protein substrates.

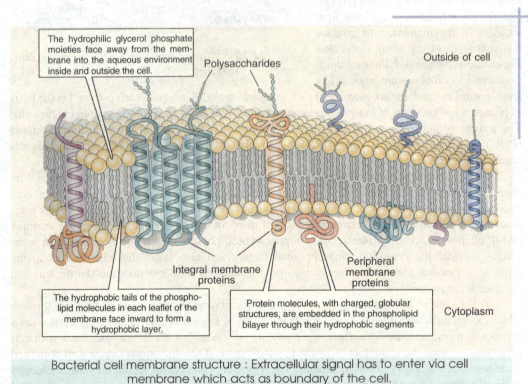

The hydrophilic glycerol phosphate moieties face away from the membrane into the aqueous environment inside and outside the cell.

Polysaccharides

Outside of cell

Integral membrane proteins

Peripheral membrane proteins

The hydrophobic tails of the phospholipid molecules in each leaflet of the membrane face inward to form a hydrophobic layer.

Protein molecules, with charged, globular structures, are embedded in the phospholipid bilayer through their hydrophobic segments

Cytoplasm

Bacterial cell membrane structure : Extracellular signal has to enter via cell membrane which acts as boundary of the cell.

(c) **The adenylate cyclase, cAMP and protein kinase A pathway:** Adenylate cyclase is activated at the membrane by interaction with the activated heterotrimeric G protein G_s. The cAMP is generated and binds to and activates protein kinase A (PKA), which phosphorylates many substrates.

(d) **Integrins, the Rho family and organisation of cytoskeletal:** The integrins are the signalling molecules that interact with the extracellular matrix on the outside of the cell and various proteins-linked to actin on the cells interior. The proteins involved include α-actin, lalin, tensin, vinculin and pavilin. A local adhesion is formed upon activation that includes focal adhesion kinase (FAK). The Src kinase is recruited and several proteins in the complex are activated by phosphorylation by Src and FAC. These signals lead to the Ras/Raf, Rho signalling pathway and to cytoskeletal rearrangement.

In eukaryotes, the central role of signalling pathway of a cell is to define its phenotype and function. The increasing novel knowledge about the components of signalling pathways and the types of genes which they interact are already being applied in new strategy to combat the cancer. For example, genetically engineered viruses are attempted to grow in such cells that lack functional p53 and kill these cells (Henderson *et al.,* 1999).

There are about 2000-5000 signal transduction proteins in mammalian cells. Bacteria have capacity to utilise eukaryotic signalling pathway during the process of infection. These findings make a line between the signalling pathways involved in infection and the other responsible for the pathology in diseases such as cancer and inflammation.

C. Cell-cell interactions and Infection

1. Bacterial Adhesion to Host Cells

Before entering inside, bacteria adhere to host cells and secrete product(s) or structural products complementary to host. Hence, bacteria are found adhered to host's epithelial cells due to direct adhesion to host cells or binding to secretory products that coat host cells or bacteria. For example, teeth are rapidly colonised by bacteria. Besides, bacteria also adher to phagocyte cells of the host and trigger immune system and may or may not be phagocytosed.

A wide variety of surfaces are available for adherence of microbial cells such as polymers of extracellular matrix (e.g. collagen, proteoglycans), bones, endothelial cells. Bacteria possess several surface molecules and structures which facilitate to adhere these surfaces. Moreover, a particular bacterium is able to adhere to only specific surface. It means that there exists tissue-tropism of bacteria (Table 24.9). Some of the bacteria (e.g. *Neisseria meningitidis*

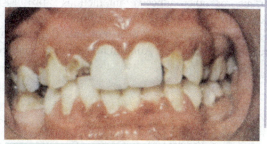

Tooth decay caused by rapid colonisation by bacteria.

and *Salmonella* spp.) encounter many types of surfaces to cause infection. But at least two facts of bacterial adhesion are very important, the physico-chemical forces facilitating invasion and the specificity of process to guide specific surfaces.

Table 24.9 : Tissue tropism of bacteria (based on Henderson *et al.,* 1999).

Bacteria	Epithelium of host tissue responsible for tropism
Bordetella pertusis	Epithelium of respiratory tract
Compylobacter jejuni	Epithelium of intestine
Helicobacter pylori	Gastric mucosa
Mycoplasma pneumoniae	Epithelium of respiratory tract
Neisseria meningitidis	Nasopharyngeal epithelium

Bacteria	Epithelium of host tissue responsible for tropism
N. gonorrhoeae	Epithelium of urethra
Salmonella typhimurium	Epithelium of intestine
Streptococcus pyogenes	Epithelium of pharynx
Vibrio cholerae	Epithelium of intestine

Bacteria and host cells interact and affect the activity of one another by the secretion of toxins, low molecular weight metabolites, hormones, enzymes and antibacterial peptides. The behaviour of eukaryotic cells is affected by LPS, peptidoglycan and membrane protein on outer wall.

2.   Basic Principles of Microbial Adhesion

(i) **Forces Affecting Adhesion:**
There are different types of forces which operate between bacteria and host cell surface before adhesion of bacterial cells (Fig. 24.13). Van der Waals and electrostatic forces apply when microbial cells are at a distance of tens of nanometers (A and B). Bacterial cell and host cell surface are mutually attracted due to Van der Waals forces. The electrostatic forces result in repulsion of these two objects. Hydrophobic interactions (C) result in adhesion of bacterial and host cells and the former is brought closer to host cell surface for other adhesive interactions such as hydrogen bonding, cation bridg-

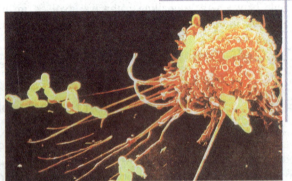

The intestinal bacteria *E. coli.* that adhere to white blood cells (phagocytic cells) of the host thus triggering immune system.

ing and specific bonding of a molecule (ligand) on the bacterial surface to a receptor molecule present on host surface. Adhesin molecules present on bacterial surface are responsible for adhesion.

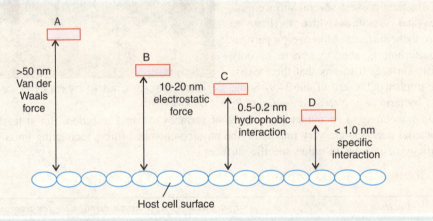

Fig. 24.13 : Various forces operating between bacteria and host surface affecting adhesion (based on Henderson *et al.*, 1999).

(ii) **Role of Bacterial Structure in Adhesion:** Bacteria possess several structures which help in adhesion of cells for example fimbriae (or pilli), fibrils, flagella, capsule and S layer. All these structures consist of adhesins. Capsule components of certain bacteria (e.g. *Streptococcus, Staphylococcus, Klebsiella, Neisseria, Haemophilus*) mediate adhesion to host cell surface. S layer consists of glycoprotein and self-assembling units (external to cell wall) which also help in adhesion. Fimbriae are present on cell surface and cause bacterial adhesion.

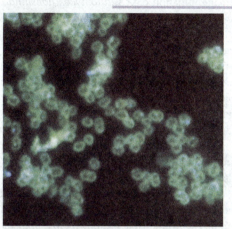

The epithelial surface secretes antimicrobial compounds (e.g. lysozyme and antibacterial peptides). The epithelium of respiratory tract is coated with mucin in which bacteria are trapped, brought to back of pharynx by ciliary action. Also, several antimicrobial compounds are found in mucin (e.g. lysozyme, lactoferrin, secretory IgA, superoxide radicals and antibacterial peptides), The urinary tract and oral cavity are always flushed by secretions of respective tissues. Even then epithelium is colonised by bacteria.

The capsule component of bacterium *Neisseria* mediates adhesion to host cell surface.

Using broad-spectrum antibiotics, normal microflora is disturbed and undesirable microorganisms may be present such as *Candida albicans, Clostridium difficile* and pseudomonads which can infect the organs. It seems that normal microflora of human exerts protective effect. How this mutualistic association between two types of bacteria got evolved ? How does host cells encourage such selective adherence ? Little is known about adherence of normal microflora of healthy tissues. Most of the information which we know are on microbial adhesion between pathogenic microorganisms and host cells. Several other facets to bacterial adhesion are given in Fig. 24.14.

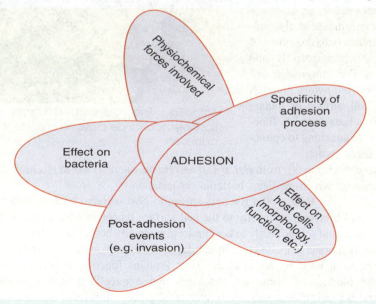

Fig. 24.14 : Facets of bacterial adhesion to host cells (based on Henderson *et al.*, 1999).

Adhesin is located at the tip or along the whole length of fimbriae. Fimbriae are widely distributed among the Gram-negative bacteria such as *Bordetella, Salmonella, Neisseria, Pseudomonas, Yersinia,* etc. Fimbriae have been classified into the five types: Type 1 (rigid fimbriae that exhibit mannose-sensitive haemagglutination e.g. *E. coli*), Type 2 (similar to type 1 but not induce haemagglutination e.g. *Actinomyces naeslundii*), Type 3 (flexible and mannose-resistant fimbriae (they are common among the enterobacteriaceae e.g. *Klebsiella pneumoniae*), Type 4 (they consist of *N*-methyl-phenylalanine in the amino terminus region of the major subunits e.g. *Pseudomonas aeruginosa*), and Type 5 (thinner than type 1, mannose-sensitive and a few in number).

(iii) **Bacterial Adhesins:** Several different types of molecules present on bacterial cell surface act as adhesins and facilitate the attaching bacteria to host cell surface. One of the most extensively explored adhesins is the lectin (glycoproteins). Lectins are present at the end of pili, capsule of Gram-negative bacteria, etc. Examples of bacteria comprising of lectins are: *E. coli* (*N*-acetyl-*D*-galactosamine), *Kleb. pneumoniae* (*N*-acetylmuraminic acid and *N*-acetyl-*D*-glucosamine), *Staph. saprophyticus* (*N*-acetyllactosamine). In Gram-negative bacteria, lipoteichoic acid (LTA) acts as an important adhesin. A glycoprotein (fibronectin) produced by many epithelial cells and other host cells act as receptor for LTA.

S. aureus produces a surface protein (210 KDa) which acts as adhesin and mediates the adhesion to fibronectin. The bacterium also attaches to the other host proteins e.g. fibrinogen, laminin. A proline-rich protein of *Mycoplasma* sp. also acts as adhesin.

Carbohydrates present on bacterial cell surfaces act as adhesins in certain bacteria. For example, *P. aeruginosa* secretes an exopolysaccharide (alginate) that acts as adhesin for attachment to tracheal cells and mucin, and binds to both.

Lipopolysaccharide (LPS) of Gram-negative bacteria play an important role in adhesion. For example, LPS of *Comp. jujeni, E. coli, Ps. aeruginosa, Sal. typhi, Shig. flexneri,* etc. mediate the bacterium to attach to host's epithelial cells.

Bacterial enzymes (e.g. glyceraldehyde 3-phosphate-dehydrogenase of *Strep. pyogenes,* gingipain R and gingipain K of *Por. gingivalis,* cell surface urease of *Hel. pylori,* glucosyl transferase of cell surface of mutant streptococci have been found to function as adhesins in attaching to epithelial cells of various tissues.

Shigella : The rod-shaped bacilli where lipopolysaccharide plays an important role in adhesion.

Molecular chaperonin 60 (from *Hel. pylori* and *Haem. ducreyi*) which is a heat–shock or stress protein are produced which mediates bacterial colonisation.

(iv) **Nature of the Host Cells:** Bacteria adhere to host surfaces in three different ways: *(a)* directly to the lipid bilayer, *(b)* directly to the cell surface receptors whose normal function is to bind host molecules, and *(c)* indirectly to host molecules already bound to the host cell surface (Fig. 24.15). The cell membrane is made up of lipid bilayer where proteins are embedded. Lipid structure is such that are recognised by bacterial adhesins. Proteins function in transport of molecules, recognition and binding of hormones, cytokines and extracellular matrix molecules, signal transduction and cell-cell interactions, carbohydrate of glycoproteins and amino acids of proteins act as receptors for bacterial adhesins.

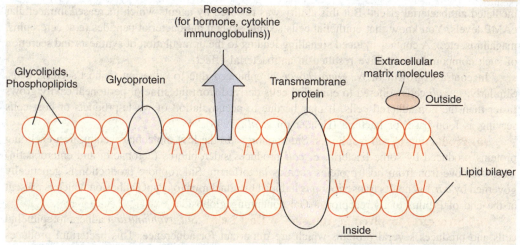

Fig. 24.15 : Host cell surface acting as receptors for bacteria adhesion.

Several receptor molecules are found on mammalian cell surfaces such as integrins, cadherins, selectins, serpentine receptors, cytokine receptors, intracellular adhesin molecules, etc.

The extracellular matrix (ECM) of host tissue consists of complex mixture of polymers such as fibronectin, fibrogen, collagen, proteoglycans and glycosaminoglycans. The ECM affects many cellular activities such as migration, proliferation and differentiation. Bacteria adhere to the ECM and arrest the activities of the host cells.

3. Effect of Adhesion on Bacteria

Before coming bacteria to their host surfaces, they reside in several environments which may not be congenial for their growth. The various environments may be soil, water, air clothing, food materials, vegetation, etc. Reaching the new environment of host, bacteria adapt through upregulation and suppression of several gene products. At least one organism uses the adhesion process to trigger the expression of gene products. These help to survive in new environment. Bacterial attachment gives a signal to initiate synthesis of structure or signalling molecules which help to infect host cell.

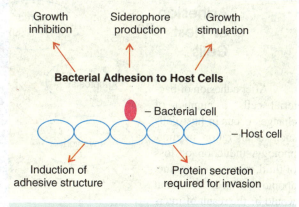

Fig. 24.16 : Effect of bacterial adhesion on host cell.

Besides, adhesion also results in phenotypic changes in few bacteria. Some of the effects of adhesion on bacteria are briefly described. The possible effects of interactions on bacteria are given in Fig. 24.16.

(i) **Effect on Bacterial Growth:** Growth of uropathogenic strain of *E. coli* is inhibited after adhesion to uroepithelial cells of human. This effect was adhesion-dependent. It may be demonstrated by separating the uropathogenic cells from bacteria by semi-permeable membrane. Ca^{++} and cAMP are involved in this process. Inhibition of Ca^{++} flux abolished the contact-mediated growth inhibition. Uro-epithelial cells of UTI (urinary tract infection) patients did not produce adhesion-

mediated antibacterial effect. But this ability was restored by agents which increased intracellular cAMP levels. You know that epithelial cells secrete several antibacterial peptides (e.g. cercropins, magainins, etc.). A contact-induced signalling leading to the upregulation of synthesis and secretion of such compounds may have resulted in antibacterial effect.

In contrast, growth of *N. gonorrhoeae* is enhanced due to adhesion to host's HeLa cells. Similarly, *E. coli* cells adhered to epithelial cells derived from intestine or peritoneal cavity grows faster than the non-adhered cells. It may be due to accumulation of waste-products of host cells serving as food base for bacteria (Henderson *et al.*, 1999).

(ii) **Production of Siderophores:** Siderophores are the low molecular weight, iron-chelating proteins produced by some bacteria. *E. coli* produces siderophores (aerobactin and enterobactin) which remove iron from host proteins such as lactofferrin. Siderophore production is genetically governed by *barA* gene. Its transcription is induced by attachment of PapG adhesion which is present at the end of P-pilus to its receptor, a Gal containing globoside.

(iii) **Structures Involved in Adhesion of Host Cells:** *Sal. typhimurium* adhers to epithelial cells and produces several proteins which are important for adherence. This bacterium produces surface appendages called **invasomes** within 15 minutes of adherence. Invasomes differ from flagella or pili being three fold thicker than the former. Invasomes mediate the internalisation of the organism by the epithelial cells. A gene present on *inv* locus governs the entry of *Sal. typhimurium* into non-phagocytic cells.

Enteropathogenic strains of *E. coli* attaching to mucosal surface form discrete micro-colonies which is known as 'localised-adherence'. *In vitro* investigation has shown that the epithelial cells of bundles of filament (i.e. bundle forming pili) induce the location-adherence.

After adhesion of *C. jujeni*, epithelial cells are stimulated. Within 60 minutes of adherence, about 14 proteins are synthesised by the bacteria.

4. Effect of Adhesion on Host Cells

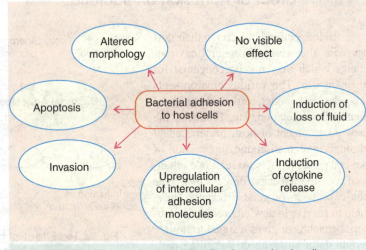

After adhesion of bacterial cells, a number of changes occur in host cells such as alteration in morphology, induction of loss of fluid, release of cytokine, apoptosis (Fig. 24.17). What would be the result of interaction ? This depends on types of host cells. Some of

Fig. 24.17 : Effect of bacterial adhesion on host cells.

the effects of bacterial invasion on host cells are discussed below:

(i) **Effect on Epithelial Cells:** Bacteria are able to colonise all types of epithelial cells of a healthy person because the microbes contain molecules (e.g. LPS, peptidoglycan, lipoteichoic acid) which harm the mammalian cells. However, some adherent microorganisms pose minimal effects. Examples of some of the bacteria are given which cause effect on epithelial cells (Table 24.10).

Table 24.10 : Effect of bacterial adhesion on epithelial cells.

Bacterial species	Effect on epithelial cells
Members of normal microflora (*Bordetella pertusis, V. cholerae*)	No visible effect
Enteropathogenic bacteria (*E. coli, H. pylori, S. pyogenes*)	Altered morphology
Oral streptococci, *E. coli, H. pylori*	Induction of cytokine release
Uropathogenic *E. coli*	Expression of intercellular adhesion molecules
Species of *Neisseria, Haemophilus, Shigella, Salmonella, Yersinia*	Invasion

(ii) **Adhesion to Fibroblast:** Fibroblasts produce materials of extracellular matrix and maintain integrity of connective tissue. Besides, they can also secrete cytokines and other inflammatory materials. Hence, they are important in host defence and other maintenance of cytokine network. A profound immunological effect and structural consequences can be obtained after interfering with their function. So far less attention has been paid on interaction between bacteria and fibroblast. Attachment of *Treponema denticola* to human gingival fibroblast induces several changes in the host cells: *(a)* retraction of pseudopods followed by rounding of cells and the formation of membrane blebs, *(b)* rearrangement of filamentous actin network into a perinuclear array, *(c)* detachment of cells from their substratum due to degradation of fibronectin by surface-associated proteases, and *(d)* apoptosis (cell death) (Ofek and Doyle, 1994).

(iii) **Adhesion to Vascular Endothelial Cells:** A continuous monolayer lining of the blood vessel walls is formed by vascular endothelial cells. These act as barrier between the blood and the vessel walls and regulate the tone of blood vessel, permeability and coagulation of blood, reactivity of leukocytes and platelets, angiogenesis and the source of vascular mediators (e.g. nitric oxide, prostacyclin, endothelin).

(iv) **Adhesion to Phagocytes:** It is important to know the mechanism of adhesion to phagocytic cells because bacteria are disposed off through this mechanism. Some bacteria adhere directly (involving the interaction between adhesins and acceptors on phagocytes) or indirectly by interacting with host components (e.g. complement or immunoglobulins) which then bind to receptors of phagocytes. Uptake of bacterium into a phagosome is induced by adhesins. Then the phagosome fuses with lysosome resulting in death of the bacterium. Moreover, there are some bacteria which survive within phagocytes, for example *M. tuberculosis, B. purtusis, Y. enterocolitica,* etc. Adhesion by some of the pathogens (*Legionella pneumophila* and *S. typhimurium*) causes the release of cytokines by the host cells. Increased expression of cytokines and chemokines occur after attachment of *L. pneumophila* or *S. typhimurium* to murine macrophages.

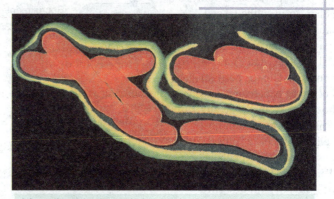

Mycobacterium tuberculosis : survives within phagocytes.

The other interesting result of bacterial adhesion to phagocytes is the **apoptosis** (programmed cell death). Apoptosis occurs when the enteric pathogens (e.g. *Y. enterocolitica*) bind to macrophages. It invades mucosa, resists phagocytosis and maintains extracellular life. It does not induce macrophages but induces death of phagocytes which show all features of apoptosis such as cytoplasm shrinkage, nuclear chromatin condensation and DNA fragmentation. Cell death occurs due to secretion of one or more Yop proteins. The bacterium involves this strategy for its survival (Ofek and Doyle, 1994).

5.   Bacterial Invasion of Host Cells

Invasion of host cells by bacteria results in several diseases as mentioned earlier. How does this occur remains questionable ? During the last decade much efforts have been made to unravel the mechanism that bacteria use for infection of host cells. Bacteria have evolved several invasive mechanisms. Most of them involve the manipulation of normal host cell cytoskeletal components such as actin and tubulin resulting in the investigation of the host cell membrane to enclose the bacterium within the vacuole. This occurs by interfering the inhibition of signal transduction or both.

In addition, *Rickettsia prowazekii* digests the part of cytoplasmic membrane. The invasive bacteria may remain viable inside the vacuole (*M. tuberculosis*), escape from vacuole and colonise the cytoplasm (*Listeria monocytogenes*) or escape from the vacuole and the cell and spread systematically (*Yersinia enterocalitica*) (Henderson *et al.*, 1999). In this section the adhesive processes leading to invasion of the host cell have been described.

(i) **Mechanism of Invasion:** Bacterial invasion of host cells is broadly classified into the three groups on the basis of involvement of microfilaments or microtubules of host cells or make entry into the cells. The type of host cells to be invaded also govern the invasion process. The ways through which bacteria invade epithelial cells, endothelial cells and macrophages are described below:

(a) **Invasion of epithelial cells:** The outer integument of our body is constituted by the epithelial cells. A varying population of microbes contributing the normal microflora colonises the epithelial interface. Thus epithelium acts as the first physical barrier for commensal microorganisms. Many bacteria enter epithelial cells of the host by inducing the rearrangement of microfilament of the cytoskeletin. Invasion of host cells by several bacteria such as *L. monocytogenes*, *Salmonella*, *Shigella*, and *Yersinia* have been most extensively studied.

Y. enterocolitica adhers to an epithelial cell. Adhesion is mediated by many adhesin between the Ail protein and Yad A protein. Close contact between bacterium and host cell is made at any point on the bacterium-host cell interface- a process described as zippering. This induces the uptake of the organism into an endocytic vacuole.

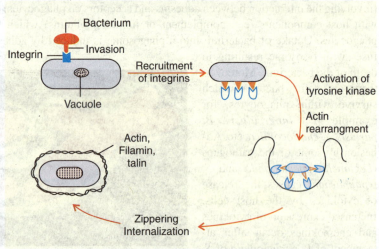

Fig. 24.18 : Different stages of invasion of epithelial cells by *Yersinia* spp.

The bacteria sink into the membrane of the epithelial cell. The host cells show a normal appearance within a few minutes of entry.

Binding of invasion of bacterium to its integrin receptor (on host cell surface) along the zippering induces the uptake of bacterium (Fig. 24.18). Clustering of integrins induces tyrosine kinase activity which is required for invasion. Cytochalasin inhibits endocytosis emplying the involvement of actin microfilaments. During early stages of internalization, clathrin lattices are soon formed beneath the bacteria. The polymerized actin and other proteins (filamin and talin) surround the vacuole. The internalized bacteria survive inside the vacuole but not reproduce.

Unlike *Yersinia*, *Salmonella* spp. invade the epithelial cells after adhesion to microvili. Within 1 minute of contact, microvilii form pseudopods extending from cell surface of host epithelium, engulf bac-

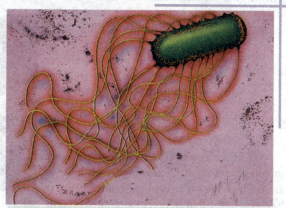

The bacterium *Salmonella* is a constitutent of the microflora which colonise the epithelial interface.

teria and internalise within a vacuole (Fig. 24.19). Mannose-specific type 1 fimbriae mediates the adhesion of *Salmonella* spp. to epithelial cells of intestine. A little information is available on the initial host signal transduction events after bacterial invasion. After invasion, intracellular level of Ca^{++} increases which polymerises actin synthesis and inhibits bacterial invasion. Fimbriae-mediated adhesion of *Salmonella* takes place.

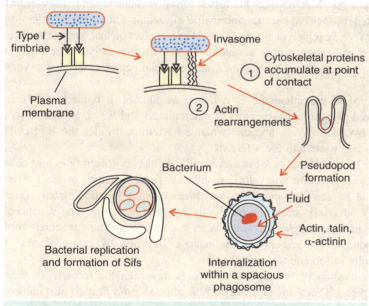

Fig. 24.19 : Different stages of invasion of epithelial cells by *Salmonella* spp.

Besides, cytoskeletal proteins (actin, talin, tubulin, tropomyosin and ezrin) accumulate in host cell at the site of bacterial adhesion. Several proteins of host cell membrane form aggregates in the vicinity of bacterial attachment. Consequently, bacterium is taken up by macropinocytosis (a process which involves intake of large quantity of extracellular fluid). Thus bacterium resides on a large fluid filled vacuole called spaceous phagosome surrounded by polymerised actin, talin and actinin. About 4-6 hours of invasion bacteria proliferate. It is accompanied by the formation of lysosomal fibrillar structures (*Salmonella*-induced filaments, Sifs) attached to phagosomes.

(b) **Invasion of epithelial cells:** There are several pathogenic bacteria which enter at a site of host cell and spread throughout the body, for example *H. influenzae, N. gonorrhoea, S. dysenteriae, S. pneumoniae, S. typhi,* etc. They enter the blood stream and cross the cellular barrier i.e. the endothelium. Invasion takes place by one of the four main routes:

- invasion followed by intracellular persistence without multiplication (e.g. *S. aureus, P. aeruginosa*),
- invasion followed by intracellular replication (*Rickettsia rickettsii*),
- traversed without cell disruption (spirochaetes), and
- invasion within phagocytes (*Listeria monocytogenes*).

E. coli causes diarrhoeae, urinary tract infection and neonatal meningitis. A little is known how does it pass blood-brain barrier. It has been found that an outer membrane protein (Omp A) of the bacterium plays a central role in invading the host cells. Bacterial Omp A

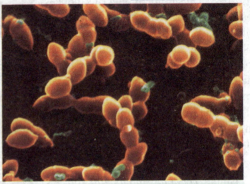

Streptococcus pneumoniae : A pathogenic bacterium involved in the invasion of epithelial cells.

protein mediates binding of bacteria to the receptors (*N*-acetylglucosamine, β1-4-*N*-acetylglucosamine epitope) present on BMEC (brain microvascular endothelial cells). The polymers of β1-4-linked *N*-acetylglucosamine prevent the entry of *E. coli* into the cerebrospinal fluid of neonatal rats. Hence, meningitis caused by it can be controlled by using receptor analogues.

(c) **Invasion of macrophages:** Macrophages are a part of our immune system. They engulf antibody or complement-coated (opsonised) bacteria, internalise in vacuole and kill them. Adhesion is mediated by interaction between Fc region of antibodies and Fcr receptors on macrophage surface. Binding causes internalisation of bacteria within a vauole. Then it fuses with lysosome to form phagosome inside which bacterium is killed by enzymes, antimicrobial peptides, reactive oxygen species and low pH.

Fimbrial protein FimH of uropathogenic *E. coli* acts as adhesin. It binds to CD48 (a glycosylphatidyl inositol-linked glycoprotein) receptors of macrophages and *E. coli* is internalised.

(ii) **Consequences of Invasion:** What happens when a bacterium invades the host cell? Certainly both the host cells and bacterium are affected.

(a) **Effect of host cells:** It is a difficult task to describe all the changes occurring in host cells due to bacterial invasion. Bacteria affect the host cells in many ways finally resulting in death. After invasion several cells respond by secreting cytokines which activate the immune system. Over production of cytokines may adversely affect the host cells e.g. *S. dysenteriae*. The diarrhoeal response to infection is mediated by postglandin which is an important regulator for secretion of gastro-intestinal fluid by inducing Cl⁻ secretion from mucosa. Infection by enterobacteria of intestinal epithelial cells results in secretion of postglandins.

Within 15 minutes on invasion of macrophages by *S. typhimurium* the former lost their phagocytic ability, induces apoptosis resulting in death of 50% cells. *Shigella flexneri* also induces induction of apoptosis in macrophages.

(b) **Effect on bacteria:** The bacteria have several options after invading the host cells. They may live within the vacuole of host cell (e.g. *M. tuberculosis, S. typhimurium, Brucella* spp., *Burkhulderia cepacia*) within the cytoplasm (e.g. *Shigella* spp., *Rickettsia* spp. *Listeria monocytogenes*) or may exit the cell and live extracellular life (e.g. *Yersinia* spp.). Bacteria living extracellularly are transported to another sites in the host.

Irrespective of above option, the bacteria adapt the new environmental conditions by expressing gene products that help their survival in new habitats. The bacterial regulatory system responds to changes of environmental factors e.g. pH, osmolarity and concentration of O_2, CO_2, micro- and macro- nutrients and antibacterial substances.

For example, *Salmonella* spp. after ingestion invades epithelial cells of mucosa. Its invasion depends on production of secreted invasion proteins (Sips) exported by Type III secretion system which is activated after contact with cell membrane of host cells. Genes located on *Salmonella* pathogenicity island (SPI-2) encode the proteins. After internalisation, it remains within a vacuole where it replicates until its exit.

Within the vacuole, expression of iron- and magnesium-regulated genes (i.e. iro A and mgt B, respectively is increased). Low pH of vacuole upregulates the expression of lysine decarboxylase gene. It suggests that O_2 and lysine are present in the vacuole at pH 6.0.

After internalisation, *Salmonella* grows and survives within the macrophage by expressing a large number of genes. The environmental conditions within the macrophage (i.e. limited nutrients, low pH and high osmolarity) force the bacteria to produce alternative sigma factor for RNA polymerase. The increased levels of sigma factor render the bacteria more resistant to adverse conditions and enable it to respond within the macrophage.

Development of molecular and genetical techniques has helped the investigation of ways in which bacteria respond within their host. These techniques include isolation of bacterial mRNA, synthesis of bacterial cDNA and probes, signature-tagged mutagenesis, differential fluorescence induction, *in vivo* expression technology, etc.

***(iii)* Bacterial Survival and Growth After Invasion:** Once the bacterium has invaded the epithelial cell, it may proliferate within the cell and come out of cell or infect deeper tissues and spread the outer sites. Thus the bacterium leads two types of life style: extracellular and intracellular life styles.

If the bacteria remains extracellular, they face secretions of blood or tissues such as complements, antibodies, antimicrobial molecules discharges by phagocytes and phagocytic cells. Bacteria have evolved various ways to deal with these antimicrobial substances and phagocytic cells through production of toxins (*see* preceding section), capsule (*S. aureus, K. pneumoniae, E. coli, S. pyogenes*), etc.

There is a large number of bacteria which lead intracellular lifestyle surviving inside the vacuole, phagolysosome or cytoplasm of infected host cells.

***(a)* Survival in phagosome:** It may be exemplified by *Coxiella burnetti* causing Q fever. Its main habitat is macrophage inside which it grows and multiplies. It exists in two phases: I and II. Phase I bacteria have a smooth form of lipopolysac-

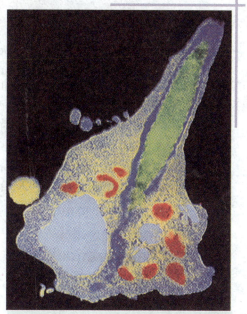

An extracellular bacterium (green) that has been phagocytosed by secretions of a neutrophil (nuclear lobes in blue).

charide and are highly virulent, whereas phase II bacteria have a rough lipopolysaccharide and have reduced virulence.

After adhesion, phase I bacteria bind to a leukocyte, response integrin (a membrane protein) and integrin-associated proteins (Fig. 24.20). The bacterium is internalised by microfilament-dependent process. Then phagosome and lysosome get fused to form apparent normal phagolysosome-containing membrane proton ATPase and lysosomal enzymes. The infected host cell is capable of asymmetric division resulting in two daughter cells, one containing the vacuole (inside which bacterium is present) and the other parasite-free cell. The vacuole containing cell is broken liberating the bacterium, whereas the other cell may be invaded by bacterium repeating the similar cycle. Little is known how the bacterium reproduces in acidic phagolysosome (pH 5.2).

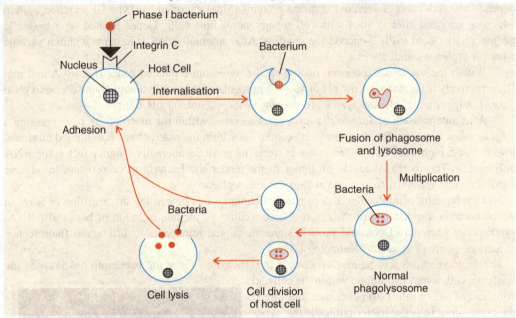

Fig. 24.20 : Stages in life cycle of *Coxiella burnetii* (based on Henderson *et al.*, 1999).

(b) **Survival within vacuoles:** *Legionella pneumophila* causing legionnaire's disease invades the macrophages and gets internalised by coiled phagocytosis (Fig. 24.21). After 15 minutes of internalisation, phagosome is surrounded by smooth vesicles and after 1 hour by mitochondria.

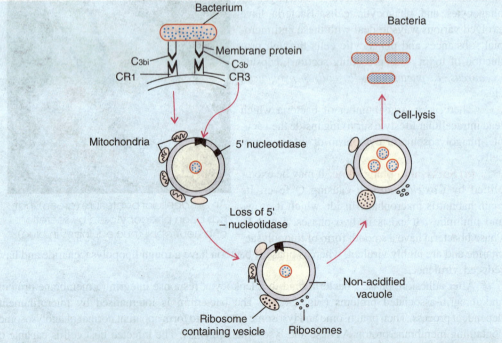

Fig. 24.21 : Multiplication of *Legionella pneumophila* inside infected macrophage.

The phagosome does not fuse with lysosome. Hence, it lacks endosomal receptor (transferrin) and endosomal/lysosomal markers (CD63). Phagosome-lysosomal fusion is prevented by the polycationic protein (Mip) of the bacterium. After 4 to 8 hours of internalisation, the phagosome is surrounded by ribosome and ribose-containing vesicles. Bacterium multiplies exponentially with the doubling time of 2 hours using bacterial cell organelles resulting in cell lysis.

(c) **Survival within the host cytoplasm:** There is a number of bacteria which escape from vacuole (after invading the host cell) and remain within the cytoplasm, for example *L. monocytiogenes, Shigella* spp., *S. aureus*, streptococci, *Rickettsia* spp., *Haemophilus ducreyi,* of these only *Rickettsia* spp. are the obligate intracellular parasite (McCrae *et al.*, 1997).

4.   Bacterial Protein Toxins- Agents of Disease

Bacterial protein toxins damage the host cells. Toxin production involves many stages such as toxin expression, transportation from bacterial cell using protein mixture, and export out side the cell. Some toxins affect entry of bacteria into host cells. Some survive the extracellular harsh conditions. Some toxins act on the cell surface, whereas the others enter the cell in a membrane-bound vesicle. The active part of toxin is taken across the vesicle membrane into the cytoplasm where it recognises its target and modifies it. Thus the key example of cellular microbiology is the target recognition and modification. The toxin research explains disease, identifies new cellular components and leads to antibacterial therapy. Protein toxin research has relevance with microbial pathogenesis, microbial genetics, receptors, membrane-translocated cell biology and structural biology. So far more than 300 toxins of Gram-positive bacteria have been characterised.

Historically, plague (caused by *Y. pestis*) and cholera (caused by *V. cholerae*) have been the most famous toxin-related diseases. Hippocrates described cholera over 2,000 years ago. By the end of 1880s, Pasteur, Koch, Roux, Kitasato and von Behring identified toxin-producing bacteria, their proteina-ceous nature and action. Development of rationally designed vaccines to combat disease could be advanced after deeper understanding of relationship between structure and function of toxins.

In 1883, *C. diphtheriae* (producing diphtheria toxin) and *V. cholerae* (producing chorale toxin) was cultured. Thereafter, Roux and von Behring developed antitoxin for therapeutic administration. In 1888, Yersin and Roux noted that the action of diphtheria toxin was enzymatic. By 1890, its antitoxin therapy (production of substances to combat

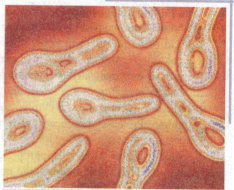

Corynebacterium diphtheriae, a club-shaped rod bacterium produces diptheria toxin, the principal virulence factor for diptheria.

diphtheria toxin by challenged animals) was recognised. Similarly, in 1890, von Behring and Kitasato prepared antitoxin. In 1925, non-toxic antigenic vaccine was developed using formaldehyde-treated *C. tetani* culture.

The α-toxin (phospholipase C that damages animal cell membrane) of *Clostridium perfringens* was such toxin whose mechanism of action was identified first. During 1960s, intracellular action of diphtheria toxin could be understood. The co-enzyme NAD played a role in the action of this toxin. Diphtheria toxin catalysed ADP-ribosylation of the translocation factor EF2. This factor is a protein which interacts with the machinery of protein synthesis. Detailed knowledge of structure-function relationship of diphtheria toxin has a broad therapeutic application.

(i) **Classification of Toxins:** Based on activity, toxins are divided into three types: type I (that act the cell membrane), type II (that attack the cell membrane), and type III (that penetrate the membrane to act inside the cell).

(a) **Membrane-damaging toxins:** This group of toxins are called type II toxins which directly act on cell membrane, form holes resulting in cell death. More than 100 toxins have been identified and categorised into several groups as given below:

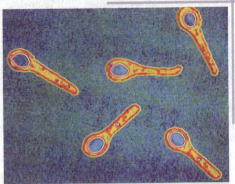

The club-shaped cells of *Clostridium tetani* the formaldehyde treated culture of which was used to develop non-toxic antigenic vaccine.

- *Pore forming toxins:* Such toxins enter into the cell membrane as oligomers and form pores. The size of pores differ with different toxins. Some pore forming toxins have cellular activity i.e. invoke cytokine production. Besides, many intracellular toxins induce pores due to translocation of their catalytic domains across the membrane into cytosol. The thio-activated cholesterol-bing toxins are produced by four genera of Gram-positive bacteria, for example species of *Streptococcus* (e.g. streptolysin O, pneumolysin), *Listeria* (listerolysin O, ivanolysin), *Clostridium* (tetanolysin, perfringolysin O, septicolysin O, histolyticolysin O, chauveolysin), and *Bacillus* (cereolysin O, alveolysin, thuringolysin O). These toxins attack cells containing cholesterol in their membrane and form pores of about 30-40 nm having 30 monomers.

  Gram-negative bacteria produce RTX toxins, for example *E. coli,* haemolysin, leukotoxins from *Pasteurella haemolytica, Proteus, Bordetella* adenylate cyclase. These toxins form pores of about 1-2 nm consisting of 7 monomers and damage the normal function of host cells. The α-toxin of *S. aureus* form pores in host cell which results in cell death through apoptosis.

- *Toxins that damage membrane enzymatically:* There are many toxins that damage host's cell membrane enzymatically. For example phospholipases produced by *L. monocytogenes, S. aureus, P. aeruginosa, B. cereus,* and *Aeromonas.* Phospholipase C (PLC) or α-toxin of *C. perfringens* has necrotic and cytolytic activity. PLC of *P. aeruginosa* damages lung surfactant in human. Proteases produced by *Porphyromonas gingivalis* is implicated in gum disease.

(b) **Membrane transducing toxins:** These type of toxins are type I toxins which damage host cells by subtle means through inappropriate activation of cellular receptors. They send wrong message into the cell which confuses the normal routes of communication. The examples are: the stable toxin (ST) of *E. coli* and emetic toxin of *B. cereus.* The ST binds to membrane receptors to stimulate guanyl cyclase and give rise to the intracellular message (cGMP). The cGMP activates protein kinase G and modulates several signalling pathways.

Bacterial superantigenic toxins directly stimulate immune response by acting as mitogens. They bind to the T cell receptor and MHC Class II antigen directly and activate one or several susets of about 5-20% of T cells.

Examples of superantigens are enterotoxins (causing food poisoning) and exotoxins (causing toxic sock) of *S. aureus,* and erythrogenic toxins of *S. pyogenes.*

(c) **Intracellular toxins:** These are type III toxins which act in most subtle way. They act in different stages (Fig. 24.22). They have to gain intracellular access, survive attack by proteases and protons and trick the cell into them to their target and destroy enzymatically. They are the most deadly group of toxins, for example botulinum and tetanus neurotoxins are lethal to human at a dose of about 0.1 ng and affect nerves.

Neurotoxins of *C. botulinum* and *C. tetani* act as protease. They block the function of peripheral nerves and cause a flaccid paralysis, stimulate adenylate cyclase (cAMP), the high concentration of which causes massive fluid accumulation in the lumen of the gut resulting in watery effect on immune function. Tetanus toxin attacks nerve cells in the CNS and its effects are more dramatic leading to muscle spasm and rigid paralysis.

The anthrax lethal factor (LF) is a zinc protease that cleaves the N-terminus of MAP (mitogen-activated protein) kinase to inactivate it.

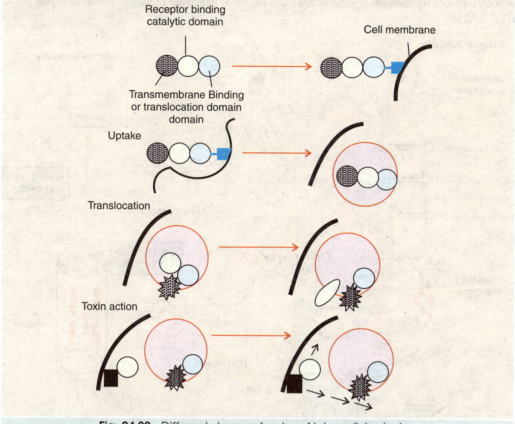

Fig. 24.22 : Different stages of entry of intracellular toxins.

D. Immune Response to Bacterial Infection

Immunity is the complex of cells and humoral factors involved toward off disease caused by pathogens. Immunity is of two types: *innate* (i.e. natural) *immunity* (utilised by all living cells) and *acquired* or *adaptive immunity* (evolved about 400 million years ago operating in vertebrates).

I. Innate Immunity

Innate immunity involves several cell populations such as epithelial cells, monocytes, macrophages, dentric cells, leukocytes, natural killer (NK) cells, lymphocyte sub-populations (e.g. CD5 positive B lymphocytes) which bridge the divide between innate and acquired immunity.

The cells arise in the bone marrow from precursor cell populations. Humoral systems are too important. These involve many types of cytokines, acute-phase proteins, enzymes like lysozymes, metal-binding proteins, integral membrane iron transporters, antibacterial peptides and the complement pathways. Immunity has been described in detail in Chapter 19.

1. Bacteria-Host Interactions at Body Surfaces

The totality of mechanisms constitutes our immunity to infection. Infection is caused rarely due to multifunctional systems of innate immunity. It is divided into those systems which *(a)* act at body surfaces (epithelium or submucosa), and *(b)* those which are present in the submucosal tissue (Fig. 24.23).

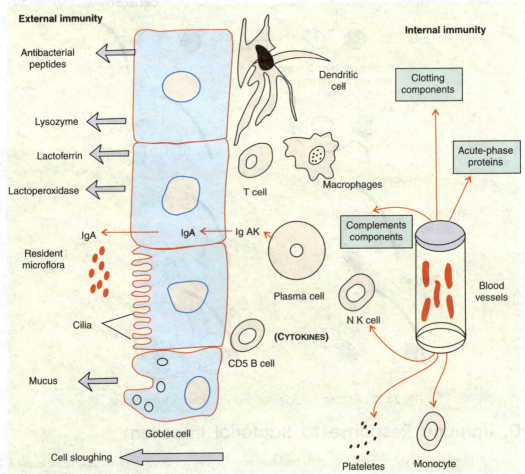

Fig. 24.23 : Innate immune systems of the epithelia and submucosal tissues (modified after Henderson *et al.,* 1999).

It can be further subdivided into those activities which act on external body surface (i.e. lining of mouth, lungs, intestine, etc.) and those which function within the body. The various mucosal surface (epithelial lining) have evolved the mechanism which inhibit the growth and infection of pathogens. At different epithelial sites, there exists site-specific mechanisms to inhibit the infection. These mechanisms include: (a) synthesis and release of antibacterial components e.g. lysozyme which degrade peptidoglycan of bacterial cell walls (or antibacterial peptides which kill bacteria), (b) constantly removing bacteria adhered to epithelium through sloughing off cells, and (c) the inhibitory effect of normal microflora on growth of pathogenic bacteria. The antibacterial mechanism employed by external epithelia are given in Table 24.11.

Table 24.11 : Mechanism utilised at various sites of epithelium and mucosa to prevent infection by pathogenic microorganisms.

Sites of epithelium/mucosa	Mechanisms operating at different sites
Bladder	Low pH, flushing action of urine, physical barrier of urethra
Colon	Mucus, resident microflora, sloughing of cells
Eyes	Tears, lysozyme, lactofferin, blinking, sIG A
Lungs	Macrophages, antibiotic peptides
Mouth	Resident microflora, lysozyme, sloughing off cells, lactoferin, sIG A, flow of saliva, antibiotic peptides
Nasopharynx	Resident microflora, lysozyme, sloughing off cells, lactoferrin, blinking, sIG A, phagoryter
Skin	Dry environment, acid conditions, resident microflora, sloughing off cells, lysozyme, lactoferrin
Stomach	Low pH, protease
Small intestine	Mucus, flow of fluid, sloughing off cells
Vagina	Low pH, resident microflora

2. Recognition of Bacteria/Bacterial Infection by Innate Immunity

The innate immune system recognises key molecular structures of pathogen i.e. the molecules that are essential for bacterial survival and unlikely evolve their structure, as such mutation would be lethal. Such molecules are called PAMPS (pathogen-associated molecular pattern).

Such molecules are called the structural components such as lipopolysaccharide and peptidoglycan, possibly DNA with high CpG content (where cytosine is unmethylated). Molecular chaperones play a role in bacterial recognition. The invertebrates and vertebrates have evolved a range of receptors referred to as pattern recognition receptors (PRRs) which are cell bound and able to recognise bacterial molecules PAMPs which are essential for cell function. Binding of PAMPs to the PRRs on the cells of vertebrate results in pro-

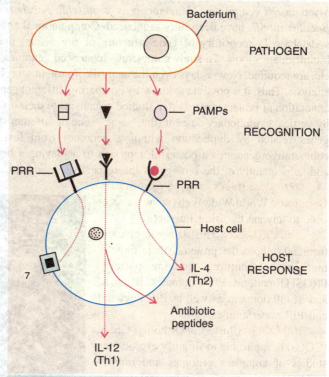

Fig. 24.24 : Recognition of bacteria by innate immune responses (based on Henderson *et al.*, 1999).

duction of antibacterial peptide, cytokines (e.g. interleukin IL-4 nad IL-12) for the induction of Th_1 and Th_2 responses and the stimulatory molecules B_7-1, B_7-2 (Fig. 24.24).

The first PRR identified was CD14 present on monocytes and plasma which binds to bacterial carbohydrates including lipopolysaccharide, peptidoglycan and lipoarabinomannan and activate cells bearing this receptor. Binding of PAMPs to their receptors is an early evolved recognition system which stimulates the production of a range of molecules required for defence against bacterial parasite.

E. Cellular Microbiology – Future Direction

1. Comparative Genomics

All the genes of an organism are not functional. In different groups of organisms the percentage of functional genes varies. For example in bacteria 3-5 genes are non-functional, whereas in humans 97 % genes are non-functional. Besides, the level of evolutionary conservation of microbial proteins is rather uniform with ~70% of gene products. Each of the sequenced genomes has homologs in distant genomes. Thus, the function of many of these genes can be predicted by comparing different genomes and by transferring functional annotation of proteins from better studied organisms to their orthologs from lesser studied organisms.

Based on the above facts, study of comparative genomics proved a powerful approach for achieving a better understanding of the genomes and, subsequently of the biology of the respective organisms. Recently, some of the genome of the microorganisms viz. *Haemophilus influenzae, Mycoplasma genitalium, Methanococcus jannaschii, Saccharomyces cerevisiae, Escherichia coli, Bacillus subtilis* have been fully sequenced. Computational analysis of complete genomes requires a database (a repository of gene structure of organisms) that store genomic informations and bioinformatics tools. To study completely sequenced genomes, analysis of nucleic acids, proteins, etc. are required. Now-a-days even the analysis of protein sets also proved as a tool to study genome analysis. Thus, it is possible to know by comparing different genomes and by transferring functional annotation of proteins from better-studied organisms to their orthologs [i.e. genes that are connected by vertical evolutionary descent (the "same" gene in different species)] as opposed to paralogs (i.e. genes related by duplication within a genome) from lesser-studied organisms. This makes comparative genomics a powerful approach to achieving a better understanding of the genomes, and subsequently of the biology of the respective organisms.

(i) **Databases for Comparative Genomics:** World Wide Web (www) is accessible to anyone by using Internet.

(a) **PEDANT:** This database gives information about the proteins, their three-dimensional structures, enzyme patterns, PROSITE patterns, Pfam domains, BLOCKS and SCOP domains as well as PIR keywords and PIR superfamilies.

(b) **COGs:** Clusters of Ortholog Groups (COGs) is applicable to simplify evolutionary studies of complete genomes and improve functional assignments of individual proteins. It comprises of ~2,800 conserved families of proteins from each of the sequenced genomes.

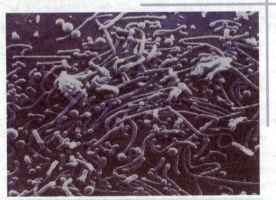

Mycoplasma the genome of which has been fully sequenced.

It contains orthologus sets of proteins from at least three phylogenetic lineages which are assumed to have evolved from an individual ancestral protein. The functions of orthologs are same in all organisms. The protein families in the COGs database are separated into 17 functional groups that include a group of uncharacterized yet conserved proteins as well as a group of proteins for which only a general functional assignment appeared appropriate. In COGs database due to storage of diverse nature of data on proteins, the similarity searches also give some information for those proteins which has no clear informations in databases. The databases also act as a tool for a comparative analysis of complete genomes.

(c) **KEGG:** Kyoto Encyclopaedia of Genes and Genomes (KEGG) Centers on cellular metabolism was proposed by Kaneshisa and Goto (2000). A comprehensive set of metabolic pathway charts both general and specific has been given for the sequenced (genome) organism. In this, enzymes identified in a particular organism are colour-coded, so that one can easily trace the pathways. It also provides the enzymes coded for the orthologus genes. These genes if located adjacent to each other, form like operons, for example comparison between two complete genomes in which genes are located relatively close or adjacent (with in five genes) can be made. This site is useful to get informations for the analysis of metabolism in various organisms.

(d) **MBGD:** Microbial Genome Database (MBGD) is situated in the University of Tokyo, Japan. This database helps to search for microbial genomes. MBGD accept the several sequences at once (~2000 residues) for searching against all of the complete genomics available displays colour–coded functions of the detected homologs, and their location on a circular genome map. This database also gives informations regarding the functions e.g. degradation of hydrocarbon or biosynthesis of nucleotides, etc.

(e) **WIT:** Similar to KEGG, WIT ("What is there" database) gives informations regarding metabolic reconstruction for completely sequenced genomes. The WIT features are to provide sequence of reactions between two bifurcations besides to include proteins from many partially-sequenced genomes. These features of WIT provide many more informations on the sequences of the same proteins/enzymes obtained from different organisms.

Bioinformatics Subgroups: The bioinformatics has more subgroups viz. networking, sequence database and alignment theories, phylogenetic analysis, secondary structure predictions and DNA analysis, biomolecular structures, dynamics and function, protein motifs, modelling analysis of 3-D structures of macromolecules, applications in the discovery of synthetic molecules to heat, human diseases, and molecular mechanisms involved with gene regulation, etc.

(ii) **Steps of Sequence Formation:** The tool of bioinformatics provides the analysis of sequence information. This process involves:

- Identifying the genes in the DNA sequences from various organisms.
- Developing methods to study the structure and/or functions of newly identified sequences and corresponding structural RNA sequences.
- Identifying families of related sequences and the development of models.
- Aligning similar sequences and generating phylogenetic trees to examine evolutionary relationships.

To know the biological and biophysical

Decoding the information of biological sequences. This flask of genetically engineered cells is the starting point for the biotechnological production of a protein targeted to cancer cells.

knowledge, conversion of sequence information is required. Informations of the biological sequence can deciphere the structural, functional and evolutionary clues encoded in the languages of biological sequences. The decoding of languages may be decomposed into sentences (proteins), words (motifs) and letters (amino acids), and the code may be tackled at a variety of these levels. A single letter change within a word can sometimes change its meaning for example, a chain codon for glutamic acid (GAA) to valine (GUA) in homozygous individuals. This minute difference results in a change from a normal healthy state to fatal sickle cell anaemia.

(iii) **Basic Requirements:** Following are some of the requirements:

- Biological research on the web.
- Sequence analysis, pair wise alignment and database searching.
- Multiple sequence alignments, trees and profiles.
- Visualizing protein structures and computing structural properties.
- Predicting protein structure and function from sequence.
- Tools for genomics and proteomics.

The well known packages (softwares) for DNA and protein sequence analysis include Staden and Gene world (for DNA and protein sequence); Gene Thesarus (access to public data and integration with proprietary data), Lasergene (for coding analysis, pattern site matching, structure and comparative analysis, restriction site analysis, PCR primer and probe designing, sequence editing, assembly and analysis, etc.), CINEMA (package provides facilities for motif identification using BLAST), EMBOSS (using nucleotide sequence pattern analysis, codon usage analysis, gene identification tools, protein motif identification and rapid database searching with sequence pattern), EGCG (for fragment assembly, mapping multiple sequence analysis, pattern recognition nucleotide and protein sequence analysis, etc.). The biological data and information storage are given in Table 24.12.

Table 24.12 : Biological data and their source of storage of informations

Subject	Source	Link
Biomedical literature	PubMed	www.ncbi.nlm.gov/entrez/query.fegi
Nucleic acid sequence	GenBank	www.ncbi.nlm.gov/entrez/
	SRS at EMBL/EMI	query.fegi?dbzNucleotide
		http://srs.ebi.ac.uk
Protein sequence	Gene Bank	www.nbrf.georgetown.edu
	SWISSPORT at	
	EXPASY PIR	

(iv) **Classification of Databases:** The databases are broadly classified into two categories: *sequence databases* (it involves both proteins and nucleic acid sequences), and *structural databases* (it involves only protein databases).

Moreover, it is also classified into three categories *(a)* primary database, *(b)* secondary database, and *(c)* composite database.

Primary databases contain information of the sequence or structure alone of either protein or nucleic acid e.g. PIR or protein sequences, GenBank and DDBJ for genome sequences. Secondary databases contain derived informations from the primary databases, for example informations on conserved sequence, signature sequence and active site residues of protein families by using SCOP, eMOTIF, etc. The composite database is obviating the need to search multiple resources. The SCOP is structural classification of proteins in which the proteins are classified into hierarchial levels such as classes, folds, superfamilies.

(v) **Comparative Modelling or Homology Modelling:** It is useful in aligning two sequences to identify segments that share similarity. It later identifies the structure of desired protein. After predicting the structure of the homology, rigid body assembly approach is applied for assembling the structure that represents the core loop regions, side chains, etc. In sediment matching procedure, coordinates are calculated from approximate position of conserved atoms of the templates. The alignment of the sequence of interest with one or more structural templates can be used to derive a set of distance constraints which gives informations on distance geometry or retrained energy minimization or retained molecular dynamics to obtain the structure.

(a) **Threading:** It is a technique to match a sequence with a protein shape in the absence of any substantial sequence identity to proteins of known structure, whereas comparative modelling requires protein sequences. Threading is followed by *scoring*, that creates a profile for each site or using a potential based pair wise interaction. Potential energy functions may be obtained from *ab initio* quantum mechanical calculations or from thermodynamic, spectroscopic or crystallographic method or by combination method.

(b) **Sequence analysis:** In order to understand the protein/nucleic acid structure and evolution, the analysis of their sequence data is required. The sequence analysis is the detection of homologus (orthologus: same function, different species) or paralogus (different but related functions within one organism) relationships by means of routine database searches. Some of the important resources are outlined in the following:

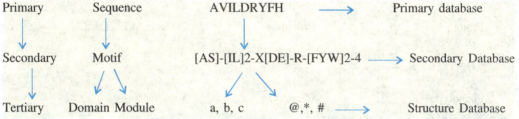

Primary	Sequence	AVILDRYFH	⟶	Primary database
Secondary	Motif	[AS]-[IL]2-X[DE]-R-[FYW]2-4 ⟶		Secondary Database
Tertiary	Domain Module	a, b, c	@,*, # ⟶	Structure Database

The primary structure of protein lies in its amino acids sequences which are stored in primary databases as linear alphabets that denote the constituent residues. The secondary structure of a protein corresponds to regions of local regularity e.g. α-helices and β-helices which in sequence alignments are often apparent as well conserved motifs. These are stored in secondary databases of patterns e.g. regular expressions, fingerprints, blocks, profiles, etc. The tertiary structural elements, which may form discrete domains within a fold (a, b, c), or may give rise to autonomous folding units or modules (such as @, *, #), complete fold, domains and modules are stored in structural databases as sets of atomic co-ordinates.

Analysing DNA sequences : Indentifying the DNA fragments on electrophoresis gel.

2. Functional Genomics

Functional genomics is to place all of the genes in the genome of an organism within a functional frame work. Actually, in every organism about 12-15% genes are structural genes which are expressed for certain characters. These are transcribed in a given cell. This is helpful in overall functioning of the cell and organism.

Functional genomics brings together genetics with gene transcripts, proteins and metabolites by analyzing genome sequencing. Functional genomics is driving a shift from vertical analysis of single genes, proteins or metabolites towards horizontal analysis of full suites of genes, proteins and metabolites. This may help in molecular participation of a given biological process. This offers the prospect of determining a truly holistic picture of life.

(i) **Functional Genomics Toolbox:** The functional genomics emerged in response to the challenges posed by complete genome sequences. To understand this process it is necessary to know the biochemical and physiological function of every gene product and their complexes. The activity of genes manifests at a number of different levels, including RNA, protein and metabolite levels and analyses at these levels can provide insight not only into the possible function of individual

gene but also the cooperation that occurs between genes and gene products to produce a defined biological outcome. The technology, involved in defining functional genomics are DNA or oligonucleotide microarray technology for determining mRNA, 2D gels and mass spectroscopy and other methods for analyzing different proteins and GC-MS or LC-MS for identifying and quantifying different metabolites in a cell. High throughput methods for forward and reverse genetics are also integral to functional genomics (Fig. 24.25).

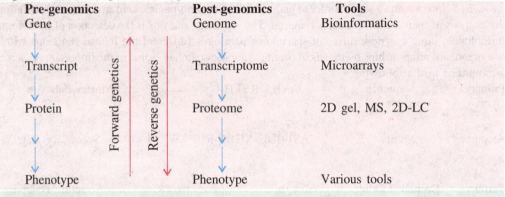

Pre-genomics	Post-genomics	Tools
Gene	Genome	Bioinformatics
Transcript	Transcriptome	Microarrays
Protein	Proteome	2D gel, MS, 2D-LC
Phenotype	Phenotype	Various tools

Fig. 24.25 : Tools of functional genomics.

(ii) **Methods in Functional Genomics:** Functional genomics lies on gene expression, profiling (mRNA) in protein expression, reverse genetics, the generation of targeted mutations in genes of interest besides forward mutation rate, the generation of random mutations in the genome for desirable mutants and bioinformatics. These criteria help in providing maximum information of a particular organism. This helps in understanding the biological process at the molecular level and also useful to identify novel genes regulating this process. To understand the gene function, it is desirable to identify genes and to understand its expression at the whole genome level. There are many prokaryotic and eukaryotic organisms whose genomes are fully sequenced. The current discovery is mapping of whole sequences of genes present in human genome. It is possible to assign functions to novel genes and proteins, and to understand biological processes at the molecular level. The integrated understanding of the control of gene expression and knowledge of signal transduction, cell signalling and overall cell function are dynamic tools to study regulation of gene expression in any given cell type. In yeast cells, transcripts associated with different parts of the cell cycle form discrete clusters. These studies allowed sequence tags encoding proteins of unknown function to be assigned to putative classes based on their clustering with genes of known function. Here, role of functional genomics will be to test those reputative functions and apply to resolve complex biological processes.

(iii) **Future Prospects:** It has good future as briefly given below.

(a) **In human pathology:** Application of gene expression profiling in understanding the human cells and tissues to disease is under way. It would be possible in future to study the modifications in gene sequence during infection. Such studies will yield fundamental insights into the etiology and pathogenesis.

(b) **In parallel sequencing:** In a number of laboratories, most of the sequences are generated using different approaches. Integrating very different datasets is not as simple as assembling a sequence itself which serves as an absolute standard. Although transcriptional profiling can be used to construct the standardized databases based on absolute RNA and protein levels, yet this is clearly not the case for relative gene expression data.

3.   Evolution of Genomes in Microorganisms

The prokaryotes contain a large single, circular dsDNA molecule, usually less than 5 Mb long.

In addition, they may also contain plasmids. The genome is arranged into **operons.** A typical prokaryotic genome contains a small amount of non-coding DNA distributed throughout the sequence. For example, *E.coli* contains only ~11% the non-coding DNA. On the other hand, the eukaryotic cells contain chromosomes present in the nucleus. Each chromosome contains a single dsDNA molecule. A small amount of DNA appears in two organelles i.e. mitochondria and chloroplasts. The genetic code of nuclear and organellar genes are translated differently. The difference in genome size of different organisms is due to different amounts of simple repetitive sequences often called as *junk* DNA.

The genomic sequences have been found useful in the study of evolution of genomes. The comparative genomics gives the information about the gene arrangement in the genome of microorganisms belonging to different classes, varying nature of proteins and its expression patterns. The function of the protein present in one class of microbe has a homology in the other. If so, the homologus proteins carry out the same function in microorganisms belonging to two different classes.

Andrade and co-workers compared the protein functions of three major domains of life. Their classification contained as major processes involving energy, information, and communication and regulation. In case of energy a number of processes such as biosynthesis of co-factors, amino acids, metabolisms, fatty acids including lipid analysis, nucleotide biosynthesis and transport have been included, whereas replication, transcription and translation processes comprised of information. The communication and regulation deal with regulatory functions, cell envelop into cell wall and cellular processes. They determine number of genes in three genera of different phyla *i.e. Haemophilus influenzae* (bacteria), *Methanococcus jannashii* (archaea) and *Saccharomyces cerevisiae* (eukarya) which contain 1680, 1735 and 6278 genes, respectively. Some of the proteins are shared but some are unique to each domain. *M. jannaschii* shares some proteins with *H. influenzae* and some with *S. cerevisiae.*

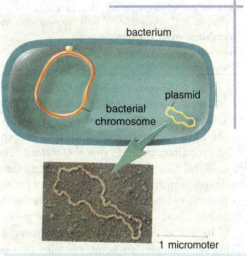

In addition to their single, circular chromosome, bacteria commonly possess rings of DNA called plasmids which may carry useful genes.

Mushegian and Konin (1996) compared the genomes of *Mycoplasma genitalium* and *H. influenzae.* From 1703 protein-coding genes of *H. influenzae*, 240 are homologues of proteins of *M. genitalium.* They concluded that these proteins must be essential, but might not be sufficient for autonomous life. Some essential functions might be carried out by unrelated proteins in the two genomes. For example, the common set of 240 proteins left gaps in essential pathways, which could be filled by adding 22 enzymes from *M. genitalicum.* A number of 250 genes proposed necessary and sufficient minimal set finally after removing functional redundancy and parasite-specific genes. According to them, the proposed minimal genome functional class includes protein synthesis, DNA replication, recombination and repair, transcription apparatus, chaperon-like proteins, glycolytic pathway, no nucleotide biosynthesis, amino acid or fatty acid biosynthesis, protein export machinery and limited functions of metabolic transport proteins. The identification of the gene complement of the common ancestor of *M. genitalium* and *H. influenzae* remained unanswered. The 71% proteins of 256 genes have been found to have recognizable homologues among eukaryotic or archaeal proteins.

Genome analysis has revealed families of proteins with homologues in archaea, bacteria and eukarya. It is assumed that these got evolved from an individual ancestral gene through a series of evolution and duplication events. This may be due to effects of horizontal gene transfer.

In case of *Aquiflex aeolicus*, 83% of the proteins have archaeal and eukaryotic homologues to *Borrelia burgdorferi* in which 52% of the proteins have archaeal and eukaryotic homologues. Archaeal genomes have 62-71% proteins higher with bacterial and eukaryotic homologues but only 35% of the proteins of yeast have bacterial and archaeal homologues. It is further necessary to know whether the common set of proteins present in different phyla have common functions. The minimal set of proteins of *M. genitalium* has around 30% homologues in all genomes. Other essential functions must be carried out by unrelated proteins. The common protein families appearing in archaea, bacteria and eukarya involved in translation process are given in Table 24.13.

Table 24.13 : Protein families involved in translation process in microbes

Functional classes of proteins	Number of families in the genomes
Cellular processes including chaperon	9
Secretion, cell division, cell wall, biosynthesis, metabolism	9
Translation (with ribosome structure)	53
Transcription	4
Replication, recombination and repair	5

The WEB resource is helpful for databases of aligned gene families.

(i) **Horizontal Gene Transfer (HGT):** The horizontal gene transfer is defined as the acquisition of genetic material by one organism, by natural means. This can take place by direct uptake as in pneumococcal transformation experiment demonstrated by Avery and coworkers (for details see Chapter 8, *Microbial Genetics*). The genome analysis revealed that HGT is a common process that includes discrepancies among the evolutionary trees constructed from different genes, and direct sequence comparison between genes from different species. For example, *E. coli* appears to have entered the genome by horizontal transfer after divergence from the *Salmonella* lineage about 100 million years ago. The HGT is more prevalent among operational genes in microbial evolution.

In case of *Bradyrhizobium japonicum*, a slow growing root–nodulating, symbiotic nitrogen fixing bacterium in soybean, contains two genes of ammonia assimilatory enzyme, glutamine synthetase. In this case one gene is similar to the other symbiotic nitrogen fixing bacteria but the second one is found identical to higher plants. Similar in the photorespiration, enzyme ribulose 1, 5-bisphosphate carboxylase/oxygenase fixes carbon dioxide at the entry of Calvin cycle (dark reaction), passes on between bacteria, mitochondria and algal chloroplast as well as undergoing gene duplication in prokaryotes. HGT involves a phenomenon called NOR in prokaryotes. Both eukarya and prokaryotes are chimeras. NOR is gene transfer necessarily limited to ancient ancestors. In the genome of *M. tuberculosis*, 8 genes appeared to human genome. There is a mixed type of informations available that numerous bacterial genes have entered the human genome. Further all informational genes from *M. jannaschii* are found to be similar to that of yeast. Similarly, eukaryotes derive their informational genes from archaea, and operational genes from proteobacteria, cyanobacteria and methanogens. The informational genes are less subject to horizontal transfer that determines the identity of the species.

(ii) **WEB Resources for Databases:** For continuous progress of genome evolution, it is further suggested to use the following WEB resources for databases of aligned gene families.

Pfam: Protein families database:

http://www.sanger.ac.uk/software/pfam

COG:cluster of orthologus groups:

http://www.ncbi.nlm.nih.gov/cog/

HOBCAGEN: Homologus Bacterial genes database:

http://pbil.univ-lyon-fr/databases/hobacgen.html

HOVERGEN: Homologus Vertebrate Genes Database:

http://pbil.univ-lyon-fr/databases/hovergen.html

TAED: The Adaptive Evolution Database:

http://www.sbc.sv.se/~liberles/TAED.html

Many other websites can also be used to procure the latest information about genomes of microorganisms.

QUESTIONS

1. What is cellular microbiology ? Write the emergence of cellular microbiology.
2. Discuss in detail the eukaryotic cell-to-cell signalling with emphasis on role of cytokines.
3. What is quorum sensing ? Write the mechanism of quorum sensing. How is it a mechanism of virulence in bacteria.
4. Give a detailed account of intracellular signalling and basic building blocks used in it.
5. Write in detail adhesion of host cells, role of bacterial structure in adhesion, nature of host cells and adhesins.
6. Write in detail bacterial invasion of host cells and the mechanism and consequences of invasion.
7. What are bacterial toxins ? Write types of bacterial toxins.
8. Discuss in detail the cell mediated and humoral immunity.
9. Give a detailed account of structural and functional genomics.
10. What is the difference between a global and a local alignment strategy?
11. How many exons are present in unknown sequence?
12. Give an account of the databases which are used in protein and nucleic acid studies?
13. Classify different databases and how can we use these databases in macromolecular study of proteins and nucleic acids?
15. Give an account about the web sources by which we can access different databases.
16. Write short notes on the following:

    (i) Bacterial diseases, *(ii)* Pathogenicity islands, *(iii)* Endocrine hormone signalling, *(iv)* Cytokine and its sub-families, *(v)* Ca^+ and CAM in signalling, *(vi)* Role of cell membrane in signalling, *(vii)* Effect of adhesion on bacteria, (viii) Bacterial survival after invasion, (ix) Innate immunity, (x) Comparative genomics.

REFERENCES

Aktories, A. 1997. *Bacterial toxins: tools in cell biology and pharmacology.* Chapman & Hall: London.

Cossart, P. *et al.* 1996. Cellular microbiology emerging. *Sci.* **271**: 315-316.

Groisman, E.A. and Ochman, H. 1997. Pathogenicity islands: bacterial evolution in quantum leaps. *Cell.* **87**: 791-794.

Hancock, J.T. 1997. *Cell signalling.* Chapman & Hall, London.

Hendeson, B. *et al.* 1999. *Cellular Microbiology.* John Wiley & Sons, New York, p. 451.

Kaprelyants, A.S. and Kell, D.B. 1996. Do bacteria need to communicate with each other for growth ? *Trends Microbiol.* **237**: 237-242.

Lee, C.A. 1996. Pathogenicity island and the evolution of bacterial pathogens. *Infect. Agents Disease.* **5**: 1-7.

McCrae, M.A. *et.al.* 1997. *Molecular aspects of host-pathogen interactions.* Cambridge Univ. Press, U.K.

Ofek, I and Doyle, R.J. 1994. *Bacterial adhesion to cells and tissues.* Chapman & Hall, London.

Pace, N.R. 1997. A molecular view of microbial diversity and the biosphere. *Sci.* **276**: 734-740.

Microbial Ecology I: Ecological Groups and Microbial Interactions

25

A. Ecological Groups of Microorganisms

B. Microbial Interactions

I t was E. Haeckel (1869) who coined the term *Oikologie* (Greek *Oikos* means house, and *logos* means study) from which the recent term, *ecology* has originated. Daubenmire (1947) defined the term ecology by using a statement attributed to Haeckel, "*the study of the reciprocal relations between organisms and their environment*". Thus, ecology is the study of organisms in their natural home or habitat. Moreover, the important feature of ecology is that it always comprehends (along with organisms) the nonliving environments and deals with the systems of levels, higher than the organisms in the biological spectrum of levels of organisation of systems of *genes, cells, organs, organisms, population and communities* (Ambasht, 1984). *System* means a unified made of regularly interacting and interdependent components. Thus ecology deals with holistic approaches to unravel the mystery of these two components and their relationships.

In the above definition two aspects are very clear: the organisms and the environments. The term *organisms* include all living components of an ecosystem *i.e.* plants and animals. It can further be subdivided into *macroscopic* and *microscopic* organisms. Discussing ecology as a whole is not the aim of the authors. This Chapter concerns only with the microscopic organisms of soil, water and air and their interactions *i.e.* a new branch of ecology called *microbial*

Bacteroids inside root nodule of a legume

ecology. With the advancing knowledge of microorganisms in different habitats and their functions, microbial ecology has emerged as a new discipline of ecology which has gained momentum only in recent years.

A. Ecological Groups of Microorganisms

Microorganisms are an important component of an ecosystem. Several ecological categories can be made on certain grounds while grouping the microorganisms.

Microbial ecology : The microscopic organisms of *soil, water and air* and their interactions constitute microbial ecology.

1. Based on Oxygen Requirement

On the basis of their ability to grow in the presence or absence of O_2 they can be divided into three distinct categories: *aerobes* (which grow in the presence of O_2), *anaerobes* (which grow only in the absence of O_2), and *facultative anaerobes* (which has capability to grow either in the absence or the presence of the air).

2. Based on Carbon Sources as Energy

Carbon is a chief source of energy and a building block of cell wall and other cell components. All living organisms are divided into the two major groups, the autotrophs and the heterotrophs.

(i) **Autotrophs (or lithotrophs):** In autotrophs the carbon source is CO_2. Energy is provided through photosynthesis in photoautotrophs (*i.e.* phototrophs) or oxidation of inorganic compounds as in chemotrophs (for detail see Chapters 12 and 13). Examples of autotrophs are cyanobacteria, microscopic algae, and photosynthetic bacteria.

(ii) **Heterotrophs (or Chemoorganotrophs):** In heterotrophs the sources of carbon and energy of biosynthetic mechanism of the cell are the dead organic materials. These materials are broken down by the heterotrophs for their carbon and energy to form their cell components and cell as well. There is a large number of heterotrophs represented by bacteria, fungi, actinomycetes, etc.

3. Based on Temperature

When temperature becomes a limiting factor *i.e.* governs the growth and activities of microorganisms, the latter becomes adapted to the new environment. On the basis of temperature regime, microorganisms may be categorised in the following groups:

(i) **Psychrophiles.** Psychrophiles are those microorganisms which grow at a very low temperature (below 20°C). Further the microorganisms growing at low temperature can be classified into two groups: (*a*) *facultative psychrophiles*, and (*b*) *psychrotrophs*. Facultative psychrophiles are

capable of growing at temperature ranging from 0°C or less to 20°C with an optimum growth at 15°C. Some isolates from Antarctic waters have been found to show maximal growth at 10°C or less. On the otherhand psychotrophs are those microorganisms which can grow at 5°C or below, irrespective of their upper or optimum growth temperatures. Singh *et al.* (1984) have reviewed the ecological aspects of microbial life at low temperatures.

In the troposphere (upto 27,000 meter altitude) where temperature remains below –40°C, bacteria, fungi and other microorganisms have been isolated. Procter and Parken (1942) have recorded the presence of *Bacillus* over polar Pacific and Atlantic oceans where the temperature remains below –20°C. Moreover, from some glacial caves (*e.g.* caves in the Arctic, Lapland the pyrenees, the Alps in Romania) where temperature ranges between –0.8°C and 5°C, certain bacteria such as *Arthrobacter, Pseudomonas* and *Flavobacterium* have been isolated. The optimal temperature required for growth by *Polaromonas vacuolata* is 4°C.

(*ii*) **Mesophiles.** The microorganisms growing optimally at temperatures between 25°C and 35°C, and have capacity to grow at about 15°C to 45°C are known as mesophiles. Most of microorganisms are mesophiles and constitute the major component of an ecosystem, for example cyanobacteria (*e.g. Nostoc, Anabaena, Oscillatoria,* etc), bacteria (*e.g. Rhizobium, Mycobacterium, Corynebacterium, Azotobacter,* etc.), fungi (*e.g. Fusarium, Trichoderma, Puccinia, Pythium,* etc.), amoebae, etc.

(*iii*) **Thermophiles (Heat-loving Microorganisms).** Microorganisms growing readily at temperatures of 45 to 65°C are known as thermophiles. The obligate thermophiles are incapable of multiplying below 40°C. Johri and Satyanarayana (1984) reviewed thermophilic fungi of paddy straw (*e.g. Aspergillus fumigatus, Chaetomium thermophila, Humicola lanuginosa, Thermoascus*

A large colony of *Nostoc* : A mesophilic cyanobacterium.

aurantiacus), coal mine soil (*e.g. A. fumigatus, C. thermophila, Torula thermophile, Paecilomyces sp.,* etc.)

(*iv*) **Hyperthermophiles and Super-hyperthermophiles.** The microorganisms that prefer to grow above 80°C to 100°C (the boiling point of water at sea level) are called *hyper thermophiles*. Thomas D. Brock of the University of Wisconsin about 30 years ago first discovered the earliest specimens of microbial life in hot springs and other waters. In 1960 Brock with his colleagues identified the first hyperthermophile capable of growing at temperature greater than 70°C. The bacterium is now known as *Thermus aquaticus*. From this bacterium an enzyme named *Taq* has been isolated which is widely used in polymerase chain reaction (PCR) technology (for detail see Chapter 6). Some users of PCR have replaced the *Taq* with *Pfu* polymerase. This enzyme is isolated from *Pyrococcus furiosus*.

In addition, this team found the first hyperthermophile in an extremely hot and acidic spring which is known as *Sulfolobus acidocaldarius*. This bacterium grows prolifically at temperatures as high as 85°C. To date more than 50 hyperthermophiles have been isolated, many by Karl O.

Stetter and his colleagues at the university of Regensburg in Germany. The most heat - resistant of these microbes is *Pyrolobus fumarii* which grows in the walls of smokers. It multiplies at temperature upto 113°C. It stops growing at temperatures below 90°C. Example of the other hyperthermophile is *Methanopyrus* that lives in deep-sea chimneys and produces methane.

Now there is a question about the upper limit for life. Do *super-hyperthermophiles* capable of growing at 200 or 300°C exist? No body knows, although current understanding suggests that the limit will be about 150°C. Above this temperature probably no life forms could prevent dissolution of the chemical bonds that maintain the integrity of DNA and the other essential molecules (Madigan and Marrs, 1997). All such organisms growing at extremes of environment are termed as *extremophiles*.

4. Based on Habitat

On the basis of habitat in broad sense, microorganisms are divided into the following groups:

(*i*) Soil Microorganisms: Microorganisms dwelling in soil subsystem are known as soil microorganisms or soil microflora, and the branch of microbiology dealing with soil microflora is known as *soil microbiology*. This branch has been separately discussed in detail in Chapter 25.

(*ii*) Aquatic Microorganisms: Microorganisms residing in water subsystem irrespective of its quality or physical or chemical nature are in broad

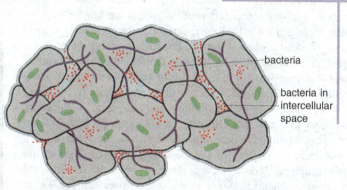

bacteria

bacteria in intercellular space

The Microenvironment. The world of Microorganisms in soil : Bacteria present as isolated microcolonies on surfaces and in pores, which are lovered by thin water films : Filamentous fungi are able to grow on and between these aggregated particles. Protozoa move in water films and graze on bacteria.

sense called aquatic (water) microorganisms. The branch of microbiology dealing with aquatic microorganisms is known as *aquatic (water) microbiology*. It has been discussed in detail in Chapter 28. Most of the human pathogens spread through water (see Chapters 20, 21 and 22).

(*iii*) Aeromicroflora: Microorganisms living in air are generally known as *aeromicroflora* or microorganisms of air, and the branch of microbiology dealing with aeromicroflora is known as *aeromicrobiology*. Several fungal and viral diseases are transmitted through air (see Chapters 20, 21 and 22). Aeromicroflora has been discussed in detail in Chapter 29, *Aeromicrobiology*.

5. Microorganisms Living in Extreme Environment i.e. Extremophiles

Extremophiles thrive under conditions that would kill the other creatures. The molecules that enable extremophiles to prosper are becoming commercially useful now a days. Of the particular interest are the enzymes that help the extremophiles to function in brutal circumstances. Psychrophiles and hyperthermophiles have been described earlier. Example of other groups of high and low pH loving microorganisms are also not uncommon. Most natural environments on the earth are essentially neutral, having pH values between 5 and 9. *Acidophiles* (acid loving) thrive in the rare habitats having a pH below 5, and *alkaliphiles* favour habitats with a pH above 9 (Madigan and Marrs, 1997). For detail see Chapter 26, Extremophiles.

6.   On the Basis of Mode of Nutrition or Habit

There are different ways by which the microorganisms derive nutrition for their growth and development.

(*i*) **Saprophytism:** Saprophytism is a phenomenon which refers to getting nutrition from dead organic materials, and such microorganisms are known as saprophytes for example *Aspergillus, Penicillium, Rhizopus, Mucor,* etc. The saprophytes are equipped with extracellular enzyme-producing capacity according to the available substrate. Some time in the presence of a living host a few saprophytes change their tendency and cause disease. Such saprophytes are called *facultative parasites.* Facultative parasites are basically saprophytes but have tendency to behave as parasite as well. Such microorganisms are very dangerous. They are also termed as opportunistic microorganisms. For example, some species of *Fusarium, Pythium,* etc. are facultative parasites.

(*ii*) **Parasitism:** Parasitism refers to deriving nutrition from a living plant or animal host, and microorganisms associated with parasitism are known as para-sites (living upon). There are certain nutrients which are not found in dead organic materials. For such nutrients the parasites have to infect the plant or animals. On the other hand it can be said that parasitism is a tendency of parasites to infect living hosts. When a parasite is very virulent and cannot live without a living host, it is called *obligate parasite* such as *Puccinia,* powdery mildews, etc. Obligate parasitism is the highest level of specialization for deriving nutrition. On the other hand a parasite, in the absence of a suitable living host can pass its life as saprophyte. It is a second mode of leading the life and survival mechanism. Such types of parasites are known as *facultative saprophytes i.e.* the parasites that have faculty to live as saprophyte in the absence of a suitable host, for example *Phytophthora infestans* (causing late blight of potato can be cultured in laboratory on oat meal agar medium), *Taphrina deformans* (a leaf curl fungus of peach, etc.), some smut fungi, etc.

Root nodules of leguminous plants have *Rhizobium* which are nitrogen-fixing bacterium.

(*iii*) **Symbiosis:** In parasitism advantages are only to the microorganisms. The hosts are the losers. Consequently, there develops disease. However, in other case both the microbes and hosts are benefited as far as nutrition is concerned. Such association of mutual benefit is known as *symbiosis.* Symbiosis can be seen in lichens, mycorrhiza, root nodules of legumes and non-leguminous plants. For detailed discussion see the preceding section, *Microbial Interactions.*

B. Microbial Interactions

Microorganisms are ubiquitous in their occurrence. However, in natural environment they interact among themselves with plants, with animals, and, moreover, with their niches. Finally, different types of interrelations are established. Reasons for microbial interactions are the competition for nutrients (including oxygen) and space in an ecological niche. Baker and Cook (1974) pointed out that a microbe may not affect the other, or may affect by one or more of the following ways: (*a*) by stimulation of growth and development of associate, (*b*) by inhibition of growth and development of the associate, (*c*) by stimulating the formation of resting bodies by the associate, (*d*) by inhibiting the formation of resting bodies by the associate, (*e*) by enforcing the dormancy of the associate, (*f*) by causing lysis of the associate, (*g*) by harming the population of plants, (*h*) by directly benefiting the plants, and (*i*) by getting influenced by its own microenvironmental factors. Some of the possible microbial interactions have been discussed in this section.

However, on the basis of relative advantage to each partner *i.e.* hosts and microorganisms, the relationships are basically of three types : (*a*) *neutralism* (where host remains unaffected by the microbe), (*b*) *mutualism* (where both partners get benefits from the association), and (*c*) *parasitism* (where one partner gets benefits and the second suffers from damages).

I. Clay-Humus-Microbe Interaction

Clay mineral (and humic substances) affect the activity, ecology and population of microorganisms in soil. Clays modify the physico-chemical environment of the microbes which either enhance or attenuate the growth of individual microbial population. After release from clays, the organic material is either degraded by microorganisms or again bind to clays. Microorganisms have a negative charge at the pH of most microbial habitats. The magnitude of electronegativity on cell walls of bacteria and fungi is regulated by pH, amino acid residues and changes in wall composition (Archibald *et al*, 1973).

Clay minerals get adsorbed and bind with proteins, amino acids, small peptides and humic substrates. Microorganisms utilize the nutrients for their growth and activity directly from clay-protein, clay-amino acids or peptides, and clay-humic substrate complexes. Moreover, high levels of clay (*e.g.* montmorillonite) soil interferes and restricts infection of banana rootlets by *Fusarium oxysporum* f.sp. *cubense*, and thus exerts natural biological control of panama disease. The clays and humic colloids influence the distribution and activity of *Streptomyces, Nocardia* and *Micromonospora* (Goodfellow and Williams, 1983). Clay particles (*e.g.* kaolinite) is known to reduce the toxicity of cadmium (Cd) on *Macrophomina phaseolina* (Dubey and Dwivedi, 1985)

II. Plant-Microbe Interactions

The above ground (foliage) and below ground (roots) portions of plants are constantly interact with a large number of microorganisms (*e.g.* bacteria, actinomycetes, fungi, amoebae, nematodes, and algae) and viruses, and develop several types of inter-relationships. Microbial interactions with both above ground and below ground parts of plants are briefly discussed in this section. Moreover, considering the result of interactions, it may develop destructive, neutral, symbiotic or beneficial association with plants.

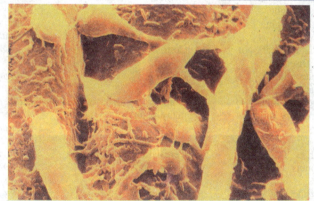

Plant-microbe interactions : Plant roots release nutrients that allow intensive development of bacteria and fungi on and near the plant root surface.

Interactions on Above Ground Parts

Microbial interactions on above ground part of plant occur in a varieties of ways where the foliage especially leaf surface (phyllosphere and phylloplane) acts as microbial niche. For detail description *see* Chapter 29.

1.   Destructive Associations (Diseases)

Plants provide a substantial ecological niche for microorganisms. However, the abundance of this potential niche with respect to any individual microbe is more apparent than real, since a few are able to grow on a wide range of plant species. Microorganisms show specificity with the hosts, organ, tissue and age of plants. The microorganisms that lead to destructive association are called pathogens. Example of some of pathogenic microorganisms is given in Table 25.1.

Disease development is governed by the resultant of three important factors: (*a*) host susceptibility, (*b*) congenial environment, and (*c*) virulent pathogen. In the presence of resistant host, unfavourable environment, or avirulent pathogen, disease will not develop.

Plant-microbe interaction occurs at molecular level. In this interaction 'gene-for-gene relationship' of H.H. Flor (1940) implies. A gene-for-gene relationship exists when the presence of a gene in one population is contingent on the continued presence of a gene in another population and where the interactions between the two genes lead to a single phenotypic expression by which the presence or absence of the relevant gene in either organism may be recognised (Person *et al*, 1962).

Table 25.1 : Some microbial pathogens causing disease on above ground parts of plants

Diseases	Hosts	Pathogen
Algal Diseases		
Red rust	Citrus, mango, sapota, cocoa, guava, tea	*Cephaleuros parasitica* and *C. virescens* coffee
Bacterial Diseases		
Stem canker and wilt	Tomato and other member of Solanaceae	*Corynebacterium michiganese*
Fire blight	Apple	*Erwinia amylovora*
Stem blight	Pea	*Pseudomonas pisi*
Canker	Citrus	*Xanthomonas citri*
Angular leaf spot	Cotton	*X.malvacearum*
Fungal Diseases		
Late blight	Potato	*Phytophthora infestans*
Early blight	Potato	*Alternaria solani*
Downy mildew	Grape	*Plasmopara viticola*
Powdery mildews	many plants	Species of *Erysiphe, Phyllactinia, Uncinula, Oidium, Podosphaera, Sphaerotheca*
Ergot	Cereals and grasses	*Claviceps purpurea*
Smuts	Many plants	*Urocystis, Sphacelotheca, Tolyposporium, Ustilago, Tilletia,*
White rust	Crucifers	*Albugo spp.*
Rust	Many plants	
	Wheat	*Puccinia graminis tritici, P.recondida*
	Linseed	*Melampsora lini*
Tikka leaf spot	Peanut and groundnut	*Cercospora personata*
Blast	Rice	*Pyricularia oryzae*
Red rot	Sugarcane	*Colletotrichum falcatum*
Leaf spot	Guava	*C. gloeosporioides*
Charcoal rot	Soybean, sunflower, etc	*Macrophomina phaseolina*
Mycoplasma Diseases		
Little leaf	Several plants *e.g.* Brinjal	*Mycoplasma* spp.
Phyllody	Sesame	*Mycoplasma* sp.
Sandle spike	Sandle wood	*Mycoplasma*

2.   Beneficial Association (Symbiosis)

The excellent example of plant-microbe interaction resulting beneficial association visualised on above ground part is the development of stem nodules. There are three known genera of legumes which are known to bear stem nodules are *Aeschynomene, Sesbania* and *Neptunia*. The stem nodules develop as a result of interaction between these plants and *Azorhizobium* species. Rhizobia develop symbiotic association with hosts, fix atmospheric nitrogen and benefit the plants. *S. aculeata* is the most popular green manure in north India which contributes about 70 kg of nitrogen and 15-20 tonnes/ha wet biomass to the soil. *A. americana* is a wild annual legume which is also used as green manure. *S. rostrata* bears both stem as well as root nodules. Detail description of root-nodule development and nitrogen fixation is given in Chapter 14. In addition, *Anabaena azollae* establishes symbiotic association with *Azolla* which is a member of pteridophyta. Species of *Nostoc* establishes symbiotic relationship with *Anthoceros* and *Blasia*, members of Bryophyta.

Interactions on Below ground Parts

Similar to above ground part, plant root-microbe interactions occur in soil as well which lead different types of associations, e.g. destructive, associative or symbiotic. One of the interesting point is that the microbe has to pass the 'rhizosphere' region before the start of interaction with plant roots. Rhizosphere and rhizoplane regions have been discussed in detail in Chapter 27, *Soil Microbiology*.

1.   Destructive Associations

Like destructive association of above ground parts, the roots also result in a destructive associations. The symptoms developed by the pathogens on root are damping off, wilt, rot, knot, scab, etc. Root diseases caused by different groups of pathogens are listed in Table 25.2.

The pathogens infect roots. Entry of pathogens takes place through wounds caused by fungi or nematodes, cracks or root hairs. In most of the cases penetration is preceded by the formation of a specific cushion like structure (appressorium) which exerts mechanical pressure on root surface. Some pathogens directly penetrate the root tissues. In *Rhizoctonia solani* multicellular cushions are seen on the roots or hypocotyl of infected plants. Nematodes directly inflict a slight mechanical injury on plant root. Their saliva is toxic for host tissues which results in cellular hypertrophy and hyperplasia, suppression of mitosis, cell necrosis and growth stimulation. Second stage larvae of *Meloidogyne* and *Heterodera* normally enter the root at or just behind the root tip. *Meloidogyne* larvae enter through the ruptures made by emerging roots, cracks on root surfaces, nodular tissues, etc. and results in development of root knots.

Crown gall : A bacterial disease caused by *Agrobacterium tumefaciens* on fruit and trees.

Table 25.2 : Some important pathogens causing diseases on roots of several plant species

Diseases	Hosts	Pathogens
Bacterial Diseases		
Scab	Potato	*Streptomyces scabies*
Crown gall	Fruit trees, dicots	*Agrobacterium tumifaciens*
Fungal Diseases		
Club-root disease	Cruciferous plants	*Plasmodiophora brassicae*
Powdery scab	Potato	*Spongospora subterranea*
Wart disease	Potato	*Synchytrium endobioticum*
Damping off	Nursery seedlings	*Pythium aphanidermatum*
Root-rot	Wheat	*P. graminicolum*
	sugarcane, corn, etc.	*P. arrhenomanes*
Wilt and foot-rot	Coconut and arecanut	*Ganoderma lucidum*
Wilt	Cotton, arhar, banana,	Species of *Fusarium*
Root rot/charcoal rot	Tomato, okra, soybean	*Macrophomina phaseolina*
Foot-rot	Barley, soybean	*Sclerotium rolfsii*
Root rot	Many plants	*Rhizoctonia solani*
Nematode disease		
Root-knots	Many plants e.g.	*Meloidogyne javanica,*
	tomato, sugarcane, etc.	*M. incognita*
Root lesions	Many plants	*Pratylenchus* spp.
		Heterodera rostochiensis

Certain wilt causing species of *Fusarium* (e.g. F. *udum*, F.*oxysporum* f. sp. *cubense, F. oxysporum* f sp. *lycopersici*, etc.) infect root, enter in vascular supply i.e. xylem bundles and produce mycelia that block the xylem vessels. These act as mechanical plug for xylem vessels. Consequently plants show wilting symptoms. Interestingly, *Macrophomina phaseolina* enters in roots and gets established in root tissues. It produces intraxylem sclerotia. Sclerotia are produced in such a high amount that impart sprinkling charcoal like symptoms. Therefore, root rot caused by this pathogen is called charcoal-rot. Certain fungi such as *Pythium, Rhizoctonia*, etc. cause damping-off of seedlings of several crop plants. *Synchytrium endobioticum* causes wart of potato tubers.

A member of actinomycetes (*e.g. Streptomyces scabies*) causes scab disease of potato. *Agrobacterium tumifaciens*, a soil-borne bacterium, causes crown gall of fruit trees including roots. Affected plants become stunted with restricted growth of plant part and poor fruit set. *Pseudomonas solanacearum* causing brown-rot and bacterial wilt of tomato, potato and other solanaceous plant is a well known pathogen. After cutting open the affected tubers, and creamy, viscous exudation from open surface is observed and the dark brown

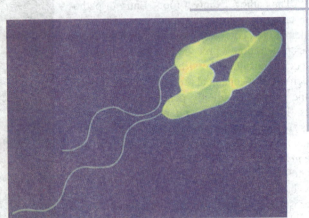

An electron micrograph of *Pseudomonas solanacenum* causing brown root of tomato, potato and other solanaceous plants.

discolouration of the vascular region becomes distinct. Consequently, tuber formation is affected and size of tubers is greatly reduced.

2. Beneficial Associations (Symbiosis)

Symbiosis is the phenomenon of living together where both the parteners are benefited. The microsymbionts derive freshly prepared food from the host plant which lack in soil. The macrosymbionts get certain nutrients from soil which are not readily available such as trace elements, nitrogen, phosphorus, etc. However, as a result of interaction of microorganisms with plant roots there may or may not develop apparent symbiotic structure. Symbiotic associations with different groups of microorganisms are discussed below:

(*i*) **Cyanobacterial Symbiosis:** The term cyanobacteria is of recent origin which includes the members of cyanophyceae. They may be both heterocystous and non-heterocystous forms. Heterocyst is the site of nitrogen fixation. The non-heterocystous forms also fix nitrogen. *Anabaena cycadae* is associated with the coralloid roots of *Cycas*. It is present in cortex in a well defined region which is known as algal zone.

(*ii*) **Bacterial Symbiosis:** Among bacteria there are two categories of symbiosis, one that does not form apparent symbiotic structure *i.e.* root nodules, and the second group which forms root nodules. However, there is a third group which enhances plant growth without entering in symbiosis.

(*a*) **Associative symbionts:** The first group includes the species of *Azospirillum* which are intimately associated with their host. These have been isolated from the rhizoplane region. As a result of infection root nodules are not formed but pictures of root hair deformation are known. Moreover, *Azospirillum* also invades cortical and vascular tissues of host, and enhances the number of lateral root hairs. This results in an increase in mineral uptake which are probably due to phytochrome production rather than N_2 fixation. Host specificity of *Azospirillum* differs from that of *Rhizobium*. Due to intimate association of *Azospirillum* with roots of several non-leguminous plants, *Azospirillum* and the other such bacteria are called `associative symbiont. The other non-nodule forming associative symbionts are *Azotobacter paspali* (found on roots of tropical grasses), *Beijerinckia* (shows host specificity with sugarcane root), *Azospirillum* (with roots of corn, wheat, sorghum), etc.

Much work has been done on *Azospirillum*. It is associated with roots in such a way that a gentle washing do not dislodge the nitrogen metabolizing activity. It has been estimated that *A. paspali* contributes 15-93 kg N/ha/annum on sugarcane root. It saves nitrogen fertilizer equivalent to 20-40 kg/ha.

(*b*) **PGPR (Plant growth promoting rhizobacteria):** The bacteria which colonize the rhizosphere of root are commonly known as rhizobacteria. The non-symbiotic beneficial rhizobacteria which affect the plant growth favourably are called PGPR. The PGPR have been discovered by Kloepper *et al.* (1980). PGPR belong to genera of *Pseudomonas, Bacillus* and *Streptomyces*, and most of them are fluorescent pseudomonads. The other types are non-fluorescent pseudomonads, *e.g. Serratia* and *Arthobacter*. The most common species of *Bacillus* are *B. polymyxa, B. circulans* and *B. macerens*.

These bacteria increase the growth of host plants. The increase in plant growth is due to (*a*) changes in balance of rhizosphere microflora producing an indirect effect on the crop, (*b*) control of pathogens and other harmful microorganisms in the rhizosphere, (*c*) production of growth hormones like gibberellin and indole acetic acid, (*d*) release of nutrients from soil, (*e*) possible production of vitamins or conversion of materials to a usable form by the host, and (*f*) possible nitrogen fixation by rhizobacteria.

(*c*) **Legume-*Rhizobium* symbiosis:** *Rhizobium*, a soil bacterium, enters in symbiosis with leguminous plants. It develops root nodules which are the site of N_s fixation. For detail discussion see Chapter 14, *Nitrogen Fixation*.

(*iii*) **Actinomycete-Non-legume Symbiosis:** From this class the species of *Frankia* are known to develop nodules which are known as *actinorhiza*. Nitrogen fixing nodulated non-legumes are:

the species of *Alnus, Casuarina, Cercocarpus, Comptonia, Hippophae, Discaria, Dryas, Elaeagnus, Myrica, Purshia, Shepherdia*, etc. These plants grow in such a condition where the concentration of nitrogen is low. One of the most extensively studied plant is the alder trees, *Alnus nepalensis* which grows in nitrogen-deficient soil. The extent of nitrogen gain by such angiosperms varies with soil types, climatic conditions and plant age. The nitrogen gain with *Alnus* is 12-200 kg/ha/annum, and with *Hippophae* 27-179 kg/ha/annum.

(*iv*) **Fungal Symbiosis (Mycorrhiza):** In 1885, it was a German Forest pathologist, A.B. Frank, who for the first time coined the term mycorrhiza to denote plant-fungus association. Mycorrhiza (fungus-root) has been defined as an apparent structure developed as a result of symbiotic association between fungi and plant roots. Mycorrhizal associations are diverse in both structure and physiological function. Garrett (1950) grouped the mycorrhizal fungi into the ecological category of root-inhabiting fungi (see Fig. 27.2). This indeed be regarded as end terms in the specialization of root-inhabitants, that is, of an ecological group that includes many important soil-borne plant pathogens.

Mycorrhizae develop as a result of symbiotic association between fungi and plant roots.

Frank classified the mycorrhizae into *ectotrophic* and *endotrophic* ones on the basis of trophic levels. However, on the basis of strictly morphological and anatomical features, mycorrhizae are divided into the three broad groups: *ectomycorrhiza, endomycorrhiza* and *ectendomycorrhiza* which correspond to the older and still commonly used terms ectotrophic, endotrophic and ectendotrophic mycorrhizae. That is literally by *outside, inside* and *outside-inside feeding*, respectively.

Harley and Smith (1983) have recognised the endomycorrhizae into five distinct types: (*a*) vesicular-arbascular (VA) mycorrhiza, (*b*) arbutoid mycorrhiza, (*c*) monotropoid mycorrhiza, (*d*) ericoid mycorrhiza, and (*e*) orchid mycorrhiza. Marks (1991) has recognised the seven forms of mycorrhiza (Table 25.3), the special features of which are briefly discussed as below:

Table 25.3 : Mycorrhizal types, their characters and distribution within the plant kingdom (after Marks, 1991)

	Mycorrhiza	*Host range*	*Types of relationships*
1.	Vesicular-arbuscular mycorrhiza	All groups of plant kingdom	Coiled intracellular hyphae, vesicle and arbuscules present
2.	Ectomycorrhiza	Gymnosperms and angiosperms	Sheath, intercellular hyphae
3.	Ectendomycorrhiza	Gymnosperms and angiosperms	Sheath optinal, inter and intra-cellular hyphae
4.	Arbutoid mycorrhiza	Very restricted, Ericales	Sheath, inter-and coiled intra-cellular hyphae
5.	Monotropoid mycorrhiza	Very restricted, Monotropaceae	Sheath, inter-and coiled intra-cellular hyphae
6.	Ericoid mycorrhiza	Very restricted, ericales	No sheath, no inter-cellular hyphae, long coiled
7.	Orchid mycorrhiza	Very restricted	Only coiled, intraclicular hyphae

(*a*) **Ectomycorrhiza:** Only 5% vascular plants develop ectomycorrhiza which predominates in family Pinaceae, Fagaceae, Betulaceae, Juglandaceae and Myrtaceae and in other tropical and temperate families. Fungi that participate in ectotrophic association include agaric Basidiomycetes, Gasteromycetes, Ascomycetes, fungi imperfecti and occasionally phycomycetes.

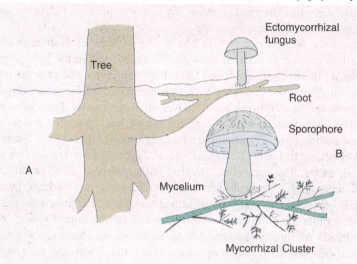

Fig. 25.1 : Ectomycorrhizal symbiosis. A, a tree root showing symbiosis of an ectomycorrhizal fungus; B, an enlarged view of a sporophore (of fungal symbiont) growing on mycorrhizal root of host tree (diagrammatic).

Strict host specificity is rare and, therefore, one plant may form mycorrhizae with several fungi simultaneously. Therefore, over 5,000 fungi of Basidiomycetes-Ascomycetes involved in forming ectomycorrhizae on 2000 woody plants. The fungi interact with feeder roots which in turn, undergo morphogenesis. The mycorrhizae may be unforked, bifurcated, nodular, multiforked or coralloid. Outside the root surface fungal mycelia form a compact and multilayered covering known as *mantle* (Photoplate 25.1A,B). It prevents the direct contact of root tissues with rhizosphere. A typical ectomycorrhizal symbiosis has been shown in Figs. 25.1 A-B.

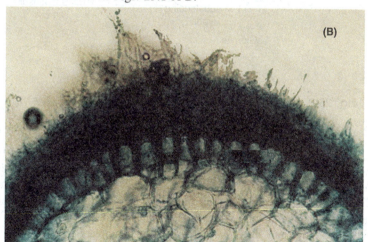

Photoplate 25.1: Ectomycorrhizal symbiosis in oak tree. A, oak seedling bearing ectomycorrhizal (ECM) lateral roots; B, a transverse section of ectomycorrizal root showing mycelial network, mantle and Hartig net (Coutesy : Dr. H.S. Ginwal, *Ph.D. Thesis*, Kumaun University, Nainital).

Thickness of mantle varies from 20-40 mm depending on mycorrhizal fungi, temperature, nutritional factors, etc. The fungus forms a network of mycelia in cortex which is known as *Hartig net*. The mycelia never enter the endodermis.

The fungi forming ectomycorrhiza are *Amanita muscaria, Boletus edulis, Cenococcus geophilus, Inocybe rimosa, Laccaria laccata, Leccinum, Lepiota, Russula spp., Pisolithus tinctorius, Suillus spp., Scleroderma citrinum, Rhizopogon* spp.,

(*b*) **Ectendomycorrhiza:** Ectendomycorrhiza shares the features of both ecto-and endo-mycorrhiza. They have less developed external mantle. The hyphae within the host penetrate its cells as well as grow within them. These are found in both gymnosperms and angiosperms. Very little is known about the fungi involved in this types of association due to little researches on them.

(*c*) **Vesicular-arbascular Mycorrhiza (VAM):** Over 90% of vascular plants of world flora form VA mycorrhiza. The mycosymbionts are widespread among both cultivated and wild plants, and found in bryophytes, pteridophytes, gymnosperms and angiosperms (Harley and Smith, 1983). The fungi forming VAM belong to family Endogonaceae of Zygomycotina. Hyphae are aseptate, inter-and intra-cellular in cortex. The intracellular hyphae either become coiled or differentiated into densely branched arbuscules. Arbuscules function as haustoria and perhaps involved in interchange of materials between plant and fungus. In addition, large, multinucleate, terminal or intercalary oil-rich vesicles may be produced on both inter-and intra-cellular hyphae. VAM are formed by about hundreads of fungal species. All of them belong to only six genera *viz.*, *Acaulospora, Gigaspora, Glomus, Entrophospora, Sclerocystis* and *Scutellospora*. Diagrams of *Glomus* and *Gigaspora* are given in Fig. 25.2.

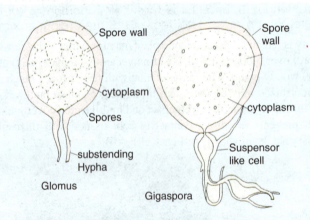

Fig. 25.2 : Spores of vesicular-arbuscular mycorrhizal (VAM) fungi (diagrammatic). *Glomus* has globose spore with constricted substending hypha without septum. *Gigaspora* has a bulbous substending hypha.

(*d*) **Ericoid mycorrhiza:** Ericoid mycorrhiza occurs throughout the fine root systems (hair roots) in the tribe Ericoidae of family Ericaceae (except tribe Arbutoidae). Many genera such as *Epachris, Leucopogon, Monotoa, Rhododendron, Vaccinum*, etc. develop ericoid mycorrhiza. Plants are woody shrubs or small trees found in open or acid peaty soil. They have usually fine roots on which the fungus established to outermost layer of cortical cells forming dense intracellular cells. The fungi may all be ascomycetes, for example *Pezizella, Clavaria* spp., etc.

(*e*) **Arbutoid mycorrhiza:** Mycorrhiza of the tribe Arbutoidae of family Ericaceae was first described from *Arbutus unedo*. The host plants are mostly woody shrubs and trees. Roots are typically herorhizic (the short roots being converted into mycorrhiza with a well defined sheath and a Hartig net), the fungus penetrates cortical cells where it forms extensive coils of hyphae. The mycosymbionts are of Basidiomycetes. Many of fungal symbionts which form symbiosis in these plant, also form mycorrhiza with conifers (Harley and Smith, 1983). It has been suggested that a transition between ecto- and endo- moycorrhizae exists in the arbutoid type of mycorrhiza, accounting for the term ectendomycorrhiza some times applied to this phenomenon.

(*f*) **Monotropoid mycorrhiza:** The family Monotropaceae which includes achlorophyllous plants (*e.g. Monotropa hypopitys*), develops monotropoid mycorrhiza. These plants completely depend on mycorrhizal fungi for carbon and energy. Roots form ball throughout which fungal mycelium ramifies enclosing the mycorrhizal roots of neighbouring green plants. The root ball is the survival organ of *Monotropa* during winter and after return of favourable conditions it gives rise to flowering shoots. With the root growth, a sheath and Hartig net are formed. From the hyphae a peg like haustoria push into epidermal and cortical cells. In the start, host cell wall invaginates to include fungal pegs, but finally pegs penetrate cell wall and emerge into cells. The structure and function of monotropoid mycorrhiza change with seasonal development of the host plants (Harley and Smith, 1983).

(*g*) **Orchid mycorrhiza:** In nature, orchids germinate only with infected endomycorrhizal fungi that subsequently colonize the host plants. The fungi are mostly the form genus *Rhizoctonia* with perfect state *Ceratobasidium, Sebacina* and *Tulasnella* occurring in Basidiomycetes (mainly Tulasnales) and Ascomycetes.

Effect of mycorrhizal fungi on their hosts

(*a*) **Mycorhizosphere effect:** Similar to rhizosphere, mycorrhizosphere (the close vicinity of ectomycorrhizae) shows increased microbial community leading to mycorhizosphere effect. The photosynthates flow into soil through roots and mycorrhizae support a diverse community of soil microorganisms, many of which influence plant growth. The mycorrhizosphere microorganisms may be facultative anaerobes, extracellular chitinase producers, phosphate solubilizers, and producers of siderophores, antibiotics, hormones, plant growth-suppressors and promotors (Lindermann, 1988).

Mycorrhizae aid plant growth : Soybeans without mycorrhizae (left) and with different strains of mycorrhizae (center and right).

(*b*) **Nutrient uptake and Translocation:** Mycorrhizae increase the absorptive surface of root resulting in increased uptake of water and nutrients from the soil. The ectomycorrhizal fungi translocate phosphorus, nitrogen, calcium and amino acids, and increase translocation of Zn, Na and other minerals to the hosts. Their hyphae extract N and transport from soil to plant due to increased absorptive surface area. Plant available phosphorus in soil is in small amount (1-15%) of total P content. The by-products of fungi dissolve several insoluble nutrients. Three mechanism of mycorrhizal activity has been discussed for weathering soil P and transport to host plants: (*a*) the interaction of mycorrhizal fungi and phosphate solubilizing bacteria, (*b*) production of phosphatases by the mycorrhizal fungi, and (*c*) production of organic acids by mycorrhizal fungi (Allen, 1991).

Translocation of P in fungal hyphae takes place by cytoplasmic streaming. P is stored in the form of polyphosphates due to polyphosphate kinase activity. Then P is transferred to host plant after break down of phosphates by phosphatases and release of inorganic phosphate. P is accumulated in mantle and Hartig net and, thereafter, transferred from Hartig net to host tissue.

Typical irregular branching of white, smooth ectomycorhizal fungi produces IAA possibly involved in longevity of pine roots.

Similarly, the VAM fungi also influence growth, exudation and nutrient uptake in host plants. Polyphosphate granules have been found in arbuscules, hyphae and vesicles of VAM fungi. Chitin appears to be the main carbohydrate-related material present in vesicle and hyphal walls. *Glomus fasciculatus* translocate P over a distance of at least 7 cm, and *Rhizopogon luteus* to 12 cm.

(*c*) **Transfer of metabolites from Host to Fungal symbiont and the other plants:** The products of photosynthesis (photosynthates) move from host to the fungal symbiont. However, host to host transfer of carbohydrate via a shared fungal symbiont also takes place.

(*d*) **Growth-hormone/antibiotic production:** Some of the ectomycorrhizal fungi produce indole acetic acid (IAA) which possibly is involved in morphogenesis and longevity of roots. Moreover, *Leucopaxillus Cerealis* var. *piceina* is known to produce growth inhibiting antibiotics.

(*e*) **Plant Protection/Biocontrol of Pathogens:** Both VAM and ectomycorrhizal fungi make the plants drought and frost resistant, increase tolerance to stress against soil temperature, soil toxins, high acidity, and heavy metal toxicity.

VAM and ectomycorrhizal fungi inhibit the infection of pathogens to plant roots. The fungal mantle acts as a passive mechanical barrier influencing either the pathogen or its spread in host tissues.

Works on mycorrhizal fungi in India

Though the study of mycorrhizal fungi has been a neglected field, yet in India works on mycorrhizal fungi was started by B.K. Bakshi at Forest Research Institute, Dehra Dun. His project report `*Mycorrhiza and its role in forestry* (1974) published by F.R.I. has served as a mile stone for the beginner. Mishra and Sharma (1981) reported the association with *Pinus kesiya* of *Amanita muscaria, Boletus edulis, Cenococcus geophilus, Inocybe rimosa, Russula roseipes, Scleroderma aurantium, Suillus bovinus* from Meghalaya of North East Himayala. Moreover, association of ectomycorrhizal fungi with some other gymnosperm (*Cedrus, Cryptomeria* and *Pinus*) has been explored by R.R. Mishra and his research group. The work on this aspect done by K.S. Thind and his research group at Chandigarh cannot be overlooked. Lakhanpal (1987) has reported 72 fungal species forming mycorrhiza in several forest trees in the Western Himalaya. At Forest Research Institute, Dehra Dun, in addition to significant research work done by B.K. Bakshi and coworkers, mycorrhizal researches are being strenghthened. *Amanita muscaria, L. laccata* and *Scleroderma citrinus* were recorded from *Pinus patula,* and *A. muscaria,* and *L. chinensis* were identified from

Eucalyptus globulus (ICFRE, 1993). Ginwal *et al* (1996) surveyed oak forests of Kumaon Himalaya and reported the genera *Agaricus, Amanita, Russula, Entoloma, Laccaria, Scleroderma, Rhizopogon* and *Tricholoma* that form mycorrhiza. A total of 27 sporophore producing ectomycorrhizal fungi were identified. Variations in genotypes of the host, biotic disturbances in the forest, tree density, soil temperature, soil pH, moisture and the presence of antagonistic microorganisms were the factors that governed the occurrence of ectomycorrhizal fungi in different oak forests (Ginwal *et al.*, 1996). The western and eastern Ghats of South India have been extensively surveyed by A. Mahadevan and his research group of Madras University. Dubey *et al.* (1998) have reviewed the influence of nutrients on formation and growth of ectomycorrhiza.

In India, B.K. Bakshi (1974) was the first to publish an account of 14 spore types of 5 genera of Endogonaceae (VAM Fungi) such as *Glomus, Gigaspora, Acaulospora, Endogone* and *Sclerocystis* from forest soils. Sharma *et al* (1986, 1987) recorded the species of *Acaulospora, Gigaspora, Glomus* and *Sclerocystis* from rhizosphere soils of different forest trees of Meghalaya. Negi (1993) recorded *Endogone* sp. in soils of *Cupressus torulosa* and VA-mycorrhizal fungi in root tissues from Nainital. Vesicles of varying colours and numbers were observed on *C.torulosa* roots throughout the year. The maximum infection was recorded in winter and minimum in summer seasons (Dubey and Negi, 1995). In Himachal Pradesh 10 districts were periodically surveyed over

a wide altitudinal range of 500 to 2500 meters. A total of ten VAM fungi of four genera were recorded (ICFRE, 1993).

Many epiphytic and terrestrial orchids are considerably dependent on mycorrhizal fungi for their carbon sources. Katiyar *et al* (1986) studied the mycorrhizal status in certain tropical epiphytic orchids from Khasi and Garo hills of Meghalaya. Diversity of mycorrhizal fungi in the Himalaya with emphasis on their forms, function and management has been reviewed by Dubey and Ginwal (1997).

A significant work on VAM fungi has been done by Sudhir Chandra and Kehri of the University of Allahabad especially on management of waste land and introduction in crop field for high yield of crops. D.J. Bagyaraj and his research group have done excellent work on VAM fungi as far as crop improvement is concerned. Reena and Bagyaraj (1990) screened several VAM fungi

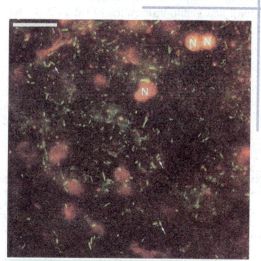

Fungal-bacterial interaction : Stained bacterial endosymbiont in unfixed spores of fungus *Gigaspora margarita.* Living bacteria fluoresce bright yellow green; lipids and fungal nuclei (N) appear as diffuse masses.

for their suitability to use as inoculants for two slow growing tree species, *Acacia nilotica* and *Calliandra calothyrsus*. They found that the inoculated seedlings had greater plant height, leaf number, stem girth, biomass and phosphorus and zinc content. Johri and Mathew (1989) successfully produced inocula of VAM fungi in bulk on *Phaseolus mungo* roots of *G. fasciculatus, G. caledonium, Gigaspora margarita,* and *G. calospora*. Of these *G. margarita* showed the highest percentage of root infection. Tata Energy Research Institute (TERI), New Delhi is contributing significant work on mycorrhizal fungi. In recent years researches on VAM fungi have gained momentum throughout the country.

III. Animal-Microbe Interactions

There are many kinds of microorganisms that interact with different groups of animals and develop a variety of relationships. Some of the relationships have been discussed in this section.

1. Destructive Associations

Pathogenic microbes interact with animals including man and cause many kinds of disease. For detail description of protozoan, bacterial and fungal (as well as viral) diseases see Chapters 21 and 22, *Medical microbiology*. However, destructive associations are also found between two microbes such as fungi and amoebae, nematode and fungi, etc. (see section *microbe-microbe interactions*).

2. Neutral Association (neutralism)

Normal microbiota of human body: There is a large number of microorganisms that normally act as the resident of different body organs of humans such as skin, nose and nasopharynx, oropharynx, respiratory tract, mouth, eyes, external ears, stomach, small intestine, large intestine (colon), and genito-urinary tract (Table 25.4). Reasons of having informations about the normal human microbiota are :

(*a*) to have an understanding of microorganisms at specific site so that greater insite into the possible infections can be provided,

(*b*) to help the physician investigator so that he can understand the causes and consequences of overgrowth of microorganisms normally absent at a specific body site, and

(*c*) to increase awareness of the role of indigenous microbiota that stimulate host immune response

Table 25.4 : Normal microbiota of human body

Body sites	Microorganisms
Ear	Coagulase-negative staphylococci, diphtheroids, pseudomonads, Enterobacteriaceae
Eye	Coagulase-negative staphylococci, *Haemophilus* spp., *Staphylococcus aureus*, *Streptococcus* spp.
Intestine (small)	Species of *Lactobacillus, Bacteroides, Clostridium,* Enterococci, *Mycobacterium*
Intestine (large)	*Actinomyces, Acinetobacter, Bacteroides, Clostridium,* Coagulase-negative staphylococci, *Escherichia coli,* Enterococci, *Lactobacillus, Klebsiella, Mycobacterium, Pseudomonas, Proteus, S. aureus,* streptococci
Mouth	Species of *Actinomyces, Candida,* coagulase-negative staphylococci, Diphtheroids, *Fusobacterium, Haemophilus, Neisseria, Porphyromonas, Prevotella, S. aureus, Streptococcus pneumoniae.*
Nose	Coagulase-negative staphylococci, *Haemophilus, Neisseria, S. aureus, S. pneumoniae,* Viridans streptococci
Skin	Coagulase-negative staphylococci, diphtheroids, *Bacillus, Candida, S. aureus,* streptococci, *Mycobacterium*
Stomach	Species of *Lactobacillus, Streptococcus, Staphylococcus, Peptostreptococcus*
Urethra	Coagulase-negative staphylococci, diphtheroids, *Bacteroides, Fusobacterium, Mycobacterium, Peptostreptococcus*
Vagina	Species of *Bacteroides, Candida, Clostridium,* diphtheroids, *Lactobacillus, Gardnerella veginalis, Peptostreptococcus,* streptococci,

3.　Symbiotic Associations

Symbiotic associations of bacteria, fungi and protozoans with insects, birds and herbivorous mammals are discussed below:

(*i*) Ectosymbiosis of protozoa, Bacteria and Fungi with Insects and Birds: Most of the animals such as insects (termites and cockroaches) cannot utilize the cellulose and lignin components of woody tissues of tree due to lack of cellulose and lignin degrading enzymes. Therefore, several insects develop ectosymbiotic association with cellulose- and lignin-decomposing microorganisms that can degrade these substrates. All termites and cockroaches that eat upon wood, harbour flagellated protozoa in their guts. These protozoa digest cellulose. In turn the protozoa develop symbiotic association with certain N_2-fixing bacteria and spirochetes which perhaps also help in cellulose degradation. In addition, during moulting season of cockroaches hormones (*e.g.* ecdysone) are secreted which induce cyst formation in symbiont protozoan.

(*ii*) Endosymbiosis of Bacteria and Fungi with Birds and Insects: Moreover, there is a group of birds belonging to the genus *Indicator* which are commonly known as honey guides. These birds are found in Africa and also in India. These birds eat upon remnants of exposed honey comb but cannot digest bees wax. Therefore, they harbour in their intestine the two microbes, *Micrococcus cerolyticus* and *Candida albicans* for carrying out the digestion of bees wax.

Except carnivorous insects, the others that live upon blood or plant sap develop symbiotic association with bacteria such as coryneforms and Gram-negative rods, and *Nocardia* (a member of actinomycetes). These microsymbiont are present in insect hosts in specialised cells. The cells that contain fungi are called *mycetocytes,* and those that contain bacteria are called *bacteriocytes.* These microsymbionts provide to the insects with some growth factors (that are lacking in insects) and some essential amino acids. Also the microsymbionts assist in breakdown of certain waste products.

(*iii*) Ruminant Symbiosis: The herbivorous mammals (*e.g.* cattles, sheep, goats, camels, etc) are known as ruminants because they have a special region of gut which is called *rumen*. These animals use plant cellulose as the source of carbohydrate which is not digested in normal gut. The cellulosic material is digested in rumen which acts as incubation chamber teeming with protozoa and bacteria. In some animals like cow, the size of rumen is very large. Some of anaerobic cellulose-digesting bacteria (*e.g. Bacteroides succinogens, Ruminococcus flavofaciens, R.albus* and *Botryovibrio fibrisolvens*) develop mutualistic symbiosis, and hydrolyse cellulose and other complex polysaccharides to simpler forms which in turn are fermented to fatty acids (*e.g.* acetic acid, propionic acid, butyric acid) and gases (methane and carbon dioxide). Some of the bacteria are capable of digesting proteins, lipids and starch as well. Lignin fraction of plant remains undigested. The rumen bacteria ferment proteins and lipids and produce hydro-

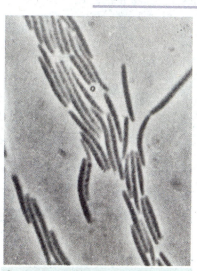

Scanning electron micrograph of bacterial colonies of *Micrococcus* on agar (× 31,000) : The bacterium is harboured in the intestine of honey guides for carrying out the digestion of bees wax.

gen and carbon dioxides gase, which in turn is converted into methane by *Methanobacterium ruminantium*. The bacteria of rumen multiply into a large population. However, most of them are passed into stomach along with undigested material where they are killed by proteases and other enzymes. The fatty acids in rumen are absorbed and gases are passed out.

IV. Microbe-microbe Interactions

Different types of beneficial and harmful interrelationships between micro-organisms, and plants/animals have been discussed earlier. Similarly, microorganisms interact themselves and lead to beneficial and harmful relationships. Some of the interactions and interrelationships have been discussed in this connection.

1. Symbiosis Between Alga and Fungus (Lichens)

Crustose (encrusting) lichens growing an a granite post.

Lichen is a thallus of dual organism *i.e.* a fungus and an alga that form a self supporting combination.. The fungal component is called *mycobiont* and the algal partner as *phycobiont*. The two groups of organisms live in close proximity and appear as a single plant. The fungus forms the thallus of the lichen, whereas the alga occupies only 5-10% mass of the thallus.

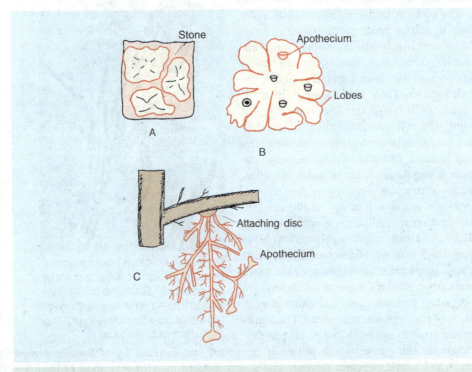

Fig. 25.3 : Lichens. A, crustose lichen; B, foliose lichen; C, fruticose lichen.

Mycelium of the fungal partner forms a close network that appears as tissue. Inside this compact mass of mycelium algal cells are embedded. Generally, fungi derive nutrition saprophytically from dead organic materials, or parasitically from a living host. But in lichen fungal mycelium derives nutrition from the alga. The algal cells form food by themselves and/or fix N_2 from the atmosphere which then are diffused into fungal hyphae. This type of mode of nutrition is called *biotrophic nutrition* which is seen in lichen. The members of algae forming lichen belong to Cyanophyta or Chlorophyta. However, it may be unicellular or filamentous forms. The genera of blue-green algae are *Nostoc, Gloeocapsa, Rivularia* and *Stigonema*. Of the green algae, species of *Trebouxia* are the most common unicellular green algae. The fungal partners forming lichen are mostly the members of Ascomycetes, and 2-3 genera of Basidiomycetes. No fungus of Phycomycetes enters into lichen formation. Symbiosis is based on the facts that alga provides food to fungus, and fungus provides shelter to alga.

(*i*) **Classification:** On the basis of nature of fungal partner and fructification types lichen are divided into two groups: *ascolichens* (in which fungal component is an Ascomycete), and *basidiolichens* (in which the fungal component is a Basidiomycete). However, on the basis of the habitat lichens are divided into three groups: *saxicolous* (growing on rocks or stones), *corticolous* (growing on leaves and bark of trees epiphytically) and *terricolous* (growing on soil).

(*ii*) **Lichen Thallus:** As in lower plant, in lichens also the plant body is known as thallus. Lichen thalli are grey, or greyish green in colour. On the basis of structure of thalli, lichens are of three main types (Fig. 25.3): (*a*) *crustose lichens* (flat thalli, without any lobe, growing on stones, rocks, bark or any hard sub-strata, and appears like crust, for example *Haemmatomma puniceum* and *Graphic scripta*), (*b*) *foliose lichens* (thalli are flat, much lobed and leaf-like appearing as twisted leaves, have distinct lower and upper surface, attached to substrate with rhizoid-like structure called *rhizinae*, for example *Chaudhuria, Cetraria, Parmelia, Peltigera, Physcia* and *Xanthoria*), and (*c*) *fruticose lichens* (thalli are most conspicuous, most complex, and slender and freely branched, the

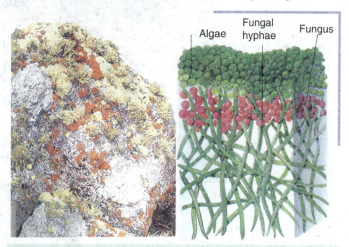

Lichens are symbootic associations : T.S. of a lichen thallus showing bottom green alga penetrated byfungal hyphae.

branches are cylindrical, flattened and form thread like tuft, thalli not differentiated into upper and lower surfaces, for example *Cladonia, Ramalina* and *Usnea*.

2. Antagonistic Interactions (Antagonism)

The composition of the microflora/microfauna of any habitat is governed by the biological balance created through interactions and associations of all individuals present in a community. However, the environmental conditions upset the equilibrium. Any inhibitory effect of an organism created by any means to the other organism(s) is known as antagonistic interaction, and the phenomenon of this activity is called *antagonism*. Antagonism is the balancing wheel of the nature.

Through this mechanism some sorts of biological equilibrium is maintained. Antagonism has three facets, *amensalism, competition*, and *parasitism* and *predation*.

(i) Amensalism (Antibiosis and Lysis): Amensalism is the phenomenon where one microbial species is adversely affected by the other species, whereas the other species is unaffected by the first one. Generally, amensalism is accomplished by secretion of inhibitory substances such as antibiotics, etc. Antibiosis is a situation where the metabolites secreted by organism A inhibits the organism B, but the organism A is unaffected (Photoplate 25.2). Metabolites penetrate the cell wall and inhibit its activity by chemical toxicity. Generally, antimicrobial metabolites produced by microorganisms are antibiotics, siderophores, enzymes, etc. The potent antagonists e.g. *Trichoderma harzianum* and *T. viride* are known to secrete cell wall lysing enzymes, β-1, 3-glucanase, chitinase, etc. Lysis of fungal mycelium occurs due to secretion of enzymes.

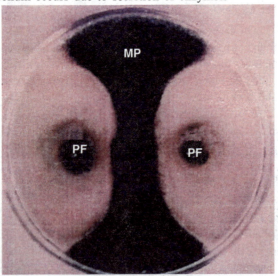

Photoplate 25.2 : Antibiosis between *Macrophomina phaseolina* (MP) and *Pseudomonas fluorescens* (PF). Arrow indicates the zone of inhibition

Siderophores: Siderophores are the other extracellular secondary metabolites which are secreted by bacteria (*e.g. Aerobacter aerogenes, Arthrobacter pascens, Pseudomonas cepacia, P.fluorescens*), Actinomycetes (*Streptomyces* spp.), yeast (*Rhodotorula* spp.), fungi (*Penicillium* spp.), and dinoflagellates (*Prorocentrum minimum*). Siderophores are commonly known as microbial iron-chelating compounds because these have a very high chelating affinity for Fe^{3+} ions and very low affinity with Fe^{2+} ions. Siderophores are low molecular weight compounds. These after chelating iron (III) transport it into bacterial cells. Kloepper *et al* (1980) were the first to demonstrate the importance of siderophore production by PGPR in enhancement of plant growth. Siderophores chelate Fe^{3+} and make Fe^{3+} deficient condition for other microorganisms. Consequently growth of mi-

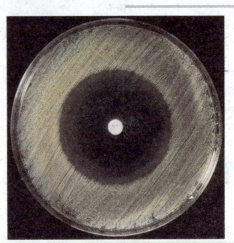

Amensalism: A negative Microbe-Microbe Interaction. Antibiotic production and inhibition of growth of a susceptible bacterium on an agar medium.

crobe is inhibited. When the siderophore producing PGPR is present on root surface, it supplies iron to plant. Therefore, plant growth is stimulated. For example, secretion of siderophore by *Pseudomonas fluorescens* and inhibition in growth of *Macrophomina phaseolina* (forming a clear zone) is shown in Photoplate 24.2. Role of siderophores in biological control of plant pathogens is of much importance in recent years.

(*ii*) **Competition:** Among the microorganisms, competition exists for nutrients, including oxygen and space but not for water potential, temperature or pH. Success in competition for substrate by any particular species is determined by competitive saprophytic ability and inoculum potential of that species (Garrett, 1950, 1956). Garrett (1950) has suggested four characteristics which are likely to contribute to the competitive saprophytic ability: (*a*) rapid germination of fungal propagules and fast growth of young hyphae towards a source of soluble nutrients, (*b*) appropriate enzyme equipment for degradation of carbon constitu-

ents of plant tissues, (c) secretion of fungistatic and bacteriostatic growth products including antibiotics, and (d) tolerance of fungistatic substances produced by competitive microorganisms.

Parasitic castration of plants by Endophytic Fungi : Stroma of the fungus *Atkinsonella hypoxylon* infecting *Danthonia compressa* and causing abortion of the terminal spikelets.

Thus competition exists for limiting resources. The inadequate quantity of readily available carbon compounds is a more likely basis for competition. At low level of carbon, the fast growers will often hold slow growers in check when both are added to sterilized soil. But there is no such check on the less active heterotroph when carbon supply is adequate (Alexander, 1997). Under these conditions, competitiveness is directly correlated with growth rate.

(*iii*) **Parasitism and Predation:** Parasitism is a phenomenon where one organism consumes another organism, often in a subtle and non-debilitating relationship. Predation is an apparent mode of antagonism where a living organism is mechanically attacked by the other with the consequences of death of the former. It is often violent and destructive relationship. These phenomena are dealt with the example of fungi, amoebae and nematodes (Table 25.5).

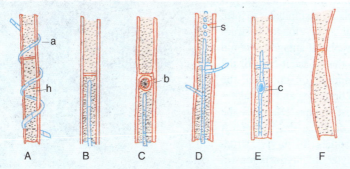

Fig. 25.4 : Mycoparasitism showing post-interaction events. A, coiling (*a*, antagonist; h, host hypha); B, penetration; C, barrier formation (*b*) by host; D, branching and sporulation (s) by antagonist; E, chlamydospore (*c*) formation; F, lysis of host hyphae (diagrammatic, after Dubey and Dwivedi, 1986).

(*a*) **Mycoparasitism (Fungus-fungus interaction):** When one fungus is parasitized by the other fungus, this phenomenon is called mycoparasitism. The parasitizing fungus is called hyper parasite and the parasitized fungus as hypoparasite (Fig. 25.4). Mycoparasitism commonly occurs in nature. As a result of inter-fungus interaction, several events take place which lead to predation viz., coiling, penetration, branching, sporulation, resting body formation, barrier formation to check the entry of pathogen, and lysis of host cell(s) (Fig. 25.4).

In coiling event (A) the hyperparasite *i.e.* antagonist (*a*) recognises its host hypha *i.e.* hypoparasite (h) among the microbial community, comes in its contact and coils around the host hypha. Host recognition by the antagonist has been discussed on molecular basis. Manocha (1985) has given the basis of host-recognition by mycoparasites. Cell wall surface of host and non-host microbes contains D-glucose and N-acetyl-D-galactosamine residues as lectins present on the cell wall, an antagonist recognises the suitable sites (lectin residues) and binds the host hypha. As a result of coiling the host hypha loses its strength. Antagonist dissolves cell wall of host and enters inside the lumen of the later (Fig. 25.4B).

Table 25.5 : Examples of predation and parasitism (Dubey, 2005).

Mode of antagonism	Plant Pathogens	Antagonists	Post-infection events
Mycophagy	Cochliobolus sativus	soil amoebae	Perforation in conidia
	Gaeumannomyces graminis var.tritici	soil amoebae	Penetration and hyphal lysis
Mycoparasitism	Botrytis alli	Gliocladium roseum	Penetration of hyphae
	Cochliobolus salivas	Myrothecium verrucaria	Antibiosis and penetration.
	Rhizoctonia solani and Fomes annosus	Trichoderma viride	Coiling, cytoplasm coagulation
	Sclerotium rolfsii	T. harzianum	Coiling, penetration lysis
Nematophagy	Heterodera rostochiensis	Phialospora hetroderae	Cyst penetration and egg killing

Some times host develops a resistant barrier (Fig. 25.4C) to prevent the penetration and proliferation inside the lumen. Host's cytoplasm accumulates to form a spherical, irregular or elongated structure, so that the hypha of antagonist could not pass towards the adjacent cells of the hypha (C). Depending on nutrition, the antagonist forms branches and sporulates (s) inside the host hypha (D). Until the host's nutrients deplete, the antagonist produces resting bodies (the survival structures), for example chlamydospores (*c*) inside the host hypha (E) due to loss of nutrients and vigour for survival (Table 25.5; Fig. 25.4F) (Dubey and Dwivedi, 1986).

Mycelia, microconidia and chlamydospores of *Histoplasma capsulatum var*-capsulation the host fungus in soil. Chlamydospores (Survival structures develop a resistant barrier.

(*b*) **Mycophagy:** Mycophagy is the phenomenon of feeding upon fungi by amoebae. Many amoebae are known to feed on pathogenic fungi. The antagonistic soil amoebae are *Arachnula, Archelle, Gephyramoeba, Geococcus, Saccamoeba, Vampyrella,* etc. These amoebae interact with fungal hyphae and make perforations. The fungi on which perforations have been observed are *Cochliobolus sativus, Gaeumannomyces graminis* var. *tritici, Fusarium oxysporum, Phytophthora cinnamomi* (Chakraborty and Warcup, 1983; Dwivedi, 1986). On the lysed hyphae of these fungi amoebae develop round cysts (Dwivedi, 1986).

Chakraborty *et al* (1983) have described the following three major steps of feeding on fungal propagules by soil amoebae:

Attachment: As a matter of chance trophozoites of amoebae attach to fungal propagules i.e. conidia, hyphae, etc. The attachment occurs by chemotaxis or thigmotaxis.

Engulfment: The fungal propagules according to its size are fully engulfed by amoebae. But the small trophozoites attached to the hyphal wall or spore make perforations on it.

Digestion: The completely or partially engulfed propagules/cytoplasm of the host fungi are digested in a large central vacuole formed inside the cysts.

(*c*) **Nematophagy:** The phenomenon of eating upon nematodes by fungi is known as nematophagy and the fungi as *predaceous fungi*. Fungi are mechanically involved in attacking and killing the nematodes resulting in consumption of nematodes. The predaceous fungi are widely distributed in the surface litter and decaying organic matter. Over 50 species of fungi are known that attack nematodes. Different developmental stages of nematodes are susceptible to attack by different types of fungi.

As early as 1869, for the first time M.S. Woronin established the fact that the predaceous fungi capture and destruct the nematodes with certain specialised trapping organs. During 1930s, C. Drechsler added greatly to the list of predaceous fungi and unravelled the mechanism of trapping. Duddington (1957) reviewed the work of fungi that attack microscopic animals and contributed significantly to the knowledge of nematophagous fungi.

Nematophagous (Predaceous) Fungi

The predaceous fungi are also termed as nematophagous fungi. The nematophagous fungi are of three main types on the basis of ecological habit: (*a*) nematode-trapping fungi, (*b*) endoparasitic fungi, and the (*c*) egg parasites.

(*i*) **Nematode-trapping Fungi:** Fungi capturing nematodes are called nematode-trapping fungi. Such fungi have evolved structural adaptations to trap or penetrate their prey. They may be predatory or endoparasites. There are varieties of ways by which fungi trap nematodes resulting in their death. These methods are discussed as below:

(*a*) **Adhesive hyphae:** The fungal hyphae form adhesive which capture nematodes. These hyphae produce adhesive at any point in response to nematode contact or the hyphae are coated with adhesive along their entire surface. At the point of hyphae where contact is made for capture, a thick and yellowish chemical material is secreted for example, *Stylopage hadra*. Thereafter, an outgrowth of hyphae similar to appressorium develops. When the nematode is trapped, it becomes inactive first and killed in the last after penetration of hyphae. After penetration, elongate, unbranched absorptive hyphae grow along the nematode body and completely exploit the contents.

(*b*) **Adhesive branches:** The nematode trapping fungi produce the most primitive and simple organ of capture, the adhesive branches, which are a few cells in height. From the main prostrate hyphae short laterals grows as erect branches on or below the substrate. Over the whole surface of branch a thin film of adhesive material is coated. Examples of adhesive branch producing fungi are *Dactylella cionopaga* and *D. gephyropaga* (Fig. 25.5 A).

(*c*) **Adhesive nets:** Nets are formed by fungal hyphae which are adhesive in nature. Nets may be in the form of a single hoop-like loop (*e.g. Arthrobotrys musiformis*) to a complex multibranched networks (*e.g.* A. *oligospora*). Upon observation with electron microscope it appears that the hyphae are coated with adhesive material. As the nematode comes in contact of hyphae it is attached at many points resulting in penetration by infectious hyphae. Initially penetration is accompanied by the formation of infectious bulb which leads to form hyphae which grow inside nematode. Hyphal growth exploit nutrients and results death of the prey.

(*d*) **Adhesive knobs:** Morphologically a distinct adhesive cell, globose to sub-globose in structure, is produced at the apex of a slender non-adhesive stalk containing 1-3 cells. A thin film of adhesive material is produced over the surface of knob. If a nematode is caught by a knob, soon it is attacked by several knobs with subsequent penetration. The immobilized nematode is destroyed thereafter. Examples of adhesive knob-producing fungi are *Dactyleria candida, Dactylella* and *Nematoctonus*.

(*e*) **Non-constricting rings:** From the prostrate creeping septate hyphae there arise erect and lateral branches which form non-constricting rings. Initially the branch is slender but widens subsequently and being curved to form a circular structure. At the point where tip of branch makes contact with supporting stalk, cell walls get fused. Thus it results in formation of three-celled ring with a stalk. A nematode enters the ring and moves forward. This results in marked constriction in cuticle. Generally rings are impossible to dislodge. During struggle the rings break from the weak point. Therefore, nematodes containing rings can move. Initially rings do not have any harmful effect on nematode but eventually nematode is penetrated and its body content is consumed. *Dactylaria can-*

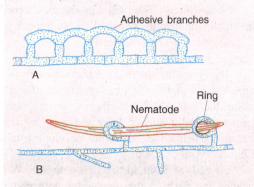

Fig. 25.5 : Predaceous (nematophagous) fungi. A, scalariform adhesive branches of *Dactylaria gephyropaga*; B, constricting rings of *Arthrobotrys dactyloides*.

dida and *D. lysipage* produce non-constricting rings, inspite of producing the adhesive knobs.

(*f*) **Constricting rings:** The constricting rings are produced similar to non-constricting rings but the supporting stalk is shorter and stouter (Fig. 25.5 B). In this case also a three celled ring is formed. It is a most sophisticated ring formed by predaceous fungi such as *Arthrobotrys anchonia, Dactylaria brachopaga* and *Dactylella* which are abundant in soil.

The nematode is captured by garroting action of the ring cell. By swallowing the ring cell graps the nematode in a single hold when a nematode enters into the ring, friction created by its body induces to swell the rings soon. The cells swell inwardly by three times greater than the original one within 1/10 of a second resulting in body of nematode deeply constricted. Struggle between nematode and fungi goes on for a few minutes. Thereafter nematode becomes still and hyphae from ring cell penetrate the body and exploit the nutrients of nematode with the consequences of death. The most potential predatory fungi are the species of *Dactylaria*

(*g*) **Mechanism of interaction:** The body of nematode consists of a low molecular weight peptide (or possibly a single amino acid) which is called *nemin* (Pramer and Kuyama, 1963). Nemin is water soluble and potential stimulant for trap-formation. It causes morphological changes in nematophagous fungi. The process of nematode-fungus interaction is accomplished through a series of molecular events resulting in nematode death. Pramer and Kuyama (1963) identified on the trap of *A. oligospora* the initiator of prey-predator recognition as *lectin*. The lectin of fungus binds especially to the sugar, N-acetyl-D-galactosamine, present on nematode cuticle. The lectin has also been purified. It is a protein of molecular weight 22,000 daltons. The trapped nematodes secrete mucilage which has been identified under electron microscope. The nematode cuticle is lysed at the point where lectin combines with N-acetyl-D-glucosamine. Within an hour fungal hyphae penetrate the prey. The enzyme collagenase is secreted by the fungus which dissolves collagen protein of nematode cuticle. The hyphae which have penetrated the nematode digest the body content and translocated to rest of the parts of hyphae for fungal growth and reproduction.

(*ii*) **Endoparasitic Fungi:** Unlike nematode-trapping fungi, the endoparastic fungi do not extensively produce mycelium external to nematode body. But they attack nematodes through many modifications brought about in conidia. The endoparasitic fungi are species of *Cephalosporium, Meria, Verticillium, Catenaria, Meristacrum*, etc.

Catenaria anguillulae produces zoospores which track down nematodes by swarming, eventually encyst near nematode body orifice (*e.g.* anus, vulva and buccal cavity), penetrate and colonize the prey. The encysted zoospores produce germ tube which penetrate nematode through orifice or by dissolving cuticle. The infectious hyphae grow well inside nematode body, digest content and lyse the prey. Zoosporangia

Fungal Nematode Interaction : The fungus *Arthrobotrys*, the nemalode strangler, traps its nematode prey in a noose like modified hypha that swells when the inside of the loop is contacted.

are produced inside body from which numerous uniflagellated zoospores are liberated.

In addition, *Meristacrum asterospermum* forms adhesive conidia which attach to the cuticle of nematode. It germinates to form the hypha which swells and acts as infectious thallus. Similarly, adhesive spores are also produced by species of *Meria, Cephalosporium* and *Verticillium*. In *M. coniospora* an adhesive bud develops at the distal end of tear-drop shaped spores. The conidia attach to nematode body, germinate and penetrate through cuticle. After penetration, an infectious hypha in body cavity of nematode is formed, the amount of which increases eventually with the result of nematode death.

(*iii*) **Egg Parasites:** There are a few saprophytic fungi which attack on nematode eggs. When a fungal hypha comes in contact of an egg, a swollen structure at terminal portion develops at the point of contact. It gets attached to the egg where from a narrow infectious tube develops that penetrates the shell of the egg. After penetration, the infectious hyphae swell up and form a post-penetration bulb which looks like appressorium. From this structure there develops numerous irregularly branched absorption hyphae that consumes egg nutrients. Examples of egg parasites are *Dactyllela oviparasitica* and *Paecilomyces lilacinus* that penetrate root-knot or cyst nematode.

QUESTIONS

1. Classify the microorganisms on the basis of carbon sources and temperature.
2. Write a brief note on extremophiles.
3. Classify the microorganisms on the basis of nutrition.
4. What do you know about plant-microbe interactions? Discuss in brief destructive associations.
5. Write an extended note on plant-microbe interactions with emphasis on *symbiosis*.

6. What do you know about symbiosis? Write in brief different types of symbiotic interactions in plants.

7. What is mycorrhiza? Write in brief different types of mycorrhiza studied by you.

8. What are VAM fungi? Discuss in detail about benefits of VAM fungi provided to plant hosts.

9. Write in brief the effects of mycorrhizal fungi on their hosts.

10. Write in brief about works done on mycorrhizal fungi in India.

11. What do you know about microbe-microbe interactions? Write in detail about antagonism.

12. What do you know about parasitism and predation? Discuss in detail with suitable examples.

13. What do you know about predaceous fungi? Write an extended note on different types of predation mechanisms with suitable examples.

14. Write short notes on the following :

(*i*) Psychrophiles, (*ii*) Hyperthermophiles, (*iii*) Super-hyper-thermophiles, (*iv*) Extremophiles, (*v*) Saprophytes, (*vi*) Symbiosis, (*vii*) Associative symbionts, (*viii*) PGPR, (*ix*) Actinorhiza, (*x*) Ectomycorrhiza, (*xi*) VAM fungi, (*xii*) Mycorrhizosphere, (*xiii*) Normal microbiota of human body (*xiv*) Rumen symbiosis, (*xv*) Lichens, (*xvi*) Antagonism, (*xvii*) Amensalism, (*xviii*) Mycoparasitism, (*ix*) Predaceous fungi.

REFERENCES

Alexander, M. 1977. *Soil Microbiology*. 2nd ed. John Wiley & Sons, Inc, New York

Allen, M.F. 1991. *The Ecology of Mycorrhiza*. Cambridge Univ. Press, London, p. 184.

Ambasht, R.S. 1984. *A Text book of Plant Ecology*. 7th ed., Students' Friends & Co., Lanka, Varanasi.

Archibald, A.R.; Baddiley, J. and Heptinstall, A. 1973. The alanine ester component and magnessium binding capacity of walls of *Streptococcus aureus* H grown at different pH values. *Biochem et Biophysica Acta*. **291**: 629-634.

Baker, K.F. and Cook, R.J. 1974. *Biological control of Plant Pathogens*. W.H. Freeman & Co., San Fransisco, p. 433.

Bakshi, B.K. 1974. *Mycorrhiza and its Role in Forestry*. PL480 Project Report, F.R.I., Dehra Dun, India.

Chakraborty, S. and Warcup, J.H. 1983. *Soil Biol. Biochem*. 15: 181-185.

Chakraborty, S.; Old, K.M. and Warcup, S.H. 1983: *Soil Biol. Biochem* 15: 17-24.

Dubey, R.C. and Dwivedi, R.S 1985. Toxicity of cadmium on the growth of *Macrophomina phaseolina* causing root-rat of soybean as influenced by kaolinite, pH, zinc and managenese. Proc. *Indian Nati.Sci. Acad.* **B51** 259-264.

Dubey, R.C. and Dwivedi, R.S. 1986. Destructive mycoparasitic behaviour of *Fusarium solani* (Mart.) Appl. and Wool against *Mucor spinosus* van Tieghem. *Microbios Letters*, **32**: 123-127.

Dubey, R.C. and Negi, C.M.S. 1995. Seasonal occurrence of VA mycorrhizal fungi in roots of cypress trees in relation to edaphic factors. *Acta Botanica Indica*. 23: 173-175.

Dubey, R.C. and Ginwal, H.S. 1997. Prospects of mycorrhizal fungi in the Himalaya: Forms, function and management. In `*Himalayan Microbial Diversity*' (eds. S.C.Sati, J. Saxena, and R.C. Dubey), pp.317-338, Today and Tomorrow's Print & Pubt, New Delhi.

Dubey, R.C.; Pandey, S. and Tripathi, P. 1998. Influence of nutrients on formation and growth of ectomycorrhiza. In '*Trends in Microbial Exploitation*. (eds. Bharat Rai, R.S. Upadhyay & N.K. Dubey), pp. 56-70, International Soc. for Cons. Natural Resources, B.H.U., Varanasi.

Doubenmire, R.F. 1947. *Plant and Environment*. John Wiley & Sons, Inc, New York.

Duddington, C.L. 1957. *The Friendly Fungi : A new Approach to the Eel worm Problem*. Faber & Faber, London, p.188.

Dwivedi, R.S. 1986. Role of soil amoebae in take-all decline of wheat. *Indian Phytopath* 39: 550-560.

Goodfellow, M. and Williams, S.T. 1983 Ecology of actinomycetes. *Ann. Rev. Microbial*. 37: 189-216

Ginwal, H.S., Dubey, R.C. and Singh, R.P. 1996. Diversity of ectomycorrhizal fungi in different central Himalayan Oak forests. *Ann. For*. 4: 65-69.

Harley, J.L. and Smith, S.E. 1983. *Mycorrhizal symbiosis*. Academic Press, London, p.461.

I.C.F.R.E. 1993. Annual Research Report for Forestry Research and Education. DehraDun, p.80.

Johri, B.N. and Satyanarayana, T. 1984. Ecology of thermophilic fungi. In `Progress in Microbial Ecology`, eds. K.G. Mukherji, V.P. Agnihotri and R.P. Singh). pp. 349-361, Print House (India), Lucknow.

Johri, B.N. and Mathew, J. 1989. Strategies for mass cultivation of vesicular-arbuscular mycorrhizal fungi. In `Plant-Microbe Interactions`. (ed K.S. Bilgrami), pp. 293-304, Narendra Publ. House, Delhi.

Katiyar, R.S. Sharma, G.D. and Mishra, R.R. 1986. Mycorrhizal infection of epiphytic orchids in tropical forests of Meghalaya (India). *J. Indian Bot. Soc.* **65** : 329-334.

Kloepper, J.W. *et al.* 1980. Enhanced plant growth by siderophores produced by plant growth promoting rhizobacteria. *Nature*. **286**: 885-886.

Lakhanpal, T.N. 1987. In `Mycorrhiza Round Table`. Proc. of workshop. J.N.U., New Delhi, pp.53-83.

Lindermann, R.G. 1988. Mycorrhizal interaction with the rhizosphere microflora: the mycorrhizosphere. effect. *Phytopath*. **78**: 366-370.

Madigan, M.T. and Marrs, B.L. 1997. Extremophiles. *Sci.Am.* (April): 82-87.

Manocha, M.S. 1985. Specificity of mycoparasite attachment to the host cell surface *Can. J. Bot 63:* 772-778.

Marks, G.C. 1991. Casual morphology and evolution of mycorrhizae. *Agric. Ecosystem Environ.* **35**: 89-104.

Mishra, R.R. and Sharma, G.D.J *Indian Bot.Soc.* **60:** 168-171.

Person, C. I. Samborsiki, D.J. and Rohringer, R. 1962. The gene-for-gene concept. *Nature.* **194:** 561-562.

Pramer, D. and Kuyama, S. 1963. *Symp on Biochemical bases of morphogenesis in fungi.* **27:** 282-292.

Procter, B.E. and Parker, B.W. 1942. In *Aerobiology* (ed. S. Moulton), pp. 48-54, AAAs Publ. No. 17, Washington, D.C.

Reena, J and Bagyaraj, D.S. 1990. *Arid Soil Res. and Rehabilitation.* **4:** 261-268

Sharma, S.K; Sharma, G.D. and Mishra, R.R. 1986. Records of *Acaulospora* spp. from India. *Curr. Sci.* **55:** 724-726.

Sharma, S.K.; Sharma, G.D. and Mishra, R.R. 1987 Endogonaceae in subtropical forests of North-East India. *J. Indian Bat. Soc.* **66:** 266-268.

Singh, V.P., Saxena, R.K. and Srivastava, Sheela, 1984. Microbial life at low temperatures: ecological aspects. In `Progress in Microbial Ecology* (eds K.G. Mukherji, V.P. Agnihotri and R.P. Singh), pp. 341-347, Print House (India). Lucknow.

Microbial Ecology-II : Extremophiles

(Acidophiles, Alkalophiles, Halophiles, Psychrophiles, Thermophiles and Hyperthermophiles, Barophiles)

26

Extremophiles : Micro organisms that can withstand the extremes of the environment.

The extremophiles are those microrganisms whose optimal growth conditions are found outside of the 'normal' environment. It is now well recognized that many parts of the world contains extreme environment such as polar regions, acidic and alkalophilic springs and cold pressurized depths of the ocean are colonized by microbes which are specially adapted to these exceptional environments. Some restricted ranges of microbes have the ability to inhabit in extreme environments. **'Extreme'** is defined as the fact that microbes not only survive but actually grow in some of unusual environment on earth. It has stimulated scientific curiosity about the mechanisms permitting the survival and growth in such surroundings. These special organisms might provide a valuable resource for the exploitation of new chemical and biotechnological processes.

Exit in hostile environment.

Most of the extremophiles are prokaryotes and archaea. They produce enzymes, antibiotics, etc. which have biotechnological importance. Thermostable enzymes for specific applications in industries are more robust than their low temperature relates and may show enhanced resistance to organic solvents. In addition, they remove and recover metals and degrade toxic pollutants.

This chapter describes the important groups of extremophiles such as acidophiles, alkalophiles, halophiles, thermophiles and hyperthermophiles, psychrophiles and barophiles.

1. Acidophiles

Most natural environments on the earth are essentially neutral, having pH between 5 and 9. Only a few microbial species can grow at pH less than 2 or greater than 10. Microorganisms that live at low pH are called **acidophiles**. Fungi as a group tend to be more acid tolerant than bacteria. Many fungi grow optimally at pH 5 or below and a few grow well at pH values as low as 2. Several species of *Thiobacillus* and genera of archaea including *Sulfolobus* and *Thermoplasma* are acidophilic.

Thi. ferroxidans and *Sulfolobus* sp. oxidize sulfide mineral and produce sulphuric acid. The most important factor for obligate acidophily is the cytoplasmic membrane of obligatory acidophilic bacteria which actually dissolves and lyses the cell wall. This suggests that high concentration of H^+ ions are needed for membrane stability.

Massive growth of the extreme acidophile *Ferroplasma* in a california mine.

Highly acidic environment is formed naturally from geochemical activities (such as the production of gases in hydrothermal vents and some hot springs) and from the metabolic activities of certain acidophile themselves. Acidophiles are also found in the debris left over coal mining. Interestingly, acid-loving extremophiles can not tolerate great acidity inside their cells, where it would destroy DNA. They survive by keeping the acid out. But the defensive molecules provide this protection as well as others that come in contact with the environment must be able to operate in extreme acidity. Indeed **extremozymes** (their enzymes providing adaptability) are able to work at pH below one, more acidic than even vinegar or juice of stomach. Such enzymes have been isolated from the cell wall and underlying cell membrane of some acidophiles.

(i) **Physiology:** Obligate acidophiles have an optimum pH for growth which remains extremely low (1 to 4). To shield the intracellular enzymes and other components from low to medium pH, the organisms maintain a large pH gradient across the membrane. Special forms of lipids are present in their membrane which may minimize the leakage of H^+ down the pH value. For instance, the presence of certain fatty acids has been reported to provide special adaptations to growth and survival at extremely low pH. Acidophiles maintain the cytoplasmic pH

The bacterium *Thermoplasma acidophilum*, that lives in hot acidic springs, survives by keeping the acid out. Its'DNA is protected from the acid by a protein coat).

around 6.5. In these organisms, the pH remains generally 1-2 which is lower in comparison to neutrophiles and alkalophiles. In acidophiles the pH is compensated by positive inside electric potential which is opposite to that present in neutrophiles. The reversed electric potential is generated by electrogenic K^+ uptake which allows the cells to extrude more H^+ and thus maintain the internal pH.

(ii) **Molecular Adaptation:** Most critical factor for obligate acidophily lies in the cytoplasmic membrane. When the pH is raised to neutrality, the cytoplasmic membrane of obligately acidophilic bacteria actually dissolve and the cells lyse. It is suggested that the high concentration of hydrogen ions are required for stability of membrane that allows bacteria to survive.

(iii) **Applications:** Potential applications of acid-tolerant extremozymes range from catalysts for the synthesis of compounds in acidic solutions to additives for animal feed which are intended to work in animal stomach. When added to feed, the enzymes improve the digestibility of expensive grains, therefore avoiding the need for more costly food. Rusticyanin proteins from acidophiles help in acid stability. Expression of heterogenous arsenic resistance genes in the iron-oxidizing *Thiobacillus ferrooxidans* has been established as biotechnological approach of bioremediation.

2. Alkalophiles

Alkalophilies live in soils laden with carbonate and in Soda lakes, such as those found in the Rift Valley of Africa and the west U.S. The first alkalophilic bacterium was reported in year 1968. Most alkalophilic prokaryotes studied have been aerobic non-marine bacteria and reported as *Bacillus* spp. Krulwich and Guffanti (1989) separated them into two broad categories: alkali-tolerant organisms (pH 7.0-9.0) (which cannot grow above pH 9.5) and alkalophilic organisms (pH 10.0-12.0). Most of the alkalophilic organisms are aerobic or facultative anaerobic. Some alkalophiles are *Bacillus alkalophilus*, *B. firmus* RAB, *Bacillus* sp. No. 8-1 and *Bacillus* sp. No. C-125 which bear flagella and hence are motile. The flagella induced motility is considered by a sodium motive force (smf) instead of proton motive force (pmf). They are motile at pH 9-10.5 but no motility is seen at pH 8. The Indigo-reducing alkalophilic bacterium (*Bacillus* sp.) isolated from indigo ball was used to improve the indigo fermentation process. Their cell wall contains acidic compounds similar in composition to peptidoglycans.

(i) **Physiology:** The cell surface of alkalophiles can maintain the neutral intracellular pH in alkaline environment of pH 10-13. The recommended concentration of NaOH for large scale fermentation is 5.2% depending upon organism. The pH should remain 8.5-11. Sodium ions (Na^+) are required for growth, sporulation and also for germination. The presence of sodium ions in the surrounding environment has proved to be essential for effective solute transport through the membranes.

In the Na^+ ion membrane transport system, the H^+ is exchanged with Na^+ by Na^+/H^+ antiport system, thus generating a sodium motive force (smf). This drives substrate accompanied by Na^+ ions into the cell. The incorporation of α-aminobutyrate (AIB) increased two fold as the external pH shifts from 7 to 9, and the presence of Na^+ ions significantly enhance the incorporation. Molecular cloning of DNA fragments conferring alkalophily was isolated and cloned. This fragment is responsible for Na^+/H^+ antiport system in the alkalophily of alkalophilic microorganisms.

(ii) **Molecular Adaptation:** Alkalophiles contain unusual diether lipids bonded with glycerol phosphate just like other archaea. In these lipids, long chain, branched hydrocarbons, either of the phytanly or biphytanyl type, are present. The intracellular pH remains neutral in order to prevent alkali-lablie macromolecules in the cell. The intracellular pH may vary by 1-1.5 pH units from neutrality which helps these organisms to survive in highly alkaline external environment.

(iii) **Applications:** Some alkalophiles produce hydrolytic enzymes such as alkaline proteases, which function well at alkaline pH. These are used as supplements for house hold detergents. For example an alkaline protease called **subtilisin** has been produced from *B. subtilis* which is used in detergent. The stone washed denim fabric is due to the use of these enzymes. These enzymes soften and fade fabric by degrading cellulose and releasing dyes (Table 26.1).

Table 26.1 : Some extremozymes and their applications.

Extremozyme	Uses
Thermozymes	Required for DNA amplification reactions and industrially important product formation
Halozymes	Proteases, alkaline phosphatases, lipases and amylases are used in industry for manufacturing of detergents
Acidozymes	Sulphate oxygenase, *Thiobacillus* dehydrogenase, rusticyanin (acid stable e⁻ carrier) and thrompsin
Psychozyme	Pectinase, lipase, cellulase, amylase for detergents, Food processing (cheese making, meat tendering, lactate hydrolysis), Biosensors (environmental applications), biotransformations and contact lens cleaning solutions
Alkalozymes	Protease (detergents), amylase (starch industry), cyclomaltodextrin glucanotransferase (chemical and pharmaceutical), pullunases (detergents), xylanses (pulp and paper industry), pectinases (paper production)

3. Halophiles

Halophiles are the Gram-negative, non-spore forming, non-motile bacteria that reproduce by binary fission. They appear red pigmented due to the presence of carotenoids but sometimes they are colourless. They contain the largest plasmid so far known among all the known bacteria.

Halophiles are able to live in salty conditions through a fascinating adaptation. Because water tends to flow from the areas of high to low solute concentrations. A cell suspended in a very salty solution will lose water and become dehydrated unless its cytoplasm contains a higher concentration of salt than its environment. Halophiles contend with this problem by producing large amounts of an internal solute or by containing a solute extracted from outside. For example, *Halobacterium salinarum* concentrates KCl in the interior of the cell. The enzymes in its cytoplasm will function only if a high concentration of HCl is present. But their cellular proteins contacting the environment require a high concentration of NaCl.

Microorganisms growing in extreme environments : salterns turn red by halophilic algae and halobacteria.

This group of bacteria lives in highly saline environment (>3.5% salt concentration) such as neutral salt lakes or artificial saline source like salted food, fish, etc. Extreme halophilic organisms require at least 1.5 M (about 9%) NaCl but most of them have optimum growth at 2-4 M NaCl (12.23%). Some examples of prokaryotic extremely halophilic bacteria occurring in nature are given in Table 26.2.

Table 26.2 : Prokaryotic genera of extremely halophilic species.

Halobacteria	Methanogens	Bacteria
Halobacterium salinarium	Methanobacterium sp.	Acetohalobium sp.
Halobacterium halobium		
Haloferax mediterranei		Actinopolyspora sp.
Haloarcula sp.		Ectothiorhodospira sp.
Halococcus acetoinfaciens		
Halococcus agglomeratus		
Natronobacterium gregoryi		
Natronococcus sp.		

(i) **Physiology:** Halophilic bacteria lack peptidoglycan in cell walls and contain ether-linked lipids and archaean type RNA polymerases but *Natrobacterium* is extremely alkalophilic as well. Former also contains diether lipids not present in other extreme halophiles. They are chemoorganotrophic bacteria that require amino acids, organic acids and vitamins for optimum growth. Some times they oxidize carbohydrates as energy source. Cytochromes *a, b* and *c* are present but membrane mediated chemiosmosis generates proton motive force. They also require sodium for Na$^+$ ions. *Halobacterium* exceptionally thrives in osmotically stressful environment and does

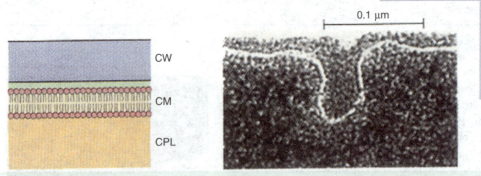

Electron micrograph of *Methanobacterium formicicum*, a typical gram positive extremely halophilic bacterium.

not produce compatible solutes. Peptidoglycan is absent in their cell wall. Aspartate and glutamate (acidic amino acids) are present. The negative charges of the carboxyl groups of these amino acids are shielded by Na$^+$ ions. The ribosomes of *Halobacterium* requires high K$^+$ ions for stability, which is a unique feature as no other group of prokaryotes requires it for internal components.

The membrane lipids of these archaea are composed of diphytanylglycerol, diether analogues of glycerophospholipids. The extreme halophiles contain high intracellular concentration of Na$^+$ and K$^+$ and their proteins seem to have adapted to this high salt concentration by having a higher fraction of acidic amino acid residues and a more compact packing of a polypeptide chain than protein from non-halophilic bacteria. In the halophilic bacteria generally a Na$^+$/H$^+$ antiporter is used to pump Na$^+$ outwards and solute uptake has been shown to be Na$^+$ coupled in several halobacterial species.

(ii) **Molecular Adaptation:** In such bacteria K$^+$ ions inside the cell is more than Na$^+$ ion outside the cell which act as its solute. Hence, the cells maintain cellular integrity. Halobacteria lack peptidoglycans in their cell walls and contain ether-linked lipids and archaean type RNA polymerases which maintain the rigidity at salty conditions. These changes in cytoplasmic membrane allow such bacteria to survive.

(iii) **Applications:** Certain extreme halophiles synthesize a protein called bacteriorhodopsin into their membrane. Some produce polyhydroxy alkanoates and polysaccharides, enzymes and compatible solutes. They are also used in oil recovery, cancer detection, drug screening and

biodegradation of residue and toxic compounds. Kushner (1985) defined the halobacteria based on utilization of optimum salt concentration for their growth. In this system, non-halophiles are those that grow best in media containing < 0.2 M NaCl, slight halophiles (marine bacteria) grow best at 0.2 to 0.5 M NaCl, moderate halophiles at 0.5 to 2.5 M NaCl, and extreme halophiles grow in media containing 2.5 M to 5.2 M (saturated) NaCl.

It is interesting to note that all extreme halophiles are archaea except for two species of the photosynthetic *Ectothiorhodospira*, one of the *Acetohalobium* and one actinomycete *Actinopolyspora*. Some actinomycete species of the genus *Methanohalobium* has been described. The bright red colour water of the salterns is now known to be due to the bacterioruberin pigments of the halobacteria. The biotechnological potential of halobacteria with commercial interest is following:

(a) **Bacteriorhodopsin:** The retinal proteins of halobacteria have been observed as integral proteins of the purple membrane, containing one of the proteins called bacteriorhodopsin. This protein is light-driven, proton translocator and converts sunlight to electricity. The bacteriorhodopsin absorbs light at 570 nm. It exists in two forms. The trans configuration after excitation converted to the cis form following the absorption of light (Fig. 26.1). In this case, ATP synthesis is prevented and the electrical potential arising from the proton gradient will be the source of electricity. It is used in optical data processing and as light sensors. A photographic film based on purple membrane displays the interesting properties as it does not require developing. Holographic films of this type are suitable for computer memory i.e. parallel processing (Rodriguez-Velera and Lillio, 1992). Recently, *biochips* have been introduced in new generation of computers (Hong, 1986). In future, robots with vision may have biosensors based on this protein. Desalination of water is also demonstrated by the application of bacteriorhodopsin.

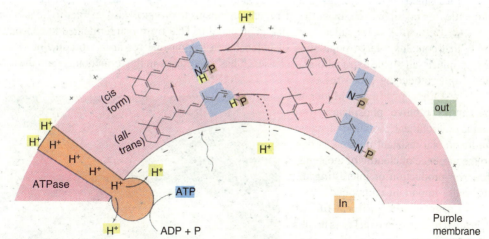

Fig. 26.1 : Bacteriorhodopsin proton pump working under the influence of light.

(b) **Bioplastic or polyhydroxy alkanoates (PHA):** This kind of heteropolymer is biodegradable. It exhibits total resistance to water and degraded in human tissues; hence it is biocompatible. It has pharmaceutical and clinical importance, including the use in delayed drug release, bone replacement and surgical sutures. Production of PHA is always higher by using *Halof. mediterranei*. In addition, these halobacteria possess high genomic stability which is a pre-requisite for industrial purposes.

(c) **Polysaccharides:** Microbial exopolysacchrides are used as stabilizers, thickness, gelling agents and emulsifiers in the pharmaceutical industries, paint and oil recovery, paper, textile and food industry. *Halof. mediterranei* produces a highly sulphated and acidic heteropolysaccharides (up to 3 g/l) which contain mannose as a major component. Such a polymer combines excellent rheological properties with a remarkable resistance to extreme of salinity, temperature and pH.

(d) **Microbially enhanced oil recovery:** Residual oil in natural oil fields can be extracted by injection of pressurized water down in a new well. The bacterial biopolymers are of interest in enhanced oil recovery because of their bio-surface activity and properties of bio-emulsifiers.

(e) **Cancer detection:** A protein (84 kDa) has been used from *Halobacterium halobium* as an antigen to detect antibodies against the human *e-myc* oncogene product in the sera of cancer patient suffering from pyrolytic leukaemia cell line (HL-60). The use of halobacterial antigens as probe for some types of cancer seems to be promising.

(f) **Drug screening:** Plasmid, pGRB-1 of *Halobacterium* strain GRB-1 used in the pre-screening of new antibiotics and anti-tumor drugs affect eukaryotic type II DNA topoisomerase and quinotone drugs which act on DNA gyrase. Such drug causes DNA cleavage of small plasmid from halophilic archaea *in vivo*.

(g) **Liposomes:** Ether-linked lipid of the halobacteria is used in liposome preparation having great value in the cosmetic industry. Such liposomes would be more resistant to biodegradation, good shelf-life and resistance to other bacteria.

(h) **Enzymes:** Proteases and amylases from *Halobacterium salinarium, H. halobium*, and lipases from several halobacteria have been reported. A site-specific endonuclease activity has been reported in *H. halobium*.

(i) **Bioremediation:** Bertrand *et al.* (1990) observed that the halobacterial strain EH4 isolated from a salt-mark was found to degrade alkanes and other aromatic compounds in the presence of salt.

(j) **Gas vacuoles or vesicles:** Some *Halobacterium* spp. produce intracellular gas filled organelles called vacuoles of gas vesicles which provide buoyancy. In the future, the genes of such properties are possible to engineer in other microorganisms to produce gas vacuoles to float in water.

(k) **In food:** A sauce called 'nam pla' is prepared in Thai from fish fermented in concentrated brine that contains a large population of halobacteria responsible for aroma production. Because they produce salt-stable extracellular proteases. It has importance in the fermentation and the flavour and aroma producing processes.

(l) **Other products:** Moderate halophiles remove phosphate from saline environment. Isolation of stable antimicrobial-resistant mutants is due to the presence of cloning of the genes for over-production of interesting industrially important compounds.

Large scale cultivation of *Spirulina platensis* in Israel uses brackish water which is unsuitable for agriculture and the *Spirulina* biomass is marketed as a healthy food. *Spirulina* grows optimally in alkaline lakes with a salt concentration ranging from 2 to 7%.

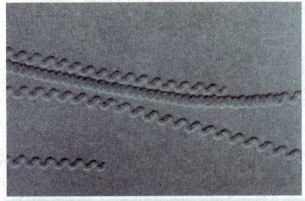

A helical, filamentous cyanobacterium *Spirulina* which is marketed as healthy food.

4.   Psychrophiles

Temperature is an important environmental factor which influences the different groups of microorganisms. Different groups of microorganism based on different temperature regime are given in Fig. 26.2.

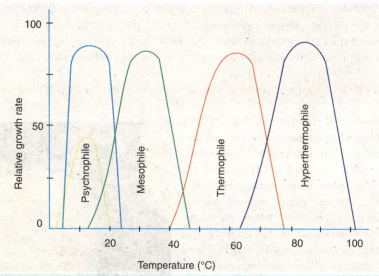

Fig. 26.2 : Different groups of microorganisms categorized on the basis of different temperature regime.

Cold environments are actually more common similar to hot environment during summer. The oceans which maintain an average temperature of 1-3°C make up our half the earth's surface. The vast land areas of the Arctic and Antarctica are permanently frozen or unfrozen for only a few weeks in summer.

James T. Staley and his colleagues at the University of Washington have shown that microbial communities populate ice ocean water of Antarctic sea that remains frozen for much of the years. These communities include photosynthetic eukarya, notably algae and diatoms as well as variety of bacteria. *Polasomonas vacuolata* obtained by Staley's group is a prime representative of a psychrophile. Psychrotolerant can be isolated from more widely distributed habitat than psychrophiles. They can be isolated from soil, water in temperate climates as well as meat, milk and other dairy products, vegetables and fruits under refrigeration.

They grow best between 20 and 40°C but cannot grow at 0°C. After several weeks of incubation their visible growth can be observed. Its optimum temperature for growth is 4°C, and 12°C for reproduction. The cold-loving microorganisms have started to interest manufacturers who need enzymes that work at refrigerator temperature such as food processors, makers of fragrances and producers of cold-wash laundry detergents.

Some psychrophiles can be dangerous organisms for man e.g. *Pseudomonas syringae, Erwinia* sp., *Yersinia enterocolitica,* etc. Most of the foods or food products are stored at freshing temperature so that the pathogenic or saprophytic microbes cease to grow.

A majority of marine microbes is psychrophiles due to their habitat (ocean). Generally, these are Gram-negative rod shaped bacteria. Among them are pseudomonads of which *P. geniculata* is the most common. The other microbes are *P. putrefaciens, P.fragi* and *P.fluorescens, Flavobacterium* spp, *Alcaligenes* spp, *Achromobacter* and a few strains of *Escherichia, Aerobacter, Aeromonas, Serratia, Proteus, Chromobacter* and *Vibrio* are psychrophilic in nature. The common psychrophilic yeasts are species of *Candida, Cryptococcus, Rhodotorula* and *Torulopsis.*

Physiologically the Gram-negative property of the bacteria and high proportion of G+C contents are present in such microorganisms. Psychrophiles contain an increased amount of unsaturated fatty acids in their lipids. Flagellum disappears after increasing the temperature.

(i) **Physiology:** Psychrophiles produce enzymes that function optimally in the cold. Its cell membrane contains high content of unsaturated fatty acid which maintains a semi-fluid state at low temperature. The lipids of some psychrophilic bacteria also contain polyunsaturated fatty acids and long chain hydrocarbons with multiple double bonds.

(ii) **Molecular Adaptation:** The active transport in such organisms occurs at low temperature. It indicates that the cytoplasmic membranes of psychrophiles are constructed in such a way that low temperature does not inhibit membrane function. The membrane contains polyunsaturated fatty acids in their lipids which maintains the rigidity at low temperature and organisms thus are able to survive.

(iii) **Applications:** Psychrophiles and their products have many applications as described below:

(a) **Source of pharmaceuticals:** Many psychrophiles such as *Streptomyces, Alteromonas, Bacillus, Micrococcus, Moraxella, Pseudomonas* and *Vibrio* have been isolated from deep-sea sediments. They grow at temperature between - 3 and -30^0C. Aquatic plants and animals are highly prone to infestation by pathogenic micro-organism. An *Alteromonas* sp. has been reported to synthesize 2, 3-indolinedione (isatin). This compound protects *Palaeoman macrodactylus* from pathogenic fungus *Lagenidium callinectes*. Similarly, another strain of *Alteromonas* sp. is intimately associated with the marine sponge *Halichondria okada* and produces a tetracycline alkaloid e.g. alterimide. There is wide scope

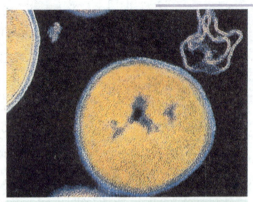

Bacillus, a psychrophile isolated from deep-sea sediments.

for the discovery of novel biologically active compounds in marine microbiology. An antitumor polysaccharide has been isolated as narinactin from marine actinomycetes. A mixture of protease and amylase isolated from *Bacillus subtilis* removes the dental plaque. Mainly lipases are used as stereo-specific catalysts and in the biotransformations of various high value compounds such as flavouring agents and pharmaceuticals. Trehalose is formed by an enzyme trehalase present in several psychrophilic bacteria.

(b) **Bacterial ice nucleating agents:** There are several uses for ice-nucleating agents (INA) produced by bacteria. They are being used in artificial snow-making, in the production of ice creams and other frozen foods. These are also used in immunodiagnostic kits as a conjugate to antibodies and as a substitute for silver iodide in cloud seeding. Among several organisms, bacterial INAs have attracted much attention due to its ability to form ice nuclei at relatively high temperature in comparison to other sources.

(c) **Fermentation industry:** Mesophilic yeasts containing unsaturated fatty acids in membranes (lipids) have been found to be resistant between -80 and $-$

Bread rises as CO_2 is liberated by fermenting mesophilic yeast (containing unsaturated fatty acids) which converts glucose to ethanol via the alcoholic fermentation pathway.

20°C. These are preferred for its storage in baking and other processing industries. Fermentation at 6-8°C reduces the inhibitory effect of ethanol on cell membrane of the yeast cells.

(d) **In microbial leaching:** Currently microbial leaching operations involve oxidative solubilization of copper and uranium ores. Leaching operation in temperate countries is carried out at very low ambient temperature. Microbial leaching operation from sulfide ores is carried out at 4-37°C.

(e) **In bioremediation:** Psychrophiles have ability to degrade various compounds in their natural habitat. They are used in bioremediation of several pollutants at low temperature. The bacterial strains were found to mineralize dodecane, hexadecane, naphthalene, toluene. It has been demonstrated in laboratory and field experiments using specific bacterial strains. A psychrophilic bacterium, *Rhodococcus* sp. strain Q15 has been studied for its ability to degrade *n*-alkanes and diesel fuel at low temperature.

(f) **Denitrification of drinking water sources:** The presence of high nitrate concentration in water has become a major problem in many countries. The most widely used practices for removal of NO_3 is the biological denitrification. Most of the denitrification processes are carried out at 10^0C. The rate of denitrification in these cases can be enhanced by employing psychrophilic bacteria isolated from permanently cold habitat.

(g) **Anaerobic digestion of organic wastes:** The obligate anaerobes which convert organic acids to CH_4 and CO_2 i.e. methanogens are highly sensitive to low temperature. The rate of methanogenesis can be increased several times by low temperature adaptation by methanogens. The process can be made possible by selective enrichment of psychrophilic methanogens through long term laboratory trials. *Methanogenium frigidum* isolated from Ace lake (Antarctica) grows optimally at 15^0C. This bacterium is found to produce methane from hydrogen and carbon dioxide.

5. Thermophiles and Hyperthermophiles

Hyperthermophilic bacteria are archaea that represent the organism at the upper temperature border of life (Stetter, 1992). Neutrophilic and slightly acidophilic hyperthermophiles are found in terrestrial solfataric fields, and deep oils reservoir. These exhibit specific adaptations to their environments and most of the bacteria are strictly anaerobic.

Various factors both abiotic and biotic, that control the growth of all living organisms are called *biotope*. The moderate thermophiles are called *extreme thermophiles* which grow optimally between 80 and 100°C. The hyperthermophiles are unable to grow below 80^0C but adapted to high temperature as they do not even grow at 80°C. Some of the examples are given in Fig. 26.3. *Thermotoga* has rod shaped cells surrounded by a characteristic sheath-like structure (the 'toga') which balloons out at the end (A). Archaeal coccoid sulphate reducers are the members of the genus *Archeoglobus* (B). *Methanopyrus kandleri* is a rod shaped methanogen (C).

The hyperthermophiles can grow in natural as well as in artificial environmental conditions. Natural sulphur-biotopes are usually associated with active volcanism. In such situation, soil and surface waters from S-containing acidic fields (pH 0.5-6.0) and neutral to slightly alkaline hot spring environment persists. Well-known biotopes (a biotope has upper and lower limits for growth for each of environmental factors) of hyperthermophiles are volcanic areas such as hot springs and solfataric fields i.e. high temperature fields located within volcanic zones with much sulphur acidic soil, acidic hot springs and boiling mud. Few of hyperthermophiles live in shallow submarine hydrothermal systems and abyssal hot vent systems called "black smokers" having temperature of about 270-380°C. The black smokers are mineral-rich hot water that makes cloud of precipitated material on mixing with sea water. Other biotopes are smouldering coal refuse piles having acidic pH and geothermally heated soil reservoirs (Fuchs, 1992).

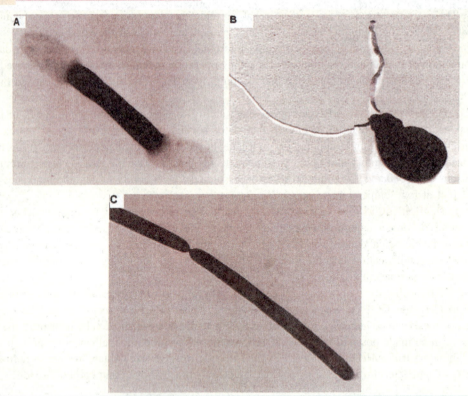

Fig. 26.3 : Cells of *Thermotoga maritima* (A), *Archaeoglobus lithotropicus* (B) and *Methanopyrus kandleri* (C).

Most of the hyperthermophiles are anaerobic due to low solubility of oxygen at high temperature and the presence of red gases. Anaerobic chemolithoautotrophic hyperthermophiles completely independent on sun, but they could even exist in other planets also. Hydrothermal vents in the bottom of the ocean have temperature of 350^0C or greater and also show the existence of hyperthermophiles. The recently discovered non-volcanic biotope embedded in deep geothermal heated oil stratification of extracted fluids evidenced for such microbial communities.

For the cultivation of such bacteria, samples are brought to the laboratory without temperature control. They are isolated by enrichment culture technique with variation in compo-

A hot spring coloured green and blue by halophilic cyano bacterial growth.

sition of substrate and control of *in situ* temperature. Agar is not suitable, hence more heat-stable polymer such as gellan gum or polysilicate gels are used for solidification.

Many taxonomic types of cultured hyperthermophiles are already known so far. They represent 52 species belonging to 23 genera and 11 orders of hyperthermophilic bactreria and archaea known in literature. The organisms whose optimum growth temperature is < 45°C are called thermophiles and those above 80°C are called hyperthermophiles.

(i) **Physiology:** The enzymes and proteins are much more stable than the other forms and these macromolecules function at high temperature. Thermophilic proteins have different amino acid sequences that catalyse the same reaction in a mesophile which allow it to fold in a different way and thereby show heat tolerant effect. All thermophiles contain reverse gyrase, a unique type 1 DNA topoisomerase that stabilizes DNA.

Heat stability of proteins from hyperthermophiles is also due to increased number of salt bridges (bridging of charges on amino acids by Na^+ or other cations) present and densely packed highly hydrophobic interior of the protein, which have membranes rich in saturated fatty acids. This allows the membrane to remain stable and function at high temperature. Most of the hyperthermophiles are archaea which do not contain fatty acids, the lipids in their membranes but instead have hydrocarbons of various lengths composed of repeating units of 5-6 compound phytans bonded by ether linkage to glycerophosphate. With increase in temperature of growth an increase in degree of saturation, chain length and/or iso-branching of the acyl chains are observed. Sometimes, special lipids (the sterol like hopanoids) are present in thermophiles. These may also affect an adaptation to life at high temperature by making the membrane more rigid.

(ii) **Molecular Adaptation:** These bacteria contain heat–stable enzymes and proteins which regulate various macromolecular functions at high temperature. The critical amino acids substituted in one or more locations in these enzymes allow them to fold in a different manner and thereby withstand the denaturing effect of heat resulting into the survival of these organisms. Further, the cytoplasmic membrane contains lipids rich in saturated fatty acids, thus allow the membrane to remain stable and functional at high temperature. The thermophilic archaea do not contain fatty acids in their lipids, neither its membrane has ester linkages with glycerol phosphate. This imparts more rigidity to its membrane systems.

(iii) **Applications:** Most of the microorganisms that thrive above the boiling point of water belong to archaea. The enzymes of thermophiles are of great interest. Hyperthermophiles have focused on thermostable enzymes from vent. The proteins (chaperons) were also discovered. These proteins are expressed under stress conditions and involved in protein foldings.

(a) **Enzymes:** New enzymes from hyperthermophiles have reduced the number of steps needed to transform starch into fructose syrup. The amylase, glucoamylase, pullunases and glucosidases are the enzymes used in starch industry. Pullunases are found in anaerobic bacteria. Amylases are widely used in textile, confectionary, paper, brewing, and alcohol industries. Similarly, glucosidases are used for hydrolyzing lactose syrup and mixtures to glucose and galactose. They may have clinical applications since there is evidence of a lactase deficiency in the population which is either inherited or is the result of ageing. Glucose isomerase is widely used in the food industry which converts glucose to fructose for use as sweetner.

A source of acid drainage from a mine into a stream. The soil and water have turned red due to the presence of precipitated iron oxides caused by the activity of extremely thermophilic bacteria such as *Thiobacillus*.

Due to the thermal stability of the enzymes, hyperthermophiles have been the subject of intensive investigation. Thermostable enzymes are more resistant to the denaturing activities of detergents and organic solvents. The amylases have been extracted from *Pyrococcus furiosus* and *Pyrococcus woessei*. Enzymes have been exploited from some archaea e.g. *Desulfurococcus mucosus, Staphylothermus marinus, Thermococcus celer* and *Thermococcus litoralis*. A toga-associated amylase has also been detected from *Thermotoga maritima*. This enzyme is active between 70 and 100°C at pH 6. *Fervidobacterium pullunolyticum* has the potentiality of producing thermotolerant enzyme optima at 90°C.

Certain bacteria and archaea such as *P. woesei, P. furiosus, Thermococcus litoralis, T. celer, F. pennavorans, D. mucosus,* etc. are reported to produce pullunase II (amylopullanase) having 90 kDa molecular weight with temperature optima 105°C and pH 6. Some of these (*P. woesei* and *P. furiosus*) also produce glucosidases with temperature optima 110-115°C. These are useful for the bioconversion of starch into various useful products of industrial significance.

A thermostable exo-4-β-cellobiohydrolase with a half life of 70 minutes at 108°C has been isolated from *Thermotoga* sp. strain FjSS3-B. Similarly, thermostable xylanases have been reported from *Thermotoga maritima, T. neoplolitiana, T. thermarum*. *P. furiosus* exhibited β-xylanosidase activity. The enzymes from *Thermotoga* sp. are extremely stable with half-life of 8h at 90°C.

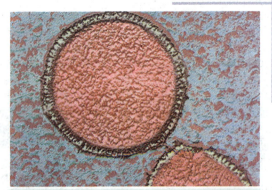

The protein hydrolyzing enzymes (proteases) have been isolated, purified and characterized from a number of thermophilic and hyperthermophilic microorganisms specially *Pyrococcus, Thermococcus, Sulfolobus, Staphylothermus* and *Desulfurococcus*. Pyrolysin, an enzyme associated with the cell envelope which is a serine-type protease has temperature optima of 110°C and a half

Thermostable enzymes in thermophilic bacterium *Staphylothermus marinus* have great potential in industry.

life of 4h at 100°C. It has been identified and characterized in *P. furiosus* and *P. woesei*. The serine-protease from *Sulfurococcus mucosus* exhibits its activity at 100°C.

A unique protease which hydrolyses keratin of chicken feather, hair and wool has been characterized from a bacterium *F. pennavorans*. A thermophilic glucose isomerase was characterized and purified from *Thermotoga maritima*. Ferredoxins from *Thermoplasma acidophilum, Sulfolobus acidocaldarius* and *Desulfurococcus mobilis* have also been investigated. Hydrogenase, having a half-life of 21 h at 80-85°C has been isolated from *Pyr. furiosus*. A thermoactive pyruvate-ferredoxin-oxidoreductase (POR) which catalyses the oxidative decarboxylation of pyruvate to acetyl-CoA and CO_2 has been detected in *D. amylolyticus, H. butylicus, Thermococcus celer, Pyrococcus woesei, P. furiosus* and *Thermotoga maritima*.

Enzymes involved in amino acid biosynthesis such as aromatic aminotransferase from *Thermococcus litoralis* and *Sul. solfataricus* have been detected. An extremely thermostable enzyme with optimum activity at 100°C has also been detected from *Methanobacterium thermoformicimum*. The purified enzyme from *P. woesei* and *P. furiosus* has molecular mass of identical subunits 45 kDa each. The enzymes have heat-stability up to 70% after heat treatment at 100°C for 1 hour.

Glutamate synthetase (GS) is responsible for the synthesis of glutamine from glutamate and ammonia. The half-life of partially purified GS is 2 hours at 100°C. Two thermo-active aromatic aminotransferases from *Thermococcus lithoralis* has been purified and characterized, which are active at 100°C temperature. The enzyme aspartate aminotransferase transferring amino group from glutamate to oxaloacetate has been detected in *Sul. solfataricus*.

Taq polymerase is very important enzyme used in molecular biology for the amplification of DNA using polymerase chain reaction (PCR). This enzyme found in *Thermus aquaticus* is active at 80°C at pH 8 (Chien *et al*, 1976). Simpson *et al.* (1990) has investigated the other DNA polymerase from *Thermotoga* sp. Certain archaea such as *Sul. acidocaldarius* and *Sul. solfataricus* consists of DNA polymerase of a single polypeptide chain with a molecular mass of 100 kDa. The DNA polymerase from *P. furiosus* has also been purified. The DNA ligase has been characterized from *Thermus thermophilus*. Topoisomerases type I purified from *Sulfolobus acidocaldarius, Desulfurococcus amylolyticus, Thermoplasma acidophilum, Fervidobacterium islandicum, Thermotoga maritima,* and *Methanopyrus kandleri,* while topoisomerase II has so far been isolated from *Sulfolobus acidocaldarius. Thermotoga maritima* contains of both gyrase and reverse gyrase enzymes. Repair of extensive DNA damage caused by ionizing-radiation at 95°C has been demonstrated in *Pyrococcus furiosus.*

(b) **Chaperons:** The *chaperons* are the proteins which express under stress conditions such as elevated temperatures. They are involved in protein folding (Ellis, 1990). These are detected in *Su. shibate* and *Su. solfactaricus*. It is called thermophilic factor which has 55 kDa molecular mass. Due to increase in high concentration of intracellular protein up to 105°C, this protein complex is called **thermosome**. The thermosome consists of a cylindrical complex of a two stacked identical rings, each unit consists of 8 subunits around a central channel. Both subunits contain 56 and 59 kDa molecular mass. They also bind the unfolded proteins similar to chaperons. A thermostable disulfide-bond forming enzyme has been isolated, characterized and purified from *Sul. solfataricus.*

6.   Barophiles

Barophiles are those bacteria that grow at high pressure at 400-500 atmosphere (atm) on 2 to 3°C. Such conditions exist in deep-sea habitat about 100 metre in depth. Many are barotolerant and do not grow at pressures above 500 atm. but some live in the gut of invertebrates (amphipods and holothurians). *Photobacterium shewanella* and *Colwelha* inhabit more rapidly. Some thermo-philic archaea are barophiles e.g. *Purococcus* spp. and *Methanococcus jannaschii*. Barophiles adapt the extreme pressure (200-600 bars) involving macromolecular structures in cells. Increasing pressure makes structures more compacts, and this tendency has been the principle of microscopic ordering.

(i) **Physiology:** There are variations in membrane structure and function. The amount of mono-unsaturated fatty acids in the membrane increases due to increase in the pressure. The organism is thereby able to circumvent the loss of membrane fluidity imposed by increasing the pressure. As the pressure decreases, membrane fluidity presumably increases and the cells respond by decreasing the level of mono-unsaturated fatty acids. It is evidenced that increased pressure decreases the binding capacity of enzymes for their substrates. Thus the enzymes must be folded in such a way as to minimize these pressures in barophiles. It is not known whether H^+, Na^+ or both are used as coupling ions in energy transduction in these organisms.

(ii) **Molecular Adaptation:** In the cytoplasmic membranes of high pressure tolerant microbes, the amount of unsaturated fatty acids is more which allows the adaptive significance. Further, the adaptativity may also be due to changes in protein composition of the cell wall outer membrane called OmpH protein, a type of *porin*. The porins are structural proteins meant for diffusion of organic molecules through the outer membrane and in to the periplasm. It is observed that OompH system is pressure–dependent and required for growth at high pressure.

(iii) **Applications:** Barophiles are the major source of unsaturated fatty acids or polyunsaturated fatty acid. The microbial barophilism is helpful in enhancing the mining. Underground mining operations usually occur at increased pressures and temperatures and barophilic thermophiles are better adapted under such situations. Recently, vacant salt mine area has been worked out as

fermenters for the biological gasification of pretreated lignite or agricultural crops based on the involvement of extremophiles endowed with adaptation to high pressure and temperature besides salinity.

QUESTIONS

1. Name few microorganisms which are exploited for extremozymes. What is their industrial significance?
2. Define extremophiles. How are these microorganisms classified based on temperature? Describe it in detail.
3. Write short notes on the following:
   (a) Acidophiles, (b) Barophiles, (c) Alkalophiles
4. Define barotolerants. What kinds of physiological and molecular adaptations allow it to grow optimally under pressure?
5. Write a detailed account on the applications of thermophilic microorganisms.
6. Write a brief account on the following:
   (a) Psychrophiles, (b) Bacteriorhodopsin, (c) Halophiles
7. What are thermophiles? How do they grow in hydrothermal vents?
8. Write an account on molecular adaptations of different group of extremophiles.

REFERENCES

Bertrand, T.C. *et al.* 1990. Biodegradation of hydrocarbons by an extremely halophilic archaebacterium. *Letters Appl. Microbiol.* **11**: 260-263.

Chien, A., Edgar, D. B., Trela, J. M. 1976. Deoxyribonucleic acid polymerase from the extreme thermophile *Thermus aquaticus. J. Bacteriol.* **127**: 1550-57

Ellis, R.J. 1990. Molecular chaperons: the plant connection. *Sci.* **250**:954-59

Fuchs, G., Ecker, A. & Strauss, G. 1992. Bioenergetics and autotrophic metabolism of chemolithotrophic archaebacteria. In *The archaebacteria. Biochemistry and Biotechnology*, Eds Danson, M.J., *et al.* pp. 23-39. Portland Press: London.

Hong, F.T. 1986. The Bacteriorhodopsin membrane system as a prototype molecular computing element. *Biosystem.* **19**: 223-236.

Krulwich, T.A., Guffanti, A.A. (1989). Alkalophilic bacteria. *Ann. Rev. Microbiol.* **43**: 435-463.

Kushner, D.J. 1985. The Halobacteriaceae. In *The Bacteria*, Vol. 8, Eds Woese, C.R. & Wolfe, R.S., pp 171-214. Academic Press: London.

Roderiguez-Valera, F. & Lillio, J. G. 1992. Halobacteria as producers of polyhydroxyalkanoates. *FEMS Microbiol. Review.* **103**: 181-186.

Simpson, H.D. Coolbear, T., Vermue, M., Daniel, R.M.1990. Purification and some properties of a thermostable DNA polymerase from a *Thermotoga* species. *Biochem. Cell. Biol.* **68**: 1292-96

Stetter, K.O. 1992. The genus *Archaeoglobus*. In *The Prokaryotes*, vol. 1, 2nd edn. Eds Balows, A. *et al.* ,pp. 707-711. Springer-Verlag: New York.

Microbial Ecology – III : Soil Microbiology

27

A. Soil as a habitat for microorganisms

B. Rhizosphere and Rhizoplane microorganisms

C. Organic Matter Decomposition

D. Biogeochemical Cycling

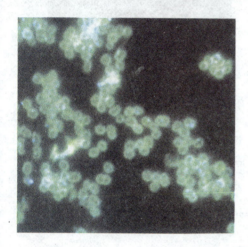

S oil is the outer region of earth-crust consisting of loose material formed by gradual weathering of rock, and gives to plant both mechanical and nutritional support. Soil can be defined as *the space-time continuum forming the upper part of the earth crust*. Thus, soil is a complex product of parental material, or geology, topography, climate, time and biological activity on anthropogenic activity (Griffin 1972). Soil formation influenced by different factors may be written as below:

$$s\int^{b,p,ct,t}$$

Where, = b, biological activity; p, parental rock; c, climate; t, time; t, topography.

Waksman (1916-1944) was the first to stress the study of soil microorganisms.

A. Soil as a Habitat for Microorganisms

Soil is a unique habitat which harbours a variety of microflora and fauna, and gives mechanical and nutritional support to higher plants on which human civilization is based. Structural components of soil and its major constituents determine soil quality.

Soil : *A medium of mechanical support cutting across diverse microflora*

1.   Soil Quality

Soil is as important as water and air. Human life could not be sustained with access to soil, because it is the source of most of our food. The economic well being of most of the nations on earth depends greatly on arable soils and the ways of maintenance of their productivity. Moreover, a good quality soil also acts as an environmental filters for cleaning air and water. Soil is the ultimate receptors and reservoir of nutrients released from organic matter. It also sends the nutrients back to plants.

Soil quality can be defined as the capacity of a soil to function within the boundaries of ecosystem to sustain biological productivity, maintain environmental quality and promote plant and animal health.

Soil quality encompasses its capacity for (i) crop productivity, (ii) food safety, and (iii) health of animals and humans. Soil quality can improve or deteriorate depending on the influencing factors. The factors that cannot be affected are; geology, topography, climate and time (Fig. 27.1) The factors which can be influenced are humus content, phosphorus status, degree of saturation, etc.

Soil formation : Torrent waterfall rolls the heavy rock masses and grinds them into finer particles.

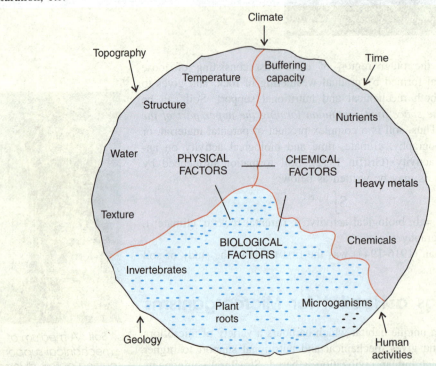

Fig. 27.1 : The complex structure of soil created by the influence of geology, topography, climate, time and human activity.

2.   Physico-chemical Properties of Soil

Physico-chemical properties of soil include soil texture, water, air, inorganic chemicals, and organic matter. The biological factors of soil are soil flora and fauna.

Mechanical composition (texture) of soil is determined on the basis of size of soil particles i.e. sand, silt and clay particles (Table 27.1). However, the ratio of soil particles governs the porosity (pore size of soil), soil water (present in pores), air temperature, pH, inorganic and organic matters and microorganisms and their community-size. The amount of each component is changed with soil types.

The sources of inorganic material in soil are the parent rocks which get changed by the physical and chemical processes of weathering. Therefore, soil consists of different sized mineral particles (Table 27.1), the ratio of which determines the characteristics of soils. On the basis of particle size soil minerals are divided into: (*a*) *sand* particles (about 50 µm diameter) which are the fragments of rock materials, (*b*) *silt* particles (2-50 µm diameter) which contian primary min-

Biological weathering.

erals (quartz), and (*c*) *clay* particles (less than 2 µm diameter) composed of secondary minerals (*e.g.* kaolinite, montmorillonite, ellite, etc.). Clay particles are negatively charged particles which are the important components of soil environment influencing the physico-chemical and biological properties of soil (Gray and Williams, 1971). Chemical nature of soil minerals differ, and it has been divided into *silicates* and *non-silicates*. The non-silicate group includes oxides, hydrides, sulphates, chlorides, carbonates and phosphates. The silicates are very complex structures but vary widely in its stability and resistance to decomposition. Among these, the most influential soil particles, as far as microbial activity is concerned, are the colloidal size clays and humic materials. It plays significant role in determining the availability of nutrients and in the interaction of extracellular enzymes and antibiotic substances produced by the microorganisms (Burns, 1983).

(*i*) **Organic matter:** The surface layer of soil consists of a relatively unchanged mass of plant/animal remains called litter. After microbial decomposition organic matter is converted into unidentifiable amorphous material which is known as *humus*.

Table 27.1 : Mechanical composition of soil and relative size of soil particles.

Soil particles	Diameter of particles (mm)
Sand	
Very coarse sand	2.00 — 1.00
Coarse sand	1.00 — 0.50
Medium sand	0.50 — 0.25
Fine sand	0.25 — 0.10
Very fine sand	0.10 — 0.05
Silt	0.05 — 0.002
Clay	< 0.002

(ii) Soil-water and Air: Soil-water and air play a significant role as it influence the metabolic activities of macro and micro biota. Soil water and air are directly related to the soil texture because the portion of pore space devoid of water is filled with gas, and water film in pores, gives a mean for the movement, germination and growth of spores therein. The detailed account of microbial activity affected by various edaphic factors associated with water regime has been discussed by Griffin (1972).

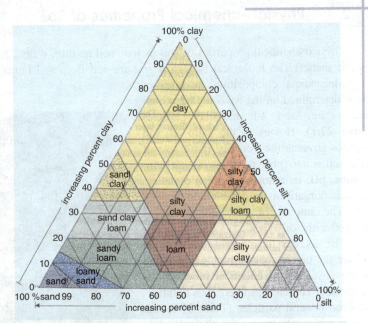

Soil texture depends on the percentages of clay, sift and sand particles in the soil.

(iii) Soil Microbes: Microorganisms which live in soil are algae, bacteria, actinomycetes, fungi, protozoa, nematodes, etc. (Fig. 27.2). A brief discription of soil microorganisms has been given below.

(a) Soil Algae: Soil algae (both prokaryotes and eukaryotes) luxuriantly grow where adequate amount of moisture and light are present. They play a variety of roles in soil. One of the important role of blue-green algae is that it has revolutionised the field of agriculture microbiology due to use of cyanobacterial biofertilizer (see Chapter 30). On the other hand they can also be used in reclamation of sodic soil *i.e.* alkaline soil, sewage treatment, etc. The prominent genera are *Anabaena, Calothrix, Oscillatoria, Aulosira, Nostoc, Scytonema, Tolypothrix,* etc (for detail see Chapter 14, *Nitrogen fixation*)

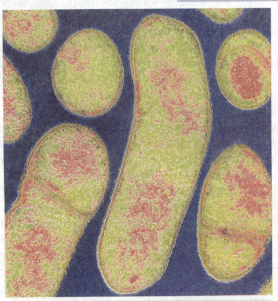

Clostridium botulinum, the anaerobic soil bacterium.

(b) Bacteria: Bacteria are the smallest unicellular prokaryotes ($0.5 - 1 \times 1.0 - 2.0$ µm), the most abundant group and usually more numerous than others, the number of which varies between 10^8 and 10^{10} cells per gram soil. However, in an agriculture field their number goes to about 3×10^9/g soil which accounts for about

3 tonnes wet weight per acre. Based on regular presence, bacteria are divided into two groups : (*a*) soil *indigenous* (*i.e.* true resident) or autochthonous, and (*b*) soil *invader* or allochthonous. Moreover, the number and types of bacteria are influenced by soil types and their microenvironment, organic matter, cultivation practices, etc. They are found in high number in cultivated than virgin land, maximum in rhizosphere and less in non-rhizosphere soil possibly due to aeration and nutrient availability (Rovira, 1965, Alexander, 1977). The inner region of soil aggregates contained higher level of Gram-negative bacteria, while the outer region contained higher level of Gram-positive bacteria. This may be due to polymer formation, motility, surface changes, and life cycle of bacteria involved.

Bacteria do not occur freely in the soil solution but are closely attached to soil particles or embedded in organic mat-
ter; even after adding the dispersing agents bacteria are not completely dis-lodged from the soil par-ticles and distributed in sus-pension as individual cell. Moreover, they play a ma-jor role in organic matter decomposition, bio-trans-formation, biogas produc-tion, nitrogen fixation, etc. Example of some of soil bacteria is *Agrobacterium, Arthrobacter, Bacillus, Alcaligens, Clostridium, Corynebacterium, Erwinia, Nitrosomonas, Nitrobacter, Pseudomonas, Rhizobium, Thiobacillus*, etc.

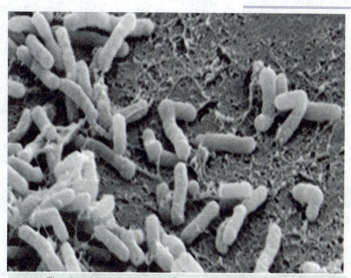

Electron micrograph of an *Agrobacterium* spp.

(*c*) **Actinomycetes:** Actinomycetes share the characters of both bacteria and fungi (see Chapter 2), and they are commonly known as "ray-fungi" because of their close affinity with fungi. They are Gram-positive and release antibiotic substances. However, the earthy odour of newly wetted soils has been found to be a volatile growth product of actinomycetes. Population of actinomycetes in soil remains greater in grass land and pasteur soil than in the cultivated land. In temperate zones the number of actinomycetes ranges from 10^5 to 10^8 per gram soil. The most limiting factor is the pH which governs their abundance in soil. Its luxurient growth is favoured by neutral or alkaline pH (6.0 to 8.0) (Garrett, 1981). The important members of actinomycetes are: *Actinomyces, Actinoplanes, Micromonospora, Microbispora, Nocardia, Streptomyces, Thermoactinomyces*, etc.

(*d*) **Bacteriophages:** Bacteriophages as well as plant and animal viruses have been observed in the soil. However, there role has not been clearly understood.

(*e*) **Protozoa:** In moist soil most of the members of microfauna remain in encysted form. The population of each group is 10^3 per gram wet soil. The role of soil protozoa is predatory, as these eat upon bacteria and thereby regulate their population. However, the number of protozoa can be correlated with plant root growth and indirectly with status of soil nutrients (Griffin, 1972; Garrett, 1981). Interaction of soil-amoebae with fungal hyphae has been discussed in section *Microbial Interactions* (see *Mycophagy*)

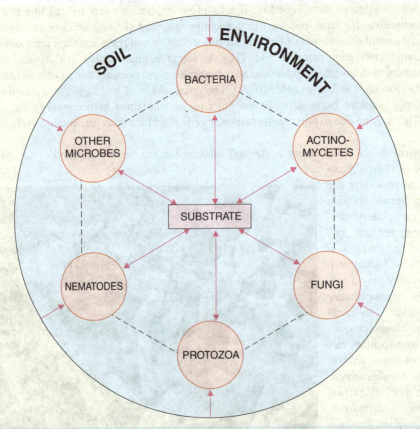

Fig. 27.2 : Microbial world of soil environment showing possible interactions between microbes and energy substrate, microbes and physical factors, and between the microbes themselves (energy substrates include inorganic chemicals, dead organics and living plant root and soil fauna (after Dubey *et al.,* 1992).

(*f*) **Nematodes:** Until the role of nematodes in soil was understood nematology was in infancy stage. In recent years the ecology of nematodes has been greatly advanced. Nematodes derive nutrients for their growth and reproduction from the cell contents and cytoplasm of protozoa, bacteria, fungi, etc. (Griffin, 1972). Some common example of protozoa is *Colpoda, Pleurotricha, Heteromita, Cercomonas, Oikomonas, Phalansterium,* etc.

(*g*) **Fungi:** In most of aerated or cultivated soils fungi share a major part of the total microbial biomass because of their large diameter and extensive net work of mycelium. However, population of soil fungi ranges from

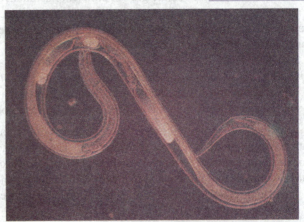

One square meter of forest soil teems with two to four million nematodes.

2×10^4 to 1×10^6 propagules per gram dry soil and its number differs according to isolation procedure and composition of media. Fungi derive nutrients for their growth from organic matters, living animals (including protozoa, arthropods, nematodes, etc.) and/or living plants establishing different types of relationships. Garrett (1950) classified the soil fungi on the basis of substrate-specialization (*i.e.* fundamental niche) and duration of parasitism as given in Fig. 27.3.

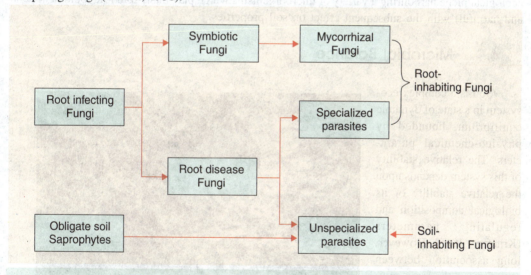

(a)   (b)

Representatives of two phyla of fungi : Ascomycetes and Basidiomycetes.

The *root inhabiting fungi* are characterized by an expanding parasitic phase on the living host with a little declining saprophytic phase after the death of the invaded host. But the *soil inhabiting fungi* are characterized by the ability to survive indefinitely as soil saprophyte. Success in competitive colonization of substrate by any particular soil fungus depends directly on its competitive saprophytic ability (CSA), inoculum potential at the surface of the substrate, and inversely as the aggregate inoculum potential of the competing fungi (Garrett, 1956).

Fig. 27.3 : Classification of soil fungi according to Garrett (1950).

Many unspecialized root infecting fungi can probably exist as competitive soil saprophytes in the absence of living host, they have characteristics that confer a high degree of competitive saprophytic ability such as (*a*) a high mycelial growth rate allied to rapid germination of resting propagules when stimulated by nutrient diffusion from a potential substrate, (*b*) a sufficient arrangement of tissue-degrading enzymes, (*c*) production of fungistatic growth products including antibiotics, and (*d*) tolerance of those produced by other microorganisms (Garrett, 1956). However, most specialized parasites have evolved in such a way as to cause minimum possible damages to the host plant functioning as a piece of biological machinery existing for support and protection of parasite. The distribution of specialized parasite is determined by its host species (Garrett, 1970). Some of the soil saprophytes are: *Alternaria, Aspergillus, Cladosporium, Dematium, Gliocladium, Helminthosporium, Humicola, Metarrhizium,* etc., and fungi associated with plant disease are : *Armillaria, Fusarium, Helminthosporium, Ophiobolus, Phytophthora, Plasmodiophora, Pythium, Rhizoctonia, Sclerotium, Thielaviopsis, Verticillium,* etc.

3.   Is 'Habitat' a Better Term for Microorganism ?

It is rather better to use the term 'niche' than 'habitat', as the habitat is the place where an organism lives, whereas niche speaks the habitat as well as the role of the organism in that habitat with respect to other organisms and the environment. Odum (1971) has used the two analogies, 'address' for habitat and `profession' for niche of the organisms. Among the active microbial species in soil, food specialization makes possible the existence of a large number of ecological niches within a given habitat. Even in such specialized and restricted habitat, as the rhizoplane (see *microbial interactions*) of an individual plant, it is possible that most of the microbial species that remain present therein are not competitive. Dwivedi and Saravanamuthu (1985) have discussed this restriction as the inherent genetic potential of a species which determines the range of tolerance to the varied physico-chemical and environmental parameters and called as *fundamental niche*. Moreover, if an ecological niche is a 'substrate' the key for understanding seems to be the realization that the scale of time and distance applicable to microorganisms are extremely small, and the soil is extremely heterogeneous on these scales (Griffin, 1972). Thus, soil is the best heterogeneous ecological niche harbouring a variety of microorganisms which play many fold role (both beneficial and harmful) with the subsequent effect on soil properties.

4.   Microbial Balance

Soil is a complex eco-system in a state of dynamic equilibrium, bounded by physico-chemical parameters. The relative stability of this system depends upon the relative stability of its biological composition and regulating parameters (Kruetzer, 1965). However, long association between organisms in the same environment brings about a kind of permanance or balance among them which is commonly known as equilibrium

A fungal mycelium growing through leaves on the forest floor in Maryland.

between needs of organisms and their numbers. The equilibrium is possible only when, during coexistence action and interaction between different microorganisms of varied potential go on. Thus, at a given time population of one species increases, while that of the others perhaps decreases due to microbial interaction in an ecological niche. In natural soil if the greater number of interacting factors, results in the more stationary microbial balance (Wilhelm, 1965). In rhizosphere the balance would be less stable due to continuous release of energy sources (host tissues/exudates) for microorganisms.

Baker and Cook (1974) pointed out that the presence of microorganisms at a given place and time is determined by (a) its having or being introduced there, (b) the existence of physico-chemical environment favourable to its development, (c) the presence of associated organisms (symbionts, hosts) favourable to its development, or organisms (host or parasite) required for its survival, and (d) the inhibition or absence of organism (disease organisms, pests, antagonists) so detrimental to it as to cause its extinction. An organism will increase until the limitations imposed by the biotic and abiotic environment just counter balance the rate of increase. Thus, in biological sense limitations check the chaos and epidemics of microorganisms.

Owing to the presence of complex substrates the ecological niches become very complex. However, physico-chemical and genetical diversity of microorganisms even in nutritionally deficient types, allow them to range over many environments with varying degree of success. For example, *Macrophomina phaseolina* is a root parasite pathogenic to many plants under high soil temperature (30-35°C) and dry conditions, but it was found to parasitize quickly the roots of weeds growing in cool environment where the fungus was not pathogenic.

Man is the supreme in living organisms in disrupting the balance of nature. This act has been exploited in the case of microorganisms what we call as biological control of soil inhibiting pathogens. In contrast, when man started cultivation innumerable microbial balances were broken resulting in an increase in parasites harmful to cultivated plants. Ordish (1967) stated "we are civilized only because we have upset the balance of nature for our advantage; unless we continue to do so we shall perish".

B. Rhizosphere and Rhizoplane Microorganisms

In 1904, L. Hiltner for the first time coined the term `rhizosphere' to denote the area of intense microbiological activity that extends several millimeter from the root system of the growing plants. Microorganisms growing under the influence of roots are often qualitatively and quantitatively different from those inhabiting remote from this influence in the soil environment. Therefore, the rhizosphere is a unique subterranean habitat for microorganisms. The rhizosphere microflora of one plant differ from the rhizosphere microflora of the other plant. Thus, rhizosphere microorganisms differ plant to plant both qualitatively and quantitatively.

The rhizosphere region can be divided into two zones: the inner rhizosphere which is in a close vicinity of root surface, and the outer rhizosphere embracing the immediate adjacent soil (Fig. 27.4). In 1949, F.E. Clark has suggested to use the term `rhizoplane' for root surface itself in studying the rhizosphere phenomenon. Balandreau and Knowles (1978) have termed the epidermis/cortex region, the *endorhizosphere* and the zone in the immediate vicinity to epidermis, the *exorhizosphere* to denote the intemacy of microbial associations. Between the rhizosphere and soil there is an area of transition in which the root influence diminishes with distance. Therefore, it is generally accepted that the term rhizosphere soil refers to the thin layer adhering to a root after the loose soil and clumps have been removed by shaking. The soil coating varies in thickness according to root types, presence of moisture and condition of soil. This certainly influences the 'rhizosphere effect'.

1. Reasons of Increased Microbial Activity in Rhizosphere

The outer epidermal walls of living root hairs and all plant roots are covered with mucilage and cuticle (see Fig. 27.4). Organic and inorganic compounds accumulated in cytoplasm of root cells are diffused out. This loss occurs probably due to unfavourable conditions external to root. The phenomenon of loss of organic and inorganic compounds from root surface is known as root exudation. In addition, the root-hairs are sloughed off during secondary thickening. All these root tissues and organic and inorganic compounds constitute a food base (for microorganisms) which are generally lacking in non-rhizosphere soil. The microorganisms colonize the rhizosphere to utilize them as food, and in turn release exudates from their own cells. Thus, they are regarded as selective sieves.

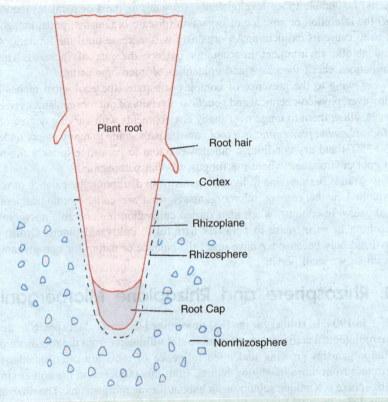

Plant root

Root hair

Cortex

Rhizoplane

Rhizosphere

Root Cap

Nonrhizosphere

Fig. 27.4 : A typical plant root showing rhizosphere, rhizoplane and non-rhizosphere regions.

2.   Composition of Root Exudates

A.D. Rovira (1965) has reviewed the plant root exudation and its influence upon soil microorganisms. She has described the nature of root exudates as below:

(a) Root exudates contain carbohydrates, organic acids, nucleotides, flavonones, enzymes, etc.

(b) Atleast ten sugars such as glucose, fructose, sucrose, xylose, maltose, rhamnose, arabinose, raffinose, and a few oligosaccharides have

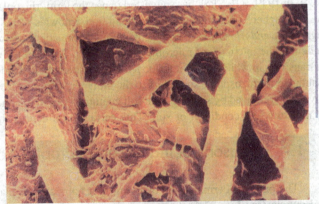

A scanning electron micrograph showing bacteria and fungi colonising the rhizosphere. Plant roots release nutrients that allow intensive development of bacteria and fungi on and near the plant surface.

been identified in the exudates of a wide range of plant.

(c) Generally organic acids, involved in uptake of nutrients by roots are found inside roots in significant leavels. However, these are exudated in the root exudates to some extent such as tartaric acid, oxalic acid, malic acid, acetic acid, citric acid, fumaric acid, etc. (Rovira, 1965).

(d) The amino acids detected in root exudates are leucine, valine, glutamine, asparagine, serine, glycine, glutamic acid, phenylalanine, threonine, tyrosine, lysine, proline, tryptophan, etc.

(e) Ten vitamins have been identified in the exudates from a variety of plants in trace amount. These may be sufficient to meet the requirements of some of the vitamin-requiring microorganisms of the rhizosphere. Vitamins found in root exudates are: biotin, thiamin, pantothenate, niacin, choline, inositol, pyridoxine, p-amino benzoic acid, m-methyl nicotinic acid, `M'-factor.

(f) The nucleotides, flavonones and enzymes are leaked from the zone of active cell division immediately behind the root cap. In this zone the nucleic acids are synthesised, and respiration occurs at the greatest level involving 3, 4-dioxyflavonone. Nucleotides, flavonones and enzymes detected in root exudates are: adenine, guanine, uridine, citidine, flavonones, phosphatase, invertase, amylase, protease, polygalacturonase, etc.

(g) A wide range of miscellaneous compounds is released from roots, and several of these are toxic to microorganisms e.g. hydrocyanic acid, glycosides, saponins, etc. The other miscellaneous compounds present in root exudates are: auxins, organic phosphorus compounds, benetonite colouring compounds, nodulating inhibitors, *Azotobacter* growth-stimulator, scopoletin (6-methoxy-7-hydroxycoumarin), etc.

3.   Factors Affecting Exudation

Root exuation is affected by many factors. Examples of some of the factors are discussed here with.

(a) Temporary wilting of plants increases the release of amino acids from roots in sand or soil,

(b) Exudation is increased under high light and temperature condition and is greatest during the first few weeks of growth,

(c) Secondary metabolites of certain bacteria causes a three fold increase in exudation of scopoletin without affecting root-growth. Microorganisms alter the patterns of amino acids in exudates,

(d) In the presence of competitive rhizobia polygalacturonase is released from the roots, polymyxin and the other polypeptide antibiotics increase the leakage of inorganic and organic materials from roots. Microorganisms affect the patterns of exudation in at least three ways: by altering the permeability of root cells, by modifying the metabolism of roots, and by modifying some of the materials released from the roots (Rovira, 1965).

4.   Rhizosphere Microorganisms

The rhizosphere region is a highly favourable habitat for the proliferation and metabolisms of numerous types of microorganisms (Alexander, 1977). The microbial community of this zone can be examined by means of cultural, microscopic and manometric techniques. Dubey and Dwivedi (1988) screened the rhizosphere and non-rhizosphere microfungi of soybean both qualitatively and quantitatively. Always, the number of rhizosphere microfungi was higher than the number of non-rhizosphere fungi (Table 27.2) The dominant fungi of rhizosphere were *Aspergillus flavus, A. fumigatus, A. luchuensis, A. niger, A. terreus, Cladosporium cladosporioides, Curvularia lunata* and *Fusarium oxysporum,* whereas the dominant fungi of rhizoplane were *A. niger, Cladosporium herbarum, F. oxysporum, F. solani, Macrophomina phaseolina, Neocosmospora vasinfecta* and *Rhizoctonia solani.* In addition, mycorrhizal fungi are also known to be present in rhizosphere soil and rhizoplane of roots.

Protozoa are relatively conspicuous particularly the small flagellates, large ciliates and amoeboidal forms. They are situated in the water films on the root hairs and on the epidermal tissue. Cysts of nematodes have also been reported in the rhizosphere region, for example *Heterodera, Pectus, Tylenchus, Acrobeles, Helicotylenchus, Meloidogyne,* etc.

Less is known about the algae except the blue-green algae present in the rhizosphere soil. This may be because of establishing symbiotic associations in certain plants such as coralloid roots of *Cycas.*

Bacteria reported from the rhizosphere and rhizoplane regions irrespective of their dominance are: *Arthrobacter, Pseudomonas, Bacillus brevis, B. circulans, B. polymyxa, B. megaterium, Agrobacterium radiobacter, A. tumifaciens, Azotobacter, Flavobacterium, Rhizobium* spp., *Cellulomonas, Micrococcus, Mycobacterium,* etc.

Actinomycetes are also important constituents of rhizosphere and rhizoplane microflora of different biosynthetic capabilities, antagonistic potentiality and taxonomic groups. Examples are *Actinomyces chromogenes, Frankia* (inside root tissues), *Nocardia* spp., *Micromonospora* spp., *Streptomyces antibioticus, S. scabies, S. griseus* etc.

Crown gall tumor of a tomato plant caused by *Agrobacterium tumefaciens* which colonises the rhizosplane and rhizosphere regions.

Table 27.2 : Average number of fungi in the rhizosphere and non-rhizosphere regions of soybean at different growth stages (source Dubey and Dwivedi, 1988).

Soil	Average number of fungi (×10⁴/g soil) Growth Stages				
	SDL	PRF	FLR	PFL	SNT
Non-rhizosphere	5.037	5.064	7.060	6.622	5.187
Rhizosphere	8.460	12.770	20.019	17.406	15.076
R:S Ratio	1.69	2.521	2.835	2.629	2.906

SDL, seedling; PRF, preflowering; FLR, flowering; PFL, post-flowering; SNT, senescent; R:S, rhizosphere: non-rhizosphere ratio.

5.   The Rhizosphere Effect

As described earlier, the rhizosphere is a zone of increased microbial community as well as microbial activities influenced by the root itself. However, this influence can be measured simply by plating technique and expressed as a rhizosphere effect (i.e. a stimulation that can be measured on quantitative basis by the use of rhizosphere : soil (R:S) ratio, obtained by dividing the number of microorganism in the rhizosphere soil by the number of microorganisms in the non-rhizosphere soil).

R:S ratio of soybean changing with different growth stages has been shown in Table 27.2. It is apparent that different types of microbes dominate at a particular growth stage. A single microbial species will have to compete all the time to become a permanent inhabitant of the rhizosphere region, which is rather impractical. This microbial selection during different growth stages and changing scenario of microbial community in rhizosphere/rhizoplane are ultimately governed by root exudates of host as well as the result of microbial interactions.

As a rule, actinomycetes, protozoa and algae are not significantly benefited by roots, and the R:S ratio rarely exceeds 2:1 or 3:1.

The bacterial count in rhizosphere soil is the maximum (R:S values 10 to 20 or more), and varies with plant species, plant age and fertilization. However, there is no selective stimulation or inhibition. Generally, bacteria of several distinctly different physiological, taxonomic and morphological groups are found to grow in rhizosphere region. However, on generic basis *Pseudomonas, Flavobacterium, Alcaligens* and *Agrobacterium* are especially common (Alexander, 1977). In contrast, total count of fungi is not altered by the root. Not total, but specific fungal genera are stimulated with plant species, plant age and soil types and environmental factors (Upadhyay and Rai, 1982).

The study of rhizosphere is very important because several pathogenic microorganisms have to pass through the rhizosphere and infect root system. Moreover, manipulation of rhizosphere environment is of recent consideration as far as control of soil-borne plant pathogens is concerned (Dubey *et al*, 1992).

6. Rhizosphere Engineering

Rhizosphere effect is established in the root region due to secretion of root exudates as compared to non-rhizospheric soil. Gene expression is altered through genetic engineering in plants; therefore, the root exudates and secretions are also changed which result in aggressive root colonisation by desirable microorganisms. Tissue-specific gene expression has great potential for the genetic engineering of the rhizosphere. The use of border cells (i.e. cells released out from root) into the surrounding soil due to their metabolically active nature supports rhizosphere process. It is now possible to manipulate the rhizosphere through changes in root exudation to alter the microbial flora to influence the micro-bial gene expression or to alter the chemistry of root region.

Most of the plant growth promoting rhizobacteria (PGPR) inhibit the deleterious plant patho-gens by involving proteins, peptides, etc. Their gene manipulation may help in engineering proteins/pep-tides which ultimately dif-fuse out in the rhizosphere. The role of **opine-concept**

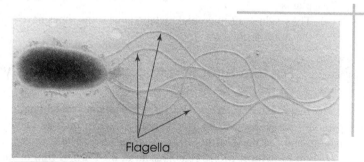

Pseudomonas fluorescens living in the rhizosphere of plant species (by involving proteins, peptides etc.) which produce antifungal agents, which ward off potential pathogens.

Flagella

in plant-microbe interactions is now evidenced by the fact that these amino acid derivative or sugar derivative molecules are not catabolised by most of the soil microorganisms. Hence, the use of opine-catabolising bacteria has definitely play a role, thus establishing the opine-concept. The modified form of this concept is to engineer plants to overproduce some common nutrients that do not allow to grow the undersirable microorganisms in the rhizosphere.

Plants might be engineered to release a small molecule inducer of bacterial gene expression into the rhizosphere. In response, an engineered microbial population would synthesise a desired organic molecule e.g. siderophores which indirectly play a role in biological control. Similarly, transgenic plants, release rhizopine (L-3-O-methyl-syllo-inosamine) into the rhizosphere which allow rhizopine catabolising bacteria to grow in their vicinity. Plants can potentially remediate the soil contaminants, accumulate metals for recovery purposes and clean the contaminated soil. On the other hand, bacteria can contribute to remediation by catabolising organic molecules, mobilising soil bound metals and excreting siderophores. Proton used in decreasing soil pH and solubilising metal cations is also an advantage of rhizosphere engineering.

7.   Effect of Microflora on Host Plants

The rhizosphere microorganisms have either beneficial or harmful effects on the development of plant. The microorganisms are intimately associated on the rhizoplane, therefore, any toxic or beneficial substance produced by them has direct effect on plant. Direct effect of host on microflora has been discussed earlier. Some of the possible effects are briefly described as below:

(a) The microorganisms catalyse the reactions in the rhizosphere and produce CO_2 and form organic acids that in turn solubilize the inorganic nutrients of plants.

(b) Aerobic bacteria utilize O_2 and produce CO_2, therefore, lower O_2 and increase CO_2 tension that reduces root elongation, and nutrient and water uptake.

(c) Some of the rhizosphere microorganisms produce growth-stimulating substances and release elements in organic forms through the process of mineralization.

(d) Plant growth regulators such as indole acetic acid, gibberellins, cytokinins, etc. are known to be produced by the rhizosphere microflora.

(e) They influence phosphorus availability to plant through the process of mineralization and immobilization. However, when plant suffers from nutrient scarcity during summer in tropical areas the microorganism release the immobilized nutrients. Therefore, they act as sink between soil and plant roots in nutrient poor systems (Singh et al., 1989).

(f) Microorganisms in the rhizosphere zone change the availability or toxicity of sulphur to plants.

(g) The products of microbial metabolisms some times have toxic effect on plants, therefore, these are termed as *phytotoxins*.

8.   Factors Affecting Microbial Community in Soil

The major external factors that influence the microbial community in soil are moisture, pH, temperature, gases, organic and inorganic fertilizers, organic matter of soil, types of vegetation, ploughing, and season (Griffin, 1972, Mueller et al, 1985, Dubey and Dwivedi, 1988)

(i) **Soil Moisture:** Moisture is present in the form of film in soil pores. The amount of water increases with increase in porosity of soil. Pore-size depends on soil texture i.e. composition of sand, silt and clays. Moreover, soil moisture is affected through irrigation, drainage or management practices like tillage or crop rotation that enhance the intake and transmission of water by soil.

(ii) **Organic and Inorganic Chemicals:** The chemicals are very important for microorganisms as these provide nutrition for growth, activity and survival of microorganisms in ecologically deficient niches in soil. The chemical factors are gases, acids, micro- and macro-elements, clay minerals, etc. In the soil solution gases (oxygen, methane and carbon dioxide), and microorganisms are dissolved. However, the dissolved components are in constantly shifting equilibrium with the solid phase, soil air, moisture as well as with soil organisms and plant root activity. It has been found that low potassium and high nitrogen favour cotton wilt by *Fusarium vasinfectum*. Soil-borne fungi are sensitive to pH. As a result of pH range for vigour and growth, they are more destructive at acid and neutral at alkaline conditions. For example, *Plasmodiophora brassicae* favours best in acid soil, and the disease produced by it is uncommon or mild in soil of pH more than 7.5. Acidophilic natives of *Trichoderma viride* increased in soil on addition of sulphur, carbon disulphide, and methylbromide due to lowering down of pH to about 4.0.

Due to excess soil moisture in peat bogs, the growing plants cannot absorb enough minerals (as their roots become waterlogged) and therefore the microbial population is lower.

(*iii*) **Soil Organic Matter:** The dead organic material of plant and animal origin serve as total soil organic matter which later is subjected to microbial colonization and decomposition. However, upon incorporation of green manures, crop residues, etc. in soil, the community size of microorganisms gets increased. At the same time application of these organic matter alters the composition of soil microflora, microfauna, and relative dominance of antagonistic bacteria, actinomycetes, fungi, amoebae, etc. (Baker and Cook, 1974).

(*iv*) **Types of Vegetation and its Growth Stages:** The dominance of one or the other groups is related to the type of vegetation and growth stages of a plant. Dubey and Dwivedi (1988) found an increased population of fungi in the non-rhizosphere and rhizosphere of soybean according to season and growth stages, respectively (Table 27.2). In the rhizosphere aspergilli, fusaria and penicillia were dominant in addition to the other fungal species. However, frequency of *Macrophomina phaseolina* and *Neocosmospora vasinfecta* increased on rhizoplane with onset of senescence. This selective action of plants has been discussed to be due to microbial response either to specific root-exudates or chemical constituents of sloughed-off tissues that undergo decomposition (Rovira, 1965). Moreover, Mueller *et al.* (1985) determined the incidence of fungi and bacteria occurring in the roots of six soybean cultivars growing in fields cropped for 3 years either with corn or soybean. Cropping history affected the recovery of *M. phaseolina*, *Phomopsis* spp. and *Trichoderma* spp. but not *Fusarium* spp. or *Gliocladium roseum*. Recovery of *Trichoderma* spp. was greater following corn than following soybean. After death of the plant soil saprophytes colonize rapidly, thus total spectrum of microflora in the rhizosphere is changed (Upadhyay and Rai, 1982).

(*v*) **Different Seasons:** The amount of plant available nutrients is governed by the number and activity of microorganisms. They remain in constant dynamic state in soil where microbial community is greatly influenced by physico-chemical and biological factors (Rovira, 1965; Griffin, 1972; Garrett, 1981). Changes in microbial community are known in soils of tropical, sub-tropical and temperate regions. Shail and Dubey (1997) have studied the seasonal changes in microbial community (bacteria and fungi) and species diversity in fungi in banj-oak and chir-pine forest soils of Kumaon Himalaya in relation to edaphic factors. Maximum number of fungal taxa, and average number of bacteria and fungi (per gram soil) were recorded in rainy season and minimum in summer season from both the soils (Table 27.3).

Table 27.3 : Seasonal changes in average number of fungi and bacteria per gram dry soil in banj-oak and chir-pine forest soils of Himalaya (source Shail and Dubey, 1997)

Forest soils	Seasons	Fungi (CFU × 10³)	Bacteria (CFU × 10⁵)
		Average number (g⁻¹ dry soil)	
Banj-oak	Summer	199	987
	Rainy	369	1408
	Winter	189	1161
Chir-pine	Summer	70	908
	Rainy	272	1344
	Winter	129	1072

CFU, colony forming units of microorganisms

C. Organic Matter Decomposition

Decomposition and photosynthesis are the two important process of an ecosystem. Litter (the organic remains of biological origin) is an organic chemical-carrier of nutrients present in different ecosystems. The rate of decomposition is the function of structural components of litter *i.e.* the structural components of litter governs the rate of decomposition. However, the plant/animal materials present a variety of substances in soil which both physically and chemically are heterogeneous.

1.   Composition of Litter

The organic constituents of plant that give structural topography generally contain six broad groups of consituents as given below:

(*i*) **Cellulose:** Cellulose is the most abundant chemical constituent of a cell. It is a carbohydrate composed of glucose units linked by β-linkages at 1 and 4 carbon atoms of sugars. Total number of glucose units varies from 2000 to 10,000. Total amount of cellulose varies from 15 to 60% in litter.

(*ii*) **Hemicellulose:** Hemicelluloses are found in close proximity to cellulose in the primary and secondary walls. The simple sugars (or monosaccharides) or uronic acids (derivatives of simple sugars) are bound together to form a large molecule which is called hemicellulose. The common sugars are xylose, arabinose, glucose and galactose, and uronic acids are glucuronic and galacturonic acids. It makes up to 30% of organic matter by weight.

(*iii*) **Lignin:** Certain plants especially woody species contribute large amount of lignin. It is found in the secondary layers of the cell wall and also in middle lamella. The lignin molecule is the aromatic compounds containing C, H and O. The basic unit of lignin is a phenylpropane type structure. It makes up 5 to 30% of plant.

(*iv*) **Water-soluble Components:** The water soluble components of plant include simple sugars, amino acids and aliphatic acids. Altogether these contribute to 5 to 30% of plant by weight.

(*v*) **Ether- and Alcohol-soluble Components:** These components of plant cell wall include fats, oils, waxes, resins and a number of pigments.

(*vi*) **Proteins :** Proteins are the main constituents of nucleic acids and cytoplasm, and found to some extent in cell wall as well. Total amount of proteins vary from 1 to 13% of total tissue weight.

2.   Carbon Assimilation and Immobilization

Microorganisms depending on substrate specificity colonize the organic matter and decompose it. However, the organic matter serves two functions for microflora. First, it provides energy for growth and, secondly it provides carbon for the formation of new cell material. During this process certain waste products are also produced by microorganisms *e.g.* organic acids, CO_2, methane, etc. (Alexander, 1977). The process of conversion of substrate to protoplasmic carbon is known as *assimilation.* About 20-40% substrate is assimilated and rest is released as CO_2 or accumulates as waste. Aerobic bacteria assimilate substrate by 5-10%, whereas anaerobic bacteria assimilate by 2-5%. Fungi assimilate carbon in forming mycelium by 30-40%. The term *substrate* is used in a wider sense for corpse of plant or animal tissues on which the microorganisms live for its food and energy.

As the carbon assimilation occurs at the same time the other inorganic chemicals *e.g.* nitrogen, phosphorus, potassium and sulphur are also taken up for the formation of new protoplasm. Assimilation of inorganic substances is an important process that takes place in microbial cells. By this process microorganisms accumulate inorganic substances in their cells and reduce the concentration of plant-available nutrients in soil. This event of accumulation of inorganic substance by the microorganisms and making the plants nutrient-deficient is known as *immobilization.* The magnitude of immobilization is proportional to net quantity of microbial cells formed. It is also related to carbon assimilation by a factor governed by the C:N, C:P, C:K or C:S ratio of newly generated protoplasm (Alexander, 1977)

3.   Organic Matter Dynamics in Soil

When organic matter is present in soil it is gradually rendered into a uniform, dark coloured amorphous mass by microorganisms which is designated as *humus.* Humus serves as a source of energy for the development of various groups of microorganisms and as a result of decomposition CO_2, NH_3 and other products are given off. Fungi along with the other microbial goroups chiefly bacteria and actinomycetes decompose the organic matter in soil and release the nutrients which are locked up in complex form in organic matter. However, the process of decomposition starts when the plants are in senescent stage. After com-

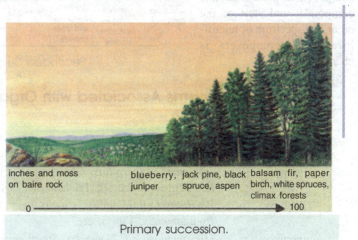

inches and moss on baire rock

blueberry, juniper

jack pine, black spruce, aspen

balsam fir, paper birch, white spruces, climax forests

0 ──────────────────────────► 100

Primary succession.

ing in the contact of soil some other groups of microorganisms colonize the litter. At this stage termites too play a role in physical breaking of the litter. Thereafter, microorganisms of different groups colonize the substrate depending upon its chemical composition. Thus microbial succession occurs on the decomposing material till it fully disappears in elemental forms. The events of sequential appearance of microorganisms on a substrate with respect to time is called *succession.* Mostly succession of fungi on decaying material has been worked out in detail. Garrett (1981) has given a general trend of fungal succession on a substrate as given in Table 27.4.

Table 27.4 : General trend of fungal succession

Senescent tissue		Dead Tissue	
Stage 1a	Stage 1b	Stage 2	Stage 3
Weak parasites	Primary saprophytic sugar fungi living on sugars and carbon compounds simpler than cellulose (mucoraceous phycomycetes)	Cellulose decomposers and associated secondary saprophytic sugar fungi sharing products of cellulose decomposition (ascomycetes and some mucorales)	Cellulose and lignin decomposers, and other associated fungi (basidiomycetes and others)

Primary colonizers   Secondary colonizers

⟶ Trend of succession ⟶

During the course of decomposition, the water-soluble components are metabolised first. Thereafter, cellulose and hemicellulose disappear gradually. The lignins disappear in the last stage of decomposition because these are resistant to decomposition. During the course of disappearance of one component and start of the other, the spectrum of microflora i.e. decomposers is changed.

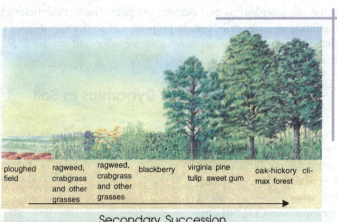

ploughed field | ragweed, crabgrass and other grasses | ragweed, crabgrass and other grasses | blackberry | virginia pine tulip sweet gum | oak-hickory climax forest

Secondary Succession.

4. Microorganisms Associated with Organic Matter Decomposition

(i) **Cellulose Decomposers:** Following are the decomposers of cellulose:

Fungi: *Alternaria, Aspergillus, Chaetomium, Coprinus, Fomes, Fusarium, Penicillium, Rhizopus, Trichoderma, Trametes, Verticillium*, etc.

Bacteria: *Bacillus, Cellulomonas, Clostridium, Corynebacterium, Cytophaga, Pseudomonas, Vibrio*, etc.

Actinomycetes: *Micromonospora, Nocardia, Streptomyces, Streptosporangium*, etc.

(ii) **Hemicellulose Decomposers:** Following is a list of decomposers of hemicelluloses:

Fungi: *Alternaria, Aspergillus, Chaetomium, Fusarium, Glomerella, Penicillium, Trichoderma*, etc.

Bacteria: *Bacillus, Cytophaga, Erwinia, Pseudomonas*, etc.

Streptomycetes: *Streptomyces*, etc.

(iii) **Lignin Decomposers:** Many of the basidiomycetes are capable of degrading lignin but the reaction is slow, because of the slow growth of microorganisms. Aerobic bacteria are also able to bring about some degradation of lignin, but little attention has been paid. Examples of some of the lignin decomposers are given below:

Fungi: *Agaricus, Armillaria, Clavaria, Clitocybe, Coprinus, Ganoderma, Phaliota, Pleurotus, Polyporus, Poria, Trametes, Ustulina,* etc.

Bacteria: Species of *Arthrobacter, Flavobacterium, Micrococcus, Pseudomonas, Xanthomonas,* etc.

5.   Factors Affecting Organic Matter Decomposition

There are many factors that govern the rate of decomposition of organic matter such as structural topography of litter, chemical constituents of litter, mechanical composition of soil (*i.e.* ratio of sand, silt and clay particles), temperature, aeration, soil pH, nitrogen contents present in soil.

(*i*) **Litter Quality :** Structural and chemical properties of litter make up its quality. However, it is less specific term reflecting decomposition rate. Dubey and Pandey (1996) have recorded the higher number of fungal taxa from decomposing litter of *Cupressus torulosa* than *Pinus roxburghii* litter decomposing in respective forests of Kumaon Himalaya. In contrast, the litter of bryophytes are decomposed at a very slow rate because of presence of lignin like complex chemicals in its thalli.

(*ii*) **Temperature:** Temperature governs the growth and microbial activity in its natural habitats. Singh *et al* (1989) studied the fungal communities associated with decomposition of leaf litter of oak in different forests along an elevational transect. At different elevations temperature differs. They found that species diversity and fungal counts on oak leaf litter were markedly affected by the environment changes brought about by the native leaf litter. The pattern of fungal species occurring on litter changed with the progress of decay of substrate.

(*iii*) **Aeration:** In the pores of soil, sufficient amount of oxygen is present which is required by aerobic flora. In water-logged conditions where O_2 becomes a limiting factor aerobic microorganisms will be absent, and only anaerobic microorganisms will grow and decompose the organic matter. Soil texture affects aeration and the later affects microoragnisms.

(*iv*) **Soil pH:** Soil pH is governed by the presence of cations and anions. Certainly these affect microbial growth. For example, actinomycetes prefer to grow above soil pH 7, bacteria below 7, and fungi between pH 5 to 6.

(*v*) **Inorganic Chemical:** The concentration of already available inorganic substances also affect the rate of decomposition of added matter on soil. In addition after decomposition from humus the elements N, P, K, Na, Mg, Ca, etc are released in soil. Some amount is taken up by the growing microorganisms and the remainder is made available to plants.

(*vi*) **Moisture:** Soil moisture varies according to water holding capacity (WHC) of a given soil. However, water is required to carry out the physiological processes. Significance of water in soil for various activities of microorganisms has nicely been discussed by Griffin (1974) in his book *Ecology of Soil Fungi.* Rai and Srivastava (1982) have found the half life of litter in a tropical mixed dry deciduous forest about 173 days. Moisture plays a critical role in determining the activity of microorganism in decomposition.

6.   Microbial Biomass as an Index of Soil Fertility

In recent years, the role of microorganisms in decomposition of organic matter and mineralization of bioelements, and more specifically the large pool of mineral nutrients in their cytoplasm and their turn over into the soil again available to plant roots, have been much studied in many countries including India. Microbial biomass, irrespective of soil types, is a part of soil organic matter which includes the living microbial components *i.e.* bacteria, actinomycetes, fungi, algae, protozoa and other microfauna. Microbial biomass is influenced by affecting the microbial activity. Singh *et al.* (1989) have advocated that the microbial biomass in nutrient poor system acts as a sink and a source of nutrients. It accumulates during dry period and release these rapidlly during monosoon period so that plant growth can be started. This accumulation of nutrients is called *immobilization.* Thus in nutrient-poor ecosystem, nutrient retention and withdrawal mechanisms are effective (Singh *et al*, 1984).

Cultivation leads to considerable loss of soil organic matter and, therefore, microbial biomass also changes (Srivastava and Singh, 1989). The turn over time of microbial biomass is estimated at 1-3 years. Since microbial biomass is the most active fraction of soil organic matter, therefore, any change in organic matter may be supposed that microbial biomass too has changed. Singh and Singh (1995) have found that microbial biomass is associated with water-stable macro and micro aggregates in forest, savana and cropland soils of a seasonally dry tropical region.

7.   Soil Fertility

Soil fertility is the sustainable capacity of a soil to produce good yields of high quality on the basis of chemical, physical and biological factors. The three factors are needed to assess soil quality or fertility. Much work has been done on physical, chemical, climatic and cultivation data. But biological soil components have never been used to assess soil quality. The parameters for estimation of biological factors are soil respiration, nitrogen mineralisation, nitrification, denitrification, nitrogen fixation, soil enzyme activity (e.g. urease, acid phosphatase and alkaline phosphatase, etc.). The level of microbial biomass of a soil denotes soil quality at a time on the basis of above parameters. Fig. 27.5 shows the integrated evaluation of biological, chemical, physical, cultivation and climatic data (Torstensson *et al.* 1998)

A fertile soil plays an important role in enhancing the yields of tomatoes.

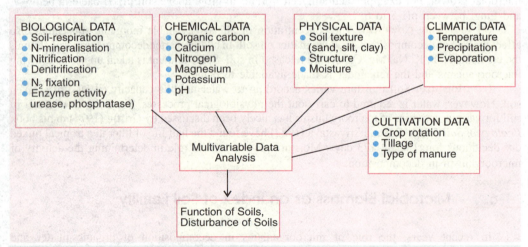

BIOLOGICAL DATA
- Soil-respiration
- N-mineralisation
- Nitrification
- Denitrification
- N$_2$ fixation
- Enzyme activity urease, phosphatase)

CHEMICAL DATA
- Organic carbon
- Calcium
- Nitrogen
- Magnesium
- Potassium
- pH

PHYSICAL DATA
- Soil texture (sand, silt, clay)
- Structure
- Moisture

CLIMATIC DATA
- Temperature
- Precipitation
- Evaporation

CULTIVATION DATA
- Crop rotation
- Tillage
- Type of manure

Multivariable Data Analysis

Function of Soils, Disturbance of Soils

Fig. 27.5 : Integrated evaluation of biological, chemical, physical, cultivation and climatic data for a soil which shows functioning of soil. These also act as indicators of disturbances arising in soil. (based on Tortensson et al., 1998).

Soil Function: The basic natural function of soil system should be: (*i*) primary production by creation a rooting substrate, (*ii*) to support above ground, (*iii*) to decompose organic material resulting in release of simple organic compounds and minerals, *(iv)* to exchange gases, solutes and organisms between the aqueous, solid and gas phases, and *(v)* to support diversity of live forms and their adaptations, in response to heterogeneous physical, chemical and biological conditions.

D. Biogeochemical Cycling

The major plant nutrients derived from soil are nitrogen, phosphorus and potassium because these are made biologically available to plants. Out of the three nutrients nitrogen stands out as the most susceptible to microbial transformations as it builts up protein, and many components of cytoplasm of microorganisms, plants and animals. Phosphorus is the second to nitrogen which is required by both plants and microorganisms. It is found in soil, microorganisms and plants in the form of organic and inorganic compounds. It play a major role in accumulation and release of energy during metabolism. Potassium is the major cation that plants obtain from soil. Its quantity in plants is often inadequate. In addition, there are many important chemicals present in plants, animals and microorganisms such as carbon, sulphur, iron, manganese, zinc, etc. Thus, biogeochemical cycling associated with microorganisms is very important for the maintenance of soil fertility. Biogeochemical cycling of carbon nitrogen, phosphorus and potassium (N,P,K) is discussed in detail.

1. The Carbon Cycle

In nature carbon exists in the form of inorganic and complex organic compounds. In atmosphere the concentration of CO_2 is only 0.32% which is less than what is required by plants for photosynthesis. The CO_2 is the main source of carbon required to build the organic world. The CO_2 returns back into the atmosphere through the respiration process by all groups of organisms. The other method of returning carbon is through degradation (decomposition) of organic matter by microorganisms (see *organic matter dynamics*). A simplified carbon cycle is given in Fig. 27.6.

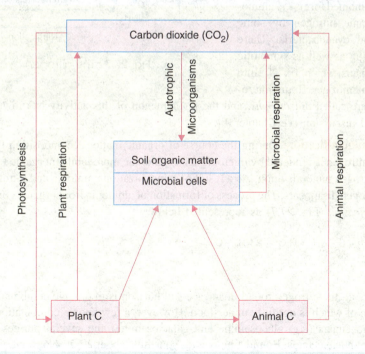

Fig. 27.6 : The carbon cycle.

2.   The Nitrogen Cycle

Nitrogen is found in the atmosphere in the highest concentration (79%). It is an essential constituent of proteins and chlorophyll found in organisms. In spite of being in such a high amount, nitrogen is not directly taken by the animals or plants from the atmosphere. In soil it remains in limited amount, and whatever concentration is found, that is governed by microbial activities. Therefore, concentration of nitrogen made available in soil directly governs soil fertility. The key processes of biogeochemical cycling of nitrogen are nitrogen fixation, ammonification, nitrification, denitrification, and nitrite ammonification (Fig. 27.7).

(i) Nitrogen Fixation: Nitrogen fixation is a process of conversion of gaseous form of nitrogen (N_2) into combined forms *i.e.* ammonia or organic nitrogen by some bacteria and cyanobacteria. There are free-living as well as symbiotic microorganisms which fix N_2 into proteins. The nitrogen-fixing microorganisms are called *diazotrophs*, and the phenomenon of this activity is known as *diazotrophy*. For detailed discussion see Chapter 14.

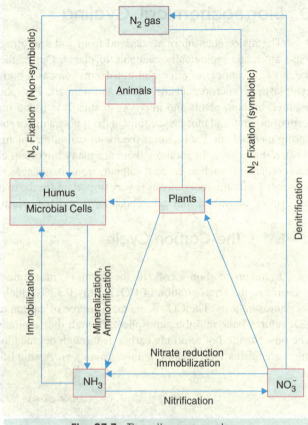

Fig. 27.7 : The nitrogen cycle.

(ii) Ammonification: During the course of organic matter decomposition the complex form of organic nitrogen is rendered by microorganisms into the more labile inorganic form. This process is called nitrogen mineralization. As a result of mineralization ammonia and nitrate are formed and organic nitrogen disappears. The process of formation of ammonia from organic compound is known as *ammonification* (Fig. 27.7) as represented below:

$$NH_2 - \underset{\underset{O}{\|}}{C} - NH_2 + H_2O \rightarrow 2\,NH_3 + CO_2$$

Mostly all the nitrogen found in surface soil horizon is in organic combination, the chemical composition of which is not fully understood. However, a few known combinations are free amino acids, amino–sugars (*e.g.* glucosamine and galactosamine) and several purines and pyrimidines derived from nucleic acids. Bound amino acids constitute about 20-50% of total nitrogen of humus, whereas 5-11% is amino-sugar nitrogen. Moreover, the net change in the amount of inorganic nitrogen (N_i) is expressed as below:

$$N_i = \text{Organic nitrogen mineralized} - (N_a + N_p + N_l + N_d)$$

where, N_a = nitrogen assimilated by microorganisms

N_p = nitrogen removed by the plants

N_l = nitrogen lost by leaching

N_d = nitrogen volatilized by denitrification

Microbiology: A diverse microflora liberates ammonia from organic nitrogen compounds. These include bacteria (e.g. *Pseudomonas, Bacillus, Clostridium, Serratia, Micrococcus*, etc.), fungi (*e.g. Alternaria, Aspergillus, Mucor, Penicillium, Rhizopus*, etc), and actinomycetes. They synthesize extracellular proteolytic enzymes for the decomposition of proteins. The ammonifying population includes aerobes and anaerobes.

Proteins are converted to peptides and amino acids by extracellular proteolytic enzymes. Ammonia is produced after deamination (e.g. oxidative or reductive deamination) or the amino acids according to reactions as below:

Actinomyces liberates ammonia from organic nitrogen compounds.

Oxidative deamination:

$$\underset{\text{Amino acid}}{R\!-\!\underset{\displaystyle \overset{\displaystyle NH_2}{\|}}{C}H\!-\!COOH} + O_2 \rightarrow \underset{\text{Keto acid}}{R\!-\!\underset{\displaystyle \overset{\displaystyle O}{\|}}{C}\!-\!COOH} + NH_4$$

Reductive deamination:

$$\underset{\text{Amino acid}}{R\!-\!\underset{\displaystyle \overset{\displaystyle NH_2}{\|}}{C}H\!-\!COOH} + 2H \rightarrow \underset{\text{acid}}{R\!-\!CH_2\!-\!COOH} + NH_4$$

NH_4 predominates in acidic and neutral environment with the increase of pH, therefore, NH_3 predominates and is released into the atmosphere.

$$NH_4 \longrightarrow NH_3 + H^+$$

Ammonium is formed slowly at water levels slightly below the permanent wilting percentage. Ammonification is not eliminated by soil submergence and the process is rapid in wet paddy fields, where O_2 level is quite low (Alexander, 1977).

Urea is found in soil which is made available as a decomoposed product of nitrogenous bases, as fertilizer and as animal excretory product. However, urea is readily hydrolysed, and transformed to ammonia.

(*iii*) **Nitrification:** The process of oxidation of ammonium ions (oxidation level= - 3) to nitrite ions (oxidation level = +3) and subsequently to nitrate ions (oxidation level = +5) is known as *nitrification*. Thus, ammonium (the most reduced form of inorganic nitrogen) acts as the starting point for nitrification. Nitrate is also produced (on addition to soil) in manure piles, during sewage processing, and marine environment. In 1877, T. Schloesing and A. Muntz for the first time gave experimental evidence that nitrification is of biological origin. In 1890, S. Winogradsky isolated the responsible microorganisms from soil. However, L. Pasteur postulated earlier that the formation of nitrate was microbiological and analogous to the conversion of alcohol to vinegar.

There are certain chemical properties of the microbiological habitat that serve to change the magnitude of the transformation. Nitrification is proportional to the cation exchange capacity of soil. In alkaline soil where the concentration of salt is high, nitrate production is declined probably due to low tolerance of nitrifiers. In environment having a near neutral reaction, formation of nitrate from ammonia is rapid, whereas in acidic soils nitrate is formed faster from organic material (Alexander, 1965).

Microbiology: The nitrifying bacteria are very important for plants because they affect nutrient availability. Nitric acid is formed from ammonia which drops the pH. This affect changes in concentration of soluble potassium, phosphate, magnesium, iron, manganese and calcium. Nitrification is an example of aerobic respiration and appears to be present in a few autotrophic bacteria. The chemolithotrophs derive energy through these two processes (conversion of ammonia to nitrite, and nitrite to nitrate). These steps are carried out by two different types of nitrifying bacteria.

(*a*) **Bacteria oxidizing ammonium to nitrite:** The nitrifying bacteria are *Nitrosomonas, Nitrosococcus, Nitrosolobus, Nitrosospira, Nitrosovibrio*, etc. The reaction characterizing the chemoautotrophic bacteria of the first step of nitrification is as below:

$$NH_4^+ + 1\tfrac{1}{2} O_2 \longrightarrow NO_2^- + 2H^+ + H_2O$$
$$(-3) \qquad\qquad\qquad (+3)$$

Here the oxidation state is changed from -3 of ammonia to +3 of nitrous acid through the removal of electrons. The complete reaction is as below:

$$NH_3 \longrightarrow NH_2OH \longrightarrow (HNO?) \; NO \longrightarrow NO_2^-$$

Ammonium is first converted to hydroxylamine (NH_2OH) which in turn is changed to undefined metabolite (possible nitroxyl, HNO). This intermediate is oxidized to nitrite by the way of NO. Here accumulation of NH_2OH may lead toxicity to cells.

(*b*) **Bacteria oxidizing nitrite to nitrate:** The example of such bacteria is *Nitrobacter, Nitrospira* and *Nitrococcus*. They change nitrogen oxidation state from +3 to +5 as given below:

$$NO_2^- + H_2O \longrightarrow H_2O.NO_2^- \longrightarrow NO_3^- + 2H^+$$
$$2H^+ + \tfrac{1}{2} O_2^- \longrightarrow H_2O$$

Mostly nitrification is carried out by autotrophic bacteria but there are some heterotrophic bacteria and fungi which also take part in nitrification. The examples are strains of *Nitrosomonas, Aspergillus flavus*, etc.

(*iv*) **Denitrification:** Nitrogen is converted into different forms by microorganisms. When nitrate is added to soil, N_2, N_2O (nitrous oxide) and NO (nitric oxide) are evolved after reduction of nitrate as given below:

$$NO_3^- \longrightarrow NO_2^- \longrightarrow N_2O \longrightarrow N_2$$

Thus, microbial reduction of nitrate and nitrite with the liberation of molecular nitrogen and nitrous oxide is called *denitrification*. These are the volatile products and, therefore, are lost to the atmosphere and fail to enter the cell structure. Thus, denitrification is essentially a respiratory mechanism in which nitrate replaces the molecular oxygen. Therefore, denitrification may be termed as *nitrate respiration*. There are three possible reactions through which volatilization of nitrogen occurs: non-biological losses of ammonia, chemical decomposition of nitrite, and microbial denitrification resulting in evolution of N_2, N_2O and NO into the atmosphere. For the purpose of crop production, nitrogen volatilization has deleterious effect because it depletes part of soil reserves of essential nutrients.

Microbiology: Fungi and actinomycetes have not been found to be associated with denitrification. It is carried out only by certain bacteria such as the genera: *Pseudomonas, Bacillus, Paracoccus* as well as occasionally *Thiobacillus denitrificans, Chromobacterium, Corynebacterium, Hyphomicrobrium* and *Serratia* species.

Denitrifying bacteria are abundant in arabic fields and count for about a million per gram soil. Their population is higher in rhizosphere soil. Denitrifying bacteria are aerobic but nitrate is used as the electron donor for their growth in absence of O_2. In addition,

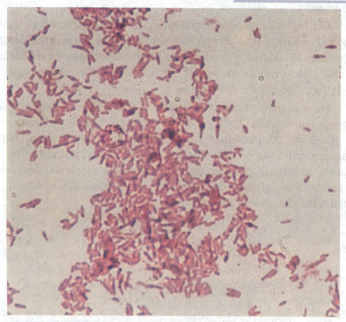

Corynebacterium, a bacterium found to be associated with denitrification.

several chemoautotrophs (*Thiobacillus denitrificans* and *Paracoccus denitrificans*) are capable of reducing nitrate to nitrogen (Pyne, 1973).

There are three microbiological reactions of nitrate : first, a complete reduction to ammonia, second an incomplete reduction and accumulation of nitrite, and third, a reduction to nitrite followed by the evolution of gases (*i.e.* denitrification). The biochemical pathway of nitrate reduction and denitrification is given below:

$$2HNO_3 \xrightarrow[-2H_2O]{+4H} 2HNO_2 \begin{cases} \xrightarrow[-2H_2O]{+2H} 2NO_2 \xrightarrow[-2H_2O]{+2H} N_2O \xrightarrow[-H_2O]{+2H} N_2 \\ \\ \xrightarrow{} ? \longrightarrow [2NH_2OH] \longrightarrow 2NH_3 \end{cases}$$

3. The Phosphorus Cycle

Phosphorus is an important constituent of protoplasm and required for metabolism of all living organisms. However, the major store house of phosphorus is the rock deposits. Agricultural crops contain 0.05 to 0.5% of phosphorus in their tissues in the form of several compounds such as phytin, phospholipids, nucleic acids, phosphorylated sugars, coenzymes and related compounds.

Several transformation processes are done by the microorganisms: (*a*) alteration of solubility of inorganic compounds of phosphorus, (*b*) mineralization of organic compounds with the release of inorganic phosphate, (*c*) immobilization of phosphorus i.e. conversion of inorganic, available ions into cell components, and (*d*) oxidation/reduction reaction of inorganic phosphorus compounds (Fig. 27.8). However, microbiologically only two processes (mineralization and immobilization) are important.

The bulk of phosphorus present in bacterial cell accounts for 1/3 to 1/2 of all phosphorus. In soil 15 to 85% of total phosphorus is organic. Soil rich in organic matter contains abundant organic phosphorus. Therefore, a good correlation exists between the concentrations of organic phosphorus, organic carbon and total nitrogen. The C:N:P ratio in carrington silt loam soil at soil depth 0-15 cm is found 98 : 8.3 : 1.

The inorganic phosphorus, which is unavailable to plants, is solubilized by many microorganisms into solution. Such bacteria are abundant on root surfaces of plants and account for 10^5-10^7 per gram soil. Phosphate solubilization commonly requires acid production, but liberation of hydrogen sulfide (a product that reacts with ferric phosphate to yield ferrous sulfide) by some bacteria also made phosphorus more available to plants. Species of *Pseudomonas, Bacillus, Flavobacterium, Mycobacterium, Micrococcus, Penicillium, Fusarium, Aspergillus,* etc. are associated with phosphorus conversion.

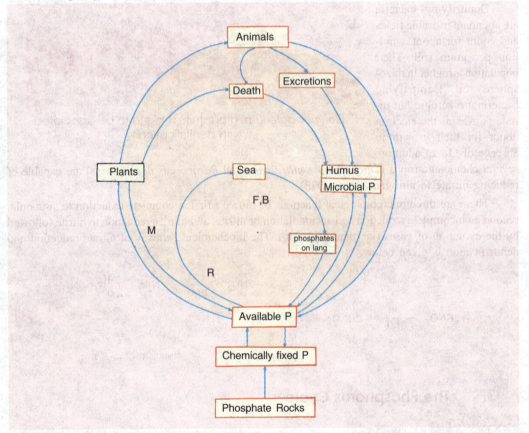

Fig. 27.8 : The phosphorus cycle I, immobilisation : M, mineralisation; R, run-off; F.B, by fishing and birds.

Soil contains a large amount of organic phosphorus which is unavailable to plants unless it is microbiologically converted into inorganic forms. This process is known as mineralization, which is achieved by decomposition. The cleavage of phosphorus from organic matter is done by enzymes which are collectively known as *phosphatases*.

The mycorrhizal fungi play a key role in making the phosphorus available to host plants by extracting it from unavailable fraction of soil reservoir. For detail description see Chapter 24.

Phosphorus exists in several oxidation forms such as -3 form of phosphine (PH_3) to the oxidized +5 form of orthosphosphate.

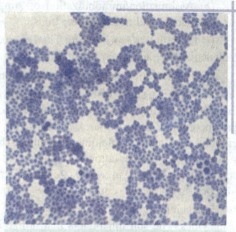

Micrococcus luteus, a bacterium which makes phosphorus more available to plants

4. The Sulfur Cycle

Sulfur is an essential nutrient of plants and animals. It is most abundant in earth crust in low concentration and is unavailable to plants. In soil sulfur enters in the form of plant residues, animal wastes, chemical fertilizers and rain water. It is taken up by the plant roots as the sulfate ions (SO_4^{-2}) which is required for their growth and development. It occurs in plant, animal and microbial proteins in the amino acids, cystine and methionine, and in B vitamins e.g. thiamine, biotin and lipoic acid. Also it is found in excretory products of animals as free sulfate, thiosulfate, thiocyanate, and taurine.

Organic and inorganic forms of sulfur is microbiologically metabolised in soil through different transformation processes (Fig. 27.9) as given below:

(*a*) Decomposition of organic sulfur compounds by microorganisms into smaller units and finally into inorganic compounds (*i.e.* mineralization),

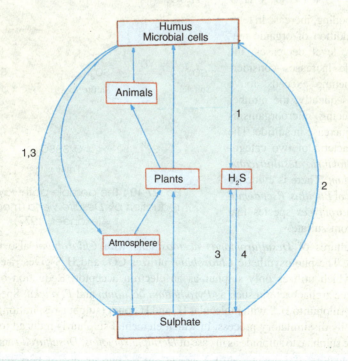

Fig. 27.9 : The sulfur cycle. 1, mineralisation; 2, immobilisation; 3, sulfur oxidation; 4, sulfate reduction.

(*b*) Assimilation/immobilization of simple sulfur compounds and their incorporation into bacterial, fungal and actinomycete cells,

(*c*) Oxidation of inorganic ions and compounds such as sulfides, thiosulfate, polythionates and elemental sulfur, and

(*d*) Reduction of sulfate and other anions to sulfide.

Plants derive sulfur from soil in the form of sulfate (some directly from atmosphere). Animals get sulfur by feeding plants, or animals. The dead parts of plants, animals and also microorganisms upon incorporation into soil are decomposed by microorganisms. Proteins are hydrolysed into amino acids and other sulfur-containing molecules. These in turn are attacked by microorganisms resulting in release and accumulation of sulfate and sulfide. Moreover, in anaerobic conditions H_2S accumulates, whereas in aerobic environment the combined sulfur is metabolized to sulfate. However, in nature the concentration of sulfate and sulfide is low.

Oxidation states of sulfur varies from -2 of sulfite to +6 of sulfate. The reaction is catalysed by enzymes. The soil inhabitants that utilize inorganic sulfur are autotrophs or heterotrophs, for example *Thiobacillus thiooxidans, T. thioparus, T. ferrooxidans, T. novellus, T. denitrificans.* The heterotrophic bacteria, actinomycetes and fungi also oxidized inorganic sulfur compounds. The examples of these microbes are species of *Arthrobacter, Bacillus, Flavobacterium* and *Pseudomonas* that generate thiosulfate from elemental sulfur. The filamentous fungi (e.g. species of *Aspergillus, Penicillium* and *Microsporum*) produce sulfate from organic substrates such as methionine, cystine, thiourea, taurine, etc.

T. thiooxidans and *T. novellus* carry out the following reaction:

$$Na_2S_2O_3 + 2O_2 + H_2O \longrightarrow 2NaHSO_4$$

T. thiooxidans oxidizes sulfur as below:

$$S + 1\tfrac{1}{2}\,O_2 + H_2O \longrightarrow H_2SO_4$$

Due to flooding, increase in temperature and addition of organic material when O_2 level decreases, the level of sulfide increases considerably, and sometime exceeds above 150 ppm. Consequently, the number of sulfate reducing microorganisms increases with increase in sulfide. The predominant bacteria of two categories: are *Desulfovibrio desulfuricans, Desulfotomaculum.* There is evidence for the presence of *Bacillus, Pseudomonas* and *Saccharomyces* species that liberate H_2S from sulfate.

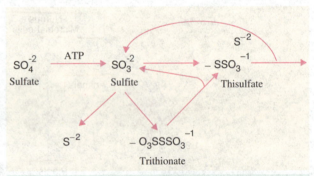

Fig. 27.10 : The possible pathways of sulfate reduction by *Desulfo vibrio* (modified after Alexander, 1977).

Mixed cultures of *Desulfuromonas acetooxidans* with *Chlorobium* also reduce elemental sulphur to H_2S. The photosynthetic *Chlorobium* utilizes CO_2 and H_2S gives rise sulphur. It is *D. acetooxidans* which utilizes only sulphur as an electron acceptor. Reduction of sulphur is also observed in some archaebacteria such as *Pyrodictium occultum* and *P. buckii.* Species of *Acidianus* can also reduce sulphur to H_2S with molecular H_2 and H donor (sulphur-respiration). Certain bacteria reduce sulphur by dissimilatory process, in which elemental sulphur is reduced to sulphide but are unable to reduce sulphate to sulphide as in case of *Desulphuromonas, Desulfurella* and *Campylobacter.*

Direct reduction of sulphate can be carried out by two groups of bacteria.

Non-acetate oxidizing bacteria	Acetate oxidizing bacteria
Desulfovibrio, Desulfomicrobium	*Desulfobacter, Desulfobacterium,*
Desulfobotulus, Desulfotomaculum,	*Desulfococcus, Desulfonema,*
Desulfomonile, Desulfobacula,	*Desulfosarcina, Desulfoarculus,*
Archaeoglobus, Desulfobulbus,	*Desulfacinum, Desulforhabdus,*
Thermodesulfobacterium	*Thermosulforhabdus*

The mechanism of H_2S formation from sulfate is not fully understood. However, possible pathways of sulfate reduction by *Desulfovibrio* is given in Fig. 27.10.

Fig. 27.10 shows the three hypothetical pathways for sulfur metabolisms: (*a*) direct reduction of sulfate to sulfite without forming free sulfur products, (*b*) formation of thiosulfate from sulfite which in turn is cleaved to form sulfide and sulfite, and (*c*) an initial production of trithionate which is subsequently converted to a mixture of thiosulfate and sulfite (Alexander, 1977).

Phylogenetically, *Desulfotomaculum* (endospore forming Gram-positive bacterium), placed with the *Clostridium* subdivision, whereas other sulfate reducing bacteria (Gram-negative) are in the delta subdivision of the purple bacterial group. Few are related to *Bdellovibrio* and gliding myxobacteria genera *Desulfovibrio, Desulfobacter*. Most of them are able to grow chemolithotrophically with H_2 as an electron donor, sulphate as electron acceptor, and CO_2 as sole C source. Many bacteria use nitrate instead of sulfate. This is the reason that some sulphate reducing bacteria are able to fix nitrogen e.g. *Desulfovibrio, Desulfobacter*.

A sulphate reducing bacterium *Desulfovibrio saprovorans.*

QUESTIONS

1. Discuss in detail that soil is a habital for microorganisms.
2. How many types of microorganisms live in soil? Describe in brief different factors affecting them.
3. What is microbial niche? How does it differ from habitat?
4. What do you know about rhizosphere and rhizoplane regions? Write in details the reasons of increased microbial activity in these regions.
5. Write an essay on factors affecting microbial community in soil.
6. What is organic matter? Write different process of decomposition.
7. What is biogeochemical cycling? Discuss the carbon cycling.
8. Write short notes on the following:
   (*i*) Microbial niche, (*ii*) Microbial balance, (*iii*) Rhizosphere microflora, (*iv*) Rhizosphere effect, (*v*) Plant litter, (*vi*) Organic matter decomposers, (*vii*) Microbial biomass, (*viii*) Ammonification, (*ix*) Nitrification, (*x*) Denitrification, (*xi*) Phosphorus cycle, (*xii*) Nitrogen cycle, (*xiii*) Sulfur cycle, (*xiv*) Sulphate reducing bacteria

REFERENCES

Alexander, M. 1965. Nitrification. In *Soil Nitrogen* (eds. W.V. Bartholonew and F.E. Clark), pp. 307-343, *Am. Soc. Agron.* Madison, Wisconsin.

Alexander, M 1977. *Soil Microbiology*. 2nd ed. John Wiley & Sons, Inc. New York.

Baker, K.F. and Cook, R.J. 1974. *Biological Control of Plant Pathogens.* W.H. Freeman & Co., San Francisco, p. 433.

Balandreau, J. and Knowles, R. 1978. The rhizosphere. In *Interactions between non-pathogenic soil microorganisms and Plants* (eds. Y.R. Dammergues and S.V. Krupa), pp. 243-268, Elserier, Amsterdom.

Burns, R.G. 1983. In *Microbes in their Natural Environment.* Symp. 34 (eds. I.H. Slater, R. Whittenbury and J.W.T. Wimpenny), *Soc. Gen. Microbiol,* Ltd. Cambridge Univ. Press.

Dubey, R.C. and Dwivedi, R.S. 1988. Population dynamics of microfungi in root region of soybean with reference to growth stages and environmental factors. *J. Indian Bot Soc.* 67:54-162.

Dubey, R.C. and Pandey, Sunita. 1996. Microbial decomposition of two coniferous leaf litter in Kumaon Himalaya. *J. Indian Bot.Soc.* **75:** 83-85.

Dubbey, R.C, Pandey, L. and Gaur, R. 1992. On some aspects of microbial activity in soil. In *Microbial Activity in the Himalaya* (ed. R.D. Khulbe) pp. 53-82, Shree Almora Book Depot, Almora.

Dwivedi, R.S and Saravanamuthu, R. 1985. The concept of niche in the ecology of soil inhabiting fungi. In *Frontiers in Applied Microbiology.* Vol I (eds. K.G. Mukherji; N.C. Pathak and V.P. Singh); pp 237-259, Print House (India) Lucknow.

Garrett, S.D. 1950. *Ecology of root-infecting fungi. Bot.Rev.* **25:** 220-254.

Garrett, S.D. 1956. *Biology of Root-Infecting Fungi.* Cambridge University Press, London, p. 292.

Garrett, S.D. 1970. *Pathogenic Root-Infecting Fungi.* Cambridge Univ. Press

Garrett, S.D. 1981. *Soil Fungi and Soil Fertility.* 2nd ed. Pergamon Press, Oxford.

Gray, T.R.G. and Williams, S.T. 1971. *Soil Microorganisms.* Longman Group Ltd., London, p.240.

Griffin, D.M. 1972. *Ecology of Soil Fungi.* Chapman & Hall, London, p. 193.

Kruetzer, W.A. 1965. The reinfestation of treated soils. In *Ecology of Soil Borne Plant Pathogens* (eds. K.F. Baker and W.C. Snyder). John Murray, London.

Mueller, J.D.; Short, B.J. and Sinclair, J.B. 1985. Effects of cropping history, cultivars and sampling dates on the internal fungi of soybean roots. *Plant Disease,* **69** : 520 - 523.

Odum, E.P. 1971. *Fundamentals of Ecology.* 3rd ed. W.B. Saunders & Co. Philadelphia, p.574.

Ordish, G. 1967. *Biological Methods in Crop pest control.* Constable, London, p.242

Pyne, W.J. 1973. Reduction of nitrogenous oxide by microorganisms. Bact. Rev. 37: 409-452.

Rai, B. and Srivastava, A. 1982. Decomposition of leaf litter in relation to microbial populations and their activity in tropical dry mixed deciduous forest. *Pedobiol* 24: 151-159.

Rovira, A.D. 1965. Plant root exudates and their influence upon soil microorganisms. In *Ecology of Soil-Borne Plant Pathogens* (eds. K.F. Baker and W.C. Snyder), pp. 170-186, John Murray London.

Shail, S. and Dubey, R.C. 1997. Seasonal changes in microbial community in relation to edaphic factors in two forest soils of Kumaun Himalaya. In *Himalayan Microbial Diversity.* (eds S.C. Sati, J.Saxena and R.C. Dubey), pp. 381-391, Today & Tomorrow's Printers and Publ., New Delhi.

Singh, S. and Singh, J.S. 1995. Microbial biomass associated with water-stable aggregates in forest, sawana and cropland soils of a seasonally dry tropical region, India. *Soil Biol.Biochem.* **27:** 1027-1033.

Singh, J.S.; Rawat, Y.S. and Chaturvedi, O.P. 1984. *Nature,* **311.** 54-56.

Singh, J.S., Raghubanshi, A.S.; Singh, R.S. and Srivastava, S.C. 1989. Microbial biomass acts as a source of plant nutrients in dry tropical forest and savana. *Nature,* **338:** 499-500.

Singh, S.P.; Pande, K.; Upadhyay, V.P. and Singh, J.S. 1990. Fungal communities associated with the decomposition of a common leaf litter (*Quercus leucotrichophora* A. Camus) along an altitudinal transect in the central Himalaya. *Biol Fertl. Soils.* **9** : 245-251.

Srivastava, S.C. and Singh, J.S. 1989. Effect of cultivation on microbial biomass C and N of dry tropical forest soil. *Biol. Fertl Soils,* **8:** 343-348.

Torstenson, L. *et at.* 1998. Need of a strategy for evaluation of arable soil. *Ambio* 27: 4-8.

Upadhyay, R.S. and Rai, B. 1982. Ecology of *Fusarium udum* causing wilt disease of pigeon-pea: population dynamics in the root region *Trans. Br. mycol Soc.* **78:** 209-220.

Wilhelm, S. Analysis of biological balance in natural soil. In *Ecology of Soil-Borne Plant pathogens.* (eds. K.F. Baker and W.C. snyder), pp. 509-517, John Murray, London.

Microbial Ecology-IV : Water Microbiology

28

A. Types of Water

B. Water Microorganisms

C. Microbiological Analysis of Water

D. Purification of Water

Water is the elixir of life. It is an essential part of protoplasm and creates a state for metabolic activities to occur smoothly. Therefore, no life can exist without water. In addition, there are thousands of microorganisms which live in water and are transported through it.

Major area (about three-fourth) of earth surface is covered by water, mainly by oceans and to some extent by lakes, rivers, streams, ponds, etc. However, water is constantly in continuous circulation. This phenomenon of movement of water is called *hydrological cycle* or *water cycle* (Fig. 28.1). Also water is present below the land which is recovered by tube wells. Water is also present in the form of ice in South and North poles of the earth, on the top of high mountains, etc. From the earth surface water is lost through evaporation, transpiration and exhalation. About 93% of evaporated ocean water cools and falls back in the form of rain on the sea, and about 900 cubic miles of water moves along with wind current over land surface and joins the cloud stock of over double this quantity of water obtained through evapo-transpiration from land (Ambasht, 1984).

Water receives microorganisms from air, soil, sewage, organic wastes, dead plants and animals, etc. It is obvious that at times almost

Water, most of it in oceans, covers three-fourths of Earth's surface.

any organisms may be found in water, those finding unfavourable conditions die, and the others finding favourable conditions grow and multiply to increase their population. In this chapter we shall briefly describe the sources of water, physico-chemical and biological properties, microbiological examination and testing of water quality as well as water treatments.

A.   Types of Water

Natural water is commonly grouped into four well-marked classes : (a) atmospheric water, (b) surface water, (c) stored water, and (d) ground water.

1.   Atmospheric Water

Rain and snow are included under the atmosphere water (Fig. 28.1). Microorganisms in the form of cells/dormant propagules and dust particles remain suspended in water and snow. Considerable number of bacteria can be isolated from rain water. After heavy rain or snow fall, the dust particles and bacteria are washed from the atmosphere. Therefore, atmosphere for some times remains free from microorgansms.

2.   Surface Water

The water present on earth surface is known as surface water (Fig. 28.1:) It is found in the form of several water bodies such as rivers, streams, ocean, lakes, etc. As soon as raindrops or snow touches the earth, it becomes contaminated by the microorganisms present in soil. Total microbial number in water depends on microbial population of soil, types of soil, types of organic materials present in soil and also on types of microorganisms and their activities. Moreover, microorganisms of water are governed by climatic, chemical and biological conditions. For example, microbial population in ponds, pools and lakes would be several

Microorganisms present in soil contaminate the surface water.

times higher than that of the rivers and streams. Also microbial population is influenced by anthropogenic activities. Frequent bathing of animals, throwing of excreta and effluents in water bodies certainly affect the total microbial load in water.

3.   Stored Water

The stagnant land water present in ponds, lakes, etc. are called stored water (Fig. 28.1) During storage, in general, the number of microorganisms gets reduced; thus it establishes to some extent the purity and stability. However, in stored water the microorganisms are affected by several factors such as sedimentation, ultra violet light, temperature, osmotic effects, food supply, and activities of other microorganisms as well.

(i) Sedimentation: Microorganisms have specific gravity slightly more than the distilled water, therefore, they slowly settle down the bottom of water bodies. Bacterial cells get attached to suspended particles that cause sedimentation. As the suspended particles settle down, the upper layer of water is made free from the microorganisms.

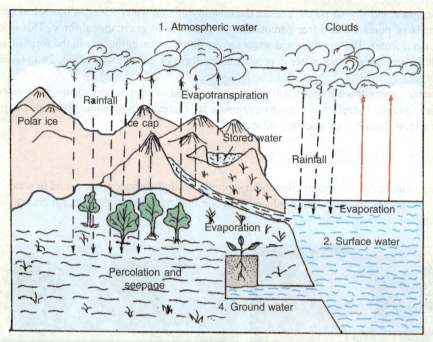

Fig. 28.1 : Hydrological cycle (based on Ambasht, 1984).

(ii) Interaction of Other Microorganisms: Predatory protozoa are present in water. They need living or dead bacteria for their food. If sufficient amount of oxygen is present, a large number of these microorganisms are engulfed by the predatory protozoa. However, when dissolved oxygen and bacteria are absent, the protozoa will not be present in water.

(iii) Light Rays: Both the vegetative cells and spores of microorganisms are killed in the presence of prolonged exposure of direct sunlight. The diffused light does not cause any effect. However, in a water supply toxicity of ultra violet light is governed by the turbidity of water i.e. it is inversely proportional to turbidity. In a clear water ultra violet light is effective to a depth of about three meters but in turbid water light rays cannot penetrate. In bright light some of the free floating cyanobacteria sink down for some time, and float again when brightness of light is diminished.

(iv) Temperature: Effect of temperature on organisms varies. Increasing temperature is harmful to microorganisms. But some pathogenic microorganisms multiply well with increasing temperature. It has been found that *E. coli* multiplies well in autoclaved water at 37°C. Raw water stored at 22°C shows increased number of bacteria, but greater number has been recorded in autoclaved water stored at 22°C as compared to 37°C (Cody *et al.*, 1961).

(v) Food Supply: The number of microorganisms is likely to increase with increased food supply in water. Several waste materials present in water act as food bases. In addition, algae growing in water act as food for microorganisms when algae are dead after completing its life cycles. In addition, toxic substances viz., acids, bases, etc. cause marked reduction in microbial community. There are certain dissolved gases such as CO_2, H_2, etc. which have harmful effect on water microflora. All together these factors bring about changes in water pH which, however, influence microbial community of water (Potter, 1960).

4.   Ground Water

Soil consists of particles of varying size (see Chapter 27), therefore, soil pore size varies. The percentage, space and size of pores regulate the quantity of rain water than can be soaked and held in soil. The large pores allow a free percolation of water due to gravitational force. This is known as *gravitational water*. The gravitational water moves down and accumulates in the form of *ground water*. A huge amount of fresh water is present in the form of ground water (Fig. 28.1) (Ambasht, 1984).

As water moves down through soil pores, bacteria and other microorganisms with suspended particles are filtered. Therefore, the microorganisms are carried only to some distance. Deep wells contain negligible number of microorganisms.

B. Water Microorganisms

A large number of microorganisms both saprophytes and pathogens are found in water which fall under the groups bacteria, fungi, algae, protozoa and nematodes. Several animal viruses are also transmitted through water.

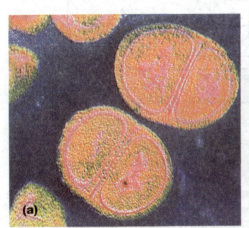
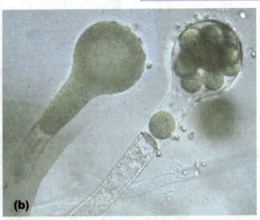

(a) The Domain Bacteria   (b) *Saprolegnia*, an aquatic fungus
Bacteria and Fungi constitute a large number of microorganisms found in water.

The majority of bacteria found in water belongs to groups: fluorescent bacteria (e.g. *Pseudomonas, Alginomonas*, etc), chromogenic rods (*Xanthomonas*, etc), coliform group (*E. coil, Aerobacter*, etc.), *Proteus* group, non-gas forming, non-chromogenic and non-spore forming rods, spore formers of the genus *Bacillus*, and pigmented and non-pigmented cocci (*Micrococcus*).

A significant work on aquatic fungi has been done at different centres in India. A large number of fungi, both saprophytes and pathogens, have been reported from a variety of water sources including potable water (Table 28.1). Works on aquatic fungi was initiated at Allahabad by Prof. R.K. Saxena, at Lucknow by S.N. Das Gupta, J.N. Rai and their associates, at Varanasi by M.S. Pavgi, Ram Dayal, R.S. Dwivedi and R.C. Mishra, at Gorakhpur by Prof. K.S. Bhargava and his coworkers, at Hyderabad by P. Rama Rao and C. Manoharachary, at Jabalpur by Prof. G.P. Agrawal and his research group, and at Nainital by R.D. Khulbe, S.C. Sati and their associates (Bhargava, 1945; Dayal, 1960, Manoharachary and Rama Rao, 1983; Khulbe, 1980, 1997; Khulbe and Sati, 1983; Mishra and Dwivedi, 1987).

Sati (1997) has reviewed the zoosporic and conidial aquatic fungi found in various water sources of Kumaun Himalaya where the water bodies are the only source of potable water. Therefore, significance of such study becomes more important than those where aquatic fungi are studied for academic and ecological importance.

Table 28.1 : Some of aquatic fungi in different water bodies
(modified after Mishra and Dwivedi, 1988)

Aquatic fungi	Sources	Reporters
Achlya americana	Tap water, Fish parasite	R.D. Khulbe & S.C. Sati, Nainital
A. androcomposita	Tank water	A. Hamid
A. debaryana	Water, fish pathogen	R.D. Khulbe & S.C. Sati, Nainital
Allomyces neo-moniliformis	Lake water	K.S. Bhargava
A. laevis	Lake water, fish pathogen	R.D. Khulbe & S.C. Sati
Blastocladia simplex	Gujar lake (Jaunpur)	R.C. Mishra & R.S. Dwivedi, B.H.U.,Varanasi
B. rostata	Water	S.N. Das Gupta, L.ucknow
Chytridium brevipes	Water	do
Cladochytrium setigerum	Water	S.K. Hasija, Jabalpur
Dictyuchus pisci	Pond water, fish parasite	Khulbe & Sati, Nainital
Isoachlya anisospora var. indica	Pond water	R.K. Saxena & K.S. Bhargava
Olpidiopsis luxurians	Lake water, fungal parasite	R.D. Khulbe, S.C. Sati, Nainital
Pythium undulatum	Pond water, fish pathogen	do
P. echinulatum	Water	C.Manoharachary, Hyderabad
Rhizophydium sp.	Gujar lake	R.C. Mishra & R.S. Dwivedi, Varanasi
Saprolegnia spp.	Tap water	R.D. Khulbe & S.C. Sati
Sapromyces indicus	Strem water	M.O.P. Iyenger, Madras

In addition, there are many water-borne protozoan, bacterial and fungal pathogens which are transmitted through water and cause serious diseases. For detailed discussion see Chapters 21 and 22. However, water-borne pathogenic microbes are given in Table 28.2.

Table 28.2 : Some of the water-borne pathogenic microbes

Microorganisms	Reservoir	Other features
Bacteria		
Aeromonas hydrophila	Free-living	Associated with cellulitis, gastroentitis, etc.
Chromobacterium violaceus	Soil run off	Saprophyte
Mycobacterium tuberculosis	Free-living, infected person	Tuberculosis, etc.
Vibrio cholerae	Free-living	Causes diarrhea
Salmonella enteriditis	Intestine of animals	Water-borne
Yersinia enterocolitica	In animals	Water-borne, gastroenteritis
Pseudomonas aeruginosa	Free living	Swimer's ear
Protozoa		
Giardia lamblia	Sheep, dogs, cats	Diarrhea
Cryptosporidium	On domestic and wild animals	Scute enterocolitis
Acanthamoeba	Sewage sludge disposal area	Corneal ulcers, granulomatous amoebic encephalitis (GAE)
Entamoeba histolytica	Water	Amoebiasis

1.   Marine Microbiology

Marine microbiology deals with microorganisms living in sea. About 97% of earth's water is present in sea. Ocean at its greatest depth is slightly more than 11,000 meters deep. Therefore, pressure in the marine environment increases by about 1 atm/10 meters in depth. The bacteria growing in vertically differentiated marine environment can be categorised in three groups: (*a*) *barotolerant* (bacteria growing between 0 and 400 atm but best at normal atmospheric pressure), (*b*) *moderate barophiles* (bacteria growing optimally at 400 atm, but still grow at 1 atm), and (*c*) *extreme barophiles* (bacteria growing only at higher pressure *i.e.* 6000 to 11,000 meter depth). The hydrostatic pressure can reach 600 to 1,100 atm in the deep sea, whereas the temperature is about 2 to 3°C. Increased pressure does not affect the barotolerants. Some bacteria living in gut of deep-sea invertebrates (*e.g.* amphipods and holothurians) are truly basophilic. These basophilic bacteria play a significant role in nutrient cycling.

In the marine system, the major source of organic mater is *phytoplanktons* (*phyto*, plant and *planktos* wandering) *i.e.* free floating plants. A common planktonic alga is *Synecococcus* which can reach to a density of 10^4 to 10^5 cells/ml at the ocean surface. *Picocyanobacteria* (very small cyanobacteria) may represent 20 to 80% of total phytoplankton biomass. In turn these act as a source of food for marine fish and other animals.

Bacteria which live in deep sea ecosystems next to hydrothermal events in the ocean floor can survive temperatures as high as 250° C.

Recently, a very interesting group of bacteria (*i.e.* archaeobacteria) has been discovered from the marine system. It has been found that about 1/3 of oceanic picoplanktons (cells > 2 μm) are archaeobacteria associated with hostile environment i.e. hot springs, deep marine "black smoker" areas, saline and thermoacidophile region). The isolation and functional role of these bacteria are still to be studied in detail.

2.   Fresh Water Microbiology

Lakes and rivers provide the major fresh water bodies which are used for potable water. However, due to growth of phytoplanktons and decomposition of organic matter nutrient status varies throughout the year. In addition, animal activities (including man) also influence the nutrient level of lakes and rivers. The nutrient-poor lakes are known as oligotrophic lakes, whereas the nutrient-rich lakes are called eutrophic lakes (Fig. 28.2). The nutrient-rich lakes contain high amount of bottom sediments containing organic matter. *Eutrophication* (enrichment of lake due to high concentration of nutrients) of lakes occur by multifarious ways where anthropogenic activities are of much importance. The eutrophic lakes support luxuriant growth of bacteria and algae. Some of the fast growing algae at optimum condition bloom well showing their maximum population. This phenomenon is known as *water blooming*, and microorganisms associated with it are called *water blooms*. The two most important lakes of India which are fully eutrophicated are the Dal Lake (Srinagar, J & K) and Naina Lake (Nainital) which are the only source of potable water.

The microorganisms growing in lakes are the genera : *Anabaena, Microcystis, Nostoc, Oscillatoria, Oedogonium, Spirulina*, diatoms, protozoa, etc.

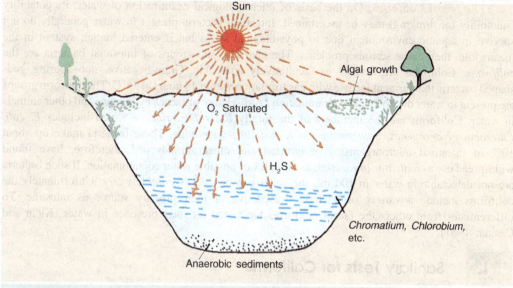

Sun

Algal growth

O_2 Saturated

H_2S

Chromatium, Chlorobium, etc.

Anaerobic sediments

Fig. 28.2 : An eutrophic lake.

The capacity of rivers and streams to process the added organic material is limited. If too much organic material is added, the water becomes anaerobic. This situation arises in those rivers and lakes which are present near the urban areas. In streams and rivers untreated or inadequately treated municipal wastes and other organic materials are discharged. These organic wastes cause changes in microbial community. Release of organic wastes from known sources is called *point source of pollution*. When the amount of organic materials added is not too high, it supports the growth of several algae. Algae produce O_2 during photosynthesis in day light and evolve CO_2 in day and night. This results in *diurnal oxygen* shifts. In the mean time oxygen level completes the self purification process. This demand of oxygen is expressed in terms of biological oxygen demand (BOD) or the chemical oxygen demand (COD).

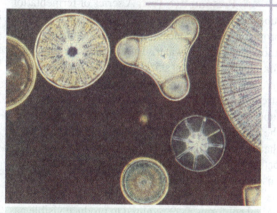

Several kinds of diatoms growing in a lake.

C. Microbiological Analysis of Water Purity

As discussed earlier, the natural water bodies such as lakes, streams, rivers contain sufficient amount of nutrients that support the growth of microorganisms. However, there are different ways by which microorganisms enter in water supply, for example broken sewer lines, congested centres, inappropriate treatment, etc. In addition, lack of awareness among people also adds in contamination of water. Unhygenic environment at public places of water collection is also one of the reasons. People suffering from communicable diseases also discharge pathogenic microbes in water through their excreta, for example amoebic dysentery, typhoid fever, bacillary dysentery *poliomyelitis*, etc. Bacteria multiply in water at 35°C and at 20°C.

The faecal Coliforms: On the basis of microbiological examination of water, its potability (suitability for drinking) may be ascertained. Intestinal bacteria present in water generally do not survive in aquatic environment due to physiological stress but if entered human system in the meanwhile, they cause serious problems. The characteristic groups of intestinal bacteria are the *coliforms*. Coliforms are defined as facultatively anaerobic, Gram-negative, non–sporing, rod-shaped bacteria that ferment lactose with gas formation within 48 hours at 35°C. The coliform groups are present in water due to faecal contamination *i.e.* discharge of faeces by humans and other animals in water. Coliforms are the members of the family *Enterobacteriaceae* which includes *E. coli, Enterobacter aerogenes, Salmonella* and *Klebsiella pneumoniae.* These bacteria make up about 10% of intestinal microorganisms of humans and other animals and, therefore, have found widespread use as *indicator organisms,* as an index of possible water contamination. If such bacteria are not detectable in water in 100 ml, the water can be said as *potable water.* Unfortunately the coliforms include a variety of bacteria irrespective of their primary source as intestine. To differentiates from others the faecal coliforms are tested for their presence in water (Klein and Casida, 1967).

1.   Sanitary Tests for Coliforms

The original test for the presence of coliform in water is done by standard multiple tube fermentation technique. This method involves the three routine standard tests: (*a*) the presumptive test, (*b*) the confirmed test, and (*c*) the complete test (Fig. 28.3)

(*i*) Presumptive Test: A series of fermentation tubes each containing lactose broth or lauryl tryptose broth of known concentration, are inoculated with known amount of water. These tubes are incubated for 24 to 48 hours at 35°C (Fig. 28.3A). Generally, five fermentation tubes containing single or double strength broth are inoculated with 10 ml water, 5 tubes with 1 ml water and 5 with 0.1 ml water. At the end of 24 hours of incubation, the tubes indicate that the coliforms are absent. These tubes are incubated for an additional 24 hour to be sure for the absence of coliforms (i.e. gas production).

(E. coli (in 3D) is a bacillus found in the human colon. The presence of *E. coli* in water may indicate contamination by sewage.

Chambers (1950) showed that 40 to 390 million per ml coliforms were required to produce visible gas in fermentation broth. The average number was 170 million per ml. However, in most of the cases 75 million coliforms per ml were required to produce the gas. This difference may be due to the ratio of coliforms to non-coliforms. The non-coliforms, if in high number, reduce the gas formation.

(*ii*) Confirmed Test: If a positive test of gas production is obtained, it does not mean that coliforms are present. The other organisms too also give false positive presumptive test because they are also capable of fermenting lactose with formation of acid and gas. The positive presumptive test is resulted due to synergism i.e. joint action of two microorganisms on a carbohydrate with production of gas which is not formed if both are grown separately. In addition, if yeasts, species of *Clostridium* and some other microorganisms are present, gas is also produced. Therefore, a confirmed test is performed for the presence of coliforms. All fermentation tubes showing gas within 48 hours at 35°C are used for confirmed test. It is of two types as described below:

The positive presumptive fermentation tube is gently shaken. A drop of its culture is transferred to brilliant green lactose bile broth fermentation tube (Fig. 28.3B). The tubes are incubated for 48 hours at 35°C. The appearance of gas within this period indicates for positive confirmed test. The

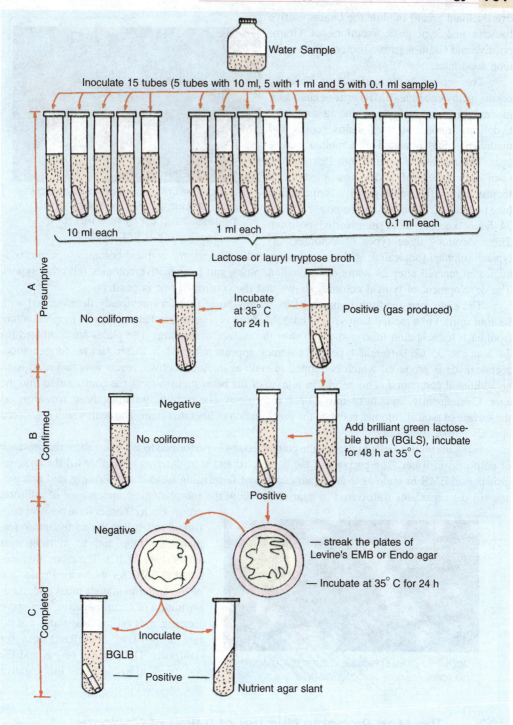

Water Sample

Inoculate 15 tubes (5 tubes with 10 ml, 5 with 1 ml and 5 with 0.1 ml sample)

10 ml each   1 ml each   0.1 ml each

Lactose or lauryl tryptose broth

A Presumptive

No coliforms ← Incubate at 35° C for 24 h → Positive (gas produced)

B Confirmed

Negative
No coliforms

Add brilliant green lactose-bile broth (BGLS), incubate for 48 h at 35° C

Positive

C Completed

Negative

— streak the plates of Levine's EMB or Endo agar
— Incubate at 35° C for 24 h

Inoculate

BGLB
—— Positive ——   Nutrient agar slant

Fig. 28.3 : The multiple tube fermentation test.

dye (brilliant green) inhibits the Gram-positive bacteria and synergistic reactions of Gram-positive and Gram-negative bacteria for a common food base.

The second confirmed test is done by eosine methylene blue (EMB) agar or endo agar method. In eosine methylene blue agar method, a definite amount of two stains (eosin and methylene blue) is added to a melted lactose agar. The medium is poured into Petri dishes. Over the surface of EMB agar medium, a loopful culture from each positive fermentation tube is streaked. Plates are incubated at 35°C for 24 hours keeping them in inverted position. There develops three types of colonies: (*a*)

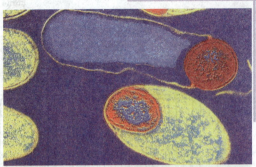

The bacterium *Clostridium* is involved in gas production like coliforms which necessitate a confirmed test for the presence of coliforms.

typical colonies (nucleated, with or without metallic sheen), (*b*) atypical colonies (opaque, non-nucleated mucoid after 24 hours of incubation, pink), and (*c*) negative colonies (all other types). The development of typical colonies shows that the confirmed test is positive.

The endo agar medium is prepared by adding basic fuchsin (previously decolourised with sodium sulfite) to a melted lactose agar base. Medium is poured into Petri dishes. A loopful culture from each fermentation tube is streaked over the surface of medium. The plates are incubated for 24 hours at 35°C. Different types of colonies appears after 24 h. After lactose fermentation, acetaldehyde is produced which is trapped in endo agar. Acetaldehyde reacts with sulfite to form an additional compound. This results in release of the basic fuchsin from the combination into the agar. Consequently, agar turns into a deep red colour. The metallic goldlike sheen appearing on the surface of typical colonies is due to the precipitation of liberated stain. The stain which is restored appears purple in colour.

(***iii***) **Completed Test:** In the last the completed test is performed to ascertain about the presence of coliforms in water. The purpose of the completed test is to determine whether (*a*) the colonies growing on EMB or endo agar are again capable of fermenting lactose and forming acid and gas, and (*b*) the organisms transferred to agar slants show the morphological appearance of coliform group. Each colonies form positive confirmed test is transferred to lactose fermentation tube and to nutrient agar slants (Fig. 28.3C). The tubes are incubated at 35°C for 48 hours. Production of gas in fermentation tubes and, demonstration of Gram-negative, non spore forming rods on the agar slants constitute a positive completed test for coliforms. The absence of gas and the rod production confirms for negative test of coliforms.

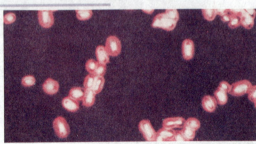

Klebsiella pneumoniae, a faecal coliform of the family, *Enterobacteriaceae*.

2. The Most Probable Number of (MPN) of Coliforms

For the first time Hoskins (1934) computed the MPN to evaluate coli-aerogenes test by fermentation tube method. Table 28.3 is based on the general formula of Hoskins (1934) for calculating the numbers of coliform microorganisms present in 100 ml of water. The figures are based on the use of five tubes as described in *presumptive test.* By referring to a MPN table (Table 28.3), a statistical range of the coliform numbers is determined by observing the number of broth tubes producing gas. Confirmed and positive tests of Fig. 28.3B are used to calculate the MPN.

(*i*) **The Membrane Filter Technique:** Goetz and Tsuneishi (1951) described a new method for enumeration of coliform microorganisms in water to which they named as *millipore filter technique.* However, it is often referred to as membrane filter technique. This technique has been accepted as the standard method for the microbiological examination of sewage, water, etc. The filtering apparatus consists of a glass or stainless steel funnel, and a flask. The funnel of stainless steel is clamped to a base containing a molecular filter. The stem of base is inserted into a fitter flask through a rubber stopper (Fig. 28.4 B).

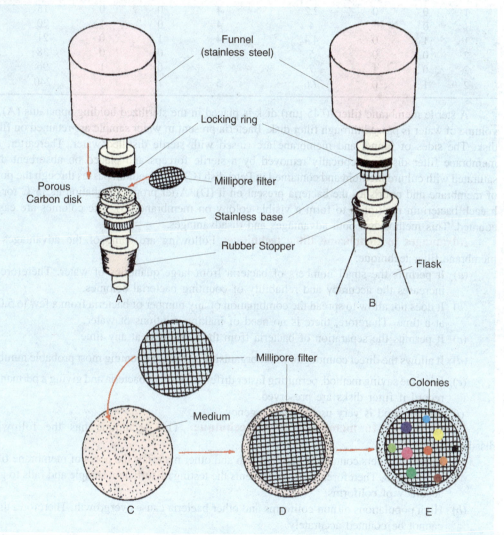

Fig. 28.4 : A millipore filter system. A, unassembled view; B, assembled view; C, plating of membrane filter containing filtered water sample on appropriate medium; D, incubation for 24 hours; E, colonies of typical coliforms growing over membrane filter.

Table 28.3 : Most probable numers (MPN) of coliform organisms present in 100 ml of a water sample

Number of positive lactose broth tubes			Number of coliform organisms	Number of positive lactose broth tubes			Number of coliform organisms
10 ml	1 ml	0.1 ml		10 ml	1 ml	0.1 ml	
0	0	0	0	3	0	0	8.8
0	1	0	2	3	0	1	11
0	1	1	4	3	1	0	12
1	0	0	2.2	4	0	0	15
1	0	1	4	4	0	1	20
1	1	0	4.4	4	1	0	21
2	0	0	5	5	0	0	38
2	0	1	7	5	0	1	96
2	1	0	7.6	5	1	0	240

A sterile membrane filter (0.45 μm) disk is placed in the sterilized holding apparatus (A). A volume of water is passed through filter disk. Bacteria present in water sample are retained on filter disk. The sides of funnel and membrane are rinsed with sterile distilled water. Thereafter, the membrane filter disc is aseptically removed by a sterile forceps and placed on absorbent disk saturated with culture medium and contained in Petri dish (C). The medium passes through the pores of membrane and nourishes the bacteria present on it (D). After proper incubation at 35°C for 24 h each bacterium multiplies to form a visible colony on membrane (D). The colonies are easily counted. This method has both advantages and disadvantages.

Advantages to membrane filter technique: Following are some of the advantages of membrane filter technique:

(a) It permits the small numbers of bacteria from large quantities of water. Therefore, it increases the accuracy and reliability of counting bacterial colonies.

(b) It does not allow to spread the combination of any number of bacteria from a few to 5,000 at a time. Therefore, there is no need of making dilutions of water.

(c) It permits the separation of bacteria from their nutrients at any time.

(d) It allows the direct counting of microorganisms instead of counting most probable number.

(e) It is time saving method, permiting faster differentiation of bacteria and giving a permanent record if filter disks are preserved.

(f) This method is very useful in emergency.

Disadvantages to membrane filter technique: This technique has the following disadvantages:

(a) In turbid waters containing algal growth and other materials the pores of membrane filter are clogged. Therefore, the filter prevents the testing of sufficient sample and fails to give accuracy of coliforms.

(b) High populations of non-coliforms and other bacteria cause overgrowth. Therefore, these cannot be counted accurately.

(c) Metals and phenols can be absorbed to membrane filter and, therefore, inhibit growth of bacteria retained on membrane filter.

2.   The Defined Substrate Test

The defined substrate test is used for both coliforms and E. coli in a single 100 ml water sample.

A specialised medium containing 0-nitrophenyl-β-D-galactopyranoside (ONPG) and 4-methylumbelliferyl-β-D-glucuronide (MUG) as the only nutrients are added with a water sample of 100 ml. The incubation bottles containing water plus both nutrients along with control are incubated at 35°C for 24 hours. If coliforms are present the medium will turn yellow within 24 hours due to break down of ONPG. When *E. coli* is present, the MUG is changed to yield a fluorescent product. The fluorescence is observed under long-wavelength UV light for the presence of *E. coli*. If the colour does not develop and give negative report for the presence of coliform, the water is suitable for drinking. If coliforms are present, *E. coli* or faecal coliforms must be detected in water.

3. IMViC Tests

E. coli and *Aerobacter aerogenes* are the most important bacteria of coliform group, the former is commonly known as faecal contaminant and the later as non-faecal contaminant. Morphologically both are similar and cannot be distinguished. Therefore, biochemical tests are being used to differentiate each of them. The tests differentiating them are collectively known as *IMViC* tests (or reactions). The abbreviation *IMViC* is prepared by using first letter of four different tests viz., indole, methyl red, Voges-Proskauer, and citrate tests. The reactions of *E. coli* and *A. aerogenes* are shown in Table 28.4.

Table 28.4 : Characteristic IMViC reaction *on E. coli* and *A. aerogenes*

Tests	E. coli	A. aerogenes
Indole	Indole is produced from tryptophan, (+)	Indole is not produced, (–)
Methyl red	*E. coli* produces acids that drop pH below 4.5. Hence methyl red is turned red	Methyl red remains yellow due to production of less acid, (–)
Voges-Proskauer	*E. coli* does not produce acetylmethylcarbinol in glucose-peptone medium, (–)	Acetylmethylcarbinol is produced by this bacterium, (+)
Citrate	Citrate as the sole carbon source does not support the growth of *E. coli*, (–)	Citrate supports the growth of this bacterium, (+)

D. Purification of Water (Water treatment)

There are several methods for purification of water the use of which depends on amount and quantity of water. For example, purification of water required for a single house-hold differs from that of a town or city. Secondly, purification of water is essential before its consumption so that disease cycle of pathogenic microorganisms can be broken. Thus, purification of water results in prevention of pathogen to reach the human body. Therefore, water purification is done with the prospect of making it satisfactory in appearance, taste, odour and free from pathogens. There are three chief methods which are used for the purification of drinking water in municipal supplies, *i.e. sedimentation, filtration* and *disinfection* (Kabler, 1962). The steps of water purification of municipal supply are given in Fig. 28.5.

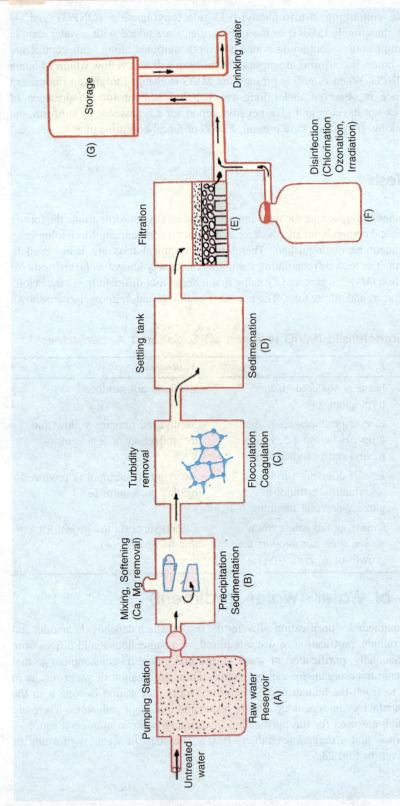

Fig. 28.5 : Steps of water purification of municipal water supplies. Depending on water quality changes in steps are possible. A, a raw water reservoir; B. sedimentation (chemical dosing, mixing, flocculation, settling, ion exchange); C, coagulation (as in B); D, sedimentation (flocs settle on bottom of sedimentation tank; E. filtration (water is filtered); F, disinfection (chlorination, ozonation, irradiation); G, Storage (chlorinated water is stored for drinking water supply).

1.　Sedimentation

Sedimentation is done when water consists of large sized organic materials such as leaves, and gravels which have run off from the soil. Suspended particles settle down depending on their size and weight and conditions of the stored water. Sedimentation is done in large reservoirs or in restricted area of settling tank. The rate of sedimentation is enhanced by adding alum, iron, salts, colloid silicates which act as *coagulants* (Fig. 28.5A-D). The suspended materials and microorganisms are entrapped by coagulants and settle down rapidly. This procedure is called coagulation or flocculation. The microorganisms remain viable for some time. Thus, sedimentation provides partial reduction of microorganisms in water due to their settling down on bottom but does not sterilize the polluted water.

The population of *Physarum* (a slime mold) is reduced as it settles down on bottom but full sterilization of polluted water does not occur by sedimentation.

2.　Filtration

It is the second step of purification. After sedimentation the water is further purified by passing to filtration unit (Fig. 28.5 E). It is the effective means of removing microorganisms and the other suspended material from the water. There are two types of sand filters which are used in water purification such as *slow sand filter* and *rapid sand filter*.

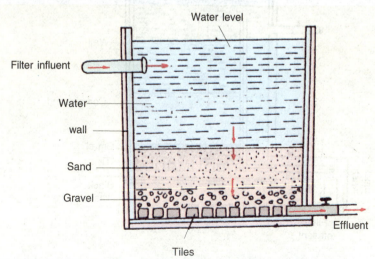

Fig. 28.6 : Diagram of the slow sand filter.

(*i*) **Slow Sand Filter:** In slow sand filtration plants the rate of filtration of water is slow, hence the plant requires a considerable area. This plant consists of a concrete floor containing drainage tiles (for collection of filtered water). The tile is covered with first coarse sand and finally 2 to 1 feet of sand at the top of plant (Fig. 28.6).

Water is passed through this plant. Water passes slowly through the filter and collected by tile drain pipes at the bottom which later on is pumped into a reservoir. If water is turbid, slow sand filters are clogged soon. Therefore, turbid water, which is to be filtered, should be clarified first by sedimentation, thereafter, passed through slow sand filters. The capacity of slow sand filter plant is to filter about 5 million of water per acre per day.

Water purification is done not by physical action but by physiological mechanisms supported by microorganisms. In the surface of layers of fine sand a colloidal material, consisting of bacteria, algae and protozoa, is attached. This mucilagenous material makes the pores more effective by closing the pores between the sand grains. Sand grains have positive charges and bacterial cell walls have negative charge. Therefore, bacteria are adsorbed on the surface of sand. Protozoa ingest bacteria. Due to intense microbial interactions, chemical concentration of water is reduced. When filtration efficiency of the plant is reduced, due to deposition of thick mucilagenous material, the plant is taken out for cleaning. Through this plant, the pathogenic microorganisms such as *Giardia* and its cysts which are not removed by any other methods can be filtered from water (Logsdon, 1991).

For the first time in 1852 (after London's severe cholera epidemic in 1849), parliament of London required that the entire water supply be passed through slow sand filters before use. This plant was installed in many countries after suffering from cholera epidemics (Logsdon, 1991).

(*ii*) **Rapid Sand Filter:** Similar to slow sand filter, the rapid sand filter is also constructed. This plant consists of layers of sand, gravel and rock (Fig. 28.7). Before filtration, water is treated with alum or ferrous sulphate in a settling tank where precipitates settle down. Then water is allowed to pass through rapid sand filter plant. This plant depends on physical trapping of fine particles and flocs or coagulants. The pores of the plants are soon clogged. It is cleaned by forcing cleaned water backward *i.e.* back washing through the beds of gravels and sands without disturbing the fine sand.

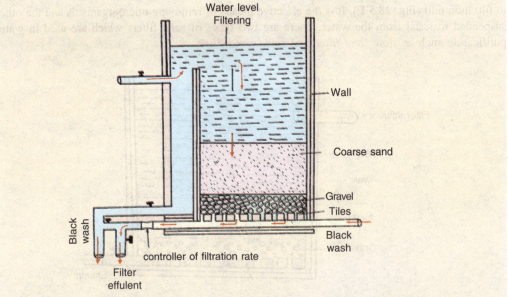

Fig. 28.7 : Diagram of the rapid sand filter.

About 99% bacteria are removed by this plant. But unfortunately the use of coagulants, rapid filtration and chemical disinfection often does not remove *Giardia lamblia* cysts, *Cryptospordium* oocysts, *Cyclospora* and viruses. More consistent removal of these pathogens is possible through slow sand filter. Therefore, water collected after filtration needs further treatment.

In addition, rapid sand filter plant operates about 50 time, faster than slow sand filter plant, and can delivers about 150 to 200 million gallons of water per acre per day. It requires less land area, less cost and less maintenance. Therefore, many plants are constructed in a chain. If one plant is being cleaned, the others are under operation.

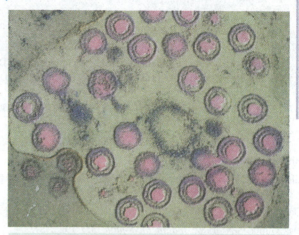

Viruses are not removed by the use of coagulants and rapid filtration.

3.   Disinfection

Some of the bacteria pass through filter even after filtration which must be killed before consumption of water. Therefore, disinfection of public water supply needs to be done. Disinfection is the final step of water purification. Solutions of sodium hypochlorite are treated in small towns but in recent years, chlorination of public water supply has become popular. Chlorination involves the release of chlorine gas in water which gets readily mixed up with water (Fig. 28.5F). The amount of chlorine required depends on organic matter and number of microorganisms present in water, and duration of time to act upon. High concentration of chlorine quickly acts upon microorganisms and *vice-versa*. Therefore, the amount of chlorine required for disinfection is called *chlorine demand*. Water is chlorinated to contain about 0.1 to 0.2 ppm of residual chlorine which reaches to this concentration after 20 minutes of its addition. However, if the concentration of chlorine exceeds its demand, peculiar odour and tastes are experienced. If action of chlorine prolongs in water containing high amount of organic matter, chloramines, are formed. Change in odour and taste of water is due to the formation of chlorophenols. In the presence of high organic matter, chlorine reacts with it and produces *halomethanes* which are a group of carcinogenic compounds.

The mechanism of action of chlorine on microorganisms is obvious. After reacting with water, chlorine is converted into hypochlorous acid which in turn quickly releases nascent oxygen. The nascent oxygen soon oxidises the cellular components of microorganisms as well as organic matter. In addition, chlorine fails to kill the microbial spores. The other gas which behaves like chlorine is the ozone. The simplest method to make water free from microbes and for consumption is boiling for 10-15 minutes.

QUESTIONS

1. Write a brief note on different types of water found in the environment.
2. Write an essay on microorganisms of water.
3. Give an account of different parameters for microbial analysis of water purity.
4. What is the most probable number (MPN)? Write in brief the different tests used to measure the MPN.

5. Write in detail the different methods of water purification.

6. Discuss in detail the sanitary tests for coliforms.

7. Write short notes on the following:

   (*i*) Sedimentation, (*ii*) Fresh water microbiology, (*iii*) Marine microbiology, (*iv*) Faecal coliforms, (*v*) Presumptive test, (*vi*) Membrane filter technique, (*vii*) Defined substrate test, (*viii*) IMViC test, (*ix*) Filtration of water, (*x*) Disinfection, (*xi*) BOD and COD.

REFERENCES

Ambasht, R.S. 1984. *A Text Book of Plant Ecology.* 7th ed. Students' Friends & Co., Lanka, Varanasi.

Bhargava, K.S. 1945. *Proc. Indian Acad. Sci.* **21B** : 344.

Chambers, C.W. 1950. Relationships of coliform bacteria to gas production in media containing lactose. *Public Health Rept.* **65:** 619.

Cody, R.M. *et al.* 1961. Coliform population in stored sewage. *J. Water Pollution Control Fed.* **33** : 164.

Das Gupta, S.N. 1982. *Indian Phytopath.* **35:** 192.

Dayal, R. 1950. *Proc Natl Acad.* Scie. **28:** 49.

Goetyz, A and Tsuneishi, N. 1951. Application of molecular fitter membranes to bacteriological analysis of water. *J.Am. Water Works Ass.* **49:** 943.

Hoskins, J.K. 1934. Most probable numbers for evaluation of coli-aerogenes tests by fermentation tube method. *Public Health Report.* **49:** 393.

Kabler, P.W. 1962. Purification and sanitary quality of water. *Ann. Rev. Microbiol.* **16**: 127.

Khulbe, R.D. 1980. Occurrence of parasitic water molds in some lakes of Nainital, India. *Hydrobiol.* **70:** 119-121.

Khulbe, R.D. and Sati, S.C. 1983. A new species of *Dictyuchus* and key to the other species. *Bibl. Mycol.* **91** : 609 - 617.

Klein, D.A and Casida, Jr. L.E. 1967. *Escherichia coli* die out from normal soil as related to nutrient availability and the indigeneous microflora. *Can. J. Microbiol* **13**: 1461.

Longsdon, G.S. 1991. Slow sand filtration. *Am. Soc. Civ. Eng.* (Thematic issue), New York.

Manoharachary, C. and Rama Rao, P. 1978. *Mysore J. Agric. Sci.* **12:** 280.

Mishra, R.C. and Dwivedi, R.S. 1987. Aquatic water molds from Gujar lake, Jaunpur-II. *J. Indian bot. Soc.* **66**: 203-208.

Mishra, R.C. and Dwivedi, R.S. 1988. Taxo-ecological studied on aquatic moulds in India. *In Perspectives in Mycology & Plant Pathology* (eds. V.P. Agnihotri, A.K. Sarbhoy, and D. Kumar), pp. 370-386. Malhotra Publ. House, New Delhi.

Potter, L.F. 1960. The effect of pH on the development of bacteria in water stored in glass containers. *Can. J. Microbiol* **6**: 257.

Prescott, L.M, Harley, J.P. and Klein, D.A. 1996. *Microbiology.* 3rd ed. Wm.C. Brown Publ.

Sati, S.C. 1997. In *Himalayan Microbial Diversity.* (eds. S.C. Sati, J.Saxena and R.C. Dubey), Today and Tomorrow's Printers & Publ., New Delhi.

Microbiology of Air (Aeromicrobiology)

29

A. Significance of Air

B. Vedic Technology for Air Purification

C. Works on Aeromicrobiology in India

D. Indoor Aeromicrobiology

E. Aeroallergens and Aeroallergy

F. Phylloplane Microflora

During 1930s the term *aerobiology* was used to denote the air-borne spores (e.g. fungi and other microorganisms) and pollen grains. Further the term was elaborated to include dispersion of insect population, fungal spores, bacteria and viruses (Jacobs, 1951). With the inception of International Biological Programme (IBP) in 1964, the term was further extended to include research work of air-borne materials of biological significance. Thus, *aeromicrobiology* deals with the study of air-borne microorganisms and viruses along with important particulate matter of air, especially smoke, dust, radionuclides, and pesticides. The important gases that affect the microorganisms are hydrogen sulphide, sulphur dioxide, carbon monoxide, chlorine, hydrogen fluoride, ozone, etc. The microbial forms are bacteria, fungi and actinomycetes (their spores), algae, spores of pteridophytes, pollen grains, micro-insects, and viruses (Table 29.1).

Pollen grains

Table 29.1 : Some common aeroallergens (based on Finkelstein, 1969)

Aeroallergens	Sources
Pollen grains	Wind-pollinated plants e.g. grasses, weeds, trees
Moulds	Saprophytic fungi multiply on dead organic substrate in the presence of optimum moisture and temperature
Danders	Feathers of chickens, ducks, hairs of cat, dogs, sheep, cattle, laboratory animals and humans.
House dust	A composite of all dust containing specific components related to mites, algae, etc.
Cosmetics	Talcs, perfume, hair tonics, lotions, bindi, etc.
Insecticides	Insecticides containing pyrethrum as a common ingredient..
Paint and varnishes	Linseed oil, organic solvents act as irritant.

A. Significance of Air

Vedas are the most ancient script of *Aryans* written between 1,500 and 8,500 BC. However, man's basic resources acting as life support system has been described as *panchbhutas i.e.* soil (*kshiti*), water (*Aap*), energy (*Agni*), space (*Aakash*), and air (*Vayu*). In addition to the description of *panchbhutas* towards spiritualism, air has been discussed as the breath of life (*Pran*). Air plays dual role *i.e.* creating the health and diseased condition.

Air as the breath of life.

The cool mild breeze coming through a garden brings fragrance and solace to humans. In such a condition one experiences comfort. Moreover, such conditions are better for control of exhalation and inhalation of air *i.e.* *Pranayam*. In addition, the stormy air (wind) becomes dangerous sometimes causing a great damage to both the natural/man-made structures on the earth (Vannucci, 1994). Moreover, air containing foul smelling and pathogenic microorganisms is not fit for human health. Such air causes many human diseases including allergy. The allergy causing air-borne agents are called **aeroallergens** that include biotic and abiotic factors. The spread of pollen grains in the air depends on the meteorological parameters.

Brutally drying stormy air (wind) has badly damaged the natural nests of penguins uprooting them.

B. Vedic Technology for Air Purification

Ours is a spiritual country. Much has been discussed about *Krimi* associated with human health in *Atharveda* (1/2/31, 4/23). The seers were aware of making their surroundings pure by using vedic technology one of which is the *Agnihotra i.e.* the smallest form of *Vedic Hom*. Thus *Agnihotra* was (and still is) the daily duty of offerings to fire, twice in a day (at morning and evening) for air purification in their houses and surroundings. Gases are produced after burning of various substances of fire oblations (*e.g.* ghee, cereals, medicinal herbs, forest herbs and vegetables) and diffused in surroundings. These gases kill/inhibit various kinds of microorganisms present in air. The action of gases is potentiated by increased temperature that causes decline in relative humidity of the atmosphere. These gases act as antimicrobial and insect-repellent. (Satya Prakash Saraswati, 1974).

Gupta and Singh (1996) at Gurukul Kangri University performed *Agnihotra* and noticed that the total microbial population of bacteria, fungi and actinomycetes decreased in the surrounding after *Yajna fire*. However, after 24 hour of *Yajna*, the surrounding environment upto 120 cm height was completely free from the microorganisms. Navneet *et al.* (1998) have discussed the results of samplings of aeromicroflora before the start of *Yajna*, during *Yajna* and after *Yajna* performed in *Prarthana Bhawan* of Gurukul Kangri University, Hardwar. They found that *Rhizopus* sp., *Fusarium* and *Alternaria alternata* were present only before the start of *Yojna*, whereas *Cladosporium cladosporioides, Penicillium cyclopium* and *Aspergillus niger* were dominant.

The total number of microbial propagules decreased when *Yajna* was over. Tilak (1982) has also reported about 70% decrease in population of aeromicroflora after *Agnihotra* in Aurangabad (Maharastra).

C. Works on Aeromicrobiology in India

1. Fungi

The first systematic work was carried out by Cunningham (1873) at Calcutta Jails which he published in the form of a book, *Microscopic Examination of Air*. Later on Prof. K.C. Mehta (1940-1952) of Agra College, Agra extensively investigated the air for the presence of wheat rust spores. Our present knowledge of rust disease of cereals is only due to his contribution. He discovered that rust spores survive on wheat and barley and self-sown plant at 2000 meter high hills. The wind currents carry the inoculum to the hills when climatic conditions of plains are unfavourable. When these are favourable, spores are brought back to plains to start reinfection on fresh wheat plants.

(*i*) **Occurrence and Epidemiology:** Work of K.C. Mehta lent support for Joshi *et al.* (1972) and Nagarajan and Singh (1973) who found that brown and black rust of wheat are disseminated from the hills of South India to Central India by "hop jump" in which a storm depression formed in Bay of Bengal/Arabian sea reaches Central India. Nagarajan and Singh (1973) found the satellite television cloud photography as a possible tool for forecasting the spread of several plant pathogens such as *Alternaria, Cercospora, Helminthosporium, Puccinia*, etc. Aeromicrobiological investigations on several crops such as rice, wheat, jwar, bajra, cotton, banana, citrus, sugarcane, and vegetables have been carried out by many Indian scientists. For a detailed discussion see the books, *Aerobiology* by Tilak (1982), and *Environmental and Aerobiology* by Jain (1998).

Studies on microbial components of air over crop fields are useful in understanding the plant pathogens and in establishing the forecasting system for disease control. A lot of work has been done on aeromycology on several crop fields. Navneet (1995-96) studied aeromycoflora over potato fields at Kurukshetra for two years and recorded 25 fungal species. The dominant fungi were *Alternaria alternata, Aspergillus flavus, A. niger, Cladosporium herbarum, Epicoccum nigrum, Penicillium citrinum, P. cyclopium* and *Trichothecium roseum*. Rain has the propound effect on aerofungi. Verma *et al.* (1998) recorded 62 aerospora over rice fied by using Tilak sampler. The most dominant fungal taxa was *Aspergillus* (75%) followed by *Penicillium* (10%), *Alternaria* (3%) and *Cladosporium* (2%). In general, aspergilli form a major component of aerospora including pathogenic species also (*e.g. A. niger*). *A niger* causes disease on jack fruit, onion, etc. *Alternaria* spp. (*e.g. A. brassicae, A. dauci, A.porri, A. solani*) cause disease on several crop plants.

(*ii*) **Aflatoxins by Aerofungi:** Aflatoxins are the secondary metabolites of some saprophytic fungi injurious for human health. Work on aflatoxin production from agricultural field to storage conditions in different parts of the country as well as in different crop fields has been done well by late Prof. K.S. Bilgrami and his associates (1975-1996) at Bhagalpur University, Bihar. The species of fungi secreting aflatoxins are *Aspergillus flavus, A. parasiticus* and others. Aflatoxin of *A. flavus* is known to cause cancer. Rati and Ramalingam (1979) surveyed *A. flavus* in air samples in outdoor air and air of poultry shed. They found that 72% of these were toxigenic, and the incidence of toxigenic strain was found to be greater in winter months. The concentration of aerospores inside poultry shed was about 10-100 times greater than that of adjacent outdoor environment. In 1960, outbreak of a lethal disease in turkey poults was resulted due to consumption of aflatoxin contaminated meals. The disease symptoms included (*a*) rapid deterioration in the conditions of birds, and (*b*) subcutaneous haemorrhages leading to death. The livers of dead birds were pale, fatty and showing extensive necrosis and biliary proliferations. Similar symptoms were also observed in duckling fed on "toxic groundnut meal", and several other animals. Organs other than liver are also affected by aflatoxins such as kidney, adrenal glands, lungs, skin, etc. (Bilgrami and Sinha, 1993). Moreover, aflatoxin B1 is known to cause mutagenicity through chromosomal aberrations and DNA breakage in plant and animal cells. Sinha *et al* (1987) observed the gross and individual types of chromosomal abnormalities and breakage in the chromosomes of bone marrow cells of mice. The breaks were more frequent in the distal regions of the longer chromosomes.

(*iii*) **Seasonal Occurrence of Aflatoxin Producing Aerofungi:** Environmental factors are responsible for occurrence of aerofungi in different regions. Choudhary (1991) serveyed the climatic conditions on incidence and severity of aflatoxin contamination of field maize crop during 1986-1990 which were cultivated as *Kharif* crop in Bihar. He noticed that temperature (during July-August) and prevalence of relative humidity (90% RH) for prolonged time are the major determinants of aflatoxin contamination. Increased level of aflatoxins in nonirrigated maize crops has been linked to higher levels of air-borne inoculum of *A. flavus* (Bilgrami and Choudhary, 1991)

2. Algae

A little work has been done on occurrence of algae in air. Generally, algae dominate upto a height of 2 meter. The most common algae found in air are the species of *Chlorella, Chlorococcum, Chlamydomonas, Aulosira, Nostoc* and *Phormidium*. Ramalingam (1971) reported some of the algal types (e.g. diatoms, *Protococcus, Spirogyra, Oscillatoria,* etc.) from Mysore. Agnihotri *et al.* (1977) have reported the lichen component of airospora with special reference to allergenic ones, from hill districts of Uttar Pradesh. The most common li-

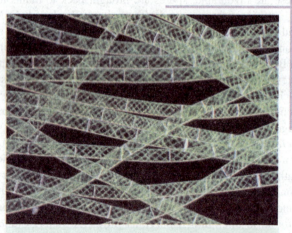

An allergenic filamentous alga, *Spirogyra.*

chens were the species of *Cladonia, Heteroderma, Parmelia, Usnea,* etc. In rainy season beauty of the world fame Taj Mahal fades due to the growth of algae on it. However, it is properly cleaned regularly. A National Laboratory for Conservation of Cultural Properties has been established at Lucknow that takes care of developing methods for conservation of monuments, archives, etc.

D. Indoor Aeromicrobiology

Indoor aeromicrobiology deals with microorganisms present in air inside the houses. There are many microorganisms which are responsible for biodeterioration of storage materials, equipment, library materials and archives. Sharma and Navneet (1996) reported aerofungi from the fermentation unit of Gurukul Kangri Pharmacy, Hardwar. Pathogenic aeromicroorganisms are found in hospital wards. Even house dust has been investigated for the presence of allergenic constituents. Some examples of indoor aerospora have been discussed below.

1. Aeromicrospora of Pharmacy

Ayurvedic drugs are prepared in pharmacy. Sharma and Navneet (1996) reported aerofungi from the fermentation unit of Gurukul Kangri Pharmacy, Hardwar. They isolated fungi during February-March, 1994. The environmental factors viz., relative humidity and temperature affected their occurrence. In the diurnal cycle, fungi showed an evening tendency. The dominant species were *Cladosporium cladosporioides, Alternaria* sp., *Penicillium cyclopium, Epicoccum nigrum*, etc.

2. Aeromicroflora of Hospitals

Hospital is an important indoor environment responsible for spread of airborne pathogens. It acts as reservoirs of pathogens which later on is transmitted to the other individuals viz., patients, hospital workers, visitor, etc. In turn it is carried over patients. Even coughing and sneezing cause the spread of microorganisms and important viruses. The hospital transmitted pathogens are *Mycobacterium tuberculosis, Staphylococcus aureus*, influenza virus, *Aspergillus flavus, A. fumigatus, Candida albicans,* etc. (Tilak, 1982).

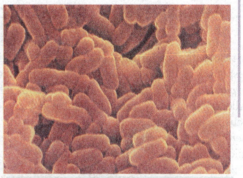

Mycobacterium tuberculosis is one of the hospital transmitted pathogens.

The species of *Aspergillus* is not a normal flora of human but it frequently causes lung infection in hospital environment. The dominant form of hospital infections is the candidiasis caused by *C. albicans*. *C. albicans* contaminates the hospital wards generally through direct contact with fingers.

Squames from the skin of persons in the operation theatre transmit pathogenic microorganisms. Squames contain many pathogenic bacteria which are transmitted to patients. Therefore, attempts must be made to check the spread of pathogens in the hospitals so that primary and secondary infections could be avoided. For detail of air-borne diseases see Chapters 21 and 22.

3. Other Houses

The indoor air never lacks spores. Even in clean rooms about 25 spores/m³ have been found. In the houses where air conditioners and coolers (humidifiers) are used. There is such chance of air-borne microorganisms because humidity and low temperature (about 25°C) create congenial environment for propagation and spread of microorganisms. Cold mist humidifiers are generally colonised by fungi (e.g. *Aspergillus, Geotrichium, Penicillium, Phialophora*, etc.), yeasts, bacteria, etc. Moreover, insects act as carrier of human pathogens and lay eggs in stagnant water (Tilak, 1982).

4. Aeromicroflora of Storage Materials

Several valuable materials are stored in houses, and personnel are appointed to look after them. However, these are deteriorated by aeromicroflora. Some examples are discussed below:

(*i*) **Library:** A library consists of thousands to millions of valuable, common and rare, printed and hand written books. The major constituent of paper is the cellulose. Therefore, cellulose-degrading microorganisms colonise and degrade the papers. The common cellulose degrading fungi are the species of *Alternaria, Aspergillus, Curvularia, Bispora, Chaetomium, Cladosporium, Fusarium, Helminthosporium, Periconia, Nigrospora, Rhizopus, Stemphilum, Trichoderma*, etc. High moisture and low temperature increase the rate of cellulose decomposition by these fungi (Tilak and Vishwe, 1975). In addition, decomposition of rexin and leather has been found very less in spite of production of high amount of enzymes.

(*ii*) **Wall Paintings:** Wall paintings are the cultural heritage in a region of a country. The world fame wall paintings at Ajanta and Ellora caves have shown the sign of biodeterioration. For the first time Tilak and Kulkarni (1972) studied the aerospora of caves. Tilak *et al.* (1972) isolated the fungal spores on the paintings and in caves of Ajanta and Ellora. Probably aerofungi have deteriorated the wall paintings. Growth of aerofungi is supported by the excreta of bats which probably serves as substrate. In addition, the meteorological factors are also responsible for growth and biodeterioration of wall paintings.

E. Aeroallergens and Aeroallergy

In 1906, Anton Van Pirquet introduced the term *allergy* to denote any altered capacity of body to react with a foreign substance. Now it is established that these phenomena are mediated by immune responses of the body. However, if an immunological cause damages the body cells, it is called *hypersensitivity*. Now-a-days, the two terms allergy and hypersensitivity are used synonymously. Allergy is caused by certain biological and abiological agents present in the atmosphere. The allergy causing agents present in air are called *aeroallergens*, and the allergy caused by them is called *aeroallergy*. Some of the prominent aeroallergens are house dusts, pollen grains, cosmetics, microbial spores or cells, etc. (Table 29.2).

Table 29.2 : Example of common aeroallergens

Aeroallergens	Examples
Algae	*Aulosira, Chlamydomonas, Lyngbya, Nostoc, Phormidium, Gloeotrichia,* diatoms, *Oscillatoria, Chlorella, Plectonema,*
Fungi	*Alternaria, Aspergillus, Candida, Chaetomium, Curvularia, Fusarium, Monilia, Penicillium, Phoma, Trichoderma*
Lichens	*Cladonia, Heterodermia, Parmelia, Usnea*
Pollen grains	*Ageratum conyzoides, Amaranthus* sp., *Argemone mexicana, Azadirachta indica, Carica papaya, Cassia fistula, Cocos nucifera, Croton, Datura metel, Catharanthus roseus, Euphorbia hirta, Ocimum sanctum, Parthenium hysterophorus, Ricinus communis, Setaria* sp., *Phoenix sylvestris.*
Others	Dust, mites, viruses, bacteria, protozoa, moss spores, fern spores, seed spiders, etc.

There are four major types of allergic reactions viz., Type I, IgE-dependent, Type-II, cytotoxic tissue specific antibody, Type-III, toxic antigen-antibody complexes with activated complement, and

Type-IV, T-lymphocyte cell-mediated hypersensitivity. The role of Type-I allergy is very common in vast majority followed by Type-III allergy.

Recently, the World Health Organisation, International Union of Immunological Societies of Allergy Nomenclature sub-committee has revised the names of allergens. Instead of italicising the names, the new names are written by using the first three letters of the genus and the first letter of the species and a number. For example, the mite allergens for *Dermatophagoides pteronyssinus* are written as Der P_1, Der P_2 and so on. Some allergens have been discussed below:

Lichens, symbiotic associations between algae and fungi, act as common aeroallergens.

1. House Dust Allergens

Generally, house dust is a mixture of hairs, moulds, bacteria, decomposed parts of cloths or furniture, small insects, mites, etc. These are inhaled by individuals who later on suffers from allergy. The most common fungal spores isolated from the allergic individuals are the species of *Aspergillus, Rhizopus, Cladosporium, Curvularia, Helminthosporium, Phoma, Fusarium, Cephalosporium, Nigrospora, Cladosporium, Epicoccum, Penicillium*, etc. A significant work on respiratory allergy has been done at V.P. Chest Institute (Delhi), and at the other centres viz., Jaipur, Kanpur, Lucknow, Kolkata and Aurangabad.

Mites are the important components of house dust and strong allergens for humans. Some mite species are *Dermatophagoides farinae, D. gallinae* (poultry mite), *D. domesticus, D. destructor, D. pteronyssinus,* etc. Among these *D.pteronyssinus* has been most extensively studied. *D. pteronyssinus* is found in large number on mattresses, blankets, pillows, etc where human scales are abundant. These are distributed in all houses equally.

D. pteronyssinus is commonly known as *house dust mite* which dominate by 88% almost in all houses. Children shed considerable amount of skin in house, which serves as a culture medium to *D. pteronyssinus*. Thus, human skin is a good source of nutrients in the house dust. A significant work on allergenic parts of mites has been done. Cunnington (1972) found that the dead mites and their different body parts, faeces and secretions are equally potent allergens as the live mites. The housewives, domestic cleaners, upholsterners and decorators are found to be victimised by houst dust mite. Torey *et al.* (1981) also reported that mite faeces are the major source of house dust allergens. They are related to the presence of mites, and radioimmunoassay revealed more than 95% of the allergen accumulating in mite cultures was associated with faecal particles.

2. Pollen Grains

The climatic conditions in India vary in different regions, therefore, variations also occur in vegetation. Due to great diversity in vegetation and climatic conditions the aeroallergens are also of different types in different regions and places. In India aeroallergenic pollen grains are contributed mainly from the plants belonging to the family Poaceae, Chenopodiaceae, Amaranthaceae and Asteraceae (Table 27). Tilak and Vishwe (1979) recorded 33 pollen types of which herb pollen grains were the highest in number. After attaining maturity, the pollen grains get disseminated and dispersed through vectors like wind, water and insects.

Occurrence and abundance of pollen grains in the environment differ according to season and vegetation types. However, most of the pollen grains are found throughout the year. Allergenic potential of pollen grains is governed by their chemical constituents. Chakraborty *et al.* (1998) have analysed and clinically tested the allergenic potential of some common pollen grains recorded in Kolkata viz., *Areca catechu, Azadirachta indica, Carica papaya, Catharanthus, Phoenix* and *Datura.* They reported 49% sensitivity by *Azadirachta* pollen grains followed by *Datura* (35%) and *Catharanthus* (26%). Chemical nature of these pollen grains showed differences in total carbohydrate, total protein and total lipid composition.

3. Cosmetics

Cosmetics are the articles that some one applies to, sprinkles on or rubs into their body to cleanse, beautify or promote attractiveness or alter their appearance; the product must not normally harm body function or structure. The examples of cosmetics of daily use are creams, gels, powders, lotions, perfumery items, oils, dyes, lipsticks, *bindi,* eye cosmetics, etc. These cosmetics show dermatological problems. Bhaduria (1998) reviewed the cosmetics and allergy caused by them. The causes of allergies are : (*a*) the quality products of cosmetics, (*b*) spoilage of cosmetics by

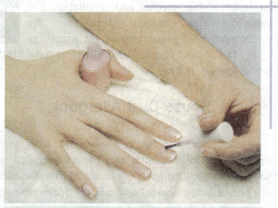

Daily use cosmetics show dermatological problems.

microorganisms viz., *E.coli, Aerobacter aerogenes, Bacillus subtilis, Alternaria, Aspergillus, Cladosporium, Fusarium, Candida, Torula, Trichoderma,* etc, (*c*) secretion of toxins by microorganisms, (*d*) chemical ingredients of cosmetics, (*e*) inadequate preservative system, (*f*) packaging materials, and (*g*) mode of use. Pasricha (1988) has reported a confirmed diagnosis and the cause of contact dermatitis by performing *Patch test* as below:

Hair dyes	:	The most common cause of contact dermatitis among the cosmetics. The paraphenylenediamine present in hair dyes is the most common sensitizer and gave positive patch test.
Hair oils	:	Mustard oil gave positive patch test.
Shampoos	:	Selson Shampoo (containing selenium sulphide and cetavlon) gave positive patch test.
Lipsticks	:	Dye used for colour causes contact dermatitis; eosin is a potent photocontact sensitizer.
Bindi	:	PVC disc used in *bindi* gave positive patch test. Adhesive material of *bindi* contains 70-80% paratertiary-butyl phenol which is a causal factor of contact dermatitis and other skin allergies after constant use of *bindis.*
Tooth pastes	:	Foaming agent is known to be chief cause of contact dermatitis.
Hair-removers	:	Barium sulphide in hair creams/lotions is a strong cauterizing agent for the skin. It can damage the skin if left in contact for a long time.

F. Phylloplane Microflora

Last (1955) studied the distribution of members of Sporobolomycetaceae on wheat and barley in Britain, and Ruinen (1956) investigated the occurrence of *Beijerinckia* (a nitrogen fixer) on the leaves in Indonesia. These two workers independently introduced the term *phyllosphere* to describe the leaf surface as the habitat of microorganisms. Ruinen referred the leaf surface as the *neglected milieu* of microorganisms. In 1958, Kerling suggested to use the term *phylloplane* for leaf surface habitat.

The leaf surface acts as the landing stage for the microbial propagules. These are deposited by impaction, by boundary layer exchange, by sedimentation under the influence of gravity and in rain and splash droplets (Pugh, 1984). The spores present on leaf surface get nutrients defused from leaf and also from pollen grains present on leaf surface. Moreover, honey bees and insects discharge excreta on leaf surface which serve as food base for microorganisms. The insects that chew upon leaves also cause secretion of nutrients. Microorganisms colonise the leaves when it remain in unfolding stage. Thereafter, microbial competition is established on leaf surface. In addition, they are constantly influenced by temperature fluctuations, ultra violet radiation, desiccation/moisture content present in atmosphere.

Some common phylloplane microflora are bacteria (*Beijerinckia* sp., *Pseudomonas trifoli*, etc.), yeasts (*Candida albicans, Cryptococcus diffluens, Saccharomyces cerevisiae, Torulopsis colliculosa*), and fungi

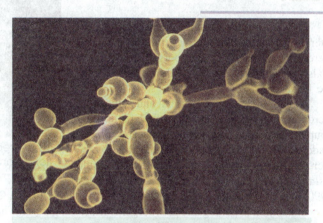

The yeast *Candida albicans*, a common phylloplane microflora.

(*Alternaria alternata, Aspergillus niger, A. terreus, Aureobasidium pullulans, Cephalosporium roseo-griseum, Cladosporium cladosporioides, Curvularia lunata, Drechslera australiensis, Epicoccum purpurascens, Fusarium oxysporum, Myrothecium roridum, Penicillium oxalicum, Phoma glomerata, Trichoderma harzianum, T. viride*, etc.).

1. Phylloplane Pathogens

The microorganisms living on phylloplane are of several types *i.e.* saprophytes, symbionts and pathogens. Some common pathogenic fungi are given in Table 29.3.

Table 29.3 : Some phylloplane pathogens

Pathogens	Disease
Albugo candida	White rust of crucifers
Alternaria brassicae	Leaf spot of crucifers
A. solani	Early blight of potato
Cercospora arachidicola and C.personata	Leaf spot or tikka disease of groundnut
Colletotrichum gloeosporioides	Leaf and fruit disease
Erysiphe graminis tritici	Powdery mildew of wheat
Helminthosporium oryzae	Brown spot of paddy

Pathogens	Disease
Melampsora lini	Rust of linseed
Peronospora parasitica	Downy mildew of crucifers
P. viciae	Downy mildew of pea, beans
Plasmopara viticola	Downy mildew of grape-vine
Pyricularia oryzae	Rice blast disease
Phytophthora infestans	Late blight of potato
Puccinia graminis tritici	Rust disease of wheat
Taphrina deformans	Peach leaf curl
T. maculans	Leaf spot of turmeric and zinger
Uromyces fabae	Rust of pea

2. Microflora of Floral Parts

Flowers produce pollen grains, nector and several chemicals that attract microorganisms. Therefore, a variety of microorganisms fall on petals, sepals, ovary, etc. Several insects too colonise floral parts.

Among those, both beneficial and pathogenic microorganisms are present. One of the pathogenic group of fungi is the smuts of graminaceous plants. In addition, fruit infection occurs some times by the microorganisms of floral parts, for example *Pestalotia psidii* causing fruit rot of guava. Pandey (1990) has reported the fungi on floral parts of guava (Table 29.4).

The insect rose aphid colonises the floral parts and sucks sugar-rich juice from them.

In addition to pathogenic microorganisms, floral parts also harbour beneficial microorganisms. For example, *dhataki* flowers (*Woodfordia fructicosa*) are used in Ayurvedic Pharmacy in the production of *amritarishta*. Roshan Lal and Maheshwari (1996) isolated bacteria, yeasts (*Saccharomyces* sp) and fungi (aspergilli and *Rhizopus nigricans*) from *dhataki* flowers, and studied the potential for production of *amritarishta*. It was found that *Saccharomyces* sp. produced *amritarishta* with 7% alcohol when grown on raw material through fermentation process.

Table 29.4 : Distribution of fungi (cm^{-3x} 10^3) on floral parts of guava isolated by dilution plate method (after Pandey, 1990)

Fungal species	Floral parts		
	Unopened bud	Petals	Sepals
Alternaria alternata	28	6	9
Aspergillus flavus	—	1	
A. niger	3	2	—
Aureobasidium pullulans	32	8	19
Candida sp.	1	—	—
Cladosporium cladosporioides	37	14	17
C. herbarum	4	2	—
Drechslera australiensis	2	—	5
Epicoccum purpurascens	10	2	12
Fusarium oxysporum	3	3	24
F. semetectum	7	—	5
Pestalotia psidii	—	—	4
White sterile mycelia	—	—	1
Yellow sterile mycelia	—	—	2

3.   Characteristics of Phylloplane Microflora

Since phylloplane microflora is exposed into the environment, it is constantly influenced by meteorological factors. Therefore, microflora develop certain characteristic features so that it can cope with environment.

(*i*) **Morphological Characteristics:** The phylloplane inhabitants possess morphological characteristics for their survival. These include the development of pigment in their mycelia, spores, pycnidia, apothecia, cleistothecia for protection against strong light and desiccation. The dark pigments act as light screen. These pigments are often referred to as *melanin*.

(*ii*) **Physiological Characteristics:** Phylloplane fungi possess a number of physiological characteristics discussed as below:

(*a*) **Nutrition:** Phylloplane microfungi have ability to decompose cellulose by producing cellulases. In addition, pectinases, cutinases proteases have also been estimated in many fungi e.g. *Alternaria alternata, Aureobasidium pullulans, Botrytis cineria, Cladosporium herbarum.*

(*b*) **Radiations:** Besides certain exceptions, high intensity light inhibits mycelial growth, while light (of normal intensity) is not harmful to phylloplane fungi. The UV portion of spectrum plays a significant role. Fungi containing melanin pigment are resistant to UV exposure as compared to those containing hyaline mycelia. Hyaline spores of *Aureobasidium* and *Sporobolomyces* are killed by exposure of UV for five minutes, whereas dark spores of *Alternaria* and *Epicoccum* survive even after exposure for 35 minutes. High intensity of UV radiation becomes lethal to microorganisms.

(*c*) **Relative humidity:** During rain and dew formation, high relative humidity on leaf surface is found. But when there is wind the microenvironment is reduced. Therefore, phylloplane fungi get advantages at low levels of relative humidity. Generally, a faster growth rate of the germ tube and mycelium of some fungi occurs at 100% relative humidity. Even some could not grow below 93% of relative humidity.

(*d*) **Temperature:** Generally, the phylloplane fungi are mesophilic growing at temperature 20-25°C. Even some fungi (e.g. *Alternaria alternata, Aureobasidium pullulans, Cladosporium herbarum, Botrytis cinerea*) can grow below 0°C.

4.   Microbial Interaction on Leaf Surface

As discussed in Chapter 25, many kinds of interactions occur on phylloplane region between the pathogens and saprophytes. Rai and Singh (1980) investigated the antagonistic activities of some phylloplane fungi (of mustard and barley) against *Alternaria brassicae* and *Drechslera graminea*. The antagonistic fungi were *Aureobasidium pullulans, Epicoccum purpurescens, Cladosporium cladosporioides* and *A. alternata*. The most significant effect was observed when the spores of leaf surface fungi or their metabolites were sprayed on leaves prior to inoculation of the pathogens. Pandey *et al.* (1993) studied the antagonistic activities of some phylloplane fungi of guava against *Colletorichum gloeosporioides* and *Pestalotia psidii*. They recorded pronouned inhibition in lesion development after application of spore suspension (3×10^5 propagules/ml) of *A. pullulans, C. cladosporioides, E. purpurescens, F. oxysporum* and *Trichoderma harzianum* against *C. gloeosporioides,* and *A. niger, A. terreus, C. roseo-griseum* and *T. harzianum* against *P. psidii*. Such studies help in biological control of plant pathogens. For a detailed discussion of mechanism of biological control (*i.e.* antagonism), see Chapter 25.

QUESTIONS

1. What is aeromicrobiology? discuss in brief the concept of air and method of air purification during *Vedic* period.

2. Write an essay on aeromicrobiology in India with special reference to fungi and algae.

3. What is indoor microbiology? Write in brief aeromicroflora of hospitals, library and storage materials.

4. What are the aeroallergens? Write different bio-chemical components of aeroallergens.

5. What do you know about phylloplane? Discuss in brief the microflora of leaf surface and floral parts of any plants.

6. What are the morphological and physiological characteristics of phylloplane microfungi.

7. Write short notes on the following:

    (*i*) Significance of *Agnihotra*, (*ii*) Aeromicrobiology, (*iii*) Aerofungi and aflatoxins, (*iv*) Indoor aeromicrobiology, (*v*) Aeroallergens, (*vi*) Cosmetics and aeroallergens, (*vii*) Phylloplane microorganisms, (*viii*) Phylloplane pathogens, (*ix*) Microflora of floral parts, (*x*) Characteristics of phylloplane microfungi, (*xi*) Microbial interactions on leaf surface.

REFERENCES

Bhaduri, R. 1998. Cosmetic allergy; A review. In. *Environmental and Aerobiology* (ed. A.K. Jain) pp. 245-256. *Research Periodicals & Book Publ. House*, Texas, U.S.A.

Bilgrami, K.S. and Choudhary, A.K. 1990. *Indian Phytopath.* **43:** 38-42.

Bilgrami, K.S. and Sinha, K.K. 1993. Mycotoxins - Chemical, biological and environmental aspects. In *Fungal Ecology and Biotechnology* (eds. B. Rai, D. K. Arora, N.K. Dubey and P.D. Sharma). pp. 295-310, Rastogi Publ. Meerut.

Chakraborty, P. *et al.* 1998. Clinical tests and chemical analysis of some common aeroallergens from Calcutta. In *Environmental and Aerobiology* (eds. A.K. Jain). pp. 239-294. Research Periodicals & Book Publ. House, Texas, U.S.A.

Choudhary, A.K. 1991. *Ph.D.Thesis.* Bhagalpur Univ. Bhagulpur (Bihar)

Cunningham, D.D. 1873. *Microscopic examination of air.* Govt. Printer, Calcutta.

Cunnington, A.M. 1972. House dust mites and respiratory allergy. A qualitative survey of species occurring in finish house dust. *J.Resp.Dis* **53**: 338.

Gupta, G.P. and Singh, K. 1996. Studies on Vedic technology for environmental health. *Natl Seminar on Biotech*: New Trends and Prospects, Gurukul Kangri Univ., Hardwar, Abstract, pp. 41-42.

Jacobs, W.C. 1991. *American Meteoral. Soc.* Boston. 1103-111.

Jain, A.K. 1998. *Environmental and Aerobiology.* Research periodicals & Book Publ. House, Texas

Joshi, L.M., Sastri, S.E. and Gera, I.D. 1972. Epidemiological aspects of *Puccinia graminis tritici* in India. *Proc. Indian Natl. Sci. Acad.* **37B:** 445-453.

Kerling, L.C.P. 1958. *Tydschr. Pl. Ziekten.* **64:** 402-410.

Last, F.T. 1955. *Trans. Br. mycol. Soc.* **38**: 221-239

Nagarjan, S and Singh, H.1973. Satellite television cloud photography as a possible tool to forecast plant disease spread. *Curr. Sci.* **42** : 273-274.

Navneet 1995-96. Aeromycoflora over potato fields *J. Natural and Physical Sci.* 9-10. 61-71.

Navneet, Chand, S. and Sharma, V.K. 1998. *Agnihotra*- the air purifier. *Aryabhatt* (Gurukul Kangri Univ. Hardwar.

Pandey, R.R. 1990. Mycoflora associated with flora parts of guava (*Psidium guajava* L). *Acta Botanica Indica.* **18:** 59-63.

Pandey, R.R. Arora, D.K. and Dubey, R.C. 1993. Antagonistic interactions between fungal pathogens and phylloplane fungi of guava. *Mycopath.* **124:** 31-39.

Rai, B. and Singh D.B. 1980. Antagonistic activity of some leaf surface microfungi against Alternaria brassicae and Drechslera graminea. *Trans Br. mycol. Soc.* 75: 363-369.

Roshan Lal and Maheshwari, D.K. 1996. Studies on the potential of microflora associated with *dhataki* flowers (*Woodfordia fructicosa* Kurz.) in the production of *Amritarishta*. Natl. Seminar on Biotechnology: New Trends & prospects, Gurukul Kangri Univ. Hardwar, Abstr. p. 30-31.

Rati, E.A. and Ramalingam, A. 1979. Toxic strains among air-borne isolates of *Aspergillus flavus* Link. *Indian J. Exp. Biol.* **17:** 97-98.

Ruinen, J. 1956. *Nature* (London). 177: 220-221.

Sharma, V.K. and Navneet. 1996. Aeromycoflora of Gurukul Kangri Pharmacy, Haridwar. National Seminar on Biotechnology: New Trends & Prospects, Gurukul Kangri Univ., Hardwar. Abstr. b. 54.

Satya Prakash Saraswati 1974. *Agnihotra*—a study from chemical stand point. *Jan Gyan Prakashan*, New Delhi, p.106.

Tilak, S.T. *Aerobiology.* Vaijayant Prakashan, Aurangabad, p.207.

Tilak, S.T. and Kulkarni, R.L. 1972. Microbial content of air inside and outside of caves at Aurangabad. *Curr. Sci* -23: 850-851.

Tilak, S.T. and Vishwe, D.B. 1975. Microbial content of air inside library. *Biovigyanam* 1: 187-190.

Tilak, S.T. and Vishwe, D.B. 1979. Aeropalynology at Aurangabad. IV th Int. Polynological Conf., Lucknow.

Tilak, S.T. *et al.* 1972. Studies in the microbiological deterioration of paintings at Ajanta and Ellora. *Studies Museumology.* **8:** 20-25.

Torey, E.R. *et al.* 1981. Mite faeces are a major source of dust allergens. *Nature.* **289:** 592.

Upadhyay, R.K and Arora, D.K. 1995-96. Sporostatic nature of neem smoke and its possible ecological influence on air fungal flora of polluted site *J.Sci. Res.* (B.H.U.) **26:** 125-129.

Vannucci, M. 1994. Ecological readings in the vedas. *D.K. Print World (P) Ltd.* New Delhi, p.116.

Verma, K.S.; Agarwal, R. and Saraf, R. 1998 Aeromycology of rice field with special reference to circadian periodicity of some dominant fungal spores. In *Environmental and Aerobiology* (ed A.K. Jain), pp. 189-

Environmental Microbiology

30

A. Waste as a Resource
B. Sewage (Wastewater) Treatment
C. Microbial Leaching
D. Biodegradation
E. Microorganisms in Abatement of Heavy Metal Pollution
F. Water Pollution Management
G. Biofiltration
H. Biodeterioration

Microbiology has spread its scope into various applied sciences like environmental engineering and other subjects which are covered in this book. On one hand, many species of microbes are responsible for producing improved quality and variety of products for betterment of life of human being, while on the otherhand, urbanization, industrialization and increase in human growth/population rate are responsible for degradation of environment. Microbes are able to degrade solid waste (lignocellulosics) into compost, some are able to degrade pesticides, waste and those generated in industries such as heavy metals present in industrial effluents and from thermal power stations, sewage sludge and in other sources. Some microorganisms are also able to degrade petrol and petroleum products entering as oil pollutants due to oil spills in the marine environment. Microorganisms have a role to play in bioleaching of ores. Thus, the study of microorganisms is of prime importance in furthering our knowledge on environment.

Microbes degrade sludge into compost.

A. Waste as a Resource

In some countries (particularly in third world) wastes (solid as well as liquid) are the major source of pollution due to generation in large quantities, and pose environmental problems. Wastes generated after biogas production in the form of slurry, used as compost for agricultural fields. The human and animal excreta are used for biogas production. The `Integrated Health and Energy System Project' for poor countries used for growing *Spirulina*, an alga rich in protein. The sewage is also used for the growth of *Spirulina*. The liquid effluents are added in artificially constructed ponds meant for cultivation of algae. In such system, $NaNO_3$ and $NaHCO_3$ are mixed so as to induce growth of *Spirulina*. The liquid coming out from factories and sewage, after treatement proved beneficial for irrigation purposes. Large quantities of organic wastes are used for cultivation of mycelial forms and other fungal forms. Such waste, if enriched in cellulose (agricultural waste) proved as a potential source for production of enzymes (cellulases and xylanases), proteins, biogas, ethanol, glucose, fructose syrup, etc.

Indiscriminate sewage discharge is destroying the coral reefs by causing dense algal growth which blocks sunlight from corals' dinoflagellates thereby depriving the corals of nutrients.

Urban (city) garbage is a problem in Indian cities. It is estimated that production of city garbage in India is about 41,000 tonnes per day, the annual production is about 15 million tonnes. Nonetheless, city garbage produced in Bombay, Chennai and Kolkata is comparable to that of developed countries. In India, garbage is dumped in the outer skirt of the cities. In some big cities, municipal solid waste compositing plants are in operation (Sharma, 1984). The Bombay and Delhi, Municipal Corporations have already established treatment plants. The Central Mechanical Engineering and Research Institute, Durgapur has established a first pilot plant to produce electricity by using city garbage. The plant has a capacity to use about 500 kg garbage/ha, as a result of which about 5 KW electricity can be generated. Process of electricity production is the conversion of garbage anaerobically into biogas and in turn into electricity. Now-a-days, municipal, agricultural and light industrial wastes are used for conversion into energy by direct burning in refuse fired energy system (Ghose and Bisaria, 1981).

The mixture of wood and bark waste burnt directly is collectively called "hog fuel". The hog fuel combustion technology has been developed in the USA. This fuel is produced in large sized boilers made up of steel. A cogeneration technology has been developed to generate electricity from hog fuel, and to use the exhaust heat in the form of process steam for manufacturing operations.

I. Organic Compost

(*i*) **Definition:** Compost is defined in the Oxford Advanced Learner's dictionary as the process of mixture of decayed organic matter, manure, etc. added to soil to improve the growth of plants. According to Biddlestone *et al.* (1973) composting is the decomposition of heterogenous organic matter by a mixed microbial population in a moist warm aerobic environment. Incomplete microbial degradation of organic waste, where the microbial processes vary from aerobic to anaerobic form are stated as compost (Crawford, 1983).

(*ii*) **Process of Composting:** For making the compost, the crop-residues are degraded in specially designed pits soak to conserve nutrients in a confined environment where cattle dungs, farm wastes (arranged in layers) and urine are allowed to remain for desired periods. The contents are either exposed to air or loosely covered with a mud pack so as to prevent water logging during rainy season. After 6-8 months incubation, the material is ready for use in the farm as an organic fertilizer (Subba Rao, 1986). Most composts deliberately consist of a mixture of pectin, hemicellulose, cellulose and lignin. The compost may contain a number of chemical fractions each with its own decomposition characteristics. Usually the process of compost formation is aerobic. It is intended to produce a product compost which can be used as a organic fertilizer and soil conditioner.

(*iii*) **Factors Affecting Composting:** These are several factors that affect composting.

(*a*) **Microorganisms:** The selection of suitable microbes depends on the type of composting process *i.e.* aerobic or anaerobic, type of raw material, etc. The efficient cellulolytic cultures, such as species of *Aspergillus, Trichoderma, Penicillium* and *Trichurus* accelerate composting for efficient recycling of dry crop wastes with high C:N ratio and reduce the composting period by about 1 month. Enrichment of partially composted crop wastes can be achieved by *Azotobacter* and phosphate solubilizers to improve the nitrogen as compared with controls.

Moreover, the presence of a mixture of anaerobic forms of microorganisms in dung or biogas slurry proved potent in making compost. Actually, the compost carries agriculturally useful microorganisms which aid in the improvement of soil fertility.

(*b*) **Soil:** The soil can be defined as a natural medium for plant growth composed of minerals, organic materials and living organisms. The biological activities and microbial metabolism in the soil contribute to its texture and fertility.

(*c*) **Organic Matter:** The amount of organic matter present in any soil determines its natural suitability for plant cultivation. The value of compost has not only in its N P K content but also in the substantial quantities of humus which are essential for main-

Soil constitutes one of the major factors affecting (composting).

tenance of soil organic matter and fertility levels in tropical and sub-tropical soils (Gaur, 1987).

(*iv*) **Role of Compost:** Composting reduces the soluble nitrogen contents of agricultural wastes. If these wastes are spread directly onto land the highly soluble nitrogen compounds can be easily washed into water courses. Composting also results in phosphorus compounds becoming bound up in new microbial cells so that run-off can be avoided. Higher crop yields have also been claimed for composted versus directly applied animal manures (Thompson, 1977). The incorporation of organic remains in the form of compost, farmyard manure, cereal residues and green manure is known to influence favourably the physical, chemical, physico-chemical and biological properties of soil (Hesse, 1984). Various workers have observed that composting is one of the oldest solid waste treatment methods known to man. Although compost making has been practised since Biblical times but modern interest probably stems from an extended visit to China, Japan and Korea in 1909 by Professor R.H. King of the U.S. Department of Agriculture. His observations were further studied by Sir Alberts Howard, a botanist employed by the Indian government, who developed the Indore composting process. The process is involved the construction of pits (30 feet long, 14 feet wide and 2 feet deep) to conceive heat and moisture, except during the monsoon season when they were built above ground. The heap was composted by building up of plant material, animal manure, dust soil, ash and moisture. The heap was turned after 16, 30 and 60 days and carted to the fields after 90 days. Subsequently, the process was improved and developed. It has been shown that, when compared to inorganic fertilizers of similar nutrient value, compost has increased crop yield by as much as 10%. Further, compost has the additional advantages that they add humus to the soil and improve aeration and water holding capacity to the soil (Biddlestone, 1973).

(v) **Vermicomposting:** Vermicomposting is the operation of composting process of organic materials by involving earthworms. It is a sustainable biofertilizer generated from organic wastes. Vermicompost is an excellent source of nutrients for vegetables, ornamentals, fruits and plantation crops. Using vermicompost one can get 10-15% more crop yield, besides improvement in quality of the products. For the first time, in 1970, vermicomposting was started in Ontario (Canada). In recent years the USA, Japan and Philippines are the leaders of vermicompost producers. So far least attention has been paid in India. But in recent years, Government and nongovernment organisations (NGOs) are trying to popularise the vermicomposting process.

Earthworm, *Eisenia* is involved in vermicomposting.

Through an NGO, Pithoragarh Municipality (Uttaranchal) has started vermicomposting by using a thermotolerant earthworm *Eisenia foetida*. Vermicomposting is also being done at Shanti Kunj (Haridwar). One kg earthworm can consume 1 kg organic materials in a day. They secrete as casting which are rich in Ca, Mg, K, N and available P. Depending on substrate quality, vermicompost consists of 2.5-3.% N,1-1.5% P and 1.5-2.0% K, useful microroganisms (bacteria, fungi, actinomycetes, protozoa), hormones, enzymes and vitamins.

Earthworms make tunnels and mix soil. Thus they aerate the soil which promote the growth of bacteria and actinomycetes. Consequently, microbial activity of soil is increased due to increase in enzymatic and biological activity of earthworms. About 500 species of earthworms are known in India and over 3,000 in the world. The most common members of the earthworm to be used in vermicomposting include: *Eisenia andrie. E. foetida, Dravida willsii, Endrilus euginee, Lamito mauritii, Lubrieus rubellus* and *Perionyx excavatus*.

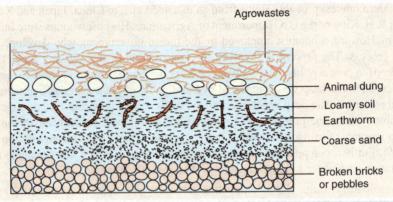

Fig. 30.1 : Diagram of a pit for operation of vermicomposting process.

Process of Vermicomposting: Process of vermicomposting can be done in pits or concrete tanks, wells or wooden crates. A pit of 2×1×1 m³ dimension (1 m maximum depth) is dug under a shade to prevent the entry of water during rain (30.1). Wooden bricks or pebbles are spread on the bottom of pit followed by coarse sand to facilitate the drainage. It is covered by a layer of loamy soil which is moistened and inoculated by earthworms. It is covered by small lumps of fresh or dry cattle dung followed by a layer of hay or dry leaves or agrowastes. Every day for about 20-25 days water is sprinkled over it to keep the entire set up moist. Until the pit is full dry and green leaves are put into the pit in each week. Vermicompost is ready after 40-45 days. Vermicompost appears soft, spongy, dark brown with sweet smelling. Then it is harvested and kept in dark. It is sieved and packed in polythene to retain 20% moisture content.

II. Biogas

Many developing countries are encouraging for the installation of biogas plants to meet out the demand of fuel. India is one of the pioneer countries in biogas technology. U.P. Government in the years 1957 and 1960 established a permanent station named 'Gobar gas Research Station' at Ajitmal in

Earthworms inoculate the loamy soil layer during vermi composting.

Etawah district. There are many other institutions where research and development programmes are carried out such as Khadi and Village Industries Commission, (KVIC), Bombay, the Gram Vikas Sansthan, Lucknow and National Environmental Engineering Research Institute, Nagpur. The Non-conventional Energy Development Agency (NEDA) of Uttar Pradesh has already installed several night soil–based biogas plants throughout the state. One of the major plants installed at Kanpur, is the Rajapurwa Biogas Plant where about 1400 kg human waste from 50 seat toilet complex is pooled through underground pipelines at one place. The waste is processed through a digester into about 55 m³ methane gas. This quantity is enough to run an 8 HP engine and pumpset and to provide fuel to some biogas lamps and cooking burners to about 3500 dwellers. The sewage plant at Okhla

(New Delhi) has 15 digesters of 5665 m³ capacity each, and produces 17,000 m³ gas per day which is equivalent to about 10,000 litres kerosene per day. National sugar Institute, Kanpur has developed methods for production of biogas from bagasse and other agricultural residues. It is estimated that five cattles generate dung to produce 2 m³ biogas plant to meet the demand of cooking and lighting for a family of 4-5 people. Biogas and H_2 based engines are successfully developed.

(*i*) **Benefits from Biogas Plants:** In Asia, biogas is used mainly for cooking and lighting purposes. In addition, there are many other advantages in installing the biogas plants. It is used in internal combustion engines to power pumps and electric generators. Sludge is used as fertilizer. The most economical benefits are minimising environmental pollution and meeting the demand of energy for various purposes.

(*ii*) **Feedstock Materials:** There are two sources of biomass i.e. plant and animal for biogas production. The biomass obtained from plants is aquatic or terrestrial in origin, while biomass generated from animals includes cattle dung manure from poultry, goat, sheep and slaughter houses, fisheries waste, etc. Cattle dung is most potent for biogas production. Besides dung (gobar), agricultural residue, apple pomace and deteriorated or dumped wheat grains are also proved to be good source for biogas production.

(*iii*) **Biogas Production (Anaerobic Digestion)** The anaerobic digestion is carried out in an air tight cylindrical tanks which is called digester. A digester is made up of concrete bricks and cement or steel. It has a side opening (charge pit) into which organic materials for digestion are incorporated. There lies a cylindrical contanier above the digester to collect the gas. A diagram of single stage digester for gobar gas plant is shown in Fig. 30.2. It is noticed that after 50 days, sufficient gas is produced in gas tank, which is used for house hold purposes. Usually, digesters are burried in soil in order to benefit from insulation provided by soil. In cold climate, digester can be heated.

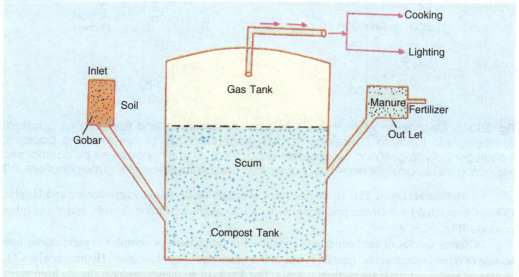

Fig. 30.2 : Single stage digester for gobar gas production.

Anaerobic digestion is accomplished in three stages, solubilisation, acidogenesis and methanogenesis.

(*a*) **Solubilisation:** It is the initial stage when feed stock is solubilized by water and enzyme. The feed stock is dissolved in water to make slurry. The complex polymers are hydrolysed into organic acids alcohols by hydrolytic fermentative methanogenic bacteria which are mostly anaerobes (Fig. 30.3A).

(*b*) **Acidogenesis:** In this stage, the second group of bacteria *i.e.* facultative anaerobic and H_2 producing acidogenic bacteria convert the simple organic material via oxidation/reduction reactions into acetate, H_2 and CO_2. These substances serve as food for the final stage. Fatty acid is converted into acetate, H_2 and CO_2 via acetogenic dehydrogenation by obligate H_2 producing acetogenic bacteria. There is another group of acetogenic bacteria which produce acetate and other acids from H_2 and CO_2 via acetogenic hydrogenation (Fig. 30.3B).

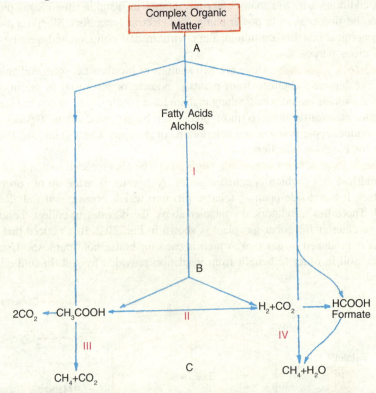

Fig. 30.3 : Mechanism of methane formation; A, hydrolytic and fermentative bacteria; B, acetogenic bacteria (I-acetogenic dehydrogenation by proton reducing bacteria, II-acetogenic hydrogenation by acetogenic bacteria), C, methanogenesis by acetoclastic methanogens i.e., acetate respiratory bacteria (III) and hydrogen oxydizinig methanogens (IV)

(*c*) **Methanogenesis:** This is the last stage of anaerobic digestion where acetate and H_2 plus CO_2 are converted by methane producing bacteria into methane, carbon dioxide, water and other products (Fig. 30.3C).

Different species of methanogens are involved in breakdown of complex organic matter into acetate or other organic acids. Acetate is one of the substrates of methanogens. Hydrogen with CO_2 in general acts as substrate for methanogenesis. The details of methanogens have already been given in Chapter 2.

All the bacteria require H_2 and formate (except *M. bryantii, M. thermoautotrophicum* and *M. arboriphilus*) for growth and methane production, whereas *M. barkerii* requires (besides H_2) methanol, methyl amine and acetate for their growth. Thus, the methanogens are either autotrophs, or utilize simple organic compounds as formate, acetate, and methyl amine and occupy the terminal position in anaerobic food chain.

(*iv*) **Mechanism of Methane Formation:** Metabolic activities of methanogens are quite peculiar. Carbon dioxide fixation, Calvin cycle, serine or hexulose pathways are absent in them.

Several coenzymes such as methyl coenzymes M, hydroxymethyl coenzyme M, coenzyme F420, coenzyme F430, component B, methanofuran or carbon dioxide reducing factor, methanopterin and formaldehyde activating factors are present.

The primary reaction in which carbon dioxide formation occurs is given below.

$$CO + H_2O \longrightarrow CO_2 + H_2$$

The secondary reaction takes place in the presence of sufficient hydrogen.

$$CO_2 + 4H_2 \longrightarrow CH_4 + 2H_2O$$

Other reactions showing methane formation from various substrates are given below:

$$4CH_3OH \longrightarrow 3CH_4 + CO_2 + 2H_2O$$
$$4HCOOH \longrightarrow CH_4 + 3CO_2 + 2H_2O$$
$$CH_3COOH \longrightarrow 12CH_4 + 12CO_2$$

(*v*) **Factors Affecting Methane Formation:** Following are the factors affecting methane production.

(*a*) **Slurry:** Proper solubilisation of organic materials (the ratio between solid and water) should be 1:1 when it is house hold type.

(*b*) **Seeding:** In the beginning, seeding of slurry with small amount of sludge of another digester activates methane evolution. Sludge contains acetogenic and methanogenic bacteria.

(*c*) **pH:** For the production of sufficient amount of methane, optimum pH of digester should be maintained between pH 6-8 as the acidic medium lowers down methane formation.

(*d*) **Temperature:** Fluctuation in temperature reduces methane formation, because of inhibition in growth of methanogens. In case of mesophilic digestion, temperature should be between 30°C and 40°C but in case of thermophilic ones, it should be between 50 and 60°C.

(*e*) **Carbon-nitrogen (C:N) ratio:** Improper C : N ratio lowers methane production. Maximum digestion occurs when C : N ratio is 30 : 1. Amendment of nitrogen or carbon substrates should be done exogenously according to chemical nature of substrate.

(*f*) **Creation of anaerobic conditions:** It is obvious that methane production takes place in strictly anaerobic condition, therefore, the digesters should be totally air tight. In Indian conditions, digesters are buried in soil.

(*g*) **Addition of algae:** Ramamoorthy and Sulochna (1989) have found an enhancement in biogas production on addition of algae, *Zygogonium* species. The amount of biogas produced from the algae was twice (344 ml/g dry algae) in comparison to cow dung (179 ml per g dry cow dung). Therefore, addition of algae holds promise to get biogas in sufficient amount.

B. Sewage (Wastewater) Treatment

Sewage is the used and wastewater consisting of human excreta, wash waters, and industrial and agricultural wastes (e.g. wastes from live stock i.e. chicken, cattle, horse, etc) that enter the sewage system. In general sewage contains about 95.5% water and 0.1 to 0.5% organic and inorganic materials. The solid remains in suspended form in water. The celluloses, lignocellulo-

The outlet pipe in the background is discharging raw sewage in water, thus the need for sewage treatment arises.

ses, proteins and fats are found in colloidal form in water. The other inorganic materials are dissolved and are found in ionic forms. From the industries disposed wastes consist of detergents, antibiotics, paints, biocides, etc. The pulp and paper industries discharge cellulosic and inorganic chemicals. Therefore, sewage composition differs with types of industrial effluents discharged into sewage systems.

(*i*) **The Sewage Microorganisms:** A variety of microorganisms are present in water for example bacteria, fungi, protozoa, algae, nematodes, amoebae and viruses. Some of the common pathogenic microorganisms are given in Table 28.2. The microorganisms are aerobes, obligate anaerobes and facultative anaerobes present in millions in one millilitre of sewage. Most of them are intestinal and soil bacteria. The common bacteria are coliforms, streptococci, micrococci, lactobacilli, clostridia, pseudomonads, etc. The other pathogens are those causing amoebic dysentery, cholera, typhoid fever, bacterial dysentery, polio, hepatitis, etc. (Bitton, 1994). These organisms are heterotrophs and thus decompose organic matter of sewage. In the beginning aerobic bacteria dominate and decompose organic material and at the end anaerobic bacteria predominate (e.g. methanogens) which produce methane (CH_4), carbon dioxide (CO_2) and hydrogen (H_2) gases. By collecting sewage now-a-days several biogas plants are being operated in Bombay, Okhla (Delhi) and Kanpur where biogas is supplied in selected villages for lighting and cooking purposes.

(*ii*) **BOD and COD:** In the initial stage of decomposition of organic materials of sewage, aerobic microorganisms are involved. These completely oxidise the organic materials under aerobic condition rendering into complete stabilization of sewage. During this process, all the dissolved oxygen is consumed for oxidation of organic materials. Consequently there develops anaerobic conditions in water which results in death of water animals and emitting foul smell due to incomplete oxidation of organic materials in sewage. This occurs because the wastes have now high *biochemical oxygen demand* (BOD) required by the aerobic bacteria which the sewage has no more oxygen to bring about oxidation process. Thus, BOD is the amount of oxygen required by organisms in water under certain standard conditions. The sewage of high organic material has high BOD value and *vice versa*. Thus, BOD is the important indication of levels of biological pollution in water, and provides an index of the amount of microbially oxidizable organic matter.

One can measure BOD in the laboratory. Test sample of water is taken in BOD bottles. Initially, the level of dissolved oxygen in the water sample is determined. Thereafter, bottles are stoppered at 20°C. After incubation for five days the amount of dissolved oxygen is again measured. The difference in oxygen values is the BOD which denotes the amount of oxygen utilized within five days. BOD is expressed as parts per million.

On the otherhand, chemical oxygen demand (COD) is the amount of chemical oxidation required to convert organic matter in water and wastewater to carbon dioxide (Montgomery, 1985).

Basically, there are two levels of sewage treatment on the basis of amount of sewage generated by humans: small scale treatment and large scale treatment. Small scale treatment of sewage is done in small homes and rural areas, whereas large scale treatment is done in towns and cities by municipal bodies.

I. Small Scale Sewage Treatment

There are several methods of sewage treatment on small scale. A few of them are described below:

(*i*) **Cesspools:** Human waste is thrown in cesspools in many homes. It is constructed in underground part with concrete in such a way that it contains wall of cylindrical rings with pores (Fig. 30.4). Its opening is near the ground level. Wastewater (sewage) enters the cesspool through the inlet pipe. The bottom of cesspool remains open. Therefore, the suspended solid material falls on the bottom of cesspool and forms sludge after getting deposited in huge amount. Water passes out through the open bottom of cesspool and through pores into the surrounding soil. The organic materials of the sludge are decomposed by anaerobic bacteria resulting in release and deposition of breakdown products on the ground. Thus the amount of breakdown products exceeds; it forms thick layers which need to be cleaned by using strong acids. Dried bacterial preparation of *Bacillus subtilis* or yeast cells should be added at intervals. These accelerate the decomposition of sludge deposited at the bottom of cesspool.

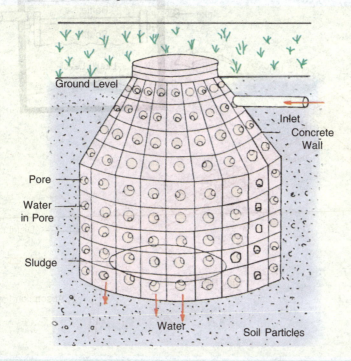

Fig. 30.4 : Diagram of cesspool. Wastes enter through inlet. Water passes into surrounding soil through pores of wall and bottom. Solid materials accumulate at the bottom as sludge.

(*ii*) Septic Tanks: In rural areas, individual family uses septic tank because of lack of public sewers. Septic tank is a metallic or concrete tank which is kept below the ground level some where near the homes. Into septic tanks all the domestic wastes flow through the inlet pipes. A family of four members needs a septic tank of 3 × 5 × 5 feets by accumulating about 750 gallons of sewage. The suspended organic materials are accumulated at the bottom of tank, whereas the water flows through outlets to a distribution box (Fig. 30.5A) which is connected with perforated pipes that open under the soil surface in the surrounding areas. Therefore, effluents from the tank is passed to underground surface of soil. Through this process, the patho-

An aerial view of a modern conventional sewage Treatment plant.

genic microbes are not eliminated. Therefore, the drinking water supply must be kept at certain distance of the pipes of septic tanks.

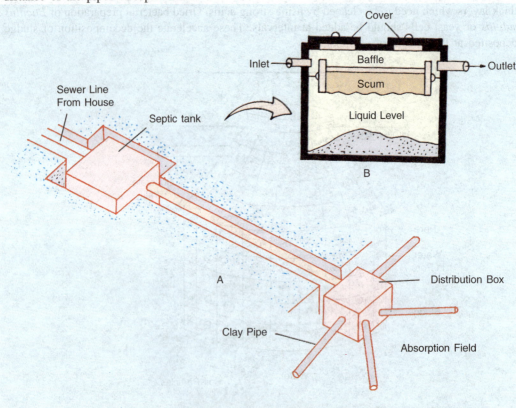

Fig. 30.5 : Installation of septic tank for sewage disposal from houses. A, overall installation including absorption field; B, a view of septic tank where water enters through inlet and solids accumulate as sludge.

The organic materials accumulated in septic tank is decomposed by anaerobic bacteria releasing into water several by-products such as sugars, alcohols, organic acids, amino acids, fatty acids, glycerols and gases (e.g. H_2, H_2S, CH_4, CO_2, etc). However, there remains undigested organic materials called sludge. *Sludge* is removed from the septic tank at certain intervals by pumping process otherwise it will block the tank and pipes. Sludge acts as a source of humus when applied in field.

In addition, in small towns sewage is collected into a large ponds which are called *oxidation lagoons*. The sewage is discharged into oxidation lagoons where organic materials are oxidized first by aerobic organisms and the sediments are decomposed by anaerobic microorganisms.

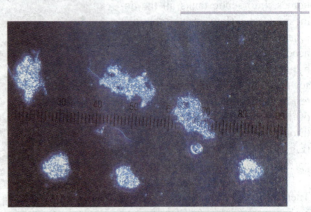

Sludge, the undigested organic material, blocking the septic tanks, needs to be removed at intervals.

II. Large Scale Sewage Treatment

Sewage treatment on a large scale of populations of city is known as large scale sewage treatment. In cities sewage and garbage are generated in massive amount per day which are treated by municipal plants. A schematic view of waste treatment by a municipal plant is shown in Fig. 30.6. Overall processes of conventional municipal sewage plant can be divided into three steps: the *primary treatment, secondary treatment,* and *tertiary treatment.* The primary treatment is viewed for physical separation of insoluble materials, to lower BOD; the secondary separation is based on microbial decomposition of organic materials in the effluents; the tertiary treatment is the chemical removal of inorganic nutrients and pathogenic microbes (Table 30.1).

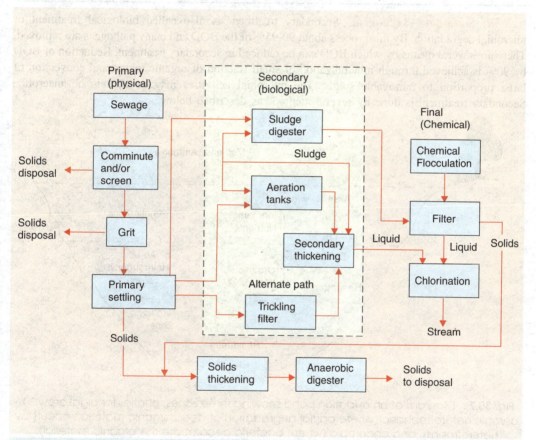

Fig. 30.6 : Flow chart of different stages of sewage treatment.

(*i*) **Primary Treatment:** Primary treatment is the physical removal of 20-30% of organic materials present in sewage in particulate form. The particulate material is removed by screening, precipitation of small particulate and settling in basin or tanks where the raw sewage is piped into huge and open tanks. The solid material (sludge) is removed and kept in landfill/composting for anaerobic digestion. The liquid portion is piped into sludge tanks. The accumulated materials in sludge tanks are subjected to aluminium sulfate or the other coagulants so that the suspended particles, organic materials and microorganisms should be trapped as in sedimentation process of water purification.

Table 30.1 : Major steps in primary, secondary and tertiary treatment of wastes (based on Prescott *et al*, 1996)

Treatment Steps	Processes
Primary	Screening and removal of insoluble particulate materials, addition of alum and other coagulants
Secondary	Biological removal of dissolved organic matter through trickling filters, activated sludge, lagoons, extended aeration systems and anaerobic digesters
Tertiary	Chemical removal of inorganic nutrients, virus removal/inactivation, trace chemical removal

(*ii*) **Secondary Treatment:** Secondary treatment is also called biological treatment or microbial degradation. By this process about 90-95% of the BOD and many pathogens are removed. There are several means by which BOD can be reduced in secondary treatment. Reduction of BOD by 90% is achieved through mineralization of small fraction of organic matter and conversion of large proportion to removable solids. The microbial activities may be aerobic or anaerobic. Secondary treatment is done by several methods as described below:

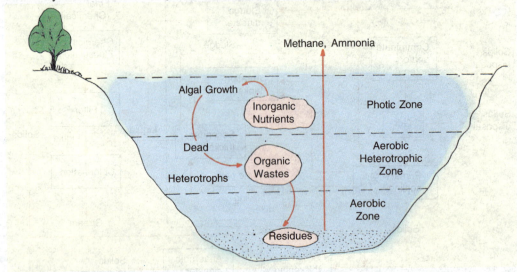

Fig. 30.7 : Diagram of an oxidation pond showing three zones; photic (for algal growth), aerobic heterotrophasic (where partial degradation of dead organic materials occurs by heterotrophs), and anaerobic (where bacteria decompose the organic material).

(*a*) **The oxidation ponds:** The oxidation ponds (also called lagoons or stabilization ponds) permit the growth of algal forms on waste-water effluent (Fig. 30.7). It is used for secondary treatment in rural areas or industrial sectors. The organic materials are degraded by heterotrophic bacteria into simpler forms that in turn support the growth of algae. Algae use these nutrients to increase their biomass. Air supplies oxygen for biochemical oxidation of organics. Oxygen evolved by algae after photosynthesis maintains the oxygen deficit created by heterotrophs. The efficiency of oxidation process can be improved by constructing shallow ponds. The algae growing in oxidation ponds are: *Chlorella pyrenoidosa, C. ellipsoides, Scenedesmus acutus, S. quadricauda, Spirulina platensis,* etc. Secondary treatment through oxidation ponds is the aerobic sewage treatment device.

(*b*) **The trickling filter:** Aerobic secondary treatment also can be carried out with a trickling filter (Fig 30.8). It is a simple sewage treatment device that consists of a bed of a crushed stone, gravel, slag, or synthetic material with drains made at the bottom of the tank. Thus the trickling filter has a pile of rocks over which sewage or organic wastes slowly trickle. A revolving sprinkler (arm) is suspended over a bed of porous material which distributes the liquid sewage over it, and collects the effluents at the bottom. Due to spraying process, sewage is saturated by oxygen.

The porous filter bed becomes coated with slimy bacterial growth mainly by *Zooglea ramigera* and other slime producing bacteria. The slime is colonised by the heterotrophic microorganisms e.g. bacteria (*Beggiatoa alba, Sphaerotilus natons, Achromobacter* spp., *species of Pseudomonas* and *Flavobacterium*), fungi, nematodes, protozoa, etc. These microorganisms form a stationary microbial culture becuase of continuous supply of nutrients present in sewage and metabolising the organic constituents into the more stable end products. Therefore, BOD of effluent is reduced by these microorganisms. The microorganisms get air through porous bed. A newly constructed bed needs a few weeks to function efficiently unless the zoogleal film is coated over it (Bitton, 1994).

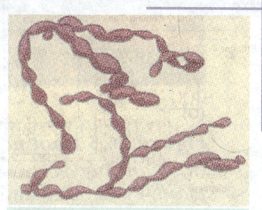

The bacterium *Pseudomonas,* one of the colonisers of bacterial slime during the sewage treatment.

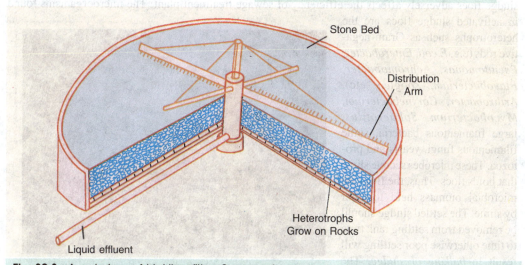

Stone Bed

Distribution Arm

Heterotrophs Grow on Rocks

Liquid effluent

Fig. 30.8 : A cut view of trickling filter. Sewage from primary treatment is spread over rock bed by distribution arm. Water is filtered through rock bed.

(c) The activated sludge: It is also one of the widely used aerobic treatment systems, for waste water in which very vigorous aeration of the sewage is done. The sewage is passed into an aeration tank from primary settling tank. Sewage is aerated by mechanical stirring (Fig. 30.9). Due to vigorous aeration of sewage floc-formation occurs. The colloidal and finely suspended matter of sewage form aggregates which are called floccules. The flocs are permitted to settle down in secondary settling tank. The particles of floc i.e. activated sludge contain large amount of metabolising bacteria together with yeasts, fungi and protozoa.

The activated sludge is introduced in primary settling tank and aeration tank just for rapid development of microorganisms and rapid exploitation of organic matter. This process is repeated i.e. addition of settled sludge to fresh sewage, aeration, sedimentation, addition of settled sludge to fresh sewage, and so on. This repeating process results in complete flocculation of fresh sewage within a few hours. Activated sludge process reduces the BOD of effluent to 10-15% as compared to raw sewage.

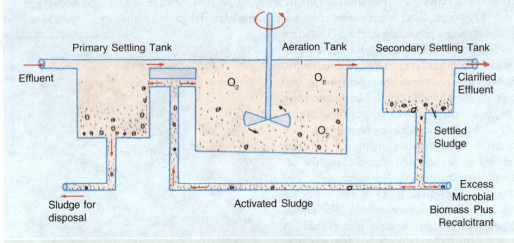

Fig. 30.9 : Aerobic activated sludge : secondary sewage treatment system.

The use of activated sludge hastens the efficiency of system. A poor settlement of activated sludge flocs adversely affects the efficiency of sewage treatment plant. The microorganisms found in activated sludge flocs are the heterotrophs such as Gram-negative rods (e.g. *E.coli, Enterobacter, Pseudomonas, Achromobacter, Flavobacterium, Zooglea,* etc), *Arthrobacter, Corynebacterium, Mycobacterium, Sphaerotilus,* large filamentous bacteria, some filamentous fungi, yeasts and protozoa. These microbes secrete slime that holds flocs. Thus, the flocs are microbial biomass held together by slime. The settled sludge should be removed from settling tank time to time otherwise poor settling will result in *bulking of sludge*. The bulking sludge is caused by massive development of filamentous

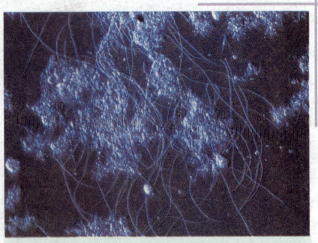

A poor settlement of activated sludge lowers the quality of the final effluent.

bacteria (e.g. *Sphaerotilus, Baggiatoa, Thiothrix*), and filamentous fungi (e.g. *Cephalosporium, Cladosporium, Geotrichum,* etc.). Thus, the settled sludge is permitted to anaerobic treatment and reinoculation of fresh sewage (Cowan *et al.,* 1995).

Advantage of using activated sludge process are : (*a*) significant reduction in BOD and suspended solids, (*b*) reduction in intestinal pathogens, (*c*) requirement of little land, and (*d*) no need of high dilution of final effluent.

(*d*) **Anaerobic digesters:** All the aerobic processes produce excess microbial biomass or sewage sludge which contains many recalcitrant organics. The sludge from aerobic sewage treatment together with the materials settled down in primary treatment are further treated in anaerobic digesters through the process of anaerobic digestion (Fig. 30.10). These digesters are used only for processing of settled sewage sludge and the treatment of very high BOD industrial effluents. Anaerobic digesters are large fermentation tanks designed to operate anaerobically with continuous supply of untreated sludge and removal of final, stabilized sludge product. However, in these tanks provisons are made for mechanical mixing, heating, gas collection, sludge addition and removal of final stabilized sludge. High amount of suspended organic materials with high number of bacterial community (10^9 - 10^{10} CFU/ml) is found. The organics are decomposed by a number of anaerobic microorganisms whose population is found to be 2 - 3 times greater than the anaerobes. Anaerobic digestion involves the following three steps.

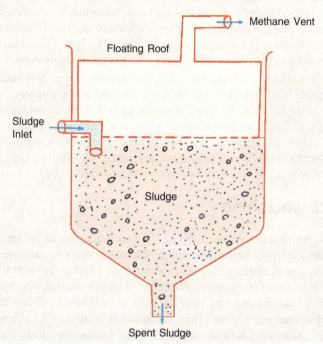

Fig. 30.10 : Anaerobic sludge digester.

Fermentation: The fermentation of sludge components to form organic acids (including acetate) from organic polymers is done by a number of bacteria such as species of *Bacteroides, Clostridium, Peptostreptococcus, Eubacterium, Lactobacillus,* etc. The organic acids are butyrate, propionate, lactate, succinate, acetate along with ethanol and H_2, CO_2, etc.

Lactobacillus, a bacterium involved in the fermentation of sludge component.

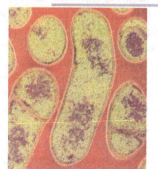

Electron micrograph of *Clostridium,* a bacterium important in sewage breakdown.

• **Acetogenic reactions:** The products (e.g. butyrate, propionate, lactate, succinate, ethanol) produced during fermentation are utilized as substrate by several acetogenic bacteria viz., *Syntrophomonas, Syntrophobacter* and *Acetobacterium*. The products produced as a result of acetogenic reactions are: acetate, H_2 and CO_2.

• **Methanogenesis:** The products produced during acetogenesis are utilized as substrate by methanogenic bacteria. Acetate is used to produce $CH_4 + CO_2$ by *Methanosarcina* and methanothrix. H_2 and HCO_3^- are used to produce methane by several bacteria e.g. *Methanobrevibacter, Methanomicrobium, Methanogenium, Methanobacterium, Methanococcus,* and *Methanospirillum*. A critical balance between oxidants and reductants is maintained during methanogenic processes. The hydrogen concentration must be maintained at a low level so that it can function most efficiently. Upon accumulation of hydrogen and organic acids, methane production is inhibited. Thus, the final product of anaerobic digestion is a mixture of gases (70% CH_4, 30% CO_2), microbial biomass and nonbiodegradable residues (e.g. heavy metals, polychlorinated biphenyls, etc.).

(*iii*) **Tertiary Treatment:** Tertiary treatment is aimed to remove non-biodegradable organic materials, heavy metals, and minerals. The salts of nitrogen and phosphorus must be removed because they cause eutrophication. By using activated carbon filters the organic pollutants can be removed, whereas by adding lime the phosphorus is precipitated as calcium phosphate. Nitrogen can be removed by stripping, volatilization as NH_3 at high pH values. Ammonia can be converted by chlorination to dichloromine which in turn is converted to N_2. Tertiary treatment is expensive, therefore, it is not employed unless very necessary (Montgomery, 1985).

C. Microbial Leaching

Algae and oligotrophous nonspore forming bacteria belonging to the genera *Pseudomonas, Corynebacterium* and *Arthrobacter* predominate in the rocks. Many of them are capable of fixing atmospheric nitrogen. Rocks disintegrate mainly along cracks and at the surface. Accumulation of organic substances, leaching of elements and formation of clayey minerals occur through weathering of rock-crust which are accompanied by large deposits of bauxites, kaolinite clays and nickel, are more complex land microenvironments. Their formation is related to the degradation of original rocks under certain oxidative/reductive conditions. The weathering crusts are characterised by general distribution of autotrophic bacteria which oxidise reduced compounds of nitrogen, sulphur and iron. Nonspore forming bacteria prevail among heterotrophs, particularly belonging to the genera *Arthrobacter, Corynebacterium* and *Mycobacterium* in a pure culture which assimilate phenols, ethanol and other alcohols besides growing on nitrogen-poor rocks. The microflora of sedimentary rocks in the zone of hypergenesis is more diverse and the geochemical activity of various bacterial groups depends upon the reduction-oxidation conditions of the medium.

It is interesting to note that microbes are involved not only in the leaching and migration of elements but also in the formation of minerals under present conditions (Table 30.2). The second source of organic matter for heterotrophs in rocks is metabolites of photo- and chemolitho-trophs. Algae and thiobacilli are known to liberate 5-50 and 20-50% of fixed carbon, respectively, as organic compounds into the medium. A possibility of growth and biomass accumulation of heterotrophs in cultures of thiobacilli and nitrifying bacteria has been proved beneficial. Hence, it is obvious that rocks contain organic substances at concentrations which considerably exceeds those required for the beginning of growth of microbial cenoses.

Metal ores are present in various forms in nature. Rocks are the rich source of various metals. Ore deposits are characterized by a considerable concentration of elements, indicating that biogeochemical process of their transformation. Of course, ore deposits are an excellent ecological niche for the activity of a specific autotrophic microflora. Thiobacilli utilising the energy of oxidation of iron, sulphur and its reduced compounds predominate in the deposits of sulphur and sulphide ores. *Leptospirillum ferrooxidans* ($Fe^{2+} \rightarrow Fe^{3+}$), *Thiobacillus organoparus* and *T. acidophilus* ($S \rightarrow SO_4$) are found in ore deposits under the favourable physicochemical conditions; elements are leached from ores and transferred with ore waters. Leaching often proceeds on such a large scale that it can be a source for obtaining considerable amount of metals.

The microbial leaching process starts with the acid reaction in which ores are percolated down, rich in the mineral, which is collected and allowed to precipitate, purified and recovered.

1.   Some Examples of Leaching

(i) **Copper Leaching:** Copper leaching is practical, feasible and in use throughout the world for many years. In this process, the leaching solution contains sulphate and iron carries the microbial nutrients in and dissolved copper out. The copper containing solution is precipitated. About 5-6% of the world copper production is obtained via microbial leaching of chalcocite, chalcopyrite or covellite. Chalcopyrite is oxidised as given below :

$$2CuFeS_2 + 8\tfrac{1}{2}O_2 + H_2SO_4 \longrightarrow 2\ CuSO_4 + Fe_2(SO_4)_3 + H_2O$$

Covellite is oxidised as below:

$$CuS + 2O_2 \longrightarrow CuSO_4$$

(ii) **Uranium Leaching:** This is an indirect process in which the microbes do not interact with uranium ore directly but act on iron oxidant. Ferric sulphate and sulphuric acid can be obtained by *Thiobacillus ferrooxidans* from the pyrite within the uranium ore. The reaction is given below:

$$2FeS_2 + H_2O + 7.5O_2 \longrightarrow Fe_2(SO_4)_3 + H_2SO_4$$

In this process, the dissolved uranium is extracted from the leach liquor with organic solvents such as tributyl phosphate and the uranium is subsequently precipitated. Another recovery process is the use of iron exchanges.

The uranium process is quite significant because of moving such a vast amounts of uranium. In the leaching process insoluble tetravalent uranium is oxidixed with hot H_2SO_4/Fe_3^+ solution to soluble hexavalent uranium sulphate. The reaction is as below:

$$UO_2 + Fe_2(SO_4)_3 \longrightarrow UO_2SO_4 + 2FeSO_4$$

Table 30.2 : Microorganisms involved in respective biogeochemical processes (rock weathering) and leaching

Microorganisms Involved	Process
Thiobacillus ferrooxidans *T. thiooxidans*, other thiobacilli	Oxidation of organic substances and pyrite, leaching of elements; formation of clayey minerals.
Nitrifying bacteria, *Thiobacillus* sp., heterotrophic microbes	Oxidation of NH_4^+, NO_2, pyrite, organic-substances, degradation of serpentines, leaching of Mg, Ni, Si, Ca and other elements.
Mycobacterium, Arthrobacter Corynebacterium, Pseudomonas, Azotobacter, nitrifying bacteria, *Desulfotomaculum, T. ferrooxidans T. thiooxidans,* thiobacilli similar to *T. thiopans.*	Oxidation of organic substance and reduced sulphur compound degradation of pigmatite,leaching of Li, Al, Si and Fe

D. Biodegradation

Biodegradation is a microbiological process where complex polymers are broken down into less harmful forms, and in turn utilized by them as a source of energy.

1.   Biodegradation of Petroleum (Hydrocarbon)

Petroleum and its products are the hydrocarbons. It is a rich source of organic matter and is oxidised if comes in contact with air and moisture. There are some microorganisms which cleave the hydrocarbons into simpler molecules. *Bioremediation* is an important process now-a-days used for abatement of pollution. In this process oil or other pollutants are utilized by microorganisms if added with inorganic nutrients. The importance of bioremediation in oil spill of marine (sea) environment is widely studied.

Fungi and bacteria are the main agents which decompose oil and oil products. Besides, cyanobacteria, yeast and algae have shown to oxidise hydrocarbons. The simplest hydrocarbon pollutant is methane. It is degraded by a specialized group of bacteria called *methanotrophic* bacteria. As you know, oil is insoluble in water and is less dense; it floats on the surface and forms slicks or oil films. Hydrocarbon–oxidising microorganisms develop rapidly in such films. Oil is present both in anoxic (absence of O_2) as well as oxic (presence of O_2) environment as is evidenced by the presence of natural oil deposits.

Many pseudomonads, different cyanobacteria, various corynebacteria and mycobacteria are able for degradation of petroleum products. Initially, the nonvolatile components are oxidised by bacteria and in the later process, certain fractions of branched-chain and polycyclic hydrocarbons are degraded slowly. Sometimes, it has an impact on fisheries.

It is important to mention that addition of inorganic nutrients such as phosphorus and nitrogen to oil spill increase bio-remediation rates significantly. The microbial production of hydrocarbon occurs in colonial alga, *Botryococcus braunci,* by secretion of long-chain hydrocarbons (C_{30} to C_{36}) that have the consistency of oil. An increasing interest has been shown in using this type of microbe as renewable sources of production of petroleum.

Biotechnological Approaches for Abatement of Pollution: In recent years experimentations on biotechnology have been done for the production of potential microbial strains capable of degrading pollutants. Dr. Anand Mohan Chakraborty (an India-borne American Scientist) succeded in producting a genetically engineered strain of *Pseudomonas putida* that utilized complex chemical compounds. It was called as *super bug*. For detail discussion see Chapter 11, *Gene Cloning in Microorganisms.*

Microbial biogeochemistry has far reaching importance in the area of bioleaching of metals. The bioleaching process allows recovery of about 70% of the mineral from low grade ores as in the case of copper. The application of *Thiobacillus ferrooxidans* population helps to recover this metal. It involved the biological oxidation of copper present in these ores to produce soluble copper sulphate. The copper sulphate can be recovered by reacting the leaching solution. Sometimes, bioleaching process requires addition of nitrogen and phosphorus, if these are low in ores. These added minerals enhance the solubilisation process. Vernadskii (1934) thought about the possibility of solubilization of silicates by soil microorganisms to liberate various cations of silicate elements. Aleksandrov and Zak (1950) isolated certain bacteria capable of decomposing alumino silicate, which were named as "silicate bacteria" such as *Proteus mirabilis*. Microorganisms interact with silicate materials either by decomposing and solubilizing of silicate materials or utilize silica in a dissimilatory fashion by incorporating it into their cells or bodies in the dissolved form, releasing it as free silicic acid. They also assimilate silica by taking it up in dissolved form. *Sarcina ureae* is known to release silicon from quartz occurred as a result of alkalification of the medium.

Significant quantities of Si, Al, Fe and Mg were reported to be solubilized by *Penicillium simplissimum* WB-28 from dunite, peridotite, basalt, granite and quartzite rocks due to the production of citric acid. The destruction of apophyllite, olivine and Ca and Zn silicate by *Pseudomonas* and other soil organisms was found to be accompanied by the releases of ketogluconic acid and other organic acids.

Solubilization of silica from diatoms has been found to be associated with bacterial activity which in many cases has been reported to be due to hydrolytic enzymes. Biodegradation of different aluminosilicates for the recovery of aluminium has been widely studied. Acidolysis, complexolysis and alkalolysis are considered to be the acting mechanisms depending upon the type of the metabolite secreted. The action of `silicate bacteria' on aluminosilicates has been connected with the formation of mucilagenous capsules as well as with the production of different metabolites such as organic and amino acids. The ability of heterotrophic bacteria, *B. mucilaginous* to degrade silicate and aluminosilicate minerals has been used to develop a technological scheme for the dressing of low grade magnesite and bauxite. *Bacillus lichenoformis*, isolated from magnesite ore deposit, uptake silica and silicon which was restricted to adsorption onto bacterial cell surface rather than an internal cell surface uptake through membrane. Recently, Halder *et al.* (1993) reported different strains of root nodule bacteria, *Rhizobium* and *Bradyrhizobium*, capable of solubilising silicates from different synthetic silicates.

In a country like India with its vast unexploited mineral potential, bioleaching assumes a great national significance.

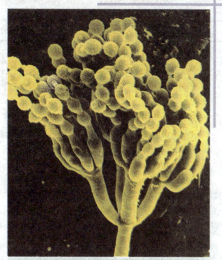

Penicillium solubilizes significant quantities of Si, Ai, Fe and Mg by the production of citric acid.

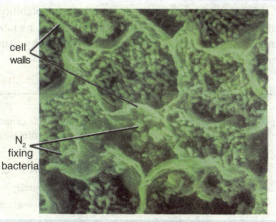

cell walls

N₂ fixing bacteria

The nitrogen fixing bacterium *Rhizobium*, the strains of which have been found capable of solubilising silicates.

2. Microbial Degradation of Xenobiotics

Xenobiotics are those chemicals which do not exist in nature. These are man-made, synthesized compound such as pesticides. Pesticides are toxic chemicals which act by interfering with microbial reactions in the target organisms. Since most of the time, these are added in soil and may affect those microorganisms which are important in maintaining soil fertility. Such organisms also detoxify pesticides in soil. Thus any chemical which seriously affects the soil microflora may harm soil fertility and crop production. Pesticide loss can also occur by volatilization, leaching or spontaneous break down.

The microorganisms that are able to metabolize pesticides and herbicides are given in Table 30.3.

Table 30.3 : Microorganisms and pesticide degradation products

Name of microbe	Chemical	Degradation product
Clostridium sporogenes	γ-BHC	1,2,3,5 - tetrachlorobenzene
Pseudomonas putida	γ-BHC	γ-BTC & α-BHC
Yeast	DDT	1, 1-dichloro-2-2 bis (p-chlorophenyl) ethane [TDE or DDD]
Proteus vulgaris	DDT	TDE
Aspergillus niger, Penicillium chrysogenum	Aldrin	Dieldrin
Trichoderma viride, Mucor plumbeus	Malathion	dyfoxon

These are diverse group of microorganisms which are able to metabolize pesticides and herbicide, including genera of both fungi and bacteria.

(i) **Characteristics of Microbial Metabolism:** Most of the metabolic activities in the microbial world are meant for production of energy. Most of the organic molecules can serve as a source of energy to atleast some microbes. However, a few groups of chemicals are foreign to microorganisms. Among insecticidal compounds, the halogen-containing chemicals must be regarded as foreign, or unusable material as such for microorganisms.

Microorganisms, if mutated may have adaptability towards chemicals that are initially toxic to them. In such cases the pesticide–degrading metabolic activities become higher. In general microbes alter the pesticide degradation process by using several mechanisms. Some of them are given below (Table 30.4).

Table 30.4 : General classification of microbial metabolism of pesticides

A. Enzymatic

    1. Incidental metabolism of pesticides which cannot serve as energy sources

    (a) Wide-spectrum metabolism involving hydrolases, oxidases, etc.

        (i)   Insecticides as substrates

        (ii)   Insecticides as electron acceptors or donors

      (b) Co-metabolism of compounds structurally very similar to the natural substrate

    2. Catabolism: Insecticides serve as energy sources.

    3. Detoxification: Serving as resistance mechanism

B. Non-enzymatic:

    1. Photosynthetic break down

    2. Contribution via pH change

    3. Production of organic and inorganic reactants

    4. Production of cofactors

(a) **Enzymatic process:** It is most important to know whether the microbe derives energy from the process or not. It is generally possible that incidental metabolism is more prevalent form of microbial metabolism when the amount of pesticide is low in comparison with other carbon sources. The catabolic metabolism could occur when the amount of pesticide is high, coupled with favourable chemical structure of pesticide that allows it to be microbially degradable and utilizable as a carbon source.

Through the studies of chlorinated aromatic pesticides, it may be possible to select a microbial strain capable of degrading the pesticides by enriching the medium with a non-chlorinated analogue of the pesticide. By such approach, even very stable and usually non-degradable pesticides might be made susceptible to microbial attack.

(*b*) **Non-enzymatic process:** Some pesticides are photochemically altered in the environment. There are two ways in which microbial products can promote photochemical reactions. In the first case, microbial products can act as photosensitizers by absorbing the energy from light and transmitted to insecticidal molecule. In another case, microbial products can facilitate photochemical reactions by serving as donors or acceptors.

Very limited informations are available on the importance of the microbial formation of organic products capable of reaching with pesticides. Such reactants can be postulated to include amino acids, peptides, alkylating agents organic acids, vitamins, etc. Insecticidal chemicals are known to react with amino acid particularly with an -SH moiety.

Finally, microbial production of cofactors which are used in both enzymatic and non-enzymatic reactions should not be overlooked. Co-factors are those which promote the overall reactions involving an organic chemical without becoming a part of the chemical reaction product derived from that chemical.

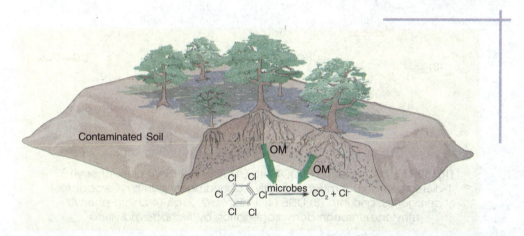

Microbes efficiently carry out metabolic processes when organic matter (OM) is released from the plant roots leading to enhanced degradation of pesticides.

(*ii*) Common Processes of Insecticidal Metabolism

(*a*) **Hydrolytic processes:** Most of the microbial activities are based on hydrolytic processes. It does not occur in other biological group. For example, major metabolic product in *Trichoderma viride* is the phenol of mexacarbate (*i.e.* the hydrolysis product) as compared with metabolism in animals which gave various oxidation products. The reason for such hydrolytic reactions being common in the microbial world is that many of the microbes secrete hydrolytic enzymes exogenously as in the case of fungi. Nearly all the exoenzymes secreted by microbial cells seen to be related to the metabolism of large molecules.

(*b*) **Reductive systems:** The major conversion product of parathion is aminoparathion. This is due to microorganisms. On the otherhand, products of oxidative reactions is such as para-oxon diethyl-thiophosphoric acid which occurs due to animal metabolism. Another important microbial reaction on insecticide is reductive dechlorination. The reaction proceeds by replacing a chlorine atom on a non-aromatic carbon with a hydrogen atom. The best known case being the conversion of DDT to TDE and DDE (Fig. 30.11).

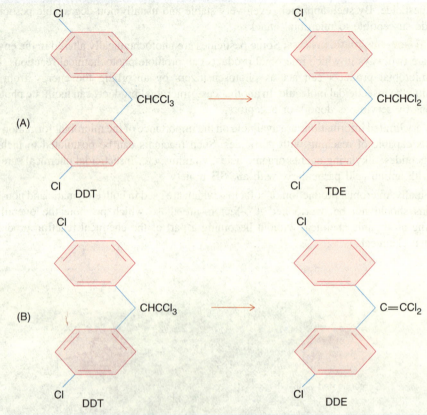

Fig. 30.11 : Degradation of DDT into (A) TDE (2, 2 bis-(4-chlor-phenyl)-1, 1-dichloroethane) through reductive dehydrogenation by *Aerobacter aerogenes*, and into (B) DDE (1, 1-dichlor-2, 2, bis-(4-chloro-phenyl) -ethylene) through dehydrogenation by *Trichoderma viride*.

Other insecticides which are known to go through such dechlorination reactions are gamma-BHC.

(*c*) **Oxidation:** There are several oxidative reactions such as epoxidation of cyclodienes such as altruis and hepatachlor to corresponding epoxides, oxidation of thioethers to sulphoxides and sulphones, oxidative dealkylation of alkylamines, aromatic ring opening, decarboxylation, etc.

The important aspect is to identify one by one the key metabolic reactions and the stable end products in order to provide the necessary informations to understand the processes, tendencies and role of microorganisms in altering the character of the important group of environmental pollutants.

E. Microorganisms in Abatement of Heavy Metal Pollution

The metals and their compounds no doubt are indispensable to the safety and economy of most of the nations but also considered as key factors in the liberation of modern civilization from hunger, disease and discomfort in human life. But after long spell of silence, it has now been realized that the liberal and indiscriminate use of metals since the beginning of civilization has resulted in ecological panic to all biotic components of the earth. Most of the heavy metals such as cadmium, vanadium, manganese, arsenic, lead, iron, mercury, zinc, nickel, chromium, copper, antimony, etc. are toxic to the biological system, not only due to their inert nature but also due to their persistence and long run cumulative effects. The occurrence of "Itai Itai" disease due to cadmium, "minamata" disease due to mercury, miscarriage in women due to lead, cancer of skin, lungs and liver due to arsenic are some of the examples of heavy metal pollution. In many cases, the causes were mostly due to the industrial discharges or irresponsible handling of the products or substrates of the industries.

(*i*) **Heavy Metal Tolerance in Microbes:** Fungi, bacteria, actinomycetes, algae etc. are the major microorganisms reported to tolerate heavy metals which universally occur in diverse ecosystems with meaning frequency.

(*a*) **Algae:** The species of *Chlorella, Anabaena inaequalis, Westiellopsis prolifica, Stigeo clonium tenue, Synechococcus* species, *Selenastrum capricornutum* etc. tolerate heavy metals.

(*b*) **Bacteria:** Bacteria resistant to heavy metals were frequently isolated from environmental sources such as soil and water. Resistance to mercury compounds being a common property of both Gram-positive and Gram-negative bacteria. *Staphylococcus aureus* a Gram-positive organism was intermediate in heavy metal sensitivity between *E.coli* and *Pseudomonas aeruginosa* which are Gram-negative organisms. The other bacterial genera such as *Bacillus cereus, Mycobacterium scrofulaceum, Streptococcus agalactiae, Streptomyces lividans, Thiobacillus ferrooxidans, Pseudomonas aeruginosa, Yersinia enterocolitica, Staphylococcus aureus,* etc. were reported to tolerate both cadmium and mercurry. Bacterial species of *Arthrobacter, Bacillus, Brevibacterium, Corynebacterium, Nocardia, Serratia,* etc. absorb mercury and lead alongwith other heavy metals in the solution (Nakajima and Sakaguchi, 1986). *Alcaligenes faecalis* tolerate zinc and cadmium when grown in nutrient broth (Prahalad and Seenayya, 1988).

(*c*) **Actinomycetes:** *Actinomyces flavoviridis* and several species of *Streptomyces* exhibited high ability to absorb mercury and lead alongwith the other heavy metals from mixed metal solution of manganese, cobalt, nickel, copper, zinc, cadmium, mercury, lead and uranium. *Actinomyces levoris,* and *S. viridochromogenes* were shown to accumulate a large amount of uranium from aqueous systems (Horikoslhi *et al.* 1981).

(*d*) **Fungi:** Heavy metal tolerance is a regular phenomenon exhibited by a number of fungal species. Studies have revealed that fungi accumulate heavy metals from dilute background concentrations. Lead and copper were more readily accumulated by fungi and actinomycetes in comparison to zinc, managenese, cobalt, nickel and cadmium which make selective accumulation of these heavy metals by fungi different from many bacteria and yeasts. Yeasts are least sensitive to heavy metals (Avakyan, 1987). *Trichoderma viride, Aspergillus niger* and *A. giganteus* tolerate nickel concentration but showed prolonged growth and inhibited spore formation and spore germination. Nakajima and Sakaguchi (1986) reported the selective absorption of mercury and lead from a mixed metal solution alongwith the other heavy metals by *A. niger, A. oryzae, Chaetomium globosum, Fusarium oxysporum, Giberrella fujikuroi, Mucor hiemalis, Neurospora sitophila, Penicillium chrysogenum, P. lilacinum* and *Rhizopns oryzae* besides yeast species of the genera *Candida, Hensenula, Saccharomyces* and *Torulopsis.*

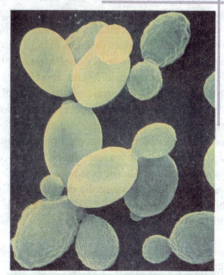

Yeast species of the genera *Saccharomyces* showing heavy metal tolerance.

Dubey and Dwivedi (1985) reported *Macrophomina phaseolina*-tolerant to 500 ppm of cadmium *in vitro*. However, toxicity of this metal was influenced by kaolinite, pH and the presence of zinc and manganese in the medium. Cd toxicity increased with increase in pH, and decreased with increase in concentration of Zn and Mn.

(*e*) **Higher plants:** Decontamination of soils polluted with heavy metals is possibly one of the most intractable of environmental problems. Soils become contaminated in this way either naturally due to proximity to metal, outcrops, or as a result of mining, industry and the dumping of waste.

There are some reports about the potential of certain plants capable of supplying a `green' solution to this problem. Plants capable of accumulating high concentrations of metals such as zinc, nickel, cadmium, lead, copper and cobalt could provide an effective and practical method of cleaning-up heavily polluted soils. There is an excellent potential for using hyperaccumulater plants to remove metals through this process. Certain types (e.g. *Thlaspi caerulescens* of family Brassicaceae) has been found to be a strong hyper-accumulator of zinc and cadmium. Further research will help to identify the faster growing and most strongly metal-accumulating genotypes; explore the possibility of genetic engineering to improve metal uptake characteristics.

Certain plant species of family Brassicaceae such as *Brassica oleracea* cv Greyhound, cabbage, *Raphanus sativus* cv French Breakfast, radish, *Thlaspi caerulescens*, have been reported as a strong Zn accumulator; *Alyssum lesbiacum* and *A. murale*, both are known as Ni hyperaccumulators from serpentine soils in Greece, and also from *Arabidopsis thahania*, a widespread species.

One aspect needed further study is the disposal of the potentially hazardous biomass produced by an effective hyperaccumulator crop. It is suggested that one option could be the controlled ashing of harvested material to yield a residue in which metals such as Zn and Ni may be concentrated by more than 10 per cent.

(*f*) **Cyanobacteria:** Several species of microalgae including the green alga, *Chlorella* (Aksu and Kutsal, 1991), blue green alga, *Anabaena* (Mallick and Rai, 1994), marine algae (Holan *et al.*, 1993), bacteria (Aksu *et al*, 1991), masses (Coloras *et al*, 1992), and macrophytes (Mallick *et al.*, 1996) have been used for heavy metal removal. However, monospecificity and good operational conditions are some of the prerequisites, difficult to maintain, that limit the practical application of these organisms. Recently, Rai *et al.* (1998) studied biosorption of Cd^{2+} and Ni^{2+} by a capsulated nuisance cyanobacterium, *Microcystis* both from field and laboratory. The naturally occurring cells showed higher efficiency for Ni^{2+} and Cd^{2+} biosorption as compared to laboratory cells.

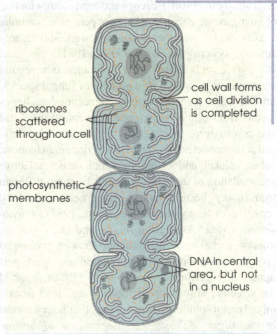

ribosomes scattered throughout cell

cell wall forms as cell division is completed

photosynthetic membranes

DNA in central area, but not in a nucleus

Anabaena, a filmentous blue-green alga, has been used for heavy metal removal.

(*ii*) **Mechanism of Heavy Metal Resistance:** There are four main physiological mechanism of heavy metal resistance: (*a*) inactivation, (*b*) impermeability, (*c*) bypass, and (*d*) altered target site(s).

On the otherhand, earlier studies (Ashida, 1965) reported that fungi tend to survive heavy metal stress either by adaptation or mutation. Not only fungi but few algae such as *Stigeoclonium tenue* can grow in a zinc–rich effluent and showed slightly greater resistance under *in vitro* conditions due to possible adaptation. Enzymic flavoprotein is believed to play a key role in the mercury detoxification system of many bacteria. Bacterial strains which additionally detoxify organomercurials possess another enzyme organomercurial lyase which cleaves the covalent carbon mercury bond to release Hg^{2+} which is then volatilized by the action of mercuric reductase.

Many plasmids responsible for heavy metal toterance were reported in bacteria. Mediation of penicillinase plasmid in importing resistance to divalent metal ions of mercury and cadmium is reported in *Staphylococcus aureus*. It has been suggested that heavy metal resistances may have been selected in earlier times, and that they are merely carried along today for selection for antibiotic resistances. In Tokyo in 1970s both heavy metal resistance and antibiotic resistance were found with high frequencies in *Escherichia coli* isolated from hospital patients, whereas heavy metal resistance determinants were found in *E. coli* from an industrially polluted river. Selection occurs for resistances to both types of agents in the hospitals, but only for resistance to toxic heavy metals in the river environment. The major recent progress has consisted of the cloning and DNA sequence analysis of determinants for mercury, arsenic, cadmium, and tellurium resistance and the initial reports of still additional resistances.

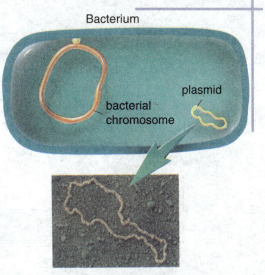

Bacterial plasmids encode enzymes for biodegradation of specific compounds.

F. Water Pollution Management

The level of water pollution is increasing in large cities in many countries. Moreover, efforts are being made to minimise water pollution levels and apply biotechnological tools to wastewater treatment. Japan has applied genetic engineering methods and improved bioreactors for wastewater and industrial waste treatment.

 ## 1. Use of Commercial Blends of Microorganisms/Enzymes in Wastewater Treatment

(i) **Bioaugmentation (Use of Blends of Microorganism):** Acceleration of biodegradation of specific compounds by inoculating bacterial cells is called **bioaugmentation.** Bacterial cells contain specific plasmid which encodes enzymes for degradation of those compounds. A variety of plasmids have been reported from *Alcaligenes, Acinetobacter, Arthrobacter, Beijerinkia, Klebsiella, Flavobacterium* and *Pseudomonas*. Several genetically engineered strains have been developed exploiting *Pseudomonas*.

Microroganisms capable of degrading herbicides/other chemicals in industrial water are isolated from wastewater, compost, sludge, etc. Some of the strains may be irradiated to enhance their ability and mutants are selected (30.12) Before their use in the environment they are tested in laboratory for their biodegradation ability. Bioassays are also used to assess the toxicity of the wasterwater for commercial preparation of microbial seeds. Selected strains are used in large fermentor to get mass culture. Then they are preserved through lyophilization, drying and freezing.

Commercial bioaugmentation products are single culture of consortia of microroganisms with certain degradative properties or their desirable characters. At present most important users are the industrial wastewater treatment plants. The selected microorganism is added to a bioreactor so that potential for biodegradation of wastes must be maintained or enhanced. Due to trade secrets information on bioformulation of mixture of microbial cultures are not scanty.

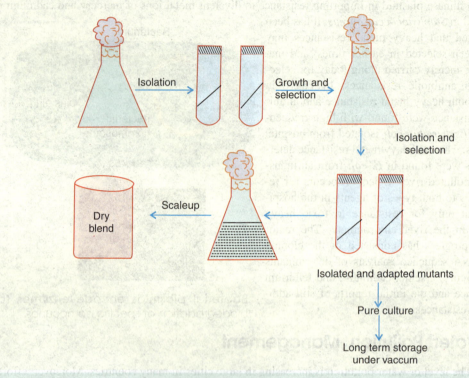

Fig. 30.12 : Isolation and purification of microbial blends used for pollution control.

Application of biaugmentation includes : *(a)* the increased BOD removal in wastewater treatment plants, *(b)* reduction of sludge volume by about 30% after addition of selected microorganisms, *(c)* use of mixed cultures in sludge digestion, *(d)* biotreatment of hydrocarbon waste, and *(e)* biotreatment of hazardous wastes.

The use of added microorganisms for treating hazardous wastes such as phenol, ethylene glycol, formaldehyde has been attempted. Bioaugmentation with parachlorophenol-degrading bacteria decomposed 96% para-chlorophenol in 9 hours. Cells of *Candida tropicalis* have been used for removal of high concentration of phenol present in freshwater. Ability of a bioreactor to dechlorinate 3-chlorobenzoate was increased after addition of *Desulfomonile tiedjei* to a methanogenic upflow anaerobic granular sludge banket. Anoxygenic phototrophic bacteria have also been considered for the degradation of toxic compounds in wastes (Bitton, 1999).

Some demerits of bioaugmentation are: *(a)* need of an acclimation period prior to onset of biodegradation, *(b)* a short survival or lack of growth of microbial inocula in the seeded bioreactors, and *(c)* some times negative or non-conclusion of some of commercial products.

(ii) **Use of Enzymes in Wastewater Treatment:** Several enzymes have been detected in wastewaters such as catalase, phosphate esterases and aminopeptidases. These enzymes can be added to freshwater to improve biodegradation of xenobiotic compounds. For example parathion hydrolases (isolated from *Pseudomonas* and *Flavobacterium*) have been used to cleanup the containers of parathion and detoxification of wastes containing high concentration of organophos-phates. A little information is available on use of enzymes in wastewater treatment plants. This technology is applied to reduce the production of excessive amount of extracellular polysaccharides during wastewater treatment because overproduction of polysaccharides results in increased water retention with reduced rate of dewatering. Addition of enzyme can degrade these expolymers.

Some specific enzymes (e.g. horseradish peroxidase) can catalyse the polymerisation and precipitation of aromatic compounds (e.g. substituted aniline and phenols). Horseradish peroxidase catalyses the oxidation of phenol and chlorophenols by hydrogen peroxide. The extracellular fungal laccases (obtained from *Trametes versicolor* or *Botrytis cinerea)* can be used for the treatment of effluents generated by the pulp and paper industry because this enzyme can be useful for dechlorination of chlorinated phenolic compounds or oxidation of aromatic compounds even at adverse environmental conditions such as low pH, high temperature, presence of organic solvents, etc. Therefore, attention has now been paid to use **extremozyme** of microorganisms that can work at extreme environments also.

2. Use of Immobilised Cells in Wastewater Treatment

There are various methods to immobilise the microbial, plant and animal cells such as *(a) entrapment* (immobilisation of cells in polymeric materials e.g. alginate, carrageenan, polyacrylamide, polyurethrane foam, etc., *(b) adsorption* (on sand beads, porous silica, porous brick or wood), *(c) covalent binding* (using hydroxymethyl acrylate). Entrapment is the most popular approach. Immobilised cells have been used for the treatment of various wastes for decontamination of water or wastewater containing natural or xenobiotic compounds and decontamination of soils and aquifers. Some of the following examples of pollution control has been discussed by Bitton (1999).

(a) **Removal of brown lignin compounds:** Brown lignin compounds are found in paper mill effluents. Immobilised white-rot fungus *(Coriolus versicolor)* can be used to remove this compounds.

(b) **Biodegradation of phenolic compounds:** Much information is available on degradation of phenolic compounds by using bacteria. Immobilised cells of *Pseudomonas, Arthrobacter* and *Alcaligens* degrade chlorinated phenols. Bioreactors containing *Flavobacterium* immobilised in calcium alginate can degrade pentachorophenol at maximum rate (i.e 0.85 mg/g beads/ hours. Immobilised tyrosinase can remove rapidly phenol, chlorophenol, methoxylphenol and cresols from fresh water. A polycation derived from chitin (chitosan) can remove efficiently the coloured products from the effluents.

(c) **Methane production by Immobilised methanogens:** Anaerobic waste treatment can be enhanced using immobilised methanogens in two-staged bioreactors.

(d) **Dehalogenation of chloroaromatics:** Immobilised cells of *Pseudomonas* sp. removes chlorides linked to aromatic compunds.

(e) **Immobilised activated sludge microorganisms:** A high treatment efficiency was achieved with a two step process consisting a reactor containing immobilised activated sludge microorganisms followed by a biofilm reactor.

(f) **Use of immobilised algae to remove micronutrients from wastewater effluent:** Immobilised *Phormidium* or *Scenedesmus* removes nitrogen and phosphorus from wastewater effluents.

(g) **Use of immobilised cells/enzymes in biosensor technology:** Biosensor is a device consisting of a wide range of biological elements and a transducer. Biological sensing elements are immobilised microorganisms, enzymes, nucleic acids or antibody which interacts with an analyte and produce a signal. This signal is transmitted to transducer which converts it into an electrical signal (Fig. 30.13). Different types of biosensors have been developed for its use in food, clinical, pharmaceutical and wastewater treatment. For details see *A textbook of Biotechnology* by R.C. Dubey.

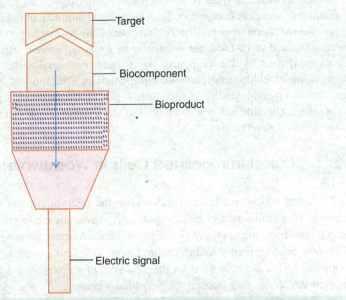

Fig. 30.13 : Components of a biosensor.

3.   Role of Microorganisms in Metal Removal

Generally physical and chemical methods are used for removal of heavy metals from wastewater such as oxidation, reduction, precipitation, ultrafiltration, etc. Use of microorganisms is an alternative to physical and chemical methods.

There are several microorganisms growing in marine water, fresh water and wastewater. *Bacillus licheniformis* and *Zooglea ramigera* have been isolated from activated sludge. They produce extracellular polymer which complex and accumulate metals such as iron, copper, cadmium, nickel or uranium. The accumulated metals are released from biomass upon treatment with Hcl. Fungal mycelia (e.g. *Aspergillus* and *Penicillium)* also remove metals from wastewater and offer a good alternative for detoxification of effluents. Biosorption has shown that *Aspergillus oryzae* can remove cadmium efficiently from solution (Table 30.5).

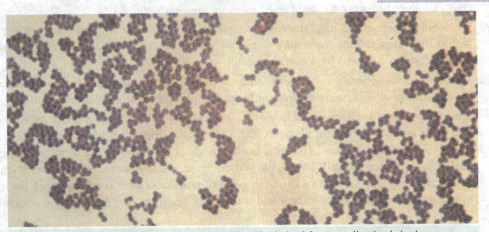

Bacillus licheniformis, a bacterium isolated from activated sludge.

Table 30.5 : Microorganisms involved in metal removal from industrial wastewater (based on Bitton, 1999).

Microorganisms	Metals removed
Aspergillus niger	Copper, cadmium, zinc
A. oryzae	Cadmium
Chlorella vulgaris	Gold, copper, mercury, zinc
Penicillium spinulosum	Copper, cadmium, zinc
Rhizopus arrhizus	Uranium
Saccharomyces cerevisiae	Uranium
Trichoderma viride	Copper
ATM-Bioclaim™	Biotechnology-based use of granulated product derived from biomass

In recent years recombinant bacteria are being investigated from removal of specific metals from contaminated water. For example, a genetically engineered *E. coli* was developed which expresses H^{2+} transport system and metallothionein (a metal-binding protein). It was able to accumulate 9 μmol Hg^{2+}/g cell dry weight. Bioaccumulation could not be affected by chelating agents, Na^+ Mg^{2+} and Cd^{2+}.

Mechanism of metal removal: Microorganisms remove metals by the following mechanisms *(a)* **adsorption** (negatively charged cell surfaces of microorganisms bind to the metal ions), *(b)* **complexation** (microorganisms produced organic acids (e.g. citric acid, oxalic acid, gluconic acid, formic acid, lactic acid, malic acid) which chelate metal ions. Biosorption of metals also takes place due to carboxylic groups found in microbial polysaccharides and other polymers, *(c)* **precipitation** (some bacteria produce ammonia, organic bases or H_2S which precipitate metals as hydroxides or sulfates. For example, *Desulfovibrio* and *Desulfotomaculum transform* SO_4 to H_2S which promotes extracellular precipitation of insoluble metal sulfides. *Klebsiella aerogenes* detoxifies cadmium to cadmium sulphate which precipitates on cell surface, *(d)* **volatilization** (some bacteria causes methylation of Hg^{2+} and converts to dimethyl mercury which is a volatile compound).

4. Application of Recombinant DNA Technology in Waste Treatment

Still this technology is at the stage of infancy due to the lack of knowledge and fear in the society for release of genetically engineered microorganisms (GEMs). But its major use is to detect the pathogens and to increase biodegradation of xenobiotics in wastewater treatment plants. The major tools of recombinant DNA technology are the nucleic acid probes and PCR to detect pathogens in effluents of wastewater plant. Some pathogens detected by PCR are *E. coli, Shigella flexneri, Salmonella, Legionella pneumophila, Pseudomonas aeruginosa, Yersinia,* Hepatitis A virus, HIV and *Giardia.* Moreover, the molecular-based technique must be validated in order to be considered by regulatory agencies.

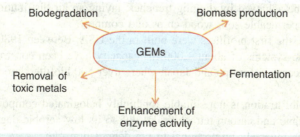

Fig. 30.14 : Area of application of genetically engineered microorganisms (GEMs).

GEMs are used in several areas of waste treatment such as biomass production, biodegradation of recalcitrant, removal of toxic metals, fermentation (methane and organic acid production), enhancement of enzyme activity, increased resistance to toxic inhibitors (Fig. 30.14). Recombinant DNA technology is involved in two steps: *(a)* searching out of microorganisms of desired function, and *(b)* transfer of character of desired function to the other microbes relevant to environment. Such microbe is called genetically engineered microorganism (GEM).

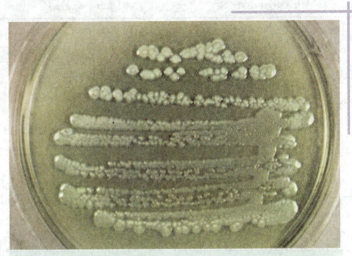

Industrial wastes have harsh environment for the growth and maintenance of GEMs. Environment is made harsh due to extremes of

Genetically engineered pseudomonas strains help to degrade components of crude oil.

temperature, pH, salinity, oxygen, redox potential and ionic composition. Special bioreactors are constructed where microorganisms are used to degrade industrial wastes. For example, biofilm reactors are preferred for this purpose because there will be less chance for potential release of GEMs into the environment. Many plasmids containing strains of *Pseudomonas* are used to degrade several components of crude oils. Using recombinant DNA technology the level of several enzymes has been increased. These enzymes are tryptophan synthetase, α-amylase, DNA ligase, benzylpenicillin acylase. These techniques help to improve enzyme stability and catalytic efficiency, increase their substrate range, to create multifunctional hybrid enzymes with improved substrate flux. This would be an exciting area of industrial waste treatment.

However, there is a fear for deliberate release of GEMs into the environment. Therefore, the society must be educated about the possible risk, if any, regarding the GEMs.

G. Biofiltration

Biofiltration is a new technology used to purify contaminated air evolved from volatile organic and inorganic compounds by involving microorganisms. It is a low cost technology gradually becoming popular due to simple operational and waste-removal efficiencies. Biofiltration is the oldest biotechnological method for removal of undesired foul gas components from air. Since 1920, biofilters were used to remove odorous compounds from wastewater treatment plants or animal farming. It could be achieved by digging trenches, laying an air distribution system and refilling the trenches with permeable soil, wood chips and compost.

In the 1960s, the first biofilters were built in the USA. Between 1980s and 1990s, about 30 large and full-scale systems of about 1,000 m^3 capacity have been constructed. Biofiltration has more industrial success in Europe and Japan where over 500 biofilters are in operation (Soccol *et al.*, 2003).

Moreover, biofiltration is not suitable for highly halogenated compounds (*e.g.* trichloroethylene, trichloroethane and carbon tetrachloride) due to its low aerobic degradation. Also the size of a biofiltration is inversely proportional to the degradation rate.

(i) **Biofilters:** Biofiltration is done by using biofilters. Biofilters are the packed-bed units in which gas is blown through bed of compost or soil covered by an active biofilm made by the natural microorganisms. Diagram of a biofilter is given in Fig. 30.15. The microorganisms consume the gaseous organic pollutants and use as source of carbon and energy. Instead, it may contain an inner support where a special pool of microorganisms is cultivated. The harmful compounds are degraded by an active biofilm covering the bed. The unwanted odorous organic compounds from gaseous phase are removed. They are absorbed or adsorbed on porous solid base of the biofilter, or dissolved into liquid phase and then oxidised by the microorganisms. Biofiltration is beneficial because it does not require large amount of energy operation.

(ii) **Microorganisms:** Different aspects have been studied regarding the microbial potential of biofilters, *(a)* isolation and characterisation, *(b)* use of pure cultures of bacteria or fungi, *(c)* mixed microbial population, *(d)* effect of enrichment culture including application of special strains, types of microorganisms and their metabolic activities, *(e)* effect of external conditions on microbial activity, and *(f)* release of microorganisms from biofilters.

Microorganisms present in biofiltration are mainly aerobic ones. Most of them present in biofilters are bacteria (mostly coryneforms and endospore formers), occasionally pseudomonads, protozoa, invertebrates and few actinomycetes (mainly *Streptomyces* spp.) and some fungi, (mostly *Alternaria, Aspergillus, Botrytis, Cladosporium, Fusarium, Mortierella, Rhizopus, Penicillium, Trichoderma*). Fungi from a large specific area which remain in direct contact of air flowing through filter.

Microorganisms are the most critical components of biofilters. Because they transform or degrade the contaminants. Naturally occurring microorganisms are available for the process because they get adapted to the contaminants. In some cases a specific microbe or genetically engineered microorganisms may be used (Schroeder, 2002).

(iii) **Biofilter Media:** The filter media must have some characteristics for performance of the biofilter. Because all the filter media allow the polluted air to interact closely with the degradative microorganisms, oxygen and water. Constitution of the physical media provides fine porous, large surface area and distribution of a uniform pore size. Which strongly defines the efficiency of biofilm. Inorganic bed material has a good flow properties and consists of a variety of metal oxides, glass or ceramics beads; PVC is commonly used as packing material.

The active microbial biofilm will adhers

Fig. 30.15 : Diagram of a biofilter.

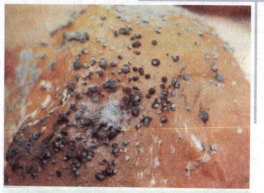

An aerobic multicellular common mold *Penicillium* (growing on an apple).

onto the biofilter media. The amount of microorganisms present depends on the availability of surface which in turn increases efficiency of biofilters. Thus the suitable biofilter media have large surface area for both adsorption of contaminants and support for microbial growth. Synthetic or inert media must be inoculated with soil, compost or sewage sludge. Because these materials have a big and complex population of microorganisms available to develop the proper microbial culture for the process. Pure culture can also be tested as inoculum. Suitable biofilter media have the ability to retain moisture to sustain biofilm layer and retain capacity of nutrient supply to microbes that form active biofilm. Some of the materials used as biofilter media are compost, peat, soil, activated carbon, wood chips or bark, perlite, vermiculite, lava rock and inert plastic material.

(iv) **Mechanism of Biofiltration:** Removal of contaminated material is multistep process. The contaminants are converted into liquid phase and transported to bacterial cell in the biofilm and transferred across the cell membrane, where the compound is degraded and used in cell metabolism (Schroeder, 2002). The treatment process depends on two mechanisms: *(a)* direct adsorption in biofilm and degradation, *(b)* adsorption on organic media and biodegradation, and *(c)* dissolution in aqueous phase and degradation. After biodegradation the contaminants are exhausted from the biofilter. The process can be expressed as below:

Pollutants + O_2 + Microorganism $\rightarrow$ Microbial cells + CO_2 + H_2O

Immobilisation of microorganisms to the bedding materials in biofilters: Immobilisation of microorganisms consists of two processes: *(a)* the self-attachment of cells to the filter bedding material, and *(b)* the artificial immobilisation of microorganisms to the bedding material. Self-attachment of microorganisms to a surface depends on the microbial culture *i.e.* secretion of glycocalyx (extracellular polysaccharide) and several forces such as electrostatic interaction, covalent bond formation, hydrophobic interaction, and partial covalent bond between microorganisms and hydroxyl groups on surfaces. Immobilisation of microorganisms at filter bedding is done by five methods such as: carrier bonding, cross-linking, entrapment, microcapsulation and membrane methods. For detail see *A textbook of Biotechnology* by R.C. Dubey.

H. Biodeterioration

It is a phenomenon of reducing the quality of products by microorganisms through secretion of enzymes and other metabolites.

1.   Biodeterioration of Stored Plant Food Material

The stored (unprocessed) plant material (fruits, seeds, etc.) are usually decayed due to post harvest attack of bacteria and fungi. This kind of spoilage is called biodeterioration. However, loss of plant materials before harvest is covered in plant pathology. The microbes can damage the plant materials partially as well as completely that definitely down the grade or the quality of the products. A large number of bacteria such as *Erwinia* sp., *Corynebacterium* sp. and fungi such as *Phytophthora* sp., *Streptomyces* sp., *Curvularia* sp., *Aspergillus* sp. etc. The storage fungi are usually developed from dormant spores/ mycelium that cause post harvest spoilage resulting into deterioration of food, fruits, seeds, etc. during storage. Now-a-days deterioration is mostly checked by altering storage conditions by chilling, using inert gases and by treatment with low dosages of gamma-radiation. The most affected products are soft fruits and salad vegetables, whereas grains, oilseeds and legumes are durable products and root vegetables and apples are of semi-perishable nature.

2.   Biodeterioration of Leather

Leather is a product of animal hides. Besides its wool and animal glues, there is other animal product which are attacked by microorganisms since leather contains keratins, animal fats and proteins. Therefore, it is rapidly deteriorated by lipolytic and proteolytic microorganisms which secrete lipases and proteases.

Leather production from animal hides involves soaking process which is carried out in water where several bacilli (*B. sublitis, B. megaterium. B. punilis*) attack it. They secrete several extracellular enzymes that may remain active long after the death of their producer organism. During leather deterioration microbial damage may cause tensile strength and colouration. Since finished leather is quite acidic, hence fungi such as *Rhizopus, Mucor, Cunninghamella* and *Aspergillus* deteriorate it quite fairly, whereas bacteria are secondary colonizers. Wool, fur, and feathers mainly contain cystein rich proteins (keratin), that is degraded by keratinophilic fungi such as *Trichophyto*n sp., *Chryosporium* sp. and *Streptomyces* sp. They damage, colour, impart odour and affect the tensile strength. Biodeterioration of these items now-a-days is checked by incorporating biocide (bromopol i.e. 2-bromo-nitropropane-1,3-diol) during processing of the above items for clothing.

3. Biodeterioration of Stone and Building Materials

Old ancient monuments, natural rocks, etc. are all attacked by various microorganisms such as large number of cyanobacteria, especially *Pleurococcus, Oscillatoria,* etc., lichens, fungi, including *Botrytis, Penicillium,* and *Trichoderma* sp. They use concrete, brick and mortar of building material as nutrients. Various biocides in washes, bleaching compounds, phenolics and organo-tin compounds are being used to save the building materials.

Actually, microbes colonize the site which may cause excessive expansion and contraction associated with wetting and drying of colonies. Entrapment of water within the colonies and crack can lead to enhanced damage. In addition, hyphal penetration into the surface layers of these materials can result in crack formation which may be promoted by excretion of corrosive metabolites. Several organic acids solubilize calcium carbonate while oxalic and citric acids solubilize silicates. *Desulfovibrio desulfuricans* reduce sulphur compounds and produces H_2S which is then oxidized to sulphuric acid by *Thiobacillus thiooxidans*.

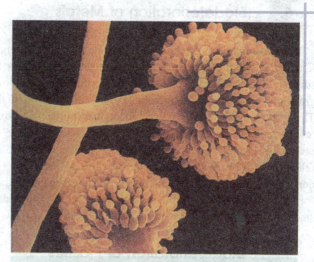

Characteristic conidiophores of *Aspergillus*, a fungus actively involved in biodeterioration of stored plant food material.

Various nitrifying bacteria, *Nitrobacter, Nitrosomonas*, also solubilise calcium associated with building material. They oxidize ammonia to nitrite leading in the formation of soluble form of calcium i.e. calcium nitrite. Besides, a large number of pseudomonads corrode steel and iron structures used in building construction.

4. Biodeterioration of Paper and Other Cellulosic Materials

Plant cells are composed of cellulose, hemicellulose, lignin and pectic substances. Therefore, the plant cell biomass derives paper and card. Besides, it is used to thicken cosmetics, paints, etc.

Although, a large number of factors are responsible for the microbial spoilage of paper, etc. Deterioration of paper occurs by various cellulolytic fungi such as *Trichoderma chaetomium* and *Aspergillus,* and bacteria such as *Cellulomonas,* etc. These microorganisms secrete cellulases (endo-β-glucanases,) exo-β–glucanases, and β–glucosidase) convert cellulose into glucose. Besides, some microorganisms also secrete enzyme xylanases that deteriorate hemicellulose present in paper. However, many other microbial activities have major effect on material strength.

5. Biodeterioration of Fuels and Lubricants

Fuels and lubricants are the products of petroleum. Fuels and lubricants are attacked only if water is present. Lubricants deteriorate more quickly as they consists of an emulsion of soil finely dispersed in water, in the presence of emulsifying agent. These hydrocarbons are shorter carbon chain length (petrol, gasoline) less susceptible to biodeterioration in comparison to those having long chain length such as diesel oil and kerosene. Further they also contain phenolic substances.

A large number of fungal species e.g. *Aspergillus, Hormoconis* (formly *Cladosporium resinae)* often found to be associated with oil sludge. These soil fungi degrade linear and branched alkanes to aromatic rings. Biodeterioration of these hydrocarbons containing materials may be controlled by adding biocides (organoboron, isothiazolones, etc.).

6. Biodeterioration of Metals

Microorganisms are known to corrode metals. They make biofilm by colonization/concentration of cells, which release corrosive metabolic products resulting in removal of hydrogen by sulphate reducing bacteria. Actually, microbial concentrations of the cells appear from an oxygen gradient that develops as a microbial colony, in contact with the metal, utilizes the available oxygen. These colonies have both oxygen-accessible (border) and oxygen-limited zone (centre), which are having negative and positive charge particles respectively. This leads to metal ion formation by producing insoluble hydroxides. Iron corrosion occurs mainly due to a bacterium, *Gallionella* (chemolithotroph) that oxidizes ferrous ions to ferric ions and forms insoluble ferric hydroxide deposits at the site of microbial attack. Various organic and inorganic acids are of microbial origin for example *H. resinae* produces sulphuric acid to sulphide. The organism, comes in contact of steel/iron resulting in corrosion and breakage/leakage of water pipe lines.

7. Biodeterioration of Plastics

These are polymeric materials made up of polyethylene, polystyrene, polyvinyl chloride (PVC) and polyesters. Many plastics are resistant to microbial attack but addition of various other materials makes the plastics prone to microbial attack. *Streptomyces rubireticuli* and *Penicillium* sp. are reported to deteriorate PVC and polyamides (nylon), respectively. Polyesters, polycaprolactone and polybutylene adipate are degraded by bacteria and fungi.

8. Biodeterioration of Cosmetics and Pharmaceutical Products

Cosmetics are being manufactured in the form of lotions, creams, liquids, solids and powder forms. Some are stickers and used externally for cleaning and decorative proposes. They consists of large quantity of water, animal, plant/mineral oils, natural gums thickening agents, carbohydrates, aroma and flavouring agents in addition to protein hydrolysates, milk, beer, egg, plant extracts, etc.

These product formulations are good sources of nutrients for microbes. Although, preservatives are added but due to complex nature of formulations, preservatives become less effective. Sometimes, creams and lotions are contaminated with pseudomonads, although with low levels of this group of organisms do not harm the individual but is applied to patients with skin infection/damage, situation may become worst.

Some of the deteriorated cosmetics impart foul odours due to production of organic acids, fatty acids, amines, ammonia, and hydrogen sulfide. Production of ammonia or acid leads to alter the pH, which may change the consistency and colour of the products by developing lumps and slime. Sometimes gas bubbles are also generated. Such products later on become unstable and form separate oil and water phases.

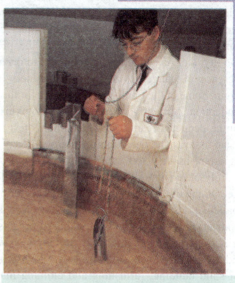

Beer, brewed under controlled conditions, is a good source of nutrients for microbes as a product formulation of cosmetics.

Various lipids (oils, fats) are susceptible to microbial attack when dispersed in aqueous formulations after degradation. They give rise to glycerol and fatty acids. The fatty acids may further break down via β-oxidation to form odorous ketones. Many other additive chemicals particularly glycerol and sorbitol are used in toothpaste, etc. These allow various microbes to grow and secrete amylases, cellulases, etc. responsible for degradation of such carbon containing microbial nutrients. Shampoo and detergents often contain sodium dodecyl sulfate which act as substrates for enzyme alkyl sulphatases. The species of *Pseudomonas*, *Citrobacter* and *Aerobacter* secrete enzymes that breakdown the finished items and may generate unpleasant odour, particularly H_2S.

Pharma products are both sterile (injectable, intravenous infusions, etc.) and non-sterile forms (tablets, capsules, powders, suspensions, syrups). Many non-sterile pharma products contain low amount of active principles besides a large quantity of additive substances. Similar to cosmetic and lotions, they many contain preservatives. The presence of low levels of pathogenic microorganisms, higher level of opportunistic pathogens besides contamination with toxic microbial metabolites lead to deterioration. Various disinfectants contain ammonia compounds, reported to be contaminated with pseudomonads, for example *Pseudomonas aeruginosa* contaminates eye drops, washes and mascara, etc. Some injectables contain Gram-negative bacteria, fungi, etc. which produced endotoxins (pyrogens) that directly induce fever. For safe practices use of ultrapure water is recommended in the manufacture of intravenous drugs and infusions.

The active ingradients may also become ineffective due to microbial contamination. It leads to breakdown of antibiotics (for e.g. penicillin is attacked by β-lactamases). Such enzymes are secreted by various groups of microorganisms. Similarly, Aspirin is broken down by *Acinetobacter iwoffii,* Atropine is degraded by *Pseudomonas* and *Corynebacterium* spp., and *Cladosporium herbarum* causes spoilage of hydrocortisone.

Safe-practices : Microbial quality control is a necessary step to check the deterioration of crude and finished materials. It is conducted to monitor microbial contamination of raw materials, to monitor and confirm the efficacy of operations such as sterilization, to control the pathogenic microorganisms (by their absence), and to evaluate the expected storage period. The assessment is carried out by total microbial count and/or estimation of groups. e.g. coliforms, or the detection of specific microorganisms. Further microbial activity within a product may also be confirmed from changes occurring to the pH, viscosity, stability, etc. As stated, spoilage may also occur due to the presence of dangerous metabolite products of microbial origin. Some of the fungi are responsible for the mycotoxins production. Similarly, presence of endotoxins (pyrogens) can be deducted readily via the Limulus test (amoebocyte lysate assay.) The detection of microbial enzymes is often important as they too can persist after cell death. Many hydrolysate enzymes (cellulase) can pose to product storage life. On the other hand, to test the efficacy of preservatives challenge test examine their effectiveness by utilizing biodeteriogens, or a group of standard organisms such as *P. aeruginosa, Staphylococcus aureus, Candida albicans* and *Aspergillus niger.* Soil burial tests are useful techniques for solid building, furnishing and clothing materials.

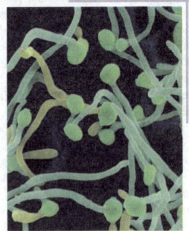

candida albicans constitutes a group of standard micro-organisms to test the efficacy of *preservatives.*

I. Microbial Plastics

It is interesting to note that certain bacterial species naturally synthesize polymers such as poly-β -hydroxybutyrate (PHB). This type of material is used as raw material for plastic. The organism used to grow under controlled conditions and if substrate is altered then PHB is obtained with different chemical properties. For example, acetate and butyrate lead to poly-β-hydroxybutyrate by certain group of bacteria which yields a polymer. *Copolymers* containing alternating repeats of various monomers can give rise to these biological products. The bacterial plastics are substitute for synthetic polymers which are not only nuisance to environmentalist

Chemical plastics persist in nature, thus the need for microbial plastics arises.

but also responsible for animal death. The use of polythene bags in our country leads to land fill. Mostly, chemical plastics are foreign to the environment in which they are dumped, hence persist in nature. Although, genetically engineered microorganisms are recently developed to tackle this problem.

QUESTIONS

1. Discuss the role of microorganisms in environmental microbiology.
2. Write the biodegradation of pesticides.
3. Which chemical processes lead to biodegradation of DDT.
4. Name few microorganisms and associated biodegradation process of pollutants.

5. Write a critical account on biodegradation of Industrial Waste containing heavy metals.

6. Describe the process of solid waste conversion process.

7. Define compost. How to yield the compost for agricultural practices?

8. Write short notes on:

   (a) Organic fertilizer   (b) Bioremediation
   (c) Xenobiotics   (d) Uranium recovery
   (e) Microbial plastics   (f) Sewage

9. Describe the process of liquid waste treatment by microorganisms.

10. Why is activated sludge a more efficient means of removing BOD than a sludge digester?

11. What are septic tanks and oxidation ponds. What is their significance?

12. Write the primary, secondary and tertiary treatment processes used in wastewater treatment.

13. Write a short account on:

   (a) BOD   (b) Sludge
   (c) Methanogens   (d) Trickling filter
   (e) Activated sludge   (f) Biogas technology
   (g) Lagoons   (h) Vermicomposting

14. Discuss the process of biodeterioration of the following:

   a) Leather   (b) Paper
   (c) Lubricants   (d) Metals
   (e) Plastics   (f) Cosmetics
   (g) Pharma products   (h) Bioaugmentation

REFERENCES

Alkesandrov, G. and G.A. Lak. 1950. *Microbiologia*, 19:97.

Biddlestone, A.L. and K.R. Gray 1973. The chemical Engineer, 210, 76.

Bitton, G. 1994. Wastewater Microbiology. Wiley-Liss, Gainsville, NY.

Bitton G, 1990. Use of Commercial blends of microoogranisms and enzyme. In *Wastewater Microbiology,* 2nd Ed. Wiley-Liss: New York.

Bartha, R. 1986. Biotechnology of petroleum pollutants biodegradation. *Microbiol. Ecol.* 12: 155-172.

Bollag, J.M. 1982. Microbial transformation of pesticides. Adv. Appl Microbiol. 18: 75-130.

Choudhry, G.R. 1995. Biological degradation and bioremediation of Toxic Chemicals, Timber Press, Portland, OR.

Cowan, R.M., Love, N.G. Sock, S.M. and K.White. 1995. Activated sludge and other aerobic suspended culture process. *Water Env. Res.* 67: 433-450.

Crawford, J.H. 1983. Composting of agricultural wastes: A review. *Process Biochem.* 1, 2: 14-18.

Dagley, S. 1987. Lessons from Biodegradation. *Ann. Rev. Microbiol.* 41.

Dubey, R.C. and Dwivedi, R.S. 1985. *Proc. Indian Natl. Sci. Acad.* B51:259-264.

Gaylarde, C.D. and H.A. Videla, 1995. *Bioextraction and Biodeterioration of Metals*. Cambridge University Press, NY.

Gaur, A.C. 1984. In *"Organic matter and Rice"*. International Rice Research Institute, Manita, 502-514.

Halder, A.K., P.K. Roy and A.K. Mitra. 1994. *Everyman Science.* XXIX: 136-139.

Halder, A.K., A. Banerjee, A.K. Mishra and P.K. Chakvarty. 1992. *Trans. Bos. Res. Inst.* 55, 27 : 1-2.

Hesse, P.K. 1984. In *Organic matter and Rice*. International Rice Research Institute, Manila, 35-43.

Kelly, O.P. Norris, P.R. and C.L. Brterley. 1979. Microbiological methods for the extraction and recovery of metals. p.p 263-308. In A.T. Bull, D.C. Ellwood and C. Ratledge In A.T. Bull, D.C. Ellwood and C.Ratledge (eds.). *Microbial Technology current state future prospects*. Cambridge University Press, Cambridge.

Lederberg, J. 1992. *Encyclopaedia of Microbiology*. Academic Press, San Diego.

Montomery, J.M. 1985. *Water treatment:* Principle and Design. John Wiley & Sons., NY.

Pooley, F.D. 1982. Bacteria accumulate silver during leaching of sulphide ore numerials. *Nature*, 296: 642-643.

Ratledge, C. 1994. *Biochemistry of microbial degradation*. Kluwer Academic, Dordrect, the Netherlands.

Rehm, H.J. and G. Reed. 1987. *Biotechnology*. vol. 8 Microbial Degradation. VCH Press.

Spani, J.C. 1995. *Biodegradation of nonaromatic compounds*. Plenum Press.

Soccol, C.R. *et al.* 2003. Biofiltration: an emerging technology. *Indian J. Biotechnol,* **2:**396-410.

Schroeder, E.D. 2002. Trends in application of gas phase bioreactor. *Env. Sci. Biotechnol,* **1:**65-74.

Subba Rao, N.S. 1986. *Soil Microorganisms and plant growth*, Oxford and IBH Publishing Co. Pvt.Ltd, New Delhi.

Sugio, T., Domatsu, C., Torno, T. and K. Imai. 1984. Role of ferrous ions in synthetic coballous sulphate leaching of *Thiobacillus ferroxidans. Appl. Env. Microbial* 48: 461-467.

Thompson, R. 1977. *Compost Science.* 6, 18.

Tuovinen, O.H. And D.P.Kelly. 1974. Use of microorganisms for the recovery of metals. *Int. Metal. Rev.* 19:21-23.

Vernadsku, V.I. 1934. A treatise on Geochemistry Gozizat (Russian).

Young, L.Y. and C.E. Cerniglia. 1995. *Microbial Transformation and degradation of toxic organic chemicals*. Witey-Liss, N.Y.

Agricultural Microbiology
(Biofertilizers and Biopesticides)

31

A. Crop Production (Biofertilizers)

B. Crop Protection (Biofertilizers)

C. Microbiology of Lignocellulose Degradation in the Rumen

Agricultural microbiology is concerned with various roles played by microorganisms in general. Although, agricultural microbiology is further classified into soil microbiology and plant pathology but the above two branches are better known as separate sciences. During the last decade, major findings were made i.e. (*a*) soluble phosphate fertilizers could be used for crop improvement, (*b*) soil fertility could be maintained by the use of organic fertilizers, and (*c*) plant diseases could be controlled by using microorganisms through the process of antagonism. These subjects have received major breakthrough in recent years. Several biofertilizers have been commercialised for increasing crop fertility. Many bacteria and fungi have been discovered for their use in controlling various plant diseases. The products are made and marketed under different trade names by both Indian and foreign companies. But the use of indigenous microbial preparations always proved better in comparison to that of introduced ones by multinational companies. Bureau of Indian Standard (BIS) has also passed norms and published a booklet. It is now obvious to learn that use of 'microbial inoculant' will be more helpful in solving present and future problems of man.

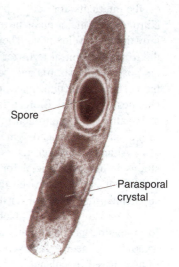

Spore

Parasporal crystal

Bacillus thruingiensis :
Source of Bacterial
Insecticide

Microbial Inoculants: A formulation that contains one or more beneficial microbial strains (or species) in an easy-to-use and economical carrier material, organic, inorganic, or synthesized from defined molecules. The desired effects of the inoculants on plant growth include nitrogen fixation, in bio-control of (mainly) soil-borne diseases, the enhancement of mineral uptake, weathering of soil minerals, and nutritional or hormonal effects.

Microbial consortium or mixed inoculants: It is a combination of microorganisms that interact with other synergistic microorganisms. All of them provide nutrients, remove inhibitory products, and stimulate each other through physical or biochemical activities. A mixture of *Azospirillum* and *Cellulomonas* in which the former provides nitrogen to the *Cellulomonas* which degrades cellulose into glucose. This glucose acts as C source to *Azospirillum*.

A. Crop Production (Biofertilizers)

1. Production of Bacterial Biofertilizer

With day-by-day increasing the population, especially in developing countries like India, the stress on agriculture is also increasing continuously. With the development, the land area under farming is not increasing but is further decreasing, this has posed extra burden on the agriculture. Therefore, the land available for agriculture should be economically utilized and maximum results be obtained.

Most of our agricultural lands are deprived of either one mineral or the other. These minerals are essential for the growth and development of plants. One of the nutrients for any type of plant is *nitrogen*. Nitrogen is a major element required by the plant for growth and development. The nitrogen is provided in the form of chemical fertilizer. Such chemical fertiliz-

Enhancing soil fertility by using organic fertilizers.

ers pose health hazards and pollution problem in soil besides these are quite expensive, bringing the cost of production much higher. Therefore, biofertilizers are being recommended in place of chemical fertilizers. **Biofertilizers** are the formulations of living microorganisms which are able to fix atmospheric nitrogen in the available form for plants (nitrate form) either by living freely in the soil or associated symbiotically with plants.

Although nitrogen fixers are present in the soil, enrichment of soil with effective microbial strains is much beneficial for the crop yields. Use of composite biofertilizers can increase soil fertility. It has been proved that biofertilizers are cost effective, cheap and renewable source to supplement the chemical fertilizers.

(i) History: In 1895, Nobbe and Hiltner applied for patents in England and the United States for a legume inoculant that was later marketed as *Nitragin*. Nitragin was produced on gelatin and agar nutrient media. However, agar based inoculants were soon replaced by peat-based ones because in agar based inoculants, mortality was very high during the dry phase. In India, the production of biofertilizers on commerical scale was started only during late 1960's when yellow seeded soybean was introduced for the first time. Recognition of Indian peat as suitable carrier for production of biofertilizer in 1969 further augmented the growth of biofertilizer industry in India. The performance of Indian peat-based biofertilizer at Indian Agricultural Research Institute, New Delhi, was found to be comparable to that obtained with imported `Nitragin' biofertilizer from the U.S.A. Since then,

the process of development of biofertilizer, specially of rhizobial biofertilizer for various crops in India has made a tremendous success.

(*ii*) **Production of Biofertilizer:** In order to meet the food requirements of ever increasing population, the nitrogen fertilizer requirement for crop production by 2000 A.D. was estimated to be about 11.4×10^6 tonnes. Biological nitro-

gen fixation can be the key to fill up this gap because of high cost and several other demerits of chemical fertilizers. For production of a good and efficient biofertilizer, first of all an efficient nitrogen fixing strain is required, then its inoculum (the form in which the strain is to be applied in fields) is produced. Packing, storing and maintenance are other aspects of biofertilizer production. While producing biofertilizers the standards laid down by BIS have also to be kept in mind for making the product authentic. Commercial production of biofertilizer is shown in Fig. 31.1.

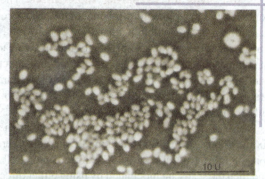

Nitrobacter, a rod shaped nitrifying bacteria, involved in the production of biofertilizer.

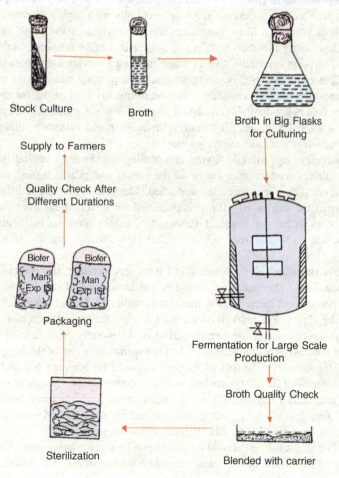

Fig. 31.1 : Commercial production of biofertilizer.

(*iii*) **Criteria for Strain Selection:** The efficient nitrogen fixing strain is evolved or selected in laboratory, maintained and multiplied on nutritionally rich artificial medium before inoculating the seed or soil. In soil, the strain has to survive and multiply to compete for infection site on roots against hostile environment in soil.

(*iv*) **Steps for Preparing Biofertilizer:** The isolated strain is inoculated in small flasks containing suitable medium for inoculum production. The volume of the starter culture should be a minimum of 1% to obtain atleast 1×10^9 cells/ml. Now the culture obtained is added to the carrier for inoculant (biofertilizer) preparation. Carriers carry the nitrogen fixing organisms to the fields. In some cases carrier is first sterilised and then inoculated, while in other cases it is first inoculated and then sterilised by UV irradiation. The inoculum is now packed with 10^9-10^{10} viable cells per gram. Final moisture content should be around 40-60%. For large scale production of inoculum, culture fermenters are used.

(*a*) **Seed Pelleting:** Direct seed coating with the gum arabic or sugary syrup and useful nitrogen fixing strains especially the coating of rhizobia over specific host legume seeds are another method for obtaining fruitful results. As before, first of all the inoculum is prepared of the desired strain and then the seeds are inoculated by using either direct coating method or slurry method. Immediately after seed coating, $CaCO_3$ is added to sticky seeds. The practice of seed inoculation dates back to 1896 when Voecher used this technique. In many soils the nodule bacteria are absent or are not adequate in either number or quality to meet the nitrogen requirements of the plants. Under these conditions, it is necessary to inoculate seeds or seedlings with highly effective rhizobia.

(*b*) **Inoculant Carriers:** Most inoculants are the mixture of the broth culture and a finely milled, neutralized carrier material. Carrier is a substance having properties such as, non-toxicity, good moisture absorption capacity, free of lump forming material, easy to sterilize, inexpensive, easily available and good buffering capacity, so that it can prolong maintain the growth of nitrogen fixing microorganisms which it is carrying. The most frequently used carrier for inoculant production is peat. However, peat is not available in certain countries such as India. A wide range of substitutes c.g. lignite, coal, charcoal, bagasse, filter mud, vermiculite, polyacrylamide, mineral soils, vegetable oils, etc. have been tested as alternative carriers.

Carrier processing e.g. mining, drying and milling are the most capital intensive aspects of inoculant (biofertiliser) production. First of all the carrier like peat is mined, drained and cleared off stones, roots, etc. Then, it is shredded and dried. The peat is then passed through heavy mills. Material with a particle size of 10–40 μm is collected for seed coating. Peat with particle size of 500-1500 μm is used for soil implant inoculant. Carriers have to be neutralised by adding precipitated calcium carbonate (pH 6.5-7.0) After this, the carriers are sterilised for use as inoculants.

(*c*) **Quality Standards for Inoculants:** Like every product, the biofertilizers should also follow certain standards. The inoculant should be carrier-based and should contain a minimum of 10^8 viable cells per gram of carrier on dry mass basis within 15 days of manufacture, and 10^7 within 15 days before the expiry date marked on the packet when the inoculant is stored at 25-30°C. The inoculant should have a maximum expiry period of 12 months from date of manufacture. The inoculant should not have any contaminant. The contamination is one of the biggest problems faced by the biofertilizer industry. The pH of inoculant should be between 6.0 and 7.5. Each packet containing the biofertiliser should be marked with the informations *e.g*, name of product, leguminous crop for which intended, name and address of manufacture, type of carrier, batch or code number, date of manufacture, date of expiry, net quantity meant for net area and storage instructions. Each packet should also be marked with ISI (BIS) certification mark.

The inoculant (biofertilizer) should be stored in a cool place away from direct heat preferably at a temperature of 15°C. The bioinoculant should be packed in 50-75 μ low density polyethylene packets.

Two main methods of inoculation are currently being used *(a)* seed inoculation and *(b)* soil inoculation. The soil inoculation is done by delivering the inoculant directly into the sowing furrow with the seeds. Seed inoculation by pelleting or coating the seed with inoculant is the most popular methods.

(v) **Green Manuring:** Green manuring is defined as a "farming practice where a leguminous plant which has derived enough benefits from its association with appropriate species of *Rhizobium* is ploughed into the field soil and then a nonlegume is sown and allowed to get benefitted from the already present nitrogen fixer". The practice of green mauuring began from time immemorial, from several century B.C. in India and China. During the course of time, availability of chemical fertilizers decreased the significance of green manuring. In recent years, due to hike in price of chemical fertilizers, the practice of green manuring is reemphasized.

Some of the cultivated legumes and annual legumes such as *Crotolaria juncea, C. striata, Cassia mimosoides, Cyamopsis pamas, Glycine wightii, Indigofera linifolia, Sesbania rostrata, Leucaena leucocephala,* etc. contribute nitrogen.

In addition to nitrogen, green manuring provides organic matter, phosphorus, potassium besides minimising the pathogenic organisms in soil (Ghai and Thomas, 1989). The reclamation of "usar lands" can also be done by green manuring.

In India besides a large number of private and semi-Government organisations, the National Biofertilizer Development Centre sponsored by the Ministry of Agriculture and the establishment of National Centres for blue green algal collections at IARI, New Delhi, the Department of Biotechnology, Govt. of India, Ministry of Science & Technology are the major developments that reflect our concern to harness biofertilizers in our agricultural economy.

2. Algal and Other Biofertilizers

Biological nitrogen fertilizers play a vital role to solve the problems of soil fertility, soil productivity and environmental quality. *Anabaena azollae,* a cyanobacterium lives in symbiotic association with the free floating water fern *Azolla.* The symbiotic system *Azolla-Anabaena* complex is known to contribute 40-60 kg N ha^{-1} per rice crop. *Anabaena azolle* can grow photoautotrophically and fixes atmosphere nitrogen. The nitrogen fixing cyanobacteria such as *A. azollae* and *A. variabilis* when immobilized in polyurethane foam and sugar cane waste have significantly increased the nitrogen fixing activity and ammonia secretion. The inoculation of cyanobacteria in rice crop significantly influenced the growth of rice crop by secretion of ammonia in flood water. The use of neem cake coupled with the inculation of *Azolla* greatly increased the nitrogen utilization efficiency in rice crop. Besides *Anabaena,* other nitrogen fixing cyanobacteria like *Aulosira, Calothrix, Hapalosiphon, Scytonema, Tolypothrix* and *Westiellopsis* have been held responsible for the spontaneous fertility of the tropic rice fields.

In addition to contributing N, the cyanobacteria add organic matter, secrete growth promoting substance like auxins and vitamins, mobilise insoluble phosphate and improve physical and chemical nature of the soil. Algalization has been shown to ameliorate the saline-alkali soils, help in the formation of soil

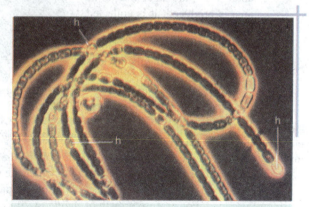

The nitrogen fixing cyanobacteria *Anabaena azollae* with heterocysts (h).

aggregates, reduce soil compaction, and narrow C:N ratio. These organisms enable the crop to utilize more of the applied nutrients leading to increased fertilizer utilising efficiency of crop plant. Most of the cyanobacteria act as supplements to fertilizer N contributing up to 30 kg N ha^{-1} season^{-1}. The increase in the crop yield varies between 5-25 percent.

(i) Mass Production of Cyanobacterial Biofertilizers: For out door cultivation of cyanobacterial biofertilizers, the regional specific strain should be used. In such practices, a mixture of 5 or 6 regionally acclimatised strains of cyanobacteria e.g. species of *Anabaena, Aulosira, Cylindrospermum, Gloeotrichia, Nostoc, Plectonema, Tolypothrix* etc. are generally used as starter inoculum. The following methods are used for mass cultivation: (*a*) cemented tank method (*b*) shallow metal troughs method, (*c*) polythene lined pit method and (*d*) field method. The polythene lined method is most suitable for small and marginal farmers for the preparation of biofertilizer. In this method, small pits are prepared in field and lined with thick polythene sheets. The mass cultivation of cyanobacteria is done by using any of the above four methods; the steps are given below:

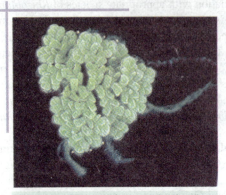

Azolla-Anabaena Symbiosis : *Azolla,* a water fern.

(*a*) Prepare the cemented tank, shallow trays of iron sheets or polythene lined pits in an open area. Width of tanks or pits should not be more than 1.5 m. This will facilitate the proper handling of culture.

(*b*) Transfer 2-3 kg soil and add 100 g superphosphate. Water the pit to about 10 cm height. Mix lime to adjust the pH. Add 2 ml of insecticide to protect the culture from mosquitoes. Mix well and allow to settle down soil particles.

(*c*) When water becomes clear, sprinkle 100 g starter culture on the surface of water.

(*d*) When temperature remains around 35–40°C during summer, optimum growth of cyanobacteria is achieved. The water level is always maintained about 10 cm during the period.

(*e*) After drying, the algal mass (mat) is separated from the soil that forms flakes. During summer about 1 kg pure algal mat per m^2 area is produced. It is collected, powdered, and packed in polythene bag and supplied to the farmers after sealing the packets.

(*f*) The algal flakes can be used as starter inoculum again.

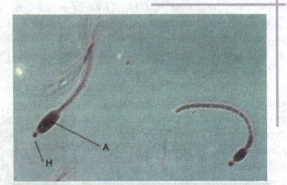

Terminal heterocysts (H) and subterminal akinetes (A) of *Cylindrospermum,* a cyanobacterium involved in mass production of biofertilizers.

(ii) Mass Cultivation of *Azolla*: The aquatic heterosporus fern contains endophytic cyanobacterium, *Anabaena azollae* in its leaf cavity. There are number of species of *Azolla,* namely *A. caroliniana, A, filiculoides, A. maxicana, A. nilotica, A. pinnata* and *A. rubra* which are used as biofertilizer especially for paddy. For mass cultivation of *Azolla,* microplots (20 m^2) are prepared in nurseries in which sufficient water (5-10 cm) is added. For profuse growth of *Azolla,* 4-20 kg P$_2$O$_5$/ha is also amended. Optimum pH (8.0) and temperature (14-30°C) should be maintained. Finally, microplots are inoculated with fresh *Azolla* (0.5 to 0.4 kg/m^2). An insecticide (Furadon) is used to check the insect's attack. After 3 weeks, the mat of *Azolla* is ready for harvest and the same microplot is inoculated with fresh Azolla to repeat the cultivation.

Azolla mat is harvested and dried to use as green manure. There are two methods for its application in field: (*a*) inocorporation of *Azolla* in soil prior to rice cultivation, and (*b*) transplantation of rice followed by water draining and incorporation of *Azolla* (Singh, 1980). However, reports from the IRRI, Manila (Philippines) revealed that growing of *Azolla* in rice field before rice transplantation increased the yield equivalent to that obtained from 30 kg/ha nitrogen as urea or ammonium phosphate.

3.   Endophytic Nitrogen Fixers

Recently, several non-leguminous and particularly graminaceous species such as rice, wheat and forage grasses have registered tremendous interest in nitrogen fixation. Isolation of a number of diazotrophic bacteria such as *Azospirillum*, *Herbaspirillum* and *Acetobacter* is reported.

The term endophyte refers to microorganisms (bacteria and fungi) that colonize root interior of plants and live most of their life inside the plant tissue. Spilitting the term endophyte into facultative and obligate was suggested to distinguish, respectively, strains that are able to colonize both the surface and root interior and to survive well in soil from those that do not survive well in soil but colonize the root interior and aerial parts.

(*i*) **Facultative Endophytic Diazotrophs:** This group is composed of *Azospirillum* spp. and considered important with non-legume plants. Although *A. lipoferum* was the first species of the genus isolated by Tarrand *et al.* (1978). *A. brasilense* among all the seven known species is the best characterized at physiological and molecular levels.

(*ii*) **Obligate Endophytic Diazotrophs:** This group includes *Acetobacter diazotrophicus* (syn. *Gluconacetobacter diazotrophicus*) a nitrogen fixing bacterium clustered in the alpha sub-class of the proteobacteria *Azoarcus* spp., *Herbaspirillum* spp. and a partially identified *Burkholderia* sp. are clustered in the beta sub-class of the proteobacteria.

(*iii*) **Other Bacteria:** *Alcaligens,* a diazotrophic member of this genus has been consistently isolated from the rhizosphere of wet rice land. *Burkholderia*, the other bacterium appears to have potential as rice inoculant. In the case of *Klebsiella,* substituted nitrogen fixation has been observed in rice inoculated with *K. oxytoca* or any other *Klebsiella* spp. that are considered as endophytes. The diazotrophic nature of some members of the genus *Pseudomonas* is still a matter of debate. Nevertheless, several bacteria within it are clearly diazotrophic such as *Pseudomonas diazotrophicus, P. flurorescens, P. saccharophila* and *P. stutzeri.* Recently, several researchers have attempted to construct an artificial association between rhizobia and rice particularly with *Azorhizobium caulinodans.*

(*a*) **Isolation and Identification of Endophytes:** For isolation and identification of natural diazotrophs from plant samples, root or stem, washed with sterile water, surface sterilized with 70% ethanol for 5 minutes and with 0.2% mercuric chloride for 30 second, washed several times using sterile water. Sterilization of root and stem will be verified by rolling them on BMS agar plates. Then homogenize the sample in a mortar and pestle in sterile phosphate buffer saline/ 1% sugar solution and serially diluted and 0.1 ml sample transfer into vials containing 5-8 ml of respective semisolid media for the targeted bacterium with respective C sources with an initial pH of 6.0. The number of diazotrophic populations is determined by the most probable number methods using a McCrady table. Vials with veil pellicles reaching the surface after incubation at 30°C with or without gas production and with positive reaction for acetylene reduction activity, show the presence of good endophytes.

(*b*) **Applications in Agriculture:** Obligate endophytes have an enormous potential for use because of their ability to colonize the entire plant interior and establish themselves niches protected from oxygen or other inhibitory factors; thus their potential to fix nitrogen can be expressed maximally. Recent studies in Brazil showed that the sugarcane varieties fix up to 80% nitrogen. It has been reported that wetland rice receives some nitrogen by endodiazotrophs. Tropical pasture grasses such as *Brachiaria, Digitaria, Panicum* and *Paspalum* spp. fix nitrogen.

4. Biofertilizers aiding Phosphorus Nutrition

Tropical soils are deficient in phosphorus. Further most of the microorganisms solubilize P and thus make it available for plant growth. It is estimated that in most tropical soils, 75% super phosphate applied is fixed and only 25% is available for plant growth. There are some fungi such as *Aspergillus awamori, Penicillium digitatum,* etc. and bacteria like *Bacillus polymyxa, Pseudomonas striata,* etc. that solubilize unavailable form of P to available form. India has 250 mt of rock phosphate deposits. The cheaper source of rock phosphate like Mussoorie rock phosphate and Udaipur rock phosphate available in our country can be used along with phosphate solubilising microorganisms (Table 31.1)

Penicillium solubilizes unavailable form of P to available form.

Vesicular-arbuscular mycorrhizal (VAM) fungi colonize roots of several crop plants. They are zygomycetous fungi belonging to the genera *Glomus, Gigaspora, Acaulospora, Sclercystis,* etc. These are obligate symbionts and can not be cultured on synthetic media. They help plant growth through improved phosphorus nutrition and protect the roots against pathogens. Nearly 25-30% of phosphate fertilizer can be saved through inoculation with efficient VAM fungi as reported by Bagyaraj (1992).

5. Production of Mycorrhizal Biofertilizer

Methods of inoculum production of mycorrhizal fungi differ with respects to their nature, depending upon types *i.e.,* ectomycorrhizal or endomycorrhizal.

(*i*) **Ectomycorrhizal Fungi:** In this case, the basidiospores, chopped sporocarps, sclerotia, pure mycelial culture, fragmented mycorrhizal roots or soil from mycorrhizosphere region can be used as inoculum. The inoculum is mixed with nursery soil and seeds are sown thereafter.

Institute for mycorrhizal Research and Development, USA and Abbot Laboratories, USA have developed a mycelial inoculum of *Pisolithus tinctorius* in a mycelial vermiculite-peat moss substrate with trade name 'MycoRhiz' which is commercially available on large quantities (Tale 31.2).

(*ii*) **VA Mycorrhizal Fungi:** VA mycorrhiza can be produced on a large scale by pot culture technique. This requires the host plant mycorrhizal fungi and natural soil. The host plants which support large scale production of inoculum are sudan grass, strawberry, sorghum, maize, onion, citrus, etc. The starter inoculum of VAM can be isolated from soil by wet sieving and decantation technique (Gerdeman and Nicolson, 1966). VAM spores are surface sterilised and brought to the pot culture. Commonly used pot substrates are sand : soil (1:1, w/w) with a little amount of moisture.

There are two methods of using the inoculum: (*a*) using a dried spore-root-soil to plants by placing the inoculum several centimeter below the seeds or seedlings, (*b*) using a mixture of soil-roots, and spores in soil pellets and spores are adhered to seed surface with adhesive.

Commercially available pot culture of VA mycorrhizal hosts grown under aseptic conditions can provide effective inoculum. Various types of VAM inocula are currently produced by Native Plants, Inc (NPI), Salt Lake City. In India, Tata Energy Research Institute (TERI), New Delhi and Forest Research Institute, Dehradun have established mycorrhizae banks. Inocula of these can be procured as needed and used in horticulture and forestry programmes (see R.C., Dubey, *A Textbook of Biotechnology*).

Table 31.1 : Some major nitrogen-fixing microorganisms and beneficiaries plants.

Name of microorganisms	Name of crop plants which receive benefits
Rhizobium spp. living symbiotically in root nodules	All grain legumes (pulses), some oil yielding (soybean, ground nut), some fodder legumes (*e.g.* clover),
Nostoc, Anabaena, Aulosira and others (free living blue green algae)	Rice
Anabaena azollae living symbiotically with the waterfern,	Rice *Azolla* spp.
Azotobacter chroococcum (free living bacterium)	Rice, maize, cotton and others
Frankia spp. (actinomycete) living symbiotically in nonlegume root nodules	*Alnus, Casuarina* and others
Azospirillum spp. (associate symbiont)	Maize, sorghum, pearl-millet, finger millet and others
Bacillus polymyxa, Clostridium spp. *Rhodospirillum* spp.	non-specific hosts

Table 31.2 : Commercially available biofertilizers and their manufacturers, beneficiary crop and associated microorganism

Product	Manufacturer's Name	Microbe used	Beneficial crop.
NitraginTM	Nitragin Sales Corpn. Wisconsin, 53209	*Rhizobium*	Soybean
Rhizocote	Coated seed Ltd., Nelson, New Zealand	*Rhizobium*	Legumes
Nodosit	Union Chemiques S.A. Belzium	*Rhizobium*	Legumes
Rhizonit	Phylaxia Allami Budapest, Hungary	*Rhizobium*	Legumes
Nitrazina	Wytwornia Walcz. Poland	*Azotobacter*	Cereals & vegetables
N-germ	Laboratoire de Microbiologie France	*BGA*	Rice
Tropical Inoculants	Tropical Inoculants Brisbane, Queensland	*Azotobacter*	Rice and wheat
Nodulaid	Agricultural Lab. New South Wales, UK	*Rhizobium*	Legumes
Azotobacterin	Tashkent laboratories Moscow	*Azotobacter*	Vegetables, cereals
Nodion	Indian Organic Chems. ltd. Mahew Mahal, Bombay	*Rhizobium*	Legumes
Azoteeka	Bacifil, 25 Nawal Kishore Rd. Lucknow	*Azotobacter*	Cereals

Product	Manufacturer's Name	Microbe used	Beneficial crop.
Agro-teeka	National Fertilizers & Chemicals 11, Ind. Area-II, Ramdarbar, Chandigarh	Azotobacter	Wheat, rice, maize, tea, Sugarcane potato
Rhizoteeka	Microbes India, 87, Lenin Savabe, Calcutta	Rhizobium	Legumes
Nitrogeron	Root Nodne Pvt. Ltd. Australia	Rhizobium	Legume

B. Crop Protection (Biopesticides)

1. Microbial Herbicides

The use of endemic or exotic plant pathogens to kill weeds is the efforts of phytopathologists. The microbial origin of herbicides is definitely a major contribution and an alternative to chemical weedicides. Some of the successes with exotic pathogens have been the use of *Puccinia chondrillina* from Southern Europe to control skeleton weed (*Chondrilla juncea*) in Australia, while the use of *Cercosporella riparia* to control *Ageratina riparia* introduced in Hawaii from Jamaica, and the use of introduced rust *Phragmidium violaceum* to control wild black berry (*Rubus* spp.). Similarly, the use of *Cercospora rodmanii* to control water hyacinth and *Colletotrichum gloeosporioides* to control *Aeschynomene virginica*, the host specific pathotype *Phytophthora citrophthora* to control milk-weed vine (*Morreria odorate*) have been established. Such type of weedicides needs a close cooperation, whereas the use of endemic microbial herbicides needs cooperation from various agencies.

2. Bacterial Insecticides

Bacteria are often associated with plant and human diseases. However, there are certain bacteria such as *Clostridium acetobutylicum* (acetone-butanol production), *Bacillus licheniformis* (antibiotic- bacitracin production), *B. megaterium* (vitamin B12 production) and *Pseudomonas fluorescens* (2-ketogluconate), are some of them which are used for industrial product formation. In addition, there are certain bacteria which are of immense importance to mankind. They are pathogenic to insects, pests and other pathogens and kill these wide range of parasitic organisms. The majority of bacteria isolated from insects can be regarded as facultative pathogens. Interests in the use of bacteria as biological

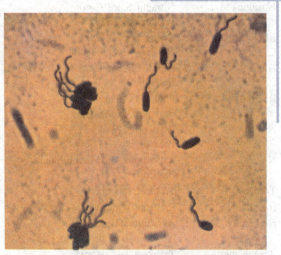

Pseudomonas fluorescens : Used for industrial product formation, as a bacterial insecticide.

control agent was stimulated by an increasing number of insects and pests. Some of the bacteria which are being used as bacterial insecticides are listed in Table 31.3.

Table 31.3 : Bacterial insecticides and their host range

Product/Antagonist	Target pathogen /host	Source (Country)
Blue circle/ Pseudomonas cepacia	Rhizoctonia solani, Pythium spp./vegetable seed treatment	Stine Microbials, Madison W1,USA
Daggerg/ Pseudomonas fluorescens	Rhizoctonia solani Pythium spp., Macrophomina phaseolina, Fusarium spp./cotton seed treatment	Ecogen Inc. Langhorne, PA USA
Epic/ Bacillus subtilis	Rhizoctonia solani, Fusarium species/cotton, legumes	Gustafson Inc., Dallas TX, USA
Bactophyll Bacillus subtitis	Various fungi/vegetables	NPO vector, Novosibirsk, Russia.

(i) **Pseudomonads as Bacterial Insecticides:** *Pseudomonas aeruginosa* is among the most frequently described pathogen causing disease in insects. It is still unknown that certain strains pathogenic to insects by feeding differ from those that are potentially pathogenic to man. *P. aeruginosa* that produces toxic enzymes, has also been used as a model organism to study the mechanism of insect pathogenicity and immunity. Non-fluorescent pseudomonads isolated from insects include *Pseudomonas alcaligens*, *P. cepacia*, *P. maltophila* and *P. acidovorans*. Aeromonads occur in laboratories devoted to the culture of insects in aquatic habitat. Some aeromonads such as *Aeromonas hydrophila* and *A. formicans* are pathogenic to insects.

Pseudomonas cepacia is known to be a versatile bacterium of soil (Sinsabaugh and Howard, 1975), as a plant pathogen (Burkhalder, 1950) and a human pathogen (Ederer and Matsen, 1972) as well as a broad spectrum antagonist to plant pathogens through the production of various types of antibiotics such as pyrrolnitrin (Janisiewicz and Roitman, 1988). *P. cepacia* is a Gram-negative bacterium that has been reported to produce siderophores. This bacterium also acts as plant growth promoting rhizobacteria (PGPR). Suppression of plant diseases may involve secretion of siderophores or antibiotics and/or aggressive root colonization by organisms that displace or exclude deleterious rhizosphere microorganisms. *Pseudomonas fluorescens* is one of the most important biological control agents of many plant disease causing organisms. These are also common PGPR that secrete siderophores. Seed inoculation with these organisms helps in inducing growth and suppression of diseases. It produces fluorescent siderophores called pyoverdine or pseudobactrin which is characteristic of the fluorescent pseudomonads. The siderophores are low molecular-mass, water soluble, high affinity Fe(III) chelators. Siderophores are secreted under iron-limiting conditions as a means to secure available iron present at low concentration in soil. The ability of certain pseudomonads to utilize a wide range of ferric siderophores as a source of metabolic iron may contribute to their competitiveness and survival in the soil.

(ii) **Bacillus Species as Bacterial Insecticides**

(a) *Bacillus thuringiensis***:** This bacterium is now widely known as Bt. It is the most important bacilli reported to kill a wide range of insects like moths, beetle, mosquitoes, flies, aphids, insects, ants, termites, midges, butterflies even some pathogenic fungi such as *Pythium ultimum* and *Fusarium oxysporum* f.sp. *lycopersici*, depending upon the host strains of the bacterium. Some of the Bt strains are pathogenic to cockroaches, snails and protozoans. Due to a wide range of host killing, Bt occupies a tremendous significance in agriculture due to exo- and endo-toxin production (Bulla *et al*, 1980) and its related ease of mass production in submerged fermentation on relatively cheap media (Faust, 1974; Ignoffo and Anderson, 1979) besides many chemical compounds designed for use in controlling economically and biomedically important insects.

The Bt bacterium was first discovered by a Japanese scientist Ishiwata in 1902 who isolated a bacterium from diseased or unhealthy silk larvae, and was named as *Bacillus satto*. In the year 1912, a German microbiologist also isolated this bacterium from infected insects from a flour mill. He named the bacterium which caused the insect disease, as *Bacillus thuringiensis*. After a gap of 50 years, the importance of this organism was realized and development of commercial formulation of Bt as biopesticide in USA was carried out in 1960. Now hundreds of Bt strains are commercially available round the world and all the major insect pests are susceptible to these strains.

The mechanism of action of Bt endotoxins on insect is quite interesting. The endotoxin is the protein. The crystalline proteins upon ingestion by the insect larvae are solubilized under highly alkaline conditions prevalent in the midgut. The toxins are digested by the enzymes, called proteases into active fragments. These active fragments bind to receptor proteins present in the gut epithelial membrane. Upon binding, the toxin molecules form pores in the gut membrane. As a result the osmotic equilibrium of the cell is disturbed, the cells swell and burst. This results into the death of insect larvae.

Now, more than 50 Bt genes have been isolated, cloned and characterized. Making use of microorganisms and genetic engineering, transgenic microbes and transgenic plants are developed. Many crop plants are colonized by harmless bacteria. Such bacteria are identified and genetically transformed by vectors carrying Bt genes. These bacterial formulations are then sprayed on the crop which provide a protective cover to the crop. Certain crop plants are also being protected against insects by genetic mediation *i.e.* transgenic plants (resistant to insects) with Bt genes for example bollworm resistant cotton, stem borer resistant rice, corn borer resistant maize, potato beetle and tuber moth resistant potato, tomato resistant to pinworm, etc. At IARI, New Delhi, Bt transgenic cabbage and cauliflower plants have been developed. Apart from crop plants, many forest tree species are also being transformed using Bt genes. Now Bt technology is being used in expressing two different kinds of Bt genes in transgenic plants or microorganisms. This technique helps in preventing insects from developing resistance to Bt toxin proteins.

Bt formulations are being used to control *Aedes* and *Anopheles* mosquitoes, black fly which spread yellow fever, malaria and blindness respectively. Now-a-days Bt genes are being expressed in aquatic bacteria and cyanobacteria so that during their propogation in lakes, the breeding of mosquito could be checked. Scientists in Anna University, Chennai are engaged in research on Bt. Bt strains are also being developed towards killing of liver flukes, tapeworms, etc. which cause disease in cattle and human.

Bt is produced commercially in the form of powder the composition of which is given in Table 31.4.

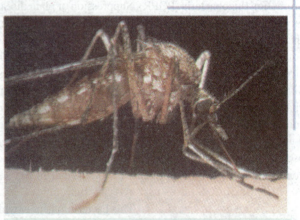

Bt formulations are used to control *Aedes* mosquito which spreads yellow fever.

Table 31.4 : B₄ medium for Production of *Bacillus thuringiensis* powder

Ingredients	Amount in g lit⁻¹
Beer	2 lit
Proflo (defatted cooked cotton seed flour)	10
Dextrose	15
Yeast extract	2
Bacto-peptone	2
$MgSO_4.7H_2O$	0.3
$FeSO_4.7H_2O$	0.02
$ZnSO_4.7H_2O$	0.02
$CaCO_3$	1.0

All the ingredients are mixed to form powder thoroughly, the powder was administered to the insects by mixing it into their diet, with larvae being allowed to feed on the powder diet mixtures. The effect of this exposure was judged by a single criterion, *death*. The calculations of potentials of dry powders of δ-endotoxins of *B. thuringiensis* is given below:

$$\frac{L_{C50} \text{ Standard}}{L_{C50} \text{ test sample}} \times \text{Potency of standard (IU / mg)}$$

(*b*) **Toxins produced by *B. thuringiensis*:** Bt produces several toxins, four of which will be considered here: α-exotoxin (heat-labile toxin); β-exotoxin (fly-factor or heat-stable exotoxins); δ-endotoxin (crystalline toxin or just crystal or parasporal body) and louse factor.

The α-exotoxin identifies as lecitherase c water soluble, heat labile and toxic to insect while, β-exotoxin defined as adenine nucleotide and ATP analogue and given the name "thuringiensin" which is water soluble heat-stable toxin. The δ-endotoxin in the crystal of Bt has a limited activity spectrum. It is produced during sporulation. In 1974, Ginrich *et al* reported that four species of mammal - biting lice were susceptible to powders containing the spore endotoxin complex of Bt-var Kurstaki (HD-1), an isolate of Bt that does not produce β-exotoxin, a toxin to which these lice are susceptible. It seems improbable that the endotoxin could be responsible for the action. Hence, the toxicity to lice was due to a new chemical which was called the "*louse-factor*".

Damaged cotton flower (left) and healthy flower (right) after *Bt* applications

(*c*) **The Other Bacteria which Control Insects:** Among the species of *Bacillus*, there are other obligate spore-forming bacteria such as *Bacillus papilliae* and *B. lentimorbus*, facultative spore former such as *B. cereus* which are pathogenic to insect. The control of Japanese beetle by inducing milky disease with *B. lentimorbus* and *B. papilliae* have been reported. *E. coli* is reported from flies as surface contaminant. The strains of *E. blattae* has been isolated from cockroach. Strains of *Enterobacter*, *Serratia* and *Klebsiella* are frequently reported as facultative insect pathogens.

Serratia liquefaciens and *S. marcescens* control lab insects in culture due to unsatisfactory nutrition and physical conditions in the maintenance. *Alcaligenes odorans* has been isolated from insects where this bacteria acts as facultative pathogen.

(d) **Bacteria associated with nematodes:** Relationship between the insect (host) and the nematode gut has been studied by Poiner (1979). Nematode larvae entering an insect body were reported to cause septicaemia by dissolving the insect gut wall and thus introducing bacteria to the haemolymph. An entomocidal (insect-killing) bacterium named *Achromobacter nematophilus* was described initially from the nematode *Neoplectana carpocapsae* and regarded as a commensal symbiont of the nematode. The nematode genera, *Neoplectana* and *Heterorhabditis* carry in its gut bacterial species named *Xenorhabdus nematophilus* and *X. fluorescens* belong to family Entero-bacteriaceae. Even one bacterium from each species is enough to kill insect.

The advantage of using bacterial insecticides are: (*i*) permanency, (*ii*) economic feasiblity due to low cost, (*iii*) environmental safety, (*iv*) absence of development of resistance.

3.   Virus Insecticides

There are viruses or their products commercially exploited in place of chemical insecticides. Viruses of the family Baculoviridae are pathogenic to arthropods. Viruses contain lipid envelope with circular, double stranded DNA genome of 80-150 Kilo base pairs depending upon the viruses (Buchanan *et al.,* 1987; Palmiter and Brinster, 1986). Naked viral DNA is infectious *per se*. Baculoviruses are restricted in their host ranges. They do not infect vertebrates, non-arthropod invertebrates, microorganisms or plants. Indeed baculoviruses infect only a few arthropod species. Their use dates back to the 19th century. But the commercial use of virus insecticides has been limited by their high specificity (limited host range) and slow action.

Viruses most frequently considered for control of insects (usually saw flies and lepidoptera) are the occluded viruses, namely NPV, cytoplasmic polyhedrosis (CPV), granulosis (GV) and entomopox viruses (EPN).

4.   Entomopathogenic Fungi

Many entomopathogenic fungi overcome their hosts only after limited growth in the haemocoel so toxins are presumed to cause host death. The importance of toxins to the virulence of an entomopathogenic fungus is difficult to evaluate. Because toxin production in the host must be preceded by following activities: (*i*) attachment of the infective unit on the cuticile, (*ii*) germination of infective unit, (*iii*) peneteration followed by formation of germ tube and appearance of infection pegs from appressoria, (*iv*) multiplication of hyphal bodies in the haemocoel, (*v*) production of toxic metabolites, (*vi*) death of the host, (*vii*) growth in the mycelial phase with invasion of all host organs, (*viii*) penetration of hyphae from the interior through the cuticle to the exterior of the insect, (*ix*) production of infective units on the exterior of the insects.

(a) *Metarhizium anisopliae:* This fungus produces destruxin B and desmethyl destruxin B in silk worm larvae. These toxins are referred to as vivotoxins. In addition to low molecular weight compounds, proteases and other enzymes are produced by entomopathogenic fungi. Injection of *Entomophthora* species cultures into *Galleria mellonella* larvae caused blackening similar to that noted in infected larvae.

Metarhizium culture filtrate is toxic to coleoptera haemocytes *in vitro* producing changes in organelles. The extract of mycelium is toxic if it comes in the contact of adult house fly. The solvent extracts of *Metarhizium* kill silk worms if it is injected into intrahaemocoelic injection. Six cyclodepsipeptides with five member amino acids viz., β-alanine, alanine, valine, isoleucine and proline have been isolated from filtrates of *M. anisopliae* cultures.

Cytichalasins are the other fungal metabolites. Their origin is from phenylalanine or tryptophan linked to C_{14} or C_{18} polyketide chain. Their activity includes inhibition of cytoplasmic cleavage in cultured mammalian cells.

(*b*) *Beauveria bassiana* **and** *B. brongriartii*: *B. bassiana* and *B. brongriartii* are the other fungal species that attack insects. Beauvericin is a depsipeptide. It comprises of a cyclic repeating sequence of 3 molecules of *N*-methyl phenylalanine alternating with three molecules of 2-hydroxy isovaleric acid. It has also been isolated from mycelium of *Paecilomyces fumosoroseus*.

Beauverolides and bassianolide are the other cyclodepsipeptides. The beauverolides H and I were isolated from *B. bassiana,* while bassianolide which is composed of 4 molecules each of L-N methyl leucine and D-α-hydroxy isovaleric acid has been isolated from *B. brongriartii.* The isorolides A, B, and C are also cyclodepsipeptides which were found in *B. brongriartii.* Two very similar pigments, tanellin and bassianin, produced *in vitro* by some strains of both *B.bassiana* and *B. brongriartii* are concentrated in the mycelium rather than released into the medium. The dibenzoquinone pigment oosporein, produced by many isolates of *Beauveria*, probably accounts for the reddish colour of infected caterpillars. *B. brongriartii* converts 20% of the original solids in a peptone medium into oxalic acid. This acid is a general poison. Oxalate crystrals have been noted on the surface of insects killed by *B. bassana*.

(*c*) *Verticillium lecanii*: *V. lecanii* is non-fastidious and can grow on all conventional media meant for culturing of fungi. The most frequently recorded hosts are scale insects and aphids. Much less reports of hosts have been recorded in other orders. *V. lecanii* sometimes hyperparasitizes phytopathogenic fungi, mostly rusts and powdery mildews.

The control of aphids and seales is possible by using *V. lecanii* conidia or blastospores suspended in phosphate buffer containing 0.02% Triton X-100 as wetting agent. After evening sprays, plants were covered with polythene blackout sheets to restrict day length for flower initiation.

The appraisal of *V. lecanii* as a microbial insecticide is based on the choice of infectious material *i.e.* between conidia and blastospores. Production of conidia on agar is too expensive and also difficult to ensure culture purity. Alternatively, the conidia can be produced on a cheap granular solid substrate. *V. lecanii* is reported to be a promising biological control agent aganist aphids of chrysenthemum in green houses.

(*d*) *Hirsutella thompsonii*: *H. thompsonii* a deuteromycetous fungus, is a potential killer of citrus rust mite (*Phyllocoptrula oleivora*). Most of the species are pathogenic to invertebrates. The fungal strains have also been reported from the citrus bud mite, *Eriophyes sheldone* and coconut flower mite, *E. guerreronis.*

In the USA, a commercial formulation of conidia has been introduced by Abbott Laboratories, North Chicago, Illinois. *H. thompsonii* can be cultured on agar media. The spore suspension is mixed in semisolid medium containing wheat bran (60 g) and distilled water (60 ml) to which 250 ppm per litre of chloramphenicol is added. The flasks are incubated at 25°C for 2 weeks to get bran-fungal mat. It is air dried and blended, finally stored at 18-20°C.

(*e*) *Nomuraea rileyi*: It is the Ascomycetous fungus for which first effort of mass-production was made for applying as entomopathogenic agent. This fungus usually is found to induce extensive epizootics in caterpillar pests on cabbage, clover, soybean and velvet beans and thus is a potential agent for use as a microbial insecticide. The natural epizootics were observed on *Bombyx mori*, *Peridroma saucia, Leptinotarsa decemlineata.* The unidentified compounds were extracted from mycelium produced in submerged culture.

(*f*) **The other fungi:** *Aspergillus ochraceus, Paecilomyces fumosoroseus, Fusarium solani Trichoderma harzianum*, *T. reesei, Gliocladium virens,* etc. have also been reported for other microbial pesticides.

C. Microbiology of Lignocellulose Degradation in the rumen

Lignocellulose (fodder) degradation in the rumen is a complex succession of reactions. A number of rumen microorganisms are cellulolytic and hemicellulolytic. These organisms act in consortium of mixed culture to hydrolyse the plant cell wall polymers. Small plant fragments are engulfed intact by the larger rumen ciliates such as *Eudiplodimuim maggii.* Large, fresh fragments are subjected to invasion cellulolytic chytridiomycetes fungal zoospores, such as *Neocallimatrix frontalis,* and motile cellulolytic bacteria namely, *Butyrivibrio fibrisolvens, Eubacterium cellulosolvens,* etc.

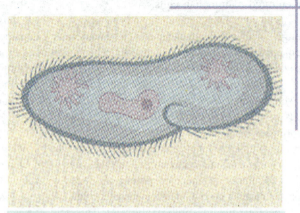

Lignoocelluloysis by ciliate protozoa : Small plant fragments are ingested and subjected to hydrolysis by intracellular enzymes.

Lignocellulolysis by ciliate protozoa may be the least complex of the rumen microorganisms. In these species small plant fragments are ingested and subjected to hydrolysis by intracellular enzymes probably of protozoal origin. The ingested plant fragments may carry adherent celluloytic bacteria, and it is possible that their enzymes may remain active insides the ciliate. Removal of fermentation products such as volatile fatty acids, through the rumen wall, and lactate, succinate and hydrogen by metabolism by other rumen bacteria is also believed to help cellulolysis in *vivo.*

1. Use of Recombinant DNA Technology in Rumen Bacteria

Genetic modification of rumen bacteria by using rDNA technology may improve the rate or extent of lignocellulolysis in the animal. The task of developing effective rDNA technology in rumen bacteria, particularly for improving lignocellulose digestion is complex. After the rate limiting steps have been defined, it is necessary to identify and clone suitable heterologous genes for insertion into a rumen bacterium. It is then necessary to develop effective vectors and transformation system to develop methods to monitor both the survival of the gene in the host cell and its survival in the rumen, and finally to measure its effect on *in vivo* digestibility of lignocellulose.

Besides rDNA technology, genetic mutation is another technique, which produces greater quantities of rate-limiting enzymes. Mutagenesis in *Ruminococcus albus* cause regulation of cellulose enzymes. Genomic DNA libraries have been constructed of fragments from the cellulolytic rumen bacteria *Bacteriodes succinogens, Ruminococcus albus, Butyrivibrio fibrisolvens* and others. Gene cloning in rumen bacteria may provide valuable insights into our understanding of the process operating in rumen.

Rumen Microflora : It is very much similar to the chemostat or a fermenter in which cellulosic biomass in the form of fodder is digested. Actually, it varies in size from cow to sheep wherein it has a capacity of approximately 30 litres to 6 litres, respectively. The temperature remains 30°C and pH 6.5. The anaerobic atmosphere persists in rumen. It runs continuously. Since rumen contains microorganisms (symbionts and parasites), the fodder is digested in this region. The ruminants animals do not have celluloytic enzymes to digest cellulose but mammals can metabolise them due to presence of cellulose decomposers namely, *Fibrobacter succinogenes, Butyrivibrous fibrisolvens, Ruminobacter albus, Clostridium lochheadii;* starch decomposers such as *Bacteriodes ruminicola,*

Ruminobacter amylophilus, Selenomonas luminatium, Succinomonas amyolyticus. Streptococcus, bovis; lactate decomposers e.g. *Selenomonas lactilytica* and *Megasphaera elsdenii,* pectin decomposers *(Lachnospora multiparus)* and lastly methanogens, *Methanobrevibacter ruminantium* and *Methanomicrobium mobile.*

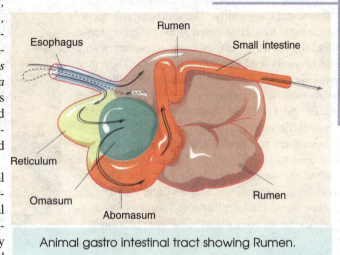

Animal gastro intestinal tract showing Rumen.

The rumen has about 10^2/ml microbial population of anaerobes. The concept of microbial consortium or biofilm is an example wherein combined activity of several bacterial species and protozoa occur. For instance, a fermentative bacterium produced hydrogen gas and another species of methanogens uses it for production of methane gas.

Rumen also consists of ciliate protozoal fauna of obligate anaerobes. They eat and ingest rumen bacteria, hence maintain the bacterial densities. Rumen also possesses some anaerobic fungi, which ferment cellulose and volatile fatty acids. They also play a role in degradation of other polysaccharides including a partial degradation of lignin, hemicellulose and pectin.

The microflora and their action depends upon the type of fodder available in rumen. For example, if an animal is fed legume hay rich in pectin, the bacterial genes *(Lachnospira multiparus)* will be more active. Certain rumen bacteria produce alcohol as a fermentation production, yet ethanol rarely accumulated in the rumen due to the fact that it can be fermented to acetate and H_2. This hydrogen never accumulates because it is quickly used to reduce CO_2 to CH_4 by methanogens. Acetate is not converted to CH_4 in the rumen because the retention time is too short for development of acetolastic methanogens, which typi-

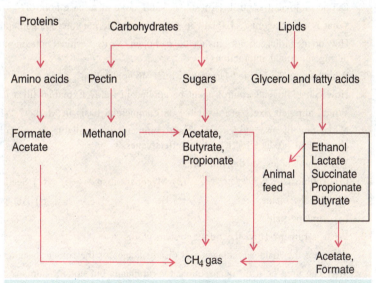

Fig. 31.2 : Mechanism of fodder degradation and production of CH_4 which serves as a hydrogen sink.

cally grow very slowly. Sometimes a change in the microbial composition of the rumen causes illness or even death of the animal. If an animal is given grain diet, a high density and population of *Streptococcus bovis* occurs. *S. bovis* produces lactate from starchy grains, which acidifies the rumen called *acidosis* (Fig. 31.2.).

Certain fungi (anaerobic) are present in the symbiotic gut community. The uniflagellates spores of these fungi have been noted as motile spores move around the rumen until they come in contact with a piece of swallowed vegetational matter. They then grow long projections called rhizoids, into the food. The rhizoids produce enzymes that digest the cell wall of the food and disintegrates in to pulp. Eventually, a sporangium develops and more flagellate spores are released. These microorganisms also seem able to use N_2, which has been secreted or ingested into the rumen in the form of urea. They use this to synthesize proteins. Rumen fungi also degrade cellulose, hemicellulose, lignin and pectate substances of woody-herbaceous plants used as fodder for animals.

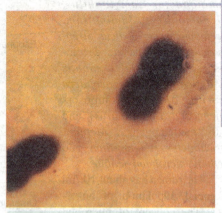

Streptococcus bovis, the causal agent of acidosis.

QUESTIONS

1. What is biofertilizer? Write note on its significance?
2. What type of microorganisms are used as bioinoculants? Write a critical note on phosphate solubilisers?
3. Write short notes on the following:
   (a) Bacterization   (b) Biofertilizers
   (c) Mycrrohizae   (d) Cyanobacterial biofertilizers
4. Define biological control. How far is it effective in disease control?
5. What is an antagonist? Explain in brief about different microbial interaction processes?
6. How do the bioinsecticides invade the pathogen? Write a critical note on morphological changes occurring during microbial interactions?
7. What is mycoherbicide? Discuss in brief its application?
8. How can biological control agent be applied in field? Explain with a few examples.
9. Which nitrogen fixers are available on commercial basis in Market? What is their significance?
10. Name few bioinsecticides which are commercially available in the market. How will you test the commercial product for its quality/effectiveness?
11. Write short notes on the following:
    (a) Antibiosis   (b) Mycoparasitism   (c) Seed inoculation
    (d) Mycoherbicide   (e) Bt   (f) Antagonism
    (g) Viral pesticides
12. What is rumen? How the rumen system works?
13. What are the main advantages and disadvantages of the rumen system?
14. Why can urea be a nitrogen source for ruminants but not for human?
15. What is the use of rDNA technology is rumen bacteria?

REFERENCES

Anonyomus 1986. *Cereal Nitrogen Fixation* ICRISAT, Hyderabad.

Anonymous 1997. TCDC International Workshop on application of Biotechnology in Biofertilizers and Biopesticides. Department of Biochem. Engg. & Biotech., IIT, New Delhi.

Bagyaraj, D.J. 1984. Biological interactions with VA mycorrhizal fungi. In *VA mycorrhiza*, C.L. Powell and D. J. Bagyaraj (Eds.). P.P. 131-153, CRC Press, Boca Raton, Florida.

Baker, K. F. and Snyder, W.C. 1965. *Ecology of soil-borne plant pathogens–Prelude to biological control* Univ. of california Press. Berkely.

Brown, M.E. 1974. Seed and bacterisation *Ann. Rev. Phytopath.* 12: 311-331.

Dreyfus, B., Rinando, G and Dommergues, Y.R. 1985. Observations on the use of *Sesbania rostrata* as green manure in paddy fields. *Mircen J. Appl. Microbiol. Biotechnol.* 1. 111-121.

Fred, E.B., I.L. Baldwin and E. Mc Coy. 1932. *Root nodule bacteria and leguminous plants*. Unit of Wisconsin Press, Madison, PP. 118.

Garrett S. D. 1982. *Soil fungi and soil fertility*. Pergamon press, Oxford.

Hobsen, H.J., N.N. Aung, P. Somasegran and U.G. Kang. 1991. Oils as adhesives for seed inoculation and their influence on the survival of *Rhizobuim* spp. and *Bradyrhizobium* spp. on inoculated seed. *World J. Microbiol. Biotechnol* 7:324-330.

Iswaran, V., A. Sen and R. Apte. 1972. Plant compost as a substitute for peat for legume inoculants. *Curr Sci.* 41: 257.

Natesh, S., Chopra, V.L. and S. Ranachandran. 1987. *Biotechnology in Agriculture*. Oxford & IBH Publ. Co., New Delhi.

Nobbe, F. and L. Hiltner. 1896. Landw Versuchosten 47 : 257.

Peters, G.A., Ito, O., Tyagi, V.V.S, Mayne, B.C., Kaplan, D. and Calvert, H.E. 1981. Photosynthesis and N_2 fixation in the *Azolla-Anabaena* Symbiosis. In *Current Perspectives in Nitrogen fixation*, A. H. Gibson and W.E. Newton. Aust. Acad. Sci. Canberra, pp. 121-24.

Powell, C.L. and D. J. Bagyaraj 1984. Effect of mycorrhizal inoculation on the nursery production of blue berry cuttings. *N.Z.J. Agrie. Res.* 27: 467-71.

Rangaswami, G. 1988. Soil-plant-microbe interrelationship. *Indian Phytopath.* 41 : 165-172.

Roughley, R. J. and J. M. Vincent. 1967. Growth and survival of *Rhizobuim* spp. in peat culture. *J. Appl. Bacteriol* 30: 362-376.

Singh, R. N. 1961. *The Role of blue green algae in nitrogen economy of Indian agriculture*. ICAR, New Delhi.

Seeniva, M.N. and D. J. Bagyaraj. 1988. *Chloris gayana* (*Rhodes* grass) a better host for mass production of *Glomus fasciculatium* inoculum. *Pl. Soil* 106: 289-90.

Somasegran, P. and H. Hoben. 1994. *Hand book of Rhizobia. Methods in legume-Rhizobium Technology*. University of Hawaii. NiFTAL Project.

Subba Rao, N.S. Tilak, K.V.B.R., Laxmi Kumari, M. and C.S. Singh. 1979. *Azospirillum*- A new bacterial fertilizer for tropical crops. *Science Reporter* 16: 690-92.

Thompson, J. A. 1980. Production and quality control of legume inoculants. P.P. 489-533. In F.J. Bergersen (ed.) *Methods for evaluating biological nitrogen fixation*. John wiley & Sons.

Venkatraman, G.S. 1972. *Algal biofertilizers and Rice cuttivation*. Today & Tomorrow's Printers & Publishers, New Delhi

Watanabe, A and Y. Yamanato. 1971. Algal Nitrogen fixation in the tropics. *Pl. Soil.* 403-413 (special vol.)

Instrumentation in Microbiology

32

Normally naked eyes cannot see the objects less than about 100 μm in diameter. This is also the resolving power of eyes. The size of bacteria varies from 1-5 μm while the size of viruses ranges from 25-350 μm. The hand lenses giving enlargement of 2 to 10 X have been known and used from ancient times. Technically these are also microscopes.

A. Microscopy

1. Beginning of Microscopy

A dutch specialist of spectacle, Zoocharia Janssen (1590) used second lens and thus the image formed by primary lens was greatly enlarged of the order of 50 to 100 X. This is the basic principle on which modern compound microscope is based. In 1610 Galileo invented some what "improved microscope". Robert Hooke (1635-1703) made and used a compound microscope in 1660s and published a book "Micrographia". The highest magnification of it was 200 X but he did not observe bacteria. A contemporary of Hooke, named Antoni van Leeuwenhoek (1632-1723) discovered unicellular, independently living microorganisms which he called

Robert Hooke

"animalcules". Obviously, they are protozoa and bacteria. Due to this achievement, he was called "father of microbiology". Professionally Lecuwenhoek was a linen merchant. During his time he used to make minute, simple but powerful lenses. This instruments were simple magnifying glasses, generally consisting of a small, single, biconvex, almost spherical lenses, but they gave magnification to about 300 X. The size of the objects which he examined was determined by comparison. For this purpose he used sometimes grains of sand, seeds of millets or mustard or blood corpuscles, etc. In all, he made 419 lenses, some mounted in gold while most in brass. C. Huygens developed two lens eye pieces, while Abber (1840-1905) developed apochromatic objectives and sub-stage condenser. The apochromatic are those which are relatively free from chromatic and spherical aberrations. The professional use of compound microscope was observed after 1820. The total magnification produced by modern objectives and ocular together, is the product of two magnifications. In addition to magnification, however, resolution, is the ability to distinguish two adjacent points as separate.

2. Types of Microscopes

A microscope may be defined as optical instrument consisting of a lens, or combination of lenses for making enlarged or magnified images of minute objects. The microscope is absolutely indispensable to the biologists in general and to the microbiologists in particular. Generally, microscopes are of two categories.

3. Light Microscope or Compound Microscope

A microscope is an instrument which makes enlarged image of minute objects near the objective lens. The image is formed as shown in the Fig. 32.1.

The compound microscope has two set lenses. One is known as objective and the other eye piece. These are mounted in a holder commonly known as body tube. The lens system nearest to the specimen is called objective which magnifies the specimen to a definite number of times. The second lens system is called eye piece which is the nearest to eye. It further magnifies the image formed by the objective. Accurate focusing is attained by a special screw appliance called as fine adjustment.

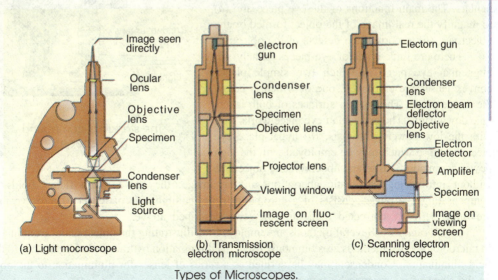

(a) Light mocroscope

(b) Transmission electron microscope

(c) Scanning electron microscope

Types of Microscopes.

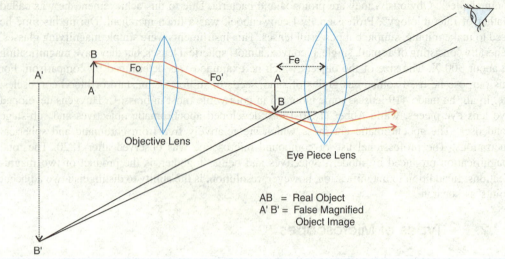

AB = Real Object
A' B' = False Magnified
Object Image

Fig. 32.1 : Image formation in compound microscope.

(i) Parts of Compound Microscope : A compound microscope consists of several parts as described below:

(a) Objective Lens: The chief characteristics of the objective lens are: (a) to gather the light rays coming any point to the object, (b) to unite the light in a point of image, (c) to magnify the image.

There are three types of objectives namely (a) *achromatic* which are simplest in construction and adequate for most purpose, (b) *flourite* are those in which the aberrations are largely eliminated by use of these objectives, (c) *apochromatic* which are most corrected with respect to aberrations.

(b) Eye piece: The eye piece is also known as oculars. The main functions of the eye piece are: (a) to magnify the real image of the object formed by the objectives, (b) to correct the defects of the objective.

There are three types of eye pieces namely: (a) huggenian eye piece in which two simple plano-convex lenses are employed, one of which is below the image plane. The convex surfaces of both layers face downwards. The huggenian eye piece works well with the low power achromats, (b) hyper-place eye piece are those which may be employed with the high power acromatic flourite and apochromatic objectives

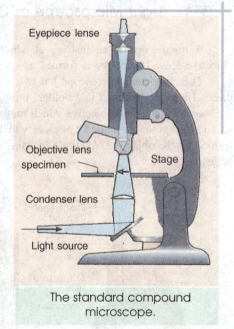

The standard compound microscope.

without introducing chromatic aberrations in the image; they give much flatter fields, (c) compensating eye piece consists of a chromatic triplet combination of lenses. These eye piece are more perfectly corrected then the other two described above.

(c) Condensers: Several methods are employed for illustrating the object under examination. In microbiology two methods are commonly used: (a) illumination by transmitted light, and (b) dark field illumination. Condensers may be defined as a series of lenses for illumination to the object on the stage of microscope by transmitted light.

(ii) Resolving Power: It is defined as the power of an objective able to separate distinctly two adjacent points. The resolving power of a microscope is a function of the wavelength of light used and the numerical aperture (NA) of the lens system. The larger the NA, the greater the resolving power of the objective and the finer the detail it can reveal.

(iii) Numerical Aperture (NA): The angle subtended by the optical axis and the outermost rays still covered by the objective is the measure of aperture of the objective. It is half-aperture angle. The magnitude of this angle is expressed as a sine value.

The sine value of the half aperture angle multiplied by the refractive index (n) of the medium (the space between the front lens and cover slip) gives the numerical aperture (Fig. 32.2).

Hence NA = n sine θ

With dry objectives the value of n is 1 because 1 is the refractive index of air. When immersion oil is used as the medium, n = 1.56 and if θ = 58°, then NA = S sine, θ = 1.56 × sine 58 = 1.56 × 0.85 = 1.33.

If objects are smaller, the smallest wavelength of visible light cannot be seen. In order to see such minute objects it would be necessary to use rays of shorter wavelength.

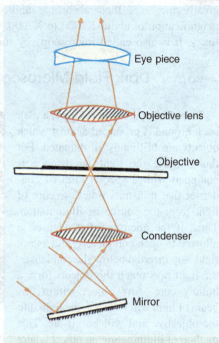

Fig. 32.2 : Functioning of oil immersion.

(iv) Limit of Resolution: It is the smallest distance by which two objects can be separated and distinguished as two separate objects. The greater resolution in light microscopy is obtained with the shortest wavelength of visible light and an objective with maximum NA. The relationship between NA and resolution can be expressed as follows:

$$d = \frac{\lambda}{2\,NA}$$

where, d = resolution

λ = wavelength of light

NA = numerical aperture

(v) Magnification: Magnification is the increase in size of the object. Magnification beyond the resolving power is of no value. It is because the larger image will be less distinct in detail and fuzzy in appearance.

4.   Bright Field Microscopy

In bright-field microscopy, the microscope field (the area observed) is bright and the microorganisms appear dark because they absorb some of the light. Normally, microorganisms do not absorb much light, but staining them with dye

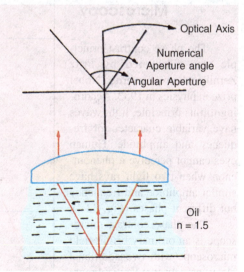

Fig. 32.3 : Schematic representation of the compound bright field microscope.

greatly increases their absorbing ability. Generally microscope of this type produce useful magnification of about X1000 to X2000. At magnification greater than X2000, the image appears fuzzy. It is also called microscopy by transmitted light (Fig. 32.3).

5.   Dark Field Microscopy

In this type of microscopy, a dark back ground is produced against which objects are brilliantly illuminated. For this purpose the light microscope is equipped with a special kind of condenser that transmits a hollow core of light from the source of illumination. Thus, if the aperture of condenser is allowed to open completely, and a dark field stop inserted below the condenser, the light rays reach the objects form a hollow core. Any object within this beam of light will reflect some light into the objective and will be visible. This method of illuminating an object where

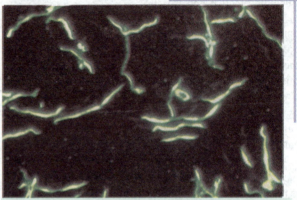

Treponema pallidum, the spirochete that causes syphilis; dark-field microscopy.

the object appears self-illuminous against a dark field, called dark-field illustration. The condensers used are Abbe condenser, paraboloid condenser and cardoid condenser. Dark field microscopy is particularly valuable for the examination of unstained microorganisms suspended in fluid wet mount and hanging drop operations.

6.   Phase Contrast Microscopy

The phase contrast principle was discovered by Fritz Zernike who was awarded Nobel prize in physics in 1953. According to this principle, light waves have variable character for frequency and amplitude. Human eyes cannot perceive a phenomenon when two light rays have similar amplitude and frequency but different phases (Fig. 32.4).

The phase contrast microscope is an ordinary bright field microscope with two additional plates, namely annular diaphragm and phase shifting plate, which enables the usage forming rings to be phase shifted with respect

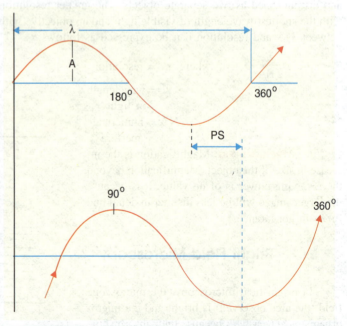

Fig. 32.4 : Light waves exhibiting similar properties of amplitude (A), and wavelength (λ) but differing from one another by being out of phase (Ps).

to others. Annular diaphragm allows only a ray of light to pass through the condenser and then to object. The phase shifting plate is placed at the rear focal plane of objective lens. This disc has a ring of optical dielectric material on which the ring of light from annular diaphragm is focused. The ring has the property of retarding or advancing the phase of light of a quarter of a wavelength traversing it.

Phase contrast microscopy.

From each translucent or transparent particle in the object, consider a single ray of incident light. From this two rays result, one, the direct (undiffracted) ray comes through the annular diaphragm passes through the objective and focussed on phase-shifting ring which either retards or advances the ray one quarter wavelength with respect to second ray. The second ray is also derived from the incident ray, but modified by being scattered or diffracted in passing around margin of the object. This ray does not pass through the phase-shifting ring but traverse the other area of transparent disc. Its wavelength is neither retarded nor advanced. Thus, there is one quarter wavelength out of phase. The contrast between the two is called phase-contrast. It is valuable device in wet mounts and hanging drops. It increases visualization of cellular structure and traverse the other area of transparent disc.

7. Fluorescent Microscopy

Many chemical substances absorb light. After absorbing light of a particular wavelength and energy, some substances emit light of larger wavelength and lesser energy content. Such substances are called fluorescent and the phenomenon is termed as *flourescence*. This is the phenomenon which is applied in fluorescent microscopy. In practice, microorganisms are stained with a fluorescent dye and then illuminated with blue light. The blue light is absorbed and green is emitted.

Blue and green fluorscence can be excited by ultraviolet radiation as well. A high intensity mercury lamp is used as light source which emits white light. Two filters are used, one is the exiter filter which transmits only blue light to pass to specimen and blocks out all other colours; the second is barrier filter which blocks out blue light (excitation radiation) and allows green light (or other light emitted by fluorescing specimen) to pass through and reach the eye (Fig. 32.5).

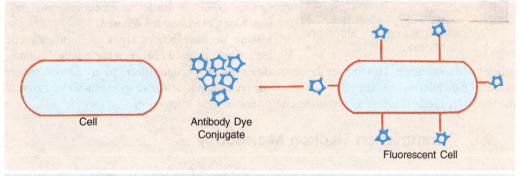

| Cell | Antibody Dye Conjugate | Fluorescent Cell |

Fig. 32.5 : Fluorescent cell.

As shown in Fig. 32.5 a high intensity mercury lamp is used as the light source and it emits white light. The excited filter transmits only blue light to the specimen and blocks out all other colours. The blue light is reflected downward to the specimen by a dichromatic mirror (which reflects light of certain colours but transmits light of other colours). The specimen is stained with a fluorescent dye: certain portions of the specimen retain the dye, others do not. The stained portion absorbs blue light and emit green light which pass upward, penetrate the dichromatic mirror and reaches the barrier. This filter allows the green light to pass to the eye and blocks out the resident blue light from the specimen which may not have completely deflected by the dichromatic mirror. Thus, the eye perceives the stained portions of the specimen of glowing green against a black background.

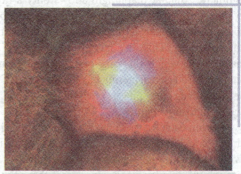

In fluoresecent microscopy, a fluorescent dye that binds to a specific cell material is stimulated by a beam of light.

Some suitable fluorescence are : acridine orange, acridine yellow; acriflavine auramine 0; fluorescence titan yellow G; rhencine A, etc.

8.   Electron Microscopy

In 1931 Knoll and Ruska, German scientists discovered electron microscopy. Von Borries and Ruska (1938) in Berlin constructed first practical electron microscope. The commercial instrument first came in around 1940.

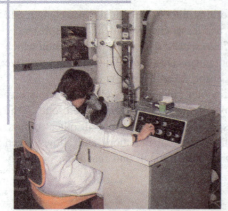

A Modern Transmission Electron Microscope.

In electron microscope the source of illumination is electron beam. The construction and principle of electron microscope are easily related to those of light microscope. The range of wave length of visible light used in light microscope is 4000 Å - 7800 Å, while with an electron microscope employing 60-80 KV electron, the wave length is only 0.05 Å. In the instrument as shown in the figure, the electron gun generates electron beam. These electrons are concentrated by other components of electron gun producing a fast moving narrow beam of electron. Electrons are focused by electromagnetic lenses. Electromagnetic lens consists of wire encased in soft iron casing. When electric current is passed through the coil, it generates an electromagnetic field through which electrons are focused.

There are three general types of electromagnetic lenses. The one is placed between the source of illumination and the specimen. This focuses the beam of electron on specimen functions in a similar manner as that of light microscope. The other two lenses are on the opposite side of specimen which magnify the image in similar fashion as objective and ocular in light microscope.

9.   Transmission Electron Microscopy

The electron source is commonly a tungsten filament of 30-150 KV potential. The electron beam passes through the centre of ring like magnetic condenser and becomes converged on the specimen. After being transmitted through the specimen (hence transmission electron microscope (TEM), the magnetic objective focuses the electron into a first (real) image of the object which is enlarged (2000 times). The magnetic projector lens then magnifies a portion of the first image producing mangnification upto 240,000 or more.

The final enlarged image can be reviewed by striking a fluorescent screen which makes it visible. The image can also be thrown upon a photographic plate for permanent record. Portions of the photographs may be enlarged four to six times giving the picture in the range of two million times as large as the object.

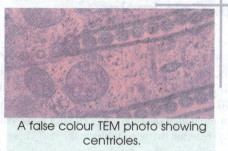

A false colour TEM photo showing centrioles.

$$0 = \frac{\text{Magnifies 2 million times}}{\text{2 mm diameter}} \longrightarrow 4 \text{ million max or 2.5 miles in diameter.}$$

Molecules in the microscope interfers with the movement of electrons. To prevent this, the interior of the microscope is kept in the state of high vacuum, around 10-4–10-6 mm Hg. It is also necessary to have specimen ultra thin. The electron beams has very poor penetrating power, therefore, only small objects or very thin sections of the specimen can be examined.

10. Scanning Electron Microscopy (SEM)

Scanning electron microscopes combine the mechanism of electron microscopy and television. SEM became commercially available in early 1960's and the researchers were Knoll, Von Ardenne, Zworytein etc.

In SEM, electrons are not transmitted through the very thin specimen from below but impinge on its surface from above. The specimen may be opaque and of any manageable thickness and size. If the specimen is an electron conductor, it needs only to be held on an appropriate support. If it is non-conductor, it is allowed to dry but if moist, freeze dried in liquid nitrogen is necessary. The specimen is then coated with metal vapour (gold) in vacuum.

A false colour SEM photo of *Paramaecium*.

The electrons originate at high energy (20,000 V) from a hot tungsten or lanthanum hexoboride cathode "gun". These electrons are sharply focused, adjusted and narrowed by an arrangement of magnetic fields. Instead of forming a broad inverted cone of rays, in SEM a needle sharp probe (about 5 - 10 mm in diameter) is made. This primary beam (probe) acts only as an exciter of image forming secondary electrons emerging from the surface of the specimen.

The probe scans the specimen like that on a blank TV screen. The probe can impinge on depth and heights with equal speed and accuracy giving great depth of field and producing images with three dimensions. Images are elicited from wherever the probe strikes the metal coated areas of the specimen. Magnification is the ratio of final image to the diameter of area scanned.

Any of the secondary electrons with sufficient energy can emerge from the surface. Those that emerge not too far from the point of impact of the probe, can be used to form an image. The useful secondary electrons are magnetically deflected to a collector or detector. Here, they produce a signal that represents at any single moment, only 5-10 mm area or spot of impingement of the probe on the specimen. The successive signals from the collector are amplified and transmitted to a cathode ray (TV) tube. The scanning beam and TV tube beam are synchronized. The image scan by the eye on TV screen is thus the sequence of signals representing in *araster pattern*, the successive areas traversed by the primary probe beam. Exposure may range from a few second to one-half hour or more. The TV image may be photographed, video taped or processed in motion on a computer.

B. Colorimetry

Most of the biochemical experiments involve the measurement of a compound or group of compounds present in a complex mixture. Probably the most widely used method for determining the concentration of biochemical compounds is *colorimetry*, which makes use of that when white light passes through a coloured solution, some wavelengths are absorbed more than others (Fig. 30.6). Many compounds are not coloured, but can be converted into coloured compounds. Some of them are not coloured even after conversion but can be made to absorb light in the visible region by reaction with suitable reagents. These reactions are often specific and in most cases quite sensitive, so that quantities of material in the region of millimole per litre concentration can be measured. The major advantage is that complete isolation of the compound is not necessary and the constituents of a complex mixture such as blood can be determined after little treatment. As discussed below the depth of colour is proportional to the concentration of the amount of light absorbed is proportional to the intensity of the colour and hence to the concentration. There are two laws on which the principle of colorimetry is based.

1.   Lambert's Law

When a ray of monochromatic light passes through an absorbing medium, its intensity decreases exponentially as the length of absorbing medium increases.

2.   Beer's Law

When a ray of monochromatic light passes through an absorbing medium its intensity decreases exponentially as the concentration of the absorbing medium increasing.

These two laws combined together are called Lambert-Beer law (Fig. 32.6).

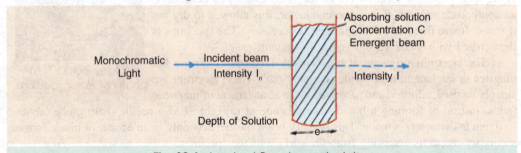

Fig. 32.6 : Lambert-Beer Law principle.

(i) Limitations of the Lambert-Beer Law: Due to increase in concentration of the solution, sometimes a non-linear plot is obtained. This might be due to the reasons : (*a*) light must be narrowed, preferably monochromatic, (*b*) wavelength of the light used should be at the absorption maximum of the solution. This also gives the greatest sensitivity, (*c*) there must be no ionization, association, dissociation or solvation of the solute with the concentration or time, (*d*) the solution is too concentrated to give intense colour.

3.   The Photoelectric Colorimeter

The basic arrangements of a typical colorimeter is shown in Fig. 32.7. In this instrument, when a white light from a tungston lamp passes through a slit, then a condenser lens, to give a parallel beam which falls on the solution under investigation contained in an absorption cell or cuvette. It is made of glass with the sides facing the beam cut parallel to each other. Generally, the cell samples are 1 cm square and will hold 3 ml liquid. The absorption cell is followed by filter, which allow maximum transmission of the colour absorbed after selection. If a blue solution is under examination, then red is absorbed and a red filter is selected. The colour of the filter is, therefore, complementary to the colour of the solution under investigation. In some instruments the filter is located before the absorption cell. The filters give narrow transmission bands and, therefore, approximate to monochromatic light. After this, the light then falls onto a photocell which generates an electric current and is direct proportion to the intensity of light falling on it. The electrical signal is increased in strength by the amplifier, and the amplified signal passes to a galvanometer, which is calibrated with a logarithmic scale so as to give absorbance readings directly. In these systems, the blank solution is first put in the colorimeter and the galvanometer is adjusted to zero extinction, which is followed by the test solution and the extinction is read directly. Now, a better method is to split the light beam, pass through the sample and the other through the blank. After this balance the two circuits give zero deflection on the galvanometer. The extinction is determined from the potentiometer reading which balances the circuit.

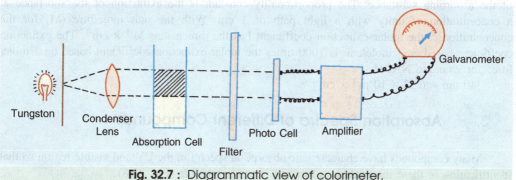

Fig. 32.7 : Diagrammatic view of colorimeter.

C. Spectrophotometry

A spectrophotometer is a sophisticated type of colorimeter where monochromatic light is provided by a grating or prism. The band width of the light passed by a filter is quite broad, so that it may be difficult to distinguish between two compounds of closely related absorption with a colorimeter. A spectrophotometer is then required, when the two peaks can be selected on the monochromator. Some compounds are absorbed in the UV region and their concentration is not possible to determine by using colorimeter. For such purposes, concentration is determined by spectrophotometer which also operates below 190 nm.

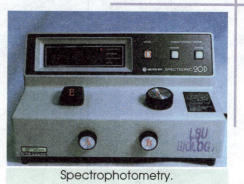

Spectrophotometry.

1. Absorption Spectra

Many compounds have characteristic absorption spectra in the UV and visible region so that identification of these substances in a mixture is possible.

Glass cuvettes are much cheaper than silica (quartz) and are used wherever possible. Their main limitation is that glass absorbs UV radiation and quartz can be used below 360 nm, so quartz cuvette are employed below this wavelength. The glass cuvette range varies from 360-800 nm, while quartz cuvette ranges from 200-800 nm. A tungston lamp produces a broad range of radiant energy down to about 360 nm. To obtain the UV region of the spectrum a deuterium lamp is used in the range 360-400 nm, then a blue filter is placed in the light beam read against a reagent blank which contains everything except the compound to be measured. The blank is first placed in the instrument and the scale adjusted to zero extinction (100% transmittance) before reading any solution. Alternatively, the extinction can be read aganist distilled water and the absorbance of the blank substracted from that of the test solution.

2. Absorption Spectrum and Extinction Coefficient

The Bonguer-Lambert-Beer law stated that

$E = c \times d$ where c = concentration and d = light path; E-molar extinction coefficient.

It stated that the extinction (E) is proportional to the light path and to concentration c of the absorbing substance. The proportionality constant is the extinction of the substance at a concentration of unity with a light path of 1 cm. With the unit mole/litre (M) for the concentration c, the molar extinction coefficient has the dimensions $M^{-1} \times cm^{-1}$. The extinction coefficient based on 1 mole/cm^3 is 1000 times the molar extinction coefficient based on 1 mole/litre; for example NADH

340 nm = $6.22 \times 10^3 m^{-1} \times cm^{-1}$

3. Absorption Spectra of Different Compounds

Many compounds have characteristic absorption spectra in the UV and visible region so that identification of these materials in a mixture is easily possible (Fig. 32.8).

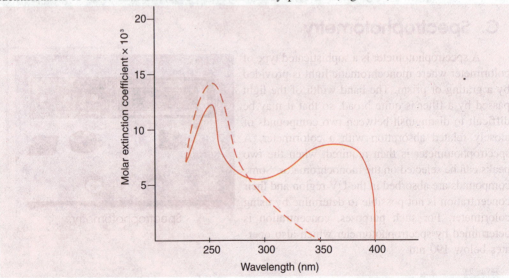

Fig. 32.8 : Relationship between wavelength (nm) and extinction coefficient of a compound.

The most frequently used wavelength in the UV region is 340 nm. At this wavelength, the reduced forms of the pyridine nucleotide coenzyme $NADH_2$ and $NADPH_2$ absorb strongly, while the oxidised forms do not (Fig. 32.9). NAD has a typical dinucleotide structure.

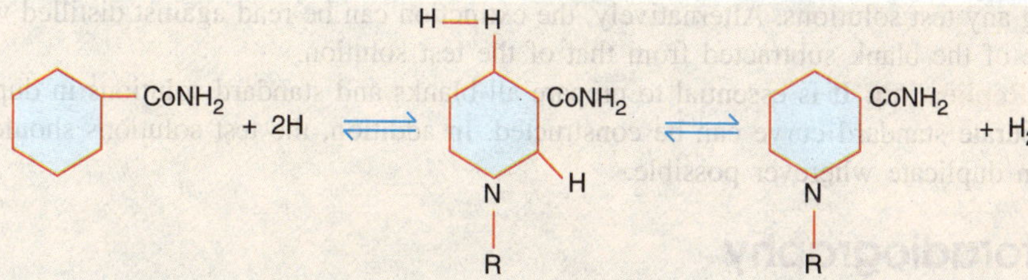

Fig. 32.9 : Reduction of nicotinamide adenine dinucleotide.

(*i*) **Proteins:** Proteins absorb strongly at 280 nm according to their content of the amino acids tyrosine and tryptophan and this provides a sensitive and non-destructive from assay. Proteins also absorb in the far ultraviolet because of the peptide bond.

(*ii*) **Nucleic Acid:** Nucleic acids and their component bases show maximum absorption in the region of 260 nm. The extent of the absorption of nucleic acids is a measure of their integrity, since the partially degraded acids absorb more strongly than the native materials. The spectra of the component bases are also sufficiently different to be used in their identification.

(*iii*) **Haemoglobins:** When haemoglobins are modified by the effect of certain drugs or carbon monoxide, characteristic shifts of their absorption maximum occur, so that the presence of these modified forms can be detected and measured.

(*iv*) **p-Triphenylphosphate:** This is a substrate used for phosphates, which catalyses the hydrolysis of the compound to p-nitrophenol and inorganic phosphate. In alkaline solution, the product p-nitrophenol gives a typical yellow colour with a maximum absorption at 405 nm. The substrate does not absorb in this region so the progress of the reaction in alkaline solution can be followed.

4.　Some Practical Points

The detailed operation of a particular instrument must, of course, be obtained by careful reading the instruction manual, but a few general points concerning the use and care of colorimeters and spectrophotometers are given below.

(*i*) **Cleaning Cuvettes:** Cuvettes are cleaned by soaking in 50% v/v nitric acid and then thoroughly rinsed in distilled water.

(*ii*) **Using the Cuvettes:** First of all, fill the cuvettes with distilled water and check them against each other to correct for any small differences in optical properties. Always wipe the outside of the cuvettes with soft tissue paper before placing in the cell holder and do not handle them by the optical faces. When all the measurements have been taken, wash them with distilled water and leave in the inverted position to drain.

(*iii*) **Absorption of Radiation by Cuvettes:** Glass cuvettes are much cheaper than silica ones and used wherever possible. Their main limitations is that glass absorbs ultraviolet radiation and they can be used at or above 360 nm, so silica cells are employed below this wavelength.

　　　　Glass cuvettes range : 360 – 800 nm
　　　　Silica (quartz) cuvettes range : 200 – 800 nm.

(*iv*) **Light Source:** A tungston lamp produces a broad range of radiant energy down to about 360 nm. To obtain the ultraviolet region of the spectrum a deuterium lamp is used as the light source. If the tungston lamp is used in the range 360-400 nm, then a blue filter is placed in the light beam.

(*v*) **Photocells:** When working at wavelengths up to 625 nm a `blue' photocell receives the emitted light and a `red' photocell receives above this wavelength. Photocells are exposed to light for the shortest time necessary to take a reading in order to avoid fatigue.

(*vi*) **Blanks:** The extinction of a solution is read aganist a reagent blank which contains everything except the compound to be measured. This blank is first placed in the instrument for the reading any test solutions. Alternatively, the extinction can be read against distilled water and absorbance of the blank subtracted from that of the test solution.

(*vii*) **Replication:** It is essential to prepare all blanks and standard solutions in duplicate so that an accurate standard curve can be constructed. In addition, the test solutions should also be prepared in duplicate wherever possible.

D. Autoradiography

The use of radioactive radiations to obtain the photographic film of the test material, incorporated with the radioactive tracers, is called autoradiography and the film obtained is called autoradiograph. After development the irradiated areas appear on the film as dark areas corresponding to the distribution of the tracer.

1. Principle

Autoradiography can be detected either directly with a scintillation counter or indirectly via their effect on photographic film. At the light microscopic level the autoradiography is based on the principle that if a photographic emulsion is brought into contact with radioactive material, the ionic radiations will convert the emulsion as dark spots of silver at certain points.

$$\text{Silver halide} \xrightarrow[\text{radiations}]{\text{ionising}} \text{Silver (black spots)}$$

Radioactive substances are introduced into the test material either in a given chemical form or tagged with certain metabolic precursors. For example, nucleic acid can be made radioactive by incorporation of radioactive phosphate during nucleic acid synthesis. The newly synthesised nucleic acid thus becomes radioactively labelled.

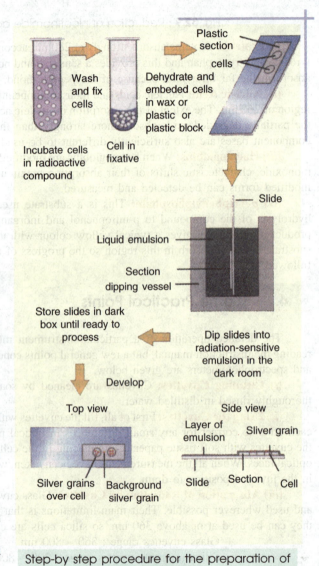

Step-by step procedure for the preparation of an autoradiograph.

2. Types of Radiation

The following three types of radiations are used in autoradiography:

(*i*) **Alpha Rays:** The alpha rays particles which consist of 2 neutrons and 2 protons and are infact charged helium atoms. Radium 226 is their source.

(*ii*) **Beta Rays:** The beta rays are electrons ejected or emitted by nuclei. The energy levels of these electrons may vary. When highly charged they are called hard beta as of ^{32}P but soft beta is emitted from ^{14}C.

(*iii*) **Gamma Rays:** The gamma rays are electromagnetic rays and resemble X-rays. ^{60}CO is gamma emitter.

3. Procedure for Continuous Labelling

(*i*) **Exposure of Cells to 3HTDR (Tritiated Thymidine):** Tritiated thymidine (specific activity 12,600 in Ci/m mol) obtained from BARC should be added to the experimental material at the last `S' phase prior to harvesting. Before preparation the material should be thoroughly washed for removing the traces of radioactive thymidine left in the medium which may cause excessive background radiation.

$$3HTDR + S \text{ phase of the organism} \longrightarrow \text{Labelled}$$

(*ii*) **Preparation of Film:** Air dried slides are to be stained with desired stain and then stripped film method may be followed:

Stripping film (AR-10 Kodak Ltd.) mounted on glass plates are to be removed from the refrigerator and allowed to cool down to room temperature. All the subsequent procedures are to be carried out in a dark room fitted with red light and low temperature (about 20°C).

Slides are marked on the specimen side and immersed in clean distilled water in coplin jar. The photographic plates are unwrapped, held with the emulsion side up and squares ($4 \times 4 \text{ cm}^2$) were cut with a sharp scalpel. Using a pair of small forceps the squares are stripped off the plate and floated with emulsion side down in a tray containing distilled water. When the squares straighten out, the slides are immersed in water with specimen side up and into a position beneath the square of the film. The slide is then lifted out of water. The film covering the specimen straightens out and the overlapping ends flooded onto the other side of slide. Slides are now allowed to dry in a tray, packed in a slide box protected from light and kept with a small amount of silica gel in cold for 21 days exposure before developing.

Developing Autoradiographic Film: All steps are carried out in the dark room fitted with lamp. The stored slides are painted on the reverse with euparol or nail polish to prevent shifting of the film and then kept for 5 minutes in developing solution (Kodak), washed thoroughly in distilled water fixed in Kodak metafix for 2 minutes. Again rinse in distilled water and dried, the film can now be observed to locate the deposition of silver grains.

4. Measurement of Radioactivity

Apart from making films of the test material, autoradiography can directly be measured by various counters like Geiger-Muller counter, Scintillation Counter, Proportional Counter, etc. While counting the back ground count should also be considered.

(*i*) **Geiger-Muller Counter:** It is made of a glass or metal tube containing a mixture of gases, an inert gas like helium or argon and an organic vapour or a gas eg. isopropanol or isobutane or cooking gas (which is a mixture of hydrocarbons). It has a thin window, usually made of mica at one end to enclose the gas. Depending upon the thickness of the window and the energy of radiation of the radio-isotopes a fraction of the particles emitted enter the counting tube and ionize the gas mixture inside. The small current generated is magnified at high voltage, and measured with a scaler calibrated to record disintegrations per unit time. The counts recorded include counts due to ionization of the gas mixture by cosmic rays passing through the tube, any radioactivity present in the surrounding area, etc. This is known as the background. These background counts have to be substracted from the counts given by the sample. Corrections also have to be applied for (*a*) half life of decay, (*b*) geometry, (*c*) coincidence loss, and (*d*) self absorption. If measurements are done at the same shelf and same position, no correction for geometry is necessary. For long lived isotopes half life of decay is unimportant. The results of Geiger-Muller counter are expressed in counts/min. The main limitation of gas counter is its *dead time*. After the initial ionization the cloud of slow moving positive ions formed potential on the cathode. Thus the counter is dead for about 300 μ sec. after each ionizing event. It is used for X-rays, γ-rays, electrons and β-particles detection (Fig. 32.10).

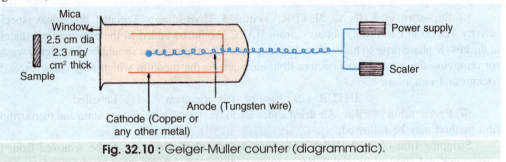

Fig. 32.10 : Geiger-Muller counter (diagrammatic).

(*ii*) **Scintillation Counter:** Certain organic and inorganic materials emit light flashes or scintillation when charged. Flashes or particles, X-rays, γ-rays pass through them. There are three stages in the operation of this counter: absorption, scintillation process, and conversion of light into electronic impulsions *i.e.* the light flashes from the scintillators fall on a photomultiplier tube and then the signal is amplified. The amplified signal is made to be proportional to the intensity of ionizing radiations (Fig. 32.11).

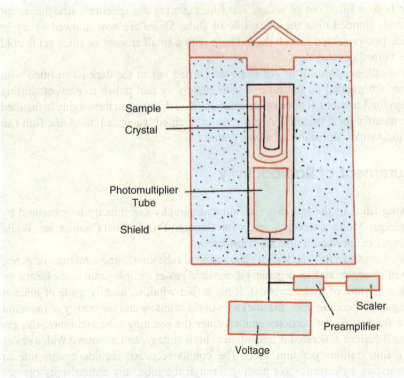

Fig. 32.11 : Scintillation counter for measurement of radiations (diagrammatic.)

E. Tracer Technique

Isotopes are used as tracers to trace the course of events. Such isotopes are stable (^{2}H, ^{15}N, ^{18}O etc.) or unstable (^{3}H, ^{14}C, ^{32}P, ^{35}S etc.). The latter emits ionizing radiations e.g. α-particles, β-particles, γ-rays, X-rays, etc. and thus can be detected easily by instruments (Geiger Muller Counter etc.) which can detect the ionization of the medium through which they pass. The former needs instruments like a mass spectrometer which with the help of a powerful magnetic field separates a heavy isotope from a lighter isotope.

The α-particles being helium nuclei are heavy and, therefore, cannot penetrate long distances, but are strongly ionizing. The β-particles are much lighter and travel distances depending on their energy. The γ-rays and X-rays are even more penetrating. The unstable isotopes decay exponentially to stable forms and the time taken for half decay is known as the half life of the radioisotope. Thus ^{13}N has a half life of 10.5 min, ^{32}P, 13.8 days, ^{35}S, 85 days. ^{3}H, 10 yrs and ^{14}C, 5688 yrs. ^{3}H emits weak γ-rays (0.018 Mev) and ^{14}C (0.156 Mev) and ^{32}P, 1.6 MeV (1Mev = 1 million ev)emit strong rays. The energy of radiation, the half life of the isotope, the solubility and the metabolic role of its compounds are the major factors which determine the extent of health hazards of a radioisotope.

With appropriate precautions the radioisotopes can be handled, processed, detected and measured with practically no risk or hazard. The movement of a radioactive element in a compound can also be traced by autoradiography. When the energetic particles emitted by a radioisotope hit a scintillating substance like NaI, ZnS, anthracene, etc. photons are ejected; these photons interact with the silver halides present in the photographic emulsion and produce the black spots as in conventional photography. This principle is also followed in scintillation counting.

The unit of ratioactivity is curie or Bequerel named after the discoverer. One Curie (Ci) is equivalent to 3.7×10^{10} disintegrations per second. One Bequerel (Bq) is 10^{10} dps. When counts are taken under identical conditions the activity in curies can be calculated directly by reference to a standard.

Working with radioisotopes involves considerable health hazard, unless adequate precautions are taken. Blood counts have to be taken periodically and film badges should always be worn when handling a radioisotope. Under no circumstances radioactive substances should come in contact with any part of the body. Care should always be taken so that no radioactive gas is inhaled. No one should eat, drink or smoke in a radio-isotope laboratory. One should handle radioisotopes, particularly β-emitters like ^{32}P or α-emitters like ^{60}CO some distance away from the source. μCi quantities are usually used in biological studies and these are less hazardous.

Washing should never be pooled down a sink. They should be dried inside a hood and kept in a bin maintained for this purpose. The contents of bin have to be buried under ground in a safe place or sent to the Bhabha Atomic Research Centre, Bombay.

Hands, aprons, table tops, etc. should be monitored regularly and decontaminated, if necessary. Headache, dizziness, abnormal blood counts, etc. are indications of radiation sickness. Work must be stopped and doctors consulted in such cases.

(i) Preparation of Samples for Radioactivity Measurement Procedure

(a) Transfer 0.1 ml of the radioactive substance in solution to the planchet. Allow it to spread; add a few drops of distilled water or ethanol (if the substance is water or alcohol soluble) and place under a heating lamp at a distance where no splattering takes place.

(b) Place the planchets inside Petri dishes, cool and count.

(ii) For Solid Samples

(a) Weigh the empty planchet.

(b) Transfer the fine powder with a spatula carefully to the planchet and spread to cover the entire surface. Pad the surface gently until it appears to be smooth.

(*c*) Weigh the planchet with the powder. Substract the initial weight of the planchet. If the difference exceeds the minimum thickness for saturation, take counts. If not more powder has to be introduced and the sample prepared again.

F. Chromatography

Tswett (1903) defined chromatography as the process of separation of coloured substances, but now-a-days it is performed on mixtures of coloured substances including gases. The common feature to all the chromatography methods is the use of two phases: (*a*) stationary phase (*b*) mobile phase. The separation of the coloured substance depends upon the two phases. On the basis of nature of the stationary phase, the classification is given below:

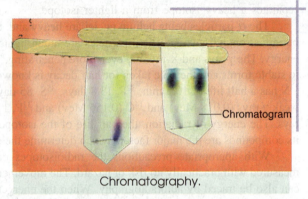

Chromatography.

If the stationary phase is solid, the process is called *adsorption* while mobile phase is liquid called *partition* chromatography. The mobile phase may be either a liquid or gas. On the basis of stationary and mobile phase, chromatography system is of four types:

(*a*) Liquid - solid   (e.g. classical adsorption chromatography, TLC and ion-exchange)
(*b*) Gas-solid   (e.g. gas solid chromatography)
(*c*) Liquid-liquid   (e.g. classical partition chromatography, paper chromatography)
(*d*) Gas-liquid   (e.g. gas-liquid chromatography, capillary coloumn chromatography)

(*i*) Classification Based on Methods: As stated earlier, the classification is based on phases (solid or liquid) called adsorption and partition. The adsorption phase may contain a mobile phase in liquid or gaseous form. Similarly the partition liquid chromatography has a liquid mobile phase or gaseous form of mobile phase called liquid-liquid chromatography and gas-liquid chromatography respectively.

All separations by chromatography depends on the fact that the substances to be separated distribute themselves between the mobile phase and the stationary phases in proportions which vary from one substance to another. The manner in which the substances are distributed is most conveniently discussed by referring to the 'sorption isotherm'.

(*ii*) Sorption Isotherm: The amount of a particular substance taken up `sorbed' by the stationary phase depends on the concentration is the mobile phase. The curve obtained by plotting the amount sorbed against concentration at constant temperature, is the `sorption isotherm'. The shape of the isotherm is one of the most important factors governing chromatography behaviour. The term `sorption' includes the *adsorption* which refers to the increase in concentration at the interface between the mobile and the stationary (solid) phase, while *absorption* is the dissolution of a substance from a mobile phase into the liquid stationary phase.

1.   Adsorption Column Chromatography

The general chromatography method was introduced into biology by Tswett in 1906 and various adsorbents such as starch, and aluminium oxide are poured as a slurry into a long glass tube. The name column chromatography was given to it because column of materials was used for adsorption of different substances.

An unknown compound is poured on the open top of the tube and identified by slowly trickling down an appropriate solvent through the long tube commonly spoken as a column. The rate at which any compound migrates through the column depends upon the balance between its affinity for the solvent and its adsorption on the particles in the column. By column chromatography the different coloured components of a mixture are separated and the separated components travel down the column at different rates. By changing receivers the different fractions are collected separately as they leave the bottom of the column. Colourless substances are frequently separated by collecting many small fractions in succession from the column, testing each fraction chemically or by other means, and combining all fractions containing single component.

Various adsorbents which are in common use are usually inert compounds like charcoal, alumina, $CaCO_3$, $Ca(PO_4)_2$, Cellulose, silica, glass kaolin (clay material). Silica gel and Kiesulguhr are most commonly used now a days. The individual adsorbent particles should be of uniform size and sufficiently hard for adsorption during the packing of the column.

Adsorption chromatography is widely used in the separation of aminoacids, lipids, steroids, sugars, and similar other compounds of relatively low molecular weight. A molecule that is adsorbed more strongly will migrate very slowly in comparison to another molecule which is poorly adsorbed.

2.   Thin Layer Chromatography

(*i*) **Principle:** The separation of compounds on a thin layer is similar in many ways to paper chromatography but has the added advantage that a variety of supporting media can be used so that separation can be by adsorption, ion exchange, partition chromatography or gel filteration depending on the nature of the medium employed. The method is very quick and separations can be completed in a hour. Compounds can be detected at a lower concentration than on paper as the spots are very compact. Furthermore, separated compounds can be detected by corrosive sprays and elevated temperature with some thin materials, which of course is not possible with paper.

(*ii*) **Preparation of Thin Layer:** The Rf value is affected by the thickness of the layer below 200 μm and a depth of 250 μm is suitable for most separations. There are several good spreaders on the market which, when carefully used, can produce even layer of required thickness.

The slurry (5ml) of alumina (made for TLC) following the directions given as per manufacturer. Take the glass slides and keep them flat, pipette out 1-2 ml of the slurry into them. By tilting the slides, spread the slurry evenly on the surface. Lining the edges with vaseline will be of help. Spot a mixture of methylene blue and cresol red, and dip the slide in a beaker jar containing Na_2CO_3 solution saturated with butanol. The slide must be handled with care. There are now a number of prepared thin layer plates that are commercially available and these may be more convenient to use than drying to prepare plates in the lab.

(*iii*) **Solvents:** Electrostatic attraction play a big role in the adsorption phenomena, and therefore, the polar nature of the solvent used influences adsorption considerably. Generally, adsorption is maximal in nonpolar solvents and decreases as the polarity is increased.

(*iv*) **Development:** It is essential to make sure that the atmosphere of the separation chamber is fully saturated, otherwise Rf values will vary widely from tank to tank. This can be ensured by using as small tank as possible. Development of the plate is usually by the ascending technique is very rapid. The spots are evaluated in the same way as in paper chromatography.

(*v*) **Precautions**

(*a*) The coated glass plates should be clean and dried before making a thin layer over it.

(*b*) Plates should be activated before use in drying over at 105°C.

(*c*) Samples should be spotted carefully.

3. Gas Chromatography (GC)

The technique of gas chromatography was used for the first time by A.T. James, and P. Martin (1952) to separate long chain fatty acids. The phenomenon of differential adsorption can be used to separate gases and vaporizable substances as well. **This technique is called Gas Liquid chromatography**. In this an inert solid base is coated with a liquid like paraffin oil or silicone oils, and packed in long tubes. This column is kept in an oven and generally kept around 150-200°C. An inert gas like N_2 is passed through as carrier. The mixture of gases to be separated are now injected into one end of the column and as they are carried through by the carrier gas, they get absorbed, released and reabsorbed. The gas with least adsorptivity emerges first out of the column. There are several methods of detection of the emerging material, primary among them are flame ionization detector, electrical resistance detector, and electron capture device.

The principle of the gas chromatography is also the same as any other type of chromatography. In this case, the substance to be analysed is partitioned between mobile phase and a stationary phase. The mobile phase in GC is a gas instead of a liquid. The stationary phase consists of a liquid material coated on an inert solid support. The equipment consists of supply of carrier gas, an injector, temperature regulated oven, column, detector and recorder. Most commonly used carrier gases are helium, argon, nitrogen or hydrogen. The material to be analysed can be injected in the form of solid, liquid or gases.

The injector helps the column temperature in order to vaporise the sample and to prevent its condensation in an injector system. The column in a GC made up of glass or metal measuring about 1.8-3.6 m long with 3-6 mm diameter. The column has an inert solid support coated with a liquid phase. Column is heated to the desired temperature by an oven around the column. The detectors in the gas chromatography are situated at a short distance from the column. Recorder is a multivoltmeter which converts the signal of the detector on graph.

G. Centrifugation

1. Sand

If the sand particles are suspended in water, the particles due to varying in their size, density and the velocity of the medium, settle down at different rates. On the otherhand, the gravitational pull is also involved which is about 980 cm/sec^2 or 1 g unit under normal condition. If the gravitational force is increased then particles called `light' will also sediment.

Actually, if we apply the centrifugal force, the gravitational force is induced. As you know that the centrifugal force acts in a direction away from the centre of the axis. The faster the speed of rotation, the greater will be the force. This can expressed as below:

Centrifugal force = (angular velocity)^2 × radius. Angular velocity is related to rotations/min (rpm) by the following formula:

$$\text{Angular velocity} = \frac{2\pi \times \text{rpm}}{60} \text{ resolution/sec.}$$

The centrifugal force is generally expressed as relative centrifugal field (RCF) in g units as:

$$RCF = [4\pi^2 \, (\text{rpm})^2]/3600 \times 980 \, g \text{ units}$$

or $\quad$ 1.11×10^{-5} rpm rg units

2. Centrifuge

This instrument is based on centrifugal forces. Basically, it has containers rotated around the central axis with the help of electric motor. Now-a-days, cooling centrifuges, high speed centrifuges and ultracentrifuges are available with the different types of rotors i.e. angle head and swinging bucket types. In the angle type rotor, the sample kept at an angle of about 30° to the horizontal whereas, in the latter, the sample while spinning are horizontal. The most common centrifuge is clinical centrifuges. In the case of ultracentrifuges, the spun rotors produce force under vacuum to reduce friction.

3. Zonal Centrifugation

For the separation of the components of a mixture, the sample in solution is centrifuged. The sedimentation takes place in a solution present in a column. The density of the solution increases as we move down the column. The solution should be inert as a result of which gradient is formed. This sample mixture is now put in a column which allows the sample to form different bands according to their rate of sedimentation. The components with high sedimentation rate are at lower end of the column. The size and shape of the molecule also affect the sedimentation, as sedimentation coefficient is a function of a mass of the particle. The bands are formed at various positions which can be separated by puncturing the tube, bands are collected separately as depicted in the figure 32.12.

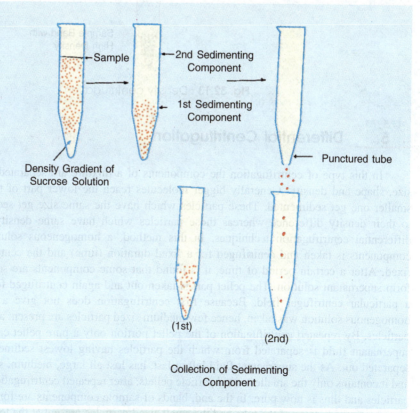

Fig. 32.12 : Zonal centrifugation.

4.   Density Gradient Centrifugation

The sample is mixed in a dense solution possessing low concentration and having fast diffusing property. The mixture is spun in centrifuge but before starting the process the sample is to be in a uniform mixture. After centrifugation, the solution forms a density gradient and the sample components occupy those positions in the density gradient which correspond to their density. The bands of sample formed are separated by puncturing the tube. The method is also called isopycnic centrifugation (Fig. 32.13).

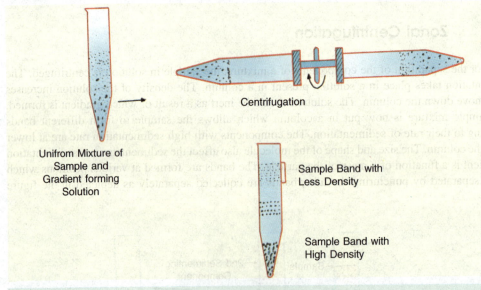

Unifrom Mixture of
Sample and
Gradient forming
Solution

Centrifugation

Sample Band with
Less Density

Sample Band with
High Density

Fig. 32.13 : Density centrifugation.

5.   Differential Centrifugation

In this type of centrifugation the components of a mixture are separated according to their size, shape and density. Generally bigger molecules reach the lower part of the coloumn before smaller one get sedimented. These particles which have the same size get sedimented according to their density difference whereas those particles which have same density are separated by differential centrifugation techniques. In this method, a homogeneous solution of mixture of components is taken and centrifuged for a fixed duration (time) and the centrifugal field is also fixed. After a certain period of time, it is found that some components are sedimented and they form supernatant solution. The pellet part is taken out and again centrifuged for fixed duration at a particular centrifugal field. Because first centrifugation does not give a pure pellet. Since homogenous solution was taken, hence few medium sized particles are present along with large size particles. By repeated centrifugation of the pellet portion only a pure pellet can be obtained. The supernatant fluid is separated from which the particles having lowest sedimentation rate can be separated out. As the supernetant fluid left at last, has lost all large, medium, small sized particles and it contains only the smallest sized particle pellets, after repeated centrifugation, only large sized particles and thus is now pure. In the end, bands of sample components are formed the large sized particles at the bottom of the tube and the small sized particles present at the top of the tube. Thus, the different components of a mixture are separated.

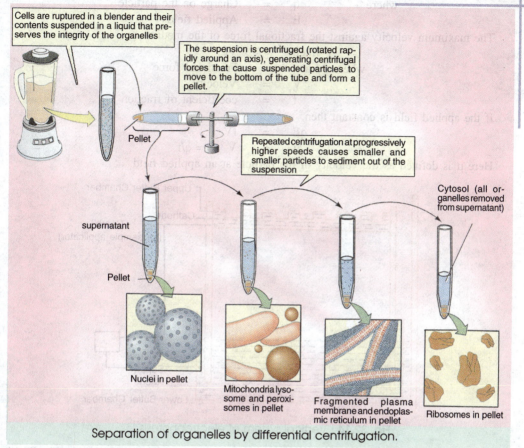

Cells are ruptured in a blender and their contents suspended in a liquid that preserves the integrity of the organelles

The suspension is centrifuged (rotated rapidly around an axis), generating centrifugal forces that cause suspended particles to move to the bottom of the tube and form a pellet.

Pellet

Repeated centrifugation at progressively higher speeds causes smaller and smaller particles to sediment out of the suspension

Cytosol (all organelles removed from supernatant)

supernatant

Pellet

Nuclei in pellet

Mitochondria lysosome and peroxisomes in pellet

Fragmented plasma membrane and endoplasmic reticulum in pellet

Ribosomes in pellet

Separation of organelles by differential centrifugation.

H. Electrophoresis

Tiseleus (1937) developed the process of electrophoresis. This is a process of separation of particles where on the basis of difference in charges and molecular size and shape the particles under the influence of an electric field migrate. Many biochemicals such as amino acids, peptides, proteins and nucleic acid possess ionizable groups.

The process takes place by dissolving the sample into a buffer solution and the supporting medium (such as starch for protein separation) used for casting the gel, also saturated with the

Separation of DNA restriction fragments by gel electrophoresis.

buffer. The isoelectric point of the molecule and the buffer decided the charge on the molecule. If the buffer is at a above pH with that of isoelectric point, the sample will be negatively charged and will migrate towards the anode and vice-versa.

This method exploits the underlying fact. On the application of electric field the electric force experienced by any particle can be written as:

$$F_{elec} = qE$$

where   q   =   Charge on the particle

E   =   Applied field

The maximum velocity against the fractional force of the medium.

F fract.   =   Vf

where f fract.   =   Fractional force

V   =   Velocity

f   =   coefficient of fraction

If the applied field is constant then,

$$qE = fV$$
$$\mu = V^{-2}/E = q/f$$

Here μ is defined as the velocity of the particle at an applied field.

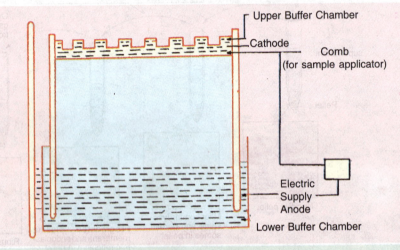

Fig. 32.14 : Vertical system of electrophoresis.

Electrophoresis can be performed by both vertical as well as horizontal methods (Fig. 32.14 and Fig. 32.15). Both horizontal and vertical systems of electrophoresis chiefly comprise of two parts. the electric supply unit and the electrophoretic unit. The difference lies in the mode of arrangement.

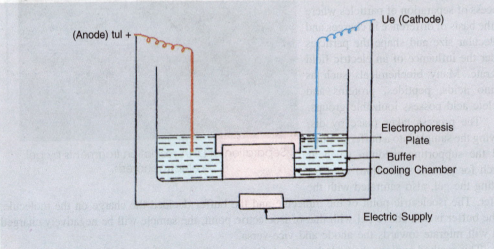

Fig. 32.15 : Horizontal system of electrophoresis.

Method: In gel electrophoresis, the agarose gel is present between two glass plates placed together by means of plastic chips. It is immersed in a buffer solution which maintains the required pH for the molecule to remain charged and thus facilitate electrophoresis. The sample is applied by means of pipetting in microquantity. The comb is removed after the sample application. The comb adds the sample in the upper buffer chamber. If the substance to be separated is a protein, the pH is maintained at pH 9 and the reason is that protein molecules are negatively charged at this pH. To ensure further uniform distrubution of charged molecules agents like SDS (sodium dodecyl sulphate) is added. After application of electric field or lower side, they are absorbed into various bands depending upon this charge and molecular weight, the small sized particles move faster and are close to the anode, the developed electrophoresis plate is dried and analysed by using various techniques.

The electrophoresis is of several types, such as paper, slab gel, disc electrophoresis, etc. This technique is specially useful in medical microbiology where in the bacterial antigens (proteins) are electrophorized and the composition helps in preparing an antibody against it.

I. Electrofocussing

This is a latest techniques for the separation or protein or you may express it as "electrophoresis in a pH gradient". The proteins migrate in an electric field, but when they reach a point where the pH is the same as its isoionic point, their further movement is prevented. This is called electrofocussing, because protein has been focussed to its isoionic point.

J. Gel Filteration

Biological macromolecules form a class of substances with special functions which are controlled *in vivo* by small changes in the environment. Changes in the pH, concentration of metal ions, cofactors etc. may have a profound effect on the molecules being studied and it is clearly necessary to have available mild separation techniques which operate independently of these factors. Gel filteration is one of these techniques. A gel filteration separation can be performed in the presence of essential ions or cofactors, detergents, urea, at high or low ionic strength, at 37°C or in the cold room.

As a solute passes down a chromatographic bed its movement depends upon the bulk flow of the mobile phase and upon the Brownian motion if the solute molecules which causes their diffusion both into and out of the stationary phase. The separation in gel filtration depends on the different abilities of the various sample molecules which never enter the stationary phase, move through the chromatographic bed fastest. Smaller molecules, which can enter the gel pores, move more slowly through the column, since they spend a proportion of their time in the stationary phase. The choice of a suitable gel for filtration is basically a case of finding the gel whose fractionation range covers the range of molecular sizes in the sample to be fractionated, since molecules of biochemical interest may range in size from a few hundred to many millions in molecular weight. The most common media used include sephadex G-types (sephadex is a bead-forming cross-linked dextran gel which swells in water and aqueous salt solution). It is stable in the pH range 2-12 and in buffers containing dissociating agents such as urea and detergents and can be sterilized by autoclaving. Besides, sephadex, sepharose is also used and it is a bead forming agarose gel stable in aqueous suspension pH range 4-9. Sepharose melts on heating and should not be used above 40°C or autoclaved. Sepharose provides an excellent medium for the fractionation of high molecular weight substances such as protein complexes and polysaccharides. Besides, sepharose has been proved to be a excellent medium for immobilisation of enzymes, antibodies hormones and other ligands for affinity chromatography.

A latest introduction of media for gel filteration is sephacryl (which is prepared for covalently cross-linking ally (dextran with N-N- methylene bisacrylamide to yield a highly stable matrix. It can be used in an aqueous buffer system in concentration urea or guanidine hydrochloride and in a number of organic solvents. The covalently cross-linked matrix cannot melt on heating and sephacryl is conveniently sterilized by autoclaving.

K. Hot Air Oven

This is a dry air type sterilizer with three walls and two air spaces. The outer walls are made up of thick asbestos to reduce the radiation of heat. The hot air steriliser is operated electrically. In this case the heater coil is either be placed at the bottom of the oven or on the side walls. A convection current travel a complete circuit through the wall space and interior of the oven. The temperature inside the oven is controlled by thermostat.

(*i*) **Principle:** The hot air steriliser is operated at a temperature of 160 to 180°C for a period of one and a half hour. If the temperature goes above 180°C there is a danger of cotton being charred. The hot air steriliser is used for sterilizing all kinds of laboratory glassware, such as test tubes, Petri dishes, pipettes, flasks, bottles, etc. Other materials which will not be burnt at high temperature may also be sterilized in hot air steriliser. Petri dishes may also be put in metal cans or wrapped with paper and placed inside the steriliser.

(*ii*) **Precaution:** It is necessary to check the proper temperature at which the materials are sterilized. Under no circumstances should the hot air oven be used to sterilize culture media, as the liquids will boil to dryness. There should be temperature controlling device for maintaining the temperature required for sterilization.

(*iii*) **Uses:** The hot air sterilizer is used for sterilizing laboratory glass ware such as test tubes, Petri dishes, pipettes, flasks, bottles and other materials which will not be burned at higher temperature.

L. Autoclave

The autoclave is a cylindrical vessel having double walls around all parts except the upper side. It is built to withstand the steam pressure of at least 30 lb per sq. inch.

(*i*) **Principle:** The principle used here is to increase the temperature of steam (gas) in a closed system that increases its temperature. The water molecules become more aggregated that increases their penetration considerably. The water boils at

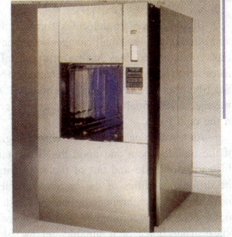

A large automatic autoclave.

100°C depending upon the vapour pressure of the atmosphere. The temperature will be increased if the vapour pressure is increased. This relationship between pressure and temperature is shown in Table 32.1.

Table 32.1 : Pressure-temperature relationship

Pressure (Pounds/inch²)	Temperature °C	Pressure (Pounds/inch²)	Temperature (°C)
0	100	20	126.5
5	109	25	130.5
10	115.5	30	135.5
15	121.5	40	141.5

The autoclave is usually operated at 15 1b/sq. inch steam pressure for 30 min., which as seen from the above table corresponds to 121.5°C. This temperature for a period of 30 min. is sufficient to kill all the spores and vegetative cells of microorganisms.

(*ii*) **Precautions:** The following precautions are to be taken: The level of water should be checked before operating. The air should be completely evacuated from the autoclave and the steam must have access to the materials to be sterilized.

For example, if a material such as cotton wool or glass beads are to be sterilized in a glass bottle closed with rubber stopper the sterilization would not be complete as steam cannot pass through the rubber stopper.

(*iii*) **Procedure:** Sufficient amount of water is placed inside the autoclave. The material is placed inside the autoclave for sterilization. The cotton plug should be covered with a piece of butter paper so that the plug does not wet. The lid of autoclave should be tightened with the help of screws, then switch on the plug. The steam outlet is kept open till we feel that the air from inside the autoclave has been evacuated and then close the steam outlet. The pressure is allowed to remain at 15 l/sq. inch for 15-30 min., is done by controlling the steam in the valve. After 30 minutes switch off the current and let the autoclave cool down and thus the pressure comes down to zero mark. Then the autoclave is cooled down, the lid is opened and taken out the materials.

(*iv*) **Uses:** The autoclave is used to sterilize usual non-carbohydrate media, broths and agar media, contaminated media, aprons, rubber tubings, rubber gloves etc. This types of sterilization is also used in the commercial canning of fruits and vegetables and also in order to manufacture sterilized milk.

M. Laminar Air Flow

Laminar air flow is an equipment having an air blower in the rear side of the chamber which can produce air flow with uniform velocity along parallel flow lines. There is a special filter system-high efficiency particulate air filter (HEPA filter) which can remove particles as small as 0.3 μm. In front of the blower, there is also peculiar mechanism from which the air blown from the blower produces uniform air velocity along parallel flow lines. These are horizontal and vertical types.

(*i*) **Principle:** Laminar flow can produce dust free air current with uniform velocity along parallel flow lines which help in transferring the culture media bacteria free. Air is passed through these special filters into the enclosure and the filters does not allow any kind of microbes to enter into the system. Due to uniform velocity and parallel flow of air current we can perform pouring, plating, slanting, streaking without any kind of contamination.

(*ii*) **Precautions:** Following precautions should be taken care of before handling the apparatus: We should put off our shoes before entering to operate the apparatus. We should wash our hands with soap and we should not talk inside the chamber while doing experiment, otherwise there will be a chance for contamination with certain bacteria or microorganisms through air of our mouth.

(*iii*) **Procedure:** Dust particles are removed from the surface of the laminar flow with the help of smooth cloth using alcohol. The UV light should be switch on for 30 minutes before performing the experiment and the front covering glass of laminar flow is opened and kept properly. The air blower is set at the desired degree so that the air inside the chamber is to be expelled because the air which is already inside the chamber may be contaminated. One should sit properly in front of the chamber and wash the hands and stage of the chamber again with alcohol to reduce contamination. All the experiments i.e. pouring, plating, streaking etc. are to be done with in the flame zone of the sprit lamp. The required materials are to be placed side by side on the stage of laminar flow.

(*iv*) **Uses:** Within the chamber of laminar flow, we can transfer any media for culturing bacteria or fungi or any microbe without any contamination. The parallel and smooth air flow blown out from inside the chamber of the laminar flow, is dust free.

N. Incubator

An incubator is an equipment that consists of copper/steel chamber, around which warm water or air circulates either by electricity or by means of small gas flame. The temperature of the incubator is kept constant by thermostat control.

(*i*) **Principle:** Incubator is operated to culture or for growing an organism in a suitable medium at proper temperature. In an incubator the variation of temperature should not be more than one degree celsius (1°C). In large incubator the variation of temperature goes up to 2 or 3° C. Small square type incubators are better than large incubators. If a lower temperature than that of the room temperature is needed, the water, before circulating around the upper chamber, is directed by the thermostat to pass through an ice chest or a small boiler according to the requirement of temperature.

(*ii*) **Precautions:** The door of the incubator should be opened only when necessary. If the tubes are to be incubated for a long time or at higher temperature,

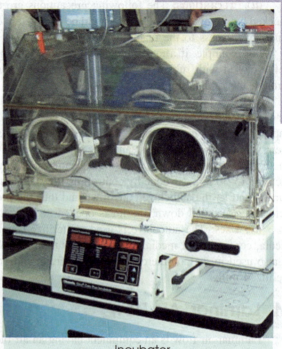

Incubator.

the medium may become too dry due to excessive evaporation. In such case, cotton plug should be pushed inside the neck of the tube and a media rubber cap should be placed to cover the plug. If the Petri dishes are to be incubated for a long time these may be placed in moist chamber with a damp sterile cotton wool at the bottom.

(*iii*) **Uses:** The method of incubation of culture depends upon the temperature and the oxygen requirements of the organism. For this purpose the incubator is used to maintain different temperature required for growth of organism in a bacteriological laboratory.

QUESTIONS

1. How is the evolution of tools linked with the development of microbiology? Compare the electron and compound microscope as to their resolving power and magnification.

2. Compare and contrast the principle, methods of operation and utilities of scanning electron microscope and transmission electron microscope (SEM vs. TEM).

3. Discuss the application of simple verses compound verses electron microscope.

4. Briefly comment upon the structural and functional characteristics and role played by the following types of microscopes.
   (*a*) Dark field microscope
   (*b*) Fluorescent microscope
   (*c*) Phase contrast microscope

5. What is the principle of centrifugation? Write different kind of centrifugation process in detail.

6. Write short notes on the following

   (a) Zonal centrifugation

   (b) Density gradient centrifugation

   (c) Differential centrifugation

7. Define the process of electrophoresis? How will you perform the process of gel electrophoresis?

8. Write short answers on the following:

   (a) Electrofocussing

   (b) Electrophoresis

9. Briefly write the fundamental laws of photometry. Compare the utilities of colorimetry and spectrophotometry.

10. Which instrument is used to measure the colour? Write in detail about the process.

11. What are two laws of colorimetry on which it works? Write in detail.

12. What is the significance of spectrophotometry?

13. What is the principle of chromatography? Write their classification and method so as to separate different amino acids from a mixture.

14. Write in short about the following:

   (A) TLC

   (B) Detectors

   (C) GLC

15. On which principle oven and incubator are based? What is their use in Microbiology?

16. Write short notes on the following:

   (a) Laminar air flow   (c) Geiger Muller counter

   (b) Autoclave   (d) Scintillation counter

17. What is the principle of autoradiography? Mention different type of radiations? How radioactivity is measured?

REFERENCES

Bair, E. J. (1962). *Introduction to chemical Instrumentation.* McGraw-Hill, New York.

Beckman, A.O. W.S. Gallaway, W.Kaye and W.F. Ulrich (1977). *History of spectrophotomer* of Beckman Instruments, Inc., Anal. Chem., 49, 280

David, D. J. (1974). *Gas chromatography detectors,* John, New York.

Grob, R.L. (1977). *Modern Practice of gas Chromatography.* John Wiley, New York.

Laub, R.J. and R.L. Pecsok (1978). *Physiochemical application of Gas chromatography,* Wiley-Inter science, New York.

O' Farell (1975). Two-dimensional electrophoresis technique. *J. Biol. Chem.*

Righelti, P.G. and J.W. Drysdale (1975). In *Laboratory techniques in Biochemistry and Molecular Biology,* 5(2) eds. T.S. Work and E.work, North Holland Pub. Co.

Seeley, H.W. Jr. and Paul J. van Demark (1972). *Microbes in Action.* W.H. Freeman Co.

Strobel, H.A. (1973). *Chemical Instrumentation,* 2nd ed., Addison-Wesley, Reading, Mass.

Willard, H.H., J.A. Dean, L.L. Merritt and F.A. Settle (1986). *Instrumental methods of analysis.* CBS Publishers & Distributors, Delhi.

INDEX

A

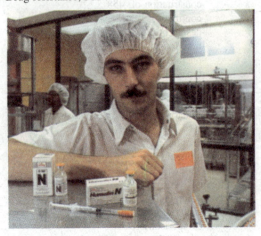

H

I

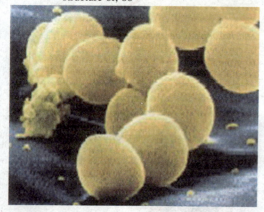

S

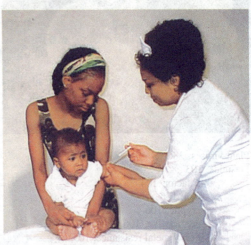

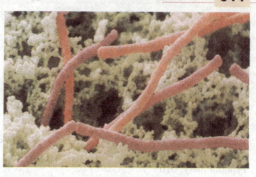

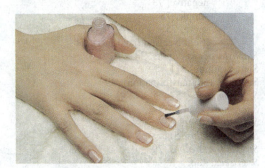